SATELLITES

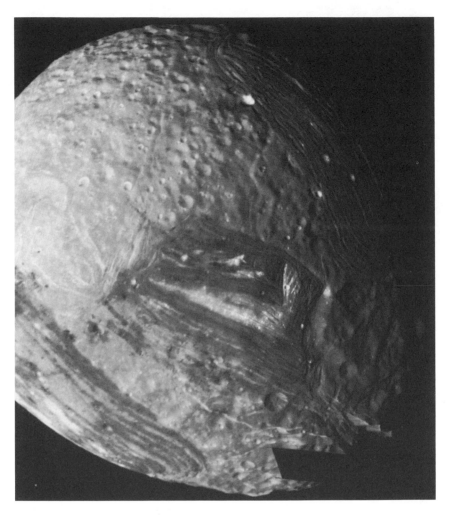

The Uranian satellite Miranda reveals a complex geologic history in this 400 km wide view acquired by Voyager 2 on January 24, 1986. This mosaic of seven images was taken from ranges between 40,000 and 30,000 km and has resolutions of 600−750 m/line pair. Because of rapid spacecraft motion, the mosaic distorts the shape of Miranda. Data lost in transmission cause the break at the lower right. About one-half of the satellite is ancient, heavily cratered terrain. Younger tectonic features make large areas of grooved and faulted terrain at upper right and lower left. A smaller rectangular area of faulting has a bright chevron-shaped albedo marking. Global fracture systems are associated with fault valleys ~10 km deep at the lower part of the terminator (photograph courtesy of NASA/JPL:P-29541).

SATELLITES

Edited by

Joseph A. Burns
Mildred Shapley Matthews

With 45 collaborating authors

THE UNIVERSITY OF ARIZONA PRESS
TUCSON

About the cover:

The front cover displays surface features on Io as obtained by Voyager 1 on 5 March 1979. The back cover shows the satellites of Uranus photographed by G. P. Kuiper on 25 April 1962 with the Struve 2-m reflector of the McDonald Observatory. Satellites, clockwise from the right, are Umbriel, Ariel, Titania, Oberon and Miranda.

THE UNIVERSITY OF ARIZONA PRESS

Copyright © 1986
The Arizona Board of Regents
All Rights Reserved

This book was set in 10/12 Linotron Times Roman.
Manufactured in the U.S.A.

Library of Congress Cataloging-in-Publication Data

Satellites.

(Space science series)
"Based on a conference, Natural satellites,
Colloquium 77 of the International Astronomical Union
(IAU) held at Cornell University, Ithaca, New York,
from July 5 to 9, 1983"—Pref.
Bibliography: p.
Includes index.
1. Satellites—Congresses. I. Burns, Joseph A.
II. Matthews, Mildred Shapley. III. International
Astronomical Union. Colloquium (77th : 1983 :
Cornell University) IV. Series.
QB401.S38   1986   523.9'8   86-19145

ISBN 0-8165-0983-2 (alk. paper)

CONTENTS

COLLABORATING AUTHORS

R. H. Brown, *836*
J. A. Burns, *ix, 1, 117*
M. H. Carr, *629*
C. R. Chapman, *492*
A. F. Cheng, *403*
R. N. Clark, *437*
S. K. Croft, *293*
D. P. Cruikshank, *836*
S. Dermott, *802*
M. J. Drake, *581*
F. P. Fanale, *437*
M. J. Gaffey, *437*
J. Gradie, *629*
P. K. Haff, *403*
A. W. Harris, *39*
J. W. Head, *581*
K. Housen, *342*
D. M. Hunten, *629*
R. E. Johnson, *403*
T. V. Johnson, *342*
W. M. Kaula, *581*
L. J. Lanzerotti, *403*
J. I. Lunine, *39*

M. C. Malin, *689*
D. L. Matson, *342*
W. B. McKinnon, *492, 718*
D. Morrison, *764*
D. B. Nash, *629*
T. Owen, *764*
E. M. Parmentier, *718*
S. J. Peale, *159*
G. V. Pechernikova, *89*
D. C. Pieri, *689*
R. T. Reynolds, *224*
E. L. Ruskol, *89*
V. S. Safronov, *89*
G. Schubert, *224*
L. A. Soderblom, *764*
T. Spohn, *224*
S. W. Squyres, *293*
D. J. Stevenson, *39*
P. Thomas, 342, *802*
J. Veverka, *342, 802*
A. V. Vitjazev, *89*
C. F. Yoder, *629*

PREFACE

Few topics of research have blossomed as rapidly as planetary science and, within this discipline, the study of natural satellites has burgeoned perhaps the most of all. In the mid-nineteen seventies, when the previous volume on planetary satellites (Burns 1977) was prepared, natural satellites (beyond the Earth's Moon) were just beginning to be studied seriously. At that time most results came from groundbased observations or were derived from theoretical speculation. Even though the investigation of satellites was in its infancy, the objects had already revealed themselves to be diverse and quite interesting places. For instance, Io—with its odd optical properties, its gas tori and its apparent control over Jupiter's decametric radiation—was already a puzzle. Since information was limited, this puzzle and many others could hardly be addressed, let alone solved. Certainly no overall understanding of planetary satellites was available, despite the general recognition that these objects contained some of the best clues about the solar system's origin.

Today the situation is quite different. The Voyager missions have provided a rich harvest of information about these moons. Meanwhile theoretical developments have kept pace. As a consequence, our view of planetary satellites nowadays is substantially more detailed and more informed than it was a mere decade ago. This newly available treasure trove of data raises questions for disciplines that had scarce interest in moons a few years past. As a single example of this, we note that many geologists and geophysicists who are now intrigued by the complex surface morphologies of moons then barely knew of the objects. The paradigms of each discipline have been modified in different ways in response to our vastly improved knowledge of satellites. But in each case—from the active volcanoes on youthful Io to Titan's dense primitive atmosphere—the features observed on the satellites have led to a deepening of our appreciation of the solar system's beauty and complexity. Some day, we hope, they will even lead us to comprehend truly how the solar system originated.

Like the other volumes in the Space Science Series, this book is based on a conference—"Natural Satellites," Colloquium 77 of the International Astronomical Union (IAU), held at Cornell University, Ithaca, New York, from July 5 to 9, 1983. The effective date of the material in this book is, nevertheless, August 1985.

We have been extremely fortunate to deal with a talented and outstanding group of authors, to whom we express our deep thanks. All chapters were reviewed by at least two scientists; this group of referees has been instrumental in maintaining the standards of the Series. The book's preparation was carried out with invaluable help from B. Boettcher, G. Podleski, and M. Talman in Ithaca, and M. Magisos in Tucson. The Map Section at the end of the book has been prepared with the help of R. Batson, B. Boettcher, J. K. Burns, K. Pierce and P. Thomas; the color plates of the Galilean satellites, including that used on the cover, were kindly provided by A. McEwen and L. Soderblom. The book's bibliography was compiled by M. Magisos. T. Gehrels and J. Veverka provided essential advice on the composition and scientific content of this book. D. Morrison, who originally started as a coeditor but was called away by other duties, played a pivotal role in developing the book's initial concept, including its title, and in selecting its topics, authors and referees. Financial support for this book was generously provided by the National Science Foundation, the National Aeronautics and Space Administration, the Voyager and Galileo projects and Cornell University. Finally we wish to express appreciation to the University of Arizona Press for their efforts in publishing this book.

Joseph A. Burns
Mildred S. Matthews

SATELLITES

1. SOME BACKGROUND ABOUT SATELLITES

JOSEPH A. BURNS
Cornell University

More than fifty moons have been discovered and studied from groundbased observatories and spacecraft. These objects range in size from collisional fragments that are a mere ten kilometers in breadth to satellites that rival the planets in size and complexity. This chapter tabulates the orbital parameters, physical properties and the discovery circumstances for the various satellites; pertinent information on the planets is also given. Satellites are classified as regular satellites, irregular satellites *or* collisional debris; *the Moon, Triton and Charon do not fit this classification scheme. The philosophy that guides the book's remainder is laid out and précis of individual chapters are provided so as to illustrate some of the links between various chapters.*

Clearly the satellites are just as interesting as the planets they orbit. But the vast growth in our information about satellites and the diversity of the objects themselves present hurdles to any full understanding of the satellites. This book is intended to remove these obstacles by summarizing our knowledge of satellites so as to put it in perspective and by pointing out the common processes that affect all satellites. It approaches this problem in a matrix fashion. The book's first half contains ten chapters which describe various processes that operate to varying degrees on all satellites. Generally these chapters first lay out some basic principles and then apply them in turn to those satellites for which available data show the processes to be pertinent. In this way the differences and similarities of the sundry satellites can be appreciated. The second half of the book is composed of seven chapters about individual satellites or classes of satellites. In these, the principles from the first part are called upon to develop a comprehensive picture of the distinguishing properties of specific satellites. This format, with extensive cross referencing between chapters, should provide the reader with an excellent overview of what is known—and what is not—about the natural satellites in the mid-nineteen eighties.

Useful background material for the book will be found in this introduction and in the end matter. This chapter contains four tables of planet and satellite data, a scheme for classifying the satellites, and a quick survey of the book including a brief description of the various satellites in the solar system as presented here and elsewhere. End matter in the book includes (i) a color section containing Mercator projections of the four Galilean satellites, taken from Voyager images, plus several color figures from the chapters; (ii) equal-area Lambert projection maps of the major satellites visited by spacecraft using material generously provided by R. Batson of the United States Geological Survey (these have been put together through the efforts of B. Boettcher, J. A. and J. K. Burns, K. Pierce, G. Podleski and P. Thomas), and sketches of the identified features on some small satellites; (iii) a glossary of frequently used terms; (iv) a complete bibliography of all the text references, which should be a useful research tool in its own right; and (v) an index.

I. SATELLITE AND PLANETARY DATA

By early 1986 some 54 natural satellites were confirmed to exist, and there was good circumstantial evidence for at least a half-dozen more objects. Table I lists all the known moons with their names as approved by the International Astronomical Union; the derivation of these names is described in Sec. IV. Table I also indicates the discoverer of each satellite and, for each planet's retinue, orders them by the years in which they were found. For obvious reasons, faint objects have only been discerned in the past few years and often these discoveries have been made by members of the Voyager spacecraft imaging team. In Sec. IV, which concerns the individual satellites, we briefly recount some discovery circumstances and mention some unsuccessful attempts to find additional satellites (cf. Kuiper 1961) as well as future plans for spacecraft exploration. Table I also provides pronunciations which are based, following Newburn and Gulkis (1973), on the unabridged *Random House Dictionary of the English Language*, where possible. These pronunciations are not presented in the belief that they are correct—alternative choices could be made and questions can certainly be raised as to, for example, whether ancient or modern Greek sounds are more appropriate in the twentieth century for the names of Hellenic gods (see Consolmagno and Reiche 1982), etc. We offer these proposals in the simple (but vain) hope of producing consistent pronunciation among practitioners.

Table II gives some properties of the planets that might prove useful in studies of satellites. Gravitational parameters (mass M and gravitational coefficients J_2 and J_4) are generally determined from the orbits of flyby space-

TABLE I
Satellite Discoveries[a]

Year	Satellite		Pronunciation	Discoverer(s)	Country	V_0
		Moon	mо̅о̅n			−12.7
1877	MI	Phobos	fō'bos	Hall	USA	11.3
1877	MII	Deimos	dī'mos	Hall	USA	12.4
1610	JI	Io	ī'ō	Galileo (Marius?)	Italy (Germany)	5.0
1610	JII	Europa	yо̅о̅ rо' pə	Galileo (Marius?)	Italy (Germany)	5.3
1610	JIII	Ganymede	gan' ə mēd'	Galileo (Marius?)	Italy (Germany)	4.6
1610	JIV	Callisto	kə lis' tō	Galileo (Marius?)	Italy (Germany)	5.6
1892	JV	Amalthea	am'əl thē' ə	Barnard	USA	14.1
1904/5	JVI	Himalia	him' əli ə	Perrine	USA	15.0
1904/5	JVII	Elara	ē' lə rə	Perrine	USA	16.6
1908	JVIII	Pasiphae	pə sif'ə ē	Melotte	England	16.9
1914	JIX	Sinope	si nō' pē	Nicholson	USA	18.0
1938	JX	Lysithea	lis' i thē' ə	Nicholson	USA	18.2
1938	JXI	Carme	kär' mē	Nicholson	USA	17.9
1951	JXII	Ananke	ə' nan kē	Nicholson	USA	18.9
1974	JXIII	Leda	lē' də	Kowal	USA	20.2
1979	JXV	Adrastea	ad'ra stē' ə	Jewitt, Danielson	USA	18.7
1979/80	JXIV	Thebe	thē'bē	Synnott	USA	16.0
1979/80	JXVI	Metis	mē'tis	Synnott	USA	17.5

TABLE I (continued)

Year	Satellite	Pronunciation	Discoverer(s)	Country	V_0
1655	SVI	Titan	tīt' ən	Huygens	Holland	8.3
1671	SVIII	Iapetus	ī ap' i təs	Cassini	France	10.2–11.9
1672	SV	Rhea	rē' ə	Cassini	France	9.7
1684	SIII	Tethys	tē' this	Cassini	France	10.2
1684	SIV	Dione	dī o' ne	Cassini	France	10.4
1789	SI	Mimas	mī' mas	Herschel	England	12.9
1789	SII	Enceladus	en sel' ə dəs	Herschel	England	11.7
1848	SVII	Hyperion	hī pēr' ē ən	W.&G. Bond/Lassell	USA/England	14.2
1898	SIX	Phoebe	fē' bē	Pickering	USA (Peru)	16.5
1966/80	SX	Janus (large coorbital)	jā' nəs	Dollfus	France	14.5
1966/80	SXI	Epimetheus (small coorbital)	ep'ə mē'thē əs	Fountain, Larson/Walker	USA	15.7
1980	SXII	Helene (Dione Lagrangian)	hə lēn'	Lecacheux, Laques	France	18.4
1980	SXIII	Telesto (leading Tethys Lagrangian)	tə les'tō	Reitsema, Smith, Larson, Fountain	USA	18.7
1980	SXIV	Calypso (trailing Tethys Lagrangian)	kə lip'sō	Pascu, Seidelmann, Baum, Currie	USA	19
1980	SXVI	Prometheus (F-ring inner shepherd)	prə mē'thē əs	Collins et al.	USA	15.8
1980	SXVII	Pandora (F-ring outer shepherd)	pan dōr'ə	Collins et al.	USA	16.5
1980	SXV	Atlas	at'ləs	Terrile	USA	18
1787	UIII	Titania	ti tā' nē ə	Herschel	England	14.0
1787	UIV	Oberon	ō' bə ron'	Herschel	England	14.2
1851	UI	Ariel	âr' ē əl	Lassell	England	14.4

TABLE I (continued)

Year	Satellite		Pronunciation	Discoverer(s)	Country	V_0
1851	UII	Umbriel	um′ brē el′	Lassell	England	15.3
1948	UV	Miranda	mi ran′ də	Kuiper	USA	16.5
1985	1985U1			Voyager 2	USA	~20.5
1986	1986U1-1986U9			Voyager 2	USA	~23–25
1846	NI	Triton	trit′ ən	Lassell	England	13.6
1949	NII	Nereid	nēr′ ē id	Kuiper	USA	18.7
1978	PI	Charon	kâr′ ən	Christy	USA	16.8

[a]This table is an updated version of one by Morrison et al. (1977); circumstances of recent satellite discoveries come from IAU records, the original papers, personal recollections and Beatty et al. (1982, p. 220). V_0 magnitudes (i.e. mean opposition visual magnitudes at zero phase angle) are those given in the *Astronomical Almanac* (1985). Pronunciation symbols include a̲ as in air and t̲h̲ as in thin or path, as well as the usual long vowels, marked with a bar, ā for example, and the short vowels which are unmarked. The schwa, written ə, has the sound of a̲ in above, of e̲ in system, of i̲ in easily, of o̲ in gallop and of u̲ in circus. Additional discussion of this material is given in the adjacent text as well as in the section on the various satellites.

craft or natural satellites (Aksnes 1977); the Uranian values are specified by the precisely defined kinematics of the rings (Elliot and Nicholson 1984; French et al. 1986). Much more elaborate expansions are available for the gravity fields of Earth and Mars (see Sjogren 1983 for references) since these planets have been orbited repeatedly by artificial spacecraft. Although specific numbers are listed for them, equatorial radii R and optical flattenings are actually somewhat ill-defined because terrestrial planets have lumpy shapes that do not form perfectly spheroidal equipotential surfaces and because values for the gaseous planets depend upon how deep an atmosphere is probed by an occultation. Nevertheless, "equatorial radii" are required to normalize distances in gravitational expansions; the ones for the giant planets that we have taken here are those adopted in planetary ring studies (see Greenberg and Brahic 1984) and are not those at the 1 bar pressure level, as found in some other books.

Densities provide the best, but still a crude, indicator of gross composition; values for the outer planets are questionable primarily because of uncertainties in radii (see just above). Escape velocities, $V_e = (2GM/R)^{1/2}$, give lower bounds to impact velocities of interplanetary projectiles onto the planet; they can also be used to compute circular orbital velocities of satellites which are equal to $0.707 \, V_e R^{*-1/2}$, where R^* is the satellite's orbital semimajor axis expressed in planetary radii (see Table III). $I/(MR^2)$ is the ratio of I (a body's moment of inertia about an axis through its center) relative to the inertia that a mass M would have if at radius R; the ratio is 0.4 for a homogeneous sphere and is proportionately less for more centrally condensed objects. It can be calculated from models of internal density distributions or computed approximately for planets that are in hydrostatic equilibrium, knowing the planet's J_2, density, flattening and rotation rate (see, e.g., Dermott 1984c; chapter by Thomas et al.).

In connection with satellite studies, the rotation period of the planet's interior is also that of its magnetic field and thus that of the inner magnetospheric plasma. Since most satellites lie near their planet's equatorial plane, the planet's obliquity (i.e., the angle between its rotation axis and its orbit plane) illustrates how the satellite orbits are oriented in space.

Blackbody temperatures provide rough, but useful, guides as to which materials are stable in the vicinities of particular planets. The subsolar blackbody temperatures listed in Table II are the equilibrium temperatures of insulated, totally absorbent surfaces held normal to the solar radiation flux; the values for a black sphere reduce those by a factor of $\sqrt{2}$. The listed effective temperatures are those of a sphere with the planet's albedo p; depending upon a satellite's actual albedo (see Table IV), its temperature will differ from these in the ratio of $(1 - p)^{1/4}$ for the two objects.

The orbital properties of the satellites ordered by the distances of the satellites from their planets, are listed in Table III. Semimajor axes are generally computed from orbital periods, knowing the planet's J_2 and J_4 (see Table II),

TABLE II
Planetary Parameters[a]

Planet	Orbital Distance (AU)	Orbital Distance (10^9 m)	Mass (10^{24} kg)	Equatorial Radius (10^6 m)	Density (10^3 kg m^{-3})	Escape Velocity (km s^{-1})	J_2 (10^{-5})	J_4 (10^{-5})	I/MR^2
Mercury	0.387	57.9	0.3303	2.439	5.43	4.25	(8 ± 6)	—	—
Venus	0.723	108.2	4.870	6.051	5.25	10.4	0.6	—	(0.34)
Earth	1.000	149.6	5.976	6.378	5.518	11.2	108.3	− 0.16	0.331
Mars	1.524	227.9	0.6421	3.393	3.95	5.02	196.0 ± 1.8	− 3.2 ± 0.7	0.365
Jupiter	5.203	778.3	1900	71.398	1.33	59.6	1473.6 ± 0.1	− 58.7 ± 0.5	(0.26)
Saturn	9.539	1427.0	568.8	60.33	0.69	35.5	1667 ± 3	− 103 ± 7	(0.22)
Uranus	19.182	2869.6	86.87	26.20	(1.15)	21.3	333.9 ± 0.3	(− 3.2 ± 0.4)	(0.20)
Neptune	30.058	4496.6	102.0	25.23	(1.55)	23.3	(430 ± 30)	—	(0.27)
Pluto	39.44	5900.1	(0.013)	(1.5)	(0.9)	(1.1)	—	—	—

TABLE II (continued)

	Rotation Period	Obliquity	Blackbody Temperatures (K)			Optical Flattening
			Sphere	Subsolar	Effective	
Mercury	58.65 d	$(2° \pm 3°)$	448	633	442	(0)
Venus	243.01 ($\pm$ 0.03) d	177°.3	328	464	244	(0)
Earth	23.9345 h	23°.45	279	394	253	0.0034
Mars	24.6299 h	23°.98	226	320	216	0.0059
Jupiter	9.841 h (equator) 9.925 h (field)	3°.12	122	173	87	0.065
Saturn	10.233 h (equator) 10.675 h (field)	26°.73	91	128	63	0.108
Uranus	17.3 h	97°.86	64	90	33	(0.024)
Neptune	(18.2 $\pm$ 0.4) h	(29°.56)	50	71	32	(0.017)
Pluto	6.387 d	(118°.5)	44	62	(43)	

[a]Orbital distances come from Allen (1973), while planetary gravity fields are from Sjogren (1983). Parameter values for individual planets that have been studied in detail come instead from the following: Mercury (Gault et al. 1977); Jupiter (Campbell and Synnott 1985); Uranus (Voyager 2 data 1986; French et al. 1986); and Neptune (Harris 1984b). Rotational parameters are taken from Allen (1973), the *Astronomical Almanac* (1985), Davies et al. (1983), and Harris and Ward (1982). Additional information on the spin and shape of Uranus and Neptune are summarized by French (1984) and Belton and Terrile (1984). The optical flattening listed for Neptune is derived from the tabulated J_2 and the rotation period (Dermott 1984c); the actually observed flattening does not yield a self-consistent theory. Temperatures come from Allen (1973) and Goody and Walker (1972). ( ) indicate possible errors of 5% or more. Additional information on this table is described in the text.

TABLE III[a]
Satellite Orbits

Planet		Satellite	Orbital Semimajor Axis 10^6 m	(Planetary Radii)	Orbital Period (days)	Rotational Period[b] (days)	Eccentricity (forced) free	Inclination (forced) free	Laplace Plane Tilt
Earth		Moon	384.4	(60.3)	27.3217	s, 1	0.05490	5°15*	23°45
Mars	MI	Phobos	9.378	(2.76)	0.319	s, 1	0.015	1°02	0°01
	MII	Deimos	23.459	(6.91)	1.263	s	0.00052	1°82	0°92
Jupiter	JXVI	Metis	127.96	(1.7922)	0.2948		<0.004	~0°	
	JXV	Adrastea	128.98	(1.8065)	0.2983		~0	~0°	
	JV	Amalthea	181.3	(2.539)	0.4981	s	0.003	0°40	
	JXIV	Thebe	221.90	(3.108)	0.6745		0.015 ± 0.006	0°8 ± 0°2	
	JI	Io	421.6	(5.905)	1.769	s	$(0.0041)–10^{-5}$	0°040	
	JII	Europa	670.9	(9.397)	3.551	s	$(0.0101)–10^{-4}$	0°470	
	JIII	Ganymede	1,070	(14.99)	7.155	s	(0.0006)–0.0015	0°195	
	JIV	Callisto	1,883	(26.37)	16.689	s	0.007	0°281	
	JXIII	Leda	11,094	(155.4)	238.72		0.148	27°*	
	JVI	Himalia	11,480	(160.8)	250.57	0.4	0.158	28°*	
	JX	Lysithea	11,720	(164.2)	259.22		0.107	29°*	
	JVII	Elara	11,737	(164.4)	259.65		0.207	28°*	
	JXII	Ananke	21,200	(296.9)	631R		0.169	147°*	
	JXI	Carme	22,600	(316.5)	692R		0.207	163°*	
	JVIII	Pasiphae	23,500	(329.1)	735R		0.378	148°*	
	JIX	Sinope	23,700	(331.9)	758R		0.275	153°*	3°12
		Halo		(1.4–1.71)				≤10°	
		Main Ring		(1.71–1.81)			~0	0°	
		Gossamer Ring		(1.81–3)				0°	
Saturn	SXV	Atlas	137.64	(2.281)	0.602		~0	~0°	
	SXVI	Prometheus	139.35	(2.310)	0.613		0.0024 ± 0.0006	0°0 ± 0°1	
	SXVII	Pandora	141.70	(2.349)	0.629		0.0042 ± 0.0006	0°0 ± 0°1	
	[SXI	Epimetheus	151.422	(2.510)	0.694	s	0.009 ± 0.002	0°34 ± 0°05	
	[SX	Janus	151.472	(2.511)	0.695	s	0.007 ± 0.002	0°14 ± 0°05	

TABLE III[a] (continued)

Planet	Satellite	Orbital Semimajor Axis 10⁶ m	(Planetary Radii)	Orbital Period (days)	Rotational Period[b] (days)	Eccentricity (forced) free	Inclination (forced) free	Laplace Plane Tilt
	SI	Mimas	185.52	(3.075)	0.942	s	0.0202	(1°53)	
	SII	Enceladus	238.02	(3.945)	1.370	s	(0.0045)	0°02	
	SIII	Tethys	294.66	(4.884)	1.888	s	0.0000	(1°09)	
	[SXIII	Telesto]	294.66	(4.884)	1.888		~0	~0°	
	[SXIV	Calypso]	294.66	(4.884)	1.888		~0	~0°	
	SIV	Dione	377.40	(6.256)	2.737	s	(0.0022)	0°02	
	[SXII	Helene]	377.40	(6.256)	2.737		0.005	0°2	
	SV	Rhea	527.04	(8.736)	4.518	s	(0.0010)–0.0003	0°35	
	SVI	Titan	1221.85	(20.25)	15.945		0.0292	0°33	
	SVII	Hyperion	1481.1	(24.55)	21.277	chaotic	(0.1042)	0°43	
	SVIII	Iapetus	3561.3	(59.03)	79.331	s	0.0283	7°52	14°84
	SIX	Phoebe	12952	(214.7)	550.48R	0.4	0.163	175°3*	26°73
	D Ring			(1.11–1.235)					
	C Ring			(1.235–1.525)					
	B Ring			(1.525–1.940)					
	A Ring			(2.025–2.267)					
	F Ring			(2.324)					
	G Ring			(2.82)					
	E Ring			(3–8)					
Uranus	1986U7		49.75	(1.90)	0.336		~0	~0	
	1986U8		53.77	(2.05)	0.377		~0	~0	
	1986U9		59.16	(2.26)	0.435		~0	~0	
	1986U3		61.77	(2.36)	0.465		~0	~0	
	1986U6		62.65	(2.39)	0.476		~0	~0	
	1986U2		64.63	(2.47)	0.494		~0	~0	
	1986U1		66.10	(2.52)	0.515		~0	~0	
	1986U4		69.93	(2.67)	0.560		~0	~0	
	1986U5		75.25	(2.87)	0.624		~0	~0	
	1985U1		86.00	(3.28)	0.764		~0	~0	

UV	Miranda	129.8	(4.95)	1.413	s	0.0027	4°.22
UI	Ariel	191.2	(7.30)	2.520	s	0.0034	0°.31
UII	Umbriel	266.0	(10.15)	4.144	s	0.0050	0°.36
UIII	Titania	435.8	(16.64)	8.706	s	0.0022	0°.14
UIV	Oberon	582.6	(22.24)	13.463	s	0.0008	0°.10
	Rings		(1.60–1.95)			$10^{-2} - 10^{-4}$	$10^{-3} - 10^{-4}$
Neptune	NI Triton	354.3	(14.0)	5.877R	s	<0.0005	159°.0 ± 1°.5
	NII Nereid	551.5	(219)	350.16		0.75	27°.6* 29°.56
	Ring-like Arc	~65	(~3)				
Pluto	PI Charon	19.1	(12.7)	6.387		~0	94°.3 ± 1°.5

[a] Most data are from the *Astronomical Almanac* (1985) except as noted below and in the text. The orbits of the outermost sets of Jovian satellites are strongly perturbed by the Sun. Thus the listed osculating orbital elements are of limited use in constructing ephemerides (see Kovalevsky and Sagnier 1977, especially their Fig. 4.3); numerical integrations are preferable. For the inclinations of the outer Jovian satellites, we tabulate the values of Kovalevsky and Sagnier (1977) since they are given relative to the ecliptic. Information on the Uranian satellite orbits is taken from Veillet (1983a), which contains the best quality and most recent observations of the system; however, the analysis of these observations does not include secular perturbations so that some orbital elements could be in error (Dermott and Nicholson 1986); results for Miranda should be correct however. Orbital elements for Nereid are taken from Mignard (1981c), those for the Saturnian ring moons from Aksnes (1985), and those for Triton from Harris (1984b). Charon's orbital elements are adopted from Tholen (1985). Satellites denoted with an * correspond to those on distant orbits dominated by solar effects rather than by the planet's oblateness (see Table 7.1 in Burns 1977b). For inclinations and eccentricities the parentheses identify values expressed in planetary radii. ( ) denote forced values; [ ] denote satellites that are in coupled orbits. R indicates a retrograde orbit. Additional discussion of this table is given in the text. Voyager 2 values for Uranian satellites are preliminary (Smith et al. 1986).
[b] s = synchronous, l = libration detected.

[11]

as described, for example, by Aksnes (1977). The tabulated semimajor axes of Saturn's coorbital pair are instantaneous ones valid at the time of the Voyager flyby; these oscillate as the satellites periodically interchange their orbits (see Peale's chapter). The values of semimajor axes given in terms of planetary radii take the numbers listed in the first column and divide by planetary radii from Table II. Information on satellite rotations are found in the *Astronomical Almanac* (1985) and Peale (1977), as well as in this book's chapters by Veverka et al. and Thomas et al.

Orbital eccentricities and inclinations may be affected by resonances, as summarized by Peale (1976b; chapter herein) and Greenberg (1977,1984c). The inner three Galilean satellites are locked in the Laplace resonance which forces their orbital eccentricities; forced eccentricities are identified in the table by parentheses, with those listed coming from Yoder and Peale (1981; cf. Table II in Peale's chapter). The Enceladus-Dione pair of Saturn are in a 2:1 simple eccentricity resonance (see Table I in Peale's chapter; also Greenberg 1977). Hyperion's eccentricity is induced by Titan through a 4:3 simple eccentricity resonance but the little satellite has meager influence on that massive moon. Rhea's eccentricity is forced by Titan through a secular interaction (Greenberg 1977). Mimas and Tethys are locked in a 2:1 mixed inclination resonance (Peale 1976b) that maintains the alignment of their pericenters.

Orbital inclinations of Table III are given relative to the local Laplace plane (defined in the glossary and Burns' chapter): for a satellite near its planet, this surface lies approximately in the planet's equator while, for distant satellites, it is in the planet's orbital plane about the Sun. The tabulated tilts of the Laplace planes, measured off planetary equators, are from Veverka and Burns (1980) for Mars' satellites, and from Ward (1981) for Iapetus, while for the outlying satellites they are simply the planet's obliquity (cf. Table II). Those satellites with tilted Laplace planes will have variable inclinations (when measured from the planet's equator) as the orbit precesses about the pole of the Laplace plane (see, e.g., Mignard et al. 1986).

The physical properties of the satellites are given in Table IV, where the satellites of each planet are grouped in order of size. Satellite radii can be accurately determined in one of two ways: with spacecraft images interpreted through geodetic techniques or from infrared measurements coupled with visual observations which together yield both the satellite's size and albedo (see discussion in the chapter by Veverka et al.; also Morrison 1977). If these observations are unavailable, radii can be estimated once the object's brightness is combined with its solar distance (see Table I) and an albedo chosen to be like those of its neighboring satellites (Morrison et al. 1977). The IAU-adopted radii of the larger satellites are tabulated by Sjogren (1983) and Davies et al. (1983). Saturn's satellite data, originally published by Smith et al. (1982), are contained in Morrison (1982a,1983) and Morrison et al. (1984). The sizes listed there have been revised slightly in our tabulation now that control networks have been developed from the Voyager images; minor changes are available for Mimas (Davies and Katayama [1983a]; Thomas and Dermott [1985] match limb profiles of Voyager images instead), Enceladus (Davies and Katayama 1983a), Tethys and Dione (Davies and Katayama 1983b), Rhea (Davies and Katayama 1983c), and Iapetus (Davies and Katayama 1984). The sizes of Hyperion (Thomas and Veverka 1985) and Phoebe (Thomas et al. 1983a) come from reanalysis of the Voyager images. Tidal distortions are detectable in the shape of Io (Davies and Katayama 1981; cf. Synnott et al. 1985) as well as those of Mimas and perhaps Enceladus (Davies and Katayama 1983a; cf. Thomas and Dermott 1985). Voyager images (Thomas et al. 1983a; see chapter by Thomas et al.) provide the sizes and geometric albedos of the small Saturnian satellites. Radii of Thebe, Metis and Adrastea were similarly evaluated by S. P. Synnott (personal communication, 1983). Estimates of the sizes of the outer Jovian satellites are selected from Tholen and Zellner (1984), who apply the measured albedos of JVI and JVII (Cruikshank et al. 1982) to the visual magnitudes of the other outer satellites; if these tiny moons are jagged (as expected), the tabulated sizes correspond to a single, not necessarily representative, cross section. The radii and albedos of all Uranian moons are drawn from infrared observations by Brown et al. (1982b), except Miranda's values which are selected from Brown and Clark (1984). These all have been modified slightly from Voyager's 1986 flyby through the system. Triton's infrared radiation was originally detected by Lebofsky et al. (1982); the presence of an atmosphere may invalidate the radius calculation and this could account for Triton's improbable density (Harris 1984b; chapter by Cruikshank and Brown). If Triton's density were in fact 2000 kg m^{-3} and the tabulated mass were correct, its radius would be ~ 2500 km. The sizes of Nereid and Charon are estimated from visual magnitudes and assumed albedos; in the latter's case, the ongoing eclipses/occultations will allow an improved size determination.

Satellites having three radii listed are irregular in shape. The tabulated axes are those of a nearest-fit triaxial ellipsoid; these numbers should not be construed to imply that the satellites are in fact triaxial ellipsoids. A satellite that has tidally evolved to its minimum energy/synchronous rotation state will on average have its long axis pointed toward its planet and its short axis normal to its orbit plane (see Peale 1977 and his chapter herein; see also chapter by Chapman and McKinnon).

Most well-determined masses of satellites listed in Table IV are derived from spacecraft perturbations; Sjogren (1983) records some values. The Moon's gravity field is accurately documented from the Lunar Orbiter and Apollo missions, while the masses of Phobos and Deimos come from a recent redetermination of the perturbations suffered by the Viking Orbiters during their remarkably close flybys of these moons (see Sjogren 1983). Galilean satellite masses, ascertained from the four Pioneer and Voyager flybys, have been computed by Campbell and Synnott (1985). The tabulated masses for Tethys, Dione, Rhea, Titan and Iapetus come from the deflections of the Voyager tra-

TABLE IV
Satellite Physical Properties[a]

Planet	Satellite	Radius (km)	Mass (10^{20} kg)	Relative Mass	Density (10^3 kg m^{-3})	Geometric Albedo	Surface Composition
Earth	Moon	1738	734.9 (±0.7)	1.23×10^{-2}	3.34	0.12	rock
Mars	MI Phobos	$(13.5 \pm 1) \times (10.7 \pm 0.7) \times (9.6 \pm 0.7)$	$1.26 (\pm 0.1) \times 10^{-4}$	2.0×10^{-8}	2.2 (±0.5)	0.06	carbonaceous?
	MII Deimos	$(7.5 \pm \frac{1}{3}) \times (6.0 \pm 0.5) \times (5.5 \pm 1)$	$1.8 (\pm 0.15) \times 10^{-5}$	2.8×10^{-9}	1.7 (±0.5)	0.06	carbonaceous?
Jupiter	JIII Ganymede	2631 ± 10	1482.3 ± 0.5	7.803×10^{-5}	1.94	0.4	dirty ice
	JIV Callisto	2400 ± 10	1076.6 ± 0.5	5.667×10^{-5}	1.86	0.2	dirty ice
	JI Io	1815 ± 5	894 ± 2	4.704×10^{-5}	3.57	0.6	sulfur, SO_2
	JII Europa	1569 ± 10	480 ± 2	2.526×10^{-5}	2.97	0.6	ice
	JV Amalthea	$(135 \pm 10) \times (82 \pm 8) \times (75 \pm 5)$	—	—	—	0.06	sulfur/rock?
	JVI Himalia	90 ± 10	—	—	—	0.03	carbonaceous?
	JXIV Thebe	$? \times (55 \pm^{10}_{5}) \times (45 \pm^{10}_{5})$	—	—	—	0.05–0.10	rock?
	JVII Elara	40 ± 5	—	—	—	0.03	carbonaceous?
	JVIII Pasiphae	~35	—	—	—	—	carbonaceous?
	JXVI Metis	$? \times (20 \pm 2) \times (20 \pm 2)$	—	—	—	0.05–0.10	rock?
	JXI Carme	~22	—	—	—	—	carbonaceous?
	JIX Sinope	~20	—	—	—	—	carbonaceous?
	JX Lysithea	~20	—	—	—	—	carbonaceous?
	JXII Ananke	~15	—	—	—	—	carbonaceous?
	JXV Adrastea	12.5 × 10 × 7.5	—	—	—	0.05–0.10	rock?
	JXIII Leda	~8	—	—	—	—	carbonaceous?
	Rings		—	—	—	~0.05	rock?
Saturn	SVI Titan	solid 2575 (±2) clouds 2775	1345.7 ± 0.3	2.36×10^{-4}	1.881	0.2	nitrogen/methane atm.
	SV Rhea	764 ± 4	24.9 ± 1.5	4.4×10^{-6}	1.33	0.65	ice
	SVIII Iapetus	718 ± 8	18.8 ± 1.2	3.3×10^{-6}	1.21	0.5/0.04	ice/carbonaceous?
	SIV Dione	559 ± 5	10.5 ± 0.3	1.8×10^{-6}	1.44	0.55	ice
	SIII Tethys	524 ± 5	7.6 ± 0.9	1.3×10^{-6}	1.26	0.80	ice
	SII Enceladus	251 ± 5	0.8 ± 0.3	1.5×10^{-7}	1.24	1.04	pure ice
	SI Mimas	197 ± 3	0.38 ± 0.01	6.6×10^{-8}	1.17	0.77	ice
	SVII Hyperion	$(175 \pm 15) \times (120 \pm 10) \times (100 \pm 10)$	—	—	—	0.25	dirty ice

	Satellite						
	SIX Phoebe	$(115 \pm 10) \times (110 \pm 10) \times (105 \pm 10)$	—	—		0.06	carbonaceous?
	SX Janus	$(110 \pm 5) \times (95 \pm 5) \times (80 \pm 5)$	—	—		0.5	ice?
	SXI Epimetheus	$(70 \pm 8) \times (58 \pm 8) \times (50 \pm 5)$	—	—		0.5	ice?
	SXVI Prometheus	$(70 \pm 5) \times (50 \pm 7) \times (37 \pm 8)$	—	—		0.5	ice?
	SXVII Pandora	$(55 \pm 8) \times (43 \pm 5) \times (33 \pm 5)$	—	—		0.5	ice?
	SXV Atlas	$(19 \pm 4) \times ? \times (14 \pm 4)$	—	—		0.5	ice?
	SXII Helene	$(18 \pm 3) \times ? \times (<15)$	—	—		0.6	ice?
	SXIV Calypso	$(15 \pm 3) \times (13 \pm 5) \times (8 \pm 3)$	—	—		0.9	ice?
	SXIII Telesto	$? \times (12 \pm 3) \times (11 \times 3)$	—	—		0.6	ice?
	Rings			$\sim 5 \times 10^{-8}$		0.2–0.6	dirty ice
Uranus	UIII Titania	800 ± 5	34.3	—	1.59 ± 0.09	0.23 ± 0.04	dirty ice
	UIV Oberon	775 ± 10	28.7	—	1.50 ± 0.10	0.20 ± 0.04	dirty ice
	UII Umbriel	595 ± 10	11.8	—	1.44 ± 0.28	0.16 ± 0.04	dirty ice
	UI Ariel	580 ± 5	14.4	—	1.65 ± 0.30	0.38 ± 0.07	dirty ice
	UV Miranda	242 ± 5	0.71	—	1.26 ± 0.39	0.22 ± 0.05	dirty ice
	1985U1	85 ± 5	—	—	—	0.06 ± 0.02	ice?
	1986U1	~40	—	—	—	~0.09	ice?
	1986U2	~40	—	—	—	~0.06	ice?
	1986U3	~30	—	—	—	~0.04	ice?
	1986U4	~30	—	—	—	~0.05	ice?
	1986U5	~30	—	—	—	~0.05	ice?
	1986U6	~30	—	—	—	~0.04	ice?
	1986U7	~25	—	—	—	~0.05	ice?
	1986U8	~25	—	—	—	~0.05	ice?
	1986U9	~25	—	—	—	~0.05	ice?
	Rings	—	—	10^{-9}–10^{-11}	—	~0.04	ice?
Neptune	Triton	1750 ± 250	1300 ± 250	$1.28\ (\pm 0.23) \times 10^{-3}$ (5)	—	~0.36	methane ice/nitrogen?
	Nereid	~200		$\sim 10^{-1}$			
Pluto	Charon	~500			(0.8)		methane ice

aMost of the references on which this table is based are given in the text. Other recent tabulations of physical properties of outer planet satellites are found in Morrison (1982b,c,1983) and Gehrels and Matthews (1984) as well as individual chapters in this book. Albedos are taken from Morrison (1983), Thomas et al. (chapter), plus Buratti and Veverka (1984). Surface compositions are general classifications of materials; see Morrison (1983) and chapters herein. Voyager 2 data for Uranus are preliminary (Smith et al. 1986; Tyler et al. 1986) jectories (Tyler et al. 1981,1982). The masses of Mimas and Enceladus, as specified by the Voyager flybys, are of lower accuracy than, and contradict, those derived from resonance theories (cf. Tyler et al. 1982b; Greenberg 1984c); we choose the latter. The masses of Janus and Epimetheus are calculable theoretically from mutual perturbations (Yoder and Synnott 1984; Peale's chapter) but current observations yield unlikely results. The classical motion; this value is questioned by Harris (1984b), who believes that it may be a large overestimate. Voyager 2 values are now included for the Uranian satellites.

Relative masses are evaluated by dividing the tabulated numbers by the planetary masses given in Table II. Values for the rings of Uranus (Elliot and Nicholson 1984) and Saturn (Cuzzi et al. 1984) are similar to those of small satellites ($R \lesssim 100$ km); the Jovian ring's mass is too poorly constrained to bother tabulating. The relative masses of the Uranian satellites are estimates based on the listed radii and a nominal density of 1300 kg m^{-3} (Dermott and Nicholson 1985); alternate masses have been derived by Veillet (1983a; cf. footnotes to Table III; chapter by Cruikshank and Brown). The relative mass of Charon assumes both it and Pluto have the same density and albedo.

Density values are computed from the tabulated radii and masses. Ideally, tidal distortions could also indicate mean density (Soter and Harris 1977; Morrison and Burns 1976; Thomas and Dermott 1985; chapter by Thomas et al.) or, if the density is known, any substantial central condensations (Dermott 1984c).

II. A GROSS CLASSIFICATION SCHEME

Based on their orbits and sizes (Tables III and IV), most known moons fall into three general classes: regular satellites, collisional shards, irregular satellites.

Regular Satellites

The regular satellites, which include virtually all the classical major satellites, comprise the first class; they form miniature solar systems about the three largest planets. Their sizes range from the largest satellites (Ganymede and Titan) to intermediate objects (most Saturnian and all Uranian moons). The orbits of these moons are evenly spaced about their planets, are nearly circular and lie close to their planets' equators; certainly in at least the Jovian case, the densities of these satellite members vary systematically with distance from their planets, much like that of the planets from the Sun. Although differences exist, the regular satellite systems are thought to originate through processes that have caused the solar system itself (chapters by Stevenson et al. and Safronov et al.), and are worth attention for this reason. They are complex bodies that in most ways are comparable to planets. In addition they allow us to survey satellites in a comparative manner, balancing the importance of distance from planet against distance from Sun. This group is composed of the four Galilean satellites of Jupiter, the eight classical satellites of Saturn (Mimas to Iapetus), and all five classically known Uranian satellites. These well-observed objects are described in this book in great detail.

Collisional Shards

Intermingled with the regular satellites are the second class, the collisional shards, most of which have been discovered since 1979, chiefly by Voyager (Aksnes [1985] summarizes their discovery circumstances and their celestial mechanics). These objects are tiny, craggy chunks that presumably are the remnants of once-larger satellites, battered and ground down by the on-going meteoroid flux (see Harris 1984a; chapter by Chapman and McKinnon). Often they are adjacent to their planet; probably this does not indicate that they originated as tidally disrupted fragments, rather that the meteoroid flux is more intense there. This debris may trace the processes whereby circumplanetary disks turn into those satellites extant today (chapter by Stevenson et al.). In the Saturnian system (Cruikshank et al. 1984b) these collisional end products are usually found in dynamically interesting orbits: the coorbital pair Janus-Epimetheus, the F ring shepherds, and the Lagrangian satellites of Tethys and Dione. At one end of the size scale (and the planetary proximity scale), they grade into ring particles. Indeed all four Jovian members (Amalthea, Adrastea, Thebe and Metis) are embedded in Jupiter's diaphanous ring system and at least two of these seem to sculpt its main ring's form (Burns et al. 1980,1984; Showalter et al. 1986). Atlas, lying within a faint ring, skirts the outer edge of Saturn's main A ring (Graps et al. 1984) but does not constrain the ring's diffusional expansion as originally thought (Cuzzi et al. 1984). There is good reason to believe that additional small satellites exist in the Saturn, Uranus and Neptune systems (see subsections below in Sec. IV). Ten such objects were found about Uranus by Voyager. The Galileo mission will probably discover more in the Jovian system.

At the larger end of the size scale the collisional shards grade into the smallest, closest regular satellites—Thebe/Amalthea, Mimas and Miranda; these four moons have comparatively the most irregular orbits amongst the regular satellites and lie across the size transition where objects disrupted by the collisional flux do not reaccumulate into spheres (see chapters by Thomas et al. and Chapman and McKinnon). In this group we place Phobos and Deimos by the process of elimination: they are too small to be considered regular satellites and dynamical evidence is accumulating that they have not been captured (Burns' chapter). The physical properties of these satellites, as well as those of the final class, are summarized in the chapter by Thomas et al.

Irregular Satellites

The irregular satellites have elongate, highly inclined (often retrograde) orbits that suggest capture of these small objects. They lie in the outer reaches

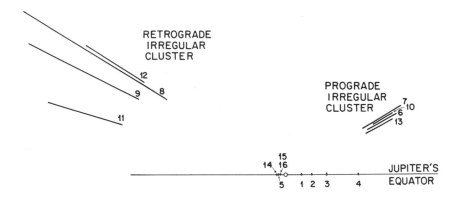

JUPITER'S SATELLITES

Fig. 1. A sketch of the Jovian system which best illustrates the orbital character of regular satellites, irregular satellites and collisional debris. Orbits are positioned according to inclination;
orbital eccentricities are indicated by showing apocenter and pericenter distances. Orbital radii
for the outer satellites are plotted at 25% of the scale used for the other satellites.

of their planet's gravitational grasp on orbits that can be strongly perturbed by
the Sun (cf. the chapter by Burns for a discussion of capture and the limiting
distance for stable orbits). Jupiter has two clusters of 4 members apiece in this
category; in fact, the Jovian system best illustrates the distinction between
regular and irregular satellites (see Fig. 1). The inner Jovian group are at
~ 160 R_J (Jovian radii) with $\langle e \rangle \sim 0.15$ and $\langle i \rangle \sim 28°$. The outer Jovian group
are at ~ 310 R_J with $\langle e \rangle \sim 0.25$ and $\langle i \rangle \sim 150°$. Fruitless searches by C. T.
Kowal for additional Jovian members require that fewer small satellites exist
than should be expected for collisional debris. Saturn's Phoebe is in this class
and Neptune's Nereid also clearly falls into this category.

Three satellites—which also happen to be (i) the largest satellites relative to their planets; (ii) those that are almost binary companions to their planets; and (iii) those that have undergone extensive tidal evolution—do not satisfy the criteria for any of the above three classes. The Earth's Moon is unique:
it is the only substantial satellite of a terrestrial planet; it has a distant, slightly
inclined and eccentric orbit; and it possesses a chemical makeup which distinguishes it from its primary (see the chapter by Kaula et al.). Its great distance
from Earth is thought to result from tidal evolution (Burns' chapter) and alone
would not rule out classifying this object as a regular satellite. It may have
originated in a singular event: great attention is currently being shown in the
possibility that the Earth's Moon may have formed from a circumplanetary
nebula that itself was produced out of the escaped ejecta of a nearly cata-

strophic terrestrial impact event (chapter by Stevenson et al.; Hartmann et al. 1986). Triton and Charon, the other two unclassifiable satellites, are distant and difficult observational targets. Yet it is known that these large satellites have methane ice on their surfaces and, like Titan, may have atmospheres containing nitrogen and methane (Cruikshank and Brown chapter); these compositions merely reflect the frigid temperatures in the outer solar system. Triton's retrograde orbit hints at an origin by capture (see Burns' chapter) but its large size means that such an origin would not have been easily accomplished (see chapter by Stevenson et al.). Charon-Pluto are better considered as a binary object rather than a satellite-parent pair; the dynamical evolution of this unique body is considered in detail in Peale's chapter.

III. AN OUTLINE OF SATELLITE PROCESSES

Following this introduction, the book proper starts with two chapters on satellite origin. These overviews are followed by eight chapters describing the various processes that determine the nature of the satellites we see today. Included in this set are a pair of chapters on dynamical evolution, one of which particularly investigates resonant orbits and unusual spin states while the other presents more general ideas of tidal evolution. The next chapter is a detailed description of thermal evolution. Processes that govern surface structure are considered in the succeeding three chapters. One concerns tectonics, another surface physics and the last magnetospheric interactions. The final two chapters in the processes section of the book review our knowledge of surface composition and cratering. Immediately below I highlight the contribution of these chapters to the book and this material to our current understanding of satellites.

The chapters concerning processes cover many of the same topics as ones in the previous volume *Planetary Satellites* (Burns 1977a), although orbital mechanics are de-emphasized herein and physics pertinent to surfaces is emphasized. Besides, the current chapters are much deeper treatments, reflecting the stimulation (and ground truth) provided to satellite science by the Voyager mission. Almost all of these chapters use the Earth's Moon, the satellite which has been visited and studied most intensively by solar system astronomers, as a touchstone for the models presented. Our difficulty in explaining various lunar features should alert the reader to treat all models with appropriate caution.

Origins

A fresh view of the ideas from physics and chemistry that are germane to the solar system's origin is presented in the chapter by Stevenson et al. Three modes of satellite origin (cf. Lissauer and Cuzzi 1985) are considered to be plausible: growth in particulate or gaseous circumplanetary disks which are a side product of the planet's accretion (the topic of a more quantitative treatment in the next chapter by Safronov et al.), formation from the debris launched following a giant impact, or capture aided by gas drag. Stevenson et al. argue that, until planetary formation itself is better understood, choosing the manner of any specific satellite's origin will be virtually impossible. Thus much of the review is concerned with planetary formation insofar as it impacts ideas on concomitant satellite formation. Then the authors invoke dynamical and cosmochemical arguments to guide their selection of possible origins for each of the satellite systems; however, they are not dogmatic in their choices. Stevenson et al. propose that the Earth's Moon formed by splash-off following an oblique impact (cf. Hartmann et al. 1986), Phobos and Deimos coformed from collisional debris which accumulated in orbit as Mars grew, the regular satellites and rings of Jupiter and Saturn developed in circumplanetary disks by accretion, the irregular satellites were captured and the collisional shards are leftover debris that has been shattered. A theme of the chapter is that more data are needed to constrain origin scenarios, and this is especially the case for the satellites of Uranus and Neptune and Pluto-Charon, where several possible origins are put forward. Throughout this chapter the authors remind us that a variety of processes contribute to produce the satellites which are extant today.

Safronov et al. in the following chapter thoroughly review the Soviet school of satellite cosmogony and concur with Stevenson et al. that natural satellites are an unavoidable consequence of planetary formation. They consider that circumplanetary swarms develop when heliocentric particles collide inelastically within a planet's gravitational sphere. Once a circumplanetary swarm forms, it is replenished through continuing collisions by additional heliocentric material. The swarm attempts to collapse onto the planet since mass but little angular momentum is added to it; this tendency is partially resisted by internal collisions which simultaneously try to spread the disk. Time scales for the swarm's processes, such as protosatellite growth, are very short compared to the swarm's lifetime. The appropriate physics is reasonably well understood for the solid-body swarms about the terrestrial planets, but gas-dust disks about the giant planets are just beginning to be investigated. Safronov et al. provide parameterized expressions for the relative velocities of particles in disks, the thickness of disks, and the accumulation or fragmentation of particles. These processes have many analogs in other celestial systems, e.g., the solar nebula, planetary rings and accretion disks.

Dynamical and Thermal Evolution

The evolution of satellite orbits due to tides is considered by Burns, who motivates the appropriate equations. These equations have been numerically integrated to track the possible past histories of the Earth's Moon and the Martian satellites. The current outward evolution of the Moon is so rapid as to require that the Moon would have been at the Earth's surface only ~ 1.5 Gyr ago. However, today's rate is shown to be anomalously high and to result from a near-resonance between the Moon's principal forcing term and the primary natural frequencies of the Earth's ocean basins, where most dissipation occurs. This indicates that lunar orbital history calculations, developed from today's conditions, are misleading guides as to the Moon's starting point (see chapters by Kaula et al. and Stevenson et al.; Boss and Peale 1986). Tides presently produce significant decreases in Phobos' orbital size and eccentricity, but scarcely affect Deimos' orbit. While integrations of Phobos' orbit into the past suggest a captured object, likely these calculations too are deceptive. This happens because recent passages by Phobos through orbital resonances have the capacity to alter its past orbit dramatically. Hence, as in the case of the Moon, orbital histories based on straightforward integrations of current conditions give wrong answers. The best guess today from dynamics is that the Martian satellites were formed *in situ*. Burns also addresses, but to a lesser extent, some dynamical evolution problems in the outer solar system: the development of resonances, the tidal evolution of the Pluto-Charon system and other synchronously locked pairs, secular perturbations of the Uranian satellite orbits, and the orbital history of retrograde Triton. The first two of these topics are considered in much greater mathematical detail by Peale. The final points made in Burns' chapter are that satellites can only be captured if energy is lost and that all extant satellite orbits are stable with respect to escape.

The consequences of tidal dissipation, especially in regard to resonances, are further addressed in Peale's chapter. Answers to important problems in satellite dynamics are presented. Peale uses a graphical scheme to demonstrate that orbital resonances can originate when the orbits of a pair of satellites evolve tidally at different rates until a commensurability is achieved. This two-body resonance solution is applied to describe how the three-body Laplace resonance among Io, Europa and Ganymede might have originated and evolved. This resonance plays a crucial role in the continuation of Io's tidal heating (see chapters by Nash et al. and by Schubert et al.) and the theory provides a link between Io's heat output and the rate of Jupiter's tidal dissipation. The motions of Janus and Epimetheus, Saturn's coorbital satellites (see chapter by Thomas et al.), are easily explicated and the relative horseshoe orbits of the pair are found to be currently very stable. Another unique satellite system considered by Peale is that of Pluto-Charon, whose mutually synchronously locked configuration is determined to be the natural outcome of tidal evolution, given the relatively large mass ratio of the pair (see Burns' chapter). The chapter closes with a portrayal of Hyperion's spin evolution where it is proven that Hyperion's highly irregular shape, when combined with its eccentric orbit, prevents the satellite from settling into synchronous rotation. This satellite is destined to tumble chaotically forever, a situation entirely consistent with current observations (Thomas and Veverka 1985).

Thermal histories, compositions and interior models are appraised by

Schubert et al. in a comprehensive review. The authors sketch the manifold thermal evolution paths that satellites may undergo, depending on their composition, heat sources and heat transfer mechanisms. Satellites at their birth are somewhat warm due to the gravitational energy released during their accretion; subsequent heating is caused by differentiation, radioactive decay, which is proportional to silicate content, and tidal heating, which is manifestly important for Io and may be (or have been) significant for Europa, Earth's Moon, Triton and Enceladus. Heat transport occurs by conduction, subsolidus convection and magma migration. Thermal histories are closely tied to interior models; both therefore become complicated and questionable for the ice-rock composition common in the outer solar system due to uncertainties in the phase structure and rheological properties of ice and ice-rock mixtures. Quite different evolutionary paths can arise depending upon whether, and how, the satellite is differentiated. Schubert et al. describe the expected interior structures and thermal histories for most satellites. Small bodies are undifferentiated, cold and dormant. Large satellites have very complex interior histories and may be still evolving thermally today. Clear evidence of extensive endogenic activity on many intermediate-sized satellites, notably Enceladus, indicates that some of them too are complex thermal entities.

Surface Processes

The icy satellites of Jupiter and Saturn display a surface architecture quite different from that observed on the rocky bodies of the inner solar system. Squyres and Croft start their chapter by describing the tectonic features seen on the three outer Galilean satellites (see also the Europa chapter by Malin and Pieri, and the Ganymede-Callisto chapter by McKinnon and Parmentier) and the Saturnian satellites (see chapter by Morrison et al.). Of these, Ganymede has the most intricate geology with its dark, hummocky, heavily cratered terrain cut through by a younger, bright terrain characterized by grooves. Additional interesting surface detail seen on other satellites is also given in this chapter. Stresses are generated on icy satellites in several ways. Global expansion, caused by thermal evolution, differentiation, phase changes or silicate dehydration, may account for much of the generally observed extensional tectonics. Global shell deformations, due to despinning and tidal distortion, may generate large-scale structures. Solid-state convection, by itself, probably does not produce any features but may contribute to local stresses. Impacts generally cause stress patterns which are localized, but which may occasionally extend as large as hemispheric bull's-eye basins like Callisto's Valhalla. These stresses find their tectonic expressions as faults, fractures, furrows and grooves. Squyres and Croft point out that most icy satellites have been resurfaced in part through water volcanism and/or diapirism (see Kirk and Stevenson 1986); their surfaces have also been modified through viscous relaxation (see the chapter by Chapman and McKinnon) and by mass wasting.

Processes that govern local surface structure are considered in the chapter by Veverka et al., who discuss the principles behind, and the results from, various remote-sensing techniques. Such methods include photometry, polarimetry and radiometry, which are the cornerstones of observational planetary astronomy, whether done from a mountaintop or from a scan platform aboard a spacecraft. Prior to presenting these subjects, the chapter provides a new discussion of regolith formation on satellites. In particular it shows how impact-generated regoliths should vary with surface composition (rocky or icy), local gravity (small or large moon), and proximity to planet (near or well outside the planet's Roche limit). A major segment of this chapter reviews the remote-sensing data available about satellites and indicates the physical characteristics which can be inferred from photometric and radiometric data. Because of limited *in situ* experience with icy materials, further laboratory work and modeling will be needed to delineate satellite surfaces fully. The final section of the chapter by Veverka et al. addresses the observed distributions of albedo and color across satellite surfaces; it concludes that no single mechanism can explain the global patterns seen on all the satellites of Jupiter and Saturn.

In the following chapter Cheng et al. lay out the interactions of planetary magnetospheres with satellite surfaces and rings. These interactions alter surfaces, generate and/or modify satellite atmospheres, and influence magnetospheric properties. Several other recent reviews on this topic, pertaining specifically to the Jovian examples, are published in Morrison (1982c) and Dessler (1983). The chapter here starts by recalling laboratory experiments in which energetic particles (protons and heavy ions) are accelerated through a near vacuum into various ice films (H_2O, SO_2, CH_4, CO_2) at low temperatures; these experiments quantify sputtering yields and delineate any structural or compositional changes. Using these measurements as applied to the icy satellites of Jupiter, Saturn and Uranus, the authors describe likely effects on erosion, albedo and chemical variations, and the production as well as maintenance of tenuous atmospheres. They argue that sputtering may have played an especially important role in darkening the Uranian rings and moons. A large portion of the chapter details the influences of the Jovian magnetosphere on Io and the pivotal part Io plays in determining Jovian magnetospheric composition and processes; while the connection between Io and the magnetosphere is undeniably crucial, its exact nature remains controversial. The Saturnian system presents two unique interactions. First, Saturn's main rings and E ring supply and modify the magnetospheric plasma; Saturnian satellite surfaces are altered just as in the Jovian case and, in addition, the main rings have an atmosphere of unknown cause, which might be connected to magnetospheric processes. Second, atomic nitrogen and hydrogen leak out of Titan's massive atmosphere into the Saturnian magnetosphere to populate toroids surrounding the planet; this process has many similarities to mechanisms suggested for restocking the clouds of Ionian material found in the Jovian magnetosphere.

Composition and Cratering

Satellite compositions, as determined mainly by reflectance spectroscopy and secondarily from thermal and morphological observations, concern Clark et al. in their chapter. Only for the Moon, whose samples are available, can an astronomical identification by remote sensing be tested; in this case, the lunar composition of basaltic and anorthositic rocks and glasses, as determined in terrestrial laboratories, is confirmed. In order to permit the interpretation of the spectra of icy satellites, the authors plot the reflectance spectra of expected minerals and ices (H_2O, CO_2 and CH_4) as well as how water ice spectra are affected by various contaminants. From diagrams such as these and the presence of S and O in the inner Jovian magnetosphere, the surface of Io is thought to be covered by allotropes of sulfur and SO_2; however these identifications are not without their problems, which Clark et al. and Nash et al. comment upon. The other Galilean satellites have surfaces predominantly of water ice; Callisto, the darkest, still contains $\gtrsim 90\%$ water ice although it may have some silicate or carbonaceous areas. Saturn's rings and the surfaces of most Saturnian satellites, like those of Europa and Ganymede, are nearly pure water ice. The leading side of Iapetus seems to be coated with a bland, dark red organic compound similar to substances found in carbonaceous chondrites and suspected on asteroids in the outer solar system. Mixtures of water ice also containing various amounts of this impurity may account for the spectra of Hyperion and the Uranian satellites. Methane ice as well as liquid and/or solid nitrogen covers Triton. Pluto-Charon seems similar to Triton but has more methane. The meager evidence available, generally broadband photometric observations, suggests that the surfaces of the small satellites of Mars, Jupiter and Saturn are, respectively, composed of carbonaceous, rocky and icy materials. These compositions are listed in my Table IV.

The cratering of planetary satellites is described by Chapman and McKinnon in the final chapter about satellite processes. Substantial progress has occurred since the late 1970s in understanding the two main aspects of this topic. The first part of this encyclopedic chapter reviews cratering mechanics and introduces nondimensional scaling parameters from which crater sizes and ejecta patterns can be estimated. This subject has advanced markedly in the last few years and is applied here especially to cratering of icy surfaces, a topic which is just receiving attention. The second half of the chapter concerns crater statistics; the Voyager observations of crater distributions at 5.2 and 9.6 AU have greatly extended the coverage of impact features to a region where cometary impacts might be expected to dominate. This section considers whether saturation cratering has occurred and concludes that probably it has. The authors illustrate how to convolve the physics of the chapter's first half with the observed crater populations to ascertain true production functions. The goal of this portion is to determine whether the same population of projectiles strike the terrestrial bodies as have collided with the icy moons in the outer solar system; if such were the case, cratering chronologies could be linked and events on outer solar system bodies could be tied to the established lunar time scale. Chapman and McKinnon conclude that 4, and possibly 5, distinct populations are needed to explain the observed crater statistics; these include cometary and asteroidal projectiles plus a planetocentric component for Saturn. Accordingly these authors maintain that an absolute interplanetary correlation of geologic time cannot yet be inferred.

IV. THE SATELLITES

The chapters in the latter half of the book detail mankind's knowledge about individual satellites or types of satellites. In contrast to the chapters in the book's first half, which present the general concepts of a particular discipline of planetary science and then apply them to all the satellites, the chapters of the second portion describe the known attributes of a satellite (or class of satellites) as a way to paint a complete picture of the object(s). As an introduction to these chapters, I now provide some very general background material on each family of satellites. Usually I will first describe a satellite system in broad terms. I will then recall the circumstances of discovery for known satellites, as well as their namings, and then I will record some unproductive attempts to locate additional satellites about their planets. As appropriate, I will mention any spacecraft exploration planned for these systems. I will also point the reader to places where these satellites are discussed in this book as well as to other reference sources that merit attention.

Moon

Earth's companion falls into a category by itself, due to its visibility from our globe and its direct exploration by humans and their machines. It is the first satellite, both in distance from the Sun and in the subjective eyes of man. However, compared to other satellites, the Moon is not especially remarkable although it presents the second largest mass ratio relative to its planet (Table III). In part, this mass accounts for the substantial tidal evolution that its orbit has undergone (Burns' chapter). The lunar orbit currently has a modest eccentricity and inclination, which are believed to give clues about the Moon's origin. The question of the lunar origin (chapters by Stevenson et al., by Burns and by Kaula et al.) is a matter of some controversy and of active research (Hartmann et al. 1986); the currently popular idea, that the Moon originated following the nearly catastrophic collision of a massive protoplanet into Earth, seems to offend the fewest parties, but it still has not yet been critically examined.

The Moon has, at most, a small, possibly molten core (see chapter by Schubert et al.) and a low-density crust, which formed by differentiation ~ 4.4 Gyr ago and which suffered major melting some time later. The lunar surface is extensively cratered with the heavy bombardment ending about 4.0 Gyr ago; the maria (lava flows which are principally found on the Moon's Earth-facing side and whose production essentially ceased about 2.5 Gyr ago) are relatively lightly cratered (chapter by Chapman and McKinnon). Compared to the Earth, the Moon is enriched in refractory siderophiles and in mafic components but depleted in iron and volatiles; its oxidation levels and silica content also seem low in comparison to terrestrial rocks. Cosmochemists are continuing to probe the mysteries contained in the samples returned by the Apollo and Luna missions.

Volumes have been written about the Moon's geology, physics and mineralogy, and we only cite some of the most recent books: Fiedler (1971), Frondel (1975), Guest and Greeley (1977), Schultz (1976), Mutch (1970), Taylor (1975,1982) and Cadogan (1981). The yearly progress made in lunar science is displayed in the annual abstract volumes and the *Proceedings of the Lunar and Planetary Science Conferences*, once published in *Geochimica and Cosmochimica Acta* and now part of the red *Journal of Geophysical Research*. Review articles about the Moon were common in the 1970s, and were carried especially by the *Reviews of Geophysics and Space Physics*. A recent survey of the Moon's geology is that by Wilhelms (1984*b*). Texts on planetary science (see, e.g., Glass 1982; Murray et al. 1981; Hartmann 1983) generally devote a large fraction of their space to the topic.

In this book, many chapters in the book's first half use the Moon to calibrate the observations/theories of other satellites; the Earth's companion is the standard by which the other satellites are measured. These chapters include the Moon's orbital evolution studied by Burns, its thermal history discussed by Schubert et al., its surface properties (especially regolith development) described by Veverka et al., its surface composition portrayed by Clark et al. and its cratering history given by Chapman and McKinnon. A chapter specifically summarizing the Moon's characteristics is given by Kaula et al. They have a difficult task, one not often faced by satellite enthusiasts, namely, to list the salient properties of an object about which so much is known. They first describe lunar geology, where the primary shaper is cratering although volcanism and tectonism also play roles. They then consider lunar petrology and geochemistry so as to try to place constraints on the Moon's bulk composition. The principal lunar elemental constituents are similar to the Earth's mantle plus crust, with the main distinction being the Moon's deficiency of volatiles. The internal structure of the Moon, as determined by seismology and restricted by thermal history, is quite simple: a megaregolith (the outcome of repeated impact) overlies a thin crustal layer (generated long ago by differentiation) which sits atop the mantle. The mantle gradually softens at depth to form the central asthenosphere; a small iron core is possible. All these disciplines help limit the Moon's origin, which these authors conclude must be due to some *ad hoc* mechanism. They caution that the Moon, probably the best understood object in the solar system by virtue of its simplicity and the detailed information available about it, unhappily may be a poor guide to the other satellites.

One does not ask which "astronomer" discovered the Moon. Other natural Earth satellites have, however, been sought unsuccessfully, with the primary site studied being the triangular libration points of the Earth-Moon system (see, e.g., Valdes and Freitas 1983). Runcorn (1983) nevertheless has long maintained that several lines of evidence point to the past existence of lunar satellites; Tombaugh et al. (1959) achieved a limiting magnitude of 13-14 for most of the lunar satellite environment. Almost certainly today, the largest objects orbiting Earth, other than the Moon, are man-made. The Soviet, American and European space programs are all studying possible lunar missions in the 1990s, but none of these will search for moons of the Earth or the Moon.

Phobos and Deimos

Mars is orbited by two diminutive, very dark, roughly triaxial satellites with heavily pockmarked surfaces. Both objects move on nearly circular, slightly inclined orbits close to Mars. Deimos is just beyond synchronous orbit while Phobos lies within the Roche limit for Mars. Much of Phobos is covered with long, linear grooves (see Map Section) whose cause is not well understood, whereas Deimos appears much smoother. Spectral observations and the low measured densities of ~ 2000 kg m^{-3} suggest that both satellites are carbonaceous and therefore from the outer solar system. This has led to the speculation that the objects are captured but, as Burns' chapter points out, such an origin is unlikely.

A. Hall of the U.S. Naval Observatory, with prodding from his wife C. A. Stickney Hall, discovered both Martian moons during a week in August 1877. He called the innermost, Phobos (fear), and its companion, Deimos (panic), after attendants of the war god Mars. Searches for additional Martian satellites have been carried out from Earth by Kuiper (1961) who placed an upper limit of 0.75 km for the radius of any object with an assumed albedo of 0.05 outside the orbit of Phobos. Studies of 19 pre-insertion Mariner 9 images were also unsuccessful (Pollack et al. 1973*b*) as were surveys made by Viking Orbiter-1 (Duxbury et al. 1982); the latter would have found any objects 500 m or larger with albedo of 0.05 or brighter inside the orbit of Phobos and within 350 km of the planet's equatorial plane. Hubble Space Telescope (HST) may be able to contribute to this search.

The dynamical characteristics of the Martian satellites were reviewed by Burns (1972); more recently, several papers in a special issue of *Vistas in Astronomy* (22, part 2, 1978) devoted to the Martian satellites on the centennial of their discovery were concerned with their celestial mechanics and dynami cal characteristics. Reviews emphasizing spacecraft observations were contained in the same volume; other surveys are by Pollack (1977) and Veverka and Thomas (1979). Veverka and Burns (1980) and Zharkov et al. (1984) give more up-to-date overviews. There is no specific treatment of these objects in this book, but the chapter by Thomas et al. describe them in some detail and Burns devotes space to their orbital evolution; other "processes" chapters cover them in passing. The Soviet Union is planning a sophisticated rendezvous mission, including geochemical analysis, in 1989.

The Jovian Satellites

Credit for discovering the Galilean satellites is given to Galileo who in 1610 first pointed his primitive telescope at Jupiter and thereby helped overturn the predominant heliocentric theory. Simon Marius found these satellites at almost the same time, and there may even be truth to the claim that Chinese astronomers made naked-eye sightings of these objects in the fourth century B.C. (Pang 1983). The discovery in 1974 of JXIII Leda, a 20th magnitude object, with the 122 cm Schmidt telescope at Palomar by C. Kowal was the last of the Jovian satellites made from the ground; a 21st magnitude object located in 1975 has not been since recovered (Kowal 1975). Searches by Voyager in the environs of Jupiter's rings have been more profitable with three detections to date; Metis was originally sighted through the shadow it cast on Jupiter's atmosphere and has been seen only in transit across Jupiter's disk (see references in Burns et al. 1984).

The 16 satellites of Jupiter come in three distinct sets. Besides the 4 well-studied Galilean moons (Io, Europa, Ganymede and Callisto), a quartet of craggy collisional shards orbit within the region of Jupiter's faint ring system and eight irregular satellites—2 clusters of 4 small satellites apiece—loop around the giant planet at great distances. As described in the previous section, the small ring-moons are likely to be remnants of once-larger objects. The irregular satellites are thought to have been captured, either through a collision (or two) near the outer limits of Jupiter's sphere of influence (see chapter by Stevenson et al.) or by gas drag (Burns' chapter).

The Galilean satellites are named for illicit lovers of Zeus (Jupiter), as are all the irregular Jovian satellites. The small exterior moons on prograde orbits have names ending in a, whereas those in retrograde orbits terminate in e. Adreastea was a Cretan nymph who nursed the baby Zeus and also fed him on the milk of Amalthea, either a nymph herself or a she-goat. Metis, Zeus' first wife, was swallowed by her husband. Thebe was a Boeotian nymph who was married to Zethus. The naming of features on the satellites, as given in the Map Section, is described by Davies (1982).

The Galilean satellites come in two pairs: (i) Io and Europa—the inner, silicate-rich lunar-sized satellites ($\rho \sim 3500$ kg m^{-3}) plus (ii) Ganymede and Callisto—the outer, ice-rock giant satellites ($\rho \sim 1900$ kg m^{-3}). This dichotomy in density, the orbital regularity possessed by the system, and the complexity of the individual satellites has caused the Galilean satellite family to be termed a miniature solar system (cf. chapter by Stevenson et al. which presents a revisionist view). Three chapters in this book are devoted exclusively to the Galilean moons which are also displayed in the Map and Color Sections. These chapters follow upon the fine reviews contained in *Satellites of Jupiter* (Morrison 1982*c*), but now five additional years have elapsed in which the Voyager data have been digested further. Since another satellite system has been surveyed, the treatment here is able to place these diverse and intriguing objects in a broader context. It is interesting that, with the approach of the Galileo mission, scientific attention is returning to the Galilean satellites after being given over to the Saturnian and Uranian moons.

The odd satellite Io is characterized by Nash et al. in their chapter. It is covered by the most youthful, bright and variegated surface of all the Jovian satellites. Io's volcanism was the surprise of the Voyager mission; at the time of the Jovian encounters, nine volcanoes falling into two classes were active. As all planetary scientists know, on the eve of the Voyager 1 encounter Peale et al. (1978) predicted that tides furnish significant heat to Io's thermal budget (cf. the chapter by Schubert et al. who modify the accepted tidal heating paradigm somewhat). Long-term heating develops through the continual flexing of Io's shape, which occurs because Io's orbit is forced to be eccentric through the Laplace resonance that links the motions of Io, Europa and Ganymede; the dynamical history of this resonance is presented in full mathematical detail in Peale's chapter and to a much lesser extent in the chapters by Burns and by Nash et al. The complex, coupled dynamical-thermal history is described in the chapters by Schubert et al. and Nash et al. Most scientists today believe that the presently observed heat flow is more than tidal heating can generate and comes from a few localized hot spots associated with volcanic calderas. Major, unresolved issues include whether the heating is episodic and whether silicates or sulfur drive the volcanism. Given these uncertainties Schubert et al. argue in their chapter that scenarios exist whereby the apparent mismatch between observed flow and supply can be explained. Io's surface geology is dominated by volcanic features which produce three kinds of surface structures (vents, plains and mountains). However, silicates must also be present in the surface in order to support some of the observed surface relief. The pastel hues which colorfully paint the satellite's surface apparently come from a palette of sulfur allotropes with SO_2 frost providing a white canvas, especially in equatorial regions on Io's leading hemisphere (Clark et al. chapter).

The bizarre inner Galilean moon also significantly interacts in a variable way with the Jovian magnetosphere (chapters by Cheng et al. and Nash et al.). The Io plasma torus, mostly ionized S and O, corotates with the Jovian magnetic field; it lies about Jupiter's centrifugal equator and is centered around Io's radius. Neutral clouds of sodium and potassium extend partially along Io's orbital path. The mechanism(s) whereby Io supplies the torus and neutral cloud are uncertain. While apparently providing most heavy components of the Jovian magnetospheric plasma, Io (and its compatriot moons) simultaneously absorb, and are coated by, the same species. Through such a process the Jovian ring, the four small inner moons and the trailing side of Europa are probably stained with sulfur compounds (see chapters by Clark et al. and Cheng et al.). Io also somehow helps trigger Jupiter's sporadic and powerful decametric radiation.

Europa is a silicate object encased in a thin ice sheet; a narrow layer of liquid water may be sandwiched between the two. The chapter by Malin and Pieri about this lunar-sized satellite emphasizes the youthfulness of its surface as indicated by the near absence of craters. Few landforms, besides long, arcuate lineaments which crisscross the satellite's surface, are visible, presumably because only poor resolution spacecraft images are available. Tectonic explanations for the cracked planet-wide lineaments involving either tidal stresses or global expansion of the satellite are preferred in this chapter as well as in that by Squyres and Croft. The ice sheathing of Europa restricts possible compositions and thermal histories (chapters by Malin and Pieri and by Schubert et al.). As with all the Galilean satellites, sputtering of the surface by magnetospheric plasma may be important in developing as well as in modifying the surface texture and makeup (see chapters by Veverka et al. and Cheng et al.).

Ganymede and Callisto, on the basis of their similar size and density, are two satellites that prior to Voyager were suspected to be twins. As demonstrated by the chapter by McKinnon and Parmentier, clearly they are not (see Map Section). Callisto, a dust-mantled ball of rock and ice, is probably the most heavily cratered object in the solar system today: everywhere its surface is covered by impact features extending up to globe-girdling multiringed impact structures (see chapter by Chapman and McKinnon). Ganymede, the largest satellite in the solar system, displays incredible tectonic variety with some ancient and some youthful surfaces (see chapter by Squyres and Croft). The chapter by McKinnon and Parmentier focuses on the dichotomy between these neighboring giants and uses it to explore various unresolved questions concerning the regolith, surface geology, interior structure and tectonics. A major issue is the relative importance of thermal history and internal differentiation versus conditions of origin in determining the surfaces seen today (see also chapters by Stevenson et al. and Schubert et al.). While McKinnon and Parmentier come to no certain conclusions as to why Ganymede and Callisto are so different, they outline a number of plausible scenarios whereby the final dissimilar states of the two ice-rock Galilean satellites could be achieved. These satellites provide extremely valuable laboratories in which to test the ruling paradigms of tectonism, volcanism and impact cratering developed from the study of the terrestrial planets—comparative planetology at its best.

The primary reference for these objects, *Satellites of Jupiter* (Morrison 1982c), was written only a dozen or so months after the startling findings of the Voyager flybys. Thus it is not the usual compilation of tried and true reviews but rather contains primary source material from the mission as well as the planetary community's first attempts at synthesis. Other original sources are the initial reports in *Science* (204:945–1006, 1979; 206:925–995, 1979) and *Nature* (280:725–806, 1979) plus special issues of the *Journal of Geophysical Research* (86:8123–8841, 1981) and *Icarus* (40:225–547, 1980). Post-Voyager review articles are by Morrison (1982c,1983) while the status of pre-Voyager science was examined by Morrison and Burns (1976) and Johnson (1978). The interactions of the Jovian satellites, particularly Io, with the magnetosphere are described at length in Dessler (1983) as well as in several chapters concerning the Io torus in Morrison (1982c). Io's atmosphere was earlier surveyed by Trafton (1981).

The summaries of the Galilean satellites presented here will be valuable stepping-stones between Voyager reconnaissance flybys and the long-lived Galileo mission. The latter is due for launch in the late 1980s; thirty months later it will start through the system on a two-year trek which is scheduled to include at least a dozen close flybys of the Galilean satellites.

Satellites of Saturn

The number of catalogued Saturnian satellites nearly doubled in 1980 when Voyager discovered the three inner satellites and Earth-based observers found the coorbital pair (separately spied by Voyager) as well as the three Lagrangian satellites. Groundbased observations were successful at this time because Saturn's rings were nearly edge-on to Earth and so they scattered little sunlight. At the time of the previous ring plane crossing in the mid 1960s, the E ring was seen clearly for the first time and at least one satellite was suspected to be present at the coorbital location. All these eight latest additions are what we call collisional shards; they join eight regular satellites and retrograde Phoebe.

All of the regular Saturnian satellites (see chapter by Morrison et al. and the Map Section) but Titan are objects with ice/rock interiors and icy surfaces; variable crater densities across the surfaces of these modest-sized moons indicate surprising endogenic activity, some of it quite recent. The regular satellites are tied together by numerous resonances which account for most orbital eccentricities. Saturn's system is dominated by Titan, the first moon discovered after the Galilean satellites and the only satellite with a very substantial atmosphere (Hunten et al. 1984). The inner satellites induce most of the understood structure in the ring system.

The discoverers of the classical Saturnian satellites included many astronomical giants—C. Huygens, J. D. Cassini, W. Herschel and W. H. Pickering—who were perhaps drawn to Saturn's environs by its exquisite rings. The rings themselves (see Greenberg and Brahic 1984) are in a sense a collection of small satellites, but they are not treated in this book except by Clark et al., who review their spectrophotometry, and to a lesser extent by

Peale and by Burns, who investigate their dynamical connection to the inner Saturnian satellites.

A 80 to 90% complete search of Saturn's environs down to 23rd magnitude has located no other satellites (D. P. Cruikshank, personal communication, 1985); Phoebe alone seems to be at a great distance from Saturn. Nevertheless, other constituents are suspected in Saturn's retinue. Almost certainly a small undetected satellite pries Encke's division open (Cuzzi and Scargle 1985; Showalter et al. 1985*a*) and at least one other may be present in the Cassini division (Marouf et al. 1985). Charged particle absorptions detected by Pioneer 11 near the F ring suggest entities there (Burns and Cuzzi 1985) and similar clumpy material or a satellite appears to lie in Mimas' orbit (Chenette and Stone 1983). Several blips on Voyager images (for which unique orbits could not be determined) hint that other collisional shards may circle the planet (S. P. Synnott, personal communication, 1985). Among the latter, a particularly strong case can be made for a Lagrangian satellite of Enceladus (Synnott 1986).

Most of the names of the classical satellites were suggested by J. Herschel at the start of the nineteenth century. In Greek mythology a Titan was one of a family of giants born to Uranus and Gaea; Tethys, Dione, Rhea and Phoebe were Titans and sisters of Saturn, while Hyperion and Iapetus were his brothers. Mimas was a giant, and Enceladus a giant or a Titan. Janus is the two-faced ancient diety of all beginnings. Atlas, Prometheus and Epimetheus are children of Iapetus and Clymene. Pandora, the wife of Epimetheus, was sent to Earth (along with her infamous box) to bring misery to the human race. Telesto, Calypso, Helene and Electra are among the three thousand nymph daughters of Oceanus and Tethys.

Several review chapters in the book *Saturn* (Gehrels and Matthews 1984) are devoted to the satellites of this planet. Greenberg (1984*c*) discusses the numerous orbital resonances, while Morrison et al. (1984) and Cruikshank et al. (1984*b*), respectively, consider the geological and optical properties of all Saturn's satellites. In the same book Titan (Hunten et al. 1984) and its magnetospheric interaction (Neubauer et al. 1984) receive the depth of treatment they deserve. The production of organic matter, so important a component of Titan's atmosphere, is described by Sagan et al. (1984). Pre-Voyager discussions of Saturn's satellites include a review by Cruikshank (1979), a series of papers in a NASA conference publication (Hunten and Morrison 1978) and a consideration of Titan's atmosphere (Trafton 1981). The primary source materials from the Voyager flybys through the Saturn system are issues of *Science* (212:159–243, 1981; 215:499–594, 1982) and *Nature* (292:675–755, 1981). Three special issues (*Icarus* 53:163–387, 1983; *Icarus* 54:159–360, 1983; and *JGR* 88:8625–9018, 1983) are given over to Saturn, its rings and satellites.

In view of the recent reviews on Saturn's satellites, only a single chapter of this book specifically addresses the Saturn system. This chapter by

Morrison et al. is based in part upon Morrison et al. (1984) and Owen (1982) and describes all the satellites, including Titan; it is especially valuable for its comparative aspects, i.e., how the Saturnian satellites fit into the general framework of other satellite systems.

Observations of the Saturnian system have significantly influenced thinking about satellites in general, and this is reflected in the coverage of these objects in the book's first half. For example, the distinct cratering record left on the surfaces of the Saturnian satellites shows that they must have been peppered by debris from one of their extinct compatriots. Combined with the craters seen on the Galilean satellites and terrestrial planets, the Saturnian craters require four and probably five distinct populations of projectiles (chapter by Chapman and McKinnon). The fact that small satellites, such as Enceladus, exhibit fresh surfaces and global tectonics has mandated a rethinking of heat sources and thermal histories (chapters by Stevenson et al., Squyres and Croft, and Schubert et al.). The chemical composition of Titan's atmosphere severely restricts modes of origin (chapters by Stevenson et al. and Morrison et al.). The presence of the E ring and its apparent association with Enceladus demonstrates the interaction of a magnetosphere with a satellite system (Burns et al. 1984; chapter by Cheng et al.). The nature (Veverka et al. chapter) and composition (Clark et al. chapter) of the icy surfaces of Saturn's moons provide valuable counterpoints to those of Ganymede and Callisto. The comparable-sized Uranian moons will furnish additional perspective; at present, the Uranian surfaces seem to be like Hyperion's. The Saturn system is a planetary dynamicist's delight with its ring's architecture (see Greenberg and Brahic 1984), numerous resonances (including Lagrangian satellites, the coorbital duo and the more mundane orbit-orbit satellite resonances; Peale's chapter), and Hyperion's spin, the largest-scale chaotic motion demonstrated in the Universe (Peale's chapter).

According to the chapter by Morrison et al., Titan with its massive and complex atmosphere, enigmatic Enceladus and ying-yang Iapetus are particularly notable amongst Saturn's progeny. As described also by Hunten et al. (1984), Titan possesses a predominantly nitrogen atmosphere denser than Earth's. More than a dozen organic species have been identified in Titan's spectrum; most of them have been produced by photochemical reactions in the atmosphere. Titan is suspected to be covered by a global ocean of ethane, intermixed with a sludge of more complex hydrocarbons. Enceladus shows the clearest evidence of recent endogenic activity. Even though presently far from adequate, tidal heating of an interior containing a low-density water-ammonia eutectic may be able to generate surface features. To a lesser degree, the other intermediate-sized satellites also indicate endogenic activity. All of them have low densities, and consist mainly of H_2O ice with 30 to 40% rock mixed in; they have quite pure water ice surfaces (Clark et al. chapter). Iapetus, with its black carbonaceous leading hemisphere and icy trailing hemisphere, provides ample proof that exogenic processes also affect the surfaces of the Saturnian satellites (cf. Mignard et al. 1986). The other mid-sized satellites also display leading/trailing asymmetries in photometric properties but are much less exaggerated. The orbital characteristics of the numerous small satellites of Saturn are limned in the section above on a classification scheme and in the chapter by Thomas et al.

The only spacecraft mission to the vicinity of Saturn currently being discussed is the joint U.S.-European Cassini mission, which in the mid-1990s would send a probe into Titan's atmosphere and simultaneously place an orbiter about Saturn to view its rings and satellites.

The Satellites of Uranus, Neptune and Pluto

More than a century after Cassini reported sighting his final Saturnian moon, W. Herschel in 1787 discovered the next satellites, the outermost pair around Uranus, the planet he had found six years earlier. Two more were located by Lassell in 1851. Miranda, hidden in the glare of the planet, escaped detection until Kuiper found it in 1948. These last two workers extended their searches to the next planet out, finding Triton in 1846, immediately after Neptune itself was discerned, and Nereid in 1949. J. Christy of the U.S. Naval Observatory noted that Pluto's image was not perfectly circular and that the bump moved with the planet's spin period; his claim of a satellite, borne out in subsequent CCD and speckle images, has received ample confirmation by the series of eclipses/occultations starting in early 1985 (Binzel et al. 1985a).

Searches for additional satellites about these three planets have been unsuccessful. Sinton (1972) looked for satellites interior to Miranda and as close as the Roche limit, using a methane filter (0.89 μm) to reduce the planet's brightness, and saw no satellite brighter than magnitude 17. Infrared CCD images from the rings out to Ariel fail to find any objects that are more than 1/2 Miranda's brightness (W. Forrest and P. D. Nicholson, personal communication, 1985); a similar survey of Neptune's neighborhood also was unsuccessful. Smith (1984) tried imaging in the deep methane absorption band at 0.89 μm with a CCD mounted on a coronagraph; the search reaches mag $\sim$ 23 well away from Uranus and Neptune but is less sensitive to the planet. The limiting radius for satellites of Uranus (Neptune) is $\sim$ 10 km ($\sim$ 40 km) at distances greater than 15 planetary radii increasing to $\sim$ 100 km ($\sim$ 2000 km) at $\sim$ 3 R_p. Cruikshank (see his chapter with Brown) has sought satellites exterior to Oberon down to 23rd mag without success.

Nevertheless theoreticians (Goldreich and Tremaine 1979) are confident that small shepherd satellites ($R \sim 10$ km) push apart the nine strands that make up the Uranian system (Elliot and Nicholson 1984). Likely the survival of the discontinuous arc-like structure about Neptune (Hubbard et al. 1986) may also involve one or more small satellites (Lissauer 1985). This latest "ring" presumably accounts for the occultation, once thought to be due to a satellite, noticed by Reitsema et al. (1982; Hubbard 1986). Voyager 2 penetrated the Uranian bull's-eye on January 24, 1986, flying near Miranda, and is scheduled to reach Neptune in August 1989. Pluto will not be visited by man's machines for many years.

Rather than being named after figures in Greek and Roman mythology, the appelations of the Uranian satellites are those of fairies from plays by Pope and Shakespeare. Ariel and Umbriel appear in the former's "Rape of the Lock," whereas Oberon and Titania are the king and queen of the fairies in "A Midsummer Night's Dream" and Miranda is Prospero's daughter in "The Tempest." Triton was a minor sea god who was the son of Amphitrite and Neptune, son of Saturn. Nereid is a sea nymph and attendant to Neptune. In Greek mythology, Charon ferries souls of the dead over the Styx to where Pluto, god of the underworld, reigns.

Little research has been carried out so far on these outer solar system denizens. The volume *Uranus and Neptune* (edited by Bergstralh 1984), written in preparation for the Voyager encounters, contains most of the recent reviews on these satellites. Cruikshank (1982) previously surveyed the Uranian satellites on the bicentennial of Uranus' discovery. Triton's atmosphere and that of Pluto is considered by Trafton (1981). A volume produced to commemorate the 50th anniversary of the discovery of Pluto discusses its satellite only a little, as it had just been identified (*Icarus* 44:1–71, 1980; cf. Whyte 1980).

The chapter in this book by Cruikshank and Brown compiles our knowledge of the satellites of the outer solar system; the same authors (Brown and Cruikshank 1985) have also prepared a popular account on the same topic. The surfaces of these satellites resemble that of Saturn's Hyperion and are composed of water ice contaminated by some dark bland material. The Uranian objects, ignoring Miranda, have been claimed to have the most regular orbits of all (Greenberg 1975a), despite their planet being tilted on its side relative to its orbit. By comparison, the two satellites of Neptune have irregular orbits: the very large Triton is on a highly inclined, and retrograde, but circular orbit while Nereid moves prograde in the most eccentric orbit of any known natural satellite in the solar system. Spectroscopy indicates methane and possibly nitrogen on Triton's surface; this large satellite has a seasonally variable atmosphere and perhaps lakes or even an ocean. The orbit and spins of the Pluto-Charon pair are tidally synchronized; the system's specifications will be clearer as the results of the ongoing eclipse/occultation season are unravelled (Binzel et al. 1985a; Tholen et al. 1985). Pluto is smaller than many satellites. The surface of the satellite cannot be analyzed separately from that of its planet; Pluto has a mottled surface of methane and is shrouded by a tenuous atmosphere of methane and perhaps other gases (cf. Brosch et al. 1986).

V. CONCLUSION

Much has been learned in the past ten years about satellites as individuals and together as a class of celestial bodies. These objects exhibit diverse morphologies, having responded in various ways to the physical processes that act on all of them. The different paths that they have traveled may indicate different environments, distinct intrinsic compositions or some chance event. They are among the foremost tracers of the scheme whereby the solar system originated.

The last decade's remarkable growth in our information base about satellites is displayed in this book. This explosion of data will continue with the imminent exploration of Uranus and Neptune by Voyager and with the advent of the Hubble Space Telescope (HST) era. Once this burst of facts is received, the explosive expansion of satellite data is unlikely to be sustained. Then will come the hard exercise of putting it all together. This next phase—to understand satellites fully—will be much more difficult. A start has already been made toward developing a coherent view of satellites, as abundantly evidenced in the next seventeen chapters of this book. While striving toward this goal of understanding, many may be nostalgic for the distant past when one could be expert about all aspects of satellites or for the present era of exploration when exciting science was simple to do. But easy or not, this phase of total comprehension is coming and the current book is a first step toward it.

Acknowledgments.   The tables provided here were originally presented at IAU Colloquium 77, for which they were assembled with help from D. P. Cruikshank, M. Davies, R. Harrington, J. Lieske, D. Morrison, P. Nicholson, S. Synnott, P. Thomas, and especially A. Harris. Improvements and corrections to the original material were kindly suggested by K. Aksnes, J. K. Beatty, A. Brahic, D. P. Cruikshank, S. Dermott, A. Dobrovolskis, R. Greenberg, A. W. Harris, J. Lissauer, W. McKinnon, F. Mignard, D. Nash, P. Nicholson, T. Owen, D. Pascu, G. Schubert, K. Seidelmann, S. Squyres, P. Thomas, J. Veverka and I. P. Williams. Preparation of this material was supported in part by NASA and NSF. The manuscript was assembled by M. Talman and G. Podleski.

Note added in Proof: The Voyager 2 observations (Smith et al. 1986; Hanel et al. 1986; Tyler et al. 1986) of the Uranian system have revolutionized our view of this set of satellites. Cruikshank and Brown discuss these results and show spacecraft images of the major satellites. Here I abstract a few preliminary findings of Smith et al. (1986).

Besides observing the known members of the Uranian system in detail, Voyager discovered ten new satellites. All are similar to the collisional shards in the Jovian and Saturnian systems; they are dark (~ 5% albedo), small and lie close to their planet's ring system in nearly circular orbits that are roughly coplanar with Uranus' equatorial plane. The largest of these satellites, 1985 UI (see Fig. A-8 of the Cruikshank and Brown chapter), is about 170 km across, roughly equidimensional and heavily cratered. A pair of satellites straddle Uranus' outermost ε ring and apparently account for its confinement; the other suspected ring shepherds, down to a limit of ≤ 4 km radius, have not been detected. The remaining eight newly found moons are located in the region between the rings and Miranda; orbits (see Table III) are close together and spaced quite evenly. These objects, like the nearby Uranian rings, are very much darker than their larger compatriots.

Masses of the previously known moons have been determined through the mutual perturbation observed by the spacecraft (cf. Tyler et al. 1986); the values are not in accord with Veillet's (1983a) calculations (cf. Dermott and Nicholson 1986). The observed radii agree quite well with previous determinations inferred from groundbased infrared measurements. Satellite densities derived from the Voyager measurements are given in Table IV; they do not show an obvious trend with orbital radius, as do the Jovian values, and they differ significantly from the lower densities of the icy Saturnian satellites. These densities are somewhat too high to be produced from an equilibrium condensation mixture of rock, water and ammonia/methane compounds, while they seem slightly low to be compatible with the estimated properties of CO-rich assemblages.

Surface albedoes of the satellites, especially the small ones, are fairly low. Photometric properties indicate a different surface microstructure (see the chapter by Veverka et al.). The spectra of all satellites, including individual bright patches on them, are quite similar and, like those of the rings, are remarkably flat. Most surfaces are thought to be intimate mixtures of ice and a dark spectrally neutral component (see the chapter by Clark et al.) with brighter areas containing less of dark contaminant. Since the dark constituent is not reddened, it must be relatively free of iron-bearing silicates and sulfur (or they are obscured by an opaque phase), nor is it like the material thought to be mixed with water ice on the surfaces of the Jovian and Saturnian satellites, including Iapetus' dark side; it also differs from the organic material believed to color the red and dark (D class) minor planets abundant in the outer asteroid belt. Phoebe and/or C class asteroids provide the best spectral match. Cheng et al. in their chapter, had predicted that such surface properties were to be expected if Uranus had a magnetosphere with properties like those detected by Voyager 2: the irradiation of methane-rich ices by energetic ions can darken surfaces and produce complex organic materials. Alternatively, the material may be primitive. In a few places, a different subsurface composition is indicated: isolated patches of very dark material are found on the floors of a few large craters of Oberon (similar features are present on Iapetus) and bright material is seen along several scarps on Titania.

Oberon and Titania (Figs. A-1 and A-2 of the Cruikshank and Brown chapter) are nearly twins in regard to size, density, color and albedo. However, Titania's surface contains many more extensive fault scarps and graben than Oberon's. Oberon's surface, covered with numerous large (100 km diameter) craters, resembles the ancient heavily cratered highlands of the inner planets. Titania is gouged instead by smaller (10 to 50 km) craters, which may be caused by planetocentric debris rather than by cometary projectiles (see the chapter by Chapman and McKinnon).

Umbriel and Ariel (Figs. A-3 and A-4 of the chapter by Cruikshank and Brown) have similar diameters, and could have comparable densities, but their surfaces differ. Umbriel's surface is almost totally coated with a uniform dark material such that no impact rays are even visible. Its surface topography, along with Oberon's, seems to be the oldest on the major satellites; it is dominated by large ancient craters much like those seen on Oberon. In contrast, very few craters larger than 50 km are apparent on Ariel, which in this regard resembles Titania. Ariel has the brightest and geologically youngest surface of the Uranian satellites. Significant portions of Ariel's outer layers have been resurfaced by extruded material through which long narrow sinuous valleys, reminiscent of lunar rilles, are cut. Fault systems dominate the landscape and imply global expansion, perhaps through differentiation (see the Squyres and Croft chapter).

Miranda's surface (Figs. A-5, A-6 and A-7 of Cruikshank and Brown's chapter and the frontispiece to the book), an amalgam of many solar system terrains, is truly bizarre. Nearly half of the satellite's visible surface is (the expected) ancient heavily cratered terrain. Three regions of younger terrain, which are rectangular to ovoid in plan, make up the remaining scene that Voyager saw; they are considered by the Voyager imaging team to be variants of the same internal process. Complex sets of parallel and intersecting scarps and ridges cover these young terrains. The ridges are notably similar to the grooved terrain recognized on Ganymede, Europa and Enceladus. Patterns that appear to be flows are seen at high resolution; one appears to have originated at a volcanic cone. Globe-girdling fault systems seem to be extensional in form. Outcrops expose bright and dark materials. Craters on Miranda have a lunar-like spectrum, as do those on Umbriel and Oberon, but the innermost Uranian satellite has three times the lunar crater density while the other two are slightly less heavily cratered than the Moon. Gravitational focusing due to Uranus can account for the higher cratering flux at Miranda; fluxes are computed to have been high enough to disrupt Miranda several times and to have shattered any parent precursor of the newly discovered satellites.

The abundant evidence for extensional tectonics and for resurfacing indicates an unanticipated large level of internal activity for the Uranian satellites. Several causes of this internal heating may be suggested. The higher densities of the Uranian satellites compared to those of the icy Saturnian satellites imply increased radiogenic heating (see the chapter by Schubert et al.). Due to low surface albedos, surface temperatures of the Uranian satellites are higher than those of the Saturnian satellites; the Voyager IRIS experiment (Hanel et al. 1986) measured brightness temperatures of $86(\pm 1$ K and $84(\pm 1)$ K for Miranda and Ariel, respectively (cf. Table II). These two effects may have allowed internal temperatures to exceed the melting point of ammonia-water mixtures (see the Stevenson et al. chapter). Satellite resonances, or near-resonances, may have also played an important part in heating Ariel and Miranda.

This synopsis of the preliminary findings by the Voyager spacecraft teams (Smith et al. 1986; Hanel et al. 1986; Tyler et al. 1986) covers in scant depth the startling wealth of information returned from the spacecraft. Nevertheless, it (and the addendum to the chapter by Cruikshank and Brown) should convey that these observations will profoundly influence our understanding of all satellites. The revelation of Uranus' retinue—and how this satellite family conforms (but also does not) to our expectations—serves to remind us that a full comprehension of planetary satellites is not yet available.

2. ORIGINS OF SATELLITES

D. J. STEVENSON
California Institute of Technology

A. W. HARRIS
Jet Propulsion Laboratory

and

J. I. LUNINE
University of Arizona

Satellites are an inevitable consequence of most plausible planetary accumulation processes. They can arise from gaseous or particulate circumplanetary disks, continuously fed during accretion of the planet or infrequently created by large impacts. They can also arise from capture, aided by gas drag. Fission (in the Darwinian sense) is highly unlikely. The relative importance of these processes depends primarily on: (a) the disposition and dynamics of the gas phase; and (b) the mass spectrum of the planetesimals which feed a growing planet. It is not possible to assess these factors with great confidence at present because of considerable uncertainty concerning the mode(s) of planetary formation. This review seeks, therefore, to assess critically the alternatives within the context of current ideas of the early solar system, guided by both cosmochemical and dynamical constraints, but unencumbered by prejudices concerning planetary growth. Topics discussed include the dynamics of both gaseous and particulate disks, the role of large impacts in creating satellite source material, the role of capture, and the thermodynamics of satellite accretion. Possible explanations for each of the satellite systems are offered.

I. INTRODUCTION

A cosmogonist living on Ganymede would probably preoccupy himself with understanding the origin and evolution of the Jovian system, maybe eventually raising his sights to a comparative study of the Saturnian system and even the solar system. Inevitably, the terrestrial viewpoint has evolved in a different way. Discussions of satellite origin have usually been afterthoughts or extensions of planetary origin scenarios. The Earth's Moon has received most attention and there has been an emphasis on scenarios that extend or mimic those invoked to describe the origin of the terrestrial planets. There has also been a preoccupation with the dynamical issues, because the cosmochemical constraints were either nonexistent or equivocal.

In fact, satellites would seem to be an inevitable and important part of any model of solar system origin. The large satellites (Moon, Io, Europa, Ganymede, Callisto, Titan, Triton) are as diverse and interesting as the planetary family and their cosmochemical makeup is becoming better understood. The intermediate-sized and small satellites (extending continuously to ring particles in size) provide essential insights into the compositional and dynamical environment of accretion and the influence of later impacts. Perhaps most importantly, there are several regular satellite systems but there is only one known solar system thus far. The perspective gained by considering the similarities and differences among these systems enables one to appreciate the diversity of circumstances and the unlikelihood of a single model which encompasses all.

Table I summarizes the observations most relevant to origin hypotheses, illustrating both pattern and diversity among satellites (for more specifics, see the tables in the introductory chapter). Perhaps the most striking cosmochemical pattern is the systematic increase in ice content with distance from the primary among the Galilean satellites. It is tempting to attribute this to a temperature gradient imposed by proto-Jupiter or a Jovian nebula, just as the compositional systematics of the solar system are frequently interpreted as a temperature gradient in the solar nebula. We shall see that, although there may be elements of truth to this analogy, the validity of a miniature solar system picture is questionable. Similar systematic trends are not excluded for Saturn and Uranus, but neither do the data demand them. In many ways, the Saturnian system is strikingly different from the Jovian system. The Uranian system may even have an inverted systematic: Veillet (1983a; cf. Dermott and Nicholson 1986) suggests that the *outer* satellites may be more dense (i.e., more rock-rich than the inner satellites). If correct, this argues for a different origin scenario. In different ways, Triton and the Earth's Moon also argue against all-encompassing generalization: Triton because of its retrograde and inclined orbit and the Moon because of its unusual composition (not readily reconciled with any primary condensation hypothesis).

TABLE I

Summary of Satellite Properties Relevant to Origin Hypothesis

Planet	Satellite	Orbital Radius[a]	Orbital Character[b]	Compositional Features[c]
Earth	Moon	60	RT	Iron poor, volatile poor, similar to Earth mantle?
Mars	Phobos, Deimos	2.8, 7	RT	Carbonaceous chondritic
Jupiter	Inner, small bodies	2–3	RT	Rock
	Io	6	RT	Rock
	Europa	10	RT	Rock + 100 km H_2O layer
	Ganymede	15	R	60% rock, 40% ice
	Callisto	27	R	60% rock, 40% ice
	Outer, small satellites:			
	Inner group	~170	$i = 25-30°$	Asteroidal?
	Outer group	~300	~150°	
Saturn	Inner, small bodies	2–3	R	Ice-rich?
	Mimas, Enceladus,			
	Tethys, Dione, Rhea	3–9	R	~40% rock, 60% ice
	Titan	20	R	60% rock, 40% ice including N_2, CH_4
	Hyperion	25	R	Rock?
	Iapetus	60	R	40% rock, 60% ice
	Phoebe	220	$i = 150°$	Carbonaceous chondritic?
Uranus	Miranda, Ariel, Umbriel	5–11	R(T?)	Icy?
	Titania, Oberon	17–23	R	Rock-ice?[d]
Neptune	Triton	14	$i = 150°$	Ice, including CH_4, N_2?[e]
	Nereid	~230	$i = 28°$	?
Pluto	Charon	~10	RT	Icy?

[a] Approximate distance in units of planetary radius.

[b] $R \equiv$ regular (low inclination and eccentricity), $T \equiv$ tidally evolved orbit.

[c] Rock and ice percentages are approximate mass fractions.

[d] Veillet (1983a); Stevenson (1984b).

[e] Cruikshank et al. (1984a).

Four patterns may have general validity, although their interpretation is uncertain:

1. In those cases where there is more than one satellite with substantial mass, a "regular" family of satellites with low-inclination prograde orbits exists. Jupiter, Saturn and Uranus conform to this rule. This argues for the existence of an equatorial disk of material from which the satellites formed. (But what happened to Neptune's disk?)

2. Regular satellite families extend out to distances that tend to scale with the size of the central planet (i.e., 20 to 50 planet radii). Notice that this size is *not* analogous to the solar system. Curiously, the reconstituted mass (total amount of cosmic composition material required to make the satellite family) is a few percent of the central planet mass; this *is* analogous to the reconstituted solar nebula mass expressed as a fraction of the Sun.

3. Large, icy satellites tend to be ice-poor relative to small, icy satellites. Ganymede, Callisto and Titan are all roughly 60% rock, 40% ice by mass whereas the small Saturnian satellites appear to be close to a cosmic mixture ($\sim 40\%$ rock, 60% ice by mass). This argues for a mass-dependent mechanism of selective accretion of rock or selective loss of ice—an important clue to the environment and time scale for the accretion of large satellites.

4. Volatile ices (e.g., CH_4, N_2) tend to be found or inferred on satellites which are distant from the Sun or from the planet or (frequently) both. Unfortunately, this is a confused pattern because there are two factors operating and we lack sufficient information to disentangle them. For example, would Ganymede be like Titan if Jupiter had formed out where Saturn is and vice-versa? (i.e., Ganymede's formation environment was probably warmer than that of Titan partly because Jupiter is more massive and, hence, had a warmer nebula and higher luminosity and partly because Jupiter is closer to the Sun. But are these effects comparably important?).

We seek here to consider both the patterns and the diversity, not by describing detailed models or even a suite of models but instead, by laying before the reader the issues and their ramifications. Major unresolved issues concerning the early solar system are readily acknowledged, unlike some existing reviews. A large part of the exposition relates to planetary formation, but from the perspective that satellites are likely to be an unavoidable by-product of the formation process. The emphasis is on ideas and approximate quantification (factors of π and $\sqrt{2}$ being dropped wherever they are inessential to the main points). Cosmochemistry is given equal billing with dynamics, in anticipation that the abundances and chemical form of satellite constituents are diagnostic of their source material (e.g., circumplanetary or circumsolar?). There is no other published review with the scope of what is attempted here, although the brief but wide-ranging recent review by Ruskol (1982a) deserves mention, and the chapter by Safronov et al. in this book provides an excellent summary of the Soviet school of planetary cosmogony.

We begin in Sec. II with a brief survey of ideas for the origin of the solar system and then proceed to the problems of planetary formation, with an emphasis on the popular (but not universally accepted) notion of an accretion disk. The important features discussed are dynamic time scales, condensation, gravitational instability and, most importantly, the mode(s) and timing of planetary formation. This leads naturally to an assessment of the ways in which satellites can form: from material that condenses within a gaseous disk emplaced around a planet, from accumulation of particulates fed from solar orbit into planetary orbit by dissipative processes, from planetary material (perhaps injected into orbit by impact), or by accumulation elsewhere in the solar system followed later by capture. The following Sec. III describes the formation of circumplanetary disks, perhaps the situation most relevant to large satellite formation. Large impacts, interplanetary collisions, and gaseous accretion are each considered. A simplified analysis of the various dynamic and thermal time scales is given for these disks in Sec. IV, primarily for the conditions appropriate to giant planets. Although this analysis has much in common with the solar nebula problem and relies on many of the ideas developed in Sec. II, some important differences arise which lead us to question the common assumption that the regular satellite systems are miniature solar systems.

Section V deals with the accretion energetics of large satellites, Sec. VI considers the role of impact disruption and ablation, and Sec. VII covers the dynamics of intact satellite capture. Accretion energetics are discussed because of their probable relevance to several geological and chemical problems (the differences between Ganymede and Callisto, the origin of Titan's atmosphere and the formation of the lunar highlands). Impact disruption is probably important for the early evolution of small satellites/rings, especially for Saturn. Satellite capture is discussed, partly for its likely role for irregular satellites, but also for its important role in gas-rich scenarios (as advocated by Hayashi and coworkers). The review concludes with an attempted synthesis in which possible explanations are advanced for the origins and properties of all the planetary satellites.

II. SOLAR SYSTEM ORIGIN AND PLANETARY FORMATION

Modern ideas of solar system formation have been greatly influenced by our improved understanding of the interstellar medium, especially the existence and nature of dense, molecular hydrogen clouds (see "Protostars and Planets," ed. T. Gehrels 1978; and "Protostars and Planets II," eds. Black and Matthews 1985). Recent work on T Tauri (Hanson et al. 1983), IRAS (Infrared Astronomical Satellite) observations (Aumann et al. 1984), and groundbased observations of nearby stars (Smith and Terrile 1984; McCarthy et al. 1985) are expected to influence further the development of cosmogonical models. Solar system origin is generally believed to involve a gaseous disk formed from gravitational collapse of a rotating, interstellar cloud of dust and gas, with the disk having far greater extent than the central condensed body because of the angular momentum budget. Gaseous disk models are in the spirit of the early (18th century) models of Kant and Laplace. Although the detailed dynamics of the collapse and disk formation (interplay of rotation and magnetic fields, role of triggering events such as supernova-induced shock waves or spiral density waves) remain controversial, the availability of appropriate conditions for collapse is not seriously in doubt.

The mass distribution (or, equivalently, total angular momentum) of the forming disk is very uncertain, especially since there is no reason to suppose that the central concentration of mass is equal to the mass of the Sun, at least in the earliest stages. This uncertainty is the primary reason for the range of models encountered in the literature. The models developed by Cameron (1978,1983a) are generally characterized as high mass models because the total mass is roughly twice the solar value. However, only about 0.1 $M_\odot$ (1 $M_\odot \equiv$ one solar mass) is in that part of the disk interior to the present orbit of Neptune. In this respect, his models of the disk are only about twice as massive as the so-called minimal mass nebula models, best represented in the work of Hayashi and collaborators (see Hayashi 1981; Hayashi et al. 1985). These operate on the pragmatic (but probably unreasonable) principle that what you can see now tells you what you had then. Minimal mass simply means taking an inventory of the total amount of involatile (rock) constituents in the present planetary system and calculating the mass of the reservoir of assumed cosmic abundance from which this material was derived. The dominant contribution (and uncertainty) in this reconstitution is the amount of heavy material at the centers of the giant planets (apparently $\sim$ 10 $M_\oplus$ per planet, where 1 $M_\oplus \equiv$ one Earth mass, but with an uncertain ice/rock ratio; see Stevenson 1982a). The biggest uncertainty, however, lies in the treatment of disk dynamics during and after accretion.

A. Accretion Disk Dynamics

Most current models are based on, or related to, the now-classic work of Lynden-Bell and Pringle (1974), some aspects of which were present in the earlier work of Shakura and Sunyaev (1973). We will develop the concepts here since they will be useful for discussions of both the solar system and satellite systems. The essential idea is a disk in which radial redistribution of mass and angular momentum are achieved by viscous stresses $\rho \nu R(d\Omega/dR)$, where ρ is the mass per unit volume, ν is the effective kinematic viscosity and $\Omega(R)$ is the angular velocity at distance R from the central axis. In this disk (schematically illustrated in Fig. 1), the viscous couple on material between R and $R + dR$ is

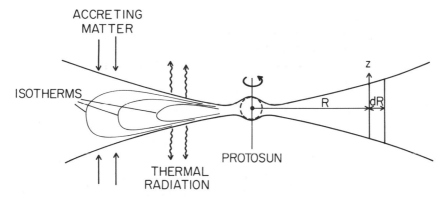

Fig. 1. Schematic illustration of the cross section of an accretion disk (thickness exaggerated by a factor of ~ 2 for clarity). The outer bowtie-shaped region is the radiating surface (optical depth unity in the infrared). Lobe-shaped isotherms are also indicated. The disk is shown connected continuously to the protosun but may actually be bounded by a shock front. On the right, a cross section of an elemental annulus is shown, the torque balance for which leads to Eq. (2) in the text.

$$ g \equiv \int_{-\infty}^{\infty} \rho\, \nu\, R^2\, \frac{d\Omega}{dR}\, 2\pi R \, dz = 2\pi R^3 \nu \sigma\, \frac{d\Omega}{dR} \tag{1} $$

where σ is the surface density of the disk (mass per unit area). Angular momentum balance requires $D/Dt\,(h\delta m) = dR(dg/dR)$ where $h = R^2\Omega$ is the specific angular momentum, δm is the mass element between R and $R + dR$, and D/Dt is the total time derivative. In steady state ($\partial/\partial t = 0$), we obtain the useful result

$$ F\, \frac{dh}{dR} = \frac{dg}{dR} \tag{2} $$

where F is the mass flux radially outward. If g goes to zero at both $R = 0$ and $R \to \infty$, then it follows that for $dh/dR > 0$ (the most likely case, e.g., as in Keplerian motion), there is mass inflow at small R and mass outflow at the disk extremities—but always angular momentum outflow. Of course, this would normally not be a steady state situation, but it helps in understanding the evolutionary trend.

If we perform dimensional analysis on Eq. (2), then we find a characteristic time scale $\tau_{\mathrm{visc}} \sim R^2/\nu$ for the redistribution of a large fraction of the disk mass. For $R \sim 10^{15}$ cm, characteristic of our solar system, this time exceeds the age of the solar system if $\nu \lesssim 10^{13}$ cm^2 s^{-1}. For comparison, the molecular kinematic viscosity of hydrogen at typical primordial solar nebula conditions is $\sim 10^5$ cm^2 s^{-1}. Clearly, significant evolution of viscous accretion disks in a time scale $\lesssim 10^9$ yr requires an effective viscosity caused by turbulence. This is perhaps the most important unsolved problem of disk dynamics. A popular choice is $\nu = \alpha c H$ where c is the sound speed, H is the scale height (essentially the disk thickness) and α is a fudge parameter, sometimes chosen to be as large as ~ 0.2 (see, e.g., Cameron 1983a) but other times as low as $\sim 10^{-2}$ (Lin and Bodenheimer 1982). Choices of α are usually either *ad hoc* or based on unverified assertions about the existence and nature of turbulence. The nearest to a predictive theory for α is the work of Canuto et al. (1984), which is based on the expectation that the disk is thermally convective in the vertical direction (meaning the direction perpendicular to the plane of the disk). A necessary, but not sufficient, condition is that the disk is optically thick. However, more work is needed before the effective viscosity can be said to be even qualitatively understood.

The energetics of the disk can be appreciated by considering a vertical cylinder of unit cross section. If both top and bottom surfaces radiate into free space at an effective ("photospheric") temperature T_e, then the energy loss is $2\sigma_{SB}T_e^4$, where σ_{SB} is the Stefan-Boltzmann constant. If this is balanced locally by viscous dissipation, then

$$\sigma \nu R^2 \left(\frac{d\Omega}{dR} \right)^2 \simeq 2\sigma_{SB}T_e^4. \tag{3}$$

In the present context, viscous dissipation simply means the local cascade of eddy motions into smaller eddies, eventually balanced by molecular viscosity. In a Kolmogorov cascade, this process depends only on the large-scale characteristics of the flow. It is important to realize that the photospheric temperature is not necessarily the temperature of the disk interior. If the disk has a vertical optical depth τ and the opacity is uniform with height, then it is easily shown from the equation of radiative transport that $T_c \simeq (1 + \tau)^{2/7} T_e$, where T_c is the midplane temperature. If the opacity is not uniform with height, then thermal convection may occur and a smaller value of T_c/T_e is possible. In any event, significant temperature differences between midplane and photosphere are possible and will be important in our later discussion. This point deserves emphasis because it is frequently overlooked in published models. In particular, it means that the concept of a single temperature at a radial position R may be meaningless.

The thickness of the disk is readily calculated, at least for the simple case where the disk is approximately isothermal and the dominant contribution to the vertical component of gravity is provided by the central mass (the protosun) of mass M_c. In this case, hydrostatic equilibrium implies

$$\frac{dp}{dz} \cong -\rho \, \frac{z}{R} \, \frac{GM_c}{R^2} \tag{4}$$

where p is the pressure and ρ is the density.

Application of the ideal gas equation of state $p = (\rho/\mu)kT$, where μ is the molecular weight, then gives

$$\rho = \rho_c \exp(-z^2/H^2) \tag{5}$$

$$H \equiv \left(\frac{2kT}{\mu} \frac{R^3}{GM_c} \right)^{1/2}. \tag{6}$$

As a simple application of the above ideas, substitute Eq. (6) into $\nu \approx cH$ and then solve Eq. (3). The startlingly simple result is $T_e \sim 10^6 \rho^{2/5}$, where ρ is in g cm^{-3} and molecular hydrogen is assumed. This result is independent of mass. If we further suppose that $\sigma \propto R^{-1}$, then application of the above equations yields $H \propto R^{1.1}$, $\rho \propto R^{-2.1}$ and $T_e \propto R^{-1}$. The latter result, in particular, enables one to see how the mass distribution in the disk is related to the temperature distribution. As an example, a nominal solar nebula model can be constructed by requiring that $T \sim 140$ K at 5 AU from the Sun, in order that water ice condense at the orbit of Jupiter. For $\sigma \propto R^{-1}$ and $\alpha = 0.1$ one then finds $T_e \simeq 700$ (1 AU/R) K, $H \simeq 7 \times 10^{11}$ $(R/1$ AU$)^{1.1}$ cm, $\rho \simeq 8 \times 10^{-9}$ (1 AU/$R)^{2.1}$ g cm^{-3} and $\sigma \simeq 1.3 \times 10^4$ (1 AU/R) g cm^{-2}. The disk mass is (0.03 M$_\odot$)(R_{outer}/1 AU), where R_{outer} is the outer radius of the disk.

One more disk property of great importance is the Kelvin-Helmholtz cooling time, τ_{cool}, assuming that the only energy source is the internal heat of the disk:

$$\tau_{cool} \simeq \frac{\sigma C_p T}{2\sigma_{SB} T_e^4}. \tag{7}$$

If the optical depth $\tau \lesssim 1$ then for the example given above, $\tau_{cool} \sim 1$ yr. However, if $\tau \sim 10^2$ (say) and $T \sim 500$ K (reasonable for the orbit of the Earth), then $\tau_{cool} \sim 10^4$ yr and almost as long as the viscous time scale $\tau_{visc} \sim R^2/\nu$.

B. Disk Instabilities

Details of nebula disk models change (the "evolution time" of Cameron's modeling effort is one or two years) but the enduring feature that distinguishes high-mass from low-mass models is the existence and nature of gravitational instabilities, a crucial issue for understanding the formation of planets and satellites. Consider a patch of the accretion disk with a characteristic linear dimension (in the plane of the disk) of $\ell \lesssim H \ll R$. We can identify three characteristic time scales: an orbital time scale $\tau_{rot} \sim \Omega^{-1}$ associated with the effects of the Coriolis force and differential rotation, a sound wave travel-time $\tau_{sound} \sim \ell/c$, and a characteristic gravitational time scale $\tau_{grav} \sim \sqrt{\ell/\sigma G}$, associated with the free-fall collapse of the patch. (Notice that all these time scales are derivable purely from dimensional analysis.) Crudely speaking, gravitational collapse occurs provided τ_{grav} is less than either of the other time scales. If $\tau_{grav} > \tau_{rot}$, then subelements of the patch separate because of the stabilizing effect of the orbital motion before they can collapse towards each other by self-gravitation. If $\tau_{grav} > \tau_{sound}$, then the collapse can be prevented by a pressure change. In general, collapse will be possible for all length scales which satisfy $\tau_{grav} < \tau_{rot}$ (i.e., $\ell \lesssim \sigma G/\Omega^2$) *and* $\tau_{grav} < \tau_{sound}$ (i.e., $\ell \gtrsim c^2/\sigma G$). Clearly, the lowest surface density σ for which both inequalities can be satisfied is $\sigma_{crit} \sim c\Omega/G$ and the corresponding instability wavelength is $\ell \sim c/\Omega$.

   In reality, the effect of orbital motion is a competition between stabilizing (Coriolis) and destablizing (shear) components. The correct dispersion relation for waves of the form $\exp(i[kR + m\theta + \omega t])$, assuming $kR \gg 1$ or m, is

$$(\omega - m\Omega)^2 = K^2 + c^2 k^2 - 2\pi G\sigma|k| \qquad (8)$$

$$K^2 \equiv 2\Omega\left(2\Omega + R\,\frac{d\Omega}{dR}\right) \qquad (9)$$

where θ is the polar angle (Lin and Shu 1966; Goldreich and Ward 1973). The lowest σ for which ω has an imaginary part (and the wave is unstable) occurs at $k = K/c$, $\sigma_{crit} = cK/\pi G$, essentially as in the crude argument except that Ω is replaced by the epicyclic frequency K. This is physically reasonable since the epicyclic frequency is the frequency of simple harmonic motion for the perturbation motion of a particle displaced infinitesimally from a circular orbit. For a Keplerian disk, where fluid elements obey $\Omega \propto R^{-3/2}$, $K = \Omega$. In the dispersion relation above, c^2 is defined as $(\sigma/\rho)(dp/d\sigma)$. In a gas, it will be the actual sound speed; in a particulate disk (e.g., Saturn's rings) c will be essentially the random component of the velocity of the particles; in a mixture of gas and particles there is no completely general result.

C. Giant Gaseous Protoplanets

   A gaseous disk with $T_e \sim 500$ K at $R = 1$ AU typically requires $\sigma \gtrsim 10^5$ g cm^{-2} for gravitational instability. Cameron's disk models satisfy the criterion $\sigma > cK/\pi G$ early in the evolution (elapsed time $\sim 10^4$ yr, typically), leading him to propose that the primordial solar nebula fragmented into a large number of giant, gaseous protoplanets (Cameron 1978; Cameron et al. 1982). The proposed instabilities may be global (i.e., depending on the structure of the entire disk and very large) as well as the local instabilities governed by Eq. (9), but the criterion for instability is likely to be similar. At first sight, this appears to be an attractive way of initiating the formation of giant planets, especially since Eq. (9) predicts that the characteristic mass of the region that goes unstable in the local instability is $\sigma_{crit} \ell^2 \sim c^2/\Omega G \sim$ Jupiter's mass. However, there are a number of difficulties with the hypothesis. First, the existence of the instability is not well established, even for the high surface densities favored by Cameron, because the instability criterion is only marginally satisfied (i.e., by a factor ~ 2). Since the critical wavelength is not significantly larger than the thickness of the disk, the thin-disk approximation breaks down. Cameron appeals to the numerical results obtained by Yabushita (1969) for Saturn's rings but these cannot be used because Yabushita omitted the sound speed (or, equivalently for Saturn's rings, the dispersion velocity), a very poor approximation for the nebula.

The second and possibly greatest difficulty with giant, gaseous protoplanets is that it is very difficult to see how to evolve from them to the present solar system. The difficulties exist for both the giant planets and the terrestrial planets. Although the gaseous protoplanets may evolve in such a way as to cause rain-out of iron and silicates to the center (Slattery et al. 1980), this is insufficient to explain the ten or so Earth mass "rock" cores of the present giant planets, unless additional (*ad hoc*) assumptions are made about feeding material from the solar nebula into the protoplanet or the size distribution of protoplanets. Although it is possible to eliminate some or all of a gaseous envelope at a later stage (by evaporation or tidal stripping or high, early solar ultraviolet radiation), it is also difficult to reconcile the terrestrial planetary properties with the properties of gaseous protoplanets. For example, these models seem to imply an almost purely iron core, incompatible with the core of the Earth which contains substantial amounts of lighter elements (Jacobs 1975). It has to be acknowledged, however, that sufficient work has not yet been done on the implications of the model that we can reach a conclusion on its capabilities. We choose to ignore this model in the remainder of this review for the pragmatic reason that its implications for satellites are completely unknown. To the extent that the evolution for the giant, gaseous protoplanet scenario eventually merges with other models discussed below, the neglect is only partial.

D. Condensation, Sedimentation, Coagulation

The starting material for the solar system is a cloud of gas and dust. Some but not all of the dust is evaporated by the high temperatures produced in the formation and evolution of the accretion disk. In the outermost part of the solar system, a lot of interstellar dust may survive. This may explain, for example, the very high D/H observed in parts of some meteorites (Yang and Epstein 1983). This could be important for understanding the properties of satellites and comets, since the interstellar material has a chemical makeup different from that predicted for thermodynamic equilibrium. For example, much of the nitrogen is in the form of N_2 rather than NH_3 and much of the carbon is in the form of CO or organics with low H/C ratios rather than CH_4 (see Herbst [1978] for a discussion of interstellar chemistry). A similar situation can arise if N_2 and CO are formed at high temperature in the solar nebula and then quenched to low temperature without reequilibrating to the preferred thermodynamic (hydrogen-rich) forms (Lewis and Prinn 1980; Prinn and Fegley 1981).

At any radial position in the disk, there will be a complicated thermal history, but the long-term trend will be a cooling one, leading to a partial condensation of material from the gas phase. It is possible to predict which materials condense and the order in which they condense, provided one assumes equilibrium chemistry (Grossman and Larimer 1974). For our purposes, it suffices to note some very general points which are common to most models. First, free iron condenses at a very similar temperature to major silicate phases. Consequently, it is implausible to explain iron-rich bodies (e.g., Mercury) or iron-poor bodies (Moon) solely on the basis of the temperature in their regions of formation. Second, water ice condenses at around 160 K at the low pressures encountered in the solar nebula and this appears to demarcate approximately the terrestrial planet zone from the giant planet zone. By virtue of its abundance, water ice is the most important condensate, probably exceeding in mass the combined total of all the other condensates in the solar system. The third point about condensation scenarios is that models seldom predict temperatures low enough for the direct condensation of abundant but highly volatile ices such as CH_4, N_2 or CO and certainly not Ne, H_2 or, of course, He. However, temperatures may be low enough (~ 40 to 60 K) to allow the formation of clathrate hydrates, compounds of the form $\sim X.7H_2O$, where X can be a mixture of CH_4, N_2, CO and other volatile molecules. Kinetic inhibition may be a problem, however, and the most likely place for clathrate formation is in dense but cold nebulae around giant planets (Lunine and Stevenson 1985a).

The general picture is an accreting disk in which the temperature first increases and then subsequently decreases (either from a Lagrangian or from an Eulerian point of view). "Meteorological" effects (upwelling and downwelling convective motions) are superimposed on these long-term "climatic" trends. In the cooling phase, nucleation of condensates occurs from the gas phase. This nucleation can be either heterogeneous (on the surfaces of preexisting refractory seed nuclei) or homogeneous. Growth of these condensation centers to micron-sized grains is by diffusion alone and rapid ($\lesssim 10^3$ yr). Subsequent growth can be slower but is aided by the relative motion between gas and grains or caused by possible sticking of grains during collisions. Condensation is likely to begin near the disk photosphere, where the temperature is lowest, and cause the formation of a cloud of micron-sized grains. Unlike the gas, which is supported by pressure gradients, the dust grains settle toward the midplane of the disk. Since the molecular mean free path is greater than the grain size, the drag exerted on the grains is ballistic rather than viscous. In the grain frame of reference, molecules impacting from upwind tend to be going faster on average than molecules from downwind by $\sim v$, where v is the grain settling velocity. The drag force is then $\sim \pi r m n c v$, where r is the grain radius, n is the gas number density, m is the gas molecular weight and c

is the sound speed. Since the vertical component of gravity is $\sim z\Omega^2$ (Eq. 4), an estimate of the sedimentation time is then

$$\tau_{\text{sed}} \sim 3 \times 10^5 \left(\frac{1\,\mu}{r} \right) \text{yr.} \tag{10}$$

A more detailed analysis (Weidenschilling 1980; Nakagawa et al. 1981) suggests short sedimentation times ($\sim 10^3$ yr) because of the growth of large (up to 1 cm sized) particles by sticking. The situation is complicated and not yet well understood because of the probable presence of turbulence.

If turbulence is weak or absent then the concentration of solid particles toward the disk midplane leads inevitably to gravitational instabilities. This can be appreciated by examining Eq. (9) and realizing that the near-midplane effective sound speed c is decreasing as sedimentation proceeds. If the gas-solid medium is thought of as a pseudogas in which most of the density ρ arises from the solid particles but most of the pressure p arises from the gas, then $c^2 \sim p/\rho \ll p/\rho_{\text{gas}}$. The precise form of the instability is likely to be more complicated than Eq. (9) and the above crude analysis of c (see Sekiya 1983) but the result, independent of details, is likely to be the clumping of solid grains into planetesimals of a few kilometers in radius. The planetesimal mass $\sim \sigma^3_{\text{solid}} G^2/\Omega^4$ can be estimated by dimensional analysis alone or by finding the mass within a patch of area $\sim \pi(\pi/k)^2$ where k is the critical wave vector predicted by Eq. (9).

If substantial turbulence is present, gravitational instabilities may not occur but planetesimals of about a kilometer in radius may still be possible by sticking of particles (Weidenschilling 1980). The only difficult situation may be strong turbulence where collisional disruption could exceed coagulation.

E. Evolution of a Planetesimal Swarm

The nebula now consists of a mid-plane layer of planetesimals in closely spaced, approximately Keplerian orbits, immersed in a gaseous disk. The subsequent evolution depends on when and how the excess gas is dispersed. Clearly, the giant planets must have formed when the nebula had not yet dispersed. The environment for terrestrial planet formation is less apparent. In any event, one needs to consider evolution of the planetesimal swarm both in the presence of and in the absence of a gaseous medium.

Both accumulation scenarios are discussed at length by Wetherill (1980a), and some aspects are discussed further by Harris and Ward (1982). The gas-free accumulation was extensively studied by Safronov (1969; see also the chapter by Safronov et al.), essentially by analytical techniques, although later numerical work has also played an important role (Wetherill 1980b; Greenberg 1982b). The gaseous accretion scenario has been analyzed primarily by Hayashi and coworkers (Hayashi 1981; Nakagawa et al. 1983; Hayashi et al. 1985). In both models, the early evolution from kilometer-sized bodies to bodies having radii several hundred kilometers is rapid ($\sim 10^4$ yr) because the orbits are so closely spaced that collisions are possible even with zero eccentricity. The subsequent evolution is invariably much slower. In the Safronov theory, a steady state eventually occurs in which the gradual increase in random velocities of planetesimals, caused by noncollisional (gravitational) encounters, is balanced by the loss of energy caused by inelastic collisions. In both the analytical theory and numerical simulations of the later stages (Wetherill 1980a,b), most of the mass is in the larger bodies and, hence, the dominant gravitational perturbers are the most massive bodies. Wetherill shows that Safronov's prediction for a steady state is then valid and the random velocity v in a swarm of bodies perturbed by a large body of mass M and radius r must be of order the escape velocity from that body

$$\text{v} = \left(\frac{GM}{r\theta} \right)^{1/2} \tag{11}$$

where θ is known as the Safronov number and is not greatly different from unity. (The random velocity is the deviation from circular Keplerian motion about the Sun.) In this discussion we are ignoring the important but difficult issue of the time evolution of the planetesimal size distribution; we shall return to this later. For the present, it suffices to mention the possibility that runaway growth may occur at early stages of the accumulation (Greenberg et al. 1984), potentially invalidating the Safronov picture in the sense that the assumed size distribution may not be attainable.

If ρ_s is the mass density of the planetesimal swarm, then it follows that in the Safronov model, a given planetesimal collides with an amount of solid material $\sim \pi r^2 \rho_s \text{v}$ per unit time. A characteristic accretion time is then

$$\tau_{acc} \sim \frac{M}{\pi r^2 \rho_s \text{v}} \sim \frac{M}{\pi r^2 \sigma_s \Omega} \tag{12}$$

where σ_s is the surface density of solids ($\sim \rho_s \text{v}/\Omega$). Notice that for $\theta \gtrsim 1$, gravitational focusing is not a major effect but it may be a big enough effect to modify the relative growth rates of the Earth and Moon, for example. The density ρ_s can be estimated from existing terrestrial planetary masses and spacing, or by assuming that M is smeared out over a toroidal volume with large radius $\sim R$ (the orbital radius of the planet) and small radius $\sim R(\text{v}/\text{v}_k)$, where v_k is the Keplerian velocity. For Earth, one finds $\tau_{acc} \sim 10^7$ yr, but increasing to $\gtrsim 10^9$ yr in the giant planet region.

The accumulation process can be significantly different in the presence of gas. The early evolution differs because of gas drag (Kusaka et al. 1970; Whipple 1972; Weidenschilling 1977) but bodies of several hundred kilometers in radius may still form in $\sim 10^4$ yr (Nakagawa 1978). Formation of bodies with radii $\gtrsim 10^3$ km may take in excess of 10^6 yr at Earth orbit and

$\sim 10^7$ yr at Jupiter's orbit but a number of factors could modify these estimates (see Nakazawa and Nakagawa 1981). Once gas drag becomes unimportant, many of the physical processes are closely analogous to the Safronov theory. Some of the problems are potentially very different. For example, ice-rich planetesimals can undergo evaporation (either thermally or by accretional energy input) and modify their gaseous environment from cosmic to steam-rich. Bodies $\gtrsim 10^3$ km in radius form bound atmospheres which can extend out to a distance $\sim GM/c^2$ and provide a much larger cross section for subsequent accretion and growth (Hayashi et al. 1977). This effect, in particular, can reduce the time scale for the later stages of accretion and allow the formation of a core of several Earth masses at Jupiter's orbit in as little as 10^7 yr. If the gaseous medium persists then the proto-Earth, for example, would have had a massive atmosphere, perhaps 5% $M_\oplus$, with a bottom temperature of up to ~ 4000 K (Hayashi et al. 1979). More detailed modeling including various assumptions on grain opacity (Mizuno and Wetherill 1984) yields lower temperatures. This model is not obviously incompatible with the present terrestrial planetary properties because there appears to be no difficulty in subsequently dissipating this massive atmosphere, using the early high ultraviolet and particle flux from the Sun (Sekiya et al. 1981). However, the model favored by the Japanese group necessarily implies a solar-like reservoir of noble gases dissolved in the primordial magma ocean of the planet and is therefore difficult to reconcile with the enormous (and nonsolar-like) variations of ^{36}Ar between the present atmospheres of Earth, Venus and Mars. The difference between Earth and Venus is particularly puzzling since the model of Hayashi et al. (1979) would predict essentially identical primordial states for these planets. The resolution of this difficulty, if any is possible, must lie in subsequent differences in the evolutions of these planets.

Variations of noble gas content may be more readily reconciled with an essentially gas-free scenario in which adsorption of noble gases onto grains occurred in the earliest stages but the nebula was dissipated before large bodies accumulated (Wetherill 1981a). The Safronov model of terrestrial planet formation thus remains as the model with the fewest difficulties. It does not, however, explain the giant planets.

F. Giant Planet Formation

The central region of rock and ice in the giant planets is about ten Earth masses or more. (Some earlier models of Jupiter, in particular, gave lower values but the most complete modeling attempts give this result; see Hubbard and Stevenson [1984].) This material is almost certainly primordial in the sense that it probably did not rain out from overlying hydrogen-rich layers after the planet formed (Stevenson 1982b). Additional enrichment of heavy material is likely in the hydrogen-rich envelope, however. Indeed, it is known that the carbon abundance in the atmosphere of Jupiter is twice solar (Gautier et al. 1982) and the enhancement may be greater in Saturn (Courtin et al. 1983).

Since it is very unlikely that methane ice or clathrate could condense at Jupiter's orbit, the Jovian enhancement suggests accretion of Uranus and Neptune zone planetesimals after formation of the planet. We return to this in Sec. VI.

These observational constraints are consistent with "nucleated collapse" of the giant planets, in which a central dense core grows until the overlying atmosphere becomes so massive that hydrostatic equilibrium is not possible. The idea was first quantified by Perri and Cameron (1974) who obtained unreasonably large estimates for the required core mass because they assumed an adiabatic atmosphere. More realistic models (Harris 1978a; Mizuno et al. 1978; Mizuno 1980) indicate a core mass of about the observed size, although the results depend on assumed accretion rates and grain opacities. Collapse, once initiated, is slow (quasi-static) rather than hydrodynamic, except for the possible complications of molecular dissociation or ionization. The sequence of events may have been as follows. The cores of the giant planets may have grown in much the same way as the terrestrial planets were formed. However, these cores, unlike those in the terrestrial zone, became massive enough when the gas phase was still present that collapse was initiated. Subsequently, the protoplanets grew to their present size by the gaseous accretion of approximately cosmic composition material. The final masses may have been determined by the timing and rapidity of this accretion more than by the total reservoir of material available (which was presumably comparable for each planet). Detailed calculations of the collapse and accretion are in progress (Bodenheimer and Pollack 1984).

The main difficulty with this model is that the formation time for the large cores may be very long ($\sim 10^7$ yr). One possible resolution of this problem is that nucleated collapse occurs at a much lower mass, perhaps $\sim$ one Mars mass, and that many such bodies subsequently merge to form the giant planet (Stevenson 1984a). In this case, the similarity of core masses for the giant planets must be viewed as being merely an indication of the amount of solid matter in the accretion zone of each planet rather than the preferred mass for nucleated collapse.

G. Implications for Satellites

Are satellites an inevitable by-product of the planetary accumulation process? The answer is almost certainly yes, but a detailed answer requires consideration of the planetesimal dynamics and size distribution during accretion, and the timing of nebula dispersal.

Consider, first, the Safronov (gas-free) accumulation scenario. The question we need to address is whether a significant amount of material is emplaced in orbit around a growing protoplanet during the accretion. Four processes can be envisaged (although the distinctions between some of them are sometimes slight):

1. As material is accreted, the protoplanet angular momentum increases until fissional instability is achieved (Darwin 1880; Wise 1963). This is conceivable, in principle, if the planetesimals arrived with a sufficiently biased angular momentum. However, one would expect that the net angular momentum of the swarm relative to the protoplanet is very small. This is supported by the observed low angular momentum of solar system bodies relative to that needed for fission (see discussion of Wood 1977) and theoretical arguments (see, e.g., Harris 1977). Although fission leading to the formation of a ring (Durisen and Scott 1984) may be more favorable than the Darwinian notion of binary fission, the basic problem of excess angular momentum remains. This theory is, therefore, not attractive, at least in the form that it is usually stated (but see (4), below).

2. Opik (1972) pointed out that in a near encounter of a large (Moon-sized) body with a protoplanet, the innermost parts of the body may be moving less rapidly than the local Keplerian velocity and could, after tidal disruption of the body, go into orbit about the planet. Numerical calculations (Wood and Mitler 1974; Mitler 1975; Kaula and Beachey 1984; Nakazawa and Hayashi 1984, described in Hayashi et al. 1985) indicate that this is possible, but only if the heliocentric orbital velocity of the body is close to that of the planet. (A difference of ≤ 2 km s^{-1} is required for Earth orbit.) An example (numerical results of Nakazawa and Hayashi) is shown in Fig. 2. These events had rather low probability in the Safronov model, where most planetesimals would have been moving

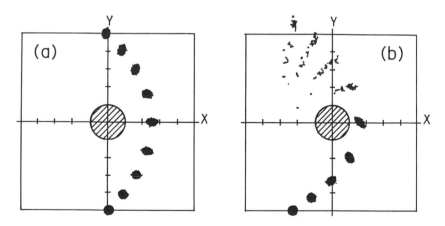

Fig. 2. Tidal disruption of a planetesimal in close encounter with a planet, according to calculations of Nakazawa and Hayashi (1984). The planetesimal is treated as a zero-strength, self-gravitating assembly of 560 elemental masses. In case (a), distances of closest approach is greater than 1.8 planetary radii, and no disruption occurs. In the closer encounter of (b), most of the fragments resulting from disruption end up in orbit about the planet.

faster, but since close encounters are much more frequent than impacts, this process has significant probability. This could be described as a "capture" scenario, although the capture is not as an intact body.

3.   Ruskol (1960), motivated by the ideas of O. Yu. Schmidt (see Ruskol 1982*a*) pointed out that when two planetesimals collide within the Hill sphere of a planet, at least some of the debris may have insufficient energy to escape the Hill sphere, yet sufficient angular momentum to avoid falling onto the protoplanet. Debris from many collisions of this kind can form a circumplanetary disk. Subsequent collisions between planetesimals and the disk can augment its growth. One or more satellites will form from aggregation within this disk at or beyond the Roche limit. This coaccretion model has been developed further by A. W. Harris and W. M. Kaula (1975) and is described in more detail in the chapter by

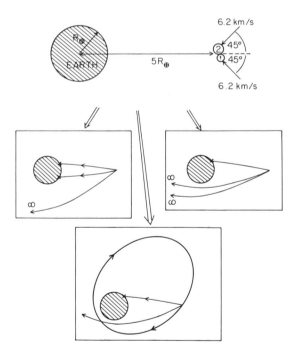

Fig. 3. An example of how collisions between planetesimals within the Hill sphere of a protoplanet can sometimes lead to orbital injection of debris. In this particular case, two bodies (one twice as massive as the other) enter the Hill sphere with velocities ~ one quarter of the escape velocity of the Earth and collide in such a way that three equal mass bodies emerge from the impact with half of the initial kinetic energy (in the center of mass frame) dissipated. Many possible outcomes were computed; the three shown are the most common for this particular example. Orbital injection occurs in a substantial fraction of cases. Impacts are also frequent and involve both bound and unbound trajectories.

Safronov et al. Some elementary aspects are discussed further below, and Fig. 3 shows examples of several possible outcomes of two planetesimals colliding within the sphere of influence of the proto-Earth.

4. Large bodies collide with the protoplanet. The resulting gas dynamical (non-Keplerian) motion of impact-generated vapor and debris leads to some orbital injection, causing formation of a disk from which one or more satellites may form. Large impacting bodies are needed because the gas dynamical effects must operate over a distance that is a significant fraction of the protoplanetary radius. The importance of large impacts was advocated by Hartmann and Davis (1975), and the physical concepts were more completely described by Cameron and Ward (1976) and Ward and Cameron (1978), for application to lunar origin. The basic principles are illustrated in Fig. 4 and a partial quantification is discussed in Sec. III.B. A variant of this scenario involves spin-out of a disk of superrotat-

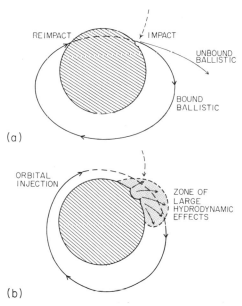

Fig. 4. (a) Impact event in which debris leaves the impact site ballistically. Kepler's laws dictate that in the absence of other collisions, every debris particle must either escape hyperbolically or reimpact the planet. (b) If a macroscopic zone of non-Keplerian motion exists, because of the pressure gradients associated with an expanding vapor cloud caused by the impact, then the periapse of some bound trajectories can be raised above the planetary surface, as indicated. This can lead to orbital injection and disk formation, provided the material has net angular momentum. The illustration is highly idealized because it localizes the non-Keplerian effects. In reality, even larger impacts lead to a planet-encircling cloud of gas and debris in which gas dynamical and collisional (viscous) processes are important even at large distances from the impact site. See text for more details.

TABLE II
Illustrative Inventories of Planetesimals Required to Form the Earth

$\alpha = 2^a$	$\alpha = 5/3^b$
1 Mars	5 Mars
10 Moons	3 Moons
10^2 Iapetus	12 Iapetus
.	.
.	.
.	.
10^9 1 km planetesimals	10^5 1 km planetesimals
TOTAL = 1 Earth Mass	TOTAL = 1 Earth Mass

[a] Equal amount of mass in each logarithmic mass interval down to 10^{-10} Earth masses.
[b] Equal amount of cross-sectional area is contributed by each logarithmic mass interval.

ing vapor and liquid formed from an impact event (in the spirit of Ringwood [1979]).

Leaving aside questions of efficiency for the moment, it is clear that coaccretion is most attractive if much of the planetesimal mass is in small bodies, and large impacts are likely to be important if most of the incoming mass is in large bodies. In many models of accretion, the mass spectrum $N(m)$ ~ $m^{-\alpha}$, with $5/3 < \alpha < 2$ (see discussion in Wetherill 1980a). Here, $N(m)\mathrm{d}m$ is defined as the number of planetesimals with masses between m and $m + \mathrm{d}m$. Table II shows an inventory of planetesimals required to form the Earth for two limiting cases, $\alpha = 5/3$ (for which each mass decade has an equal amount of surface area) and $\alpha = 2$ (for which each mass decade has an equal amount of mass). Each line in Table II refers to roughly a decade in mass (so that "10^2 Iapetus" means "roughly one hundred bodies each with a mass within a factor three of the mass of Iapetus." Several caveats and explanations are necessary before assessing the implications of these distributions. First, it is often said that the Safronov model predicts a second biggest body (i.e., the largest body accreted by a particular planet) which is smaller than the planet by a large factor ~ $(2\theta)^3$ ~ 10^3. This is an overinterpretation of the Safronov calculation, because the model artificially isolates a preferred embryo planet in a specified zone of accretion and does not, therefore, make any pretense of explaining why there are 4 (rather than, say, 10^2) terrestrial planets. Models which avoid this artificiality do not have a large discrepancy between a planetary mass and the largest accreted planetesimal (Wetherill 1980b). Second, the low mass cutoff in Table II is somewhat arbitrary but plausible; it corresponds to the planetesimal formed by gravitational instability. Third, the accretion is hierarchical, so the largest bodies in the list should arrive toward the end of accretion. Fourth, and perhaps most important, a power-law spectrum with $5/3 < \alpha < 2$ may have considerable physical justification for steady state populations but there are some indications that it might not actually be achieved. Greenberg et al. (1984, and references therein) argue for an early runaway accretion in which a few large bodies form but most of the mass remains in small bodies. Unfortunately, these "particles in a box" calculations can only describe earlier stages of accretion, since they do not include Keplerian dynamics. No calculation has ever been made for the evolution all the way from kilometer-sized planetesimals to fully grown planets; as a consequence, no firm justification for the spectra assumed in Table II exists.

To the extent that the tabulated planetesimal inventories represent reality, they provide support for more than one process of satellite origin if α is near 2 since there is then significant mass in both small and large bodies. Even in this limit, very close encounters and impacts with large bodies are inevitable. If α is significantly < 2, then coaccretion becomes less attractive and the role of late, large impacts assumes greater importance.

We turn now to gas-rich accretion scenarios of solid planets (Hayashi et al. 1985). The four possibilities listed above are then still possible, but much less likely. Fission is still highly unlikely and tidal disruption, though as likely as before, may lead to small debris that undergoes rapid orbital decay because of gas drag and collides with the protoplanet. Likewise, coaccretion and large impact will not work if they lead to material that undergoes rapid loss of energy and angular momentum. The time scale for this loss depends on whether the gas within the Hill sphere is supported entirely by pressure gradients or partly by orbital motion. In the former case, assuming high Reynolds number for the flow past the body, this time scale is of order the time it takes for the body to encounter a mass of gas equal to its own mass, roughly $\rho_0 r / \rho_g v$ (ρ_g = gas density, ρ_0 = density of the body of radius r moving at velocity v). For reasonable choices of $\rho_g \sim 10^{-6}$ g cm^{-3}, $\rho_0 \sim 1$ g cm^{-3}, v $\sim 10^6$ cm s^{-1}, this time scale is $\sim (r/1$ cm) in seconds, a short time except for very large bodies. However, gas drag offers the possibility of intact capture of large bodies, the hypothesis favored by Hayashi and coworkers for lunar origin (Nakazawa et al. 1983). We discuss this in Sec. VII much later in the review since, unlike other models discussed here, the satellite is formed in exactly the same way as the planet (although elsewhere in the solar system).

One very important mode of satellite formation remains, appropriate to giant planets. As gas accretes onto the protogiant planet, it may have sufficient angular momentum to form a disk. Subsequent evolution of this disk may have some analogies to the solar system, but also some striking differences. We analyze the formation and dynamics of these disks further immediately below (Sec. III).

III. DISK FORMATION

Clearly, most of the favored satellite formation processes involve formation of a circumplanetary disk of solids or gas. In this section, we examine the nature and efficiency of the formation of these disks.

A. Coaccretional Models

We wish to estimate the amount of material placed in orbit about a protoplanet by collisions of planetesimals within the Hill sphere (Ruskol 1960; see Fig. 3). Consider a sphere of radius Ar_p centered on the protoplanet, radius r_p, with $A \gg 1$. This sphere represents the region within which interplanetesimal collisions can produce bound orbits. ($A \sim 50$, typically.) Suppose that all the planetesimals are equal in size, with radius r and number density n. The mean distance that a planetesimal travels before hitting another planetesimal is $\lambda \simeq 1/\pi r^2 n$. Consider a particular planetesimal entering the sphere of radius Ar_p. The probability that it will hit another planetesimal whilst traversing this sphere is $\sim Ar_p/\lambda$. The probability that it will hit the protoplanet is $\sim 1/A^2$. If we suppose that almost all of the planetesimals eventually accrete onto the protoplanet, then clearly the total mass of the material placed in orbit is at most

$$m_{\text{disk}} \simeq \frac{(Ar_p/\lambda)}{(1/A^2)} \, m_{\text{planet}} \tag{13}$$

which is $\sim 10^{-4} (10 \text{ km}/r) \, m_{\text{planet}}$ for parameter values appropriate to Earth accretion, allowing for the (small) effect of gravitational focusing. The result is labeled as a disk mass because subsequent collisions among fragments will tend to cause coplanarity of orbits. The formula is not meaningful for $r \lesssim 1$ km, however, because such a planetesimal swarm would tend to have formed larger planetesimals at an early time. It is an upper bound because some of the material involved in these collisions will either escape or collide with the planet.

Many planets have satellites with mass $\lesssim 10^{-4}$ of the planet mass. However, it is not correct to assume that the disk mass estimate is necessarily the mass of a final satellite, if any. For example, Harris and Kaula (1975) showed that, if a Moon embryo $\gtrsim 10^{-5} \, M_\oplus$ (1 $M_\oplus \equiv$ one earth mass) is present when the proto-Earth is $\sim 0.1 \, M_\oplus$, then the embryo can grow by accretion to its present mass ($\sim 10^{-2} \, M_\oplus$), being fed by the same planetesimal swarm that feeds the proto-Earth. We will return to aspects of this coaccretion model later.

If $r = 10^2$ km, the m_{disk} can be formed by a single interplanetesimal collision. Clearly, the probability of this singular event drops rapidly as the planetesimal size increases beyond ~ 100 km. We conclude that a circumplanetary swarm is inevitable by the mechanism envisaged by Ruskol (1960), but may not be important unless most of the planetesimal mass is in bodies $\lesssim 100$ km radius. In fact, this criterion may not be satisfied, as we discussed above (see Table II).

B. Impact Generation of Disks

We first discuss solid body impacts (relevant to terrestrial planets). The giant planets are considered at the end of this subsection (III.B). As Fig. 4 illustrates, an impact cannot lead to orbital injection unless there are substantial non-Keplerian contributions to the motion of the debris (vapor, liquid and solid). These effects can arise from pressure gradients or viscous stresses in the expanding cloud and must operate over a substantial distance (meaning a significant fraction of one orbit) so that the Keplerian tendency for reimpact of negative energy debris is overcome. We consider, first, the ability of impacts to generate sufficient vapor to provide large hydrodynamic effects, and second, the distance over which these effects operate.

The heat of vaporization for silicates is $\sim 10^{11}$ erg g^{-1}. As Zeldovich and Raizer (1967) discuss, complete vaporization in the pressure release following an impact requires that the projectile have a specific energy content that is about a factor of ten greater than the heat of vaporization. This suggests that the mass of vapor produced is of order (one projectile mass) $\times$ ($v_{impact}/14$ km s^{-1})^2, a result first obtained by Stanyukovich (1950). Subsequent, more detailed calculations (Ahrens and O'Keefe 1972; Rigden and Ahrens 1981; O'Keefe and Ahrens 1982) predict a somewhat lower vapor production. If both the target and the projectile are cold and have Earth-mantle composition, then the production of one projectile mass of vapor (MgO, SiO, and O$_2$ molecules primarily) requires $v_{impact} \approx 20$ km s^{-1}. If both target and projectile are hot (2000 K), a plausible state for their interiors because of accretional heat (Kaula 1979,1980; Stevenson 1981), then $v_{impact} \approx 12$ to 14 km s^{-1} is required. In each case "one projectile mass of vapor" actually means that, upon expansion to an ambient pressure of 1 bar, a projectile mass of vapor remains uncondensed. Further adiabatic expansion causes the formation of silicate rain. Conversely, one projectile mass of vapor remains uncondensed at 1 kbar pressure for a lower velocity, $v_{impact} \approx 10$ to 12 km s^{-1} (material initially hot). This is a significant factor because gravity may prevent the expansion of the impact-generated gas cloud to low pressure, if the mass of vapor produced is very large. (0.1 Moon mass of vapor enveloping the Earth gives a basal pressure of 1 kbar.) It is significant that these velocity estimates are a little greater than the escape velocity from Earth and, therefore, comparable to the planetesimal impact velocity on Earth predicted by the Safronov theory, toward the end of accretion. This implies that the effects discussed here are applicable to only Earth and Venus among the terrestrial planets, and not to Mercury and Mars where impact velocities were lower unless a significant amount of very high-velocity material arrives later (e.g., Uranus and Neptune zone planetesimals; see Sec. VI). The detailed implications of these results require a computer simulation (O'Keefe and Ahrens 1982) but since the most highly shocked (hence, vaporized) material is among the material which leaves the impact site first and fastest, it is reasonable to suppose that for the above velocity estimates, hydrodynamic effects will be very important for at least the leading edge of the expanding gas cloud.

The conclusions thus far apply equally well to projectiles with a radius of one meter and with a radius of one thousand kilometers. However, the expanding plume from a small impact event is eventually modified by the back pressure of a preexisting atmosphere, as in nuclear detonation or even the 10 km radius bolide of the Cretaceous-Tertiary extinction event (Jones and Kodis 1982). Under these circumstances, little material escapes the vicinity of the impact and any that does escape is on a Keplerian trajectory for all but the first few kilometers of its motion and, hence, unlikely to end up in orbit. We can quantify this as follows: suppose that a particle is accelerated by gas dynamics up to a radial velocity v_r and tangential velocity v_θ at a height h above the surface of a planet, and subsequently moves only under the action of planetary gravity (i.e., Keplerian trajectory). We assume, for definiteness, a spherical planet with the same mass $M_\oplus$ and radius $R_\oplus$ as the present Earth. We express the initial velocity components at $r = R_\oplus + h$ in the form $v_r = \delta(2GM_\oplus/R_\oplus)^{1/2}$ and $v_\theta = (1 - \varepsilon)(2GM_\oplus/R_\oplus)^{1/2}$ and assume δ, $\varepsilon \ll 1$ and $h \ll R_\oplus$. It is simple to show that the subsequent trajectory is bound provided $\varepsilon > h/2R_\oplus$ and that for these trajectories, the periapse is above the planet surface provided $\varepsilon < 2/3(h/R_\oplus)^{1/2}$ and $|\delta| < (h/R_\oplus - 9/4\ \varepsilon^2)^{1/2}$. The fraction of possible radial and tangential velocities which lead to orbital injection are each of order $(h/R_\oplus)^{1/2}$ and the probability of injection is thus of order the product of these, i.e., $\propto (h/R_\oplus)$. In other words, the probability of orbital injection is proportional to the linear dimension of the zone in which the hydrodynamic effects are very important (other factors being equal).

We estimate this zone as follows. According to the Navier-Stokes equation, the change in kinetic energy of a gas element expanding from $r = R_\oplus + h$ to $r = \infty$ (neglecting gravity) for steady flow is given by

$$\int_{R_\oplus + h}^{\infty} u\ \frac{\partial u}{\partial r}\ dr = \frac{1}{2}(u^2|_{r = \infty} - u^2|_{r = R_\oplus + h})$$

$$= \int_{R_\oplus + h}^{\infty} -\frac{1}{\rho}\ \frac{\partial p}{\partial r}\ dr = \frac{\gamma p}{(\gamma - 1)\rho}\bigg|_{r = R_\oplus + h} \tag{14}$$

where $p \propto \rho^\gamma$ is assumed for the equation of state for adiabatic gas expansion. If we require that the change in $\frac{1}{2}\ u^2$ is 10% of the gravitational energy, then, in the case of the Earth, we need to define h as the height at which $p \approx 10^3$ bar, $\rho \approx 0.1$ g cm^{-3} ($T \approx 5000$ K, $\gamma \approx 1.2$; this assumes SiO + MgO + $\frac{1}{2}O_2$ vapor). For a projectile mass M_{proj} and a vapor production of

$\sim M_{proj}(v_{impact}/14 \text{ km s}^{-1})^2$, it follows that an appropriate choice of the zone dimension h is

$$\frac{h}{R_{\oplus}} \simeq 3\left(\frac{M_{proj}}{M_{\oplus}}\right)^{1/3}\left(\frac{v_{impact}}{14 \text{ km s}^{-1}}\right)^{2/3} \qquad (15)$$

(provided $h \ll R_{\oplus}$). This suggests that impacting bodies with mass $\gtrsim 10^{-3}$ $M_{\oplus}$ are capable, in principle, of injecting a significant fraction of their own mass in orbit. Of course, detailed quantification of how much escapes and how much reimpacts requires a numerical simulation (see, e.g., Cameron 1985), including the important effects of oblique impact. As a general rule, impacts at velocities greatly in excess of escape velocity will lead to hyperbolic escape rather than orbital injection (see Sec. V) (see Benz et al. 1986).

We turn, now, to orbital injection from impacts into gaseous bodies. In some respects, this is easier because the target material is already in a high entropy state. However, the impacting body may plunge deep (many times its own radius) into the planet before it encounters a mass of gas comparable to itself. Much of the shocked gas may be prevented from entering orbit by the presence of unshocked, overlying gas. (Although the projectile punches a cylindrical hole in the atmosphere which remains open for a significant time, shocked gas that escapes out through this opening may not have the right trajectory for orbital injection.) This problem requires numerical simulation. It has been suggested that impact on Uranus could create a disk from which the satellites formed (Cameron 1975; Stevenson 1984b).

C. Gaseous Accretion

If the giant planets formed by a nucleated instability, as discussed in Sec. II.F, then gas flowing into the Hill sphere can form a compact disk (radius $\ll$ Hill sphere radius) about the protoplanet. One possible estimate of the disk radius r_d can be obtained from the analysis of Cassen and Pettibone (1976) by assuming gas elements conserve specific angular momentum Γ once they enter the Hill sphere, and equating centrifugal to gravitational forces: $\Gamma^2/r_d^3 \sim GM/r_d^2$, where M is the protoplanet mass. For a plausible value of Γ $\simeq r_H^2 \Omega/4$ (see Ruskol 1982a), where r_H is the Hill sphere radius and $\Omega \equiv (GM_{\theta}/r_p^3)^{1/2}$ is the Keplerian angular velocity at the planetary orbital radius r_p, it follows that

$$r_d \simeq r_H/48. \qquad (16)$$

This predicts a disk radius ~ 20 times the present radius of a giant planet, typically, coincidentally about the same size as the regular satellite systems of the giant planets, as we discussed in the introduction. Numerical simulations of gas accretion (Sekiya et al., described in Hayashi et al. [1985]) indicate streamlines which lead to prograde gas flow about the protoplanet (as required to explain the observed satellite systems) and an angular momentum comparable to or less than that used in the above example. However, direct application of Cassen and Pettibone (1976) may not be appropriate because of the importance of tidal effects. A complete description may also require inclusion of the processes (e.g., eddy viscosity) whereby the resulting disk redistributes angular momentum. This problem requires more work.

The current angular momentum of the giant planets is much less than that implied by the above discussion. It is clear that enormous amounts of angular momentum must be lost. (Notice that this is a very different dynamical situation from that applicable to gas-free terrestrial planet accretion, where the random velocity of planetesimals is more important than the Keplerian velocity difference across a Hill sphere and the average net angular momentum of accreting planetesimals is very low.) However, it is likely that processes exist to remove this angular momentum, including tidal interaction of the disk with the solar nebula and possibly hydromagnetic processes later in the evolution (invoked by Cameron [1975]). Nevertheless, more work is needed on this to assess whether this is quantitatively acceptable.

IV. DYNAMICS OF A PROTOSATELLITE DISK

Now that we have demonstrated that a circumplanetary (protosatellite) disk is likely to be formed during planetary accretion, it is necessary to examine its dynamics. Several dynamical processes are present in a particulate or gas/solid disk in Keplerian motion which tend to dissipate it or rearrange its component parts. In order to evaluate the relative importance of these processes in the solar nebula vs. a protosatellite disk, it is useful to compare the dimensional time scales of each process. In Table III we list approximate values of the properties of a low-mass solar nebula in the Jupiter/Saturn zone, compared to a low-mass protosatellite disk, estimated by smearing out the satellite masses (e.g., the Galilean satellites or Saturnian satellites), augmented to planetary abundances by gas, over the relevant orbital dimensions. If the process of satellite formation is inefficient, or if the protosatellite disk were of solar rather than planetary composition, then a more massive disk would be inferred (cf. Weidenschilling 1982). The next entry in Table III is the size of gravitational instability products which would arise if the solid particulate component of the disk were to fragment into discrete bodies (Goldreich and Ward 1974; Ward 1976). Following that, we list time scales of several dynamical processes which tend to alter the disk. Dimensional equations for each of these processes were summarized by Harris (1984a), along with references to the original works. We here recall only the nature of each process.

Viscous spreading of the gas has already been mentioned (Sec. II.A). The lower limit tabulated assumes a fully turbulent disk ($\alpha \sim 0.1$ in the previous discussion); if turbulence is not well developed, the time scale of spreading is much larger. The case for turbulence in a protosatellite disk is much

TABLE III
Comparison between Solar Nebula and Protosatellite Nebula Properties and Time Scales

	Solar Nebula	Protosatellite Nebula
Properties		
Temperature	200 K	200 K
Orbit radius	$\sim 10^{14}$ cm	$\sim 10^{11}$ cm
Orbit frequency	$\sim 10^{-8}$ s^{-1}	$\sim 10^{-5}$ s^{-1}
Surface density, solids	3 g cm^{-2}	3×10^3 g cm^{-2}
Surface density, gas	200 g cm^{-2}	3×10^4 g cm^{-2}
Scale height of gas disk	10^{13} cm	3×10^{10} cm
Gas density	2×10^{-11} g cm^{-3}	10^{-6} g cm^{-3}
Size of gravity instability products	100 km[a]	100 km[a]
Time Scales in Years		
Viscous spreading (gas)	$> 10^5$	$> 10^2$
Gas drag, disk	10^7	10^3
Gas drag, large particles	10^8 $(r/1$ km$)$	10^3 $(r/1$ km$)$
Drag from infall of circumsolar matter, disk	—	m_s/m_p (τ_{planet}) [b]
Drag from infall of circumsolar, large particles	—	r_s/r_p (τ_{planet}) [b]
Coagulation/fragmentation	2×10^5 $(r/1$ km$)$	0.1 $(r/1$ km$)$
Gravitational instability	10	0.03
Cooling/condensation	—	10^6

[a] Could be as little as 1 km. See Greenberg et al. (1984).

[b] τ_{planet} is the time scale for accreting the planet, m_p = mass of planet, m_s = mass of satellite, r_s = radius of satellite, r_p = radius of planet.

stronger than for it in the solar nebula, since the higher surface density of the former assures an optically-thick, and thus likely convective, disk.

Aerodynamic drag between the solids in the disk and the gas will likely exist due to the partial support of the gas by the radial gradient of the gas pressure, leading to a slightly non-Keplerian orbital velocity of the gas component. For an optically-thick disk of small particles, the drag can be estimated in terms of boundary layer flow (see, e.g., Goldreich and Ward 1973), or for larger particles where the gas penetrates the solid layer, one can estimate the drag forces on individual particles (see, e.g., Weidenschilling 1977,1982). Both time scales are listed in Table III. The crossover occurs at an optical depth of the solids of order unity, or at a mean particle size of ~ 3 cm in the solar nebula, or ~ 30 m in the protosatellite disk.

In the case of a satellite disk, matter may be falling onto the disk from circumsolar orbit over a time long compared to the other evolutionary time scales of the satellite disk. Such might be the case for the rate if giant planet formation were controlled by the time scales of coagulation and/or viscous spreading in the solar nebula (cf. Table III). Matter arriving from circumsolar orbit, subject to gravitational forces alone, would impinge on the disk with much less than orbital angular momentum; hence, it would cause a drag (Harris and Kaula 1975; Harris 1978b). The result of this is that a satellite, or a disk capable of stopping the matter falling on it, will decay into the planet upon encountering a mass comparable to its own from solar orbit. The time scales of this drag on a disk or on a protosatellite are estimated by Harris (1978b), and summarized in Table III. Values are not tabulated for the solar nebula, since most dynamical studies agree that the initial collapse of the protostellar cloud to form the protosun and solar nebula disk occurred more rapidly than the subsequent cooling, condensation and accumulation of the planets.

The time scale of growth of solid bodies by collisional aggregation in the absence of gas (Safronov 1969) is listed next. Rather similar time scales may apply in the presence of gas (Weidenschilling 1982). The same time scale exists for particles destroying each other by collisional fragmentation. If both aggregation and fragmentation are occurring, the time scale of change of particle size may be much longer or the size distribution may even be in equilibrium.

The next time scale listed is that of gravitational instability, which is of the order of the orbit period. Inside the Roche limit of the protoplanet, these instabilities still occur (Ward and Cameron 1978) but lead to rapid viscous spreading rather than formation of large bodies. A similar process can occur in a liquid-gas mixture (Thompson and Stevenson 1983).

The last time scale listed is the time required for Jupiter or Saturn to cool sufficiently to allow ices to condense in the zones where ices are seen, assuming the planets formed more rapidly than that (Pollack et al. 1976; Bodenheimer et al. 1980).

The conclusion which we draw from the various time scales in Table III is that, in general, the disruptive processes which are expected in a proto-satellite disk act on time scales much shorter than the formation and/or condensation time scales of such a disk, but much longer than the gravitational instability or coagulation time scales. Thus a gaseous disk which formed along with a giant gaseous protoplanet might be dissipated even before the solids could condense. A disk about a slower-forming planet would be subject to rapid decay as a result of further infall of circumsolar material; however, the solid components of such a disk might coagulate even more rapidly. Hence, we envision a sequence of formation in which the gaseous component of the disk is rapidly evolving, with the solid particles coalescing into protosatellites on an even shorter time scale. These intermediate-sized bodies still evolve quite rapidly compared to the growth rate of the planets; hence, the present satellite systems may be only the last few of a much larger suite of satellites formed during the growth of the planets, and may have evolved substantially from the orbital positions at which they formed. This is strikingly different from the evolution of the solar nebula.

Several problems remain in understanding disk evolution, primarily centered around the redistribution of angular momentum at resonances. (The same processes play an important role in the dynamics of Saturn's rings.) As a satellite forms in a disk, it may clear a gap (tunnel) in the surrounding material, modifying the accretion rate and accretion environment (Lin and Papaloizou 1979a; Papaloizou and Lin 1984). The implications of this for protosatellite disks are still poorly understood (see Appendix of Lunine and Stevenson 1982).

V. SATELLITE ACCRETION

The accretion of a satellite from smaller components provides the initial source of energy for surface evolution, internal differentiation and, in some cases, creation of a primordial atmosphere. The mode and end state of accretion for a given-sized satellite are a strong function of conditions in the region of formation, i.e., presence or absence of gas, ambient temperature/pressure, and time scales of dynamic versus thermal processes. Voyager images of Galilean and Saturnian satellites have prompted comparative models of satellite formation in an attempt to constrain modes of accretion (see, e.g., Schubert et al. 1981; Coradini et al. 1982b; Lunine and Stevenson 1982). The results so far have been ambiguous but further progress both in observations and theory may elucidate the role of accretion in subsequent satellite evolution. In this section we review current ideas on satellite accretion processes, as well as what can be learned about such processes by looking at the present states of these bodies. Closely tied to accretional processes is the subsequent thermal evolution of satellites. Although we touch on this where necessary to illustrate the coupling between formation and present state, we do not attempt

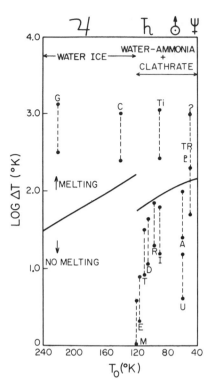

Fig. 5. Crude estimate of surface temperature rise ΔT during accretion (Eq. 17) vs. ambient nebular temperature T_0 in formation region. Range in accretion temperature reflects uncertainty in specific heat of material into which accretional energy is deposited. Solid lines indicate threshold temperature rise for which water ice melting occurs; for $T_0 < 120$ K melting of the ammonia-water eutectic defines the threshold. Symbols: G (Ganymede), C (Callisto), Ti (Titan), M (Mimas), E (Enceladus), T (Tethys), D (Dione), R (Rhea), I (Iapetus), A (Ariel), U (Umbriel), ♇ (Pluto), TR (Triton). The mass of Triton is very uncertain.

a review of thermal models (at the risk of neglecting some work) since this topic is covered in the chapter by Schubert et al.

We concentrate on the largest satellites, which are massive enough to have undergone substantial melting (and hence evolution) during accretion. We also focus on giant planet satellites because the only other large satellite (the Moon) is covered in a separate chapter by Kaula et al. Figure 5 plots the surface temperature rise due to accretional energy input for the icy outer-planet satellites. The maximum accretional temperature rise ΔT due to planetesimals encountering the satellite at the surface escape velocity is

$$\Delta T = \frac{GM}{rC_p} \tag{17}$$

where M and r are mass and radius of satellite (here the present-day values), C_p is the specific heat at constant pressure of the material in which the energy is deposited, and G is the gravitational constant. The range in accretional temperature rise in the figure reflects a range of C_p values, depending on whether the energy is deposited in the growing body itself or in a gaseous envelope surrounding the body (the composition of which also changes C_p). Equation (17) assumes zero initial planetesimal velocity, and all energy is buried in the near-surface or atmosphere with none lost to space by radiation or redistribution by conductive or convective processes *in* the body. It also ignores latent heat effects caused by melting of the icy component of accreting material. Nevertheless, it provides a useful crude scheme for gauging the relative importance of accretion amongst the satellites. Also plotted is the threshold surface temperature rise above which the bulk icy constituents of the satellite would melt. Melting is considered a crucial stage of accretion since rock-ice differentiation is initiated, which (as discussed below and in chapters by Schubert et al., by McKinnon and Parmentier, and by Squyres and Croft) may have profound effects on the further evolution of these bodies. Also, the vapor pressures of the volatile constituents are sufficiently high at the melting point that the existence and composition of a primordial atmosphere then becomes an issue. Planetesimals are assumed to contain water ice as the main volatile constituent in the Jovian system, and ammonia water plus clathrate hydrate in the Saturn system and beyond (Lewis 1974). The possible existence of a strong temperature gradient in the Jovian nebular environment has been argued above; a similar gradient in the Saturnian nebula is somewhat less compelling in view of the apparent lack of a simple compositional gradient in the regular satellites of that system (Tyler et al. 1982).

Figure 5 shows that Ganymede, Callisto, and Titan are sufficiently massive that large-scale melting of these bodies during accretion was possible; modifications to the models which might limit Callisto's melting will be discussed below. Io and Europa also could have experienced large temperature increases; they are not plotted because they are rocky objects which may not have incorporated much water ice during accretion (Pollack and Reynolds 1974) and hence may not have melted (although silicate dehydration could have occurred). The intermediate-sized Saturnian and the Uranian satellites (using masses from Veillet [1983a], for the latter) are not massive enough to have undergone accretional melting; Rhea, Oberon, and Titania may be marginal cases. Triton and Pluto are truly intermediate in that some modest melting of these bodies is likely.

Figure 5 thus suggests that we may profitably restrict ourselves to looking at the Galilean satellites, Titan and perhaps Triton and Pluto, if we wish to understand the effect of accretional energy on the internal structure of satellites and hence subsequent satellite evolution. The intermediate-sized Saturnian satellites may provide a record of the incorporation of highly volatile substances such as methane clathrate hydrate; however, these satellites accreted cold enough that they were likely primarily homogeneous mixtures of rock and ice at the end of formation (see, e.g., Stevenson 1982a; Ellsworth and Schubert 1983). Given the restricted group of satellites we are considering, what might the present surface state of these bodies tell us about their origins? We outline the relevant observations here; they are examined in light of possible accretion models discussed later in this section.

Perhaps the most dramatic diagnostic, and yet one which has so far eluded satisfactory explanation, is the striking difference in surface features between Ganymede and Callisto as revealed by Voyager (Smith et al. 1979a; see the chapter by McKinnon and Parmentier). Ganymede is characterized by two terrains: an old, heavily cratered, low-albedo terrain with estimated age in excess of 4 Gyr and a younger (~ 3.5 Gyr), smoother, higher albedo terrain characterized by grooves (long, linear features, a few hundred meters in height) of unknown tectonic origin. Callisto's surface consists entirely of the old, heavily cratered terrain. Either Callisto did not undergo the event or events responsible for the partial resurfacing of Ganymede, or underwent this process so early in its history that the evidence has been obliterated. The former interpretation is most plausible. Although some models explain the presence of resurfaced terrain on Ganymede (and lack thereof on Callisto) in terms of the different masses and hence radiogenic heating of these bodies (Cassen et al. 1980a), the similarity in bulk properties has led others to propose a more fundamental difference in conditions surrounding the bodies during accretion and hence in the degree to which their interiors were initially differentiated (Schubert et al. 1981; Lunine and Stevenson 1982). Because of the large amount of accretional heating and melting which both Ganymede and Callisto could have undergone, it has not been possible to conclude with certainty that the end states of their formation were different; however, qualitative differences for certain accretional models are possible and are described below.

Most recently Voyager and groundbased data on the composition of Titan's atmosphere have opened the possibility that clues to the formation physics of this body may be found in the abundances of key atmospheric gases. Titan's atmosphere is primarily N_2, but with a small amount of CH_4 and possibly Ar (see review by Hunten et al. 1984 and the chapter by Morrison et al.). Other species (e.g., CO) are present in small amounts and may have implications for accretion, but the most important issues are as follows. In what form was nitrogen supplied to Titan? What is the total reservoir of methane? Was all the carbon supplied to Titan originally in the form of methane? What is the noble gas reservoir? Although work on these issues is in its early stages (see Owen 1982; Lunine and Stevenson 1982a; Lunine et al. 1983; Lunine 1985) it is becoming clear that both the trapped gas composition of incoming planetesimals (i.e., clathrate hydrate) as well as the mode of formation of the satellite may have strongly influenced the current state of the atmosphere and surface. Triton's and Pluto's surface ice and gas composition may eventually provide similar constraints on their modes of accretion; both observation and theory need to be advanced before such constraints can be applied. Both Triton and Pluto have methane (Apt et al. 1983) and it is possible that Triton has condensed nitrogen (Cruikshank et al. 1984a).

Models of satellite accretion may be divided into gaseous and gas-free, depending on the ambient nebular gas pressure in the region of formation. The two cases are distinguished primarily in the way that infalling planetesimals deposit their energy in and around the growing satellite. Those two extremes in turn may be divided depending upon accretion time scales, size of growing body, and opacity in a gaseous envelope surrounding the body. In one gas-free limit, large planetesimals impacting the surface deposit a portion h of their energy in the near surface layers. We write the energy balance as (Safronov 1969; Kaula 1979):

$$h\left(\frac{v^2}{2} + \frac{GM}{r}\right) \frac{dm}{dt} = C_p \, \Delta T \frac{dm}{dt} \tag{18}$$

where v = initial planetesimal velocity and dm/dt = growth rate of satellite.

Models of Jovian and Saturnian satellite accretion based on this process have been constructed by Coradini et al. 1982; Schubert et al. 1981; Ellsworth and Schubert 1982, and lunar accretion by Kaula (1979) and Ransford and Kaula (1980). A major uncertainty which plagues this model is the choice of h, the fraction of energy of the impacting body retained as heat in the accreting satellite. Estimates of h in the literature range from 0.1 to 0.9 (Schubert et al. 1981; Kaula 1979). Kaula estimated h to be 0.8 to 0.9, based on computer modeling of energy partitioning in the initial stages of projectile impact (O'Keefe and Ahrens 1977). As he notes, however, a substantial fraction of this retained heat energy may be radiated away in jets of material flung up by the impact process, since the flight time of some of the material can be minutes. Because of uncertainties in calculating the radiative loss, an estimated h value of 0.1 to 0.2 was used in Kaula's studies. Coradini et al. (1982b) calculated the partitioning of energy in large impacts for the case of the icy Galilean satellites, and found $h \sim 0.1$ to 0.2. So long as the impactor size is greater than the thermal skin depth of the surface, significant energy retention is expected.

An alternative model applicable to accretion of small planetesimals, or accretion in a gaseous but optically-thin medium is

$$T_s = \left[T_0^4 + \frac{\left(\dfrac{GM}{r} + \dfrac{v^2}{2}\right)\dfrac{dm}{dt}}{4\,\pi\,R^2\,\sigma} \right]^{1/4} \tag{19}$$

where σ = Stephan-Boltzmann constant, T_0 = ambient nebular temperature, T_s = accretional surface temperature. This expression, from Safronov (1969),

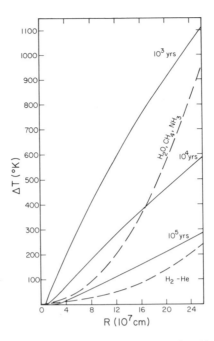

Fig. 6. Surface temperature rise during accretion vs. radius for objects with radiative (solid lines) and convective (dotted lines) envelopes. The former corresponds to the limiting case of small planetesimals in gas-free accretion (Eq. 19). Accretion times in years indicated for radiative cases; results for two different envelope compositions are shown for convective case.

corresponds to the limiting gas-free case in which the surface is bombarded by small planetesimals which do not deposit any energy in the interior of the satellite but instead radiate it away at the surface, and should thus apply crudely to the situation where planetesimals are slowed and broken up in an optically-thin gas and ultimately strike the satellite surface as small objects. Hanks and Anderson (1972) have applied this model to the case of gas-free lunar accretion by small planetesimals. Figure 6 plots temperature rise from Eq. (19) vs. accreting satellite radius for zero initial relative velocity and various choices of accretion time scales (which have been discussed above). Note that for the shorter plausible time scale (10^2 to 10^4 yr) huge surface temperatures are achieved in the radiative case.

If one considers higher gas densities which may be obtained in optically-thick Jovian and Saturnian nebulae, the accretion process is different. When the growing core of a satellite exceeds a mass of roughly 10^{25} g, it gravitationally dominates a large region of the surrounding gas. The extent of this region, the Hill sphere, is defined as the distance to the inner Lagrangian point of the satellite (Mizuno 1980) and substantially increases the cross section of the satellite for planetesimal capture. The region in which capture is likely is given by $\sim GMc^{-2}$ where c is the sound speed. Planetesimals spiral through and break up in the Hill sphere, depositing a portion of their kinetic energy in the envelope. As shown in Lunine and Stevenson (1982), the energy deposition rate for accretional time scales of $\lesssim 10^4$ yr is sufficiently high to force a convective temperature structure in the gas envelope, even when only gaseous opacity is considered. In the limit that planetesimal material equilibrates completely with the atmosphere and rains out gently on the surface, the surface temperature is determined by the atmospheric adiabat. One may then calculate the temperature and pressure through the envelope down to the surface. The adiabat is tied to the pressure and temperature in the nebula which is assumed to be continuous across the Hill sphere boundary. In fact, the interface between nebula and satellite envelope is subjected to gas flow processes (Miki 1982) which may introduce pressure and temperature perturbations; the possible implications of this flow for satellite accretion conditions have not as yet been investigated in the literature.

To verify that sufficient energy is deposited to maintain an adiabat, one can calculate the surface temperature rise from energy deposited by incoming planetesimals impacting the surface using a modification of Eq. (17), and verifying that this value is higher than the temperature obtained at the base of a convective atmosphere. For Ganymede, Callisto and Titan this criterion is satisfied throughout much of the accretion, even when latent heats of melting and vaporization of various icy constituents are accounted for. For Io and Europa, latent heat of dehydration of silicates may inhibit convection during portions of the accretion process.

Figure 6 plots surface temperature rise vs. surface radius of the satellite for the convective envelope case. Here a simple dry adiabat is used. The lower dotted curve is for a hydrogen-helium atmosphere, the upper for one dominated by water, ammonia or methane vapor. The latter case is relevant to late stages of accretion when substantial volatilization of ices has occurred.

To compute the envelope adiabat properly, we must account for the condensation processes as planetesimals disseminate their ices. The saturated adiabat, in a form generalized somewhat from Houghton (1977, p. 19), is given in Lunine and Stevenson (1982a). It differs significantly from the dry adiabat once the melting point of the ice is exceeded and can be considerably different in the limit where the vapor dominates the gas pressure. An equally important consideration is the condition at the envelope-nebula interface, which must thus be chosen carefully. For example, in the Galilean satellite system a constraint on the nebular temperature is provided by the bulk densities of the satellites, which imply a hydrated silicate composition for Europa vs. ice-silicate for Ganymede. A plausible radial temperature gradient (see Sec. II) of $T \propto R^{-1}$, where R is distance from Jupiter, constrains temperatures at Callisto once the temperature near Ganymede is fixed to be just below the condensation point of water ice (see also Pollack et al. 1976). This yields 240 K at Ganymede and 130 K at Callisto, taking into account solar insolation which lessens the temperature gradient near Callisto. Note that optically-thin nebulae possess temperature gradients $\propto R^{-1/2}$ (Pollack and Reynolds 1974); however, this is physically implausible since protoplanetary nebulae are optically thick in molecular H_2 alone. Optically-thick nebular models provide the greatest temperature contrast between the formation regions of Ganymede and Callisto, and hence between the surface temperatures of the accreting satellites themselves.

We compare the results of the above models of accretion with what is observed at present on large satellites and focus first on gas-free studies of lunar formation. A major challenge for lunar accretion models has been to account for near-surface melting and differentiation of the Moon, implied by isotopic evidence (Wasserburg et al. 1977). Accretion by small bodies, in which the impact energy is radiated away at the surface, can result in melting only for very short accretion time scales, $< 10^4$ yr in the models of Hanks and Anderson (1972). The plausibility of such time scales for the Earth-Moon system is discussed elsewhere in this chapter (see Sec. II.E). Kaula (1980) considered the case of large impactors and explicitly calculated depth of impact energy deposition and conduction of the impact heating through the interior to construct an accretional thermal profile. He found that melting could be achieved with reasonable heat retention factors ($h \gtrsim 0.2$). However, unless the planetesimals were $< 10^{-3}$ times the mass of the Moon, and of uniform mass, the melt region was too deep in his model interior to be consistent with the geochemical data. The host of issues associated with lunar formation, centered in particular on geochemical evidence, are treated in the chapter by Kaula et al. devoted to the Moon. The point we wish to make here is that evidence exists for early melting of at least a portion of the lunar interior, and has motivated models for accretion-induced melting in a gas-free environment. The apparent uniqueness of the Earth's Moon, however, makes it difficult to constrain the relevant accretional processes; one feels more comfortable examining a suite of bodies with similar or systematically varying characteristics. It is also possible that lunar formation involved very different processes from what we discuss here (see Sec. VIII). To this end we focus in the remainder of this section on the Galilean satellites and Titan.

The strikingly regular density gradient and orbital characteristics of the Galilean satellite system has encouraged attempts to constrain models of accretion using the observed satellite properties. The densities of the inner two Galilean satellites imply that they are primarily silicate bodies, with the silicates probably hydrated to some degree initially (Lewis 1971a). The present states of Europa and Io are consistent with either gaseous or gas-free accretion, in that either type of scenario provides enough energy input in principle to produce some dehydration of the silicates during the later stages of accretion. However, the surfaces of both satellites show evidence for subsequent modification by internal melting or dehydration which erased their post-formational states. For Io the primary driving process is probably tidal dissipation (Peale et al. 1979); the role of tidal heating for Europa is less clear but could be important (Squyres et al. 1983b; cf. chapter by Schubert et al.). Therefore, their presently observed surfaces may not be diagnostic of accretion processes.

Ganymede and Callisto, on the other hand, may provide strong constraints on the physical mechanisms operating during accretion, largely because their inferred similar bulk physical properties and ice-rock compositions are overlain by contrasting surfaces. Schubert et al. (1981) proposed that Callisto shows no evidence for tectonic activity because it is an undifferentiated body which removes radiogenic heat by subsolidus convection. Ganymede on the other hand, was partially differentiated after formation with an ice upper mantle and rock lower mantle overlying an undifferentiated core. This unstable situation, it is claimed, would lead to core overturn and subsequent resurfacing as a result of the redistribution. Although the precise mechanism for resurfacing is likely more complex and still controversial (see Squyres [1980a] and more recently Kirk and Stevenson [1983] as well as the chapter by McKinnon and Parmentier), the implication that Ganymede is differentiated while Callisto is not remains attractive. Both simple gaseous and gas-free models predict substantial melting and hence differentiation for these two bodies, despite the fact that ambient nebular temperatures for the two bodies differ by almost a factor of two in the optically-thick case. Differences in asymptotic approach velocities of planetesimals striking Ganymede and Callisto in principle would have produced some gradation in melting between the bodies if such differences were large and the velocities themselves comparable to the escape velocities from the satellite surfaces. Neither of these criteria is likely to have held for Joviocentric planetesimals (see previous sections). Solar orbiting bodies would accrete onto both satellites with similar velocities many times the satellite escape velocities (based on numbers in Shoemaker and Wolfe [1983]). If such material constituted a substantial fraction of the accreting mass, substantial amounts of melting would occur in *both* Ganymede and Callisto. We thus find it unlikely that differences in planetesimal approach velocities could have produced substantial melting on Ganymede but not Callisto. The gas-free accretion scenario can produce modest melting in Ganymede and essentially none in Callisto if a sufficiently low value of the heat retention factor ($h < 0.1$) is chosen; the different ambient temperatures in the two regions would then put Ganymede and Callisto on opposite sides of the melting curve. It is difficult to justify on physical grounds the *particular* choice of h that simultaneously would achieve these states, although low values of h may not be unreasonable (Schubert et al. 1981). Another approach is the model of Coradini et al. (1982b) which calculated thermal profiles within Ganymede and Callisto during accretional heating, taking account of heat conduction/convection and the depth to which the projectile initially deposits energy. Both bodies then possess substantial internal melted regions capped by a solid crust, with Ganymede's crust only half as thick as Callisto's. (Solid crusts are produced in the model because at a given location, the radiative cooling time of the impact-melted zone is less than the time between impacts.) Although the authors suggest that Ganymede's thinner crust could favor groove formation on its surface as opposed to Callisto's, no explicit mechanism dependent on the different thicknesses is proposed.

It is possible that the optically thick models lead to a large difference in degree of internal melting. Lunine and Stevenson (1982a) consider the consequences of planetesimal breakup due to gas drag and find that much of the material which rains down onto the surface is small enough to have low terminal velocity yet large enough that it has not equilibrated with the hot atmosphere by conduction. These meter-sized icy fragments cool the surface and allow Callisto to form without substantial melting. Ganymede is nevertheless predicted to have a substantial primordial water ocean.

Another interesting but less well-understood aspect of accretion in a gaseous medium concerns the extent to which the large icy satellites may evaporate H_2O and thereby increase their rock/ice ratio to the observed enhanced value of around $0.6:0.4$ by mass (the solar value is $0.4:0.6$). This would be a hydrodynamic process, in which part of the infall energy is made available for outflow of H_2O vapor. The process is energetically conceivable because for the large icy satellites $GM/rL \sim 1$, where L is the latent heat of vaporization for water ice.

Before leaving Ganymede and Callisto, it is worth noting that if their differences are attributed to varying degrees of differentiation of their interiors, then this favors homogeneous over heterogeneous accretion, since the accretion of a rocky body prior to the addition of an icy veneer will produce a differentiated structure independent of surface accretional temperatures. In any case, the short dynamical time scales indicated above make implausible models in which rock accretes into a satellite core prior to the condensation and accretion of water ice (Pollack and Reynolds 1974).

Titan offers us an example of a body for which the incorporation of highly volatile material, i.e., ammonia-water ice and clathrate hydrate, as well as the accretional process itself, may have played a key role in producing the presently observed atmosphere. It is remarkable that some of the small Saturnian satellites show clear evidence of resurfacing, but Callisto (which is much larger) does not. This argues for the presence of a more volatile ice in the Saturnian satellites and its *absence* in Callisto (and, by implication, Ganymede). Current understanding of how a H_2O-NH_3 satellite would evolve (Stevenson 1982a; Hunten et al. 1984) strongly suggests that resurfacing would take place. Thus, it is likely that bulk quantities of ammonia and more volatile clathrate hydrate did not condense in any system closer to the Sun than Saturn's. The possibility of the incorporation of methane clathrate hydrate in Titan was suggested by Miller (1961). There are no published quantitative accretion models of Titan. A rough gas-free accretion calculation along the lines of Schubert et al. (1981), using a similar h value, produces substantial melting of the ammonia-water component of the body. Whether an initial atmosphere can be formed and retained in a gas-free scenario is at present unclear; possible outcomes of Titan accretion scenarios are tabulated in Hunten et al. (1984). Two of the authors (Lunine and Stevenson) have calculated a gaseous accretion case assuming planetesimals contained water ammonia ice and (primarily) methane clathrate hydrate initially. The high volatility of these constituents results in the satellite's initially hydrogen- and helium-dominated atmosphere quickly becoming ammonia- and methane-dominated. The resulting low atmospheric specific heat coupled with the lower atmosphere running off the clathrate saturated adiabat late in the accretion, produces high temperatures, up to 600 K, and high methane and ammonia pressures, of order 0.5 kilobar total at the close of accretion. Most of the water-ammonia component of Titan then comprises a liquid mantle which extends to the surface, in contact with the ammonia-methane atmosphere. Most of the mass of chemical species more volatile than ammonia and originally sequestered in the planetesimal clathrate (e.g., methane and the noble gases) will have been released into the atmosphere during accretion. This reservoir provides a potential source for the present-day atmosphere. Discussion of the as yet speculative processes of escape of the primordial atmosphere, subsequent conversion of the NH_3 to N_2 by photolysis (Atreya et al. 1978) and incorporation of some of the gases into a near-surface clathrate reservoir are beyond the scope of this review. It is interesting to note, however, that the current atmospheric pressures of N_2 and CH_4 (Lindal et al. 1983) are close to what would be expected if these gases had been in equilibrium with clathrate at 180 K, the freezing point of an ammonia-water liquid. (If a hydrocarbon ocean is present today, the CH_4 value will be determined by ocean composition and hence may be misleading; see Lunine et al. [1983].) This suggests that the present atmosphere evolved from some much hotter state. It is not clear whether such an atmosphere could have evolved in a gas-free accretion scenario. Our ability to constrain modes of accretion to be compatible with the observed present conditions on Titan is as yet not satisfactory; in particular, accretion in a CO-N_2-rich as opposed to a CH_4-NH_3-rich nebula (Lewis and Prinn 1980; Prinn and Fegley 1981) has yet to be fully explored.

One problem which should be of concern in gaseous accretion models is the question of the stability of the gas envelope to hydrodynamic collapse. The envelopes around the accreting bodies are assumed to be in hydrostatic equilibrium; however, as the solid body grows, it reaches a critical mass at which the envelope is unstable to hydrodynamic collapse forming a Jupiter-like planet. Bodies accreting in a hydrogen-helium environment reach $10 \, M_\oplus$ before collapse ensues (Mizuno 1980). However, Stevenson (1982b) showed that the critical core mass goes as $\sim \mu^{-12/7} t^{-3/4}$ where μ is the molecular weight of the atmosphere and t is the accretion time. For Titan in particular, the gas envelope in the late stages of accretion is dominated by high molecular weight volatiles such as water, ammonia, and methane. So long as the accretion time is short ($< 10^8$ yr), the atmosphere is stable.

At the close of accretion, however, it is conceivable that the rate of growth of the satellite slows drastically *before* the nebular gas is dissipated. If accretion is shut off completely prior to nebula dissipation, the thermal luminosity of the atmosphere is at best only marginally sufficient to prevent collapse. It may be necessary to invoke continued accretion of planetesimals as the nebula was removed, or tidal effects such as tunneling (Weidenschilling 1982), to isolate the envelope from the nebula as accretion tailed off. Neither of these possibilities as yet poses any serious dilemma for the gaseous accretion story.

Perhaps the greatest current challenge for accretion models lies in Triton and Pluto, two icy bodies with volatile-rich surfaces. Because of the unique dynamical conditions of each of these bodies at present, application of accretion theories of regular satellite systems may be questionable at best. It is even possible that Triton is a captured body (McKinnon 1984; see also Secs. VII and VIII). As Fig. 5 shows, bodies the size of Triton or Pluto can undergo substantial heating during accretion, perhaps leading to enhanced loss of the more volatile ices. Nevertheless, both bodies have methane (Apt et al. 1983) and liquid nitrogen has been suggested for the surface of Triton (Cruikshank et al. 1984). Perhaps the most interesting issues concern comparison of Titan and Triton. How do their reservoirs of CH_4 and N_2 compare? Were they obtained in similar ways (e.g., does N_2 form from photochemical processes in both cases)? Until further observational data are available, it will not be possible to understand better their formation environments.

VI. IMPACT DISRUPTION AND ABLATION

Satellite accretion is usually a rapid process ($\lesssim 10^4$ yr) but much can happen later to change the satellite families that we see. Impacts by Sun-orbiting bodies were probably frequent during the first few hundred million years and continue up to the present time. These impacts can disrupt small satellites and ablate volatile material from larger satellites. We first discuss disruption.

A. Impact Disruption

It has been suggested (see, e.g., Shoemaker and Wolfe 1982) that some of the small satellites of the solar system (e.g., Phobos, Deimos, the small inner satellites of Saturn) may have been disrupted by large meteoritic impacts and then reaccumulated again. The time scale of such reaccumulation is very short, even for very small satellites (cf. Table III). Soter (1971) estimates the time for sweepup of debris from circumplanetary orbit to be ~ 10 yr for Phobos and $\sim 10^4$ yr for Deimos. Similarly short times can be derived for small Saturnian satellites (Burns et al. 1984). The most compelling evidence in favor of this scenario is that the largest craters seen on several satellites (e.g., Phobos, Mimas) are so large that the satellites were nearly disrupted by the collisions which left the craters. It would seem unlikely that the bombardment flux in the past was just enough to provide such nearly-catastrophic collisions, but no more. Furthermore, by estimating the bombardment flux from the crater record on larger satellites, which were probably not disrupted, it appears likely that the smaller neighbors were subjected to fluxes at least several times that required to disrupt them (Shoemaker and Wolfe 1982; Smith et al. 1983). An important corollary of this hypothesis is that rings may be the result of such disruptions of satellites within the Roche limit, where reaccumulation would be prevented by the planet's tidal forces (Harris 1984a).

Important questions raised by the above scenario are how much energy is required to disrupt a body of ~ 100 km size, and what are the likely consequences of such an event? The energy necessary to disrupt a large body can be divided into that necessary to overcome the material strength of the body S (energy/unit mass), and that necessary to overcome gravity. The mass of projectile m colliding at velocity v necessary to cause disruption of a body of mass m_s and radius r_s is thus obtained (see chapter by Chapman and McKinnon) from

$$\frac{1}{2} m\mathrm{v}^2 \gtrsim m_s S + \frac{3}{5} \frac{G m_s^2}{\gamma r_s} .$$ (20)

The factor γ is introduced in order to account for the inefficiency of conversion of impact kinetic energy into kinetic energy of the fragments. Laboratory experience on small targets suggest $S \simeq 10^8$ erg g^{-1} for rock and ~ 10^6 erg g^{-1} for ices, and $\gamma \sim 10^{-1}$ (see, e.g., Davis et al. 1979; Fujiwara et al. 1977; Greenberg et al. 1977). For the Saturnian satellites, the impact velocity is essentially that arising from the infall of the projectile into the Saturnian gravity field. For the Martian satellites, the impact velocity is dominated by the rms crossing velocity of projectiles with Mars: ~ 5 km s^{-1} for asteroids or ~ 30 km s^{-1} for comets. For the above constants, the projectile/target mass ratio required to disrupt an icy satellite about Saturn is

$$\frac{m}{m_s} \gtrsim 10^{-7} + 30\left(\frac{r_s}{r_p}\right)^2 .$$ (21)

For satellites more than a few km in diameter, the second term (gravitational energy) dominates. For satellites as small as ~ 10 km, $m/m_s \sim 3 \times 10^{-7}$; hence a meteoroid only ~ 100 m in diameter might disrupt the satellite. For satellites ~ 100 km in radius, $m/m_s \sim 10^{-4}$; thus a projectile of ~ 10 km diameter would be required for disruption. For the case of Martian satellites, using the values of $S \sim 10^8$ dyne g^{-1} and $\gamma \sim 0.1$, we find $m/m_s \sim 10^{-3}$ for an asteroidal impact or ~ 10^{-4} for a cometary impact. In each case, the material strength dominates gravity. In a subsequent breakup, where the material strength might be negligible, the mass ratio required to disrupt the Martian satellites would be $\sim 10^{-4.5}$ for asteroidal impacts or $\sim 10^{-6}$ for a cometary impact. The present asteroidal flux near Mars suggests that the largest collision expected with Phobos in 4.5×10^{-9} yr is $m/m_s \sim 10^{-4}$ (Soter 1971). This suggests that if Phobos is a solid rocky body, it could have avoided disruption by the present flux of asteroids. However, if it was once disrupted, perhaps by a higher flux in the past, then it probably has been re-disrupted repeatedly, perhaps every billion years or so, by the present asteroidal flux. Collision mass ratios similar to those for the Martian satellites apply for hypothetical satellites of asteroids, which have been proposed in the same size range.

It is interesting to note that, because of the high velocities involved, collisions energetic enough to disrupt a body nonetheless impart very little change in its momentum. Note that for $m/m_s \sim 10^{-3}$ and $v \sim 5$ km s^{-1}, the change in the target's velocity is only $\Delta V \sim 5$ m s^{-1}. Hence a disrupted satellite, once reaccumulated, returns to an orbit hardly different from before, with the exception that any eccentricity or inclination of the original orbit should be lost.

The dispersion velocity of the fragments upon disruption might be greater than the above ΔV, however. The velocity should be of order $S^{1/2}$, owing to elastic recoil as the body yields, or of order of the escape velocity from the satellite, whichever is larger. Hence, fragments may scatter with velocities of ~ 100 m s^{-1}. Such a dispersion velocity would leave the fragments very tightly confined in a narrow zone about Jupiter or Saturn (e, $i \sim 0.003$), and rather well confined even about Mars (e, $i \sim 0.03$). However, the fragments of a disrupted satellite of an asteroid would be much more poorly confined, and indeed such a disruption would probably result in the loss of a substantial fraction of the satellite's mass from the system. Davis et al. (1979) estimate that asteroids smaller than ~ 50 km in diameter have mean collision lifetimes less than the age of the solar system. If this is correct, then probably only the very largest asteroids ($\gtrsim 300$ km in diameter) could retain satellites smaller than ~ 50 km in diameter since only such large bodies could retain the debris of a disrupted satellite in order for it to reaccumulate in orbit about its primary.

B. Impact Ablation

The implied presence of impact disruption of small satellites also requires that large satellites were hit by high velocity (cometary) bodies. In these circumstances, there is a net loss of material rather than accretion, because much of the vapor, liquid, and solid debris leaves the impact site with greater than escape velocity. Numerical simulations of impacts suggest that net loss occurs from large Galilean satellites or Titan for bodies impacting at velocities $\gtrsim 10$ km s^{-1} (but depending on projectile composition and density;

see Figs. 22 and 23 of O'Keefe and Ahrens 1982*b*). This may be an under-estimate of the ablation if a large amount of vapor is produced, since the computer calculations do not assess pressure gradient acceleration effects after the impact. Suppose that the methane enhancement in Jupiter (Courtin et al. 1983) is attributed to late infalls of high-velocity Uranus and Neptune zone planetesimals—a reasonable hypothesis because methane cannot condense at Jupiter's orbit as pure ice or clathrate and insufficient carbon is present in carbonaceous chondrites. The amount of material required to accrete onto Jupiter is several Earth masses. Allowing for gravitational focusing, it is straightforward to show that Io would have been subjected to collisions involving a total mass of material ~ 10 to 50% of its own mass. Of course, some (perhaps most) of the ablated material would reaccrete if it were rock; but, if it were ice (H_2O), then it might be lost forever (swept away by magnetospheric processes; see the chapter by Cheng et al.). It is conceivable, therefore, that the compositional gradation among the Galilean satellites is a consequence of impact. This is an extreme hypothesis but it serves to illustrate that ablation may have been important.

VII. SATELLITE CAPTURE

In order for a prospective satellite to be captured from interplanetary space into a bound orbit about a planet, it must lose a great deal of kinetic energy while inside of the planet's sphere of influence. To date, scenarios which have been proposed appear to be dynamically unlikely or even impossible, or require conditions which were unlikely to have existed. Nonetheless, certain of the present satellites invite capture scenarios to explain their origin, either because of compositional differences from their primaries (Moon, Martian satellites) or because of their orbital characteristics (outer Jovian satellites, Phoebe, Neptunian satellites). (See also the chapter by Burns.) Several of the processes that we have already discussed could be described as capture: tidal disruption of a large planetesimal passing within the Roche limit; co-accretion and impact. However, we wish to focus here on intact capture. The only plausible way to achieve this involves gas drag. Hunten (1979) proposed that the Martian satellites may have been captured as a result of aerodynamic drag with an extended protoatmosphere about Mars. Pollack et al. (1979) suggested a similar mechanism for capture of the irregular satellites of Jupiter, Saturn, and Neptune. Recently, Nakazawa et al. (1983) have even suggested such an origin for the Earth's moon. They point out that if a body enters the Hill sphere of the proto-Earth with low energy (corresponding to a velocity at infinity ≲ 2% of escape velocity of the Earth), then it undergoes many (usually prograde) orbits around the Earth before eventually escaping if there is no gas drag (see also Hayashi et al. 1977; Heppenheimer and Porco 1977). Higher energy injection leads to a small number of (equally prograde and retrograde) revolutions before escaping. The low-energy injections are very favorable for capture, even if the gas density is low, because of the many orbits in which to reduce the energy. The eventual almost circular orbit can be stabilized provided the gas is gradually escaping. Aside from the difficulties with the Hayashi model that have already been mentioned, the very low energy of encounter together with the unusual chemical makeup of the Moon make this model unattractive. For the outer planets, an extended gas envelope suitable for trapping satellites could exist during the pre-collapse phase of a giant gaseous protoplanet. However, capture could also occur in the outer part of the gaseous disk that may form in the nucleated collapse scenario of giant planet formation. (In the outer regions, the disk is so fat that the damping of captured satellite inclination may not be severe.)

VIII. SPECIFIC ORIGIN SCENARIOS

A. Mercury, Venus

If either of these two planets once had large, close-in satellites, they would have been removed by the effects of tidal friction (Burns 1973; Ward and Reid 1973; Counselman 1973; see the chapter by Burns). Hence, the present lack of satellites cannot be used to constrain theories of satellite origin about terrestrial planets.

B. The Moon

In spite of the wealth of knowledge returned by lunar exploration and the decade of careful study following the Apollo project, the origin of the Moon remains an enigma. The similarity of isotopic abundances seem to require formation from an essentially common source of matter, yet the depletion in iron and volatiles requires some differentiation of that source material which ended up in the Moon. These constraints would seem to be best satisfied by a scenario of fission or splash-off by oblique impact. The latter is potentially satisfactory from a dynamical standpoint, and may satisfy the chemical constraints, but more work is needed. The coformation model of Harris and Kaula (1975) is satisfactory on dynamical grounds, but apparently fails to account for the iron depletion of the Moon. As we discussed in Secs. II.G and III, the relative importance of large impacts and coformation may depend primarily on the mass spectrum of the bodies from which the Earth formed. The impact origin hypothesis leads to very high temperatures for the protomoon (Thompson and Stevenson 1983; Cameron 1985) whereas coformation may lead to lower temperatures. See the chapter by Kaula et al. for a current summary of constraints on lunar origin models, and the chapter by Burns for comments on orbital evolution. There will be a book on the *Origin of the Moon* (eds. Hartmann, Phillips, and Taylor), anticipated to be published in 1986.

C. Phobos and Deimos

Yoder (1982) has shown that the present eccentricity of Phobos' orbit can be accounted for by resonance passages during the tidal decay of its orbit. Thus the back-integration of the orbit of Phobos which leads to a high eccentricity (see, e.g., Singer 1971) is probably a misleading result and should not be taken to suggest a capture origin. This is discussed in some detail in Burns' chapter.

The recent popularity of a capture origin for Phobos and Deimos (see, e.g., Hunten 1979; Lambeck 1979; Cazenave et al. 1980) can be traced to the Viking results indicating low mean densities of ~ 2 g cm^{-3} and dark surfaces suggestive of carbonaceous chondrite composition. Since asteroids which resemble the Martian satellites are commonly found in the outer part of the asteroid belt, the satellites must have come from there, so the argument goes. We note, however, that within our limited ability to distinguish one material from another by remote sensing, Phobos and Deimos may be of essentially the same composition as inner main belt asteroids, or for that matter of Mars itself, except that they appear to be undifferentiated. The higher albedo asteroid types which make up the bulk of the inner asteroid belt are generally considered to be differentiated bodies. Although the heat source which caused the differentiation remains a matter of debate, it is commonly assumed that differentiation would occur only in large asteroid parent bodies. Furthermore, mutual collisions among the asteroids should cause bodies smaller than ~ 50 km in diameter to have been catastrophically disrupted in a time shorter than the age of the solar system (Davis et al. 1979). Thus the present population of small asteroids is not the primordial one; the primordial bodies have been destroyed and replaced by fragments of larger asteroids. Thus the fact that the present population of asteroids near Mars in the size range of its satellites are not of carbonaceous composition should not be taken to indicate that the planetesimals from which Mars accumulated were not of carbonaceous composition.

We propose, then, that the Martian satellites coformed in orbit around Mars from collisional debris which accumulated in orbit as Mars grew (see, e.g., Ruskol 1960,1963,1972c; Harris and Kaula 1975; Harris 1978b). Since that material was never incorporated into bodies larger than the present satellites, it never became differentiated. Furthermore, since they have not been melted, we expect that they have remained loose aggregates of debris throughout, thus explaining their low mean densities. Dobrovolskis (1982) has shown that the satellites need no more internal strength than lunar regolith to retain their present figures. It is possible that the satellites have been repeatedly disrupted by collisions over the age of the solar system. Unlike asteroids, the Martian satellites are able to reaccumulate following disruption. Thus they may be even more primitive than asteroids of the same size, which are likely collisionally evolved from much larger bodies, which in turn may have been differentiated before disruption.

D. Regular Satellites and Rings of Jupiter and Saturn

The time scale of aggregation of solid material about the giant planets is so much shorter than the time scales of dissipative processes, that we are led to conclude that satellites formed and migrated inward to destruction in large numbers during the course of the planets' growth. The present systems are thus only the tail end of the process, and the ring systems are probably satellites left within the Roche limit at the conclusion of planet growth, later disrupted by meteoritic bombardment (Harris 1984a). Whether the satellites were seeded by infalling planetesimals captured by the planet or grew from an accretion disk surrounding the planet depends upon the scenario of planet formation one believes in. The existence of satellites is not strongly diagnostic of either scenario.

The similar masses and regular density progression of the Galilean satellites argues, however, for a common accretional origin. The progression of densities in particular suggests (but does not require) that these satellites accreted within a strong ambient temperature gradient, and that orbital evolution inward during formation excluded Europa from regions where water ice condensate was stable but was not so rapid that Ganymede crossed out of the ice stability threshold. The striking surficial differences, implying qualitatively different interiors of Ganymede and Callisto, have seriously challenged accretion models to produce two very different bodies out of the same budget of materials. Accretion in a gaseous nebula seems even marginally able to produce simultaneously a primarily unmelted Callisto and a substantially melted Ganymede. Gas-free accretion may also achieve this, but with the choice of an adjustable parameter that cannot be verified. The vaporization of water ice and its transport out of the satellites' envelopes late in the accretion process can in principle explain the modest ice-to-rock depletion in Ganymede and Callisto, as well as in Titan. Some process common to the four Galilean satellites and Titan which operated at the end of accretion to arrest their growth at roughly the same rock mass is as yet unidentified. Since Ganymede, Callisto and Titan have very similar *total* masses, it is possible that the efficiency of their terminal accretion was drastically reduced by the process of volatile vaporization. The existence of Io and Europa, however, suggests some external agent such as disk dissipation or tidal tunneling (suggested by Weidenschilling [1982] for small satellites) caused accretion to tail off. The former seems an unlikely explanation unless the accretions of the Galilean satellites and Titan were precisely coeval.

The inclusion of Titan in treatment of the large satellites should be approached with trepidation, since it is a unique member of the Saturn system. No satisfactory explanation for the size distribution of the regular Saturn satellites exists that can simultaneously account for Titan-, Rhea,- and Mimas-class objects. Collisional disruption of the inner Saturnian satellites has likely occurred, and the rings may be composed of satellites disrupted within the

Roche limit. All the Saturnian satellites with the exception of Titan likely formed undifferentiated, with subsequent resurfacing caused by the radiogenic heating of a low melting point fluid such as ammonia hydrate.

Titan's present atmosphere may be understood as being derived in part from substantial accretional heating. Plausible ambient temperatures during formation would have allowed accretion of methane clathrate hydrate and ammonia hydrate, producing a massive hot primordial ammonia and methane atmosphere overlying an ammonia-water liquid mantle, extending as an ocean to the surface. Cooling of the atmosphere-ocean system, conversion of NH_3 to N_2 (Atreya et al. 1978) and eventual sequestering of atmospheric components in the ocean as dissolved gases and clathrate are plausible subsequent processes leading to the present state (Lunine and Stevenson 1982b). The current abundance of gaseous N_2 can then be understood as that remaining when the surface froze over at ~ 175 K, terminating the liquid mantle-atmosphere contact. Outgassing of some fraction of the CH_4 trapped in the interior, most likely in one or a few catastrophic episodes would provide the budget needed for photochemistry and a postulated hydrocarbon ocean in the contemporary era. Less likely, because of the resulting excessive water ice depletion relative to rock and overabundance of CO relative to N_2 as measured at present, is a scenario in which Titan accretes in a CO-N_2-rich nebula.

E. Irregular Satellites of Jupiter and Phoebe

These satellites were likely captured, in some sense. We favor a collisional capture to accomplish the initial transformation from heliocentric to planetocentric orbit. In the context of the nucleated growth scenario of planet formation, considerable further inward evolution could be expected from drag with heliocentric orbiting gas, and as the planet gains mass. Indeed, these outer satellites may be the seeds which would have evolved inward and grown larger to become the next generation of regular satellites, had accumulation continued.

F. Iapetus

This satellite is often regarded as irregular; however, it should be noted that its orbit is essentially circular, and inclined only $\sim 8°$ to the Laplacian plane. Ward (1981) has suggested that Iapetus may have formed in its Laplacian plane, but was left in a somewhat inclined orbit by the rapid dissipation of a protosatellite disk around Saturn.

G. Satellites of Uranus and Neptune

One is tempted to assume by analogy that the Uranian satellites formed in the same way as the Jovian satellites. However, the formation of the two planets may have been very different, since the gas component of Uranus and Neptune is substantially less than the mass of ice and rock. The fact that the Uranian satellites are prograde and equatorial in spite of the planet's retrograde spin is noteworthy. This, and Triton's retrograde orbit, lead us to conclude that the angular momentum of the satellite systems was not dictated by heliocentric motion. In the case of Uranus, the satellite disk appears to have been in communication with the planet's spin; in the case of Neptune, Triton's orbit appears quite unconnected with the planet's spin. One possibility is that Triton was captured (McKinnon 1984). Another possibility is that Triton had no neighbors to interact with, whereas Uranian satellites may have evolved into the equatorial plane through a series of collisions between initially inclined satellites. An impact origin for both the obliquity of Uranus and a disk from which the satellites formed is one possibility (Cameron 1975; Stevenson 1984b). Although such an event excites the eccentricity of Uranus' orbit, this can subsequently be damped by a sufficient number of smaller impacts. (A similar effect is observed in the numerical simulations of Earth accretion by Wetherill [1980b].)

Although the masses of the Uranian satellites are too low to have produced substantial accretional heating, the eventual reconnaissance of their surfaces may reveal much about the chemistry of their feeding zones. In particular, the high densities of Oberon and Titania (Veillet 1983a; cf. Dermott and Nicholson 1986) are consistent with formation in a CO-rich, rather than CH_4-rich gas. This might occur by impact heating (Stevenson 1984b). This raises the possibility that these satellites accreted primarily N_2 rather than NH_3. Since the former does not form a eutectic fluid with H_2O, resurfacing may not occur on these bodies as it does on the Saturnian satellites of similar size. The comparatively low albedos of the Uranian satellites (Brown et al. 1982b) is at least consistent with this proposal.

Triton is probably massive enough to have produced some melting and volatilization during accretion. This is compatible with (but not yet diagnostic of) the presence of a volatile-rich veneer. The crude density determination hints at formation in a CO-N_2-rich feeding zone, although the apparent coexistence of CH_4 (Apt 1983) and N_2 (Cruikshank et al. 1984a) on its surface could argue for a more complex suite of C and N condensates. It is unlikely that Triton was ever hot enough as a result of accretion to permit significant photochemical conversion of NH_3 to N_2 (unless the mass has been underestimated). Pluto exhibits a CH_4-rich surface and atmosphere; based on its mass determination, modest melting and formation of an atmosphere could have occurred during accretion. Neither Pluto nor Triton can as yet constrain effectively accretion models; better mass, radius and surface composition information are desperately needed to test our concepts of formation of these bodies.

H. Pluto-Charon System

The present configuration of the Pluto-Charon system is a tidal end state (see the chapters by Peale and by Burns). Without knowing the mass ratio of the pair, it is impossible to know exactly the angular momentum content of the system and thereby trace back the history of Pluto's spin rate versus the orbital separation of the pair. The mass ratio estimated by assuming equal densities and albedos of the two bodies implies that the initial rotation period of Pluto may have been as short as a few hours, thus suggesting that the system may have formed by fission (Lin 1981). Unlike the fission hypothesis for lunar origin, the mass ratio of Charon to Pluto falls in the expected range of a pair formed by fission following rotational instability, and the present angular momentum of the system may be compatible with rotational instability of a single body. However, the mechanism of spinup to the point of rotational instability remains unspecified (see the comments in Peale's chapter). If collisional spinup were the cause, then the actual fission event may be more accurately described by the oblique collision, splash-off scenario, since the collision which triggered the fission would no doubt have altered the outcome from the result of smooth transition from Jacobi ellipsoid to Poincare figure to fission (cf. Wise 1963).

I. Asteroidal Satellites

The very existence of the Pluto-Charon system, the smallest planet with the largest satellite/planet mass ratio, lends tantalizing credence to the possibility that asteroids may have satellites. Both Pluto and the asteroids appear to have formed in environments where the planetesimal swarm was stirred up by more massive neighbors in such a way that the collision velocities considerably exceeded the surface escape velocities of the bodies in question. Harris (1979) has noted that this condition should lead to collisional overspinning of the bodies involved. Although the details of what happens when a body is collisionally overspun have not been worked out, it is clear that if a large collision event dissipates too much energy to allow matter to escape back into space, and at the same time delivers enough angular momentum to render the single body rotationally unstable, something must happen. Weidenschilling (1981) has compared theoretical spin rates of equilibrium figures of rotation, up to and including fissioned binaries, and finds the observed distribution of asteroid rotation rates suggestive of the possibility that some are fissioned binaries. It is also noteworthy that several asteroids with rotation periods of 2–6 days are known, and they may be tidally-evolved binaries, similar to the Pluto-Charon system.

IX. CONCLUDING COMMENTS

At this point, some readers may feel frustration at the lack of a unifying picture of satellite origin. We have endeavored to show, on the contrary, that unification (in the sense of identifying a dominant process) would be a false goal. Despite the patterns described in the introduction, the diversity of planets and the satellite systems suggest a variety of contributing processes. In a sense, there is an underlying optimism: satellites are not difficult to make. On the contrary, we seek reasons why some planets do *not* have satellites; why Mars does not have a large satellite; why Neptune lacks a regular satellite system. If there is a sense in which frustration is warranted, it lies in the lack of relevant data rather than theoretical ideas.

The role of observations is great, especially cosmochemical clues on planetary and satellite origin: detailed knowledge of Titan's atmosphere; noble gas abundances in giant planets; composition and structure of Ganymede and Callisto; identification of volatile ices in small satellites. In most of these areas, we have barely scratched the surface.

Theoretical advances are also needed: orbital injection by impact must be further quantified; gas accretion and disk formation for giant planets must be modeled. In fact, all the tasks required for improving our understanding of planetary formation are part of a better understanding of satellite origins.

It was once thought that we could unravel the mysteries of the solar system by going to the Moon. Planetary scientists are no longer so naive. A more realistic expectation, perhaps, is to seek in the properties and nature of the satellites some clues to how planets were put together. It is hoped that this review has provided some guidelines on how to proceed toward this goal.

Acknowledgments. The authors gratefully acknowledge support by several grants from the National Aeronautics and Space Administration. We thank J. Burns and J. Pollack for their helpful comments on an earlier version of this review.

3. PROTOSATELLITE SWARMS

V. S. SAFRONOV, G. V. PECHERNIKOVA, E. L. RUSKOL
and A. V. VITJAZEV
USSR Academy of Sciences

The formation of natural satellites is considered now to be a by-product of the formation of the planets. In this chapter two types of protosatellite swarms will be considered—swarms of solid bodies around terrestrial planets and gas-dust disks around giant planets. Circumplanetary swarms are formed due to the inelastic collision and subsequent capture of particles by the planet's gravitational field. We consider these processes as well as the dynamical evolution of swarms in detail. During later stages a swarm is replenished by new material coming mainly from erosional collisions of high-velocity external bodies with members of the swarm. The characteristic time scales of kinetic processes in the swarms are orders of magnitude shorter than the lifetime of swarms, determined by the accumulation time for the planets. The velocities of bodies in a swarm stirred by impacts of external bodies are estimated. They dominate the main mass of the swarm during most of the accumulation. Equations are developed for the surface density distribution in the swarm as it accretes matter from the zone of the planet. Two processes dominate the evolution: contraction of the swarm due to the small angular momentum of the accreted material and diffusive expansion of the swarm. The origin and evolution of gas-dust disks near giant planets is discussed in connection with the general problem of the origin of these planets. The formation of Jupiter and Saturn was characterized by violent accretion of gas remaining in the solar nebula. The powerful energetics of this process in the Jovian disk caused the observed decrease in density of its satellites with the distance from the planet. Some of these features are common for circumplanetary and circumstellar disks as well as for planetary rings. A comparative study of their dynamics should provide a better understanding of their evolution.

The formation of natural satellites is usually considered as a by-product of the formation of the planets themselves. The often mentioned structural similarity of satellite systems near giant planets and the planetary system implies a similar mode of formation. However, the various satellites differ considerably from one another and this presumably reflects different conditions of planetary formation. For example, the rotational slowing of Mercury and Venus by solar tides led to the loss of all possible close satellites for these two planets (Burns 1973; Ward and Reid 1973). The Earth-Moon system, owing to the smaller tidal influence of the Sun but having strong interaction within the system, had a different evolution. The zone of Mars was greatly influenced by bodies invading from the zone of Jupiter. Satellite swarms of Jupiter and Saturn formed in the presence of gas while it was rapidly accreting onto the planets. Uranus presumably was subjected to a catastrophic collision (megaimpact) which inclined its rotational axis by 98°. Its protosatellite swarm was positioned exactly in its equatorial plane. In contrast, more massive Neptune with a rotation axis in a more normal position has no regular satellite system appropriate to giant planets but has a massive retrograde satellite. Very small Pluto has a relatively large satellite. Recently, several satellites of asteroids may have been discovered.

Not all peculiarities of satellite systems can be explained in a systematic way. The retrograde orbits are probably the result of a capture, i.e., a sporadic phenomenon. Therefore, the classification of satellites as regular and irregular seems to be justified also from a cosmogonical point of view. The protosatellite swarms considered in this chapter relate mainly to regular satellites. In this respect, the Moon can be treated as a regular satellite if we assume it formed from an evolved protosatellite swarm, but as an irregular satellite if other ways of formation are considered. The comparatively large mass of the Moon and its different chemical composition from the Earth gave birth to catastrophic (or exotic) hypotheses of its origin (see chapter by Kaula et al.). Among them the capture hypotheses (Urey 1962; Alfvén 1963) and the fission hypotheses (Darwin 1908; Bullen 1967) may be mentioned. The hypotheses of the origin of the Moon in a swarm formed around the Earth as a result of a catastrophic (nearly grazing) collision of a very large body with about the mass of Mars (Mitler 1975; Wood 1977; Kaula 1977) occupies an intermediate position. In order to make a choice between the regular and irregular method of the protolunar swarm formation, both mechanisms need further investigation.

The idea of the origin of satellites in circumplanetary swarms of solid bodies and particles, captured by a planet's gravitational field during mutual collisions was put forward in a qualitative form by O. Yu. Schmidt (1957). Its quantitative development was undertaken by Ruskol (1960,1963,1975), while Harris and Kaula (1975a) considered it in regards to the lunar origin. The gas-free accumulation of satellites was addressed in general by Safronov and Ruskol (1977) and satellites of gas-accreting giant planets by Ruskol (1981, 1982b) and Weidenshilling (1982).

No doubt the similarity between the structures of satellite systems and planetary system is related to the similarity of the accumulation process of solid bodies in protosatellite swarms and in protoplanetary disks. This provides an additional argument against the formation of planets by contraction of gaseous condensations, because such a process would be impossible in satellite systems. Because the theory of accumulation of planets is a "primary problem," it has been developed more broadly than that for satellites; hence it is useful to mention briefly the main results. The evolution of a system of preplanetary bodies is determined by the variation with time of its main parameters, the distribution of masses and velocities of bodies (Safronov 1969). It can be shown that the distribution of mass m approaches an inverse power law

$$n(m) = c\, m^{-q}, \quad m < m_1 \tag{1}$$

where the largest mass m_1 increases with time. Only the largest bodies violate this law. If bodies coalesce in collisions, the power index is $q \approx 1.6$. Fragmentation enlarges q to ≈ 1.8. Note that for $q < 2$, most mass is concentrated in the largest bodies. The random velocities, v, of bodies with respect to Keplerian circular motion increase due to mutual gravitational perturbations at encounters and decrease due to inelastic collisions. For a power law in the mass distribution according to Eq. (1) with $q < 2$, these velocities rather quickly tend to an equilibrium value

$$v = \left(\frac{G\, m}{\theta\, r} \right)^{1/2} \tag{2}$$

where m and r are the mass and the radius of the largest body in the zone of its interaction with other bodies, G is the gravitational constant and θ is a dimensionless parameter. In the absence of gas, θ is equal to $3 - 5$; gas slows the velocities of bodies so that θ increases by half an order of magnitude for km-sized bodies but by an order of magnitude for meter-sized ones.

Larger bodies grow relatively faster than smaller ones, becoming so-called planet embryos; the number of embryos decreases as they grow by sweeping out surrounding material. Only the largest bodies survive to form finally the present planetary system. In the terrestrial planet zone, the process lasts $\sim 10^8$ yr.

The problem of satellite formation can be divided into two parts: (1) the development of the protosatellite swarm, and (2) its evolution to form satellites. While the second part has much in common with the problem of evolution of the protoplanetary cloud and accumulation of the planets, the first one substantially differs from the problem of the cloud's origin. Only accretional gas-dust disks around Jupiter and Saturn were similar to the protoplanetary accretional disk. It is necessary to point out that the time scale of formation of the protoplanetary disk ($10^6 - 10^7$ yr) is much shorter than its subsequent evolution ($\sim 10^8$ yr), but the time scales for formation and evolution of proto-satellite swarms are about the same (see chapter by Stevenson et al. Table I).

First, we shall consider the formation of circumplanetary swarms of solid material in the absence of gas; this is a satisfactory approximation for the terrestrial planets and perhaps also for Uranus and Neptune.

I. FORMATION AND GROWTH OF THE PROTOSATELLITE SWARM

Mutual inelastic collisions of particles with bodies in the vicinity of a planet decrease relative velocities making it possible for these objects to be captured by the gravitational field of the planet, thereby transferring them from heliocentric to planetocentric orbits. At first (in the absence of a swarm), collisions occurred between particles gravitationally not bound to the planet (free-free collisions); but, once the density (ρ_2) of the circumplanetary swarm so formed exceeded the density (ρ_1) in the planetary zone, new particles were captured mainly by collisions with the particles of the swarm (free-bound collisions). (Any values related to the zone of the planet, we shall denote by the subscript 1, while those related to the swarm will have the subscript 2.) Particles could be captured by the swarm following single collisions with particles of comparative size or greater, or by numerous collisions with small particles having a total mass of about the mass of the captured particle. An order of magnitude estimate of the captured mass is obtained by considering only collisions between particles comparable in size. For target particles in the mass interval Δm between m/χ and χm (where $\chi \sim 2$), the mathematical expectation of a collision of a particle crossing the swarm along a normal trajectory equals

$$M(m, \Delta m) = \xi \pi r^2 H_2 n_2(m) \Delta m = (\chi - \chi^{-1}) \xi \pi r^2 c_2 H_2 m^{1-q_2}. \quad (3)$$

Here H_2 is the swarm thickness. For collisions of comparable bodies, the cross-section factor $\xi \approx 2$. The value $c_2 H_2$ is determined by the surface density σ_2 of the swarm

$$\sigma_2 = \frac{c_2 H_2 m_2^{2-q_2}}{(2 - q_2)}. \quad (4)$$

As the number of bodies impacting a cm^2 of the swarm's surface in the mass interval m to $m + dm$ equals $v_1 n_1(m) dm$, the mass distribution of particles captured into the swarm is also an inverse power law (Eq. 1) with the exponent

$$q_2' = q_1 + q_2 - \frac{5}{3} \quad (5)$$

(see Ruskol 1975). The usually accepted values of $q_1 \approx q_2 \approx 11/6$ give $q_2' \approx 2$. This has led to the conclusion that mainly small particles of the planetary zone were selectively captured into the swarm. The mass inflow I_{cs} of bodies participating in such "comparable-size" collisions can be found by multiplying Eq. (3) by $v_1 n_1(m)$ and integrating over m. For $q_1 + q_2 = 11/3$ we get

$$I_{cs} = \int_{m_0}^{m_2} M(m, \Delta m) v_1 n_1(m) m \, dm \approx \frac{1}{10} \frac{\rho_1 v_1 \sigma_2}{\delta (r_1 r_2)^{1/2}} \log (r_2/r_0) \quad (6)$$

where δ is the particle density, m_0 and m_2 (with corresponding radii r_0 and r_2) are the lower and upper limits for the distribution $n_2(m)$.

The condition for capture of a particle m by particles smaller than m,

$$\pi r^2 H_2 \int_{m_0}^{m} m n_2(m) dm = \frac{\pi r^2 c_2 H_2}{2 - q_2} m_2^{2-q_2} \geq m = \frac{4}{3} \pi \delta r^3 \quad (7)$$

gives the maximum size r_c for captured particles

$$r_c^{3q_2 - 5} = \frac{3\sigma_2}{4\delta r_2^{6 - 3q_2}}. \quad (8)$$

With $q_2 = 11/6$

$$r_c = \left(\frac{3\sigma_2}{4\delta} \right)^2 \frac{1}{r_2}. \quad (9)$$

All particles with $r < r_c$ are captured into the swarm by smaller particles. The inflow of such particles for $q_1 = 11/6$ equals

$$I_{sp}(11/6) = \frac{3\rho_1 v_1 \sigma_2}{4\delta(r_1 r_2)^{1/2}} = \rho_1 v_1 \tau_{z2} \left(\frac{r_2}{r_1} \right)^{1/2} \quad (10)$$

where $\tau_{z2} = 3\sigma_2/4\delta r_2 = \pi r_2^2 \sigma_2/m_2$ is the optical thickness of the disk along the z-axis when all bodies equal the largest body m_2. Comparing Eq. (6) with Eq. (10), we see that $I_{sp} \sim I_{cs}$.

It is important also to evaluate capture by large bodies. By analogy with Eq. (3), one can find the expected number of collisions of the body m with all bodies $m' > m$ when it crosses the swarm

$$M(m) = \int_{m}^{m_2} M(m, m') \, dm' \approx \int_{m}^{m_2} \pi r'^2 H_2 c_2 m'^{-q_2} \, dm'$$

$$= \left(\frac{3}{4\pi\delta} \right)^{2/3} \frac{\pi H_2 c_2}{q_2 - 5/3} m^{5/3 - q_2} \left[1 - \left(\frac{m}{m_2} \right)^{-5/3} \right]. \quad (11)$$

Multiplying this expression by $mn_1(m)dm$ and integrating over all m, we get the flow of matter captured by the swarm in collisions of bodies with larger ones

$$I_{lb} \approx \frac{(2 - q_2)\rho_1 v_1 \tau_{z2}}{\left(\dfrac{11}{3} - q_1 - q_2\right) r_1^{3(2-q_1)} r_2^{3q_1-6}} \tag{12}$$

for $q_1 + q_2 > 11/3$, $r_c \ll r_2$, and

$$I_{lb} \approx \frac{(2 - q_1)(2 - q_2)3\tau_{z2}\rho_1 v_1 \ln(r_2/r_c)}{(q_2 - 5/3)r_1^{6-3q_1} r_2^{5-3q_2}} \tag{13}$$

for $q_1 + q_2 = 11/3$. When $q_1 = q_2 = 11/6$

$$I_{lb} = \frac{\rho_1 v_1 \tau_{z2}}{2} \left(\frac{r_2}{r_1}\right)^{1/2} \ln(r_2/r_c). \tag{14}$$

The lower limit of integration should be taken equal to r_c, because as shown above, all bodies smaller than r_c are captured into the swarm, providing the flow I_{sp} is according to Eq. (10).

Comparing I_{lb} with I_{sp} and I_{cs}, we see that for $q_2 = 11/6$, capture by larger bodies is several times more effective than capture by comparable and smaller ones. As in the case of comparable bodies, the frequency of capture of bodies $m \ll m_2$ is proportional to $m^{5/3 - q_2}$ (cf. Eqs. [3] and [11]). Therefore, the mass distribution of the captured bodies has the same exponent (Eq. 5), so our conclusion on the predominance of capture of smaller particles into the swarm still holds. With $q_2 > 11/6$ and $r_c < r_2$, capture of particles by smaller particles in the swarm sharply increases. I_{lb} grows slowly and becomes even slightly less than I_{sp}.

These relations for I_{cs} and I_{lb} are obtained assuming a "transparent" swarm, in which the capture of bodies proceeds throughout the swarm's whole volume, and the mathematical expectation M of a collision is equal to its probability. In contrast, I_{sp} corresponds to the model of an opaque swarm, in which all particles with $r < r_c$ are captured. The first model was applied by Ruskol (1975) for all bodies and the second one by Harris and Kaula (1975) and Kaula and Harris (1975) (also for all bodies). It is clear that the first model is certainly valid for larger bodies. It becomes incorrect for $r < r_c'$ where r_c' satisfies the condition $M(r_c') \approx 1$ in Eq. (11). We find

$$\left(\frac{r_c'}{r_2}\right)^{3q_2 - 5} = \frac{2 - q_2}{q_2 - 5/3}\frac{3\sigma_2}{4\delta r_2}\left[1 - \left(\frac{r_c'}{r_2}\right)^{3q_2 - 5}\right]$$

$$= \frac{(2 - q_2)\tau_{z2}}{(q_2 - 5/3) + (2 - q_2)\tau_{z2}} \tag{15}$$

with $q_2 = 11/6$, $r_c' \lesssim r_c$ (cf. Eq. [8]). It can be shown that for $q_2 > 11/6$, $r_c' < r_c$. Therefore, the lower limit used in the derivation of I_{lb} provides the swarm's "transparency" for all $r > r_c$.

The total flux of bodies captured into the swarm by large and small bodies for $q_2 = 11/6$ is

$$I\,(11/6) = I_{sp} + I_{lb} = \rho_1 v_1 \tau_{z2} \left(\frac{r_2}{r_1}\right)^{1/2} (1 - \ln \tau_{z2}). \qquad (16)$$

The second term should be put to zero for $\tau_{z2} \geq 1$, i.e., for $3\sigma_2 \geq 4\delta r_2$ because then $r_c > r_2$. Thus, in order to determine I, it is necessary to correctly evaluate r_2.

If the accumulation of bodies in the swarm is not stopped by the destructive action of bodies incoming from the zone of the planet, and if r_2 is not very small, then the second term in Eq. (16) is on the order of a magnitude higher than the first one. The amount of solids captured into the swarm considerably increases with the increase of q_2. For example, if $q_2 = 23/12$ (the average of $q_1 = 11/6$ and 2),

$$I\,(23/12) = \frac{5\rho_1 v_1 \tau_{z2}^{2/3}}{3} \left(\frac{r_2}{r_1}\right)^{1/2}. \qquad (17)$$

For $\sigma_2 \sim 10^2$ and $r_2 \sim 10^6$, Eq. (17) gives about a five times higher flow for I than does Eq. (16).

This simple consideration does not take into account some substantial features of the process. First, it ignores that the probability of capture decreases with distance from the planet, being zero at the boundary of the Hill's lobe (R_{L_1} being its radius). In collisions, bodies lose some part (roughly one half) of their kinetic energy. But before approaching the planet they possess a random velocity (Eq. 2) and this energy loss may be insufficient to allow their capture into the external parts of the swarm. Ruskol (1975) proposed an approximate dependence of the capture probability on the distance R_2 from the planet (p)

$$p\,(R_2) = \frac{\theta_1}{1 + 2\theta_1} \left(1 - \frac{R_2}{R_{L_1}} \right) \qquad (18)$$

where $l = R_2/r_p$, $R_{L_1} = R_1 (m_p/3M_\odot)^{1/3}$, and m_p is the planet's mass. A factor of this type should be added to Eq. (16) for the mass flow captured into the swarm. Without this factor, Eq. (16) characterizes the flow of particles scattered rather than captured by the swarm.

Second, we did not consider that the larger bodies crossing the swarm can sweep out smaller bodies encountered on the way. Using the same procedure as in the derivation of Eq. (14), it is easy to find that if such sweeping occurs, the corresponding mass loss from the swarm is also determined by Eq. (14), with the only substitution being the logarithm of r_2 instead of r_1. However, the result of the collision of bodies substantially depends on the impact velocity. When bodies rebound after collisions and do not coalesce, the mass loss by the swarm is considerably less. During the final stage of planetary accumulation, when the velocities of bodies become high enough, they disintegrate in collisions and the removal of matter from the swarm becomes again less effective. Even the opposite can be expected; bodies crossing the swarm are subjected to erosion through impacts of smaller bodies and leave a part of their mass in the swarm. Because the mass lost by a body during this stage is much greater than the impacting mass, and because a considerable part of it remains in the swarm, erosion becomes a very effective mechanism for feeding the swarm.

When a body m is struck by a small body m', a crater is formed on the body m and it loses a mass $\gamma m'$. According to Marcus (1969), $\gamma = K v_i^2$, where v_i is the impact velocity, and K is a parameter which depends on the rock strength. If v_i is expressed in km s^{-1}, K can be taken as ~ 10. For a collision with bodies larger than m/γ', where $\gamma ' > \gamma$, the body m disintegrates. Its larger fragments are subjected to further erosion, and smaller ones are captured into the swarm. Bodies of radii smaller than a certain radius r_e, almost completely evaporate while crossing the swarm. The mass loss experienced by the body m while traversing the distance $s = H_2$ can be found by integrating the expression

$$dm \approx - \gamma \pi r^2 \rho_2 \left(\frac{m}{\gamma_1 m_2} \right)^{2 - q_2} ds. \tag{19}$$

For $q_2 = 11/6$ we get the maximum size of a body that completely vanishes;

$$r_e \approx \left(\frac{\gamma \sigma_2}{8\delta} \right)^2 \frac{1}{\gamma_1^{1/3} r_2} = \frac{\gamma^2}{36 \gamma_1^{1/3}} r_c. \tag{20}$$

If we do not consider catastrophic collisions, then $\gamma_1 = \gamma'$; if catastrophic collisions are considered, $\gamma_1 \approx 1$. On the other hand, deceleration of the body during erosion leads to a slightly less effective value of γ. From the same Eq. (19) we find that the larger body erodes a mass (for $q_2 = 11/6$)

$$\Delta m = m \left\{ 1 - \left[1 - \frac{\gamma \tau_{z2}}{4} \left(\frac{r_2}{r} \right)^{1/2} \right] \right\}^6. \tag{21}$$

Integrating over all m from m_e to m_1 and adding the erosion of smaller bodies with $m < m_e$, we find the total erosional loss of mass by bodies entering the swarm from the planet's zone:

$$v_1 \int \Delta m \, n_1(m) dm \approx \gamma \ln \frac{(r_1/r_c)^{1/2}}{\gamma} I_{sp}. \tag{22}$$

Presumably more than one-tenth of this quantity (the particles thrown out from the crater in directions close to the swarm's plane) should remain in the swarm. Therefore, the erosion due to high-velocity impacts ($\gamma \sim 10^2 - 10^3$) provides a more effective replenishment of the swarm than collisions at small velocities

$$I_e \approx \gamma_2 \log \frac{(r_1/r_c)^{1/2}}{\gamma} I_{sp} \tag{23}$$

where $\gamma_2 \gtrsim 0.1\,\gamma$.

A. Relative Velocities of Bodies in the Swarm

The velocities of bodies v_2 inside the swarm are important parameters because they determine the thickness of the swarm H_2 and its volume density ρ_2, when the surface density σ_2 is known. But even more important is that the velocities of bodies determine the character of evolution of the swarm: at low velocities there is effective accumulation of bodies, while at high velocities, they fragment. As in the case of the accumulation of bodies in the protoplanetary disk, the velocities of bodies in protosatellite swarms can be evaluated by balancing the mechanical energy acquired by the swarm from some sources to the energy lost in inelastic collisions of bodies. There are two such sources:

1. Energy brought in by bodies of the planet's zone as they enter the swarm;
2. Gravitational interactions among the swarm's bodies.

In the early stages, the bodies of the swarm are small, as are their mutual perturbations. The first source then dominates but it becomes less effective with time as ρ_1 decreases. The second source in contrast becomes more important as the bodies in the swarm grow until it dominates in the later stages. Then the velocities of bodies are determined by Equation (2), with m and r close to the mass and radius of the largest body at a given distance R_2 from the center of the swarm. Below we evaluate the first source—the energy brought by bodies entering the swarm from the planet's zone.

Consider the change of kinetic energy in a swarm member m'' after its inelastic collision with another body m'. Let the normal relative velocity v_n of the bodies reduce after the impact by a factor ε while the tangential one v_t does not change. Then their velocities v' and v'', relative to the Keplerian velocity V_{K2} reduce after the impact to u' and u'', respectively. For u'' we have (cf., e.g., Yavorsky and Detlaf 1968)

$$u_n'' = \frac{m'(1 + \varepsilon)v_n' + (m'' - \varepsilon m')v_n''}{m' + m''} \tag{24}$$

$$u_t'' = v_t'' \tag{25}$$

where

$$v_t' = v' \cos\varphi; \quad v_n' = v' \sin\varphi; \quad v_t'' = v'' \cos\psi; \quad v_n'' = v'' \sin\psi. \tag{26}$$

The change in kinetic energy of m'' per unit mass is

$$\zeta(\varphi,\psi)v''^2 = u''^2 - v''^2 = u_n''^2 - v_n''^2. \tag{27}$$

Averaging $\zeta(\varphi,\psi)$ over all directions and denoting $m'/m'' = x$ and $v'/v'' = y$, we obtain

$$\zeta = \frac{f_1(\varepsilon,y)x^2 - f_2(\varepsilon)x}{2(1 + x)^2} \tag{28}$$

where

$$f_1(\varepsilon,y) = (1 + \varepsilon)^2 y^2 + \varepsilon^2 - 1; \quad f_2(\varepsilon) = 2(1 + \varepsilon). \tag{29}$$

The expected change of energy of m'' per unit time due to its collisions with swarm bodies is

$$\zeta_2(m'') = \xi\pi v_2 c_2 \int_0^{m_2} \zeta(m'',m')(r'' + r')^2\, m'^{-q_2}\, dm'$$

$$= \frac{1}{2}\, \xi\pi v_2 c_2 r''^2 m''^{1 - q_2} \int_0^{m_2/m''} \frac{f_1(\varepsilon,y)x^2 - f_2(\varepsilon)x}{(1 + x)^2} \times$$

$$(1 + x^{1/3})^2 x^{-q_2} dx \tag{30}$$

where v_2 is the average relative velocity in the swarm, $q_2 < 2$ and $c_2 = (2 - q_2)\rho_2 m_2^{q_2 - 2}$. Assuming a power dependence of v_2' on their masses $v_2' = v_{02}(m'/m_2)^\alpha$, we have $y = v_2'/v_2'' = (m'/m'')^\alpha = x^\alpha$.

Similarly the energy change of the body m'' during collisions with bodies of the planet's zone per unit time is

$$\zeta_1(m'') = \frac{1}{2}\, \xi\pi v_1 c_1 r''^2 m''^{1 - q_1} \int_0^{m_1/m''} \frac{f_1(\varepsilon,\bar{y})x^2 - f_2(\varepsilon)x}{(1 + x)^2} \times$$

$$(1 + x^{1/3})^2\, x^{-q_1}\, dx \tag{31}$$

where $c_1 = (2 - q_1)\rho_1 m_1^{q_1 - 2}$, $\bar{y} = v_1/v_2'' = (v_1/v_{02})(m_2/m'')^\alpha$ with v_1 the average velocity of bodies in the planet's zone relative to the circular Keplerian velocity in the swarm $V_{K2} = (Gm_p/R_2)^{1/2}$:

$$v_1^2 = v_{10}^2 + 3V_{K2}^2; \quad v_{10}^2 = \frac{Gm_p}{\theta_1 r_p}. \tag{32}$$

The "equilibrium" value of the velocity v_2 of the body m'' can be found from the condition:

$$\zeta_1(m'') + \zeta_2(m'') = 0. \tag{33}$$

Substituting Eqs. (30) and (31) into Eq. (33), we obtain the equation for v_2

$$\rho_2 v_2 (2 - q_2) m_2^{q_2 - 2} \int_0^{m_2/m''} \frac{f_1(\varepsilon, y)x - f_2(\varepsilon)}{(1 + x)^2} (1 + x^{1/3})^2 x^{1 - q_2} \, dx +$$

$$\rho_1 v_1 (2 - q_1) m_1^{q_1 - 2} m''^{q_2 - q_1} \int_0^{m_1/m''} \frac{f_1(\varepsilon, \bar{y})x - f_2(\varepsilon)}{(1 + x)^2} (1 + x^{1/3})^2 x^{1 - q_1} \, dx$$

$$= 0. \tag{34}$$

Integration of Eq. (34) over x gives an expression for $v_2(m'')$. From tedious transformations (omitted here), one can find that $\alpha \approx (q_2 - q_1)/2 < 0.1$. We see that with the assumptions mentioned above there is no equipartition of energy in the system. The exponent α is small and is not important. It will be neglected in further considerations.

Integrating Eq. (34) over all m'' from m_0 to m_2 and putting $\rho_2 v_2 = 4\sigma_2/P_2$, $\rho_1 v_1 = \rho_{10} v_{10} [(1 + 2\theta_1/l)(1 + 3\theta_1/l)]^{1/2} = 4\sigma_1[(1 + 2\theta_1/l) (1 + 3\theta_1/l)]^{1/2}/P_1$ (P_1 and P_2 are the periods of revolution around the Sun and the planet, respectively), we obtain an equation for the energy balance in collisions of bodies of the swarm among themselves and with the external bodies per unit mass of all bodies. Again omitting complicated transformations, we get

$$\left(\frac{v_1}{v_2}\right)^2 = \frac{3 - \varepsilon}{1 + \varepsilon}$$

$$+ \frac{C_1 - C_2 \varepsilon}{1 + \varepsilon} \frac{\sigma_2}{\sigma_1} \frac{P_1}{P_2} \left(\frac{m_1}{m_2}\right)^{2 - q_1} \left[\left(1 + \frac{2\theta_1}{l}\right)\left(1 + \frac{3\theta_1}{l}\right)\right]^{-1/2} \tag{35}$$

where C_1 and C_2 are coefficients depending on q_1 and q_2, which are obtained by integration of Eq. (34) over x, and then over m''. From Eqs. (35) and (32) we find

$$v_2^2 = \tag{36}$$

$$\frac{(1 + 3\theta_1/l)(1 + \varepsilon) v_{10}^2}{3 - \varepsilon + (C_1 - C_2 \varepsilon) \dfrac{\sigma_2}{\sigma_1} \dfrac{P_1}{P_2} \left(\dfrac{m_1}{m_2}\right)^{2 - q_1} \left[\left(1 + \dfrac{2\theta_1}{l}\right)\left(1 + \dfrac{3\theta_1}{l}\right)\right]^{-1/2}}$$

The first two terms in the denominator of Eq. (36) are important only near the outer boundary of the swarm where σ_2 is small and the third term becomes less than $3 - \varepsilon$. For all the rest of the swarm, these terms may be neglected, and then

$$v_2^2 \approx \left(1 + \frac{3\theta_1}{l}\right)^{3/2} \left(1 + \frac{2\theta_1}{l}\right)^{1/2} \frac{1 + \varepsilon}{C_1 - C_2 \varepsilon} \frac{\sigma_1}{\sigma_2} \frac{P_2}{P_1} \left(\frac{m_2}{m_1}\right)^{2 - q_1} v_{10}^2. \tag{37}$$

With $q_1 = q_2 = 11/6$, we obtain $C_1 \approx 2$, $C_2 \approx 1.2$; with $q_1 = 11/6$ and $q_2 = 23/12$, $C_1 \approx 3.4$ and $C_2 \approx 1.2$, respectively. Thus, when the exponent q_2 for the mass distribution of bodies in the swarm is higher, relative velocities are lower; then for $\varepsilon^2 = 0.5$

$$v_2\left(q_2 = \frac{23}{12}\right) / v_2\left(q_2 = \frac{11}{6}\right) \approx \frac{2}{3}. \tag{38}$$

When $v_2 \lesssim v_e(m_2)$, gravitational interaction of bodies in the swarm should be taken into account as well as viscous stirring by mutual collisions. These sources of energy were important in the part of the swarm nearest to the planet and at the late stage of accumulation when σ_1 became small. The estimates can be made in the same way as were done by Safronov (1969) and by Goldreich and Tremaine (1978).

     Large bodies with high velocities could also create in the swarm flows of small particles resembling large-scale turbulent motions in a continuous medium with a mixing length $\sim H_2$ and $v_2 < v_{turb} < v_1$, which considerably increases the viscosity of the swarm. Such motions could influence the evolution of the swarm (see Sec. I.B below).

B. Thickness of the Swarm

     Using the value of v_2 found above, the thickness of the swarm $H_2 = P_2 v_2/4$ can be estimated. For the inner parts of the swarm (near the planet) one can take $v_1 \approx (3Gm_p/R_2)^{1/2}$. Then for $q_1 = 11/6$

$$\frac{H_2}{R_2} \propto \left(\frac{\sigma_1 P_2}{\sigma_2 P_1}\right)^{1/2} \left(\frac{r_2}{r_1}\right)^{1/4} \propto R_2^{3/4 + b/2} \tag{39}$$

where we assume $\sigma_2 \propto R_2^{-b}$. For $R_2 \sim 10 r_p$, $H_2/R_2 \sim 10^{-2}$. At greater distances, H_2/R_2 increases faster, approaching unity already halfway to the edge of the Hill lobe. Therefore, the swarm is very thin near the planet and almost spherical on the periphery.

C. Volume Density of the Swarm

It is interesting to compare the volume density $\rho_2 = \sigma_2/H_2$ with the critical density $\rho_{cr} \approx 2\rho^* = 2(3m_p/4\pi R_2^3)$ for which the swarm becomes gravitationally unstable (Safronov 1969). For small distances from the planet, using Eq. (39), we have

$$\rho_2/\rho^* \approx \frac{G}{R_2}\left(\frac{r_1}{r_2}\right)^{1/4}\left(\frac{P_1}{\sigma_1}\right)^{1/2}(\sigma_2 P_2)^{3/2} \propto R_2^{5/4 - 3b/2}. \tag{40}$$

At $b > 5/6$ the most favorable conditions for instability are near the planet. The specific result depends of course on the mass of the swarm and the index b. In the Earth's zone, for $m_p = m_\oplus$ and the distance $R_2 = 3r_p$ near the Roche limit, the density ρ_2 reaches the critical value $2\rho^*$ at $\sigma_2 \approx 2 \times 10^4$ g cm^{-2}, i.e., at a swarm mass of 2×10^{24} g for $b = 2$ and at a swarm mass 3×10^{25} g for $b = 1$. For even less mass, the "classic" gravitational instability in the swarm is impossible. When such a value of σ_2 is reached, it can only be known by studying the evolution of the protosatellite swarm.

D. Accumulation and Fragmentation of Bodies in the Swarm

Small relative velocities of bodies in the swarm ("normal" values of θ_2 in Eq. 2) favored accumulation by collisions. In contrast high-velocity impacts of external bodies entering the swarm destroyed them; but before any body disintegrated, it succeeded in growing to a certain size. An average maximum size r_2 of bodies can be found from the rate of growth of bodies and their average lifetimes. Only a few bodies occasionally avoided destructive collisions for a long period and reached larger sizes.

For a body of a constant radius r, its lifetime before the catastrophic collision with a nonswarm body of mass larger than m/γ' is found by the integration of the mass distribution function (Ruskol 1975; see her Eq. 4.80)

$$\tau_f = \frac{5}{3}\frac{P_1 \delta (r_1 r)^{1/2}}{(\gamma')^{5/6}\sigma_1}. \tag{41}$$

The body also undergoes an erosional mass loss due to impacts by external bodies smaller than m/γ'. The characteristic time for erosion $\tau_e = m/\dot{m}_e$ is comparable to its lifetime τ_f (for $q = 11/6$, $\tau_e/\tau_f = \gamma'/5\gamma$). Nevertheless, erosion has practically no influence on the body's lifetime because the mass continuously lost by erosion is almost three orders of magnitude less than the mass gained through accumulation in collisions with the swarm's bodies whose velocities are low (see Eq. 37)

$$\frac{\dot{m}_e}{\dot{m}_a} = -\frac{\gamma}{(\gamma')^{1/6}(1 + 2\theta_2)}\frac{\sigma_1}{\sigma_2}\frac{P_2}{P_1}\left(\frac{r_2}{r_1}\right)^{1/2}. \tag{42}$$

The lifetime of small bodies in the swarm is very short. Before joining larger bodies, they only double their mass. The largest bodies grow most effectively. Their characteristic time of accumulation

$$\tau_a = \frac{m}{\dot{m}_a} = \frac{P_2 \,\delta\, r}{3(1 + 2\theta_2)\,\sigma_2}\left(\frac{r_2}{r}\right)^{1/2} \tag{43}$$

is two or three orders of magnitude less than the lifetime τ_f given by Eq. (41). Thus, before they can be destroyed by impacts with external large bodies, they have time to sweep out most of the material in their local feeding zones and to become embryo satellites in the same way as embryo planets formed. The width of the feeding zone (and the related embryo mass m_2) were determined by the velocities of bodies in the zone, but after the time τ_f elapsed, the embryo is destroyed by an impact of an external body larger than m_2/γ', and another largest body becomes a new embryo. This process was repeated many times over, because the growth of the planet and replenishment of the proto-satellite swarm lasted for $\sim 10^8$ yr, while the lifetime between catastrophic collisions of the embryo is $< 10^6$ yr. The width of the embryo feeding zone is $2\Delta R_2 \sim 2R_2 v_2/V_{K2}$. Therefore the embryo mass can be approximately taken equal to

$$m_2 \sim 4\pi R_2^2 v_2 \frac{\sigma_2}{V_{K2}}. \tag{44}$$

Substituting in Eq. (44) the value v_2 from Eq. (37), we find for $q_2 = 11/6$

$$\left(\frac{r_2}{r_1}\right)^{11/2} \approx 9\,\frac{1 + \varepsilon}{C_1 - C_2\,\varepsilon}\,\frac{\theta_1 r_p^4}{\delta^2 r_1^6}\,\frac{\sigma_1 \sigma_2 P_2}{P_1}\left(3 + \frac{l}{\theta_1}\right)^{3/2}\left(2 + \frac{l}{\theta_1}\right)^{1/2} l^3. \tag{45}$$

The radii r_2 of the largest bodies depend weakly on σ_2. In a wide interval of possible values of σ_2 from 1 to 10^3 g cm^{-2}, they change only 3.5 times, namely from 15 km to 50 km for $r_p = 5000$ km, $r_1 = 1000$ km, $\sigma_1 = 5$ g cm^{-2} and $R_2 = 10\, r_p$. More pronounced is the dependence on R_2; the embryo sizes are roughly proportional to the distance from the planet. This order of magnitude estimate of m_2 does not take into account that

1. Efficiency of the accumulation of m_2 at $\theta_2 < 1$ can be lower than that assumed (coalescence in every collision);
2. When m_2 sweeps up most of the material in its zone, new bodies enter

the zone due to diffusion from neighboring zones (see below) allowing a further slower increase in m_2.

The factors work in opposite ways giving us an excuse to neglect both of them.

Equations (37) and (45) for v_2 and r_2 permit us to write the velocities of presatellite bodies in the form of Eq. (2): $v_2^2 = Gm_2/\theta_2 r_2$ and to evaluate the parameter θ_2

$$\theta_2 \approx \left[\left(\frac{C_1 - C_2 \, \varepsilon}{1 + \varepsilon} \, \frac{P_1}{\sigma_1 \theta_1} \right)^8 \frac{9}{\delta^2} \left(\frac{r_1 G}{\pi} \right)^4 \frac{\sigma_2^{14} \, l^{19}}{r_p^{10} (3 + l/\theta_1)^{12} (2 + l/\theta_1)^4} \right]^{1/11} .$$

(46)

This relation shows a considerably larger θ_2 (a smaller v_2) in the inner zone of the swarm and also of course at the late stage of accumulation when $\sigma_1 \to 0$. However, in this case we should take into account viscous stirring by mutual collisions and gravitational perturbations which would decrease θ_2 because actually $\theta_2^{-1} = \theta_{21}^{-1} + \theta_{22}^{-1}$, where θ_{21} is related to the perturbations of external bodies and θ_{22} to the mutual interactions of the swarm's bodies. For large θ_{21}, $\theta_2 \approx \theta_{22}$, that is, the velocities are moderate and favor the accumulation of bodies. In the outer part, $\theta_2 \ll 1$, and the considerable random velocities of bodies can lead to their erosion and fragmentation during collisions. The intermediate values $\theta_2 \sim 1$ occur for distances $R_2 \approx 10 \, r_p$ with values $r_p = 5000$ km, $r_1 = 1000$ km and $\sigma_2 \sim 10^2$ g cm^{-2}.

Once r_2 is found, the only unknown quantity σ_2 remains, which is determined by the inflow of captured material from the planet's zone, and by the character of radial motions inside the swarm. The evolution of the swarm will be discussed in Sec. II. Here we shall estimate only the increase σ_2 of the swarm in the absence of radial redistribution of its material. In this case, $\partial \sigma_2 / \partial t = 1$, so that using Eq. (16), we get an almost exponential growth of σ_2 with the characteristic time scale

$$\tau_\sigma \sim \frac{\delta (r_1 r_2)^{1/2}}{3 \, \sigma_1 \, \ln \tau_{z2}^{-1}} \, P_1$$

(47)

which is $\sim 10^5$ yr. A higher value of the flow I during the erosive stage yields almost γ_2 times less τ_σ.

II. EVOLUTION OF PROTOSATELLITE SWARM

The value of τ_σ given by Eq. (47) does not characterize directly the real change of σ_2 nor that of the total mass of the swarm. The increase in the planet's mass as a result of accumulation leads to a proportional decrease in the swarm's dimensions. The flow of mass I onto the swarm also leads to its contraction owing to the small angular momentum of the accreted material.

The redistribution of the swarm material can be described by hydro-dynamic equations of motion, similar to those used in the study of evolution of circumstellar accretional disks. Let us consider a thin axisymmetric disk rotating with approximately Keplerian velocity around the z axis, which passes through the planet. We average its characteristics over z. Then the continuity equation for the surface density of the disk can be written in the form

$$\frac{\partial \sigma}{\partial t} + \frac{1}{R} \frac{\partial}{\partial R} (R \, \sigma \sum_i v_{Ri}) = I \qquad (48)$$

where I is the flow of material onto the disk per cm^2 per s, and v_{Ri} is the radial velocity of the disk material caused by some mechanism i. All values here are related to the swarm, but we shall omit the subscript "2" unless it would cause a misunderstanding. Initially this equation was developed only considering the orbital contraction due to the planet's mass increase (Ruskol 1963). Conservation of the body's angular momentum gives the Jeans' invariant $R \cdot m_p = $ const. Therefore

$$v_{R1} = - \frac{R \, \dot{m}_p}{m_p} . \qquad (49)$$

Later, the swarm's contraction due to the low specific angular momentum k_J of the accreted matter in comparison with the Keplerian one $k = \omega R^2$ was separately considered (Ruskol 1972a,b)

$$v_{R2} = - \frac{2R(1 - k_J/k) \, I}{\sigma} . \qquad (50)$$

In accordance with the hypothesis by Artemjev and Radzievskij (see Ruskol 1972a,b), k_J was assumed equal to $\omega_1 R^2$. Harris and Kaula (1975), considering the orbital evolution of a satellite which accumulates simultaneously with the planet from the same source of matter of the planet's zone (not from the circumplanetary swarm), have taken $k_J \approx 0.2 \, k$. They also took into account the tidal removal of the satellite from the planet.

Pechernikova et al. (1984) have numerically investigated the evolution of circumplanetary swarms taking $k_J \sim R^{3/2}$ on the basis of their earlier study of planetary rotation (Vitjazev and Pechernikova 1981). In Eq. (48) both terms v_{R1} and v_{R2} were taken into account. From calculations a power law $\sigma_2 \propto R_2^{-b}$ has been found, with $b \gtrsim 2$ for distances $3r_p \lesssim R_2 \lesssim 50 \, r_p$ but $b \lesssim 2$ for $R_2 \gtrsim 50 \, r_p$. For the calculated models, maximum values of $\sigma_2 \lesssim 10^2$ g cm^{-2} (near the Roche limit) were found (Fig. 1). The total mass of the swarm during the active stage of the Earth's growth and up until the final stage did not exceed $10^{-5}m_p$. A formally increased efficiency of capture into the swarm by a factor of 5 or 10 was not accompanied by a similar growth in the swarm's mass;

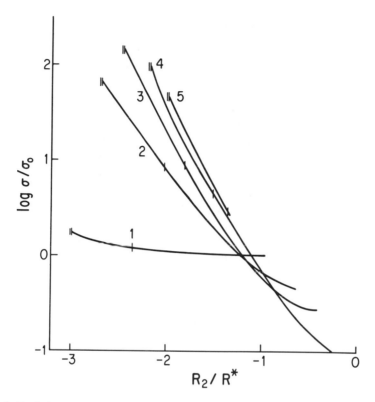

Fig. 1. Evolution of surface density distribution in the circumterrestrial disk according to Pechernikova et al. (1984). R_2 is normalized to $R^* = R_{L1}$ for the present Earth, and σ to $\sigma_0 = 1$ g cm^{-2}.

N	1	2	3	4	5
t(yr)	6.2×10^6	1.3×10^7	2.3×10^7	5.0×10^7	1.4×10^8
$(m_p/m_\oplus)$	0.001	0.010	0.043	0.30	0.94

The Roche limit is indicated on the curves by (‖) and $R_2 = 10r_p$ by (|).

the amount of material passing through the swarm and falling onto the Earth only increased by the same factor of 5 or 10. Hence, the models considered could only explain the formation of satellite systems with a mass 10^{-5} or 10^{-4} of the planet's mass.

The next mechanism causing radial motion in the disk is connected with the angular momentum transfer which is described usually in terms of viscous transport or diffusion. In investigations of accretional gaseous disks turbulent kinematic viscosity $\nu = v_t \lambda / 3$ (where v_t is the turbulent velocity and λ is the mixing length) is widely used. From equations of continuity and of angular

V. S. SAFRONOV ET AL.

momentum conservation one finds (see, e.g., Lynden-Bell and Pringle 1974)

$$v_{R3} = -\frac{3}{\omega R^2 \sigma} \frac{\partial}{\partial R} (\sigma v \omega R^2). \tag{51}$$

The same relation can be used for the rotating swarm of bodies with small mean free path. However, when the time between collisions is larger than the period of revolution around the planet, the viscosity should be determined differently taking into account that radial displacement $\Delta R \approx \lambda$ before collision or gravitational encounter varies between 0 and $2eR$ (Safronov 1969)

$$v = \lambda^2/3\tau \approx e^2 R_{\frac{1}{2}}^2/\tau \tag{52}$$

where $\tau^{-1} = \tau_s^{-1} + \tau_g^{-1}$ and τ_s is the average time interval between collisions of the body moving along the orbit with eccentricity e, τ_g is the time between its effective gravitational interactions with other bodies (relaxation time). A more general expression for the viscosity at any mean free path of particles in Saturn's rings was used by Goldreich and Tremaine (1978). Considering only collisions of particles, they evaluated the stress tensor and found that the hydrodynamic equations can be used to describe the motion of particles with large mean free path with the value of v close to that of Eq. (52) (without τ_g). Stewart and Kaula (1980) have found from the Boltzmann equation for the case of collisions almost the same relation (Eq. 51). However, for the case of gravitational interactions, the energy generated by viscous stress interactions of bodies is only about one tenth of the energy found with the viscosity given by Eq. (52). Nakagawa et al. (1983) have used Eq. (51) for planetesimals, considering the process as a diffusional one and simply replacing v by the diffusion coefficient D which is defined by the same relation (Eq. 52) except that τ is taken equal to the time of dissipation of random motion in the gas. We assume that at the late stage of accumulation of the terrestrial planets, when presatellite swarms were formed, the gas almost completely dissipated from their region and therefore we do not take it into account.

The protosatellite swarms of Jupiter and Saturn were initially embedded in gas. The gas rotated around the planet more slowly than the solid bodies and offered resistance to their orbital motion through the drag force f_φ, that diminished orbital radii at a rate

$$v_{R4} = \frac{2f_\varphi}{\omega}. \tag{53}$$

Neglecting a possible difference between coefficients of the viscosity (v) and diffusion and considering the motion of bodies with large mean free path ($\tau > P/2$) we have from Eq. (52)

$$\nu = D \approx v_2^2 R_2^2 \left(\frac{1}{\tau_s} + \frac{1}{\tau_g} \right) / V_k^2. \tag{54}$$

The second term is important mainly for comparatively low velocities of bodies when $\theta_2 > 1$ (near the planet); the first term comes into play when $\theta_2 < 1$. At $l \gtrsim 10$, the term with τ_s dominates. For bodies of radius r (Safronov 1969)

$$\tau_s = \frac{(3q_2 - 5)\delta P_2 r}{3\sqrt{2}\,\xi\,\sigma_2} \left(\frac{r_2}{r} \right)^{3(2 - q_2)} \tag{55}$$

where $\xi \approx 2$. Substituting this value into Eq. (54) and neglecting the τ_g term, we find that for almost the whole swarm (except only a small central part) when $q_2 = 11/6$

$$D(r) = D(r_2)\left(\frac{r_2}{r} \right)^{1/2} \approx \left(\frac{l}{\theta_1} + 3 \right) \frac{12\sqrt{2}\,\sigma_1 R_2^2}{\delta P_1 (rr_1)^{1/2}} \tag{56}$$

when Eq. (37) is taken into account. For smaller bodies the diffusion is more rapid than for larger ones because, while they have similar orbital eccentricities, they collide more frequently. However, Eq. (56) is not applicable to small particles with $r < r^* = \sigma_2^2/\delta^2 r_2 \sim 10^{-3}$ cm for which $\tau_s < P_2/4$, $\Delta R < eR_2$ and $D \propto r^{1/2}$. The total mass of such particles is negligibly small. Hence, the average value of D for all bodies can be found by integrating over the mass spectrum from r^* to r_2

$$\bar{D} \approx \frac{1}{2} \ln\left(\frac{r_2}{r^*} \right) D(r_2) = \ln\frac{\delta r_2}{\sigma_2} D(r_2) \approx 10\, D(r_2). \tag{57}$$

With $\sigma_1 = 5$ g cm^{-2}, $r_1 = 1000$ km, $l = 10$ and $r_2 = 25$ km from Eqs. (56) and (57), we find $D(r_2) \approx 10^7$ cm^2 s^{-1}, $\bar{D} \approx 10^8$ cm^2 s^{-1}. This gives a characteristic diffusion time $\tau_D \sim R_2^2/\bar{D} \sim 10^4$ yr. Then, because $v_{R3} \sim \bar{D}/R_2$, the diffusional term in Eq. (48) happens to be almost an order of magnitude greater than the term with v_{R2} and still increases with the increase of R_2— their ratio is $\sim (3 + l/\theta_1)/p(R_2)$. Then one can expect that the evolution of the swarm proceeds in the direction of the decrease of v_{R3}; i.e., according to Eq. (51) $\sigma_2(R_2) \rightarrow \text{const}/(DR_2^{1/2})$ (cf. Vitjazev and Pechernikova 1982; Cassen and Summers 1983). However, the theory of mass transport in rotating swarms of bodies is not yet completed and its results have been obtained with various simplifying assumptions (isotropic velocities, etc.). Hence, definite conclusions about the behavior of $\sigma(R)$ would be premature at present. We note that for the swarm of protoplanetary bodies diffusion was not so significant.

III. GAS-DUST PROTOSATELLITE DISKS AROUND GIANT PLANETS

The accumulation process of the giant planets and their protosatellite swarms was more complicated than the process in the terrestrial zone owing to the active participation of gas in their formation, to the merging of their feeding zones and to the ejection of bodies from the solar system during the final stage. The chemical composition of Jupiter and Saturn is not much different from that of the Sun. Consequently, gas dominated their formation, a view totally accepted; but, insofar as the mode of formation goes, opinions still diverge. In the model of a massive solar nebula ($M \approx 2$ $M_\odot$) by Cameron (1978), the planets originate from giant contracting gaseous condensations that separate at an early stage in evolution of the solar nebula. The evolution of such isolated nonrotating condensations of constant mass up to the present state of these planets has been calculated in a sequence of papers (see, e.g., Pollack and Reynolds 1974; Pollack et al. 1977; Bodenheimer et al. 1980). The main characteristics of the process have been revealed, and the physical properties of objects estimated. However, the model of the massive nebula itself meets with serious difficulties when its evolution is considered. In particular the following questions remain:

1. Why did the gravitational instability not provide gaseous condensations more massive than Jupiter?
2. Why did numerous condensations with masses of about the same as Jupiter's not coalesce into more massive ones?
3. Why and through what process were more than 99% of those condensations that rotated around the Sun afterwards removed from the solar system, so that a regular planetary system of much lower total mass remained?

A modification of the idea on gaseous protoplanets is the Prentice and ter Haar hypothesis (1979a) in which a continuously rotating mass fragments by turbulence into a system of Laplacian rings which later would evolve into planets and satellites.

In a low-mass nebula, gaseous protoplanets cannot form (Ruskol 1960a; Cassen et al. 1981). In such a nebula the formation of giant planets is considered as a two-stage process. First, solid bodies accumulate in the same way as in the zone of terrestrial planets, and nuclei of future planets form; then, when they reach a critical mass (2–3 $M_\oplus$), the accretion of gas by these nuclei begins. This scheme is based on ideas of gas accretion by gravitating bodies (Bondi and Hoyle 1944) and of hydrodynamic instability of massive atmospheres of bodies (Perri and Cameron 1974). It was developed later by several authors (see, e.g., Safronov 1969; Harris 1978a; Mizuno 1980; Safronov and Ruskol 1982; Weidenschilling 1982).

It has been found that gas is not uniformly accreted by the planet from its

TABLE I[a]
Parameters for Different Stages of Accretion of Giant Protoplanets

Stage	Jupiter Duration (yr)	Jupiter Mass Increase ($M_\oplus$)	Jupiter Mass Escaped ($M_\oplus$)	Saturn Duration (yr)	Saturn Mass Increase ($M_\oplus$)	Saturn Mass Escaped ($M_\oplus$)
1	3×10^7–10^8	1–3	800–3000[b]	2×10^8	2–3	1200[b]
2	10^5–10^6	10	15	10^6	8	6
3	2×10^2	40	0			
4	10^4	85	0	6×10^4	20	0
5	10^6–10^7	120	20	10^6–10^7	35	15
6	10^8	60	120	3×10^8	30	90

[a] From Safronov and Ruskol 1982.
[b] Gases.

feeding zone and that the accelerated accretion did not take place according to the classic formula $dm/dt \propto m^2$, except for a restricted intermediate stage. The duration of different stages for Jupiter and Saturn is given in Table I.

A. Reconstruction of Protosatellite Disks

The nuclei of the giant planets were surrounded by swarms of solid material. When the gas began accreting the swarms turned into accretionary gas-dust disks. Increase in the central mass led to a proportional decrease of the disk's dimensions according to the Jeans invariant. Hence almost all material entering the disk early fell onto the planet. It is interesting to consider the later stage when the major part of the disk no longer fell onto the planet but evolved into a satellite system. Unfortunately, the theory of gaseous planets' accretion is not yet developed to such a degree that would allow us to determine all the properties of the disk. We must reconstruct the surface density and the mass of the disk from the present-day total mass of satellites by diluting it with light gases (Weidenschilling 1982; Lunine and Stevenson 1982a). Estimates of the volume density of the disk material and of the surface temperature of the growing giant planet then allows us to determine P-T conditions at a given distance in the central plane of the disk where the regular satellites formed. Figure 2 shows non-smoothed "reconstructed" surface densities in the satellite systems of Jupiter and Saturn. A high value of the surface density in Io's zone ($\sim 3 \times 10^6$ g cm^{-2}) is obtained from its present-day value (10^4 g cm^{-2}) increased 300 times to account for H + He. Such large values of σ for the gas and the solid material are the highest in the solar system. They also provide a very high volume density in the disk as can be calculated from the relation for a disk of a small mass rotating around a central body (see, e.g., Safronov 1969):

$$\rho = \rho_0 \, e^{-\pi z^2/H^2} \tag{58}$$

where $\rho_0 = \sigma/H$ is the density in the central plane, $H = Pv_T/4$ is the scale height of the disk, and v_T is the mean thermal velocity. For a broad temperature interval in the disk from 1000 K down to 100 K, the thickness of the disk H at Io's distance varies between one and two Jupiter radii (R_J), and the volume density between 2×10^{-4} g cm^{-3} and 4×10^{-4} g cm^{-3}. It is interesting to note that this top value for disks is nevertheless several times less than the density $\rho^* = \rho_J (R_J/R)^3$ corresponding to that in a sphere of Jupiter's mass with the diameter of Io's orbit. This result is a serious argument against gravitational instability in circumplanetary gaseous disks. However, it does not forbid such an instability in thin dust layers where a higher density can be reached (Coradini et al. 1982).

B. Dimensions of Accretion Disks

The reconstructed disks of Jupiter and Saturn in Fig. 2 have radial dimensions of at least ~ 30 R$_J$ and ~ 30 R$_S$ because the feeding zones of Cal-

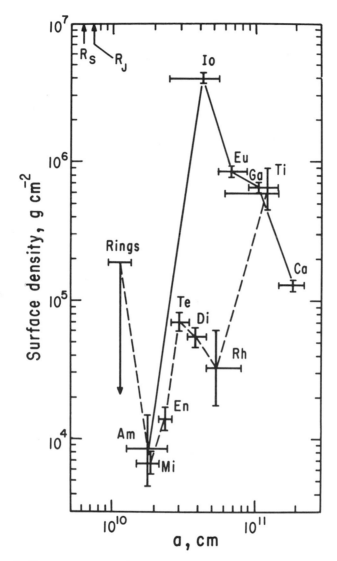

Fig. 2. The reconstructed surface density in the disks of Jupiter and Saturn according to Weidenschilling (1982).

listo (26 R_J) and of Titan (20 R_S) extended somewhat beyond their orbits. What was the size of the accretion gaseous disk? The complicated problem of nonstationary gas accretion onto a growing planet in a system differentially rotating around the Sun has not yet been solved. Even in a simplified form (stationary accretion, constant mass, slow rotation at infinity), extended numerical calculations are needed. Figure 3 shows the velocity field in the re-

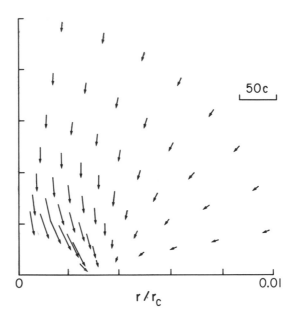

Fig. 3. Velocities of accretional gas flow into the disk according to Cassen and Pettibone (1976).

gion of formation of an accretionary disk (Cassen and Pettibone 1976). The accretion is supersonic for $R_2 < 0.9 \times (M/M_J)^{2/3} R_{L1}$, and disk dimensions are determined by the specific angular momentum Γ_∞ of gas far from the planet. The value $\varepsilon = (\Gamma_\infty c/GM)^2$ (where c is the velocity of sound) is a small parameter which determines the extension of the region where the matter is partly accreted as well onto the disk forming around the planet. The radius of the disk is approximately equal to εr_c, where $r_c = GM/c^2$. This result by Cassen and Pettibone can be applied to accretional disks of Jupiter and Saturn, because the matter contained in their Hill's lobes has a positive planetocentric angular momentum. It is of the order of $\Gamma_\infty = 1/4 \, \omega_1 R_1^2 (M/3M_\odot)^{2/3}$. The value of ε is then $< 10^{-2}$, and the radius of the disk is only a small part of the radius of Hill's lobe:

$$R_d \sim \varepsilon r_c = \frac{R_{L1}}{48}. \tag{59}$$

For Jupiter $R_d \approx 16 \, R_J$, for Saturn $\approx 23 \, R_S$. These values can be considered as lower limits because the sizes of the disks increased due to viscous forces. Consequently, the process of gaseous accretion, which created giant planets, was able at the same time to create circumplanetary accretion disks necessary for the formation of the regular satellites. The initial composition of the disks could be about the same as that of the growing planets. The disks were heated mainly by radiation received from the planet and additionally by the accretional gas flowing onto the disk. The contribution of solar radiation during accretion was much less. Unfortunately all estimates of energetics of the process give intervals too wide for determining possible temperatures in the disk. Hence, some smoothed disk models are usually assumed in which the temperature drops with distance roughly as R^{-1}, with the condensation point of water vapor (230 K: after Lunine and Stevenson 1982a) at the boundary between rocky and rocky-and-icy satellites.

In order to understand the energetics of the process, it is desirable to estimate the temperature in the disk without any additional assumptions. Already the simplest approximate estimates reveal substantial differences between the disks of Jupiter and Saturn (Ruskol 1981). Let a total energy of accretion, equal to half the potential energy of spherical homogeneous planets with radii one to three times the present ones, be uniformly emitted into space within time intervals of active accretion ($3 \times 10^4 - 10^5$ yr for Jupiter and $4 \times 10^5 - 3 \times 10^6$ yr for Saturn). The temperature in the disk can be taken as the blackbody temperature for a transparent disk ($T \propto R^{-1/2}$) or for an opaque disk ($T \propto R^{-1}$). Then we obtain a wide range of temperatures (see Fig. 4). Nevertheless, the results show a considerable heating of the inner region of Jupiter's disk (orbits of Amalthea, Io and Europa) while Saturn's disk remains rather cold at all distances. This difference, which correlates with the variations of chemical composition of satellites with distance from the planet, results from

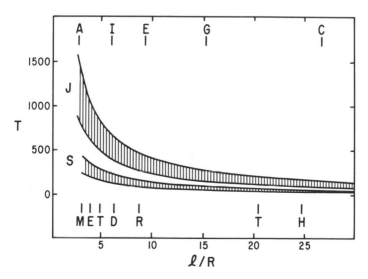

Fig. 4. Temperature in the accretionary disks of Jupiter and Saturn at the present distances of their satellites (figure from Ruskol 1981).

the fact that the accretional energy of Jupiter is an order of magnitude higher while its time scale of accretion is an order of magnitude less than that of Saturn. In this scheme the anomalous composition of Titan compared to other Saturn's satellites still has no explanation.

The assumptions made above are too rough; the nonuniform character of the process should be taken into account (see Table I). The evolution of proto-satellite swarms of giant planets is first of all the evolution of their condensable component from which the satellites were formed. Thus the detailed consideration of processes in the swarms of the terrestrial planets described in Secs. I and II has a direct relation to the swarms of giant planets. However, there was an additional strong interaction of solids with the gas. A separate evolution for each component has been actively studied (although many unclear problems remain); some effects of interaction between gas and solids were also studied, but no satisfactory consideration of their common evolution has been given. There have only been attempts to give a qualitative description of the scenario (see, e.g., Coradini et al. 1982; Weidenschilling 1982; Lunine and Stevenson 1982a). The main points of the process can be described as follows.

Protosatellite swarms of solid matter began to form around the solid nuclei of giant planets even prior to the beginning of gas accretion (nuclei masses < 2 or $3 \, M_\oplus$). The presence of gas diminished the velocities of bodies and particles and by that favored the accumulation of bodies. Gas drag led to a contraction of their orbits, but, in the outer parts of the Hill lobe, new bodies appeared due to the long duration of the process. During the subsequent growth of the planet's mass, all bodies of the inner zone of the swarm fell onto the planet, but the bodies from the periphery did not reach the planet and instead remained in the inner part of the disk. They certainly underwent an intense impact transformation; in particular, they may have lost all pristine volatile elements.

During the gas-accretion stage, a gaseous disk formed around the planet with small solid bodies suspended in the gas. Simultaneously, large particles continued to be captured in the disk and they moved in the planet's zone independently of the gas. The flow I_{sp} of captured particles was at least two orders of magnitude higher than in the terrestrial zone, due to the higher surface density of the disk with the addition of the gas (according to Eq. [10], $I_{sp} \propto \sigma_2$). Larger bodies could be captured through collisions with large bodies of the disk. During the stage of active accretion a sufficiently high temperature was maintained in the disk which was heated by the radiation of the growing planet penetrating mainly from higher z-coordinates and the energy of gas falling onto the disk. Usually only the first energy source is considered and it is assumed $T_d \approx C R^{-1}$, with the value of C chosen so that ice condenses at the present distance of Ganymede. The second source presumably can also be significant, particularly in the outer part of the swarm, because its energy decreases with R as $R^{-3/4}$, i.e., slower than for R^{-1}; but, a reliable quantitative estimate of it requires a knowledge of the disk configuration and the direction of the gas flow.

Radial motions are produced in the disk by the interaction of gaseous and solid components. Various effects work in opposite directions. Gas drag, owing to a difference in angular velocities of bodies and of the gas, forces bodies to move toward greater gas density, i.e., predominantly toward the planet. Due to another effect (tidal torque), a large body (a "shepherd" in Goldreich's terminology) pushes the gas away from its orbit. The gas inside the orbit loses a part of its angular momentum because of faster rotation and moves inwards while the gas outside the orbit rotates slower and moves outwards. A tunnel of rarefied gas appears around the orbit. All the other solid bodies and particles, because they tend to denser gas regions, move away from the tunnel. The minimum mass of the body creating the tunnel was computed at first to be $\sim 10^{17}$ g (Weidenschilling 1982), i.e., of the same order as the mass of planetesimals formed by gravitational instability (Coradini et al. 1982), and later, by a different method, to be $\sim 10^{25}$ g, i.e., almost the mass of Galilean satellites (Lunine and Stevenson 1982a). The presence of numerous large bodies with crossing orbits makes the probability of formation of stable tunnels very low. It can be shown, according to Eqs. (44) and (46) that such a situation took place almost at the end of accumulation. If the turbulence is maintained in the gas (by continuing gas accretion onto the disk, or perturbations in the gas created by passing large bodies, etc.), then density variations (including tunnels) should be smoothed considerably. Besides, with a high-turbulent viscosity, the disk should disperse: gas is removed from the inner region toward the planet, and from the outer regions—out to the Hill sphere. Gas carries away smaller solid particles, leaving some of them at a "safe" distance from the planet. Diffusion of solid bodies in the swarm (as described in Sec. II) works in the same direction.

It is interesting to note that at a rather late stage of decaying accretion, when the planet's luminosity falls but its mass is still growing (see Table I), the temperature in the vicinity of a body approaching the planet is almost constant, according to Eq. (49) and to a simple estimate of temperature. Thus, each satellite accumulates at its own almost isothermal conditions.

It is important to compare the rate of accumulation of solid bodies with the rate of accretion of gas onto the disk. There is an opinion that turbulence hinders the growth of particles and their precipitation toward the central plane, and that this process begins only after the gas accretion stage. According to another point of view, cohesion and growth of particles (not solid spheres but porous aggregates of low density) took place also at an earlier stage. When these particles became sufficiently large (meter-sized), turbulence in the gas no longer could maintain their random velocities above the critical value. As a result, the instability rose in the solid component and the formation of larger bodies was facilitated.

IV. CONCLUSION

This chapter has concentrated on the dynamics of circumplanetary disks and protosatellite swarms. Many questions remain unanswered. Some of these are discussed in other chapters of this book. The time has come for a synthesis of all earlier attempts to explain the origin of satellites (capture, fission, binary accretion). Such a complex approach is particularly desirable in the study of the Moon's formation. Furthermore, dynamical considerations should be more closely connected with chemical studies. Finally, we believe that the theory of circumplanetary protosatellite disks should be a part of a general theory of disk-like structures in astrophysics. The origin and evolution of galaxies, accretional disks in double-star systems, protoplanetary disks around some stars, swarms of bodies and particles around planets (in particular, planetary rings) now attract the steadily growing attention of specialists, both theorists and observers.

Acknowledgments. The authors are grateful to J. A. Burns for his remarks and for considerable improvement of the text, and to A. W. Harris and an anonymous referee for a thorough reading of the manuscript and for many useful critical remarks.

4. THE EVOLUTION OF SATELLITE ORBITS

JOSEPH A. BURNS

Cornell University

This chapter summarizes how the orbits of the natural satellites have evolved; it emphasizes results obtained in the last decade. In the mid-1970s most orbital history calculations were concerned with the Moon or the Martian satellites. Nowadays the relevant processes acting on these bodies are much better understood, although the origins of these moons are still uncertain. Accordingly, the scientific focus has shifted to the outer solar system, where difficult problems lie, particularly in the development of resonances. The observed outward evolution of the lunar orbit is surprisingly rapid and seems to result from a near-resonance between the natural frequency of the Earth's oceans and the Moon's principal forcing term; this realization eliminates a previous conflict between the Moon's age and its evolution time scale. Tidal evolution has caused important changes in Phobos' orbital semimajor axis and eccentricity but not its inclination, nor in the orbit of Deimos; passages through several resonances may have produced part of Phobos' eccentricity. Current orbits of the Martian moons do not necessarily suggest capture. The synchronous configuration of the Pluto-Charon system is expected from tidal evolution; binary asteroids, if they exist and have the appropriate mass ratio, should also be driven to a similar state. The final end-state for retrograde satellites and hypothetical satellites of tidally despun planets is loss from the system. Resonances in the outer solar system can develop due to differential tidal expansion. The nearness of satellites to the edge of Saturn's rings precludes the rapid expansion predicted from ring torques, unless either the satellites or rings are young, or the phenomenon is incorrectly described. The orbital eccentricity and inclination of any Uranian satellite are likely to be secularly forced by other satellites. The circular orbit of Triton, Neptune's massive retrograde satellite, is caused by tidal action. The irregular satellites of Jupiter, Saturn and Neptune were probably captured as the solar system formed. New criteria have been developed to characterize stable three-body orbits.

Despite the popular notion that the heavens are unchanging, the solar system has evolved, albeit slowly, and continues to do so today. Energy loss is the primary cause of these inexorable changes. This chapter describes the evolution of satellite orbits, particularly as caused by tides within the planet or the satellite itself; orbits of the Saturnian satellites can also be modified through interactions with the ring system.

I will pay special attention to the substantial advances made since my previous reviews on satellite evolution (Burns 1977*b*,1982). For example, now the processes governing the Moon's evolution are better comprehended (Lambeck 1977,1980; Hansen 1982; Brosche and Sündermann 1982); despite this, it appears that calculations of the past lunar orbital history are likely to be less valuable in constraining the Moon's origin than once hoped (Boss and Peale 1986). Significant improvements have been made in understanding how the current orbits of the Martian satellites could have arisen (Yoder 1982; Szeto 1983). Tidal evolution is believed to be responsible for the resonant structure of the Jovian and Saturnian satellite systems (see Peale's chapter); the probability of capture into some of these resonances can now be calculated explicitly. The evolution of the Laplace resonance and its connection to Io's tidal heating has been delineated (Yoder 1979*b*; Yoder and Peale 1981; cf. Greenberg 1981,1982*a*; chapters by Peale, by Schubert et al. and by Nash et al.) although major issues remain unresolved. The orbital evolutions of the Uranian satellites (Dermott 1984*a*; Dermott and Nicholson 1986; Squyres et al. 1985), Triton (McKinnon 1984) and Pluto-Charon (Peale's chapter and below) are beginning to be addressed as well.

Capture from heliocentric orbits has been suggested for years as a way to account for a few planetary satellites, particularly the irregular ones (see, e.g., Burns 1977*b*); explicit calculations for specific drag models (Pollack et al. 1979; Hunten 1979) have been performed. On the subject of capture, much effort has been devoted to developing a new criterion to identify when an object is trapped stably (Markellos and Roy 1981). This topic is no longer an academic one since unseen satellites of asteroids (van Flandern et al. 1979) could threaten those spacecraft scheduled to pass nearby; the Galileo spacecraft is scheduled to scrutinize some asteroid on its way to Jupiter and several asteroids are to be visited by the CRAF (Comet Rendezvous Asteroid Flyby) mission in the early 1990s.

I. TIDAL EVOLUTION

A. Qualitative Remarks

Tides in a planet fundamentally arise because gravity is a function of radial distance. Since parts of the planet located at somewhat different distances from the perturbing satellite experience slightly different gravitational forces, various portions of the body are accelerated differently. But these parts are not entirely free to move separately since the planet retains its integrity. Instead the planet deforms in response to the varying attraction of gravity; to do so, internal strains (and concomitant stresses) develop and supply the necessary forces to make all parts accelerate similarly. The planet is stretched along the line which connects its center to that of the satellite (see Fig. 1a): the satellite's attraction at the planet's sub-satellite point is more than the planet's central acceleration, while at the supra-satellite point the satellite's attraction is less than the requisite acceleration at the supra-satellite point (see Burns 1982). Since the distorting force depends on the difference in gravity across the body, the force (and the consequent deformation) varies as R/r^3, where R is the planet's radius and r is the distance from the planet to the satellite.

The distortion described here occurs only ideally: in truth, the alignment along the line of centers is inexact due to energy loss. Whether the planet is solid (Mars), liquid (Jupiter) or part solid/part liquid (Earth), some dissipation will always occur whether through inelasticity, viscosity, breaking waves or whatever process. For any of these, as long as resonant phenomena do not occur and the processes are not too dissipative, the planet's response is accurately represented by simply time-delaying and reducing the ideal response. The linear system composed of a mass, spring and dashpot is a useful heuristic analog.

The orbital consequences of this tidal distortion are now described qualitatively for the usual case when the planet's spin ω is faster than the orbital angular velocity n; a quantitative calculation follows. Any time delay allows the tidal bulge to rotate from beneath the satellite that generated it (Fig. 1b). That is to say, as with the mass-spring-dashpot, the largest deformation occurs after (in time) the maximum force: for $\omega > n$, later in time means that the bulge *leads* in angle. The gravitational attraction between the satellite and the asymmetrically placed bulge gives rise to a torque that slows the planet's rotation, and ultimately would cause this spin to be synchronously locked with the satellite's orbital rate (as has apparently transpired with Pluto and Charon); the counterpart of this process causes satellite rotations to be synchronously locked with the orbits of their "satellites," namely the planets themselves (Peale 1977). Simultaneously, so as to conserve the system's total angular momentum, the same forces augment the orbit's angular momentum by the amount that has been lost from the spin. Since a component of the bulge's attraction is in the direction of the satellite's motion, energy is simultaneously added to the orbit, causing it to expand.

Because the tidal force felt by the satellite results from the potential of a slightly distorted planet, it varies $\sim r^{-4}$; it is also proportional to the time delay, which denotes the magnitude of the asymmetry. Hence, combined with the distortion itself, the tidal force is $\sim r^{-7}$ as long as the satellite is not too close to the planet, in which case higher-order terms must be included. Accordingly, once even modest eccentricities are present, forces at pericenter dominate the evolution. If the evolution is idealized as due solely to a pericenter impulse, then pericenter distance $a(1 - e)$ must remain constant while

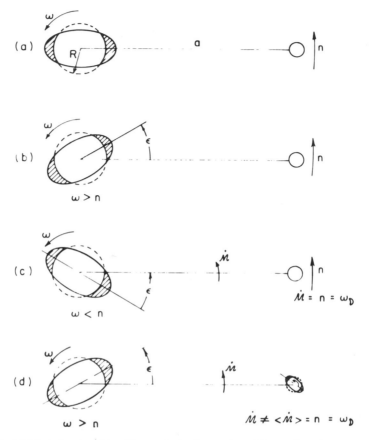

Fig. 1. (a) Ideal planetary tides. The gravitational attraction of the satellite distorts the planet, which responds immediately, assuming perfect materials. No angular momentum nor energy is transferred to the satellite's orbit since the gravitational field of the planet is symmetric about the line of centers. (b) Planetary tides for the usual case in the solar system, $\omega > n$. The presence of friction delays (in time) the tidal bulge which precedes by ε the perturbation that caused it. Angular momentum and energy are given from the planet's rotation to the satellite's orbit: the planet's spin slows while the orbit grows. (c) Planetary tides when $\omega < n$, including dissipation. This situation holds for Phobos, Metis, Adrastea and the retrograde satellites. The time-delayed bulge now lags behind the satellite position. Angular momentum and energy flow from the satellite's orbit to the planet's spin: the orbit collapses, circularizes (for $2\omega < 3n$) and the planet's rotation speeds up. (d) Tides in a satellite near pericenter. While, on the average, $n = \omega_{\text{moon}}$, the satellite's orientation librates about the line of centers. To maintain synchronism as the orbit evolves outward, the satellite's long axis is permanently offset. The torque by the planet on the satellite tends to shrink the orbit but its effect is quite small.

orbital semimajor axis a increases for $\omega > n$: hence this model indicates that orbital eccentricity e generally grows due to the action of planetary tides (Jeffreys 1961). One slight modification should be mentioned. The history of the orbital eccentricity is largely determined by the variation in the tidal amplitude as the satellite approaches and recedes from the planet. Hence, de/dt has the sign of $(2\omega - 3n)$ whereas da/dt, governed mainly by the satellite's mean semidiurnal tide (see Fig. 1b), has the sign of $2(\omega - n)$. The orbital inclination I will also be altered because, in general, the planet's rotation carries the tidal bulge out of the satellite's orbit plane. This produces a torque normal to the orbit plane that usually slowly aligns the planet's spin angular momentum vector with the satellite's orbital angular momentum vector. (This evolution of the rotation axis' orientation does not apply when the spin rate $\dot{\psi}$ $\geq 2n$; e.g., a rapidly rotating Venus would be driven toward an obliquity between $0°$ and $90°$ as it slowed down by tides. Only when $\dot{\psi} < 2n$ is the equilibrium obliquity $0°$ [Goldreich and Peale 1970].) The Moon's orbit at present contains most of the angular momentum stored in the Earth-Moon system as a whole; so as the Moon evolves outward, its orbital inclination must decrease if angular momentum is to be conserved (see Kaula 1964; Lambeck 1975; Burns 1977b).

It is worthwhile to note that specific tidal components dominate the evolution of particular orbital elements (cf. Lambeck 1980, pp. 112–117; see Sec. I. E below). For a constant tidal lag angle, the M_2 (lunar semidiurnal) tide, that sketched in Fig. 1b, determines four-fifths of the evolution of a and I since the tidal bulge due to this component continually adds energy to the orbit and produces a torque on the orbit plane; it does not, however, simultaneously alter the eccentricity much because its effect does not vary around the orbit. Eccentricity changes are almost entirely caused by the N_2 (lunar elliptic semidiurnal) tide, which has scarce influence on a and I; as mentioned above, the large tide near pericenter means that the apocenter grows while the small tide at apocenter has negligible effect on pericenter height. The consequences for I of the K_1 (lunisolar diurnal) tide and the O_1 (lunar declination) tide are nearly as large as that of the M_2 tide but they almost cancel one another.

The qualitative discussion given above for fast planetary rotations reverses for slow ones (Fig. 1c): orbits shrink (for $\omega < n$), circularize (for $2\omega < 3n$) and lift off the equator.

Tides within the satellite's interior also cause some orbital evolution. Despite the spins of all regular satellites being synchronously locked to their orbital motions (Peale 1977), the tidal bulge on the satellite is not perfectly aligned, nor is it static, and each of these produces evolution. Consider a perfectly circular satellite orbit with $\omega > n$. As described above, the satellite orbit will evolve outward due to planetary tides. For synchronism to be maintained throughout the orbital expansion, the satellite's spin must slow down in pace with the satellite's lessened orbital rate. This is accomplished by the long axis of the Moon's tidally deformed shape being slightly offset from the line of centers (see Fig. 1d), which allows the planet to produce the appropriate decelerating torque on this asymmetric mass distribution (Yoder 1982; Mignard 1982*b*).

Tides on a satellite moving along an elliptical path have two additional features. First are "push-pull" (radial) tides in which the size of the tidal bulge varies in step with changes in the planet-satellite separation. Energy is dissipated within the satellite through this continual flexing. Hence, due to a time delay with similar cause to that described earlier for the planetary bulge, the tidal bulge on the satellite reaches a maximum shortly after pericenter passage. Second, the instantaneous orbital angular velocity, which varies with orbital longitude in order to satisfy Kepler's second law, oscillates about its mean value, which is the satellite's spin rate for a synchronously locked satellite. Hence the tidal bulge is generally misaligned with the planet-satellite line, sometimes leading it and sometimes trailing it. Again, energy is lost as the tidal bulge rocks back and forth across the satellite's face. As a consequence of all these effects, energy is dissipated while orbital angular momentum changes relatively little. This circularizes orbits (Goldreich 1963; Burns 1976*b*,1977*b*) and competes with planetary tides that, as we saw above, usually tend to increase orbital eccentricities. Lambeck (1980) estimates that the lunar tidal effect on the Moon's orbit today is less than one-sixth that of the Earth's (see below), but no general conclusion can be drawn for other epochs or other planet-satellite systems. Nevertheless, the fact that most major satellites move on nearly circular orbits presumably testifies that satellite tides have dominated the evolution of orbital eccentricities (Goldreich 1963; Burns 1977*b*) or that the total tidal expansion of orbits has been limited.

B. Tidal Evolution Equations

To describe the past and future histories of satellite orbits, we need differential equations that express the evolution of the orbit form under the action of tides. The orbit form is characterized by the *osculating orbital elements* which are six quantities that describe the orbit at any time t: for an ellipse one possible set is a (orbital semimajor axis), e (eccentricity), I (inclination), Ω (longitude of ascending node) and ω_p (argument of pericenter) as well as the mean anomaly $\mathcal{M}$ (angular position on the orbit) which allows an epoch (i.e., starting time at pericenter) to be defined (see, e.g., Danby 1962). The Gauss and Lagrange orbital perturbation equations, respectively, then provide time rates of change of these osculating orbital elements in terms of perturbing forces and disturbing potential functions; these perturbations or disturbing functions are the quantities above and beyond the point mass force or potential. Since tides are most easily represented by a potential, we will employ the Lagrange equations.

The tidal disturbing function at $\mathbf{r}$, due to the tides raised by satellite (mass m^* at $\mathbf{r}^*$) on a planet of radius R, is

$$\Delta U(\mathbf{r}) = \left(\frac{Gm^*}{r^*}\right) \sum_{\ell=2}^{\infty} k_\ell \left(\frac{R}{r^*}\right)^\ell \left(\frac{R}{r}\right)^{\ell+1} P_{\ell 0}(\cos S) \tag{1}$$

where $P_{\ell 0}(x)$ is the Legendre polynomial of argument x, and S is the position angle between $\mathbf{r}(r,\phi,\lambda)$ and $\mathbf{r}^*(r^*,\phi^*,\lambda^*)$, given by spherical trigonometry as

$$\cos S = \sin \phi \sin \phi^* + \cos \phi \cos \phi^* \cos(\lambda - \lambda^*) \tag{2}$$

(Lambeck 1980). In Eq. (1) k_ℓ is the ℓth degree Love number, defined as the fractional increase (over the imposed tidal potential) in the ℓth degree potential at the planet's (displaced) surface arising solely from the planet's deformation; for a homogeneous elastic body of rigidity μ and density ρ, the second-order Love number is

$$k_2 = \frac{3/2}{1 + 19\mu/(2g\rho R)} \tag{3}$$

where g is surface gravity (Love 1944). The Earth's k_2 is observed to be ≈ 0.25 which implies $\mu \approx 10^{12}$ dyne cm^{-2}, similar to that measured seismically and that of steel. For small objects, such as satellites, the term in parentheses is $\ll \mu$, so that $k_2 \approx (4\pi/19)G\rho^2 R^2/\mu$; the Moon's $k_2 \approx 0.02$ to 0.03. The factors in Eq. (1) are, respectively, the point mass potential, the appropriate Love number so as to include only the part of the potential due to the planet's deformation, the radial dependence of the ℓth potential at the planet's surface, the radial dependence of the potential effect back at the point of interest and finally the angular distribution across the surface.

To use Eq. (1) in the orbital perturbation equations, which are written as functions of orbital elements, we must express the positions of the tide-raising body (identified here by asterisks) and the perturbed satellite (without asterisks) in terms of their orbital elements rather than in spherical polar coordinates. Kaula (1964; see also Lambeck 1980) provides the appropriate transformation as well as the addition theorem for spherical harmonics that is needed. With these we can write

$$\Delta U(\mathbf{r}) = \sum_{\ell=2}^{\infty} \sum_{f=0}^{\ell} \sum_{p=0}^{\ell} \sum_{q=-\infty}^{\infty} \sum_{j=0}^{\ell} \sum_{g=-\infty}^{\infty} \Delta U_{\ell fpqjg} \tag{4}$$

with

$$\Delta U_{\ell fpqjg} = \frac{Gm^*}{a^*} k_\ell \left(\frac{R}{a^*}\right)^\ell \left(\frac{R}{a}\right)^{\ell+1} (2 - \delta_{0f}) \frac{(\ell - f)!}{(\ell + f)!}$$
$$\times F_{\ell fp}(I^*)F_{\ell fj}(I)G_{\ell pq}(e^*)G_{\ell jg}(e)\cos(v^*_{\ell fpq} - v_{\ell fjg} + \varepsilon_{\ell fpq}) \tag{5}$$

where δ_{0f} is the Kronecker delta, while $F_{\ell fp}(I^*)$ and $G_{\ell pq}(e^*)$ are polynomials in sin I^* and e^* listed in Table I; the angle $v^*_{\ell fpq} = (\ell - 2p)\omega_p^* + (\ell - 2p + q)\mathcal{M}^* + f(\Omega^* - \theta)$ with θ denoting sidereal angle. The lag angle

$$\varepsilon_{\ell fpq} \approx [(\ell - 2p + q)n - f\dot{\theta}]\Delta t \qquad (6)$$

where an overdot signifies a time derivative and Δt is the tidal delay time, which is negative and is specified for a particular interior model or determined *a posteriori* once an orbit's evolution is measurable.

The Lagrange perturbation equations (see, e.g., Brouwer and Clemence 1961) contain partial derivatives of the potential (Eq. 4) with respect to the orbital elements of the perturbed body. Of course, in the typical case, the tide-raising body and the body affected by the tides are the same satellite so that, after the derivatives are taken, the orbital elements of the two can be equated. Other simplifications are also possible. First, since only long time "secular" changes are important, we can ignore terms that are periodic in ω_p, Ω or $\mathcal{M}$. Furthermore, even though, following Kaula (1964), we have allowed higher-order terms each to have their own phase lag $\varepsilon_{\ell fpq}$, it is common (given our understanding of planetary interiors and dissipative mechanisms) to set the phase lags equal for all solid-body tidal components, following Darwin (cf. Lambeck 1977). In addition, continuing the same pragmatic philosophy, the above development (which is strictly valid only for solid-body tides) may also be applied (cautiously) to oceanic tides. The two types of tides are certainly different, since boundary conditions and resonances may play an important (and variable) role for oceanic tides, but equivalent phase lags and Love numbers can be defined for the ocean case (see Lambeck 1980, pp. 293–295). More importantly, for many histories, we only need worry about second-degree ($\ell = 2$) terms, since the other terms drop off fast with R/a, generally a small quantity. Indeed, once higher-order terms are important contributors, tidal amplitudes are likely to be so large that the rheological properties of the solid Earth are no longer linear. In particular, most of the oceanic dissipation occurs in the M_2 tide, whose equilibrium shape is well represented by a displaced ellipsoid. In the past this approximation did not necessarily hold; this important complication will be discussed later. Lastly, since the $G_{\ell pq}(e)$ are proportional to $e^{|q|}$ (see Table I), the summation over q in Eq. (5) need only be carried out over a small number of terms, say $|q| \leq 2$.

With these simplifications, we can arrive at the secular changes in the elements by substituting Eq. (4) into the Lagrange perturbation equations (Brouwer and Clemence 1961; Danby 1962). For the three most descriptive orbital elements, Lambeck (1980, p. 292; cf. Burns 1982, p. 451) gives

$$\dot{a}_{\ell fpq} = 2K_{\ell f}[F_{\ell fp}(I)]^2 [G_{\ell pq}(e)]^2 (\ell - 2p + q) \sin \varepsilon_{\ell fpq} \qquad (7)$$

TABLE I

Orbital Functions $F_{\ell fp}(I)$ and $G_{\ell pq}(e)$ for $\ell = 2$ and $-2 < q < 2$ to order e^2 [a]

ℓ	f	p	$F_{\ell fp}(I)$
2	0	0	$-\frac{3}{8}\sin^2 I$
2	0	1	$\frac{3}{4}\sin^2 I - \frac{1}{2}$
2	0	2	$-\frac{3}{8}\sin^2 I$
2	1	0	$\frac{3}{4}\sin I\,(1 + \cos I)$
2	1	1	$-\frac{3}{2}\sin I\cos I$
2	1	2	$-\frac{3}{4}\sin I\,(\cos I - 1)$
2	2	0	$\frac{3}{4}(1 + \cos I)^2$
2	2	1	$\frac{3}{2}\sin^2 I$
2	2	2	$\frac{3}{4}(1 - \cos I)^2$

ℓ	p	q	p	q	$G_{\ell pq}(e)$
2	0	-2	2	2	0
2	0	-1	2	1	$-\frac{1}{2}e + \cdots$
2	0	0	2	0	$1 - \frac{5}{2}e^2 +$
					$\cdots$
2	0	1	2	-1	$\frac{7}{2}e + \cdots$
2	0	2	2	-2	$\frac{17}{2}e^2 + \cdots$
2	1	-2	1	2	$\frac{9}{4}e^2 + \cdots$
2	1	-1	1	1	$\frac{3}{2}e + \cdots$
2	1	0	1	0	$(1 - e^2)^{-3/2}$

[a]Table from Kaula (1964, p. 664) and Lambeck (1980, p. 114).

$$\dot{e}_{\ell fpq} = K_{\ell f}[(1 - e^2)^{1/2}/ae][F_{\ell fp}(I)]^2[G_{\ell pq}(e)]^2$$

$$\times [(1 - e^2)^{1/2}(\ell - 2p + q) - (\ell - 2p)] \sin \varepsilon_{\ell fpq} \qquad (8)$$

and

$$\dot{I}_{\ell fpq} = K_{\ell f}\frac{(\ell - 2p) \cos I - f}{a(1 - e^2)^{1/2} \sin I}[F_{\ell fp}(I)]^2[G_{\ell pq}(e)]^2 \sin \varepsilon_{\ell fpq} \qquad (9)$$

where the coefficient

$$K_{\ell f} \doteq \frac{Gm^*k_\ell}{[G(M + m^*)a]^{1/2}}\left(\frac{R}{a}\right)^{2\ell + 1}\frac{(\ell - f)!}{(\ell + f)!}(2 - \delta_{0f}). \qquad (10)$$

Caution is required in using Eq. (9) for $\dot{I}$ since, for the inclination, in contrast to the other orbital elements, one must specify what reference plane is being used. Mignard (1979) surmounts this problem by deriving an equation equivalent to that given above, but doing it relative to an invariable plane. This approach is particularly useful when the reference plane changes during the evolution.

For the long time evolution of the satellite's orbit, we only care about secular terms, those with zero frequency, since the others average to zero. So $\dot{v}^*_{\ell fpq} - \dot{v}_{\ell fjg} = 0$ in Eq. (5) or $p = j$ and $q = g$. The lag angle $\varepsilon_{\ell fpq}$ depends on the nature of the material, as well as on the tidal magnitude and frequency (Burns 1977b).

Satellite and solar tidal effects can also be calculated within this same formalism. For satellite tides it is merely a matter of interchanging the planetary and satellite parameters in Eq. (5), and multiplying by the mass ratio $M_\odot/m^*$ (see Kaula 1964). For a given $2fpq$ component, the ratio of satellite-to-planet tidal effects for perturbations of a, e, or I is

$$A = \frac{k_2^* \sin \varepsilon^*_{2fpq}}{k_2 \sin \varepsilon_{2fpq}}\left(\frac{M}{m^*}\right)^2\left(\frac{R^*}{R}\right)^5 \qquad (11)$$

(Kaula 1964; Lambeck 1980; Mignard 1981b,d,1982b); for the Earth-Moon system $A \approx 0.67 \sin \varepsilon^*/\sin \varepsilon$, where the ε's can depend on tidal component (e.g., $\varepsilon^*_{2200} = 0$).

Solar tides raised on the planet can also be included in this framework by substituting solar parameters to replace the satellite values in Eq. (8) and multiplying by $M_\odot/m^*$. They imperceptibly alter the Earth's orbit today. However, they do slow the Earth's spin somewhat at present and, although they will never dominate the evolution during the Sun's lifetime, they will become relatively more important in this regard as the Moon withdraws from Earth. The removal of system angular momentum by the Sun will then be crucial in the system's evolutionary history (cf. Ward 1982); this might also have been true for any satellites that once orbited Mercury or Venus (Burns 1973; Ward and Reid 1973; see Sec. II.A below).

C. Inclination

A striking feature of the current solar system's structure is that satellites (at least those near their primaries) have very small orbital inclinations. A process must be present to reestablish continually these low inclinations because the Sun's torques on the planets cause planetary rotational axes to precess at a rate $\dot{\eta}_\odot$. Hence, without some mechanism to induce satellite orbit planes to follow the precessing planetary equators, a satellite's orbit would soon drift off its planet's equator.

It is well established in celestial mechanics that a close satellite, such as an artificial spacecraft, will maintain a fixed average orbital inclination off the planet's equator (Danby 1962, p. 261); this orbit precesses about the planet's rotation pole at a steady rate $\dot{\eta}_s$ which depends on the planet's oblateness and the orbit's size, shape and inclination (Danby 1962). Furthermore, whenever $\dot{\eta}_s \gg \dot{\eta}_\odot$, the satellite's inclination relative to the planet's precessing equator will remain essentially constant even as the planet precesses (Goldreich 1965a); this calculation neglects mutual and solar perturbations of the satellite, and ignores any tidal evolution of the satellite orbit. This process then keeps inclinations small if they start out that way.

Another classical result of celestial mechanics is that the mean orbit plane of a distant satellite, such as the Earth's Moon, precesses steadily around the pole of the planet's orbit plane as a result of solar perturbations (Danby 1962, pp. 283–284). This (and the close satellite case of the preceding paragraph) can be understood in elementary dynamical terms as the steady precession of the orbital angular momentum vector produced by the average gravitational torque acting on the orbit. In each of the two cases, the orbit's angular momentum vector sweeps out a circular cone around a fixed direction in space. For satellites which are neither distant nor close, we can also define a mean pole about which the pole of the satellite's orbit plane precesses on a nearly circular path. The plane normal to this precession pole is called the Laplace plane. From the above discussion, the Laplace plane is the planet's equator for close satellites, but is the planet's orbit for distant moons: it transfers smoothly between the two for intermediate-size orbits (Table III of the introductory chapter lists the Laplace plane locations in the neighborhood of the various satellites).

The division between close satellites (which maintain fixed orbital inclinations relative to their primary's equator) and distant satellites (which keep constant orbital inclinations relative to their planet's orbit plane about the Sun or "ecliptic") occurs in the region where the torque applied to the satellite orbit by the planetary oblateness is comparable to that due to the Sun. This criterion implies that the transfer to solar control happens close to the distance

$$a_{crit} \approx (2J_2 R^2 a_\odot{}^3 M/M_\odot)^{1/5} \tag{12}$$

where $a_\odot$ is the planet's orbital semimajor axis and J_2 is the planet's oblateness coefficient, listed in Table II of the introductory chapter (Goldreich 1965a). For outer satellites whose orbits circumscribe closer satellites, the "oblateness" corresponding to these inner bodies (which gravitationally act like an equatorial bulge of the planet) should be included in J_2 (Goldreich 1966). Circular orbits are assumed by Eq. (12), and a_{crit} is increased when eccentric satellite orbits are considered instead (Mignard 1981d).

With few exceptions, the division given by Eq. (12) accurately reflects the transition between regular satellites and irregular ones (Goldreich 1966; Burns 1977b, 1982). The regular satellites (which, as described in the introductory chapter, include almost all the larger satellites) lie within a degree or two of their planet's equator (except for Miranda which is at 4°) and have nearly circular orbits; presumably they formed by accretion out of the circumplanetary disks which surrounded their planets and which lay in the planets' Laplace planes (equatorial planes) (see chapters by Safronov et al. and by Stevenson et al.). The irregular satellites generally are small with highly inclined and eccentric orbits; they are believed to be captured objects (see Sec. III). The only satellites for which this division into regular and irregular satellites fails are those where the concept might be suspect. For instance, the Earth's Moon with an orbital inclination of 5°.1 to the ecliptic may have tidally evolved across the transition zone between near and far (see below); a_{crit} is 10 $R_\oplus$ today but, including the Earth's additional flattening which corresponds to the planet's faster rotation when the Moon was closer, a_{crit} would have been 17 $R_\oplus$, if the Moon were ever at 10 $R_\oplus$. Iapetus has an orbital inclination of 7°.5 relative to the local Laplace plane, which itself is tilted by 8° relative to Saturn's equator; this satellite lies almost at the transition point given by Eq. (11). Ward (1981) believes that Iapetus' unique orbit (high inclination but small eccentricity) may be a signature from its accretion history and may imply typical accretion time scales. Triton, like the Earth's Moon, may have tidally evolved through the transition zone on its inward spiral (McCord 1966; McKinnon 1984).

Through $F_{\ell fp}(I)$ the orbital evolution (Eqs. 7–9) involve the inclination, which from the above discussion has a relatively simple history for close or distant satellites; therefore the coupling of I into all the evolution equations (Eqs. 7 and 9) can usually be handled in a straightforward manner. For example, Goldreich (1966) numerically averaged his lunar orbital evolution equations over a precessional period and then considered this averaged equation for a; for the Moon this is acceptable only in to about 10 $R_\oplus$, the distance given by Eq. (11). Cazenave et al. (1980) and Mignard (1981*d*) developed schemes to follow Phobos' orbit through the transition zone. They found, not surprisingly in retrospect, that for slow enough evolutions the orbital inclination remains fixed relative to the evolving Laplace plane. Accordingly, a gradually evolving satellite that originates in its planet's equatorial plane and tidally moves outward would end up in the planet's orbit plane. The transfer problem for the Moon has been addressed by Mignard (1981*b*), who patched two separate solutions together in the transition region, and Conway (1982), who developed an analytical solution to the linearized precessional equations.

D. Calculations of the Lunar Orbital History

For obvious reasons, the orbital evolution equations described above have been applied most systematically to the tidal evolution of Earth's companion. The dynamical history of the Earth-Moon pair is more complex than most other systems because:

1. Both ocean tides and solid-body tides affect the evolution;
2. Solar tides significantly brake the primary's rotation;
3. Much angular momentum has been transferred from the Earth's spin to the lunar orbit due to the large relative mass ratio m/M;
4. The Sun sensibly perturbs the lunar orbit;
5. The Moon's orbit presently has a nonnegligible eccentricity and inclination.

On the other hand, one has the advantage in treating this problem that various techniques have been used to measure several aspects of the actual evolution; that is, more so than for any other case, the theory is testable. Although the primary motivation for comprehending the lunar orbital history is to constrain the Moon's origin (or more ideally to specify it exactly), here we will not emphasize schemes for lunar origin since they are discussed in the chapters by Kaula et al., Schubert et al. and by Stevenson et al. Most reviews about the dynamical aspects of lunar origin (see, e.g., Kaula 1971; Kaula and Harris 1973,1975; Öpik 1972) were written some time ago, during the Apollo era; only the article by Boss and Peale (1986) was prepared recently after the paradigm of catastrophic collision for lunar origin had become popular.

A general admonition for orbital histories is in order. Almost all dissipation models that have been used in orbital histories so far are linear. This means that, when the Moon gets close to the Earth, say within 10 $R_\oplus$, the model fails because tidal amplitudes will be large. Accordingly, any calculations within this distance become very questionable. The bottom line is that orbital history computations cannot clearly distinguish between capture, fission or binary accretion (Boss and Peale 1986).

I now briefly chronicle previous studies of lunar orbital evolution. In the mid-eighteenth century I. Kant suggested that lunar tides account for the observed slowing of the Earth's rotation. About a hundred years later J. R. Mayer, arguing from conservation of energy, surmised that therefore the Moon must evolve outward from the Earth. The first quantitative attempts to trace the Moon's orbital evolution as driven by tidal friction were begun more than a century ago by Darwin (Darwin [1908] compiled these pioneering studies). Using reasonable tidal models, combined with graphical and rough numerical solutions, he demonstrated that the Earth and Moon could have been quite near one another long ago and he postulated that the fissioning of proto-Earth had given birth to the two objects we know today. Gerstenkorn (1955, 1969) was the first to allow a large inclination in his solution. Integrating backwards in time, he showed that at close distance ($\sim 3\ R_\oplus$) the lunar orbit flipped over the Earth's pole and then the orbit receded away: he proposed that the Earth had tidally captured the Moon from an independent heliocentric orbit along a retrograde hyperbolic path. Kaula (1964) dealt with a problem similar to Darwin's but with a vastly improved numerical treatment after first developing the complete tidal perturbation equations. Other solutions in this period were made by G. J. F. MacDonald, S. F. Singer, L. B. Slichter and N. A. Sorokin who each investigated different tidal models (Hansen [1982] and Boss and Peale [1986] summarize these works).

Goldreich (1966) wrote a benchmark paper on the Moon's orbital history, including both lunar and solar tides on the Earth but neglecting explicit oceanic tides as well as tides in the Moon. He considered an inclined but circular orbit, arguing that the Moon's small eccentricity of today could only have been less in the past (Goldreich 1963). He accounted for the Earth's forced precession and the lunar orbital precession due to the Sun plus the Earth's oblateness, and correctly averaged his equations over the appropriate precession periods. Goldreich (1966) plotted lunar orbital semimajor axis a versus lunar orbital inclination (relative to both the ecliptic and to the Earth's equator), as well as the Earth's obliquity versus a. On such plots, one is unsure how far back to carry the integration in a because that depends on the rate at which the path is traversed; the trajectories are specified but one never knows where the Moon enters the path. The reason that time is not taken as the variable is that the evolution rate along a particular curve depends linearly on the lag angle ε, which in 1966 was (and, to a considerably lesser extent, still is) poorly constrained by the available data. As we shall learn below, today's apparent lag angle is atypical and should not be used to determine the time scale of the evolution.

The plots from Goldreich (1966) have been reprinted often (see, e.g., Kaula 1971; Burns 1977b,1982; Lambeck 1980) so we will only describe their attributes; instead we will present Mignard's (1982b) results below. The lunar evolution histories of Goldreich (1966) show that, as the Moon evolves inward going toward the past, the Earth's obliquity Φ monotonically drops from today's value of 23°.5 to about 18° near 30 $R_\oplus$ (cf. Ward 1982). Over the same region, the lunar orbital inclination I relative to the ecliptic slowly climbs from $5°1$ to more than $6°$. The inclination i relative to Earth's equator oscillates between two limits because the lunar orbit precesses about the pole of the ecliptic plane and not about the Earth's rotation pole; this means that sometimes I should be added to Φ but other times subtracted: i varies between $\Phi + I$ and $\Phi - I$. At about 30 $R_\oplus$ the Laplace plane (see Sec. I.C) starts to diverge from the ecliptic and move toward the Earth's equator, which it reaches when the Moon is near 7 $R_\oplus$. Until the equatorial Laplace plane is attained, the previous trends continue: Φ and i decrease while the maximum I increases. The new feature in the transition region is that all three quantities vary between limits during the precession of the lunar orbit about the displaced Laplace plane. At very close distances, when Goldreich's neglect of eccentricity becomes suspect, i starts to increase sharply, reminiscent of Gerstenkorn's finding of large inclinations at closest approach. The overall nature of this lunar orbital history is remarkably similar, whether the phase lag is constant or linearly dependent on rate.

The large i found by Goldreich (1966) and Gerstenkorn (1955,1969) at all close distances is difficult to reconcile with any simple theory for lunar origin that is based on fission or binary accretion. This profoundly important conclusion is unaffected even when arbitrary changes are made in the Earth's obliquity; it also does not seem to be significantly influenced by an amplitude dependence or frequency dependence in the lag angle (Goldreich 1966). Pathological rheologies for the Earth material can lessen the final inclination (cf. Rubincam 1975) but it is far simpler to invoke a different origin scheme or to claim that a post-origin impact into the Moon is responsible for the nonzero starting inclination required by today's lunar orbit.

The most recent computations that discuss lunar orbital evolution without considering ocean tidal models explicitly are those by Mignard (1979, 1980,1981b) and Conway (1982). The latter author used a frequency-dependent model of tidal friction for solar-plus-lunar tides; inclined and eccentric orbits were considered and several terms in the Earth's gravity field were included. The variational equations were analytically averaged over the relative precession period, in place of Goldreich's numerical attack. Results similar to Goldreich's were generally realized.

In a series of papers Mignard (1979,1980,1981b,1982b) studied the evolution of a planet-satellite system under the influence of tides. Only the last of these papers incorporates solar perturbations, which we have learned are important just when $r \gtrsim 15$ $R_\oplus$; at close distances the simpler approach of Mignard (1980) is adequate. Both solar and lunar tides on the Earth as well as Earth tides on the Moon are considered to be frequency dependent. The ratio of lunar dissipation to terrestrial dissipation is given by A as defined in Eq. (10); possible values of A are listed in Secs. I.B and I.E. Mignard averages the Gauss perturbation equations over the mean anomaly and over the lunar orbital precession period. He numerically integrates a set of differential equations for five parameters that describe the state of the system; they are semi-

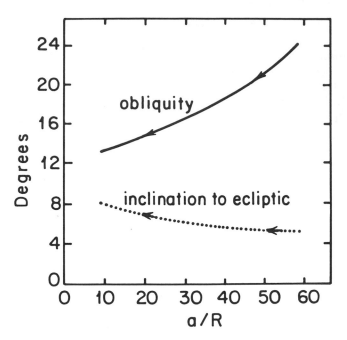

Fig. 2. The variation of the Earth's obliquity and the lunar orbital inclination I to the ecliptic in the case of a Moon with no dissipation (figure from Mignard 1982b).

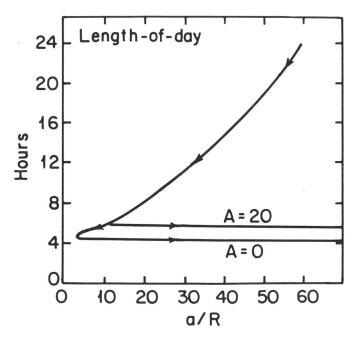

Fig. 3. The history of the length of the day in terms of lunar orbital semimajor axis. $A = 0$ corresponds to no lunar dissipation (see Eq. 10) while $A = 20$ implies high lunar dissipation (figure from Mignard 1982b).

major axis, eccentricity, orientation of the Moon's orbit plane, plus the Earth's obliquity and spin rate. The primary difference from Goldreich's treatment is the generalization of the problem to include lunar eccentricity. The variation of the Earth's obliquity and the Moon's orbital inclination with respect to the ecliptic are shown in Fig. 2. The Earth's rotation period is plotted in Fig. 3 versus lunar distance. As the Moon moves inward, angular momentum is transferred from its orbit back to the Earth's rotation. The primordial spin period of the Earth was closer to those typical of other solar system bodies than is its spin period today (Burns 1975).

The past history of the Moon's eccentricity is plotted in Fig. 4. For both a Moon with no dissipation ($A = 0$) and one with large dissipation ($A = 20$), the lunar eccentricity was less in the recent past than its current value while it was very large back at the time the Earth is neared. The major consequence of lunar dissipation is that the distance of closest approach moves from $\sim 3\,R_\oplus$ for negligible ($A = 0$) and intermediate ($A \lesssim 3$) lunar dissipation to $\sim 10\,R_\oplus$ for high dissipation. If one interprets the plotted history as indicating capture

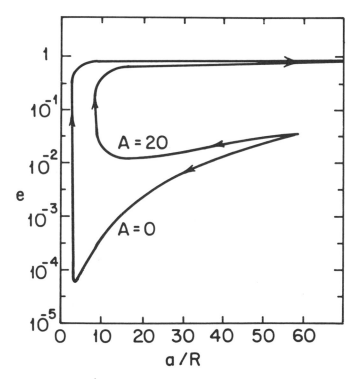

Fig. 4. The history of the lunar orbital eccentricity in terms of the semimajor axis. A is the ratio of lunar to terrestrial tidal effects given by Eq. (10) (figure from Mignard 1981b,1982b).

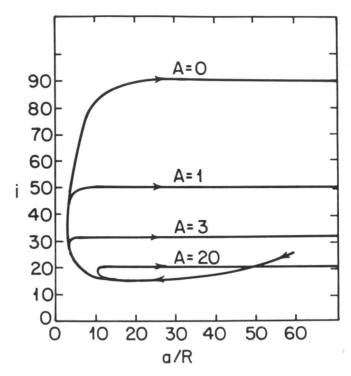

Fig. 5. The history of the lunar inclination i relative to Earth's equator in terms of the semimajor axis. A is the ratio of lunar to terrestrial tidal effects given by Eq. (10) (figure from Mignard 1981b,1982b).

of the Moon by Earth, this event is allowed to take place beyond the Roche limit in the high-dissipation case (see discussion in Boss and Peale 1986).

The mean lunar inclination traces a path relative to Earth's equator as shown in Fig. 5. This history is effectively independent of lunar dissipation until about 10 $R_\oplus$. Large changes in i, greatest for the least lunar dissipation, occur near closest approach. Mignard (1981b) claims that these results suggest an origin by capture for the Moon but, in contrast to Gerstenkorn (1955,1969), that prograde capture is likely since $A \neq 0$ (see Fig. 5). However, we should recall that any linear dissipation model is suspect at these close distances so that histories should not be believed.

As described next, however, a problem is that, following capture, the orbit evolves much too rapidly. Of course, this serious time scale discrepancy would no longer be a concern if the Earth's propensity for dissipation were to be time dependent, such as might occur with oceanic tides. This neglect of explicit, time variable ocean tides in most previous plots of lunar orbital history has vitiated any conclusions about lunar origin drawn from previous orbital histories.

E. The Lunar Time Scale: A Quandary and its Resolution

No single number defines the rate of lunar orbital evolution under tides. Each orbital element satisfies a separate perturbation equation (see Eq. 8), responds to different tidal components (rates are given in Sec. I.A and at the end of this section; Lambeck 1980, Table 10.1), and varies in its own manner as the orbit character (a,e,i) changes. Nevertheless, I will describe a's rate today as *the* lunar evolution rate since changes in a are most easily measured and since the Moon's distance at $t = 4.6$ Gyr ago might best constrain origin. An alternative argument, which allows the Moon's location at its origin to be specified by the current lunar shape, was originally proposed by Jeffreys more than seventy years ago. It was recently reconsidered by Lambeck and Pullan (1980) who argue that the Moon's current shape includes a fossil tidal bulge from 4.0 Gyr ago when the Moon was at ~ 25 $R_{\oplus}$.

The characteristic time for lunar orbital evolution can be easily calculated. I consider here only the evolution caused by the M_2 tide ($\ell fpq = 2200$), the lunar principal tide, which generates four-fifths of the secular change in a (Lambeck 1980). According to Eq. (8), this tide produces

$$\dot{a}_{2200} = 3m^*k_2 \left(\frac{G/a}{M + m^*}\right)^{1/2} \left(\frac{R}{a}\right)^5 \sin \varepsilon_{2200} \tag{13}$$

which is easily integrated back in time from today's starting conditions ($t = t_0$, $a = a_0$):

$$a^{13/2} - a_0^{13/2} = (39/2)m^*k_2 \left(\frac{G}{M + m^*}\right)^{1/2} R^5(t - t_0) \sin \varepsilon. \tag{14}$$

Although the subscript is now dropped, the reader should be aware that ε is twice the observed lag angle $[(n - \dot{\theta})\Delta t]$ (see Eq. 6). Note the strong dependence on a and that small satellites move proportionately more slowly than large ones.

Kepler's third law states that $n^2a^3 = G(M + m^*)$, so that

$$\dot{n}_{\ell fpq} = - (3/2)(n_{\ell fpq}/a_{\ell fpq})\dot{a}_{\ell fpq}. \tag{15}$$

Hence, Eq. (14) can be rearranged to

$$(a/a_0)^{13/2} = 1 - (13/3)(\dot{n}_0/n_0)(t - t_0). \tag{16}$$

One can measure $\dot{a}$ directly through lunar laser ranging or, more accurately but less directly, find it through the intermediary of $\dot{n}$, the lunar "secular acceleration," after first removing the "nontidal" acceleration (i.e., that caused by the Sun's gravitation and first explained by Laplace). Alternative ways include, (1) equating the loss of the Earth's angular momentum (as indicated by the deceleration of Earth's spin) to that gained by the lunar orbit, after first incorporating an estimate of the solar tidal contribution, and (2) detecting the actual tidal effects on artificial satellites which define various $\varepsilon_{\ell fpq}$ and k_ℓ and then substituting them into the perturbation equations.

Representative values from these various methods are listed below. Note that all of these basically provide d^2L/dt^2 with L the Moon's mean longitude; at one time just one-half this quantity was tabulated. Direct laser ranging of the Moon yields an acceleration of $-25.3(\pm 1.2)$ arcsec/century^2 (Yoder et al. 1984; Dickey et al. 1983; Ferrari et al. 1980; Mulholland 1980; see Cazenave 1982). The lunar laser ranging value is based upon the most accurate and homogeneous data set applicable to the problem (W. Kaula, personal communication, 1985). These data have been scrutinized thoroughly and the original results from the Jet Propulsion Laboratory team received an independent check from R. W. King and associates at Massachusetts Institute of Technology. The accuracy of $\dot{n}$ should improve as a function of time as $t^{-5/2}$ and so that the standard deviation in $\dot{n}$ should fall below $0.5''/cy^2$ by the end of 1988 (Mignard 1985). Several other schemes are available to specify $\dot{n}$. The classical value of $-22.4(\pm 7.0)''/cy^2$ (Spencer Jones 1939; the estimated error is from Lambeck 1980, p. 303) is based on telescopic observations from 1677 to 1936 of stellar occultations by the Moon, of solar longitudes and of transits by Mercury and Venus across the Sun's face. More recent observations of transits of Mercury yield $-26(\pm 2)''/cy^2$ (Morrison and Ward 1975). The number which derives from lunar occultations positioned in atomic time is -21.4 $(\pm 2.6)''/cy^2$ (van Flandern 1981). From the response of artificial satellites to tidal bulges produced by the Moon, Cazenave (1982) believes that $\dot{n} = -26.1(\pm 1.6)''/cy^2$. Ancient records of lunar and solar eclipses provide evidence for changes in the Earth's spin rate $\dot{\psi}$ and the lunar acceleration; a summary of this data (Lambeck 1980, pp. 303–318) concludes that $\dot{n} = -28(\pm 2)''/cy^2$ and $\dot{\psi} = 1100(\pm 100)''/cy^2$ averaged over the past several millenia (see also Mignard 1985). One might be led from this quick summary to infer that these numbers are narrowly specified. This is true as of 1985, but just ten years ago it was not, with measured numbers varying by more than a factor of three; it is still entirely possible that unmodelled systematic errors might pervade all solutions. Caution is demanded.

While the above values of $\dot{n}$ and $\dot{\psi}$ pertain only to the present time, growth lines on bivalves, corals and stromatolites may indicate corresponding numbers from 10^8 to 10^9 yr ago. Even though scrutinized for more than twenty years, these biological measures of tidal evolution are controversial and difficult to interpret clearly. Ideally, however, they should provide the number of solar days per year and per synodic month as well as the number of synodic months per year; from theoretical considerations the year should have a nearly fixed length. Admirable surveys of this literature are by Scrutton (1978) and Lambeck (1980, pp. 360–390). The latter estimates $\dot{n} = -20.7$

$(\pm 2.1)''/cy^2$ and $\dot\psi = -1040(\pm 40)''/cy^2$, comparable to the current tidal accelerations of the Moon and Earth. In addition, the data unequivocally demonstrate that the Earth's spin varied through the geologic record; it is uncertain, however, whether it changed smoothly or erratically. In addition, over a similar interval, sedimentary records suggest that tidal heights in the Cretaceous period (about 65 Myr ago) through the late Precambrian era (> 570 Myr ago) were comparable to those today.

The above summary indicates that we can be reasonably confident in the current value of the lunar secular acceleration. If we are, a check may be made: following the introductory arguments given in connection with Fig. 1, the additional lunar orbital angular momentum can only be present because the Earth's rotation has slowed down. An excellent test of the theory then should be "Is the observed $\dot n$ consistent with the loss of spin angular momentum as manifested in $\dot\psi$?" The once-troubling answer to this question was that an additional "nontidal" secular acceleration of about $300''/cy^2$ was needed (Mignard 1985). The explanation for this puzzle comes from the Lageos spacecraft, the acceleration of whose node indicates that the Earth's flattening is changing (Yoder et al. 1982, perhaps in response to the recent deglaciation (Peltier 1985). This decrease in J_2 causes a proportional lessening of the Earth's moment of inertia and an accompanying acceleration of the spin rate of $250(\pm 50)''/cy^2$.

Using Eqs. (14) or (16), it seems as though it should now be trivial to compute *where* the Moon started and thereby to settle *how* the Moon originated. Since the current evolution rate ($\dot n_0$ or $\dot a_0$) is known, it can be substituted in Eq. (16) and a computed for $t - t_0 = 4.6$ Gyr ago. Unhappily this strategy fails because Eq. (16) indicates that the Moon arrives at Earth ($a = 0$) when $t - t_0 = (3/13)(n_0/\dot n_0)$, or 1.5 Gyr ago for $\dot n_0 = -25''/cy^2$. That the Moon was captured only relatively recently was once an acceptable answer—indeed Darwin himself had suggested an evolution time of $\sim 10^8$ yr following capture. However, the dating of rocks returned by the Apollo and Luna missions requires that the Moon be relatively quiescent for at least the past 3, and probably 4, Gyr; even lunar lava flows, which are confined to a brief interval around 3.5 Gyr ago, seem to come only from shallow depths (see the chapter by Kaula et al.) and so were not produced during a capture event. Similarly, thermal histories of the Earth rule out any recent heat pulse of the magnitude that would accompany lunar capture (cf. Lambeck 1980, pp. 391–394).

Clearly, the implicit assumption which is made when integrating Eqs. (13) or (15) (namely, that only a and t are variable) is incorrect: the present evolution must be abnormally rapid. This implies that the Love number k_2 or, much more likely, the effective lag angle ε (i.e., the tidal energy dissipation) must be unusually large today.

Energy loss in various oscillatory systems, whether electric circuits or free vibrations of the Earth, is conveniently described by the specific dissipation function Q^{-1}, which is defined by

$$Q^{-1} = (1/2\pi)\Delta E/E \qquad (17)$$

where E is the peak elastic energy stored in the system over a cycle and ΔE is the amount of energy dissipated per unit cycle. For slow motions of weak, linearly dissipative systems (i.e., large Q), the lag angle between stress and strain is

$$\sin \varepsilon_{\ell fpq} = Q^{-1}. \qquad (18)$$

Equivalent definitions can be given for materials having particular constitutive laws such as a Maxwell body or a Kelvin-Voight solid (Lambeck 1980, pp. 15–21; Ross and Schubert 1986; van Arsdale 1985). Burns (1977b) describes how one might expect Q^{-1} to vary with frequency and amplitude, and how it depends on the material and its state (pressure and temperature).

At seismic frequencies between 3×10^{-4} Hz and 1 Hz, the Earth's Q lies within a factor or two of 400 (Lambeck 1980, pp. 18–20). Typical values for Q of a few hundred are also measured in the laboratory when seismic waves with frequencies up to 10^6 Hz pass through various types of rocks (Burns 1977b). The picture is less clear in the case of very low frequencies; for example, the Chandler wobble (at 3×10^{-6} Hz) seems to decay with $Q \approx 100$. However, this observation is difficult to make, especially since the cause of the wobble remains uncertain (Smith and Dahlen 1981).

The energy dissipation implied by the orbital evolution rate is so sizeable as to suggest that something in the problem's formulation is amiss. From Eqs. (13) and (15) we have $\sin \varepsilon_{2200} \approx 5°$, or that the effective $Q \approx 12$ from Eq. (18). This is very much lower than typical values for Q, and indicates from Eq. (17) that much tidal energy is not recovered. We will see shortly that the anomalous Q for Earth tides is due primarily to the oceans.

Values of energy loss in the Moon are also required in order to select the past lunar evolution from the suite of possible solutions (see Figs. 4 and 5). Lunar rotational motions, as probed by lunar laser ranging (Ferrari et al. 1980; Cappalo et al. 1981; Yoder et al. 1984), yield $k_2 \approx 0.02$ to 0.03 (Cazenave 1982), as expected by Eq. (3) with $\mu = 10^{12}$ dyne cm^{-2}, but $Q \approx$ 10 to 30. It is particularly difficult to explain such a low Q in the Moon since it has no oceans (discounting its maria); for solid dissipation the damping of lunar seismic waves indicates $Q \approx 10^3$ to 10^4. The low Q for the bulk Moon may instead result from a viscous coupling between the lunar core and mantle (Yoder 1981b). If the lunar $Q = 10$ and the lunar $k_2 = 0.02$, then $A \simeq 1$ from Eq. (11); Mignard (1982b) selects $A = 20$ as an upper bound by considering the case when only solid body tides ($Q \simeq 300$) operate on Earth.

Likely choices for the Q's of other solar system members are described as my discussion of their evolution requires them (see also Goldreich and Soter 1966; Burns 1977b).

The possibly crucial role played by oceanic dissipation in determining lunar orbital evolution has been long recognized. While, in his late nineteenth century orbital evolution calculations G. H. Darwin treated the solid Earth as a viscous fluid, in several passages he doubts whether it would behave as such, and suggests instead that the dissipation might be occurring in the oceans. G. I. Taylor in 1919 and H. Jeffreys a year later demonstrated that significant energy was in fact being lost in the shallow seas. Over the next half century, it was acknowledged that oceanic dissipation might be important and might change as continents moved about. Nevertheless, the lunar orbital histories which were developed in the Apollo era always expressed the past Q of the Earth as constant and often as due to solid-body effects. Perhaps this was to allow simple calculable histories and to avoid ephemeral evolutions that would depend upon ancient, unknown coastal outlines (see Lambeck 1980, p. 289).

Lambeck (1975,1977,1980) was the first to use available global ocean tidal models, which incorporate terms for ocean loading and gravitational self-attraction into the Laplace tidal equations, to demonstrate quantitatively the influence of the principal diurnal and semi-diurnal tides. It was his achievement to show that oceanic dissipation is controlled largely by global processes rather than by local ones, since the latter are unlikely to influence greatly the ocean tidal bulge unless shallow seas determine the phase of the global ocean tides. In this regard there may be some hope of correctly modeling past oceanic dissipation (see below). Cazenave (1982) tabulates the lunar accelerations expected from several global tidal models developed toward the end of the 1970s. To illustrate the controlling influence of oceanic tides on lunar orbital history, we consider the results of Lambeck (1980, p. 335) who evaluates secular changes in the lunar orbital elements (a,e,I) resulting from a best theoretical model of eight separate ocean tidal components. He finds that $\dot{n} = -30.5(\pm 0.31)''/\text{cy}^2$, of which $-27.4''/\text{cy}^2$ is caused by the M_2 tide; equivalently $\dot{a} = 1.44(\pm 0.15) \times 10^{-9}$ m s^{-1}. The total $\dot{e}$ is $+5.73(\pm 1.75) \times 10^{-19}$ s^{-1}, where the N_2 tide supplies $+5.82 \times 10^{-19}$ s^{-1}. For $\dot{I}$ Lambeck computes $-4.17(\pm 0.47) \times 10^{-19}$ s^{-1} with the M_2 tide accounting for -3.46×10^{-19} s^{-1} and the K_1 tide providing -1.38×10^{-19} s^{-1}. Obviously these tiny $\dot{e}$ and $\dot{I}$ are not measurable. The supposition that oceanic tides largely govern lunar evolution was certified when the effects of the M_2 ocean tide were detected in the orbital perturbations of the artificial satellites Geos 3 and Starlette (Cazenave 1982). The M_2 ocean tidal parameters derived from these observations would induce $\dot{n} = -21.9(\pm 1.6)''/\text{cy}^2$ for the Moon, or nearly all the observed evolution. Further refinements of tidal parameters are to be expected from this technique particularly with the forthcoming Topex mission. In theory, the evolution rate caused by the oceans can be subtracted from the total observed value to infer that portion of the evolution caused by solid-body tides. The remaining rate derived in this way indicates a Q for the Earth's mantle between 200 and 400, with great uncertainty; we note this higher value agrees with other measures of the Earth's solid-body dissipation.

F. Oceanic Dissipation in the Past

The realization that oceanic tides dictate most of today's lunar orbital evolution is a major advance in our understanding of how the solar system evolves. It automatically implies that the past evolution was different. In concordance with this new understanding is the fact that ancient oceans are computed to produce significantly less energy loss and transfer of angular momentum between Earth and Moon than do corresponding models of today's seas (Sündermann and Brosche 1978; Krohn and Sündermann 1982). Hence, to trace even qualitatively the past lunar track so as to address the overriding question of lunar origin, we must understand the specific cause of the large dissipation today. Is it merely the presence of a large fractional covering of liquid or might it depend on the contemporary configuration of ocean basins/continental shelves?

The energy loss in oceanic tides can be calculated through three separate approaches (see Brosche and Sündermann 1978,1982; Lambeck 1980, pp. 319–334). The first evaluates the rate at which work is done by the Sun and Moon on the deformed ocean surface. This requires either a globally accurate tidal model or world-wide observations, but it does not involve poorly known tidal currents nor any true understanding of the nature of the energy sink. The second method assumes that bottom friction is the dominant dissipation mechanism; since the loss rate varies as the cube of the tidal velocity, only velocities in shallow seas (where particularly strong currents exist) are important but they must be accurately specified. Jeffreys (1920) originally showed that most of the Earth's rotational energy loss could be produced in this way. The third method is based upon the second in that energy loss in the shallow seas can be derived from an energy balance across the entrances to these waters; the advantage in this treatment over the second is that the precise mechanism of energy loss need not be stipulated. Both Lambeck (1980) and Hansen (1982) tabulate about a dozen previous estimates of tidal energy loss. All of these methods find energy loss rates between 1 and 4×10^{19} erg s^{-1}, a remarkably narrow range, given the different assumptions and measurements; the upper limit is nearly the amount suggested by astronomical observations. In addition to tides, the bending of ice sheets, percolation through sediments, and precessional torques on the core may absorb some power (see Hansen 1982).

Questions as to the precise dissipation mechanism or its location need not concern us here. Whereas energy is dissipated in local processes that depend on higher harmonic terms, the orbital evolution of the Moon is primarily determined by second-degree harmonics. Hence we are principally interested in the amplitudes and the phase lags of these second-degree terms, rather than those of any other components.

Resonance is thought to play a fundamental role in the exceptional oceanic dissipation observed today. It is well known that an underdamped, linear, forced harmonic oscillator dissipates energy most effectively when driven near its natural frequency (see Fig. 4 in Hansen 1982). One oscillatory mode for an ocean is that of free gravity waves, whose period for a rectangular basin is approximately

$$T_g = \frac{2L}{N\sqrt{gD}} \qquad N = 1, 2, 3. \ . \ . \tag{19}$$

where L and D are the basin's length and depth (Sündermann 1978). Oceans 4 km deep (close to the actual mean ocean depth of 3770 m) with lengths of 3000, 5000 and 9000 km have natural sloshing periods of 8.4, 14.0 and 25.2 hr, respectively (see Table 6.3 in Lambeck 1980). Surely the natural modes of such basins would be well coupled to the diurnal (23.93 to 25.82 hr) and especially the semi-diurnal (11.97 to 12.66 hr) tides of today. Tidal gauge observations of actual sea levels have confirmed that tidal forcing frequencies fall within the measured spectrum of major oceanic normal modes; for example, normal modes of 14.8 and 9.3 hr have been inferred for the North Atlantic from tidal data taken at Bermuda and the Azores. Platzman et al. (1981) find normal mode periods of 13.0 and 14.1 hr in the North Atlantic and 15.5 hr for the Pacific Ocean using realistic bathymetry and an ocean tidal model (cf. references in Sündermann 1982).

The importance of today's near-resonance condition on lunar orbital histories has been demonstrated by Hansen (1982) and Webb (1980,1982a,b). The former author gives a heuristic demonstration of the process by considering the lunar evolution to be governed by oceanic energy losses which respond like a damped harmonic oscillator that is near resonance. Once the ratio of the system's natural frequency σ to the tidal forcing frequency ω is selected for this analog model, the current loss rate fixes a coefficient, which effectively determines a phase lag. (Two possible solution trajectories exist; Hansen [1982] argues from present tidal models that the weak damping case is the more appropriate choice.) The energy loss (and the associated tidal torque which drives the evolution) is known then for all positions since it depends only on this coefficient, the Earth-Moon separation and ω/σ, which varies in a given way as the Earth's rotation changes in response to the same tides. Different past oceanic configurations, particularly variable areas of shallow seas, could conceivably alter the phase lag, in which case the past evolution rate could be undetermined for yet another reason.

According to this model, the history of the Earth-Moon separation depends critically on the particular choice for the starting value of ω/σ (Fig. 6). If the system is presently being forced at $\omega/\sigma = 0.75$, then as the Earth's rotation increases in the past, the system moves closer to resonance and energy

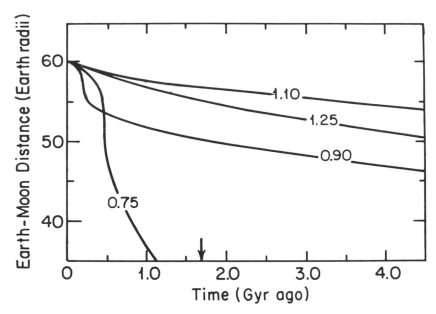

Fig. 6. The history of the Earth-Moon distance for a harmonic oscillator model with weak damping. Shown are results starting with several different initial values for the ratio of tidal forcing frequency ω to ocean natural frequency σ. The arrow indicates the time of closest approach for the case $\omega/\sigma = 0.75$ (figure from Hansen 1982).

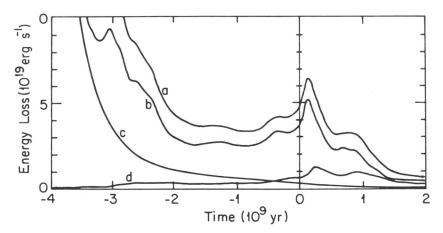

Fig. 7. The power dissipated in the oceans and the solid Earth plotted as a function of time from the present: (a) total power dissipated; (b) oceanic losses due to the M_2 tide; (c) solid-body losses in the M_2 tide for $Q = 150$; (d) oceanic losses due to the S_2 tide (figure from Webb 1982a).

losses are more severe; the orbit evolves swiftly, particularly in the vicinity of resonance, where the evolution curve's slope is steep. On the other hand, if the system is presently above resonance (e.g., $\omega/\sigma = 1.25$), the Earth's faster paleorotation means that in the past the system was even farther away from resonance. The evolution therefore is slow; even the radial dependence in the tidal equations is not enough to counteract the increasing distance to the resonance. Figure 7 makes the same point in a more sophisticated, but more subjective, way. Here oceanic tidal dissipation is calculated from bottom friction in an average hemispherical ocean whose tidal response is described by Laplace's tidal equations. Resonances with the M_2 tide produce most of today's loss. Tidal dissipation was less in the recent past as the effects of the resonance gradually disappear; in the more remote past the influence of the Moon's closer distance may eventually overwhelm the absence of resonance.

Lunar orbital histories, which include the changing response of oceanic tidal models, have been presented by Webb (1982a,b) and Hansen (1982). In each, the dynamical model is simplified by considering a circular lunar orbit and ignoring precession due to solar torques. Ocean-wide distributions of tidal amplitude are computed by numerical solutions to Laplace's tidal equations with dissipation by linear bottom friction. Hansen (1982) calculates the oceanic tidal torque from these tidal heights and uses it to integrate the orbital evolution equations backward; Webb (1982a,b) instead models the energy lost. Each treats several idealized distributions of land versus sea. Webb (1982a,b) looks at a hemispherical ocean and averages its response over all possible orientations; Hansen (1982) considers a smaller spherical cap located either along the equator or at the pole.

All of these models find that the energy dissipation, and hence the rate of lunar evolution, was much less in the past than at present. A typical orbital history is shown in Fig. 8. In these simulations the Sun is ignored: hence the system's total angular momentum L_T (which is the sum of Moon's orbital L_0 plus the Earth's rotational L_s) is fixed in direction and magnitude throughout the evolution. Ocean tides induce changes in the Earth's rotation period, I' (the lunar inclination relative to L_T), γ (the terrestrial obliquity relative to L_T) and i (defined as before). While specific histories depend on the particular continent model chosen and of course on the validity of the particular assumptions in the model (which are not discussed in detail by either author), the general conclusion that the past coupling between Earth and Moon was weaker than today's is almost surely true and is relatively insensitive to the continental configuration. As described earlier, this occurs because, with the faster paleorotation, the semi-diurnal tidal frequencies would be resonant only with lesser normal modes rather than the global modes; but it is the latter that determine the tidal torque at large a/R. Nevertheless, two points should be made: first, the calamitous consequences of a close approach between Earth and Moon have almost surely been averted, and second, just as importantly, past lunar orbital histories are poor guides to constrain lunar origins (Boss and

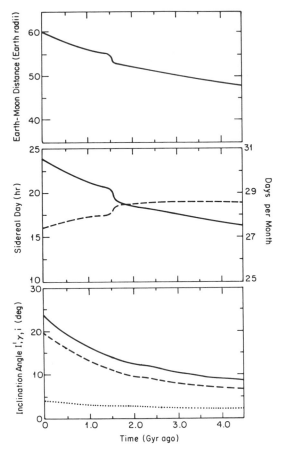

Fig. 8. Lunar orbital history if the Earth has all its land mass in a polar continent and the weak friction model prevails. The top panel shows the Earth-Moon separation. The middle panel displays the length of the sidereal day (solid) and the number of days per sidereal month (dashed). The bottom panel gives the relative lunar inclination I' (dotted), the relative terrestrial obliquity γ (dashed) and the relative inclination i (solid) (figure from Hansen 1982).

Peale 1986). Thus, even though there is renewed interest in the question of the Moon's origin (Boss 1986), partly as a result of a recent conference (Hartmann et al. 1986) on that topic which highlighted the idea of a catastrophic birth for the Moon in a giant impact, orbital evolution calculations are likely to assist little in refining ideas.

G. The Martian Satellites

Phobos and Deimos circle Mars on low-eccentricity orbits that lie nearly in Mars' equatorial plane. While Deimos ($a = 6.90\ \mathrm{R}_\delta$) is just outside the synchronous orbit ($a = 6.02\ \mathrm{R}_\delta$), Phobos at $a = 2.76\ \mathrm{R}_\delta$ is well within it. The orbital evolution of the Martian pair has been an active research topic since the late 1970s for two reasons. First, besides the Moon, Phobos is the only other satellite actually observed to be secularly accelerating (see Burns 1978); along with other close satellites, its orbital evolution should be detectable by astrometric telescopes planned for Space Station. Second, based on close flybys of the Viking Orbiters in 1977, the Martian moons are characterized as asteroidal instead of terrestrial; they have craggy, dark carbonaceous surfaces and low densities of ~ 2 g cm^{-3} (Veverka and Burns 1980). These properties suggest that they may have originated in the outer solar system rather than *in situ*. Dynamicists have been urged to find ways whereby satellite capture could be achieved (see Burns 1978; Pollack et al. 1979; Hunten 1979; Lambeck 1979; Cazenave et al. 1980). For general reviews of the orbital evolution of the Martian satellites see Burns (1972,1978), Veverka and Burns (1980), Pollack (1977), Szeto (1983) and Zharkov et al. (1984).

Many mechanisms, including some quite bizarre ones, have been proffered to explain Phobos' secular acceleration of $\sim 10^{-3}$ deg yr^{-2}; Burns (1972) lists many of these. It is generally agreed now that solid-body Martian tides are entirely adequate to generate the observed acceleration (Burns 1972, 1978; Pollack 1977); the inferred solid-body Q for Mars is between 50 and 150, and quite plausible. Since Phobos lies within synchronous orbit, its tidal evolution is *toward* the planet while tides drive Deimos, beyond synchronous orbit, yet farther away from the planet (see Figs. 1b and 1c). Accepting that solid-body tides dominate the present evolution, Phobos' orbit will collapse to the Martian surface in $\sim 5 \times 10^7$ yr from now, according to Eq. (16).

Planetary tides also modify the satellite's orbital eccentricity just as they do that of the Moon. Phobos' orbit is circularized by tides in both Mars and the satellite itself; for the latter, push-pull radial tides are less important than tides driven by the forced libratory (rocking) motion (Yoder 1982). If one takes Phobos' current eccentricity at face value and naively integrates Eq. (8) into the past (see below), its orbit could have been quite elliptical long ago (Goldreich 1963; Singer 1968); the effect of the Martian tidal torque alone would mean that Phobos' primordial e was ~ 0.1 to 0.2 about 4.6 Gyr ago. Tidal dissipation in the satellite would induce Phobos' original e to be substantially higher. Indeed the orbit becomes so elliptical that, in the past, its pericenter is closer to Mars even as the orbit size expands. This hints that tidal evolution might have proceeded yet faster earlier than it does now. The implication of these calculations then is that some time < 4.6 Gyr ago Phobos was on a large elongate orbit; namely, it was captured (Singer 1968; Lambeck 1979; Cazenave et al. 1980; Mignard 1981*d*; Szeto 1983; but see below). Typical evolution histories are shown in Fig. 9.

On the other hand, Deimos, because of its greater distance from Mars, is scarcely influenced by tides: its ancient orbit remains nearly circular and expands hardly at all away from the synchronous position (Fig. 10). Hence the dynamical evidence is firmly against Deimos' capture. However, as we have just discussed, it appears that Phobos could have been acquired by Mars. If it

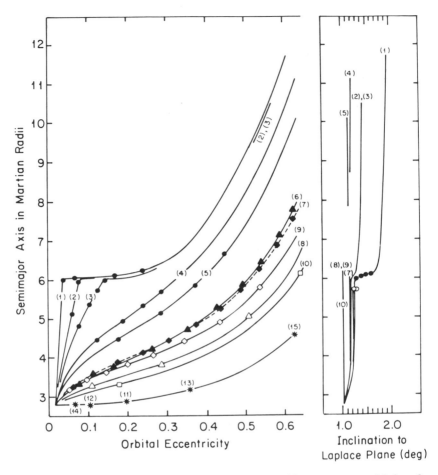

Fig. 9. (a) Phobos' orbital history in (a,e) space, considering fifteen separate possible laws for Q's dependence on frequency for the satellite-planet pair. Dots on curves (1) to (5) are time markers at 1 Gyr intervals. Symbols on curve (6) to (10) occur 50 Myr apart, while * for the Q models (11) to (15), corresponding to constant angular momentum, each denote Phobos' position 60 Myr ago. (b) Phobos' orbital evolution in (a,i) space, which shows little change relative to the Laplace plane (figure from Szeto 1983).

had been, then its elongate orbit would have crossed Deimos' for an extended period prior to evolving inward by tides. During this interval when the orbits of the Martian satellite pair were interlaced, the characteristic time for mutual collisions is much briefer ($\lesssim 10^3$ to 10^6 yr) than the evolution time itself, indicating that a collision was inevitable (Lambeck 1979; Cazenave et al. 1980; Szeto 1983). Therefore Deimos represents a barrier, effectively limiting the amount that Phobos could possibly have evolved. At the simplest level, there seems to be no way around this problem. Given that Deimos' orbit today is so regular, it is difficult to accept either that the satellites were captured sequen-

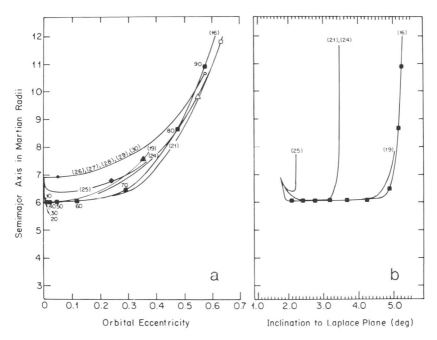

Fig. 10. (a) Deimos' orbital history in (a,e) space. Curves (16)–(30) use the same Q laws as (1)–(15), respectively, in Fig. 9. Numbers alongside symbols on curve (16) mark billions of years ago. Time markers are shown for curve 19 (filled triangle) at 1000 Gyr, curve 21 (open square) at 60 Gyr, curve 24 (open triangle) at 700 Gyr and curve 25 (filled diamond) at 500 Gyr. The two circles on the single curve (26–30) give different time markers for the different laws: (26), 180/320 Gyr; (27), 380/620 Gyr; (28), 360/570 Gyr; (29) 3300/4200 Gyr and (30) 680/990 Gyr. (b) Deimos' past inclination history for selected Q laws. Time markers are 10 Gyr apart. The evolution corresponding to those Q laws that are not shown lie between curves (16) and (25) except for runs (26) to (30) that show no evolution whatsoever (figure from Szeto 1983).

tially (with Phobos evolving first) or that they originated together as the largest fragments of a catastrophic collision (Yoder 1982).

The most probable resolution of this contradiction (that Phobos' current orbit suggests a capture which is not permitted according to Deimos' orbit) is that Phobos' present orbital eccentricity may result from relatively recent events rather than being the tidally damped remnant of a much larger past value. Yoder (1982) has shown that Phobos has passed through three gravitational resonances (see Burns 1978) at 2.9, 3.2 and 3.9 R$_\delta$. He demonstrates that the orbital eccentricity and inclination undergo jumps whose precise magnitudes depend on Mars' eccentricity and obliquity at the moment of passage through each resonance. If such jumps have taken place, the presently measured orbital eccentricity and inclination may be much higher than their primordial values. Accordingly, if current numbers are taken as starting points for any numerical integration, misleading results will be generated. If the satellite orbits are to have never crossed, Phobos' $\mu Q \gtrsim 10^{12}$ dyne cm^{-2}, indicating that the satellite has substantial internal strength (Yoder 1982); a similar bound arises from the supposition that Phobos' observed e results from resonant interactions alone.

A comparable evolution calculation shows that Deimos should have encountered a 2 : 1 resonance with Phobos. This resonant interaction should have similarly excited an eccentricity $e_D \approx 0.002$, or more than six times the observed number. Subsequent tidal dissipation seems inadequate to damp this eccentricity, unless Deimos' forced libration is exceptionally large.

The inclinations of the Martian satellite orbits also weigh against a capture origin. Both Cazenave et al. (1980) and Mignard (1981d) developed schemes to follow the orbits' evolutions relative to the Laplace plane even for large eccentricities (see also Szeto 1983). They found that the orbits stay close to the Laplace plane so that the current near-equatorial paths, when evolved back into history, switch gradually onto the Martian orbital plane. At first glance, this coincidence of a possible ancient plane for the Martian satellites with a plane near that in which most solar system bodies lie might seem to argue for capture. Actually the reverse is the case. As viewed by an observer on the planet, the orbits of asteroids passing close to Mars are not generally near Mars' orbital plane because only the relative motions matter; if asteroids were to be captured by Mars, they would not lie in the planet's orbital plane but rather would be randomly oriented. By energy considerations, capture could only occur at very low approach speeds since little energy can be dissipated at great distances from the planet. Hence Mars and the protosatellite-cum-asteroid must have had almost identical orbits, one more improbable circumstance required by capture.

Much like in the case of the Earth's Moon, we now have a substantially clearer understanding of the processes that must be included when discussing the orbital evolution of the Martian satellites. Unfortunately, however, the past orbits are less certainly known than we thought they were in the mid-1970s. To summarize, our understanding of the long-time dynamical history of the three satellites which orbit terrestrial planets has improved markedly. With this progress has come the slow and sad realization that unpredictable changes in each system—modifications in the Moon's evolution as the Earth's ocean basins wax and wane, and uncertain changes in Phobos' eccentricity due to resonance passages—remove each system from the realm of deterministic dynamics.

II. THE OUTER SOLAR SYSTEM

A. The Jovian and Saturnian Systems

The Voyagers' journeys through the region of the giant planets, and in particular their observations of planetary rings, have reawakened the curiosity of the dynamical community about the outer solar system. The problems of interest there are often quite different from those previously described about the Moon and Phobos. For example, since exploration is ongoing and data are yet to be digested, discoveries crying for explanation are being made almost monthly. Moreover, the fact that relatively large satellites are located so close to the major planets (see Table III in the introductory chapter) is *prima facie* evidence that tidal evolution in the outer solar system is relatively sluggish (see below; Goldreich and Soter 1966) unless satellites somehow recently originated (see Borderies et al. 1984). Lastly, linkages, such as resonances, between various components of the systems are common and of fundamental importance. Resonance phenomena, which are thought to be established through differential tidal expansion (Borderies and Goldreich 1984), are discussed extensively in Peale's chapter and will not be addressed here.

Tidal evolution in the outer solar system is driven by different physics than that for the terrestrial planets. This is easily seen by the much larger Q's found for the giant planets. Io's proximity to Jupiter after 4.6×10^9 yr fixes the planet's averaged Q to be $> 6 \times 10^4$ (Yoder and Peale 1981), including Io's transfer of angular momentum to Europa and Ganymede as stipulated by the Laplace resonance; lower bounds that are similar or somewhat smaller can be placed on the Q's of Saturn and Uranus from the nearness of Mimas and Ariel to their respective planets (Goldreich and Soter 1966; slightly different values are given in the chapter by Schubert et al., Sec. III). The absence of a noticeable secular acceleration in Triton's orbit means that $Q_N > 2 \times 10^3$ (Harris 1984b). If resonances are instituted by differential tidal expansion (see Peale's chapter), $Q \leq 10^6$ for Jupiter and 7×10^4 for Saturn (Goldreich and Soter 1966) to allow a reasonable probability for enough evolution to transpire that resonances should commonly be seen. These Q's are several orders of magnitude larger (i.e., energy loss is very little) than those of the Earth, Moon and Mars; nevertheless, attempts to identify potential sources of dissipation for the gaseous giants have generally been in vain (see references in Burns 1977b and in Hubbard and Stevenson 1984). Dissipation in fluid planetary exteriors is inadequate by many orders of magnitude. Inelastic dissipation in the rocky cores of giant planets could, however, provide Q's as small as $\sim 10^5$ to 10^6 depending on the particular properties of the cores (Dermott 1979a). Another potentially important contributor to energy dissipation is the irreversible entropy losses associated with the formation of helium raindrops as tidal perturbations pass overhead (Stevenson 1983).

The usual tidal process transfers angular momentum from the planet's rotation outward to a satellite's orbit through the intermediary of the displaced tidal bulge (Fig. 1). In the outer solar system, torques also develop between planetary rings and moons (Goldreich and Tremaine 1979); through this process, ring edges are restrained from spreading and narrow ringlets, such as Saturn's F ring and presumably all the Uranian rings, are shepherded. These torques arise by the gravitational interaction between the satellites and the rings' mass distribution, which is axisymmetric due to the density and bending waves that the satellites themselves induce. While a general picture of satellite torques is starting to emerge, a complete understanding of their operation is yet to be achieved (cf. Borderies et al. 1984; Shu et al. 1985).

The torque exerted by a satellite (mass m^*, relative distance Δa) on a narrow ring of mass m_r is estimated to be (Goldreich and Tremaine 1982)

$$T \sim \frac{G^2 m^{*2}}{n^2 (\Delta a)^4} m_r. \tag{20}$$

The time scales for orbital expansion under these torques can be remarkably short (Goldreich and Tremaine 1982; Borderies et al. 1984), especially for the Saturnian ring-moons (Atlas, the F ring shepherds and the coorbital satellites), because they lie so close to the rings. For example, the shepherds as individuals would recede from the outer edge of the A ring to their present location in only $\sim 10^7$ yr (Goldreich and Tremaine 1982). The same process should have moved ring material very substantially (Lissauer et al. 1984), causing most of the A ring to collapse into the B ring. Two approaches to try to resolve this time scale problem have proven unsuccessful: first, the inclusion of nonlinear effects for density waves lengthens time scales but does so by less than an order of magnitude (Shu et al. 1985; Longaretti and Borderies 1986); and second, resonances between the ring-moons and more distant satellites, which could allow the ring moon to serve merely as a transfer point for the ring's angular momentum on its way out, have not been found (Borderies et al. 1984). Either the rings and satellites are surprisingly young, or the process is incompletely described. Neither alternative is appealing.

Planetary tides generally increase orbital eccentricities when orbits expand as described early in this chapter (Sec. I.A); in competition with this, orbits are circularized by the satellite's tidal energy losses, which are driven by the eccentricity. The eccentricity of any orbitally evolving moon thus reaches equilibrium at a value where the satellite's internal energy loss equals that added by tidal evolution. It is generally acknowledged that the active volcanism observed on Io is caused by this tidal heating (Peale et al. 1979; see chapters by Schubert et al. and by Nash et al.). The measured heat flux restricts Jupiter's Q to be $< 4 \times 10^4$ (Yoder and Peale 1981; Greenberg 1981, 1982a). Greenberg's (1981,1982a) alternative view, elaborated on in Sec. VII of the chapter by Schubert et al., is that Io did not start out in resonance with Europa and Ganymede. He suggests that Io's melting may be recent and/or episodic; in this case $Q_J \gtrsim 10^7$. If the smooth plains of Enceladus also record past tidal heating episodes of that tiny satellite, $Q_S \lesssim 10^5$ (Yoder 1981a); as an alternate explanation for Enceladus' surprising form, Lissauer et al. (1984) have proposed that Janus and Enceladus were once locked in a 2:1 resonance in order to enhance Enceladus' e enough to produce melting but that the resonance has since been disrupted.

Other research topics in orbital evolution/dynamics for the Saturn system that have been stimulated by Voyager discoveries include:

1. The linked motion of the coorbital satellites Janus-Epimetheus (Dermott and Murray 1981a,b; Goldreich and Tremaine 1982; Yoder et al. 1983; Lissauer et al. 1985a; Peale's chapter);
2. The libration of the Trojan satellites first glimpsed in 1980 by Voyager and groundbased observers in the orbits of Dione and Tethys;
3. The chaotic trajectories of material in the vicinity of Hyperion (Farinella et al. 1983);
4. Hyperion's chaotic rotation (Wisdom et al. 1984; Peale's chapter);
5. The loss of debris from satellites or ring particles near the Roche limit (Dobrovolskis and Burns 1980; Davis et al. 1981; Weidenschilling et al. 1984; Papadakos and Williams 1983);
6. Circumplanetary dust dynamics (Burns et al. 1979; Mignard 1982a);
7. The putative coating of Iapetus by Phoebe dust (Mignard et al. 1986);
8. The orbital evolution caused by Titan's possible ocean (Sagan and Dermott 1982).

Most attention about the Saturnian system's dynamics has been deservedly devoted to the connection between satellite orbits and ring features (gaps, waves, edges, kinks and so forth) but, since that topic is extensively covered by another book in this series (Greenberg and Brahic 1984), I will not address it here.

B. The Uranian Satellites

The moons of Uranus do not have any orbit-orbit resonances like those that permeate the satellite systems arrayed about Jupiter and Saturn. But a Uranian trio lies near a Laplace-type resonance, such as that which ties the inner three Galilean satellites together (see Peale's chapter); the mean motions of Miranda (UV), Ariel (UI) and Umbriel (UII), the three innermost satellites, satisfy the following relation

$$n_V - 3n_I + 2n_{II} = -0.078°/\text{day}. \tag{21}$$

This near-resonance is not, however, stable; nevertheless it can be, and has been, used to determine the products of the masses (Greenberg 1976,1984a).

As Table III in the introductory chapter indicates, the Uranian satellite system is perhaps the most regular of the regular satellites; i.e., orbital eccentricities and inclinations (except Miranda's 4°) are quite small, and the satellite spacing is fairly uniform. However, from Veillet's (1983a,b) observational synopsis, eccentricities are clearly nonzero. This is surprising for a system that one might suspect would be tidally evolved. We can compute a typical circularization time by returning to Eq. (8) and adding the effects of M_2 tides both in the satellite and in the planet. For a homogeneous satellite, we find

$$\frac{de}{dt} = \frac{3}{2}\left[\frac{19}{4}\frac{k_{2U}}{Q_U}\frac{\rho}{\rho_U}\left(\frac{r}{a}\right) - 7\frac{k_2}{Q}\frac{\rho_U}{\rho}\left(\frac{R}{a}\right)\right]\left(\frac{R}{a}\right)^2\left(\frac{r}{a}\right)^2 ne \tag{22}$$

where r, ρ, k_2 and Q pertain to the satellite rather than to Uranus (subscript U). The first term of the bracketed expression, that due to the tides in Uranus, is negligible compared to the satellite tidal portion. With reasonable choices for the various parameters in Eq. (22), Squyres et al. (1985; cf. Dermott 1984b) compute eccentricity decay times on the order of 10^9 to 10^{10} yr for the outer two satellites; this implies that the present eccentricities of these Uranian satellites might be primordial. In contrast, the calculated decay times for the inner three satellites are quite short, of order 10^8 yr. Yet the eccentricities of all Uranian satellites—those suspected to be damped and those undamped— are comparable. Squyres et al. (1985) have listed possible ways out of this dilemma. The true explanation is more likely that of Dermott and Nicholson (1986) who have pointed out that mutual secular perturbations, similar to those governing planetary orbital histories, operate between the Uranian moons, particularly the outer pair; much earlier Greenberg (1975b,1984a) had realized that secular perturbations may influence this system's motion. This means that the orbital eccentricities of the various satellites, as well as their inclinations, are coupled together, and all (with their respective precession rates) vary considerably over time. Typical periods are 10 to 10^3 yr, so the tabulated orbital elements (which are determined by fitting observations over decades) are probably not representative of average values. This recognition of the importance of mutual interaction calls into serious question Veillet's (1983a,b) analysis of past observations which was used to choose orbital elements and which, from the nodal precession rates, had given the somewhat odd set of satellite densities not confirmed by Voyager 2. Dermott and Nicholson (1986) do note that the eccentricities of the inner satellite pair are probably forced; their continual damping will produce an appreciable heat source, particularly for Miranda and Ariel (Dermott 1984b; Squyres et al. 1985). Voyager 2 images (Smith et al. 1986) show that these two satellites, especially the former, have had fluids flow across their surfaces and have undergone massive internal evolution.

C. The Satellites of Neptune, Pluto-Charon, and Hypothetical Satellites

By the standard tenets of satellite origin, both massive retrograde Triton and Nereid with its highly elliptical orbit must be captured satellites (chapter by Stevenson et al.). An alternative point of view is that of Lyttleton (1936) who, noting that Pluto's elliptical heliocentric orbit passes well within Neptune's, hypothesized that Pluto was an escaped satellite of Neptune. Lyttleton's original proposal suggested that Pluto's ejection from the system followed a close interaction with Triton, whose orbit was made retrograde during the event. Based upon numerical experiments, Harrington and van Flandern (1979; cf. Dormand and Woolfson 1980) advocated instead that passage of a ponderous (2 to 5 Earth masses) planetesimal through the Neptune satellite system placed Triton on its odd orbit and ejected Pluto. Farinella et al. (1980) take the middle ground and propose that Triton, subsequent to its capture, gradually modified Pluto's orbit to the point of escape. These scenarios are untenable because (1) Pluto's mass is far too little to reverse Triton's orbital direction (Farinella et al. 1980; McKinnan 1984); and (2) the Pluto-Charon pair would be disrupted in any energetic interaction (Lin 1981; Peale's chapter).Much more plausible is the possibility that Pluto (and Charon?), Triton and Nereid are all remnant planetesimals; the capture of the last two by Neptune, perhaps through gas drag (Pollack et al. 1979), accounts for that planet's entire retinue, while Pluto-Charon were not swept up by Neptune because their 3:2 resonance with Neptune's orbit prevents close encounters with the planet.

The orbital evolution of retrograde Triton was first studied by McCord (1966; cf. Banfi 1984), who recognized that tidal forces contract the orbit (see Fig. 1c). As already noted, this collapse has not been detected ($\dot{n} < 2°$/century[2]) so that lower bounds are $Q_N > 2 \times 10^3$ and the final crash time $> 3 \times 10^7$ yr. These lower bounds can be raised an order of magnitude if a prograde satellite is associated with the material that produces occasional stellar occultations at about 3 R_N (see Hubbard 1986). McCord (1968; cf. Burns 1973; Ward and Reid 1973) generalized these results to identify those conditions (limiting satellite masses and initial orbital radii) under which hypothetical satellites will be lost over the age of the solar system, showing that such losses would be commonplace.

Triton's negligible orbital eccentricity ($e \lesssim 0.005$) can be readily explained as a consequence of tidal decay. From Eq. (22), with planetary tides ignored once again, the decay time is

$$t_0 = \frac{2}{21}\left(\frac{m_T}{m_N}\right)\left(\frac{a}{r_T}\right)^5\left(\frac{Q}{k_{2_T}}\right)n^{-1}. \tag{23}$$

For $Q_T \approx 10^2$ and $k_{2_T} = 10^{-1}$ to 10^{-2}, $t_0 = 10^7$ yr, a very short time, this implies that Triton's orbit is now circular, and has been for aeons (Harris 1984b). The orbital circularization took place so long ago that tectonic evidence of it on Triton's surface has surely vanished. Nevertheless, the enormous heat pulse associated with this process might have released clathrated gases to the satellite's surface (McKinnon 1984) and might be ultimately responsible for Triton's unusual surface composition (chapter by Cruikshank and Brown).

The Pluto-Charon pair is unusual, and possibly even unique, amongst the solar system's dynamical entities. By dint of its exceptional mass ratio, this system has evolved to the state of dual synchronous rotation in which it is found today (the properties of this system should be refined following the tracking of its eclipses/occultations that started in 1985 [Binzel et al. 1985]). The only equivalent solar system members may be binary asteroids, which have never been unequivocally detected but which have been hypothesized to exist as a natural outgrowth of collisional events in the asteroid belt (Farinella et al. 1982); searches for them have been instituted by Cellino et al. (1985)

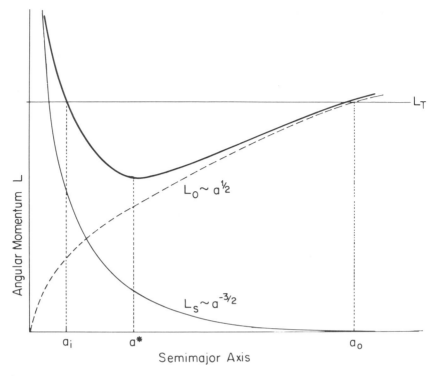

Fig. 11. The angular momentum for a synchronously rotating satellite plotted against semimajor axis. L_s is the angular momentum in the primary's spin while the satellite's orbit contains L_0. a_i is the inner, and a_0 the outer, synchronous spin state corresponding to a specified total angular momentum L_T. The minimum total angular momentum is achieved at a^*, where $L_0 = 3 L_s$. See Fig. 9 in Peale's chapter.

among others. Of course, close binary stars are usually tidally evolved and their evolutionary histories (Shu and Lubow 1981) are a useful analog for Pluto-Charon, if we overlook mass exchange.

The evolutionary history of the Pluto-Charon duo is described in Peale's chapter whose results will be summarized here, and then extended to address an associated problem for any satellites that might have once been about Mercury and Venus. Peale shows that in general there are two solutions for the semimajor axis at which the total angular momentum (contained in the planet's spin plus the satellite's orbit) achieves a given value for the dual synchronous rotation state. He also demonstrates that the inner state is unstable. Accordingly, Pluto-Charon presumably resides in the stable outer state now, and always will since the system is effectively isolated from external torques. For a self-consistent set of the parameters in Peale's expression, the outer synchronous state is at $a/R = 13.0$.

The presence of a third body naturally alters the dynamics of the synchronously locked pair that we have been considering up until now (see references in Burns 1977b). This situation pertains to any one-time satellites of Mercury or Venus, satellites of other satellites, or binary asteroids, rather than to the case of Pluto-Charon which is too far from the Sun to be affected by it. The consequence of the third body is to withdraw angular momentum from the pair, specifically by acting on the primary's spin. Figure 11 illustrates that any reduction in the system's total angular momentum moves the inner synchronous state outward and the outer state inward.

Dermott (1978), among others, has suggested that escape would occur if the inner synchronous state (at a_i) were ever driven beyond the planet's sphere of influence. However, the question is moot because, as we have mentioned above, the inner synchronous state is unstable and so the satellite will not follow a_i on its outward path. On the other hand, satellites can escape if, when pushed outward by tides, they reach the sphere of influence before the system achieves the synchronous rotation that we have been describing, whether at a_i or at a_0, the outer synchronous state's semimajor axis. By way of illustration, one could imagine slight parameter changes that would allow the Earth's Moon, which is currently evolving outwards toward the outer dual synchronous state, to escape. Kumar (1977; cf. Donnison 1978) invoked this process to account for the absence of satellites about Mercury and Venus. Singer (1970) claimed that the loss of a satellite and its subsequent recollision into Venus could explain that planet's retrograde rotation as well as its hot atmosphere. The evolution of a tidally altered three-body system is discussed in the context of Mercury and Venus by Burns (1973) and Ward and Reid (1973), who each severely limited the size and location of possible past moons. Reid (1973), Runcorn (1983) and Conway (1986) consider satellites of satellites.

For a satellite trapped at a_0, the withdrawal of angular momentum from the system causes the satellite to accelerate as it spirals toward the primary. This may seem contradictory. The external torque tries to slow the primary's spin but, because of the synchronous lock, in fact withdraws the angular momentum from the satellite's orbit (in Fig. 11 notice how L_0 and L_s change as L_T is reduced near a_0) while the planet spins *faster*. Prior to its destruction on the planet's surface, the satellite may break the synchronous lock (if it reaches a^*, where our argument on the stability of a_0 fails) or, if flimsy enough, it may fragment near the Roche limit (cf. Dobrovolskis 1982; Boss and Mizuno 1985) to form a ring system (Harris 1984a). A sufficiently strong satellite would plunge intact to the surface where its impact should generate an elongated surface feature (Burns 1976a). Such equatorial surficial scars have not been discerned on either Mercury or Venus; Schultz and Lutz-Garihan (1981) have also unsuccessfully sought evidence for such grazing impacts on Mars.

III. GRAVITATIONAL CAPTURE/ESCAPE

Circumstances under which satellites might be tidally eliminated from orbit by crashing onto their planet's surfaces or by escape have just been described. An associated topic is whether some of the most distant satellites, say the outer cluster of Jupiter's irregular satellites, might someday leak away to their own independent heliocentric orbits under the action of strictly point-mass gravity. Since Newton's equations of motion are time-reversible, an equivalent (and an interesting) question concerns whether these same satellites might be objects captured in the distant past from their own paths about the Sun. Morrison and Burns (1976), Burns (1977b), Pollack et al. (1979) and Pollack and Fanale (1982) provide historical references on the problem.

Several criteria have been put forward to quantify how far out a satellite can go before escape (i.e., transfer from an orbit controlled by one primary to an orbit about another) becomes probable (Burns 1982). The most elementary yardstick assumes that transfer occurs when the gravitational forces from the two primaries (planet and Sun) balance on the satellite. This is incorrect because it ignores the fact that the satellite almost shares the planet's orbit, and therefore accelerates by almost the same amount. According to a second criterion, orbital transfer will occur when the solar perturbation felt by the particle (as viewed in a planetocentric frame of reference) is the same fraction of the direct planetary attraction as is the ratio of the planetary perturbation (viewed in a heliocentric reference frame) to the direct solar attraction. The radius of this *sphere of influence* (Plummer 1918, p. 235) is

$$\rho_I \lesssim \mu^{2/5} \, r \tag{24}$$

where μ is the mass ratio $M/M_\odot$ and r is the planet's orbital semimajor axis; this distance is used in preliminary spacecraft mission plans as the point where elliptic orbits about the Sun begin to be treated as hyperbolic trajectories past the planet. The last alternative returns to the classical circular restricted three-body problem (Szebehely 1967) and says that, only if a Hill (or zero-velocity) curve is open, can transfer occur; since initial conditions specify the zero-velocity curve forever (as long as the problem's idealization is valid and there is no energy change in the meantime), any captured object can always escape and vice versa. In this framework, bound objects lie closer than the distance to the inner Lagrange equilibrium point, or

$$\rho_H \lesssim (\mu/3)^{1/3} \, r. \tag{25}$$

This distance is intimately connected to the Roche limit, when the planet is taken to fill uniformly a sphere of this radius. This viewpoint ignores the fact that satellites are in orbit and so have a velocity that enters into the computation of the Jacobi constant which fixes the particular Hill curve. More complicated expressions for the maximum orbital radius of stable satellites are provided by Keenan (1981); they lie close to Eq. (25). Stability criteria are also available for more general problems, including unrestricted mass ratios (Pendleton and Black 1983) or hierarchical three-body systems (Roy et al. 1984).

The latter paper calls upon integrals of energy and angular momentum to construct a stability criterion (Saari 1984).

Numerical experiments, which occasionally include the perturbation of other objects and more often account for the ellipticity of the primaries' orbits, demonstrate that the Hill sphere is a useful guide for distinguishing permanently captured objects from temporarily retained ones. However, orbits only out to $\sim (1/3$ or $1/2)\rho_H$ are found to be stable in the long term (see, e.g., Keenan 1981; Huang and Innanen 1983; or papers referenced by Burns 1977b). Also, computationally, retrograde orbits are observed to be more firmly grasped than are prograde ones. Huang and Innanen (1983) profer an explanation for this observation. It is comforting to see that the solar system also recognizes this fact; the outermost known Jovian satellites are retrograde, as is Phoebe, Saturn's most distant moon.

Numerical simulations of "satellite" capture by Jupiter are the subject of a series of papers by Carusi and Valsecchi (e.g., 1979). Such captures, since energy is not dissipated, are always temporary and they are also extremely rare, requiring precise timing of capture conditions.

Despite the difficulty of snaring satellites, capture is the favored origin for Phoebe, Nereid, the irregular Jovian satellites and perhaps even massive Triton (Kuiper 1951; Pollack et al. 1979). The comparatively small size of most of these satellites, their elongate and inclined orbits at great distances from their planets (Morrison et al. 1977), and their possible association with D-type asteroids (very dark, unusually red asteroids that are commonly found beyond the outer edge of the main asteroid belt; chapters by Thomas et al. and by Clark et al.) all speak for an origin distinct from the regular satellites. Byl and Ovenden (1975) consider capture in terms of the restricted three-body problem. They calculate that the Jacobi constants (i.e., the energy measured in a coordinate system rotating with Jupiter's orbit) of the Jovian irregular satellite pairs need only be altered by less than a percent each for escape to be allowed. Capture has also been suggested as a way to emplace the Martian satellites (Hunten 1979; Pollack et al. 1979; Veverka and Burns 1980).

In order to dissipate the energy necessary to effect permanent capture, two processes are serious contenders; Burns (1977b) has dismissed several others. Colombo and Franklin (1971) suggest that two objects (asteroids, satellites, or one of each) collided within Jupiter's sphere of influence, and the detritus from their impact are the extant satellite clusters. Stevenson et al. in their chapter endorse this viewpoint. A variant of this process, collisions within the Hill sphere which leave at least one object captured, has been called upon to transfer small particles from heliocentric orbits into a circumplanetary disk, as was first proposed by Ruskol in a scenario for the Moon's origin; this mechanism is considered quantitatively in the chapters by Stevenson et al. and especially by Safronov et al. Gas drag capture, in which solar-orbiting planetoids are slowed sufficiently during their passage through a primordial circumplanetary nebula to transfer to a planetocentric orbit, is preferred by

Pollack et al. (1979) and Pollack and Fanale (1982). Numerical integrations demonstrate that, for capture to happen while traversing the nebula, the planetoid's speed must be slowed by at least tens of percent. This in turn means that the planetoid, during its flight across the nebula, must interact with gas that has mass equal to tens of percent, defined as the fraction β, of its own mass. Since β is inversely proportional to the planetoid's radius, small objects are preferentially acquired. Nevertheless, for expected nebula models (Pollack et al. 1977), objects up to 10 to 10^2 km can be slowed enough to be captured near the edges of the nebula; deeper penetrations of the nebula are necessary to ensnare a satellite like Triton. The abrupt deceleration that transpires in the capture event will likely cause fragmentation. The relative speed between members of either Jovian cluster are of the same order as escape velocities off the largest member; they hint that each cluster was born when a primordial object barely split apart.

Once capture by gas drag is accomplished, orbital evolution proceeds swiftly because the acquired planetoid continues to sweep through the circumplanetary envelope. The characteristic time for evolution is $\sim P/\beta$ (where P is the satellite's orbital period) or ~ 10 yr. This rapid evolution has an important implication: post-capture orbital evolution can quickly turn irregular orbits into fairly circular, uninclined ones that ultimately crash into the planet. Indeed some process must halt the evolution (i.e., remove the nebula) if any satellites are to be found in orbit. Fortunately models of nebular evolution do undergo such a collapse phase. If this scenario is correct, the current outer satellites are only the last of many captured objects, their predecessors having fallen into the planet proper.

IV. CONCLUSION

Even though the solar system itself has not changed much in the decade since the last review of satellite evolution (Burns 1977b), our understanding of it has. I foresee a similar evolution over the next decade, with resonances and ring-satellite interactions better comprehended. The advent of supercomputers should help sort out the complex interactions of many-body systems. Just as surely, new and equally interesting puzzles will arise to be solved.

Acknowledgments. I sincerely thank F. Mignard, A. R. Dobrovolskis, R. Greenberg, W. M. Kaula, K. Lambeck and S. J. Peale for reading, correcting and improving an earlier version of the manuscript. I appreciate the technical help provided by B. Boettcher, T. Martin, G. Podleski and C. Snyder, and the forbearance of the co-editor.

5. ORBITAL RESONANCES, UNUSUAL CONFIGURATIONS AND EXOTIC ROTATION STATES AMONG PLANETARY SATELLITES

S. J. PEALE
University of California at Santa Barbara

Several examples of satellite dynamics are presented where significant progress has been made in understanding a complex problem, where a long-standing problem has finally been solved, where newly discovered configurations have motivated novel descriptions or where an entirely new phenomenon has been revealed. The origin of orbital resonances is shown in the demonstration of the evolution of a pair of planetary satellites through a commensurability of the mean motions by a sequence of diagrams of constant energy curves in a two-dimensional phase space, where the closed curve corresponding to the motion in each successive diagram is identified by its adiabatically conserved area. All of the major features of orbital resonance capture and evolution can be thus understood with a few simple ideas. Qualifications on the application of the theory to real resonances in the solar system are presented. The two-body resonances form a basis for the solution of the problem of origin and evolution of the three-body Laplace resonance among the Galilean satellites of Jupiter. Dissipation in Io is crucial to the damping of the amplitude of the Laplace libration to its observed small value. The balance of the effects of tidal dissipation in Io to that in Jupiter leads to rather tight bounds on the rate of dissipation of tidal energy in Jupiter. Motion in the relative horseshoe orbits of Saturn's coorbital satellites is described very well by a simple expansion about circular reference orbits. The coorbitals are currently very stable, and their relative motions can be used for the determination of the masses of both satellites. Pluto and its relatively large satellite Charon form an unusual system where the relative size and proximity of Charon lead to a most probable state where both Pluto and Charon are rotating synchronously with their orbital motion. The normal tidal evolution of a satellite spin toward synchronous rotation is frustrated in the case of Saturn's satellite Hyperion where gravitational torques on the large permanent asymmetry cause it to tumble chaotically. Observations of Hyperion's lightcurve are consistent with the chaotic rotation but do not verify it with certainy.

I. INTRODUCTION

The planetary satellite systems have provided a long list of puzzles involving origins and evolutions of various configurations that have slowly yielded to solution over the years. Much more detailed information about these systems from recent spacecraft observations has motivated a flurry of dynamical analysis—some of which does not even depend much on the new observations. We describe several examples of this recent activity, where the emphasis is on resonances and the consequences of the dissipation of tidal energy. By resonance we mean a situation in which the mean angular velocities of two satellites have a ratio near that of two small integers.

The dynamical evolutions of various satellite configurations in the solar system include the effects of energy dissipation on both the orbits and the spins. We discuss first (Sec. II) the origin and evolution of the two-body orbital resonances among the satellites, where significant progress has been made in simplifying the rather complex and diverse mathematical descriptions. It is now possible to understand the origin and evolution of orbital resonances in terms of a few simple principles. We begin here with a heuristic discussion of a two-body orbital resonance as an introduction to the description of the mathematical development. This development starts with first principles, and will require the reader to be familiar with graduate level classical mechanics but only the more commonly known jargon of celestial mechanics. The reward for the persevering reader will be a knowledge of how the relatively complex system is reduced to a single degree of freedom to yield the simplified Hamiltonian used in the analysis and of the approximations used in getting it. We show how the evolution of a two-body orbital resonance due to differential tidal expansion of the respective orbits can be followed through a sequence of diagrams showing curves of constant Hamiltonian in the two-dimensional phase space. The trajectory of the system in this phase space is essentially one of the curves of constant Hamiltonian in each diagram. Since the Hamiltonian is not conserved as the system evolves, the trajectory corresponding to the motion in successive diagrams is identified by the adiabatically conserved action. The conditions for capture into an orbital resonance as tides push the system toward a commensurability of the mean motions and the subsequent evolution within the resonance are easily seen in the sequence of diagrams. The model also allows analytic determinations of the probability of capture into the resonance when the system reaches a commensurability of mean motions.

This analysis has greatly eased the understanding of the establishment and current configurations of orbital resonances among the satellites. We indicate how it can give reasonably accurate quantitative descriptions of real resonances in some cases, but many existing resonances require a more accurate analysis—often with much more elaborate models including dissipation within the satellites and, in the case of Saturn's satellites, strong interactions with the rings.

In Sec. III we show how the analysis of the two-body orbital resonances described in Sec. II suffices for the description of the two-body resonances among the Galilean satellites of Jupiter. This description works in spite of the existence of the three-body Laplace resonance and the simultaneous libration of more than one resonance variable at the same commensurability of the mean motions. The two-body resonances form the basis for understanding the evolution to, and capture within, the more complex Laplace orbital resonance involving the inner three Galilean satellites. The subsequent analysis describes the evolution of the Galilean satellite system to its current configuration of a very small libration amplitude for the three-body Laplace resonance, where a high rate of dissipation of tidal energy in the satellite Io is crucial to the damping of the libration. This theory for the origin of the Laplace resonance, especially the understanding of the almost zero amplitude of libration of the Laplace angle, must rank as one of the major accomplishments in the study of dynamical evolution in the last decade. A perhaps unexpected bonus from this work is the establishment of rather tight constraints on the rate of dissipation of tidal energy in Jupiter. As in Sec. II, sufficient mathematical detail is presented here to enable one to follow the development without mystery concerning the source of the conclusions and the approximations involved. In Sec. IV we describe the coorbital satellites of Saturn, where the evolutionary aspects of this system apply more to the future than to the past. The interest in this system lies in its uniqueness and in the very simple modification of the analysis of the restricted three-body problem which suffices for its description. The relative horseshoe orbits for the satellites are shown to be currently very stable. In principle, the masses and densities of both coorbitals can be found by inserting observational parameters into the analysis, but accurate values will require more precise observations from a spacecraft orbiting Saturn. From an emphasis on orbital configuration and evolution, we turn (Sec. V) to a description of the Pluto-Charon system, where both planet and satellite appear to rotate synchronously with the orbital motion. This isolated system has thus reached the ultimate endpoint of tidal evolution, and its recent discovery, its uniqueness and the interesting dynamics of the dual-synchronous rotation state warrants its inclusion here. Two possible states of dual-synchronous rotation for a given total angular momentum are shown to exist—an unstable state at close separation and a stable state at a more distant separation. The Pluto-Charon system occupies the latter. The tidal evolution of the spins of the satellites discussed to this point have been a relentless retardation by tides toward the currently observed synchronous rotation. Except for the Pluto-Charon example, this evolution has not been elaborated, since it has been long understood. Hyperion, on the other hand, is distinguished from other satellites of its size by being very nonspherical. In addition, it occupies an orbital resonance with Titan which maintains its very large orbital eccentricity of about 0.1. We show in Sec. VI that this combination of circumstances forces Hyperion to evolve not toward an orderly synchronous or other commensurate spin-orbit state, but to a region in phase space where it will most likely tumble chaotically, essentially forever.

II. ORBITAL RESONANCES

If two satellites orbiting the same primary have mean orbital angular velocities (mean motions) which have a ratio very near that of two small integers, the satellite motions are said to be commensurate and to define an orbital resonance. Orbital resonances among the satellites of the major planets are of interest because many more such resonances exist than can be accounted for by a random distribution of orbits (Roy and Ovenden 1955). The inner three Galilean satellites of Jupiter, Io, Europa and Ganymede, are locked in the famous Laplace relation where $n_1 - 3n_2 + 2n_3 = 0$ with $\lambda_1 + 3\lambda_2 + 2\lambda_3$ librating about 180° with very small amplitude. The subscripts on the mean motions and mean longitudes (n_i, λ_i) number the satellites sequentially from the closest to the farthest from Jupiter. In addition, $\lambda_1 - 2\lambda_2 + \bar{\omega}_1$ and $\lambda_2 - 2\lambda_3 + \bar{\omega}_2$ librate about 0° and $\lambda_1 - 2\lambda_2 + \bar{\omega}_2$ librates about π, where the $\bar{\omega}_i$ are the longitudes of periapses. At Saturn, Enceladus-Dione, Mimas-Tethys and Titan-Hyperion are locked in orbital resonances with $n_i/n_j = 1/2$, $1/2$ and $3/4$, respectively. For Mimas-Tethys, $2\lambda_1 - 4\lambda_3 + \Omega_1 + \Omega_3$ librates about 0° (Ω_i are the longitudes of ascending nodes), for Enceladus-Dione, $\lambda_2 - 2\lambda_4 + \bar{\omega}_2$ librates about 0° and for Titan-Hyperion, $3\lambda_6 - 4\lambda_7 + \bar{\omega}_7$ librates about π. Voyagers 1 and 2, as well as contemporaneous groundbased observations, revealed several examples of stable commensurabilities in the Saturn system where two satellites' periods were nearly identical. Some were librating about the Trojan points 60° in front of or behind a large satellite whereas one pair, the so-called coorbitals, have horseshoe-shaped orbits relative to each other. We shall address the analysis of this last 1:1 resonance in Sec. IV and consider here only the resonances where the mean orbital motions are near the ratio $j:j + k$ where j and k are nonzero integers.

The hypothesis of the formation of orbital resonances by the differential tidal expansion of initially randomly distributed orbits was suggested by T. Gold (personal communication, 1960). Goldreich (1965) showed the resonances to be stable against continued tidal expansion of the orbits and Allan (1969) demonstrated how the age of a resonance follows from the amplitude of libration if the rate of tidal expansion of the orbits is assumed. Greenberg et al. (1972) and Greenberg (1973a) demonstrated the automatic capture of a Titan-Hyperion type resonance and the subsequent evolution within the resonance, and Sinclair (1972) calculated capture probabilities for the resonances involving Mimas-Tethys and Enceladus-Dione, respectively, as the resonance

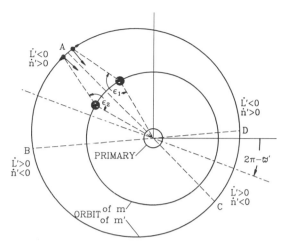

Fig. 1. Schematic diagram of resonant orbits demonstrating stability. Arbitrary positions of repetitive conjunctions are at points A, B, C and D. L and n are angular momentum and mean motion, respectively.

variable of each system passed from circulation to libration. Yoder (1973, 1979*a*) developed the first analytic theory which described the complete origin and evolution of an arbitrary two-body resonance including analytically determined capture probabilities. The field was reviewed by Peale (1976*b*) and by Greenberg (1977). Heuristic descriptions of several resonance properties and processes are given in both reviews. One of these descriptions of stability and some evolutionary aspects are given here to illuminate the physical basis for the mathematical development to follow.

The stability of two-body orbital resonances was understood at the time of Laplace and follows from the ensuing argument. Consider two satellites of masses $m \gg m'$ in coplanar orbits about a primary. The inner satellite m is assumed to be in a circular orbit and is so much more massive than m' that perturbations by the latter can be ignored. The mean motions are assumed to be nearly commensurate, and m' to be in an eccentric orbit. The orbits are shown schematically in Fig. 1, where $\bar{\omega}'$ is the longitude of the outer orbit's pericenter. Four arbitrary positions of repetitive conjunctions are indicated by dashed lines, and the relative positions of each satellite just before and just after a conjunction are also shown along with the radial and tangential components of the perturbing force by m on m'. A dot over a symbol indicates time differentiation.

During the period from opposition to conjunction, m removes angular momentum from m' via the tangential component of the perturbing force and, from conjunction to the next opposition, it adds angular momentum. If the conjunction occurs exactly at pericenter or exactly at apocenter, the effects of the tangential forces integrate to zero, and there is no net transfer. Repetitive conjunctions at any other point destroy this symmetry. If we assume that the line of apsides is fixed and also assume precise commensurability of the mean motions, a conjunction at point A in Fig. 1, for example, would be followed by successive conjunctions at the same point for noninteracting satellites. However, the tangential component of the perturbing force is larger prior to conjunction than after (for $\varepsilon_1 = \varepsilon_2$ in Fig. 1) because the orbits are diverging. In addition, the angular velocity of m' is closer to that of m prior to conjunction, as m' is slowing down as it approaches apocenter. This means m catches up with m' more slowly than it recedes after conjunction, so the larger tangential force opposing the motion of m is also applied for a longer time than the smaller tangential force in the opposite sense after conjunction. Hence, a conjunction at A leads to a net loss of angular momentum by m' over an entire synodic period. The resulting increase in the mean orbital angular velocity n' means that the next conjunction is closer to apocenter.

Similarly, a conjunction after apocenter (point B in Fig. 1) results in a net gain of angular momentum by m', and a tendency for the next conjunction to be again closer to apocenter. The conjunctions thus librate stably about the apocenter of m', preserving the commensurability. Allowing a secular variation of $\tilde{\omega}'$ does not change this conclusion as the ratio n/n' is adjusted such that conjunctions still librate about the apocenter.

The same arguments applied to a conjunction at points C or D near pericenter show that conjunctions are again driven toward apocenter. The pericenter conjunctions thus correspond to an unstable equilibrium configuration like that of a pendulum near the top of its support. The stable point of the analogous pendulum corresponds to the apocenter conjunction.

Now suppose conjunctions occur repetitively at apocenter with no libration and that the inner orbit is being expanded by tidal interactions with the primary. The orbital period of m will increase, and successive conjunctions will occur slightly after apocenter on the average. Angular momentum will thus be secularly transferred in just the right amount to preserve the commensurability against the tendency of the tide to disrupt it (Goldreich 1965).

Two other characteristics of a stable commensurability can now be understood. First, if conjunctions always occur at apocenter of the outer satellite in this example, the radial force of m on m' accelerates m' toward the primary, and m' follows a trajectory slightly inside the trajectory it would have followed if m were not there. This means m' will reach its closest point to the primary slightly sooner than normal and the line of apsides will have rotated in a retrograde sense. If m is sufficiently massive, this regression of the line of apsides due to the resonant perturbation (conjunction always at apocenter) can dominate the normal prograde motion due to the oblateness of the primary and the secular perturbations from other satellites. This is actually realized in the Titan-Hyperion resonance where the line of apsides of Hyperion's orbit regresses about $19°$ yr^{-1}.

The second characteristic is the secular increase of the eccentricity in

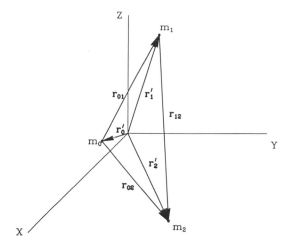

Fig. 2. Definition of variables used in orbital resonance analysis.

this type of orbital resonance. Recall that a tidal expansion of the inner orbit causes the conjunctions of the stable resonance to occur slightly after apocenter. A radial impulse force anywhere between apocenter and pericenter causes the orbiting body to fall closer to the primary, thereby increasing the eccentricity e' of its orbit. With conjunctions now occurring slightly after apocenter, the maximum of the radial perturbative force tends to increase e' secularly.

The stability of a resonance to tidal expansion of the orbits and the growth of the eccentricity of that satellite orbit whose longitude of apoapse or periapse is always near the conjunctions is thus understood from simple ideas. These properties of orbital resonances emerge naturally in the description below, and they play fundamental roles in the evolutionary process. Additional heuristic descriptions are found in Peale (1976b) and in Greenberg (1977). See also Greenberg (1973b) for a lucid description of the Mimas-Tethys resonance, which involves the inclinations of *both* orbits to Saturn's equatorial plane with libration of conjunctions about the mean of ascending node longitudes.

The relatively complex descriptions of origin and evolution of orbital resonances by differential tidal expansion of the orbits have recently been simplified enormously (Henrard 1982a; Henrard and Lemaître 1983). An elegant description of the capture process (Henrard 1982a) has led to a second, much simplified analytic evaluation of capture probability (Borderies and Goldreich 1984) applicable to the most common orbital resonances. However, a modification of these capture probabilities may be necessary when the chaotic nature of the separatrix (the curve in phase space separating circulation from libration) is taken into account (Wisdom et al. 1984) (see Sec. VI).

We shall develop here the Hamiltonian appropriate to two-body resonances from first principles and justify various approximations. Henrard and Lemaître (1983) use this Hamiltonian together with the adiabatic invariance of the system action in a description in which most of the resonance phenomena are easily followed and understood as the system evolves into and within libration due to tidal expansion of the orbits. The success of this simplified theory in explaining the origin and current states of some orbital resonances, as well as cases in which better approximations are necessary, will be demonstrated by examples.

We describe the motion of two interacting satellites orbiting a primary as a system of three point masses $m_0 \gg m_1, m_2$ with position vectors $\mathbf{r}_i'$ from a fixed origin at the center of mass (Fig. 2). Relative positions are indicated by $\mathbf{r}_{ij} = \mathbf{r}_j' - \mathbf{r}_i'$. The equations of motion are

$$m_i \ddot{\mathbf{r}}_i' = \sum_{i \neq j} \frac{G\, m_i m_j\, \mathbf{r}_{ij}}{r_{ij}^3} = -\nabla_i V \tag{1}$$

where

$$V = -\sum_{i \neq j} \frac{G\, m_i m_j}{r_{ij}} \tag{2}$$

with G being the gravitational constant and ∇_i indicating differentiation with respect to the coordinates of the ith mass. By referring the position of m_1 to m_0 and that of m_2 to the center of mass of m_1 and m_0, the potential can be separated into a central term for the elliptic motion of the m_i about their respective centers (i.e., m_0 and the center of mass of m_0 and m_1) and a common disturbing potential which appears in the equation of motion of each m_i ($i > 0$). This system of coordinates was found by Jacobi (see, e.g., Brouwer and Clemence 1961, p. 588). There results

$$\frac{d^2 \mathbf{r}_i}{dt^2} + \frac{G\, m_0 m_i \mathbf{r}_i}{r_i^3 m_i'} = -\nabla_i \frac{\Phi}{m_i'} \tag{3}$$

where

$$\mathbf{r}_i = \mathbf{r}_i' - \sum_{k=0}^{i-1} m_k \mathbf{r}_k' \Big/ \sum_{k=0}^{i-1} m_k \tag{4}$$

locates m_i relative to their respective centers,

$$m_i' = \frac{m_i \sum_{k=0}^{i-1} m_k}{\sum_{k=0}^{i} m_k} \tag{5}$$

and

$$\Phi = -G\left[\frac{m_1 m_2}{r_{12}} + m_0 m_2\left(\frac{1}{r_{02}} - \frac{1}{r_2}\right)\right] \tag{6}$$

is the disturbing potential for the two masses m_1 and m_2, which is just V less the part that produces the second term in Eq. (3). From Eq. (6) one sees how Φ and hence Eq. (3) can be generalized to any number of masses (see Eq. 54 in Sec. III below).

The left-hand side of Eq. (3) is just the expression for two-body motion with $G(m_0 + m_i)$ replaced by $\mu_i = Gm_0 m_i/m_i'$. The Cartesian coordinates and velocities can thus be transformed into a canonical set, (see, e.g., Plummer 1918, pp. 142–153) and subsequent canonical transformations used to yield

$$L_i = m_i' \sqrt{\mu_i a_i} \qquad , \quad \lambda_i = \varepsilon_i + \int^t n_i dt$$
$$W_i = L_i (1 - \sqrt{1 - e_i^2}) \qquad , \quad w_i = -\bar{\omega}_i$$
$$Z_i = L_i \sqrt{1 - e_i^2}(1 - \cos I_i) \quad , \quad z_i = -\Omega_i \tag{7}$$

$$\frac{dL_i}{dt} = -\frac{\partial H}{\partial \lambda_i} , \qquad\qquad \frac{d\lambda_i}{dt} = \frac{\partial H}{\partial L_i}$$

$$\frac{dW_i}{dt} = -\frac{\partial H}{\partial w_i} , \qquad\qquad \frac{dw_i}{dt} = \frac{\partial H}{\partial W_i}$$

$$\frac{dZ_i}{dt} = -\frac{\partial H}{\partial z_i} , \qquad\qquad \frac{dz_i}{dt} = \frac{\partial H}{\partial Z_i} \tag{8}$$

where a_i, e_i, I_i, λ_i, $\bar{\omega}_i$, Ω_i, ε_i are, respectively, the semimajor axes, eccentricities, inclinations of the orbit planes to some reference plane (here the equator of m_0), mean longitudes, longitudes of the periapses, longitudes of the ascending nodes of the orbit planes on the reference plane and mean longitudes at epoch, and $n_i \equiv \sqrt{\mu_i/a_i^3}$ are the orbital mean motions. H is the Hamiltonian given by

$$H = -\sum_{i=1}^{2} \frac{m_i' \mu_i^2}{2L_i^2} + \Phi. \tag{9}$$

The two terms in Φ are called the direct and indirect terms, respectively. From Eq. (4)

$$\mathbf{r}_{02} = \mathbf{r}_2' - \mathbf{r}_0' = \mathbf{r}_2 + K_1 \mathbf{r}_1 \tag{10}$$

where $K_1 = m_1/(m_0 + m_1)$ and we see that the indirect part of Φ is expandable in a rapidly converging series in $K_1(r_1/r_2)$:

$$\frac{1}{r_{02}} - \frac{1}{r_2} = \frac{-m_1}{m_0 + m_1} \frac{r_1}{r_2^2} \cos S_{12} + 0\left(\frac{m_1}{m_0}\right)^2. \tag{11}$$

We neglect terms of order $(m_1/m_0)^2$ and higher, and also let $m_0 + m_1 \approx m_0$ which means $m_i' \approx m_i$. With these approximations, we can expand Φ to the form (Allan 1969)

$$\Phi = -\sum \frac{Gm_1 m_2}{a_2} C I_1^{|\ell - m - 2p_1|} I_2^{|\ell - m - 2p_2|} e_1^{|q_1|} e_2^{|q_2|} \cos \phi_{\ell m p_1 p_2 q_1 q_2} \tag{12}$$

where

$$\phi_{\ell m p_1 p_2 q_1 q_2} = (\ell - 2p_1 + q_1)\lambda_1 - (\ell - 2p_2 + q_2)\lambda_2 - (\ell - m - 2p_1)\Omega_1$$
$$+ (\ell - m - 2p_2)\Omega_2 - q_1\tilde{\omega}_1 + q_2\tilde{\omega}_2 \tag{13}$$

and where we assume $a_2 > a_1$. In Eq. (12) C is a series in a_1/a_2, e_1^2, e_2^2, I_1^2, I_2^2 whose lowest order term is of order $(a_1/a_2)^\ell$; that is, the lowest order term in C contains neither e nor I. The summation indices have the range $2 \leq \ell < \infty$, $0 \leq p < \ell$, $0 \leq m < \ell$, $-\infty < q < \infty$. An important property of the series is the equality of the coefficient of $\tilde{\omega}$ with the lowest power of e in the coefficient of the cosine and the equality of the coefficient of Ω with the lowest power of I. When Φ is expressed in terms of the canonical variables, L, W, Z are confined to the coefficients of the cosines and λ, w, z are confined to the arguments. Rotational invariance requires that the sum of the coefficients in each argument vanish.

Near a low order commensurability of the mean motions of two satellites, the frequency of some of the terms becomes very small. Often the combination of a large coefficient and a small frequency makes the perturbations due to a single term completely dominant. Keeping only this term in the expansion of Φ is equivalent to averaging over high-frequency terms and ignoring all the remaining terms with small coefficients. This is a much more severe approximation than neglecting the terms which are higher order in the masses, because terms which have small frequencies but which are not the lowest order in e or I may have other factors in the coefficient of the cosine that make them comparable in magnitude to the lowest order term. Still, keeping only the single term in Φ is usually a good approximation if e (or I) is not too large. It is this model we shall investigate.

For the purpose of illustration, we shall choose a simple eccentricity type resonance where a single term in Φ dominates. The disturbing potential becomes

$$\Phi = \frac{Gm_1 m_2}{a_2} C_1 e_1^{|k|} \cos(j\lambda_1 - (j + k)\lambda_2 + k\bar{\omega}_1). \tag{14}$$

If $j = k = 1$, Eq. (14) would be appropriate for the Enceladus-Dione resonance.

The secular motions of the nodes and periapses due to the oblateness of m_0 and the presence of the other satellites are important in separating the frequencies of several variables having the same ratio of the coefficients of λ_1 and λ_2 (e.g., $\lambda_1 - 2\lambda_2 + \bar{\omega}_1$, $\lambda_1 - 2\lambda_2 + \bar{\omega}_2$). Only because the secular variations of $\bar{\omega}_i$ and Ω_i are sufficiently different for the two satellites, can we consider all but one resonance variable to be high frequency and hence ignorable. (We shall see later that two resonances with nearly identical frequencies can still be treated independently provided the respective resonance variables contain no common longitudes of periapses or common longitudes of ascending nodes.) We can include these secular motions by adding to Eq. (14) the terms from Eq. (12) with $\phi = 0$, the zero frequency terms from the disturbing functions involving the other satellites and the Sun as well as the secular terms from expansion of m_0's nonspherical field. Rather than carry this exercise out in detail, we note that

$$\left.\frac{dw}{dt}\right|_s = -\dot{\bar{\omega}}_s = \frac{\partial H}{\partial W} \tag{15}$$

which we can obtain by adding the term $-W\dot{\bar{\omega}}_s$ to H. Similarly $-Z\dot{\Omega}_s$ and $L\dot{\lambda}_s$ account for the secular motion of the node and the addition to the mean motion. The subscript s denotes secular. Then

$$\begin{aligned}
H = &-\frac{G^2 m_0^2 m_1^3}{2L_1^2} - \frac{G^2 m_0^2 m_2^3}{2L_2^2} - \frac{G^2 m_0 m_1 m_2^3}{L_2^2} C \left(\frac{2W_1}{L_1}\right)^{|k|/2} \\
&\times \cos(j\lambda_1 - (j + k)\lambda_2 - kw_1) - W_1\dot{\bar{\omega}}_{s1} - Z_1\dot{\Omega}_{s1} \\
&+ L_1\dot{\lambda}_{s1} - W_2\dot{\bar{\omega}}_{s2} - Z_2\dot{\Omega}_{s2} + L_2\dot{\lambda}_{s2}
\end{aligned} \tag{16}$$

is the Hamiltonian for the two-body, simple eccentricity type resonance, where from Eq. (7) $W_1 \approx L_1 e_1^2/2$ has been used.

There are two sets of variables L_i, W_i, Z_i, λ_i, w_i, z_i corresponding to 6 degrees of freedom, but the above approximate form of H allows us to reduce this to 4 degrees of freedom. Since $z_i = -\Omega_1$ does not appear in Φ (i.e., z_i is cyclic), the Z_i are constants of the motion and $\partial H/\partial Z_i = -\Omega_{si}$ are independent of the other variables. Next, the angle variables appear only in the combination $j\lambda_1 - (j + k)\lambda_2 + k\bar{\omega}_1$ which suggests we make the further change of variables (recall $w_i = -\bar{\omega}_i$)

$$
\begin{aligned}
\theta_1 &= j\lambda_1 - (j + k)\lambda_2 - kw_1 \\
\theta_2 &= j\lambda_1 - (j + k)\lambda_2 - kw_2 \\
\theta_3 &= \lambda_1 \\
\theta_4 &= \lambda_2.
\end{aligned}
\tag{17}
$$

The momenta Θ_i conjugate to θ_i are obtained from the differential relation (Brouwer and Clemence 1961, p. 539)

$$
\sum_{i=1}^{4} \Theta_i d\theta_i = \sum_{j=1}^{2} (L_j d\lambda_j + w_j dw_j)
\tag{18}
$$

which yields

$$
\begin{aligned}
j(\Theta_1 + \Theta_2) + \Theta_3 &= L_1 \\
-(j + k)(\Theta_1 + \Theta_2) + \Theta_4 &= L_2 \\
-k\Theta_1 &= W_1 \\
-k\Theta_2 &= W_2.
\end{aligned}
\tag{19}
$$

Since θ_2, θ_3 and θ_4 do not appear in H, Θ_2, Θ_3 and Θ_4 are constants of the motion, and the problem is reduced to one degree of freedom with only θ_1 and Θ_1 as variables.

Now $\Theta_2 = -W_2/k = -L_2(1 - \sqrt{1 - e_2^2}) \approx -L_2 e_2^2/2$ and since it is constant, we can set it equal to zero or include its value in Θ_3 and Θ_4 in Eq. (19) with negligible change in the resonance behavior. Then in the new variables

$$
\begin{aligned}
H = &-\frac{G^2 m_0^2 m_1^3}{2(\Theta_3 + j\Theta_1)^2} - \frac{G^2 m_0^2 m_2^3}{2[\Theta_4 - (j + k)\Theta_1]^2} \\
&- \frac{G^2 m_1 m_2^3 m_0 C(-2k\Theta_1)^{k/2} \cos\theta_1}{[\Theta_4 - (j + k)\Theta_1]^2[\Theta_3 + j\Theta_1]^{k/2}} \\
&+ \frac{\Theta_1}{k}\dot{\bar{\omega}}_s + \frac{\Theta_2}{k}\dot{\bar{\omega}}_{s2} + (j\Theta_1 + \Theta_3)\dot{\lambda}_{s1} \\
&+ [-(j + k)\Theta_1 + \Theta_4]\dot{\lambda}_{s2}.
\end{aligned}
\tag{20}
$$

From Eqs. (19) and (7) we see that $|\Theta_1| \ll \Theta_3$, and $|\Theta_1| \ll \Theta_4$; hence, we can expand the first two terms in H to second order in Θ_1/Θ_3 and Θ_1/Θ_4

consistent with the magnitude of the terms kept in Φ. Since the coefficient of the cosine is a factor m_1/m_0 smaller than the first two terms, we need not expand the denominator of this term in Θ_1. We carry out the expansions, absorb all the constant terms into H, use W_1 as a variable instead of Θ_1, replace the angle variable θ_1 by $\theta_1' = \theta_1/k$ and drop the prime. Then with $H' = -H$ and $dW_1/dt = -\partial H'/d\theta$, $d\theta_1/dt = \partial H'/dW_1$, we arrive at the Hamiltonian used by Henrard and Lemaître (1983).

$$H' = \alpha W_1 + \beta W_1^2 + \varepsilon(2W_1)^{k/2} \cos k\theta_1 \qquad (21)$$

where

$$\alpha = (jn_1^* - (j+k)n_2^* + k\dot{\omega}_{s1})/k$$

$$\beta = \frac{3}{2k^2}\left[\frac{j^2}{m_1 a_1^2} + \frac{(j+k)^2}{m_2 a_2^2}\right]$$

$$\varepsilon = C(n_1^2)^{1-\frac{k}{4}}\frac{a_1}{a_2}^{3-k}\frac{m_2}{m_0}m_1^{1-\frac{k}{2}} \qquad (22)$$

where $n_i^* = n_i + \dot{\lambda}_{si}$ with $\Theta_3 \approx L_1$ and $\Theta_4 \approx L_2$ being used. This Hamiltonian is the same form used by Yoder (1973,1979b) in his original analytic solution.

The reason for transforming the angle variable to θ_1/k from the old θ_1 is to have $k\theta_1$ as the argument of the cosine in Eq. (21). We shall use a canonical transformation below similar to $x = (2W_1)^{1/2}\cos\theta_1$ and $y = (2W)^{1/2}\sin\theta_1$ where the Hamiltonian in the new variables x and y is analytic at the origin for $k > 1$ only with $k\theta_1$ as the cosine argument. This property of the disturbing potential is called the d'Alembert characteristic. Although we shall explicitly treat only the $k = 1$ case in detail, this proper choice of canonical variables is an important consideration in the analysis of higher-order resonances.

We have assumed C to be constant without expressing it in terms of the Θ_i. This is justified as follows. The lowest-order term in C involves only a_1/a_2. Then $\delta C/C \sim \delta a/a \sim \delta L/L$. Since $\Theta_3 = L_1 + jW_1/k$ is a constant of the motion, $\delta L_1 = -j\,\delta W_1/k$. But $\delta e_1/e_1 \sim \delta W_1/W_1 = 0(\delta L_1/L_1)/e_1^2$ or $0(\delta C/C)/e_1^2$. So the variation in the coefficient of the cosine during the conservative oscillations is determined almost completely by the variation in e_1, and keeping C constant is a good approximation.

All three coefficients and the "constants" absorbed in H' will be slowly varying as tides raised on m_0 cause n_1^* and n_2^* to decrease differentially. However, α is small near the resonance and the fractional change in α due to the changes in n_i^* will generally be much larger than the fractional change in β or ε. The exception to this condition occurs when the mass m_2 is sufficiently larger than m_1 that $\langle|jdn_1^*/dt - (j+k)dn_2^*/dt|\rangle$ is very small, where $\langle\rangle$ indicates averaged value. In this case substantial changes in the a_i would occur for significant change in α. Since $\delta\beta/\beta \sim \delta a/a$ and $\delta\varepsilon/\varepsilon \sim \delta a/a$, the fractional change in β and ε would not be small compared to $\delta\alpha/\alpha$. However, for the existing resonances among the satellites we can consider β and ε approximately constant compared to α. The approximation is least applicable to the Mimas-Tethys resonance where $m_2/m_1 \approx 17$, but the physical processes involved remain the same and are adequately described by the model.

The approximate constancy of β and ε compared to α is a necessary condition for the applicability of the Henrard-Lemaître model in which the Hamiltonian is reduced to a dimensionless form with a single variable parameter. We consider only the case $k = 1$, which applies to all the two-body resonances among the satellites except the Mimas-Tethys resonance and even this can be approximated as a $k = 1$ resonance because m_2/m_1 is so large. Henrard and Lemaître (1983) simplify the Hamiltonian by the following change in variables:

$$R = \left(\frac{2\beta}{\varepsilon}\right)^{2/3} W_1$$

$$\phi = \theta_1 + \pi \qquad \text{if } \varepsilon\beta > 0$$

$$\phi = \theta_1 \qquad \text{if } \varepsilon\beta < 0$$

$$\tau = \left(\frac{\beta\varepsilon^2}{4}\right)^{1/3} t. \tag{23}$$

The Hamiltonian becomes

$$K = -3(\delta + 1)R + R^2 - 2\sqrt{2R} \cos \phi \tag{24}$$

where

$$\delta = -\left(\frac{4\alpha^3}{27\beta\varepsilon^2}\right)^{1/3} - 1 \tag{25}$$

and

$$\frac{dR}{d\tau} = -\frac{\partial K}{\partial \phi} \quad , \quad \frac{d\phi}{d\tau} = \frac{\partial K}{\partial R} . \tag{26}$$

In Eq. (24) the variations due to tides are confined to the single parameter δ.

Since K is a constant of the motion in the absence of tides, the nature of the motion can be ascertained by plotting R vs. ϕ for various values of K. These curves are called the level curves of the Hamiltonian or constant energy curves. It is more convenient to do this in Poincare variables, with

$$x = \sqrt{2R} \cos \phi$$
$$y = \sqrt{2R} \sin \phi \ . \tag{27}$$

With $\sqrt{2R}$ used instead of R in the definitions of x and y, the transformation is canonical with

$$\frac{dx}{dt} = -\frac{\partial K}{\partial y}, \qquad \frac{dy}{dt} = \frac{\partial K}{\partial x} \tag{28}$$

and

$$K = -\Delta \frac{(x^2 + y^2)}{2} + \frac{(x^2 + y^2)^2}{4} - 2x \tag{29}$$

where $\Delta = 3(1 + \delta)$. Also $\sqrt{2R} \propto e_1$, and the radius from the origin to the trajectory measures the instantaneous eccentricity.

The stationary points follow from $\partial K/\partial x = \partial K/\partial y = 0$ and are the roots of

$$x^3 - \Delta x - 2 = 0,$$
$$y = 0. \tag{30}$$

The parameter δ has been constructed such that for $\delta < 0$ ($\Delta < 3$) there are no negative real roots of Eq. (30) and one positive real root, whereas for $\delta > 0$ ($\Delta > 3$), there are two negative real roots and one positive real root.

A typical set of level curves is shown in Fig. 3, where $\Delta \approx 9$. Curves for positive and negative circulation and for libration are indicated. We define libration here as any state where ϕ is bounded. The curve marked separatrix separates regions of circulation from those of libration in Fig. 3, although it is possible to have a librating system, according to our definition, without the trajectory being inside the separatrix. There are three stationary points apparent in Fig. 3 (two stable and one unstable corresponding to the roots of Eq. 30), and a trajectory near the stable point on the negative x axis is not within the separatrix but would correspond to libration about $\phi = 180°$.

Henrard (personal communication, 1985) is careful to point out that libration of a resonance variable ϕ does not necessarily imply much dynamical significance. In our example, ϕ could be librating even though n_1 and n_2 are far from commensurability when e is very small and $\tilde{\omega}$ varies rapidly. In this case, the libration has little or no effect on the system evolution unless it is maintained until n_1 and n_2 approach the commensurability. There is thus some argument for reserving the term "resonance" for those systems described by a level curve contained within the separatrix (Henrard and Lemaître 1983). However, we shall retain our definition of libration, but keep

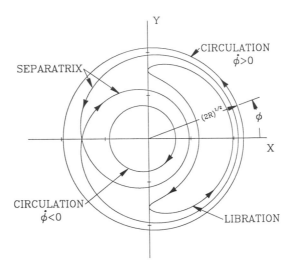

Fig. 3. A typical set of level curves of the Hamiltonian appropriate to a two-body orbital resonance. Positive and negative circulation and libration curves are indicated as well as the curve
(separatrix) separating circulation from libration.

in mind that the "resonance" is significant only when n_1/n_2 is relatively close
to commensurability. (See the qualifications below for more discussion on this
point.)

The character of the set of level curves depends on the value of Δ (or δ).
For example, there is only one stationary point and no separatrix for $\Delta < 3$,
and both inner and outer curves of the separatrix expand as Δ increases above
3. As Δ is only slowly varying, the actual trajectories of the system in the xy
plane are very close to the trajectories for constant Δ and the system can be
thought to evolve slowly through a set of level curve diagrams as Δ varies. To
follow the behavior of the system as the set of possible level curve trajectories
changes with Δ, we need one more important principle.

As long as Δ does not change very much during a period of the motion,
the action

$$J = \oint R\mathrm{d}\phi = \oint x\mathrm{d}y \tag{31}$$

is an adiabatic invariant. This has been proven for periodic systems with a
slowly varying parameter by Gardner (1959), Lenard (1959), and explicitly
for nonlinear oscillations appropriate to orbital resonance by Henrard (1982a;
see also Landau and Lifshitz 1960; Yoder 1979b). The exception to the adiabatic conservation of J occurs when a trajectory crosses the separatrix since
the period on the separatrix is infinite, due to asymptotic approach to the unstable equilibrium point. Hence, we expect the action to be conserved until a

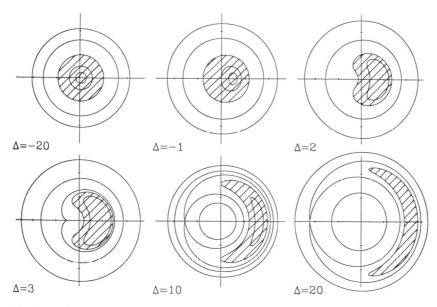

$\Delta=-20$ $\Delta=-1$ $\Delta=2$

$\Delta=3$ $\Delta=10$ $\Delta=20$

Fig. 4. Tidal evolution of two bodies toward orbital resonance as a series of level curves of the Hamiltonian. The particular curve appropriate to the motion encloses the shaded area which is the conserved system action determined by initial conditions far from resonance. The sense of motion about the curves is shown in Fig. 3. This series illustrates automatic capture into libration.

separatrix is crossed at which time the action will change in the transition to a librating or oppositely circulating state, but thereafter remain constant through further evolution following the transition. From Eq. (31), we see that the action is just the area within a trajectory and is determined by initial conditions far from resonance. Hence, we can identify the trajectory traversed by the system at any stage of the evolution as that trajectory which encloses an area equal to the initial action integral or that area enclosed by one or both curves of the separatrix (not the initial action) after a transition through that separatrix. The series of level curve diagrams coupled with the adiabatically conserved action allow us to understand in a simple way all of the major characteristics of the origin and evolution of orbital resonances. We do this by considering the consequences of several initial conditions far from resonance and two directions of approach to the resonance.

Figure 4 shows the trajectory evolution on a series of xy phase plane plots for the case where Δ increases from negative values as the resonance is approached. From Eqs. (22) and (25), a negative $\delta(\Delta < 3)$ corresponds to positive α and the resonance is being approached as the inside satellite catches up to the resonance configuration by the more rapid expansion of its orbit; that is, n_1/n_2 is decreasing. We have chosen a relatively small initial eccentricity far from resonance so the action (area inside trajectory) is small. Maintaining the area inside the trajectory corresponding to the motion as Δ increases through the series of diagrams shows that the system evolves smoothly into libration about $\phi = 0°$. We have included the separatrix trajectory in the diagrams after its appearance at $\Delta = 3$. For the small eccentricity chosen as an initial condition, the capture into resonance is certain. Subsequent evolution within the resonance is seen to be a continued increase in the forced eccentricity (determined by the separation from the origin of the stationary point enclosed by the trajectory) as the amplitude of libration is reduced.

Figure 5 shows a similar evolution over the same change in Δ but starting with a much larger eccentricity far from resonance. Now the separatrix forms inside the circulation trajectory and expands as Δ increases until the outer curve of the separatrix is nearly the same as the trajectory. At this point, the trajectory must cross the separatrix and a transition from the state of positive circulation must occur. This is the resonance encounter and the transition from positive circulation can have but one of two outcomes. After following a trajectory very close to the separatrix, the system can either reverse its direction of motion and trace the crescent shaped trajectory of libration (capture into resonance) or continue its negative circulation (escape). The action is not conserved across the transition, but assumes a new essentially constant value, which is equal to the area inside the crescent-shaped separatrix at the time of transition into libration (Fig. 5, $\Delta = 6$) or to the area of the inside curve of the separatrix for transition into negative circulation. The evolution beyond transition can now be followed as before with the proper trajectory once again being determined by the conserved area within. In libration, the initial amplitude is 180°, and it is reduced as the forced eccentricity increases, as shown in Fig. 5. Unlike the situation for a small eccentricity far from resonance where capture into libration was certain, here the capture is probabilistic. A procedure for estimating that probability will be discussed below.

Comparison of Figs. 4 and 5 shows that the criterion for certain capture into resonance is that the initial action far from resonance be less than the area inside the separatrix first formed when $\Delta = 3$ ($\delta = 0$). Since the initial trajectories far from resonance are essentially circles centered at the origin, the criterion for certain capture is

$$J_0 = \pi(x_0^2 + y_0^2) = 2\pi R_0 < J_c = 6\pi \tag{32}$$

where J_0 is directly proportional to e_0^2 and J_c is the area inside the critical separatrix for $\Delta = 3$. The integral for J_c is most easily done using the first form of Eq. (31) involving R and ϕ, where Eqs. (24) and (26) are used to change the integration variable from ϕ to R and $dR/d\tau = 0$ and $d\phi/d\tau = 0$ at $R(\phi = 180°) = R^*$ are used to factor terms in the integrand. From Eq. (32) the criterion becomes

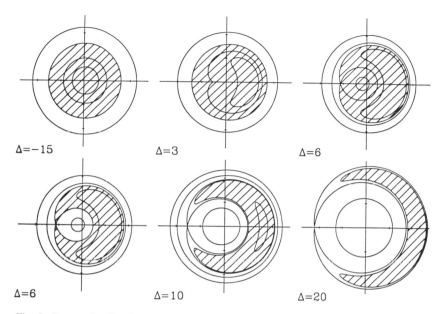

$\Delta=-15$ $\Delta=3$ $\Delta=6$

$\Delta=6$ $\Delta=10$ $\Delta=20$

Fig. 5. Same as for Fig. 4 except now the separatrix forms before the resonance is reached because of the higher initial eccentricity. The series shows capture into libration which is now probabilistic. There is a discontinuous change in the action (shaded area) upon transition from circulation to libration.

$$e_{10}^2 < \frac{6}{m_1\sqrt{\mu_1 a_1}} \left(\frac{\varepsilon}{2\beta} \right)^{2/3} \tag{33}$$

for certain capture.

The age of a resonance can be estimated by comparing the current value of Δ and the value of Δ in the past when the area inside the separatrix was equal to the current action. If the current action is less than the area of the critical separatrix at $\Delta = 3$, the transition into resonance was automatic, and the value of Δ when this occurred corresponds to the trajectory passing through the origin of the xy plane. With current and initial values of Δ determined, their difference gives the change in α from Eq. (25), and the age $T = \alpha/\dot{\alpha}$ where $\dot{\alpha}$ follows from the rate of change of $j\dot{n}_1^* - (j + k)\dot{n}_2^*$ due to tides raised on m_0. This age is qualified in one of the applications below where the implicit assumptions on which it is based are questioned.

Before illustrating the calculation of the capture probability, we show the consequence of approaching the resonance from the opposite direction with Δ initially large and positive and decreasing. This corresponds to the case where the orbits are initially too close together for the resonance, but $m_2 \gg m_1$ such that m_2's orbit expands faster than m_1's even though it is farther away. This evolution is shown in Fig. 6 where the system starts in negative circula-

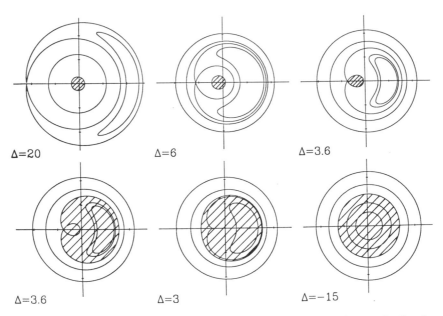

Fig. 6. Same as for Fig. 4, except now the resonance is approached from the opposite direction (orbits initially too close for resonance). This evolution illustrates temporary inverted libration but eventual certain escape from the resonance.

tion with relatively small eccentricity and action. As Δ decreases, the separatrix shrinks and eventually the area enclosed by the inner curve of the separatrix is equal to the initial action (Fig. 6, $\Delta = 3.6$). Again the system has no choice but to traverse a path first very close to the inside curve of the separatrix and then close to the outside curve. This time, however, it must remain in positive circulation as the separatrix continues to shrink. The shrinking area enclosed by the separatrix means the action is only invariant after the transition if the circulating trajectory is left behind as the separatrix shrinks away from it. Capture into the resonance libration from this direction of approach is impossible. This was noted by Sinclair (1972) as well as Yoder (1973) and is discussed in the review by Peale (1976b). However, it is more simply understood in this approach due to Henrard and Lemaître (1983). For the initial conditions assumed for Fig. 6, we see that the system is temporarily trapped in a state of inverted libration about $\phi = \pi$. A bound on e_0 for such a temporary inverted libration follows from the condition that the initial action be less than the area of the inner separatrix when that separatrix passes through the origin. The inverted libration is always unstable to continued tidal evolution of the system.

The major consequence of passing through the resonance with decreasing Δ is the substantial jump in the free eccentricity. The new action is the area within the outer separatrix curve at the time of transition. If J_1 and J_2 are the areas within the inside and outside curves of the separatrix, respectively, $J_1 = 2\pi R_1$, $J_2 = 2\pi R_2$, where R_1 and R_2 are the nearly constant values of R on far opposite sides of the resonance, respectively. The initial eccentricity determines J_1 which in turn determines the value of $\Delta = \Delta_T$ at transition. Given Δ_T, J_2 (the area of the outside separatrix) is determined by an integration described earlier, and we find

$$J_1 + J_2 = 2\pi(R_1 + R_2) = 2\pi\Delta_T \tag{34}$$

and from Eq. (25) with $\Delta_T = 3(\delta_T + 1)$,

$$e_{1i}^2 + e_{1f}^2 = -\frac{2\alpha_T}{\beta L_1} \tag{35}$$

with e_{1i} and e_{1f} being initial and final values of e_1 far from resonance and α_T the value of α at transition. Note that $\Delta_T > 3$ ($\delta > 0$) for this transition, requiring $\alpha_T < 0$ (Eq. 25).

Capture Probability

Sinclair (1972,1974) determined capture probabilities for several orbital resonances encountered by the satellites of Saturn with numerical techniques, and Yoder (1973) gave the first analytic determination of the probabilities for general resonances (cf. Peale 1976b). We shall outline a method developed by Henrard (1982a) which led to a second, much simpler analytic determination of the capture probability by Borderies and Goldreich (1983).

A capture probability only applies to the case where $\Delta > 3$ ($P_c = 1$ for $\Delta_T < 3$ at resonance encounter) and a separatrix exists as the system approaches resonance. Transition occurs when the separatrix has expanded to the point where the system makes a last positive circulation very close to the outside curve of the separatrix, and the next motion is a traverse in the opposite sense along a trajectory very close to the inside curve of the separatrix. Let N be the value of the Hamiltonian $K(R, \phi, \Delta)$ (Eq. 24) relative to its value $K(\bar{R}, \phi, \Delta)$ on the separatrix where $\bar{R}(\phi)$ corresponds to the separatrix. Since $K(\bar{R}, \phi, \Delta) = K^*(R^*, \phi^*, \Delta)$ where R^* and ϕ^* are the values at the unstable equilibrium point on the separatrix, we can write

$$N = K(R, r, \Delta) - K^*(R^*, r^*, \Delta). \tag{36}$$

As K and K^* are fixed if Δ is fixed, the total variation in N must be due to the variation in Δ:

$$\frac{dN}{dt} = \left(\frac{\partial K}{\partial \Delta} - \frac{\partial K^*}{\partial \Delta} \right) \frac{d\Delta}{dt} . \tag{37}$$

As the separatrix for which $K = K^*$ approaches the circulation trajectory with Hamiltonian K, necessarily N approaches zero as both K and K^* change. From Eq. (24) a larger circulation trajectory for given Δ when R is large yields a larger K, so N is initially positive and $dN/dt < 0$.

For the last circulation trajectory near the outer separatrix curve before the separatrix is crossed, let

$$B_1 = \oint_{c_1} \frac{dN}{dt} dt = \dot{\Delta} \oint_{c_1} \left(\frac{\partial K}{\partial \Delta} - \frac{\partial K^*}{\partial \Delta} \right) dt = \dot{\Delta} \oint_{c_1} \left(\frac{\partial K}{\partial \Delta} - \frac{\partial K^*}{\partial \Delta} \right) \frac{d\tilde{R}}{\dot{R}} < 0 \tag{38}$$

be the change in the relative Hamiltonian, where $dt = dR/\dot{R} = d\tilde{R}/\dot{\tilde{R}}$, since the trajectory is very near the separatrix and c_1 is the outside curve. The next motion will be a traverse of the inside separatrix curve in the opposite sense. Call the change in the relative Hamiltonian for this traverse

$$B_2 = \dot{\Delta} \oint_{c_2} \left(\frac{\partial K}{\partial \Delta} - \frac{\partial K^*}{\partial \Delta} \right) \frac{d\tilde{R}}{\dot{\tilde{R}}} \tag{39}$$

where c_2 refers to the inside separatrix curve. K also increases away from the inside separatrix curve so, if B_2 is also < 0, capture is certain. If $B_2 > |B_1|$, the system gains more relative energy in traverse of inner separatrix than it lost in tracing the outer separatrix and escape is certain. If $B_2 > 0$ but $B_1 + B_2 < 0$ and we assume the possible values of N over the range $0 \leq N \leq -B_1$ at the start of the last outside circulation to have a uniform probability distribution, the capture probability is

$$P_c = \frac{B_1 + B_2}{B_1} = \frac{2}{1 + \dfrac{\pi}{2 \sin^{-1} \left[\dfrac{R_{max} + R_{min} - 2R^*}{R_{max} - R_{min}} \right]}} \tag{40}$$

where we have used Eqs. (24) and (26) to write

$$B_i = -2\dot{\Delta} \int_{R^*}^{R_i} \frac{d\tilde{R}}{\sqrt{(\tilde{R} - R_{min})(R_{max} - \tilde{R})}} \tag{41}$$

where $R_1 = R_{max}$, $R_2 = R_{min}$, the maximum and minimum values of R on the separatrix. Equation (40) was first obtained by Yoder (1973).

We need but evaluate R_{max}, R_{min} and R^* on the separatrix in terms of Δ. From Eq. (27), $x^* = -\sqrt{2R^*}$ where x^* is the most negative root of Eq. (30). The extremes in R occur on the x axis, so from Eq. (29) with $x_{max} = \sqrt{2R_{max}}$, $x_{min} = \sqrt{2R_{min}}$, x_{max} and x_{min} are roots of $x^4 - 2\Delta x^2 - 8x - 4K$

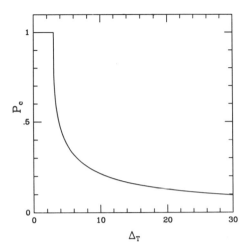

Fig. 7. Probability of capture into a two-body resonance with $n_1:n_2 = j:j + 1$ as a function of the value of $\Delta = \Delta_T$ at the time of transition.

and x^* is a double root. K can be expressed in terms of Δ and x^* with the help of Eq. (30), $(x - x^*)^2$ factored from the quartic and the remaining quadratic solved for x_{max} and x_{min}. There results the expression obtained by Borderies and Goldreich (1984):

$$P_c = \cfrac{2}{1 + \cfrac{\pi}{2 \, \sin^{-1}(sz)^{-3/2}}} \tag{42}$$

where $s = \sqrt{\Delta_T/3}$, $z = \cos(\xi/3) + \sqrt{3} \sin(\xi/3)$ with $\cos \xi = (\Delta_T/3)^{-3/2}$. $P_c(\Delta_T)$ is known and we can evaluate P_c as a function of the initial eccentricity far from resonance by determining the area of the outside separatrix curve as a function of Δ. Then Δ_T is the value of Δ corresponding to an area of the outside separatrix curve equal to the initial action $J = 2\pi R_0$ where $R_0 \propto e_0^2$ by Eqs. (23) and (7). P_c is given as a function of Δ_T for the $j:j + 1$ orbit-orbit resonance in Fig. 7.

An Application

The resonance between Enceladus and Dione is of the type discussed in the illustrative example above with $j = k = 1$. Table I gives the relevant parameters for the two satellites which, when substituted into Eq. (22), yield

$$\alpha = 0°072/\text{day} = 1.45 \times 10^{-8} \text{ rad s}^{-1}$$
$$\beta = 4.05 \times 10^{-44} \text{ g}^{-1} \text{ cm}^{-2}$$
$$\varepsilon = -3.347 \times 10^9 \text{ g}^{1/2} \text{ cm s}^{-3/2} \tag{43}$$

TABLE I
Parameter Values for the Enceladus-Dione Resonance

	Enceladus	Dione
n^*	262°732/day	131°535/day
a	238,040 km	377,420 km
e	0.0044	0.0022
$\dot{\bar{\omega}}_s$	0°410/day	0°084/day
m/m_0	1.5×10^{-7}	1.9×10^{-6}

where $C = -1.19$ is determined from the coefficient of $\cos(\lambda_1 - 2\lambda_2 + \bar{\omega}_1)$ in the expansion of Φ for $a_1/a_2 = 0.63$ corresponding to $n_1/n_2 = 2$ (see, e.g., Brouwer and Clemence 1961, p. 490). From Eq. (25), $\delta = -2.0$, and Enceladus and Dione are in a state where no separatrix has formed and the capture into resonance is certain as noted by Sinclair (1972), Yoder (1973) and others. In fact, from Eq. (33) capture would be certain for any $e_{10} < 0.017$ which is much larger than the current value of 0.0044.

The amplitude of libration of $\lambda_1 - 2\lambda_2 + \bar{\omega}_1$ about $0°$ is near $1°$ (Sinclair 1972). From Fig. 4, we see that the trajectories for $\delta \approx -2$ are nearly circular, and the libration amplitude yields the radius of the circle as $\rho = \sqrt{2R_{eq}}$ $\tan 1°$ with R_{eq} being the value of R at the stationary point within the trajectory. From Eqs. (23) and (7), $e_1 = 7.38 \times 10^{-3} \sqrt{2R}$ from which $\sqrt{2R} = 0.603$ when $e_1 = 0.0044$ and $\rho = 0.0105$. The action is just $\pi\rho^2$ and if this is conserved, the transition into resonance occurred when the circle of radius ρ passed through the origin, or the stationary point was at $x = \rho$. From Eq. (30), this yields $\delta = -62.5$ at transition and the age of the resonance is the time for δ to go from -62.5 to -2.

From Eqs. (25) and (22), $d\delta/dt \propto d\alpha/dt = dn_1^*/dt - 2dn_2^*/dt + d\dot{\bar{\omega}}_{1s}/dt$, where

$$\frac{dn^*}{dt} = \frac{-9}{2} k_2 \frac{m}{m_0} \left(\frac{R_s}{a}\right)^5 \frac{n^2}{Q_0} \tag{44}$$

with $k_2 = 0.34$ (Gavrilov and Zharkov 1977), $Q_0 > 1.6 \times 10^4$ (Goldreich and Soter 1966) and R_s being the Love number for Saturn, the dissipation function for Saturn and the radius of the satellite, respectively. Substitution of the values of the remaining parameters from Table I yields $d\alpha/dt = -1.7 \times 10^{-23}$ rad s^{-2} and an age of the resonance of 1.6 Gyr.

Qualifications

The Henrard-Lemaître formulation thus describes the origin and evolution of orbital resonances in an orderly and easily understood manner.

The somewhat diverse approaches to the study of such resonances by Allan (1969), Sinclair (1972,1974), Greenberg (1973a) and Yoder (1973), for example, are hereby reduced to probably the simplest form and provide a basis for any study of orbital resonances. As is often the case, however, this most elegant and simple model must be applied with caution in interpreting the current configuration and inferred histories of observed orbital resonances. Let us begin with the above example of Enceladus-Dione which appears on the surface to satisfy all the criteria for the model approximations.

Perhaps the most critical assumption is the isolation of a single resonance term in R to simplify the Hamiltonian. The terms nearest in frequency all have $n_1/n_2 \approx 2$, and those frequencies corresponding to the lowest order coefficients are

$$
\begin{aligned}
(2n_1^* - 4n_2^* + 2\dot{\Omega}_1) &= -1\overset{\circ}{.}496/\text{day} \\
(2n_1^* - 4n_2^* + \dot{\Omega}_1 + \dot{\Omega}_2) &= -1\overset{\circ}{.}170/\text{day} \\
(2n_1^* - 4n_2^* + 2\dot{\Omega}_2) &= -0\overset{\circ}{.}844/\text{day} \\
n_1^* - 2n_2^* + \dot{\bar{\omega}}_2 &= -0\overset{\circ}{.}254/\text{day} \\
2n_1^* - 4n_2^* + \dot{\bar{\omega}}_1 + \dot{\bar{\omega}}_2 &= -0\overset{\circ}{.}254/\text{day} \\
n_1^* - 2n_2^* + \dot{\bar{\omega}}_1 &= 0.
\end{aligned}
\tag{45}
$$

The frequencies are separated by the different secular motions of Ω_i and $\bar{\omega}_i$. The identity of two frequencies depends on the existence of the last zero frequency, and this identity would vanish with the resonance. The separation of the frequencies in Eq. (45) is sufficient that all but the resonant term can be considered high-frequency terms whose negligible influence justifies selection of only the single term in R for analysis of the *existing* configuration. However, we are interested in the origin and evolution of the system and the assumption of only a single term in R always dominating the dynamics is not necessarily correct.

First, Sinclair (1972) noticed that the existing resonance would normally be the last one encountered among those in Eq. (45) as n_1 was reduced. The current orbital inclinations are quite small and uncertain. Sinclair (1974) used Struve's (1933) values of $I_1 = I_2 = 4.1 \times 10^{-4}$ rad to determine the values of I_i before an assumed transition from positive to negative circulation. Notice from Figs. 4 and 5 that such a transition reduces the action and hence the eccentricity in the example and would similarly reduce the inclination for an inclination type resonance. So, having escaped capture into the series of inclination resonances implies larger inclinations in the past for the orbits of Enceladus and Dione. These larger inclinations insure a nonzero probability of escape for all the inclination resonances.

We can perform the same analysis as that above using the Henrard-Lemaître formulations for the resonance with $\theta_2 \ (= \lambda_1 - 2\lambda_2 + \bar{\omega}_2)$ as the librating variable. The value of $\varepsilon = 2.87 \times 10^8$ g$^{1/2}$ cm s$^{-3/2}$, $\alpha =$

$-0°254/\text{day}$ from Eq. (45) and β has the same value of 4.05×10^{-44} g^{-1} cm^{-2} as it had for the $\theta_1 \, (= \lambda_1 - 2\lambda_2 + \tilde\omega_1)$ libration. With $e_2 = 0.0022 = 7.7 \times 10^{-4} \sqrt{2R}$, the current action $J = 25.65$, which is the area of the inside separatrix curve at transition. This determines $\Delta_T = 13.0$ and a capture probability $P_c = 0.18$ which was also found by Sinclair (1974). From nonzero probability of escape for all the resonances, Sinclair (1974) showed the consistency of the current Enceladus-Dione resonance with the tidal hypothesis of origin under the assumption that the resonances are encountered in the order Eq. (45). However, from Eqs. (16) and (7) with $d\tilde\omega_1/dt = \partial H/\partial W_1 = \dot{\tilde\omega}_{s1} - (C'/e_1)\cos\theta_1$, a small e_1 induces a large negative contribution to $\dot{\tilde\omega}_1$ when θ_1 is small. For the value of $e_1 \approx 7 \times 10^{-5}$ when libration was established consistent with conserved action, $\dot{\tilde\omega}_{1\,\text{res}} \approx -4°5/\text{day}$ so the resonance in which we find Enceladus-Dione would be the first one encountered instead of last as given by Eq. (45) (Sinclair 1974; Peale 1976b).

The decrease in $n_1^* - 2n_2^*$ within the resonance means each of the other resonances in the 2:1 set were encountered (resonance angles passed through zero frequency) while the system was librating in the existing resonance. The justification of the approximation of keeping a single term in R by other terms being high frequency or having small coefficients is not necessarily valid. However, Sinclair (1983) has pointed out the independence of some of the resonances clustered around the 2:1 commensurability in the sense that each can evolve without significantly affecting the states of any of the others. This follows from the fact that the orbital parameters affected by a given resonance are often essentially unique to that resonance. Such independence has been demonstrated numerically by Wisdom (personal communication, 1983) in a study of the two first-order eccentricity resonances. Second-order resonances involving the sum of the two node longitudes or the sum of the two longitudes of pericenters do affect the same orbital parameters as other first- and second-order 2:1 resonances, so one expects some coupling here. Wisdom's (1983) calculations indicate a weak effect on the evolutions of two overlapping resonances in this case, but the resonances essentially still evolve independently for Enceladus-Dione. The current occupancy of only the first-order eccentricity resonance after apparently having passed through the other 2:1 possibilities thus depends only on the nonzero probability of escape from each of these other resonances, a condition found to be satisfied by Sinclair (1974). The selection of a single term in the disturbing function for the study of a particular resonance is valid even when the frequency is not separated from that of another nearby resonance, provided that the only common angle variables in the two descriptions of the resonances are the mean longitudes.

Another qualification on the use of the simple model is that the eccentricity not be too large or the satellite orbits not be too close together. This is illustrated by the application of the model to the Titan-Hyperion resonance where $3\lambda_1 - 4\lambda_2 + \tilde\omega_2$ librates about π with an amplitude of 36° and with $C = 3.26$ in Eq. (16). We use the mean eccentricity of Hyperion's orbit of 0.104

to determine $R = 779e_{\frac{3}{2}}^2 = 8.43$ or $x = \sqrt{2R} = 4.11$ as the location of the stationary point (Eqs. 7, 21, 22, 23). From Eq. (30), $\delta = 4.458$ or $\Delta = 3(\delta + 1) = 16.37$. From Eqs. (24) and (26), the zero of $d\phi/d\tau$ yields the value of $R = R_m = 8.383$ at the extreme of the libration at $\phi_m = 36°$. Substitution of R_m into Eq. (24) yields the Hamiltonian $K = -73.580$. Equation (29) can now be solved for the extreme values of x with $y = 0$, and from Eqs. (27), (23) and (7), the extremes $0.096 \lesssim e \lesssim 0.111$ with the stationary point at $e = 0.104$. This is almost twice the amplitude of 0.004 found for the fluctuation in e during the 1.75 yr libration of the resonance variable by Woltjer (1928; see also Taylor 1984). [Most of the observed variation in Hyperion's eccentricity $(0.08 \lesssim e \lesssim 0.13)$ results from Titan's large eccentricity and the relative motion of the pericenters.]

Much of the discrepancy can be traced to the expansion in Laplace coefficients and the neglect of higher harmonics of the resonance variable. Although the factor e^n appears in the coefficient of the harmonic $\cos[n(3\lambda_1 - 4\lambda_2 + \tilde{\omega}_2)]$, the factors $C(a_1/a_2)$ grow so rapidly with n that the coefficients are comparable in magnitude to that of the first harmonic. This slow convergence of the expansion is also evident in the determination of the resonance driving of $\tilde{\omega}_2$'s retrograde motion. Henrard (1982b) includes terms in the disturbing potential up to the sixth harmonic of the resonance variable and finds $0.103 < e < 0.115$, which is not a sufficient reduction of the discrepancy. The simple theory where a single resonance term is picked from the expansion of the disturbing function in Laplace coefficients is clearly inadequate to describe the Titan-Hyperion resonance, although its application has been dominant in the literature (see, e.g., Peale 1976b).

It is noteworthy that the very small amplitude of libration for the current Enceladus-Dione resonance is not necessarily an indication of a small free eccentricity at transition—at least not as small as we indicated above. Dissipation of energy in the variation of tides raised on Enceladus by Saturn necessarily reduces the free eccentricity, and, as we shall see shortly, limits the growth of the forced eccentricity (Sec. III). Tidal energy is dissipated in a synchronously rotating satellite at the rate (Peale and Cassen 1978)

$$\frac{dE}{dt} = -\frac{21}{2} m_0 n^3 a^2 \left(\frac{R_s}{a}\right)^5 \frac{k_2^s}{Q_s} e^2 \tag{46}$$

where k_2^s, Q_s are the potential Love number and dissipation function for the satellite. The eccentricity is the source of the variation of the amplitude and orientation of the tidal bulge on the satellite leading to the dissipation (see Burns' chapter). The satellite cannot reduce its spin rate in synchronous lock so the energy dissipated must come from the orbital energy $-Gm_0m/2a$, and a must decrease. But orbital angular momentum $m\sqrt{Gm_0a(1 - e^2)}$ is conserved (no transfer from spin), so $dE/dt = (Gm_1m_0/2a^2)da/dt$ leads to

$$\frac{de}{dt} = -\frac{21}{2}\frac{m_0}{m}\left(\frac{R_s}{a}\right)^5 n\frac{k_2^s}{Q_s}e. \tag{47}$$

The amplitude of libration vanishes with the free eccentricity, so satellite dissipation could have been the dominant cause of the small libration amplitude. If this is true, the action is no longer a conserved quantity and the area of the trajectory in the xy plane increases as we go back in time. There is also a retardation of the rate of change of $n_1^* - 2n_2^*$ due to satellite dissipation (see Sec. III below). However, if we ignore this for the time being, the larger area enclosed by the circular trajectory in the past means it will intersect the origin a shorter time ago, and the resonance age would be less than the 1.7 Gyr determined above. If satellite dissipation were large enough, even the order in which the resonances in Eq. (45) were encountered could be altered, since a larger free e_1 in the past means $\dot{\omega}_1$ was less negative at encounter.

It is also possible that the departure of the Enceladus-Dione pair from the history inferred from the simple tidal model is much more drastic than the modification implied by the satellite dissipation alone. This follows from the fact that Enceladus is a small, icy satellite and tidal dissipation for the current eccentricity is limited (Yoder 1981a). Yet parts of this satellite's surface are almost crater-free and are therefore geologically young (Smith et al. 1982; chapter by Morrison et al.). This requires considerable internal heating to provide the necessary activity for smoothing the surface. Since the current forced eccentricity of 0.0044 is too small for significant tidal heating of a solid Enceladus, and since tidal heating appears to be the only viable means for providing the energy (see the chapter by Schubert et al.), Enceladus' eccentricity must have been considerably higher in the past. A proposed earlier orbital resonance with Tethys has difficulty in generating sufficient eccentricity (Yoder 1981a), whereas a fairly recent 2:1 resonance with Janus (one of the coorbitals), although capable in principle of driving the eccentricity to sufficiently large values, is precariously weak because of Janus' small mass.

The unique thing about the latter hypothesis (Lissauer et al. 1984) is that Janus is driven out by torques from density waves generated by Janus in Saturn's A ring at the 7:6, 6:5, etc. orbital resonance positions. The rate of angular momentum transfer is so large that, were Janus locked into the 2:1 resonance with Enceladus, the system would be driven deep into the resonance pushing Enceladus' eccentricity to a large value determined by the dissipative properties in Enceladus (see Sec. III). Enceladus would have approached the 2:1 resonance with Dione far more rapidly than inferred earlier from tides raised on Saturn. Accordingly, the value of $n_1 - 2n_2$ would have been much smaller and the eccentricity much larger than those compatible with the simple 2:1 resonance existing today between Enceladus and Dione. The interaction with Dione would most likely have destroyed the 2:1 resonance between Enceladus and Janus because of this incompatibility and Dione's far

dominant mass. If this happened, Enceladus would have entered the 2:1 resonance with Dione with an eccentricity much too large to be sustained by tides
raised on Saturn in the face of the high rate of dissipation in Enceladus tending to reduce it. The eccentricity would have damped down to the current
value in only 17 Myr if the breakup of the 2:1 resonance with Janus coincided
roughly with the establishment of the 2:1 resonance with Dione. This time is
determined by the current separation of Janus from the 2:1 resonance position and the rate at which the ring torques are believed to be transferring angular momentum. This 17 Myr age of the Enceladus-Dione resonance, and
the evolution from high eccentricity to low is a drastically different history
from the 2.7 Gyr evolution inferred from the simple model. Perhaps a more
astounding result which is independent of the possible past existence of a
Janus-Enceladus resonance is that the ring torques on Janus and other close
ring moons are so large that the rings themselves are in a state of rapid
change. The rings have insufficient angular momentum to transfer at the inferred rate for very long and may therefore be less than 200 Myr old (Lissauer
et al. 1984).

Although the simple model with a single term in the disturbing potential
forms a basis for the discussion of the origin and evolution of the orbital resonances among the major satellites of Saturn, circumstances require much
more elaborate models to adequately describe the Titan-Hyperion and the
Enceladus-Dione resonances. The former requires at least the inclusion of
higher harmonics of the resonance variable to describe the librations or an
expression not involving Laplace coefficients, whereas the latter almost certainly requires inclusion of dissipation within Enceladus which in turn may
imply a totally different history from that deduced from the simple model.
The Mimas-Tethys 2:1 resonance is reasonably well described by the simple
model, although Tethys' large mass, which results in a rate of expansion of
its orbit near that of Mimas, means the approach to resonance and evolution within the resonance is relatively slow. In this case, the reduction of
the Hamiltonian to a form containing a single variable parameter is less
appropriate.

The profound effects of dissipation within Enceladus on the dynamical
history of the satellite systems of which it is or may have been a part are still
speculative. In the Jupiter system, on the other hand, the dissipation of tidal
energy within the satellite Io (Peale et al. 1979) has made this body the most
thermally active solid object in the solar system (Smith et al. 1979a; chapters
by Schubert et al. and Nash et al.). We can use the simple resonance model
again as a basis for discussing the resonances among the Galilean satellites,
but now dissipation in the satellites must be included explicitly. The inclusion
of this dissipation allowed an understanding of the origin and evolution of the
three-body system of resonances among the Galilean satellites (Yoder 1979b),
which had eluded scientific minds for 300 yr. It is this history of the Galilean
satellites which we describe in the next section.

III. THE GALILEAN SATELLITE SYSTEM

The possible importance of dissipation of tidal energy in the satellites to the dynamical evolution of the Galilean satellite system is supported by the effect of that dissipation on the thermal evolution of Io, the innermost member. The value of the forced eccentricity can be found within the framework of the theory developed in Sec. II, even though more than a single term must be kept in the disturbing function for the general analysis we develop below. The effect of the libration of $\lambda_1 - 2\lambda_2 + \bar{\omega}_2$ is to force Europa's eccentricity but not Io's, since conjunctions of Io and Europa occur near Europa's apocenter and e_2 is forced to that value to keep $\bar{\omega}_2$ in step with the conjunctions (see, e.g., Peale et al. 1979). The libration of $\lambda_2 - 2\lambda_3 + \bar{\omega}_2$ also forces Europa's eccentricity, and this additional forced eccentricity for Europa leads to a decrease in $n_1 - 2n_2$ to maintain $\lambda_1 - 2\lambda_2 - \bar{\omega}_2$ in libration. The continued libration of $\lambda_1 - 2\lambda_2 - \bar{\omega}_1$ forces e_1 to a larger value, but its magnitude is determined completely by the value of the two-body parameter $n_1 - 2n_2$; that is, only the resonance term with argument $\lambda_1 - 2\lambda_2 + \bar{\omega}_1$ is used for determining the forced value of e_1 and we can use the theory of Sec. II above.

Table II (Yoder and Peale 1981) lists the important parameters for the Galilean satellites. In addition, $n_1 - 2n_2 = n_2 - 2n_3 = 0°739/$day from which we obtain from Eq. (22) for the libration of $\lambda_1 - 2\lambda_2 + \bar{\omega}_1$

$$\alpha = 0°900/\text{day} = 1.82 \times 10^{-7} \text{ rad s}^{-1}$$
$$\beta = 3.716 \times 10^{-47} \text{ g}^{-1} \text{ cm}^{-2}$$
$$\varepsilon = -1.989 \times 10^{12} \text{ g}^{1/2} \text{ cm s}^{-3/2}. \tag{48a}$$

From Eq. (25)

$$\delta = -2.823 \tag{48b}$$

and the positive real root of Eq. (30) is $x = 0.358$, which yields the eccentricity corresponding to the stable stationary point of $e_1 = 0.0043$ from Eqs. (7), (23) and (27). This is the mean eccentricity which must be maintained for Io's orbit given the observed value of α in Eq. (48). Substitution of this value of e_1 into Eq. (46) yields a rate of tidal heating of Io of (Peale et al. 1979)

$$\frac{dE_1}{dt} = \frac{1.9 \times 10^{21}}{Q_1} \text{ erg s}^{-1} \tag{49}$$

where a solid, homogeneous body with Love number $k_2 \approx 3 \, \rho g R / 19\mu$ is assumed with rigidity $\mu = 6.5 \times 10^{11}$ dyne cm^{-2} (that of the outer layers of the Moon [Nakamura et al. 1976]), and the remaining parameters are from Table

TABLE II

Orbital Parameters for the Galilean Satellites[a]

	Io	Europa	Ganymede	Callisto
$M/M_J \times 10^5$	4.684 ± 0.022	2.523 ± 0.025	7.803 ± 0.030	5.661 ± 0.019
$n(°/\text{day})$	203.4890	101.1747	50.3176	21.5711
$\dot{\tilde{\omega}}_s(°/\text{day})$	0.161	0.048	0.007	0.002
$\dot{\Omega}_s(°/\text{day})$	-0.134	-0.033	0.007	-0.002
$a(\text{km})$	422,000	671,400	1,071,000	1,884,000
e_{forced} (2:1)	0.0041	0.0101	0.0006	
e_{free}	$(1 \pm 2) \times 10^{-5}$	$(9.2 \pm 1.9) \times 10^{-5}$	0.0015	0.0073
$\sin I_{\text{free}}$	$(7.0 \pm 1.9) \times 10^{-4}$	0.0082	0.0034	0.0049
$R(\text{km})$	1816 ± 5	1569 ± 10	2631 ± 10	2400 ± 10
$\rho(\text{g cm}^{-3})$	3.53	3.03	1.93	1.79

[a]Table after Yoder and Peale (1981).

II. Io's tidal heating is also considered in the chapters by Schubert et al. and Nash et al.

The value of Q_1 for rocks on Earth is typically near 100 (Knopoff 1964), and the value for Mars for tides raised by Phobos is comparable (Shor 1975; Smith and Born 1976; cf. Burns' chapter). Substitution of $Q_1 = 100$ in Eq. (49) yields a rate of dissipation in Io which is three times that in the Moon from radiogenic sources and a rate per unit volume in the center about 10 times the lunar average. As the Moon appears to be molten or nearly molten in the center (Nakamura et al. 1976; Ferrari et al. 1980; Yoder 1981*b*; Stevenson and Yoder 1982), it is likely that the center of Io was melted by tidal dissipation. Once melted in the center, a less rigid Io suffers a greater amplitude of tidal flexing leading to a higher dissipation per unit volume in the surrounding shell. The total increase in dissipation exceeds the loss due to the reduction in the volume of solid material in which the dissipation is occurring. Peale et al. (1979) demonstrated that solid-state convection would be unable to rid the satellite of this heat as fast as it was generated by the tides and therefore could not prevent a thermal runaway from an initial state of melting in the center to a state where only a relatively thin shell of solid material remained near the surface. (Although it is likely that the inner core would solidify eventually [Schubert et al. 1981 and chapter herein], the high rate of dissipation in the shell persists as long as it is decoupled from the solid core by a liquid layer [Cassen et al. 1982].) That Io had indeed been heated extensively by tides was dramatically verified by the observation of extensive active volcanism in Voyager 1 images (Smith et al. 1979*a*).

Although the stability of the Galilean resonances, including the libration of $\lambda_1 - 3\lambda_2 + 2\lambda_3$ about 180°, was understood at the time of Laplace (this latter resonance bears his name), the almost immeasurably small amplitude of libration of the Laplace angle in particular defied explanation in the framework of the assembly of the resonances by differential tidal expansion of the orbits (see, e.g., Sinclair 1975). In a major dynamical feat, Yoder (1979*b*) showed that inclusion of the dissipation of tidal energy in Io made almost all the observational constraints on the system consistent with an origin and evolution of the resonances from tides raised on Jupiter. (This analysis was later elaborated and extended [see Yoder and Peale 1981].) The dissipation of tidal energy in Io is completely dominant in rapidly reducing the amplitudes of libration to the small values currently observed, and one need no longer infer special damping conditions at the time of origin of the satellite system. An added bonus is the establishment of rather tight constraints on the rate of tidal dissipation in both Io and Jupiter. The upper bound on the Q of Jupiter has led to new dynamical processes inferred for the interiors of the giant planets (Stevenson 1983).

We can develop a formulation in canonical variables for the interaction of the four satellites just as we did for two (see, e.g., Yoder and Peale 1981), but it is more expedient to change variables to a noncanonical set of orbital elements a, e, I, $\bar{\omega}$, Ω, λ. If α_i and β_i are conjugate coordinates and momenta used in Sec. II for a single satellite and the new variables are represented by γ_j, we have for Hamiltonian H

$$\frac{d\gamma_j}{dt} = \sum_{k=1}^{6} [\gamma_j, \gamma_k]_{\alpha, \beta} \frac{\partial H}{\partial \gamma_k} \tag{50}$$

where

$$[\gamma_j, \gamma_k]_{\alpha, \beta} = \sum_{i=1}^{3} \left(\frac{\partial \gamma_j}{\partial \alpha_i} \frac{\partial \gamma_k}{\partial \beta_i} - \frac{\partial \gamma_j}{\partial \beta_i} \frac{\partial \gamma_k}{\partial \alpha_i} \right) \tag{51}$$

are the Poisson brackets. There follows (see, e.g., Plummer 1918, p. 142):

$$\frac{da}{dt} = -\frac{2}{m} \sqrt{\frac{a}{\mu}} \frac{\partial H}{\partial \lambda}$$

$$\frac{de}{dt} = \frac{\sqrt{1-e^2}}{me\sqrt{\mu a}} \frac{\partial H}{\partial \bar{\omega}} - \left(\frac{(1-e^2) - \sqrt{1-e^2}}{me\sqrt{\mu a}} \right) \frac{\partial H}{\partial \lambda}$$

$$\frac{dI}{dt} = \frac{\tan \frac{1}{2} I}{m\sqrt{\mu a(1-e^2)}} \left(\frac{\partial H}{\partial \lambda} + \frac{\partial H}{\partial \bar{\omega}} \right) + \frac{1}{m \sin I \sqrt{\mu a(1-e^2)}} \frac{\partial H}{\partial \Omega}$$

$$\frac{d\lambda}{dt} = \frac{2}{m} \sqrt{\frac{a}{\mu}} \frac{\partial H}{\partial a} + \frac{(1-e^2) - \sqrt{1-e^2}}{me\sqrt{\mu a}} \frac{\partial H}{\partial e} - \frac{\tan \frac{1}{2} I}{m\sqrt{\mu a(1-e^2)}} \frac{\partial H}{\partial I}$$

$$\frac{d\omega}{dt} = \frac{-\sqrt{1-e^2}}{me\sqrt{\mu a}} \frac{\partial H}{\partial e} - \frac{\tan \frac{1}{2} I}{m\sqrt{\mu a(1-e^2)}} \frac{\partial H}{\partial I}$$

$$\frac{d\Omega}{dt} = \frac{-1}{m \sin I \sqrt{\mu a(1-e^2)}} \frac{\partial H}{\partial I} \tag{52}$$

where

$$H = -\sum_{i=1}^{4} \frac{Gm_0 m_i}{2a_i} + \Phi \tag{53}$$

with

$$\Phi = -\sum_{\substack{i,j=1 \\ i<j}}^{4} \frac{Gm_i m_j}{r_{ij}} - \sum_{k=2}^{4} Gm_0 m_k \left[\frac{1}{r_{0k}} - \frac{1}{r_k} \right] \tag{54}$$

following from the generalization of Eq. (6) from two to four satellites. Appropriate subscripts may be placed on the variables in Eq. (52) for each satellite. Equation (54) is expanded to lowest order in m_i/m_0 and subsequently to a series like Eq. (12), from which the important terms may be selected. To the disturbing potential Φ must be added the contribution of the nonspherically symmetric parts of Jupiter's gravitational field leading to conservative secular motions. The effects of tidal dissipation in both Jupiter and in the satellites must also be included. Perturbations by the Sun are small and have been omitted although, as we shall point out later, they may have an important influence on the interpretation of the libration amplitude of the Laplace angle.

The selection of the low-frequency (resonant) terms in the expanded version of Φ eliminates Callisto completely except for its contribution to the conservative secular motions. Next, the current values of $n_1 - 2n_2 = n_2 - 2n_3$ $= 0^\circ.739/\text{day}$ coupled with the secular motions, $\dot{\Omega}_{si} \geq -0^\circ.134/\text{day}$ (the subscript s indicates secular) means all the inclination resonances associated with the $2:1$ commensurability (see Eq. 45) have not been encountered and are sufficiently far away to be considered high frequency. Further simplification is obtained by keeping only those resonant terms in Φ which are first order in the eccentricity, which is reasonable since these terms are also multiplied by one factor of m_i/m_0 and the eccentricities are relatively small. This last simplification eliminates the terms with argument $\lambda_1 - 4\lambda_3 + 3\tilde{\omega}_1$, since the coefficient has a factor e_1^3. The infinite number of terms in Φ has now been reduced to only those terms involving the arguments $\lambda_1 - 2\lambda_2 + \tilde{\omega}_1$, $\lambda_1 - 2\lambda_2 + \tilde{\omega}_2$, $\lambda_2 - 2\lambda_3 + \tilde{\omega}_2$, $\lambda_2 - 2\lambda_3 + \tilde{\omega}_3$ and the secular terms.

The general assumption is that the orbits of the Galilean satellites were originally in a nonresonant configuration with Io being pushed away from Jupiter most rapidly such that it approaches the $2:1$ resonance with Europa. Those terms in Φ involving λ_3 are thus initially high frequency and are not included here. Although two resonant terms are important in Φ, the librations $\lambda_1 - 2\lambda_2 + \tilde{\omega}_1$ and $\lambda_1 - 2\lambda_2 + \tilde{\omega}_2$ are semi-independent, since most of the variation of each resonance variable is in the respective $\tilde{\omega}_i$, and we may consider the capture of each variable into resonance whether or not the other variable is librating at the time. (See the discussion of the Enceladus-Dione resonance in Sec. II.) The fact that the δ's for both resonance variables are now negative (Eq. 48b) and were more negative in the past, shows that both variables were captured into their respective librating states with certainty. Hence, we can consider the subsequent evolution of the Io-Europa system with both resonance variables librating.

From Eqs. (52) and (53) with $n_i \equiv \sqrt{\mu_i/a_i^3}$

$$\frac{\mathrm{d}n_i}{\mathrm{d}t} = \frac{3}{m_i a_i^2} \frac{\partial \Phi}{\partial \lambda_i}$$

$$\frac{de_i}{dt} = \frac{1}{m_i e_i \sqrt{\mu_i a_i}} \frac{\partial \Phi}{\partial \tilde{\omega}_i}$$

$$\frac{d\tilde{\omega}_i}{dt} = \frac{-1}{m_i e_i \sqrt{\mu_i a_i}} \frac{\partial \Phi}{\partial e} \tag{55}$$

where higher-order terms in the equations of variation are omitted, consistent with the order of the terms kept in Φ, $\partial\Phi/\partial\Omega = \partial\Phi/\partial I = 0$ and the equations for λ and I are not needed. With

$$\Phi = \frac{-Gm_1 m_2}{a_2} \{C_1 e_1 \cos(\lambda_1 - 2\lambda_2 + \tilde{\omega}_1) + C_2 e_2 \cos(\lambda_1 - 2\lambda_2 + \tilde{\omega}_2)\} \tag{56}$$

for the resonance part with $C_1 = -1.19$ and $C_2 = 0.428$ for $a_1/a_2 = 0.63$ (see, e.g., Brouwer and Clemence 1961, p. 490), we need but add the effects of tidal dissipation and the secular motions.

The disturbing potential from which the tidal effects are found is given by (Kaula 1964; see Burns' chapter)

$$\Phi_T(m, b) = \frac{-k_2 Gm^2 R^5}{a^{*3} a^3} \sum_{m'=0}^{2} \frac{(2-m')!}{(2+m')!} (2 - \delta_{0m'})$$

$$\times \sum_{p=0}^{\ell} \sum_{q=-\infty}^{\infty} [F_{2m'p}(I)][F_{2m'p}(I^*)][G_{2pq}(e)][G_{2pq}(e^*)]$$

$$\times \cos(v^*_{2m'pq} - \epsilon_{2m'pq} - v_{2m'pq}) \tag{57}$$

where m is the tide-raising body, k_2 and R are the Love number and radius, respectively, of the body on which the tide is raised and

$$v_{2m'pq} = (2 - 2p + q)\lambda - q\tilde{\omega} - (2 - 2p - m')\Omega - m'\psi \tag{58}$$

with ψ being an angle defining the rotation of the tidally distorted body. The starred variables refer to m as the body raising the tide and the unstarred variables refer to m as the body reacting to the tidal distribution of mass. Generally, secular tidal effects arise only when disturbed and disturbing bodies are the same, so starred and unstarred coordinates are equated after the derivatives relative to the unstarred variables are taken. There is a secular contribution to the perturbations of one satellite by the tide raised on m_0 by another if the orbital periods are commensurate, but these are small and will be neglected. The angle ϵ_{2mpq} is the phase lag of the response of the tidally distorted body relative to the phase of the tide-generating potential and is numerically equal to $1/Q$. The $F_{2mp}(I)$ are closed functions of $\cos I$ and $G_{2pq}(e)$ are infinite series, which are both tabulated in Kaula (1964) and in Table I of

Burns' chapter. For our purposes here we can set $I = 0$ in which case $F_{201}(0)$ $= -1/2$, $F_{220}(0) = 3$ and all remaining F's are zero. The $d\Phi_T/d\lambda$ term must be retained in the de/dt equation since those terms in Eq. (57) with $q = 0$ are not factored by e.

Substitution of Eqs. (56) and (57) into Eq. (55) yields

$$\frac{dn_1}{dt} = 3n_1^2 \frac{m_2}{m_0} \frac{a_2^2}{a_1^2} (C_1 e_1 \sin \theta_1 + C_2 e_2 \sin \theta_2)$$

$$- c_1 n_1^2 [1 - (7D_1 - 12.75)e_1^2]$$

$$\frac{dn_2}{dt} = -6 \frac{m_1}{m_0} n_2^2 (C_1 e_1 \sin \theta_1 + C_2 e_2 \sin \theta_2)$$

$$- c_2 n_2^2 [1 - (7D_2 - 12.75)e_2^2]$$

$$\frac{de_1}{dt} = \frac{m_2}{m_0} n_1 \frac{a_1}{a_2} C_1 \sin \theta_1 - \frac{c_1 n_1}{3} (7D_1 - 4.75)e_1$$

$$\frac{de_2}{dt} = \frac{m_1}{m_0} n_2 C_2 \sin \theta_2 - \frac{c_2 n_2}{3} (7D_2 - 4.75)e_2$$

$$\frac{d\bar{\omega}_1}{dt} = n_1 \frac{m_2}{m_0} \frac{a_1}{a_2} \frac{C_1}{e_1} \cos \theta_1 + \dot{\bar{\omega}}_{1s}$$

$$\frac{d\bar{\omega}_2}{dt} = n_2 \frac{m_1}{m_0} \frac{C_2}{e_2} \cos \theta_2 + \dot{\bar{\omega}}_{2s} \tag{59}$$

where $\theta_1 = \lambda_1 - 2\lambda_2 + \bar{\omega}_1$ and $\theta_2 = \lambda_1 - 2\lambda_2 + \bar{\omega}_2$ and

$$c_i = \frac{9}{2} \frac{k_2^J}{Q_J} \left(\frac{R_J}{a_i}\right)^5 \frac{m_i}{m_0}$$

$$D_i = \frac{k_2^i}{k_2^J} \left(\frac{m_0}{m_i}\right)^2 \left(\frac{R_i}{R_J}\right)^5 f \frac{Q_J}{Q_i} . \tag{60}$$

The Love numbers $k_2^i \approx 3\rho_i g_i R_i / 19\mu_i$ for the satellites are those for a homogeneous, incompressible sphere, and f is an enhancement factor to account for added dissipation if a satellite interior is partially molten. Equations (59) represent the variations with dissipative effects included, both in Jupiter and in the satellites. The D_i are measures of the ratio of the dissipation within the satellites to that in Jupiter. The satellite dissipation tends to increase n_i and decrease e_i. The numerical constant in the coefficient of e_i corresponds to the lowest-order term in eccentricity for dissipation in Jupiter, and it leads to an increase in e_i and a decrease in n_i. The $c_i n_i^2$ are just the rates of change of n_i from tidal torques from m_0 for m_i in a circular orbit.

We can simplify the dissipative terms by noting that $k_2^{(1)} \approx 0.036$ for μ

$= 5 \times 10^{11}$ dyne cm^{-2} and that $k_2^f = 0.38$ (Gavrilov and Zharkov 1977) from which

$$D_1 > 0.478 \frac{Q_J}{Q_1} \tag{61}$$

since $f > 1$. But $Q_J > 6 \times 10^4$ from the proximity of Io to Jupiter after 4.6 Gyr (Goldreich and Soter 1966), so with $Q_1 = 0(100)$ by analogy with terrestrial rocks, $D_1 > 300$ (we shall argue later that $D_1 \approx 4000$), and $7D_1 \gg$ the numerical constant in the coefficients of e_1. This means the effect of dissipation in Jupiter in changing e_1 and in the e_1 effect on n_1 is small compared to the effect of dissipation in the satellite and may be neglected. Next $c_2 n_2^2 \approx 0.026 \ c_1 n_1^2$ and we also make little error in neglecting the dissipation in Europa altogether.

Equations (59) are most easily solved by changing variables first to $h_i = e_i \sin \tilde{\omega}_i$, $k_i = e_i \cos \tilde{\omega}_i$ to eliminate the singularity when $e_i = 0$ and finally to $p_i = k_i - \iota h_i$, $q_j = k_i + \iota h_i$, where $\iota = \sqrt{-1}$. The dissipative terms are written in terms of the new variables as follows:

$$\left. \frac{dh_1}{dt} \right|_{diss} = \frac{de_1}{dt} \sin \tilde{\omega}_1 + e_1 \cos \tilde{\omega}_1 \frac{d\tilde{\omega}_1}{dt}$$

$$= -\frac{7}{3} c_1 n_1 D_1 e_1 \sin \tilde{\omega}_1 + \dot{\tilde{\omega}}_{s1} e_1 \cos \tilde{\omega}_1 \tag{62}$$

where only the secular $\dot{\tilde{\omega}}_{s1}$ enters here since the resonance-controlled motion is accounted for by terms with argument θ_i. There results

$$\frac{dp_1}{dt} = -\iota n_1 \frac{a_1}{a_2} \frac{m_2}{m_0} C_1 \exp(\iota V_1) - \left[\frac{7}{3} c_1 n_1 D_1 + \iota \dot{\tilde{\omega}}_{s1} \right] p_1$$

$$\frac{dq_1}{dt} = \iota n_1 \frac{a_1}{a_2} \frac{m_2}{m_0} C_1 \exp(-\iota V_1) - \left[\frac{7}{3} c_1 n_1 D_1 - \iota \dot{\tilde{\omega}}_{s1} \right] q_1$$

$$\frac{dp_2}{dt} = -\iota n_2 \frac{m_1}{m_0} C_2 \exp(\iota V_1) - \left[\frac{7}{3} c_2 n_2 D_2 + \iota \dot{\tilde{\omega}}_{s2} \right] p_2$$

$$\frac{dq_2}{dt} = \iota n_2 \frac{m_1}{m_0} C_2 \exp(-\iota V_1) - \left[\frac{7}{3} c_2 n_2 D_2 - \iota \dot{\tilde{\omega}}_{s2} \right] q_2 \tag{63}$$

where $V_1 = \lambda_1 - 2\lambda_2$. The equations for n_1 and n_2 are unchanged except for dropping the small numerical constant in the coefficients of e_i.

The equations are solved by successive approximations justified as follows. Before the system entered the resonances, the free eccentricities of Io and Europa were damped at rates given by Eq. (47) with time constants of

$$\tau_1 = 3.3 \times 10^4 \, Q_1 \text{ yr}$$
$$\tau_2 = 1.4 \times 10^6 \, Q_2 \text{ yr.} \tag{64}$$

With $Q_i = 100$ both eccentricities must have been extremely small by the time the 2:1 commensurability was approached. The automatic transitions into libration as described in Fig. 4 occurred when the δ's were very negative with the trajectory in the phase plane being a tiny circle. There could have been only a negligibly small fluctuation in $n_1^* - 2n_2^*$ even when the libration in θ_i was 90° immediately after libration began since the librations are accommodated by the variations in $\dot{\tilde{\omega}}_i$ (see, e.g., Peale 1976b). As the forced eccentricity grows (the small circular trajectory moving to the right of the origin in Fig. 4), librations could begin to cause fluctuations in $n_1^* - 2n_2^*$, but the librations are reduced to negligible amplitude by this time as can be perceived from Fig. 4 with a tiny circular trajectory about the stationary point on the positive x_1 axis. Hence, throughout the history of the evolving two-body resonance $\dot{V}_1 = n_1^* - 2n_2^*$ ($n_i^* = n_i + \dot{\lambda}_{si}$; Eq. 16) is only slowly varying, and we can use the zero order solution $n_i^* = $ constant in Eqs. (63).

Equations (63) now separate from the n_i equations and, with $\nu_1 = \dot{V}_1 = n_1^* - 2n_2^*$, have the particular solutions

$$p_1 = -n_1 \frac{a_1}{a_2} \frac{m_2}{m_0} \frac{C_1}{(\nu_1 + \dot{\tilde{\omega}}_{s1})} \exp \iota(\nu_1 t + \xi_1)$$

$$p_2 = -n_2 \frac{m_1}{m_0} \frac{C_2}{(\nu_1 + \dot{\tilde{\omega}}_{s2})} \exp \iota(\nu_1 t + \xi_2) \tag{65}$$

where q_i are the complex conjugates of Eq. (65) and

$$\xi_i = \frac{7}{3} \frac{c_i n_i D_i}{\nu_1 + \dot{\tilde{\omega}}_{si}}. \tag{66}$$

We have used $1 + \iota\xi_i \approx \exp \iota\xi_i$ and neglected ξ_i^2 in the denominator. The solution of the homogeneous equation

$$p_1 = p_{10} \exp\left[\frac{-7}{3} c_1 n_1 D_1 - \iota\dot{\tilde{\omega}}_{s1}\right] t \tag{67}$$

is a transient with the very short time constant given by Eq. (64), and it can be ignored. Note that the damping of the free eccentricity represented by Eq. (67) coincides with the damping of the libration amplitude of the resonance variable. The stationary solution in terms of the forced eccentricity and the center of libration follows from Eq. (65), since $p_1 = e_1 \cos \tilde{\omega}_1 - \iota e_1 \sin \tilde{\omega}_1$. We obtain

$$e_{12} = -n_1 \frac{a_1}{a_2} \frac{m_2}{m_0} \frac{C_1}{\nu_1 + \dot{\tilde{\omega}}_{s1}}$$

$$e_{21} = n_2 \frac{m_1}{m_0} \frac{C_2}{\nu_1 + \dot{\tilde{\omega}}_{s2}}$$

$$\tilde{\omega}_1 = -(V_1 + \xi_1)$$

$$\tilde{\omega}_2 = -(V_1 + \xi_2) + \pi \tag{68}$$

where e_{ij} indicates the value of e_i forced by m_j. The effect of satellite dissipation is to cause conjunctions to occur slightly past the pericenter of Io's orbit and past the apocenter of Europa's orbit. That is, $V_1 + \tilde{\omega}_1 = \lambda_1 - 2\lambda_2 + \tilde{\omega}_1 = -\xi_1$ so when $\lambda_1 = \lambda_2$, $\lambda_1 = \tilde{\omega}_1 + \xi_1$. There is an additional phase shift in the same direction from the secular decrease in n_1 causing it to arrive late at the conjunction point. This latter phase shift can be determined from the condition that $(d/dt)(n_1 - 2n_2 + \dot{\tilde{\omega}}_i) = 0$ but the effect is relatively small compared to the satellite dissipation and will be neglected.

Now $e_i \cos \tilde{\omega}_i$ and $e_i \sin \tilde{\omega}_i$ obtained from Eq. (65) are known functions of time which, when substituted into the equations for dn_i/dt yield

$$\frac{dn_1}{dt} = -c_1 n_1^2 (1 - 14D_1 e_{12}^2) + 7 c_2 n_1^2 \frac{a_1}{a_2} \frac{m_2}{m_1} D_2 e_{21}^2$$

$$\frac{dn_2}{dt} = -c_2 n_2^2 (1 + 7D_2 e_{21}^2) - 14 \frac{m_1}{m_2} \frac{a_2}{a_1} n_2^2 c_1 D_1 e_{12}^2. \tag{69}$$

We neglect the dissipation in Europa compared to that in Io and combine these to give

$$\frac{d\nu_1}{dt} = \frac{dn_1}{dt} - 2 \frac{dn_2}{dt} = -c_1 n_1^2 (1 - 34.5D_1 e_{12}^2) + 2c_2 n_2^2. \tag{70}$$

We have used Eqs. (66) and (68) to obtain Eqs. (69) and (70). In Eq. (70) we see that $d\nu_1/dt$ vanishes when $e_{12}^2 = (34.5D_1)^{-1}$, a consequence of the dissipative effects in Io tending to reduce e_1 balancing the effect of tides raised on Jupiter tending to increase e_1 by driving n_1 closer to the exact commensurability. Once $\dot{\nu}_1 = 0$, e_{12} and e_{21} have equilibrium values determined by Eq. (68) which change only on the slow time scale of the orbit expansion. Transfer of angular momentum between m_1 and m_2 is not efficient until n_1 is reasonably close to $2n_2$ and e_{12} is close to its equilibrium value. The time required for e_{12} to approach equilibrium is thus comparable to that for n_1 to approach $2n_2$ from some unknown initial value. The value of $D_1 \approx 4200$ determined below leads to $e_{12} = 0.0026$ and from Eq. (68) $e_{21} = 0.0014$ at equilibrium.

After equilibrium is reached, the system expands with the outward acceleration of Europa maintaining its mean motion at a fixed factor near 0.5 times the mean motion of Io. This stable state is maintained until Europa encounters the 2:1 resonance with Ganymede. Just as dissipation in Io tends to repel Europa by the secular transfer of angular momentum, dissipation in Europa would tend to transfer angular momentum to Ganymede in the 2:1 resonance. However, if Europa were acting alone, the equilibrium eccentricity for the Europa-Ganymede interaction would be three times the current value of 0.01. Long before this can happen, the frequency $\nu_2 = n_2^* - 2n_3^*$ of the outer pair approaches $\nu_1 = n_1^* - 2n_2^*$ of the inner pair. The vanishing of the difference $\nu_1 - \nu_2$ describes the presently observed three-body Laplace relation.

Once the 2:1 resonance with Ganymede is approached, we must add the appropriate resonance terms in Φ for the Europa-Ganymede interaction. Corresponding Io-Ganymede terms are omitted as they are third order in e, so Io and Ganymede interact almost entirely by using Europa as an intermediary.

The disturbing function becomes

$$
\Phi = \frac{Gm_1 m_2}{a_2} [C_1 e_1 \cos(\lambda_1 - 2\lambda_2 + \tilde{\omega}_1) + C_2 e_2 \cos(\lambda_1 - 2\lambda_2 + \tilde{\omega}_2)]
$$

$$
+ \frac{Gm_2 m_3}{a_3} [C_1 e_2 \cos(\lambda_2 - 2\lambda_3 + \tilde{\omega}_2) + C_2 e_3 \cos(\lambda_2 - 2\lambda_3 + \tilde{\omega}_3)]
$$

$$
(71)
$$

and the variations in n_i, e_i, $\tilde{\omega}_i$ follow from Eqs. (55) and (57). The procedure is identical to that used before, and we find

$$
e_2 \cos \tilde{\omega}_2 = - n_2 \frac{m_1}{m_0} \frac{C_2 \cos(V_1 + \xi_2)}{\nu_1 + \dot{\tilde{\omega}}_{s2}} - n_2 \frac{m_3}{m_0} \frac{a_2}{a_3} \frac{C_1 \cos(V_2 + \xi_3)}{\nu_2 + \dot{\tilde{\omega}}_{s2}}
$$

$$
e_2 \sin \tilde{\omega}_2 = n_2 \frac{m_1}{m_0} \frac{C_2 \sin(V_1 + \xi_2)}{\nu_1 + \dot{\tilde{\omega}}_{s2}} + n_2 \frac{m_3}{m_0} \frac{a_2}{a_3} \frac{C_1 \sin(V_2 + \xi_3)}{\nu_2 + \dot{\tilde{\omega}}_{s2}}
$$

$$
e_3 \cos \tilde{\omega}_3 = - n_3 \frac{m_2}{m_0} \frac{C_2 \cos(V_2 + \xi_4)}{\nu_2 + \dot{\tilde{\omega}}_{s3}} + e_{30} \exp[-Kt] \cos \dot{\tilde{\omega}}_{3s} t
$$

$$
e_3 \sin \tilde{\omega}_3 = n_3 \frac{m_2}{m_0} \frac{C_2 \sin(V_2 + \xi_4)}{\nu_2 + \dot{\tilde{\omega}}_{s3}} + e_{30} \exp[-Kt] \sin \dot{\tilde{\omega}}_{3s} t \qquad (72)
$$

where $V_2 = \lambda_2 - 2\lambda_3$, ξ_i is given for $i = 3, 4$ by Eq. (66) with $\nu_2 = n_2 - 2n_3$ replacing ν_1, and e_{30} is an initial free eccentricity. The expressions for $e_1 \cos \tilde{\omega}_1$ and $e_1 \sin \tilde{\omega}$, are still given by the real and imaginary parts of Eq. (65), and we have retained the (not so) transient solution for m_3 with $K = 7/3$ $c_3 n_3 D_3$ as the inverse of the time constant also given by Eq. (47). This time constant is 2.5×10^{-5} $\mu_3 Q_3$ yr which perhaps should have led to a more damped e_{30} if the rigidity μ_3 were that of ice (4.8×10^{10} dyne cm^{-2}). But e_{30} is observed to be about 0.0015 compared to a forced $e_{32} \approx 0.0007$ and $\lambda_2 - 2\lambda_3 + \bar{\omega}_3$ is not librating; that is, the circular trajectory of radius $\propto e_{30}$ in Fig. 4 encloses the origin as the center of the circle is a distance $\propto e_{32} < e_{30}$ away.

If we neglect ξ_i and $c_i n_i^2$ for $i > 1$, substitute Eq. (72) into the new expressions for dn_i/dt and write the forced values of the eccentricities as

$$e_{12} = -n_1 \frac{m_2}{m_0} \frac{a_1}{a_2} \frac{C_1}{v_1 + \dot{\bar{\omega}}_{s1}}$$

$$e_{21} = n_2 \frac{m_1}{m_0} \frac{C_2}{v_1 + \dot{\bar{\omega}}_{s2}}$$

$$e_{23} = -n_2 \frac{m_3}{m_0} \frac{a_2}{a_3} \frac{C_1}{v_2 + \dot{\bar{\omega}}_{s2}} \tag{73}$$

where $e_2 = e_{21} + e_{23}$ is now forced by both Io and Ganymede, we find

$$\frac{dn_1}{dt} = -3n_1^2 n_2 \frac{m_2 m_3}{m_0^2} \frac{a_1}{a_3} \frac{C_1 C_2}{v_2 + \dot{\bar{\omega}}_{s2}} \sin(V_1 - V_2) - c_1 n_1^2 (1 - 14 D_1 e_1^2),$$

$$\frac{dn_2}{dt} = n_2^3 \frac{m_1 m_3}{m_2^0} \frac{a_2}{a_3} C_1 C_2 \left(\frac{6}{v_2 + \dot{\bar{\omega}}_{s2}} + \frac{3}{v_1 + \dot{\bar{\omega}}_{s2}} \right) \sin(V_1 - V_2)$$

$$- 14 C_1 n_2^2 \frac{m_1}{m_2} \frac{a_2}{a_1} D_1 e_{12}^2,$$

$$\frac{dn_3}{dt} = -6n_3^2 \frac{m_1 m_2}{m_0^2} \frac{C_1 C_2 n_2}{v_2 + \dot{\bar{\omega}}_{s2}} \sin(V_1 - V_2). \tag{74}$$

With $\varphi = \lambda_1 - 3\lambda_2 + 2\lambda_3 = V_1 - V_2$, we can write from Eq. (74)

$$\frac{d^2 \varphi}{dt^2} + A n_2^2 \sin \varphi = -c_1 n_1^2 (1 - 44.8 D_1 e_{12}^2) \tag{75}$$

$$\frac{dv_1}{dt} + (A_1 - 2A_2) n_2^2 \sin \varphi = -c_1 n_1^2 (1 - 34.5 D_1 e_{12}^2) \tag{76}$$

where

$$A = A_1 - 3A_2 + 2A_3$$

$$A_1 = 3 C_1 C_2 \frac{a_2}{a_1} \frac{m_2 m_3}{m_0^2} \frac{n_2}{v_2 + \dot{\bar{\omega}}_{s2}}$$

$$A_2 = -3C_1C_2 \frac{a_1}{a_2} \frac{m_1 m_3}{m_0^2} \left(\frac{2n_2}{\nu_2 + \dot{\omega}_{s2}} + \frac{n_2}{\nu_1 + \dot{\omega}_{s2}} \right)$$

$$A_3 = 6C_1C_2 \left(\frac{a_2}{a_3} \right)^2 \frac{m_1 m_2}{m_0^2} \frac{n_2}{\nu_1 + \dot{\omega}_{s2}}. \tag{77}$$

The left-hand side of Eq. (75) is that obtained by Laplace and reveals the pendulumlike stability of the libration of φ. The right-hand side of Eq. (75) includes the effects of dissipative tides raised on Jupiter by Io and of tidal dissipation in Io only. Inclusion of dissipation in the remaining satellites has only a small effect on the outcome.

The encounter of Europa with the 2:1 commensurability with Ganymede has introduced fluctuations into ν_1 as indicated by Eq. (76). It is these fluctuations which hold the key to the capture of φ into libration. We can eliminate $\sin \varphi$ between Eqs. (75) and (76) to yield

$$\frac{d\nu_1}{dt} = \frac{A_1 - 2A_2}{A} \frac{d\dot{\varphi}}{dt} - 0.32 \, c_1 n_1^2 (1 - 12.6 D_1 e_{12}^2). \tag{78}$$

Integration of Eq. (78) gives

$$\nu_1 = 0.68 \, \dot{\varphi} + \nu_{10} \tag{79}$$

where $\langle \nu_1 \rangle = 0.68 \langle \dot{\varphi} \rangle + \nu_{10}$ is slowly varying.

At resonance $\langle \dot{\varphi} \rangle = 0$ and $\dot{\varphi}$ represents fluctuations about zero with $\langle \nu_1 \rangle = \nu_{10}$. For the approach to resonance we can define $\delta\dot{\varphi}$ as the periodic part of $\dot{\varphi}$ which is a fluctuation about some mean value. The fluctuating part of ν_1 is thus

$$\delta\nu_1 = 0.68\delta\dot{\varphi}. \tag{80}$$

Equation (75) contains ν_1 and ν_2 through A and through e_{12}. It is the fluctuations in A and e_{12} through ν_1 and ν_2 which define the capture scheme. The fluctuations in ν_2 follow from the definition of $\dot{\varphi} = \nu_1 - \nu_2$

$$\delta\nu_2 = \delta\nu_1 - \delta\dot{\varphi} = -0.32\delta\dot{\varphi}. \tag{81}$$

Now $\delta A/A \approx 0.04 \, \delta\dot{\varphi}/(\nu_2 + \dot{\omega}_{s2})$ whereas $\delta(e_{12}^2)/e_{12}^2 = -1.36 \, \delta\dot{\varphi}/(\nu_1 + \dot{\omega}_{s1})$, so we can neglect the former and write

$$e_{12}^2 = \langle e_{12}^2 \rangle \left[1 - \frac{1.36\delta\dot{\varphi}}{\nu_1 + \dot{\omega}_{s1}} \right] \tag{82}$$

and Eq. (75) becomes

$$\ddot{\varphi} + An_2^2 \sin \varphi = -c_1 n_1^2 (1 - 44.8 D_1 e_{12}^2) - \sigma \delta\dot{\varphi} \tag{83}$$

where

$$\sigma = 60.9 \frac{c_1 n_1^2 D_1 e_{12}^2}{\nu_1 + \dot{\omega}_{s1}} \tag{84}$$

and where e_{12}^2 is now the average value with the fluctuations explicitly displayed by $\delta\dot{\varphi}$.

Initially $\dot{\varphi} = \nu_1 - \nu_2 < 0$ since $\nu_2 = n_2 - 2n_3$ must be large. So $\langle\dot{\varphi}\rangle = 0$ can only be approached if $-c_1 n_1^2 (1 - 44.8 D_1 e_{12}^2) > 0$ or $e_{12} > (1/44.8 D_1)^{1/2}$. But this is assured if the equilibrium in the Io-Europa 2:1 resonance is reached before encounter with Ganymede since $e_{12} = (1/34.5 D_1)^{1/2}$ in that case.

Equation (83) is that of a pendulum with an applied torque good for either circulation or libration of the Laplace angle φ. As the system passes through resonance, the mean value of $\dot{\varphi}$ is essentially zero and $\delta\dot{\varphi} = \dot{\varphi}$. Replacing $\delta\dot{\varphi}$ by $\dot{\varphi}$ in Eq. (83) yields a form identical with that derived for spin-orbit coupling by Goldreich and Peale (1966), and we can use their expression for the capture probability to obtain (Yoder 1979b; Yoder and Peale 1981)

$$P_c = \frac{2}{1 + \dfrac{\pi}{4} \dfrac{\ddot{\varphi}_T}{\sigma |An_2^2|^{1/2}}} = \frac{2}{1 + \left(\dfrac{D_1}{3700}\right)^{3/4}} \tag{85}$$

where $\ddot{\varphi}_T = -c_1 n_1^2 (1 - 44.8 D_1 e_{12}^2)$ is the secular rate of change in $\langle\dot{\varphi}\rangle$ from the tidal expansion of the orbits and $e_{12} = (1/34.5 D_1)$ at the time of capture is used to obtain the final form with Eqs. (73) and (77) giving $\nu_1 + \dot{\omega}_{si}$ in terms of D_1 in A and σ. Capture into the resonance is certain if $D_1 < 3700$.

The $\dot{\varphi}$ term will damp the librations in φ and we see directly how the dissipation in Io can provide the previously elusive means of accounting for the extremely small amplitude of the Laplace libration. The tides will continue to reduce n_1 and, with the addition of Ganymede to the resonant system, Eq. (78) shows that ν_1 is also further reduced as e_{12} increases toward a new equilibrium value of $(12.6 D_1)^{-1/2}$. In libration $\langle d\dot{\varphi}/dt\rangle = 0$ in Eq. (78). If the current value of $e_1 = 0.0041$ is the equilibrium value, $D_1 = 4600$. A larger equilibrium e_{12} would result in a smaller D, and we see from Eq. (85) that

$$P_c \geq 0.9. \tag{86}$$

Henrard (1983) has refined the mathematical model used here and finds $P_c = 1$, with similarly slight changes in other parameters. That the value of e_1 is indeed currently at equilibrium follows from the analysis of the damping of the libration in φ.

We pointed out in the discussion of Enceladus that the action, here represented by

$$J = \oint \dot\varphi \, d\varphi \qquad (87)$$

is no longer conserved when there is dissipation in the satellites, and we see from Eq. (87) that J vanishes with the amplitude of libration. This suggests we follow the decrease in J to analyze the tidal damping of the libration in φ. This approach is especially convenient because it avoids altogether any problems with the infinite period on the separatrix which has to be dealt with if we monitor φ_{max} directly. If we write the libration energy as

$$E = \frac{1}{2} \dot\varphi^2 - An_{\frac{3}{2}}^2 \cos \varphi \qquad (88)$$

then $J = J(E, A)$, and

$$\frac{dJ}{dt} = \left[\frac{dE}{dt} \frac{\partial}{\partial E} + \frac{dA}{dt} \frac{\partial}{\partial A} \right] \oint \dot\varphi \, d\varphi$$

$$= \left[\left(\frac{\partial E}{\partial t} + \frac{\partial E}{\partial A} \frac{dA}{dt} \right) \frac{\partial}{\partial E} + \frac{dA}{dt} \frac{\partial}{\partial A} \right] \oint \dot\varphi \, d\varphi \qquad (89)$$

where $\partial E/\partial t$ is with A held constant and follows by multiplying Eq. (83) by $\dot\varphi$,

$$\frac{dJ}{dt} = \left[\left(-\sigma\dot\varphi^2 + \ddot\varphi_T\dot\varphi - n^2 \cos \varphi \frac{dA}{dt} \right) \frac{\partial}{\partial E} + \frac{dA}{dt} \frac{\partial}{\partial A} \right] \oint \dot\varphi \, d\varphi. \qquad (90)$$

Solving Eq. (88) for $\dot\varphi$, we find $\partial\dot\varphi/\partial E = 1/\dot\varphi$ and $\partial\dot\varphi/\partial A = n^2 \cos \varphi/\dot\varphi$. Making these substitutions in Eq. (90) and averaging over a period τ yields

$$\left\langle \frac{dJ}{dt} \right\rangle = -\sigma J \qquad (91)$$

and

$$J(t) = J(0) \exp \int_0^t -\sigma(t) dt \qquad (92)$$

where $\tau = \oint \dot\varphi^{-1} d\varphi$, σ and $\ddot\varphi_T$ are assumed constant during the averaging over the libration period.

An explicit solution of Eq. (91) follows if we change the integration variable from t to $z = \gamma_1/\overline{\gamma}_1$ where $\gamma_1 = \nu_1 + \dot{\omega}_{s1}$ and $\overline{\gamma}_1$ is the value of γ_1 at the equilibrium $e_{12} = (12.6D_1)^{-1/2}$ (Yoder and Peale 1981). There results

$$\left|\frac{A(t)}{A(0)}\right|^{1/2} \frac{\pi}{4} \sin^2 \frac{\varphi_m}{2} = \left[\frac{z_0^2(1 - z^2)}{z^2(1 - z_0^2)}\right]^{7.6} \tag{93}$$

for small libration amplitude φ_m, where the Laplace resonance was established at $t = 0$ when $z = z_0 = 1.63$. From Eq. (77) $A(t)/A(0) = [z_0 - (\dot{\omega}_{s1} - \dot{\omega}_{s2})/\gamma_1]/[z - (\dot{\omega}_{s1} - \dot{\omega}_{s2})/\gamma_1]$ and Eq. (93) yields $z = 1.047$ for a current amplitude of $\varphi_m = 0°.066$ (Lieske 1980). From a solution of Eq. (78) with $\langle \dot{\varphi} \rangle = 0$,

$$t = \frac{\overline{\gamma}_1}{0.32c_1n_1}\left[z_0 - z + \frac{1}{2}\ln\left(\frac{1 + z}{1 + z_0}\frac{1 - z_0}{1 - z}\right)\right] \tag{94}$$

which yields an age of the Laplace resonance of (Yoder and Peale 1981)

$$t = 1600 \, Q_J \text{ yr} \tag{95}$$

for $z = 1.047$. If $\varphi_m = 0°.066$ is not a remnant amplitude of the damping process, Eq. (95) is not applicable and the age of the resonance cannot be determined in terms of Q_J. The amplitude may be forced by a solar perturbation (Yoder and Peale 1981).

The more important result of the damping analysis is that $\varphi_m \to 0$ as $z \to 1$, i.e., as the system approaches equilibrium. The value of $\varphi_m = 0°.066$ and $z = 1.047$ with $e_{12} = 0.0041$ gives $e_{12} = (13D_1)^{-1/2} = 0.0043$ at equilibrium from which $D_1 = 4200$. If $e_{12} = 0.0041$ is the equilibrium value ($\varphi_m = 0°.066$ not a remnant amplitude), $D_1 = 4600$. Even if $\varphi_m = 0°.066$ is only an upper bound on the free amplitude, Eq. (95) places an upper bound on Q_J of

$$Q_J \lesssim 3 \times 10^6 \tag{96}$$

since $t < 4.6$ Gyr.

The upper bound on Q_J given in Eq. (96) is also supported by the observation of the extensive thermal activity on Io. Surely the tidal dissipation in Io is at least as high as the energy generated within from radioactive decay, since, for example, the Moon most likely has a similar content of radioactive elements as Io, yet shows no current thermal activity. Cassen et al. (1979b) have estimated the lunar heat source from radioactivity to be 6.9×10^{-8} erg g^{-1} s^{-1} at the present time with nearly four times this amount 4.6 Gyr ago. This implies a current deposition of radiogenic heat of 6×10^{18} erg s^{-1}, and from Eq. (49) $Q_1 \lesssim 300$ if tidal dissipation is to exceed this. From Eq. (60) with $D_1 = 4200$, $k_2^{(1)} = 0.036$, $f = 1$, $Q_J \lesssim 2.6 \times 10^6$.

This upper bound on Q_J is considerably below some estimates of Q_J from first principles (Goldreich and Nicholson 1977; Hubbard 1974; Burns' chapter), and has led to the suggestion that tidal dissipation in Jupiter may not have caused the assembly of the resonances. The resonances would then have to be assembled by unspecified processes at the time the Jupiter system itself was formed (see, e.g., Greenberg 1982a). However, the observed dissipation in Io would result in an increase in ν_1 and an eventual destruction of the resonances if there were no transfer of angular momentum from Jupiter tending to decrease ν. (Only the term containing D_1 would remain in Eq. 78.) The only way a primordial origin could be compatible with observation is for the system to have started deeper in the resonance with larger e's and smaller ν's, and we would now be watching the system relax toward smaller values of eccentricity (Peale and Greenberg 1980).

However, a linear stability analysis (Yoder and Peale 1981) has shown the Laplace resonance to be unstable at the current stationary values of the angles for $e_1 > 0.012$, so the system could probably not have started any deeper in the resonance than this. An origin deeper within the resonance might have been possible if the stationary values of the angles change and oscillations about these angles remain stable for $e_1 > 0.012$ (Greenberg 1984b,1985), and schemes for storage of the system on the other side of the resonance can be contrived. However, the latter schemes are highly implausible and none has been shown even to be possible (Yoder and Peale 1981). An initial $e_1 > 0.012$ would in any case be rapidly reduced to values less than this bound by the expected high rate of dissipation in Io. If 0.012 is assumed to be a firm upper bound on e_1, the total energy available for heating Io is somewhat less than the change in the orbital energy between the initial configuration with $e_1 = 0.012$ and the present configuration with $e_1 = 0.0041$ under the condition of conserved angular momentum. This latter energy is 5.5×10^{35} erg compared with about 2×10^{36} erg available from radioactive decay in Io (Yoder and Peale 1981). As tidal dissipation must exceed the radiogenic source of heat today and would have been larger in the past, one can infer that the tidal dissipation has been considerably more than that necessary to relax the system from its most extreme resonance configuration to that we see today. The only way to accommodate this dissipation is for Jupiter to supply sufficient torque to retard the relaxation. Otherwise the system would have relaxed to smaller eccentricities than we now observe (Peale and Greenberg 1980; Yoder and Peale 1981). The current eccentricity is therefore close to the equilibrium value whether or not the resonance was primordial and the upper bound $Q_J \lesssim 3 \times 10^6$ still applies.

Stevenson (1983) has found a mechanism of dissipation of tidal energy in Jupiter involving a phase change of helium which yields Q_J within the bounds imposed by the dynamics of the satellite system.

The lower bound $Q_J > 6.6 \times 10^4$ established by the proximity of Io to Jupiter after 4.5 Gyr is apparently in conflict with the measured heat flow from the satellite, which has been estimated to be near 1500 erg cm^{-2} s^{-1} averaged over the surface (Matson et al. 1981a; Sinton 1981; Morrison and Telesco 1980; chapter by Schubert et al.). One can calculate the dissipation in Io as a function of Q_J independent of the properties of the satellite provided that an equilibrium configuration is assumed (Lissauer et al. 1984). For a given Q_J, the rate at which the torque does work on Io is known and the rate of increase in the orbital energy in the preserved configuration can be calculated. The rate of work done exceeds the rate of increase in the orbital energy by the rate at which energy is dissipated in Io. For minimum Q_J this dissipation corresponds to a surface flux density of about 800 erg cm^{-2} s^{-1} or only about half of that estimated from observation. It has been pointed out by Johnson et al. (1984) that the observations of Io in the infrared from which the heat flow had been estimated were done while Io was eclipsed by Jupiter and were all therefore of the same hemisphere. Since the estimates of total heat flux had been made on the assumption of a uniform distribution of sources, a lack of sources on the far hemisphere would reduce these estimates to values near the above maximum *average* heat flux determined by the dynamical constraints. Further observations of Io at a variety of phases do show a concentration of hot spots in the previously observed hemisphere (Io eclipsed when viewed from Earth) (Johnson et al. 1984), but continuing heat flux estimates from all sources show time fluctuations about 1500 erg cm^{-2} s^{-1} with no apparent trend in the mean value (McEwen et al. 1985). This implies that the heat flux observed over the last six years must considerably exceed the long-term average flux.

Yoder's (1979b) solution of the long-standing problem of the origin and subsequent damping of the Laplace libration is one of the most outstanding accomplishments of dynamical analysis during the past decade. It was probably not anticipated that the solution of this particular problem in dynamical evolution would also lead to such narrow bounds on the dissipative properties of a giant gaseous planet.

The understanding of the origin and evolution of the resonances among the Galilean satellites has been a fascinating exercise with the introduction of the dissipation of tidal energy in the satellites providing a solution to a long-standing enigma about the route to the currently completely damped configuration. We turn now to a system in which dissipation appears to be more important in drastically altering the dynamical configuration in the future than it has been in establishing the current state. The interest in the coorbital satellites of Saturn discussed in the next section is generated by their being bodies of comparable mass locked in 1:1 orbital resonance and having such a large amplitude of libration that one satellite approaches quite close to the other, alternately in front and behind. An elegantly simple modification of the restricted three-body problem provides an adequate description of the system and allows a determination of the masses and ultimately the densities of both satellites.

IV. THE COORBITAL SATELLITES OF SATURN

The 1:1 orbital resonance has received an enormous amount of attention in the literature as part of a more general study of the restricted three-body problem (see, e.g., Szebehely 1967) with application to the Trojan asteroids. These asteroids librate about the stable stationary points of the restricted three-body problem, the L_4 and L_5 points 60° ahead and behind Jupiter with average mean motions identical to that of Jupiter. Three more Trojan-type objects were found among the satellites of Saturn by groundbased observers at the time of the Voyager 1 and 2 flybys—one librating about Dione's L_4 point (leading by 60°) and one each librating about Tethys' L_4 and L_5 points.

The coorbital satellites, Janus and Epimetheus, orbit at a mean distance of 151,000 km from the center of Saturn. This latter example of a 1:1 resonance is unique in the sense that the satellite masses are comparable ($M_1/M_2 = 1/5$ to $1/3$), and the amplitude of libration is so large that the satellites can come quite close to each other. The orbit of Epimetheus in a frame rotating with Janus has the shape of a horseshoe enveloping both the L_4 and L_5 Lagrange points instead of a Trojan-like path confined to a region near one or the other point. Reduction of groundbased observations and Voyager 1 and 2 orbit determinations yield $(M_1 + M_2)/M = (3.9 \pm 1.2) \times 10^{-9}$ and $M_2/(M_1 + M_2) = 0.216 \pm 0.009$, where M is the mass of Saturn (C. F. Yoder and S. Synnott, personal communication, 1985). Estimates of the volumes of the satellites by P. Thomas (unpublished, 1984; see chapter herein) then give densities of 0.85 ($\pm$ 0.3) g cm^{-3}. The small mass ratio allows a simple analytic approximation which adequately describes the motion of the coorbital pair and can be used to infer some aspects of their dynamical evolution (Yoder et al. 1983).

Following Yoder et al. (1983), we can write the equations of motion of Janus as perturbed as Epimetheus as

$$\frac{d}{dt}\left(r_1^2 \frac{d\theta_1}{dt}\right) = \frac{\partial F_{12}}{\partial \theta_1} \tag{97}$$

$$\frac{d^2 r_1}{dt^2} - r_1\left(\frac{d\theta_1}{dt}\right)^2 = -\frac{GM}{r_1^2} + \frac{\partial F_{12}}{\partial r_1} \tag{98}$$

where

$$F_{12} = GM_2\left[\frac{1}{\Delta} - \frac{r_1^2}{r_2^2}\cos(\theta_1 - \theta_2)\right] \tag{99}$$

$$\Delta^2 = r_1^2 + r_2^2 - 2r_1 r_2 \cos(\theta_1 - \theta_2). \tag{100}$$

The subscripts 1 and 2 correspond, respectively, to Janus and Epimetheus. The polar coordinates r and θ are referred to the center of mass of Saturn, and the satellite orbits are assumed coplanar. The perturbations from other satellites, the rings, and the Sun are neglected for the time being. The influence of the oblateness of Saturn does not qualitatively change the character of the motion and would unnecessarily clutter the analysis. A similar set of equations is appropriate for the effect of Janus on Epimetheus, where F_{21} is not simply proportional to F_{12} because of the indirect term resulting from the use of a noninertial frame. We can write the equations as perturbations from circular reference orbits with a_0 and n_0 representing the mean distance from Saturn and mean motion, respectively, and δn_i and δr_i defined by

$$\frac{d\theta_i}{dt} = n_0 + \delta n_i \tag{101}$$

$$r_i = a_0 \left(1 - \frac{2}{3} \frac{\delta n_i}{n_0} \right) + \delta r_i. \tag{102}$$

In Eq. (102) the variation in r_i is separated into that part resulting from Kepler's third law with the orbits remaining circular and the additional increment related to an induced eccentricity. The advantage of this separation will be evident below. Substitution of Eqs. (101) and (102) into (97) and (98) and expansion to first order in δr_1 and δn_1 yields

$$-\frac{1}{3} a_0^2 \frac{d(\delta n_1)}{dt} + 2n_0 a_0 \frac{d(\delta r_1)}{dt} = \frac{\partial F_{12}}{\partial \theta_1} \tag{103}$$

$$\frac{d^2(\delta r_1)}{dt^2} - \frac{2}{3} a_0 n_0^{-1} \frac{d^2(\delta n_1)}{dt^2} - 3n_0^2 \delta r_1 = \frac{\partial F_{12}}{\partial r_1}. \tag{104}$$

The relative magnitudes of the terms in Eqs. (103) and (104) for the variations in δr_1 and δn_1 due to the mutual interactions depend on the small parameter $\varepsilon = [(M_1 + M_2)/M]^{1/2}$. Although the satellites interact only when they are close to each other, the average rate of change of the increments δn_i or δr_i is the magnitude of the change divided by the time between interactions, which is half the libration period. At the time of the Voyager 1 observations, Epimetheus was gaining on Janus by $0°.254/\text{day}$, leading to a time between encounters of about 4 yr. In terms of ε, this time is of order $2\pi/n_0\varepsilon$ with ε^2 being of order M_1/M given above. Using this representation of the time in Eqs. (103) and (104) and noting that F_{12} is of order $\varepsilon^2 a_0^2 n_0^2$, one finds that δn is of order εn_0 and δr is of order $\varepsilon^2 a_0$. Hence, a formulation correct to order ε^2 is obtained by omitting δr in Eq. (103) and replacing both r_1 and r_2 by a_0 in F_{ij}. In this approximation $M_1 F_{12} = M_2 F_{21}$ and we can write the equation for the variation of the difference angle $\phi = \theta_1 - \theta_2$,

$$-\frac{1}{3} \frac{d^2\phi}{dt^2} = \varepsilon^2 n_0^2 \frac{\partial}{\partial \phi} \left[\frac{1}{2\left| \sin \dfrac{\phi}{2} \right|} - \cos \phi \right] \tag{105}$$

which has the first integral

$$E = -\frac{1}{6}\left(\frac{d\phi}{dt}\right)^2 - \frac{\varepsilon^2 n_0^2}{2x}(1 + 4x^3) \tag{106}$$

with $x = |\sin \phi/2|$.

Bounds on the motion of ϕ are determined by $d\phi/dt = 0$ in Eq. (106) and correspond to the roots of

$$R(x) = x - A(1 + 4x^3) \tag{107}$$

where $A = -\varepsilon^2 n_0^2/2E > 0$, since E is negative definite. With $\sin \alpha = (3A)^{3/2}$, the three roots of Eq. (107) are

$$x_1 = \frac{\sin(\alpha/3)}{(3A)^{1/2}}$$

$$x_2 = \frac{\sin(\alpha/3 + 120°)}{(3A)^{1/2}}$$

$$x_3 = \frac{\sin(\alpha/3 - 120°)}{(3A)^{1/2}} \tag{108}$$

where x is confined between the two positive roots x_1 and x_2. For $A = 1/3$, $x_1 = x_2 = 1/2$ which means ϕ remains fixed at either $\pm 60°$. These are the stationary solutions of the L_4 and L_5 Lagrange points corresponding to $d^2\phi/dt^2 = d\phi/dt = 0$. The latter conditions also yield an unstable stationary point at $\phi = \pi$. For A slightly less than 1/3, Eq. (105) can be expanded about the stationary solution to demonstrate harmonic motion about the stable Lagrange points at a frequency of $(27/4)^{1/2}n_0\varepsilon$. Decreasing A further yields the classic tadpole-shaped orbits (in the rotating frame) which encircle the Lagrange points. At $A = 1/5$, the root $x_2 = 1$ and the tails of the two tadpole-shaped orbits meet at the unstable L_2 Lagrange stationary point at $\phi = \pi$. The closest approach of the two satellites is $23°\!.9$ in this case. The root x_2 exceeds 1 for $A < 1/5$ and only x_1 provides a real bound on the variation of ϕ, leading to the horseshoe-shaped relative orbit where ϕ oscillates between the limits of $\phi_{\min} = \pm 2 \sin^{-1}(x_1)$. Figure 8 shows the orbits of the two satellites in a frame rotating at the mean motion n_0 for several values of A.

The relative horseshoe orbit is easy to understand from a physical point of view. As M_2 approaches M_1 from behind, the mutual interaction causes M_2 to gain angular momentum and lose angular velocity as its orbital radius is increased while M_1 loses angular momentum and gains angular velocity. This causes the relative velocity to decrease and eventually change sign, and M_1 is left with the larger orbital angular velocity and smaller semimajor axis after the encounter. M_1 then proceeds to chase M_2, which is now moving more

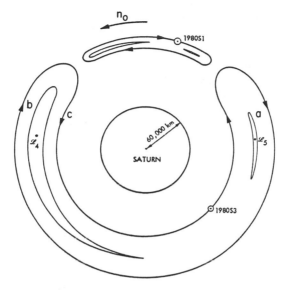

Fig. 8. Orbits of Saturn's coorbital satellites in a frame rotating with the average mean motion n_0. (a) Small librations about the Lagrange L_5 point with $A = 0.30$ and $\dot{\phi}_{max} = 0°025/$day. (b) Limiting tadpole orbit with $A = 0.2$ and $\dot{\phi}_{max} = 0°10/$day. (c) Horseshoe orbit with $A = 0.06$ and $\dot{\phi}_{max} = 0°26/$day. The positions of 1980S1 and 1980S3 in October, 1980 are shown (figure after Yoder et al. 1983).

slowly. The two satellites again exchange angular momentum as M_1 approaches M_2 from behind, and the relative motion is again reversed. By this means the satellites repeatedly bounce off each other as they orbit at approximately the same distance from Saturn.

It is interesting that most of classic motions of the restricted three-body problem follow in a modified form from the simple analysis of Yoder et al. appropriate when two of the masses are comparable yet small compared with the third. Information concerning the other two Lagrange stationary points of the classic problem (L_1 and L_2) is lost in the approximation. The nearly spherical surface about Janus containing the L_1 and L_2 points has a radius of $a_0(\varepsilon^2/3)^{1/3}$ (see, e.g., Szebehely 1967), and if Epimetheus is inside this sphere, the mutual attraction between the two satellites is comparable to that between one satellite and Saturn. Hence, the approximation is not valid if the satellites approach this separation. But this sphere of influence is only about 200 km in radius for the estimated value of ε^2, whereas the estimated distance of closest approach for the observed relative angular velocity of $0°254/$day is more than 20,000 km (Yoder et al. 1983). The approximation thus represents a good description of the motion of Saturn's coorbitals.

The minimum separation of the satellites corresponding to the root x_1 follows from the value of $A = -\varepsilon^2 n_0^2/2E$ where E is determined from the observed $d\phi/dt = 0°254/$day at the observed value of x corresponding to a separation of $117°$ (Synnott et al. 1981). As we have only guessed at the value of ε from the assumed density, it is desirable to turn the above analysis around and determine ε more precisely by measuring the distance of closest approach. Dermott and Murray (1981a,b) have discussed extensively the observational determination of the masses of the coorbitals and have considered the observational problems and accuracies of several procedures. In addition to the sum of the masses, the mass ratio can be determined from the ratio of the amplitudes of longitude variation for the two satellites ($\Delta\theta_1/\Delta\theta_2 = M_2/M_1$), and each mass can be found separately. It is likely that the error in the volume determination for these irregularly shaped satellites will ultimately be the limiting factor in determining the densities for the foreseeable future. Images from many aspects are necessary for an accurate volume, and these must await a spacecraft in orbit about Saturn.

All of our discussion so far has been based on the assumption of circular, coplanar orbits, whereas Janus and Epimetheus have orbital eccentricities of 0.007 and 0.009 and inclinations of $0°14$ and $0°34$, respectively (Synnott et al. 1981). The effects of the nonzero eccentricities and inclinations as well as perturbations by other nearby satellites (including resonances) and Saturn's oblate figure have been carefully evaluated by Yoder et al. (1983). Mutually induced variations in eccentricity and inclination lead to only a 50 km short-period variation in the orbital positions, but the nonzero values of the e's and i's cause an increase in the minimum separation of 2.1% and a 0.13% asymmetry in the leading and trailing close approaches. Long-period variations in the e's and i's from the mutual interaction are shown to bring the satellites no closer than 15,000 km, which means that the satellites will never pass nor collide if mutual interaction alone is considered.

Stability over the relatively short time scale of 200 yr has been demonstrated by Harrington and Seidelmann (1981) by a numerical integration which included the perturbations of the other major satellites. Torques from Saturn's nearby A ring increase the stability of the libration as the libration amplitude is *reduced* by any secular expansion of the orbits whose rate decreases with increasing orbital radius (Yoder et al. 1983; Lissauer et al. 1985a). The effects of near resonances between the satellite mean orbital motions and those of Mimas and Enceladus (4:3 and 2:1, respectively) are currently negligibly small, but it appears unlikely that the 1:1 resonance will survive the *future* passage through 4:3 resonance with Mimas—at least with current values of the orbital and physical parameters of both the coorbitals and Mimas.

In summary, the horseshoe orbits of Saturn's coorbital satellites are currently very stable, and their dynamics allows a determination of each of the satellite masses. Corrections to the circular, coplanar orbit analysis due to the nonzero eccentricities and inclinations and to Saturn's oblateness can be included in the above analysis. Since observations from Earth are limited to those times when the Earth is in Saturn's ring plane, it appears that determinations of the properties of the coorbital satellites more accurate than those quoted above must await the observations from a long-lived Saturn orbiter.

The coorbitals, as well as the satellites considered in the earlier sections, are rotating synchronously with their orbital motion—a natural result of tidal evolution. The discovery of Pluto's satellite Charon (Christy and Harrington 1979) revealed a planet-satellite system which may be unique in the solar system. Charon is not much smaller than Pluto and the two bodies are sufficiently close to each other that most probably as much angular momentum has been transferred from the spins to the orbit as can be. The system represents therefore a true endpoint of tidal evolution whose details are treated in the next section.

V. THE PLUTO-CHARON SYSTEM

Farinella et al. (1979) and Dermott (1978) have given the most extensive analyses of the Pluto-Charon system, where both primary and satellite appear to be rotating synchronously with their orbital motion. In this state of dual synchronous rotation, the total angular momentum is given by

$$L_o + L_s = m\sqrt{\mu a} + m_1 R_1^2 \left(\alpha_1 + \alpha_2 \frac{m_2}{m_1} \frac{R_2^2}{R_1^2}\right) \sqrt{\frac{\mu}{a^3}} \quad (109)$$

where L_o and L_s are orbital and spin angular momenta, respectively, $m_{1,2}$ are primary and satellite masses, $m = (m_1 m_2)/(m_1 + m_2)$, $R_{1,2}$ are the respective radii, $\alpha_{1,2}$ measures of the central condensation ($\alpha = 0.4$ for homogeneous sphere), $\mu = G(m_1 + m_2)$ with G being the gravitational constant and a is the orbital semimajor axis. If we fix the body parameters but let a vary while maintaining the synchronous state, $L_o + L_s$ has large values for very large a and for very small a as shown in Fig. 9. Hence, for sufficiently large and conserved angular momentum, there are in general two values of a for which the dual synchroneity of the rotations with the orbital motion is possible. A self-consistent set of parameters for Pluto-Charon, $R_1 = 1500$ km, $R_2 = 700$ km, $m_1 = 0.0022$ M$_\oplus$, $m_2 = 0.0002$ M$_\oplus$, $a = 19,470$ km, $P =$ orbital period $= 6.3867$ day (Harrington and Christy 1980; Hege et al. 1982) yields $a/R_1 = 13.0$ for the current outside synchronous state. Using this value of a/R_1 to determine L with $\alpha_1 = \alpha_2 = 0.4$, we find from Eq. (109) and Fig. 9 the inner synchronous state to be at $a/R_1 = 1.5$ with primary and satellite nearly touching. As the second term in Eq. (109) dominates for the inner state, most of the total angular momentum is in the spins of the two bodies, whereas the dominance of the first term in the outer state means the angular momentum is mostly orbital (Fig. 9).

The inner synchronous state is unstable (see, e.g., Lin 1981), which can be understood as follows. The contribution of m_2 to the spin angular momentum is only 2% of that due to m_1 so $L_s \approx \alpha_1 m_1 R_1^2 \dot{\theta}$ where $\dot{\theta}$ is the spin an-

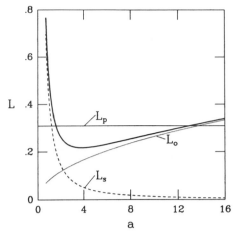

Fig. 9. The dependence of the angular momentum of the Pluto-Charon system in dual synchronous rotation on orbital semimajor axis is shown as the sum of the spin (L_s) and orbital
(L_o) angular momenta. The semimajor axis a is in units of the radius R_1 of Pluto and the
angular momenta are in units of $m_1\sqrt{\mu R_1}$. L_p is the current angular momentum with the system occupying the outer dual synchronous state at $a = 13$ (cf. Fig. 11 of Burns' chapter).

gular velocity of the primary (initially $= n =$ mean orbital angular velocity).
Then if we perturb the system

$$\Delta L_s = L_s \frac{\Delta\dot\theta}{\dot\theta}$$

$$\Delta L_o = \Delta(m\mu^{2/3}n^{-1/3}) = -\frac{1}{3} L_o \frac{\Delta n}{n} \qquad (110)$$

and from conservation of $L_o + L_s$, we have

$$L_s \frac{\Delta\dot\theta}{\dot\theta} = \frac{1}{3} L_o \frac{\Delta n}{n} . \qquad (111)$$

From Eq. (109), $3L_s = L_o$ corresponds to the value of a for which $d(L_s +
L_o)/da = 0$ (Dermott 1978). At the inner synchronous state $3L_s > L_o$ and
from Eq. (111) $|\Delta\dot\theta|/\dot\theta < |\Delta n|/n$. Moving the satellite farther away means $\Delta n
< 0$. $\Delta\dot\theta$ is also < 0, since angular momentum is transferred but its magnitude
is less, so the primary is rotating faster than the orbital motion and the system
expands outward to the outside synchronous state through the tidal transfer of
angular momentum. Moving the satellite in increases n and $\dot\theta$ but the former
by a greater amount. The primary would now be rotating more slowly than the
orbital motion and the satellite would spiral into the primary. This instability
cannot be frustrated by permanent deviations from axial symmetry because

librations would grow in amplitude and the system would be freed from libration on one or the other side of the synchronous state. Since $3L_s < L_o$ for the outer synchronous state, all the above arguments are reversed and this outer state is stable. It is likely that the system has evolved fully, and Pluto-Charon occupies the outer dual-synchronous state at present.

Since Pluto-Charon would be inside the mutual Roche limits in the inner dual synchronous state and since the total-system angular momentum is comparable to that for rotational instability of the combined masses, Lin (1981) and Mignard (1981a) have proposed fission of a rapidly rotating object for the origin of Charon. The problem with this proposal is the impossibility of forming a solar system body which is initially rotationally unstable or of finding a means of accelerating the body to instability after formation. Lin offers no acceleration mechanism and Mignard suggests a rotational acceleration during a close encounter with Neptune's satellite Triton as Pluto was thrown out of that system. To accelerate Pluto from a minimum 10 hr period (it is presumed to have a synchronous rotation period if orbiting Neptune at the Roche limit) to the 4 hr period required for instability requires a transfer of energy of about 10% of its gravitational binding energy. Pluto would have been shattered. Also, the insertion of Pluto into the stable 3/2 orbital resonance with Neptune from an escape trajectory from this planet has probability zero. Perhaps one of the processes currently being discussed for the origin of the Moon (Boss and Peale 1986) will suffice for the origin of this interesting system. Additional discussion of the Pluto-Charon system and associated problems for binary asteroids is contained in Burns' chapter.

From this state of complete tidal relaxation, we go now to a satellite for which such a relaxation is impossible. Saturn's satellite Hyperion is almost certainly tumbling in a state of chaos that will most probably persist for the lifetime of the solar system. Reasons for this bizarre rotation state and how it came about are outlined in the next section.

VI. THE ROTATION OF HYPERION

When rotation histories of the planetary satellites were last reviewed (Peale 1977), observations of as yet undetermined spins were not expected to yield any surprises except for a possible nonsynchronous spin-orbit resonance for Hyperion (Peale 1978). However, images of Hyperion by the Voyager spacecraft (Smith et al. 1982) revealed a shape so asymmetric that earlier analyses of satellite and planetary spins (involving averaging over high-frequency terms) are not applicable. The possible spin state where Hyperion is rotating stably at an angular velocity which is 1.5 times its orbital angular velocity as suggested by Peale (1978) does not even exist. It is most likely that Hyperion will always tumble in a chaotic way with the magnitude and direction of its spin vector exhibiting large variations on time scales comparable to the orbital period (Wisdom et al. 1984).

This chaotic rotation is best understood by recalling some of the details of the analysis of spin-orbit coupling (Goldreich and Peale 1966; Colombo and Shapiro 1966) and showing where that analysis fails for Hyperion. We thus consider first the case where the spin axis of a planet or satellite remains perpendicular to its orbit plane. In the absence of a tidal torque, the equation of motion for the spin is (Goldreich and Peale 1966)

$$\frac{d^2\theta}{dt^2} = -\frac{3}{2}\frac{Gm}{r^3}\frac{(B-A)}{C}\sin 2(\theta - f)$$

$$= -\frac{\omega_0^2}{2}\sum_{k=-\infty}^{\infty} H\left(\frac{k}{2}, e\right)\sin(2\theta - k\ell) \tag{112}$$

where the spin axis coincides with the axis of maximum moment, θ is the angle between the axis of minimum moment (in the orbit plane) and a line from the primary to the orbit periapse, f is the true anomaly, $A < B < C$ are the principal moments of inertia, m is the primary mass, r is the primary-satellite separation and $\omega_0^2 = 3(B - A)n^2/C$ with $n = \sqrt{Gm/a^3}$ being the orbital mean motion. The series expansion in the mean anomaly ℓ follows from the periodicity of $(a^3/r^3)\sin 2f$ and $(a^3/r^3)\cos 2f$. $H(k/2,e)$ are series in e each with factor $e^{2(k/2-1)}$ which are tabulated by Cayley (1859) and, for fewer values of k, by Goldreich and Peale (1966). A spin-orbit resonance occurs when $\dot\theta = pn + \dot\gamma$ for some half integer p, where $\dot\gamma/n \ll 1$. Since $n = \ell$, the term in the infinite series with $k = 2p$ is slowly varying. With $\theta = p\ell + \gamma$, γ is the angle between the long axis of the satellite and the direction to the primary when the satellite is at periapse. With this substitution Eq. (112) becomes

$$\frac{d^2\gamma}{dt^2} + \frac{\omega_0^2}{2} H(p, e)\sin 2\gamma = -\sum_{\substack{k=-\infty \\ k \neq 2p}}^{\infty} H\left(\frac{k}{2}, e\right)\sin[(2p - k)\ell + 2\gamma]$$

$$\tag{113}$$

and $\omega_0\sqrt{H(p,e)}$ is seen to be the frequency of small oscillations in γ.

Every term on the right-hand side of Eq. (113) contains ℓ and therefore oscillates at a high frequency compared to the term containing only γ. If $(B - A)/C \ll 1$, γ does not change much during an orbit period and we can average the right-hand side over this period holding γ constant. The resulting averaged equation of motion is Eq. (113) with the right-hand side equal to 0. This averaging procedure is not valid if $(B - A)H(p,e)/C$ is so large that γ varies substantially over an orbit period. For the application of the theory to a nearly spherical Mercury by Goldreich and Peale (1966), the averaging procedure is valid and the resulting pendulum equation in Eq. (113) demonstrates the stable libration of the spin angular velocity about $\dot\theta = pn$ for several values of p in addition to the observed resonance at $p = 3/2$.

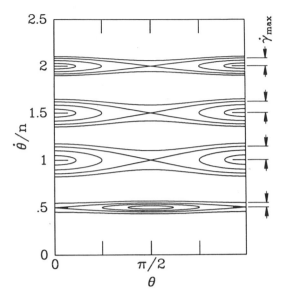

Fig. 10. Phase trajectories near spin-orbit resonances for averaged equations of motion (figure after Peale 1984).

Without the right-hand side, Eq. (113) is integrable but is valid only near a resonance, since otherwise γ is not approximately constant during the averaging process. The first integral

$$\frac{\dot{\gamma}^2}{2} - \frac{\omega_0^2}{4} H(p,e) \cos 2\gamma = E \tag{114}$$

can be used to make a phase plane plot of $\dot{\gamma}$ vs. γ which shows the character of the motion. A unique trajectory in the phase plane results for each value of E, and we can map all possible motions by choosing a range of E's. For $-\omega_0^2 H(p,e)/4 < E < \omega_0^2 H(p,e)/4 = E_c$, the motion is a libration. For $E > E_c$, the motion is either a positive or negative circulation. For $E = \omega_0^2 H(p,e)/4$, the trajectory is the infinite period separatrix separating libration from rotation. In Fig. 10, we have added pn to $\dot{\gamma}$ and plotted representative phase trajectories about several resonances. As 2γ appears in the cosine, we have restricted the phase plane plot to $0 \leq \gamma \leq \pi$ and replaced γ by θ on the abscissa with the understanding that this is the value of θ when the satellite is at periapse ($\ell = 0$). Large values of $\omega_0 = 0.15n$ and $e = 0.2$ were used to better illustrate the trajectories in Fig. 10 although these are outside the range of validity of the averaging procedure.

If librations are completely damped, $\dot{\gamma} \equiv 0 (\dot{\theta} = pn)$ and the phase plane plot reduces to a single point which we may place at either 0 or π ($\pi/2$ for $p =$

1/2). If $E < E_c$, libration curves fall inside the separatrix as closed curves. (0 and π are equivalent and the curves are closed if we wrap Fig. 10 around a cylinder.) For $E > E_c$, the curves stretch across the diagram representing positive or negative relative rotation. The curves are still closed on the cylinder, but θ increases without bound. The intersection of the separatrix with the line $\dot{\theta}/n = p$ ($\dot{\gamma} = 0$) corresponds to the unstable pendulum equilibrium and $|\dot{\gamma}|$ reaches a maximum value of $|\dot{\gamma}_{max}|$ when $\gamma = 0$ or π on this curve. This value of $|\dot{\gamma}_{max}|$ on the separatrix is the half-width of the resonance since, if $|\dot{\gamma}|$ were any larger at this point, $\dot{\theta} - pn = \dot{\gamma}$ would circulate instead of librate. For Mercury with $(B - A)/C = 10^{-4}$ and $p = 3/2$, $|\dot{\gamma}_{max}| = 0.014n$.

As plotted in Fig. 10, each phase diagram represents trajectories corresponding to exact integrals of an approximate equation, and any trajectory is traversed smoothly for given initial values of γ and $\dot{\gamma}$. However, the resonance equations are only valid when $\dot{\gamma}$ is small ($\dot{\theta}$ near pn) so we cannot plot points far away from a resonance. Nevertheless, the form of the phase plane plot suggests the following procedure. We can check the validity of the approximation and plot rotation trajectories between resonances by integrating the exact equation of motion Eq. (112) numerically (equivalent to keeping the high-frequency terms in Eq. 113) and plotting the values of $\dot{\theta}$ and θ each time the satellite passes periapse. The resulting phase plane plot is called a surface of section and consists of discrete points instead of continuous curves.

For the application to Mercury, the values of $\dot{\theta}$ and θ at periapse fall on smooth curves indistinguishable from those obtained from the approximate equation. The integrals of the motion still appear to exist although analytically obscure. We can now experiment with the surface of section by increasing $(B - A)/C$ in Eq. (112). The discrete points $(\theta, \dot{\theta})$ at periapse are still confined to a closed curve with $\langle \dot{\theta} \rangle = pn$ when initial conditions are near one of the resonances and to a smooth curve of circulation when initial conditions are far from resonance. However, initial conditions near a separatrix lead to points $(\theta, \dot{\theta})$ confined to a definite region surrounding the separatrix but randomly distributed over that region. The motion has become chaotic in the sense that infinitesimal changes in initial conditions lead to drastically different trajectories in phase space and the definite region around the separatrix over which the points $(\theta, \dot{\theta})$ are scattered is called a chaotic zone in the surface of section.

Another consequence of increasing $(B - A)/C$ is that the widths of the resonances are increased. In Fig. 10 this corresponds to a spreading of the separatrix curves as the chaotic zone about each separatrix also increases in area. Continued increase in $(B - A)/C$ eventually leads to an overlap of nearby resonances in the sense that the separatrices of two adjacent resonances would now be close to each other at $\theta = 0$ or π if each resonance were treated as if no other resonances existed. In this case, a widespread chaotic zone fills the region between the two resonances and surrounds islands of stable libration that shrink in area as $(B - A)/C$ is further increased. In a physical sense, initial values of $(\theta, \dot{\theta})$ could be consistent with libration about *either* of two adjacent resonant angular velocities if the resonances were isolated. However, one resonance perturbs the other so strongly that libration in either resonance is impossible for those initial conditions and widespread chaotic motion follows.

Chirikov (1979) has proposed a resonance overlap criterion which states that when the sum of two unperturbed half-widths equals the separation between resonance centers, large-scale chaos ensues. For the $p = 1$ and $p = 3/2$ resonances this criterion becomes

$$\omega_0\sqrt{H(1,e)} + \omega_0\sqrt{H\left(\frac{3}{2}, e\right)} = n/2 \qquad (115)$$

with $H(1,e) \approx 1$ and $H(3/2,e) \approx 7e/2$, $\omega_0 \approx n(2 + \sqrt{14e})^{-1}$. The chaotic region that exists around the separatrix for values of ω_0 considerably less than that necessary for resonance overlap in Eq. (6) can also be inferred to occur because of the overlap of other resonances. The period P of the motion becomes very large as a trajectory gets very near the separatrix and the corresponding frequency $2\pi/P$ becomes arbitrarily small. If there is a nearby resonance with frequency ω_1, we see that as we ease up to the separatrix, there are an infinity of resonances with $2\pi N/P = \omega_1$, where N is an integer. Each resonance has a width and the extent of the chaotic zone around each separatrix is estimated by how near the trajectory must be to the separatrix (how small $2\pi/P$ must be) for two of these resonances to overlap (Chirikov 1979).

For Hyperion the mean value of e is 0.1 and the value of ω_0 determined from Voyager 2 images is $\omega_0 = 0.89 \pm 0.22$ (T. C. Duxbury personal communication, 1983). However, the value of ω_0 for widespread chaotic motion near $\dot{\theta} = n$ and $3n/2$ from Eq. (115) is 0.31 for $e = 0.1$, which implies that there is a large chaotic zone surrounding these resonances, and probably others, for Hyperion. Figure 11 shows the surface of section for Hyperion with $\omega_0 = 0.89$ verifying the large chaotic zone which in fact engulfs all states from $p = 1/2$ to $p = 2$. The island of quasi-periodic librations for $p = 1/2$, is reduced to the small remnant in the lower center of the sea of chaos ($H(1/2,e) < 0$, so the satellite librates about $\theta = \pi/2$ for $p = 1/2$), and the 3/2 state has disappeared altogether. This is understood since the half-width of the unperturbed $p = 1$ state is larger than the separation of the $p = 1$ and $p = 3/2$ resonances. Quasi-periodic libration is also shown by the points falling on smooth curves for $p = 1, 2, 9/4, 5/2, 3$ and $7/2$ states. (The 9/4 resonance results from a second-order mixing of frequencies.) The centers of these islands of stable libration are substantially displaced from the mean values $\langle\dot{\theta}\rangle \equiv pn$ on the surface of section because of the large forced libration with the period of the orbit.

If Hyperion's spin axis were to remain normal to its orbit plane, tidal dissipation would most likely bring the spin to the chaotic zone, since capture

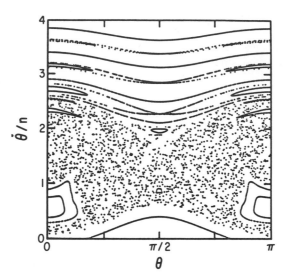

Fig. 11. Surface of section appropriate for Hyperion ($\omega_0 = 0.89n$, $e = 0.1$). A widespread chaotic zone surrounds islands of librational stability at all spin-orbit states from $p = 1/2$ to $p = 9/4$. There is no stable libration about the $p = 3/2$ state (figure after Wisdom et al. 1984).

into any of the higher-order resonances has negligibly small probability (see Goldreich and Peale 1966). Once within the chaotic zone, all points within it are repeatedly accessible and the rotation could be trapped into any of the resonances represented by the islands of quasi-periodic motion in the chaotic sea in Fig. 11. Old ideas about capture probabilities have to be revised, however, since now the satellite would have many chances at each resonance which was accessible instead of just one as in the Mercury case. In fact, even in the situation where the chaotic zone is limited to a narrow region about the separatrix, the capture probabilities would remain as before only if the tides could drag the spin across the width of the chaotic zone in a time less than the cycle time of γ (Wisdom et al. 1984). If Hyperion were to be trapped into an island, continuing dissipation would drive the trajectory to the island center representing the forced periodic libration. We might have therefore expected to find Hyperion librating near the synchronous island in Fig. 11.

But this last expectation is only possible if the spin axis remains near the orbit normal during the periodic motion represented by an island center in Fig. 11. In fact, Floquet theory (see, e.g., Kane 1965) applied to the Euler equations shows the axis orientation normal to the orbit plane to be unstable at the $p = 1$ and $p = 1/2$ states for nearly all the values of ω_0 within the uncertainties of the observations, but stable at the $p = 2$ state (Wisdom et al. 1984). The Euler equations for rigid body motion (see, e.g., Goldstein, 1980, p. 158) are solved in terms of angles like the Euler angles to specify the body orientation. Inertial axes are fixed in the orbit with the x axis directed from

Saturn to Hyperion's orbit periapse, the y axis in the direction of orbital motion and the z axis parallel to the orbital angular momentum. If a, b, c are principal axes of the satellite which start coincident with the xyz axis, θ is a rotation about c, φ a rotation about the new position of a, and ψ is a rotation about the last position of b instead of a rotation about c to avoid a coordinate singularity at $\varphi = 0$, the point about which we wish to test the stability. The notation for θ and φ is reversed from that usually seen in order to coincide with our earlier definition of θ as rotation about the spin axis when the axis is normal to the orbit plane.

With the components of the angular velocity along the body axes and the direction cosines of the vector to Saturn relative to these axes expressed in terms of θ, φ, ψ, $\dot{\theta}$, $\dot{\varphi}$, $\dot{\psi}$, we can write the Euler equations in terms of these variables, solve for $\ddot{\theta}$, $\ddot{\varphi}$, $\ddot{\psi}$ in terms of angles and angular velocities and set down the six equations to be solved as

$$\frac{d\dot{\theta}}{dt} = f_1\,(\theta,\,\varphi,\,\psi,\,\dot{\theta},\,\dot{\varphi},\,\dot{\psi}) \quad ; \quad \frac{d\theta}{dt} = \dot{\theta}$$

$$\frac{d\dot{\varphi}}{dt} = f_2\,(\theta,\,\varphi,\,\psi,\,\dot{\theta},\,\dot{\varphi},\,\dot{\psi}) \quad ; \quad \frac{d\varphi}{dt} = \dot{\varphi}$$

$$\frac{d\dot{\psi}}{dt} - f_3\,(\theta,\,\varphi,\,\psi,\,\dot{\theta},\,\dot{\varphi},\,\dot{\psi}) \quad ; \quad \frac{d\psi}{dt} = \dot{\psi}. \tag{116}$$

As the motion is periodic at the island centers, the coefficients in these equations are periodic, which satisfies the necessary condition for the application of Floquet theory. If p_θ, p_φ, p_ψ are momenta conjugate to θ, φ, ψ, then, φ, ψ, p_φ, p_ψ are zero if the axis is started perpendicular to the orbit plane and they remain zero if the axis is undisturbed. In this case, θ is identical to the θ used earlier in this section. A trajectory near the periodic trajectory is specified by $\theta' = \theta + \delta\theta$, $\varphi' = \varphi + \delta\varphi$, $\psi' = \psi + \delta\psi$, $p_\theta' = p_\theta + \delta p_\theta'$, $p_\varphi' = p_\varphi + \delta p_\varphi$, $p_\psi' = p_\psi + \delta p_\psi$. A linear transformation between the values of the variables at $t = 0$ and $t = \tau$ (one period later) is generated by the following. Assume all the initial increments in the canonical variables are zero except one, which is given a small value. Find the corresponding initial values of $(\dot{\theta},\,\dot{\varphi},\,\dot{\psi},\,\theta,\,\varphi,\,\psi)$, numerically integrate Eq. (116) over one period to find the corresponding increments $\delta p_\theta(\tau)$, $\delta p_\varphi(\tau)$, $\delta p_\psi(\tau)$, $\delta\theta(\tau)$, $\delta\varphi(\tau)$, $\delta\psi(\tau)$ and repeat this process by assuming a different variable to be incremented initially to find a new set of increments for all the variables one period later until all variables have been so treated. This gives six sets of increments at $t = \tau$ corresponding to the six independent initial conditions. Normalize each set by dividing each member of the set by the initial increment which generated it. If we form a 6×6 matrix with the normalized sets of increments as columns, the matrix represents a linear transformation which maps an arbitrary initial set of increments into their values one period later. As the coefficients in the equations have returned to their original values, we may apply the same linear transformation repeatedly to step the solution one period at a time. If any of the eigenvalues of this transformation (called Floquet multipliers) has a modulus > 1, the periodic motion is unstable. By choosing canonical variables in a Hamiltonian system each eigenvalue with modulus > 1 is matched by another modulus < 1 such that the product of all the eigenvalues is 1 (Poincaré 1892). A necessary condition for stability is thus that all the eigenvalues of the transformation matrix have modulus 1. The imaginary exponents of the eigenvalues yield the characteristic frequencies of the perturbed motion.

Floquet theory only tests for the linear stability of the axis orientation. To check that the spin axis is not stabilized slightly displaced from a direction normal to the orbit plane by nonlinear effects or that it exhibits a large amplitude, periodic variation, two nearby trajectories in phase space are integrated. If these trajectories separate linearly in time, the motion is quasiperiodic; if they separate exponentially, the motion is chaotic—again in the sense that infinitesimal changes in initial conditions lead to drastically different trajectories in phase. A measure of the exponential separation of two nearby trajectories are the Lyapunov characteristic exponents (see, e.g., Wisdom 1983) defined by

$$\lambda = \lim_{t \to \infty} \frac{\ln\left(\dfrac{d(t)}{d(t_0)}\right)}{t - t_0} \tag{117}$$

where

$$d(t) = \sqrt{\delta\theta^2 + \delta\varphi^2 + \delta\psi^2 + \delta p_\theta^2 + \delta p_\psi^2 + \delta p_\varphi^2} \tag{118}$$

is the ordinary Euclidean separation between two nearby trajectories. The conjugate variables obey the same equations as those derived from the Euler equations but the motion need not be periodic. A nonzero value of λ indicates a chaotic motion.

Numerical determination of the λ's for nearby trajectories for the periodic solution shows that the motion of the spin axis away from the orbit normal direction is fully chaotic for the $p = 1$ and $p = 1/2$ states. There is no nonlinear stabilization and the motion is not periodic with large amplitude. In addition, the attitude of the satellite is unstable everywhere in the chaotic zone surrounding the islands of stability (Wisdom et al. 1984). Hence, capture into the $p = 1$ or $p = 1/2$ state as tides dissipate rotational energy is not possible because these states are attitude unstable and Hyperion must tumble in an essentially random manner if tides have reduced the spin to the chaotic zone.

Capture into one of the attitude-stable islands ($p = 2$ or 9/4) is extremely improbable as Hyperion must arrive there with its spin axis close to the orbit normal and remain near the island long enough for a weak tidal dissipation to capture it from a strongly chaotic region ($\lambda = 0.1$).

One series of observations of Hyperion's lightcurve by Thomas et al. (1984) yields a least-squares fit to a sinusoid with 13 day period. The 14 data points were obtained from the Voyager 2 images over a two-month interval as the spacecraft approached Saturn. Variations in the image shape and orientation correlated with the apparent image brightness determined that the axis of rotation during at least part of the observation period was nearly in the orbit plane. Both the 13 day period and the axis orientation are consistent with chaotic rotation, but not with any other state of rotation expected from evolution via dissipative processes. It is easy to get a 13 day least-squares period from a lightcurve generated numerically from a chaotically rotating Hyperion when that curve is sampled only 14 times over a 60 day period (Wisdom and Peale 1984; Peale and Wisdom 1984). Many other periods were also obtainable by changing the 60 day window of sampling, as would be expected from the chaotic nature of the tumbling. Finding Hyperion's spin axis in its orbit plane could only occur if it were tumbling.

The autocorrelation of the numerical lightcurve is essentially zero for all intervals exceeding about 1.5 day. This implies that observations must be made at least once per day to define the chaotic curve. No current data set has this necessary sampling frequency, but a set of unpublished, groundbased observations by J. Goguen (personal communication, 1984) has short segments over a given observing run which match segments of the numerically generated curve in regions of rapid change. Although these two sets of observations do not completely verify the chaotic rotation of Hyperion, they are entirely consistent with and even suggestive of such a rotation. A definitive tracing of the lightcurve must wait for a nearly daily sampling over several Hyperion orbit periods. We expect to find Hyperion as the only example of continuously observable chaotic motion in the solar system (see Binzel et al. 1985b).

VII. CONCLUSION

We have discussed a simplification of the rather complex problem of the origin and evolution of orbital resonances where the evolution through a resonance by differential tidal expansion of the orbits can be followed through a sequence of diagrams of level curves of a simplified Hamiltonian. The motion corresponds to tracing a given level curve trajectory in the two-dimensional phase space, where the particular trajectory corresponding to the motion is determined far from resonance by initial conditions. As the system evolves, the motion trajectory is identified by the adiabatically conserved action as the Hamiltonian changes in the dissipative process. Most of the features of capture or escape from a resonance and evolution within a resonance are easily interpreted by following the motion through the series of diagrams. This description of resonance evolution, which does so well in simplifying many concepts, must be applied with caution to real situations in the solar system as criteria for the approximations may not be satisfied or a more complex model may be necessary.

The simple description of the two-body orbital resonance serves as a good basis for the development of the analysis of the more complex Galilean satellite resonances. It is used in the solution of the 300 year old puzzle of the origin of the three-body Laplace resonance with its negligibly small amplitude of libration. The use of Io's equilibrium orbital eccentricity to relate the rate of tidal energy dissipation in Io to that in Jupiter, with bounds on the former placing rather restrictive bounds on the latter, was a surprising bonus of the dynamical analysis.

The next two topics discussed were more novel descriptions of newly discovered, unique dynamical configurations rather than breakthroughs in analysis or solutions of old problems. The nature of the motion of the coorbital satellites of Saturn followed from a modification of the analysis for the restricted three-body problem, where now the two satellites have comparable masses but are very small compared to Saturn. The current relative horseshoe orbits are very stable but probably will not survive a future 4:3 orbital resonance with Mimas. The masses of each of the two coorbital satellites can be determined from their relative motions, but precise values of the masses along with densities must await the multiple observations from an orbiting spacecraft.

The large relative size of Pluto's newly discovered satellite Charon and its proximity to Pluto led to the inference that the system must have reached the ultimate endpoint of tidal evolution, where both primary and satellite are rotating synchronously with their orbital motion. Two dual-synchronous states are possible with a given angular momentum: an unstable state with the bodies almost touching and a more distant stable state which Pluto-Charon is thought to occupy. In spite of the existence of the inner state and sufficient angular momentum to cause rotational instability of a combined object, an origin of Charon by fission is not likely.

Finally, we have discussed the exotic rotation state of Saturn's satellite Hyperion which appears doomed to tumble chaotically indefinitely. Tidal evolution will retard the spin to a value where gravitational torques (resulting from Saturn's field acting on the strikingly asymmetric satellite in an eccentric orbit) initiate and maintain the chaotic motion. Observations of Hyperion's lightcurve support, but do not absolutely verify, the inferred motion. This verification awaits daily sampling of the lightcurve over several Hyperion orbit periods or frequent imaging from a spacecraft orbiting Saturn, but at present we have no reason to doubt the chaotic motion.

It is hard to predict what new configurations will be discovered in the future among the planetary satellites, whose dynamics, origin, and evolution remain to be described, or what long-standing dynamical problems will be solved or simplified. But the satellite system around Uranus is well populated with objects, and we have just had a close look.

Acknowledgments.   The author appreciates very much the careful reviews by J. Henrard, C. Yoder and J. Lissauer. Several errors and omitted references were pointed out. Thanks are also due J. A. Burns and A. T. Sinclair whose comments on the first draft of the manuscript were very helpful. This work was supported in part by a grant from the NASA Planetary Geology and Geophysics program.

6. THERMAL HISTORIES, COMPOSITIONS AND INTERNAL STRUCTURES OF THE MOONS OF THE SOLAR SYSTEM

GERALD SCHUBERT
University of California at Los Angeles

TILMAN SPOHN
Institut für Planetologie der Westfaelischen-Wilhelms Universität

and

RAY T. REYNOLDS
NASA Ames Research Center

Satellites evolve thermally as heat sources that compete with energy transfer processes that tend to cool it. Sources of heat in satellite interiors include gravitational energy released during accretion and differentiation, thermal energy produced by radioactive decay, and frictional energy created by tidal flexing. Heat can be transferred by conduction (radiation), subsolidus convection, and magma migration. The composition of a satellite fixes its complement of radioactive elements and sets the thermal, mechanical and rheological properties that control heat transport. Due to diverse compositions, heat sources and heat transfer mechanisms, the various satellites evolve differently.

Accretional energy influences a satellite's initial thermal state and controls the extent of early differentiation; larger satellites are heated more. The fraction of gravitational energy made available during accretion and retained as heat in a satellite's interior is highly uncertain. Accretional heating may have differentiated at least the outer layers of the Moon, the rocky Galilean satellites Io and Europa, and the icy Galilean satellite Ganymede. The heavily cratered primordial surface of Callisto suggests that this moon might never have differentiated. Accretional heating is probably insufficient to have differentiated the interiors of the small icy Saturnian satellites Mimas, Tethys, Dione, Rhea and Iapetus. Differentiation of a rocky satellite refers to the separation of a light crust or a heavy metallic core; differentiation of an icy satellite refers to the separation of ice (water) and rock. Radiogenic heating can lead to differentiation after formation (accumulation) but the process is mitigated against by sub-solidus convective heat transport. Major questions remain about the differentiation of the satellites. Do the Moon and Io have iron cores? What is the degree of ice-rock separation in the icy outer-planet satellites? Probably Ganymede has extensive regions of undifferentiated ice-rock mixture while Callisto and the icy Saturnian satellites are largely undifferentiated. The energy released upon differentiation is generally inconsequential for later thermal evolution.

Radiogenic heating by the decay of U, Th and K is usually the dominant post-accretional heat source in satellite interiors; the radiogenic volumetric heating rate is proportional to the silicate mass fraction of the body. Since radiogenic heating is proportional to volume and cooling is proportional to surface area, larger moons are usually more thermally active. However, other factors, e.g., silicate mass fraction, and internal thermal activity, especially if gauged by endogenic surface modification, are not simply controlled by satellite radius.

Tidal heating is probably the major energy source for Io and it might also be, or might once have been, important for Europa, Enceladus, the Moon, and possibly other satellites. The energy to drive Io's volcanism is generated by dissipation in an essentially rigid outer shell and in the less rigid, more fluidlike viscous layer that must underlie the shell. Heat transfer by magma migration near Io's surface must be important; Io's lithosphere would have to be too thin to conduct all the heat observed to be escaping from the satellite.

A satellite's rheological properties govern how hot it must get to transport internally generated heat to its surface. If enough heat is generated in a satellite, it will warm nearly to the solidus of its major constituent, and convection will supersede conduction (radiation) as the major heat transfer mechanism. Temperatures can thus rise relatively high in rocky satellites but they must remain below about 273 K in the portions of the low-density outer planet satellites where the rheology of ice controls the heat transfer. Subsolidus convection is generally efficient enough to prevent radiogenic heating from producing extensive partial melting in mantles of rocky satellites or from melting icy satellites. In some satellites (e.g., Mimas) even conduction is enough to preclude the partial melting of the major constituent. At present, there should be no extensive liquid water region in any of the outer planet satellites except perhaps for a near surface layer under the ice crusts of Europa and Enceladus (kept liquid by tidal heating in the crust) and thin layers containing NH_3 or salts (kept liquid by the reduced freezing points of the impure water).

While the rheology of water ice probably controls the thermal evolution of the icy satellites, the presence of low melting point constituents, such as ammonia hydrate, may be crucial to their resurfacing. Internal melting of these components and their upward migration and extrusion onto the surface could possibly occur even when the satellite as a whole is in a relatively cold, conductive state. One may be able to understand the difference between Callisto's cratered surface and the endogenically modified surfaces of some of the smaller icy Saturnian satellites such as Dione and Rhea in this way. Subsolidus convection should not now be occurring in these bodies, and if it occurred in the past, then it did so for only limited periods of time.

I. INTRODUCTION

The satellites of the solar system have followed a number of different paths in their thermal evolutions. For example, the Martian moons Phobos and Deimos are so small—a few tens of kilometers across—that they have been cold, thermally inactive bodies ever since their formation or last major collisional event. They are essentially isothermal, irregularly shaped chunks of rock that might not even be "natural" satellites of Mars. On the other hand, the Moon—the only other satellite of a terrestrial planet—has been thermally active, at least for the first 1 to 1.5 Gyr of its evolution. During this period the Moon differentiated its highland crust and poured lavas out over its surface to form the lunar maria. According to Runcorn (1978), the Moon might even have formed a core and generated its own magnetic field in this early phase of its thermal evolution. For the last 3 Gyr, however, there has been no major endogenic modification of the Moon's surface. Insofar as its outward appearance is concerned, the Moon has been a thermally lifeless body for the last several billion years.

The major satellites of the outer solar system comprise a large and diversified group whose members provide stark contrasts among themselves and with the Moon. The discoveries of the Voyager spacecraft (Smith et al. 1979a,b,1981,1982) opened these new worlds to our scrutiny and thereby redefined the entire subject of planetary thermal evolution. Although there are apparently inactive bodies with ancient cratered surfaces like Jupiter's satellite Callisto and Saturn's satellite Mimas, there are highly active ones like Jupiter's Io which probably is the most volcanically active body in the solar system. Europa, another satellite of Jupiter, and Enceladus, a small moon of Saturn, have surfaces that have been strongly modified by geologically recent internal processes. The surfaces of many of the other Jovian and Saturnian satellites have also been modified by endogenic activity that took place after the heavy cratering episode early in the evolution of the solar system. The surface of Titan, the largest Saturnian moon, is obscured by an atmosphere having a column mass 50% greater than Earth's atmosphere even though the satellite is only about 40% of the Earth's radius and 2% of its mass. Very little information is available on the satellites of Uranus and Neptune which are awaiting a visit by the Voyager 2 spacecraft. Even less is known about the single moon of Pluto.

It is generally accepted that outward heat transport from the deep interior of a planet or satellite by subsolidus convection controls the dynamic activity of the body and the extent of endogenic surface modification. The level of activity is determined by the rate at which heat is converted to mechanical work and eventually lost through the surface. The vigor of thermal convection in the satellite's interior and, accordingly, the rate of heat loss are strongly dependent on the rate of internal heating, the interior viscosity, and the size of the body (see Sec. V on heat transfer below). The smaller the planet or satellite, the more rapidly it will cool and the more thermally evolved it will be at any given time.

The differences in thermal activity between the Moon and various other satellites cannot be attributed to size, however, since some active satellites are relatively small (Enceladus' radius is only about 15% of the lunar radius) while another, Io, is comparable. In a similar vein Ganymede, Jupiter's largest moon, is comparable in size to Mercury (its radius is about 110% as large), yet the grooved terrain of the satellite's surface indicates either more vigorous endogenic activity on that body than ever occurred on the innermost planet or activity that persisted longer. The greater activity of some of the Jovian and Saturnian satellites is due not to size but to different compositions and modes of heating. The Moon and the terrestrial planets are heated mainly by the radioactive decay of uranium, thorium and potassium. Io is strongly tidally heated because of its close proximity to Jupiter and the forced eccentricity of its orbit. Enceladus might have undergone significant tidal heating in the past if its eccentricity was once much higher than it is now. None of the bodies of the inner solar system are significantly heated by tidal dissipation whereas all outer solar system satellites undoubtedly also possess the radiogenic heat source of the Moon and terrestrial planets. Strong tidal heating is only important for a few of the outer solar system satellites; it probably accounts for Io's extensive volcanism, may be involved in producing Europa's odd surface, and might be responsible for Enceladus' highly modified surface.

The densities of Ganymede and Callisto (Table I) suggest that water ice is a major component of these bodies. The fraction of water ice in the satellites of Saturn, which have densities close to 1000 kg m^{-3} (with the exception of Titan; see Table I), is even larger. Small amounts of condensed volatiles with low melting points, such as ammonia hydrate, might also be incorporated in the icy satellites. The compositions of these bodies, especially the presence of a low melting point constituent, may explain why some of them have highly modified surfaces, compared to the Moon, for example, while others do not. Differences in surface modification among the icy satellites and the Moon may also depend on the relative ease of deformation of ice and ice-rock mixtures compared with rock at temperatures near to, but below, the solidus of water ice.

Chapters in this book dealing with particular satellites will necessarily touch on the thermal evolution of these bodies. The purpose of this chapter is to discuss thermal evolution models of satellites in a more general way by emphasizing the fundamental physical processes common to all planetary bodies and by comparing the different evolutionary tracks followed by individual satellites. We begin by considering the compositions and structures of satellites (Sec. II); we then go on to discuss heat sources (Sec. III), and the relevant thermal and mechanical properties of the materials in satellite interiors (Sec. IV), and finally heat transfer mechanisms (Sec. V). Thermal history scenarios are presented next for most of the satellites of the solar system

TABLE I
Satellite Physical Properties[a]

Satellite	Distance to Planet (10^3 km)	Radius (km)	Mass (10^{19} kg)	g (m s^{-2})	Density (10^3 kg m^{-3})	Approximate Central Pressure (MPa)[b]	Approximate Silicate Mass Fraction f		
							Anhydrous vs. Hydrated Silicates	Undifferentiated vs. Completely Differentiated Satellite Models	Thermally Evolved Undifferentiated Models
Moon	384, Earth	1738	7,349 ± 7	1.62	3.34	4,710	1		
Io	422, Jupiter	1815 ± 5	8,920 ± 40	1.80	3.55	5,800	1		
Europa	671, Jupiter	1569 ± 10	4,870 ± 50	1.32	3.01	3,180	0.94–1.0	0.85–0.90	
Ganymede	1070, Jupiter	2631 ± 10	14,900 ± 60	1.44	1.93	3,610	0.58–0.73	0.43–0.49	
Callisto	1880, Jupiter	2400 ± 10	10,750 ± 40	1.25	1.83	2,700	0.52–0.66	0.42–0.48	
Mimas	185, Saturn	196 ± 3	4.55 ± 0.54	0.079	1.44 ± 0.18	10.9	0.47–0.55		0.47
Enceladus	238, Saturn	250 ± 10	7.40 ± 3.6	0.079	1.13 ± 0.55	12.6	0.22–0.26		
Tethys	295, Saturn	530 ± 10	75.5 ± 9	0.18	1.21 ± 0.17	59.0	0.29–0.35		0.34
Dione	377, Saturn	560 ± 5	105.2 ± 3.3	0.22	1.43 ± 0.06	89.6	0.46–0.54		0.46
Rhea	527, Saturn	765 ± 5	249 ± 15	0.28	1.33 ± 0.09	145	0.39–0.46	0.34	0.36
Titan	1222, Saturn	2575 ± 2	13,457 ± 3	1.35	1.881 ± 0.004	3,280	0.55–0.70	0.42–0.48	
Iapetus	3561, Saturn	730 ± 10	188 ± 12	0.24	1.15 ± 0.08	100	0.24–0.28		0.25

[a] Data for the Jovian satellites are from Morrison (1982a). Saturn satellite data are from Tyler et al. (1982) and Smith et al. (1982). Compare with Table IV of the introductory chapter.

[b] Computed from $p = (2/3)\pi G \rho_{sat}^2 R_{sat}^2$ with p = pressure, ρ_{sat} = satellite density, R_{sat} = satellite radius, G = universal gravitational constant.

(Secs. VI–XIII) and these are related to the constraints imposed by surface geology. Section XIV concludes with our assessment of how well our basic ideas of planetary thermal evolution explain the different evolutionary paths followed by the satellites of the solar system.

II. COMPOSITION AND STRUCTURE

Satellite thermal evolution, composition, and structure are topics that are fundamentally intertwined. Thermal evolution is strongly influenced by composition and structure and vice versa. As an example of how temperature affects composition and structure, consider the melting and differentiation of a homogeneous satellite and how it depends on the competition between heating and cooling rates. As examples of how composition and structure affect thermal evolution, consider how differentiation concentrates and redistributes heat sources and influences convective heat transfer by the imposition of chemical compositional discontinuities. We will therefore begin our discussion of how satellites evolve thermally with a look at what they are made of and how their components are arranged. The Moon and the larger satellites of Jupiter and Saturn, listed in Table I, will be discussed in some detail. Insofar as composition is concerned, our main interest will be in the relative amounts of silicate or rock and ices of water, ammonia and methane. The rock or silicate fraction is defined in this context as consisting primarily of the silicates of magnesium and iron with smaller amounts of other elements (Ca, Al, Na, etc.) in roughly solar or cosmic proportions. Iron and perhaps sulfur and their compounds could also be present in significant amounts.

Composition

The main constraints on the compositions and structures of satellites are the observations of their masses and radii. These are listed in Table I together with the calculated densities of the satellites. The densities of Io and the Moon indicate that they are composed largely of rock, presumably incorporating different mass fractions of iron. The lower densities of the other satellites indicate that they contain substantial quantities of H_2O, and perhaps smaller amounts of NH_3 and/or CH_4 ices. The question of which ices are present in particular icy satellites may be crucial to our understanding of them. The silicate mass fraction f of a satellite is of particular importance for its thermal evolution because the amount of internal radiogenic heating is directly proportional to f (chondritic abundances of radioactive elements are generally assumed to be representative of the silicate fraction) and the viscous resistance to convective heat transfer depends on the relative amounts of rock and ice. Unfortunately this basic compositional parameter is not well constrained even though a satellite's mass and radius might accurately estimate its overall density.

An approximate silicate mass fraction can be computed from

$$f = \frac{1 - (\rho_{ice}/\rho_{sat})}{1 - (\rho_{ice}/\rho_{sil})} \qquad (1)$$

where ρ is density and the subscripts sat, sil, and ice refer to the entire satellite and its silicate and ice components, respectively. Equation (1) tacitly assumes that ρ_{sil} and ρ_{ice} are known constants. While ρ_{sat} is reasonably well determined for all the listed satellites, ρ_{sil} and ρ_{ice} are not well constrained. For example, not only is the specific silicate composition unknown, but even more importantly, we do not know whether the silicates are hydrated or dry. Thus, ρ_{sil} could be anywhere between about 2500 kg m^{-3} (hydrated silicate) and 3500 kg m^{-3} (anhydrous silicate and iron). The effect of this uncertainty in ρ_{sil} on inferred values of f can be quantified by using Eq. (1) to calculate f for the extreme values of ρ_{sil}. The results given in Table I were obtained assuming $\rho_{ice} = 1200$ kg m^{-3}, for Ganymede, Callisto and Titan, and $\rho_{ice} = 950$ kg m^{-3} for all other satellites. (The higher pressures in the deep interiors of the larger satellites [see Table I] result in denser forms of ice.)

The unknown compositions and phases of the ices also add uncertainty to f, as does the undefined temperature T and pressure p dependences of ρ_{sil} and ρ_{ice}. Phase changes in ice are particularly important because ρ_{ice} can vary between about 1600 kg m^{-3} (dense phases of water ice) and 930 kg m^{-3} (water ice I). A more rigorous approach to the determination of f requires the simultaneous solution of

$$\rho(r) = \left[\frac{1 - f}{\rho_{ice}(T,p)} + \frac{f}{\rho_{sil}(p)} \right]^{-1}, \; f \text{ assumed constant} \qquad (2)$$

$$\frac{dp}{dr} = -\rho g \qquad (3)$$

$$g = \frac{GM(r)}{r^2} \qquad (4)$$

$$\frac{dM}{dr} = 4\pi r^2 \rho \qquad (5)$$

$$\rho_{ice} = \rho_{ice}(T,p) \qquad (6)$$

$$\rho_{sil} = \rho_{sil}(T,p) \qquad (7)$$

where r is the radial distance from the satellite's center, g is the acceleration of gravity, and $M(r)$ is the mass of a satellite internal to r. The solution of Eqs. (2–7) requires a self-consistent determination of $T(r)$, the satellite's temperature profile, and an assumption about the distribution of rock in the interior.

The distribution of rock inside a satellite affects the estimate of f because it determines the maximum pressure that the ice experiences and, accordingly, whether dense high-pressure ice phases occur in the interior. If the interior is a homogeneous ice-rock mixture (undifferentiated model), ice will be placed under higher pressure than if the interior is a rock core surrounded by an ice mantle (completely differentiated model); the pressure at the center of a homogeneous satellite model is generally higher than the pressure at the rock-ice interface of a completely differentiated model. The effect of rock distribution on estimates of f is illustrated in Table I, using results from Lupo (1982). The tabulated values of f for some of the larger satellites were computed with an anhydrous value for ρ_{sil} and water ice densities for ρ_{ice}. Realistic variations in ρ_{ice} with temperature and pressure, including phase changes in the ice, were accounted for although temperature profiles were *ad hoc* and were not the same for the differentiated and undifferentiated models. These value of f are lower than the simple estimates derived above using an anhydrous value of $\rho_{sil} = 3500$ kg m^{-3}. There are two reasons for this. Lupo's (1982) models are based on a larger value of ρ_{sil} ($3661 + 0.061p$(MPa) kg m^{-3}) and they contain a large amount of ice in high-pressure phases with densities in excess of 1200 kg m^{-3}. The undifferentiated models have lower silicate mass fractions because they have larger regions of high-pressure, high-density ice.

Uncertainty in f due to lack of knowledge of a satellite's internal temperature can be just as significant as uncertainties due to ρ_{sil} and rock distribution. Ellsworth and Schubert (1983) have calculated values of f between 0.32 and 0.40 for Rhea and 0.21 and 0.26 for Iapetus, using undifferentiated water ice-rock models with different temperatures and an anhydrous ρ_{sil}. The low values of f are for cold isothermal models in which the ice I-ice II phase transition occurs. The high values of f are for warmer convective models that do not contain ice II. Self-consistently derived values of f from the thermally evolved models of Ellsworth and Schubert (1983) are also listed in Table I for several of the Saturnian satellites.

Even considering the uncertainties in f, it appears that the larger outer solar system satellites, Ganymede, Callisto and Titan, are somewhat enriched in silicates relative to a nominal cosmic abundance of 40% anhydrous silicate and 60% water ice. The smaller satellites, Enceladus, Tethys and Iapetus, are most probably depleted in silicates relative to the cosmic abundance. Dione and Rhea are, respectively, probably more and less silicate-rich than cosmic abundance, a conclusion which does not fit very well with the general observation based on satellite size. Small Mimas has an overabundance of silicates if the Voyager determination of its mass by Tyler et al. (1982) is correct (see the discussion of Table IV in the introductory chapter). The fact that the rock/ice ratios of the large icy satellites are similar but different from the cosmic abundance is a problem whose resolution may provide important constraints on either the composition and differentiation of the protoplanetary and protosatellite nebulae or the process of satellite formation.

The ice's composition generally has a small influence on f because other likely ices, i.e., $NH_3 \cdot H_2O$ and $CH_4 \cdot nH_2O$, (where $n \simeq 7$) have densities comparable to water ice and concentrations that are small compared to water ice. However, almost no data exist on the temperature and pressure dependences of the densities of these other ices at the conditions encountered in satellite interiors. Cosmic abundances suggest that ammonia could constitute at most 18 ± 6 molar percent of a $NH_3 \cdot H_2O$-H_2O system (after accounting for oxygen incorporated into rock [Stevenson 1983a]).

Structure

There are few observational constraints on the structures of satellite interiors. Mass, radius, moment of inertia factor, surface composition, and surface geology all help to focus theoretical speculation, but the range of acceptable interior models is broad. Dermott (1984c) has discussed how the shapes of satellites that are close to their primaries can be used to provide information on their states of internal differentiation. Some measurements of the shapes of Io, Mimas and Enceladus are available (Davies and Katayama 1983a), but the uncertainties in their shapes and densities preclude definitive conclusions about their internal density distributions. We will summarize possible structural models here, reserving until later much of the discussion required to substantiate the viability of the models.

There are more data on the Moon (see chapter by Kaula et al.) than any other satellite so that the only major uncertainty about its internal structure is whether or not it possesses a small metallic core (Fig. 1). Any core within the

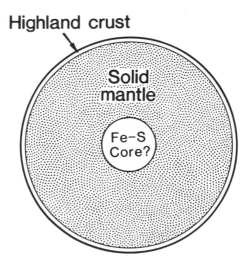

Fig. 1. Sketch of the internal structure of the Moon. The major uncertainty is whether or not the Moon has a small metallic core.

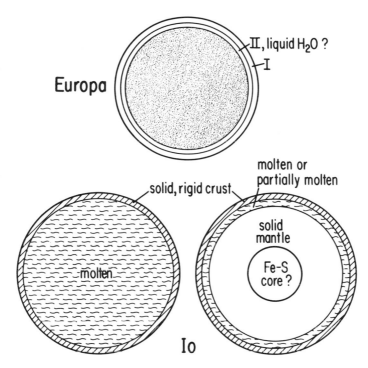

Fig. 2. Possible interior structures of Europa and Io. The shaded region inside Europa consists of hydrous or anhydrous silicates. The extent of dehydration of the deep interior is uncertain. Roman numerals I and II refer to different phases of ice.

Moon would have to be so small, however ($<$ 400 km radius), that it has little influence on the course of the Moon's thermal evolution. However, the presence or absence of a lunar core may be crucial for understanding the origin and initial thermal state of the Moon. The relatively thick lunar highlands crust is evidence that the outer several hundred kilometers of the Moon were hot enough to differentiate this crust immediately upon accumulation of the Moon some 4.5 Gyr ago.

Io is very much like the Moon in size and mass (Table I) but, as Fig. 2 shows, it may be very different internally. Io's high surface heat flow of 1 to 2 W m^{-2} (Morrison and Telesco 1980; Matson et al. 1981a; Sinton 1981; Pearl and Sinton 1982; cf. chapter by Nash et al.) and its active sulfur volcanism (Morabito et al. 1979) show that its interior is being heated at a rate much larger than any other satellite or planet. This high rate of internal heat production has led to widespread acceptance of a model in which Io is totally molten except for a thin solid and rigid crust (Peale et al. 1979; Johnson 1981; Kieffer 1982; Morrison 1982a; Pollack and Fanale 1982). The molten Io model has fundamental difficulties that will be enumerated later (Sec. VII) and they lead us to advocate the alternative structure shown in Fig. 2. This model, first suggested by Schubert et al. (1981), has a solid mantle, a molten or partially molten asthenosphere, and a rigid crust (see also Cassen et al. 1982). Io should have an iron-rich core if its interior were largely molten, but it is not certain whether such a core would exist if the mantle were solid.

Europa, though smaller than the Moon and Io, is mostly rock and there is little question that its interior has a structure similar to the one shown in Fig. 2. The major uncertainties with Europa (see the chapter by Malin and Pieri) are whether the silicates are hydrated (Ransford et al. 1981; Finnerty et al.

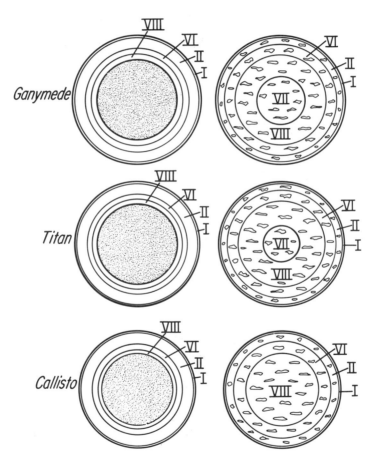

Fig. 3. Fully differentiated (left) vs. undifferentiated (right) models of Ganymede, Titan, and Callisto. The differentiated models have silicate cores (shaded) and ice mantles. The undifferentiated models have ice-rock mixtures throughout. The ice phases are designated by Roman numerals. The ice VII cores in the undifferentiated models of Ganymede and Titan require a temperature that increases with depth essentially all the way to the satellite's center. Such an increase in temperature could occur, for example, if the subsolidus temperature profile at great depth is parallel to the melting curve.

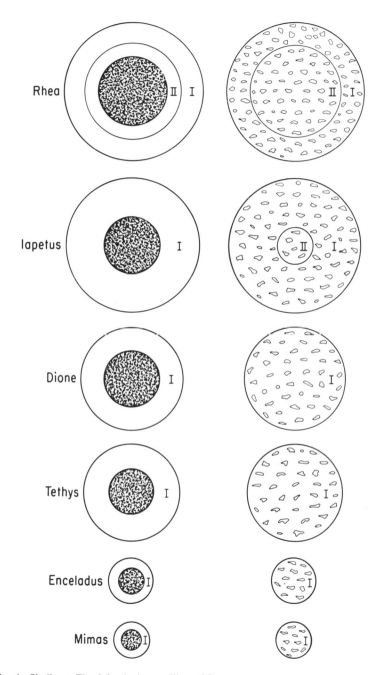

Fig. 4. Similar to Fig. 3 for the icy satellites of Saturn.

1981) or dry and whether there exists any liquid water beneath its ice crust. There should be enough heat production in Europa to have substantially dehydrated any water-rich silicates that might have originally constituted the interior.

Structural models of the other icy satellites are shown in Figs. 3 and 4. They represent two limiting cases—fully differentiated structures with rock cores surrounded by ice mantles and undifferentiated structures of homogeneous ice-rock mixtures. The rock core-ice mantle structure could be a consequence of heterogeneous accretion, or it could form by internal differentiation of a satellite that accumulated initially as a homogeneous ice-rock mixture. The silicate core-ice mantle model of the icy outer-planet satellites has received much support in the literature. Early models of these satellites had rock cores surrounded by liquid water mantles and ice crusts (Lewis 1971a,b; Consolmagno and Lewis 1976,1977; Fanale et al. 1977a; Johnson 1978). Reynolds and Cassen (1979) pointed out that the liquid mantles would rapidly freeze and more recent variations of the differentiated satellite model have ice mantles (except perhaps for Europa and Enceladus) (Cassen et al. 1979a, 1980a,b,1982; Parmentier and Head 1979; Johnson 1981).

Support for rock core-ice mantle models of the icy satellites of Jupiter and Saturn has been extensive because of the widely held view that, even if these bodies initially formed as homogeneous silicate-ice mixtures, subsequent heating of their interiors would melt the ice and separate the rock and water. Lewis (1971a,b), Consolmagno and Lewis (1976,1977,1978), and Fanale et al. (1977a) considered radiogenic heating in the silicates to be adequate to melt the ice, but more recent papers (Parmentier and Head 1979b; Thurber et al. 1980; Schubert et al. 1981) have shown that subsolidus convection in the ice could transport this heat to the surface and preclude melting of an icy satellite of any size by this heat source. This conclusion may not be valid if the satellite's silicate mass fraction exceeds a critical value above which the ice-rock mixture is considerably more resistant to deformation than pure ice. It has been proposed by Friedson and Stevenson (1983) that Ganymede differentiated because its silicate mass fraction is above the critical value.

Cassen et al. (1982) have argued that large icy satellites could be differentiated by the gravitational potential energy released upon accretion of the bodies. There is enough accretional energy (see Table II) available from the formation of Ganymede, Callisto and Titan that these satellites could be substantially differentiated. There is so little accretional energy associated with the accumulation of the small icy Saturnian satellites Mimas, Enceladus, Tethys and Dione that these bodies are almost certainly undifferentiated ice-silicate mixtures if they initially formed that way and if they have not subsequently been melted by yet another source of energy like tidal dissipation (Ellsworth and Schubert 1983).

When account is taken of the fraction of accretional energy actually re-

TABLE II
Satellite Heat Sources

Satellite	Total Accretional Energy Per Unit Mass[a] (MJ kg⁻¹)	Total Accretional Energy[a] (10^{27} J)	$\Delta T \equiv \dot{E}_{acc}/c$ [b] (K)	Energy of Differentiation in % of E_{acc}	Radiogenic Heat Production Rate[c] (10^{11} W)	Homogeneous Tidal Heating Rate[d] (10^{11} W)
Moon	1.69	124	1400		6.79	
Io	1.97	176	1600	6	4.87	17
Europa	1.24	60.5	1000	5	2.50–2.66	0.7
Ganymede	2.27	338	1300	12	4.72–5.93	
Callisto	1.79	193	1000	13	3.05–3.87	
Mimas	9.29×10^{-3}	4.23×10^{-4}	5		$(1.17–1.37) \times 10^{-3}$	
Enceladus	1.19×10^{-2}	8.77×10^{-4}	15		$(0.89–1.05) \times 10^{-3}$	6.2×10^{-4}
Tethys	5.70×10^{-2}	4.31×10^{-2}	40		$(1.19–1.44) \times 10^{-2}$	
Dione	7.52×10^{-2}	7.91×10^{-2}	50		$(2.64–3.10) \times 10^{-2}$	
Rhea	0.130	0.325	110		$(5.30–6.25) \times 10^{-2}$	
Titan	2.09	282	1200	12	4.04–5.14	
Iapetus	0.103	0.194	130		$(2.46–2.87) \times 10^{-2}$	

[a] Homogeneous accretion is assumed.

[b] Assumes specific heat $c = 1.2$ kJ kg⁻¹ K⁻¹ for the Moon, Io, and Europa; $c = 1.8$ kJ kg⁻¹ K⁻¹ for Ganymede, Callisto, Titan, and Mimas; $c = 1.5$ kJ kg⁻¹ K⁻¹ for Tethys and Dione; $c = 1.2$ kJ kg⁻¹ K⁻¹ for Rhea; $c = 0.8$ kJ kg⁻¹ K⁻¹ for Iapetus and Enceladus.

[c] The lunar heating rate is 9.2×10^{-12} W kg⁻¹ and the chondritic heating rate is 5.45×10^{-12} W kg⁻¹. The ranges of values for the icy satellites correspond to the ranges of silicate mass fractions in Table I.

[d] Estimate for tidal heating in a completely solid satellite for Io and Europa by Cassen et al. (1982) and for Enceladus by Squyres et al. (1983b).

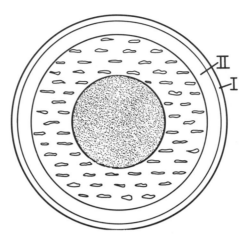

Fig. 5. An alternative structural model for either Ganymede or Callisto consisting of a silicate core (shaded) surrounded by an undifferentiated ice-rock lower mantle and an ice upper
mantle. The evolutionary steps in the establishment of this structure are described in the text.

tained in an accumulating body (a highly uncertain parameter) and the radial
distribution of this heat deposition, it is seen that accretional melting may
have been confined to only the outer regions of the larger ice-rock satellites
(Schubert et al. 1981). Figure 5 shows an alternative structural model for either Ganymede or Callisto based on this possibility. The silicate core in the
model accumulates from rock released upon melting of the outermost layers
which subsequently refreeze. The model contains a primordial undifferentiated rock-ice mixture in a spherical shell between the rock core and the ice
near the surface (Schubert et al. 1981). The undifferentiated mixture was once
nearer the center of the satellite; it has been displaced outward by the formation of the rock core. The ancient heavily cratered surface of Callisto and the
more endogenically modified surface of Ganymede may reflect different degrees of ice-rock separation in the interiors of these satellites.

The same processes that could have limited the differentiation of Ganymede and/or Callisto, i.e., retention of only some of the accretional energy
and preferential deposition of this heat in the outer layers of an accumulating
satellite, make it highly unlikely that accretional heating could have separated
the ice and rock in the middle-sized icy Saturnian satellites Rhea and Iapetus
(Ellsworth and Schubert 1983). We conclude that all the icy satellites of Saturn, except for Titan and perhaps Enceladus, have most probably not undergone extensive bulk differentiation. However, the surfaces of Saturn's icy satellites indicate that some differentiation has occurred in these bodies. The
largest icy satellites Ganymede, Callisto and Titan are partially to fully differentiated. Based on appearance, Callisto is the least differentiated of the three,
and could even be undifferentiated. An icy satellite can be resurfaced by the

melting of a minor ice component (e.g., ammonia hydrate) even though the major water ice component is not melted and the bulk of the satellite is not differentiated.

III. HEAT SOURCES

In addition to accretional energy, the interiors of the satellites could have been heated by the decay of radioactive isotopes, tidal dissipation, or differentiation and core formation. Heating by short-lived radionuclides such as ^{26}Al (Runcorn 1977) and solar wind-driven planetary electrical induction currents (Sonett et al. 1975; Herbert et al. 1977b) are other possible energy sources.

Accretional Heating

The accretion process and the conditions in the protoplanetary nebula are poorly understood (see chapter by Stevenson et al.). In any formation process, however, the total gravitational energy available is $\chi GM_s/R_s$, where G is the universal gravitational constant, M_s is the satellite mass, R_s is its radius, and χ is a constant of order unity which depends on the details of the accretion process; χ is $3/5$ if accretion is homogeneous. (This is discussed in detail also by Safronov et al. in their chapter.) The value of the total available gravitational potential energy, per unit mass, assuming homogeneous accretion, ranges from a high of 2.3 MJ kg^{-1} for Ganymede to a low of 9.3 kJ kg^{-1} for Mimas (Table II). If there were no losses, the heating could be large, of order 1000 K, for the Moon, the Galilean satellites and Titan. It would be of little significance for the smaller satellites (Mimas, Enceladus, Tethys and Dione) but it could be around 100 K for Rhea and Iapetus. However, only a fraction of this gravitational energy can be retained by the growing satellite; the rest is reradiated into space or removed by convection in the surrounding nebula. The latter would have occurred if the satellites accreted in an optically thick dense gas nebula as has been suggested for the Galilean satellites by Lunine and Stevenson (1982a). The conventional accretion model for the terrestrial planets by Safronov (1969) assumes a gas-free environment (cf. chapters by Stevenson et al. and Safronov et al.).

In any case, the fraction h of heat retained during accretion is very poorly known. In addition to its dependence on the rate of heat transfer in the nebula and on the nebula's temperature, h will depend on the size distribution of the accreting particles and planetesimals, on their relative velocity (here the proximity to the primary planet is a factor), and on the time scale of the accretion process. For models of the accreting Moon, Kaula (1980) finds that about 50% of accretional energy is retained. Coradini et al. (1982b) and Federico and Lanciano (1983) argue that, for the icy satellites, the fraction of energy retained is $< 20\%$ and may even be $< 10\%$. A similar percentage would apply

to a satellite accreting in a dense nebula. In terms of h, we can approximate
the accretional temperature T_a at radius r within an accumulating satellite as

$$T_a(r) = \frac{hGM(r)}{cr}\left[1 + \frac{ru^2}{2GM(r)}\right] + T_e \qquad (8)$$

where c is the specific heat, T_e is the ambient temperature during accretion,
$u^2/2$ is the approach kinetic energy per unit of mass of planetesimals forming
the satellite, and $M(r)$ is the mass of the satellite internal to r. For small
bodies with nearly uniform density $M(r) = 4/3\pi r^3 \rho_{sat}$.

Schubert et al. (1981) and Lanciano et al. (1981) have calculated accretional temperature profiles for Ganymede and Callisto for a wide range of values of h and an ambient temperature of $\simeq 100$ K. Ellsworth and Schubert
(1983) have done similar calculations with $h = 1$ for Saturnian satellites ranging in size from Mimas to Rhea. According to these calculations (Fig. 6), at
least the outer layers of Ganymede and Callisto (and by analogy Titan) could
have been melted, while accretional heating could not have melted the smaller
Saturnian satellites. The small moons of Saturn could have been heated enough
for solid-state convection to have occurred in their interiors. The dense cloud
scenario of Lunine and Stevenson (1982a; chapter by Stevenson et al.) also
results in a molten outer layer on Ganymede but melting could have been
avoided for Callisto if the incoming planetesimals were in the size range of 1
to 7 m in radius. In the gaseous accretion models, large planetesimals are not
likely to reach the surface because of hydrodynamic breakup (see the chapter
by Safronov et al.).

Energy of Differentiation

Some of the gravitational energy deposited in a planet or satellite during
accretion can become available for heating if the body differentiates into a
core and mantle. In contrast to accretional energy, all the energy of differentiation is retained within the planet. The average densities of the icy satellites of Jupiter and Saturn are consistent with rock fractions between about 20
and 70% by mass (Table I). After homogeneous accretion, the silicates in
these bodies could, in principle, separate to form rock cores. Similarly, the
average density of Io allows for an Fe-FeS core to exist (Cassen et al. 1982).

The gravitational energy made available by differentiation of an initially
homogeneous satellite can be written

$$U_H - U_D = 2\pi \int_0^R [\rho_H(r)V_H(r) - \rho_D(r)V_D(r)]r^2 dr \qquad (9)$$

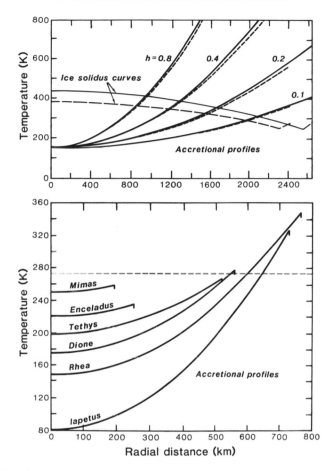

Fig. 6. Accretional temperature profiles for Ganymede and Callisto (upper plot), with different assumptions regarding the value of h (after Schubert et al. 1981), and for the Saturnian satellites (lower plot), with $h = 1$ (Ellsworth and Schubert 1983). The parameter h is the fraction of impact energy retained as heat in an accumulating satellite.

where U is the gravitational potential energy, V is the gravitational potential, R is the satellite radius, ρ is density, r is radial distance from the center of the satellite, and subscripts H and D refer to the homogeneous and differentiated states, respectively. The potential energy of a homogeneous satellite is

$$U_H = -\frac{16}{15}\pi^2 GR^5\rho_{sat}^2 \qquad (10)$$

where $\rho_H = \rho_{sat}$. The potential energy of a two-layer model of the differentiated satellite is

$$U_D = -\frac{16}{15}\pi^2 GR^5 \left[\rho_m^2 + \frac{5}{2}\rho_m(\rho_{sat} - \rho_m) \right.$$

$$\left. + \left(\frac{3}{2}\rho_m - \rho_c\right)(\rho_m - \rho_c)\left(\frac{\rho_{sat} - \rho_m}{\rho_c - \rho_m}\right)^{5/3} \right]$$

$$(11)$$

where ρ_c and ρ_m are the densities of the core and mantle, respectively. Conservation of mass is assumed in deriving Eq. (11). Thus the energy of differentiation, as a percentage of the energy of accretion, is

$$\frac{U_H - U_D}{-U_H} = -1 + \frac{5}{2}\frac{\rho_m}{\rho_{sat}} - \frac{3}{2}\frac{\rho_m^2}{\rho_{sat}^2}$$

$$(12)$$

$$+ \left(\frac{3}{2}\frac{\rho_m}{\rho_{sat}} - \frac{\rho_c}{\rho_{sat}}\right)\left(\frac{\rho_m}{\rho_{sat}} - \frac{\rho_c}{\rho_{sat}}\right)\left(\frac{\rho_{sat} - \rho_m}{\rho_c - \rho_m}\right)^{5/3}$$

The energies of differentiation of the major satellites of Jupiter and Saturn are given in Table II. For the icy satellites, ρ_c is the density of rock ($\rho_c \simeq$ 3000 kg m^{-3}; however, $\rho_c = 3500$ kg m^{-3} for Europa because the density of Europa requires that a rock density > 3000 kg m^{-3} be assumed) and ρ_m is the density of ice ($\rho_m \simeq 1000$ kg m^{-3}). For Io, ρ_c is the density of an Fe-FeS alloy ($\rho_c \simeq 5760$ kg m^{-3} for a eutectic composition at 5.5 GPa [Usselmann 1975]) and ρ_m is the density of rock. The fraction of the total accretional energy that could be dissipated during differentiation is substantial, $\simeq 12\%$, for the large icy satellites Ganymede, Callisto and Titan. The energy fraction would amount to only about 7% if the rock density were 2500 kg m^{-3}, representative of hydrated rock. Even in this case, the energy of differentiation would be sufficient to provide the latent heat of melting of water ice per unit mass of the rock-ice mixture. Friedson and Stevenson (1983) have suggested the possibility of a runaway differentiation of Ganymede partly on this basis. The energy of differentiation for Io is insignificant in comparison to the latent heat of rock or the latent heat of an iron-sulfur alloy.

Radiogenic Heating

Heat production by radioactive decay of the long-lived isotopes [238]U, [235]U, [232]Th and [40]K is the major source of energy in the post-accretionary evolution of the Moon and the terrestrial planets. It probably contributes to the heat budget of Io and Europa and to that of the icy satellites in proportion to their rock component. The rock in the icy satellites and in Io and Europa is presumed to be close to chondritic in composition. Present-day concentrations C_0 of U, Th, and K and specific heat production rates in ordinary

TABLE III
Concentration, Energy Production and Decay
Constants of Radiogenic Elements

	Concentration in Ordinary Chondrites[a] (ppm)	Concentration in Lunar Interior[b] (ppm)	Decay Energy[c] $(10^{-5}\ W\ kg^{-1})$	Decay Constant[d] λ $(10^{-10}\ yr^{-1})$
U	0.012	0.046	9.75	1.551
Th	0.040	0.170	2.60	0.495
K	840	92.0	3.52×10^{-4}	5.543

[a] Mason (1971).
[b] Langseth et al. (1976).
[c] Calculated from Birch's (1954) data.
[d] Steiger and Jaeger (1977).

chondrites are summarized in Table III. Past concentration C can be calculated from

$$C = C_0 e^{\lambda t} \tag{13}$$

where λ is the decay constant (also given in Table III) and t is time before the present. The concentrations in the Moon are estimated from the assumption of a steady-state balance between radiogenic heat production in the lunar interior and heat flow through the surface and the constraint of the abundance ratios K/U = 2000 and Th/U = 3.7 in lunar surface rocks. Since the cooling of the Moon should also contribute to the surface heat flow (Schubert et al. 1980), the lunar concentrations in Table III are likely to be overestimates. Recent geochemical studies suggest that only 50% to 75% of the Earth's surface heat flux is balanced by radiogenic heat production (O'Nions et al. 1979). It should also be noted that the lunar surface heat flow has been measured at only two locations (21 mW m^{-2} at the Apollo 15 landing site and 16 mW m^{-2} at Apollo 17). Actual heat production rates due to radioactive decay of long-lived isotopes in the Moon, Io, Europa and the icy satellites of the outer solar system can be estimated with the help of Table III and the satellite masses and silicate mass fractions from Table I. These estimates are given in Table II.

The decay of short-lived isotopes such as ^{26}Al (Fish et al. 1960) might have contributed to the heat budget in the earliest phase of the evolution of satellites because the half-life of ^{26}Al (7 × 10^5 yr) could be comparable to accretion times. However, the contribution of ^{26}Al to the heat budget for the later evolution may be ignored, except insofar as initial conditions are concerned.

Tidal Heating

Although tidal dissipation is presently a negligible heat source for the Moon (Kaula 1963; Peale and Cassen 1978; cf. Van Arsdale and Burns 1979), it is an important source of energy in some of the moons of Jupiter and Saturn because of the large masses of these planets and the small orbital distances and forced orbital eccentricities of some of these satellites. Just prior to the arrival of Voyager 1 at Jupiter, Peale et al. (1979) proposed that Io should be intensely heated by the tides raised by Jupiter. Their prediction was confirmed by the observation of satellite-wide volcanic activity (Morabito et al. 1979; Smith et al. 1979a,b; Masursky et al. 1979) and the subsequent recognition that infrared measurements indicate a large internal heat source (Morrison and Telesco 1980; Matson et al. 1981a; Sinton 1981; Pearl and Sinton 1982; see the chapter by Nash et al.). Tidal heating has also been proposed as the cause of the extensive resurfacing of Saturn's small satellite Enceladus (Yoder 1979b; Squyres et al. 1983b). Dissipation of tidal energy may also contribute to the heat budget of Europa (Cassen et al. 1979a,1980b) and may have been an important heat source in the early thermal history of Ganymede (but probably not in the early evolution of Callisto) (Cassen et al. 1980b).

The theory of tidal dissipation goes back to the work of Darwin (1898, 1908) and was developed by Gerstenkorn (1955,1969), Kaula (1963,1964), Kaula and Yoder (1976), Peale and Cassen (1978), Yoder (1979b), and Yoder and Peale (1981) (see the chapter by Peale). The gravitational field of the planet causes the satellite to deform into a prolate spheroid with its long axis pointing toward the planet. If the satellite does not rotate synchronously with its orbital period or if the satellite's orbit is eccentric, the satellite will experience a periodic forcing with a large part of the deformational energy being dissipated as heat. Internal friction causes angular momentum to be transferred between the planet and its satellite, forcing the rotation and/or eccentricity to be damped (see the discussion in the chapter by Burns). Thus, the Galilean satellites today rotate synchronously with their orbital periods as does the Earth's Moon (cf. Greenberg and Weidenschilling 1984). Eccentricities of the orbits of the satellites would also relax were it not for orbital resonances in the satellite system. Two principal tides are raised on a synchronously rotating satellite moving in an eccentric orbit: one is radial and is due to the varying distance of the satellite to the planet; the other, the dominant tide (Yoder 1979b) moves back and forth across the mean subplanetary point and is due to the libration of the satellite which results from variations of the orbital angular velocity along the elliptic orbit (see Fig. 1 in Burns' chapter).

The total rate of tidal dissipation $\dot{E}_T$ in a satellite of uniform density ρ and uniform rigidity μ, whose orbit has a small eccentricity e ($e \ll 1$), negligible obliquity and a semimajor axis a large compared to the satellite's radius R_s, is (to order e^3) given by

$$\dot{E}_T = \frac{21}{2} k_2 \frac{GM_p^2 R_s^5 e^2 n}{a^6 Q} \qquad (14)$$

(Peale and Cassen 1978; Cassen et al. 1982; chapter by Peale). In Eq. 14 M_p is the mass of the primary planet, n is the satellite's mean orbital motion, Q is its dissipation factor, k_2 is its potential Love number of second degree

$$k_2 = \frac{3/2}{1 + \dfrac{19\,\mu}{2\,\rho g R_s}} \qquad (15)$$

and g is the satellite's surface gravity. The most uncertain parameters in Eqs. (14) and (15) are Q and μ because they depend on the interior structure and the composition of a satellite. For example, if Io's interior is melted, then its effective Q will be significantly smaller than values estimated for the Moon, while the relevant value of μ will be the rigidity of Io's outer shell. Estimates for tidal heating rates in solid homogeneous models of Io, Europa and Enceladus from Peale et al. (1979), Cassen et al. (1979a,1980b) and Squyres et al. (1983a,b), are given in Table II.

Some additional constraints can be placed on $\dot{E}_T$ from orbital dynamics and the physical properties of the primary planet. The rate of tidal transfer of the planet's rotational energy to the satellite's orbital energy is (see, e.g., Officer 1974)

$$\dot{E}_p = Ln = \frac{3}{2} k_{2p} \frac{GM_s^2 R_p^5 n}{a^6 Q_p} \qquad (16)$$

where L is the tidal torque exerted on the satellite and the subscripts p and s refer to the planet and the satellite, respectively. Love numbers of the giant planets have been determined by Gavrilov and Zharkov (1977) from explicit calculations of tidal effects on the giant planets. They find k_{2p} equal to 0.379 for Jupiter, 0.341 for Saturn, 0.104 for Uranus, and 0.127 for Neptune. A lower bound on Q_p can be found from the present orbital distance of a satellite to its primary (Goldreich and Soter 1966). If the planet's rate of rotation is larger than the satellite's rate of revolution, transfer of the planet's rotational energy to the satellite's orbital energy will result in an expansion of the orbit at a rate proportional to $\dot{E}_p$. An integration over the age of the solar system of the rate of change of orbital distance gives a lower bound for Q_p. Goldreich and Soter (1966) estimated lower bounds of 10^5 for Jupiter's Q from Io's present orbital distance, 6×10^4 for Saturn's Q from Mima's present orbital radius, and 7×10^4 for Uranus from Miranda's present orbital distance. These estimates depend on k_{2p}, however, for which Goldreich and Soter (1966) used the "fluid" value of 1.5. With the values of k_{2p} calculated by Gavrilov and Zharkov (1977), the minimum Q_p values are significantly reduced to 2.5 ×

10^4 for Jupiter, 1.4×10^4 for Saturn, and 5×10^3 for Uranus. The rate of energy transfer to Io's orbit, for example, is then 1.7×10^{14} W. This is further considered in the chapter by Burns.

The ratio

$$\frac{\dot{E}_T}{\dot{E}_p} = 7 \, De^2 \tag{17}$$

with D the tidal scale factor

$$D \equiv \frac{k_{2s}}{k_{2p}} \left(\frac{R_s}{R_p} \right)^5 \left(\frac{M_p}{M_s} \right)^2 \frac{Q_p}{Q_s} \tag{18}$$

is also constrained by orbital dynamics (see the chapters by Peale and by Nash et al.). For Io in an equilibrium resonant orbit with Europa and Ganymede, Yoder (1979b) finds

$$D \leq (13e_f^2)^{-1} \tag{19}$$

where $e_f \simeq e$ is the eccentricity forced by the resonance. Thus, a bound on the planet's dissipation can be translated into a bound on the dissipation in the satellite. For Io the upper bound is 9×10^{13} W. This value is about three times larger than the one quoted by Cassen et al. (1982), who assumed $k_{2J} = 0.5$ and $Q_J = 10^5$, and is comparable to Io's surface heat flow.

IV. RHEOLOGY

The evolution of a planet or satellite depends to a large extent on the rheology of its interior. If subsolidus creep rates are sufficiently large, as for the Earth's mantle, then solid-state convection can readily remove the heat produced inside the body. The thermal state of the interior will self-regulate (Tozer 1965) because subsolidus creep is strongly temperature dependent. If creep rates are too small for subsolidus convection to occur, heat will be transported by conduction. This may lead to melting and internal differentiation, as has been discussed in Sec. II, if the heat production rate is sufficiently large.

Because creep rates slow rapidly with decreasing temperature, the convecting interiors of satellites are covered by rigid conductive shells. As a satellite cools, its conductive outer shell thickens with time. The Moon's rigid shell is about 600 km thick at present (Schubert et al. 1979). Strong tidal heating in Io's shell prevents it from thickening beyond a point where heat conduction balances tidal dissipation (see Sec. VII on Io).

A large amount of work has been published on the rheologies of rock; for reviews see Stocker and Ashby (1973), Weertman and Weertman (1975), Carter (1976) and articles in Kelly et al. (1978); and for ice I see Goodman et al. (1981), Hooke (1981) and Weertman (1983). The rheology of ices in satellites has been discussed recently by Poirier (1982) and Croft (1984) and is approached in a different context in the chapter by Squyres and Croft. The most thoroughly studied rocks and minerals include dunite and olivine. Most of the data on ice have been obtained at temperatures close to the melting point and very little is known about the rheology of phases other than that of ice I. Only very recently (Durham et al. 1983) have experiments been carried out at temperatures of interest for the small satellites of Saturn. The effect of hydrostatic pressure on ice is not well established either (Weertman 1983).

Flow mechanisms for both rock and ice include diffusion creep and dislocation glide and climb. Diffusion creep occurs through the diffusion of point defects in grains (Nabarro-Herring creep) and grain boundaries (Coble creep). Other less well-understood mechanisms include dynamic recrystallization and grain boundary melting (Ashby and Verrall 1978; Goodman et al. 1981). Additional complications arise from the presence of volatiles. Water in trace amounts usually softens rock while NH_3 in solution hardens ice (Baker and Gerberich 1979). The dominant creep mechanism, which depends on the shear stress relative to the shear modulus, the temperature relative to the melting temperature, and the grain size, can be determined from deformation maps (Ashby and Verrall 1978; Goodman et al. 1981).

The generalized flow law governing the subsolidus creep of rock and ice is

$$\dot{\varepsilon} = A\,\sigma^n\,\exp(-F/RT) \tag{20}$$

where $\dot{\varepsilon}$ is the strain rate, A and n are constants, σ is the deviatoric stress, F is the activation enthalpy of the rate-limiting microscopic transport process, R is the gas constant, and T is the temperature. If diffusion creep is dominant, $n = 1$ and the viscosity η

$$\eta = \frac{\sigma}{2\dot{\varepsilon}} = (2A)^{-1}\exp(F/RT) \tag{21}$$

is independent of stress. If dislocation creep is dominant, n is around $3-5$ and the viscosity is proportional to σ^{n-1}.

The maximum stress due to buoyancy forces in a convecting layer is of the order of $10^{-5}\,\rho g\alpha\Delta T d$(MPa), where g is surface gravity, α is thermal expansivity, ΔT is the temperature difference across the layer that drives convection, d is layer thickness, and 10^{-5} is a numerical constant chosen so that stresses of 1 to 10 MPa are calculated for the Earth's mantle. For $g \simeq 1.6$ m s^{-2}, $\alpha \simeq 10^{-5}$ K^{-1}, $\Delta T = 10^2$ K, and $d \simeq 10^6$ m, numbers appropriate for Io and the Moon, we find that the stress level in the convecting regions should be less than about 0.02 MPa. If the deformation maps for the Earth's upper mantle established by Ashby and Verrall (1978) are applicable to Io and the Moon, diffusional flow should be dominant. Ellsworth and Schubert (1983) have estimated the maximum stress level in Mimas, Enceladus, Tethys, Dione, Rhea and Iapetus to be of the order of 10^{-2} MPa. If the rheology of pure ice I_h is representative of these satellite interiors and if the average ice grain size is not much larger than 1 mm, diffusion creep should also dominate convective flow in these bodies. The stress level in the convecting interiors of Ganymede and Callisto (Schubert et al. 1981) and Titan is of the order of 0.1 MPa which puts the interiors of these satellites close to the transition between diffusion creep and dislocation creep in ice I_h. However, for modeling the thermal evolution of a satellite, the stress dependence of the viscosity may not pose a serious problem. Christensen (1984) has shown that its effect on thermal convection can be parameterized through a reduction of the activation enthalpy F.

The activation enthalpy F is given by

$$F = E_a + pV_a \tag{23}$$

where E_a is the activation energy, p is the pressure, and V_a is the activation volume. A representative value of E_a for olivine is 500 kJ mol^{-1} (Goetze and Kohlstedt 1973). For polycrystalline ice I_h at temperatures below 265 K, E_a is 60 kJ mol^{-1} (Weertman 1983); at $T \simeq 265$ K, E_a is 130 kJ mol^{-1} or more. The recent data of Durham et al. (1983) suggest that E_a may be as low as 31 kJ mol^{-1} for $T \lesssim 195$ K. Representative absolute values of activation volume for rock and ice are between 10^{-5} and 2×10^{-5} m^3 mol^{-1} (Sammis et al. 1977,1981; Ross et al. 1979; Kohlstedt et al. 1980; Weertman 1983). Activation volumes of this magnitude produce changes of 10 to 20% in activation enthalpy at the pressures encountered in the deep interiors of the largest satellites (Table I). The activation volume of ice I_h is negative as is the slope of the melting curve.

The pressure dependence of F is often conveniently modeled through the pressure dependence of the melting temperature T_m

$$F = E + pV = \bar{g}RT_m \tag{24}$$

(Weertman and Weertman 1975). Values of $\bar{g}$ for silicate rocks (compiled by Weertman and Weertman 1975) are between 20 and 30. For ice, the value of $\bar{g}$ corresponding to $E_a = 60$ kJ mol^{-1} is 26. It could be as small as 13 if the data of Durham et al. (1983) are representative or as large as 56 at temperatures close to the melting temperature.

The viscosity is often written in terms of the viscosity η_m at a reference temperature conveniently taken to be the melting temperature

$$\eta = \eta_m e^{\bar{g}\left(\frac{T_m}{T} - 1\right)}. \tag{25}$$

In most cases η_m is not the actual viscosity at T_m but rather a value extrapolated from data at temperatures well below the melting temperature. Croft (1984) presents η_m values for ice collected from the literature. Most of the data are between 10^{13} Pa s and 10^{14} Pa s, and are compatible with the value of Goodman et al. (1981) for dislocation creep mechanisms. For diffusion creep of ice, as might be relevant in the icy satellites, the data and theoretical arguments of Goodman et al. (1981) suggest $\eta_m \simeq 10^{16}$ Pa s and $\bar{g} = 25.2$ at temperatures above 200 K. At temperatures below 150 K, $\eta_m \simeq 10^{19}$ Pa s and $\bar{g}$ equals 16.5.

The rheology of high-pressure ices and of the low-temperature polymorph I_c of ice I_h has been discussed by Poirier (1982). On the basis of its crystal structure, cubic ice I_c might be even stronger than ice I_h which is already one of the strongest solids (Goodman et al. 1981). Dislocation creep in ice I_h is controlled by proton rearrangement mechanisms which would be absent in ice II where the protons are ordered. Therefore, as Poirier (1982) argues, the viscosity of ice II might be smaller than that of ice I_h. However, unpublished preliminary data on creep of ice II and ice III obtained by Echelmeyer and Kamb and briefly described by Poirier (1982) suggest that ice II is more viscous than ice I_h at the same temperature. Durham et al. (1983) observed that ductile strength drops dramatically near the transition pressure of ice I_h to ice II. According to the data of Echelmeyer and Kamb, ice III is 3 orders of magnitude less viscous than ice I_h at the same temperature. In preliminary experiments on ice II and III, Kirby et al. (1985) have found that ice III is weaker than ice I_h while ice II is stronger than ice I_h. Both ice II and III are strengthened by increasing pressure, in contrast with ice I_h which is softened with increasing pressure. The effective viscosity of ice VI at room temperature, at a pressure around 1.1 GPa, and at 90 MPa shear stress has been measured by Poirier et al. (1981) who found a viscosity of 10^{13} Pa s. This suggests a viscosity of 10^{19} to 10^{22} Pa s at a shear stress of 0.1 MPa (the expected shear stress in convecting regions of Ganymede and Callisto; Schubert et al. 1981) if a power of three to four is used for the extrapolation. Accordingly, ice VI could be much more viscous than ice I_h. Finally, ice VII is also expected to be more viscous than ice I on the basis of a comparison of their crystal structures (Poirier 1982).

Thus far we have been dealing with clean ice. Rock particles distributed in the ice will increase the ice viscosity since rock particles can be considered rigid in comparison with ice. In general, the effective viscosity of a fluid suspension will be larger than the viscosity of the pure fluid because the suspended particles exert a drag force on the fluid. In ice deforming by dislocation creep, the rock particles force the dislocations to climb around them thereby increasing the apparent activation energy for motion. The effective viscosity will increase dramatically with increasing volume concentration of solid particles once the concentration is so large that the particles form an interconnecting matrix. Roscoe's (1952) theory predicts that the critical concentration is around 60% solid particles by volume.

Hooke et al. (1972) measured the creep rate of an ice I_h-sand mixture with sand grain size ($\simeq 60$ μm) much smaller than ice grain size (300 to 500 μm) and found the creep rate to decrease exponentially with increasing volume fraction of sand. Croft (1984), by interpolating their data, concluded that viscosity could be increased by an order of magnitude over that of clean ice for silicate volume fractions appropriate to the icy satellites. The silicate volume fractions can be calculated from the mass fractions listed in Table I by multiplying the silicate mass fraction by the ratio of the satellite density to the rock density. If we assume that the rocks in Ganymede are hydrated, the volume fraction of rock would be 56% (close to the critical 60%) while for Callisto it would be about 48%. Thus the rheology of rock could have determined the interior viscosity of an undifferentiated Ganymede in which case early melting of the ice and differentiation into a rock core and ice mantle would have been inevitable. For dry silicates, however, the rock volume fraction of Ganymede is only 32%.

Friedson and Stevenson (1983) have considered the thermal histories of icy satellites with varying silicate volume fractions. They have used a generalized formula by Moshev (1979) to calculate viscosity enhancement as a function of silicate mass fraction. They further determined critical mass fractions above which convective heat loss would balance radiogenic heat production in a satellite for interior temperatures larger than the melting temperature of ice. Friedson and Stevenson (1983) conclude that Ganymede has more than the critical mass fraction of silicates if the rocks are hydrated while Callisto's rock fraction is marginal. For dry rock they find Callisto to have far less than the critical silicate fraction while Ganymede's rock content is marginal.

V. HEAT TRANSFER

The important heat transfer mechanisms in the interiors of the satellites are conduction and convection. Heat transfer by electromagnetic radiation is important on a local scale only and its contribution to the heat transport can be accounted for by an appropriate modification of the thermal conductivity. Thermal convection in a fluid spherical shell heated from below will occur, as linear stability analysis (Chandrasekhar 1961) demonstrates, if the Rayleigh number Ra

$$Ra = \frac{\alpha g d^3 \Delta T}{\kappa \nu} \tag{26}$$

is larger than a critical Rayleigh number Ra_{cr} which is of the order 10^3. In Eq. (26) α is the volumetric thermal expansion coefficient, g is surface gravity,

ΔT is the temperature difference across the layer in excess of the adiabatic temperature difference ΔT_a, d is layer thickness, κ is the thermal diffusivity, and $\nu \equiv \eta/\rho$ is the kinematic viscosity. For an internally heated sphere ($d = R_s$) or spherical shell the appropriate temperature difference is

$$\Delta T = \frac{Hd^2}{k} - \Delta T_a \tag{27}$$

where k is the thermal conductivity and H is the rate of internal energy production per unit volume.

The definition of the Rayleigh number shows that the onset of convection in a satellite depends critically on its size, its viscosity, and its thermal properties. Consider the case wherein a satellite accreted cold. Its temperature-dependent viscosity might then be large enough to inhibit convection. If the thermal conductivity is too small to allow the heat generated to be removed by conduction, the satellite's interior will warm until eventually the viscosity is sufficiently small for convection to set in. As the satellite cools and radiogenic heat production decreases with time, convection will eventually cease once there is not enough energy generated to power the flow against dissipative losses. If the satellite accreted hot, the interior will cool by thermal convection from the beginning of its thermal history. Ellsworth and Schubert (1983) have calculated thermal histories of Saturn's small satellites that illustrate these points (see discussion in Sec. X and Figs. 11–15).

The time rate of change of internal energy due to heat conduction in a sphere or spherical shell whose material parameters depend on radius r only is given by

$$\rho(r)c(r)\frac{\partial T}{\partial t} = \frac{\partial}{\partial r}\left(k(r)\frac{\partial T}{\partial r}\right) + H(r) \tag{28}$$

where $c(r)$ is the heat capacity as a function of radius. At the onset of convection the heat flux q from the sphere or shell equals the conductive heat flux q_c. The value of the Nusselt number

$$Nu \equiv q/q_c \tag{29}$$

which is the ratio of the steady-state values of q and q_c, is then unity. With increasing vigor of convection, measured by an increasing Rayleigh number, the Nusselt number increases at a rate proportional to Ra^β. For $Ra \gtrsim 10Ra_{cr}$, the Nusselt number is related to Ra by (McGregor and Emery 1969; Schubert et al. 1979)

$$Nu = b\left(\frac{Ra}{Ra_{cr}}\right)^\beta \tag{30}$$

where $\beta \simeq 0.3$ and b is a constant of order unity. Using Eq. (30) as a parameterization of the convective heat transport, a thermal history of a fluid shell with volume V bounded by the surfaces S_1 and S_2 ($S_1 > S_2$, $S_2 = 0$ for a sphere) can be calculated from the balance equation

$$\int_V \left[\frac{d}{dt} (\rho c T) - H \right] dV = \int_{S_2} q_b \, dS - \int_{S_1} q \, dS \qquad (31)$$

where q_b is the basal heat flux into the shell.

The strongly temperature-dependent rheology of silicates and water ice leads to the formation of cold rigid outer shells that shield the convective interiors of most satellites. The temperature within these shells is lower than some temperature T_ℓ below which the satellite material is rigid on a geologic time scale. Heat transfer across these shells is by heat conduction. Schubert et al. (1979) have pointed out that the cooling of the terrestrial planets Mercury and Mars and the Moon occurs through the thickening of their rigid shells or lithospheres. This is also the case for most outer solar system satellites. However, it is not true for Io whose lithosphere experiences strong tidal heating. The tidal heating rate largely determines the thickness of Io's lithosphere. If the convecting interior of a satellite is surrounded by a rigid, conducting shell, the time rate of change of the shell's thickness ℓ is given by (Schubert et al. 1979)

$$\rho c (T - T_\ell) \frac{d\ell}{dt} = -q + k \left(\frac{\partial T}{\partial r} \right)_{r = \ell} \qquad (32)$$

where $T - T_\ell$ is the temperature difference between the convecting interior and the boundary $r = \ell$, q is the heat flux from the interior, and $(\partial T / \partial r)_{r = \ell}$ is the heat flux from the boundary into the rigid shell.

VI. MOON

More is known about the Moon than all the other satellites combined. It is the only extraterrestrial body that has been sampled *in situ*. As a consequence, thermal models of the Moon have received much attention and are more highly developed than for any body other than Earth. In fact, the Moon has often served as the test case for the development of planetary thermal modeling techniques since there is sufficient information available to evaluate the validity of some numerical predictions. Although a great deal of detailed information is available for the Moon (see chapter by Kaula et al.), there are many areas critical to thermal modeling calculations wherein data are yet insufficient or interpretations are ambiguous and controversial.

The mean density of the Moon (3340 kg m^{-3}) implies a much lower iron content than that of the Earth, Venus or Mercury, and it is even depleted relative to Io. The question of the existence of a lunar iron core is not settled. An upper limit on the radius of a possible metallic core of $\simeq 400$ km has been estimated from seismic (Nakamura et al. 1974) and electromagnetic (Wiskerchen and Sonett 1977; Hood et al. 1982) data. A near-side crust about 60 km thick and detailed upper mantle structure have been deduced from the Apollo lunar seismic experiment (Nakamura et al. 1982). A thicker (or lower-density) crust is inferred for the lunar far side from gravity and topographic data (Kaula et al. 1974; Thurber and Solomon 1978). Remote-sensing measurements and surface samples (Adler et al. 1973) suggest a lunar crust composed primarily of an anorthositic gabbro. Except for the mare basalt deposits, it has been proposed that this feldspar-rich crust was formed from the crystallization and differentiation of an early magma ocean (Wood et al. 1970; Smith et al. 1970). The nature of the heat source required to produce such an ocean, its characteristics and, indeed, even its existence have been hotly debated.

Large impact basins were excavated prior to some 4 Gyr ago and the volcanic events which flooded these basins with mare basalts have been dated with Rb-Sr and ^{39}Ar-^{40}Ar techniques as occurring between 3.8 to 3.1 Gyr ago. All of the lunar material is characterized by very low volatile content and low oxygen fugacities. Relative to the Earth, the Moon is richer in refractory and ultramafic components such as Al_2O_3, FeO, MgO and TiO_2. While most of the lunar surface morphology is due to impact-related phenomena, some features such as linear rilles have been interpreted as due to crustal extension (Scott et al. 1975) while mare ridges may indicate surface compression (Lucchitta 1976). The extent and ages of these features can be used to place bounds upon the time variation of the lunar mean radius, and hence upon possible thermal histories (Solomon and Chaiken 1976). Lunar samples are weakly magnetized (Cisowski et al. 1983) and local magnetic anomalies have been detected from orbit (Schubert and Lichtenstein 1974; Hood et al. 1981) and on the lunar surface (Dyal et al. 1974). There is no current agreement as to the source of this magnetization. If, however, it is due to the existence of an early lunar dynamo (Runcorn 1977), rather than shock magnetization or some other process, the dynamo would have been operative from 4 to 3 Gyr ago, implying early melting of the bulk of the Moon. These as well as many other measured and inferred properties (see, e.g., Basaltic Volcanism Study Project [1981] as a source of discussion on many of these topics) are important to an understanding of thermal evolution. Viable thermal history models must be built on a self-consistent set of assumptions regarding these facts and interpretations.

The present thermal state of the Moon is loosely constrained by a number of geophysical observations. The surface heat flow has been measured at two sites and Langseth et al. (1976) give 18 mW m^{-2} as the most likely value for the mean lunar heat flux. Seismic (Nakamura et al. 1976a,c), electrical conductivity (Dyal et al. 1976; Sonett et al. 1972; Sonett 1982; Hood et al. 1982), and gravity (Pullan and Lambeck 1980) data are all consistent with a lunar temperature profile that rises steeply with depth in the outer 500 km to temperatures within several hundred degrees of the solidus. The temperature could be at the solidus or remain below it at depths greater than about 1000 km. The lack of apparent S wave penetration below 1000 km depth (Nakamura et al. 1982) suggests the possibility of a partially molten, deep lunar mantle.

Any comprehensive and definitive calculations of the thermal evolution of the Moon must begin with a well-defined set of initial conditions. It is here that a paradoxical situation is encountered. Although the Moon is better characterized and more thoroughly studied than any of the other satellites of the solar system, its origin has been the source of more proposed mechanisms and controversy than that of any other body (see Wood [1986] for a comprehensive review). The theories of origin have been historically characterized in terms of three general categories: capture, fission, and accretion from smaller bodies near the Earth (see chapter by Kaula et al. and that by Burns).

The capture hypothesis involves the formation of the Moon in another part of the solar system with subsequent capture by the Earth. The evolution of the lunar orbit backward in time from the present has been calculated by many authors, including Gerstenkorn (1955,1969), MacDonald (1964), Kaula (1964), Goldreich (1966), Singer (1968), Mignard (1979,1980,1981b) and Conway (1982) (it is described in detail in the chapter by Burns). These calculations indicate that while capture is possible it requires very special conditions: a small approach velocity and hence an independent orbit very similar to that of the Earth. Other problems, such as the pre-capture history of the body, a seemingly required close approach to within the Roche limit, and the question of where a body of this composition could form, remain to be satisfactorily addressed. Variations of this hypothesis include a postulated collision with a smaller moon of the Earth, gas drag from a presumed primordial atmosphere within the Hill sphere (Nakazawa et al. 1983), and capture of a portion of a planetesimal disrupted by a close approach (Öpik 1972; Wood and Mitler 1974; Smith 1974).

The fission hypothesis assumes that the Moon was originally part of a rapidly rotating Earth. The Earth then became rotationally unstable and the Moon was ejected. These theories date from the early work of Darwin (1880) and have been defended in various modifications by others, including Wise (1969), O'Keefe (1970), Ringwood (1960,1970), Binder (1974,1982), and O'Keefe and Sullivan (1978). The theories propose an explanation of the disparity in mean density between the Earth and the Moon by supposing that the Moon was derived from the Earth's mantle. Geochemical arguments have been heavily relied on by the proponents (see, e.g., Ringwood 1979) but have been questioned by others (see, e.g., Newsom and Drake 1982b). The fission theories have been severely criticized on dynamical grounds, including the problem of the removal of excess angular momentum and the high inclination of the Moon's orbit (Goldreich 1966; Conway 1982). However, calculations of Durisen and Scott (1984) lend support to the idea that sufficient angular momentum could be ejected during the fission process to account for the fact that the present angular momentum of the Earth-Moon system is only about half of what is necessary to cause dynamical instability (Kaula 1971).

The third hypothesis suggests that the Moon formed by accretion from smaller bodies in the vicinity of the Earth (Schmidt 1950; Ruskol 1967; Ganapathy et al. 1970; Kaula and Harris 1975). Usually this process is considered to have occurred at the same time as the formation of the Earth or slightly after. Some variations (e.g., multi-moon coagulation; see Ruskol 1972) could provide for a delayed formation of the Moon. The major difficulty with this theory is the problem of explaining the large difference in uncompressed mean densities of the Earth and the Moon (excepting the large body impact hypothesis, which has other problems). There must be a strong lunar depletion in iron relative to the Earth, and no adequate physical mechanism for this fractionation process has been proposed.

There has recently been growing interest in a lunar formation scenario in which the Moon accretes rapidly from a circumterrestrial disk of particulate matter. The disk is formed catastrophically from material ejected from the Earth during collision with a large ($\sim 0.1\ M_\oplus$) planetesimal late in the Earth's accretion. This hypothesis combines elements of the three earlier ones, i.e., capture (impact with the Earth), fission (material ejected from the Earth), and accretion (accretion from a disk surrounding the Earth).

The idea that the impact of such a large planetesimal is not only possible but probable has been expressed by Hartmann and Davis (1975), Wetherill (1976) and Greenberg (1979b). Hartmann and Davis (1975) further pointed out that such a large collision would eject more than enough debris from the Earth to make a Moon. Cameron and Ward (1976) suggested a way to avoid the return of the material to Earth (since the debris is typically accelerated to less than escape velocity). They proposed that a large fraction of the debris would be vaporized. This would permit hydrodynamic forces to accelerate at least some of the material beyond simple ballistic trajectories and into Earth orbit. This model has been further developed by Ward and Cameron (1978), Ringwood (1979), Cameron (1983b,1985), Thompson and Stevenson (1983), and Stevenson (1985a,b); it is described in the chapter by Stevenson et al. Due to the fact that the more complex sequence of events in this hypothesis permits a greater range of possibilities, many of the serious difficulties which have counted so heavily against the three simpler hypotheses may, in principle, be avoided. Many serious questions of detail remain, however, and much more work needs to be done before this hypothesis can be accepted.

The fission hypothesis implies a molten or near-molten Earth at the time of fission; most capture theories involve considerable tidal heating during capture. Collisional ejection and reaccretion models also predict a high-temperature origin (Thompson and Stevenson 1983). Binary accretion models do not necessarily imply an origin at high temperature (Safronov 1978; Kaula 1979; Horedt 1980) although certain versions, especially with many large-body impacts, would imply elevated temperatures in the outer layers of the bodies.

Time-dependent temperature profiles for a solid, conducting Moon heated radioactively have been calculated by a number of workers including Urey (1951), MacDonald (1961), Mayeva (1965), and Phinney and Anderson (1967). These calculations indicate that the interior of the Moon should have melted if the Moon were conducting and had chondritic abundances of radioactive elements. Levin's (1962) computations included the effect of heat of fusion. Fricker et al. (1967) introduced heat transfer by liquid convection and the redistribution (in space and time) of radioactive heat sources into a self-consistent model calculation. Similar numerical schemes were independently derived by Lee (1968) and Ornatskaya et al. (1977). Numerous efforts have been made to include additional physical effects within the framework of spherically symmetric conduction models for thermal evolution. Examples of such physical effects include: heat transfer by igneous processes (McConnell et al. 1967), accretional energy (Hanks and Anderson 1972; Kaula 1979), core formation (Siegfried and Solomon 1974), electrical heating (Sonett et al. 1975), planetary thermal expansion and thermal stress (MacDonald 1960; Solomon and Chaiken 1976) and crustal formation by crystal-liquid fractionation (Solomon and Longhi 1977; Minear and Fletcher 1978; Herbert et al. 1978).

The validity of calculations that consider only conductive heat transport (and liquid convection) was brought into question by the suggestion that subsolidus convection might be important in the lunar interior (Runcorn 1962). Convective processes were advocated by Tozer (1967) and Turcotte and Oxburgh (1969). Schubert et al. (1969) demonstrated the likelihood that the Moon and the other terrestrial planets were unstable to subsolidus convection. Turcotte and Oxburgh (1970) and Schubert et al. (1977) have considered the significance of convection for the present state of the lunar interior while Cassen and Reynolds (1973,1974) discussed the implications for lunar thermal history. Although there are no observational data that specifically require that subsolidus convection has occurred in the Moon (Cassen et al. 1979b; Phillips and Ivins 1979), the theoretical arguments in favor are impressive.

There have been several attempts at calculating global thermal history models for the Moon which incorporate convective heat transport together with the effects of melting and differentiation of radioactives. Toksöz et al. (1978) performed a restricted set of calculations by direct integration of the hydrodynamic equations. Cassen et al. (1979b) utilized a parameterized convection scheme to study a variety of initial temperature distributions and viscosity-temperature relations. They calculated the Rayleigh numbers for potentially unstable regions and applied a parameterized heat transport relation

(Nusselt number proportional to the 1/5 to 1/3 power of the Rayleigh number) to those regions in which the critical Rayleigh number was exceeded. Chacko and DeBremaecker (1982) have utilized a finite-element numerical approach to solve the hydrodynamic equations for a single axisymmetric case. The two direct integration approaches, while conceptually more appropriate, suffer from the disadvantage of long integration times and the requirement for large computational resources. This is especially serious because the results are highly dependent upon parameter values and boundary conditions, thus requiring the calculation of numerous cases. The parameterized model has been calculated for a variety of conditions. Where comparable, the results from the three different sets of calculations are generally consistent. Since the models incorporate a large number of variable parameters, it is not surprising that they can be constructed to match successfully most of the appropriate constraints.

One such parameterized convection model (Cassen et al. 1979b), shown in Fig. 7, has all the following features: the outer layers are initially molten; convection commences about 0.5 Gyr after formation; the average tempera-

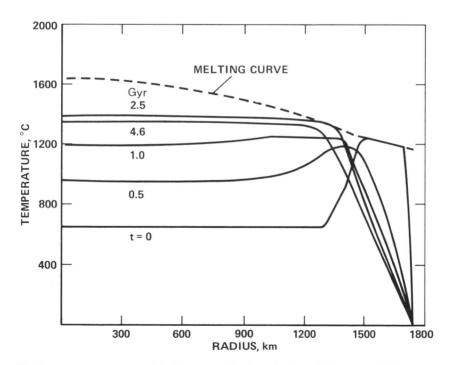

Fig. 7. Average temperature of the Moon vs. radius from the thermal history model of Cassen et al. (1979b). The indicated times are elapsed times after the onset of the thermal evolution (4.6 Gyr refers to the present).

ture rises steeply in the outer layers to near-melting at depths of about 500 km; the change in radius since 3.5 Gyr ago is < 1 km; and the present-day heat flux is between 14 and 21 mW m^{-2}. The model thus matches many of the observed or inferred constraints on lunar evolution. Cassen et al. (1979*b*) draw a number of conclusions from their calculations:

1.   Solid convection does not necessarily drive the Moon to a quasi-steady thermal balance between internal sources and loss to the surface. The thick conductive lithosphere can cause the surface heat flux to lag behind the internal energy generation rate so that at any instant of time the heat flow through the surface is larger, by 25% or more, than the total rate at which heat is produced internally (this conclusion was also reached by Schubert et al. 1980).

2.   The present thermal state is primarily determined by the rheology and melting properties of the rocks. Uncertainties in these properties preclude the determination of a specific value for the central temperature. Model calculations indicate that this temperature exceeds 1400 K. The exact numerical computations of Schubert et al. (1977) give a deep lunar temperature of 1500 to 1600 K corresponding to a viscosity of 10^{17} to 10^{18} m^2 s^{-1}.

3.   The initial temperature profile of the Moon has little or no effect upon the thermal state of the deep interior but it could affect the outer layers by determining the onset time of convection and the extent of heat source differentiation.

4.   The limit on the change in lunar radius of $\simeq 1$ km proposed by Solomon and Chaiken (1976) would, if true, preclude the possibility of an initially molten Moon since the outer layers will cool and shrink, thus requiring heating and expansion of the interior.

5.   Solid convection could have begun within the first billion years in the outer layers and mare flooding could have resulted from pressure-release melting in the upwelling region, as proposed by Turcotte and Oxburgh (1970).

6.   The later cooling history of the outer parts of the Moon is sensitive to the efficiency of heat source differentiation. An efficient differentiation process is consistent with the seismic data.

Until the mode of origin is understood, the early thermal history of the Moon will remain unknown. General outlines for the lunar thermal history of the last 4 Gyr or so are beginning to emerge in the form of models that can match many of the important observational constraints. A definitive thermal history calculation will not be possible until such major questions as the source of the lunar magnetization, the validity of the magma ocean concept, the existence of a lunar core, the precise value of the present-day lunar temperature profile, among others, are understood. Much more work on the subject of lunar thermal history remains to be accomplished.

VII. IO

Jupiter's satellite Io is a most remarkable solar system body. Though similar in size and mass to the Moon, it is the most volcanically active satellite in the solar system. Judging from its surface volcanic activity and heat flux, it is even more thermally active than any terrestrial planet. The rate of heat generation through tidal dissipation most likely sets Io apart from the other satellites.

Io's surface (see chapter by Nash et al.) is characterized by: (a) brilliant colors (white, yellow, orange, red, brown and black) which are thought to represent different allotropic forms of sulfur (Soderblom et al. 1980; Sill and Clark 1982; Fanale et al. 1982; chapter by Clark ct al.; cf. Young 1984 and Plates 4 and 5 in the Color Section; (b) a large number (> 300) of volcanic vents and surface flows, up to 9 km high mountains, and up to 2 km high steep scarps (Masursky et al. 1979; Schaber 1982); (c) a high surface-heat flux of 1.5 ($\pm$ 0.3) W m^{-2} [(6 $\pm$ 1) $\times$ 10^{13} W] determined from infrared measurements (Morrison and Telesco 1980; Matson et al. 1981a; Sinton 1981; Pearl and Sinton 1982).

No impact craters have been observed at resolutions > 1 km (Masursky et al. 1979). This suggests that Io's surface is young. It is currently resurfaced at a rate probably > 10 mm yr^{-1} (Johnson et al. 1979). During the Voyager encounters 9 active volcanic plumes were observed (Morabito et al. 1979; Strom et al. 1979). The maximum height of these plumes was about 300 km and pyroclastic deposits are spread up to 600 km from the vents. Kieffer (1982) has given a thorough discussion of the thermodynamics of the Ionian plumes. She concludes that the observational constraints allow a wide variety of reservoir conditions. The plumes could be SO$_2$ as has been proposed by Smith et al. (1979a) whose model has a sulfur-SO$_2$ pyroclastic crust that overlies a sulfur ocean. The volatile SO$_2$ is driven into the plumes by heat from the sulfur ocean which in turn is heated and replenished by silicate intrusions. The model also allows for the flow of liquid sulfur onto the crust. The minimum reservoir temperature required is 393 K.

Sulfur plumes have been proposed by Consolmagno (1979), Hapke (1979) and Reynolds et al. (1980). Hot silicate magma could vaporize the sulfur on intruding a sulfur crust. The reservoir temperature required is larger than 600 K. Such a large temperature of an Ionian hot spot has been observed by Witteborn et al. (1979).

There is probably also silicate volcanism on Io although it may not be the immediate cause of the spectacular plumes. For one thing, SO$_2$ or sulfur volcanism requires silicate intrusions to furnish the energy for the volatile plumes. The observed high surface heat flow (it is dominated by heat carried by volcanism) can hardly be delivered to a sulfur ocean or crust by heat conduction alone. Also, Io's high mountains may be of volcanic origin (Masursky et al. 1979). They would be hard to reconcile with a very thin (only a few kilometers thick), passive shell that could conduct the surface heat flux.

If the current resurfacing rate of > 10 mm yr^{-1} is also representative of the past, Io's interior could have been processed many times in 4.5 Gyr. Therefore, and because the surface heat flux suggests a hot interior, one expects Io to be differentiated. The satellite's mean density of 3.55×10^3 kg m^{-3} would also allow for an Fe-FeS core. Consolmagno and Lewis (1980) and Consolmagno (1981b) have developed a chemical-thermal evolution model that provides for a sulfur-rich crust and allows an Fe-FeS core. The model starts with an initial composition similar to that of a C2 or C3 chondrite. It is assumed that radiogenic and tidal heating are sufficient to heat Io within the first billion years to temperatures high enough for convection and outgassing of volatiles at significant rates. Io then first loses methane and carbon dioxide followed by the escape of hydrogen and water. Oxygen in the volatiles is used to produce FeO from Fe and FeS and sulfur is freed. An Fe-FeS core is possible if the deep interior is kept from losing its volatiles before the core forms. If all sulfur (5% by mass) in a dehydrated chondritic Io were in its elemental form it would make a 50 km thick crust (Consolmagno 1981b). The surface relief does not seem to be compatible with such a thick sulfur crust (Clow and Carr 1980). Significant quantities of sulfur could be present as FeS in the core.

Dissipation of tidal energy is the most likely source of Io's surface heat flow, as was originally proposed by Peale et al. (1979). The required heating is too large to be explained by the decay of radioactive isotopes (Table III) or by the ohmic dissipation of currents flowing between Jupiter and Io (Ness et al. 1979; Drobyshevski 1979; Gold 1979; Colburn 1980; Goertz 1980; Neubauer 1980; Southwood et al. 1980). The infrared data suggest that the heat flow is about as large as the absolute upper bound on tidal heating (9×10^{13} W) that applies if Io is in an equilibrium resonance orbit (see Sec. III). The current amplitude of libration of the resonance variable $\phi = \lambda_1 - 3\lambda_2 + 2\lambda_3$, where the λ_i are the mean longitudes of Io (1), Europa (2), and Ganymede (3), is $0°.066$ (Lieske 1980). It is close enough to zero to suggest that the satellite system is in a three-body resonance (see discussions in Yoder [1979b], Yoder and Peale [1981], Greenberg [1982a] and the chapter by Peale). However, there could be a problem if Jupiter's tidal dissipation factor Q_J is much larger than the minimum value of $\simeq 2.5 \times 10^4$ (Sec. III) used to establish the upper bound on Io's heat flux. According to Yoder and Peale (1981), capture of the Galilean satellites into resonance from orbital configurations far away from commensurability requires $Q_J \leq 2 \times 10^6$. These authors find that the resonance would then be 1600 Q_J years old. A similar limit on Q_J of $\leq 5 \times 10^6$ derives from scenarios starting deep in the resonance and relaxing to the present orbits (Peale and Greenberg 1980). Some independent estimates of Q_J

which are not based on satellite orbit data suggest $Q_J > 10^7$ (Greenberg 1982a; cf. Stevenson 1983). This led Greenberg (1982a) to propose scenarios in which Io is not in an equilibrium configuration with Europa and Ganymede. He points out that runaway melting due to tidal heating in an originally solid Io, as proposed by Peale et al. (1979), would suddenly decrease Io's Q and may throw it out of equilibrium. Similarly, refreezing would again change Q to induce further changes of orbital conditions. In these cases large values of Q_J could be consistent with large tidal dissipation rates in Io.

Comparison of Io's observed surface heat flow with the theoretical upper bound on the rate of tidal dissipation should take into account a few additional points (cf. chapter by Nash et al.). Io's surface heat flux is spatially highly nonuniform. Most of the heat escapes from just a few hot spots, prominent among which is Loki (Johnson et al. 1984). Thus, estimates of Io's global heat loss are uncertain because of possible hotspots at high latitudes and as yet unobserved longitudes; in addition, the component of heat flow conducted through Io's lithosphere has not been measured (Johnson et al. 1984). The most recent estimate of Io's total heat loss, $(7.5 \pm 2) \times 10^{13}$ W (McEwen et al. 1985), relies on the observed correlation between hot spots and low albedo features to extend heat flow estimates to regions of the surface not yet observed in the infrared. The current rate of heat loss from Io may not be representative of the long-term average tidal dissipation rate if Io's internal thermal activity is variable with time. That this might be the case is suggested by the association of Io's heat loss with just a few hot spots. However, it is to be emphasized that the theoretical upper bound to Io's rate of tidal energy dissipation is comparable to the estimated rate of heat loss and there is no fundamental problem in reconciling the two quantities.

The dissipation rate in a solid Io, estimated from Eq. (14) with $Q = 100$ and $\mu = 65$ GPa (representative values of the tidal Q for the terrestrial planets and of the modulus of rigidity for the outer part of the Moon [Cassen et al. 1982]) is $\dot{E}_T = 1.7 \times 10^{12}$ W. It would be up to 11 times larger (Fig. 8), or about a third of Io's surface heat flux, if the dissipation occurred in a thin shell surrounding a molten interior (Peale et al. 1979). This is because of larger strains in the thin shell for a given tidal force. The increase of the dissipation rate with decreasing shell thickness is limited by the decrease of the volume in which the tidal strains occur. Eventually, the decrease in shell volume dominates and the tidal heating rate tends to zero as the shell thickness goes to zero. The tidal heating rate can be increased above that in a thin-shell Io by enlarging the tidally strained volume. This is possible if Io's interior is only partially molten or has a partially molten asthenosphere as in one of the models illustrated in Fig. 2. The latter model was first proposed by Schubert et al. (1981). A partially molten interior with a modulus of rigidity more than 100 times less than that in the shell will allow tidal strains in the shell comparable to those in the case of a totally fluid interior, while adding to the volume in which tidal energy is dissipated. It is even possible that dissipation in the as-

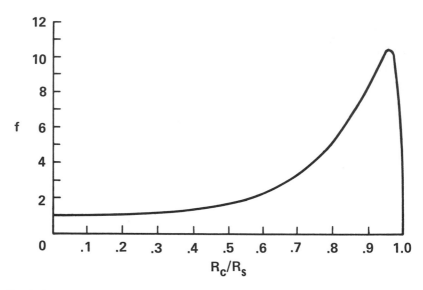

Fig. 8. The dependence of tidal dissipation rate in a model of Io on the inner radius R_c of an elastic shell that is distorting under the action of the tidal force. The outer radius of the shell R_s is the radius of Io, and the shell is assumed to be decoupled from the core. The factor f is the ratio of the tidal dissipation rate for arbitrary R_c to its value for $R_c = 0$ (figure after Peale et al. 1979).

thenosphere is larger than in the shell (Ross and Schubert 1986; Carr 1985) because Q of partially molten rock is only 10 (Murase and McBirney 1973). A modulus of rigidity of 5 GPa, only about an order of magnitude less than the "solid" value quoted above, would then allow tidal heat production at a rate comparable to the observed surface heat flow if the asthenosphere comprises all or most of Io's mantle. Io's large density allows for an Fe-FeS core. Since the eutectic Fe-FeS melting temperatures measured by Usselman (1975) are about 500 K below silicate melting temperatures at pressures expected in Io ($\lesssim 6$ GPa; Table I), at least the outer part of the core would be molten (see also Stevenson et al. 1983). Thus, as has been pointed out by Cassen et al. (1982), any solid silicate mantle layer below the asthenosphere would be a shell and experience enhanced tidal heating also. In addition, as we will show below, past tidal heating in a solid or partly solid Io might contribute to the present heat flux.

A discussion of Io's evolution into its present thermal state is difficult because the thermal evolution is coupled to the orbital evolution. The simplest case is where Io's orbital resonance is ancient. Here, the tidal heating rate could have been approximately constant throughout most of Io's thermal history and could continue essentially indefinitely. If the Galilean satellites started deeper in the resonance than they are today, Io's forced eccentricity would have been larger in the past but it might have relaxed quickly to its present near-equilibrium value if the tidal dissipation factor was small (Peale and Greenberg 1980; Greenberg 1982a). This is likely because Q decreases exponentially with increasing temperature and the combined effect of accretional, early radiogenic, and tidal heating may have heated Io's interior to temperatures close to the solidus. If the resonance is recent (some 10^8 yr old), as would be required if a surface heat flux > 1 W m^{-2} is to be explained by Yoder's (1979b) theory of capture of Io, Europa and Ganymede into resonance, then Io's present thermal state is probably also recent. If Greenberg (1982a) is correct, rapid heating could have decreased Io's Q enough to render its orbit unstable. The new equilibrium orbit would have a smaller forced eccentricity and hence the tidal heating rate would decline. In Greenberg's (1982a) scenario, oscillating forced eccentricities would lead to oscillating heating rates. Oscillatory thermal histories for Io have been derived by Ojakangas and Stevenson (1985) on the basis of a model which couples thermal and orbital evolutions through the forced eccentricity. According to this model, Io vacillates between high-heat flow regimes lasting 0(10 Myr) and low-heat flow states lasting much longer, 0(100 Myr).

Two runaway mechanisms have been proposed that could rapidly heat Io. In a solid Io, tidal heating would increase towards the center of the satellite. Peale et al. (1979) have suggested that melting at the center might trigger runaway melting since the heating rate increases with the radius R_c of the liquid interior (Fig. 8) and heat transfer in the solid part of Io might not be able to remove the tidal heat. The radius of the liquid region could even increase beyond the radius that maximizes the heating rate in the shell. The final thickness ℓ_f of the shell would be determined by a balance of heat production in the shell, heat flow into the base of the shell and heat loss through the surface of the shell. Peale et al. (1979) estimate that the shell is 20 km thick for equilibrium of heat production with heat conduction across the shell. This estimate is an upper limit because it ignores the heat flow from below due to cooling of the interior. It has been derived assuming $\mu = 65$ GPa, $k = 4$ W m^{-1} K^{-1} and a temperature difference across the shell of 1300 K. The shell could be thicker if allowance is made for volcanic heat transfer across it (O'Reilly and Davies 1980). In any case, it would probably be thinner than the thickness ℓ_m of $\simeq 100$ km that maximizes the shell's tidal heating.

Schubert et al. (1981) have questioned this model and argued that runaway melting might never occur because subsolidus convection would remove the heat generated by tidal dissipation. Since the viscosity of rock is strongly temperature dependent, Io will adjust to an increase in the heating rate by reducing its effective viscosity and increasing its vigor of convection before starting to melt. Schubert et al. (1981) find that tidal heat in a solid Io can be removed by subsolidus convection if Io's kinematic viscosity is $< 10^{17}$ m^2 s^{-1}. However, to remove the actual surface heat flow requires a viscosity of 10^{12} m^2 s^{-1}, about the minimum solid-state viscosity of rock (Tullis 1980).

Schubert et al. (1981) propose a runaway scenario that is based on the positive feedback between tidal dissipation and temperature. Since Q is strongly temperature dependent, an increase in temperature will lead to a decrease in Q which will increase the tidal dissipation rate. Of course, the rate of temperature change is limited by the self-regulating convective heat transfer. Schubert et al. (1981) give a stability analysis and show that a low $Q \simeq 10$ state would be stable and steady. Their model of Io's present interior (Fig. 2) has a low overall Q because it has a partially molten asthenosphere. If the newly formed Io was hot and if the tidal heating is ancient, Io would forever stay in the high-dissipation state. If strong tidal heating is recent, Io could evolve rapidly from a low- to a high-dissipation state (the reverse evolution from a high- to a low-dissipation state is also allowed). Because tidal energy is dissipated nonuniformly in the solid Io, the interior would first weaken near the center. The low Q interior would subsequently increase in radius similar to the growth of the melted region during the Peale et al. (1979) runaway.

Even if runaway melting occurred, tidal heating would not be able to keep the interior molten for a long time. We will argue below that Io would evolve toward the structure illustrated in Fig. 2 in a few 100 Myr. If the shell thickness ℓ_f for an equilibrium between heat removal from the shell and tidal heat production in the shell is smaller than ℓ_m, the shell thickness will be prevented from increasing above ℓ_f. Any increase in shell thickness above ℓ_f would cause enhanced tidal heating and thinning. Thus, a molten Io cannot cool by lithosphere thickening. Rather, after differentiating an Fe-FeS core, it will (partially) freeze from the inside out by removing the heat that was deposited during the runaway and liberated during core formation (Stevenson et al. 1983). The heat flux from the interior will keep the lithosphere as thin as necessary to remove the heat at the rate at which it is transferred by efficient liquid-state convection. Convection will be driven by the latent heat of the silicates and by the difference between the adiabatic temperature rise and the melting temperature rise with depth across the mantle.

A simplified energy balance similar to Eq. (31) for a homogeneous liquid interior of radius R_c is

$$\frac{4}{3} \pi R_c^3 \bar{\rho} c \frac{\mathrm{d}}{\mathrm{d}t}\left(T - \frac{L}{c}\right) = -4\pi R_c^2 q_i \tag{33}$$

where $\bar{\rho}$ and c are the average density and heat capacity of the interior, T is the average interior temperature, L is the latent heat which we take to be liberated uniformly throughout the interior volume, and q_i is the heat flux from the interior into the shell. Equations (33), (29), and (30) give the time scale

$$\tau_0 = (1 - \chi)\left(\frac{Ra_{\mathrm{cr}}}{Ra_0}\right)^{1/3} \frac{R_c^2}{\kappa} \tag{34}$$

for cooling and latent heat removal (solidification). Ra_0 is the Rayleigh number based on T equal to the melting temperature and

$$\chi \equiv -\frac{(L/c)}{R_c \dfrac{d}{dr}(T_m - T_a)} \tag{35}$$

where $T_m(r)$ and $T_a(r)$ denote the melting temperature and the adiabatic temperature, respectively, as functions of the radius r. Assuming nominal values of the model parameters $\alpha = 2 \times 10^{-5}$ K^{-1}, $g = 2$ m s^{-2}, $\Delta T = 100$ K, $\kappa = 10^{-6}$ m^2 s^{-1}, $\nu = 10^{12}$ m^2 s^{-1}, we find $\tau_0 \lesssim 300$ Myr. This time scale is smaller than or approximately equal to the age of the resonance, even for scenarios of recent capture (Yoder 1979b). As convection cools the interior and removes the latent heat, the molten layer will thin and eventually become subliquidus leaving a differentiated structure with a core, a solid mantle and an asthenosphere. The vigor of convection in the asthenosphere will decrease until it comes into equilibrium with tidal heat production. During this time, energy removal from the interior contributes significantly to the surface heat flux. In fact, most of the heat deposited in Io during runaway melting will now be removed. This illustrates the fact that tidal heat production and heat loss do not necessarily have to be in equilibrium. A similar storage of tidal heat may occur during the Schubert et al. (1981) runaway.

In summary, Io is a differentiated satellite extensively heated by tidal dissipation. Its interior most likely has an Fe-FeS core, a partially molten mantle, and a thin lithosphere shell. It may still be cooling by removing tidal heat deposited during a runaway weakening of its interior. More quantitative modeling of Io's thermal and orbital evolution is necessary to deepen our understanding of this fascinating satellite.

VIII. EUROPA

Europa (see chapters by Malin and Pieri and by Squyres and Croft), the smallest of the Galilean satellites of Jupiter, has a mean density of 3040 kg m^{-3}, a value intermediate between those of the icy satellites, such as Ganymede and Callisto, and the rocky satellites, Io and the Moon. The mean density is consistent with a formation temperature intermediate between that of Io and Ganymede (Pollack and Reynolds 1974; Cameron and Pollack 1976). A simple mixture of ice with an anhydrous rocky material characteristic of Io or the Moon gives a bulk H_2O content of at least 5% by mass (Table I). If all the H_2O were concentrated in a surficial layer, it could be > 100 km deep. Spectroscopic measurements show that Europa's surface is predominantly composed of H_2O ice with considerably fewer impurities than are present on Ganymede and Callisto. The very low surface relief (at most a few hundred meters) suggests that the H_2O layer is at least several kilometers thick. This would be necessary to cover the type of silicate relief features observed on bodies like the Moon and Io.

The surface is characterized by two basic geologic units: generally bright and smooth plains units, and hummocky mottled terrain units with a complex assemblage of spots and patches of darker material. The surface is transected by many dark linear markings and narrow bright ridges (Smith et al. 1979a,b). The linear markings appear to be the result of fractures and have been classified into several types (Pieri 1981). Some five small circular features, 10 to 30 km in diameter, have been identified as impact craters (Lucchitta and Soderblom 1982).

The present imaging resolution of surface features on Europa is generally inadequate to identify positively the processes that produced them. It is sufficient, however, to present a wealth of detail that greatly exceeds the explanatory capabilities of the simplistic theoretical models currently proposed. Consequently, thermal models must be evaluated in terms of the broad outlines of thermal evolution with the goal of developing a framework for future, more detailed investigations.

On the basis of models that included radioactive heat sources and heat transport by conduction, Consolmagno and Lewis (1976) and Fanale et al. (1977a) suggested that Europa would differentiate within the first 500 Myr. These authors obtained H_2O mantles of order 100 km deep with liquid water layers of order 50 km thick underlying an ice crust; exact thicknesses depend considerably on the assumed satellite mass fraction of H_2O. A study of the stability to subsolidus convection of such an ice crust over a liquid layer (Reynolds and Cassen 1979) showed that the ice layer would be unstable and that the increased convective heat transport would freeze a liquid water layer in a small fraction of the age of the satellite (for calculations pertinent to Europa, only the properties of ice I need be considered; water ice will generally exist only near the surface of Europa and the quantity available is such that it is unlikely that any significant mass fraction of high-density ices will form).

Peale et al. (1979) calculated the magnitude of tidal energy dissipated in a homogeneous Europa and found that, while much smaller than for Io, it was not negligible with respect to the radioactive heat source (see Table II). Electromagnetic heating should not be important for Europa's thermal history (Colburn and Reynolds 1985). Cassen et al. (1979a) calculated the rate of tidal heating in an ice shell above a liquid layer in order to examine the consequences of the assumption that Europa's H_2O mantle had once been melted. The computed shell heating rate was large enough to keep the ice crust stable against thickening. Thus a liquid layer could be maintained under these conditions and some of the consequences of such a configuration were discussed. Fracture of a thin crust by tidal or other forces would result in vigorous boiling of the exposed liquid. (Helfenstein and Parmentier [1985] conclude that nonsynchronous rotation could provide a fracturing mechanism for current activity.) During the short interval before refreezing, water vapor would be released and spread widely over the surface. This could result in active re-surfacing of the satellite. A bright albedo feature was reportedly observed on the surface of Europa during the Voyager encounter (Cook et al. 1982). The liquid layer configuration is not the only one permitted. For an initially thick ice layer that is unstable to subsolidus convection, melting could not occur, even for much higher values of the internal heat source. Thus the question of the existence of a liquid layer depends sensitively upon both initial conditions and material parameters. A correction to the original calculation (Cassen et al. 1980b) resulted in a downward revision of the heat source term and a corresponding decrease in the predicted likelihood of a liquid water layer.

These results are inherently inconclusive, however, because the calculated structure depends so strongly upon assumed initial conditions and poorly known parameters. In the case of Io, the tidal heat production rate is very high, indicating high internal temperatures. For Ganymede and Callisto, the tidal heating sources are much smaller than the radioactive heat sources and subsolidus heat transport processes dominate, resulting in solid interiors. For Europa, however, relatively small changes in assumptions and parameter values can result in very different final thermal states, because the heat production and subsolidus heat removal processes are comparable in magnitude.

Ransford et al. (1981) have proposed a model in which almost all the water in Europa is considered to be incorporated in the interior in the form of hydrated silicates. They argue that the presence of water of hydration will sufficiently lower the creep resistance of the silicates to permit subsolidus convection to occur at temperatures below the dehydration temperatures (500 to 700°C). Thus they conclude that there is only a very thin ice crust of perhaps only a few kilometers depth. Finnerty et al. (1981) modified the hydrated silicate model to propose mechanisms for the origin of some of the observed surface features. They assume that some dehydration in the deep interior would result in a completely hydrated layer some 270 km thick and also in an overall expansion of the satellite. Further, small amounts of dehydration would then allow water to migrate toward the surface and result in the extrusion of ice and ice-silicate breccias from fractures. Finnerty et al. (1981) further suggest that the cuspate ridges originate as compressional tectonic features in a region of upwelling in the anhydrous silicate interior near the anti-Jovian point. They predicted the ice crust to be thinner than 25 km and perhaps only a few km thick.

Squyres et al. (1983a) have reconsidered the tidally heated Europa model with updated parameters and with tidal heating in the silicate core as well as in an ice shell. They point out that hydrated silicates do not lose appreciable strength until dehydration occurs and liquid water is released. They assume that the water will be free to move toward the surface. With the additional tidal heat sources, they calculate the thickness of an outer hydrated silicate layer as only some 60 km and conclude that the balance of the water would be released into an overlying shell many tens of kilometers thick. They find that under these conditions an ice shell of thickness $\lesssim 30$ km would be stable to subsolidus convection. Tidal dissipation calculations indicate a mean ice thickness of 16 km and a total surface heat flow of 52 mW m^{-2}. These calculations require the formation of a liquid layer at some time in the past. A liquid layer could have formed initially during accretion or later beneath an ice cover as a water shell grew during the period of heating and dehydration of the deep interior.

While the results of these calculations support the concept of a liquid water layer between an ice crust and a silicate interior, they are highly model dependent. Squyres et al. (1983a) have summarized observational lines of evidence that provide independent support for this structure. First the fact that there are so few features which resemble impact craters, combined with the plentiful evidence for impact cratering on Ganymede and Callisto, places some constraint on the internal temperature gradient (see also Thomas and Schubert 1985). Shoemaker and Wolfe (1982) calculate a mean retention time for 10 km craters of only 3×10^7 yr, based upon cometary impact fluxes (see the chapter by Chapman and McKinnon). In order for 10 km craters to relax this rapidly, low viscosities and high subsurface temperature gradients are required. The results of Shoemaker and Wolfe (1982) and Passey and Shoemaker (1982a) imply a conducting shell < 30 km thick to provide the necessary thermal gradient. Relaxation of the smaller wavelength crater rims requires still higher temperature gradients and hence even a thinner ice shell (Parmentier and Head 1981). Thus the absence of a significant number of 10 km craters implies that the melting temperature of ice would be exceeded at depths less than the critical depth for convective instability.

Squyres et al. (1983a) estimate that the temperature gradients implied by the viscosity arguments are too high to simply be maintained by the calculated heat flow in a crust with the thermal conductivity of pure ice; a more insulating material is needed. Fracturing of an ice crust could provide the insulation by causing frost to be deposited on the surface. A frost layer would have very low density and thermal conductivity; low pressures within it would prevent compaction until temperatures reached well above 150 K (Smoluchowski and McWilliam 1984). Consequently, crustal ice thicknesses could be much less than the mean value of 16 km calculated for conduction in a solid ice layer. There is some evidence for a frost layer. The photometric function of Europa implies much more homogeneous scattering than that produced by ejecta deposits on Ganymede and Callisto (Buratti and Veverka 1983). This difference is consistent with a more frostlike and less dense structure than that typically produced by impact processes. However, on the basis of a recent radiative transfer model calculation, Buratti (1985) concludes that Europa has a compact regolith, in conflict with thermal inertias derived from eclipse measurements (Morrison and Cruikshank 1973; Hansen 1973).

Observation of sulfur on Europa's surface provides further indirect evidence for a frost layer. An ultraviolet absorption feature detected on the trailing hemisphere of Europa by Lane et al. (1981) has been identified as resulting from S—O bonds, and a mean sulfur atom density of 2×10^{14} cm^{-2} has been inferred. The sulfur deposit is attributed to sulfur atoms expelled from Io, ionized, and diffused outward to the orbit of Europa (see chapter by Cheng et al.). These ions are then deposited onto the trailing hemisphere of Europa by the relative velocity of the Jovian magnetic field which rotates much faster than the satellite mean motion. Measured ion densities in Jupiter's magnetosphere, combined with rotation rates, imply that the observed column density would be deposited in only ≈ 7 yr. One suggested explanation for these observations involves a dynamic equilibrium between the deposition of S ions and the removal of SO$_2$ molecules from Europa by charged-particle sputtering (Eviatar et al. 1981). However, such a mechanism requires a sulfur number density higher than that indicated by the interpretation of the ultraviolet observations. It would also seem to imply that S should be observed on the leading hemisphere as well. SO$_2$ has been shown to be more easily sputtered than H$_2$O (Lanzerotti et al. 1982) and, since it has a lower escape probability, it would be expected to accumulate preferentially on the leading hemisphere. Calculated sputtering rates for Europa can be very substantial (Sieveka and Johnson 1982) and sputtering would presumably be the dominant redistribution process if there is no internal source of material. The rates are calculated for a smooth surface, however; they would be considerably less for a frost-covered surface (Hapke and Cassidy 1978).

An alternative explanation (Squyres et al. 1983a) of the observed S column density is the simultaneous addition of H$_2$O molecules to the surface from internal sources at a much higher rate than the sulfur is implanted. The process discussed above for frost deposition could provide such a source. A minimum H$_2$O deposition rate can be estimated from the flux of incoming sulfur ions, the column density and the depth of column (≈ 0.5 μm) sampled by the ultraviolet measurement. Assuming that both rates are uniform in space and time, the mean water deposition rate would be expected to depend strongly upon the distribution, size and frequency of fracture events. A detailed study of sputtering, redistribution and escape processes by Eviatar et al. (1985) supports the concept of the boiling of liquid water from the interior and gives a water deposition rate of 0.04 μ yr^{-1}. Squyres et al. (1983a) suggest that particulate matter (perhaps salts or organic material) could be trapped in the rising vapor column emanating from fractures and would be preferentially deposited near the fracture sites. Such material (perhaps darkened by radiation) would tend to lower the albedo near a recent fracture and might explain some of the observed albedo features on Europa.

These two possible models for Europa's structure are summarized in Fig. 9. The Ransford et al. (1981) model has an ice crust a few kilometers thick (perhaps as much as ≈ 20 km in the Finnerty et al. [1981] version); the rest of the H$_2$O would be retained as water of hydration in the interior. Allowance is made for some dehydration of the inner silicate core with a dehydrated sili-

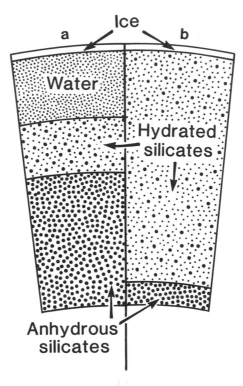

Fig. 9. Alternative structural models of Europa's interior: (a) Subsurface liquid water layer above a largely dehydrated silicate interior (Squyres et al. 1983a); (b) thin ice layer above a largely hydrated silicate interior (Ransford et al. 1981).

cate shell some 270 km thick. No tidal heating is considered and low temperatures are maintained in the outer layers by subsolidus convection. The Squyres et al. (1983a) model assumes a higher heat flux due to both tidal dissipation and radiogenic heat. This model has a larger dehydrated silicate core, a thinner hydrated silicate shell, and a much thicker outer layer of water (between 50 and 100 km thick). Two thermal states are possible in this model. One would maintain low temperatures throughout a solid ice mantle by means of subsolidus convection. The other would have high enough temperatures for a liquid shell of water to exist beneath a thin crust of ice. Active resurfacing is a feature of this model which requires that local inhomogeneities and a variable thickness frost layer produce a very thin ice crust in some regions.

Each of these models has some advantages and disadvantages in attempting to explain the observed surface features. The cold ice layer model has difficulty in explaining why there are so few craters. The thin active crust model can easily get rid of craters but the presence of even a few 10 to 30 km diameter craters is a problem. The cuspate ridges are difficult to explain by either theory. Rather than being centered on the anti-Jovian point as the Finnerty et al. (1981) model assumes, they may be ubiquitous on the surface (Lucchitta and Soderblom 1982). The liquid shell model might offer an explanation in terms of constructional features. Neither model is convincing in explaining the great range of albedo features.

It is apparent that, although the state of theoretical modeling can and should be improved, until sufficient observational data are available to permit positive identification of surface features and processes, no theoretical treatment is definitively going to explain Europa. The Galileo mission will provide high-resolution imaging of much of Europa's surface, as well as other observational data. First-order objectives are to determine the existence or nonexistence of a liquid water layer, the extent of internal differentiation, and the age of the surface. Higher-resolution images may enable the processes responsible for specific surface features to be identified and understood.

IX. GANYMEDE AND CALLISTO

Though similar in size and mass, the largest of the Jovian satellites Ganymede and Callisto are a contrast in appearance (Smith et al. 1979a,b; see chapters by McKinnon and Parmentier and by Squyres and Croft, and maps). Callisto has a very old, heavily cratered surface with two large prominent ring structures, Valhalla and Asgard (Passey and Shoemaker 1982a). Ganymede, on the other hand, has a surface of varied character including old heavily cratered terrain and younger grooved terrain with complex tectonic patterns (Shoemaker et al. 1982). The ancient heavily cratered terrain on Ganymede has a mean crater density only half that of its counterpart on Callisto (Passey and Shoemaker 1982). Topographic relaxation of craters has occurred on both Ganymede and Callisto perhaps as a consequence of viscouslike creep in the predominantly water ice surfaces of these bodies (Johnson and McGetchin 1973; Passey and Shoemaker 1982a). Crater retention ages are older on Callisto than they are on Ganymede (Passey and Shoemaker 1982a). Except for crater degradation, Callisto's surface shows no sign of endogenic modification or internal thermal activity. On the other hand, Ganymede's grooved terrain is evidence that the satellite was thermally active beyond the period of early heavy bombardment when emplacement of this terrain took place (Shoemaker et al. 1982a). In terms of structure and thermal evolution, the major challenge for Ganymede and Callisto is to understand why they are apparently so different.

Ganymede is the larger of the two satellites (2631 km in radius; Table I) and Callisto is only somewhat smaller (2400 km in radius). The low densities of these bodies (Table I), the cosmic abundances of the elements, and the spectroscopic identification of water ice as the major component of both satellites' surfaces (Sill and Clark 1982) leave little doubt that water constitutes approximately 50% by mass of Ganymede and Callisto (Table 1). Ganymede probably contains somewhat more rock than Callisto because of its higher density (Table I), but as noted earlier, there is considerable uncertainty in estimating the silicate fractions of these satellites.

The distribution of rock and ice in the interiors of Ganymede and Callisto is unknown and possible structures ranging from undifferentiated through fully differentiated have been illustrated in Figs. 3 and 5. The extent of ice-rock differentiation in these satellites is uncertain because basic questions about the production and transfer of heat in their interiors cannot be answered unequivocally. Radiogenic heating in the silicates is unlikely to melt the ice fraction of a homogeneous ice-rock interior with the viscosity of pure ice (Parmentier and Head 1979b; Thurber et al. 1980; Schubert et al. 1981). Radiogenic heating could lead to ice melting for viscosities of the ice-rock mixture greater than the pure ice viscosity (Friedson and Stevenson 1983; Croft 1984). Accretional heat is capable of melting the ice in the outer regions of Ganymede and Callisto, but the thickness of the melted layer depends crucially on the fraction of gravitational potential energy retained during the accumulation process (Fig. 6; Schubert et al. 1981). Tidal dissipation may have been a significant heat source in the early evolution of Ganymede but it almost certainly has not been important for Callisto (Cassen et al. 1982).

The observational data that can be brought to bear on the question of satellite differentiation are the compositions and appearances of the surfaces. Water makes up about 90 wt. % of Ganymede's surface but only 30 to 90 wt. % of the surface of Callisto (Sill and Clark 1982; cf. chapter by Clark et al.), consistent with at least some ice-rock differentiation of Ganymede and little, if any, ice-rock separation in Callisto. More important are the differences between the ancient surface of Callisto and the modified surface of Ganymede. The most straightforward explanation of these differences lies in the degree of differentiation of the satellites—Ganymede is at least partly differentiated while Callisto is an undifferentiated ice-rock mixture (Schubert et al. 1981). The answer to why Ganymede and Callisto are different lies then in understanding why Callisto never differentiated while Ganymede did. Though appealing, this interpretation of the geology is not compelling. It could be argued instead that while both satellites differentiated, only Ganymede remained thermally active until after the early period of intense cratering (Cassen et al. 1980a).

Friedson and Stevenson (1983) have recently suggested a plausible explanation, based on the rheology of ice-rock mixtures, for why Ganymede differentiated while Callisto might not have. As discussed in Sec. IV, the viscosity of an ice-rock mixture increases with the percentage of rock. Ganymede probably has a larger silicate fraction than Callisto; Ganymede's interior should therefore be more viscous. Not only does Ganymede have a larger rock fraction and higher viscosity than Callisto, but the percentage of rock in Ganymede may be comparable to the critical fraction (60% by volume) at which the viscosity of the mixture increases dramatically to a value determined by the silicates rather than the ice. Such a creep resistant interior would be unable to convect out the radiogenic heat of the silicates without melting the ice. A runaway differentiation, fueled by the gravitational potential energy release of sinking rock, might even be possible. If the differences between Ganymede and Callisto reflect the duration of thermal activity rather than the degree of differentiation, the larger rock fraction of Ganymede might again provide the explanation through its enhanced radiogenic heat production (Cassen et al. 1982).

The thermal evolution of an undifferentiated Callisto would be analogous to the model thermal histories of Rhea computed by Ellsworth and Schubert (1983). If the initial temperature profile resembles one of the accretional profiles in Fig. 6, then the early thermal evolution involves the conductive heating of the satellite's interior. After this period of heating, which lasts on the order of 100 Myr, the interior temperatures are high enough for subsolidus convection to occur beneath a rigid, conducting lithospheric shell. The subsequent evolution consists of cooling the interior by solid-state creep and thickening the lithosphere. Subsolidus convection would persist until the present beneath a very thick (500 to 1000 km) lithospheric shell. This scenario assumes, of course, that the viscosity of the ice-rock mixture is essentially the viscosity of ice.

The undifferentiated model of Fig. 3 shows that ice would exist in different phases in Callisto's interior; ice I is found near the surface, ice II lies below the thin ice I shell, ice VI lies at greater depth and ice VIII constitutes the deep interior. The ice II-ice VI phase transition may reduce the efficiency of subsolidus convection (Thurber et al. 1980), but it should not strongly inhibit the large-scale solid-state flow inside a homogeneous Callisto (Schubert et al. 1981). Convective flow through one of these phase transitions creates a mixed phase region in which the temperature profile coincides with the equilibrium or Clapeyron boundary separating the phases on a pressure-temperature diagram of ice (Schubert et al. 1975). The temperature difference across such a mixed phase region is approximately L/c_p (L is the latent heat of the phase transition) and the thickness of the region is about $(L/c_p)/[\rho g (dT/dp)_c]$, where $(dT/dp)_c$ is the slope of the Clapeyron curve for the phase change. Thus, in contrast to the idealized situation shown in Fig. 3, there would be no sharp phase transition in the convecting regions of a homogeneous Callisto (see Zuber and Parmentier [1984] for a similar conclusion regarding an undifferentiated model of Ganymede). In fact, two-phase regions once established by convective flow would retain their structures even upon the cessation of convection.

In single-phase ice regions of an undifferentiated Callisto, subsolidus convection would establish an adiabatic thermal state. The adiabatic temperature gradient ($\alpha T/\rho c_p$; α = thermal expansivity $\simeq 10^{-4}$ K^{-1}, $T \simeq 200$ K, $\rho \simeq 2000$ kg m^{-3}, $c_p \simeq 2$ kJ kg^{-1} K^{-1}) in Callisto is only about 10^{-2} K MPa^{-1}. Adiabatic compression in Callisto could heat its interior by no more than a few tens of degrees (central pressure in Callisto is about 2700 MPa; Table I).

If subsolidus convection is unable to remove the radiogenic heat, or if most of the accretional heat is retained, then fully differentiated models of Ganymede (and perhaps Callisto) are relevant. Melting of the ice and formation of a rock core surrounded by a liquid water mantle occur in the earliest stages of evolution (Lewis 1971a,b; Consolmagno and Lewis 1976,1977; Fanale et al. 1977a). A thin ice crust would enclose the liquid water mantle as a consequence of the low surface temperature. Figure 10 sketches the mantle temperature profile at this stage in the evolution of a fully differentiated Ganymede model (Cassen et al. 1982). The temperature is at the melting temperature of ice I at the base of the thin crust and rises along an adiabat in the water mantle until at about 500 km depth it coincides with the melting curve of ice VI (which has a steeper slope than the adiabat).

The post-accretional liquid water mantle is a transient feature of a differentiated Ganymede model. Freezing would take place both upwards from the core-mantle boundary and downwards from the crust-mantle interface. Subsolidus convection in the outer ice shell would enhance its rate of thickening (Reynolds and Cassen 1979) and the mantle would freeze solid in only several hundred million years (Cassen et al. 1982). The last part of the mantle to so-

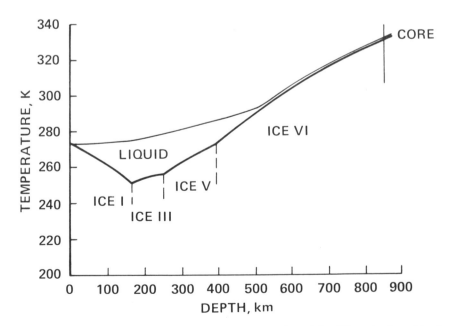

Fig. 10. Post-accretional temperature profile (thin curve) in a fully differentiated model of Ganymede (after Cassen et al. 1982). The heavy curve is the melting curve of ice.

lidify would be at the minimum in the ice solidus temperature at about 150 km depth (Fig. 10). This occurs at the ice I-ice III transition. However, a thin liquid layer might exist at 150 km depth if Ganymede incorporated any NH_3 or salts during its accumulation.

In its subsequent thermal evolution, the heat generated in Ganymede's rock core by the decay of radioactive elements would be transferred to the surface by solid-state convection in the ice mantle. The silicate core of a fully differentiated Ganymede would be larger than the Moon and subsolidus convection could be the primary heat transport mechanism in the rock core as well as in the ice mantle. The temperature profile in the mantle would lie along an adiabat in the convecting single phase regions of ice I, II, VI and VIII (Fig. 3). The transitions between adjacent phases of ice would be spread out (as in the undifferentiated Callisto model just discussed) and the temperature profile would coincide with the Clapeyron curve in the mixed phase regions. Conductive thermal boundary layers would exist in the ice at the surface and at the core-mantle interface. A conduction-dominated layer would also be present in the rock core at the core-mantle boundary. A differentiated Ganymede would thus have two rigid lithospheres, an ice lithosphere at its surface and a rock lithosphere at the core-mantle boundary.

The early evolution of a partially differentiated Ganymede (Fig. 5) involves the melting of ice in the outer layers, the collection of silicate material above the undifferentiated and cold central region of the satellite, and the rapid refreezing of the water in the outer layer (Schubert et al. 1981). Although this silicate layer would be more dense than the material below it, a large-scale Rayleigh-Taylor instability could not develop at this stage because of the very high viscosity of the cold, central material. With time, the undifferentiated central region would heat up due to downward heat conduction and internal radiogenic heating. Eventually, its viscosity would become sufficiently low that the silicate layer could become gravitationally unstable. The time scale of descent of a silicate particle through the central region τ_{RT} can be estimated from the usual Rayleigh-Taylor theory (Chandrasekhar 1961)

$$\tau_{RT} = a_c \rho \nu / g \lambda^2 \Delta\rho \tag{36}$$

where a_c is the radius of the undifferentiated region, λ is the size of a silicate-rich inhomogeneity, ν is the kinematic viscosity of the ice-rock mixture, ρ is the density of the mixture, g is the acceleration of gravity, and $\Delta\rho$ is the silicate-mixture density difference. For $\Delta\rho \simeq 1000$ kg m^{-3} and for $\nu \simeq 10^{16}$ m^2 s^{-1} any inhomogeneity with $\lambda \gtrsim 1$ km will reach the center of the satellite in a time of several billion years. Only smaller-scale inhomogeneities or inhomogeneities in the outermost thermal boundary layer (where $\nu \gg 10^{16}$ m^2 s^{-1}) would persist to the present day. Upon establishment of the structure in Fig. 5, the satellite would evolve thermally in a manner similar to that discussed above for the undifferentiated Callisto model and the fully differentiated Ganymede model. The possibility of a runaway differentiation commencing upon the melting of the outer layers of Ganymede is an alternative to the above scenario (Friedson and Stevenson 1983).

Viable thermal history models of Ganymede and Callisto must be capable of explaining the surface geology of these bodies (see the chapter by McKinnon and Parmentier). In the case of Callisto, a model with an undifferentiated interior undergoing subsolidus convection beneath a thick rigid lithosphere provides a straightforward explanation of the ancient heavily cratered surface. In the case of Ganymede, the model must explain the grooved and bright terrain. These features are generally believed to have formed by the foundering of old crust and the extrusion of liquid water onto the surface (Shoemaker et al. 1982). Crustal rifting and surface flooding require an extensional near-surface tectonic regime that is most likely realized relatively early in Ganymede's evolution (see chapter by Squyres and Croft). Complete differentiation of an initially homogeneous Ganymede would cause the satellite to expand; surface area could increase by as much as 5 to 7% (Squyres 1980a). Global expansion accompanies differentiation because the high-density ices that exist throughout the interior of the undifferentiated body melt and rise nearer the surface where the density of compressed liquid water is less than that of the originally deeply buried ice. Partial differentiation of only the outer regions of Ganymede would have a similar though smaller effect (Schubert et al. 1981). Refreezing of a liquid water mantle would produce only a small global contraction because the densities of the ices in a differentiated Ganymede are not much larger than the liquid water density (Squyres 1980a). Density increases upon freezing for all the ices except ice I, and the other phases of ice dominate the net volume change upon solidification of a differentiated Ganymede. Kirk and Stevenson (1983) have noted the possibility of a convective overturn of the relatively cold ice I that lies above the relatively warm ice III at the end of mantle refreezing. The overturn instability might occur because of the phase change; ice I displaced downward from just above the interface would convert to ice III and be slightly heavier than its surroundings due to its cold temperature (the thermal gradient would have to be superadiabatic at the interface for this instability to take place). Kirk and Stevenson (1983) suggested that this overturn could fracture the crust and resurface the satellite with clean material. This provides an explanation for the grooved and bright terrain that does not involve resurfacing by liquid water.

Satellite expansion could also result from the gradual radiogenic heating of a silicate or silicate-ice core during the first several hundred million to one billion years of Ganymede's evolution (Cassen et al. 1982; Zuber and Parmentier 1984). Degassing of any hydrated silicates in the deep interior during this early phase of heating would also contribute to a global volume increase. Shoemaker et al. (1982) have proposed mantle convection systems on two different scales (satellite-wide and shallow) to explain the structural complexities of the grooved and bright terrain.

The existence of grooved and bright terrain on Ganymede is strong evidence for its early and at least partial differentiation. The absence of these geologic units on Callisto is consistent with it being undifferentiated.

X. MIMAS, TETHYS, DIONE, RHEA, IAPETUS

The surfaces of the Saturnian satellites reflect an apparently wide range of internal activity (Smith et al. 1981,1982; also see Morrison et al. 1984 and the chapter by Morrison et al.). Dione and Rhea have extensively modified surfaces with smooth plains, degraded craters and lineaments. Tethys has areas of light cratering, a large trench extending 270° around its circumference, and some heavily cratered areas. Portions of Iapetus are relatively heavily cratered while its leading hemisphere is covered with dark material. Mimas has a uniformly heavily cratered surface that is basically unmodified. These differences among the surfaces of the Saturnian satellites can be generally reconciled in terms of the duration of subsolidus convection in their interiors and/or the occurrence of minor constituent (e.g., ammonia hydrate $NH_3 \cdot H_2O$) melting. Dione and Rhea are large enough and contain sufficient radiogenic heat sources that subsolidus convection occurs throughout a major portion of their thermal histories. This is at least consistent with the extensive endogenic modification of their surfaces. Mimas on the other hand is too small and contains too few heat sources to be thermally active except briefly early in its thermal development. This is also consistent with the preservation of its primordial surface throughout geologic time. Tethys and Iapetus are somewhere in between, with solid-state convection lasting for a significant period of time but with most of their thermal histories dominated by conductive cooling. Enceladus is a special case that we will deal with separately.

The above conclusions are based on the thermal history models of Ellsworth and Schubert (1983). The essential features of these models are that all the Saturnian satellites discussed above are undifferentiated, homogeneous ice-rock spheres heated by radioactive decay in the silicate fraction and cooled by either conductive or subsolidus convective heat transport. The structural assumption that none of these satellites could have differentiated from an initially homogeneous ice-rock mixture is founded on reasonably convincing arguments. Accretional heating is insufficient to melt any of the water ice in either Mimas or Tethys (Fig. 6). Heat of accumulation could conceivably have melted some water ice in Dione, Rhea and Iapetus, but only in the outer 10 to 100 km of these bodies and only under the most favorable of circumstances (e.g., the complete retention of accretional heat) (Ellsworth and Schubert 1983; Fig. 6). Radiogenic heating is incapable of raising the interior temperature of Mimas to the melting point of water ice. Although this source of heat is adequate, in principle, to melt the water ice in the other Saturnian satellites, convective heat transfer prevents this from occurring. These satellites contain so few silicates that the ice component should control the viscosity of the ice-rock mixture. The conclusion that these satellites are undifferentiated is not altered by the possibility of $NH_3 \cdot H_2O$ melting because ammonia hydrate is only a minor component of these bodies (at most 20% of the total ice mass [Lewis 1972]). Ammonia hydrate could have melted throughout the outer few hundred kilometers of Tethys, Dione, Rhea and Iapetus during accretion. Even more extensive melting of $NH_3 \cdot H_2O$ in the interiors of these satellites could have occurred later in their evolutions (Ellsworth and Schubert 1983). Consolmagno (1985) has proposed that Dione, Rhea and Iapetus could have been extensively differentiated by the melting of $NH_3 \cdot H_2O$. However, as stated above, we do not believe that the partial melting of a minor constituent could differentiate the bulk of the ice-rock mixture and this issue is not discussed by Consolmagno (1985).

Model thermal histories for Mimas, Tethys, Iapetus, Dione and Rhea are presented in Figs. 11–15, taken from Ellsworth and Schubert (1983). Two extreme thermal evolutions are shown for each body: (a) illustrates a hot thermal history and (b) depicts a cold one. The differences in the two thermal

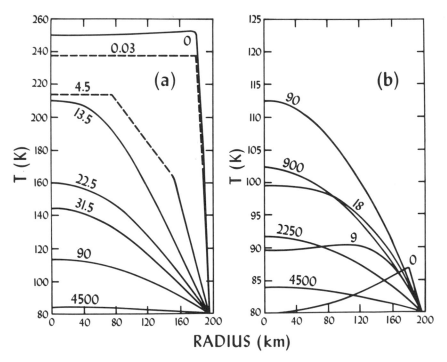

Fig. 11. Model temperature profiles for Mimas at different times (Myr) during its thermal evolution: (a) hot initial formation; (b) cold start. Solid lines indicate portions of the satellite in which heat is transferred by conduction while dashed lines denote regions of convective heat transfer (figure after Ellsworth and Schubert 1983).

histories for a particular satellite arise solely from the assumed initial temperature profiles. If ammonia hydrate is present in the interiors of these satellites, then the hot initial temperatures are inappropriate and the thermal histories should resemble the cold evolutions depicted in (b) of the figures.

Figure 11 shows that Mimas' model thermal history is basically one of
pure conductive cooling independent of initial conditions. If Mimas starts out
sufficiently hot, it can undergo subsolidus convection for a brief period (13
Myr) early in its evolution; from 13 Myr on, however, it simply cools steadily
by conduction. Following a cold start, Mimas never convects. Instead, its
deep interior warms up slightly due to radiogenic heat production in its silicate fraction and then it cools by conduction throughout the bulk of its thermal history. At the present, Mimas is a cold, nearly isothermal body with an
internal temperature of about 84 K. Mimas' thermal history during the last
4.3 Gyr is essentially independent of its initial thermal conditions.

The thermal evolution models of Tethys and Iapetus, shown in Figs. 12
and 13, respectively, are quite similar despite the 200 km difference in the
satellite radii. Because Iapetus is the larger satellite, its early convection lasts
1.8 Gyr compared with 1.4 Gyr for Tethys. Convection is also more vigorous

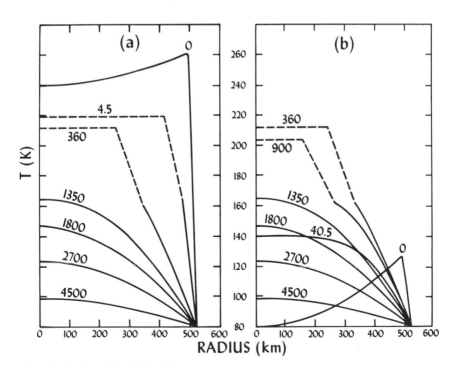

Fig. 12. Similar to Fig. 11 for Tethys.

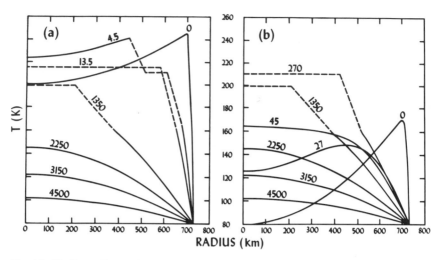

Fig. 13. Similar to Fig. 11 for Iapetus.

in Iapetus than in Tethys. For a hot start, convection begins in both satellites almost immediately, while for a cold start convection is delayed 130 Myr for Tethys and 60 Myr for Iapetus. Iapetus and Tethys are presently cold, nearly isothermal bodies with internal temperatures of 102 and 99 K, respectively. They have been cooling conductively for the last 3 Gyr, independent of their initial states. Their entire thermal histories for about the last 4 Gyr are also insensitive to initial conditions.

It is apparent from Figs. 14 and 15 that the model thermal histories of Dione and Rhea exhibit many similar features, with solid-state convection lasting 2.6 Gyr in Dione and 3.3 Gyr in Rhea independent of initial conditions. If these bodies start hot, convection begins almost immediately. Even if they start cold, only 40 Myr is required for their conductive interiors to warm up sufficiently by radioactive decay in the silicates to begin convecting. In this case, their interiors continue to warm up and the vigor of convection increases until about 140 Myr for Rhea and 180 Myr for Dione, when their thermal states are essentially the same as those for a hot start. The thermal histories for the remaining 4.3 Gyr are insensitive to the initial conditions. Rhea and Dione also end up as cold, conductive bodies with internal temperatures of 124 and 114 K, respectively. As was the case with Iapetus and Tethys, Rhea also has a longer and more vigorous convective thermal history than Dione because of its size.

The cold-start models discussed above are appropriate for delineating regions of possible $NH_3 \cdot H_2O$ melting in the satellite interiors (hot-start temperatures are above the condensation temperature of ammonia hydrate). The cold-start model of Mimas never becomes hot enough to melt ammonia hy-

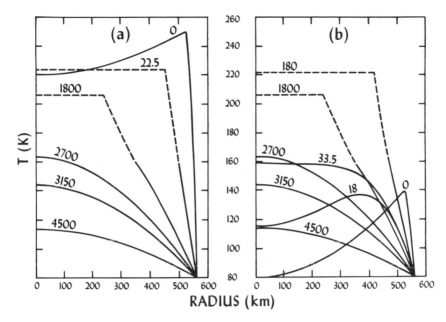

Fig. 14. Similar to Fig. 11 for Dione.

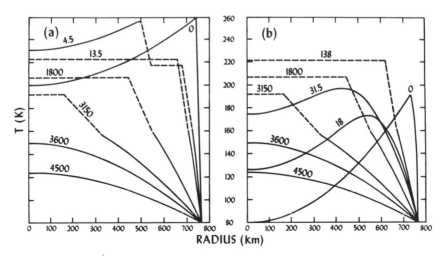

Fig. 15. Similar to Fig. 11 for Rhea.

drate anywhere in its interior. In the models of the other satellites, melting of $NH_3 \cdot H_2O$ first begins at the centers of the bodies after 85, 50, 40 and 20 Myr for Tethys, Iapetus, Dione and Rhea, respectively. Melting rapidly extends throughout most of the satellites' interiors except for regions near the surface which never partially melt. These near-surface regions are about 200 km thick on Tethys and Iapetus, and 100 km thick on Dione and Rhea. Liquid ammonia hydrate might exist closer to the surface than these limits indicate, but the liquid $NH_3 \cdot H_2O$ would have to rise through ice that is nearly rigid over geologic time and is colder than the freezing temperature of ammonia hydrate. The depth interval of $NH_3 \cdot H_2O$ partial melting reaches its maximum possible extent almost immediately after melting just begins in all the satellites. Thereafter, the melting regions decrease relatively slowly in thickness. The last location of liquid $NH_3 \cdot H_2O$ is near the satellite's center. All liquid $NH_3 \cdot H_2O$ has frozen after 1.4, 1.8, 2.8 and 3.3 Gyr in Tethys, Iapetus, Dione and Rhea.

Expansion and contraction of the satellites can occur at different stages of their thermal evolutions. During an early phase of heating of the deep interior, a satellite can expand. Later cooling will be accompanied by contraction. Growth of an ice II core during cooling can also lead to a reduction in radius and detailed thermal history calculations (Ellsworth and Schubert 1983) show that this could be important for Rhea and Iapetus. Despite the fact that most of the satellites' thermal histories are spent in cooling and contraction, the geologic evidence (faults and lineaments on the surfaces of Tethys, Dione and Rhea interpreted as graben by Tyler et al. [1981,1982]) indicates that extensional tectonism has been predominant in modifying the satellite surfaces (see the chapters by Squyres and Croft, and by Morrison et al.). These extensional features must date from an early period of satellite warming and expansion. Other possible explanations for the tensional surface features involve the freezing of liquid water if there were substantial amounts of water present early in a satellite's history and solidification of ammonia hydrate if there were enough $NH_3 \cdot H_2O$ present and if $NH_3 \cdot H_2O$ expands upon freezing (considered unlikely by Consolmagno [1985]). If melting of $NH_3 \cdot H_2O$ could differentiate the entire satellite (Consolmagno 1985), then the separation of rock and ice and the subsequent refreezing would result in planetary expansion.

The evidence of surface extension on Rhea is difficult to reconcile with the growth of a large ice II core in the models of Ellsworth and Schubert (1983). The formation of a large ice II core should lead to considerable contraction (15 km) and large compressional stresses (20 to 60 MPa). These stresses should produce prominent surface compressional features because they exceed the 2 to 3 MPa compressional strength of ice (Hobbs 1974). The absence of such features on Rhea argues against the existence of an ice II core and could imply either that Rhea is warmer than the models indicate, that it has a rock core preventing the ice II transition, or that extrapolation of the ice I and II phase boundary is incorrect. To maintain an interior warm enough

($\simeq$ 160 K vs. 129 K) to prevent the transition requires significant additional heat sources, such as tidal heating. Crude (constant-density) calculations for a totally differentiated satellite indicate that the pressure at the ice-rock interface would only be about 110 MPa, which can prevent the ice II transition if the temperature at the ice-rock boundary exceeds 130 K. Clearly, it is important to establish experimentally the pressure of the ice I-ice II phase transition at the low temperatures ($\leq$ 160 K) expected in the interiors of the icy Saturnian satellites.

Silicate fraction plays an important role in the Saturnian satellites' thermal histories as can be seen by comparing the evolutions of similar-sized bodies—Dione and Tethys, Rhea and Iapetus. The silicate-rich bodies convect nearly twice as long as the ice-rich satellites, with the length of convective activity approximately proportional to the total silicate mass. On the other hand, the size of the satellite is also important since Rhea has about twice the total silicate mass of Dione and yet they both convect for about the same length of time. In this case, the length of convective activity is approximately proportional to the silicate mass fraction times the radius (total silicate mass divided by the satellite's surface area). Thus, variations in the silicate mass fraction for the same-sized body as well as variations in the size of a satellite for a given silicate mass fraction will significantly affect the thermal evolutions.

In all of the Saturnian satellites discussed here, the melting temperature of ice I is never exceeded, but the melting of a minor component such as ammonia hydrate could be quite extensive in the four largest satellites. Without such a volatile, the endogenically modified surfaces of the Saturnian satellites are difficult to explain when compared to the ancient heavily cratered surface of Callisto, since Callisto is larger and contains a greater silicate mass fraction than even Rhea, the most thermally active of the Saturnian satellites we have considered. The melting of $NH_3 \cdot H_2O$ at depth, the migration of this liquid to the base of the lithosphere and its possible penetration of the lithosphere to form flows on the surface is considered the most likely process to account for the resurfacing of these satellites (Squyres 1981c; Stevenson and Anderson 1981; Stevenson 1982c). If the liquid ammonia hydrate does not easily penetrate the lithosphere, tidal heating of the solid shell above a liquid region could also explain the endogenic surfaces of these satellites.

XI. ENCELADUS

Enceladus (see chapter by Morrison et al. and Morrison et al. 1984) is unique among the satellites of Jupiter and Saturn in that it is the smallest satellite to show definite evidence of endogenic surface modification long after the period of satellite formation. Crater densities and surface morphology vary greatly over the surface of Enceladus. Six separate terrain types were identified by Smith et al. (1982). The sharp boundaries between different geologic units indicate different surface ages. The ages estimated from cratering statistics range from 3.8 Gyr for the most heavily cratered areas to a few hundred million years for the smooth plains on which no craters have been observed (Plescia and Boyce 1983; cf. chapter by Chapman and McKinnon). Some areas of Enceladus show grooved terrain very similar to areas on Ganymede. Squyres et al. (1983b) interpret these grooves as open extension fractures (cf. chapter by Squyres and Croft).

The surface of Enceladus has an extremely high and quite uniform albedo (Smith et al. 1982). The low mean density [1130 ($\pm$ 550) kg m^{-3}; Table I] indicates that the ice mass fraction is well over 70%. Water ice has been detected spectroscopically and no nonice contaminant has yet been identified (Clark et al. 1983a; chapter by Clark et al.). The near-surface material is thus differentiated and is almost pure ice. No NH$_3$ has been detected on Enceladus; Clark et al. (1983a) have estimated that there is < 10% by weight of constituents other than water on Enceladus' surface.

For a body as small as Enceladus (250 km radius; Table I) to remain thermally active over such an extremely long time requires an extraordinary heat source. Accretional heat and long-lived radionuclides would most probably be insufficient to bring Enceladus to the molten state early in its evolution (Ellsworth and Schubert 1983). Certainly, these heat sources could not prevent a body of the size of Enceladus from cooling rapidly to temperatures well below those necessary to resurface the satellite. Yoder (1979b) suggested that tidal heating might be important for Enceladus because it is in a 2:1 resonance with Dione. Tidal heating appears to be the only viable heat source for Enceladus. There are, however, serious problems with this mechanism as well.

The first attempt to quantitatively estimate the amount of tidal heating in a homogeneous Enceladus (Peale et al. 1980) resulted in a calculated total dissipation rate of 25 MW (for $Q = 100$), a value that is less than the energy produced by the decay of radioactive isotopes ($\simeq$ 100 MW; Table II). Peale et al. (1980) concluded that tidal heating would not be important. Yoder (1981) used a Q value of 4, but even with the 25 times higher heating rate, came to the same conclusion. Poirier et al. (1983) have estimated tidal heating rates for viscoelastic models of Enceladus. Though still higher maximum heating rates were obtained, they also concluded that tidal heating was not sufficient to melt Enceladus.

Nevertheless, the evidence for the resurfacing of Enceladus remains strong and, to add further complication to the problem, it was discovered (Baum et al. 1981) that the tenuous E ring of Saturn peaks sharply in density at the orbit of Enceladus. This implies that Enceladus is a source for E ring particles and, since the lifetime of these micron-sized particles to sputtering is estimated to be < 10^4 yr, it appears that the interior of Enceladus could even be a currently active source of particles (Morfill et al. 1983b).

The mechanism by which the E ring particles are ejected from Enceladus is of crucial importance to the deciphering of the satellite's internal state. Particles could originate in the interior of Enceladus and be propelled into space by geyserlike activity, or they could be surface fragments produced by meteoritic impact. Production as a result of internal activity would require a currently molten region in the interior (Squyres et al. 1983b). A single large impact (McKinnon 1983) would yield a broad size distribution of fragments and would be a fortuitous event indeed, producing an ephemeral E ring which would disappear in $< 10^4$ yr. Haff et al. (1984) have estimated the flux of particles produced by the hypervelocity impacts of small meteorites. They conclude that the flux of such particles is probably too low to sustain the E ring and note that the size distribution so generated would be rather broad and the fragments irregular. Furthermore, no such peak is observed at the orbit of Mimas, which has both a higher meteorite flux and a lower escape velocity. Pang et al. (1984b) have attempted to deduce the size distribution and shape of the E ring particles from Voyager star-tracker observations. They argue that a very narrow size distribution of particles (2 to 25 μm) and a nearly spherical shape is consistent with their observations. If confirmed, this result would favor a molten origin for the particles. Herkenhoff and Stevenson (1984) have a model for deriving E ring particles from volcanism which predicts nearly spherical particles of about the right size by crystal growth in an expanding vapor cloud.

Squyres et al. (1983b) have reexamined the tidal dissipation calculations for Enceladus with a view toward determining what conditions would be necessary to produce and maintain a molten interior. Stevenson (1982c) has suggested one ameliorating circumstance. If Enceladus incorporated NH_3 as a hydrate during the formation process, internal melting at the eutectic temperature of $NH_3 \cdot H_2O$ (173 K) could produce a liquid layer at temperatures below those for which subsolidus convection in ice should be important. Also, the density of the $NH_3 \cdot H_2O$ eutectic is less than that of ice, permitting the flow of liquid to the surface. If, for some reason (such as a higher eccentricity in the past; Lissauer et al. 1984) melting had occurred in Enceladus and a liquid layer had formed, there would have been a considerable enhancement of the tidal heating rate due to the greater flexure of a viscously decoupled shell (Peale and Cassen 1978). Squyres et al. (1983b) calculated the eccentricity enhancement, above the present value, needed to maintain a molten layer beneath an ice crust. For a Q of 20 and an ice crust 10 km thick the eccentricity enhancement is a factor of 5 for an $NH_3 \cdot H_2O$ eutectic liquid layer. To melt a layer by tidal dissipation would require a considerably higher value of Enceladus' initial eccentricity (up to a factor of 20 for a liquid H_2O layer). Stability calculations indicate that under these conditions an ice crust less than some 30 km thick would be stable to subsolidus convection.

Since an unforced eccentricity would damp out rapidly (on the order of 10^7 yr) resurfacing from a molten interior would thus require that the eccentricity of Enceladus had been forced to much higher levels than present, either continuously, or intermittently over a period ranging from at least 3.8 Gyr to some few hundred million years ago (Lissauer et al. 1984). Current melting would require that the eccentricity decrease had taken place quite recently, since Enceladus would freeze very rapidly (on a time scale of 50 Myr) on being damped to the present eccentricity. However, these conclusions are based on a crust having the thermal conductivity of pure ice. If the thermal conductivity of the ice layer were lower by an order of magnitude, it would be possible to maintain a liquid layer under the above conditions with the present eccentricity. Such a decreased thermal conductivity for the outer layers of Enceladus is certainly possible. Clathrates (of substances such as methane) can have conductivities that are factors of 5 to 10 below that of pure ice (Ross and Anderson 1982; Cook and Leist 1983). Amorphous ice has a low thermal conductivity and an ammonia-ice hydrate should be a factor of 2 less conducting than pure ice (Stevenson 1982c). Further, if recent resurfacing has taken place on Enceladus, the surface would consist largely of condensed frost. An open liquid surface in a vacuum will boil rapidly while the surface refreezes. Most of the water vapor would escape from Enceladus (carrying with it condensed ice particles or frozen droplets resulting from the boiling process) since the thermal rms velocity of an escaping water molecule is more than twice the escape velocity. Molecules with speeds below escape velocity would be widely distributed over the surface, however, and some could be reaccreted from orbit. The exceptionally high albedo of Enceladus is consistent with a fresh frost layer (see the chapter by Veverka et al.). Since condensed frosts have very low densities and thermal conductivities, a frost layer would be a very effective insulator. A substantial frost layer would greatly decrease the crustal thickness necessary to maintain a liquid interior.

Even if the suggestions of Squyres et al. (1983b) regarding the possibility of maintaining a currently molten interior were accepted, there remains the problem of a mechanism for initiating such melting. Lissauer et al. (1984) have noted that dynamical evolution time scales within the Saturnian system are short compared with the age of the system. In particular, the large rate of exchange of angular momentum between Saturn's rings and Saturn's newly discovered inner satellites implies substantial evolution within the past 10 to 100 Myr. Lissauer et al. (1984) demonstrated that most of the angular momentum and energy transferred outward from the rings to Janus, through density wave interactions, could have been passed on to Enceladus by a stable 2:1 Janus-Enceladus orbital commensurability. They hypothesized that this resonance existed until it was rather recently disrupted, possibly by catastrophic impact or upon establishment of the 2:1 Enceladus-Dione resonance 10^7 to 10^8 yr ago. The amount of angular momentum and energy transferred at the rate calculated by Lissauer et al. (1984) would provide Enceladus with an orbital eccentricity sufficiently large to initiate partial melting within Enceladus and to maintain a molten layer until quite recently. Heating by means of the present Dione 2:1 resonance could perhaps maintain a molten region in equilibrium or at least delay refreezing sufficiently so as to maintain a partially molten interior until the present time.

The confidence engendered by a hypothesis which involves as many special assumptions as the foregoing is certainly less than satisfactory. In addition, there are other difficulties, such as the question of why Mimas, which is closer to Saturn and has a higher eccentricity, shows no sign of internal activity since its formation. Further, the eccentricity, even in the absence of a resonance, has not yet damped to a low value. Several arguments may be presented to address this problem. First, the orbital history may have differed and not involved an earlier enhanced eccentricity. Second, Mimas may have formed sufficiently near to Saturn in its early high-temperature phase to have formed without the inclusion of volatile compounds such as NH_3, and thus it failed to achieve partial melting. Another difficulty, which has yet to be addressed, is the fact that the surface activity on Enceladus has been decidedly non-symmetric while all the theoretical investigations have treated only a symmetric model.

At least one aspect of the Squyres et al. (1983b) hypothesis is reassuringly consistent. The mass and energy requirements to generate an E ring from Enceladus are quite modest. Estimates of the total mass of the E ring ($< 10^9$ kg of 10 μm particles; Baum et al. 1981), together with estimates of the lifetimes of the particles to magnetosphere sputtering, yield equilibrium mass fluxes for maintenance of the E ring of about 10^6 kg yr^{-1}. Over the age of the solar system this would amount to the removal of a H_2O layer from Enceladus only 6 m thick. The energy flux associated with the removal of this much water is also very low, approximately 4 orders of magnitude less than the radioactive heating rate.

The evidence is compelling that Enceladus has undergone a much different and more extensive thermal evolution than the other icy Saturnian satellites. Theoretical calculations indicate some possible processes whereby such an evolutionary history could have occurred. Many uncertainties beset this interpretation, however, and much more observational evidence will be needed before the thermal history of Enceladus will be understood.

XII. TITAN

Mass and radius values for Titan are intermediate between those of Ganymede and Callisto (Table I). Thus, on the basis of size and mean density alone, Titan might be expected to have a similar bulk composition and an analogous thermal history to that of the two large icy Jovian satellites. In contrast to these bodies, however, Titan has a substantial atmosphere which totally obscures its surface (see Hunten et al. 1984; chapter by Morrison et al.). Consequently no clues from surface morphology observations are available. Furthermore, the atmosphere consists primarily of N_2 (and perhaps Ar), with some CH_4, indicating that the body may have incorporated a significant amount of volatiles (such as CH_4, N_2, NH_3, etc.) during its formation. The volatiles would presumably be present in a predominately H_2O matrix in the form of hydrates (NH_3) and clathrates (CH_4, N_2). These constituents would radically change the internal properties compared to a satellite in which water is the only volatile. Melting points of the hydrates and decomposition temperatures of the clathrates are much lower than the melting temperature of H_2O (Miller 1961), and the thermal conductivities of some of these compounds may be lower than that of H_2O ice by an order of magnitude (Ross and Anderson 1982). Values for a number of important parameters (such as viscosity) have not been measured. Until hydrate and clathrate properties are better known (under appropriate pressure and temperature conditions), thermal history calculations based on assumed compositions can provide little useful information. This is especially true for Titan because there is no surface morphology data against which to test hypotheses.

The atmosphere contains some H_2 and traces of a number of molecular species incorporating H, C and N (Hanel et al. 1981). Aerosols obscure the entire surface of Titan (Smith et al. 1981) and can be explained in terms of photochemically produced polymers (see, e.g., Podolak et al. 1979). Since such polymers would tend to fall out upon the surface, either there was initially a sufficiently large reservoir of methane to persist to the present, or additional methane has been released from the interior over time. The possibility of a methane ocean on Titan has been proposed (Tyler et al. 1981) but further interpretations of the data have not supported this hypothesis (Eshleman et al. 1983; Flasar 1983). More recent work (Lunine et al. 1983) suggests the existence of an ethane ocean (with a dissolved fraction of methane).

There is certainly a strong possibility that Titan has experienced internal melting and extended thermal activity. The estimated chondritic heat generation rate within Titan corresponds to a surface heat flux of 4 mW m^{-2} at the present time. Tidal heating (Peale et al. 1980) could add a nonnegligible increment to this, although the uncertainty in internal properties strongly influences any estimate of the magnitude. A low Q and internal strength could be expected if there is internal melting, but too large a value for tidal dissipation would present another problem. Since Titan's eccentricity is not maintained by an orbital resonance, a high dissipation rate should have reduced the eccentricity below the present value (Peale et al. 1980).

The question of the thermal state of Titan underscores the need for more experimental data on the properties of clathrates and hydrates. Surface observations are also a basic requirement and spacecraft missions involving radar mapping, as well as atmospheric and surface probes, have been proposed. Titan may eventually prove to have one of the most complex and interesting thermal histories of the satellites of the solar system.

XIII. SATELLITES OF URANUS, NEPTUNE AND PLUTO

Only Earth-based observations are available for the 8 known satellites of these planets (see chapter by Cruikshank et al.). Their inaccessibility to observation has insured that insufficient data are presently available to permit realistic thermal evolution calculations. However, this situation should soon change drastically. A Voyager spacecraft is scheduled to fly by Uranus in January 1986 and Neptune in August 1989. Success in these encounters will greatly increase our knowledge of these planetary systems. For Pluto and Charon, new information is even more imminent. The plane of Charon's orbit passes through Earth's orbit during 1985. This means that a study of mutual occultations will permit much more accurate determinations of Pluto's and Charon's mass and radius. Such observations are already in progress (Binzel et al. 1985a).

Although detailed speculation on the internal structure and thermal history of these bodies is probably premature in light of expected developments, we make here some general remarks. Charon may be large enough (Table IV), and have a sufficient amount of very low melting point volatiles (CH_4, clathrates, etc.) that it has been thermally active at least in the first aeon or so of its history. Nereid (satellite of Neptune) was probably never significantly tidally heated and it is probably too small to have other than a cold inactive interior.

Triton should be quite another case. Although it is probably somewhat smaller than Titan (Tables I and IV), it should contain an even larger complement of low melting point materials than Titan. This expectation is contradicted, however, by the only mass determination currently available (see, e.g., Greenberg 1984a) which, together with current radius estimates, yield a density of about 6000 kg m^{-3} (Table IV). This density happens to be unrealistically high and a new determination of the mass is urgently needed. If a density consistent with other outer solar system satellites is assumed, accretional and radiogenic heating should produce internal melting and a complex

TABLE IV
Properties of Triton and Charon[a]

	Radius (km)	Mass (10^{20} kg)	Density (kg m^{-3})
Triton	1320 ± 250[b]	1750 ± 250[c]	5,900 $^{+\ 15,800}_{-\ 2,700}$
Charon[d]	12	700 (assumed)	900 (assumed)

[a]Table after R. H. Brown (1984).
[b]Alden (1943).
[c]Cruikshank (1984).
[d]Christy and Harrington (1980).

thermal history. The possibility that Triton has been tidally captured and has undergone a substantial orbital evolution, introduces the potential for large amounts of internal heating (McKinnon 1984). Currently, tidal heating should be unimportant since Triton's eccentricity is < 0.005 (Harris 1984). Methane on Triton's surface has been identified (see, e.g., Apt et al. 1984). Recent measurements have indicated the presence of liquid nitrogen and an atmosphere dominated by N_2 (Cruikshank et al. 1983b). Extensive and evolved atmospheres (Trafton 1984) are often characteristic of bodies that have undergone substantial internal evolution. Triton could well be one of the most interesting objects remaining to be explored in the solar system.

Somewhat more information is available for the Uranian satellites, although not enough to provide a base for comprehensive calculations. No surface detail is visible, of course, but the satellites appear to be covered with H_2O ice and some material of neutral spectral characteristics. They range in size (Table V) from that of Mimas to objects somewhat larger than Rhea (Brown et al. 1982b; Brown and Clark 1984). Mass estimates of the satellites (Table V) have been derived from observations of their mutual gravitational perturbations (Veillet 1983a; cf. Dermott and Nicholson 1986). The densities calculated from these observations for Ariel and Umbriel are similar to those of the smaller Saturnian satellites. Miranda is so small and so close to Uranus that uncertainties in the measurements preclude the determination of a meaningful density. For Titania and Oberon, the calculated densities are about 2000 to 3000 kg m^{-3}. These values are quite surprising since bodies so far out in the solar system (and so remote from their primary) would be expected to be more volatile-rich. Both mass and radius measurements are difficult, however, and the densities (Table V) should not be considered definitive (see Dermott and Nicholson 1986). Indeed, the uncertainties in the densities are such that they allow the possibility that all the satellites could have a uniform density of

TABLE V
Properties of Uranian Satellites[a]

	Radius (km)	Mass (10^{20} kg)	Density (kg m^{-3})	Visual Geometric Albedo
Miranda	250 ± 110	1.7 ± 1.7	~ 3000	~ 0.5
Ariel	665 ± 65	15.6 ± 3.5	1300 ± 500	0.30 ± 0.06
Umbriel	555 ± 50	10.0 ± 4.2	1400 ± 600	0.19 ± 0.04
Titania	800 ± 60	59.0 ± 7.0	2700 ± 600	0.23 ± 0.04
Oberon	815 ± 70	60.0 ± 7.0	2600 ± 600	0.18 ± 0.04

[a]Masses and densities are from Veillet (1983a; cf. cautionary remarks in the introductory chapter). Radii and albedos are from R. H. Brown et al. (1982b) except for the radius and albedo of Miranda which are from R. H. Brown and R. N. Clark (1983).

about 2000 kg m^{-3}. A comparison with the Saturnian satellites of similar size suggests that internal activity in the Uranian satellites would be at least as extensive as that observed on Saturn's moons. The lower temperatures at Uranus' orbit would presumably result in the retention of even more volatiles. However, higher densities might indicate higher formation temperatures and hence less volatiles. A higher rock/ice ratio implies more radioactive heating within a body. Stevenson (1984b) has considered six classes of "generic satellites" to describe the Uranian moons. These include homogeneous and differentiated satellites with different postulated compositions. The classes yield a wide variety of surface morphologies and evolutionary paths. It will be interesting to see if even this wide range of possibilities encompasses the actual situation to be revealed by Voyager (see Table IV of introductory chapter).

Squyres and Reynolds (1983) have estimated tidal heating rates for the Uranian satellites from their observed eccentricities. It is possible that Miranda, Umbriel and, especially, Ariel are tidally heated because their current eccentricities are higher than expected. (However, Dermott and Nicholson [1986] question these eccentricities.) These satellites are not presently forced by orbital resonances that could maintain high eccentricities. The inner three satellites are, however, very near to a Laplace-type commensurability. (The four outer satellites are even closer to a commensurability relation, although not the type associated with a stable resonance [Greenberg 1975a].) Squyres et al. (1985) have calculated the satellite orbits backward in time and concluded that low-order resonances could have formed in the past. These resonances seem to be too far back in time, however, to be reconciled with Ariel's short eccentricity damping time. Dermott (1984b) points out the possibility of resonance overlap among these satellites and observes that in this circumstance the satellites could undergo chaotic motions that are capable of producing large changes in the orbital elements.

Ariel has a higher albedo than the other satellites and this suggests that its thermal history may have been significantly influenced by tidal heating (Squyres and Reynolds 1983).

Expectations are high that in the next few years we will make new and exciting discoveries about the outer planet satellites that will rival those from the Voyager explorations of the Jovian and Saturnian systems.

XIV. CONCLUSION

Knowledge of the thermal evolutions of natural satellites can provide important information on a number of fundamental problems including the conditions and processes that obtain within protoplanetary nebulae. Thermal history studies also enable us to explain the endogenic processes that have modified the surfaces of satellites and stamped them with characteristic morphologies. Finally, investigations of satellite thermal evolution give us insight into what the internal structures of these bodies might be. The type and extent of satellite thermal evolution is seen to depend primarily upon heat sources and thermal transport processes as influenced by the compositions, material properties, internal structures, dimensions, and orbital parameters of the bodies. Variations in these processes and properties have produced a very wide range of different satellite evolutionary paths. Internal activities can encompass processes from quiescent heat conduction and large-scale subsolidus convection in bodies well below their melting points, to melting and upward migration of magma culminating in such violently active surface features as the volcanoes of Io. Internal melting (as opposed to impact melting) depends not only upon the quantity of internally generated energy available, relative to the amount necessary to melt a sizable fraction of the material, but also strongly upon the rheology of the material. If a significant fraction of a given volume of material can melt at temperatures below that at which the bulk melts, then it may be possible to differentiate the body.

Small bodies with radii ≲ 100 km (e.g., satellites of Mars, Mimas) have the least complex histories; no significant thermal activity occurs long after formation or a major impact event. The large satellites with radii ≳ 1000 km (e.g., Moon, Io, Europa, Ganymede, Callisto, Titan, Triton) have experienced a range of very complex histories with activity continuing until the present. Intermediate-sized satellites (e.g., Rhea, Iapetus, Dione, Tethys, Umbriel, Titania) have experienced various degrees and durations of thermal activity with such activity having ceased by the present time (with some possible exceptions, e.g., Enceladus). Tidal heating can amplify and extend the thermally active period of any size satellite.

While a number of relevant processes have been described and outlines given of some possible satellite thermal evolutions, much more remains to be done. The properties of the satellites of the outer three planets are still almost unknown. Many first order problems remain in understanding the other bodies; these include the origin of the Moon, the existence of metallic cores in the Moon and Io, the internal structures of Europa and Enceladus, the degree of melting in Io, the nature of the difference between Ganymede and Callisto, the origin and evolution of the atmosphere of Titan, etc. Within the near future, close range spacecraft (Voyager flybys of Uranus and Neptune and Galileo Jupiter orbiter) and Earth-based observations (including Space Telescope) should greatly improve our knowledge of some of these satellites. Laboratory experiments on ices and clathrates under conditions of temperature and pressure relevant to the interiors of the outer solar system satellites will provide much needed information on the material properties of these substances. Intensified theoretical investigations utilizing the improved data base should greatly increase our understanding of these satellites and consequently of the origin and evolution of the solar system.

Acknowledgments. This work was supported in part by the National Aeronautics and Space Administration.

7. THE TECTONICS OF ICY SATELLITES

STEVEN W. SQUYRES
NASA Ames Research Center

and

STEVEN K. CROFT
University of Arizona

The icy satellites of the outer solar system display unusual and, in some cases, unique geologic features that are a result of the behavior of ice as a geologic material. Observations and theoretical calculations suggest that Europa is primarily a silicate body, but that the silicate interior is overlain by an icy crust and, perhaps, a liquid water ocean. The surface of Europa is laced with a complex pattern of fractures that have resulted from stresses in the icy crust. Ganymede shows widespread evidence for extensional deformation, suggesting a volumetrically small global expansion early in the satellite's history. Processes affecting the surface of Ganymede include faulting that produced a complex pattern of topographic grooves, and resurfacing that may have involved the emplacement of liquid water or warm, mobile ice. Callisto shows no tectonic features, and its contrast with Ganymede is one of the most important unresolved problems in the Jupiter system. Saturn's satellite Enceladus also shows evidence for geologic activity, including resurfacing and groove formation. Enceladus' very high albedo and its association with Saturn's tenuous E ring provide some evidence that it may have been active very recently. The other satellites of Saturn show evidence for varying degrees of resurfacing and extensional tectonism.

The solid planets and satellites in the solar system may be divided into two fundamental classes based on their composition. One class consists of bodies composed primarily of rock and, in some cases, metal. Bodies in this class include Mercury, Venus, the Earth, the Moon, Mars, and Io. The other class consists of bodies made up largely of ice. Important bodies in this class are Jupiter's satellites Ganymede and Callisto, Saturn's satellites Mimas, Enceladus, Tethys, Dione, Rhea and Titan, and probably also the satellites of Uranus and Neptune, as well as Pluto and its satellite Charon. Jupiter's satellite Europa is an unusual case, as it is primarily rocky in bulk composition, but has a surface geology apparently dominated by ice. Rocky solar system objects have been investigated in some detail. Until recently, however, almost nothing was known about the geology of icy objects.

With the acquisition of Voyager data, we have obtained a first look at the satellites of Jupiter and Saturn. A number of icy satellites display evidence for geologic activity on their surfaces. In nearly all cases this activity has created features different from those observed on the rocky bodies of the inner solar system. These features owe their appearance to the unique behavior of ice as a geologic material. In this chapter we summarize some aspects of our current understanding of the tectonics of icy satellites. We use the term "tectonic" here in the sense of pertaining to any geologic features formed by forces originating wholly within the body. We begin in Sec. I by describing the tectonic features observed on the icy satellites of Jupiter and Saturn, and, in the cases of Europa and Enceladus, tantalizing but inconclusive evidence that these bodies are presently active. We then discuss the processes that can lead to stresses within icy satellites (Sec. II), including volume changes, convection and shell deformation, and finally in Sec. III we discuss the faulting, fracturing and resurfacing that have resulted from these stresses.

I. TECTONIC FEATURES ON THE SATELLITES
OF JUPITER AND SATURN

Europa

Voyager images of Europa (Smith et al. 1979a,b) show some of the most convincing evidence for tectonic activity in the solar system (Fig. 1; see also the chapter by Malin and Pieri). The style of the tectonism, moreover, is in many ways unique among the planets and satellites observed to date. Unlike the other satellites which are discussed in this chapter, Europa is not made substantially of ice. It has a density of 3.03 g cm^{-3}, indicating a primarily silicate composition. If the density of the silicates is the same as for Io, however, Europa is roughly 6% H$_2$O by mass. The high albedo of the surface and the strong water absorption bands observed in its reflection spectrum (Pilcher et al. 1972; Clark and McCord 1980a) indicate that the surface is primarily H$_2$O ice. If Europa were completely differentiated, a 6% H$_2$O content would be sufficient to cover the silicate interior with a layer of ice (or water overlain by ice) ~100 km thick. The factor determining whether or not the tectonism

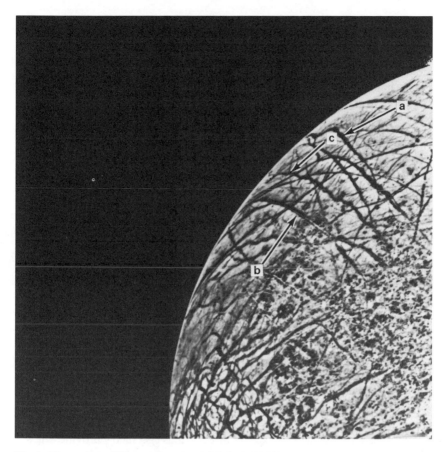

Fig. 1. The surface of Europa. Features visible include dark bands (a), and triple bands (b). A number of dark bands and triple bands show a distinctive repeating cycloid pattern, e.g. (c).

observed at the surface of Europa is dominated by ice is therefore the degree to which it has differentiated.

One of the most noteworthy observations of Europa is that its surface is nearly devoid of impact craters at Voyager resolution (~2 km/pixel). Only a handful of probable impact craters have been identified, mostly in the range of 10 to 20 km in diameter (see, e.g., Lucchitta and Soderblom 1982). The paucity of craters indicates that global resurfacing of Europa has taken place during geologically recent times. This resurfacing could have taken place by eruption of liquid water through fractures, by complete destruction of the crust and replacement by liquid water from below, or perhaps by an elevation of subsurface temperatures sufficient to erase all topography via glacier-like viscous flow of the warm ice. The last possibility would require temperatures much warmer than the solar equilibrium surface temperature of Europa, and probably could only occur if a thermally insulating surface material were present.

Two major geologic units have been mapped on the surface of Europa: mottled terrain and plains (Lucchitta and Soderblom 1982). The mottled terrain is hummocky on a lateral scale of several km, with relief that probably does not substantially exceed 100 m. It is light brown or gray in color, with numerous small circular or irregularly shaped dark spots (see Plate 10). The plains are generally brighter, smoother, and show fewer of the dark spots. Contacts between these two major units may be fairly abrupt, but are not marked by any topographic feature other than a general change in surface texture.

Both the mottled terrain and the plains are transected by a very complex intersecting pattern of linear features (see chapter by Malin and Pieri). These features show a bewildering variety in their appearance and geometry, and a number of classification schemes have been devised (see, e.g., Pieri 1981; Lucchitta and Soderblom 1982). For our purposes it is sufficient to consider three basic classes: dark bands, triple bands and bright ridges.

Dark bands are the most prominent of the linear features on Europa. (The term "dark" is actually inaccurate, as they typically have albedos of 0.5 to 0.6, and are only slightly darker than the bright plains, which have a mean albedo of 0.71 [Buratti and Veverka 1983].) They may be curving or straight, range in width from a maximum of roughly 100 km down to the resolution limit of the images, and range in length from tens of km to a few thousand km. Even at the lowest illumination angles near the terminator, no relief is observed to be associated with the dark bands. They intersect one another in a complex fashion, dividing the surface into a mosaic of bright polygons separated by dark bands. Several aspects of the bands' geometry provide clues to their formation. First, while dark bands intersect quite commonly, they are not observed to be truncated or offset laterally where they intersect other bands (Y-shaped intersections, however, are common). There is therefore no convincing evidence of transcurrent (strike-slip) motion along the dark bands. Second, many of the dark bands are wedge-shaped, tapering to sharp points at one or both ends. In this respect, they look very much like fractures that have propagated across Europa's surface. In a number of instances, the margins of wide dark bands could be matched up by closing together the blocks of bright plains material on either side while rotating them slightly (Schenk 1984). It appears that the blocks may have once been joined, and have been spread apart and rotated, with darker material filling in between. Many of the patterns formed by dark bands are similar to those formed by leads separating floes of sea ice on Earth. It is not clear why the dark bands differ in albedo from the rest of Europa. Possibilities include the presence of organic material (Schonfeld 1982) or radiation-darkened salts (see, e.g., Fanale et al. 1977a) brought to the surface from an underlying liquid layer.

Triple bands are similar in many respects to the dark bands, but differ in

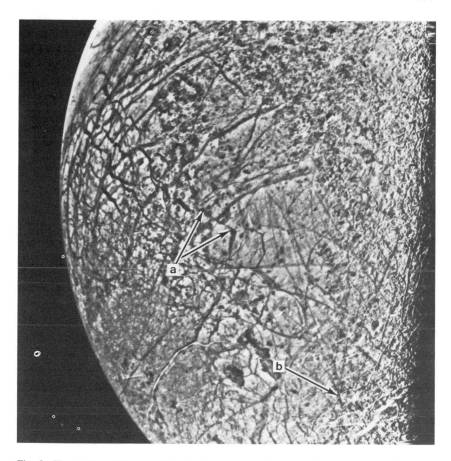

Fig. 2. The surface of Europa. Note that the geometry of some dark bands suggests the separation and rotation of crustal blocks, with new material between them forming the dark band (a). Bright ridges near the terminator (b) are approximately 10 km wide, and probably no more than 100 m high. Many display the same repeating cycloid pattern exhibited by some dark bands and triple bands. Their origin is uncertain.

that they have a distinct bright stripe running down the center. Triple bands are generally longer and straighter than most dark bands, and form many of the global-scale lineations visible in low-resolution images of Europa. Some triple bands, as they become narrower near their ends, appear to lose their central bright stripe. In a few instances, the central bright stripe of triple bands appears to be raised slightly above the surrounding plain (Malin 1980; Golombek and Bruckenthal 1983).

Bright ridges are the most prominent topographic features on Europa (Fig. 2). They have very regular widths of about 10 km, heights of at most 100 m, and lengths of hundreds of km. Like the dark bands and triple bands, they intersect commonly without transcurrent offsets. The bright ridges observed by Voyager are concentrated in a fairly narrow strip along the satellite's terminator. It is very likely, however, that they are common over much of the satellite but invisible elsewhere due to their low relief and high illumination angles. Many of the ridges have a distinctive repeating cycloid pattern.

A number of the dark bands and triple bands observed far from the terminator show this same cycloid pattern (Fig. 1). It is possible then that many of the ridges are in fact the central stripes of triple bands whose albedo contrasts are lost near the terminator. We may really only be seeing one basic type of tectonic feature on Europa: a fracture-like band that is slightly darker than its surroundings; that has a photometric contrast against the brighter background that is variable and depends on lighting geometry; that may or may not have a visible low bright ridge down the center; and that may be roughly straight or show a cycloid pattern. Galileo images to be obtained with a range of lighting geometries will help to test this idea. The origin of the cycloid pattern exhibited by many ridges and dark bands is enigmatic.

There is evidence that at least small amounts of H_2O frost have been deposited on the surface of Europa very recently (Squyres et al. 1983a). The evidence is provided by the observed strength of an ultraviolet absorption feature, attributed to S—O bonds, on the trailing hemisphere (Lane et al. 1981). From the strength of the absorption, a sulfur column density of 2×10^{16} cm^{-2} is inferred. Estimated values for the flux of S ions trapped in the Jovian magnetosphere into Europa's trailing hemisphere (Eviatar et al. 1981) predict that this column density would be deposited in only ~ 7 yr. Assuming instead that the observed column density is a result of dilution of S ions by ongoing deposition of H_2O frost, an estimate of the minimum present global mean H_2O deposition rate can be obtained. For an ultraviolet measurement depth of 0.5 μm, this rate would be ~ 0.1 μm yr^{-1}. More detailed calculations of sputtering, redistribution and escape by Eviatar et al. (1985) have supported this conclusion, but have suggested a deposition rate about half as large.

A reasonable interpretation of this observation, along with Europa's bright surface, intricate fracture pattern, smooth topography, and paucity of impact craters is that it has been active very recently, and presently has a thin ice crust underlain by an ocean of liquid water (Squyres et al. 1983). Simple theoretical calculations of tidal and radiogenic heating on Europa imply a heating rate that may be sufficient to maintain a crustal thickness of ≤ 20 km.

Ganymede

Ganymede has two principal geologic units, one dark and one bright (Smith et al. 1979a,b; chapter by McKinnon and Parmentier). Each unit covers $\sim 50\%$ of the surface. The dark terrain occurs in roughly polygonal regions that are cut through and separated by bands of bright terrain. Both units are overlain by a very thin bright frost cap near the poles.

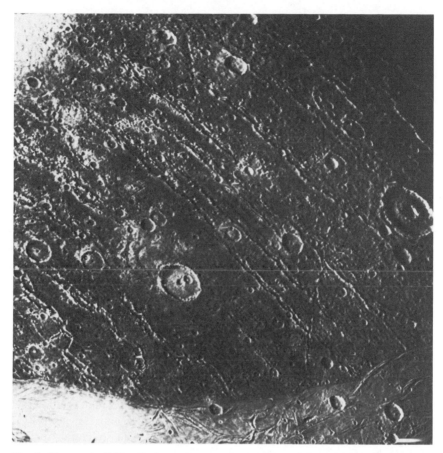

Fig. 3. Furrows in Galileo Regio on Ganymede. They are typically 10 km wide and spaced 50 km apart, and are the oldest tectonic features on Ganymede. Their origin is uncertain. The scale across the image is ~650 km.

High-resolution images of the dark terrain show it to be hummocky and heavily cratered. Some polygons of dark cratered terrain have a distinct linear texture consisting of shallow, subparallel linear depressions termed furrows (Fig. 3). The furrows are observed in a number of regions of dark cratered terrain, but are nowhere more prominent or well-developed than in Galileo Regio. In this area they are typically 10 km wide and up to hundreds of km long (Zuber and Parmentier 1984a). They have raised rims, edges that are scalloped in plan view, and depths of perhaps a few hundred m. The furrows in Galileo Regio are typically spaced 50 to 100 km apart. In a few instances the parallel furrows are cut at large angles by individual furrows that may be younger (Cassachia and Strom 1984). All of the furrows, however, are very old. They are truncated by fresh as well as highly degraded craters (palimpsests) but no craters larger than 10 to 20 km in Galileo Regio are cut through by furrows. The furrows may in fact be the oldest geologic features observed on Ganymede, tectonic or otherwise (McKinnon and Melosh 1980).

Furrows are found in a number of segments of dark cratered terrain other than Galileo Regio. For example, another well-developed set lies in Marius Regio, just southwest of Galileo Regio. These furrows are morphologically very similar to those in Galileo Regio, though somewhat narrower on the average (Zuber and Parmentier 1984a). Although the furrows in Galileo and Marius Regios are separated by only one narrow band of bright terrain, the orientations of the furrows in the two regios are quite different (Zuber and Parmentier 1984a). Less well-developed furrows are also observed in Nicholson Regio, Barnard Regio and Perrine Regio.

The origin of the furrows is unclear. One possibility is that they are extensional tectonic features formed during a very early episode of deformation. Another is that they are the remnants of a large ring structure formed by a very large impact (McKinnon and Melosh 1980). This hypothesis is motivated by the gross similarity of the furrows to concentric ring structures associated with large impacts on Callisto (see below). Unfortunately, there is no unambiguous record of such an impact on Ganymede. If such a structure ever existed, it must have been enormous, covering at least half of Ganymede. Detailed mapping shows that the centers of curvature of the furrows in Galileo Regio and Marius Regio apparently do not coincide (Zuber and Parmentier 1984a). This nonconcentricity may either imply that there was some rotation of the crustal blocks after furrow formation, or that the furrows are not related to an impact and were never concentric originally. The amount of rotation that would be required is very small. An argument against the impact hypothesis is the observation of furrows that cut across the dominant structural trend at high angles (Cassachia and Strom 1984). Their appearance is not readily explained by the radial/concentric stress field that would be generated by an impact. Without any clear evidence for or against an impact, it is unlikely that the problem of the furrows' origin will be resolved.

The bright terrain has a lower crater density than the dark cratered terrain, indicating a somewhat younger age. There is a variation in crater density of a factor of a few from place to place within the bright terrain, indicating that the period of resurfacing was of significant duration, perhaps on the order of 500 Myr (Shoemaker et al. 1982). The densities are high enough to indicate that most of the resurfacing was probably completed by 3 Gyr ago. Most of the bright resurfaced terrain is characterized by a complex pattern of long, narrow topographic depressions called grooves (Fig. 4). They are generally a few km wide and often hundreds of km long. They are morphologically distinct from furrows in several respects. Their edges are quite smooth rather than scalloped, they are heavily concentrated in the younger, bright resurfaced terrain, and they commonly occur in parallel with one another, with typical spacings between grooves of only 4 to 15 km.

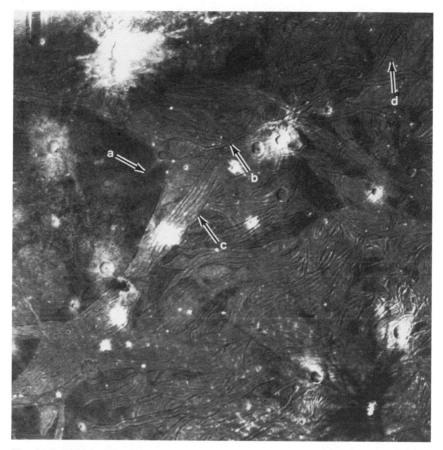

Fig. 4. Grooves on Ganymede. Features visible include single grooves in the dark, heavily cratered terrain (a), groove pairs (b), groove sets (c), and smooth, groove-free areas of bright resurfaced terrain (d). The scale across the image is ~1100 km.

For descriptive purposes, it is useful to consider three ways in which
grooves occur: single grooves, groove sets and complex grooved terrain.
Single grooves are just what their name implies, having no other nearby
grooves that lie parallel to them. Single grooves are quite common in some
areas of bright resurfaced terrain, but are also occasionally observed in dark
cratered terrain. Groove sets consist of two or more grooves that parallel one
another closely over long distances. Sets of two parallel grooves, or groove
pairs, have been cited as particularly common (Parmentier et al. 1982).
Groove sets may be up to several hundred km wide and over a thousand km
long. The grooves in groove sets may lie quite close together, producing
roughly sinusoidal topography, or may be more widely separated, with distinct flats between the grooves. The term "complex grooved terrain" is used
here to denote regions superficially similar to groove sets, but with consider-

ably less regular topography. In complex grooved terrain, no simple pattern of regularly spaced grooves can be recognized. Instead, the topography consists of unevenly spaced, occasionally asymmetric undulations that cannot clearly be classified as parallel grooves or ridges. A continuum of forms exists between the most regular groove sets and the most irregular complex grooved terrain. Groove sets and complex grooved terrain are very heavily concentrated in the bright resurfaced terrain.

The topography of the grooves is difficult to determine in detail due to the limitations of Voyager resolution and coverage. Photoclinometric profiles across the grooves at a maximum resolution of ~600 m/pixel show gentle U-shapes, with a rms slope of 5°5 and maximum slopes of only ~20° (Squyres 1981b). At scales smaller than the resolution limit, however, the topography could be considerably more rugged. Voyager resolution is insufficient to reveal the details of the topography of groove walls.

Some areas of the bright resurfaced terrain have no grooves, and at the maximum Voyager resolution appear quite smooth even under very low angle illumination. The smooth regions may occur as irregularly-shaped groove-free patches of bright terrain that are surrounded by a complicated pattern of crosscutting grooves and groove sets. They also occur as long curvilinear swaths with dimensions similar to those of groove sets. Many smooth swaths have very sharp boundaries. These boundaries may be marked only by an abrupt transition to the hummocky texture and lower albedo of the dark cratered terrain, or may be marked by single grooves. A few smooth regions have very diffuse, indistinct boundaries transitional to groove sets or to dark cratered terrain.

Single grooves, groove sets, smooth swaths and regions of complex grooved terrain may exhibit intricate crosscutting relationships. Where one groove set crosscuts another, no trace of the cut groove set is generally observed within the crosscutting set. Single grooves or groove sets lying entirely within the bright resurfaced terrain commonly lie along bright terrain/dark terrain boundaries. Other grooves and groove sets within the bright resurfaced terrain may lie at large angles to the boundary grooves or groove sets, and be truncated against them.

The complex organized patterns of the grooves clearly indicate a tectonic origin. There are three possible stress regimes that could account for the observed structure: shear, extension and compression.

The clearest evidence for shear is lateral offset of recognizable features along a linear structure. In several instances, groove sets appear offset along narrow shear zones (see, e.g., Lucchitta 1980). In each case the offset is small, generally <100 km. Another observation that could perhaps be interpreted as evidence for shear is that well-developed groove sets commonly strike into and are truncated against narrow groove sets or single grooves. It could be argued that this situation might have arisen from large amounts of shear along the groove set or single groove against which the truncation occurs. However, it is always impossible in such cases to identify the "other half" of the truncated structure along the far side of the hypothetical shear zone. The clear evidence for shear is therefore limited to a few instances of small offset.

There is considerable evidence that the bright bands and the grooves resulted from extension perpendicular to their structural trend. Many bright bands have sharp terminations at one or both ends, suggesting initiation by a fracture in the lithosphere propagating along its length. In several cases, strike-slip faults cut grooved bright bands obliquely (Squyres 1980a). If the stress field causing the inferred transcurrent faulting had the same general orientation as that forming the grooves, the fault trends and offsets in these cases indicate a maximum extensional stress that lies perpendicular to the grooves. Finally, the negative relief of the single grooves and widely spaced grooves within sets suggests very strongly that they are extensional features, perhaps long, narrow grabens or open extension fractures comparable to glacial crevasses.

There is very little obvious evidence for compressional deformation on Ganymede. No features resembling terrestrial thrust faults or subduction zones, lunar mare ridges or Mercurian lobate scarps have been observed. The topographic symmetry of most grooves argues against a thrust faulting origin, and the negative relief of single grooves and widely spaced grooves within sets argues against formation by folding. In some regions of complex grooved terrain, of course, the complexity of the topography makes it impossible to rule out a compressional origin for the features present. However, the continuum of forms that exists between simple groove sets and complex grooved terrain strongly argues for a common extensional origin for the vast majority of the features.

The interpretation of the tectonic features on Ganymede as primarily extensional in nature leads to some interesting interpretations regarding the observed truncation relationships. Many groove sets abruptly truncate against a single groove or narrow groove set. The groove or groove set against which the truncation occurs would actually mark a transform fault in these cases, with extension in the truncated set occurring on only one side of the transform. In fact, one must be careful in interpreting any features as due to strike-slip disruption of bright bands, as it is possible that the "strike-slip fault" separating two parts of an apparently offset bright band in fact predates the apparently offset feature (see, e.g., Golombek and Allison 1981). An especially interesting example of differing deformation on opposite sides of a transform fault occurs in Tiamat Sulcus, where more extension has apparently taken place on the south side of the fault than on the north side (Fig. 5).

A point of critical importance in understanding the tectonics of Ganymede is the observation that there is not a one-to-one correspondence between the presence of bright terrain and the presence of grooves. Grooves are occasionally observed in the dark cratered terrain, and some areas of bright resurfaced

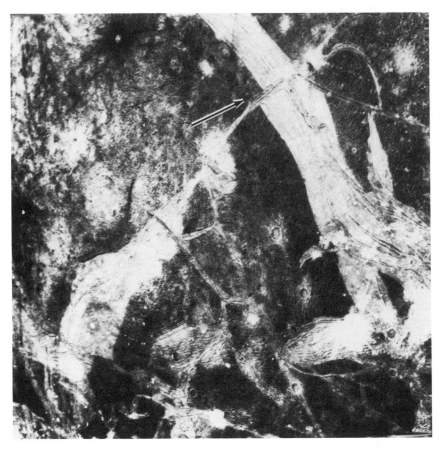

Fig. 5. Apparent transform faulting in Tiamat Sulcus on Ganymede. A broad band of grooves is
cut through by an apparent shear zone (arrow). There has apparently been more extension north
of the shear zone (north is at the top of the image) than south of it. The scale across the image is
~1200 km.

terrain are completely free of grooves. Two separate processes have therefore
operated at Ganymede's surface: one that produced bands of bright resurfaced
terrain, and one that produced grooves. The groove formation process operated most effectively, but certainly not exclusively, in the bright resurfaced
terrain. Because groove sets both crosscut and are crosscut by smooth swaths,
we may infer that resurfacing and groove formation operated concurrently
during the period of geologic activity. Resurfacing and groove formation will
be discussed individually below (Sec. III).

Callisto

Callisto, although it is similar in size and density to Ganymede, has no
clearly tectonic features (Smith et al. 1979a,b; chapter by McKinnon and

Parmentier). Most of the surface of Callisto is similar to Ganymede's dark cratered terrain, but is still darker, older and more heavily cratered. The most noteworthy geologic features on Callisto are several very large ringed impact structures. The rings surrounding the impact points are numerous and regularly spaced, and are in some respects similar to the furrows on Ganymede (McKinnon and Melosh 1980). When observed in detail, however, the rings on Callisto appear to consist largely of asymmetric ridges (Remsberg 1981). Callisto has no bright polar frost deposits like those on Ganymede. One of the most puzzling problems of the Jovian system is why Ganymede underwent intense tectonic activity while Callisto has remained dormant since the end of heavy bombardment (cf. discussion in chapters by Schubert et al., and McKinnon and Parmentier).

Enceladus

Enceladus shows the most extensive evidence for tectonic activity in the Saturn system (Smith et al. 1982; Morrison et al. 1984; chapter by Morrison et al.) (Fig. 6). It is a much smaller body than Ganymede, yet its tectonic features are in many ways quite similar. The surface of Enceladus shows a large range in impact crater density (chapter by Chapman and McKinnon). Some regions have high crater densities suggesting that they date from early in Enceladus' evolution. Other regions have no craters at all at Voyager resolution ($\sim$1 km/pixel), and several intermediate crater densities are observed as well. Enceladus has apparently undergone at least several episodes of resurfacing spread out over a substantial portion of its history. Based on a model impact flux, the youngest resurfaced regions on Enceladus probably are not much more than 10^9 yr old, and may be substantially younger (Smith et al. 1982). Enceladus differs from Ganymede in that areas of differing crater density are not marked by differing albedos. The albedo of Enceladus is nearly uniform across its entire surface, and is the highest known in the solar system (Buratti et al. 1982; chapter by Veverka et al.). This lack of contamination by nonicy material also suggests that geologic activity on Enceladus has taken place fairly recently.

The boundaries separating the oldest heavily cratered terrain on Enceladus and the various younger regions are both diffuse and sharp. Most of the boundaries in the limited Voyager coverage are of the diffuse sort, with no clearly recognizable feature marking the transition from one crater density and age to another. In at least one instance, however, a resurfaced band strikingly similar to the resurfaced bands on Ganymede cuts with very sharp margins through a region of heavily cratered terrain. The band appears to lie at a slightly lower elevation than the cratered terrain, and several craters are abruptly truncated at the boundary.

The strongest similarity that Enceladus bears to Ganymede is that it too has grooves. As on Ganymede, they are curvilinear depressions a few km wide, a few hundred m deep, and up to more than a hundred km long. Most of

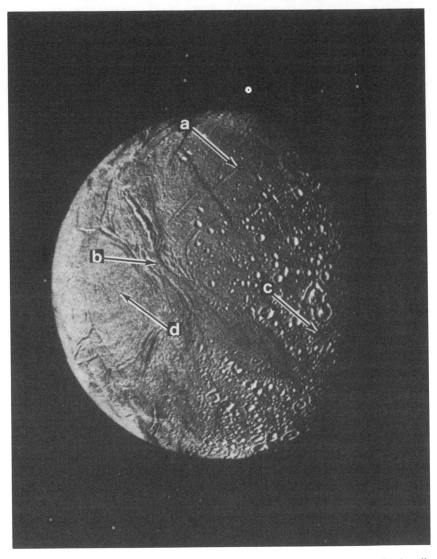

Fig. 6. Enceladus. Features visible include straight single grooves (a), groove sets (b), heavily cratered terrain (c), and smooth resurfaced terrain (d).

the grooves observed are true negative relief features, lying wholly lower than their surroundings. In one case, a ridge seems to lie significantly higher than its surroundings, from which it is separated by grooves on either side. As on Ganymede, grooves are observed singly and in sets of subparallel grooves typically spaced 5 to 10 km apart. The grooves are strongly concentrated in the resurfaced regions. In the one resurfaced band with very sharp margins, the grooves within it lie parallel to the band margins. There are also areas that appear perfectly smooth, having been resurfaced but possessing no grooves.

Grooves on Enceladus show a variation in morphology that appears correlated with the age of the terrain in which they lie. In areas with very few or no craters, the grooves are curvilinear and frequently occur in subparallel sets, much like those on Ganymede. In moderately cratered regions, grooves are less common, much straighter, and occur only singly. In the most heavily cratered regions, no grooves of any kind are observed. It must be stressed, however, that the high-resolution coverage of Enceladus is spatially limited, and the features observed are less certain than those on other bodies to be representative of those on the body as a whole.

Earth-based telescopic observations provide evidence that Enceladus may presently be geologically active. Saturn has a very diffuse, fine-grained ring, known as the E ring, far out beyond the main rings. Observations show that this ring has a radial density profile that is strongly peaked at the orbit of Enceladus (Baum et al. 1981), suggesting that it is the source of the particles. Moreover, the observed thickness of the ring suggests that little dynamical flattening has taken place, and therefore that it is a young feature. Studies of the lifetimes of micron-sized particles in the E ring suggest that sputtering is the most efficient removal mechanism. The sputtering lifetime of such particles has been estimated to be of the order of only 10^2 to 10^4 yr (Morfill et al. 1983b; Burns et al. 1984). This evidence indicates that the E ring is either a very transitory chance occurrence or is maintained by some active process. McKinnon (1983a) has pointed out that the E ring could be composed of ejecta from a very recent impact on Enceladus. In light of the clear evidence for geologically recent resurfacing presented by the low crater densities, however, an attractive interpretation of the E ring is that Enceladus is an active source of these particles (see, e.g., Squyres et al. 1983b; Pang et al. 1984a).

Other Saturnian Satellites

Several other Saturnian satellites show evidence of previous tectonic activity (Smith et al. 1981,1982; Morrison et al. 1984; chapter by Morrison et al.). Mimas, Dione and Rhea all have fractures on their surfaces that manifest themselves as narrow, shallow troughs. On Mimas these troughs are fairly straight, roughly 10 km in width and up to about 100 km long. Some lie roughly radial to the large impact crater Herschel, and it is possible that their formation was related to the Herschel event. The troughs on Dione and Rhea are of similar widths, but are longer and more irregular in plan, and there is no evidence suggesting a relationship to cratering. Troughs are less common on Rhea than on Mimas and Dione. On all three satellites they are relatively isolated features, and do not dominate the surface morphology in the way the grooves on Ganymede and Enceladus do.

The surface of Tethys is cut by an enormous trough, Ithaca Chasma, that extends at least 270° around the satellite (Fig. 7 and Map Section). It has a

Fig. 7. Tethys. The surface is cut by Ithaca Chasma, a deep tectonic trough that extends for at least 270° around the satellite.

width of up to 100 km, a depth that may be as much as ~3 km in places, and it makes up nearly 10% of the satellite's surface. It has at least one smaller trough that branches off from the main trough. The trough walls show tantalizing evidence of complex structure, including internal terraces and grooves, but the resolution of existing images is insufficient to reveal their morphologic details. Ithaca Chasma partially traces a rough great circle across Tethys. Near one pole of this great circle lies the very large impact crater Odysseus, suggesting a possible link between the two features (Moore and Ahearn 1983).

The highest-resolution images of Dione show a few very subdued curvilinear ridges (Plescia 1983; Moore 1984). They are at most a few hundred km long, and have extremely gentle slopes and low relief. Because of their subdued relief they are only visible in images near the terminator, so it is not presently possible to determine their global distribution. These features may provide evidence for compressional deformation.

None of the Saturnian satellites other than Enceladus show evidence for recent geologic activity. Mimas, Tethys, Dione and Rhea all show evidence for resurfacing early in their histories, however (Plescia and Boyce 1982). Rhea is the most thoroughly studied; it has two areas of reduced crater density, one near the north pole, and one near 30° longitude near the equator. Similar low crater-density areas are observed near the south pole of Mimas, on the trailing hemisphere of Tethys, and on the leading hemisphere of Dione. All of these resurfaced regions have fairly diffuse margins, compared to the very sharp curvilinear margins common on Ganymede and parts of Enceladus.

Saturn's outermost regular satellite, Iapetus, was imaged at lower resolution than the rest of the major satellites (best resolution ~8 km/pixel); it shows a heavily cratered surface with no conclusive evidence for tectonic activity. Iapetus is remarkable, however, in that its leading hemisphere is extremely dark, while the trailing hemisphere and the poles have high albedos more like those of the other Saturnian satellites (Morrison et al. 1975). Most of the leading hemisphere is so dark that it is impossible to identify any topography there. It has been suggested that this dark material may have been volcanically erupted to the satellite's surface (Smith et al. 1982). However, Iapetus is darkest exactly at the apex of its orbital motion and becomes continuously brighter with increasing distance from the apex (Squyres et al. 1984), supporting the idea that the albedo pattern is instead somehow related to asymmetry in the flux of impacting material (Cook and Franklin 1970; Soter 1974; Cruikshank et al. 1983a; Squyres and Sagan 1983).

Saturn's largest satellite, Titan, is completely shrouded by clouds, so nothing is known about its surface geology or tectonic history. It is similar in size and density to Ganymede, however, and must be considered a strong candidate for a tectonically active history.

II. SOURCES OF STRESS

We now consider the processes that can lead to stresses in the materials of icy satellites. It should be noted that in this chapter we use the term "crust" to refer to the entirety of a solid outer shell of ice that overlies a layer of liquid water and the term "lithosphere" to refer to the outer solid portion of a satellite that deforms in a brittle rather than a ductile fashion, regardless of whether it overlies water or warmer, more ductile ice.

Volume Changes

From the descriptions in the previous section, it is evident that several icy satellites, particularly Ganymede and to a lesser extent Enceladus, Tethys, Dione and Rhea, have surfaces shaped by extensional tectonism with little or no recognized evidence for compression. On the other hand, the low ridges on Dione may be evidence for a period of compression on that satellite. The stresses necessary to produce such features may have resulted from global changes in volume. There are a number of possible processes that could result in volume changes for icy satellites.

Differentiation. One process that can cause considerable expansion early in the history of large, dominantly icy satellites is differentiation (see chapter by Schubert et al.). The volume increase that would take place is a result of the complex physical chemistry of H_2O. Ice has a number of polymorphs (Fig. 8), only one of which, ice I, has a density less than 1.0 g cm^{-3}. The higher density polymorphs are stable at higher pressures and over a wide range of temperatures. If a satellite accretes homogeneously, silicates will initially be intermixed with these various ice phases (Consolmagno and Lewis 1976). Much of the ice in the deep interior will be ice VII or ice VIII, which have densities of ~ 1.67 g cm^{-3}. As the body is heated by radionuclides in the silicates it will tend to differentiate, with the silicates being concentrated in a core. Differentiation could take place either by complete melting of some of the ice in the interior, allowing the silicates simply to fall toward the center, or by downward flow of dense, very silicate-rich material through solid but warm and mobile ice (see, e.g., Schubert et al. 1981). Liquid water or silicate-poor ice would be forced nearer to the surface, where the pressures are lower. Rapid removal of heat from the body by near-surface solid state convection would lead to fairly rapid freezing of any large internal water layer (Reynolds and Cassen 1979). However, whether or not any large amount of liquid was generated, the net result would be that ice that was originally in the deep interior and at a high density would be transported to shallower levels and transformed to considerably less dense phases. This decrease in the mean density of the satellite's ice would require an increase in volume.

Of the satellites for which we have geologic data, only Ganymede and Callisto could contain significant quantities of the high-density ice phases necessary for expansion by differentiation (see Table I of Schubert et al. chapter). The maximum expansion possible is large. For complete differentiation, beginning with a homogeneous distribution of silicates and a temperature of 100 K, and ending with all the silicates in a core, the total increase in surface area could be as much as 6.5% for Ganymede or 5.5% for Callisto (Squyres 1980a).

A number of factors would tend to make the total expansion actually expressed by tectonic features less than the maximum possible. First, much of

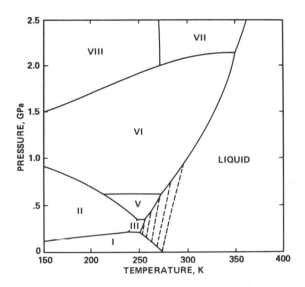

Fig. 8. The pressure-temperature phase diagram of ice. Roman numerals denote the ice polymorphs stable in each stability field. The dashed lines in the liquid field are adiabats. They give
the approximate temperature gradient that would exist in a convecting liquid layer

the differentiation may have taken place before solidification of the oldest observed crust. It is unreasonable to expect that the satellites would accrete cold.
Accretional heating would probably melt at least the outer portion of the body,
so that some of the differentiation would have taken place before solidification
of the oldest observed crust. Calculations of gas-free accretional heating suggest that melting of the outer several hundred km could take place, but that the
deep interior, where the very high density ice phases would exist, would be
largely unaffected by accretional heat (Schubert et al. 1981). Formation in a
warm nebula, on the other hand (Lunine and Stevenson 1982a), might have
led to nearly complete differentiation before formation of the oldest observed
surfaces. Second, differentiation might not have proceeded to completion. If
the internal heating was not sufficient to drive all the ice from the deep interior, the total expansion would be less.

Finally, if differentiation and expansion took place slowly enough, some
fraction of the expansion could be accommodated by ductile extension of the
near-surface ice rather than fracturing. As we shall discuss in Sec. III on fracturing, the stresses attained when an icy crust is subjected to strain are very
sensitive to the strain rate. The characteristic time for stresses to relax in a
viscoelastic material (the Maxwell time) is given by the viscosity divided by
the shear modulus. For reasonable estimates of the viscosity near Ganymede's
surface (10^{25} to 10^{27} poise; see Sec. III on viscous relaxation), the Maxwell
time is roughly 10^7 to 10^9 yr. If the time for the stresses to build up significantly exceeds the Maxwell time, then ductile deformation will take place.

We do not know the rate at which expansion would have taken place, and it seems improbable that the rate would be constant during the complete history of the expansion. The range of likely time scales for expansion is similar to the range of likely Maxwell times; surface deformation may then have been brittle only when the expansion was most rapid and the stresses highest, and ductile when strain rates were lower.

Two attempts have been made to place observational constraints on the amount of expansion Ganymede has undergone. McKinnon (1981b) notes that the large expanse of unfractured dark cratered terrain in Galileo Regio puts a limit on the amount by which its curvature could have changed due to expansion. He calculated a maximum surface area increase of ~2%, using an elastic expression for the response of the lithosphere to strain. The total expansion could, of course, have been greater if some of it were at a slow enough rate for the deformation to be ductile. Golombek (1982) also calculated a maximum expansion of 2% based on a measurement of the fractional surface area covered by grooves and the assumption that the grooves are grabens. This calculation considers deformation that has occurred since the most recent resurfacing of each grooved region, plus enough extension and downdropping beneath the resurfaced deposits to accommodate a 1 km thickness of bright material (see the discussion of deep crustal faulting below). Of course, if on the average more than 1 km of downdropping has occurred beneath the resurfaced deposits, this would underestimate the amount of brittle extensional deformation that has taken place. Nonetheless, the observable surface extension must surely be less than the amount of extension theoretically possible as a result of differentiation. If the tectonic activity on Ganymede was driven by differentiation, then the lack of tectonic features on Callisto may reflect less complete or slower differentiation of that body (cf. discussion in chapter by Schubert et al.).

Silicate Dehydration. Another process that could lead to expansion is dehydration of silicates, which has been suggested for Europa (Finnerty et al. 1981). If the satellites accreted at moderately low temperatures, they would initially contain large amounts of hydrated silicates. Dehydration reactions at low pressures typically involve significant overall volume increases in the system (i.e., the net volume of the dehydrated rock plus water exceeds the volume of the hydrated rock). One such reaction is the dehydration of serpentine to form forsterite, talc, and water:

$$5\,Mg_3Si_2O_5(OH)_4 \rightarrow 6\,Mg_2SiO_4 + Mg_3Si_4O_{10}(OH)_2 + 9\,H_2O. \quad (1)$$

The total volume change possible in such reactions is a strong function of pressure. At low pressures near the surface the volume change resulting from dehydration can be as much as several tens of percent. At pressures greater than about 1 GPa (found below depths of a few hundred km on the large satellites) it becomes small (Finnerty et al. 1981). This characteristic, plus the high temperatures required for dehydration (500 to 700° C), limit the possible bodies for which dehydration expansion could be important. The small Saturnian satellites, where internal pressures are very low, probably never became hot enough for silicate dehydration to take place (Ellsworth and Schubert 1983). For Ganymede and Callisto, dehydration expansion could not be large, because high temperatures near the surface would first melt all the ice in an ice-rock matrix, causing the rocks to fall to deeper levels where the pressures preclude substantial expansion. On Europa, however, dehydration expansion could have contributed significantly to the tectonic history.

Freezing of Ice I. Another obvious possibility for a process that could lead to expansion is freezing of liquid water (density $\rho = 1.00$ g cm^{-3}) to form ice I ($\rho = 0.92$ g cm^{-3}). While significant volume increases are possible in principle for some satellites, it is not clear that much freezing of pure liquid water has taken place in these satellites.

If complete melting and refreezing of the small Saturnian satellites could take place, surface area increases of several percent are possible on refreezing (Smith et al. 1981). Thermal models of these satellites (chapter by Schubert et al.) suggest, however, that radiogenic melting is very unlikely if they consist of only H_2O and silicates (Stevenson 1982c; Ellsworth and Schubert 1983; but see Friedson and Stevenson 1983). While radionuclides in the silicates may cause significant warming in at least Tethys, Dione, Rhea and Iapetus, solid state convection will begin at temperatures well below the melting temperature of pure H_2O, removing the heat rapidly and preventing further rise in temperature. The resurfacing observed on some of these satellites clearly shows that some internal melting has taken place, but a likely explanation is that the Saturnian satellites also contain NH_3, present with the H_2O as ammonia hydrate. The ammonia-water system has a peritectic melting point at 173 K, a temperature that could probably be reached within these satellites. While formation of a small amount of low-density ammonia-water peritectic melt is likely and could account for the observed resurfacing, the volume change involved is very small. Another possible source for heat to cause initial melting is accretion, although a model of accretional heating of the Saturnian satellites suggests that accretional heat may have been insufficient to produce significant melting (Ellsworth and Schubert 1983; chapters by Safronov et al. and Schubert et al.).

Freezing of liquid water to form ice I is not a viable means of causing significant expansion of Ganymede or Callisto, because of their large sizes and the influence of other ice phases (Squyres 1980a). Adiabats in the liquid field of the H_2O phase diagram give the temperature gradients that would be maintained by liquid convection in a liquid water layer in either satellite (Fig. 8). A liquid layer would be bounded by ice both above and below because a minimum temperature exists on the solidus curve at the ice I − ice III boundary. As cooling occurred, the adiabatic gradient would have shifted toward lower temperatures. Because both ends of the adiabats intersect the solidus, cooling would have resulted in simultaneous freezing from top downward (forming ice I) and from the base upward (forming ices VI, V, and finally III). The expansion from freezing of low-density ice I would in general have been more than offset by the contraction from freezing of the high-density ices. Only during the simultaneous freezing of ice III and ice I might a small (~0.1%) increase in surface area be possible. The net result of freezing of a thick water layer on Ganymede or Callisto would in fact be a contraction of 0.1 to 1.0%, depending on the overall H_2O mass fraction.

Of the Jovian satellites, only Europa is capable of undergoing a significant expansion as a result of freezing, because a liquid layer there would be sufficiently thin that the silicate interior, rather than dense ice phases, would lie at the base. If Europa had once had a liquid layer that has since frozen, a surface area increase of about 1% is possible (Cassen et al. 1979*a*). A liquid layer on Europa formed by accretional, radiogenic, and/or tidal heating would be expected to freeze due to cooling by solid-state convection in the overlying ice crust if the crustal thickness ever exceeded ~30 km and convective instability set in (Squyres et al. 1983*a*). Calculated tidal heating rates and observations of Europa suggest, however, that a substantial liquid layer may still be present.

Thermal Evolution in the Solid State.   A final class of processes that can cause volume changes in icy satellites involves response to temperature changes entirely in the solid state, with no melting or freezing, and no large-scale transport of material. These processes fall into two general classes. One involves simple heating expansion or cooling contraction in a single material. The other involves transition of ice phases. In a body of sufficient size for high-pressure polymorphs to occur, of course, both processes can operate simultaneously.

The small satellites of Saturn (with the probable exception of Enceladus, which may be tidally heated) have been cooling for most of their thermal histories (see, e.g., Ellsworth and Schubert 1983; see Figs. 11–15 in chapter by Schubert et al.). A small amount of global contraction can take place as a result of cooling contraction of ice I. In addition, a much greater contraction can occur if cooling leads to formation of a "core" of ice II. For all the Saturnian satellites, formation of ice II can only occur if the satellites accreted homogeneously, creating an intermixture of ice and silicates throughout the body. If substantial internal melting and differentiation took place, silicates would occupy the center and only ice I could presently exist. Assuming a homogeneous structure, Ellsworth and Schubert (1983) calculated the maximum possible contractions due to cooling of the Saturnian satellites. Their results imply maximum surface area reductions of ~1% for Mimas, Tethys and Dione, 1.5% for Iapetus, and 4% for Rhea (which can contain a substantial amount of ice II). Only Dione appears to show any possible evidence for compressional deformation (although the photographic coverage of the satellites is quite incomplete). One possible explanation is that these bodies are in fact not homogeneous, but either accreted inhomogeneously or differentiated very early in their histories due to accretional heating. Such a structure would prevent formation of ice II and would limit cooling contraction to <1%. Another possibility is that the bodies are in fact much warmer than calculated, perhaps due to near-surface thermal conductivity lowered by regolith formation or presence of NH_3 (Stevenson 1982c; Squyres et al. 1983b). In this case, temperatures low enough to permit the transition to ice II would not have been reached. For either the undifferentiated or differentiated case, the maximum extension possible through warming entirely in the solid state is limited to about 1%. This is several times less than the expansion possible by refreezing if the satellites were somehow melted initially.

For Ganymede and Callisto, transitions among the ice phases and thermal expansion entirely within the solid state can lead to volume changes. The phase transition most likely to have been significant is transition of ice V to ice II at depth as the satellites cooled (Shoemaker et al. 1982). The maximum increase in surface area from this transition is about 1%. Zuber and Parmentier (1984b) have considered the post-formation thermal evolution of both differentiated and undifferentiated models of Ganymede and Callisto. They considered heat produced by radionuclides and phase transitions as well as heat transport by conduction and solid state convection, and calculated the volume changes resulting from the thermal expansion. Over the history of the body, they found that the total surface area increase due to thermal expansion entirely within the solid state could be roughly 0.3 to 0.5%. For an undifferentiated body most of the expansion occurs in the first 500 Myr after formation, while in the differentiated case the concentration of the heat-producing silicates in the core causes the expansion to be more gradual, occurring over 2 to 3 Gyr. Their results are calculated for an initial internal temperature of 120 K throughout the body. For higher initial temperatures, which are possible due to accretional heating, the potential thermal expansion is less.

Local Thermal Stresses. These stresses could perhaps be generated on some satellites by volume changes in individual geologic units. In particular, relatively large thermal stresses could be generated by surface emplacement of thick, warm ice deposits. The plane thermal stress is given by $E\alpha_l\Delta T/(1 - \nu)$, where E is Young's modulus, α_l is the linear coefficient of thermal expansion, ΔT is the change in temperature as the body cools, and ν is Poisson's ratio. The value of α_l is a decreasing function of T, but is ~3×10^{-5} at $T =$ 200 K. Nominally, ΔT for a fresh, warm ice deposit will be on the order of 100°, and the resultant tensile thermal stresses will be several tens of MPa— sufficient to cause extensional fracturing. Total linear contraction would be on the order of 10^{-3}, so that a fresh, warm surface deposit 100 km in lateral extent could form thermal fractures with a cumulative width of ~100 m. The geometry and spacing of the fracturing are difficult to predict, but might resemble those of cooling joints common in terrestrial lavas.

Solid State Convection

On the basis of theoretical calculations, thermal convection is predicted to have occurred or be occurring in most of the icy satellites (see the chapter by Schubert et al.). The stability of an ice crust of thickness d to solid state convection depends on the Rayleigh number Ra, given by

$$Ra = \rho g \alpha_v d^3 (\Delta T / \kappa \eta) \tag{2}$$

where α_v is the volume coefficient of thermal expansion, g is gravity, ΔT here is the temperature drop across the convecting layer, κ is the thermal diffusivity, and η is the effective viscosity. Convection will occur if Ra exceeds about 10^3. Reynolds and Cassen (1979) showed that for surface gravities and heat flows characteristic of the large icy satellites, ice slabs thicker than about 30 km are unstable to solid state convection. Convection is therefore expected to be the primary mode of heat transport in the interiors of large satellites like Ganymede, Callisto and Titan. Ellsworth and Schubert (1983) investigated the extent of convection in the smaller Saturnian satellites and found that convection occurred only in the deep interior and early history of these small objects. An exception is Enceladus, which may still be undergoing convection due to ongoing tidal heating (Yoder 1979b; Squyres et al. 1983b).

The magnitude of any surface tectonic manifestations of convection will be directly related to the vigor of the flow. The vigor of convective flow may be characterized by the Rayleigh number, the characteristic flow velocity u, and the maximum stress σ in the convecting ice. The Rayleigh number is a nondimensional constant describing the relative strengths of the thermal forces driving convection and the viscous forces resisting it. The characteristic convective velocity (and the resultant stresses) may be estimated from simple parameterized convection models such as that of Sharpe and Peltier (1978). Typical values for these parameters are $Ra = 10^5$, $u = 10$ cm yr^{-1}, and $\sigma = 0.1$ to 1 kPa for the small Saturnian satellites (when convection was taking place), and $Ra = 10^6$, $u = 10$ to 100 cm yr^{-1}, and $\sigma = 100$ kPa for the large Jovian satellites. For comparison, typical values for the Earth are $Ra = 10^6$, $u = 1$ to 10 cm yr^{-1}, and $\sigma = 10$ to 50 MPa. The stresses generated in the Earth's convective flow are similar to typical tensile strengths of crustal rocks, hence widespread tensile fracturing associated with convection are both expected and observed. In contrast, the maximum stresses in the convective flows of the major icy satellites are roughly an order of magnitude smaller than the minimum tensile strength expected for the icy lithospheres of these bodies. Tectonic features due solely to convective stresses are therefore expected to be rare or nonexistent. Certainly on the Saturnian satellites, where convective stresses are three or four orders of magnitude smaller than the expected tensile strength of the lithosphere, no convection-related tectonic features are expected at all.

The apparent lack of large-scale compressive tectonic features on Ganymede is consistent with the idea that convective flow did not dominate surface tectonism there and that nothing analogous to terrestrial plate tectonics has taken place. However, even if the primary source of extensional stress was global expansion, faulting would tend to occur above convective upwellings for two reasons: (1) tensile stresses are greatest over the rising convective plumes, contributing locally to the net tensile stress; and (2) the net heat flow is higher above the rising plume, raising the local heat flow and reducing the thickness of the lithosphere being fractured. In addition, resurfacing material, whether in the liquid or solid state, would be most likely to reach the surface above the rising convective plumes. The local intensification of the extensional stress field favors formation of cracks and conduits for fluid flow, and the increased temperatures and substantially lowered viscosities would reduce the rise time for ice diapirs of a given size. Finally, the nature of the tensile stress field set up by an elongated convective upwelling (Hafner 1951) favors formation of faults parallel to the axis of the upwelling, producing (presumably) long, quasi-linear faults outlining the rising convective plumes at the surface. While the extension responsible for the tectonic features on Ganymede was probably the result of global expansion, it may have been localized by convective flow. The complex tectonic patterns on Ganymede may therefore provide information about the structure of internal convective cells at the time of tectonic activity.

Shell Deformation

A variety of processes can induce stresses in the lithosphere of a satellite by changing the satellite's shape. These processes include despinning, tidal deformation and orbital recession. The first involves relaxation of an equatorial bulge due to a reduction in rotation rate; the second involves variations in the amplitude of the tidal bulge due to orbital eccentricity; and the third involves gradual relaxation of the tidal bulge.

Despinning.   All of the satellites discussed here are in synchronous rotation; that is, their rotation period is equal to their orbital period. At the time of their formation, however, it is unlikely that the satellites rotated synchronously. The synchronous state was reached as a result of tidal interaction. Each satellite has a tidal bulge raised on it as a result of gravitational attraction of the planet about which it orbits. For an ideal material that could respond instantly, the bulge would point directly toward the planet. In reality, however, frictional dissipation of the tidal energy within the satellite results in a time lag for raising and lowering the bulge (see chapter by Burns). At any moment, then, the tidal bulge of a nonsynchronously rotating satellite is not pointed directly toward the planet. This nonalignment results in a torque upon the satellite that causes it to evolve toward synchronous rotation. Because most satellites were probably initially spinning more rapidly than they are at the present, tidal interaction would have caused a decrease in rotation rate.

Despinning of a satellite causes a reduction in its oblateness. The stresses that can be induced in a satellite's lithosphere by changes in oblateness have been studied by Melosh (1977a). The flattening f of a satellite's shape is defined as $(r_{equatorial} - r_{polar})/r$, where r is the mean radius. For a satellite of mass M rotating at angular velocity ω, the equilibrium hydrostatic flattening is given by

$$f = \frac{\text{floated}}{1 + \left\{ \frac{5}{2} \left[1 - \frac{3}{2} \, (C/M \, r^2) \right] \right\}^2} \tag{3}$$

where G is the gravitational constant, and C is the polar moment of inertia (Jeffreys 1952). Making the simplifying assumption of uniform density, Eq. (3) reduces to

$$f = \frac{5}{4} \frac{\omega^2 r^3}{G M}. \tag{4}$$

If a satellite has only a thin lithosphere that is underlain by warmer, more mobile material, it will maintain a nearly hydrostatic shape. If, on the other hand, the satellite has strength through much of its interior, the strength of the material will cause the satellite to resist shape changes. If a homogeneous, incompressible, elastic satellite initially has a hydrostatic shape, and undergoes a spin change from ω to ω', its flattening will change by

$$f - f' = \frac{\frac{5}{4} (\omega^2 - \omega'^2) r^3 / G M}{1 + (19/2)(r \, \mu / \rho G \, M)} \tag{5}$$

where μ is the satellite's shear modulus (Melosh 1977a). Note that for large icy satellites, in which $r \, \mu / \rho G \, M \ll 1$, this expression reduces to the hydrostatic change in shape. For most Saturnian satellites, however, a somewhat smaller shape change can be expected.

The magnitude of the stresses induced by these shape changes depends of course on the rheologic properties of the lithosphere. If the stresses are sufficiently large, fracturing of the crust will result. The dominant fracture pattern expected to be produced by despinning is northeast and northwest trending strike-slip faults in the mid-latitudes (Melosh 1977a). Also expected are east-west trending normal faults in the polar regions, and, in some cases, north-south trending thrust faults or folds near the equator.

For many plausible rheologies and spin changes, despinning would be expected to cause faulting on icy satellites. It is not clear, however, that faults providing geologic evidence for despinning would be preserved to the present. For most satellites, despinning would take place very early in their history. The rate of change of spin for a circularly orbiting satellite with zero obliquity is

$$\frac{d\omega}{dt} = - \frac{3k_2 G M_p^2 r^5}{C a^6 Q} \tag{6}$$

where M_p is the mass of the planet, a is the satellite's orbital radius, Q is the satellite's dissipation function, and k_2 is the Love number of the second degree for the satellite, defined

$$k_2 = \frac{3/2}{1 + (19\mu/2\rho gr)} . \tag{7}$$

Peale (1977) has tabulated maximum despinning times for all the major satellites in the solar system. Of the satellites considered here, the longest despinning time is that for Iapetus, $8.7 \times 10^8 \, Q$ yr. The Galilean satellites range from 50 Q yr for Europa to $2.2 \times 10^5 \, Q$ yr for Callisto, and the other Saturnian satellites from 140 Q yr for Mimas to $1.5 \times 10^4 \, Q$ yr for Rhea. The appropriate values to use for Q are uncertain, but are probably on the order of 100. For all the major icy satellites except Iapetus, then, despinning is geologically rapid. The Galilean satellites were probably already locked in synchronous rotation before formation of their present surfaces, so no tectonic features attributable to despinning are to be expected. The smaller Saturnian satellites would also despin rapidly, but may be marginally small enough to be respun by large impacts. It is conceivable, then, that some evidence of spin-related fracturing, if it ever took place, could be preserved. It would be extremely difficult, however, for impact respinning to generate the high spin rates necessary to produce tectonically significant stresses (Lissauer 1985).

It has been reported that linear features on Dione, such as the subdued ridges, show preferred northeast and northwest trending orientations (Moore 1984). It was suggested that this pattern may be related to spinning and despinning stresses, although the interpretation of the ridges as compressional features would require that they be formed by compressional reactivation of zones of weakness originally created by strike-slip faulting. The stresses that could be generated by despinning Dione may be roughly demonstrated by using Eq. (5), taking $\mu = 4 \times 10^{10}$ Pa (Cassen et al. 1979a) and assuming, e.g., an initial rotation period of 6 hr. The maximum stress at the surface is of the order of $\mu(f' - f)$, and would be ~5000 kPa in this case. For comparison, the shear strength of ice near the melting point is about 10^4 kPa, and it increases with decreasing temperature (Gold 1977; Haynes 1978; Hobbs 1974). Although despinning may have caused some limited deformation during the very early history of the icy satellites, its influence on the present geology has been minimal.

   Tidal Deformation.   The tidal bulge raised by a planet on a satellite causes a distortion of the satellite's shape. The lengths of the three principal semiaxes of a tidally distorted satellite (neglecting the small oblateness caused by slow rotation) are

$$r_1 = (1 + 7h\gamma/6)r \tag{8a}$$

$$r_2 = (1 - 2h\gamma/6)r \tag{8b}$$

$$r_3 = (1 - 5h\gamma/6)r \tag{8c}$$

where $h = 5k_2/3$, and $\gamma = M_p r^3/Ma^3$ (Jeffreys 1952). The term a in these equations is the instantaneous distance of the satellite from the planet. If the satellite is in an eccentric orbit, this distance changes continuously, producing a corresponding continuous variation in the size of the tidal bulge. The energy dissipated by this flexing of the satellite is the source of tidal heating. (These ideas are addressed in the chapter by Schubert et al.) Tidal flexure also of course introduces stresses in the satellite's lithosphere. Of the satellites considered here, the only one for which potentially significant tidal stresses are developed is Europa. A reasonable approximation for the behavior of Europa's lithosphere over tidal periods is that it behaves like a thin elastic shell. In this case, the maximum stress introduced by tidal flexure is given by

$$(\sigma_n - \sigma_t)_{max} = \frac{24E(\Delta r_1/r)}{7(5 + \nu)} \tag{9}$$

where σ_n and σ_t are the principal stresses normal and tangent to stress trajectories, respectively (Helfenstein and Parmentier 1983); Δr_1 is defined below.

   For the case where Europa has a thin outer shell with no appreciable strength, the shape it will assume due to tidal forces is simply the hydrostatic shape. Assuming zero strength (that is, $\mu = 0$, so that $h = 5/2$), the mean height of the tidal bulge on Europa given by $r_1 - r$ is roughly 2.3 km. The height of the bulge varies, however, due to Europa's orbital eccentricity of 0.0094. Again assuming zero strength, the variation Δr_1 of the height of the tidal bulge over an orbital period would be $\sim$130 m. This value is an upper limit, however, because over the time scale of an orbit the icy shell will have some flexural strength and resist the tidal forces. Using this limiting value for the tidal flexure, $E = 5$ GPa (Haynes 1978), and $\nu = 0.35$ (Gold 1958), Eq. (9) gives a maximum tidal stress for Europa of 2700 kPa. For comparison, a uniaxial tensile strength of 2600 kPa has been measured for polycrystalline ice at 250 K by Hawkes and Mellor (1972). This comparison is of course crude, because of the assumptions involved in calculating the stress levels and because the strength of Europa's crust may be greatly different from that of homogeneous laboratory samples at warm temperatures. It does suggest, however, that tidal stresses on Europa may be marginally capable of fracturing its ice shell.

The fracture pattern expected due to tidal flexure is complex. Helfenstein and Parmentier (1983) studied the correlation between predicted tidal fracture patterns and the orientation of actual linear features on Europa. Near the anti-Jovian point on Europa is a pattern of short reticulate dark bands with no preferred orientation. Based on their geometry, these have been interpreted as extensional fractures formed in a horizontally isotropic stress field (Pieri 1981). Tidal flexure should cause maximum extension near the sub-Jovian and anti-Jovian points; Helfenstein and Parmentier concluded that tidal stresses were a likely cause for these fractures (cf. chapter by Malin and Pieri). However, they also concluded that many of the other dark bands on Europa have orientations more consistent with stresses generated by orbital recession.

Orbital Recession. Any satellite in a prograde orbit with an orbital angular velocity at periapsis smaller than the planet's rotation frequency will gain angular momentum through tidal interaction with the planet. As described in Burns' chapter, this causes the satellite's orbit to expand outward, with the rate of change of its semimajor axis given by

$$\frac{da}{dt} = 3k_{2p}\left(\frac{G}{M_p}\right)^{1/2} \frac{r_p^5}{Q_p} \frac{M}{a^{11/2}} \tag{10}$$

where k_{2p}, r_p, and Q_p are the Love number, radius, and dissipation function of the planet, respectively. The change in the semimajor axis of the orbit of course causes a relaxation of the tidal bulge, which leads to stresses in the crust.

Helfenstein and Parmentier (1983) noted that a number of long dark bands on Europa have orientations consistent with an origin as strike-slip faults due to orbital recession stresses. Assuming elastic properties for Europa's outer shell, they calculated that about 2×10^5 km of recession would be required to build up shear stresses on the order of a few thousand kPa. In this case, however, the strain rate is so low that an elastic approximation is probably not valid. Gavrilov and Zharkov (1977) have calculated a value of k_{2p} for Jupiter of 0.379. The minimum possible value of Q_p is set by assuming Io began its history at the surface of Jupiter and has receded to its present position over the age of the solar system, and is 2.5×10^4. The present rate of recession of Europa, from Eq. (10), is therefore no more than 6.79×10^{-6} km yr^{-1}, requiring a time substantially in excess of the age of the solar system to recede 2×10^5 km. Even starting Europa at the surface of Jupiter and receding it to its present position over the age of the solar system requires a mean recession rate of only 1.3×10^{-4} km yr^{-1}, or 1.5×10^9 yr to move 2×10^5

km. At such low strain rates an elastic approximation cannot be applicable, and most of the very small amount of deformation that took place would be taken up by creep of the ice.

It seems very unlikely, then, that tidal recession has played any role in causing fracturing on Europa or any other icy satellite. The preferred orientation of dark bands reported by Helfenstein and Parmentier (1983) is of course consistent with any process that reduces the tidal bulge of the satellite. A small reduction (<130 m) occurs once every tidal period, but it is not clear why a corresponding pattern that should be produced by the increase in the bulge that also takes place each orbit is not evident.

Impact

A final source of local stress not of internal origin but which should be included to appreciate the total tectonic state of an icy satellite is impact stress and fracture (discussed in detail in the chapter by Chapman and McKinnon). Both scaling theory (Croft 1981b) and observation of fracture systems around large lunar basins (Mason et al. 1976) indicate that dense lithospheric fracturing extends outward from an impact point to about twice the rim diameter of the observed crater. In addition, laboratory impacts in brittle rocks (see, e.g., Curran et al. 1977) and ice (Croft 1981c) show less dense concentric and radial fracturing to extend outward as much as 10 crater radii from the impact point. Since impact craters are found on all the icy satellites, impact fractures should permeate the lithospheres of all these bodies to varying degrees. The icy composition of the satellites should make them more susceptible to thermal annealing of fractures at depth, but near the surface the fractures should persist for long periods. Just as volcanic activity on the Moon is largely associated with the fracture systems of large basins, conduits along which extrusion has taken place on the icy satellites may also be associated with impact fracture systems.

An even more important result of impact may be the stresses that result from gravitational modification of a large impact basin following its formation. McKinnon and Melosh (1980) have shown that a large icy satellite would not be able to support the stresses generated by the topography of the original cavity, and that prompt inward flow of material would take place. This inward flow, concentrated in the deeper, warmer, more mobile part of the crust, would result in large radial extensional stresses near the surface. Such stresses were probably responsible for formation of the ring systems observed on Callisto. Further, penetrative deformation in response to these stresses, like the stresses generated during the impacts themselves, may imprint large areas of a satellite with radially and concentrically arranged zones of weakness.

III. STYLES OF TECTONIC ACTIVITY

Faulting and Fracturing

Deep Crustal Faulting.   Based on geologic evidence, a reasonable hypothesis for the formation of the bands of resurfaced terrain on Ganymede and Enceladus is that it took place by extrusion of relatively silicate-free material to the surface in regions of lithospheric extension. There are at least three possible styles of lithospheric extension that could lead to formation of resurfaced bands (Parmentier et al. 1982) (Fig. 9). These are:

1.   Spreading analogous to that which takes place at mid-ocean ridges, with separation of lithospheric blocks and creation of large amounts of new lithosphere between them;
2.   Formation of downdropped, normal fault-bounded rift zones, similar to the East African Rift on Earth, which are later filled in by extrusion;
3.   Extrusion of silicate-poor material to the surface through narrow fissures or extension fractures, with no associated normal faulting.

The amount of extension implied by these three models varies considerably, from 100% of the band width for case 1 to only a very small fraction for case 3.

The nature of boundaries between older cratered and younger resurfaced terrain enables a selection among these possible styles. On Ganymede, contacts between bright and dark terrain are usually gently curving and very sharply defined. Craters and furrows in the dark terrain are abruptly truncated at the contacts. A similar situation exists in some areas on Enceladus. These characteristics suggest that the material making up the resurfaced bands is structurally confined. In some instances the boundary between the terrain types is diffuse, consistent with surface flooding through extension fractures or fissures (case 3 above). In general, however, the sharp boundaries of the resurfaced regions argue strongly for structural confinement of the deposits in nearly all instances on Ganymede and in some instances on Enceladus.

Both cases 1 and 2 would involve structural confinement of the resurfaced deposits. They differ, however, in several other respects. First, if lithospheric spreading had taken place, two parts of a feature such as a crater cut apart by formation of a band should appear on opposite sides of the band. This is not observed on either Ganymede or Enceladus. Second, the opposite sides of a band should "fit" back together as, for example, the continental margins of South America and Africa do. While this appears at first observation to be true in some cases, careful examination shows that the fit is generally quite poor (Allison et al. 1980). Finally, lithospheric spreading involves substantially larger amounts of extension than does deep normal faulting and rift formation. The surface of Ganymede is roughly 50% bright terrain, so its formation through lithospheric spreading cannot be explained by global expansion alone. Lithospheric spreading would require large-scale destruction of surface material and associated compressional deformation. It is difficult to

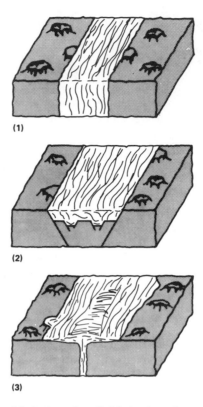

Fig. 9. Three possible models for formation of bright bands on Ganymede: (1) spreading apart of crustal blocks, with intrusion of fresh material in between; (2) formation of downdropped, normal fault-bounded rifts that are filled in with fresh material; and (3) extrusion through narrow fractures and surface flooding. The geologic evidence favors model (2). (Figure after Parmentier et al. 1982.)

reconcile this requirement with the apparent lack of compressional features. For these reasons, we reject lithospheric spreading as an important tectonic process on Ganymede or Enceladus. Formation and subsequent flooding of broad downdropped rift zones appears to be the most reasonable mechanism for formation of bright bands on Ganymede and for the sharply bounded resurfaced bands on Enceladus. It is possible that extrusion through extension fractures has also been important on Enceladus. It also is possible, however, that all the major resurfacing there involved deep normal faulting, and that in some instances the extruded deposits overflowed the rift zone margins.

Work by Schenk and McKinnon (1985) supports the hypothesis that the bright bands are rifts filled with relatively clean ice. They examined the distribution of dark halo craters in the bright terrain on Ganymede, which they interpret as craters that have excavated through the bright material into underlying darker material, producing dark ejecta deposits. By determining the minimum size of dark halo craters and calculating their depth of excavation, they concluded that the thickness of the bright material in Uruk Sulcus is roughly 1 to 2 km. The darker material into which the craters have excavated is interpreted as the downdropped old cratered terrain material, now forming the floor of the rift.

Deep normal faulting appears, then, to have been an important process on at least Ganymede, Enceladus and Tethys (where Ithaca Chasma formed but was not filled in). Additional understanding of the process may be gained by simple quantitative consideration of extension of an icy lithosphere. The major problem involved is that the lithospheres of these bodies have rheologies and strengths that are dominated by the behavior of ice. Ice as a geologic material has been extensively studied at temperatures near its melting point, but further experimental work at temperatures appropriate for the outer solar system, such as that reported by Durham et al. (1983), is sorely needed. We discuss here a very simple model of extension and deep faulting of an icy crust that is based on some measured and assumed properties of ice, and consider its qualitative and quantitative implications.

A detailed description of the rheological properties of ice is beyond the scope of this chapter (for a review see Weertman [1983]). For strain applied over very short time scales, the mechanical behavior of ice is approximately elastic, while for very long time scales the behavior is nearly viscous. We therefore consider as a reasonable first-order approximation of ice subjected to strain, a simple Maxwell viscoelastic behavior. In such a material, the stress σ is related to the applied strain ε by

$$\dot{\varepsilon} = \frac{\sigma}{\eta} + \frac{\dot{\sigma}}{E} \tag{11}$$

where η is viscosity. Taking $\sigma_{t=0} = 0$, the solution of this equation is

$$\sigma(t) = e^{-tE/\eta} E \int_0^t \varepsilon(t') e^{t'E/\eta} \, dt' \tag{12}$$

where t' is an integration variable. Consider the simplest possible case, where a horizontal slab of this material is subjected to uniaxial extensional strain. Because the viscosity of ice is strongly temperature dependent, the thermal gradient in the crust will result in a decrease in viscosity with depth. The dependence of effective viscosity on temperature is given approximately by

$$\eta = \eta_0 \exp[\beta(T_m/T - 1)] \tag{13}$$

where T is temperature, T_m is melting temperature, η_0 is the viscosity at T_m (10^{14} to 10^{15} poise), and β is a constant with a value of about 26 (Weertman 1973). Using Eq. (12), which gives the nonlithostatic stress as a function of viscosity, strain history and Young's modulus, the stresses at any depth resulting from a given uniaxial strain history may be estimated. Using failure criteria based on expected uniaxial tensile strengths for ice at low temperature and Griffith failure theory, one may calculate the depth and style of faulting resulting from these stresses. A simple model of this type has been described by Squyres (1982).

The most important result of such a model is that failure is limited to an upper cold region of the crust. This brittle zone, or lithosphere, is underlain by a ductile region in which all deformation is taken up by creep of the ice. The local thickness of the lithosphere depends on at least three factors. One is the surface temperature. All other conditions being equal, the lithosphere will be thickest where the surface is coldest. The term "surface temperature" in this context of course does not refer to the temperature given by, for example, the thermal infrared brightness of the satellite's surface. It refers instead to the temperature of material beneath the diurnal and seasonal thermal skin depth and immediately beneath any superficial frost layer or regolith with a thermal conductivity substantially lower than that of the lithosphere as a whole. The second important factor is the local geothermal gradient. For steep gradients, warm temperatures and low viscosities lie closely beneath the surface, so that the lithosphere is thin. Note, then, that as the geothermal gradient changes with time, the thickness of the lithosphere will change as well. Finally, lithospheric thickness depends on the rate at which strain takes place. For high strain rates, the elastic term of the viscoelastic equation dominates, so that stresses will be high and failure by fracturing can occur to a substantial depth. For low strain rates, however, the behavior is more nearly viscous and the lithosphere thinner. The thickness of the lithosphere of icy satellites may therefore be expected to vary with position, with time, and with the nature of the deformation taking place.

Such a model is useful for understanding the qualitative behavior of an icy crust subjected to extension. However, the quantitative results must be interpreted with a great deal of caution due to a number of large uncertainties in material properties. In performing model calculations, tensile strength was assumed to be 2600 kPa at 250 K, and to increase with decreasing temperature in a fashion similar to the behavior shown by the uniaxial compressive strength (i.e., proportional to $\exp(-0.0133\ T)$; Parameswaran and Jones 1975). A range of surface temperatures, geothermal gradients, and strain histories was considered. In order for deep normal faulting to occur, the confining stress at the base of the lithosphere must be sufficiently large to prevent formation of open extension fractures. For most geologic materials, this means that the confining stress must be greater than roughly three times the tensile strength. Calculations made using the above assumptions imply that this condition would be met for lithospheres thicker than ~25 km. A lithospheric thickness of >25 km was found to require a geothermal gradient of no more than several K km^{-1} for a wide range of surface temperatures and strain histories. If the assumptions used in the calculations are reasonable, then the implication is that the deep normal faulting involved in creation of resurfaced bands initiated when the geothermal gradient was relatively low. A depth of faulting of ≥ 25 km is certainly consistent with the large widths of the resurfaced bands observed.

A critical assumption of this calculation is that the material at depth will behave as a cohesive slab, and that its strength is not diminished by pervasive joints or microfractures. This assumption is probably valid for the depths under consideration here, due to the high temperatures of the materials involved relative to their melting point. Assuming a temperature at the base of the regolith of 140 K and a fairly modest geothermal gradient of $2°$ km^{-1} (Passey and Shoemaker 1982), the homologous temperature (T/T_m) at a depth of 25 km would be ~ 0.7. At homologous temperatures this high, thermal annealing will be fairly rapid (for comparison, rocks at a homologous temperature of 0.7 are at actual temperatures of $\sim 750°$ C). As an example of the mobility of H$_2$O molecules at such temperatures, the sublimation rate from a free surface of ice at 190 K is roughly 2 m yr^{-1} (Lebofsky 1975). The assumption of cohesion is probably not valid, however, at the colder temperatures near the surface (see below).

There are a number of serious uncertainties in this calculation. The most important are the uncertainties in the tensile strength and viscosity of the lithosphere of Ganymede. It is not clear that the tensile strength of ice near the melting point can be extrapolated to lower temperatures in the same fashion as the compressive strength. If the actual tensile strength of the deep lithosphere is less than that estimated above, then normal faulting would occur at shallower depths, and higher geothermal gradients at the time of deformation would be possible. A real need exists for extensional failure tests of ice at very low temperatures. It is also not clear what effects would result from inclusion of silicates that may be present in the ice matrix of the lithosphere. In general, inclusion of particulate material in an ice matrix tends to increase significantly both its tensile strength (see, e.g., Shvaishtein 1973) and its viscosity (Friedson and Stevenson 1983). Without knowledge of the silicate fraction of the lithosphere, the importance of these potentially significant effects is impossible to evaluate.

Because the large grabens that may have been formed by deep normal faulting on Ganymede and Enceladus have apparently been later filled in, we know little about their structure. In this regard, higher resolution imaging of Ithaca Chasma on Tethys, the one large unfilled graben on the icy satellites, would be particularly useful. Considering the great width of some bright bands, and even of some groove sets, it seems unlikely that master faults at their margins intersect at the base of the lithosphere, as has been suggested for some terrestrial rifts (see, e.g., Illies 1970). Taking reasonable dips for the master faults and observed band widths of up to several hundred km, an unreasonably low geothermal gradient ($\ll 1°$ km^{-1}) would be required. It seems likely, instead, that the rifts are formed by many roughly parallel normal faults, perhaps resulting in a complex graben and horst structure for the floor of the rift.

Groove Formation. The dominant structural features on Ganymede and Enceladus are the grooves. Grooves are also observed with variations in morphology on Mimas, Dione and Rhea. As mentioned above, the gross topographic form of the grooves strongly suggests an extensional origin. On Ganymede and Enceladus, grooves are strongly concentrated in the bands of young resurfaced terrain. In these cases they are probably the result of much shallower deformation than produced the resurfaced bands themselves which, we have argued, may be broad fault-bounded rifts filled with extruded material. Groove formation on Ganymede and Enceladus, then, is primarily a process that has modified material near the surfaces of the resurfaced bands, although a few grooves are observed in the older cratered terrain on both bodies. The details of groove morphology beyond width, approximate depth, mean slope, and spacing, however, are unclear at Voyager resolution. The geologic structures underlying the observed landforms cannot be unambiguously determined from Voyager images. Based on geologic experience from silicate bodies, two possible explanations for the grooves are very long, narrow grabens, or open extension fractures analogous to glacial crevasses. In either case, the original landforms may have undergone significant modification since their formation.

If one naively were to adopt a model such as that described in the previous section for the formation of grooves, then one would be led to the conclusion that grooves originated solely as open extension fractures. Considering the small widths of grooves, groove formation as grabens would require normal faulting, and therefore confining stresses several times the material tensile strength, at depths of only a few km. Even on Ganymede, where the gravity is highest and the confining stresses at depth are largest, this would not be the case under the assumptions of the model.

The problem with use of this model for understanding shallow fracturing or faulting, however, is that experimentally determined strengths for ice are probably not a good approximation for the strength of the upper few km of an icy lithosphere. Several sources of penetrative fracturing would be expected to decrease the strength of the upper lithosphere to values well below that measured for homogeneous laboratory samples. One source of fracturing would of course be impacts. For example, the lunar surface material seems to be largely composed of unconsolidated rubble for the upper ~1 km (Kovach et al. 1973). Impact processes would only have been important in cratered terrain units, as groove formation in the resurfaced terrain on Ganymede and Enceladus took place too soon after resurfacing for significant cratering and brecciation to occur. Other processes could operate primarily in the resurfaced terrain, such as the thermal contraction upon cooling, discussed above.

Finally, if the resurfaced deposits were emplaced in a liquid state, then a significant volumetric expansion of the deposits would occur on initial freezing. A freshly erupted deposit of water would freeze from the top down and the bottom up simultaneously. As the inner part of the deposit froze and expanded, it would exert extensional stresses that might intensively fracture the upper surface.

At depths of tens of km or more, the use of experimentally determined ice strengths for the strength of an icy lithosphere is probably more appropriate, as relatively high pressures and temperatures would promote annealing of minor fractures. Near the surface, however, the lithospheric strength may drop abruptly due to small-scale fracturing and loss of cohesion. If this indeed is the case, many grooves may be grabens. Near-surface tensile strengths of 1000 to 2000 kPa or less are required for the grooves on Ganymede to be grabens, and strengths of <100 kPa are required for smaller Saturnian satellites such as Enceladus. Considering the variety of groove morphology observed, it seems possible that both grabens and open extension fractures are present. Until much higher resolution images of grooves are available, the question of their underlying structure must remain open.

An important but complicating factor in understanding the topography of the grooves is that deformation in many cases may not be confined to the resurfaced deposits. With continued extension after initial formation of resurfaced bands, there may be continued motion on deep normal faults in the underlying material. This motion would cause normal faults to propagate to the surface, and the displacement that took place would substantially disrupt any simple pattern of parallel grabens or fractures confined to the upper few km. Where single grooves or regular groove sets are observed they probably owe their origin to shallow deformation alone, but the irregular topography of complex grooved terrain may reflect disruption by continued motion on deep normal faults in the underlying material.

An important question relating to grooves on Ganymede and Enceladus concerns the regularity of groove spacing. One attractive explanation is that the regular spacing is a result of a necking instability in the crust (Fink and Fletcher 1981). A quantitative model of development of a necking instability has been presented by Fletcher and Hallet (1983) as an explanation of regularly spaced deformation in the Basin and Range Province of the western U.S. In their model, a plastic surface unit overlies a ductile unit with an effective viscosity that decreases exponentially with depth. This configuration is similar to that which would exist on an icy satellite with a geothermal gradient. Denoting the thickness of the upper layer as H and its shear strength as τ, they find that a necking instability will develop in the upper layer if

$$\frac{\rho g H}{2\tau} < \xi H. \tag{14}$$

Here ξ is the reciprocal of the viscosity decay length within the ductile substrate; taking the viscosity relationship of Eq. (13), this will be given by

$$\xi = \frac{\beta(T_m - T_T)}{T_T^2} \frac{dT}{dz} \qquad (15)$$

where T_T is the temperature at the brittle/ductile transition.

Appropriate values for τ and $\dot{\imath}_T$ are not known, but selection of reasonable limits suggests that an instability should be possible for both Ganymede and Enceladus. Taking $\beta = 26$, a minimum shear strength of 2000 kPa, and allowing for a value of T_T as high as 200 K, we find that an instability should be possible for a geothermal gradient larger than 7° km^{-1} on Ganymede and 0°.3 km^{-1} on Enceladus. Higher shear strengths would of course allow still lower geothermal gradients. Existence of a necking instability would cause any brittle deformation that took place in the lithosphere to be concentrated at regular intervals along the axis of extension. An important result of the model is that the spacing of deformation is roughly 3 to 4 times the thickness of the brittle layer. If the regular groove spacings observed on Ganymede and Enceladus are a result of a necking instability, they imply lithospheric thicknesses in the material of the resurfaced bands at the time of groove formation on the order of only 2 km. This lithospheric thickness is less than that estimated by Golombek (1982) (4 to 6 km) on the basis of the assumption that grooves are grabens whose bounding faults intersect at the base of the lithosphere. It is also significantly less than that inferred from earlier deep faulting that may have initially produced the resurfaced bands themselves.

Variations in groove spacing that exist from place to place may reflect spatial and temporal variations in lithospheric thickness at the time of deformation. In a statistical study of groove spacing on Ganymede, Grimm and Squyres (1985) found that, while groove spacing tends to remain fairly constant within a given groove set, it can vary substantially from one set to another within a given geographic region. The implied local variations in lithospheric thickness at the time of deformation may reflect either variations in geothermal gradient or strain rate. One possible explanation, then, is that the variations reflect variations in the time of groove formation over a period when the global geothermal gradient was changing, although this would require unexpectedly rapid changes in global heat flow. A more likely possibility is that the global heat flow and geothermal gradient remained roughly constant, but the strain rate varied significantly from one groove set to the next. Finally, if the extension is concentrated above convective upwellings, the variations may reflect the different heat fluxes produced by convective upwellings of differing strength.

Resurfacing

A number of the icy satellites show evidence of resurfacing. In the case of some of the smaller Saturnian satellites, the resurfaced areas may be largely produced by erosion and deposition associated with large impact events. In particular, some of the straight grooves on Mimas are roughly radial to the large impact crater Herschel. The morphology of the grooves and the texture on the associated smooth plains is reminiscent of the "Imbrium sculpture" of some terrains on the Moon, thought to result from gouging and flow of ejecta from the Imbrium basin (Wilhelms 1980, p. 45). Similar areas are associated with the crater Odysseus on Tethys and perhaps with some of the larger craters on Dione and Rhea. However, many of the resurfaced areas on these satellites and virtually all the smooth plains on the larger satellites are not obviously associated with impacts. They are also commonly curvilinear or irregular in shape, and exhibit features inexplicable in terms of impact mechanics. For these surfaces, the most reasonable explanation is that the process was endogenic: due to extrusion of material to the surface.

The emplaced material may have been liquid water (or brine), slush, or warm ice at the time of its extrusion. Because of the extremely large variation in viscosity—and hence time scales for deformation and flow—between water ($\sim 10^{-2}$ poise) and ice ($\sim 10^{15}$ poise near melting), the rate of ascent and geometric form of ascending magmatic bodies, the mode of extrusion, and the mode of emplacement depend strongly on whether the magma was liquid or solid.

Liquid Volcanism.   If the extruded material were liquid water or an ice/water slush, the most likely form of ascent would be as a fluid-filled buoyancy-driven crack. Theoretical studies of such cracks (Weertman 1971; Stevenson 1982b,c,d) indicate their approximate dimensions and rates of ascent. Treating ice as an elastic solid through which the cracks are ascending, typical crack lengths vary from ~ 10 km (Ganymede) to ~ 100 km (Enceladus), widths from 10^2 to 10^3 cm, and rates of ascent typically from 1 to 10 cm s^{-1}. The sizes of the cracks and rates of ascent are inversely proportional to the surface gravity; hence the volume erupted is larger and the time interval between eruptive events is longer on smaller satellites (Stevenson 1982d). The interaction of fluid-filled cracks with the satellites' surface is volcanic in nature, with fast-moving flows of liquid issuing from surface fissures. Because of the low viscosity of water, the material would flow quickly away from the eruption site, preventing the construction of large conical structures typical of volcanoes on the terrestrial planets. According to a model by Wilson and Head (1983), however, progressive ice buildup in a fast-moving water or slush flow may lead to non-Newtonian behavior that produces observable flows of significant thickness on nonnegligible topographic slopes.

No such flows are obvious, but their recognition may be prevented by the relatively low resolution of Voyager images.

Some freezing of liquid water due to decompression would occur as it ascended to the surface (Parmentier and Head 1979a). This is in contrast to the behavior exhibited by silicates, which increase in melt fraction with decompression; however, the effect is small. An ascent from the minimum solidus temperature point at a pressure of 200 MPa (a depth of $\sim$150 km on Ganymede) to the surface would cause adiabatic decompression crystallization of about 10% of the liquid.

In the case of eruption of pure H_2O water through pure H_2O ice, no explosive activity would occur, because the vapor pressure of water near its melting point is quite low. On Ganymede, for example, gas bubbles could only form in an ascending water body at depths of less than $\sim$1 m below the surface. If clathrates incorporating other volatiles are present, however, explosive events may be possible. Heat from a large body of water in this case could provide enough energy to force the clathrate matrix to decompose locally into water and gas, providing pressure in excess of lithostatic (Stevenson 1982c).

A unique feature of liquid water volcanism is the fact that, in pure form, the melt is more dense than the solid. If melting occurs in pure ice I, the water will tend to migrate downward rather than upward, and extrusion simply will not occur. Because resurfacing has clearly taken place on many icy satellites, some factor must be present to circumvent the expected negative buoyancy of the extruded material. One possible method of generating positive buoyancy for liquid water is simply inclusion of silicate material in the ice matrix of the crustal material. For a crust composed dominantly of ice I, inclusion of about 15% silicates by mass would allow liquid water present to be positively buoyant. This effect might have been important in driving the resurfacing on Ganymede. Another means of increasing magma buoyancy would be inclusion of NH_3 in the melt (Stevenson 1982c; Squyres et al. 1983b; Ellsworth and Schubert 1983). The density of a $H_2O - NH_3$ peritectic melt at low pressure is only $\sim$0.87 g cm^{-3}, significantly less than that of cold ice I ($\sim$0.94 g cm^{-3}). Partial melting of a body containing NH_3 would therefore lead to production of a buoyant magma that could be forced to the surface. This effect may have been important in driving the resurfacing observed on the Saturnian satellites.

Upon reaching a satellite's surface, much of the liquid in a large eruption will flow across the surface and freeze. A fraction, however, will boil upon exposure to the vacuum of space. This vaporization has two important effects. First, the heat of vaporization of water is large. Under some circumstances the heat loss from the water due to vaporization will significantly exceed both radiative and conductive cooling (Cassen et al. 1979a). The energy for vaporization in this case comes primarily from freezing of the nearby water. Because the heat of vaporization of H_2O is 7.6 times the heat of fusion, evaporation of 1 g will cause freezing of 7.6 g. This freezing will result in much more rapid cessation of the eruption than would be calculated from radiative and conductive cooling alone.

Second, the vapor may travel for considerable distances. On large bodies like Europa or Ganymede, molecules released upward from the surface at the mean thermal speed for a temperature of 273 K can travel a few hundred km from the point of release before reimpacting the surface. A large volcanic eruption will therefore create a large bright halo of condensed frost surrounding the eruption. Satellite-wide activity would lead to a very high uniform albedo across the body. On a small body like Enceladus, the escape velocity is so low that even localized activity could completely coat the satellite with a thin layer of high albedo frost. Vapor and frost that escaped from a small satellite would not escape from the planet, however, and would remain in orbit, at least briefly, as a tenuous ring.

The high albedos of Europa and Enceladus, the evidence from sulfur distribution on Europa for recent frost deposition, and the presence of Saturn's E ring associated with Enceladus all suggest ongoing liquid volcanism on those two bodies. The bright polar caps on Ganymede also may argue for a period of liquid water volcanism early in that satellite's history. Calculations of the thermal migration of H_2O molecules across Ganymede's surface indicate that migration rates are far too low to account for complete formation of the caps by migration of water from lower latitudes (Purves and Pilcher 1980). Instead, the caps can be explained as the remnants of a former global cover of bright H_2O frost with a thickness of at most a few m (Shaya and Pilcher 1984), although other models have also been suggested (Johnson 1985). This frost coating, which now exists only near Ganymede's poles, may have formed as a result of vaporization and recondensation associated with widespread water volcanism and resurfacing early in Ganymede's history. It is noteworthy that Callisto, which shows no evidence of volcanic activity, has no comparable polar caps (Spencer and Maloney 1984).

Diapirism. The case where the "magma" extruded to the surface is in fact warm ice rising through either a denser ice/silicate mixture or simply colder clean ice is best characterized as diapirism (Parmentier and Head 1979; Kirk and Stevenson 1983). In diapirism, the viscosity of the rising material is large enough and time scales of deformation long enough that the material of the penetrated layer behaves viscously rather than elastically. Under these conditions, the rising material usually tends to form circular domes (O'Brian 1968; Whitehead and Luther 1975; Ramberg 1981). The ascent velocity v of the diapir is given by

$$v = r_d^2 g \Delta\rho / 3\eta \qquad (16)$$

where r_d is the diapir radius, $\Delta\rho$ the density difference between the penetrated medium and the diapir, and η is the effective viscosity of the penetrated me-

dium. The ascent time of a diapir is dominated by the viscosity of the penetrated medium. Because viscosity is exponentially dependent on temperature,
the rise time is strongly controlled by the satellite's thermal profile. In thermal
models where convection is explicitly included (Ellsworth and Schubert
1983; Croft 1984), the thermal profile typically consists of a nearly isothermal
layer—the core of the convection zone—topped by a convective boundary
layer where the temperature drops sharply into a conductive zone that continues with a steep thermal gradient to the surface. The viscosity structure
therefore has a thick layer of roughly constant viscosity extending through
much of the interior, capped by a "lid" where the viscosities increase enormously. Diapirs, which presumably would initiate within or below the convecting zone, would rise relatively quickly through the core of the convective
zone, but would slow to a virtual standstill somewhere in the high viscosity
lid. The thickness of the lid for which the rise time will exceed the age of the
solar system depends on gravity, $\Delta\rho$, the size of the diapir, and the surface
temperature. On large satellites like Ganymede, it is about 10 to 20 km thick,
and on the smaller Saturnian satellites, it is about 50 to 100 km.

In the case of a pure ice diapir rising through an ice/silicate mixture, the
buoyant force exists as long as the diapir does. In the case of a diapir of warm
clean ice rising through cool clean ice, buoyancy is lost with time as the diapir
cools. There is an upper limit to the height h_d to which a warm ice diapir of
radius r_d can rise in an isothermal ice layer before cooling to the temperature
of the layer. For a spherical diapir, h_d is approximately

$$h_d = \frac{g\alpha_v\rho\Delta T r_d^4}{3\kappa\eta} \tag{17}$$

where ΔT here is the initial temperature difference between the warm diapir
and its cooler surroundings. Assuming the properties of pure ice, $\Delta T = 30$ K,
and a viscosity of 10^{18} poise (typical of the viscosity deep in an ice I layer
early in a large icy satellite's thermal history), diapirs of radius smaller than
about 15 km rise $\lesssim 100$ km before stopping.

The observable morphology that would result from diapirs reaching the
surface of an icy satellite is not clear. In the case of terrestrial salt and mud
diapirs (Gussow 1968; Morgan et al. 1968; Murray 1968) and scale models
(Whitehead and Luther 1975; Ramberg 1981), diapirs tend to rise as quasi-
spheres that pierce the surface to produce radial and concentric fractures that
may become radial canyons and circumferential horst-and-graben structures.
No features of this sort are observed on the icy satellites; there are many circular features, but all seem attributable to impact phenomena. Elongated
quasi-linear diapirs do occur in some places on Earth (the "salt walls" of
Northern Germany and the Gulf of Mexico), but are usually associated with a
field of more conventional axisymmetric diapirs.

Liquid Volcanism vs. Diapirism.   The application of the possible modes of resurfacing to specific icy satellites is not necessarily straightforward. Liquid volcanism dependent on buoyancy-driven fluid-filled cracks can only take place if the density of the fluid is less than that of the crust. Solid state diapirism, on the other hand, can only be important if internal viscosity is low enough to permit diapirs to reach the surface. In the case of the Saturnian satellites, liquid volcanism driven by inclusion of NH_3 or explosive interaction with clathrates is likely because these satellites probably formed at cool enough temperatures to include the necessary compounds. In contrast, diapiric processes on these smaller satellites are unlikely to be significant precisely because the satellites are small; surface gravities are low, producing low buoyant forces, and internal temperatures either began low or dropped rapidly, producing thick high-viscosity lids on short time scales. We therefore expect that the resurfacing of the Saturnian satellites involved primarily liquid extrusion.

The situation on Europa is less clear. Europa shows strong evidence for recent deposition of at least small amounts of condensed vapor on its surface, yet inclusion of silicates in the crust does not seem reasonable as means of increasing fluid buoyancy. Further, the Galilean satellites formed at warm enough temperatures that significant quantities of NH_3 in any underlying liquid are not expected. One possibility is that some fractures in the ice crust are so wide that vapor can escape to the surface even though the exposed liquid does not. Some of the dark bands are certainly wide enough to allow this if their width represents the amount of liquid area actually exposed in a single fracture event, but it is very difficult to imagine a source of stress that could accomplish this. Another possibility is that narrow fractures open very rapidly, so that the upward momentum of the rising water carries it past its equilibrium height and onto the surface (P. Cassen, personal communication). The mechanics of such a process remain to be worked out, however, and the details of the resurfacing that has taken place on Europa remain unclear.

The resurfacing mechanism on Ganymede is also a difficult problem. It may be argued that the near-surface material on Ganymede contains enough silicate material to drive liquid water or slush to the surface. This mode is attractive from a morphologic standpoint; liquids flowing out of large fissures in long quasi-linear rift valleys would fill them with smooth, clean ice deposits. One argument for liquid volcanism on Ganymede is the presence of a remnant polar frost deposit and the lack of a similar deposit on Callisto.

A difficulty with the liquid hypothesis is that one might expect to see examples of the process at all stages of completeness, yet no dry or partially filled rifts are observed. A more serious drawback is that Ganymede is probably largely differentiated, so that sufficient near-surface silicate material to drive buoyantly eruptions of liquid water may have been lacking at the time of resurfacing. Crater densities on Ganymede's oldest surfaces are lower than densities on Callisto by a factor of about 3 to 1 (Casacchia and Strom 1984).

This implies a global thermal event distinct from the episode of bright material emplacement in both time and morphologic expression. It is likely that this early thermal event, which was sufficient to eliminate all topography prior to furrow formation, involved differentiation of at least the outer layers. On the basis of the prevalence of bright ray craters and bright-floored craters in the bright terrain on Ganymede, it has been argued that impacts in the dark terrain have excavated into relatively silicate-rich material (Squyres 1980b; Shoemaker et al. 1982). Only a very small mass fraction of silicate material is required to darken ice substantially (see, e.g., Clark and Lucey 1984), however, and this argument does not demonstrate that there is enough silicate material present to drive volcanism buoyantly. If early differentiation near the surface did indeed take place, then the present silicate component of the dark terrain is merely a light dusting of meteoritic debris that would not significantly increase the density of the crust.

The warm ice diapir model of Kirk and Stevenson (1983) was proposed to overcome the difficulty of extruding liquids to the surface of Ganymede. They recognized that if the outer part of Ganymede had melted, convective thermal boundaries in the ice layers on both sides of the refreezing liquid layer could cause a large temperature difference in the convective cores of the ice layers that might drive significant diapiric flow upon closure of the liquid layer. This model has the advantage of being able to deliver fresh material to the surface through a pure ice I barrier. As demonstrated above, diapirs would have to be in excess of 15 to 20 km in diameter to penetrate to the surface. The major problem with this model is morphological: it is not at all clear that the glacier-like flow of diapirs could form the sharp edges and long, uniform, very smooth deposits of bright terrain observed. The lobate flow fronts likely to form are not apparent. No isolated axisymmetric diapirs are observed. The characteristic radial and concentric fractures that typically accompany diapiric extrusion are also not seen.

At present, the volcanic mode of bright terrain emplacement on Ganymede seems to fit the observed morphology but cannot be clearly understood theoretically, while the diapiric emplacement mode is theoretically possible but appears inconsistent with morphology. Clearly much work remains to be done.

Surface Modification

The tectonic structures on some of the icy satellites are probably several billion years old, and have existed long enough that one might expect modification of the original topography in response to gravity and exposure to the environment of space. In particular, the observed topography of grooves on Ganymede is sufficiently different from that expected for either pristine grabens or extension fractures that either original landform would require modification to produce the observed topography. It is therefore of interest to consider various possible modification processes.

Viscous Relaxation.   Topographic relaxation caused by viscous creep of ice was anticipated to be an important process for modification of the surfaces of icy satellites long before images of the satellites were available (Johnson and McGetchin 1973). Interpretation of Voyager images of the icy satellites has prominently invoked viscous relaxation as the primary cause of the subdued relief and unusual morphology of many structures, particularly impact features (see, e.g., Smith et al. 1979a,b; Passey and Shoemaker 1982a,b; Shoemaker et al. 1982).

Although numerical schemes have been devised to simulate viscous relaxation of topographic features such as craters (see, e.g., Parmentier and Head 1981), the approximations necessary to perform such calculations generally make them inadequate to deal with the very steep surface slopes that could exist for some primary tectonic landforms. For this reason, Parmentier et al. (1982) and Squyres (1982) modeled the relaxation of topography of extension fractures and grabens using a high viscosity polymer. The time required for a given amount of relaxation to take place was calculated using simple scaling laws. As is well known from theory, large-scale topographic features relax more rapidly than fine-scale features. It was found that extension fractures relax to form single linear depressions with a width that gradually decreases relative to the initial width of the fracture. Because deviatoric stresses are highest near the bottom, the fracture closes progressively upwards, resulting in a sharp-bottomed depression that shallows with time. Grabens, on the other hand, develop raised rims and domed floors. In neither case is a feature strikingly similar to most grooves on Ganymede or Enceladus produced, although Parmentier et al. (1982) suggested that relaxed graben bore at least a superficial similarity to groove pairs. Narrow grooves with widths near the Voyager resolution limit could perhaps be viscously relaxed extension fractures, but the case is certainly not compelling.

A substantial problem with viscous relaxation of grooves arises, however, when one considers the time scales required. Based on the observed topography of impact craters, Passey and Shoemaker (1982a) calculated an effective surface viscosity for Ganymede of 1×10^{26} poise. This value may actually be a lower limit, because fresh craters formed in ice may already be shallower before any modification than their counterparts on silicate bodies (Croft 1981d, 1986). Using the expression for viscosity given in Eq. (13) and a typical mean equatorial surface temperature for Ganymede of 120 K yields a viscosity of 2.5×10^{28} poise. Even allowing for an insulating regolith that elevated near-surface temperatures by 20 K, a viscosity lower than 10^{25} poise seems difficult to achieve. Model experiments show, however, that viscosities of $\leq 10^{24}$ poise are required for any significant relaxation of grooves of any original form to have taken place on Ganymede over the age of the solar system. If the global surface viscosity were low enough to cause substantial relaxation of the grooves (i.e., $\leq 10^{23}$ poise), all craters larger than about 10 km in diameter would have virtually disappeared over the $\sim 3 \times 10^9$ yr since formation of the grooves. Such low viscosities therefore appear inconsistent with observations. These low viscosities would also make deep normal faulting impossible for any reasonable geothermal gradient, lithospheric strength, and strain history.

If these inferences are correct, the only way in which viscous relaxation could have substantially affected groove morphology would be if the relaxation took place very rapidly after emplacement of the resurfaced deposits, while they were still relatively warm. The cooling time of a layer of ice of thickness l is $\sim l^2/\kappa$, or about 10^4 yr for a layer 1 km thick and only 10^2 yr for a layer 100 m thick. Conceivably, viscous relaxation of topography could take place before cooling and stiffening of the material if groove formation occurred over these very short time scales. However, in cooling enough to permit any brittle deformation to take place, the material would also be increasing exponentially in viscosity. Further, this process could not have caused degradation of grooves observed in older cratered terrain units.

The conditions for viscous relaxation of tectonic features on the Saturnian satellites are even less favorable; surface gravities and temperatures are significantly lower than on Ganymede. Even allowing for the possible effects of inclusion of NH_3 in the ice (Stevenson 1982d), features on the Saturnian satellites are less prone to viscous relaxation than those on the Jovian satellites, and no significant relaxation of tectonic features is to be expected at present surface temperatures.

Mass Wasting.   As we have discussed above, viscous relaxation by itself does not produce topographic forms strikingly like most observed grooves from likely original tectonic landforms. Mass wasting is another process that could be important in modifying tectonic landforms after their formation. Mass wasting would tend to reduce initially steep slopes to angle of repose, filling in and widening initial groove forms and rounding the tops of sharp prominences. Because of the large amount of time available for the process, impacts would be a major contributor to breakup of slope material. Thermally induced volume changes could also be important. Mass wasting on a large scale seems to be the only process capable of generating the presently observed groove topography from initial extension fractures. It could also perhaps produce the observed topography from initial grabens, although in the latter case viscous relaxation (if possible) could also play a part in making the grooves more shallow. Unfortunately, it is not presently possible to evaluate the effectiveness of mass wasting in modifying tectonic landforms on the icy satellites. The images presently available are of insufficient resolution to allow discrimination of mass wasting deposits in the grooves. If mass wasting has indeed played a major role in transforming grabens or extension fractures into the present groove forms, it has operated significantly more effectively than mass wasting on the Moon, where many ancient grabens are well preserved. If higher resolution images of the grooves do not show substantial evidence for slope modification by mass wasting, then the search for other possible mechanisms of groove formation should be intensified.

IV. SUMMARY AND CONCLUSIONS

Europa

Observations and theoretical calculations indicate that Europa is a primarily silicate body, but suggest that the silicate interior is overlain by a liquid water ocean, which is in turn overlain by a relatively thin ice shell. Geologic activity has taken place relatively recently. The complex pattern of fracture-like lineaments clearly has a tectonic origin. The details of the formation of the lineaments, particularly the bright ridges and triple bands, are not understood. A reticulate pattern of short dark lineaments near the anti-Jovian point may owe its origin to extensional fracturing caused by tidal stresses. Global expansion due to silicate dehydration may have taken place early in the satellite's history, but its influence on the observed features is unclear, as they appear to be fairly recent. Stresses due to orbital recession are not likely to have been geologically important. Some linear features have orientations consistent with origin by tidally-induced strike-slip faulting, but other expected orientations are absent, and no strike-slip offset along the features is observed at Voyager resolution. The origin of the stresses that caused many of the features, especially those whose geometry implies rotations of tens of km, is not understood. The nature of the resurfacing that has taken place is also poorly understood, although it probably involves eruption of only small amounts of liquid water to the surface.

Ganymede

There is widespread evidence on the surface of Ganymede for extensional deformation, limited evidence for shear, and little or no evidence for compression. Nothing analogous to terrestrial plate tectonics has taken place on Ganymede or on any other icy satellite observed. The bands of bright resurfaced terrain on Ganymede are probably broad grabens formed by global expansion and filled with deposits of ice fairly early in the satellite's history. A number of processes could have caused global expansion, with the largest expansion possible being that due to internal differentiation. Tectonic extension may have been concentrated above zones of convective upwelling. The resurfacing may have taken place either as volcanic eruption of liquid water to the surface or as slow rise of clean warm ice diapirs. The volcanic model is more consistent with the observed morphology, but magma buoyancy considerations require a silicate component of the lithosphere whose presence cannot be verified and is not expected according to some models. Grooves within the bands are thought to be extensional features, perhaps extension fractures or long, narrow grabens, that were formed during the same episode of global extension. The regular spacing between grooves within a given groove set may be due to a necking instability in the lithosphere, and the spacing is probably an indicator of the thickness of the brittle lithosphere at the time of deformation. Lithospheric thickness depends on a number of factors, including surface temperature, geothermal gradient, and strain rate. Some modification of the grooves since their formation is likely, but significant modification by viscous relaxation requires very rapid groove formation immediately following a resurfacing event, and cannot account for modification of grooves in the older cratered terrain. Ganymede underwent a period of global resurfacing and furrow formation very early in its history, the nature of which is poorly understood.

Enceladus

Enceladus has been the most geologically active of the Saturnian satellites for which we have data. The activity has probably been driven by tidal heating, and the association of Enceladus with Saturn's E ring suggests that activity has continued up to the present. Geologically, Enceladus appears similar to Ganymede, with broad resurfaced regions and sets of parallel grooves. Resurfacing on Enceladus probably involves eruption of liquid to the surface, with vaporization of the liquid coating the satellite with bright frost and also forming the E ring. Eruption of liquid to the surface may be aided by inclusion of NH_3, which substantially lowers both the initial melting temperature of the ice and the density of the melt produced. Convection is unlikely to be a significant source of stress. A possible source of stress is freezing of an internal liquid water component to form ice I.

Other Saturnian Satellites

Most of the other Saturnian satellites show evidence for partial resurfacing early in their histories. As on Enceladus, this resurfacing may have been aided by the presence of NH_3. The satellites show a variety of apparently extensional features, the most impressive of which is the large graben Ithaca Chasma on Tethys. Significant global expansion could be caused by freezing of internal liquid water, but it is not clear that any mechanism exists to cause substantial melting in the satellites initially. Dione shows a pattern of very subdued ridges that may provide evidence for a minor global contraction.

The geologic evolution of icy satellites was still an unexplored field of research in 1980. The years since then have seen an explosion in the quantity of data and ideas, and significant growth in real knowledge. We have attempted to summarize here our understanding and opinions of both current knowledge and many current ideas. The next decade will hopefully bring Voyager encounters with the satellites of Uranus and Neptune, voluminous and extremely high-quality data from the Galileo mission to the Jupiter system, and perhaps a mission that will reveal the first details of the geology of Titan.

Our most certain conclusion is that many ideas and some accepted facts will change in the near future.

Acknowledgments. We are grateful to W. McKinnon, M. Golombek, and J. Spencer for helpful reviews and comments. This work was supported by the NASA Planetary Geology Program.

8. THE PHYSICAL CHARACTERISTICS OF SATELLITE SURFACES

J. VEVERKA and P. THOMAS
Cornell University

T. V. JOHNSON and D. MATSON
Jet Propulsion Laboratory

and

K. HOUSEN
Boeing Aerospace

Both exogenic and endogenic effects have been proposed to explain the major observed characteristics of satellite surfaces. The current view is that the basic properties of most surfaces result from the intrinsic composition of a body and its geologic history. Exogenic effects have, however, played a role in modifying the appearance of nearly all surfaces. The most important exogenic effect is impact cratering, one manifestation of which is the production of micrometeoroid gardened regoliths on airless bodies. On large, silicate bodies the micrometeoroid bombardment can produce an optically mature, dark agglutinate-rich soil; the nature of regoliths on predominantly icy satellites remains uncertain. Direct accumulation of infalling material does not appear to play a major role in modifying most surfaces. Solar wind radiation effects have not altered greatly the optical properties of solar system objects; magnetospheric charged particles may have modified the optical properties of some outer planet satellites (e.g., sulfur ion bombardment in the case of some of the satellites of Jupiter). Other effects, such as aeolian and liquid/solid chemical weathering, may be important on satellites with atmospheres like Titan and Triton.

I. INTRODUCTION

In the mid 1970s, it was usual to consider the surfaces of satellites to be ancient, their characteristics dominated by extended histories of impact cratering. Indeed, the few satellites that had been investigated in detail, our own Moon and the two small satellites of Mars, reinforced this general impression. Then, as Voyager ventured into the trans-asteroid regions of the solar system, this perspective changed dramatically. The properties of Io's surface are determined primarily by ongoing volcanic activity; even on icy satellites, such as Europa and Enceladus, internal processes are important. In addition, we now have to deal with the properties and formation of regoliths on icy as well as rocky satellites. The evolution of icy regoliths probably is the more complex process, in that sublimation/condensation and sputtering due to charged particles must be considered.

While it has been usual to restrict discussions to regoliths on airless bodies, we now have to deal with Io and its thin, variable atmosphere. We also know that Titan and probably Triton have substantial atmospheres. Suggestions even exist that these two satellites may in part be covered by so-called seas. Since substantive data on these putative seas are lacking, this review will deal only with solid surfaces.

Since the last major reviews of this subject in 1977, many developments have occurred, the most important of which have been preliminary spacecraft investigations of most satellites in the solar system; in 1977 only the Moon, Phobos and Deimos had been studied directly. The bulk of the data then available was from Earth-based observations. *Planetary Satellites* (Burns 1977a) contained reviews of satellite photometry and polarimetry (Veverka 1977a,b), radiometry (Morrison 1977), and compositional information derived from Earth-based color and spectral measurements (Johnson and Pilcher 1977). Since then, specific review articles dealing with natural satellites have appeared, summarizing our accumulating knowledge of the surface characteristics of specific objects. Examples include reviews of Phobos and Deimos (Veverka and Burns 1980), of the Galilean satellites of Jupiter (Morrison 1982), of the outer satellites of Jupiter (Cruikshank et al. 1982), of the small inner satellites of Jupiter (Veverka et al. 1982b), of the satellites of Saturn (Cruikshank 1979; Cruikshank et al. 1983a) and those of Uranus (Cruikshank 1984; Brown 1984). This chapter is intended to bring the situation up to date and touch on all relevant investigations (photometry, imaging, radiometry, etc.). Complementary chapters in this book include those on the composition of satellite surfaces (Clark et al.), on cratering (Chapman and McKinnon) and on the interaction of charged particles with satellite surfaces (Cheng et al.), as well as those dealing with individual satellites.

A discussion of the physical properties of satellite surfaces is almost synonymous with a discussion of satellite regoliths—those outermost layers which are most readily amenable to external studies by remote sensing techniques. Regolith is a general term for a surficial layer of loose material of any origin (cf. Bates and Jackson 1980). Planetary scientists usually associate the term with impact-generated debris, often very finely comminuted. But a satellite such as Io has a regolith largely of volcanic origin; on other satellites loose material can be generated by ice volcanism, sputtering, or other nonimpact processes. Typical photometric techniques are sensitive to the topmost hundred microns of a surface layer, thermal measurements to no more than the topmost few centimeters, whereas at microwave and radio frequencies we can probe to some tens of centimeters below a surface. All such investigations are therefore superficial in the geologic sense.

The only satellite for which we have both remotely sensed data on its regolith, in addition to *in situ* information and samples, is our own Moon (see the chapter by Kaula et al.). A crucial task for future planetary exploration must be the sampling of other regoliths to calibrate how representative the lunar one really is in terms of its general physical properties. To what extent are the properties of icy and icy/silicate regoliths similar to those made of rocky material alone? Does surface gravity play a dominant role in regolith development? Is regolith development on bodies such as Phobos and Amalthea affected by the fact that these satellites exist deep in the potential wells of their parent planets? Are the regoliths of some satellites influenced by interactions with magnetospheric particles? (See the chapter by Cheng et al.) Are the regoliths of some satellites contaminated with material derived from neighboring objects? Our studies have reached a new level of complexity; we must consider the evolution of regoliths made out of a variety of materials ranging from very volatile to very refractory, and regoliths which are probably affected by a whole suite of processes (Fig. 1). We must try to understand how the processes act on various satellites and how they determine the observable properties of regoliths.

Our review is divided into four parts. We begin (Sec. II) with a brief survey of what we know about the regolith of the Moon, both from remote sensing and *in situ* observations. One purpose is to summarize the extent to which one can decide whether the major observed characteristics of the lunar surface are due to internal or external processes and then to apply the lessons learned to other satellites. Section III synthesizes ideas about regolith formation and evolution, starting with a discussion of cratering by external objects and leading to a review of how such impact-generated regoliths are modified by internal processes. Specifically, we discuss how impact-generated regoliths are expected to vary with composition (rocky vs. icy), gravity (small satellite vs. large satellite), and environment (proximity to a large planet, etc.). Section IV reviews remote sensing data on satellite surfaces and focuses on physical characteristics determined from photometric and radiometric data. Finally, in Sec. V we address a class of problems dealing with the observed albedo distributions on satellites. We stress that no single mechanism will explain the observed lightcurves of the satellites of Jupiter and Saturn.

PROCESSES

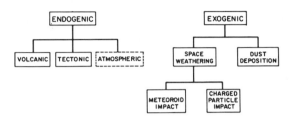

Fig. 1. Processes affecting satellite surfaces.

II. THE MOON AS A REFERENCE STANDARD

A. Lunar Regolith and Its Evolution

The Moon, described in the chapter by Kaula et al. remains the only satellite for which we have direct data on surface characteristics to test our theories and our inferences from remote sensing observations. It is also the only satellite for which an absolute chronology exists. Our ideas about the role and nature of impact cratering, regolith formation and space weathering on airless bodies are strongly influenced by our experience with the Moon.

Several decades ago, it was common to ask whether the observable characteristics of the lunar surface were determined by internal (volcanic) or external (impact) processes. Following detailed studies of Meteor Crater in Arizona by Shoemaker (1963), little doubt remained that the major features seen in telescopic and spacecraft images of lunar craters were consistent with impact cratering; field work by Apollo astronauts only reinforced this conclusion. However, as studies of lunar samples quickly showed, volcanism also played a major role in shaping the lunar surface early in its history (Head 1976). In the Moon's case, neither the inactive body affected only by exogenic battering envisioned by some, nor the volcanically active object seen by others, proved to be correct (Fig. 2).

One extreme pre-Apollo view of the lunar surface held that the regolith is dominated by foreign meteoritic material. But when returned lunar material was examined, $\leq 1\%$ meteoritic component was detected geochemically (see, e.g., Papike et al. 1982). Yet, the soil was found to be more complex than mere comminuted lunar rock (see discussion in the chapter by Kaula et al.). Simply described, it is a mixture of fine rock grains and dark, glassy agglutinate materials. The current consensus is that such soils are the result of exposure to, and continued reworking by, micrometeoroid impacts. In these impacts, most of the meteoroid material is vaporized or ejected at high speed from the initial impact region, preventing the buildup of a significant exogenic component. Melting of the endogenous material and the production of opaque,

MOON

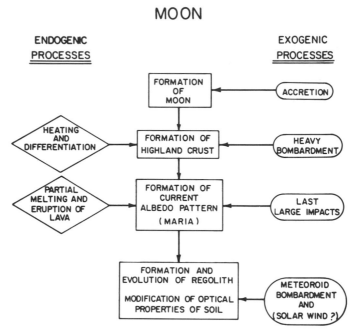

Fig. 2. Schematic representation of the interaction of endogenic and exogenic processes in determining the present characteristics of the lunar regolith.

glassy agglutinates from opaque phases of iron- and titanium-bearing minerals produce the dark soils which dominate most of the lunar surface (Taylor 1975).

The physical characteristics of the lunar regolith (Fig. 3) are consistent with a long exposure to micrometeoroid impacts (Lindsay 1976). From returned samples, we know that the outermost layer of the Moon consists of a very fine powder dominated by particles smaller than 100 μm in diameter. The surface is very porous; typical densities of the upper 1 cm are ~ 1.4 g cm^{-3}. The regolith is known to be vertically inhomogeneous and layered. Principal constituents include rock fragments, glasses, agglutinates, and about 1% foreign components from meteorites. In terms of texture (compaction and particle size), there is little difference between the maria and the highlands; the albedo difference between the two terrains is due largely to a difference in composition (as demonstrated by distinct spectral curves; see the chapter by Clark et al.) rather than to differences in particle size or in darkening by the solar wind. Solar wind darkening, once thought by some to be a major cause of albedo modification on the Moon, is now believed to play, at most, a minor role (Matson et al. 1977).

Early polarimetric (Lyot 1929) and the photometric (Minnaert 1961) measurements of the Moon were correctly interpreted to imply a very porous

Fig. 3. The lunar regolith: boulders and fine debris at Hadley Rille. This scene is the result of slow exogenic modification of an ancient volcanic terrain. The boulders are approximately 1 m high.

texture, later referred to as a "fairy castle" texture (see, e.g., Hapke and van Horn 1963). Infrared observations showed that the surface responds very rapidly to changes in insolation, indicating a low thermal inertia that is characteristic of very porous surfaces in which very poor intergrain contact exists (cf. Wesselink 1948). Eclipse cooling and heating curves gave evidence of the importance of large blocks on the surface, especially in association with the youngest craters (Saari and Shorthill 1963; Buhl 1971; and others). Radio and radar measurements confirm that the blocky texture extends down to the subsurface (Ostro 1983; Kheim and Langseth 1975).

Estimates of regolith characteristics at depth in the maria and uplands have been attempted by various methods (crater morphology, seismometry, etc.; Lindsay 1976). For many places in the mare, the regolith may extend to depths of 10 to 30 m; in the uplands much higher values are suggested. Mega-regoliths of 10 or more km have been proposed by some investigators (see, e.g., Hartmann 1972a and the chapter by Chapman and McKinnon).

The Moon provides several immediate lessons for understanding the regoliths of other bodies. For instance, there is the question of why lunar rays have photometric characteristics which are distinct from those of the background material. One convincing explanation is that advanced by Adams and McCord (1973) who suggested that initially rays consist of abundant pact-generated glass which is subsequently mixed with the adjacent and underlying regolith (but see also the more recent work of Pieters et al. [1985]). Such ideas are readily extended to other satellites. For example, Shoemaker et al. (1982) proposed that rays on Ganymede are removed by micromixing of clean ejecta ice with a dirtier regolith below, and that one can date fresher craters in a relative way by the prominence of their ray patterns, as is done on the Moon.

In terms of explaining albedo distributions on other satellites, the Moon provides two useful lessons. The first is that impacts are not very efficient in redistributing surface material over a satellite as large as the Moon. The past 4 Gyr of impacts have not been sufficient to blur the contacts between the maria and highlands to any significant degree. This fact has been amply confirmed by both X-ray, γ-ray and optical measurements (Metzger et al. 1973; Arnold et al. 1972; Johnson et al. 1977a,b) and by theoretical calculations, a lesson which should be remembered in dealing with other large or medium-sized satellites.

A second lesson learned from the Moon is that albedo and color patterns on satellites are likely to represent the result of a complex history of interaction among different processes. It may be too simplistic to ask: Is the unusual albedo dichotomy observed on Iapetus the result of strictly internal or external processes? On the Moon, the distribution of albedo is inhomogeneous because there is more mare material on the Earth-facing side. Mare basalts are the results of internal heating; yet the maria fill basins that were excavated by large impacts. It has been postulated that the reason that more mare basalts reached the surface on the Earth-facing side (although statistically there are just as many basins on the farside) is due to an asymmetry in the thickness of the lunar crust (Kaula et al. 1974; chapter by Kaula et al.). If such a crustal asymmetry exists (Wasson and Warren 1980), it could be due to internal processes, external processes (e.g., related to the existence of an early impact-generated magma ocean), or a combination of both.

The Moon also serves to remind us of the crucial importance of large-scale roughness in determining local surface temperatures and in influencing the total amount of scattered sunlight from a given area at a given phase angle. The influences of blocky ejecta on the thermal properties (Buhl 1971) as well as that of crater and other topography (Winter and Krupp 1971; Golden 1979) on local temperatures are well known. From lunar images we see shadows cast by roughness elements of various sizes and can appreciate their importance on the photometric properties (Schonberg 1925)—a lesson that must be borne in mind in interpreting the photometry of satellites for which less detailed imaging is available.

B. Other Regoliths: Similarities and Differences

With the exception of Io, and to a lesser extent Europa, all the satellite surfaces studied so far exhibit evidence of significant meteoroid bombardment. Most have very heavily cratered areas which must date back to an early period when impact fluxes of large bodies were much greater than at present, based on our understanding of the lunar chronology plus current comet and asteroid flux levels (Woronow et al. 1982; Passey and Shoemaker 1982; chapter by Chapman and McKinnon). However, many of these satellites also show evidence of significant resurfacing by endogenic activity. The nature of these resurfacing processes is not well understood, particularly for the icy bodies. The very existence of geologically recent resurfacing on Io, Europa and Enceladus has required the development of new concepts concerning tidal heating of satellites by their planets (Cassen et al. 1982; chapters by Schubert et al., Nash et al. and Peale et al.).

The microevolution of surfaces subsequent to their formation by major geologic events has been the subject of much debate. Our planet's surface is dominated by a variety of weathering processes which may have few analogs in the rest of the solar system, if one excludes certain climatic episodes in the history of Mars. Unlike the Earth, most solid bodies in the solar system, particularly the planetary satellites, are airless and have no liquid water on their surfaces under present conditions. Thus, weathering processes should be closer to lunar processes than to anything with which we are familiar on our planet.

The general principles which govern the production of lunar soil should be applicable to other bodies. Mercury appears to exhibit the spectral characteristics of mature mare soils (Vilas and McCord 1976). Matson et al. (1977) have argued that asteroid spectra suggest far less retention of impact-reworked, glass-rich agglutinate material than is the case for the Moon, a result consistent with the weaker gravity of these smaller bodies. In their regolith properties, Phobos and Deimos are those likely to be similar to smaller asteroids, as also may be the smaller, presumably captured satellites of Jupiter.

Io and Europa are the only two large satellites with bulk compositions of rock. Io's surface is so dominated by volcanic processes that impact-produced effects are probably negligible; Europa's rocky body is sheathed in ice, making it more like the other outer planet satellites in terms of surface processes. Outer planet satellites generally have surfaces consisting of varying mixtures of water frost and dark silicate or carbonaceous material. All of these bodies are expected to have particulate, impact-generated regoliths. Sharp boundaries between bright and dark albedo areas are common, suggesting that impact generation of regoliths need not involve global mixing of debris.

Modification of albedo markings by infalling material has often been proposed in the case of some satellites, especially Iapetus (see Sec. V, below). In addition, a strong case has been made for the alteration of satellite surfaces by trapped particles around Jupiter, Saturn and possibly Uranus (chapter by Cheng et al.). Many specific issues remain unresolved; for example, the extent to which sputtering erodes surfaces of the Jovian satellites (Matson et al. 1974; Lanzerotti et al. 1983) and the possible effects of sulfur ions from Io in modifying the surfaces of the other Jovian satellites. Gradie et al. (1980) have proposed that the unusual optical properties of Amalthea (very dark albedo and very red color) may result from the implantation of sulfur on a surface composed primarily of carbonaceous material, and a number of studies suggest that Europa may also show effects of sulfur implantation from Io (Lane et al. 1981; Johnson et al. 1983; Nelson et al. 1983,1985; chapter by Malin and Pieri). Weaker evidence of such an effect may exist even for Ganymede (Nelson et al. 1985).

Lanzerotti, R. E. Johnson, Haff and their coworkers have stressed the likely importance of sputtering by magnetospheric particles of the surfaces of the icy satellites of Saturn inward of, and including, Rhea. Cheng (1984b) has discussed the potential formation of a dark residue when a water/methane ice mixture is irradiated by protons, and has argued that such a process might explain the relatively low albedos of the satellites of Uranus (see the chapter by Cheng et al.).

III. MODELS OF REGOLITH GENERATION AND EVOLUTION

A. External Processes

The densely cratered surfaces observed on many solar system bodies attest to the importance of impact bombardment in their geologic evolution. A natural consequence of continual impacts is the generation of regoliths, i.e., surficial layers of broken, fragmental debris. For example, the heavy bombardment prior to mare formation extensively fractured the lunar crust and produced a debris layer thought to be several kilometers deep (Horz et al. 1976; Cashore and Woronow 1984). The less heavily cratered mare surfaces are blanketed by a debris layer typically several meters in depth (Quaide and Oberbeck 1968; Shoemaker et al. 1969). Several models of lunar regolith evolution have been constructed; comprehensive reviews of these models and of regolith properties are given by Langevin and Arnold (1977), Taylor (1975), and Lindsay (1976).

While the existence of the lunar regolith has been known for some time, the idea that smaller bodies would retain enough ejecta to develop regoliths has gained acceptance only recently. From early polarimetric observations (see, e.g., Dollfus 1961,1971; Veverka 1971*b*) the surfaces of many asteroids and satellites were interpreted as having at least thin dusty coatings, although thicker accumulations were not ruled out (Hapke 1971). A different insight into the nature of asteroid surfaces arose from the studies of brecciated and gas-rich meteorites. Comparisons of these meteorites to lunar breccias (Wilkening 1971; MacDougall et al. 1974; Rajan 1974) and observations of their abundances (Anders 1975) implied substantial (perhaps kilometer deep) accumulations of debris on some meteorite parent bodies. This surprising result prompted theoretical studies. A preliminary model by Housen (1976) and subsequent, more detailed, calculations (Housen et al. 1979; Langevin and Maurette 1980) showed that regoliths several kilometers deep could be expected on the larger asteroids and that even some bodies as small as 10 km might develop meter-scale debris layers. A review of asteroidal models is given by Housen and Wilkening (1982).

It is reasonable to assume that planetary satellites should also develop significant regoliths. While Shoemaker et al. (1982) used a model based on crater saturation to estimate regolith depth on Ganymede, very few calculations have been made which directly pertain to satellites other than the Moon.

In this section we present results of Monte Carlo computer simulations of regolith evolution on satellites. The simulations differ from earlier studies in that *observed* crater distributions are used. Hence, uncertainties associated with speculative crater scaling laws are avoided. Furthermore, recent advances in the understanding of crater ejecta are included. In order to lay a foundation for the simulations we give a brief discussion of crater formation, ejection velocities of excavated debris and ejecta blankets (see also the chapter by Chapman and McKinnon).

The growth of a regolith depends in part on the mechanical properties of surface materials, which are poorly known for most solar system bodies. Therefore, much of the following discussion refers to three generic materials: ice, competent rock and porous, nearly cohesionless soil. Laboratory analogs for these prototypes are pure water ice, basalt, and quartz sand.

1. Crater Formation. An impact on a satellite produces a roughly hemispherical shock wave resulting in comminution of subsurface materials and the initiation of material motions. The initial shock, in concert with rarefactions generated at the free surface, results in an orderly flow of material characterized by an expanding bowl-shaped cavity (see Fig. 1 in the chapter by Chapman and McKinnon). Growth of the crater is eventually halted by the strength or viscosity of the cratered medium or by gravity. As described by Chapman and McKinnon in their chapter, the relative importance of these factors depends on their respective magnitudes and on the size of the impact event. For example, for relatively small craters (a term quantified below) formed in competent media, such as rock, crater growth stops when the peak shock stress decays to a level comparable to material strength. Alternatively, mixtures of ices and silicates may behave as viscous fluids when subjected to high stresses associated with hypervelocity impact (Greeley et al. 1982; Fink et al. 1984). Finally, for sufficiently large craters in any material, or even for small impacts in low-strength or low-viscosity media, crater size is determined by gravitational body forces. These three cases are generally referred to as the strength, viscosity and gravity regimes, respectively.

The boundaries separating these cratering regimes can be shown to depend on the values of appropriate nondimensional parameters (Gault and Wedekind 1977; Housen et al. 1983; Fink et al. 1984; chapter by Chapman and McKinnon). For example, the relative importance of strength and gravity is determined by the ratio $Y(\rho gD)^{-1}$, where Y is an appropriate measure of the strength of the cratered medium, ρ is the density, g is the local gravity and D is crater diameter. If the material strength is assumed to be rate- and size-independent, then for a fixed material, the transition between the strength and gravity regimes is determined by the value of gD. For hard rock, explosion experiments indicate that the strength/gravity transition occurs roughly when $gD \simeq 2 \times 10^7$, in cgs units. Comparisons of centimeter-scale impact craters and meter-scale explosion craters in ice and in rock (Livingston 1960; Moore and Gault 1963; Lange and Ahrens 1981; Kawakami et al. 1983) suggest that the strength of ice is less than that of rock by roughly a factor of 10 or perhaps as much as 100. Hence, the strength/gravity transition in ice might occur when $gD \simeq 2 \times 10^6$.

For cases where viscosity plays a role, the relative importance of viscosity and gravity can be shown to depend on the ratio $\eta(\rho g^2 D^3)^{-1/2}$, where η is the viscosity of the stress-fluidized material (not the viscosity pertinent to long-term crater relaxation). Large and small values of this parameter, respectively, denote the viscous and gravity regimes. Impact experiments in several viscous materials (Fink et al. 1984) imply that the transition occurs when $\eta(\rho^2 gD^3)^{-1/2} \simeq 3 \times 10^{-2}$ or, equivalently, when $D \simeq 10 (\eta^2 g^{-1}\rho^{-2})^3$. Values of η appropriate to icy satellites are difficult to estimate. However, Fink et al. (1984) have noted that values of η up to $\sim 10^5$ dyne cm^{-2} s^{-1} have been observed for terrestrial flows of water, soil and rock and that the presence of ice might raise η by perhaps several orders of magnitude.

Plausible values of the various transition crater diameters are shown in Fig. 4 as a function of g for icy and rocky surface materials. As indicated there, and noted by McKinnon (1983b), material strength effects are probably not important for cratering on icy satellites, except for rather small craters. For example, on the Galilean satellites ($g \simeq 10^2$ cm s^{-2}), strength effects should not be important for craters with diameter larger than a few hundred meters. On the larger Saturnian satellites ($g \simeq 20$ cm s^{-2}), the transition to gravity scaling should occur at a crater size of roughly 1 km. Furthermore, unless the viscosity exceeds roughly 10^7 dyne cm^{-2} s^{-1}, viscous effects will also be small compared to gravity effects.

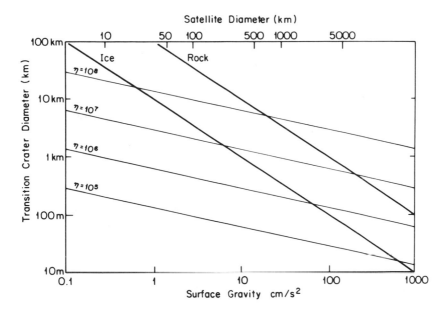

Fig. 4. Estimates of crater diameters for transitions between the strength, viscosity and gravity regimes. Heavy lines indicate a transition between the strength and gravity regimes. Light lines indicate a viscosity/gravity transition. In all cases, crater diameters above a line lie in the gravity regime.

   2. Escape of Impact Ejecta.   In an impact event, material is ejected primarily at the edge of the expanding cavity. The ejection velocities decrease monotonically as the crater grows, because the peak stress associated with the shock front decreases as it expands outward. While some of the faster debris may escape a satellite, at some point the ejection velocity may decrease below the escape velocity. Any debris ejected thereafter is returned to the surface where it forms the crater ejecta blanket. However, even so some ejecta which have temporarily escaped may orbit the satellite's primary and eventually reimpact the satellite (see, e.g., Soter 1971; Burns et al. 1984). Reaccretion of material is not considered in our Monte Carlo simulations; we address only the initial escape of debris.

   Housen et al. (1983) used the concept of a source coupling parameter to derive expressions for the volume fraction f of excavated debris ejected with velocity exceeding an arbitrary value v. In particular, they showed that, except for the slowest debris, which is ejected near the rim of the final crater, f should vary as a power of v. For example, in the strength regime

$$f \propto [v(\rho/Y)^{1/2}]^{-b} = C_1 v^{-b} \qquad (1)$$

where b and C_1 are positive constants for fixed material type. In this case f is independent of the size of the impact event. On the other hand, in the gravity regime, f depends on the crater diameter D. In particular,

$$f = C_2[v/(gD)^{1/2}]^{-b} \tag{2}$$

where C_2 is a constant for fixed material. Notice that, in this case, the fraction of ejecta with velocity exceeding v increases as crater size increases. This is a direct result of the fact that, in the gravity regime, as the size of the event increases, the crater rim forms relatively closer to the impactor (where ejection velocities are higher). Impacts in a variety of materials suggest that b depends largely on the target porosity and that b is ~ 1.7 and ~ 1.2 for nonporous and porous materials, respectively.

The fraction f_{esc} of ejecta which escapes can be found from Eqs. (1) and (2) by setting v equal to the escape velocity. In the strength regime,

$$f_{esc} = C_3 D_s^{-b} \tag{3}$$

where $C_3 = C_1(2\pi\rho G/3)^{-b/2}$. G is the gravitational constant and D_s is the satellite diameter. In the gravity regime Eq. (2) yields

$$f_{esc} = C_2[D/D_s]^{b/2}. \tag{4}$$

Rough estimates of the constants C_1, C_2 and C_3 are shown in Table I. The value of C_1 for rock was determined from the velocity measurements of Gault et al. (1963). The gravity regime constant (C_2) was estimated by matching the velocity distributions for the strength and gravity regimes at the strength/gravity transition. C_2 for ice was approximated by the value for rock. The value for dry soil was derived from cratering experiments in sand (Stoffler et al. 1975). See Housen et al. (1983) for details.

Equation (4) and the values of C_3 shown in Table I imply that for small rocky bodies, where strength scaling predominates, $f_{esc} \simeq 1$ when $D_s \simeq 20$ km (cf. Burns et al. 1984). Hence, satellites larger than these critical values should be capable of developing regoliths. Much larger satellites retain most excavated debris. For example, if the properties of the surface materials on Ganymede lie somewhere between those of ice and rock described above, then a strength/gravity transition may occur near $D \simeq 0.1$ to 1 km. For craters below the transition, only a tiny fraction (of order 0.1%) of the ejecta should escape. Above the transition, progressively larger craters lose a larger fraction. However, even craters as large as ~ 200 km on Ganymede should lose only about 10% of their ejecta.

Even very small bodies (Fig. 5) should retain a significant amount of ejecta if the surface materials are akin to cohesionless soil. For example, on a Phobos-size body, only about 25% of the ejecta from a crater as large as Stickney would escape. The fraction of ejecta which actually escapes during im-

Fig. 5. Phobos regolith. Grooves (perhaps of internal origin) and impact craters dominate the Phobos regolith (Viking Orbiter image 246A06). Vertical scale = 5 km.

TABLE I
Constants in Ejecta Velocity Distributions
(cgs units)

Material	b	C_1	C_2	C_3
Rock	1.7	10^6	0.6	4×10^{11}
Ice	1.7	1.4×10^5	0.6	5×10^{10}
Porous soil	1.2	—	0.4	—

pacts on Phobos probably differs from these estimates, for two reasons. First, tidal effects due to Mars cause the escape speed to vary by roughly a factor of 4, depending on position on the surface and launch direction (Davis et al. 1981). Second, the surface materials on Phobos are probably more cohesive than the dry sand upon which our ejecta velocity estimates are based. Increasing cohesion should result in a larger fraction of escaping ejecta.

3. Ejecta Blankets.   Housen et al. (1983) have shown that, for cratering in the gravity regime, ejecta blankets are geometrically similar; that is, the depth of ejecta at a given range r from the crater is independent of crater size, if the range and depth are expressed in units of the crater radius. Housen et al. also demonstrated that, when r is greater than a few crater radii, the depth of ejecta is proportional to $r^{-\beta}$, where β is a constant that depends largely on the porosity of the cratered material. For porous and nonporous materials, respectively, β is expected to be ~ 2.6 and ~ 2.8. At ranges less than a few crater radii the simple power law does not strictly hold. If a power law is fit locally at range r, then one finds that β increases as r decreases, but is typically in the range $3 < \beta < 4$.

Geometrical similarity of ejecta does not occur outside the gravity regime. For example, in the strength regime, ejecta are desposited relatively farther (in terms of crater radii) from the crater rim for progressively smaller and smaller craters. As a result, in the strength regime the ejecta deposits of small craters tend to be rather tenuous and discontinuous, whereas larger craters exhibit ejecta deposits more like those in the gravity regime, i.e., a significant deposit near the rim and a smooth decrease in ejecta depth with increasing range (at least out to a few crater radii).

At this point it is interesting to note that while well-developed ejecta blankets are commonly observed on Ganymede and Callisto, they are rarely seen on the Saturnian satellites (Smith et al. 1981; Moore et al. 1984). One explanation might be that the observed craters on Ganymede and Callisto were gravity-dominated while those on the Saturnian satellites were strength-dominated. However, as noted earlier, Fig. 4 implies that craters $\gtrsim 1$ km on the larger Saturnian satellites should be formed in the gravity regime. This discrepancy raises the possibility that Fig. 4 underestimates transition diameters for the Saturnian satellites.

4. Monte Carlo Algorithm.   In Monte Carlo simulations of regolith evolution, craters are randomly selected from a prescribed continuous size distribution and positioned randomly on the surface; such a simulation is described in the appendix to the chapter by Chapman and McKinnon. The effects of each crater are recorded for each point in a square grid. A trial in the simulation ends when a prescribed number of craters have been generated. At the end of a trial, the grid of points is used to construct a distribution of regolith depth. An overall distribution is constructed by combining the results of many independent trials.

The volume of nonescaping ejecta is distributed between the crater rim and cut-off range r_c. The range r_c is adopted as $5 D$ to accommodate nearly all ejecta. As noted above, the depth of ejecta varies as $r^{-\beta}$ in the far field, where $\beta \simeq 2.6$ to 2.8. Closer to the crater the exponent increases to about 3 or 4. In the interests of computational efficiency, we adopt a power law ejecta profile with an average exponent $\bar{\beta}$. A nominal value of $\bar{\beta} = 3$ is adopted, but variations from this value are considered.

The crater production functions used in our simulations are shown in Fig. 6. The distributions are presented as "relative plots," in the sense defined by Chapman and McKinnon in their chapter. Note, in these plots $R = D^3 dN/dD$, where dN is the number of craters (per unit surface area) in a diameter interval dD. The distributions in Fig. 6 generally represent the most heavily cratered terrains observed on each satellite. The observed distributions were approximated by fitting straight lines over limited ranges of crater diameter.

The use of observed crater distributions has a major advantage in that it obviates the need for speculative estimates of impactor mass frequency distributions and crater scaling laws. There are, however, two disadvantages in using observed distributions. First, the smallest crater used in a simulation is often limited by the smallest crater observable in images of a satellite surface. Second, as emphasized in the chapter by Chapman and McKinnon, processes such as crater overlap, debris fill, viscous relaxation and endogenic activity may cause the observed crater distribution to differ from the production function.

The first problem can be addressed by considering the effects of reasonable extrapolations of the distributions below the minimum observed diameter. This is discussed in more detail below. Woronow (1978) and Woronow et al. (1982) have addressed the problem of crater saturation and equilibrium on the lunar highlands and the Galilean satellites. They conclude that the observed populations are not the result of saturation. In particular, while some craters may have been destroyed, the obliteration process is such that the shape of the size-frequency distribution is preserved. If a significant number

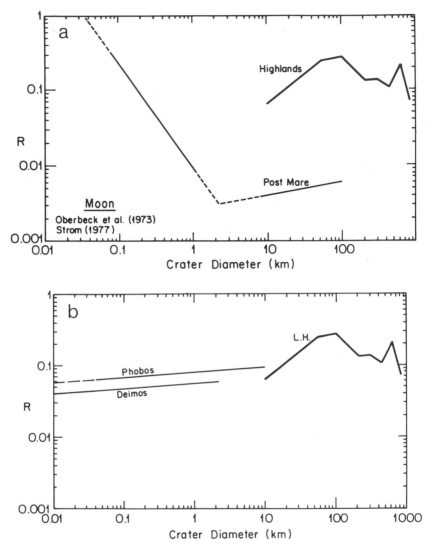

Fig. 6. Four plots which give the crater distributions used in the Monte Carlo simulations of regolith depth (Fig. 7). The distributions are shown as relative plots, as recommended by the Crater Analysis Techniques Working Group (1979). A power-law distribution with a cumulative slope of -2 would plot as a horizontal line. Similar plots are presented in the chapter by Chapman and McKinnon. L.H. in plot b represents the crater distribution of the lunar highlands, also plotted in the other three figures.

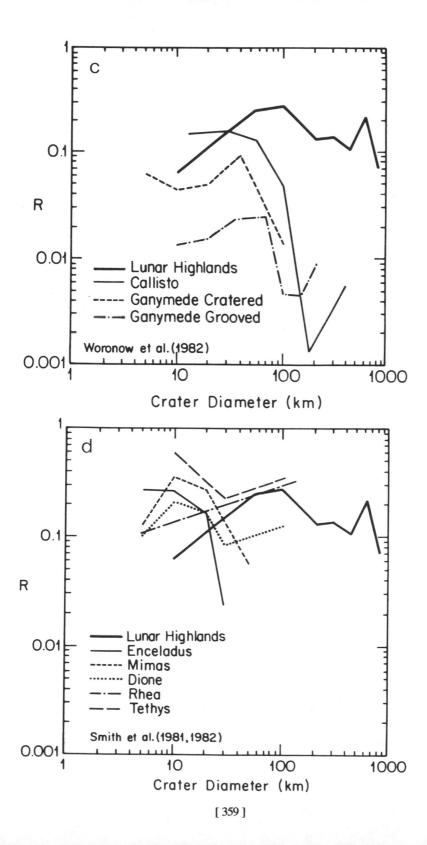

c

Lunar Highlands
Callisto
Ganymede Cratered
Ganymede Grooved

Woronow et al. (1982)

R

Crater Diameter (km)

d

Lunar Highlands
Enceladus
Mimas
Dione
Rhea
Tethys

Smith et al. (1981, 1982)

R

Crater Diameter (km)

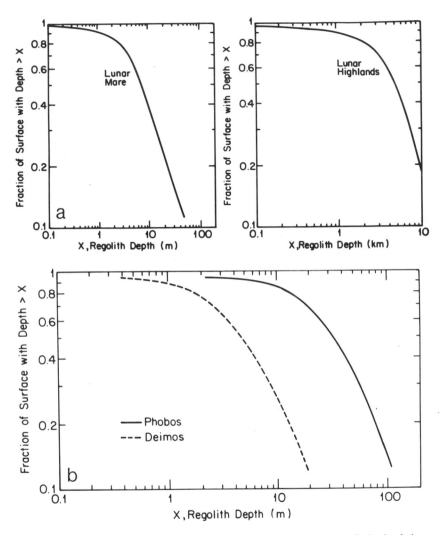

Fig. 7. Five plots showing regolith depth distributions generated in the Monte Carlo simulations, as described in the text and based on the crater distributions given in Fig. 6.

of craters have been obliterated, then the actual regolith may be somewhat deeper than we calculate. This situation could be the case if Hartmann's (1984) observations of satellite crater densities are correct. He argues that the general similarity of maximum crater densities (factor of ~ 2) on many satellites indicates saturation effects on observable crater densities. This issue is dealt with in detail in the chapter by Chapman and McKinnon.

The results of our Monte Carlo simulations are summarized in Fig. 7 and Table II. Figure 7 shows regolith distributions as the fraction of surface area

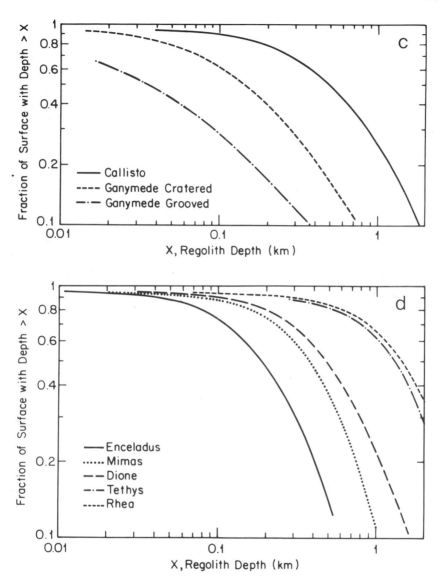

where the regolith depth exceeds a given value. Table II lists the median, mean and standard deviation of the regolith depth distribution for each case. These results are based on the nominal values $r_c/D = 5$ and $\bar{\beta} = 3$ and on the assumption that all escaping ejecta (calculated from Eq. 4) are permanently lost. The effects of reasonable variations in r_c and $\bar{\beta}$ were investigated in one case (lunar highlands) and found to produce fluctuations of only $\pm 20\%$ in the calculated regolith depth. The effects of reaccumulation of debris are discussed below (Sec. III.A.5).

TABLE II
Summary of Regolith Depth Calculations

Case	Median	Mean	Standard Deviation
Lunar highlands	5.5 km	7.1 km	6.2 km
Lunar maria	9 m	50 m	230 m
Phobos	35 m	54 m	56 m
Deimos	5 m	10 m	13 m
Callisto	540 m	870 m	1100 m
Ganymede (cratered)	160 m	320 m	460 m
Ganymede (grooved)	45 m	170 m	440 m
Mimas	390 m	500 m	430 m
Enceladus	200 m	270 m	250 m
Tethys	1400 m	1600 m	1300 m
Dione	500 m	740 m	770 m
Rhea	1500 m	1900 m	1600 m

Figure 7 shows that the regolith on the lunar highlands should typically be several kilometers deep, although large variations in the depth can be expected. For example, the regolith depth should be > 1 km over 90% of the surface and > 10 km over 20% of the surface. These estimates are consistent with the findings of Cashore and Woronow (1984). On mare surfaces, the regolith is much thinner than on the highlands, due to the relatively low crater densities (see Fig. 6). According to our simulations, the median depth on the maria should be about 9 m, consistent with seismic determinations, but about a factor of 3 larger than the depth calculated by Oberbeck et al. (1973). The difference arises because Oberbeck et al. did not include craters $\gtrsim 1$ km in their simulations; one can show that exclusion of such craters reduces the median regolith depth by about a factor of 3.

For Phobos and Deimos the calculated median regolith depths are 35 m and 5 m, respectively. The regolith on Deimos is expected to be thinner than on Phobos because both the largest crater and the overall crater density are smaller on Deimos.

From images of the satellites, Veverka and Thomas (1979) found that the regolith on Phobos may reach depths of ~ 100 m, at least locally. Our simulations imply that such depths should occur over only $\sim 10\%$ of the surface. However, if all ejecta are reaccumulated by Phobos (see Sec. III.A.5), then the median depth increases to ~ 40 m and 20% of the surface is covered by at least 100 m of debris. Note that the regolith may be even deeper than indicated in our simulations if a significant number of large craters have been obliterated.

The regolith on Callisto is calculated to be much thicker than that on Ganymede. At first thought, one might guess that the reverse would be true, because Ganymede is located closer to Jupiter, where the impacting flux should have been higher. However, the crater density is actually observed to be higher on Callisto. As noted by Woronow et al. (1982), Ganymede must have either started recording cratering events later than Callisto and/or experienced a resurfacing event more recently (see the discussion in the chapter by McKinnon and Parmentier).

Shoemaker et al. (1982) used an approximate analytic model and estimated the median regolith thickness on the polar grooved terrain of Ganymede to be 29 m. This is somewhat lower than our estimate, due in part to differences in the assumed crater size distribution. However, even if the Shoemaker et al. crater population is used, our Monte Carlo algorithm predicts a median thickness of ~ 65 m.

Calculated regolith depths vary by about a factor of 3 among the Saturnian satellites, reflecting differences in the crater populations and, to a much lesser extent, the sizes of the satellites. Rhea, Tethys and Dione should have relatively thick regoliths due to the relative abundance of large craters, i.e., the so-called population 1 craters identified by Smith et al. (1982). The low crater density on Enceladus, evidently due to recent resurfacing, gives it the thinnest regolith of the five Saturnian satellites considered. Note that we have simulated the most heavily cratered regions on Enceladus. The crater density on the smooth plains is roughly a factor of 10 to 50 less than on the heavily cratered regions shown in Fig. 6d (Smith et al. 1982). In these areas, the regolith may only be a few tens of meters thick.

5. Reaccumulation of Debris. In order to investigate the effects of reaccumulation of debris, a series of simulations was performed, assuming no ejecta escaped. The results are summarized in Table III. In general, the effect of retaining all debris is not large, primarily because in nearly all cases shown in Fig. 7 and Table II, only a small amount of ejecta (≤ 10 to 20%) escaped.

One might argue that assuming all ejecta are directly retained is not equivalent to assuming that debris are reaccumulated. For example, consider the scenario discussed by Davis et al. (1981) where the time required to reaccumulate debris is much longer than the time between cratering events. Under such circumstances, the regolith might be thicker than in the case where all ejecta are immediately returned because, delaying the return of ejecta means there will be less debris to shield the surface from further excavations into bedrock. However, this effect will be important only if the regolith in the nominal cases is deep enough to shield the surface from large impacts, which are generally the major contributors to regolith growth. For example, consider Phobos, where the median depth was calculated to be ~ 35 m. The buffering effect of the regolith will be small for all craters larger than a few hundred meters in diameter. These craters represent $> 95\%$ of the volume excavated by all sizes of craters. Hence, the growth of the regolith on Phobos does not depend on whether or not debris are returned immediately or only after a long delay. Furthermore, this holds true for all the cases studied.

TABLE III
Regolith Depth when All Debris are Retained

Case	Median	Mean	Standard Deviation
Lunar highlands	6.4 km	8.6 km	7.9 km
Phobos	41 m	66 m	73 m
Deimos	5.5 m	10 m	15 m
Callisto	550 m	890 m	1200 m
Ganymede (cratered)	160 m	320 m	460 m
Ganymede (grooved)	46 m	170 m	450 m
Mimas	410 m	530 m	460 m
Enceladus	200 m	280 m	260 m
Tethys	1400 m	1700 m	1400 m
Dione	520 m	780 m	810 m
Rhea	1600 m	2000 m	1700 m

B. Volcanic, Tectonic and Other Related Processes

Io provides the most striking example of a satellite where internal processes clearly dominate external ones in determining the nature of the present regolith.

Volcanic eruptions on Io (chapter by Nash et al.) apparently include both lavas and volatile-rich plumes. McEwen and Soderblom (1983) have suggested that the plumes can be divided into two categories: larger, shorter-lived plumes with ejection velocities ~ 1 km s^{-1} which form dark deposits over areas up to 1500 km in diameter and smaller, longer-lived plumes with lower eruption velocities, which deposit bright material over areas up to 500 km in diameter. Plume activity on Io may be sufficient to obliterate larger impact craters (Schaber 1980), but the morphology or eventual thicknesses of plume deposits remain unknown. Much of the topography of Io appears related to lavas and pyroclastic flows rather than plume deposits. A significant fraction of these materials must be silicates to support the observed topography (Clow and Carr 1980). The lava flows (described by Schaber [1980] and by Carr et al. [1979]) extend up to 200 km; the rates of eruption must be substantial, given that they have not been obliterated by plume deposits or impact craters. Although sulfur flows may show distinctive surface folding and diapirs, such diagnostic features occur at scales far too small to be detected in Voyager images (Greeley et al. 1984). The flows (whatever their composition) can be expected to be rough and variegated.

Ice volcanism (whatever the eruption style) has significantly resurfaced Europa, Ganymede and Enceladus. Europa has experienced the most extensive activity, yet no eruption centers or flows have been detected in the Voyager images (chapter by Malin and Pieri). However, Cook et al. (1983) have presented evidence that could be interpreted as indicating ongoing plumelike activity. The age of Europa's surface derived from crater densities is $\sim 10^8$ yr, but the actual resurfacing age may be considerably older (Lucchitta and Soderblom 1982). The erupted material is unlikely to be pure water and contaminants should be expected.

The resurfacing of Ganymede, associated largely but not exclusively with grooved terrain, dates to a much earlier period than that on Europa and involves only some 40% of the satellite's area (chapters by Squyres and Croft and by McKinnon and Parmentier). The resurfacing is intimately associated with extensional tectonics, now inactive. Some low, smooth areas appear to be true flows, presumably of water ice. The grooved terrain has experienced a history of faulting, infilling, more faulting, viscous relaxation, and some modification by subsequent impacts (Shoemaker et al. 1982; chapter by Squyres and Croft).

Enceladus too has experienced considerable resurfacing (Smith et al. 1982; chapter by Morrison et al.) but the resolution of Voyager images is not sufficient to reveal details of the mechanism; presumably some form of water volcanism is involved (see the chapters by Schubert et al. and by Peale). There is strong evidence that, in the case of Enceladus, resurfacing processes are still active (Haff et al. 1983; Pang et al. 1984a; Buratti et al. 1983).

Tectonic processes provide the means whereby new magmas (rock or ice) can reach the surface, and generate the topography on which mass wasting can operate. There is considerable evidence of tectonic activity on Io, and extensional tectonics are common features of many icy satellites. The causes are apparently quite different on Ganymede, Europa and smaller icy satellites such as Tethys (chapter by Squyres and Croft).

Mass wasting and downslope transport must occur on all satellites; its significance may vary widely, depending on surface composition and environmental conditions such as temperature and gravity. The process is certainly of major importance on Deimos, where many craters have been partly filled by debris and large albedo features have been formed by movement of regolith downslope. Amalthea may have similar processes (Thomas and Veverka 1982), and at least one manifestation of mass movement in a crater wall on Phobos was noted by Thomas and Veverka (1980b). In most instances, image resolutions better than those achieved by Voyager for the Galilean satellites of Jupiter and the satellites of Saturn are needed to detect signs of mass wasting. It is very likely that Galileo will detect evidence of mass wasting on some of the pronounced topography on Io. Mass wasting is also expected to have modified groove topography on Ganymede (Shoemaker et al. 1982).

Satellite surfaces must be inhomogeneous in vertical section. Certainly, on satellites such as Io, where internal processes are dominant, one must expect the outer surface to be very layered. This will also be true on satellites with impact-dominated regoliths, since such regoliths will consist of interleaved ejecta deposits in which the effects of the largest craters will be domi-

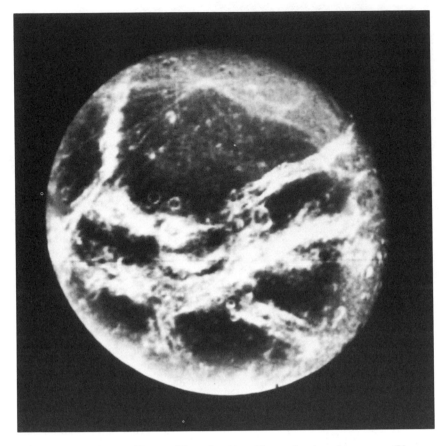

Fig. 8. Wispy albedo markings on Dione showing evidence of endogenic processes (Voyager
image 118S1-001).

nant (see chapters by McKinnon and Parmentier and by Chapman and
McKinnon). Evidence of regolith layering is abundant on the Moon (Lindsay
1976). On Phobos, at least one crater wall shows conspicuous layering. On
Deimos, the evidence is less direct, but quite conclusive. It involves the downslope motion of a brighter albedo material over a darker substrate, as well as
the presence of small dark halo craters which are explained best, as in the case
of their lunar analogs (Bell and Hawke 1982), in terms of a layered regolith
(Thomas and Veverka 1980b). Significant layering at depth is evidenced by
the dark ejecta blankets of some craters on Ganymede (Shoemaker et al.
1982; chapter by McKinnon and Parmentier). Here Galileo will provide new
important data by allowing the mapping of small craters with bright or dark
ejecta blankets on the satellite. On Callisto there is some suggestion in the
albedo patterns associated with craters that subsurface ice may be cleaner

than that at the surface. Albedo patterns on Dione suggest some form of endogenic activity (Fig. 8).

For most satellites we can ignore the role of atmospheric effects on the evolution of regoliths. Titan, Io and possibly Triton are exceptions. In the case of Io, it is almost certain that sublimation/condensation of SO_2 frost affects the texture of the regolith in some areas (Matson and Nash 1983). There is also evidence that the regolith of the polar regions on Ganymede may have been modified by a net poleward migration of water molecules (Shaya and Pilcher 1983).

IV. REMOTE SENSING OF SATELLITE REGOLITHS

We now summarize some of the principal developments in the remote sensing of satellite regoliths. Imaging data, which provide the bulk of our information on surface morphology and at present the most direct clue to the satellites' geologic evolution, are not treated explicitly. Rather, we concentrate on reviews of photometric measurements, polarimetric observations, radiometry, and microwave and radar data. Inferences about surface composition from spectral reflectance and related measurements are dealt with in the chapter by Clark et al.

A. Photometry of Satellite Surfaces

Several important breakthroughs that have occurred during the past decade are reviewed in this section. First is the introduction of various photometric functions that are suitable for dealing with the bright surfaces of satellites in the outer solar system (Sec. IV.A.1, below). Second is the fact that spacecraft exploration has made it possible for the first time to study the photometric properties of individual areas on many satellites and to extend our observations to very high phase angles not achievable from Earth. Spacecraft observations have also yielded accurate determinations of satellite radii, data that are essential for determining accurate albedos from available measurements of brightness (Sec. IV.A.2).

As a result of spacecraft exploration and a concentrated effort on the part of Earth-based observers, we now have crucial photometric data on many more satellites than was the case in the mid 1970s. For example, through the work of Degewij et al. (1980), we now have data on the photometric characteristics of the small outer satellites of Jupiter, while Brown (1982,1984) has provided an extensive survey of the photometric characteristics of the Uranian satellites. From spacecraft we now have data on satellites that are difficult to observe from Earth, either due to their proximity to the planet (e.g., Mimas and Amalthea) or because of their faintness (e.g., the small satellites of Saturn) (Sec. IV.A.3).

Another area of recent progress involves the investigation of photometric functions through laboratory measurements. Attempts have been made to test various photometric models directly against the measured scattering properties of samples of known physical characteristics. Extensive work has been done on sulfur and on candidate asteroid and Mars surface materials. Mostly due to experimental difficulties in making and preserving frost surfaces of specific textures (Sec. IV.A.1, below), little effort has been devoted to ice-rich surfaces.

1. Photometric Functions.   It has been long realized that information on the physical characteristics of a surface is contained in the manner in which the surface scatters incoming sunlight. The albedo of the surface is fundamentally determined by the optical properties of the particles that make it up, while the angular characteristics of the scattering provide information on the small-scale texture of the surface as well as on the degree of large-scale roughness. For example, if we let $\Phi(\alpha)$ be the disk-integrated brightness of a satellite expressed as a function of the phase angle α, it is well known that the behavior of $\Phi(\alpha)$ near opposition ($\alpha \sim 0°$) is a good indicator of porosity, whereas the behavior at very large α depends strongly on the importance of shadows, and therefore on the degree of large-scale roughness. Thus, the scattering properties of surfaces contain information on their physical characteristics.

If we consider a small element of surface area (Fig. 9), three angles are relevant in discussions of how this element scatters light. They are i, the angle of incidence; ε, the angle of emission; and α, the phase angle. Given a certain amount of incident sunlight, the aim is to write down an expression for $I(i,\varepsilon,\alpha)$, the intensity of scattered sunlight under the conditions shown in Fig. 9. Such a description of the scattering is called the photometric function of the surface. A fundamental aim of photometry is to provide a simple analytical description of the photometric function (or scattering law) in terms of the three geometric parameters i, ε and α. The use of such a photometric function is necessitated since it is out of the question to observe any particular surface over the full ranges of each of the three parameters. From Earth, for instance, in the case of satellites other than the Moon, only disk-averaged observations are possible (averaging over i and ε at constant α) and usually the coverage in phase angle is very restricted. For example, for the satellites of Jupiter, α observable from Earth never exceeds 12°. Spacecraft make it possible to obtain disk-resolved measurements and to cover a much wider range of phase angles. Yet, even here there are restrictions, especially for flyby missions such as Voyager. Thus, for practical purposes it is essential to have a manageable photometric function to interpolate between, and extrapolate from, the observations to other combinations of the parameters i, ε and α.

Fortunately, there have been important developments in this area. Some twenty years ago, in an era dominated by studies of the Moon, photometric functions were developed by Hapke (1963) and Irvine (1966) which correctly described the scattering from dark, porous surfaces like that of our Moon.

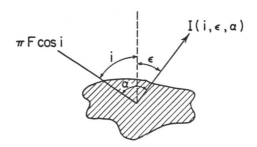

Fig. 9. Definition of photometric angles. The solar flux incident at an angle i with respect to the normal (dashed) is conventionally denoted by $\pi F \cos i$. The intensity of light scattered at an emission angle ε (phase angle α) is $I(i,\varepsilon,\alpha)$.

Recently, workable scattering laws have been developed for surfaces of arbitrary albedo.

On the empirical side, Buratti (1983) has shown that a simple function using the single parameter A of the form

$$I(i,\varepsilon,\alpha) \simeq \frac{A \cos i}{\cos i + \cos \varepsilon} f(\alpha) + (1 - A) \cos i \qquad (5)$$

can be used to fit the observed photometric properties of the icy satellites of Jupiter and Saturn rather well (Fig. 10). Here $f(\alpha)$ is an arbitrary function of the phase angle, and the parameter A determines the relative contributions of the two parts of the empirical scattering law. For $A = 1$ the scattering is lunar-like, while for $A = 0$ it follows Lambert's law. Equation (5) provides a useful fit to the icy satellite data (Fig. 10), but is empirical and not based on a rigorous derivation (see also Meador and Weaver 1975). Buratti has found that there is a close relationship between the value of A and the albedo (B_0) of an icy satellite (Fig. 11). For Enceladus, for example, $A = 0.3$, whereas for a satellite even as bright as Ganymede, A is close to 1. In general, Buratti's results show that even for icy satellites with geometric albedos as high as 0.5, A is approximately unity in the above equation and the scattering follows a lunar-like law.

Scattering laws for particulate surfaces of arbitrary albedo have been derived from radiative transfer principles by several authors (Lumme and Bowell 1981; Goguen 1981; Hapke 1981). Hapke's scattering law can be written in the form

$$I(\mu_0,\mu,\alpha) = F \frac{\tilde{\omega}_0}{4} \frac{\mu_0}{\mu + \mu_0} [S(\alpha,h)P(\alpha,g) + H(\mu)H(\mu_0) - 1] \qquad (6)$$

where πF = incident solar flux at $i = 0°$, $\mu = \cos \varepsilon$, $\mu_0 = \cos i$, $\tilde{\omega}_0$ = single particle albedo, $S(\alpha,h)$ = shadowing function, and $P(\alpha,g)$ = single particle

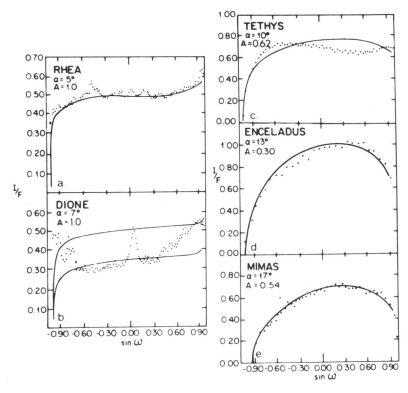

Fig. 10. Scans along the photometric equators of the five medium-sized Saturnian satellites at the lowest phase angle observed by Voyager. Fits of the empirical photometric law described by Eq. (5) are shown and the derived values of the parameter A indicated (figure from Buratti 1984).

scattering function. For convenience, the single particle scattering function can be approximated by a Henyey-Greenstein phase function

$$P(\alpha, g) = \frac{1 - g^2}{(1 + g^2 + 2g \cos \alpha)^{3/2}} \qquad (-1 < g < +1) \qquad (7)$$

where g is the asymmetry parameter $\langle \cos \alpha \rangle$. The parameter h is called the compaction parameter, and is related to the porosity of the surface (Hapke 1981).

Hapke and Wells (1981) as well as Gradie and Veverka (1984a) have shown that Equation (6) fits laboratory data well. Goguen (1981) successfully fitted his version of a similar equation to the scattering properties of a series of particulate surfaces with reflectances ranging from 0.03 to 1.0. Buratti (1983,1985) and Helfenstein et al. (1984) have shown that Hapke's equation provides a good fit to Voyager observations of the icy satellites of

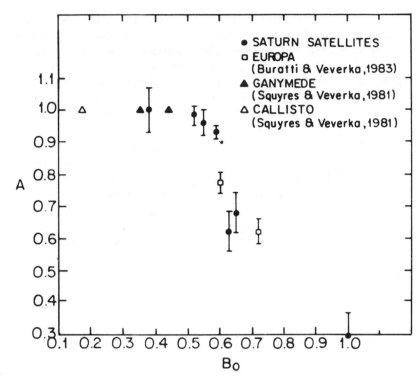

Fig. 11. A plot of the parameter B_0 at the lowest phase angle for which Voyager data are available against the parameter A (Eq. 5) for selected satellites of Saturn (Mimas, Enceladus, Tethys, Dione and Rhea) and for Europa, Ganymede and Callisto. For Europa, Dione and Rhea, data for dark and bright regions are plotted separately (figure from Buratti 1983).

Jupiter and Saturn. One interesting conclusion is that rocky regoliths and icy regoliths can have approximately similar compaction parameters (Buratti 1985).

Due to its simplicity and ease of application, a few words should be said about Minnaert's "law" according to which

$$I(\mu_0,\mu,\alpha) = FB_0(\alpha)\mu_0^{k(\alpha)}\mu^{k(\alpha)-1} \tag{8}$$

where B_0 and k are the two Minnaert parameters which are functions of α and of the wavelength λ (see Minnaert 1961). It is well known that Eq. (8) is only a crude approximation to the scattering properties of real surfaces (see, e.g., Goguen 1981), and that its indiscriminate application can lead a novice photometrist to confusion and specious results. The problem is that when Eq. (8) is fitted to the scattering properties of real surfaces at phase angles in excess of about 30°, the value of k derived is not unique, but depends on what subset of points on the planet's disk is used. For example, Goguen (1981) discusses a case in which, at a phase angle of 90°, points along the photometric equator lead to a value of $k = 1.0$, whereas points along a photometric meridian give $k = 1.2$. This problem of nonuniqueness disappears as the phase angle approaches zero, and Minnaert's law can be used as a convenient approximation (except in the immediate neighborhood of the limb) to handle small phase angle data. For instance, it has been used successfully by both McEwen and Soderblom (1984) and Simonelli and Veverka (1983) to analyze Voyager observations of individual areas on Io.

The ability to fit observed photometric data with simple functions serves several practical purposes. For example, such a procedure is necessary in determining geometric albedos and normal reflectances for limb-darkened objects such as Europa, Io and Enceladus. Such functions can also be used to interpolate between observations, or to extrapolate data to large phase angles, as in the case of determining the radiometric albedos of the white regions on Io from Voyager measurements (Simonelli and Veverka 1984b). Fitting observations to a photometric function provides a powerful method of comparing the photometric characteristics of the surfaces of various satellites, as well as intercomparing the characteristics of different regions on an individual satellite. Such comparisons can be carried out in terms of the fitted parameters on which the photometric function depends, without necessarily attaching any precise interpretation to the parameters themselves. Thus, one can compare the Minnaert limb-darkening parameter k near opposition for different satellites (Buratti 1983), or the multiple scattering parameter Q defined by Lumme and Bowell (1981), or the parameter A which appears in Eq. (5) (cf. Fig. 11). Such comparisons provide important insights even though one may not know how to connect precisely a given model parameter to a measurable physical characteristic of the regolith.

By far the most important use that could be made of photometric data would be to infer actual physical parameters for the regoliths observed. There has been much debate on how photometrically derived model parameters relate to the actual characteristics of a surface, but very little conclusive work has been accomplished to date. Most past attempts to infer regolith parameters from photometry are based on fitting to disk-integrated data. Caution has been voiced by Veverka (1977a) and by Gradie and Veverka (1984a), among others. While such procedures might be valid for homogeneous objects, we now know that most satellites have photometrically varied surfaces. For example, it is far from clear what meaning can be attached to parameters derived by fitting disk-integrated observations of a heterogeneous object like Io.

Another fundamental issue is how one deals with surface roughness on scales larger than a wavelength. Hapke (1984) and others have attempted to deal theoretically with this crucial roughness problem in an approximate way; some laboratory simulations have also been made, e.g., by Buratti (1983). While it is easy to demonstrate that large-scale roughness has a significant effect on the photometric properties of a surface, the inverse problem of the unique detection of its signature remains. In general terms, it is clear that to an acceptable approximation, the effects of large-scale roughness should be independent of wavelength, and that they should be especially important at large phase angles. Whether one can extrapolate meaningfully fits to data valid at small and moderate phase angles to large phase angles (as is necessary in the evaluation of phase integrals) without modeling large-scale surface roughness remains to be established. Some initial laboratory work has been done; for example, Buratti (1983) has shown that shadowing effects due to large-scale roughness are much less important for bright surfaces than they are for dark ones, as might be expected given that, on bright surfaces, shadows are more diluted by multiple reflections.

2. Photometry Using Spacecraft Observations. Photometry of Phobos and Deimos obtained during the Viking missions to Mars was analyzed extensively by Klaasen et al. (1979), and reviewed by Veverka and Burns (1980) and French (1980). Using observations ranging in phase angle from $0°5$ to 122°, Klaasen et al. confirmed the earlier conclusion (based on Mariner 9 data) of Noland and Veverka (1977*a,b*) that Phobos and Deimos scatter light in essentially a lunarlike fashion. These authors also established the values of the phase integrals for the two satellites: $q = 0.27$ for Phobos and $q = 0.32$ for Deimos. Radiometric Bond albedos of about 0.02 were derived. Goguen et al. (1978) analyzed the photometric properties of dark markings seen on the floors of many craters on Phobos, and concluded that this dark material had a similar composition but a much rougher texture than the average surface of the satellite.

Voyager photometry of the Galilean satellites has extended observations to large phase angles and determined the scattering properties of individual areas on the satellites. For example, Pang et al. (1981) studied Voyager photopolarimeter observations of Io and Ganymede out to phase angles of 40°. Simonelli and Veverka (1984*a*) analyzed disk-integrated Voyager observations of Io between phase angles of 2° and 159°, determining directly a radiometric Bond albedo of 0.5 ± 0.1, a value consistent with, but slightly below, previous estimates (Fig. 12). In general, Simonelli and Veverka found good agreement between Voyager and Earth-based observations at small phase angles where the two data sets overlap. A similar conclusion was reached by Buratti and Veverka (1983) in the case of observations of Europa (Fig. 13). The Voyager data for Europa, extending in phase angle from $2°9$ to 143°, yielded a phase integral of 1.09 ± 0.11, and a radiometric Bond albedo of 0.62 ± 0.14 (Buratti and Veverka 1983). Squyres and Veverka (1981) analyzed Voyager data for Ganymede and Callisto between phase angles of 10° and 124° and estimated phase integrals of 0.8 and 0.6 for Ganymede and Callisto, respectively; the corresponding values of the radiometric Bond albedos were derived to be 0.35 and 0.13.

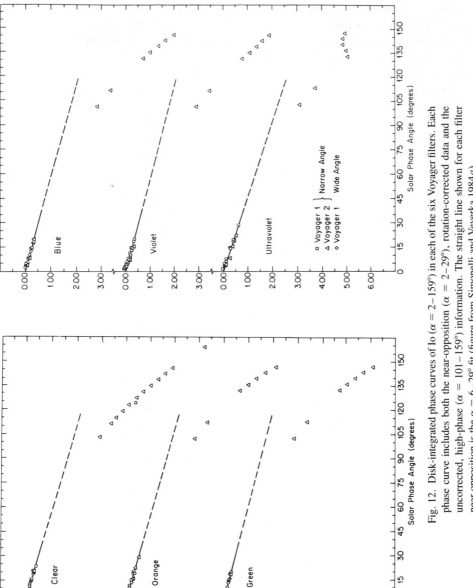

Fig. 12. Disk-integrated phase curves of Io ($\alpha = 2-159°$) in each of the six Voyager filters. Each phase curve includes both the near-opposition ($\alpha = 2-29°$), rotation-corrected data and the uncorrected, high-phase ($\alpha = 101-159°$) information. The straight line shown for each filter near opposition is the $\alpha = 6-29°$ fit (figure from Simonelli and Veverka 1984a).

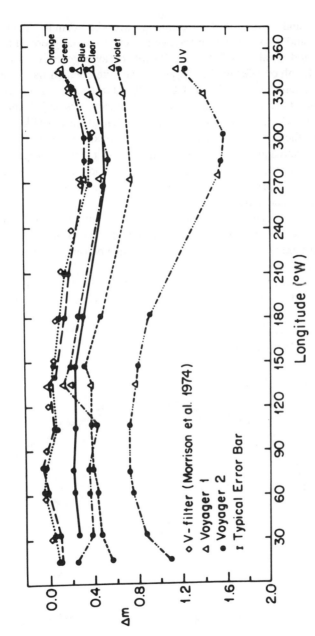

Fig. 13. Six-color Voyager observations of Europa as a function of orbital longitude (corrected for variations in solar phase angle). Groundbased V-filter observations have been included for comparison (figure from Buratti and Veverka 1983).

Squyres and Veverka (1981) estimated average normal reflectances of
0.35 ± 0.07 and 0.44 ± 0.10 for the average cratered and grooved terrain on
Ganymede, and a value of 0.18 ± 0.04 for the average cratered terrain on
Callisto. (All values refer to the Voyager clear filter, which has an effective
central wavelength near 0.47 μm.) On both satellites, the brightest craters
were estimated to reach reflectances of up to 0.7. Squyres and Veverka found
that for most surface areas on Ganymede and Callisto a lunarlike scattering
law is a satisfactory approximation. A distinctly different conclusion was
reached by Buratti and Veverka (1983) in the case of the much brighter surface
(average normal reflectance about 0.7) of Europa; here a lunarlike scattering
law is quite inappropriate and a value of the parameter A (in Eq. 5) less than 1
must be used to match the observation. Another indication of this departure is
that Europa, unlike Ganymede and Callisto, is significantly limb-darkened
near opposition. For example, Buratti and Veverka (1983) derived a value of
the Minnaert parameter $k = 0.7$ for Europa at 2°9 phase and a wavelength of
0.58 μm (Fig. 14).

The analysis of disk-resolved Io data has been a much more complex
process, since this satellite has a much more heterogeneous surface than any
of its Galilean neighbors. Disk-averaged limb-darkening parameters near opposition were derived by McEwen and Soderblom (1983), while Clancy and
Danielson (1981) obtained preliminary values for the major types of surface

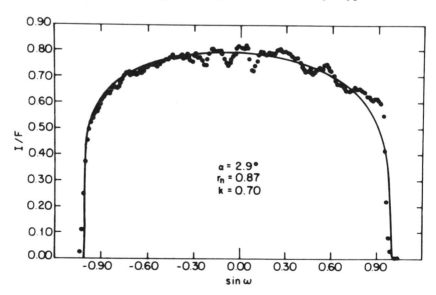

Fig. 14. A photometric scan across Europa (Voyager 1 orange filter) at a small phase angle ($\alpha =$
2°9) showing a significant degree of limb darkening. The fitted curve corresponds to a limb
darkening parameter $k = 0.70$. For a lunarlike surface, there would be almost no limb darkening ($k = 0.5$), while a Lambert surface would give $k = 1$ (figure from Buratti and Veverka
1984).

TABLE IV
Previous Determinations of the Limb Darkening of Io[a]

Solution Technique	Type of Regions	k	Reference and Phase Angle
Single-image	white	0.6 ± 0.1[b]	Clancy and Danielson (1981) $\alpha \approx 10°$
	yellow orange red black	0.5 ± 0.3	
	brown	0.8 ± 0.1	
Two-image	global averages	$0.35\ \mu m: 0.44 \pm 0.009\alpha$[c] $0.41\ \mu m: 0.49 \pm 0.006\alpha$ $0.48\ \mu m: 0.52 \pm 0.005\alpha$ $0.54\ \mu m: 0.53 \pm 0.005\alpha$ $0.59\ \mu m: 0.55 \pm 0.005\alpha$	McEwen and Soderblom (1984) $\alpha = 3–15°$

[a] Table from Simonelli and Veverka (1985).
[b] No variation in k with wavelength was observed for any color class by Clancy and Danielson.
[c] In the $k(\alpha)$ formulas of McEwen and Soderblom, α is in degrees.

areas. McEwen and Soderblom's values, reproduced in Table IV, are all less than 0.57, values which are surprisingly low given that Io's reflectance reaches 0.6 to 0.7 near 0.6 μm (Simonelli and Veverka 1984b). Gradie and Veverka (1984a) have shown that large-scale surface roughness could provide an explanation for such low values and for the fact that the k's show little wavelength dependence despite Io's reflectance changing strongly with wavelength.

Since there is continuing interest in matching the spectral reflectance properties of areas on Io to those of samples measured in the laboratory (Nash and Nelson 1979; Soderblom et al. 1980; Sill and Clark 1982; McEwen and Soderblom 1983; Gradie and Veverka 1984a; chapters by Nash et al. and by Clark et al.), it is important to determine the limb-darkening parameters for individual areas on Io (as a function of wavelength) as accurately as possible. In general, the relationship between the geometric albedo p and the normal reflectance r_n is given by

$$r_n = (k + 1/2)p \qquad (9)$$

so that unless k is exactly 0.5, the two quantities will be different (Veverka et al. 1978).

A particularly intriguing aspect of Io disk-resolved photometry is the possibility of temperature-driven brightness changes. Ordinary sulfur has been suggested as a significant constituent of Io's surface and it is known from laboratory measurements that the reflectance spectrum of ordinary sulfur is temperature-dependent, especially in the region between 0.4 and 0.5 μm.

Gradie and Veverka (1984a) have shown that this effect could be detectable at a level of a few percent in violet filter images obtained by Voyager. It has been difficult to demonstrate conclusively the presence or absence of this phenomenon, given the rudimentary state of our understanding of the photometric functions of individual areas on Io. The effect should lead to a slight increase in brightness of a region near the limb, relative to its appearance near the central meridian, due to the slightly lower temperatures at the edges of the disk. Similar considerations suggest that, if ordinary sulfur were a major surface constituent, then Io should show a slight anomalous brightening when it reappears from eclipse. The Voyager scientists searched for, but did not find, any such post-eclipse brightening of Io's disk-integrated light (Veverka et al. 1981a).

Voyager observations of Amalthea are analyzed by Veverka et al. (1981b), and reviewed by Thomas and Veverka (1982) and by Veverka et al. (1982b); the latter also includes a summary of Voyager's observations of the other three known small inner satellites (Metis, Adrastea and Thebe). Consistent with previous Earth-based observations (see below), Amalthea was found to be a very dark, very red object. The phase integral does not exceed 0.3, and the radiometric Bond albedo is less than 0.02.

The phase curves of the larger icy satellites of Saturn were derived from Voyager observations by Buratti and Veverka (1984). The phase angle coverage extends typically to about 70°, and in the case of Mimas to 133°. The phase integrals range from about 0.7 for Rhea to 0.9 for Enceladus. For Tethys and Enceladus, the radiometric Bond albedos derived from photometry agree well with those derived from Voyager IRIS radiometry (Table V); in the case of Rhea, an as yet unexplained discrepancy exists. The very high radiometric albedo of Enceladus (~ 0.9) is noteworthy.

In many cases, Buratti and Veverka (1984) found good agreement between Voyager disk-integrated data at low phase angles and Earth-based observations. But the geometric albedo for Mimas, $p_v = 0.77 \pm 0.15$ (corresponding to a mean opposition magnitude of $V_0 = + 12.5$) is definitely higher than the previously generally accepted value. However, the Voyager result is in excellent agreement with results by Franz and Millis (1983), which gave $V_0 = + 12.4$.

Buratti (1983,1984) found that for the brightest satellites a lunarlike scattering law does not hold, and that considerable limb darkening is involved at low phase angles (see Fig. 15 and Table VI). Buratti found that Eq. 5 was adequate to fit observed brightness profiles in all cases. She also fitted Hapke's equation (Eq. 6) to her data for Mimas, Enceladus, Rhea and Europa and showed that the textures of some icy regoliths are similar to that of the surface of the Moon—the texture being inferred from the value of Hapke's compaction parameter (Table VII).

One of the most startling results of the Voyager photometry of Saturn's icy satellites concerned Enceladus. Not only has this satellite a very high al-

TABLE V
Bond Albedos for Some Outer Planet Satellites

Satellite	From Voyager Photometry[a]	From Voyager Infrared (IRIS)[b]
Io	0.50 ± 0.1	—
Europa	0.62 ± 0.1	—
Ganymede	0.35	—
Callisto	0.13	—
Rhea	0.45 ± 0.1	0.65 ± 0.03
Dione	0.45 ± 0.1	—
Tethys	0.60 ± 0.1	0.73 ± 0.05
Enceladus	0.90 ± 0.1	0.89 ± 0.02
Mimas	0.60 ± 0.1	—

[a] Data for the Saturn satellites are from Buratti and Veverka (1984); those for Ganymede and Callisto are from Squyres and Veverka (1981); for Io, from Simonelli and Veverka (1984a); for Europa, from Buratti and Veverka (1983).
[b] After Cruikshank et al. (1983a).

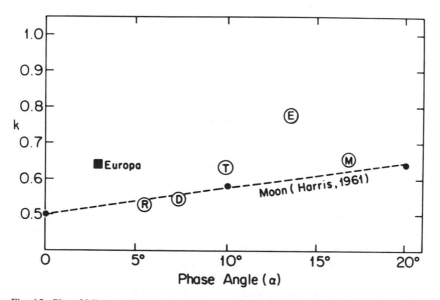

Fig. 15. Plot of Minnaert k's of the Saturnian satellites and Europa compared with the lunar k. Bright satellites have values for k significantly higher than the Moon's, whereas the moderately bright Rhea and Dione closely follow lunar behavior (figure from Buratti 1984).

TABLE VI
Minnaert Parameters for the Saturn Satellites[a]

Object	α		k	B_0
Rhea	5.5	Bright areas	0.54 ± 0.01	0.59 ± 0.01
		Dark areas	0.52 ± 0.01	0.52 ± 0.01
Dione	7.1	Bright areas	0.56 ± 0.02	0.55 ± 0.01
		Dark areas	0.53 ± 0.02	0.38 ± 0.03
Tethys	9.9		0.63 ± 0.02	0.64 ± 0.02
Enceladus	13.4		0.78 ± 0.04	1.06 ± 0.06
Mimas	16.8		0.65 ± 0.07	0.63 ± 0.02

[a] Voyager clear filter; effective passband centered near 0.47 μm. Table from Buratti (1984).

TABLE VII
Values of $\bar{\omega}_0$, h, g, and Mean Slope Angle for Hapke's Theory[a]

Satellite	Single Particle Scattering Albedo $\bar{\omega}_0$	Hapke Compaction Parameter h	Asymmetry Factor g	Slope Angle
Moon	0.25 ± 0.02	0.4 ± 0.1	-0.25 ± 0.02	
Europa	0.97 ± 0.01	1.0 ± 0.2	-0.15 ± 0.04	$23°$
Mimas	0.93 ± 0.03	0.7 ± 0.2	-0.30 ± 0.05	$30°$
Enceladus	0.99 ± 0.02	0.4 ± 0.2	-0.35 ± 0.03	
Rhea	0.76 ± 0.03	0.4 ± 0.1	-0.35 ± 0.05	

[a] Table from Buratti (1983,1985).

bedo, but the albedo, color and photometric function properties are uniform to a very high degree over the entire face imaged by Voyager 2 (Buratti and Veverka 1984), in spite of the fact that terrains of geologically very different ages are involved. The situation suggests that the observed photometric properties result from a thin, ubiquitous layer of geologically fresh frost. This conclusion is consistent with suggestions that Enceladus may be geologically active and that this activity is generating frost particles that not only cover the surface but also replenish the E-ring (Seidelman et al. 1981; Pang et al. 1984a; Haff et al. 1983; chapter by Morrison et al.).

Voyager photometric observations of Iapetus were analyzed by Goguen et al. (1983a) and by Squyres et al. (1984). Squyres et al., from observations ranging in phase angle from 8° to 90°, defined the satellite's photometric function and constructed an albedo map. They found that the reflectance of the dark material is lowest near the apex of the leading hemisphere, where normal reflectances as low as 0.02 and 0.03 are reached. Reflectances on the brighter

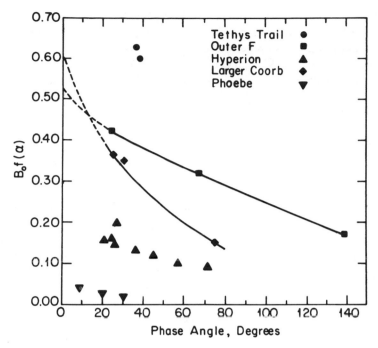

Fig. 16. Phase curves of Saturn's small icy satellites compared with those of Phoebe and Hyperion (Voyager data from Thomas et al. 1983a). Note that in this figure the parameter $B_0F(\alpha)$ is the same as $B_0(\alpha)$ in Eq. (5). Also "Larger Coorb" refers to Janus, "Outer F" to 1980S26 (Pandora), "Tethys Trail" to Calypso.

hemisphere were found to lie commonly between 0.3 and 0.4, the highest values being near 0.6.

Voyager's observations of the satellites of Saturn included both disk-integrated and disk-resolved photometric data. Diameters, phase curves and reflectances of the small satellites were analyzed by Thomas et al. (1983a; see also the chapter by Thomas et al.). These tiny bodies, which range in mean diameter from about 20 km to 190 km, were found to have geometric albedos varying from 0.4 to at least 0.8. Representative phase curves, also showing data for Phoebe and Hyperion, are given in Fig. 16. In the case of Phoebe, an average diameter of about 220 km, and a geometric albedo of about 0.05, were determined (Thomas et al. 1983b). A prograde rotation with a period of 9.4 ($\pm$ 0.2) hr was derived from the disk-integrated lightcurve and by tracking individual surface features. Voyager color data agree with earlier groundbased spectra (Tholen and Zellner 1983) showing that Phoebe has a flatter spectrum than does the dark side of Iapetus.

Voyager's observations of Hyperion have been analyzed by Thomas et al. (1984) and by Thomas and Veverka (1985). This very irregular satellite (diameters: 350 km $\times$ 240 km $\times$ 200 km) has a normal reflectance of 0.25

± 0.02 (cf. chapter by Cruikshank and Brown); at the time of the Voyager 2 flyby in 1981, it was spinning with a period of ~ 13 d about an axis nearly parallel to the satellite's orbital plane, a situation consistent with the chaotic spin state suggested by Wisdom et al. (1984) and described in Peale's chapter.

The estimates of the mean opposition magnitudes of the small satellites given by Thomas et al. (1983a) were scaled to a mean opposition magnitude of Mimas of $V_0 = + 12.9$. This scaling was carried out to effect the conversion from the Voyager filter system to the conventional V system. It now appears that Mimas is actually brighter ($V_0 = + 12.5$) so that all the magnitudes in Table 8 of Thomas et al. (1983a) should be decreased (the satellites are brighter) by 0.4 mag (Table VIII).

3. Earth-based Photometric Observations of Satellites. Photometric observations of planetary satellites made prior to 1975 using groundbased telescopes were reviewed by Veverka (1977a). Since then, virtually all the published data have dealt with the satellites of the outer planets.

An effort to monitor the magnitudes of the Galilean satellites in order to refine information on disk-integrated lightcurves and phase curves continues at the Lowell Observatory (see Lockwood et al. 1980). An interesting question concerns the magnitude of Europa's opposition effect. No opposition surge below 10° phase is evident in the Voyager data (Buratti and Veverka

TABLE VIII
Opposition Magnitudes of Selected Small Satellites of Saturn
Estimated from Voyager Observations

	Cross-Sectional Diameter at Elongation $A \times C$ (km)	Geometric Albedo[a]	+ Δm (Relative to Mimas)	V_0
Mimas	390 × 390	0.8	0	+ 12.5
Inner F Ring Shepherd	140 × 75	0.49–0.68[b]	3.1–2.7[b]	+ 15.6–15.2
Outer F Ring Shepherd	110 × 65	0.52	3.6	+ 16.1
Janus	220 × 160	0.63	1.6	+ 14.1
Epimetheus	140 × 100	0.42–0.60	3.0–2.6	+ 15.5–15.1
Telesto	24 × 22	0.53–0.75	6.3–5.9	+ 18.8–18.4
Calypso	30 × 15	0.81–1.15	6.0–5.6	+ 18.5–18.1
Dione Lagrangian	35 × 29	0.46–0.66	5.7–5.3	+ 18.2–17.8

[a] We have assumed that the geometric albedo equals the normal reflectance (i.e., no limb darkening at $\alpha = 0°$, or $k = 0.5$). If $k = 0.6$ were more appropriate, the geometric albedos in the table should be decreased by about 10% and the values of V_0 increased by 0.1 mag.

[b] Where a range is given, the two estimates correspond to procedures *a* and *b* in Table 6 of Thomas et al. (1983a).

1983) in accord with telescopic results published by Millis and Thompson (1975), but at variance with other results which indicate a small opposition effect (cf. Morrison et al. 1974). Buratti and Veverka (1983) argue that the absence of a strong opposition effect cannot be explained away as simply due to Europa's high albedo; rather it must indicate an unusual texture for the satellite's surface (see also Squyres et al. 1983b).

An additional development since 1975 has involved photometric observations of Jupiter's small, faint satellites. Millis (1978) published careful magnitude and color estimates of Amalthea which are in excellent agreement with spacecraft results (Veverka et al. 1981a). Cruikshank et al. (1982) and Degewij et al. (1980) review studies of the outer satellites. Radiometric observations of JVI and JVII have yielded albedos of about 0.03 for each and diameters of 185 and 75 km, respectively. Cruikshank et al. suggest that at least JVI and JVII have colors similar to C asteroids, whereas JVIII and JIX are more comparable to RD objects, but more recent measurements by Tholen and Zellner (1984) indicate that JVI, JVII, JVIII, JIX and JX all look like C objects (see the discussion of these objects in the chapter by Clark et al.). Tholen and Zellner also found that JXI (Carme) has an unusual spectral reflectance curve which rises at ultraviolet wavelengths. It is suspected that all of the outer satellites (with the possible exception of JXI) are low-albedo objects, and that they have short, nonsynchronous spin periods. Degewij et al. (1980) present a lightcurve for JVI which indicates a spin period of 9.2 to 9.8 hr.

In the case of the satellites of Saturn, the edge-on presentation of the rings in 1979 permitted the detection from Earth of several new small satellites (Smith et al. 1980), all of which were later imaged by Voyager. Such observations in the mid 1960s also led to the discovery of the E ring (Seidelman et al. 1981), the existence of which may be closely connected with ongoing internal activity with Enceladus (Burns et al. 1984).

Photometric observations of Saturn's larger satellites have continued at the Lowell Observatory (see, e.g., Franz and Millis 1975; personal communication 1983). For Dione and Tethys, these observers reported significant opposition surges which cannot be verified using Voyager photometry, since there are no spacecraft observations of these two satellites at phase angles below 8 to 10°. The orbital lightcurves for Rhea and Dione determined at Lowell agree with the spacecraft data (Buratti and Veverka 1984); in each case the leading hemisphere is slightly brighter. For Dione, the Voyager data show a well-defined lightcurve of amplitude about 0.6 mag, with the leading hemisphere about 1.8 times brighter than the trailing one. Such a large amplitude is consistent with some Earth-based data (see, e.g., Blair and Owen 1974), but most observers have reported smaller amplitudes. For example, extensive observations by Franz and Millis (1975; personal communication, 1983) have led to a well-defined lightcurve with a total amplitude of 0.4 mag.

The most recent photometry of Iapetus from Earth is summarized by Millis (1977) and by Cruikshank et al. (1983a). From his systematic observations covering five apparitions, Millis deduced correctly the existence of a bright area near the southern pole of the satellite. Cruikshank et al. present summary phase curves for the dark hemisphere which clearly show a pronounced opposition effect indicative of a very porous regolith texture. Additional valuable photometric and color observations of Iapetus and of Hyperion and Phoebe were published by Tholen and Zellner (1983). Using their photometry along with radii determined by Voyager imaging, Tholen and Zellner derived geometric albedos of 0.07 for the leading hemisphere of Iapetus, 0.06 for Phoebe, and limits of 0.19 to 0.25 for Hyperion, results which agree very well with spacecraft determinations (see Sec. IV.A.2, above). The value for Iapetus is higher than the lowest albedos found by Squyres et al. (1984) and by Goguen et al. (1983a) on the dark hemisphere of the satellite, but is fully consistent with the Voyager albedo map, which shows the presence of some icy material near the edges of the leading hemisphere. In the case of Hyperion, Tholen and Zellner quoted a range of values for the normal reflectance, since they could not be sure of the projected area of the satellite at the time of their observations; as already mentioned, Hyperion seems to be in a chaotic, tumbling spin state. Several groups (Hawaii, Arizona, Texas and MIT) have been monitoring the lightcurve of this 14th-magnitude object, but detailed results remain to be published. Available reports confirm, at least, that the satellite is not in a synchronous spin state.

A major addition to our knowledge of the Uranian satellites was achieved by Brown, who in his Ph.D. dissertation (Brown 1982) and subsequent work determined radii, albedos, spectral reflectance curves and photometric properties; this work is reviewed also in the chapter by Cruikshank and Brown. The satellites have sizes ranging from that of Enceladus (Miranda) to that of Rhea (Titania and Oberon) (see Table IX); their surfaces consist of dirty water ice. Judging from the relatively low albedos and the details of the spectral reflectance curves, the surface layers must be contaminated with a dark material (Brown 1984). The nature of this dark material remains unknown, but Cheng (1984b) and Cheng and Lanzerotti (1978) suggest that it may be derived from the polymerization of methane ice by charged particles in the Uranus magnetosphere. At least three of the satellites (Titania, Oberon and Ariel) show very strong opposition effects below phase angles of 1°, suggesting unusual surface textures (see Fig. 17). No detailed model of the surges has been published, but various possibilities are being pursued (see Pang and Rhodes 1983). Perhaps the very intricate texture implied by the observations is related to charged particle damage (R. E. Johnson personal communication, 1984).

Due to the large apparent tilt angle of the satellite system as seen from Earth in recent years (close to 90° in the mid 1980s), there are no reliable lightcurves for the satellites of Uranus.

Thermal radiometry of Neptune's satellite, Triton (see Sec. IV.C) indicates a radius for Triton of about 1750 km and a geometric albedo of 0.4

TABLE IX
Physical Properties of Uranus Satellites[a]

	Radius (km)	Mass (10^{23} g)	Density (g cm^{-3})	p_v
Miranda	250 ± 110	1.7 ± 1.7	~3	~0.5
Ariel	665 ± 65	15.6 ± 3.5	1.3 ± 0.5	0.30 ± 0.06
Umbriel	555 ± 50	10.0 ± 4.2	1.4 ± 0.6	0.19 ± 0.04
Titania	800 ± 60	59.0 ± 7.0	2.7 ± 0.6	0.23 ± 0.04
Oberon	815 ± 70	60.0 ± 7.0	2.6 ± 0.6	0.18 ± 0.04

[a] Masses and densities are from Veillet (1983a), but see also criticism by Dermott and Nicholson (1986); radius and visual geometric albedo (p_v) from Brown et al., except the radius and albedo of Miranda, which come from Brown and Clark (1984). Table is from Brown (1984).

(Cruikshank 1984). The relatively high albedo is consistent with the presence of ices, and possibly even liquid nitrogen as inferred by Cruikshank and his coworkers (see the chapter by Cruikshank and Brown). Available photometry and color measurements are discussed by Cruikshank et al. (1979). There is some indication that Triton shows small light variations in synchronism with its orbital period, but there is disagreement as to the details. Data summarized by Degewij et al. (1980) suggest a slightly brighter trailing hemisphere, whereas the opposite trend is reported by Franz (1978).

There are essentially no photometric data on Pluto's satellite Charon, other than an approximate estimate of its magnitude relative to Pluto (1.7 ± 0.1; Ables and Thomsen 1979). The series of mutual eclipses which began in 1985 (Binzel et al. 1985) should provide additional information.

B. Polarization Measurements

While polarization observations of asteroids have continued to provide information on albedos and hence diameters (Dollfus and Zellner 1979), supplementing those obtained by infrared radiometry (Morrison and Lebofsky 1979), there has been little activity in terms of polarimetry of satellites during the past decade. Most satellites in the outer solar system are too far from Earth to obtain a sufficiently large excursion in phase angle to determine albedos by the techniques that are employed for asteroids (Veverka 1977b). Instead, albedos and diameters for many of these objects are now known through direct spacecraft imaging and through radiometry.

One major application of polarimetry to satellite work has concerned the atmosphere of Titan. Using Pioneer 11 polarization measurements of Titan, Tomasko (1980) and Tomasko and Smith (1982) obtained important constraints on the particle size distribution of aerosols in the satellite's upper atmosphere (Hunten et al. 1984). Goguen et al. (1985) have developed an infrared polarimetric technique to obtain spatial data on Io's hot spots from

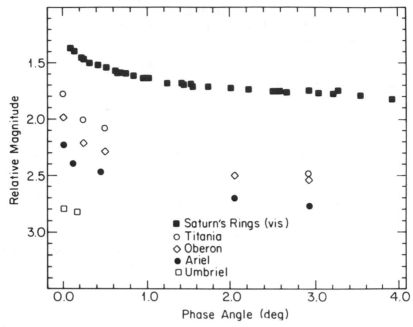

Fig. 17. Near-infrared opposition brightness surges of Ariel, Umbriel, Titania and Oberon. Also plotted are the visual opposition surge data for Saturn's rings from Franklin and Cook (1965). The data for the Uranian satellites contain an arbitrary offset to facilitate comparison to the data for Saturn's rings (figure from Brown and Cruikshank 1983; also Fig. 6 in the chapter by Cruikshank and Brown).

Earth-based observations. Dollfus and Geake (1975) have employed polarimetric observations of Callisto to argue that the trailing hemisphere of that satellite has a very unusual texture; but the argument, especially in light of other evidence (Pang et al. 1983), is not compelling.

C. Infrared Radiometry

Thermal observations of solar system objects date back more than a century to measurements of the Moon's heat made by Lord Rosse (1869). Much of the early work in this century (reviewed by Pettit [1961] and by Sinton [1962]) was carried out by Pettit and Nicholson at Mt. Wilson (see Pettit and Nicholson 1930). At wavelengths beyond several microns, the observed light from satellites is dominated by thermal emission. In contrast with the scattering of reflected sunlight in the visual, thermal emission is fundamentally different since the intensity of emitted photons is controlled primarily by temperature. Because various parameters influence the temperature, many of these can be determined from remotely sensed spectra of infrared emission. One basic parameter is the bolometric Bond albedo, since it determines the fraction of incident sunlight energy which is scattered. Surface temperatures also depend on the rotation vector which determines the nature of the diurnal cycle and the seasons, and upon the eccentricity of the planet's orbit. The amount of heat that can be retained in the surface and transported by rotation depends upon the thermal inertia of the soil or regolith. Emissivity also affects the spectral distribution of the emitted radiation. This effect tends to be small over broad wavelength reaches, even though on occasion it can be pronounced in narrow spectral intervals where electronic transitions occur.

At infrared wavelengths the observed emission can be modeled as if it came from the surface, and is primarily a function of the surface temperature. Microwave and radio wavelengths sense deeper into the surface; the emergent flux from each depth depends upon the temperature there and upon the opacity of the overlying material to microwave and radio radiation.

During the past decade, spacecraft and Earth-based measurements of infrared properties of satellites focused on three major areas. The first involved the monitoring of volcanic activity on Io, observations which emphasize the rapid time scale on which the regolith of Io is being modified. Second were observations dealing with better determinations of surface textures. Included here is the analysis of Viking measurements of Phobos and Deimos, observations of eclipse cooling curves for Iapetus, and various measurements made by Voyager concerning the surface textures of such satellites as Rhea and Io. Voyager measurements have also made it possible to refine the Bond albedos of some satellites in the outer solar system. Finally, the radiometric method of radius and albedo determination has been applied to the satellites of Uranus by Brown (1984), and to Triton by Lebofsky et al. (1982) and Cruikshank (1984). Infrared observations of Titan are reviewed by Hunten et al. (1984).

The observational coverage of planetary satellites in the thermal infrared is summarized in Table X. In this table, spatial resolution is broken down subjectively according to whether the satellite has been observed as a point or an extended source. Available temporal resolution can be divided subjectively among "scattered observations" (referring to one or a few measurements), "coverage as a function of solar phase angle" and "eclipse observations" referring to high time resolution observations of satellites as they enter or emerge from eclipse.

With the discovery of volcanism and hot spots on Io, it has become necessary to distinguish between passive and active surfaces. Passive surfaces are those heated entirely by absorbed insolation. In modeling such surfaces, energy can be assumed to be conserved: the emitted plus reflected radiation equals the insolation. Active surfaces (currently Io is the only satellite example) have significant internal heat sources. For such bodies a prime remote sensing objective is the measurement of the excess power, or heat flow from the interior.

Following the organization of Table X, passive satellites will be discussed first, focusing on the Galilean satellites, the satellites of Mars and the Moon. Then we turn our attention to Io, presently the only satellite in the active category.

TABLE X
Summary of Infrared Observations

	Disk-integrated (Limited Coverage)		Disk-integrated (Coverage as Function of Orbital Longitude)	Disk-resolved (Coverage of Specific Areas)	
Scattered observations	Titan[c] Rhea[a] Dione[a] Ariel[i]	Umbriel[i] Titania[i] Oberon[i] Triton[j]	Iapetus[b] Europa[d] Ganymede[d]	Io[e] Europa[e] Ganymede[e] Phobos[f]	Moon
Coverage as function of solar phase angle	Deimos[f,g] Phobos[f]		Io		Moon
Eclipse observations	Deimos[f] Phobos[f] Io[d] Europa[d] Ganymede[d] Callisto[d]		[h]		Moon

[a] Morrison et al. 1974.
[b] Morrison et al. 1975.
[c] Many observations made as function of wavelength to constrain atmospheric composition and structure (cf. Hunten et al. 1984).
[d] Morrison 1977, p. 276.
[e] Voyager IRIS.
[f] Viking IRTM.
[g] Mariner 9, Gatley et al. 1974.
[h] Possible for nonsynchronous satellites only.
[i] Brown 1984.
[j] Lebofsky et al. 1982.
[k] Johnson et al. 1984.

1. Thermal Emission Powered by Insolation.   The minimal case in Table X consists of a single radiometric measurement, and an accompanying visual magnitude. Knowledge of these two fluxes, visual and infrared, allows a solution for size and albedo, even when the satellite is so small that its angular size cannot be resolved directly. This radiometric method, pioneered for asteroids by Allen (1971) and Matson (1971), is reviewed by Morrison and Lebofsky (1979), Matson et al. (1978b) and Brown et al. (1982b). At a given distance from the Sun the amount of sunlight reflected by a satellite is proportional to its cross sectional area S and albedo A, while the thermal emission depends on S and $(1 - A)$. Thus, S and A can be determined from two observations. The albedo needed is the bolometric Bond albedo and the spectral distribution of the thermal emission must be modeled. The model for reradiation usually employed is a sphere with each surface element in instantaneous thermal equilibrium with sunlight. The resulting spectrum for the whole satellite is not that of a blackbody, but the sum of a series of blackbody spectra. In more detailed thermophysical modeling, it is usual to assume that the satellite's regolith is similar to that of the Moon, although very little is known about the actual thermophysical parameters appropriate to most satellites. If observations in more than one infrared bandpass are available, then the effective radiometric emissivity e can be computed. So far, such results have not differed significantly from $e \sim 0.9$, a value often assumed in the application of the radiometric method.

Brown (1982b,1984) and Brown et al. (1982) used the radiometric method to determine the radii and albedos of four satellites of Uranus: Ariel, Umbriel, Titania, and Oberon. Miranda's radius, shown in Table IX, is an indirect determination based on an estimate of the satellite's albedo from its spectral reflectance curve (Brown and Clark 1984). Thermal radiation from Triton was detected by Lebofsky et al. (1982) (cf. Morrison et al. 1982), making it possible to derive a radius of 1750 ($\pm$ 250) km, and a geometric albedo of about 0.4 (Cruikshank 1984). Thus, contrary to the opinion of many archaic textbooks, Triton is not the largest satellite in the solar system. Angular diameter measurements are now available for many satellites. With diameter known, the measured thermal emission can be used to constrain other parameters. For instance, Morrison (1977) invoked this approach to derive the photometric phase integrals q for the Galilean satellites.

Observations as a function of orbital longitude give the first order information on the heterogeneity of the surface. For example, partly on the basis of such measurements, Morrison et al. (1975) deduced that the first-order albedo distribution of Iapetus is patterned like the two halves of a baseball cover, a result subsequently confirmed by Voyager imaging (Smith et al. 1982).

Observations as a function of solar phase angle are useful for determining how fast a surface cools after sunset, and conversely how fast it warms after sunrise. Observations of the global thermal emissions of Phobos and Deimos were made with the Infrared Thermal Mapper (IRTM) aboard the Viking Orbiter at wavelengths of 11 and 20 μm covering phase angles between 0° and 120° (Lunine et al. 1982). The phase data are consistent with a thermal inertia of between 0 and 0.002 cal cm^{-2} s$^{-0.5}$ K^{-1}. The corresponding value for the Moon is ~ 0.001 in the same units. The similarity suggests that the uppermost regoliths on Deimos and Phobos are comparable to that of the Moon. The thermal skin depths on Phobos and Deimos can be estimated from Eq. 11 (below) to be about 10 mm for Phobos and 20 mm for Deimos.

Spatially resolved thermal emission data are available for the Moon, Phobos and Deimos, and for some satellites of Jupiter and Saturn. For Europa, Ganymede and Callisto, surface temperatures measured by the Voyager infrared interferometer spectrometer (IRIS) experiment (Hanel et al. 1979) are consistent with those expected from previously determined values of radiometric albedos and thermophysical parameters (reviewed by Morrison 1977). For Ganymede and Callisto (for which Earth-based data are best), a two-layer model of the uppermost regolith has been derived from eclipse observations. The model suggests a very thin upper layer (perhaps 1 mm in depth), with a very low thermal inertia (about 2×10^{-4} cal cm^{-2} K^{-1} s$^{-1/2}$) overlying a layer of higher thermal inertia (Morrison and Cruikshank 1973).

Temperature maps have been constructed for about half of the Moon, and for parts of Phobos and Deimos. Such maps can be compared with predictions from models to locate thermal anomalies. On the Moon, two major effects produce thermal anomalies. One anomaly is caused by inhomogeneous radiometric properties. The temperature response of a surface to variations in insolation depends on the thermal inertia I defined as

$$I = \sqrt{k\rho c} \qquad (10)$$

where k is the thermal conductivity, c is the specific heat capacity of the material and ρ is the bulk density of the surface layer. Anomalies in local values of I are most often caused by the presence of boulders. The second type of anomaly is due to topography: slopes and especially negative relief inside craters. The inside of craters near the equator, particularly the centers of small craters (less than a few km in diameter) can show strong positive anomalies (warmer). The center of such a crater is subjected to both direct insolation and to light scattered from the walls. Thermal emission is also received from these walls. The result is that the total energy absorbed by the floor can be substantially greater than the bolometric Bond albedo would suggest.

Mapping temperatures throughout an entire lunation provides a powerful technique for recognizing thermal anomalies, and for distinguishing between the different types (see Saari and Shorthill 1963; Shorthill and Saari 1965). Fitting a model to observations over a lunation also allows sensing the subsurface thermal structure. The thermal skin depth d, defined as the depth at which the amplitude of the temperature wave is reduced to $1/e$ of its amplitude at the surface, is:

$$d = \sqrt{Pk/\pi\rho c} \qquad (11)$$

where P is the rotation period, and k, ρ and c are defined above. Since this depth is a function of period, the length of the lunation yields the deepest probing (about 9 cm). The shallowest is offered by a lunar eclipse (~ 0.5 cm). Thus eclipse observations are useful for identifying the near-surface thermal inertia anomalies. In the case of shallow, buried rocks, anomalies would be seen in lunation data but not in eclipse data. At the right phase some anomalies can be quite strong. For example, Allen (1971) has successfully identified many boulder fields by observing regions of the Moon a few hours after local sunset when the rocks were still radiating profusely.

In some cases regolith porosities can be estimated from thermal inertias since I constrains the relative values of k, ρ and c. Choosing appropriate laboratory values for c and k determines the bulk density. The ratio of ρ to the crystalline density of the surface material yields porosity. Using this procedure for Io, Matson and Nash (1983) found a porosity of ~ 85%. On the Earth such high porosities occur in fresh, very fine grain, volcanic ash deposits (Wechsler et al. 1972), but similarly high values have been deduced for the Moon and for the satellites of Mars.

Phobos was observed during eclipse by the Mariner 9 infrared radiometer (Gatley et al. 1974). Additional eclipse data were obtained for both Phobos and Deimos by the Viking Orbiter's Infrared Thermal Mapper (IRTM) (Lunine et al. 1982). Both data sets indicated low thermal inertia (see above) for the average regoliths. In addition, Lunine et al. concluded that a higher inertia component (rock) covers about 5% of the surface of Phobos. Data for Deimos were not detailed enough to search for evidence of a blocky component.

Voyager IRIS observations of Saturn satellites (other than Titan) are reviewed by Cruikshank et al. (1984b). Important results include determinations of the radiometric Bond albedos of Rhea, Tethys and Enceladus (see Table V in Sec. IV.A.2), and of cooling during eclipse of the surface of Rhea. An analysis of the eclipse data indicates that about half the surface of Rhea cools to near 75 K, while the remainder cools to a temperature below 55 K. The higher thermal inertia material (faster cooling) may be modeled by ice blocks about 10 cm in size; the low inertia component is most likely finely textured frost (Pearl, personal communication, 1984).

The only small satellite of Jupiter or Saturn for which IRIS data are available is Amalthea. Voyager 1 IRIS observations (Hanel et al. 1979) were initially reduced to give an unusually high temperature of 185 ($\pm$ 5)K, which strongly suggested the influence of a nonsolar heat source; model calculations by Simonelli (1983) show that a maximum temperature of about 160 K is expected, assuming that Amalthea is a blackbody and that solar insolation is the heat source. Recent, more refined reductions of the IRIS data give a temperature of only 164 ($\pm$ 5)K, consistent with Simonelli's calculations and with earlier Earth-based measurements by Rieke (1975), which gave a temperature of 155 ($\pm$ 15)K. Thus, the thermal discrepancy between model and observed temperatures mentioned by Thomas and Veverka (1982) has been resolved by an improved reduction of the IRIS data.

2. An Active Satellite: The Special Case of Io. Prior to Voyager, Io was regarded as a quiescent lunar-sized body with unusual spectral properties. The prevailing compositional interpretation, the evaporite hypothesis (Fanale et al. 1974,1977a,b), relegated internal activity to a distant past. Groundbased observations at infrared wavelengths yielded peculiar results that were not fully appreciated at the time. Chief among these were the lack of agreement between brightness temperatures observed at 10 μm and 20 μm and the high level of thermal emission measured when Io was eclipsed by Jupiter's shadow. Most of these data were obtained in the early 1970s when the calibration of infrared photometry was more uncertain than it is now; the discrepancies were generally attributed to differences in calibration and technique between observers. The eclipse data were interpreted as indicating that Io had a very thin regolith and that excess emission during eclipse was due to heat conducted from bedrock a few millimeters below the surface. Voyager's discovery of volcanism on Io was the key to better models, as described here and in the chapters by Nash et al. and Schubert et al.

The original infrared photometry of Io was carried out by Gillett et al. (1970), Hansen (1972), as well as Morrison et al. (1972) and reviewed by Morrison (1977). The brightness temperatures determined are summarized in Table XI. Significant are the higher brightness temperatures at shorter wavelengths, especially at 8.4 μm. For comparison, data for Europa (from the same sources) are also included. The Europa data also show a slight wavelength dependence of brightness temperature. This dependence and similar data for Ganymede and Callisto led to the suspicion that the Io effect might be due entirely to calibration and other technical problems. Yet, no single set of correction factors for the various bandpasses could simultaneously rectify all the observations.

While the detection of anomalously high 5 μm radiation from Io by Witteborn et al. (1979) provided an early indication of Io's peculiar thermal state, the reason for the significant variation of brightness temperature with wavelength on Io becomes evident only after Voyager revealed the presence of hot spots on the satellite's surface (Fig. 18).

During the extensive eclipse observations of the Galilean satellites carried out by Morrison and Cruikshank (1973) and Hansen (1972,1973), Io distinguished itself by adjusting its temperature more rapidly to changes in insolation than any other solar system body and by having an excessively high level of emission during eclipse. Furthermore, data from eclipse disappearances were not fully consistent with those obtained for reappearances. The sharp fall in Io's flux observed as Io enters eclipse indicates a very low thermal inertia for its surface. But the flux does not continue to decrease as expected; instead, it abruptly levels off and remains approximately constant during the remainder of an eclipse. Io's brightness temperatures outside of eclipse are approxi-

TABLE XI
Brightness Temperatures of Io and Europa[a]

Effective Wavelength $\lambda(\mu m)$	Mean Brightness Temperatures (K)	
	Io	Europa
8.4	149	134
10.6	138	130
21	127	120

[a] Values as adopted by Matson et al. (1981a).

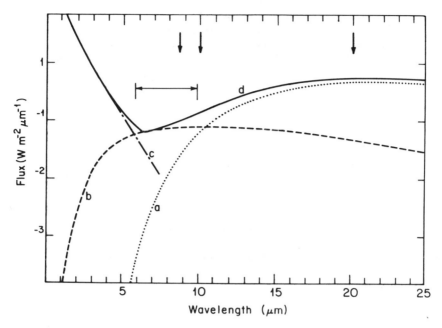

Fig. 18. Io's spectral radiance and its components as a function of wavelength: (a) thermal emission from the passive, insolation-heated surface of Io (taken as an average disk temperature of 129 K); (b) thermal emission from the hot spots (disk centered at longitude 300°W, calculated from observed Voyager hot spots as discussed in the text; (c) reflected sunlight; and (d) the sum of the components. The horizontal arrows mark the wavelength range where emission from the hot spots dominates the spectrum. The vertical arrows indicate the effective wavelengths for the data (figure from Johnson et al. 1984*a*).

mately 138 K and 127 K for 10 and 20 μm, respectively (Table XI). In eclipse, however, the emission levels at these two wavelengths correspond to brightness temperatures of 126 K and 102 K—much higher than expected. This discrepancy is now known to be caused by emission from hot spots, which is unaffected by eclipses.

The other anomalous feature of the Io eclipse data is that the observed thermal level outside of eclipse differs between disappearances and reappearances. This difference is now known to be due to the nonuniform distribution of hot spots over Io's surface (Johnson et al. 1984*a*).

In addition to the above anomalies, Hansen obtained 5 μm data in the late 1960s (O. Hansen, unpublished) that indicated excess emission well beyond what could be explained by insolation alone. As already noted, the first reported 5 μm outburst from Io was that observed by Witteborn et al. (1979). From a spectrum of the event, they were able to derive a color temperature of ~ 600 K for the excess emission, which led them to suggest volcanic activity as one possible explanation. The next step was provided by theoretical tidal

TABLE XII
Hot Spots on Io Observed by Voyager IRIS[a]

No.	Latitude	Longitude		Equivalent Radius (km)	Temperature (K)
1	30	210[b]	NW Colchis Regio	9	385
				55	165
2a	27	119[b] ⎫	Amirani/Maui[b]	13	395
b	19	122 ⎭		49	200
3	13	310	Loki Patera	⎰ 21	450
				⎱ 121	245
4	−19	257	Pele	⎰ 6	654
				⎱ 80	175
5	−40	272	Babbar Patera	⎰ 5.3	322
				⎱ 72	175
6	−41	288	Ulgen Patera	⎰ 14	355
				⎱ 78	191
7	−48	267	Svarog Patera	30	221
8	−52	344	Creidne Patera	20	231
9	−80	320	Flow in Nemea Planum	17	225

[a] Table from Pearl and Sinton (1982).
[b] Positions uncertain due to lack of concurrent imaging.

dissipation work of Peale et al. (1979), who concluded, immediately before Voyager 1's encounter with Jupiter, that Io was likely to be volcanically active.

Coupled with Voyager's discovery of active volcanoes on Io was that of the existence of hot spots (Hanel et al. 1979; also see review by Pearl and Sinton 1982). These spots, possibly lava lakes, are associated with dark markings on the surface (Table XII; also see the Map Section). A representative IRIS spectrum is shown in Fig. 19. In 1979, the largest of these spots was at Loki Patera.

Three different techniques can be used to estimate Io's heat flow. They are:

1. Eclipse observations;
2. Spectral separation;
3. Spatial resolution.

The first two methods have been applied, while the third will be employed during the Galileo mission.

The most direct way to measure Io's heat flow is to observe the satellite when it is eclipsed by Jupiter. Io's surface cools sufficiently during eclipse that the observed emission is dominated by that from hot spots. Observations at different wavelengths can be used to infer the temperatures of the hot spots;

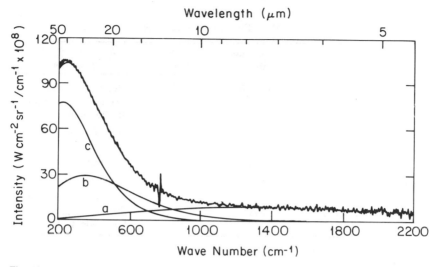

Fig. 19. Voyager IRIS spectrum of the region of Io containing Pele. The ringing centered near 760 cm^{-1} is caused by interference. The smooth curve through the data is a least-squares fit using three weighted blackbody spectra. The lower curves are the three components; from left to right, their temperatures (and fractional areas) are (a) 114 K (0.899); (b) 175 K (0.100); and (c) 654 K (5.77 × 10^{-4}) (figure from Pearl and Sinton 1982).

the observed intensities can be used to determine their effective areas (Matson et al. 1981a; Sinton et al. 1980; Morrison and Telesco 1980). The advantage of the eclipse method is that no assumptions about emission from the passive surface heated by insolation are required. The disadvantage is that Io rotates synchronously and all of the measurements are made on effectively the same hemisphere, centered on longitude 0°W.

Since the hot spots are significantly warmer than the ambient surface, their emitted radiation is shifted toward shorter wavelengths. Observations in the spectral region where hot spots dominate are the key data to the second approach of estimating Io's heat flow. In implementation this method is more difficult than the eclipse method because the emission from the passive portion of Io's surface must be modeled accurately. The brightness temperature data of Table XI provided a quick indication that traditional lunarlike modeling fails in Io's case. Matson et al. (1983) have argued that the problem may be due to unmodeled surface roughness, but the resolution remains unclear.

The spectral separation method was first employed by Hanel et al. (1979). While the IRIS observations had spatial resolution on Io, none of the footprints were small enough to completely fit inside a hot spot. Thus, Hanel et al. used the spectral separation method to obtain hot spot temperatures and areas. A sample result is shown in Fig. 19 where the observed spectrum of the Pele hot spot is broken down into components of different temperatures and

TABLE XIII
Io Heat Flow Determinations

Source	Applicable Date	Heat Flow (W m^{-2})	Remarks
Matson et al. (1981a)	ca. 1970[a]	2 ± 1	Methods a and b (see text); data from the literature
Morrison and Telesco (1980)	1979	1.5 ± 0.3	Eclipse observation
Sinton (1981)	1974/1979[a]	1.8 ± 0.6	Spectral separation
Pearl and Sinton (1982)	1979	~2.0	IRIS data
Johnson et al. (1984)	1983	~1.2	Spectral separation, longitude averaged

[a] Parts of data set are taken at different epochs.

areas. The outstanding advantage of the spectral separation method is that it can be applied to all longitudes of Io.

Observing hot spots with high spatial resolution is the best approach to measuring their emissions with high accuracy. Such observations are currently being attempted by Howell et al. (1985) using speckle imaging techniques and by Goguen and Sinton (1985) by observing the polarization of the thermal emission. The spatial resolution approach will be used by the Near Infrared Mapping Spectrometer (NIMS) if the Galileo spacecraft ever encounters Io.

The heat flow determinations made by the above methods are relatively consistent with each other (Table XIII). These values are about two orders of magnitude larger than that for the Moon (0.02 W m^{-2}), a body of the same approximate size and density as Io. The only process which can yield power at this level is the tidal dissipation proposed by Peale et al. (1979; see also the chapter by Peale). Given the high value of the heat flow one can conclude that Io's interior must be molten on a global scale (Matson et al. 1981a; cf. chapters by Schubert et al. and Nash et al.).

D. Radio and Radar Observations

Passive microwave observations of satellites other than the Moon are difficult to make from Earth in part due to confusion with emissions from the planet. To date, no spacecraft has carried a microwave radiometer to investigate satellites or other small bodies. Microwave observations of the Moon have been made over a wide range of frequencies to investigate the lunar regolith and to constrain the temperature gradient and subsurface structure (Kheim and Langseth 1975; Gary and Kheim 1978; and others). An early result of such observations were indications of the low porosity of the upper-most layers (Troitsky 1965), values which seemed unusually low, but which were ultimately confirmed by direct measurements (see Lindsay 1976). Since the penetration depth of microwave wavelengths can reach approximately ten times the wavelength, penetration depths at microwave frequencies involved depths of centimeters to tens of meters.

The only satellites that have been positively detected at microwave frequencies are Titan and the Galilean satellites of Jupiter. Microwave observations of Titan have been used to determine surface temperatures and to search for trace constituents in its atmosphere (Hunten et al. 1984). In the case of Ganymede and Callisto, Berge and Muhleman (1975) and Ulich and Conklin (1976) obtained brightness temperatures which are consistent with lunarlike regoliths and commonly quoted albedos. Recently de Pater et al. (1984) observed Io, Europa, Ganymede and Callisto at 2 and 6 cm using the VLA (Very Large Array). The radio temperatures were found qualitatively consistent with predictions for fast-rotating bodies with high radio reflectivity (low emissivity) as determined from radar measurements by Ostro (1982).

While a voluminous body of data exists on the radar properties of the Moon at various wavelengths (see Ostro [1983] for summary), the only other satellites that have been successfully observed by radar to date are the Galilean satellites of Jupiter. Various attempts to detect Titan have not been successful. The failure to detect Titan places no useful limit at present on its cross section; such a limit might make it possible to distinguish a liquid from a solid surface.

Radar observations of the Galilean satellites have been thoroughly reviewed by Ostro (1982); since that time, few observations have been made, because these bodies are inaccessible to the Arecibo radar until the late 1980s. Several interesting features in the radar observations are worth emphasizing. The first is that the radar cross sections of these objects bear a striking correlation with their optical properties: high radar reflectances corresponding to high optical albedos. For example, the geometric albedo of Europa at $\lambda = 13$ cm is 0.65, a value comparable to the satellites' optical albedo. A second feature emphasized by Ostro is the unusual scattering properties of the icy satellites; icy satellites do not scatter radar like more familiar rocky objects, an indication that the textures of their surfaces and/or the structures of their regoliths are quite distinct from those of bodies like the Moon. Specifically, the radar results indicate that the icy satellites have extremely rough surfaces at some scales greater than 13 cm.

A puzzle remains concerning the unusual radar polarization properties of the icy Galilean satellites, especially Ganymede (Ostro 1982; see the chapter by McKinnon and Parmentier). The observations clearly point to an unusual regolith texture. Ostro originally interpreted the result in terms of a contrived and geologically unlikely model in which the surface is covered with hemispherical craters which are smooth at the wavelength of observations. It is difficult to understand how such a situation would be preserved under normal geologic conditions. On the other hand, Goldstein and Green (1980) have proposed an explanation in terms of the subsurface structure of the regolith. They suggest that "the upper few meters of Ganymede is ice, crazed and fissured and covered by jagged boulders." The essential part of the model is the large number of ice-vacuum interfaces so that the surface approximates a diffuse reflector. In the conclusion to his review paper, Ostro agrees that the anomalous circular polarization inversion can be explained by randomly oriented subsurface facets. Recently, Hagfors et al. (1985) have proposed a different model involving refraction scattering rather than reflection scattering. Refraction scattering is the process whereby a wave which penetrates into a surface has its propagation vector bent as a result of small variations in the index of refraction. For sufficiently large variations, the wave may be deflected sufficiently to emerge again from the surface. Hagfors et al. show that such a process could account for observed radar properties of the icy Galilean satellites—what is required is a regolith with "convex regions of excess refractive index," perhaps a patchwork regolith made of clumps of a denser (higher index of refraction material) component surrounded by mantles of gradually decreasing density (increasing porosity and hence decreasing index of refraction). Whatever has caused the unusual regolith texture demanded by either model seems to occur on satellites as different (in terms of the relative importance of endogenic and exogenic processes) as Callisto and Europa; thus, the mechanism involved is not clear. Radar imaging of the Galilean satellites from Earth is impossible, but Ostro (1982) has obtained radar lightcurves of these bodies, and identified anomalous features in the echoes from specific longitudes. In general, given the low resolution of the data, a definitive interpretation (or interrelation with the Voyager images) is not possible at this time.

In the case of the Moon, it has proved possible to make radar maps of surface features at different frequencies and to use them for a wide variety of purposes. For example, in a series of recent papers, Zisk and his coworkers have combined radar data with other remote-sensing information in comprehensive studies of specific areas on the Moon (Zisk et al. [1977] and related references summarized in Ostro [1983]). Lunar radar images have also been used to study the radar characteristics of crater ejecta blankets and other morphologic units for eventual comparison with radar data on Venus, where the question of how one distinguishes an impact crater from other possible circular features has arisen (see, e.g., Thompson et al. 1980).

V. DISCUSSION

One critical test of how well we understand the development of regoliths is our ability to explain observed albedo and color distributions on satellites. Four general classes of patterns have been discussed in the past:

1.   Alleged systematic variations of albedo and color with satellite size;

2. Variation of albedo and color with orbital location within a system of satellites;
3. Albedo/color patterns related to leading side/trailing side hemispheric asymmetries;
4. Albedo variations with latitude on a satellite's surface.

Dealing with these points in turn, it must be emphasized that there is no established correlation between albedo and satellite size. For instance, it is not true that small icy satellites tend to be systematically darker because they have escaped internal differentiation processes.

There is much more evidence that in some cases color/albedo may be connected with orbital position. It has been suggested that there is a systematic variation in redness with distance from Io, presumably related to the ejection of sulfur from Io (see Veverka et al. 1982b). The idea may have some merit as far as Io's immediate neighbors (Amalthea and Europa) are concerned, but the connection is much more difficult to establish in the cases of Ganymede and Callisto. A less likely case involves an alleged relationship between high albedos and proximity to Enceladus in the Saturn satellite system (see Pang et al. 1984a). Whatever the details of such a mechanism might be, it is difficult to find empirical evidence that albedos of Saturn satellites bear a strong relationship to their distance from Enceladus.

That most satellites show definite color/albedo variations with longitude is a long-established fact. A venerable tradition holds that there is a trend for inner satellites to have brighter leading hemispheres and darker trailing hemispheres, and that this trend reverses by the time outer satellites such as Callisto and Iapetus are reached. Early theories which concentrated on external impact mechanisms are reviewed by Veverka (1977a). More recently, Passey and Shoemaker (1982) have suggested strong leading side/trailing side asymmetries in cratering rates and hence in regolith development, on the satellites of Jupiter and Saturn, and Hartmann (1980b) has discussed a model in which cratering lowers the albedo in the case of some ice/rock surfaces. In Hartmann's view prolonged micrometeoroid bombardment might darken such a surface as more and more ice is removed and a lag deposit of rocky material is left behind. Hartmann's mechanism, while providing a framework for interesting speculations, does not seem to apply to the Saturn satellite system; there is no correlation between albedo and crater density (a measure of surface age) on the surfaces of Saturn's icy satellites. Hartmann's scheme might be applicable to geologically simple bodies such as inactive comet nuclei far from the Sun, but in its simplest form it does not account for such effects as the role of large impacts in not only overturning, but also in melting icy regoliths. It is also true that the results of Passey and Shoemaker (1982) which typically suggest regolith depths of hundreds of meters near the apexes of motion and much thinner regoliths on the trailing faces and in polar regions cannot be tested directly at present. Voyager data do not have adequate spatial resolution, but at least the observed distribution of mappable craters ($D > 10$ km) does not display the predicted asymmetry clearly.

For three of the four Galilean satellites, we know that impact cratering has not determined the observed albedo/color asymmetry. On Io the pattern is determined to a large degree by the distribution of bright, sulfur dioxide-rich areas, a pattern which in turn seems to reflect the distribution of geologically recent volcanic activity (chapter by Nash et al.). On Ganymede, the trend is caused by the relative distributions of old cratered terrain and more recent grooved areas (see, e.g., Squyres 1980b; chapter by McKinnon and Parmentier); thus, the pattern is basically due to an endogenic process. For Europa, Johnson et al. (1983) have shown convincingly that at least the color variation may reflect the implantation of sulfur from Io—certainly an external process, but not one involving cratering (see also the chapters by Malin and Pieri and by Cheng et al.).

On the icy satellites of Saturn, Voyager data show that both endogenic and impact processes have played roles in determining the observed albedo patterns, but the relative roles are in most cases uncertain. One significant problem has to do with the limited spatial resolution achieved by Voyager and the fact that in many cases less than complete coverage was obtained at resolutions better than even 10 km/line pair. For Iapetus, strongly divergent opinions remain as to what processes are responsible for the albedo patterns. The symmetrical distribution of dark material on the leading hemisphere has led to suggestions that material spiraling in from Phoebe, Saturn's outermost satellite, is responsible for the observed pattern (see Bell et al. 1985; Mignard et al. 1986). However, because the spectrum of the dark side of Iapetus is very different from Phoebe's, even the proponents of this idea suggest that the effect must involve a secondary modification of the existing surface rather than a simple buildup of infalling debris (Squyres and Sagan 1983); other researchers believe the pattern to be primarily of endogenic origin (Smith et al. 1982). Some of these questions could be resolved if the dark hemisphere of the satellite were imaged at sufficient resolution to search for the possible presence of small bright craters. If the low-albedo material represents an externally derived coating it must be stratigraphically thin; hence, one would expect recent craters to expose cleaner ice at deeper levels.

A less dramatic case from the Jovian system is that of Callisto: here too we lack a convincing explanation of what causes the lightcurve variation (see the chapter by McKinnon and Parmentier). Adding to the uncertainty is the suggestion (cf. chapter by Cheng et al.) that sputtering by magnetospheric particles may have affected the observed albedo patterns on some Jupiter and Saturn satellites.

Latitude-dependent albedo patterns are evident on Io, Europa, Ganymede, possibly Iapetus and certainly Phoebe. Aside from internal processes, possible explanations involve the mobilization of a volatile by solar heating, charged particle sputtering or micrometeoroid gardening.

On bodies such as Ganymede and Callisto, one calculates that water ice is much more stable at the poles than at the equator, due to thermal considerations alone (Shoemaker et al. 1982). Therefore, if thermal effects were the most important ones, the equatorial regions of such satellites should develop a dark, nonvolatile lag relative to high-latitude areas. The polar caps on Ganymede have been interpreted by some as evidence of such thermal migration (Squyres 1980b). Shaya and Pilcher (1983) argue that such migration is inefficient under present conditions and instead must date back to an early epoch when fluid water erupted to the satellite's surface. Whatever the correct explanation, the absence of similar polar caps on Callisto is a strong constraint on any model. Squyres (1981a) and others have dealt with the related problem of the relative preservation of crater rays as a function of latitude on Ganymede and Callisto.

Temperatures in the Saturn system are significantly lower so that thermal migration of water should not be an important factor (see Lebofsky 1975b). Even for albedo differences as pronounced as those on Iapetus, relative sublimation rates of water ice should be insignificantly different. The latitude dependence of the albedo pattern on Iapetus (brighter poles) has been interpreted by some as consistent with an external model for producing the overall albedo dichotomy on the satellite (Squyres and Sagan 1983; see also Bell et al. 1985; Mignard et al. 1986). In the case of Phoebe, the polar regions are also slightly brighter, but it is very unlikely that any thermal migration of ice is involved (Thomas et al. 1983b).

The conspicuous brown polar regions on Io most likely reflect the pattern of volcanic activity on the satellite, but have also been cited as evidence for the efficacy of sputtering of SO_2 frost in the polar regions (see, e.g., Lanzerotti et al. 1983). Thermal migration of SO_2 frost on the surface of Io appears to be an inevitable factor (Kumar 1979), and detailed calculations of the motion of Io's temporary atmosphere are beginning to be carried out (Ingersoll et al. 1986). Here the mobilizing factors are localized sources of volcanic heat as well as differences in solar heating due to variations in insolation with latitude, topography (shadowing), and surface albedo.

There has been much discussion in the past few years as to whether charged particle sputtering plays an important role in determining the observable properties of icy satellites (e.g., albedo patterns). The efficiency of sputtering of ices by charged particles has been demonstrated thoroughly by detailed laboratory investigations (chapter by Cheng et al.). In many cases charged particle impacts should be competitive in mobilizing ices with processes such as thermal migration and micrometeoroid impact mobilization, but relative efficiencies remain to be determined.

Throughout this chapter, we have emphasized that the development of most satellite regoliths must involve a complex interplay of diverse processes. We have also stressed the degree to which our Moon, as the only satellite for which we have *in-situ* regolith data and absolute chronologies, has played a pre-eminent role in shaping our ideas about regolith evolution. In dealing with regoliths whose physical/chemical characteristics differ significantly from those of the lunar one, and which have evolved under processes whose relative roles differ markedly from those that obtained in the case of the Moon, our understanding is much more limited. Specific observations such as the anomalous radar polarization properties of some Galilean satellites, and the extremely large opposition surges of some of the satellites of Uranus should remind us that we know very little about ice and ice/rock regoliths. Important new information will come from future spacecraft exploration (for example, Galileo and Cassini) as well as from continued laboratory and theoretical work; but many decades will pass before *in-situ* regolith measurements and absolute dating on an outer planet satellite are achieved. It will be important during this interval not to get too smart too fast by pretending that all satellite regoliths are just like that of our Moon and that we understand their characteristics and fundamental processes.

Acknowledgments.   We wish to thank J. A. Burns, L. Soderblom, and an anonymous referee for numerous helpful comments. Parts of this effort were supported by a NASA grant to Cornell University.

9. INTERACTIONS OF PLANETARY MAGNETOSPHERES WITH ICY SATELLITE SURFACES

A. F. CHENG
Johns Hopkins University

P. K. HAFF
California Institute of Technology

R. E. JOHNSON
University of Virginia

and

L. J. LANZEROTTI
Bell Laboratories

When natural satellites and ring particles are embedded within magnetospheric plasmas, the charged particles interact with the surfaces of these solid bodies. These interactions have important implications for the surface, the atmosphere of the parent body, and the magnetosphere as a whole. Significant erosion of the surface by sputtering, as well as redeposition of sputter ejecta, can occur over geologic time. The surface can also be chemically modified. Sputter ejecta can make important contributions to the atmosphere; sputtering provides a lower limit to the atmospheric column density even for arbitrarily cold satellite surfaces. Sputter ejecta escaping from the parent body can form extensive neutral clouds within the magnetosphere. Ionization and dissociation within these neutral clouds can be dominant sources of low-energy plasma. The importance of these processes is discussed for the satellites and magnetospheres of Jupiter, Saturn and Uranus.

I. INTRODUCTION

All of the natural satellites in the solar system reside within a photon and particle radiation environment. The photon environment which affects satellite surfaces is primarily visible and ultraviolet radiation from the Sun. The particle environments can be either the magnetospheres of the parent planets or the solar wind. In the case of a few satellites, such as the Earth's Moon, Titan and Iapetus, both environments can be important, depending upon solar wind parameters and the orbital position of the satellite. At some times energetic ions and electrons emitted from solar flares (solar cosmic rays) can also impinge on a natural satellite not shielded by a planet's magnetic field. All of these photon and particle radiations can interact with satellite surfaces and/or satellite atmospheres and produce significant modifications.

The solar wind, composed primarily of protons and electrons with an admixture of a few percent helium ions, continually flows away from the Sun. At Earth, the average wind density is approximately 5 to 10 particles cm^{-3}. Eruptions on the Sun produce occasional outbursts of solar cosmic rays with energies from tens of thousands to tens of millions of electron volts per atomic mass unit. Planets with magnetic fields, like the Earth, Jupiter, Saturn and Uranus, can trap electrons and ions in their radiation belts. The intensities, compositions and time dependencies of such radiation are quite complicated and highly dependent upon the particular magnetosphere. Figures 1 and 2 contain, as illustrative examples, spectra of ions measured by the

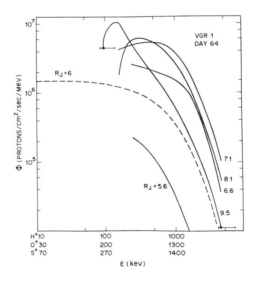

Fig. 1. Energetic ion fluxes from the LECP experiment on Voyager 1 in the inner Jovian magnetosphere (Lanzerotti et al. 1982). A proton composition is assumed. Ion fluxes at energies below 220 keV are uncertain owing to instrument saturation.

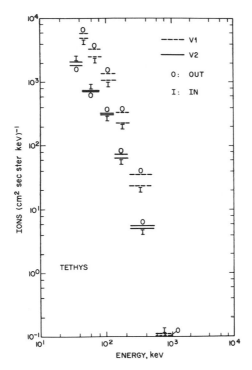

Fig. 2. Energetic ion fluxes from the LECP experiment on Voyagers 1 and 2 near Tethys. A proton composition is assumed. The dashed lines are for Voyager 1 and the solid lines for Voyager 2. The letters I and O designate inbound and outbound trajectory measurements (Lanzerotti et al. 1983).

Low Energy Charged Particle (LECP) experiment (Krimigis et al. 1977) on Voyager 1 near Io in Jupiter's magnetosphere and near Tethys in Saturn's magnetosphere. Such spectra may be time variable, a factor often of considerable importance for quantitative calculations. Further, spacecraft measurements from a single flyby cannot readily separate spatial and temporal effects on particle distributions.

It is well known that energetic particles incident on solids can eject atoms and molecules from the surface. Indeed, the impact of particle radiations on materials can result in physical and chemical alterations of the substance as well as the physical removal of material. While most laboratory studies of the surface sputtering of materials have focused on conducting or semiconducting substances, the satellite surfaces of the outer solar system are often covered with ices such as H_2O (for the icy Galilean satellites and for the Saturnian satellites and rings) and CH_4 or mixtures of H_2O and CH_4 (for the outermost planets and their satellites). A particularly interesting physical situation therefore exists for icy bodies embedded within particle radiation environments. Ices in general are insulators and should respond to irradiation differently from conductors. Studies of the particle irradiation of ices is therefore a natural extension of research, initiated during the Apollo era, on the irradiation of lunar-type materials by ions at solar wind energies (see, e.g., Wehner et al. 1963; Maurette and Price 1975; Bibring et al. 1975; Switkowski et al. 1977).

During the last several years, extensive laboratory experiments have been initiated in order to gain an understanding of the effects of particle radiation on ices of various compositions (W. L. Brown et al. 1978,1982a,b; Lepoire et al. 1983; Pirronello et al. 1981). These laboratory experiments have provided data essential for understanding the surfaces and atmospheres of natural satellites. The laboratory data are then used with satellite-based measurements of radiation fluxes in the vicinity of planetary satellites in order to assess quantitatively the resulting physical and chemical alterations of satellite surfaces and atmospheres.

An atom or molecule sputtered from the surface of a grain or a satellite can suffer a variety of fates, depending upon the satellite's physical properties and the eroded species. The physical properties of some satellites of Jupiter and Saturn are listed in Table I (more precise values are given in the introductory chapter). Some possible fates for sputtered species are illustrated in Fig. 3. The sputtered atom or molecule can escape directly or can impact the sur-

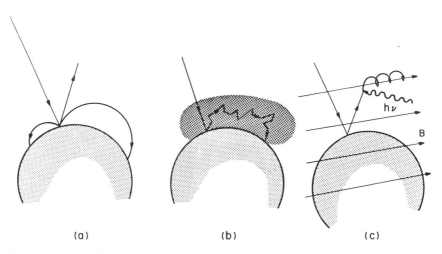

(a)   (b)   (c)

Fig. 3. Schematic illustration of possible fates of material eroded from a surface, depending on the gravitational attraction and external environment of the object: (a) an eroded atom or molecule can escape or enter into a ballistic trajectory, depending on its ejection energy and the gravitational attraction of the object; (b) an atmosphere can cause scattering and reimpact of the ejected species on the surface; or (c) ionization by charged particles or solar photons (hν) of an ejected species in the presence of an externally imposed magnetic field can cause a loss of the species. In practice, various combinations of these three conditions can also exist.

TABLE I
Satellite Data

Satellite	Radius (km)	Mass (10^{23} g)	Density (g cm^{-3})	Escape Energy (eV amu^{-1})	Escape Velocity (m s^{-1})	Distance From Planet	Corotating Ion Impact Energy (eV amu^{-1})	Energy of O$^+$ with Gyroadius Equal to Satellite Radius (keV)
Io	1820	891	3.5	0.035	2580	5.95 R$_J$	17	3.6×10^4
Europa	1500	487	3.0	0.022	2080	9.47 R$_J$	57	1.5×10^3
Ganymede	2640	1490	1.9	0.040	2790	15.1 R$_J$	170	280
Callisto	2420	1070	1.8	0.029	2390	26.6 R$_J$	550	8.5
Mimas	195	0.37	1.2	0.00013	150	3.08 R$_S$	1.4	55
Enceladus	250	0.72	1.1	0.00022	190	3.97 R$_S$	3.8	21
Tethys	525	6.10	1.0	0.00083	400	4.91 R$_S$	7.2	26
Dione	560	10.3	1.4	0.0013	500	6.29 R$_S$	14	6.8
Rhea	765	24.4	1.3	0.0023	660	8.78 R$_S$	32	1.7

face again (Fig. 3a), depending on the gravitational attraction of the body and the sputtered particle energy. If a satellite has a sufficiently thick atmosphere, the outgoing sputtered particle will equilibrate with atmospheric constituents, thereby altering the composition of the atmosphere, and eventually will be reabsorbed on the surface (Fig. 3b) or lost to space from the exobase. A particle sputtered from a satellite can be ionized by solar or stellar photons or by an external plasma (Fig. 3c). Such ionized species can escape from the satellite, especially if there is an externally imposed magnetic field (such as a planetary or interplanetary magnetic field). For the major satellites of Jupiter, Saturn and Uranus, the rotation speed of the planetary magnetic field is greater than the orbital speeds of the major satellites (Johnson 1978). Ionized species that find themselves on the planetary field lines are thus quickly swept away (Matson et al. 1974).

II. LABORATORY EXPERIMENTS

Ion bombardment of ices has been studied by several groups using thin films of ice condensed from vapor onto a cold metallic substrate. The Bell Laboratories (see, e.g., W. L. Brown et al. 1978) and Caltech (see, e.g., Lepoire et al. 1983) experimental arrangements are sketched in Figs. 4 and 5.

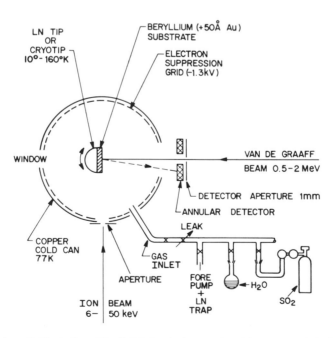

Fig. 4. Schematic illustration of the Bell Laboratories experimental arrangement, including cold finger, copper cold can, gas manifold, annular solid state detector, and ion beams from Van de Graaf and electrostatic accelerators.

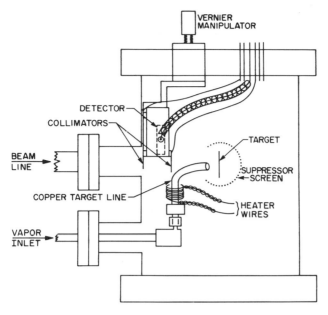

Fig. 5. Schematic illustration of the Caltech ultra-high vacuum chamber, showing the target formation system and the beam collimator/deflector.

The target ice film in both cases is grown on a beryllium substrate covered
with a thin gold "marker" layer. The Bell target (Fig. 4) is cooled by a
cyrostat capable of varying the temperature continuously from ~ 10 K. The
Bell target can be rotated to face either the vapor input or the incoming ion
beam. The backscatter detector measures particles ejected at 180° from the
incoming beam. The Caltech target (Fig. 5) is fixed with respect to the beam,
with the backscatter detector measuring particles ejected at 150°. Upstream
collimators are used to define the beams (aperture sizes ~ 1 to 2 mm^2) and
verniers precisely move the target to the selected erosion spot.

The molecular thicknesses of the films after deposition and at successive
stages of ion bombardment are monitored by Rutherford backscattering, usually with 1.5 MeV He$^+$ incident ions. Such backscattering occurs because of
the low but precisely known probability for an incident ion to collide with a
nucleus in a nearly head-on collision. The energy of the backscattered ion depends on the mass of the nucleus with which it collided and the amount of
material it traversed (losing energy) before and after the collision. The most
sensitive measure of ice film thickness is provided by the energy of the backscattered helium ions from the gold marker layer under the ice. The backscattered helium ions also directly reveal any heavy-atom constituents of a film,
such as carbon, oxygen and sulfur, providing an independent measure of the

molecular thickness of the film and a measure of its stoichiometry at various stages of erosion (W. L. Brown et al. 1980*a*,*b*; Lepoire et al. 1983).

The sputtering yields Y (molecules lost per incident ion) at normal incidence are studied as a function of beam current. As Y is found to be independent of the current in the range of the experimental parameters, the measured erosion occurs as a result of individual ions of the beam, not through a macroscopic heating of the film (or substrate).

Quadrupole mass spectrometry has been used with the Bell Laboratories apparatus as well as with experimental setups at other institutions to measure

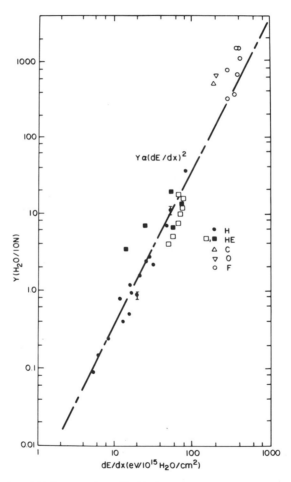

Fig. 6. Sputtering yield for water ice at liquid nitrogen temperature, showing that the yield Y scales as the square of the electronic stopping power $(dE/dx)^2_e$, for several different incident ions. ($\bullet$, $\blacksquare$, $\triangle$ and $\triangledown$) data from W. L. Brown et al. (1980*a*,*b*); ($\square$) data from Ollerhead et al. (1980); ($\circ$) data from Cooper and Tombrello (1983) and Seiberling et al. (1982).

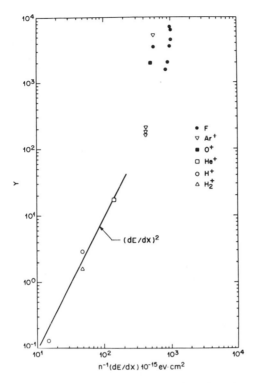

Fig. 7. Sputtering yield for SO_2 ice, at low temperature < 60 K, for various incident ions. Fluorine data are from Melcher et al. (1982); other data from Lanzerotti et al. (1982).

the composition of the sputter ejecta during ion bombardment. It has also been used to examine the fragments and new molecules formed in the films by ion bombardment and released from the films during subsequent thermal heating in the absence of ion bombardment (Pirronello et al. 1982).

Water (H_2O) and sulfur dioxide (SO_2) ices, as well as methane (CH_4) ice, are of considerable interest for present-day studies of natural satellites, and these ices are concentrated upon here. A summary of the H_2O and the SO_2 sputtering yields measured in laboratory experiments are shown in Figs. 6 and 7 for various ion species incident on a thin ice film. In these plots, the stopping power dE/dx is the energy deposited per unit path length by the ion in the film. The stopping power is a tabulated or calculated quantity (see, e.g., Anderson and Ziegler 1977; Ziegler 1977).

The representation of $Y(H_2O)$ in Fig. 8 is particularly useful for astrophysical calculations. The points are the available data for incident protons and O^+-like ions (Johnson et al. 1984c). The dashed lines represent extrapolations of the yields as determined from conventional sputtering theory (Sigmund 1969) at low energies and according to the square of the electronic

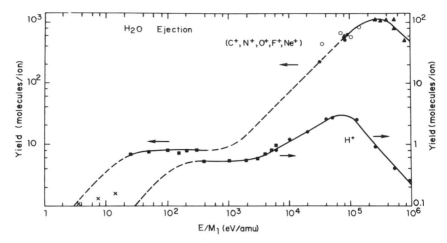

Fig. 8. Sputtering yield for H_2O ice at low temperature < 110 K for various incident ions. The dashed lines are extrapolations based on physical models. Data points were taken by various groups (see Johnson et al. 1984c).

stopping power (W. L. Brown et al. 1982a,b) at higher energies. Two distinct physical mechanisms contribute to sputtering. The nuclear (collisional) stopping power determines the yield at low energies and the electronic stopping power determines it at high energies. The two mechanisms are of comparable importance in the ~ 1 keV/amu energy region.

All of the data of Figs. 6 and 7 are taken at sufficiently low temperature that the yields are temperature independent. The dependence of the erosion yield on temperature is of particular importance for those satellites whose local daytime temperatures are close to the sublimation temperature. Figure 9 shows the temperature dependence of the erosion yield of H_2O, CO_2 and SO_2 for MeV incident He^+ ions. In these measurements, sublimation is subtracted and therefore the highest measured temperature in each case is limited by the sublimation rate in the absence of ion bombardment. As experimental measurements typically require hundreds to thousands of seconds, a practical upper limit for the temperature corresponds to an equilibrium vapor pressure of $\sim 10^{-7}$ torr, at which pressure sublimation would result in a loss of ~ 0.1 monolayer per second. All the yields show a temperature-independent region at low temperatures. At higher temperatures, two different types of behavior are evident. The H_2O and SO_2 curves show a temperature-independent region at low temperatures with a continuous rise to the sublimation limit. For CO_2 there are two sharp steps upward in Y at about 40 K and 50 K. The low-temperature region has been associated with a direct ejection process (W. L. Brown et al. 1982b), while the temperature-dependent regimes for H_2O, CO_2 and SO_2 are associated with diffusion of, or reaction among, molecular frag-

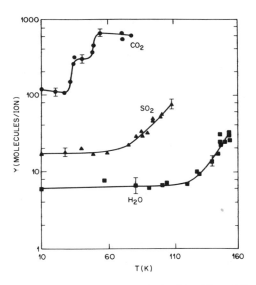

Fig. 9. Temperature dependence of sputtering yields for CO_2, SO_2 and H_2O ices, under bombardment by 1.5 MeV He^+ ions (figure from W. L. Brown et al. 1982b and Lanzerotti et al. 1982).

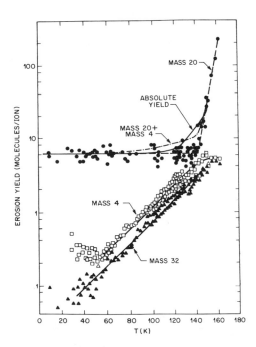

Fig. 10. Temperature-dependent sputtering and dissociation of D_2O ice under bombardment by 1.5 MeV He^+ ions. At low temperatures, sputtering of D_2O molecules dominates dissociation. As the temperature increases, the yields of dissociation products D_2 (mass 4) and O_2 (mass 32) increase rapidly. Dissociation products comprise a large fraction of the total yield above 130 K (figure from W. L. Brown et al. 1982b).

ments formed in the ice films as a result of intramolecular bond breaking due to the ionization produced by the incident ions (W. L. Brown et al. 1980b, 1982b). In the case of H_2O ice, the major constituents involved in the high-temperature increase of Y are H_2 and O_2 molecules. The yields of the constituents are shown as functions of temperature in Fig. 10, where the experiments were performed with D_2O ice in order to increase the signal-to-noise ratio in the mass spectrometer.

It is well known that organic materials (such as methane) can be polymerized by irradiation. Indeed, irradiation is a step in the fabrication process of many polymers, including the polyethylene used in cable sheathing. Cheng and Lanzerotti (1978) used initial Bell Laboratories results on the effects of particle irradiation on methane ice to propose that darkening on the rings of Uranus could be produced by irradiation of initially methane ice rings by ions and electrons trapped in the planet's magnetosphere. Laboratory experimental results have been reported by Calcagno et al. (1983a,b) on the irradiation, by 600 to 1500 keV protons, of a methane atmosphere and on the polymerization of benzene ice. Residues have also been reported from the irradiation of other ices and ice mixtures (see, e.g., Lepoire et al. 1983; Moore 1984; Moore et al. 1983; Haring et al. 1983; Foti et al. 1984).

Some Bell Laboratories results on the polymerization process as a function of fluence for 1.5 MeV helium ions incident on CH_4 ice (at 10 K) are shown in Fig. 11. The Rutherford backscatter technique was used to measure both the overall change in film thickness and the overall decrease in the number of carbon atoms in the film. These determinations can then be used to deduce the gross chemical composition of the residue that is formed, i.e., the

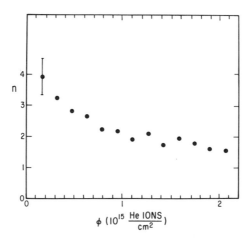

Fig. 11. Production of residual polymer with a hydrogen to carbon ratio equal to n from CH_4 ice as a function of 1.5 MeV helium ion fluence (Lanzerotti et al. 1985).

amount n of hydrogen in the polymer CH_n. The results indicate a roughly exponentially decreasing content of H in the film. It is not yet clear whether a nonzero equilibrium value is attained or whether H will slowly approach zero with increasing fluence. Irradiation by other ions at a range of incident energies is also under investigation.

III. ION IRRADIATION OF THE ICY SATELLITES

The penetration depth vs. ion energy for incident H^+, O^+ and S^+ on H_2O and SO_2 ices are shown in Fig. 12. These numbers are reliable to about 30%. The number densities 3.2×10^{22} mol cm^{-3} and 1.8×10^{22} mol cm^{-3} are used for H_2O and SO_2, respectively, to convert column densities to depths. The penetration depths are for perpendicular incidence; for nonperpendicular incidence, the values should be divided by the cosine of the angle between the surface normal and the incident direction. The values are also average penetration depths. Whereas at high energies the distribution in penetration depths (straggling) about the average is small, the ratio of the rms deviation to the penetration depth becomes of order unity at the lowest energies shown. Finally, at the lower energies, for light ions in particular, a significant fraction of the incident ions will backscatter.

As is evident from Fig. 12, the penetration effect of ions into an icy material is very much a surface phenomenon. Even for protons of 1 MeV the penetration depths are only of the order of tens of microns. Therefore, even though sputtering may remove significant amounts of material from a surface over long periods of time, the physically and chemically modified region at

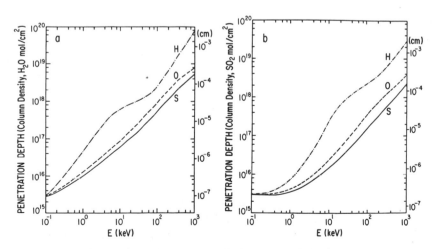

Fig. 12. (a) Penetration depth of hydrogen, oxygen, and sulfur ions into H_2O ice (Johnson et al. 1983b). Axes are given as column densities of water molecules and as cm of solid water ice at density of 3×10^{22} mol cm^{-3}. (b) As in (a) for SO_2 ice (Lanzerotti et al. 1982).

any time is a very narrow layer at the surface. However, such depths are enough to modify the albedo of a satellite (see the chapter by Veverka et al.).

In some cases incident ions cannot reach the satellite surface. Below a certain energy range, ions can be excluded by the existence of a satellite atmosphere or an intrinsic satellite magnetic field. In the former case, a small atmosphere could exist from sublimation (Kumar 1984) or from the sputtering process itself (Johnson et al. 1981; Watson 1981; Sieveka and Johnson 1985). In the latter case, one has the intriguing possibility of a self-limiting sputtered atmosphere (Lanzerotti et al. 1978).

Albedo Effects

Changes in the albedo of a satellite by particle bombardment can occur from chemical alterations of the surface material, from selective erosion effects, and/or from erosion and redeposition of the eroded species on the surfaces (see, e.g., Mendis and Axford 1974). One of the first experimental reports of the alteration of ices was one of visual observations (W. L. Brown et al. 1978) that the irradiation of clear amorphous water ice produced a highly scattering surface and a dendritic growth, altering the appearance of the ice. Such effects are produced only by energetic penetrating ions; this has been suggested as a possible source of a polar frost on Ganymede (Johnson 1985).

Changes in the reflectance spectrum can also be produced by changes in the chemistry of the surface layer. We concentrate in this chapter on charged particle effects on satellites. However, we note that Squyres and Sagan (1983) have proposed ultraviolet processing of CH_4 ice as important for producing the dark material on Iapetus, and Greenberg and his collaborators (see, e.g., Greenberg 1982) have carried out extensive laboratory experiments on ultraviolet irradiations of ice mixtures. Chemical changes have been observed in mixtures of ices irradiated by ions (Moore and Donn 1982). Formation of formaldahyde has been detected following bombardment of an equal mixture of H_2O and CO_2 (Pirronello et al. 1982). Sulfur ions which penetrate ices containing oxygen form SO bonds, giving rise to new features in the reflectance spectra. Lane et al. (1981) observed a band in the reflectance spectra of Europa's trailing hemisphere which they have identified as being produced by SO bonds, presumably caused by the bombardment of Europa's surface by magnetospheric sulfur ions. As noted in Fig. 10 for the case of D_2O ice, chemical activity can depend upon surface temperature. In addition, some laboratory experiments with SO_2 ice have shown an exposure time dependence on the formation of SO_3 (Johnson et al. 1984a; Moore 1984), a result that may be relevant to the case of ion bombardment of Europa and Io.

As discussed above, volatile compounds are generally more susceptible to sputter erosion than refractory ones, particularly at higher energies. This suggests that ion bombardment of an inhomogeneous target consisting of particles of different volatility may produce a surface layer enriched in the nonvolatile component(s). This process could be important for considerations of altered albedos because inhomogeneous surfaces are expected to characterize the regolith of many of the icy bodies in the solar system. Predominantly icy matrices are likely to be contaminated with bits of more refractory debris— remnants of meteoroid impact and/or of incomplete differentiation of the satellite crust.

Sublimation, if it is dominant, discriminates most effectively between ice and refractory grains; ice grains can eventually vanish completely, while silicate or other refractory grains will remain unaltered. Thermal evaporation of H_2O molecules is effective only where surface temperatures are high enough, as is the case in the equatorial zones of the Jovian satellites (Purves and Pilcher 1980; Sieveka and Johnson 1982). On the Saturnian satellites, sublimation is negligible.

Sputtering by energetic heavy ions (e.g., O^+ near or above 1 MeV) may also be an effective silicate enrichment mechanism. Ices tend to sputter much more rapidly in this regime (W. L. Brown et al. 1982b; Cooper and Tombrello 1983) than refractory solids (Qiu et al. 1983). The ratio of silicate to water ice sputtering yields may be as small as $\sim 10^{-3}$ for these energetic heavy ions. For protons below ~ 10 keV (as in the solar wind), the ratio is $\lesssim 0.1$. These processes have been considered in some detail by Haff (1980), who pointed out that such preferential sputtering could eventually produce a mineral "armor" over the underlying ice, ultimately decreasing the net erosion yield. Conca (1981) suggested that such a process might be relevant as an explanation of the dark-ray craters found on Ganymede.

Considerations of possible armoring should be tempered by the absence of relevant laboratory data and by other processes that may be operative, including intrinsic geologic activity, gardening by meteoroid bombardment, and redistribution of material. On satellites like Europa and Enceladus, the existence of relatively uncratered regions of large extent suggests the occurrence of some geological resurfacing mechanism (Lucchitta and Soderblom 1982; Smith et al. 1981). Stirring or gardening of the surface by meteoroid bombardment is a continuing process that will compete with an armor formation mechanism. No laboratory data exist as yet on the removal rate of small quantities of refractory or other less volatile atoms embedded in icy matrices. It is possible that enhanced yields of less volatile material could be produced from such a system, if this material is mixed into the ice on the molecular level. This may be relevant in terms of removal of such materials as sodium and potassium from Io (see Sec. IV). Finally, the creation of nonvolatile residues from volatile materials can have the same effect as preferential erosion, creating a less volatile, sputter-resistent surface layer. This may be occurring in the polar regions of Io and on the satellites and rings of Uranus.

The characteristics of redistributed materials on a surface depend on the ratio of the velocity of the ejected species to the satellite's escape velocity and on the nature of any anisotropies in the bombarding particle fluxes. Sieveka and Johnson (1982) have studied redistribution for the Jovian and Saturnian

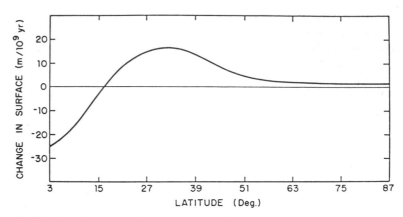

Fig. 13. Erosion and net accumulation rates on the leading hemisphere of Europa from charged
sputtering of H_2O ice.

satellites. An example for the case of Europa is shown in Fig. 13 (solid line)
for low-energy plasma-ion and order 100 keV-energy ion bombardment from
fluxes measured by the Plasma Science and Low Energy Charged Particle
(LECP) instruments on Voyager 1. The calculation includes loss from sublimation based on an albedo of 0.6 and assumes a sine-squared pitch angle
distribution for the order 100 keV-energy ions. Over the entire trailing hemisphere, a net loss of material occurs at all latitudes from direct sputter ejection
and redistribution across the satellite's surface. Over most of the leading hemisphere, a net accumulation occurs as shown in Fig. 13. This combination of
erosion and net accumulation can account for the observations by Lane et al.
(1981) of SO from S^+ implantation in water ice only on Europa's trailing
hemisphere. Alternatively, Eviatar et al. (1981) have suggested that geologic
resurfacing may account for the SO observation.

When the velocities of the incident ions are high enough for electronic
sputtering to be important (above a few keV amu^{-1}), dissociation of target
molecules will occur. For ices like H_2O, at a sufficiently high temperature,
the dissociation products escape and the target composition remains H_2O
(Fig. 10). On the other hand, bombardment of carbon-bearing ices typically
leads to chemical alteration of the target. Bombardment of an H_2O/CO_2 ice
mixture by 1.5 MeV He$^+$ leads to formation of formaldehyde (H_2CO) which
remains trapped within the ice mixture (Pirronello et al. 1982). As noted
earlier, energetic ion bombardment of benzene and methane ices leads to
formation of a black carbon-enriched, polymerized residue. Darkening and
polymerization of methane ice surfaces at Uranus from energetic ion bombardment may explain the very low albedo of the Uranian ring particles
(Cheng and Lanzerotti 1978; Pang and Nicholson 1985).

Production of Atmospheres

The production and maintenance of an atmosphere on a satellite can occur via sublimation and/or sputtering. Both processes depend on the satellite and eroded species as discussed above. The fluxes of molecules sputtered from water ice surfaces for some of the icy satellites of Jupiter and Saturn are listed in Table II. The sputtering is the estimated total yield for fast ions, assuming that the order 100 keV ions are all protons or all oxygen and that all ion fluxes are isotropic. (In the above, fast ions are distinguished from slow ions in that they have a gyrospeed large compared to the corotation speed.)

The relative importance of sputtering and sublimation for producing an atmosphere depends on the velocity distributions of the molecules leaving the surfaces for these two processes. The mean speeds for certain characteristic sputter velocity distributions are given by Johnson et al. (1983a). For the Galilean satellites of Jupiter, the characteristic velocities of sputter ejecta are typically lower than the escape velocity but higher than thermal velocities at surface temperatures (Sieveka and Johnson 1982). This has two ramifications: the scale height of a sputtered atmosphere will be larger than that of a sublimed atmosphere (Sieveka and Johnson 1985), and the sputtered molecules will make larger angular excursions across a satellite's surface. However, once sublimation dominates sputtering, the number of molecules involved increases rapidly. Therefore, sublimation has the effect of creating an exobase, from which molecules can be sputtered (Haff and Watson 1979; Sieveka and Johnson 1985). As this new surface is at a greater distance from the center of the satellite, the lower energies for gravitational escape can enhance the sputter ejection of molecules from the exobase (Haff et al. 1981).

On the icy satellites of Saturn and Uranus, a large fraction of the sput-

TABLE II
Bombardment by LECP-Measured Ions by Voyager 1

Satellite	Escape Flux $(H_2O \ cm^{-2} \ s^{-1} = 10^{-11} \ \mu m \ yr^{-1})$	
	Proton Incident	Oxygen Incident
Jupiter		
Europa	5×10^6	2×10^9
Ganymede	2×10^6	6×10^8
Callisto	2×10^5	7×10^7
Saturn		
Tethys	5×10^6	4×10^8
Dione	6×10^6	5×10^8
Rhea	2×10^6	2×10^8

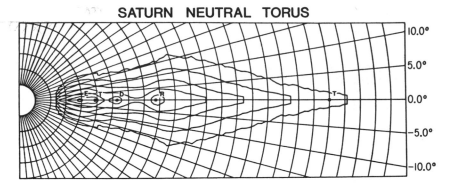

Fig. 14. Density contours for heavy atoms (molecules: H_2O, O, OH) produced by sputtering of satellite surfaces by primarily O^+ with energies of ≥ 20 keV. The outside contour corresponds to 10^{-1} atoms cm^{-3}; succeeding contours are increases by factors of $10^{1/2}$ in the densities. The letters E, T, D, R, T from left to right along the mid-line correspond to the satellites Enceladus, Tethys, Dione, Rhea and Titan. The Titan torus is not shown.

tered molecules can escape directly (Lanzerotti et al. 1983). These molecules are eventually dissociated and ionized and add to the magnetospheric particle population. Using the incident ion fluxes measured by Voyager, Johnson et al. (1984a) have calculated the sputtering loss of surface materials from these satellites. Electron impact ionization and photoionization were included in a Monte Carlo program (Barton 1983) which performed two-body orbit calculations to find the neutral density in the Saturnian magnetosphere (Fig. 14). For these results, an average escape velocity was used and the ejections from the satellites were assumed to be isotropic. The neutral density in the E ring region produced by the satellites is nonnegligible. This is the region where Pioneer and Voyager experiments reported an oxygen plasma (Frank et al. 1980; Bridge et al. 1982) and where there also appears to be an energetic oxygen plasma (Krimigis et al. 1983).

IV. INJECTIONS OF PLASMAS INTO MAGNETOSPHERES: THE SPECIAL CASE OF IO

The discovery of abundant sulfur and oxygen ions in the Jovian magnetosphere was one of the great surprises of the Voyager mission. Sulfur and oxygen ions appear to dominate the total mass and positive charge densities in the magnetosphere, and even at MeV energies their number densities may be comparable to the proton number density at similar energy per charge. This dominance by sulfur and oxygen ions makes the Jovian magnetosphere unique among astrophysical plasmas known at present. With the discovery of SO_2 volcanism on Io and the discovery of SO_2 frost features in Io's infrared absorption spectra (see the chapters by Nash et al. and by Clark et al.), it soon became obvious that Io must be the ultimate source of S and O ions in the magnetosphere. Even before Voyager, Hill and Michel (1976) and Lanzerotti et al. (1978) proposed that oxygen ion injection from water ice-covered Galilean satellites may be dynamically important in Jupiter's magnetosphere. This section discusses the influences of the Jovian magnetosphere upon Io and the importance of Io for determining magnetospheric processes.

Io's Atmosphere and Surface

While the physical and chemical properties of Io's surface and atmosphere are discussed in considerable detail in the chapter by Nash et al., it is useful to summarize here some of the more important facts relevant to magnetosphere-satellite interactions. The Pioneer 10 radio occultation experiment discovered an ionosphere at Io (Kliore et al. 1975), indicating the presence of an atmosphere. Gaseous SO_2 was identified near the volcanic plume Loki by the Voyager infrared experiment (Pearl et al. 1979), consistent with an SO_2 atmosphere in local vapor pressure equilibrium with the surface or with transient flow from Loki (Kumar and Hunten 1982). While the existence of an SO_2 atmosphere is well established, its nature remains a matter of considerable debate.

The existence of SO_2 frost on Io's surface is also established, but the physical properties of the surface (e.g., roughness; see chapter by Veverka et al.) and the chemical composition (see chapter by Clark et al.) are controversial. A consensus view is that infrared and ultraviolet reflectance measurements indicate coverage of 20% to 50% of Io's surface by optically thick SO_2 frost, while a comparable fraction is covered with elemental free sulfur (Soderblom et al. 1980; Nash et al. 1980; Fanale et al. 1982). Significant sodium and potassium must be present in some form, and silicates are also probably present at or near the surface (Fanale et al. 1982).

Two types of Io atmosphere models have been proposed: "thick" models, meaning those with an exobase well above Io's surface, and "thin" models. Kumar (1979,1980,1982) has proposed photochemical models of thick atmospheres, with the day side lower atmosphere mainly SO_2 in vapor pressure equilibrium with the surface, and the upper atmosphere mainly O and S owing to photodissociation and diffusive separation (Kumar and Hunten 1982). The dominant escape mechanism for neutrals in this case is sputtering by corotating ions from the top of the atmosphere (Haff et al. 1981), with relatively small contributions from thermal escape of O plus local ionization and pickup (Kumar 1982). For thin atmosphere models, sputtering from the surface of Io would be an important escape mechanism (Matson et al. 1974; Haff et al. 1981; Johnson et al. 1983). A thin atmosphere can be maintained by sputtering (Lanzerotti et al. 1982; Lanzerotti and Brown 1983; Sieveka and Johnson 1985), or it may be in equilibrium with a very cold subsurface regolith (Matson and Nash 1983).

Plasma Injection into the Io's Plasma Torus

Estimates of the sulfur and oxygen escape fluxes required to explain observations of Io's plasma torus are now discussed. Sulfur and oxygen, in the form of neutral atoms or molecules, typically escape from the vicinity of Io before being ionized, thereby forming extended neutral clouds within the Io plasma torus (Cheng 1980; R. A. Brown and W.-H. Ip 1981; R. A. Brown 1981; R. A. Brown et al. 1983a). When a neutral atom or molecule becomes ionized, it is immediately picked up by Jupiter's rotating magnetic field, acquiring a guiding center velocity equal to the corotation velocity and a gyrovelocity equal to the difference between the corotation and Kepler orbit velocities. The energy source for this pickup process is the electric field induced by the rotation of Jupiter's magnetic field, so the ultimate energy source is Jupiter's rotational kinetic energy. The energy input to ion gyromotion, if transferred to electrons, would be adequate to maintain the sulfur and oxygen ultraviolet line emissions measured from Io's plasma torus, assuming ionization of $\sim 10^{30}$ amu s^{-1} (Broadfoot et al. 1979). Additional energy is extracted from Jupiter's rotation as plasma is transported outward, and this energy release, for injection of 10^{30} amu s^{-1}, is roughly adequate to power Jupiter's ultraviolet aurora (Eviatar and Siscoe 1980; Dessler 1980). While these are only order of magnitude estimates, and the microscopic physics leading to optical and ultraviolet emissions is still unclear, the overall mass and energy budgets appear to be established (Hill et al. 1983a).

R. A. Brown (1981) has detected 6300 Å forbidden line emissions from neutral oxygen in the plasma torus far from Io, directly verifying that neutrals escape the vicinity of Io. A neutral sulfur cloud has also been observed in the plasma torus (Durrance et al. 1982). The measurement of 6300 Å [O I] brightness has led to independent estimates of the plasma injection rate to the Io torus. R. A. Brown (1981) noted that the inferred oxygen density implied an ionization of $\sim 10^{28}$ O s^{-1} in a uniform torus model.

Maintenance of the neutral oxygen cloud requires oxygen escape from Io, in the form of neutral atoms or oxygen-bearing molecules or both. Cheng (1982) proposed that escape, ionization, and dissociation of SO_2 from Io could maintain neutral oxygen and sulfur clouds. The SO_2 electron impact ionization rates have been measured (Orient and Srivastava 1984), but the electron impact dissociation rates remain poorly known. Cheng (1984a) estimated that escape of 1 to 4×10^{28} SO_2 s^{-1} would maintain the observed oxygen cloud and yield an average neutral sulfur density of 9 cm^{-3}. A sulfur density of 2 to 6 cm^{-3} was predicted by Johnson and Strobel (1982) in a nonuniform model. On the other hand, atomic oxygen and sulfur may also escape from Io. Smyth and Shemansky (1983) estimated that escape of 4×10^{27} O s^{-1} would maintain the oxygen cloud. Finally, the population of sulfur and oxygen ions in the plasma torus can be maintained by ionization of 7×10^{27} O s^{-1} and 3.5×10^{27} S s^{-1} (R. A. Brown et al. 1983a).

Thus, it can be concluded that $\sim 10^{30}$ amu s^{-1} in the form of S and O ions are injected into the plasma torus, and that neutrals mostly escape from Io before ionization, forming neutral clouds. The Voyager upper limit of $\sim 10^{27}$ O and S ionizations per second in the vicinity of Io (Shemansky 1980) is consistent with these conclusions, as the majority of ionizations occur far from Io in the neutral clouds. However, it remains unclear whether the sulfur and oxygen escape from Io mainly as molecular SO_2 or as atomic O and S. Independently of whether sulfur and oxygen escape separately or together in molecules, significant escape of SO_2 from Io must also occur (Cheng 1984a). This is because ions of mass/charge $= 64$, identified as SO_2^+, have been observed in the plasma torus far from Io, e.g., at 5.3 R$_J$ (Belcher 1983). S_2^+ would be a less likely identification if S_2 has relatively low sublimation and sputtering yields. However, these have recently come into question, and Linker et al. (1985) have suggested an S_2^+ identification. The mass 64 ions, whether SO_2^+ or S_2^+, cannot have been injected at Io since their lifetime against electron impact is very short compared to the radial transport time from Io's orbit. Thus they must have been injected locally, implying the presence of the parent molecule in the plasma torus.

Escape from Io

Thin Atmosphere. The incident ion fluxes important for sputtering from Io can be obtained from Voyager 1 measurements made in the plasma torus. Important contributions to the sputtering arise from corotating magnetospheric ions, whose energy of random motion is much less than that of magnetospheric corotation relative to Io (17 eV amu^{-1}), and from more energetic ions with energies above a few tens of keV. The corotating ions are incident only on Io's trailing hemisphere, whereas the more energetic ions can be incident over the entire surface, although not necessarily uniformly (Sieveka and Johnson 1982). An intrinsic Io magnetic field would cause exclusion from the surface of some low-energy ions and preferential impact areas for higher energies (see above).

The incident fluxes of corotating ions and energetic ions are given in Table III (Cheng 1984a). The corotating ion flux is obtained either from the isothermal model or the common thermal speed model of Bagenal and Sullivan (1981). Table III assumes the sputter yield for corotating SO_2^+ to be the same as for sulfur ions. The energetic ion fluxes are obtained from Voyager 1 LECP measurements (Krimigis et al. 1981a; Lanzerotti et al. 1982). The relative abundances of protons and heavy ions are unknown for the energetic ions. The energetic ion fluxes of Table III are valid for energies of 0.22 to 4 MeV (assuming protons) or for 0.37 to 4.4 MeV (assuming oxygen ions).

The escape rate of sulfur and oxygen driven by sputtering in the thin atmosphere case can be found by neglecting collisions with atmospheric neutrals for both the incident ions and the sputter ejecta. With the assumptions that 20% of Io's surface is covered by SO_2 frost and that 3% of the sputtered

TABLE III

Incident Ion Fluxes and Sputtering Yields at Io

	Incident Flux	Sputter Yield	Direct Escape Rate of Sputter Ejecta[a]
Corotating oxygen ions	$1.6 \times 10^9 \ cm^{-2} \ s^{-1 \ b}$	50	$5 \ \times 10^{25} \ SO_2 \ s^{-1}$
Corotating sulfur ions	$5.1 \times 10^9 \ cm^{-2} \ s^{-1 \ b}$	80	$2.6 \times 10^{26} \ SO_2 \ s^{-1}$
Corotating SO_2^+	$4 \ \times 10^8 \ cm^{-2} \ s^{-1 \ b}$	80?	$2 \ \times 10^{25} \ SO_2 \ s^{-1}$
Corotating oxygen ions	$6.2 \times 10^9 \ cm^{-2} \ s^{-1 \ c}$	50	$2.0 \times 10^{26} \ SO_2 \ s^{-1}$
Corotating sulfur ions	$3.6 \times 10^9 \ cm^{-2} \ s^{-1 \ c}$	80	$1.8 \times 10^{26} \ SO_2 \ s^{-1}$
Energetic ions (if all H^+)	$1.77 \times 10^5 \ E^{-1.64} \ cm^{-2} \ s^{-1} \ sr^{-1} \ MeV^{-1}$	~ 1	$9.7 \times 10^{21} \ SO_2 \ s^{-1 \ d}$
Energetic ions (if all O^+)	$2.58 \times 10^5 \ E^{-1.80} \ cm^{-2} \ s^{-1} \ sr^{-1} \ MeV^{-1}$	$1330 \ E$	$1.4 \times 10^{25} \ SO_2 \ s^{-1 \ d}$

[a] Assuming 20% SO_2 coverage and 3% escape fraction. The total loss and escape rate, including contributions from the sputter corona, can be several times the direct escape rate.

[b] Isothermal model of Bagenal and Sullivan (1981).

[c] Common thermal speed model of Bagenal and Sullivan (1981).

[d] Only energetic ion fluxes above 220 keV (for H^+) or 370 keV (for O^+) are included here. The text considers extrapolations to lower energy. The particle energy is E in MeV.

SO_2 can escape Io's gravity (Johnson et al. 1984a), the corotating ion fluxes of Table III, if incident uniformly on one hemisphere of Io, yield an escape rate of $\sim 3 \times 10^{26}$ SO_2 s^{-1}. While the SO_2 coverage on Io may exceed 20%, two other effects could reduce the estimated SO_2 sputtering rate substantially. First, if Io's surface is rough or exhibits a fairy-castle structure (Matson and Nash 1983), the sputter yields may be reduced below the values given in Table III by up to an order of magnitude (Hapke and Cassidy 1978). Second, a corotating ion flux as intense as that in Table III would be incident on an entire hemisphere of Io only if there is no intrinsic magnetic field and effects of high electric conductivity are neglected (Schulz and Eviatar 1977).

Sputtering of SO_2 at Io by energetic ions can likewise be estimated using the data of Table III. An isotropic pitch angle distribution is assumed at 6 R_J (Lanzerotti et al. 1982; Cheng et al. 1983). A 3% escape fraction is assumed; this is probably an overestimate because electronic sputtering (the dominant mechanism of energetic ion sputtering) usually gives softer energy distributions for sputter ejecta than does collision cascade sputtering (Johnson et al. 1983a).

The SO_2 escape rate driven by corotating ion sputtering is greater than that driven by energetic ion sputtering, even if the energetic ions are assumed to be all oxygen. Extrapolation to lower energies of both the energetic ion fluxes and the sputter yields, according to the power laws of Table III, does not change this conclusion, even for a lower-energy limit of 30 keV for protons or 84 keV for oxygen ions (Cheng 1984a). Lanzerotti et al. (1982) also found corotating ion sputtering rates to be greater than energetic ion sputtering rates. However, it is cautioned that the actual energetic ion fluxes below 0.22 MeV (for protons) or below 0.37 MeV (for oxygens) may be much greater than the values predicted by extrapolation of the power laws in Table III.

In addition to the direct escape of sputter ejecta, Watson (1981; see also Sieveka 1983) has noted that a sputter corona will form around a satellite, consisting of sputter ejecta still gravitationally bound to Io. SO_2 in the sputter corona can be liberated, ionized and/or dissociated by interaction with solar ultraviolet photons and magnetospheric plasma. The rates for these processes are estimated as follows (Cheng 1984a). The average flight time of SO_2 in the sputter corona is 1.2×10^3 s (Johnson et al. 1984), while the lifetime against ionization is about 2×10^4 s. Hence a fraction $\sim 5\%$ of the SO_2 injected into the sputter corona will be ionized, implying ionization of about 6×10^{26} SO_2 s^{-1}. A somewhat smaller number $\approx 9 \times 10^{25}$ SO_2 s^{-1} will be scattered into the plasma torus by collisions with corotating ions. However, the rate of SO_2 dissociation in the sputter corona is high and may yield 0.3 to 3×10^{27} O s^{-1} supplied to the atomic cloud.

Thus, the total SO_2 loss and escape rate in the thin atmosphere case, including direct escape from the surface, and ionization and dissociation in the sputter corona, is marginally adequate to maintain the plasma torus in the most optimistic case (Cheng 1984a). However, it is likely that the sputtering rates given here are overestimated by an order of magnitude because effects of surface roughness and Io's ionospheric conductivity have been neglected. Thus the thin atmosphere model faces severe difficulty in achieving adequately high escape rates of sulfur and oxygen, but it is not yet ruled out.

Thick Atmosphere.   If Io has a thick atmosphere, corotating ions can sputter atmospheric neutrals from the exobase region (Haff and Watson 1979; Haff et al. 1981). Such atmospheric sputtering may be similar to collision cascade sputtering in solids (Sigmund 1969). If so, the sputtering yield for a given atmospheric species would be roughly proportional to U_0^{-1}, where U_0 is the gravitational binding energy at the exobase. The sputtering yield would then be predicted to increase as U_0^{-1}, as the exobase altitude increases. If exobase altitudes comparable to Io's radius are feasible, then atmospheric sputtering can yield escape rates far greater than those for sputtering at Io's surface (Haff et al. 1981).

Kumar (1982) has analyzed models for thick atmospheres in photochemical equilibrium near the day side of the evening terminator. In these models, SO_2 is the dominant constituent near the surface, but with photodissociation and diffusive separation, O is dominant in the upper atmosphere. On the night side, O_2 is proposed as the major atmospheric constituent near the surface, but again O is dominant at high altitudes (as in the Earth's atmosphere). In these models, atmospheric sputtering is expected to yield very high escape rates for O, but much lower escape rates for either SO_2 or S, because these are only minor constituents near the exobase and because they have greater mass.

Kumar (1982) scaled the Haff et al. (1981) surface sputtering yield according to U_0^{-1} in order to estimate yields for an exobase altitude of 1400 km. After allowing for the relative abundances of S and O near the exobase, he found escape fluxes $\sim 3.5 \times 10^{11}$ cm^{-2} s^{-1} for O and $\sim 3.3 \times 10^9$ cm^{-2} s^{-1} for S (assuming a chemically inert surface; the oxygen escape rate remains roughly the same in a model with a chemically active surface, but the sulfur escape rate decreases to $\ll 3 \times 10^9$ cm^{-2} s^{-1}). If these thick atmosphere models apply over Io's entire trailing hemisphere, the oxygen escape rate would be $\sim 10^{29}$ O s^{-1} for either model, while the sulfur escape rate would be $\sim 10^{27}$ S s^{-1} for the inert surface model, and much less for the active surface model.

Thus, in the thick atmosphere models with photochemical equilibrium and diffusive separation, the atomic oxygen escape rate can be far in excess of that needed to maintain the plasma torus, which is $\sim 7 \times 10^{27}$ s^{-1} (R. A. Brown et al. 1983a) or $\sim 4 \times 10^{27}$ s^{-1} (Smyth and Shemansky 1983). However, the sulfur escape rate is two orders of magnitude lower than the oxygen escape rate, whereas comparable escape rates are required. The SO_2 escape rate will be negligible in these models, which leads to difficulty in understanding the observation of mass 64 ions in distant regions of the plasma torus (Bagenal and Sullivan 1981; Cheng 1984a). The oxygen escape rate can be reduced from 10^{29} s^{-1} to a value more consistent with plasma torus observations by adjusting the models to reduce the exobase altitude (Kumar 1984). However, such an adjustment would also reduce the escape rates for S and SO$_2$, leaving these far below the required rates.

It is concluded that a large ratio of oxygen-to-sulfur escape rates in the thick models with diffusive separation leads to severe difficulties in understanding the plasma torus. Cheng (1984a) has suggested that if comparable escape rates for sulfur and oxygen are required, then diffusive separation in the upper atmosphere must be limited, perhaps by rapid mixing. Volcanic venting, interactions with corotating plasma, and extreme SO$_2$ vapor pressure differences between day side and night side are expected to drive strong winds in the atmosphere. Hunten (1985) has also noted that mutual drag effects may be important in Io's atmosphere, and that these would tend to equalize the scale heights of O, S and SO$_2$.

Sodium and Potassium. An atomic sodium cloud near Io was reported in sodium D lines by R. A. Brown (1974) and studied by several investigators, including Trafton et al. (1974), Mekler and Eviatar (1974) and Lanzerotti et al. (1975). This emission comes from resonant scattering of sunlight, as opposed to sulfur and oxygen emissions from the plasma torus, that are maintained largely by electron impact excitation. The sodium cloud near Io is mainly inside of Io's orbit and extends forward from Io about 70° in orbital longitude (Matson et al. 1978a; Murcray and Goody 1978). Detections of sodium at much greater distances from Io have been reported but remain controversial (see review by Brown et al. 1983a). Maintenance of the near-Io sodium cloud, excluding the immediate vicinity of Io, requires escape of $\sim 2 \times 10^{25}$ Na s^{-1} from Io (Smyth and McElroy 1978). However, maintenance of populations remote from Io, if present, may require sources of $\sim 10^{26}$ Na s^{-1} (Brown and Schneider 1981), and inclusion of ionization losses very near Io may raise the required sodium escape rate to $\sim 10^{27}$ Na s^{-1} (Brown et al. 1983a). The Io potassium cloud, while generally similar to the sodium cloud, has not been as well studied (Trafton 1975b,1981).

Matson et al. (1974) originally suggested sputtering by magnetospheric ions as a mechanism for removing sodium from Io. Sputtering from Io's surface and atmospheric sputtering remain the most widely favored hypotheses for driving sodium escape from Io, but quantitative tests of these hypotheses are not yet feasible. Not only are the required escape rates somewhat uncertain, but the physical and chemical nature of substances containing Na and K at Io, as well as the relative abundances, remain largely unknown. Laboratory sputtering data are also generally unavailable for complex minerals containing Na and K, possibly mixed with sulfur, silicates and volatiles.

If Io's atmosphere is thin, sputtering by corotating ions is expected to be the dominant escape mechanism for Na (Haff et al. 1981; Kumar and Hunten 1982; Fanale et al. 1982). As seen in Table III, up to 10^{27} corotating heavy ions impact Io per second if an intrinsic magnetic field and effects of electric conductivity are neglected. Thus, an average sputter yield of ~ 1 Na above 2 km s^{-1} (Io's escape velocity) per corotating heavy ion impact may be required to achieve an escape of 10^{27} Na s^{-1}. A yield of ~ 0.02 Na ejected at speeds of several km s^{-1} per heavy ion impact would supply $\sim 2 \times 10^{25}$ Na s^{-1} to the near-Io sodium cloud. Such yields are not inconceivable.

However, if Io's atmosphere is thick, the corotating ions will not be able to reach the surface, and sputter ejecta from the surface will not be able to escape Io. In this case, efficient sputtering occurs from the exobase. Kumar and Hunten (1982) have suggested that Na and K atoms or compounds can be incorporated into Io's atmosphere and then ejected from Io by atmospheric sputtering. One possibility is that fine dust from volcanic vents may contain Na and K. Another possibility is that Na and K may be liberated from Io's surface by energetic ion sputtering (Summers et al. 1983), if the atmospheric column density is not much in excess of 5×10^{17} cm^{-2}. Observation of an energetic Na atom cloud with strong directional characteristics (Pilcher et al. 1984) can be explained by collisional ejection from an exosphere (Sieveka and Johnson 1984). This would also require a mixed atmosphere, as suggested above.

The spatial distribution of the atomic sodium cloud is asymmetric about Io, with the asymmetries very different for energetic Na (~ 10 to 50 eV) and low-energy Na (< 5 eV). This asymmetry can result from asymmetries in either the source or loss mechanisms, or both. Summers et al. (1983) have discussed source asymmetries, while modeling (Smyth 1983) demonstrates the importance of spatial asymmetries in ionization (loss) rates (see review by Brown et al. 1983a).

V. SATURN'S MAGNETOSPHERE AND RING MATERIAL

The rings of Saturn are bombarded by particles from the magnetosphere and by cosmic rays. These particle-ring interactions can significantly affect the ring material and can provide a source of neutral and ionized atoms to the magnetosphere. Considered below are the E ring and the inner, more substantial A and B rings.

E Ring

The E ring, lying between ~ 3 R$_S$ and ~ 8 R$_S$ (Baum et al. 1981; Burns et al. 1984), is an optically thin object (normal optical thickness $\tau \sim 4 \times 10^{-7}$ at Enceladus) composed apparently of ice grains a few micrometers in diameter (Pang et al. 1982,1984a). The outermost and most tenuous of the Saturnian rings, it is spatially associated with the icy satellite Enceladus, whose orbit at 4 R_S corresponds to the densest part of the ring (Baum et al. 1981). Bridge et al. (1982) reported the observation of a dense (30 cm^{-3}) heavy ion plasma (probably O) near the ring plane in the inner magnetosphere, confirming earlier observations by the Pioneer 11 spacecraft (Frank et al. 1980). At higher energies (> 50 keV) Krimigis et al. (1983) have identified significant oxygen densities. In the absence of a magnetic field or atmosphere associated with Enceladus, these particle populations will bombard the surface of that satellite directly, as well as the ice motes comprising the E ring. The incident flux of corotating heavy ions at 5 R_S is about 1.5×10^8 cm^{-2} s^{-1}, and the energetic ion flux there is about 10^6 cm^{-2} s^{-1}.

The ion fluxes and laboratory-derived sputtering rates can be used to estimate the lifetime of E ring grains in the magnetosphere. Under bombardment by the energetic ions measured by the LECP instrument on Voyager, the E ring particles have a lifetime estimated to be $\sim 10^4$ to 10^5 yr (Cheng et al. 1982), a value much less than the lifetime of the solar system. Haff et al. (1983) estimated that under bombardment by the low-energy plasma, the lifetime might be as short as a century or so. Since a significant fraction of the ions measured near the E ring by the LECP experiment on Voyager are heavy ions, grain lifetimes of the order of 10^3 yr are appropriate (Table II; Johnson et al. 1985a). Radial transport of E ring grains may also be rapid (Morfill et al. 1983b; review by Burns et al. 1984).

Thus, sputtering and transport processes limit the ring lifetime to geologically short times, unless a source of fresh ring material is available. One possibility is that the E ring is embedded within and maintained by a huge cloud of neutral H_2O molecules. Such a cloud would be maintained by H_2O ejection from the icy Saturnian satellites (see Fig. 14) and from the E ring itself, with the main H_2O ejection mechanisms being sputtering and perhaps meteoroid bombardment (see the discussion in the chapters by Schubert et al. and by Squyres and Croft). Loss mechanisms from the cloud would include solar ultraviolet photodissociation, electron impact ionization and dissociation, charge exchange, and sweeping by E ring grains. Ionization and dissociation of such a neutral H_2O cloud will maintain the heavy ion plasma torus near Tethys and Dione (Cheng et al. 1982; Johnson et al. 1984c). Otherwise there is no direct evidence for the existence of such a cloud. A key parameter would be the probability that an ejected H_2O molecule will be accreted onto an E ring grain rather than ionized or dissociated. If this probability is an appreciable fraction of unity, the accretion rate of H_2O onto the E ring grains may balance the loss rate to sputtering, and the E ring and its H_2O cloud may coexist in a steady state. Considering only Enceladus as a source, Haff et al. (1983) suggest a probability of ~ 0.02, in which case even optimistic estimates of micrometeoroid impact rates on Enceladus would not suffice to balance ring losses from sputtering. However, a more recently derived micrometeoroid impact flux (Morfill et al. 1983a) may require reevaluation of this source mechanism. McKinnon (1983a) has suggested that the E ring may

be the remnant of a catastrophic impact on Enceladus, in which case the ring
is truly an ephemeral object.

A and B Rings

A cloud of atomic hydrogen is associated with the main rings A and B of
Saturn. Such a ring atmosphere was predicted by Blamont (1974) and Denne-
feld (1974), and it was first detected by Weiser et al. (1977), who observed a
Lyman-α emission rate of 200 ($\pm$ 50%) Rayleigh (Ra). The existence of a
ring atmosphere was subsequently confirmed by Pioneer (Judge et al. 1980)
and Voyager (Broadfoot et al. 1981). Unfortunately, the spatial extent and geometry of the ring atmosphere are still not well defined by observations. The
Pioneer finding that the ring atmosphere emissions are brightest near the B
ring was not confirmed by Voyager. Earth-based observations, such as from
the International Ultraviolet Explorer (IUE), do not have adequate spatial
resolution to map the ring atmosphere; *in situ* observations have to contend
with separation of ring atmosphere emissions from those of the magnetosphere and the interplanetary medium. If the ring atmosphere is approximated
as a uniform sphere containing the rings, then Voyager observations (Broadfoot et al. 1981) imply an average density of ~ 600 H cm^{-3} and a total H
content $\sim 5 \times 10^{33}$. These parameters are consistent with the observations of
Weiser et al. (1977).

Recent IUE observations (Clarke et al. 1981) yielded the surprising result that the Lyman-α emission from Saturn's disk and/or ring atmosphere was
spatially nonuniform and variable on time scales of (Earth) days. Bursts of
emission at up to 1000 Ra were seen in either the northern or southern hemisphere above a relatively constant disk brightness of 700 to 800 Ra. These
bursts may be auroral phenomena or may originate from the ring atmosphere;
they are not understood.

The source mechanisms for the ring atmosphere include charged particle
sputtering of ice surfaces followed by dissociation, photosputtering, meteoroid bombardment and collisions between ring particles. Sublimation of
H_2O followed by dissociation is not a viable mechanism for ring particle
temperatures < 80 K (Cheng et al. 1982). Loss mechanisms for the ring
atmosphere include ionization, by electron impact or solar ultraviolet, and absorption by ring particles. The latter occurs because an H atom created near
the rings will pass through them once or twice per Kepler orbit, possibly encountering a ring particle; a single ring plane encounter is possible for orbits
of large apoapsis.

Carlson (1980) has estimated a lifetime of 3×10^5 s for H atoms against
absorption by ring particles, assuming unit optical depth and a sticking probability of 0.22 for atomic H on pure H_2O ice at low temperature. If correct,
this would be the dominant loss mechanism and would imply that a source of
$\sim 1.7 \times 10^{28}$ H s^{-1} is needed to maintain the ring atmosphere. However, this
estimate should be regarded with caution since it would imply a flux of $\sim 10^8$

H atoms cm^{-2} s^{-1} sticking to ring particle surfaces. This flux would suffice to cover the ring particles completely with H after a period of months. The sticking probability for H must be reevaluated, and the possibility of H atom recombination to form H_2 should be considered. A lower limit to the required H source rate is obtained by considering photoionization alone as a loss mechanism; this yields a required source of at least 3×10^{25} H s^{-1} to maintain the ring atmosphere against photoionization.

Charged particle sputtering of H_2O followed by dissociation was suggested as a source mechanism for the ring atmosphere by Cheng and Lanzerotti (1978) and was reexamined following Voyager by Cheng et al. (1982); they noted that this suggestion encounters severe difficulties. The total sputtering source of H_2O in Saturn's magnetosphere is about 10^{26} s^{-1}, dominated by ice sputtering on Enceladus, Tethys, Dione and Rhea. Sputtering of icy grain surfaces in Saturn's main rings and in the E ring is comparatively negligible. The rate of H_2O sputtering by energetic ions also exceeds sputtering by corotating ions (Johnson et al. 1985a). Charged particle sputtering of ice surfaces can maintain an H_2O cloud extending from 3 to 11 R_S in Saturn's magnetosphere, but does not contribute significantly to the ring atmosphere.

Carlson (1980) suggested photosputtering as a source mechanism for the ring atmosphere. Here solar ultraviolet dissociates H_2O near the ice surface, allowing H to escape. However, Haff et al. (1983) have suggested that the photosputtering rate estimate should be greatly reduced to incorporate the ultraviolet absorption coefficient of bulk ice rather than free H_2O, according to the original method of Harrison and Schoen (1967). This would yield a photosputtering source of $\sim 6 \times 10^{26}$ H s^{-1} when the ring plane is inclined 26° to the Sun. However, laboratory measurements of photosputtering are needed before this estimate can be accepted with confidence. Some observational evidence suggests that photosputtering may not be the dominant source mechanism. The photosputtering rate should decline significantly when the rings are nearly edge-on to the Sun (Carlson 1980), e.g., in 1980 during the Voyager and IUE observations. However, the ring atmosphere brightness at these times was greater than, or at least comparable to, the brightness during the 1975 observations when the ring plane was inclined $\sim$ 26° to the Sun.

Finally, meteoroid bombardment remains a possibly viable source mechanism (Morfill 1982; Morfill et al. 1983a; Ip 1983). Unfortunately, the estimates are uncertain by orders of magnitude (see the chapters by Veverka et al. and by Chapman and McKinnon).

Summarizing, the ring atmosphere of Saturn remains puzzling. The loss rate of H atoms lies in the range of $\sim 3 \times 10^{25}$ s^{-1} to $\sim 2 \times 10^{28}$ s^{-1}. The source of the ring atmosphere is not well understood at present and may be the atmosphere of Saturn itself.

VI. TITAN: THE CASE OF A SATELLITE'S ATMOSPHERE IN A MAGNETOSPHERE

Atomic hydrogen (and nitrogen) can be introduced to Saturn's magnetosphere from the atmosphere of the satellite Titan. Titan, during its orbit, is always embedded within Saturn's magnetosphere on the night side of the planet. On the day side, Titan's orbit can be either inside (as Voyager 1 found) or outside (as measured by Pioneer 11 and Voyager 2) the magnetosphere.

Indeed, the Saturn system, out to beyond Titan's orbit, is immersed in a huge cloud of hydrogen atoms gravitationally bound to Saturn. The existence of such an atomic hydrogen "ring," maintained by hydrogen escaping from Titan, was originally predicted by McDonough and Brice (1973). Voyager ultraviolet spectrometer observations of Lyman-α emission from the Saturn system (Broadfoot et al. 1981; Sandel et al. 1982) have shown that this hydrogen cloud, which can be called the Titan torus, extends radially from ~ 25 R_S inward to ~ 8 R_S, with a latitudinal extent of ~ 6 R_S on either side of the equator. The mean density of this cloud is ~ 20 H cm^{-3}, so the total content is $\sim 10^{35}$ H. The lifetime of atomic H in the cloud is determined mainly by photoionization, and by electron impact ionization and charge exchange from magnetosphere and solar wind plasmas. This lifetime is typically $\sim 3 \times 10^7$ s in the torus (Bridge et al. 1982), implying a source strength of $\sim 3 \times 10^{27}$ H s^{-1} escaping from Titan (Sandel et al. 1982; Bridge et al. 1982). Such escape rates may be compatible with models of the H escape rate from Titan's upper atmosphere (Hunten 1977; Podolak and Bar-Nun 1979; Strobel and Shemansky 1982; Eviatar and Podolak 1983).

Protons resulting from the ionization of hydrogen in the Titan torus may make an important contribution to the total positive charge density of the Saturnian magnetosphere (Eviatar and Podolak 1983); thus the Titan torus may play a role somewhat analogous to the neutral clouds in the Io plasma torus at Jupiter. However, the mass and energy budgets for Saturn's magnetosphere are less well characterized than those for Jupiter. Atomic nitrogen, at a density of ~ 0.4 cm^{-3}, is also predicted in the Titan torus (Strobel and Shemansky 1982; Eviatar and Podolak 1983).

VII. SATELLITES AND RINGS OF URANUS

The IUE spacecraft has recently detected intense, variable H Lyman-α emissions from Uranus (Durrance and Moos 1982; Clarke 1982). The disk-averaged Lyman-α brightness ranges from 690 Ra to 2200 Ra. This brightness is too great to be accounted for by resonant scattering, Rayleigh scattering, or Raman scattering of solar ultraviolet photons; furthermore, these mechanisms would not explain the observed variability (a factor of two in one 24-hr period) in the emissions. The Lyman-α emission from Uranus has therefore been interpreted as auroral emission, implying the existence of a magnetosphere at Uranus. However, *in situ* confirmation of a magnetosphere at Uranus must await the Voyager 2 Uranus encounter in 1986.

The magnetosphere of Uranus may have unique properties. At the time of the Voyager 2 encounter, the planetary rotation axis will point nearly toward the Sun, approximately in the ecliptic plane. Since the magnetic dipole moment is expected to be roughly aligned with the rotation axis, a pole-on magnetosphere may result, in which the magnetic tail may have a cylindrical configuration (Siscoe 1971). The magnetic field lines in the tail may become twisted owing to interaction with the solar wind, driving a field-aligned current system which would flow into and out of the sunlit pole (Hill et al. 1983*b*).

Divergent views have been presented concerning the nature of plasma sources and plasma populations at Uranus. Both the solar wind source and the ionospheric source of plasma should be much weaker, on the average, at Uranus than at Saturn or Jupiter, per unit volume of the magnetosphere. Voigt et al. (1983) have therefore predicted that the Uranian magnetosphere would be relatively devoid of plasma. Hill et al. (1983*b*) have suggested that in this case the Uranian aurora may be powered by the field-aligned current system associated with the twisting of field lines in the magnetotail. On the other hand, Cheng (1984*b*) has advocated a different view which emphasizes the possible importance of interactions between the magnetosphere and the five large, icy satellites of Uranus. In this view, the magnetosphere of Uranus contains a heavy ion plasma torus, radiation belts, and neutral particle clouds, similar to those at Saturn. In addition, interactions with magnetospheric charged particles may lead to important modification of satellite surfaces.

If the magnetic moment of Uranus obeys the same empirical scaling law (the "Magnetic Bode's Law") as those of the other magnetized planets, the surface magnetic field would be in the range of 1 to 4 Gauss. The magnetopause radius would then be 40 to 60 planetary radii. The five known Uranus satellites all lie within 22.6 planetary radii, so all five are expected to lie well within the magnetosphere. Spectroscopic observations show that all five are covered with water ice mixed with some dark, probably carbonaceous matter, and that three of the satellites are larger than Dione, while two are larger than Rhea (R. H. Brown et al. 1982*b*). It is then natural to consider the importance of particle impacts on the ices.

Cheng (1984*b*) has shown that for a Uranian rotation period of 16^h 10^m and a radius of 2.6×10^9 cm, the total rate of corotating ion sputtering for the icy satellites of Uranus is nearly equal to that for the icy satellites of Saturn, given comparable heavy ion densities. The sputter ejecta are mainly H_2O molecules which mostly escape from the Uranian satellites but remain gravitationally bound to Uranus, forming a large neutral cloud. Ionization and dissociation of this cloud then maintains the heavy ion population in the magnetosphere responsible for the sputtering in the first place.

Cheng (1984b) suggested that the inner magnetosphere of Uranus can find an equilibrium state similar to that of Saturn's inner magnetosphere. In this case, the energetic ion fluxes of $\gtrsim 50$ keV at Uranus may also be comparable to those at Saturn, so that the Uranian aurora may be powered by energetic ion precipitation into the atmosphere as a result of particle interactions with plasma waves. This latter mechanism would predict comparable auroras over the dayside and nightside poles, a prediction which can be tested by the Voyager 2 spacecraft.

Furthermore, if energetic ion fluxes at Uranus are comparable to those at Saturn, then energetic ion impacts can produce dark, carbonaceous matter without methane ice spectral features on the satellites of Uranus and yield a very low albedo for Uranian ring particles. The low albedo of Uranian ring particles was first noted by Sinton (1977) and more recently discussed by Pang and Nicholson (1985). Blackening of the Uranian ring particles by charged particle impacts was first proposed by Cheng and Lanzerotti (1978). If methane ice is present on Uranian satellites and rings, any exposed methane ice surface would be rapidly decomposed and polymerized by energetic ion impacts, forming a black, carbonaceous residue which would prevent observation of methane ice spectral features (Lanzerotti et al. 1985; Cheng 1984b; Strazzula et al. 1985). The blackened surface layer would have a depth of $\sim 10^{-4}$ g cm^{-2}, which would be opaque in the visible and near infrared, but transparent in the far infrared. Of course, the presence of methane ice on the Uranian satellites and rings is not yet established.

Laboratory experiments (Lanzerotti et al. 1985) have shown that for 500 keV proton and 1.5 MeV alpha bombardment, a fluence of $\sim 10^{16}$ cm^{-2} will transform a CH_4 ion surface to CH_n, a polymer with a ratio H/C ~ 2 (Fig. 11). Similar results are obtained with benzene ice (Calcagno et al. 1983a). The incident energetic ion flux observed by Voyager 1 outbound in Saturn's magnetosphere near Dione was 9×10^5 cm^{-2} s^{-1} above 80 keV, yielding a fluence of 10^{16} cm^{-2} in ~ 400 yr. The ion flux above 80 keV measured by Voyager 2 near 2.75 R_S was 3×10^5 cm^{-2} s^{-1} which would yield 10^{16} cm^{-2} in 1000 yr.

Cheng (1984b) concluded that any exposed methane ice surface at Uranus would be blackened and polymerized within 10^3 yr, if energetic ion fluxes there are comparable to those at Saturn. No such phenomenon would occur for H_2O ice surfaces. Energetic ion bombardment of water ice does create dissociation products O_2 and H_2, but these escape and leave behind a surface which is still H_2O ice. A pristine methane ice surface would initially be eroded by sputtering but would also be decomposed and polymerized. Fresh methane ice surfaces may be exposed continually at the Uranian satellites and rings by micrometeoroid impacts.

VIII. SUMMARY AND CONCLUSIONS

When the solid surfaces on ring particles and satellites are embedded within planetary magnetospheres, interactions occur between magnetospheric charged particles and the solid surface. These interactions have important implications for the nature of the surface, its parent body, and the magnetosphere as a whole. Impact of magnetospheric ions causes ejection of atoms and molecules from the surface, and significant erosion can result over geologic time. Furthermore, the relative abundance of various species in the surface will be altered if their erosion rates differ. Chemical alteration of the surface can result in two additional ways. First, there is ion implantation, followed by subsequent chemical reactions with surface constituents. Sulfur ion implantation may have been observed on Europa. Second, high-velocity ion impact causes dissociation of target materials, with formation of free radicals which may escape or react with surface constituents.

Atoms and molecules ejected from the surface by sputtering will also have important consequences for the parent body. If this parent body is sufficiently massive, the majority of the sputter ejecta remains gravitationally bound to the body and falls back to the surface or becomes incorporated into an atmosphere. This situation applies to the Galilean satellites of Jupiter. There can be significant redistribution of surface matter over geologic time, resulting from sputter ejecta reimpacting the surface far from where the original sputtering event occurred. At Europa, there may be a net accumulation of ~ 10 m of water ice over geologic time in much of the leading hemisphere, with comparable net erosion over the trailing hemisphere.

Sputter ejecta can make a significant contribution to the atmospheres of the Galilean satellites (except Io, where sublimation and volcanic venting of SO_2 may be dominant). Charged particle sputtering provides a lower limit to the atmospheric column density even for arbitrarily cold ice surfaces or subsurface regoliths. Particularly on the night side of the Galilean satellites, sputtering may dominate sublimation as a source for the atmosphere.

For less massive parent bodies, such as ring particles or the icy satellites of Saturn and Uranus, sputter ejecta are generally not gravitationally bound to the parent body. In this case, sputter ejecta escape directly from the satellite surface to form neutral clouds within the planetary magnetosphere. Ionization and dissociation of these neutral clouds can be significant sources of plasma for the magnetosphere. Indeed, the heavy ion plasma torus near the orbits of Dione and Tethys may be maintained by ionization and dissociation of a vast neutral H_2O cloud formed by sputter ejecta from icy surfaces in Saturn's magnetosphere.

At Io, the presence of an atmosphere complicates the processes of sputtering and escape of neutrals. Sputtering at Io, either from the surface (thin atmosphere) or the exobase (thick atmosphere), is the predominant source of sulfur and oxygen for the Io plasma torus and the Jovian magnetosphere. Neutrals escaping from Io probably include SO_2 as well as atomic O and S, and neutral clouds are formed in the Io plasma torus. Ionization and dissociation of neutrals in these clouds is the dominant source of mass for the Jovian magnetosphere. The detailed mechanisms for sulfur and oxygen escape from Io depend on the nature of Io's atmosphere. If the atmosphere is thin, the fraction of sputter ejecta able to escape Io's gravity is small enough ($\sim 3\%$) that the total SO_2 loss and escape rate may be too low to maintain the plasma torus, even when contributions from a bound sputter corona are included. If Io's atmosphere is thick, sputtering off the top of the atmosphere can yield sufficiently high escape rates of neutrals, but in this case some mechanism such as mixing in Io's upper atmosphere must be invoked in order to achieve comparable escape rates for oxygen and sulfur. Sputtering remains the most promising mechanism for maintaining the Io sodium and potassium clouds. However, the required escape rates of Na and K as well as the abundance of these substances at Io are very unclear. If Io has a thick atmosphere, Na and K must be incorporated within it in some form.

The rings of Saturn and their associated neutral clouds pose fascinating problems. The source of the atomic hydrogen cloud around Saturn's A and B rings is not well understood. Photosputtering and charged particle sputtering both encounter severe difficulties, and micrometeoroid bombardment rates are uncertain. Another major uncertainty is the rate at which H atoms are lost to sweeping by icy ring particles. The E ring also remains enigmatic. The very short lifetimes of icy grains against sputtering erosion and radial transport imply that the E ring could disappear within a time interval as short as 100 yr, unless a continual source of ring matter is available. One possibility is that the E ring may coexist with a neutral H_2O cloud in Saturn's magnetosphere; alternatively the E ring may truly be ephemeral.

The existence of a magnetosphere at Uranus has been inferred from observations of intense, variable H Lyman-α emissions from the planet. The first *in situ* observations of the magnetosphere by the Voyager 2 spacecraft in 1986 are expected to settle the issue of whether it is similar to Saturn's inner magnetosphere or, alternatively, whether it is relatively devoid of plasma. In the former case, sputtering of water ice surfaces on the Uranian satellites would maintain vast neutral clouds within the magnetosphere. Ionization and dissociation of these neutrals would be a possibly dominant plasma source. Also, any exposed methane ice surfaces would be rapidly decomposed and polymerized by charged particle impacts forming a black, carbonaceous residue. This process may explain the very low albedo of Uranian ring particles and the presence of dark, carbonaceous matter, without methane ice spectral features, on the Uranian satellites.

Acknowledgments. A. F. Cheng, P. K. Haff and R. E. Johnson acknowledge the support of the National Aeronautics and Space Administration during the preparation of this chapter. R. E. Johnson also acknowledges the support of the National Science Foundation.

10. SURFACE COMPOSITION OF NATURAL SATELLITES

ROGER N. CLARK
U.S. Geological Survey, Denver

FRASER P. FANALE
University of Hawaii

and

MICHAEL J. GAFFEY
Rensselaer Polytechnic Institute

The compositions of planetary surfaces have been largely determined by study-
ing the visible and near-infrared reflectance spectra of probable analog materi-
als and comparing these with spectra of the planets obtained from Earth-based
telescopes and spacecraft. From the position, shape and depth of absorptions in
the spectrum of a surface, the materials and their approximate abundances may
be found. Additional constraints on composition come from thermal (radio-
metric) and morphological (imaging and radar) observations. The surface com-
positions of planetary satellites are diverse. We know more about the surface of
the Moon than any other satellite because of the vast information gained by the
Apollo and other lunar missions. The surface of the Moon is primarily com-
posed of basaltic and anorthositic rocks and glasses with no abundant volatiles.
The surfaces of the satellites of Mars most likely contain hydrated and opaque
materials. The surfaces of the outer solar system satellites are rich in ices, with
rock-type material being a minor constituent. Io appears to have abundant allo-
tropes of sulfur and SO_2 as a frost or as molecules adsorbed on mineral grains
(possibly sulfur). Silicate materials may also be present on Io, but they are not
major constituents of the optical surface. The other Galilean satellites have
abundant H_2O ice on their surfaces, with Europa having the purest icy surface
and Callisto the least pure. Callisto also has some areas on its surface which

appear ice-free and probably consist of silicate and carbonaceous material. The satellites of Saturn (Enceladus, Tethys, Dione and Rhea) appear to have H_2O ice surfaces similar in purity to Europa or Ganymede, but with finer ice grains. Hyperion shows an H_2O ice surface but dirtier, containing a reddish colored material. Iapetus is very unusual in that one side displays a bright, H_2O ice surface of purity similar to Rhea, while the other side is very dark with a spectral signature of some phase similar to the organic compounds present in carbonaceous chondrites. The Uranian satellites all have abundant H_2O ice on their surfaces with impurities similar to those on Hyperion. Beyond the Uranian system, water ice has not been unambiguously found, but methane ice occurs on Triton. Liquid and/or solid nitrogen may also be present on Triton. The Pluto/Charon system appears to have a surface composition similar to Triton, but with more methane. Water ice may also be present on Triton and Pluto/ Charon, but the strong methane bands tend to mask the water absorptions. Pluto/Charon may also have abundant nitrogen, similar to Triton, but the strong methane absorptions mask detection of the nitrogen absorption.

I. INTRODUCTION

The best way to determine the surface composition of a planetary surface is to collect samples with a space probe and to return them to laboratories on Earth for sophisticated analyses. Lunar samples were treated in this way and it is technologically feasible to sample all other satellite surfaces in the solar system. Another method is to send a laboratory to the surface, as was done with the Viking landers to Mars. Unfortunately, insufficient funds at present rule out sample return missions to any other surface in the solar system. Instead, we must rely on remote sensing methods from a flyby or orbiting spacecraft, from Earth orbit or from the surface of the Earth. From an orbiting spacecraft, a combined approach using radiometric methods to map elemental composition and reflectance spectroscopy to map mineral distribution is highly effective. From Earth or Earth orbit, the most powerful approach yet developed is to analyze the reflected or emitted photons from a sunlit surface as a function of viewing geometry and photon wavelength. Analysis of this light can, at a minimum, lead to a qualitative determination of the surface composition or, in some cases, quantitative determinations of abundances of specific components.

Reflectance Spectroscopy

This technique determines the fraction of light reflected from a surface as a function of wavelength and viewing geometry. Absorption bands in the spectrum give an indication of the minerals present, and sometimes an idea of how much of a mineral might be in the surface (see Hunt 1977, and references therein). The change in the amount of reflected light as a function of viewing geometry gives an indication of the microstructure of the surface (see, e.g., chapter by Veverka et al.; Veverka 1977*a*). Modeling the scattering of photons in a particulate surface is extremely complex, so theoretical interpretations combined with laboratory studies of the reflectance properties of minerals are

required to derive mineral abundances from a reflectance spectrum of a planetary surface. Because of the importance of reflectance spectroscopy in determining surface compositions of solar system surfaces, this topic is discussed
in detail in Sec. II.

Polarization

The reflected light from a planetary surface is partially polarized. The
degree of polarization depends on the composition of the surface material,
and on its texture. The polarization of light from planetary surfaces is reviewed by Veverka (1977b; see also chapter by Veverka et al.), so the subject
is only briefly discussed here. The theoretical polarization properties of a surface are treated by Wolff (1980). In general, the polarization from a surface is
negative (polarization lying in the plane defined by source, surface and observer) when the phase angle (angle between the source, surface and observer)
is small, e.g., less than about 20°. At greater phase angles, it is positive. The
magnitude of the polarization correlates inversely with the reflectance of the
surface; lower reflectance surfaces tend to have a higher polarization. It also
correlates with texture; as the surface becomes more complex in texture, the
negative polarization becomes greater. The negative polarization (called the
negative branch) has received the most attention because all of the satellites in
the solar system, except the Moon, can be observed only at low phase angles.
The polarization of two or more surfaces (e.g. different areas on a satellite)
can be used to indicate differences in the surface. However, because of the
complexity of the polarization, it is presently not possible to interpret the
meaning of the differences in a unique manner. Perhaps, with further laboratory and theoretical studies, polarization may become a more diagnostic tool.

Radiometry

Thermally emitted photons from a planetary surface can be used to deduce the surface thermal properties. Monochromatic fluxes at one or two
wavelengths can be used to derive model dependent surface temperatures,
while measurements at many wavelengths allow detailed thermal modeling.
Thermal observations combined with visual observations can be used to derive the albedo of the surface (see, e.g., Brown et al. 1982a) which can help
constrain surface composition. Measurements obtained as a function of time
as the surface passes into the night side, or through an eclipse, can be used to
derive the thermal conductivity (see, e.g., Morrison 1977). The thermal conductivity depends on the microstructure as well as the surface material,
providing another constraint on surface composition.

The thermal emission spectrum of a surface can include both absorption
and emission bands which may be diagnostic of the surface composition.
Emission spectra are very complex functions of thermal gradients in the surface, particle type, size, and the packing density of grains, and are less understood than reflectance spectra (see, e.g., Hunt and Vincent 1968; Conel

1969). For example, as the grain size decreases, an absorption band can change to an emission band. Because of the complexity of the spectra, and the difficulty of making thermal infrared observations from Earth, little emission spectroscopy of planetary surfaces has been published. Most of the work done has been from spacecraft such as Voyager (see, e.g., Hanel et al. 1979,1981) but few features have been seen. Because the features seen in emission spectra of planetary surfaces are often weak, emission spectroscopy for use in remote sensing has received only minor attention in recent years, although with new instrumentation and better technology for making observations, emission spectroscopy may eventually prove to be a very important diagnostic tool.

Gamma-Ray Spectroscopy and X-Ray Fluorescence

Both gamma-ray spectroscopy and X-ray fluorescence can be used to sense remotely individual elements in a surface if an intervening thick atmosphere is not present. This method requires a close spacecraft approach or a lander. Applications of this technique in the U.S. planetary program have been limited thus far to the Moon, where detailed maps of the distribution and abundance of major elements (silicon, aluminum and magnesium using X-ray spectrometers, and uranium-238, thorium-235, and potassium-40 using gamma-ray spectrometers) have been obtained (see, e.g., Adler and Trombka 1977). However, a similar but improved instrument is slated for inclusion on the planned Mars Observer mission and for other orbital missions where a thick atmosphere is not encountered. For short observational periods (e.g., flybys), the natural radioisotopes of uranium, thorium and potassium can be detected and their abundance estimated on a hemispheric scale. For extended orbital missions (or station-keeping in the case of the smallest bodies) the abundance and distribution of a number of important rock-forming elements, such as iron, magnesium, silicon, calcium, aluminum and titanium can also be characterized.

Other Techniques

Surface materials can sometimes be detected in the space around the planet or satellite if they are being sputtered by high-energy particles. Detection is accomplished by remote studies of line emission spectroscopy from neutral or ionized particles or *in situ* measurements of the composition of the sputtered particles by spacecraft. Even from Earth, such studies can yield detailed maps of particle distribution and indicate the sources and mechanisms of cloud supply (see, e.g., Pilcher and Strobel 1982; chapters by Nash et al. and Cheng et al.). However, analyses *per se* are best accomplished by particle detectors and mass spectrometers aboard a spacecraft (see, e.g., Krimigis et al. 1979).

Discussion

In some cases, initial guesses can be made of surface composition by using solar system condensation models (see, e.g., Lewis 1972,1973), bulk density, and knowledge of possible endogenic processes (e.g., magmatic differentiation) or exogenic processes (e.g., meteor bombardment). In many cases the most useful measurements are those of temperature. The surface temperatures of frosts and ices as well as the temperature of clouds provide a strong constraint for identification. Even for nonices, thermal constraints can sometimes be useful. For example, the relationship between occurrence of hot spots and the spectrally bland high-temperature black regions on Io leads to a convincing, although not definitive, identification of molten sulfur. Likewise, the Voyager *in situ* analyses of ion abundances peaking near Io provides a "cross constraint" leading to a web of circumstantial evidence regarding elemental composition. This allows compositional inferences even though the spectra of sulfur allotropes are so bland that one can only say that the spectral data are *consistent* with the occurrences of several of them on the surface. Unfortunately, such inferences often come to be recognized as definitive after being repeated several times in the literature without the caveats.

This chapter reviews current knowledge concerning the surface composition of solar system satellites. Except for direct sampling, and spectroscopy, the methods discussed above only provide constraints on surface composition. Spectroscopy, whether gamma-ray or near-infrared, is the only remote sensing method which can provide direct evidence for elemental, or mineralogical composition. In all cases but the Moon, there are no direct samples, so the surface composition is an interpretation of complex remote sensing data. The most commonly used method for determining surface composition is reflectance spectroscopy, so that subject dominates the discussions in this chapter. Unfortunately, the reflectance spectrum of a particulate surface is a complex function of the optical constants of the individual materials composing the surface. Often, the spectrum of a surface composed of several materials may be dominated by the features of a constituent that is not abundant compared to the other materials in the surface. Thus, what is spectrally dominant, and what material is actually abundant may be different. In determining surface composition, spectral features identifying a material must be found, and how strong those features are relative to those of other materials will indicate actual material abundance. This chapter will try to distinguish between the interpretations that give solid indications of composition and those that only indicate spectral dominance in the data biased by a certain material possibly not abundant on the surface.

For some satellite surfaces, only broadband spectral colors have been measured and the composition is totally unknown, although speculation (often appearing very reasonable) is given. We will try to distinguish such reasonable speculation from hard data, because many times in the history of planetary science, reasonable speculation has been shown to be wrong. Dominant in the determination of surface composition is analysis of the reflectance spectrum. We begin with a discussion of the spectral properties of minerals (Sec. II) before discussing each satellite.

II. SPECTRAL ANALYSIS

One must be very careful not to overinterpret results from spectral analyses. The wording of results and conclusions should be examined closely. Quite often one will find that a particular material is spectrally dominant, and no other materials are seen in the spectrum. However, spectral dominance does not necessarily mean that the material on the surface is the dominant weight fraction. How then can the composition of a surface be determined, as opposed to finding the presence of materials that are spectrally dominant?

Traditionally, materials have been identified from specific absorption bands caused by a number of mechanisms such as molecular vibrations, electron charge transfer, conduction bands in semiconducting materials and electronic transitions within an orbital, typically the d orbital in Fe^{2+} and Fe^{3+} ions. An excellent discussion of the causes of absorption bands in minerals is given in Hunt (1977, and references therein). Each material has its own spectral signature (also seen in Hunt [1977]). Examples of absorption bands in minerals are shown in Figs. 1 and 2. When studying a particulate surface, the grain size of the material will influence the path length of photons within a material and thus affect the apparent absorptions as seen in Fig. 3 for H_2O. Such effects are not limited to H_2O ice, but are seen in all particulate materials (see, e.g., Hunt and Salisbury 1970,1971; Hunt et al. 1971a,b,1972, 1973). The change in absorption strength with grain size is caused by the change in scattering. When the grains in a surface are small, the photons encounter more index of refraction boundaries per unit path length compared to surfaces with larger grains, so the scattering is greater, and the photons have less probability of being absorbed. Without an understanding of the scattering effects, the strength of the absorptions cannot be used to determine material abundance.

Real planetary surfaces are composed of several to many different minerals, and the spectra of mineral mixtures are complex nonlinear functions of grain size, material abundance, and material opacity as well as viewing geometry and lower-order effects such as packing. Deconvolving a reflectance spectrum to mineral abundance in an unambiguous way is quite difficult. Two ways exist to solve the scattering problem: empirically and theoretically, and both are needed. The scattering is so complex that only recently have theories by Hapke (1981), Lumme and Bowell (1981), and Gougen (1981) begun to model the reflectance with reasonable precision, and it is not yet known whether these theories are sufficient. Empirical studies, such as measurements of actual mineral mixtures, are needed to check and refine the theories as well as finally to be employed in cases where the theories have been shown to be inadequate.

The reflectance of a surface at a given wavelength λ is a function of the angles of incidence i, emission e, phase angle g, and the single scattering albedo w of the grains in the surface. The equation for the bidirectional reflec-

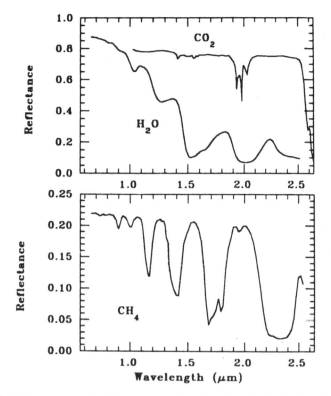

Fig. 1. Spectral reflectance of solid methane, water and carbon dioxide. Band positions identify the material. The strength of an absorption depends on the fundamental strength of a transition, and the mean photon path length through the material. The mean photon path depends on the material abundance and the grain size, thus both affect the appearance of the resulting spectrum. The CO_2 spectrum is from Smythe (1979), and the H_2O is from Clark (1981a). The CH_4 spectrum is from Brown et al. (1985).

tance of a surface can be expressed by the following equation, modified from Hapke (1981):

$$r(w,\lambda,\mu,\mu_0,g) = \frac{w}{4(\mu + \mu_0)} [F(g) + H(w,\mu)H(w,\mu_0) - 1] \qquad (1)$$

where $F(g)$ is a description of the shadowing between grains and the single particle phase function, μ is the cosine of the angle of emitted light e, μ_0 is the cosine of the angle of incident light i, and H is the Chandrasekhar H-function (Chandrasekhar 1960). This equation is equivalent to Eq. (37) of Hapke (1981) and is called the bidirectional radiance coefficient, but is more commonly called the bidirectional reflectance. The single scattering albedo of

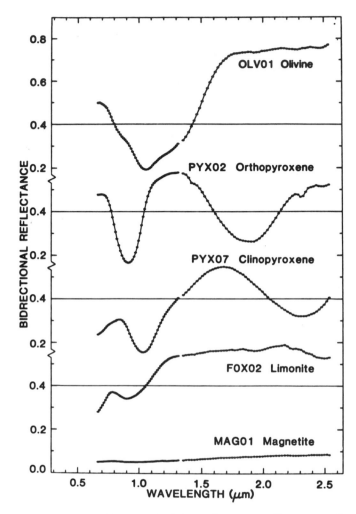

Fig. 2. Reflectance spectra of the minerals olivine, orthopyroxene, clinopyroxene, limonite, and magnetite. The absorptions are due to Fe^{2+} in the top three minerals, to Fe^{3+} in limonite, and to both in magnetite. The particular crystallographic site and ion determines the position and fundamental strength of an absorption, while the grain size influences the mean optical path affecting the observed band depths (figure from Singer 1981).

a particle is related to the complex index of refraction of the material composing the grain, as well as the grain size, and the scattering boundaries within the grain caused by imperfections such as cracks. The reader is referred to Eq. (24) of Hapke (1981) for a mathematical description of the single scattering albedo of a grain.

The reflectance of a surface composed of more than one material was described by Hapke (1981) by the equation:

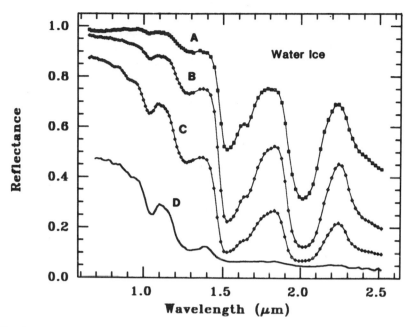

Fig. 3. The near-infrared spectral reflectance of (A) a fine-grained ($\sim 50\ \mu$m) water frost, (B) medium-grained ($\sim 200\ \mu$m) frost, (C) coarse-grained (400–2000 μm) frost, and (D) an ice block containing abundant microbubbles. The larger the effective grain size, the greater the mean photon path that photons travel in the ice, and the deeper the absorptions become. Curve D is very low in reflectance because of the large path length in ice. Frost data are from Clark (1981a); ice data are from Clark and Lucey (1984).

$$w(\text{average}) = \frac{\sum\limits_{j}(M_j w_j/\rho_j D_j)}{\sum\limits_{j}(M_j/\rho_j D_j)} \tag{2}$$

where subscript j denotes particle type, M is the mass fraction, ρ is the material density, D is the grain size, and w_i is the single scattering albedo at a given wavelength for particle type j. This equation states that the single scattering albedos add linearly with the projected surface areas of the different particle types in a surface. Because of the H-functions, the reflectance is a nonlinear function of the single scattering albedo, and a dark material will dominate a reflectance spectrum of an intimate mixture of dark material and light material. For example, consider a surface composed of two materials, each having the same density and grain size, but one material is bright with a single scattering albedo of 0.99997 and the other has a single scattering albedo of 0.20. The reflectances of the pure surfaces would be about 0.99 for the bright material and 0.03 for the dark material. If the two materials were mixed in equal proportions, then the average single scattering albedo for the surface would be 0.599985, and the reflectance would be about 0.13, which is still quite dark. If the reflectances of the pure materials were added, one would obtain a reflectance of 0.51 which is much too high.

The above illustration shows the nonlinear nature of the reflectance spectra of intimate mixtures. Because the single scattering albedo changes as a function of wavelength, the material that will tend to dominate the spectrum of a mixture is dependent on the wavelength. The material that is darker at a given wavelength will tend to dominate the spectrum. Thus, regions of the spectrum where materials have absorption bands, may be dominated by that material, but at another wavelength, a different material may dominate the spectrum.

The variation in the spectral properties of mineral mixtures is illustrated in Figs. 4, 5, and 6, where water ice, a common mineral found on many satellites in the solar system, was mixed with other materials. Depending on the reflectance of grains in the mixture at a particular wavelength, a given material can dominate the reflected light. For example, water ice is quite absorbing at 2 μm and dominates most of these spectra; however, it is very transparent at visual wavelengths and other materials present at only a fraction of a weight percent can dominate the spectrum. An analogous example will be given below when Io's spectrum is considered as a mix of SO_2 ice, which has a strong infrared absorption, plus other ultraviolet-absorbing constituents.

Clark and Roush (1984) showed that the mean photon path length as a function of wavelength can be computed theoretically. An example is shown in Fig. 7. Here a laboratory spectrum of H_2O ice is compared to a theoretically derived spectrum and the mean photon path. When the grains are small compared to the inverse of the absorption coefficient of the material, they are transparent, and photons are scattered from the surface before many can be absorbed. Absorption bands will be weak in this case. As the grains become larger, absorption increases resulting in a decreasing reflectance. Because some light is reflected from the surfaces of grains before it enters any grain, the reflectance can never reach zero for a surface consisting of randomly oriented grains. As the grain size increases, the reflectance in an absorption band reaches a minimum. But the wings of the absorption band continue to decrease as the grain size increases until the only light reflected from the grains is that due to the first surface Fresnel reflection. The absorption band depth D_B is defined as

$$D_B = (R_c - R_b)/R_c \qquad (3)$$

where R_c is the reflectance of the continuum at the absorption band center, and R_b is the reflectance at the band bottom (Clark and Roush 1984). Because the reflectance at the absorption band center decreases as the grain size increases, the absorption band depth initially increases with increasing grain size. However, as the reflectance at the band center becomes constant with

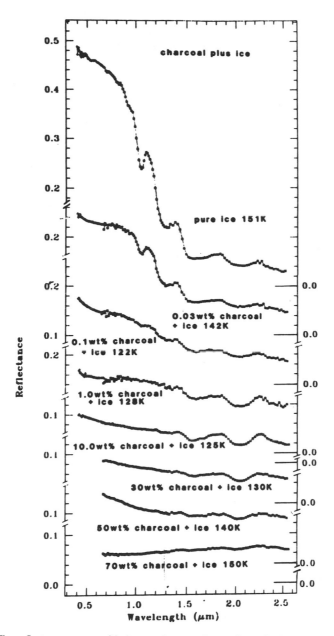

Fig. 4. The reflectance spectra of intimate mixtures of water ice and a low albedo particulate material are shown for different weight fractions of ice and particulates. Examine the ice absorptions at 1.04, 1.25, 1.6 and 2.0 μm and note the large variation in spectral properties of the overall reflectance level and of the appearance of the absorption band depths. The darker the particulate, the less photons are scattered, thus the mean photon path is reduced, resulting in weaker absorptions. The reflectance never reaches zero because specular reflection and scattering from fine structure on the surface of the samples and from microbubbles in the sample limit the photon path length in the ice (figure from Clark and Lucey 1984).

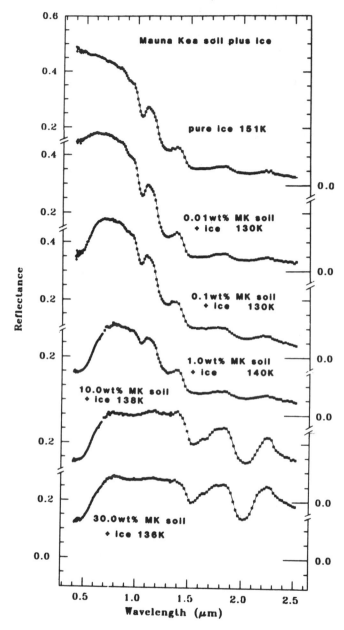

Fig. 5. The same type of ice-particulate mixture as in Fig. 4 is shown, except, for a medium-albedo particulate. Note that for the same weight fraction of particulate material mixed in the ice, the ice absorption is stronger than with charcoal (shown in Fig. 4). This is because the single scattering albedo of the particulate material (Mauna Kea soil) is higher than for charcoal grains, so that photons scatter more times before being absorbed, which results in a longer mean photon path in ice (figure from Clark and Lucey 1984).

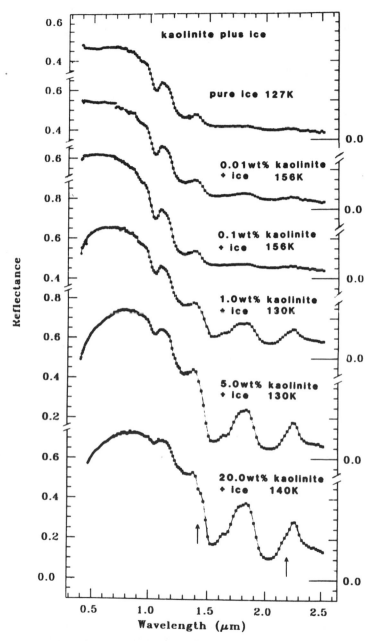

Fig. 6. The reflectance spectra of intimate mixtures of water ice and kaolinite (a high-albedo material). Spectra such as these, along with similar data shown in Figs. 4 and 5, can be used to derive trends such as that in Fig. 9; these then provide a calibration to abundance from observable spectral parameters (figure from Clark and Lucey 1984). Arrows indicate positions of absorption bands in Kaolinite.

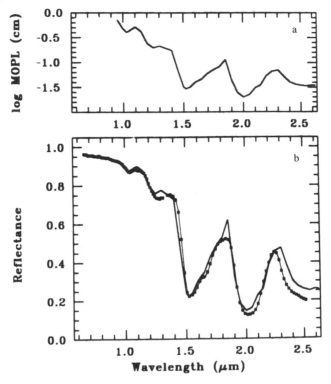

Fig. 7. A laboratory reflectance spectrum of pure water ice (points) in the near-infrared from
Clark (1981a) compared to a theoretical computation using the method described by Clark and
Roush (1984) and shown in Clark and Lucey (1984). The laboratory spectrum is that of a
medium-grained frost with equivalent grain diameters around 200 μm estimated visually with a
microscope during measurement, and with photographs whose resolution was about 25 μm.
The theoretical spectrum was for a grain size of 70 μm (with the angle of incidence and emission in the Clark and Roush [1984] Eq. [15] both equal to 60°), indicating that smaller grains
and/or crystal imperfections were present in the sample which were not seen visually or photographically. In any case, the agreement is good, matching both band depths and overall reflectance levels. In figure (a) the mean optical path length (MOPL) is plotted as a function of wavelength. It is seen that the more light absorbed by the ice, the shorter is the path length that
photons travel in the particulate surface. It is interesting that the log mean optical path length
mimics the reflectance spectrum.

increasing grain size because of the first surface reflection, the reflectance in
the wings of the absorption continue to decrease and the reflectance of the
continuum continues to decrease. The band depth, that was first increasing
with grain size now reaches a maximum and then decreases. The change in
band depth as a function of grain size or abundance is commonly referred to
as a curve of growth. This effect is mathematically described for spectra of
particulate surfaces in Clark and Roush (1984) and illustrated in Clark and

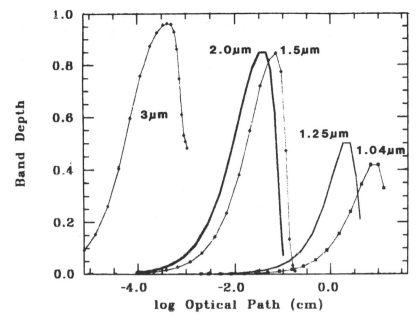

Fig. 8. The ice absorption band depth as a function of the base 10 logarithm of the mean optical path in ice is shown for 5 ice absorption features. These curves were computed from many spectra like that shown in Fig. 6 by varying the effective grain size of each spectrum. The optical path used is at the center of the absorption. At other wavelengths, the optical path is greater (figure from Clark and Lucey 1984).

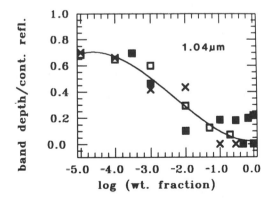

Fig. 9. The 1.04 μm band depth for water ice divided by the continuum reflectance of the ice-particulate mixture spectrum shown for the data in Figs. 4, 5 and 6 where the solid boxes are the ice-charcoal data, the crosses are the ice-Mauna Kea soil data, and the open boxes are the ice-kaolinite data. This presentation shows a crude calibration of observable spectral parameters to material abundance. Other ice absorption bands are sensitive to different amounts of impurities. Some scatter in the charcoal data is due to near-zero band depths and very low continuum reflectance, thus approaching zero divided by zero (figure from Clark and Lucey 1984).

Lucey (1984). An example of the curves of growth for several H_2O ice absorptions is shown in Fig. 8. In H_2O ice, the different absorptions will allow photons to penetrate different distances; thus each absorption will be affected differently by various grain sizes and by the presence of other minerals (see Figs. 3, 4, 5, and 6). By measuring spectra of mineral mixtures such as those shown in Figs. 4, 5, and 6, curves of growth for an absorption band can be constructed. The absorption band depth divided by the continuum reflectance at the center of the absorption has been shown, by Clark and Lucey (1984) for the case of water ice-particulate mixtures, to be a crude calibration to abundance (Fig. 9). In the visual part of the spectrum, water ice is very transparent, so the addition of small amounts of dark materials can substantially change the visual reflectance and color of the surface (Fig. 10).

In some cases, reflectance spectra of planetary surfaces do not show uniquely diagnostic features allowing identification of a particular material and other methods must be used to help constrain composition. Thermal measurements may indicate too high a temperature for a particular volatile to exist. Morphologic imaging of surface features can give clues to composition. For example, the observed surface relief on Io is inconsistent with a predominantly sulfur crust, but is consistent with a silicate crust containing several percent sulfur (Clow and Carr 1980). Another example is the presumed water flows on icy satellites (Lionel and Head 1983). These other methods, however, may add evidence for (or against) a particular material, but cannot always determine how much of a given material is on the surface. These comments are intended to point out a current technical limitation of remote phase assemblage analysis. However, it should be recalled that the limitation is somewhat tangential to the main mission of the technique; it is often the mere presence of mineral phases which most directly reflects the environmental or formational conditions rather than the precise abundance of those phases. The abundance may vary arbitrarily and stochastically, laterally and vertically (remember that most remote sensing techniques are near-surface techniques), whereas the presence of a phase—simple identification which depends only on the qualitative validity of the method—reflects environmental conditions in a systematic manner. Thus, although the quantitative abilities of optical and infrared remote sensing should not be overstated, the importance of the quantitative limitations should not be overstated either.

With these thoughts in mind, the current understanding of the surface compositions of solar system satellites is now reviewed.

III. THE MOON

More is known and has been written about the Moon than any other satellite in the solar system. By far, the bulk and most detailed of our knowledge of the Moon has come from the six Apollo missions as well as the Ranger,

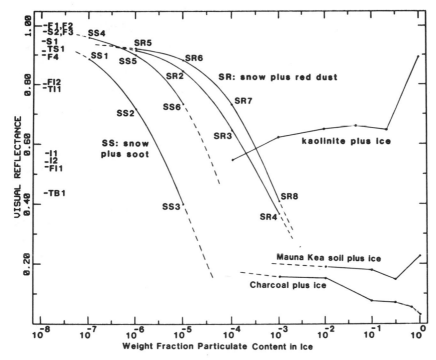

Fig. 10. The visual reflectance of selected spectra of ice, frost and mixtures of ice with particulate materials (from Clark 1982). The dotted lines indicate the expected reflectance trends for unmeasured ice-particulate values. The charcoal plus ice, kaolinite plus ice, and the Mauna Kea soil plus ice visual reflectance values were measured from samples of water plus particulates frozen rapidly from room temperature and atmospheric pressure; thus, the ice contained abundant microbubbles (data from Clark and Lucey 1984). This resulted in an increasing reflectance as the particulate concentration decreased, since the scattering by bubbles became more dominant, except in the kaolinite. In the kaolinite plus ice, the reflectance decreases as the kaolinite content decreases because the pure kaolinite is higher in visual reflectance than the pure ice. Clear ice would contain fewer or no bubbles, thus the only scattering centers would be the particulate contaminants and the resulting reflectance levels (low for dark particulates) could be easily maintained to below parts-per-million or less contaminants. The codes used are also given in Tables I and II of Clark (1982). Generally SS indicates snow plus soot; SR indicates snow plus red dust; F are frosts; S, snows; I, laboratory ices; TI, terrestrial ices; and TB, terrestrial blue ice. It is important to note that the visual color of ice plus particulate contaminants changes with grain sizes. For instance SR3 and SR7 have the same weight fraction of red dust but the finer grain size (SR3) results in a more reddish color in the snow plus dust. Similar results occur with grain-size differences of the frost alone.

Surveyor, Lunar Orbiter missions and the Soviet Luna sample-return missions. The Apollo missions returned a total of 382 kg of rocks and soil from the Moon, and three unmanned Soviet landers brought back 250 g. Geochemical data were obtained from the Apollo Orbiters using X-ray and gamma-ray spectrometer instruments. These missions have provided a very detailed look at the surface composition of the Moon. Sorting through the tens of thousands of pages published about the Moon, and synthesizing the results to a few paragraphs is a difficult task to say the least. We refer the reader to several collections which we have studied to write this section: Taylor (1982), Lindsay (1976), King (1976), Short (1975), Basaltic Volcanism Study Project (1981), and the Proceedings of the Lunar and Planetary Science Conferences. We wanted to present the composition of the lunar surface in a similar perspective to the other satellites in this chapter, but had great difficulty in collecting the proper information. Most studies of the lunar samples have focused on small details. A few studies have presented average compositions, but again, not of the lunar surface as a whole, but of small parts, such as the 20 to 90 μm sieve fraction of the soil. Apparently, the average composition of the lunar regolith (regolith taken to mean everything from the smallest submicron grain to the large boulders) has not been published in one work but is scattered through 15 years of lunar literature. It turns out, however, that the average grain size of the lunar regolith, as discussed below, is about 40 to 70 μm so that any error in examining only grain sizes encompassing that range is probably small.

The lunar surface consists of two basic geologic units: the bright highlands and the dark maria (Fig. 11; see Map Section and the chapter by Kaula et al.). The oldest regions on the Moon are the heavily cratered highlands. Large impacts formed the basins, the youngest of which are Imbrium and Orientale. There are probably at least 30 basins on the near and far side with diameters > 300 km, and many have been flooded with mare basalts. Basin distribution is even on both the near and far sides (e.g., see chapter 2 of Taylor 1982) and the apparent anisotropy in the maria distribution is caused by the paucity of mare basalt flooding on the far side. The mare basalt flows cover about 17% of the surface.

The surface of the Moon is covered by a regolith produced by intense cratering (the processes that produce the regolith are described in the chapter by Veverka et al.). The average regolith thickness is around 10 m in the highlands, while it is only 3 to 6 m at the Apollo 11 landing site (Shoemaker et al. 1970), only 2 m at the Apollo 12 site, and 8.5 m at the Apollo 14 site (Watkins and Kovach 1972). Beneath this regolith lies a thick, 1 to 3 km, zone of brecciation and fracturing (megaregolith) due to the intense early cratering of the highland crust. Over 60% of the samples returned from the highlands are breccias, with much of the rest of the material being impact melt rocks (e.g., see chapter 5 of Taylor 1982).

The most widespread component of the lunar surface is the highland ma-

Fig. 11. A photograph of the Moon from Lick Observatory. Note the sharp contrasts between the dark mare and the brighter highlands. Compare this image with that of Rhea in Fig. 15. The visual impression gives little compositional information.

terial. The most abundant mineral in the highland rocks is plagioclase feldspar followed by low-calcium pyroxene and possibly small amounts of olivine and clinopyroxene. Mafic minerals are present only as minor components. The feldspars are very calcium-rich, with compositions typically in the range of An 95 to 97. The rare earth element (REE) abundances of the average lunar highland soils are all enriched relative to chondrites. (See Sec. 5.4 of Taylor [1982] and references therein for more details.)

Table I, derived from Labotka et al. (1980), shows the soil compositions at the Apollo 11, 12, 14, 15 and 16 landing sites. The Apollo 16 samples are probably the most representative of the highland regolith, while the Apollo 11

TABLE I
Surface Composition of the Lunar Regolith from Grain Count Modal Data

Sample[a] Size fraction (μm)	10084 90-20	 20-10	12001 90-20	 20-10	12033 90-20	 20-10	14163 90-20	 20-10
Lithic clasts	1.5	—	2.9	—	0.2	0.2	6.7	1.1
Agglutinates + dark matrix breccias	35.2	12.6	21.3	4.7	22.1	7.2	36.7	11.0
Pyroxene	27.5	28.1	27.5	31.0	30.8	24.4	10.2	29.3
Plagioclase	13.1	20.0	17.4	18.1	17.9	26.6	17.2	18.3
Olivine	1.3	1.0	5.3	10.2	3.0	3.2	3.5	5.2
Silica	0.4	0.7	0.8	1.5	0.8	1.2	0.2	0.3
Ilmenite	4.0	11.9	1.2	2.7	3.0	5.2	0.5	2.0
Mare glass	9.8	16.1	18.4	24.1	4.4	11.5	2.3	4.4
Highland glass	5.1	9.5	4.7	6.4	17.1	18.5	20.1	21.7
Others	2.1	—	0.6	1.2	0.4	1.0	2.6	6.7
Total	100.0	99.9	100.1	99.9	99.7	100.0	100.0	100.0
Number of points	472	285	512	403	497	500	796	639

Sample[a] Size fraction (μm)	15221 90-20	 20-10	15271 90-20	 20-10	64501 90-20	 20-10	67461 90-20	 20-10
Lithic clasts	—	—	5.5	1.4	0.5	—	5.6	4.8
Agglutinates + dark matrix breccias	37.3	34.6	28.3	5.9	10.3	2.6	15.7	8.2
Pyroxene	24.4	13.6	21.1	31.9	7.6	12.2	6.6	9.2
Plagioclase	21.1	27.9	19.6	21.2	52.2	50.6	61.2	62.6
Olivine	4.2	4.5	2.9	3.5	3.4	3.0	3.1	3.8
Silica	0.4	0.6	0.6	0.7	—	0.6	—	—
Ilmenite	0.2	1.0	0.3	2.1	0.4	0.7	0.3	0.2
Mare glass	6.1	10.6	13.0	14.6	1.7	3.9	—	1.2
Highland glass	2.8	5.9	2.9	11.5	23.8	24.8	5.8	9.0
Others	3.7	1.4	5.8	7.3	0.1	1.7	1.8	1.0
Total	100.2	100.1	100.0	100.1	100.0	100.1	100.1	100.0
Number of points	574	509	346	288	984	1017	768	500

[a] Samples are surface soils from the Apollo 11 (10084), Apollo 12 (12001 and 12033), Apollo (14163), Apollo 15 (15221 and 15271), and Apollo 16 (64501 and 67461) lunar landing sit The Apollo 11 sample is probably the most representative of mare surface materials. The Apc 16 samples are probably the most representative of highland surface materials among those p sented in the table. From Labotka et al. (1980).

samples are probably the most representative of mare regolith. The Apollo 12, 14, 15, and 17 samples probably contain more of a mixture of highlands and mare regolith than Apollo 11 and 16 samples (Apollo 17 samples are not shown in Table I). If we assume that Apollo 11 and 16 samples are representative of the mare and highland regolith, respectively, then we see that pyroxene is more abundant than plagioclase in the mare, while the reverse is true in the highlands. Glasses and agglutinates plus breccias make up large portions of the regolith. Note that the dark mineral ilmenite is greater in the mare regolith than in the highlands.

The average compositions of Apollo soil samples in terms of weight-percent oxides are given in Table II. It can be seen that the mare soils have a greater iron and titanium content than the highlands, and that the highlands have a greater aluminum content. A list of the minerals that have been found on the Moon is given in Table III.

Because there has been so much bombardment of the lunar surface, one might expect to find some material from the impacting bodies. Probably the best indicators of meteoritic material are the abundances of trace elements. Baedacker et al. (1972) determined the abundances of Ni, Ge, Ir, and Au in lunar soils minus the abundances in crystalline rocks relative to the abundances in C1 meteorites and concluded that about 1 to 1.5% of the lunar soils are of meteoric origin.

Volatile elements in the lunar regolith are highly depleted. There are minor amounts in the highland breccias: Cl, Pb, Br, Zn, Rb, Ag and Tl. Fra Mauro basalts are the most volatile-rich of lunar samples containing Br, Cl, F,

TABLE II
Average Composition in Weight Percent Oxides of
Lunar Soil Samples ($<$ 1 mm fraction)[a]

	Apollo 11	Apollo 12	Apollo 14	Apollo 15	Luna 16	Luna 20	Apollo 16
SiO_2	42.04	46.40	47.93	46.61	41.70	45.40	44.94
TiO_2	7.48	2.66	1.74	1.36	3.38	0.47	0.58
Al_2O_3	13.92	13.50	17.60	17.18	15.33	23.44	26.71
FeO	15.74	15.50	10.37	11.62	16.64	7.37	5.49
MgO	7.90	9.73	9.24	10.46	8.78	9.19	5.96
CaO	12.01	10.50	11.19	11.64	12.50	13.38	15.57
Na_2O	0.44	0.59	0.68	0.46	0.34	0.29	0.48
K_2O	0.14	0.32	0.55	0.20	0.10	0.07	0.13
P_2O_3	0.12	0.40	0.53	0.19	—	0.06	0.12
MnO	0.21	0.21	0.14	0.16	0.21	0.10	0.07
Cr_2O_3	0.30	0.40	0.25	0.25	0.28	0.14	0.12
Total	100.30	100.21	100.22	100.13	99.26	99.91	100.17

[a]Table from Taylor (1975, p. 62).

TABLE III
Lunar Mineralogy[a]

Abundant minerals:

Pyroxenes	$(Mg,Fe,Ca)_2(Si_2O_6)$
Plagioclase	$(Ca,Na)(Al,Si)_4O_8$
Olivine	$(Mg,Fe)_2(SiO_4)$

Accessory minerals:

Ilmenite	$FeTiO_3$
Chromite	$FeCr_2O_4$
Ulvospinel	Fe_2TiO_4
Spinel	$MgAl_2O_4$
Cr-pleonaste	$(Fe,Mg)(Al,Cr)_2O_4$
Perovskite	$CaTiO_3$
Dysanalyte	Ca,REE,TiO_3
Rutile	TiO_2
Nb-REE[b]-rutile	$(Nb,Ta)(Cr,V,Ce,La)TiO_2$
Baddeleyite	ZrO_2
Zircon	$ZrSiO_4 + REE,U,Th,Pb$
Quartz	SiO_2
Tridymite	SiO_2
Cristobalite	SiO_2
Potash feldspar	$KAlSi_3O_8 + Ba$
Apatite	$Ca_5(PO_4)_3(F,Cl) + REE,U,Th,Pb$
Whitlockite	$Ca_3(PO_4)_2 + REE,U,Th$
Zirkelite	$CaZrTiO_5 + Y,REE,U,Th,Pb$
Amphibole	$(Na,Ca,K)(Mg,Fe,Mn,Ti,Al)_5Si_8O_{22}(F)$
Iron	Fe
Nickel-iron	(Fe,Ni,Co)
Copper	Cu
Troilite	FeS
Cohenite	Fe_3C
Schreibersite	$(Fe,Ni)_3P$
Corundum	Al_2O_3
Goethite	$HFeO_2$

New minerals:

Armalcolite	$(Fe,Mg)Ti_2O_5$
Tranquillityite	$(Fe,Y,Ca,Mn)(Ti,Si,Zr,Al,Cr)O_3$
Pyroxferroite	$CaFe_6(SiO_3)_7$

[a] Table from Marvin (1973).
[b] REE refers to rare earth elements.

Bi, Cd, Pb, Tl and Zn which appear mostly as surface coatings. Some $HFeO_2$ (akaganeite) has been found, but is probably due to oxyhydration of $FeCl_2$ with terrestrial water vapor. $HFeO_2$ cannot exist for any length of time under the lunar vacuum. (See chapter 5 of Taylor [1982] and references therein for more details.)

The orbital geochemical data obtained on Apollo 15 and 16 with the X-ray and gamma-ray instruments have helped to show whether the landing sites were representative of the lunar surface. The X-ray instruments provided data on the elements Si, Al and Mg. The gamma-ray instruments provided data on the elements Th, K, U, Fe, Mg and Ti. Twenty percent of the lunar surface was mapped by the gamma-ray instruments, and 10% was mapped by the X-ray instruments. The mare have a greater Th, K and U abundance than the highlands. The ratio of K to Th is less than that of terrestrial basalts and chondritic meteorites. The highest concentrations of Fe occur in the mare, consistent with the samples at the Apollo sites. Separating Ti from Fe has proven difficult with the gamma-ray data, but this shows that low-Ti basalts predominate over high-Ti basalts (see chapter 2 of *Basaltic Volcanism Study Project* [1981] and references therein). The X-ray instruments provided data that show the Al/Si values have a sharp boundary between the highlands and the mare. This result places severe constraints on the lateral movement of fine-grained material from the highlands to the mare (e.g., see chapter 5 of Taylor 1982, and references therein). The Mg/Si data show minimal differences between mare and highlands, but are slightly higher in the mare. Some craters may have penetrated through the mare basalt, exposing subsurface basalts with a higher Mg composition. An example is Messier A in Mare Fecunditatis (see chapter 2 of *Basaltic Volcanism Study Project* [1981] and references therein).

Reflectance spectra can be obtained with high accuracy from the Earth for areas as small as about 1 km over the near side of the Moon. Now that the Apollo missions have ended, spectroscopy provides a tool that can presently be used to map regions on the Moon not sampled by the Apollo missions. Reflectance spectra of the lunar surface generally increase in reflectance at longer wavelengths, and exhibit absorptions near 1 and 2 μm due to pyroxene (Fig. 12). As the lunar surface is continually bombarded by large meteoroids as well as micrometeoroids, the state of the surface is modified. A newly exposed rock will be slowly broken into smaller pieces with the bombardment process. However, this process does not continue forever, making smaller and smaller pieces because the micrometeoroids cause melting on a small scale producing glass and cementing larger pieces together. This process maintains the regolith with an average grain size in the 40 to 70 μm range. The glass produced is often dark, so as the surface undergoes this process, called maturation, it darkens and, because the glass prevents photons from encountering mineral grains, the reflectance spectrum of the surface will have weaker absorption bands. Comparison of reflectance spectra of the lunar surface with

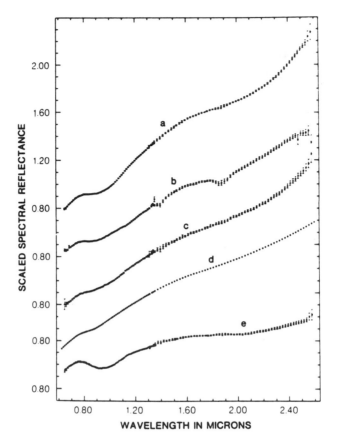

Fig. 12. Representative lunar spectra from McCord et al. (1981). Each spectrum is of a 10 to 20 km diameter area: (a) Mare Serenitatis 2; (b) Herschel; (c) Apollo 11 site; (d) Apollo 16; and (e) Aristarchus. Terrestrial atmospheric water absorptions that were not completely removed are evident for Herschel near 1.4 and 1.8 μm. The broad absorptions near 1 and 2 μm are due to pyroxene in the lunar regolith. The spectra are scaled to 1.0 at a wavelength of 1.02 μm.

spectra of returned Apollo soil samples shows that the lunar surface is covered with mature soil. Regional differences in soil types, especially in pyroxene compositions, can be determined over all lunar regions (see, e.g., Pieters 1977; McCord et al. 1981). Spectra of the ejecta of small fresh craters have been interpreted to indicate discrete lithologies of major basin-forming events (Bell and Hawke 1984).

IV. PHOBOS AND DEIMOS

The Martian satellites are very faint and close to Mars as viewed from the Earth, so there are no Earth-based spectra of these objects. All our information regarding surface composition comes from spacecraft observations from the Mariner 9 ultraviolet spectrometer, Viking lander imaging system, and the Viking Orbiter Infrared Thermal Mapper (IRTM). Pang et al. (1978) derived the reflectance spectrum of Phobos from 0.2 μm to about 0.85 μm using the Mariner 9 ultraviolet spectrometer (resolution 0.015 μm covering the 0.212 μm to about 0.35 μm region) and the three broadband Viking lander imaging filters centered near 0.45, 0.67, and 0.87 μm. These data show a flat spectral response from 0.42 to 0.87 μm with a geometric albedo of 0.05, and a decrease in albedo at shorter wavelengths, reaching 0.01 at 0.212 μm. Pang et al. found that the spectrum of Phobos matched closely with that of asteroid (1) Ceres. The bulk density of these bodies (~ 2 g cm^{-3}) is indicative of hydrated silicate assemblages and the analogy with Ceres strongly suggests an aqueously altered material (Gaffey 1978; Feierberg et al. 1981) probably with a carbonaceous chondrite precursor. An abundant opaque phase (either carbonaceous material or magnetite) with an iron-poor phyllosilicate assemblage agrees best with the spectral albedo data. This model is not unique, however, because diagnostic spectral features have not been seen. Pang et al. (1978) cite this composition as the most likely considering the spectral match with Ceres and the abundance of carbonaceous chondritic material in the solar system.

Lunine et al. (1982) derived the thermal properties of Phobos and Deimos from the Viking IRTM at wavelengths ranging from 6 to 20 μm. They found surface material of low thermal conductivity comparable to that of the Moon. They concluded that the surfaces were covered with a vertically uniform layer of finely divided material at least several centimeters thick, and on Phobos, about 5% of the surface is covered by a high inertia or blocky material.

V. THE SATELLITES OF JUPITER

Io

The surface of Io (shown in Fig. 13; see also Color Plate 1) is the most geologically active solid planetary surface in the solar system, and the phases which are dominant on Io's optical surface are dominant on no other surface (see chapter by Nash et al.). It is believed that among the most abundant of these are elemental sulfur and sulfur dioxide.

It has long been known from Earth-based hemispheric spectra that Io has one of the highest albedos of any object in the solar system and yet lacks detectable H_2O ice bands in its near-infrared spectrum (Pilcher et al. 1972; Sill and Clark 1982; also see review of early work by Johnson and Pilcher 1977). These characteristics, together with a very sharp drop in reflectance between 0.6 μm and 0.3 μm, shown in Fig. 14, suggested the possible presence of sulfur (S) on the surface (Wamsteker et al. 1974; Fanale et al. 1974). How-

Fig. 13. Voyager photo of each Galilean satellite (Smith et al. 1979; see also Color Plate 1). Clockwise from top, left are: Io, Europa, Callisto and Ganymede. Note the visual contrast between satellites, and compare the reflectance levels near 0.5 μm with the spectra in Fig. 14. (Color plates and maps of each satellite are given in the backmatter.)

ever, because S has no sharp infrared absorption features, such an identification could not be positively confirmed. Later, Cruikshank et al. (1978) and Pollack et al. (1978a) discovered a deep sharp absorption feature at 4.06 μm, which Fanale et al. (1979) and Smythe et al. (1979) identified as the $\nu_1 + \nu_3$ absorption of solid SO_2.

The 1979 Voyager 1 and 2 flybys provided imaging of Io in five broad spectral bands covering the visible and near-ultraviolet. Io's surface was found to exhibit a melange of regions of orange, yellow, white and black. In general this coloration was regarded as strongly supporting the Earth-based conclusion that S was a major surface constituent (Soderblom et al. 1979; Sagan

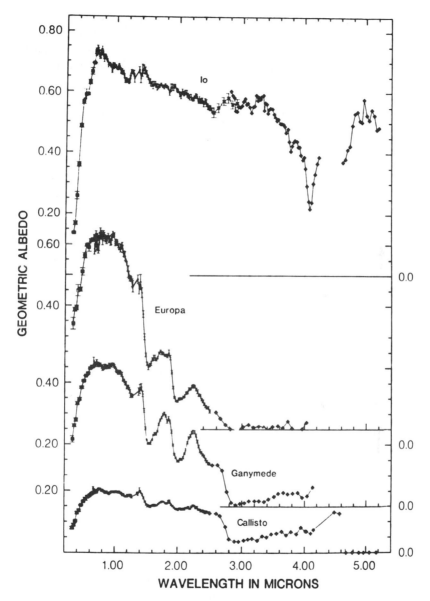

Fig. 14. Spectra of Galilean satellites derived from Clark and McCord (1980a). The spectrum of Io is of the leading side, except that the data beyond 3 μm are a mixture of both sides. The spectrum of Europa is of the trailing side, and the spectra of Ganymede and Callisto are of the leading sides.

1979). These studies concluded that virtually all the colors seen on Io could be accounted for by S allotropes or mixtures of them. These include black S, stable at 600–700 K, brown at ~ 500 K, red (S_3 and S_4) at ~ 450 K, and the familiar S_8 which is stable below the melting point (388 K) and which can be yellow or white (Sagan 1979; Soderblom et al. 1979; Sill and Clark 1982). Moreover, specific features in Io's spectrum at 0.40 and 0.55 μm (Nelson et al. 1980a) have been identified by Sill and Clark as corresponding to features in the spectra of S_3 and S_4. Particularly compelling evidence for sulfur is the relationship between occurrence (as revealed by color and morphology) and temperature (as indicated by thermal spectra). Deep calderas and fissures seem to be filled by the highest temperature material which appears to revert successively, as the flow propagates, to red, orange, yellow and white (Soderblom et al. 1979; McEwen and Soderblom 1983). Again we point out that S and O dominate the magnetosphere in the region around Io. Other elements can be only minor in flux compared to S and O. This, in essence, is the powerful, but admittedly circumstantial, case for S on Io. Recently, however, Young (1984) has concluded that the colors on Io are not consistent with elemental sulfur.

The Voyager spacecraft also contributed indirect evidence supporting the presence of frozen SO_2 on the surface, as suggested by Earth-based observations. SO_2 gas was detected near a large plume close to the subsolar point by the Infrared Spectrometry experiment (Pearl et al. 1979). The question of whether the SO_2 observed was in equilibrium at subsolar temperature with the allegedly ubiquitous frost or whether it indicated the presence of the nearby plume is too complex to be discussed here, as is the entire issue of the interaction between Io's surface and atmosphere (see discussions by Kumar and Hunten [1982], Fanale et al. [1982], and the chapter by Nash et al.). In any case, the identification of gaseous SO_2 clearly provides at least strong indirect support for the identification of SO_2 on the surface. Another way in which Voyager results support SO_2 as a major constituent is that almost all the Voyager imaging data are too bright in the ultraviolet to represent a simple mixture of the S allotropes, which would otherwise explain their colors (Soderblom et al. 1979). In one striking case—the brightest regions—even white S_8 would require an admixture of about 50% of a component as bright in the ultraviolet as is SO_2 to explain its ultraviolet brightness (Soderblom et al. 1979). Normally one could turn to ultraviolet-bright alternatives such as other ices. However, virtually all other alternatives that are bright in the ultraviolet, like H_2O ice, are eliminated on grounds of absence of infrared features in bland regions of Io's infrared spectrum. Again, this is strong supportive evidence of the presence of solid SO_2 as a major surface component. One problem with the use of S_8 as the "other" main component of the bright areas is that the predicted temperature dependence of the brightness (Wamsteker et al. 1974) was not observed in Voyager pixels (Veverka et al. 1981).

In addition to sulfur and oxygen, sodium and potassium have been detected in the Jovian magnetosphere with Io's surface as the source (see Pilcher and Strobel 1982, and references therein). Previously, Fanale et al. (1979) had eliminated nitrates and carbonates on the basis of the absence of their ancillary bands shortward of 4 μm in Io's spectrum as well as a mismatch of band positions near 4 μm. However, these workers were unable to analogously rule out sulfate owing to a breakdown in the quality of the data at 4.1 μm compounded by the steep thermal rise which begins there. If later improved spectra confirm the presence of sulfates, then several problems would be solved. First, the presence of thenardite (Na_2SO_4) in the bright areas would provide a source of the Na. Second, such a bright phase could provide an initial bright, hence cold, backdrop needed to concentrate SO_2 in the bright areas and lower the atmospheric pressure (Fanale et al. 1982). Third, the problem of using S_8 for this purpose could be avoided. Sulfate also could conceivably be formed by the action of oxygen freed by SO_2 dissociation (cf. Kumar and Hunten 1982).

Despite the generally harmonious nature of all these major observations and arguments, there are numerous uncertainties, some major, in the previously discussed indications for the composition of Io's surface. We briefly review only a few of these, together with suggestions for future work. First, the identification of S is not based on a particular absorption band and therefore is not positively confirmed. Additionally, Clow and Carr (1979) have argued that high, steep slopes observed in imaging of some volcanic caldera walls require strength in excess of that possible for S at the supposed high subsurface temperatures. This objection could be overcome if igneous silicate flows were efficiently coated to an optical depth with more volatile sublimates, including S and SO_2, which would tend to dominate the spectra—not an unreasonable possibility.

Additionally, although the SO_2 identification is specific, being related to the band center position of an identified vibrational absorption band, there is still considerable uncertainty as to how much of the feature is due to each of three forms of sulfur dioxide: SO_2 frost, adsorbed SO_2 and dissolved SO_2 (Fanale et al. 1979; Smythe et al. 1979; Nash and Nelson 1979; Nash et al. 1982; chapter herein). Adsorbed SO_2 should be expected, if conditions even approach saturation, and condensation must be preceded by extensive adsorption. Estimates of the amount of expected adsorbed SO_2 based on laboratory studies are given by Fanale et al. (1982). Nash (1982) has calculated that the adsorbed SO_2 might be able to produce a deep band, but the calculation assumed effective scattering by grains much smaller than the wavelength of the photons involved. Another argument for adsorbed SO_2 as the source of the band is that the wavelength of the Io feature may match that of adsorbed SO_2 as seen in transmission spectra better than it matches SO_2 frost or ice (Nash 1982). However, recent work by Howell et al. (1984) has eliminated previous ambiguities in the wavelength calibration of the telescopic data and now shows the telescope band position for Io and that observed for SO_2 frost with the same equipment to be identical within 0.005 μm. Lacking diffuse reflectance spectra of adsorbed SO_2, it is still not possible to put an upper limit on the contribution from that species. Another argument for adsorbed SO_2 versus frost is that the observed whole disk gas abundance is too low to be in equilibrium with Io's surface, given ubiquitous SO_2 frost (Butterworth et al. 1982). However, since Voyager data clearly show a concentration of SO_2 in the bright areas, Fanale et al. (1982) suggested that these might be able to serve as cold traps, thereby lowering the atmospheric pressure to or below the observational upper limit, while still supporting extensive surface frost. Alternatively, Matson and Nash (1983) have suggested that the effective cold trap is below the surface and at a temperature always far below the temperature of the surface near midday. This would lower the gas abundance by orders of magnitude, but then there might be almost no frost on the optically accessible surface. Again, this would require attributing virtually all the spectral feature to adsorbed SO_2 and would leave the overall spectrum of the white areas a mystery. Nash et al. further argued that Io is simply not bright enough overall in the ultraviolet to allow postulating enough frost on the surface to account for the 4.06 μm band. Thus, they favor a role for adsorbed SO_2 on this basis. However, Howell et al. (1984) have suggested a simple alternative solution: in intimate mixtures of two components, one of which is dark in the ultraviolet (like S) and the other of which is bright in the ultraviolet and dark in an infrared absorption band (like SO_2), the infrared absorber will dominate the infrared absorption region and the ultraviolet dark material will dominate the ultraviolet spectrum, each in greater proportion than their abundances (see Sec. II).

Yet another problem is that neither SO_2 nor any form of S can explain the significant flux of sodium into the magnetosphere from Io, and the suggestion of Na_2SO_4, as above, is only a speculation at present. Whether magnetospheric species are derived from Io's surface or atmosphere, S, O and Na fluxes from Io are dominant. Therefore, there must be sodium-containing phases on the surface. One alternative to the Na_2SO_4, considered by Fanale et al. (1979) and Nash and Nelson (1979) is the sulfide salt Na_2S. This compound has a spectrum so close to that of Io that substantial amounts could be mixed into a mixture of S and SO_2 without producing any noticeable deviations from Io's whole disk spectrum (Fanale et al. 1979; Nash and Nelson 1979). Furthermore, there is a good chance that the presence of Na_2S, perhaps in solution in the S, could enhance the absorption strength of the S, accounting for the steep spectral slopes mentioned earlier (the melting point of pure Na_2S is 1100° C). This mixture might subvert the ability to brighten on cooling, thus explaining the Voyager data considered by Veverka et al. (1981). The spectral data, even in conjunction with other data, still leave room for considerable variation in surface modeling.

In summary, major Io surface compositional issues that still need resolution include:

1. The purity of the surface S;
2. The question of whether silicate rocks underlie a thin S rich crust and the composition of such igneous rocks;
3. The physical state of the SO_2;
4. The chemical state of the Na on the surface;
5. The general problem of the composition of the bright areas;
6. The nature of any temporal variations in surface composition. Voyager 1 and 2 data show temporal color variations (see Smith et al. 1979b);
7. The way in which surface SO_2, in whatever form, controls the atmosphere.

Among the many important inputs to our knowledge in future years will be Galileo mission studies of surface composition including: (a) moderate (~ 20 km) spatial resolution mapping of much of the disk in 200 spectral bands from 0.7 to 5.0 μm to reveal small domains of unusual composition, which will allow the major components, confused in whole disk spectra, to be disentangled, and allow surface composition to be related to regional geologic features; (b) high spatial resolution (~ 1 km) Galileo compositional measurements across geologic boundaries such as calderas, vents, etc.; and (c) low spatial resolution (~ 80 km) Galileo synoptic observations scattered over a period of two years to detect any temporal variations (especially expected for SO_2, but also for other components). These comments, except for the synoptic observations, apply to JII, JIII and JIV as well, and it has even been suggested that JII might be exhibiting current activity (Cook et al. 1982), in which case synoptic observations of JII could be of interest. Laboratory measurements of spectra of adsorbed and dissolved SO_2 and continued study of the optical and physical properties of allotropes of S and of a solution of Na in S, are very important for understanding Io's surface composition. Ultraviolet observation of both frost and SO_2 gas abundance as well as their spatial and temporal variations (from Earth orbit) are also important.

Europa, Ganymede and Callisto

There is a considerable volume of material published on the icy satellites of Jupiter, and as with many of the other objects discussed in this chapter, a chapter on surface composition could be written on each one. We refer the reader to Sill and Clark (1982) for a previous review of the surface compositions of these objects; here we will concentrate on more recent studies. Europa, Ganymede, and Callisto (Fig. 13; see also Color Plate 1) have spectra which show water ice absorptions at 3.0, 2.0, 1.5, 1.25 and 1.04 μm (see, e.g., review by Sill and Clark 1982, and references therein; Clark et al. 1980; Clark 1982). The observed water ice absorption band depths are shown in Table IV for all solar system satellites where surface ice has been detected and

TABLE IV
Band Depths of Water Ice for Solar System Satellites [a]

	1.04μm	1.25μm	1.52μm	2.02μm	3μm
Europa (L)	0.036	0.096	0.57	0.63	~1.0
(T)	0.045	0.104	0.47	0.63	~1.0
Ganymede					
(L)	0.024	0.092	0.452	0.46	~1.0
(T)	0.044	0.106	0.381	0.31	~0.9
Callisto (L)	0.024	0.04	0.14	0.13	~0.7
(T)	0.027	0.09	0.21	0.10	~0.7
Saturn's rings	0.02	0.05	0.56	0.71	~1.0
Enceladus					
(T)			~0.54	0.60	
Tethys (L)	<0.05	~0.04	0.52	0.67	
(T)	<0.007	<0.02	0.42	0.59	
Dione (L)	<0.003	0.06	0.37	~0.45	
(T)	0.02	<0.1	~0.3	~0.35	
Rhea (L)	0.025	0.066	0.48	0.63	
(T)	0.016	0.06	0.38	~0.50	
Iapetus (L)	<0.007	<0.01	~0.06	~0.14	~0.8
(T)	<0.01	0.09	0.48	0.73	
Hyperion (L)	<0.1	<0.1	~0.38	0.5	
Miranda				~0.6	
Ariel	<0.06	<0.05	0.3	0.5	
Umbriel	<0.1	<0.05	~0.1	0.2	
Titania	<0.05	<0.05	0.2	0.4	
Oberon	<0.03	<0.05	0.2	0.2	

[a] All Galilean satellite data are from Clark et al. (1980), except those for the 3μm band, which are from Pollack et al. (1978a). The Saturnian satellite data are from Clark et al. (1984) except those for Enceladus, which are from Clark et al. (1983a) and the 3μm band depth of Iapetus which is from Lebofsky et al. (1982a). The Saturnian ring data are from Clark and McCord (1980) and Puetter and Russell (1977). Data for most Uranian satellites except Miranda are from Brown (1982) and Brown and Cruikshank (1983), and the Hyperion data are from Cruikshank and Brown (1982). Miranda data are from Brown and Clark (1984). Some 1.04μm Uranian satellite data (to be published) which were obtained on the same observing runs in 1983 as the Saturnian satellite data are included. L refers to leading side, T refers to trailing side. This table is from Clark et al. (1984).

TABLE V
Visual Reflectance and Water Ice Band Depths
of Ice and Ice-Particulate Mixtures[a]

	0.55μm vis. refl.	1.04μm	1.25μm	1.52μm	2.02μm
Fine frost	1.00	0.013	0.038	0.399	0.557
Medium frost	1.00	0.024	0.070	0.587	0.698
Med. coarse frost	0.98	0.034	0.105	0.670	0.740
Coarse frost	0.90	0.103	0.210	0.737	0.708
Pure ice block	0.47	0.254	0.390	0.534	0.176
Pure ice block	0.47	0.216	0.342	0.371	0.274
20 wt% kao + ice	0.65	0.050	0.098	0.616	0.691
5 wt% kao + ice	0.67	0.084	0.176	0.751	0.764
1.0 wt% kao + ice	0.65	0.148	0.217	0.508	0.528
0.1 wt% kao + ice	0.62	0.236	0.424	0.245	0.216
0.01 wt% kao + ice	0.55	0.232	0.369	0.346	0.302
30% wt% mks + ice	0.15	0.000	0.015	0.274	0.465
10 wt% mks + ice	0.18	0.000	0.041	0.336	0.432
1.0 wt% mks + ice	0.19	0.122	0.262	0.343	0.222
0.1 wt% mks + ice	0.42	0.162	0.280	0.273	0.265
0.01 wt% mks + ice	0.47	0.219	0.399	0.444	0.318
70 wt% chcl + ice	0.060	0.012	0.030	0.050	0.054
50 chcl + ice	0.14	0.000	0.019	0.063	0.101
30 chcl + ice	0.072	0.013	0.034	0.116	0.250
10 chcl + ice	0.073	0.013	0.068	0.237	0.363
1.0 chcl + ice	0.155	0.015	0.015	0.129	0.169
0.1 chcl + ice	0.160	0.058	0.109	0.151	0.144
0.03 chcl + ice	0.23	0.134	0.245	0.195	0.147

[a]Frost data from Clark (1981a); all other data from Clark and Lucey (1983); kao = kaolinite; mks = Mauna Kea Soil; chcl = charcoal. This table is from Clark et al. (1984).

the band depths derived from laboratories are shown in Table V for comparison. Representative spectra of the Galilean satellites are shown in Fig. 14. The many absorptions of water ice allow us to determine the abundance of the ice relative to other materials.

Early studies of the composition of the Galilean satellites were limited by data quality. Infrared instruments were just becoming sensitive enough to detect the satellites from Earth-based telescopes. Pilcher et al. (1972) were the first to identify H_2O on the surfaces of the satellites. Kieffer and Smythe (1974) compared the spectra of the Galilean satellites with CH_4, CO_2, H_2O, H_2S, NH_3 and NH_4SH frost spectra in the first detailed study to quantify the abundances of materials in the surface. Limited by data quality, they found H_2O as the dominant component on Europa and Ganymede, and limits of 5 to 28% of the other frosts assuming the frosts were optically isolated (areal mixtures) on the surface. They concluded that other materials must be present on Callisto. Pollack et al. (1978a) obtained new spectra of the satellites and performed detailed modeling of the spectra to derive quantitative abundances of minerals on the surfaces. They determined that the fractional amounts of water ice cover (in an areal mixture) on the trailing and leading sides of Ganymede and the leading side of Europa were 50 ± 15%, 65 ± 15%, and ≥ 85%, respectively, and that the 2.9 μm band of Callisto was due primarily to bound water. Again, the Pollack et al. study was limited by data quality as the data were insufficient to show the 1.04 μm ice absorption except on Europa, and its importance at that time was not fully understood.

The strength of each ice absorption in a reflectance spectrum is a direct indication of the mean optical path length through ice, and an indicator of ice grain size and/or ice purity. From higher precision data than that obtained by Pilcher et al. (1972) and Pollack et al. (1978a), Clark (1980) performed a new study of the composition of the icy Galilean satellites. He found that Europa's spectrum is dominated by deep-water ice absorptions (Fig. 14 and Table IV), which from detailed laboratory studies, can have only a few weight-percent impurities. Ganymede has more impurities in the water ice than Europa but is still probably 90 wt % or more ice. Callisto has even more impurities. However, even that study was limited by data quality because the 1.04 μm absorption on Callisto was not found. Clark et al. (1980) obtained higher signal-to-noise data on all the Galilean satellites and found the 1.04 μm band on Callisto. The 1.04 μm ice band on Callisto is 2.4% deep, a strength that indicates abundant water ice.

In contrast to the 1.04 μm band, the 3 μm water absorption on Callisto has a band depth of only about 70%. Water ice is a very strong absorber at 3 μm, and a surface layer only 1 μm thick would be essentially totally absorbing. For a surface of randomly oriented grains, the specular reflection from the grain surfaces is given by Fresnel's equations. At 3 μm, ice reflects about 1% of the light incident on its surface, so if there was abundant ice on Callisto's surface as indicated by the 1.04 μm band, then the ice band depth at 3 μm should be close to 99%. Because the 3 μm band is much weaker than this value, one of two conditions are indicated: there are ice-poor regions on the surface, or the ice surface has a fine submicron structure that scatters some photons before they are absorbed. A very fine (< 1 μm) surface structure on an ice block, or fine grains coating larger grains, might show a 3 μm absorption only 70% in depth, while still allowing photons at other wavelengths to penetrate deeper into the surface resulting in relatively strong ice absorptions. The grains would have to be considerably smaller than 1 μm and a total optical path in ice of about 1 μm would be required to give a 70% depth (see Clark and Lucey 1984).

Because of the significant vapor pressure of ice, water molecules will constantly be migrating on the surface. The result is that fine grains will grow larger with time, and fine surface structure will be filled in. The kinetics of such grain growth has been known for several decades, and application to planetary surfaces has been discussed by Clark et al. (1983a, and references therein). That study showed that at the subsolar temperature of 167 K on Callisto, small grains would grow to about 1 μm in size in only a few years, and to 1 mm in 10 million years. It is unlikely that any process such as micrometeoroid bombardment could maintain a submicron grain-size surface on such a time scale and still show the strong 1.04 μm ice absorption in the spectrum. Thus, Callisto apparently has some relatively pure water ice patches (indicated by the strong 1.04 μm ice band) and some patches of the surface that are ice-free.

Clark et al. (1983a) showed that sputtering could cause an effective growth of ice grains on an icy surface because the smaller grains would be sputtered away faster due to their larger surface-to-volume ratio compared to grains of greater size. Magnetospheric particle bombardment affects the trailing side of each satellite more strongly than the leading side because the corotational velocity of magnetospheric ions is greater than the orbital velocity of the satellites (see chapters by Cheng et al. and Veverka et al.). At the average temperature of Europa, 124 K, the sputtering rate is similar to the sublimation rate (see Johnson et al. 1981). At Ganymede, the sputtering rate is about 40 times less than the sublimation rate for an average temperature of 138 K. At Callisto, the sputtering rate is much less than the sublimation rate at an average temperature of 149 K. The sequence from Europa to Callisto would imply the largest bias in grain size from leading to trailing hemisphere at Europa and essentially no bias at Callisto. Examination of the 1.04 μm water ice absorption band depths (see Table IV) for the leading-to-trailing hemisphere indicates differences in the photon mean optical path length in ice, which is related to grain size and composition. If we assume that the basic composition of each hemisphere on a particular satellite is approximately the same, but that the satellites differ, then the mean optical path lengths are indicative of grain size differences in the ice. The sequence of ice band depths is that expected if sputtering were influencing the surface grain-size distribution (Clark et al. 1983a): JII (T) ~ JIII (T) > JII (L) > JIII (L) ~ JIV (T) ~ JIV (L), where JII is Europa, JIII is Ganymede, JIV is Callisto, L is the leading and T is the trailing hemisphere. Nelson et al. (1985) analyzed Voyager 4-channel visible data for Europa along with the hemispheric Earth-based spectra of Clark et al. (1980) and concluded that there is probably no significant compositional difference from leading to trailing hemisphere, and that the albedo and ice band depth differences were mostly due to a larger ice grain size on the trailing side.

Although no diagnostic absorptions have been found in the visible and near-infrared which are attributable to the dark material on Europa, Ganymede or Callisto, each satellite displays a reflectance spectrum which decreases from about 0.7 μm to at least 0.3 μm. The shape of this decrease is suggestive of minerals containing Fe^{3+} (cf. Clark 1980). Because the infrared portions of the spectra appear to indicate essentially pure water ice, the non-icy components must be spectrally bland with no significant slope to the spectra. These properties are also consistent with minerals containing Fe^{3+}, such as some carbonaceous chondrite meteorites (Clark 1980).

In the case of Europa, Lane et al. (1981) found an absorption in the ultraviolet which is attributable to a sulfur-oxygen bond. This absorption is seen only on Europa's trailing side and is apparently of magnetospheric origin. The question remains, however, whether SO or SO_2 is being implanted in the surface, only S is being implanted, or whether S is already present in the surface and the S—O bond is somehow created in the bombardment process.

The Small Satellites of Jupiter

The small outer satellites of Jupiter form two clusters with similar orbital elements, one group in prograde orbits (JVI Himalia, JVII Elara, JX Lysithea, JXIII Leda) and the other in retrograde orbits (JVIII Pasiphne, JIX Sinope, JXI Carme, JXII Ananke) twice as far from Jupiter (see the chapter by Thomas et al.). The *UBVRI* colors of several of these bodies (JVI, JVIII; Degewij and Van Houten 1979; Degewij et al. 1980a,b) fall generally in the domain of C-type asteroid colors (Bowell et al. 1978). Thermal infrared observations by Cruikshank (1977) indicate albedos of about 3% for JVI and JVII which are similar to those of the darkest asteroids. Cruikshank et al. (1982) concluded from the spectral and albedo data that JVI and JVII (in the inner prograde group) were spectrally similar to many of the dark C-type asteroids present in the outer asteroid belt, while JVIII and JIX (in the outer retrograde group) may be compositionally different, with spectra more similar to the dark, reddish D-type Trojan asteroids (Gradie and Tedesco 1982). Tholen and Zellner (1984) presented 8-color (subvwxpz filters, 0.336–1.042 μm) photometric data of JVI, JVII, JVIII, JIX, JX and JXI. The data for JIX was noisy and incomplete, but with the exception of JXI, the spectra of all these small outer satellites of Jupiter resemble those of C-type asteroids, and therefore presumably resemble carbonaceous chondrites in general mineralogy. Tholen and Zellner (1984) suggested that the absence of D-type spectra among the small outer Jovian satellites indicated that these objects were not derived from the same population as the Trojan asteroids or those of the outer belt. It is plausible that the two groups of outer Jovian satellites actually represent captured asteroidal bodies, perhaps subsequently disrupted. Pollack and Fanale (1982) concluded, in conflict to Tholen and Zellner, that the outer Jovian satellites came from the outer belt or the Trojan groups. However, the newer data of Tholen and Zellner (1984) appear to rule out this possibility; further study is warranted.

The largest of the small inner satellites of Jupiter, Amalthea (JV), exhibits a low albedo ($\sim 6\%$) and a very red color with scattered brighter green-ish spots (Gradie et al. 1980; Veverka et al. 1981,1982b). The dark red sur-face is generally similar in color and albedo to D-type asteroids, but Gradie et al. (1980) argue that these properties of Amalthea probably result from the extreme environment in which the satellite resides and from contamination of the surface material with sulfur from Io. The isolated greenish spots occur in association with local slopes (Veverka et al. 1982b) and may represent regions where mass wastage exposes subsurface units and/or prevents high levels of contamination by sulfur.

Three additional smaller inner satellites of Jupiter were discovered on Voyager images. Two of these, Adrastea and Metis lie at the outer edge of the Jupiter ring. Thebe lies slightly outside the orbit of Amalthea. The albedos of these three satellites appear to be similar to that of Amalthea (Veverka et al. 1982b), but direct determinations are lacking.

Because there are no data on masses or bulk densities for the small satel-lites, nothing is known about bulk composition. Actual surface composition is also not well known. Since only broadband colors have been measured, no diagnostic spectral features have been observed. At Jupiter's distance from the Sun, water ice is stable on small bodies; thus the small satellites may contain substantial amounts of water in their interiors or even on their surfaces.

Adrastea and Metis may simply represent the largest of many satellites intimately associated with Jupiter's ring. Burns et al. (1980) have suggested that erosion of these larger fragments provides the flux of small particles needed to maintain the ring. While this is a reasonable assumption, no defini-tive spectral albedo measurements yet exist to verify this connection (Veverka et al. 1982b).

VI. SATELLITES OF SATURN

Mimas, Enceladus, Tethys, Dione, Rhea and Hyperion

Water ice was first shown to exist on Tethys, Dione, Rhea and the trailing side of Iapetus with near-infrared photometry (Johnson et al. 1975; Morrison et al. 1976b) and with spectroscopy by Fink et al. (1976). Cruikshank and Brown (1982) found water ice features on Hyperion. Cruikshank et al. (1984b) have reviewed the optical properties of the satellites of Saturn. Ex-cept for some portions of Enceladus, all of these satellites appear to be heavily cratered (e.g., see Rhea in Fig. 15 and the Map Section).

The Fink et al. (1976) study showed that the 2.0 and 1.5 μm ice absorp-tions were stronger on Rhea and Tethys than on Dione and the trailing side of Iapetus. They also concluded that the fractional frost coverage must be greater than 80% for the parts of the satellites that were observed. Clark and Owensby (1982) analyzed a spectrum of the leading side of Rhea which showed the 1.04 and 1.25 μm water ice absorptions, as well as the 1.5 and 2.0 μm fea-tures observed by Fink et al. (1976). They concluded that the leading hemi-

Fig. 15. A photomosaic of Rhea, from Voyager I (Smith et al. 1981). The surface of Rhea is mostly water ice, compared to the rocky Moon, despite the similar visual impression of the lunar highlands and Rhea.

sphere of Rhea is probably at least 90 to 98 wt % water ice. Other possible constituents are particulate materials and/or clathrates.

More recently, Clark et al. (1984) measured the spectra of the leading and trailing sides of Tethys, Dione, Rhea and Iapetus, and one side of Hyperion (because its rotation is not synchronous, which hemisphere was observed is not known). Representative spectra of Saturn's satellites are shown in Figs. 16, 17 and 18. Water ice absorptions at 2.0, 1.5 and 1.25 μm are observed in spectra of all five objects (except that the 1.25 μm band was not detected in the spectrum of Hyperion). The weak 1.04 μm ice absorption was detected on the leading and trailing sides of Rhea, and the trailing side of Dione. The observed water ice band depths (or the upper limits) are given in Table IV.

The strengths of the ice absorptions on Tethys, Dione, Rhea and Iapetus (trailing side) indicate nearly pure water ice with probably less than about 1 wt % particulate minerals averaged on a global scale. Hyperion probably contains slightly more dark material than the other icy satellites. The ice could be in the form of clathrates with as much as a few weight percent trapped gases. Clark et al. (1984) also set upper limits on the surface abundance of ammonium hydroxide. The abundance upper limit depends on the precision of the data in each satellite spectrum, and the data's precision was limited by telescope observing time. They found that the NH_3 equivalent weight fraction is $\lesssim 1\%$ on Rhea (L), $\lesssim 3\%$ on Tethys (L) and Iapetus (T), $\lesssim 5\%$ on Rhea (T), and $\lesssim 10\%$ on Dione (L,T) and Tethys (T), where L refers to the leading and T refers to the trailing hemisphere.

Clark et al. (1984) showed that the water ice absorptions are stronger on the trailing sides of Tethys, Dione and Rhea than on their leading sides, similar to that seen on the Galilean satellites. This implied surface modification by charged particle bombardment from Saturn's magnetosphere on the trailing hemispheres of the satellites (cf. presentations in the chapters by Cheng et al. and Veverka et al.). Alternative explanations such as micrometeoroid bombardment might also cause some of the observed asymmetry; however, meteoroids have not been shown to impact preferentially on one hemisphere. At present, magnetospheric-induced surface modification such as is believed to occur on Europa and Ganymede is also a reasonable explanation for the asymmetries in the Saturn satellite system.

Only two spectra of Enceladus have been presented, and there are no spectral data indicative of composition for Mimas. Cruikshank (1980b) established the presence of water ice on Enceladus from a low-precision (low signal-to-noise) spectrum. Clark et al. (1983a) measured the spectrum of Enceladus at 5% resolution in the 1.5 to 2.6 μm region and found prominent water ice absorptions. They concluded that the surface of Enceladus is at least as pure ice as Rhea's surface and that the ammonium hydroxide NH_3 content has an upper limit of ~ 10 wt %. Because Mimas is flanked on one side by the water ice rings and by the icy Tethys on the other, and it has a high geometric

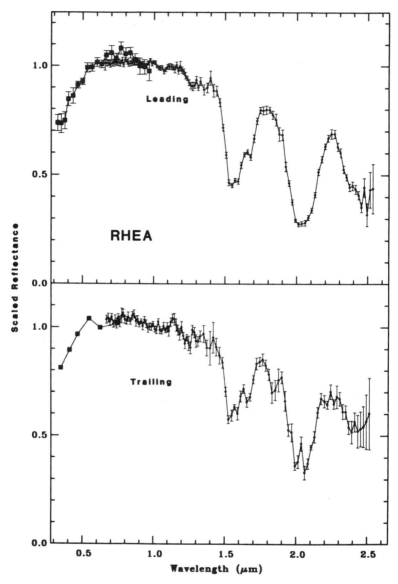

Fig. 16. Spectra of the leading and trailing sides of Saturn's icy satellite Rhea (from Clark et al. 1984). The spectra are scaled to 1.0 at a wavelength of 1.02 μm.

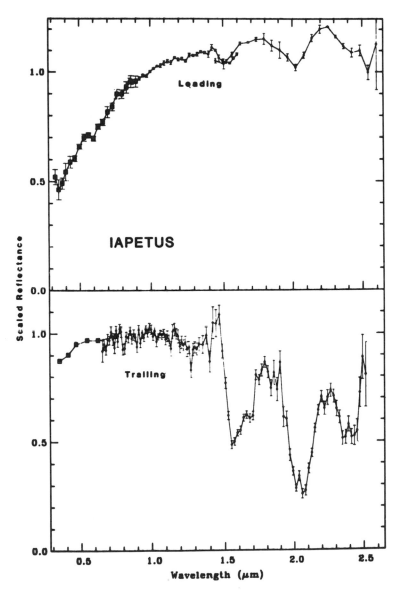

Fig. 17. Spectra of the leading and trailing sides of Saturn's satellite Iapetus (from Clark et al. 1984). The spectra are scaled to 1.0 at a wavelength of 1.02 μm.

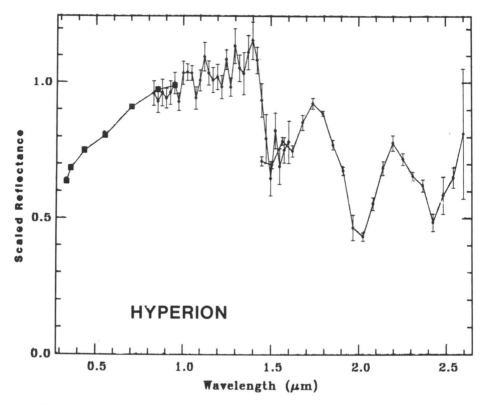

Fig. 18. Spectrum of Saturn's icy satellite Hyperion (from Clark et al. 1984). The spectrum is scaled to 1.0 at a wavelength of 1.02 μm.

albedo (0.8) as well as a low bulk density (1.4 ± 0.3: see, e.g., Smith et al. 1982), the surface of Mimas is probably similar to Tethys' nearly pure water ice.

The Strange Satellite Iapetus

The brightness variations which result from the major albedo differences between the leading and trailing hemispheres of Iapetus were first noted more than three centuries ago. Morrison et al. (1975) correctly interpreted the distribution of high- and low-albedo units, as was dramatically confirmed by the subsequent Voyager flyby of the Saturn system. Near-infrared data (Fink et al. 1976; Morrison et al. 1976; Clark et al. 1984) indicate that the brighter trailing hemisphere is primarily composed of water ice.

The dark leading hemisphere (albedo ~ 0.05: Murphy et al. 1972; Morrison et al. 1975; lowest albedo within the hemisphere ~ 0.02–0.03: Squyres et al. 1984) was studied at visible, H and K wavelengths by Veeder and Matson (1980) who maintained that the high infrared reflectance was not consistent with a carbonaceous chondrite or C-type surface material. Soifer et al. (1979) concluded from low-resolution 2.0 to 2.5 μm spectrophotometry that the flat reflectance curve of the dark side of Iapetus closely resembled that of the dark asteroids. Low-resolution photometry in the 3 μm region (Lebofsky et al. 1982a) indicates a strong absorption feature of water, possibly bound in minerals, on the dark side. Cruikshank et al. (1983a) interpreted moderate-resolution visible and near-infrared spectrophotometric data (0.3–1.0 μm, 1.4–2.5 μm and 3.0–3.8 μm) to indicate that the primary dark component of the leading hemisphere is a carbonaceous material similar to the organic fractions of the carbonaceous chondritic meteorites. These investigators also concluded that the carbonaceous phase is intrinsic to Iapetus and is concentrated by the selective removal of water ice due to the impact of high-velocity dust spiralling in from Phoebe onto the leading hemisphere, rather than an externally added component. Tholen and Zellner (1983) also concluded that the spectrum of the dark-side material was not consistent with simple addition of dust from Phoebe to the leading hemisphere of Iapetus, unless the spectral properties of Phoebe's material were substantially modified.

The Cruikshank et al. composite spectrum contained gaps, and the relative scaling between segments was somewhat uncertain. Clark et al. (1984) presented a more complete composite spectrum from 0.3 to 2.5 μm (Fig. 17) which indicates that the surface material of the dark side of Iapetus cannot be composed solely of a mixture of water ice and the organic compounds present in carbonaceous chondritic meteorites. Bell et al. (1985) produced a composite spectrum of Iapetus similar to that of Clark et al. (1984) and showed that an organic-rich mixture with a moderately iron-rich, clay mineral assemblage provides a very good match to the spectral properties of the low-albedo component of Iapetus' dark side.

Clark et al. (1984) showed that the abundance of water ice in the surface material of Iapetus' dark side could range from less than a few weight percent to as much as 95 percent by weight. Other than the 3 μm absorption, no water features have been detected in the Iapetus spectrum, which could indicate only a trace of water and essentially no ice present in the surface material. However, dark material can be extremely effective in suppressing the spectral features of ice. If the dark grains are submicron, then the dark regions of Iapetus could be at least 95 wt % ice and still not show any ice absorptions in the spectrum except at 3 μm. If the grains are larger (10 to 100 μm), then the surface could be as much as 70 wt % water.

In the absence of diagnostic absorption features, the interpretation of the dark material as an organic-rich clay cannot be regarded as definitive. While the spectral albedo properties of such materials match those of Iapetus' dark-side material, there may well be other assemblages of cosmically realistic species which could produce a similar spectral match, although none have been studied to date. Additional laboratory studies as well as improved observational data (higher data precision, increased spectral resolution, and greater wavelength coverage) would contribute to developing a unique interpretation of the dark side material on Iapetus.

The Small Satellites of Saturn

Saturn has many small satellites (see chapter by Thomas et al.): SIX (Phoebe), SX (Janus), SXI (Epimetheus), SXII (1980 SVI), SXIII (Telesto), SXIV (Calypso), 1980 SXXVI (Pandora), 1980 SXXVII (Prometheus) and SXV (Atlas). All these satellites except Phoebe have orbits within or at the orbit of Dione; the most recently discovered of these is Atlas, just outside the A ring of Saturn. Phoebe, orbiting retrograde, lies outside the orbit of Iapetus and will be discussed separately. Cruikshank et al. (1984b) reviewed the optical properties of the small satellites, which are generally consistent with the presence of water ice as the major surface component.

Most of the small satellites were observed by Voyager and, because of their intrinsic faintness and proximity to Saturn, very little Earth-based data exists on these bodies. Smith et al. (1982) and Thomas et al. (1983a,b) discussed the Voyager data on these satellites. Thomas et al. (1983a) derived the normal reflectances, colors and cross sectional radii, as well as phase information. Many of the satellites are irregular in shape and probably heavily cratered. The normal reflectances range from about 0.4 to at least 0.8. The reflectances and colors in the visual part of the spectrum span the range observed for the larger icy satellites of Saturn. This led Thomas et al. (1983a) to suggest that all the small satellites are icy with surfaces contaminated by small amounts of a dark, opaque material. This is reasonable considering the amount and purity of water ice on the larger satellites in the vicinity and on the rings. However, no spectral data have been obtained on these objects to prove the presence of ice.

Beyond Titan, Saturn has two satellites, Iapetus (discussed above), and Phoebe, the outermost in the system. Degewij et al. (1980) and Cruikshank (1980b) summarized the *UBVJHK* colors for Phoebe and concluded that this satellite had colors similar to carbonaceous chondritic material. The low surface albedo (~ 0.06) of Phoebe is consistent with a C-type material. Tholen and Zellner (1983) presented 8-color (subvwxpz filters, 0.336 to 1.042 μm) data of Phoebe and also concluded that this spectrum was similar to that of C-type asteroids, but with a pronounced absorption near 1 μm. From low-resolution Voyager images, Thomas et al. (1983b) indicate that Phoebe has a diameter of ~ 220 km with a 0.47 μm albedo varying from 0.046 to 0.06 across the surface. There are bright regions at high northern and southern latitudes, but they concluded that these did not appear to be polar caps. The Voyager color data is consistent with the earlier Earth-based spectra. Cruikshank et al. (1983a) argue that dust spiralling in from Phoebe provides the asymmetric impact flux onto the surface of Iapetus to remove selectively the icy component but does not in itself represent a significant surface contaminant on Iapetus.

Phoebe, like the dark side of Iapetus, could be predominantly icy (e.g., 95 wt % water ice with very dark, opaque, small grains, intimately mixed with the ice), and still be a dark object with no 1.5 or 2.0 μm ice bands (the 3 μm ice band would show, however). Without adequate spectral data, the existence of any ice on Phoebe is speculation. Various investigators (see, e.g., Pollack et al. 1979) have discussed the possibility that Phoebe was captured by Saturn from some other source, and therefore its composition could be completely different from the other materials found in the Saturn system.

The Rings: A Myriad of Small Satellites

Infrared spectral data (1 to 2.5 μm) of the rings of Saturn (Kuiper et al. 1970) was used to establish that water ice is a major component of the ring material (Pilcher et al. 1970; Fink et al. 1976). The decrease in reflectivity from the visible to the ultraviolet indicates that something other than pure water is present (Lebofsky et al. 1970; Irvine and Lane 1973). The reflectance spectrum further into the infrared (2 to 4 μm) also indicates water ice (Puetter and Russell 1977).

Pollack et al. (1973a) performed theoretical calculations for water frost in the 1.3 to 2.5 μm spectral region and determined a mean particle radius of between 25 and 125 μm from the shape of the 1.5 and 2.0 μm ice bands. Their theoretical spectra seemed anomalously high shortward of 1.4 μm, but provided a good match at other wavelengths.

Lebofsky (1973) and Lebofsky and Fegley (1976) found that contaminated water frosts (e.g., NH_3 + $10H_2O$, H_2S + $10H_2O$, NH_4SH + H_2O) which have been irradiated with ultraviolet light also show a decrease in reflectivity in the visible and ultraviolet. Most of these frosts have a strong absorption around 0.6 μm which may be present only weakly in spectra of Saturn's rings in the wing of a much stronger absorption at visible wavelengths. Their irradiated H_2S + $10H_2O$ frost spectrum matches Saturn's rings and Ganymede reasonably well in the visible. However, their H_2S + $10H_2O$ and NH_4HS + H_2O spectra show a very sharp decrease in reflectance beyond about 0.9 μm, and this is inconsistent with spectra of Saturn's rings.

Clark and McCord (1980b) obtained high signal-to-noise reflectance measurements of Saturn's rings whose quality approaches that attainable on laboratory samples. They identified new absorption features in spectra of Saturn's rings due to water ice at 1.25 and 1.04 μm (tentative); weak absorptions were also seen in the 0.8 to 0.9 μm region which may be due to silicates. The Clark and McCord (1980) data are combined with the 2.5 to 4.08 μm data of Puetter and Russell (1977) in Fig. 19. Due to lack of water frost spectra of many grain sizes, Puetter and Russell (1977) were uncertain about the cause of the rise in the ring spectrum at 3.6 μm, and said they could not rule out CH_4 clathrate. Clark and McCord (1980) noted that a better match might be obtained with the addition of some mineral particulates. This will tend to

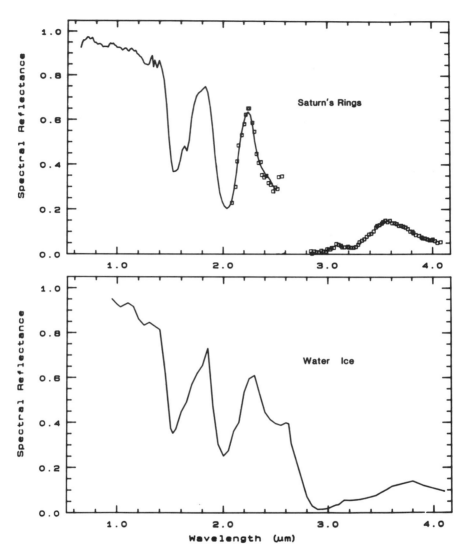

Fig. 19. Composite spectrum of Saturn's rings (top) from Clark and McCord (1980*b*) compared to a theoretical spectrum of water ice (bottom). The theoretical spectrum was generated by the methods described in Clark and Roush (1984) using the complex indices of refraction data for water ice from Irvine and Pollack (1968), a grain size of 30 μm and the angle of incidence and emission both equal to 60°. The ice absorption coefficient is maximum at 3.075 μm (14008 cm^{-1}) and the index of refraction reaches a minimum of 1.125 at 2.95 μm. These optical parameters are responsible for the reflection minimum at 2.9 μm and the small peak at 3.15 μm due to the first surface reflection from ice grains.

raise the reflectance of the mixture at any wavelength where the ice is darker than the mineral.

We have computed a theoretical spectrum of water frost of 30 μm grains using the complex index of refraction data of Irvine and Pollack (1968) and the methods described in Clark and Roush (1984). This spectrum, displayed in Fig. 19, shows a close similarity to the ring spectrum, including the first surface reflection peak at 3.15 μm and the peak near 3.6 μm. The entire 1 to 4 μm spectrum of Saturn's rings can be matched quite closely with spectra of water ice by choosing the appropriate grain size distribution. The 1.65 μm absorption is temperature dependent, becoming stronger at lower temperatures. The water ice absorption coefficients are for temperatures closer to 273 K, so the theoretical spectrum does not show the feature as strong as in the spectrum of Saturn's rings. There have been attempts to calibrate the 1.65 μm absorption to use as a temperature indicator, but in reflectance, the apparent strength is also controlled by the ice grain size and the calibration of the band is very difficult (Clark 1981a).

Clark (1980) analyzed the 0.3 to 2.5 μm spectrum of Saturn's rings and concluded that any contaminants must be spectrally neutral (no significant absorptions) in the 1 to 2.5 μm range (so the ring spectrum would appear like pure water ice) and display a decrease in reflectance shortward of 0.7 μm. There also appears to be a weak absorption near 0.8 to 0.9 μm, although this may be due to a calibration problem. Minerals containing Fe^{3+} have the required spectral features for the ice impurities. Many carbonaceous chondritic meteorites show reflectance spectra with the required characteristics. Clark (1980) concluded that the portions of Saturn's rings that are probed by visible and near-infrared photons are at least 90 wt % water ice. If the contaminant material is dark, like typical carbonaceous chondritic material, then the studies of Clark (1982) and Clark and Lucey (1984) would imply greater than 98 wt % water ice. It is possible that the ice could contain as little as a few parts-per-million small dark particles in the optical surface.

Clark (1982) showed that small changes in the ice grain size and contaminant level will affect the color and albedo in the visual part of the spectrum (Fig. 10). Thus, the radial variation in color and brightness in the rings observed by Voyager (Smith et al. 1981,1982) could be due to some or all of these effects.

Observations in the centimeter wavelength range, both active and passive, can be used to further constrain the composition of the ring particles and their size distribution. Observations at these wavelengths indicate that the rings are composed predominantly of particles in the 0.1 to several meter range (see, e.g., Cuzzi et al. 1980; Epstein et al. 1984; Marouf et al. 1983, and references therein). The ring particles are also constrained to have a silicate content of no more than about 10 wt % (Epstein et al. 1984), in close agreement with the near-infrared results. There could still be a few large

rocky boulders in the rings that could provide almost any ice/silicate ratio if
the boulders were large enough that the surface area seen by radar, optical and
near-infrared wavelengths were small compared to the surface area of ice
grains. However, repeated weak collisions among ring particles might tend to
mix any rocky material with the ice so that rocky material may not be easily
hidden from remote sensing observations.

VII. URANIAN SATELLITES

Due to the faintness of the Uranian satellites (see chapter by Cruikshank
and Brown), it has been possible to obtain near-infrared spectrophotometry
only in the last few years. The 1.5 and 2.0 μm water ice absorptions in the
spectra of Titania and Oberon were first discovered by Cruikshank and Brown
(1980), then in spectra of Ariel and Umbriel (Cruikshank and Brown 1981),
while Brown and Clark (1984) discovered the 2.0 μm ice absorption in the
spectrum of Miranda. Soifer et al. (1981) confirmed the previous results for
Umbriel, Titania and Oberon. More recently, Brown (1982, 1984) and Brown
and Cruikshank (1983) obtained higher signal-to-noise data on the four
brighter satellites (Miranda is the faintest), and these data are shown in Fig.
20. The reflectance data for Miranda is limited and runs from 1.6 to 2.5 μm;
groundbased measurements of Miranda at other wavelengths will prove exceedingly difficult because of Miranda's proximity to Uranus, and because
Uranus is very bright relative to Miranda at shorter wavelengths. The spectra
of the Uranian satellites are shown in Fig. 20.

Brown et al. (1982) derived the geometric albedos of Ariel, Umbriel,
Titania and Oberon from radiometric observations; the values are in the 0.2 to
0.3 range. Other icy objects in the solar system with similar surface albedos
are Hyperion and Callisto. In fact, the infrared spectra of Hyperion and Ariel
(Figs. 18 and 20) are quite similar, which suggests relatively pure icy surfaces
despite the low albedo. Brown (1982,1984) tried matching the albedo level
and the spectral features using laboratory spectra of frost and ice-particulate
mixtures (from Clark 1981a; Clark and Lucey 1984) numerically convolved
to the spectral resolution of the telescopic data. His best match was a linear
combination (representing areal patches on the satellite surfaces) of a spectrum of pure, fine-grained water frost and a spectrum of an intimate mixture
of 70 wt % water ice plus 30 wt % charcoal grains (grain size = 15 μm). In
analyzing spectra of Hyperion, Clark et al. (1984) assumed only intimate
mixtures and found that probably less than 1 wt % of the surface could be
small dark (nonicy) grains. Although these solutions are not unique, they constrain the surfaces of the Uranian satellites to be predominantly water ice.
Furthermore, Brown (1982,1984) established that the nonicy components in
the surfaces were spectrally bland, of low albedo, and have only a slight reddish color. Observations are needed to define the 3 μm water ice band depth;

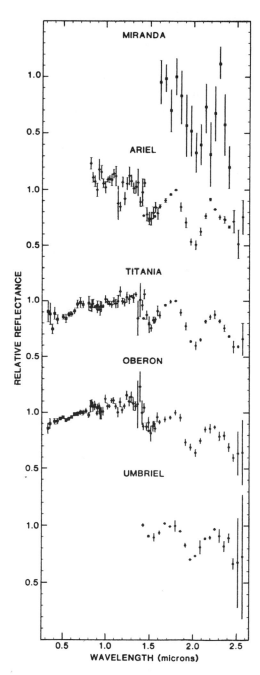

Fig. 20. Reflectance spectra of the five Uranian satellites are shown in approximate order (top to bottom) of the strength of the 2.0 μm water ice absorption. The Miranda data are from Brown and Clark (1984); the 0.3 to 1.1 μm data are from Bell et al. (1979); the 0.8 to 1.6 μm data for Ariel, Umbriel and Titania are from Brown (1982,1984); and the 1.4 to 2.6 μm data for Ariel, Umbriel, Titania and Oberon are from Brown and Cruikshank (1983).

if that band is not near 100% in depth, then the surfaces have patches free of water ice.

The water ice on the Uranian satellites may contain trapped gases of CH_4 or CO or ammonium hydroxide ice. These materials, if present in small amounts, have spectral features which would not be apparent at the spectral resolution and data precision of the current data on the Uranian satellites. Trapped gases in the form of clathrates cannot be present in amounts greater than a few wt % (see previous discussion in Sec. VI on the Saturnian satellites). The gas weight fraction is limited by the clathrate structure and the molecular weight of the gas; for a CH_4 clathrate, the gas weight fraction is limited in practice to about 10 wt %. NH_3 in the form of ammonium hydroxide could not be present in amounts greater than about 10 wt % (see Clark et al. 1984; Brown et al. 1984) or the 2.2 μm region would be significantly depressed. The spectra of both Ariel and Hyperion appear unusual in the 2.2 μm region because the spectrum appears peaked, not rounded like the other satellites, and might indicate the presence of another material besides water ice. Higher spectral resolution and higher signal-to-noise data are needed to confirm this possibility and, even if indicative of clathrates or ammonia, the abundance is small. Thus, the surfaces of the Uranian satellites are predominantly water ice.

VIII. THE SATELLITES OF NEPTUNE

Neptune has two known satellites: Triton and Nereid. Because Nereid is extremely faint, nothing is known about its surface composition, size, or albedo. As described in the chapter by Cruikshank and Brown, Triton ranks with Io as being one of the most unusual satellites in the solar system. The visible spectrum of Triton is flat from about 0.55 to 1.0 μm with a weak absorption at 0.89 μm due to methane (Apt et al. 1983). Shortward of 0.55 μm to about 0.3 μm, the reflectance decreases by a factor of 2 to 2.5 (Cruikshank et al. 1979). Apt et al. (1983) have reviewed the visible data on Triton.

In the infrared, the spectrum of Triton displays methane absorptions, reported first by Cruikshank and Silvaggio (1979), and in more detail by Cruikshank and Apt (1984). Triton, however, displays a spectral feature not yet seen for any other solar system surface. The feature occurs at 2.16 μm and has been attributed to liquid nitrogen (and/or possibly solid nitrogen) by Cruikshank et al. (1984a). Molecular nitrogen has a density-induced (2−0) absorption band at 2.16 μm. The absorption has been observed in the laboratory for nitrogen gas and liquid, but not yet in solid form. At the temperature of Triton (near 60 to 65 K), an atmosphere sufficient to account for the observed absorption could not occur. Thus, if the 2.16 μm absorption results from nitrogen, the nitrogen must be liquid, solid, or a combination of both. Cruikshank et al. (1984a) derived the absorption coefficient of liquid nitrogen as a function of wavelength and modeled the spectrum of Triton with com-

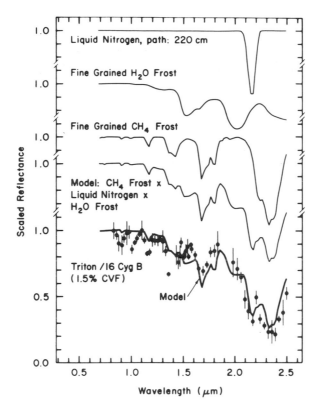

Fig. 21. The spectrum of Triton with a model spectrum, both having a resolution of 1.5%, is shown with the individual components used in the calculation. In trying to achieve a fit of the model to the data, emphasis was placed on the spectral region 1.8 to 2.5 μm (figure from Cruikshank et al. 1984a).

binations of nitrogen, solid methane, and water ice (Fig. 21). Water was needed because neither methane or nitrogen absorb, at 1.5 and 2.0 μm, and the spectrum of Triton was lower than expected at these wavelengths when water ice absorption was not included in the model.

Cruikshank et al. (1984a) estimate that Triton must have a layer of nitrogen at least tens of centimeters deep over much of its surface. Such a quantity is plausible in terms of the cosmic abundance of nitrogen and by comparison with Saturn's satellite, Titan, where a massive atmosphere of nitrogen exists. At the calculated subsolar temperature of Triton ($\sim$ 63 K), the vapor pressure of nitrogen implies a surface atmospheric pressure in the range of 0.13 to 0.30 atm.

Cruikshank et al. (1984a) speculated that dark photochemically-produced particles occur as a minor contaminant of the ices or liquid. These contaminants might account for the red slope to the Triton spectrum in the 0.3 to

0.55 μm range. However, because of the large optical paths in nitrogen (ice or liquid) needed to give the observed absorptions, any dark contaminants could be present only in very small amounts, probably less than parts-per-ten-thousand (by analogy with water ice and the studies of Clark [1982] and Clark and Lucey [1984]).

In addition to the detection of methane ice on the surface of Triton, Cruikshank and Apt (1984) have shown that the methane band strengths vary with the orbital position of Triton. Thus, the methane appears to be nonuniformly distributed around the satellite's surface.

IX. PLUTO AND CHARON

The Pluto-Charon system (also covered in the chapter by Cruikshank and Brown) is treated here as one object, because all spectral studies thus far published have not been able to separate the two. As viewed from the Earth, Charon is less than one arc-second from Pluto at maximum elongation and all spectral studies have used several arc-second apertures (atmospheric turbulence makes it impractical to obtain spectral data of each body separately). At visual wavelengths, the light from Charon is only about 25% that of Pluto, thus the following discussion applies more to Pluto than Charon. It is not known how much light Charon reflects in the infrared. Some separation of spectral parameters might be possible when the viewing geometry of the Pluto-Charon orbit changes to result in occultations (which must occur twice per Plutonian year). Since Pluto is a small object similar to several satellites of the giant gas planets, it is appropriate to discuss its surface composition as well.

The first clues to the surface composition of Pluto were obtained by Cruikshank et al. (1976) using broadband *JHK* and two narrowband (1.55 and 1.73 μm) filters. The reflectance of Pluto at these wavelengths ruled out H_2O and NH_3 ices as major spectral components. Apt et al. (1983) have reviewed the observations in the 0.3 to 1.0 μm region and the implications for methane ice on the surface. In the visual region of the spectrum, Pluto is slightly red, with the reflectance decreasing by about a factor of two from 0.7 μm to 0.4 μm. In the infrared, strong methane absorptions are observed (Fig. 22). Figure 22 is a composite of observations taken by three different teams at different times. If all sides of Pluto and Charon have basically the same features, then this spectrum reasonably represents the spectrum of Pluto. It is readily seen that the methane absorptions (indicated by the comparison methane ice spectrum) are very strong compared to those of Triton.

Currently, there is some controversy over whether the absorptions indicate a methane gas or an ice. Certainly if methane ice is present, gas must be present at a level determined by the ice temperature and the vapor pressure of methane. Fink et al. (1980a) observed several CH_4 absorptions in the spectrum of Pluto and derived a gas abundance of 27 m-amagat based primarily on

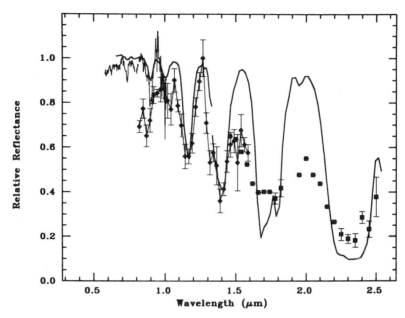

Fig. 22. A composite spectrum of Pluto (points) is compared to a laboratory spectrum of methane ice (heavy line). The disagreements at 1.5 and 2.0 μm indicates that another material, perhaps water ice, is present on Pluto that absorbs at those wavelengths but is relatively transparent at other wavelengths. The Pluto data from 0.6 to 1.0 μm (thin line) are from Fink et al. (1980a), those from 0.9 to 1.6 μm (plus symbols) are from Cruikshank and Brown (personal communication, 1983) and those from 1.5 to 2.5 μm (solid squares) are from Soifer et al. (1980). The methane ice data are from Brown et al. (1985).

the 0.89 μm band. Apt et al. (1983, and personal communication) have argued that this value must be an upper limit because the 1.15 and 1.4 μm CH_4 bands are much stronger in spectra of Pluto than that of CH_4 gas at the temperatures and pressures which occur on Pluto. The methane absorption features have also been shown to vary with Pluto's rotational phase, providing further evidence for a methane frost explanation (Buie and Fink 1984).

Pluto's subsolar temperature is probably about 58 K, in which case the average temperature would be about 52 K, and only ~45 K at 70° latitude. This would seem to be too low for Pluto to have an equilibrium atmosphere as massive as 27 m-amagat of CH_4. However, Fink et al. argued that a 10 cm photon path length in frost would be required to explain the apparent 0.89 μm band depth on Pluto, and that such a long path is unlikely. The high vapor pressure of CH_4 at Pluto temperatures would allow large, ice grains to grow so that a many-centimeter optical path is not unreasonable (Clark et al. 1983a). For such a long path, however, the CH_4 ice must be very pure with probably less than parts-per-thousand particulate contaminants.

We note that because of the very strong methane absorptions in Pluto's spectrum, a nitrogen absorption at 2.16 μm cannot be detected, and it is possible that Pluto has a substantial nitrogen atmosphere, liquid nitrogen ocean, or even nitrogen ice. Methane might be a minor constituent (even considering the 10 cm photon path) compared to nitrogen and the Pluto-Charon system might be quite similar in surface composition to Triton.

The Pluto spectrum shown in Fig. 22 agrees well with the methane ice spectrum, except near 1.5 and 2.0 μm. These two regions are where water ice has significant absorption bands. Small amounts of water ice were required by Cruikshank et al. (1984a) in the modeling of Triton's spectrum, and adding water ice to the methane ice would result in a spectrum that more closely matched the Pluto spectrum. Thus, the same constituents that are on Triton may also be present on Pluto: methane and possibly nitrogen and water.

X. CONCLUSIONS

The most highly exploited technique for remotely sensing satellite surface composition from Earth or orbital and flyby spacecraft is the identification of absorption bands in diffuse reflectance spectra which are caused by combinations of vibrational overtones, electronic transitions, charge transfers and conduction bands. Key supporting information has come from radiometric thermal measurements, atmospheric compositional measurements and magnetospheric measurements. Another powerful remote technique is gamma-ray and X-ray fluorescence analysis from orbit, which has only been fully exploited in the case of the Moon but is now proposed for several planned missions. Reflectance spectroscopy provides direct phase identifications with implications for elemental compositions while the reverse is true for the gamma ray and X ray. Reflectance studies relate directly only to the optical surface, while the gamma ray and X ray integrate to greater depths. Thus, the two techniques are highly complementary.

Spectral identification of phases based on single or multiple band identifications tend to be secure and definitive; those based on continuum shape alone can also be very useful as components of a web of independent lines of evidence, including morphological studies via imaging, thermal measurements, etc.

Frontier areas of research in remote reflectance spectroscopy include grain size determination and understanding the spectra of mixtures in order to identify component phases. Considerable progress has been made in determining quantitative abundances of phases in mixtures. However, there are limits to the reliability of the quantitative assays allowed by this approach. The primary scientific value of this remote sensing technique for surfaces of satellites or other bodies is still the identification of the presence of a given phase.

So far, high spectral resolution data of satellites in the infrared have been obtained only via Earth-based telescopes (except for lunar samples in Earth laboratories), and only as whole-disk spectra, except in the case of the Moon. The mineralogical domains on the whole-disk spectrum of a satellite are blended into one spectrum, and no correlation with surface features has been possible. This will change with the Galileo mission to Jupiter where the Near Infrared Mapping Spectrometer (NIMS) will map the surfaces of the Galilean satellites, providing a 0.7 to 5 μm spectrum for each resolution element on the surface.

Overall, water ice is abundant throughout the solar system, particularly from Jupiter to Uranus, but variety in composition is also common. There are many questions to answer. Why are dark material and dark objects present, as well as very bright objects, such as Rhea? Why do we not see signatures of ammonia or other volatiles, such as have been proposed on some satellites (e.g., that which produces the apparent icy flows on Enceladus)? What is the material responsible for the darkening and reddening of satellite surfaces? (We know about how much is present, but we cannot identify it, although many minerals containing Fe^{3+} can give such colors.) We are likely to answer some of these questions with higher signal-to-noise and higher spectral resolution data in the visual and infrared combined with more laboratory studies.

11. CRATERING OF PLANETARY SATELLITES

CLARK R. CHAPMAN
Planetary Science Institute

and

WILLIAM B. McKINNON
Washington University

In recent years, data concerning craters and cratering populations on the satellites of Jupiter and Saturn have led to great advances in understanding the bombardment history of the solar system. The first part of this chapter summarizes concurrent advances in cratering mechanics (Sec. II), with special emphasis on cratering on icy surfaces and on energy and momentum scaling of crater diameters and other cratering phenomena. The last part of the chapter treats cratering statistics (Sec. III), with emphasis on issues of crater saturation and the identification of similar or dissimilar production functions (and, by application of scaling relationships, projectile populations [Sec. IV]) on the different satellites and various terrestrial bodies. We conclude that there are 4 (possibly 5) distinct projectile populations implied, although various forms of evolved comets may be responsible for many of them. There is as yet no good link between cratering of the Moon and terrestrial planets and that of outer-planet satellites. Thus the cratering chronologies for the satellites of Jupiter and Saturn remain essentially unknown.

I. INTRODUCTION

In the early years of the age of planetary exploration by spacecraft, the study of impact cratering was a major topic. After all, extraterrestrial geology began with the Moon and the Moon's dominant landforms are craters. Many of the planets and satellites subsequently studied over the ensuing two decades have much more varied terrain, but analysis of crater morphologies, spatial distributions, and size-frequency distributions remains a major tool for understanding the geologic histories of planetary bodies. The dominant goal remains the one enunciated by Shoemaker et al. (1963), the "interplanetary correlation of geologic time." The relative frequencies of craters on different geologic units on a planetary surface provide the major method for relative dating of the units. In addition, since craters are formed by projectile populations that are frequently common to several planets and/or satellites (e.g., "new" comets in eccentric orbits can hit any planetary body), there is the potential for linking the relative chronologies from body to body in an absolute sense, tied to the absolute chronologies established by geochronological techniques applied to dating rocks from the Earth and Moon.

Other important things can be learned from studying craters. To various degrees, the morphology of a freshly formed crater, its ejecta blanket, and other influences on neighboring terrain reflect inherent differences in the structure and rheology of the surface layers in which the crater was formed. The morphology of older craters—the way they are progressively modified from their pristine appearance—reflects the exogenic and endogenic geomorphological processes operating on that part of the planetary surface. Cratering impacts may penetrate to subsurface layers otherwise unobservable; indeed, large enough impacts into a small satellite may completely disrupt it and expose its interior to view. Thus, in a variety of ways, the study of differences between craters on different geologic units, or on different bodies, can provide a calibrated measure of the geological characteristics and processes in the different locations.

A final goal of planetary cratering studies is to learn about the characteristics of the impacting projectile populations. Craters are formed by asteroids, comets and circumplanetary debris. Some of these populations may be representative of primordial planetesimal populations, or at least the fragmental remnants thereof. Of particular interest is the possibility that a record of how these populations evolved over the aeons in different parts of the solar system may be recorded in the observable cratering record on the surfaces of planets and satellites.

The purpose of this chapter is to review this field, with particular emphasis on the relatively recent evidence acquired by spacecraft imagery of other planetary satellites, especially of Jupiter and Saturn. The crater populations observed on these satellites show a variety of differences from those already well studied on the Moon, Mercury, Mars and Phobos. Thus the entire subdiscipline of cratering mechanics has been spurred in recent years. The first part of this chapter attempts to summarize our present understanding of cratering mechanics (Sec. II), with special emphasis on scaling and ejecta. We pay particular attention to the differences between cratering in rock and ice.

The second part of the chapter (Sec. III) considers evidence from crater statistics about the cratering histories on the various planetary satellites, and attempts to link this recent work with what is known about the cratering record on the terrestrial bodies. Finally, in Sec. IV, we consider the source populations and attempt to summarize and critique the various scenarios that have been offered for linking the cratering records on the various planets and satellites. Early Voyager Uranus results seem compatible with our summary.

II. MECHANICS OF CRATERING IN ICE AND ROCK

This section takes a broad look at the cratering mechanics relevant to natural satellites, and attempts to blend observational, experimental, and theoretical viewpoints. In Sec. II.A, we summarize the traditional description of the sequence of events that create an impact crater. The richness of detail inherent in such a complex physical process as crater formation is developed in the following subsections. Section II.B gives an up-to-date discussion of scaling (more or less from first principles), with special attention given to the observable aspects of the phenomenon (e.g., resulting diameters) that may be expected to vary for satellites of different sizes and different surface properties (e.g., ice vs. rock). Next is a brief and mainly theoretical discussion of impact melting and vaporization (Sec. II.C). Another major observable (and scalable) aspect of cratering is ejecta (Sec. II.D). Here we describe the various forms of ejecta and their distribution, emphasizing new observations of icy satellites. Section II.E, morphology, coherently organizes the large number of observations, also emphasizing icy satellites, but physical interpretation is less well advanced. Having concluded the cratering mechanics of individual craters themselves, we address the important large cratering events, which affect satellite geology and geophysics, in Sec. II.F; unfortunately, the mechanics of breakup are still poorly understood. Finally, the results of the scaling and morphology subsections are combined in a discussion of inter-satellite diameter scaling (Sec. II.G), which provides a link to the subsequent sections on crater and projectile populations (Secs. III and IV). A reader so inclined may go directly to Sec. III without a loss in continuity.

A. Overview

Excavation of an impact crater begins with the first contact between the cosmic object, or *projectile*, and the satellite surface, or *target* (Fig. 1, left). Atmospheric effects will be discussed later. Typical impact velocities in the solar system are on the order of 10 km s^{-1}, resulting in a high-pressure shock wave propagating into the target, accelerating target material forward, and a shock propagating back into and decelerating the projectile (though not re-

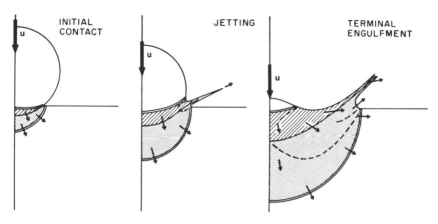

Fig. 1. The *compression* stage of impact mechanics, from Melosh (1980) and after Gault et al. (1968). Shock-wave propagation begins at initial contact between the projectile (moving at velocity *u*) and the target (the left panel shows the geometry just after initial contact). Shocked projectile and target material is lined and stippled, respectively. Arrows denote particle velocities. Rarefaction waves (dashed lines) unload the shocked zone from available free surfaces, causing jetting. *Compression* ends when the backward propagating shock reaches the rear of the meteorite, now quite distorted. Jetting merges into the beginning of formation of the ejecta curtain.

versing its motion). Jumps in pressure across the shock are on the order of 10^{2-3} GPa (1 – 10 Mbar). The actual shock pressures, densities and velocities are determined by the material properties of both projectile and target (Shoemaker 1963; Gault and Heitowit 1963; O'Keefe and Ahrens 1975; Kieffer and Simonds 1980). As the shock-compressed wad of projectile and target travels inward from the satellite surface, the forward shock expands hemispherically, engulfing increasing amounts of target mass. Eventually, the backward propagating shock reaches the rear surface of the projectile. This ends what Gault et al. (1968) termed the *compression stage* (Fig. 1). Approximate time scales range from 10^{-4} s for a 1 m projectile to ~ 1 s for a 10 km diameter projectile.

The influence of the sides of the projectile, as it is progressively engulfed by the shock and as it penetrates the target, is critical to the compression stage. Even as the shocks begin to propagate, they intersect a free surface initially formed at the triple junction of projectile, target, and void (planetary and satellite atmospheres are essentially vacua with respect to the pressures encountered in cratering). The shocked region in contact with the growing annulus-shaped free surface immediately begins unloading. The unloading or *rarefaction* wave propagates inward, and the material undergoing rarefaction is subjected to a high pressure gradient, decompressed (along an isentrope), generally melted and vaporized, and accelerated upward and outward in a jet.

Initial contact, jetting, and terminal engulfment (the end of the compression stage) are all illustrated schematically in Fig. 1. The degree of heating and volatilization depends on peak shock pressure.

At the end of the compression stage, the projectile-target interface has moved not quite one projectile diameter into the target (and possibly much less) because the particle velocities in the shocked region are slower (sometimes by more than a factor of two) than the speed of the shock through the projectile. The shocked region has been substantially unloaded from the sides, and the acceleration of the material is deviating from the strictly downward direction. An induced upward and outward motion begins to form the *ejecta curtain* (Fig. 1, right). The projectile itself is badly distorted, flattening and flowing to the sides. By the end of the compression stage, the projectile has transferred roughly one half its total energy to the target, partitioned into kinetic and internal energy (Gault and Heitowit 1963; Ahrens and O'Keefe 1977; Bryan et al. 1978; Kieffer and Simonds 1980). Momentum transfer is slower because the remaining energy in the projectile is mainly kinetic energy (Thomsen et al. 1979), and the projectile velocity u does not decrease as fast as u^2.

Momentum and energy transfer continue during the *excavation stage* (Gault et al. 1968), illustrated in Fig. 2. The shock from the initial contact has reached the back of the projectile, and the shocked region is now unloaded by a rarefaction wave propagating from this free surface. It travels at the speed of sound in the shocked material and, as this must be greater than the shock speed, eventually catches up and combines with the original hemispherical shock in the target. The combination of rarefactions from the rear and lateral boundaries of the projectile and the free surface of the target comprises a complex series of unloading events or waves whose overall effect is to rotate the velocity vectors of the flow field, basically radially outward from the point of initial contact, toward the original target surface (panels a and b of Fig. 2).

As some projectile remnants are flattened in the *transient cavity* (the bowl of the developing crater) and the original hemispherical shocked region is unloaded, an expanding shell of high pressure propagates deeply into the target as a shock wave detached from the region near the projectile. Pressures are still high behind the detached shock, though, and the projectile and target have been subjected to peak pressures far in excess of their mechanical strengths. [Typical shear strengths for intact rock and ice specimens, under static and dynamical loading, range between 10^{-2} and 1 GPa (0.1–10 kbar) (Handin 1966; Lindholm et al. 1974; Gold 1977; Lange and Ahrens 1983; Durham et al. 1983).] That which is not vaporized or melted is severely comminuted and its original mechanical integrity is destroyed. Theoretical and experimental bulk viscosity estimates for this pressure regime are low, 10^2 to 10^4 Pa s (10^3–10^5 P) (Jeanloz and Ahrens 1979); shear viscosities should be even less. Therefore, the shock-accelerated material flows in a hydrodynamic fashion.

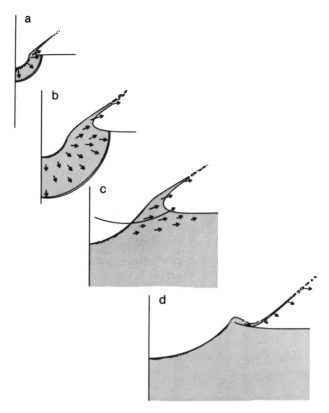

Fig. 2. Four instantaneous views of the *excavation* stage, after Gault et al. (1968), Maxwell (1977), and Melosh (1980). The shock front and flow field velocity vectors are denoted by double lines and arrows, respectively. In this example, projectile remains are indicated by the dark lining of the crater surface. Growth of the transient cavity is nearly hemispherical with material in roughly the upper 1/3 of the crater (above the streamline illustrated in panel c) being ejected. Eventually maximum dynamical depth is reached, but lateral expansion (shearing along the crater wall) continues for craters in nonliquids. Stagnation of the entire crater flow results in the classic bowl shape, raised rim and overturned flap of ejecta.

Most of the growth of the crater occurs well behind the propagating shock, and the flow field initiated by the impulsive stresses is approximately steady state and incompressible (Maxwell and Seifert 1974; Maxwell 1977; Cooper 1977; Orphal 1977). Crater growth is accomplished by true excavation (the physical lofting of material) and by the displacement of material in the near-crater region (Dence et al. 1977). The original shock eventually decays in amplitude as it diverges and is weakened by rarefactions merging with it. As the shock speed declines to transonic values, elastic precursors smear out the pressure and velocity jump, transforming the shock wave to a plastic wave. Eventually pressures decline into the elastic regime.

The transient cavity continues to grow in a nearly hemispherical manner until a maximum depth is reached. Expansion continues laterally, shearing material from the crater wall (Orphal 1977; Piekutowski 1977; see Fig. 2, panel c). However, lateral expansion of craters in water, and presumably other liquids, is minor (Gault and Sonett 1982). This description must also be modified for small craters in competent materials. For other than these small craters, the ultimate limitation on the size and shape of the crater is the dissipation of flow field kinetic energy, either viscously or frictionally, or its conversion to gravitational potential energy. The flow remains hydrodynamic so long as the stagnation pressure ρv^2 (where ρ = density and v = flow velocity) is much greater than Y, $\rho g \ell$ and $\eta v / \ell$ (where g is the acceleration of gravity, ℓ is a relevant distance scale, and Y and η are, respectively, a representative post-shock yield strength and viscosity) (Maxwell and Seifert 1974; Melosh 1980). The cratering flow field is ultimately arrested when some combination of strength, gravity or viscosity effects becomes important.

Material propelled away from the transient cavity lip forms a conical ejecta sheet, or plume, opening outward to the void. It is conical in shape because the ejection angle remains remarkably constant during cavity growth, with 35° to 55° from the horizontal bracketing laboratory impact results for sand (Stöffler et al. 1975; Oberbeck and Morrison 1976). The angle of the apparent plume, which is distinct from the ejection angle, is often near 45°. Significantly steeper ejection angles are observed for water, leading to nearly vertical plume angles (85 to 90°) (Greeley et al. 1980; Gault and Sonett 1982). The deposited ejecta forms both a continuous and discontinuous blanket and rays; if the target is competent (solid), discrete ejecta fragments can form secondary craters. The last material which is lofted at low velocity stagnates at the transient crater rim. It can form an overturned flap (Shoemaker 1962). Because the crater shape at the end of the excavation stage (the *transient crater*) is partly due to displacement, material is excavated from only a fraction ($\sim 1/3$ to $1/2$) of the total transient depth (Stöffler et al. 1975; Croft 1980; Grieve et al. 1981; Hörz et al. 1983).

Crater dimensions at the end of the excavation stage are typically one to two orders of magnitude greater than that of the original projectile. Because late-stage flow velocities are comparatively low, the duration of the excavation stage is many orders of magnitude longer than for the compression stage. In the laboratory, a 10 cm crater may form in 10^{-1} s, with compression lasting ~ 1 μs. Experiments in sand indicate that formation time scales as the square root of the diameter (Gault and Wedekind 1977; see Sec. II.B below). Thus, a 1 km diameter crater may form in ~ 10 s, and a 100 km diameter crater in ~ 100 s. Higher gravity renders excavation more difficult, resulting in smaller craters and shorter time scales.

The transient crater may also be the final one, but for craters larger than a threshold diameter (which varies from satellite to satellite), and for all craters in liquids, the transient crater is not in mechanical equilibrium and promptly collapses under the influence of gravity. This *modification stage* results in an increase in the final crater diameter with respect to the transient value, and the creation of such morphological features as slump terraces, central peaks, central pits, peak rings and multiple rings (see chapter by Kaula et al.).

B. Scaling: General Considerations

We seek to understand how crater size and shape depend on initial conditions, including those that vary from satellite to satellite. The projectile is characterized by any two of the following four quantities: mass m, velocity u, momentum $M = mu$, and kinetic energy $E = 1/2mu^2$; plus its density δ, and incidence angle θ (see Fig. 3). The target is characterized by its density ρ, material strength Y, viscosity η, and g. Additional material properties, such as the angle of internal friction ϕ may be important. In our treatment projectile shape and strength are neglected; so long as the projectile is roughly equant, there are no effects, and projectile strength does not influence the early high-pressure phase, after which mechanical integrity is moot.

We have discussed the "classical" phases of cratering above: compression, excavation, and modification. An alternative grouping of physical events is suggested by recent experiments and theory (Schmidt 1980; Holsapple and Schmidt 1982; Schmidt and Holsapple 1982; Holsapple and Bjorkman 1983):

1. Coupling phase: energy and momentum are transferred, and the cratering flow field is set up (this takes somewhat longer than the classical compression stage);
2. Power law growth stage: the flow field expands, with transient cavity dimensions expressible as simple power law functions of time (this includes a major part of the excavation stage);
3. Late stage: some combination of gravity, strength and viscosity limits cavity growth, initially causing a deviation from power law behavior and

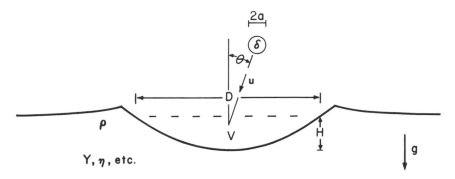

Fig. 3. Impactor of radius a and density δ strikes target of density ρ at an angle θ from the vertical. A crater of apparent diameter D, depth H and volume V results.

eventually determining the size and shape of the final (transient) crater (for larger craters the remainder of the excavation stage plus the modification stage can be considered part of the late stage, because both phases are gravity-dominated).

Early scaling relationships, based on explosion craters, assumed that the final diameter depended solely on projectile energy (cf. Shoemaker 1962; Cooper 1977). We recognize now that this is not strictly true; it depends on energy *and* momentum, i.e., of two projectiles with the same energy, the one with the greater momentum will make a larger crater (cf. Oberbeck 1977). Scaling is still relatively simple, however, because a single variable still characterizes the scaling. This variable, the *coupling parameter*, may be considered intermediate in dimensionality between energy and momentum; its exact form is determined by material properties and can be evaluated by experiment. (One way to write the coupling parameter is $mu^{6\alpha/(3-\alpha)}$. As α varies from 3/7 to 3/4, the parameter changes from mu to mu^2. We choose this particular form as an example because α turns out below to be the gravity-scaling exponent; see Eq. 3.)

Numerous cratering experiments have been performed, some using a centrifuge to impose high gravities (up to ~ 500 G) that simulate physical scales much greater than that of the laboratory. If suitable dimensionless variables are chosen, the results of such experiments on *simple materials* reveal a power law dependence between the variables or behavior asymptotic to such a dependence. A simple material is one whose material properties are neither scale- nor rate-dependent: dry sand and water appear to be simple materials under the range of impact conditions studied. The standard choice of dimensionless variables for these materials are *cratering efficiency*,

$$\pi_V = V\rho/m \tag{1}$$

where V is the excavated volume below the original ground plane (the *apparent* volume; see Grieve et al. 1981), and *gravity-scaled size*,

$$\pi_2 = 3.22 \ ga/u^2 \tag{2}$$

where a is the spherical equivalent radius of the projectile (Fig. 3). Strength effects are not expected since dry sand and water are effectively cohesionless, so a dimensionless variable incorporating Y is unnecessary. A third variable ρ/δ may be included in the set, but its influence on scaling is problematic and will be discussed below. Viscosity effects and nonvertical incidence angles (also discussed below) are also neglected.

The experimental scaling relationships for water and dry sand are illustrated in log-log form in Fig. 4 (craters formed in water are measured at maximum volume). The cratering efficiency is high for small values of π_2, meaning high velocity, low gravity, or small size; conversely, low velocity, high gravity,

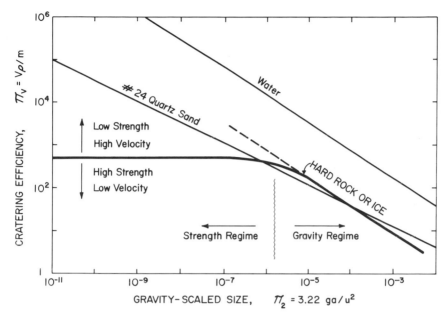

Fig. 4. Material-strength model for impact crater volume. Scaling relationships for two simple materials, water and a dry quartz sand, are taken from Schmidt (1980) and are based on data from Gault and Wedekind (1977), Oberbeck (1977) and Gault (1978). The strict power law relationship observed between the dimensionless variables translates into a linear relationship in log-log space. The simple scaling relationship for competent geologic targets is described in the text. The absolute level of the cratering efficiency (i.e., the constant A in Eq. 3) for this model in the gravity regime is that suggested by Schmidt and Holsapple (1982) and is 20 times less than that of water. The relationship between the cratering efficiencies in rock and sand is complex and nonintuitive.

or large projectiles mean low cratering efficiency. Water always craters more efficiently than sand because it has no internal friction. There is also a difference between the slopes of the two scaling relationships, probably due to porosity (Schmidt and Holsapple 1982). Porosity increases the amount of irreversible heating during shock passage, and hence decreases the coupling of the projectile's kinetic energy into the kinetic energy of the cratering flow field. The more efficient the coupling is, the steeper the slope α of the scaling relationship and the closer the relationship to pure energy scaling (i.e., the dimensions of the coupling parameter are more nearly those of energy).

The simple power laws determined by experiment imply the existence of a physically meaningful coupling parameter. Conversely, the form of the coupling parameter ensures that scaling relationships will be simple power laws. These power laws can be written as

$$\pi_V \pi_2^{\alpha} = A \tag{3}$$

TABLE I
Crater Scaling Constants[a]

	α	A	τ	T
Dry Sand	0.50 ± 0.01	0.43 ± 0.05	0.60 ± 0.01 $(0.58 \pm 0.001)^{b}$	1.64 ± 0.19
Water	0.65 ± 0.03	1.93 ± 1.45	0.59 ± 0.01 $(0.61 \pm 0.01)^{b}$	2.10 ± 0.59

[a] Sources: Schmidt (1980,1981)
[b] Theoretical values determined from $\tau = (3 + \alpha)/6$

where α and A are experimentally determined. Values for water and dry sand are given in Table I. The upper limit on α, corresponding to energy scaling, is 0.75. Useful algebraic rearrangements of Eq. (3) include calculating apparent crater volume from initial conditions

$$V = \frac{4}{3} \pi a^3 \left(\frac{\delta}{\rho} \right) A \left(\frac{3.22ga}{u^2} \right)^{-\alpha} \tag{4}$$

and determining projectile radius from apparent crater volume when other impact conditions are specified

$$a = \left[\frac{V(\rho/\delta)}{\frac{4\pi A}{3}} \right]^{1/(3 - \alpha)} \left(\frac{3.22g}{u^2} \right)^{\alpha/(3 - \alpha)} \tag{5}$$

These relationships are stated in terms not of crater diameter but of crater volume, which is more physically fundamental; the diameter can be obtained by assuming a plausible crater shape (such as a hemisphere [$V = \pi D^3/12$] or paraboloid of revolution [$V = \pi HD^2/8$, where H is apparent depth]). In laboratory experiments, craters in water are hemispheres and craters in sand are shallower bowls, and only a small change in depth/diameter ratio H/D is observed over a large range in gravity-scaled size (the craters become deeper at high π_2; Schmidt 1980). Whether the transient forms of real planetary craters deviate significantly from geometric similarity is a contentious issue (cf. Settle and Head 1979; Pike 1980b; Grieve et al. 1981). By restoring the slump terraces of collapsed lunar craters to their original positions, Settle and Head concluded that similarity of the transient crater is maintained for lunar craters of up to 70 km final diameter. Grieve et al. essentially argue the same for all terrestrial craters (the diameter of the largest being 140 km), based on shock distribution and on uplift and excavation stratigraphy.

Crater formation time T also exhibits a power law dependence on gravity-scaled size

$$\pi_T \pi_2^{\tau} = B \tag{6}$$

where π_T is the nondimensionalized formation time $T(V/\sqrt{2}\, a)$ and τ and B are experimentally determined constants; values for water and dry sand are given in Table I. Formation times for these two dissimilar materials are much more similar than their respective cratering efficiencies (Schmidt 1981). As seen in Table I, the slopes (τ values) are also quite close. Schmidt (1981) shows that this is due to the existence of a single, unique coupling parameter; in addition, he demonstrates that the coupling parameter implies formation time should be proportional to $g^{-1/2}V^{1/6}$, which provides a basis for understanding the experimental results of Gault and Wedekind (1977).

Rock (or ice) vs. Sand. The obvious question at this point is how to relate cratering experiments in sand or water to more complex geological materials such as rock or ice. A *simple material* model for the failure of rock (or ice) is that of a Mohr-Coulomb solid, incorporating an unconfined yield strength, or cohesion Y, and an angle of internal friction, ϕ (see Fig. 5). (Consideration of non-simple material behavior, specifically a scale- and rate-dependent strength, will be discussed below.) Sand, or a thoroughly fractured mass of rock or ice, possesses internal friction, but has no strength in tension. The energy required to fracture rock and create new fracture surfaces increases in proportion to the volume fractured ($\propto D^3$). The energy required to excavate a crater is proportional to the volume lifted ($\propto D^3$) times its vertical displacement ($\propto D$), totaling a D^4 dependence. These arguments in terms of "energy" are crude, but the basic point is that, whatever the respective proportionality constants, when D is very small the term $\propto D^4$ is much less than $\propto D^3$, so cratering energy goes into overcoming the strength of the target; but when D is very large, the converse is true, so cratering energy goes into overcoming gravity. The gravity dependence is more complex in frictional materials; in addition to gravitational potential energy, cratering energy goes into overcoming the "strength" provided by internal friction. This strength is proportional to overburden pressure, however, and hence to D (stress levels during the late stage are rather low, near lithostatic), again totaling a D^4 dependence for the volume fractured. For cratering in real frictional materials (generally, $\phi \sim 30$ to $50°$), which includes, for example, very small craters in sand, the energy dissipated by friction is actually more important than that directly transformed into gravitational potential energy (Holsapple 1979).

In terms of scaling cratering events in rock or ice, gravity dominates late-stage flow at large π_2; at small π_2 strength dominates and there is no gravity dependence. The dominance of either variable defines two physical regimes: the *gravity regime* and the *strength regime*. (The possible dominance of vis-

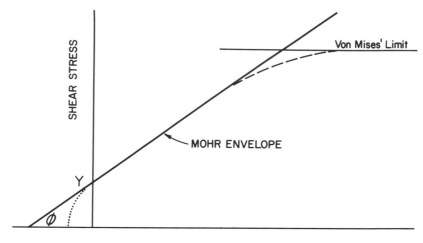

Fig. 5. Mohr envelope for failure of a Mohr-Coulomb solid (symmetric about the shear stress axis; only the top half is shown). Triaxial compression tests of real materials are described amazingly well by this failure criterion, though at high confining pressures internal friction decreases (dashed line) as asperities on sliding surfaces yield plastically. The von Mises' limit (or plateau) represents an ultimate strength; at this point, the material is in pervasive ductile failure. Strength in tension is also inadequately described. Brittle materials fail at much lower stresses (dotted line), while granular, noncohesive media have no strength in tension at all. Y is yield strength and ϕ is the angle of internal friction.

cosity will be discussed below.) Scaling relationships in each regime exhibit different asymptotic dependences (Fig. 4). The strength asymptote for a Mohr-Coulomb solid is actually a family of horizontal lines in a $\pi_V - \pi_2$ plot such as Fig. 4, depending on the specific values of strength (cohesion) and velocity. For example, the horizontal branch shown for "hard rock or ice" would lie higher for higher velocity and lower (unconfined) strength. The transition to gravity scaling takes place over at least a decade in π_2. The slope of the asymptote in the gravity regime is assumed to be the same as for water, consistent with the hypothesis that all nonporous media have the same slope α (confirmed for saturated sand by Schmidt [1984]).

The absolute level of the cratering efficiency in the gravity regime depends on the internal friction; rock or ice should crater less efficiently than water. The plausible range of internal friction angles for rubble, regardless of composition, is rather restricted (Lambe and Whitman 1979) so the gravity asymptotes for all geological materials relevant to planetary satellite surfaces may lie close together. Assuming that these asymptotes *are* coincident leads to a simple crater-scaling model: all materials fall on a single line in the gravity regime on a $\pi_V - \pi_2$ plot; a family of horizontal strength regime asymptotes branch off at levels determined by the velocity of the projectile and the cohesive yield strength (unconfined strength) of the target. The cratering efficiency in the gravity regime at large π_2 is probably uncertain by an order of magnitude but the difference between ice and rock is probably much less than an order of magnitude.

Dimensional analysis can also be applied to the strength regime. Cratering efficiency π_V can then be compared with a nondimensional strength,

$$\pi_c = Y/\delta u^2 \tag{7}$$

essentially the ratio of target yield strength to projectile kinetic energy density. Again, the existence of the coupling parameter implies a power law relationship between π_V and π_c in the strength regime

$$\pi_V \pi_c^\beta = C \tag{8}$$

where β and C are determined experimentally. Moreover, the exponent in such a relationship is related to α, the exponent in the gravity regime:

$$\beta = \frac{3\alpha}{3 - \alpha} . \tag{9}$$

The functional dependence of π_V on π_2 and π_c can be displayed in a 3-dimensional logarithmic space (Fig. 6). The strength and gravity regimes for simple

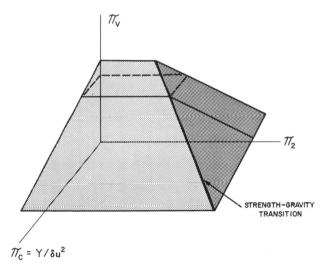

Fig. 6. Cratering efficiency prism in log π-space. Plots such as Fig. 4 are slices at constant π_c; different slices generate the family of asymptotes in the strength regime. An analogous series of diagrams could be generated by holding π_2 constant.

materials are planes in such a space. If there are no auxiliary variables, or if such variables are considered to be constant (e.g., the internal friction angle), then the 2 planes describe a universal scaling model for all geological materials. Their join defines the *strength-gravity transition*. This transition is not sharp, of course, but is smooth and continuous.

The fact that the slopes of the 2 planes in Fig. 6 are related allows the strength-gravity transition to be uniquely defined. D_t, the transition crater diameter (assuming a simple bowl-shaped geometry) between the two regimes, scales as

$$D_t \propto Y g^{-1} \delta^{-\gamma} \rho^{\gamma - 1} \tag{10}$$

with $\gamma = 2/3$ when π_c is defined as in Eq. (7) (McKinnon 1983b). It is independent of projectile velocity. Because of an ambiguity in the definition of π_c with respect to mass density, γ could be different from $2/3$, but it has not yet been determined experimentally.

Equation (10) is plotted in Fig. 7 as a function of planetary gravity for a carbonaceous chondritic projectile of $\delta = 2.2$ g cm^{-3}. The absolute values for the rock and ice curves are approximate. The hard-rock curve relies on an extrapolation from TNT events in basalt (Schmidt 1980; McKinnon 1986); solid water ice is assumed to have an unconfined strength an order of magnitude less than rock, consistent with experiments (Lange and Ahrens 1983). All the transition diameters are well below the simple-to-complex crater transition on known natural satellites (see Sec. II.E), justifying the bowl-shaped geometry assumption.

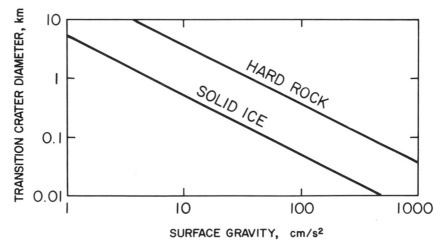

Fig. 7. Transition crater diameter between the strength and gravity regimes as a function of gravity for hard rock (approximated by basalt) and solid water ice.

Incidence Angle. The incidence angle θ is an independent nondimensional parameter. Limited data imply that the cratering efficiency for quartz sand varies as $\cos \theta$, while it varies as $\cos^2 \theta$ for granite (Gault et al. 1978). We hypothesize that it is the vertical component of velocity $u \cos \theta$ that is relevant in crater formation. If so, $\cos \theta$ should scale as u in Eqs. (3) and (8). In the gravity regime, $\pi_V \propto u^{2\alpha}$ and since $\alpha \approx 0.5$ for sand (Table I), π_V should be nearly proportional to u for these materials, as is observed. For granite in the strength regime, $\pi_V \propto u^{2\beta}$ and, since β should be approximately 0.65 for a nonporous target, the cratering efficiency (using Eq. 9) should be proportional to $u^{1.7}$. In fact, $(\cos \theta)^{1.7}$ gives a somewhat better fit to the granite data. In addition, the reduction in cratering efficiency by $\sim 65\%$ for a 45° oblique impact into water, noted by Gault and Sonett (1982), is consistent with $(\cos \theta)^{2\alpha}$ scaling when $\alpha = 0.65$. On this basis, we propose the following modifications to Eqs. (3), (6) and (8):

$$\pi_V = A\pi_2^{-\alpha} (\cos \theta)^{2\alpha} \qquad (11)$$

$$\pi_T = B\pi_2^{-\tau} (\cos \theta)^{2\tau} \qquad (12)$$

in the gravity regime, and

$$\pi_V = C\pi_c^{-\beta} (\cos \theta)^{2\beta} \qquad (13)$$

in the strength regime.

Atmospheres. Of the natural satellites, only Titan and Triton possess atmospheres that could affect the cratering process. Melosh (1981*b*) has shown that projectiles can become severely flattened or even dispersed by the differential pressure encountered during atmospheric passage when

$$a \lesssim \left(\frac{\rho_a}{\delta}\right)^{1/2} L \qquad (14)$$

where ρ_a and L are the atmospheric surface density and scale height. The crater form made by flattened or dispersed projectiles is an interesting question, but at least it should be much shallower than usual (cf. O'Keefe and Ahrens 1982*b*; Schultz and Gault 1983). The critical projectile radius, from Eq. (14), for a carbonaceous chondrite striking Titan is ~ 1 km. Knowledge of Triton's atmosphere is poor (Cruikshank et al. 1983*b*, 1984*a*), but the critical radius there is probably similar.

An atmosphere may reduce cratering efficiencies at laboratory scales somewhat (Holsapple 1979) but is probably a minor influence on large-scale impacts; an appropriate dimensionless ratio, $P/\rho ga$, where P is atmospheric pressure, would be orders of magnitude less for large-scale craters as opposed to laboratory-scale ones.

Target/Projectile Density Ratio. The role of ρ/δ in the gravity regime is intrinsic: the cratering efficiency π_V incorporates the ratio, while there is no dependence of the gravity-scaled size π_2 on either mass density term. Moreover, no dependence of π_V on ρ/δ is observed experimentally for sands; a broad range of projectile densities has been tested, as well as "pathological" targets such as iron grit (Schmidt 1980,1983).

In the strength regime, however, the kinetic-energy-density-scaled strength π_c depends on δ and it is appropriate to consider a generalization of the mass density term. Some experiments show a pronounced but poorly-defined dependence of π_V on ρ/δ in addition to π_c (Holsapple and Schmidt 1982). More strength-regime experiments are needed.

Holsapple and Bjorkman (1983) have attempted to incorporate a δ dependence directly into the coupling parameter, and they predict an explicit dependence of π_V on ρ/δ in the gravity regime. As noted above, however, the dependence for porous targets must be very slight. The dependence in the gravity regime for nonporous targets such as water, rock or ice has not actually been experimentally constrained; in particular, all the cratering experiments in water (Gault 1978; Gault and Sonett 1982) have used solely pyrex projectiles.

The Meaning of Strength. So far we have discussed the strength regime in terms of simple materials; examples are clay, alluvium and metals. Targets of crystalline materials such as rock or ice are not simple materials. Shock wave passage, by activating inherent flaws and fractures and causing them to grow and coalesce, reduces cohesion and converts a portion of the target into cohesionless "sand." Because the damage done by the shock depends both on shock pressure *and* duration, the zone of markedly reduced cohesion is spatially restricted and scale-dependent (Curran et al. 1977; Croft 1981a). The gradient of increasing strength away from the point of impact thus affects the expanding late-stage flow field, as it becomes necessary to further fracture increasing volumes of target material. Because the yield strength of rock and ice is inherently strain-rate dependent (Y increases with increasing strain rate) and strain rates must decrease with increasing crater size (i.e., $\dot{D}/D \sim T^{-1}$), the effective yield strength decreases with increasing scale (Gaffney 1984). This could lead to lower transition diameters from the strength to gravity regimes than shown in Fig. 7 (see McKinnon 1986). Large-scale geology, being structurally heterogeneous and flawed, is known to be significantly weaker than intact laboratory samples (see Handin and Logan 1981), and this discussion further illustrates how difficult it is to define the effective strength of a target based on laboratory measurements alone.

Limitations of the Model. In the gravity regime at large gravity-scaled size (π_2), the possibility exists that the (near-lithostatic) confining pressure during the late stage could reach levels associated with the von Mises' limit

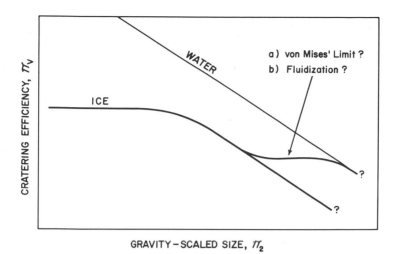

Fig. 8. Schematic cratering efficiency of water and ice, a representative competent material, as a function of gravity-scaled size (see text and Fig. 4). Cratering efficiency may increase at large π_2 if internal friction is limited by the von Mises' plateau or by little-understood fluidization mechanisms. The cratering efficiency would not, however, become greater than that of water.

(Fig. 5). Accordingly, if the angle of internal friction declines (asymptotically to 0°), then the cratering efficiency should rise and approach that of water (Fig. 8). However, the absolute values of the appropriate confining pressures are so large (~ 0.1 GPa or 1 kbar for ice [Durham et al. 1983] and of the order of 1 GPa or 10 kbar for crystalline rock [Handin 1966]) that only enormous craters ($\gtrsim 100$ km depth on Ganymede or $\gtrsim 200$ km depth on the Moon) could conceivably be affected by this limitation.

A more serious possibility is fluidization of the fractured target mass in the gravity regime. It has been recognized that the collapse of transient, simple bowl-shaped craters to more complex forms under the influence of gravity requires specific material properties in the near-crater region: cohesions under a few MPa for various planets and satellites (Melosh 1977b, 1982a) and internal friction angles $< 2°$ in general (McKinnon 1978). Various agents have been proposed to account for such very low friction angles: for silicate targets, *molten rock* acting as a lubricant and through pore pressure reduction of normal stress on fault planes (Dence et al. 1977); for ice targets, *liquid water* acting through pore pressure reduction (Passey and Shoemaker 1982; McKinnon 1982a); and *acoustic noise* for either type of target (Melosh 1979,1982a). If any combination of these is operating during the late-stage expansion of the crater, then the cratering efficiency should increase for larger gravity-scaled sizes, and approach that of water; also, the depth/diameter ratios of the transient craters should approach 1/2 as they become nearly hemispherical.

If the rheology of the material in the late-stage cratering flow can be considered fluid, its effective viscosity may be high enough to dominate strength and gravity in determining the cratering efficiency. In this case, the crater is said to have formed in the *viscous regime*. In this regime, cratering efficiency can be compared with an inverse Reynolds number,

$$\pi_\eta = \eta/(\rho a u) \tag{15}$$

where η is the target viscosity. The existence of the coupling parameter ensures that a power law relationship exists between π_V and π_η in the viscous regime,

$$\pi_V \pi_\eta^\nu = F \tag{16}$$

where ν and F are experimentally determined constants, and

$$\nu = 2\alpha/(1 - \alpha) \tag{17}$$

Note that when $\alpha = 0.65$ (as for water), $\nu = 3.7$, so control by viscosity in the viscous regime is quite pronounced. There may be an additional ρ/δ dependence, since the choice of the mass density term in Eq. (15) is ambiguous, as it was in Eq. (7).

A cratering efficiency prism analogous to Fig. 6 could be constructed, but with planes representing the viscous and gravity regimes, and the viscous-gravity transition could be defined. Unfortunately, without knowledge of the appropriate viscosities to use, transition crater diameters (as in Fig. 7) cannot be determined. We can, however, calculate when viscous scaling may be important. Experiments by Fink et al. (1984), using silicon oil targets and assuming $\alpha = 0.6$, show that viscous effects begin to affect the cratering efficiency when

$$\eta \gtrsim 0.02 \rho a^{1.2} u^{0.6} g^{0.2} \tag{18}$$

and significant effects occur when η is an order of magnitude larger. Note that the gravity dependence is comparatively weak, a consequence of the different slopes α and ν. Also, these experiments were carried out at fixed ρ/δ, so the mass density dependence in Eq. (18) is probably incomplete.

Consider a medium-sized crater on Ganymede created by a 1 km radius projectile moving at 20 km s^{-1}. With $\rho = 1$ g cm^{-3} and $g = 142$ cm s^{-2}, the effective viscosity of the cratered material would have to exceed $\sim 3 \times 10^7$ Pa s (3×10^8 P) to affect the cratering flow at all. It is problematic whether such high viscosities are achievable by the fluidization mechanisms discussed above. We believe that the strength-gravity model for impact crater volume (Fig. 4) represents the best current understanding of crater scaling for ice and rock targets, and we shall use it fruitfully, in the following sections, in intercomparisons of craters on different satellites. Further work, however, is necessary to characterize effective strengths and viscosities and the role of fluidization.

As a final note, all of the scaling relationships above require completion of the coupling phase. Thus, at very high π_2, when the cratering efficiency is of order unity (which from Fig. 4 implies the most catastrophic collisions among large satellites and planets), the final crater size becomes less than or equal to the volume of the coupling zone, and we cannot expect the scaling laws to remain valid.

C. Impact Melting and Vaporization

Shock wave passage is a thermodynamically irreversible process. Hence temperatures may increase so much that target (or projectile) material is fused or vaporized; moreover, such phase changes may remain even after the target unloads from high pressure to its original density. These processes are difficult to study experimentally, so most of our knowledge comes from analytical models and large-scale computations.

A prominent model is due to Gault and Heitowit (1963; see also the extensive discussion in Kieffer and Simonds [1980]). The model is physically curious—one-half the projectile kinetic energy is partitioned into the kinetic energy and one-half into the internal energy of a spherical volume of high-pressure target material, centered at depth, whose boundaries continually expand as the shock wave expands. At any time step, as a spherical shell of target material is added to the compressed sphere, the internal energy increase which the shell will eventually experience when it reexpands to its original density (so-called waste heat) is subtracted from the total energy of the sphere as a whole. (This somewhat unphysical treatment results in an overly rapid decrease in the total energy.) In any case, the model pressure within the sphere and the peak pressure experienced at any point on its surface decrease as the sphere volume increases, due both to geometrical spreading and to waste heat deposition.

The total energy E_t within the sphere is given at any time by

$$E_t = \frac{4 \pi \rho}{3} r^3 P(\mathcal{V}_0 - \mathcal{V})$$ (19)

where $\mathcal{V}_0$ is the initial specific volume of the target and P and $\mathcal{V}$ are the pressure and specific volume within the sphere when it expands to radius r. At very high shock pressures, the specific volume change $\mathcal{V}_0 - \mathcal{V}$ is nearly constant (see Fig. 9). Therefore, if there are no losses to waste heat, $E_t \simeq$ constant, and $P \sim r^{-3}$. Shock-pressure decay rates would be even higher if irreversible losses were accounted for.

Gault-Heitowit calculations are not in good agreement with numerical

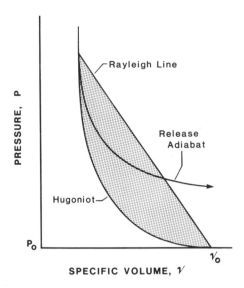

Fig. 9. Schematic thermodynamic path for shocked material. The initial state (P_0, V_0) is connected to the shocked state (P, V) by the Rayleigh line. The locus of possible shock states defines the Hugoniot. Shocked material unloads along a release adiabat. In the Gault-Heitowit model, $Pd V$ work done upon release to V_0 is approximated by the area between the Rayleigh line and the Hugoniot (shaded).

computations of high-pressure shock wave attenuation (Ahrens and O'Keefe 1977; Orphal et al. 1980) which generally give $|dlnP/dlnr| < 3$. Two factors account for this. First, a better representation of shock wave geometry is an expanding spherical shock of finite thickness (see discussion of detached shock above). The second factor concerns the waste-heat calculation. The $Pd V$ work done by the shock and subsequent unloading is the area under the Rayleigh line (the line connecting the shocked and unshocked states) minus the area under the release adiabat (Fig. 9). Release adiabats depend on the particular shock state, so the Gault-Heitowit model makes the simplifying assumption that all release adiabats follow the Hugoniot (locus of shock states). This is reasonable for low pressures but not at all for high pressures when vaporization is involved (O'Keefe and Ahrens 1977; Kieffer and Simonds 1980); it greatly overestimates the amount of waste heat and hence the rate of shock pressure decay.

The decay for an expanding, spherical, high-pressure shock of finite thickness and uniform thermodynamic state can be written as

$$dlnP/dlnr \sim -[2 + (r_0/h - 1)(f/2)] \qquad (20)$$

where r_0 is the initial radius into which the projectile kinetic energy is partitioned, h is the shock thickness, and f is a factor (< 1) that relates the actual waste heat to that predicted by Gault-Heitowit. The actual decay rates depend critically on the release adiabats used; examples would be Tillotson equations of state for gabbroic anorthosite and water ice (O'Keefe and Ahrens 1982b) and the Bakanova equation of state for water ice (Ahrens and O'Keefe 1984). A consequence of the less-steep attenuation of shock pressure, compared with Gault-Heitowit and depending on impact conditions (O'Keefe and Ahrens 1975), is that in reality there is a greater amount of impact vapor and/or melt. This vapor and melt will also be cooler (i.e., have lower specific internal energies).

Scaling. Calculations of the amount of gabbroic anorthosite target melted and vaporized by water ice, anorthosite, and iron projectiles (O'Keefe and Ahrens 1977,1982b) are presented in nondimensionalized form (Bjorkman and Schmidt, personal communication, 1983) in Fig. 10. The fusion and vaporization efficiency (target mass melted or vaporized divided by projectile mass) is plotted against normalized projectile kinetic energy. The energy densities used in the normalization, ε_v and ε_m, are the critical specific internal

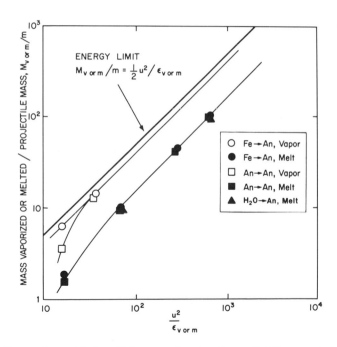

Fig. 10. Vaporization and fusion efficiency as a function of normalized impact velocity. ε_v and ε_m are critical specific energies for vaporization and melting (see text); u is impact speed. Melt data for water ice, anorthosite, and iron projectiles striking anorthosite, from O'Keefe and Ahrens (1977,1982b), appear to follow energy scaling. Extension of curves through vaporization data points is speculative.

energies in the shocked state corresponding to complete vaporization or melting upon release to a reference state; 1 bar is used here. (Another reference state could be $\mathcal{V}_0$, but note that zero pressure would be impractical, since the equilibrium phase for all materials at nearly zero pressure is solid plus vapor.)

Fusion efficiency appears to scale with projectile specific kinetic energy—vaporization efficiency may also. Both fall under a theoretical limit, where the total projectile kinetic energy is equated to the internal energy of a mass shocked to either critical value. This is not a hard upper limit, of course, as material is not generally melted or vaporized simultaneously, and internal energy densities always fall during unloading. Nevertheless the curves in Fig. 10 may be useful guides for other types of targets. For gabbroic anorthosite, ε_v and ε_m are 57.3 MJ kg^{-1} and 3.37 MJ kg^{-1}, respectively. For water ice, ε_v and ε_m are near 60 MJ kg^{-1} and 2 MJ kg^{-1} (Ahrens and O'Keefe 1984; Kieffer and Simonds 1980). Therefore the vaporization efficiencies for both types of target may be similar for velocities sufficiently above a threshold value that varies for each projectile-target combination (see Table II). Fusion efficiency by this logic would be ~ 60% greater for water-ice targets. Threshold velocities for various phase changes, from 1-dimensional impedance-match calculations, are given in Table II. For the same impactor type, the threshold velocities are uniformly lower for water-ice targets. The onset of complete vaporization does not occur at strikingly dissimilar velocities for the two targets; this is a consequence of their similar critical specific internal en-

TABLE II
Onsets of Phase Changes for One-Dimensional Impact

Target	State[a]	Shock Pressure (GPa)	Internal Energy[b] (MJ kg^{-1})	Velocity (km s^{-1}) for Impactor Type		
				Ice	GA	Iron
Gabbroic Anorthosite[c]	IM	43	2.6	7.4	4.6	3.3
	CM	52	3.4	8.3	5.2	3.8
	IV	102	7.5	12.4	7.8	5.9
	CV	590	57.3	32	21.4	16.6
Water Ice[d]	IM	7.8	1.6	3.6	2.5	2.0
	CM	10.4	2.1	4.1	2.9	2.3
	IV	13.3	3.0	4.9	3.5	2.8
	CV	165	62.0	22.3	16.3	13.9

[a] IM = incipient melting; CM = complete melting; IV = incipient vaporization; CV = complete vaporization; all upon isentropic release to 1 bar.
[b] Gabbroic anorthosite (GA) with respect to its low-pressure phase at standard temperature and pressure (STP); ice with respect to ice I_h at standard pressure and 263 K.
[c] Hugoniot for high-pressure phase and release adiabats to STP from Ahrens and O'Keefe (1977).
[d] Hugoniot for high-pressure phase from O'Keefe and Ahrens (1982b); release adiabats to 6 mbar and 263 K from Ahrens and O'Keefe (1984).

ergies. Incipient melting, complete melting, and incipient vaporization, however, occur at significantly lower velocities for water-ice targets. In particular, a water-ice target will begin to vaporize before a gabbroic anorthosite target begins to melt.

Bjorkman (1984) postulates that the coupling parameter determines the amount of impact melt and vapor, which makes sense only if coupling is complete before formation of the detached shock. Following the arguments in Sec. II.B, vaporization and melting efficiencies should scale as $(u^2/\varepsilon_v)^\beta$ and $(u^2/\varepsilon_m)^\beta$. $\beta = 0.83$ for nonporous targets, from Eq. (9); $\beta = 1.0$ for energy scaling. Using data from O'Keefe and Ahrens (1977) and Orphal et al. (1980), plus new calculations of aluminum impacting aluminum, Bjorkman finds that impact-melt-depth cubed scales with $\beta < 1$, but that melt volume scales with $\beta = 1$, consistent with Fig. 10. The reason for this discrepancy is not understood.

It is important to understand possible differences in scaling for vapor, melt and crater volumes so that we can confidently assess the relative roles of projectile size and velocity in the formation of individual terrestrial impact structures (Grieve and Cintala 1981), the role of volatilization in crater morphology (Croft 1983), and so on. It seems clear, though, in comparing the scaling of melting and vaporization efficiency with that of cratering efficiency in the gravity regime (Eq. 4), that as gravity-scaled size (π_2) increases at fixed velocity, the relative proportion of impact melt and vapor increases—possible fluidization effects become more likely as well. The rates of increase are the crucial questions.

Some Effects of Vaporization. The role of impact vaporization is enhanced on icy satellites, as opposed to silicate ones, because incipient vaporization occurs at the relatively low shock pressure of ~ 13.3 GPa (Table II). Expansion of the impact vapor cloud mixed with substantial impact melt (flow streamlines cross shock isobars) has many interesting effects. For example, solid ejecta can be accelerated by gas drag to several km s^{-1}, far in excess of what is possible by direct shock acceleration without melting (McKinnon 1981c). Diameters of these high-velocity ejecta fragments are limited to a few tens of meters, however, because larger blocks, which are less readily accelerated, are crushed by the Bernoulli pressure (Singer 1983; Vickery 1984). The vapor cloud with entrained melt and solid ejecta can form a satellite-wide frost deposit (see also Smoluchowski 1983); if the satellite is small, most of the vapor escapes into planetocentric orbit, condensing to form dust-size ring particles (McKinnon 1983a). A vapor cloud may also provide a transient pressure boundary condition for spallation (Sec. II.D below).

D. Ejecta

Ejecta, the material actually thrown out during crater formation, are usually easy to observe on satellite surfaces. Hence, many studies of this material have been made (more than of impact melt, for example). From a cratering mechanics viewpoint, however, the form, nature, distribution and provenance of ejecta remain incompletely understood.

Each phase in the cratering process (see Sec. II.B) contributes material to the total ejected mass. Early-time material, including jetted material, is ejected before *coupling* is complete and is the most highly shock-processed. Most ejecta leave during *power law growth* of the transient cavity and form most of the *continuous ejecta blanket* exterior to the crater. In the *late stage*, with cavity growth limited by strength and/or gravity, the slowest moving ejecta form the rim of the (transient) crater. An important additional form of ejecta is the spallation fragment. Such fragments come from near the free surface of the target, and are generally the most lightly shocked ejecta; they may, however, reach substantial velocities. Below we attempt to relate the different types of ejecta deposits (continuous and discontinuous blankets, fields of secondary craters and rays) to the various cratering phases or ejection regimes, and to understand the fate of impact melt and vapor and the differences from satellite to satellite.

Ejecta Blanket.   Continuous ejecta deposits on the Moon extend about one crater radius from the rim, which means that the ratio of the radius of the outer edge of the deposit R_{ce} to the crater radius R is ~ 2. Moore et al. (1974), chiefly using mapped data, found that R_{ce}/R is ~ 2.4 for craters between 0.3 and 100 km radius. Gault et al. (1975) studied a smaller number of specific craters and determined that R_{ce}/R slowly decreases with increasing crater size, declining to ~ 1.6 at 125 km crater radius. The thickness of ejecta deposits decreases rapidly away from the crater rim; the oft-quoted study of McGetchin et al. (1973) synthesized explosion- and impact-crater experiments with theory to derive the variation in ejecta thickness σ on radial range r:

$$\sigma/R = (0.14 \text{ meter}/R)^{0.26}(r/R)^{-3}. \qquad (21)$$

Both the slow variation in R_{ce}/R and the fact that normalized ejecta thickness σ/R is almost solely a function of r/R imply that continuous ejecta deposits of lunar craters tend to be geometrically similar, which has long been a guiding rule-of-thumb.

Housen et al. (1983) have elegantly shown that ejecta blankets in the *strength regime* should *not* be geometrically similar and should scale such that the deposits of larger craters are more radially restricted. Moreover, Housen et al. find that blankets *should be* geometrically similar in the *gravity regime*, applicable to most of the above studies. Note, however, that the scaling results of Housen et al. apply rigorously only to the transient crater. It is not surprising that R_{ce}/R is observed to decrease slowly with increasing scale because large craters collapse, forming slump terraces, and because still larger craters must collapse and widen to an even greater extent in order to maintain the nearly constant, limiting crater depth that is observed (see Sec. II.E). It is important to have a clear understanding of the definition of continuous ejecta deposit. Is its limit a geometrically fixed fraction of the ejecta thickness: σ/R = constant? Or is a continuous ejecta deposit only where we can recognize complete blanketing, perhaps a fixed thickness that depends on terrain roughness and secondary cratering efficiency? Morrison and Oberbeck (1978) believe that what is frequently mapped as a continuous ejecta blanket is mostly *in situ* material gardened by secondary impacts; the relatively small additional contribution of true ejecta from the primary crater adds little to the topographic thickness of the blanket. Mass motions within ejecta blankets are a further complication. Oberbeck and coworkers (see, e.g., Oberbeck et al. 1974; Oberbeck 1975; Oberbeck and Morrison 1976) have argued convincingly that secondary impacts from large lunar craters form at sufficiently high velocity to create many times the secondary ejecta mass in *tertiary ejecta*. The tertiary ejecta will have a component of momentum radially away from the primary crater, since secondary impacts are oblique. If this momentum is sufficient to overcome frictional resistance in the debris (or if a mechanism such as acoustic fluidization [Melosh 1979; also see Voight 1978] lowers the friction), then a turbulent debris flow may form and spread beyond what would be expected for a ballistically emplaced continuous ejecta blanket. Hörz et al. (1983) have proven such processes were responsible for generating the Bunte breccia of the Ries Crater ($\sim$ 13 km transient diameter) in Germany; for example, limestone blocks evidently were transported as much as 10 km in moving debris. In view of these considerations, the factor $R^{-0.26}$ in Eq. (21) must be viewed with some skepticism.

Near a crater, the topography of the ejecta blanket merges into that of the crater rim. There are difficulties in estimating rim heights and the functional form of the radial decay of rim topography, but advances have been made beyond the old Schröter's Rule equating rim volume with crater volume (cf. Pike 1974b; Settle et al. 1974; Settle and Head 1977). Pike (1977) gives the best current estimate for the variation of the rim height h_r of large lunar craters ($D > 15$ km) as a function of crater radius:

$$h_r/R = (0.31 \ \mathrm{km}/R)^{0.6}. \tag{22}$$

Observed rim heights for small craters include a component of structural uplift, but this contribution is small or negligible beyond $r \simeq 1.25R$, based on the experiments of Stöffler et al. (1975) and observations at Meteor Crater (Roddy et al. 1975). Structural uplift is even less important for larger craters due to slumping. Estimates of the degree of slumping by Shoemaker (1962), Mackin (1969) and Settle and Head (1979) imply that structural uplift can be ignored for $D \gtrsim 100$ km on the Moon. Rim height should be geometrically similar for transient craters and a slowly decreasing function of R for slumped final craters. The reduction of rim and blanket topography away from the rim crest can be approximately represented by a power law with an exponent e_r. Far from the rim crest ($r \gg R$), Housen et al. (1983) derive

$$e_r = \frac{6 + \alpha}{3 - \alpha} . \tag{23}$$

This exponent is rather insensitive to the gravity scaling exponent α, varying between 2.60 and 2.80 for $\alpha = 0.5$ to 0.65 (Table I). This range of e_r compares well with that derived from impact experiments in sand (Stöffler et al. 1975; discussed in Housen et al. 1983), impact experiments in basalt powder similar to coarse sand ($e_r = 2.4$ to 2.6; Hartmann 1985), and theoretical calculations ($e_r = 2.7$; O'Keefe and Ahrens 1976; e_r time variable but < 3; Ahrens and O'Keefe 1978). Near the rim the effective exponent is greater. Again, these relationships would be modified by any large-scale mass motions in the ejecta blanket, such as debris flows discussed above.

If craters are to remain geometrically similar in the gravity regime, then ejection velocities must increase with increasing scale or gravity. In particular, at a geometrically fixed launch point within the crater, the ejection velocity v_e must scale as $\sqrt{gR}$ for constant launch angle ψ, equivalent to the requirement that the ballistic range (equal to $v_e^2 \, g^{-1} \sin 2\psi$, when ψ is measured with respect to the horizontal) scales as R. [A similar relationship, v_e varying as $(gR)^{0.6}$, was derived by Schultz and Mendell (1978) and Schultz and Gault (1979), using earlier scaling laws.] During the power law growth phase, when gravity is not important, the launch velocity decays as a power law function of the launch position x defined in the ground plane (Housen et al. 1983):

$$\frac{v_e}{\sqrt{gR}} = K \left(\frac{x}{R} \right)^{-e_x} \tag{24}$$

where K is a constant and e_x is given by

$$e_x = \frac{3 - \alpha}{2\alpha} . \tag{25}$$

e_x varies between 2.50 and 1.81 for a range in α of 0.5 to 0.65. (Schultz and Gault [1979] had earlier adopted a power law decay of ejection velocity, with $e_x = 3.0$; they were motivated by the Maxwell Z model of the cratering flow field [see Maxwell 1977; Orphal 1977].) The power law decay of ejection velocity on launch position is not valid in the late stage, when x/R is close to 1. Rather than approaching $\sqrt{gR} \, K$ when $x = R$, v_e will decline toward zero as gravity slows cavity growth. Experimental estimates of K are imprecise, but values of order unity may be appropriate for large craters in the gravity regime.

Secondaries.   Chains, loops and clusters of secondary craters are seen in abundance around large primary craters on the Moon. The close proximity of many secondaries, and the interaction of their nearly simultaneously formed ejecta curtains, lead to distinctive morphologies such as concentric dunes, intercrater septa, radial ridges and herringbone patterns (Oberbeck and Morrison 1974; Morrison and Oberbeck 1975). Isolated secondary craters are also numerous.

There may be no limit to the sizes of secondaries and therefore to the fragments that formed them. Orientale and Imbrium secondaries may range up to 20 and 30 km in diameter, respectively (Wilhelms et al. 1978). (Those of older, larger proposed basins such as Procellarum [Whitaker 1981] have yet to be recognized.) Some apparently lie at ranges exceeding 2000 km and hence require ejection velocities up to ~ 1.5 km s^{-1}. That large, multikilometer-wide fragments could be launched to such velocities by shock is surprising. If directly accelerated, the shock pressures implied are sufficient to crush rock. In contrast, the bulk of continuous and discontinuous blanket facies around large craters are emplaced at much lower velocities and yet are relatively comminuted (Schultz and Mendell 1978). This suggests a different provenance for secondary fragments; the logical region is the near surface, where pressures, but not necessarily accelerations, are low. This hypothesis was first offered by Shoemaker (1962) and has been recently quantified by Melosh (1984).

These fragments form as surface spalls; that is, they are due to the superposition of the compressive shock wave initiated by the impact and its tensile reflection from the free surface. At a certain depth the sum of the stress waves becomes more tensile than the dynamic tensile strength of the target, and a spall plate is torn loose (see Rinehart [1968] for a discussion of stress-wave interactions). Formation of spalls requires a competent surface. Melosh (1984) derives an approximate expression for the spall thickness z_s:

$$z_s \simeq 2(\sigma_t/\rho c_L)(d/v_s) \tag{26}$$

where σ_t and c_L are the dynamic tensile strength and P-wave speed in the target, d is the penetration depth of the projectile, and v_s is the spall ejection velocity. Melosh estimates d from the Birkhoff et al. (1948) jet penetration formula

$$d \simeq 2a(\delta/\rho)^{1/2} \tag{27}$$

and v_s is shown to be approximately

$$v_s \simeq 2v_p(r)\, d/r \tag{28}$$

where $v_p(r)$ is the peak particle velocity induced by the shock. A shock decay model is then needed to evaluate ejection velocities and spall thicknesses. (Note that both Eqs. (26) and (28) are valid only for $r \gg d$.)

The spallation theory has a couple of advantages. Spall plates suffering fragmentation during ejection or while in flight can form clusters and chains of similarly sized craters, the characteristic fragment size being the spall thickness. Also there is a natural lower limit to spall thickness; spalls close to the crater are crushed by high horizontal compressive stress. The observed decrease in secondary size away from the primary crater is consistent with the spallation theory since larger spalls are slower. The narrow distribution of fragment sizes predicted by the spallation model may explain the large negative power law exponents of the incremental diameter frequency relations for secondaries, -4.6 to -5.0 (Shoemaker et al. 1965; Wilhelms et al. 1978), if the latter are interpreted as the large-diameter wing of a unimodal size distribution. The model of Melosh also predicts multikilometer spalls for basin-scale impacts, which may make large, distant secondaries physically plausible.

Spall thickness is generally proportional to dynamic tensile strength (Eq. 26), although joints, flaws, planes of weakness and inhomogeneous structure may reduce it. The tensile-strength dependence means that spalls form in the strength regime, even though the primary crater forms in the gravity regime (spall volume $\ll$ crater volume). Thus, the distribution of spall-derived secondaries "strength-scale," so they are not geometrically similar to the crater: as either crater radius or gravity increases, the secondary field contracts in relative range and increases in areal density.

It remains uncertain what the relative contributions are, as a function of radial distance, of spallation ejecta and that due to the normal, orderly process of (power law) crater growth. Equal contributions would occur at ranges that are scale- and gravity-dependent, again raising questions about the definition of the edge of the continuous ejecta blanket.

Rays. Of all the observable ejecta, ray ejecta travel the greatest distance. Rays are composed of diffuse patches and filamentary streaks oriented radially or approximately radially to the crater, often forming streamers that approximate great circles on the satellite, although loops and other geometries are known (Shoemaker 1962; Baldwin 1963). On the Moon, rays are generally characterized by their relatively high albedo, compared with both highlands and maria (Fig. 11). The albedo change is probably due to an effective increase in the crystalline silicate fraction, compared with the darker, glassy agglutinates (glass-welded aggregates), in the optical surface layer (cf. Adams and McCord 1973; Pieters 1978). The fracturing and comminution of regolith soil particles creates rough new fracture surfaces and exposes the brighter interiors of dark-coated particles; also deeper, brighter, less optically mature (glass-rich) regolith layers may be exposed. As optically immature ray material "space weathers" and darkens (see chapter by Veverka et al.), its crys-

Fig. 11. Portion of Earth-based photograph of the full Moon. The ray system of Tycho (98 km diameter, ~ 100 Myr old) is prominent; dark halo surrounding the crater rim is probably due to the presence of impact melt (Smrekar and Pieters 1984). In contrast, the older (~ 800 Myr) ray system of Copernicus (93 km diameter, upper left) has a considerably lower albedo, and may actually be photometrically mature (B. R. Hawke, personal communication, 1984), consistent with evidence for excavation of highlands material (Pieters et al. 1982; see text). Both Tycho and Proclus (28 km diameter, upper right) have bilaterally symmetric ejecta, characteristic of oblique impacts of incidence angles > 45°.

talline and glassy particle distributions will reach equilibrium. The albedo of the optically mature ray depends on the composition of the primary ray material and on the composition of the *in situ* regolith to a depth where the gardening time scale approximates the maturation time scale (~ 10 cm; Gault et al. 1974). Rayed craters are associated with the morphologically freshest class of craters (Baldwin 1963) and are deemed an indicator of extreme freshness. The fact that rays seem to fade and disappear so rapidly suggests that the overall contribution of primary ray ejecta to the observable ray material may be small, but not necessarily insignificant (see below).

The original arguments as to whether ray material is depositional or due to secondary cratering now seem misdirected. Except for small primaries, ray-forming ejecta (of any size) impact at velocities sufficient to induce vigorous secondary cratering. In fact, the more comminuted the ejecta is, the more efficiently it forms gravity-scaled craters in the regolith, although the maximum depth of excavation drops. Oberbeck (1971) has argued that ray-forming secondary impacts would have to penetrate the regolith (thickness ≤ 10 m on the maria; Quaide and Oberbeck 1968) in order to excavate fresh, bright blocks. While there is little doubt about the close association between observable secondaries and rays (Arvidson et al. 1976; Lucchitta 1977b; Maxwell and El-Baz 1978), those rays not associated with resolvable secondaries (Allen 1977) imply that secondary cratering on very small scales can also form rays.

The association of rays with small, distant secondaries raises the possibility that ray ejecta are derived from the crushed spall layer near the projectile. Alternatively, ray ejecta may be the high-velocity fraction lofted during power law crater growth, and hence may include material from deeper stratigraphic horizons. Evidence exists to support both views. Looped rays and associated secondaries at Copernicus appear related to near-surface structure (Shoemaker 1962). Conversely, Pieters et al. (1982) find spectrophotometric evidence for highlands material, presumably derived from great depth beneath the Imbrium basalts in a Copernicus ray. The detection of primary ejecta in the distant ray demonstrates, in the context of the above discussion of ray albedo, that primary ray ejecta form a nonnegligible volumetric fraction of the observable ray material (see Fig. 11). The dark, ropy KREEP glass in Apollo 12 soils identified as Copernicus ray ejecta (cf. Marvin et al. 1971; McKay et al. 1971) cannot contribute to the observable bright ray. But, of course, this highly shocked (melted) material could be derived from Copernicus right along with the lightly shocked secondary-forming fragments that do contribute to the observable ray. The ejection of materials with grossly dissimilar shock histories to similar ranges is seen in the calculations of Ahrens and O'Keefe (1978).

One thing ray-forming ejecta do *not* appear to be is high-velocity, early-time material launched before coupling is complete. The ejection velocities of this material are (at least) a fair fraction of the original impact velocity (O'Keefe and Ahrens 1975,1977,1982b; Orphal et al. 1980) and thus, for primary impacts, such material is very unlikely to be retained on any of the natural satellites (escape velocities < 2.75 km s^{-1}).

The radius of ray systems on the Moon geometrically scales crudely with crater radius (Baldwin 1963). This is consistent with ray formation being controlled by the coupling parameter (which would only not be true if rays were associated with early-time material). The azimuthal asymmetry in ray patterns, however, requires instabilities in the flow field after coupling is achieved. (Andrews [1977] correlated instabilities in the fireballs of large explosion craters with subsequent ray formation.) Unfortunately, observations of ray lengths may be severely compromised by age effects. The questions of ray scaling and provenance require additional work.

Oblique Impact. The chief paper describing experimental results on ejecta distribution as a function of incidence angle θ (see Fig. 3) is that by Gault and Wedekind (1978). Little difference from vertical impact is observed for incidence angles under 45°. At more oblique angles, the ejecta curtain tilts toward the downrange direction, and a forbidden zone in which little ejecta is deposited, develops in the uprange direction. Examples can be seen in Fig. 11. For even shallower trajectories ($\theta > 80°$), a second forbidden zone develops downrange. For impacts approaching the horizontal, ejecta travels at nearly right angles to the projectile trajectory, and the ejecta pattern resembles wings of ever-narrowing azimuthal spread. A discrete downrange jet of material is produced as well, possibly related to the jetting enhancement in oblique impacts studied by Kieffer (1977). If the vertical component of velocity is low enough, the projectile may actually bounce or ricochet.

These inferences were developed mainly from ~ 6.5 km s^{-1} impacts into pumice dust, which appears to be particularly sensitive to the effects of oblique trajectories. Thus it is of some interest to investigate these effects with different materials (such as water) and as a function of impact velocity. The angle ranges for phenomena given above should be used with caution.

Schultz and Gault (1979,1982) examined atmospheric effects on ejecta patterns (aerodynamic sorting, base surges, etc.); their findings may be applicable to Titan and Triton.

Melt, Vapor and Accretional Efficiency. The ejection velocities of shock-melted and vaporized material are usually considered to be high, but the correlation between shock level and ejection velocity is actually far from exact. Consider the trajectories in the cratering flow field: they are generally concave upward, but many initially have downward velocity components. Hence some highly shocked material may not leave the cavity until late in the growth of the crater, when ejection velocities are comparatively low (Grieve et al. 1977). Some material, lying approximately within a downward-opening cone (surrounding the stagnation trajectory) beneath the effective center of energy deposition, may not leave at all (Croft 1980).

An essential characteristic of the cratering flow is that material trajectories or streamlines cross the grossly spherical pattern of shock isobars. Shearing of the isobar pattern is not generally orderly. Reynolds numbers in the flow are given by $\rho \ell \, v/\eta$, where ℓ is the characteristic scale length over which the viscosity changes by a factor of e. Even for very rough estimates of $\ell \sim 10$ m and v ~ 1 km s^{-1}, Reynolds numbers are enormous ($\sim 10^5$ to 10^{11}) for silicate liquid and gas viscosities (Clark 1966). Thus, shear flow in the vaporized and melted portions of the target is highly unstable; the resulting turbulence should homogenize melt and vapor over a range of shock levels.

The homogenization of target rocks in terrestrial impact-melt sheets has been known for some time, and turbulent mixing is generally agreed to be responsible (see, e.g., Grieve et al. 1977; Phinney and Simonds 1977; Kieffer and Simonds 1980). The range of shock levels mixed could be sizable. For example, large-scale mixing of melt with partially melted and unmelted rock is so extensive that portions of the Manicouagan impact melt sheet were quenched within ~ 100 s (Onorato et al. 1978; Simonds et al. 1978). Most inclusions are observed to have been shocked to ≪ 20 GPa (Phinney and Simonds 1977; Simonds et al. 1978); hence few (if any) surviving inclusions were originally partially fused by shock. Those that were highly shocked were apparently digested by the melt. We assume that these inferences apply to other massive (but less well-studied) terrestrial melt sheets resembling the one at Manicouagan.

The degree of mixing of vaporized, partially vaporized, and melted material is unknown. Grieve et al. (1980) cite chemical and petrographic evidence for condensation of silicate vapor on molten iron droplets during the E. Clearwater impact. Possibly much of the vapor produced upon impact into silicate targets is intermixed and quenched by melt (Kieffer and Simonds 1980).

These inferences can be extended to impacts in ice, with a major distinction. Compared with impacts in rock, the phase change that dominates impacts in ice, for typical projectile velocities (~ 20 km s^{-1}), is vaporization rather than melting (see Table II). Water ice begins to vaporize at shock pressures an order of magnitude less than those that cause rock to vaporize. The extremely low viscosities of water ($\eta \sim 10^{-1}$ Pa s) and steam increase the level of shear turbulence during expansion of the flow field and greatly enhance the potential for mixing. Although liquid water produced by shock may mix with solid ice clasts to form a slush-like melt sheet, more vigorous mixing with copious quantities of superheated impact-generated steam may assimilate most of the water, much of the partially melted ice, and perhaps significant amounts of warm and cold ice (cf. Kieffer and Simonds 1980). Mixing of hot, high-entropy steam with colder, low-entropy water and ice may be so extensive that the ejection of volatiles from an ice impact may thermodynamically resemble the expansion of a boiling mass of water more than a condensing mass of steam (McKinnon 1983a). Furthermore, while mixing may be efficient, it does not necessarily follow that all the water or partially melted ice is expelled from the crater; melt in downward driven trajectories, especially those near the stagnation trajectory, may not be expelled when the transient cavity reaches maximum depth. Such remaining melt or slush should become volumetrically more significant at larger scales, by the arguments of Sec. II.C.

The distribution of vapor-charged ejecta will differ from that described above for ejecta blankets, secondaries, and rays. Launch angle behavior is completely different. While code calculations generally exhibit larger angles (more nearly vertical) than are observed in laboratory experiments in sand (cf. O'Keefe and Ahrens 1977; Orphal et al. 1980), when vaporization is involved the transient cavity back-fills with expanding vapor that exits the crater region isotropically (O'Keefe and Ahrens, 1976,1977,1982*b*). Such an expanding mass of vapor-charged ejecta may distribute a diffuse deposit at considerable ranges from the crater, with most material moving too rapidly to be retained on any satellite.

If the amount vaporized is considered alone, then the *accretional efficiency*, or the projectile mass minus the mass of the escaped ejecta (both expressed in units of projectile mass), is strongly negative for primary impacts on icy satellites (McKinnon 1981*c*). O'Keefe and Ahrens (1976,1982*b*) calculated that the accretional efficiency of gabbroic anorthosite targets for projectile velocities of 15 km s^{-1} was near zero when the escape velocity of the target planet or satellite was ~ 2.5 km s^{-1}. No doubt an ice target would result in much lower calculated efficiencies. On the other hand, extensive mixing of steam with less shocked phases could reduce the expansion velocities so that more volatilized material would be retained on the larger, icy satellites than would otherwise be calculated.

Observations of Ejecta on Icy Satellites. So far we have discussed ejecta in terms of mechanics and with reference to experiments and observations of the Moon and terrestrial planets. We now interpret recent observations of the icy satellites.

Continuous ejecta blankets should be observable around craters on icy satellites, so long as the craters form in the gravity regime. The expected value of the ratio R_{ce}/R, the continuous ejecta radius/crater radius, is not clear, for several reasons: (1) craters of a given size on different satellites may have collapsed and widened to different extents; (2) the edge of the continuous ejecta may be a poor indicator of scaling, as discussed above; and (3) while R_{ce}/R is a constant when referenced to the transient crater, it is not established that the same constant applies to all materials.

No systematic study of ejecta blankets on Ganymede and Callisto has yet been made, but certain aspects have been investigated. A subset ($< 5\%$) of fresh craters on Ganymede have pedestal ejecta deposits, defined by sharp, almost scarplike terminations (Strom et al. 1981; Horner and Greeley 1982). This ejecta form is thought by Horner and Greeley (1982) to result from ejecta flow, as it resembles that of certain ejecta-flow craters on Mars. They determined that the pedestal radius to crater radius ratio is ~ 2.3. Craters with dark halo deposits surrounding their rims have been identified in grooved terrain by Schenk and McKinnon (1985). They argue that those deposits represent the incorporation into the ejecta blanket of darker, cratered terrain material excavated from underneath grooved terrain material. The dark halo radius to crater radius ratio is ~ 2.8.

Do either of these studies measure the extent of the continuous ejecta

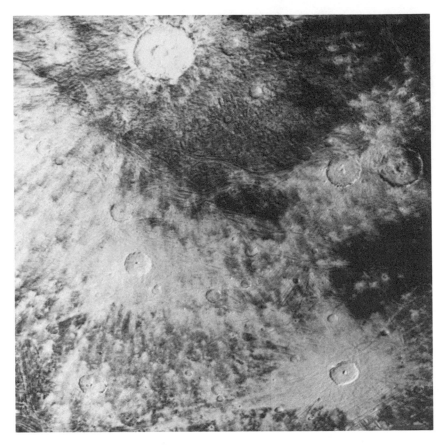

Fig. 12. Cluster of large fresh craters on Ganymede (45 to 105 km diameter), exhibiting continuous ejecta, secondaries and rays. The albedo of thin continuous ejecta and rays depends on the composition of the underlying target surface (i.e., light grooved terrain vs. dark cratered terrain), attesting to the importance of mixing of ejecta and local material by secondary cratering. (Voyager 2 frame FDS 20637.59, centered near 44°S, 163°W.)

blanket? A clue may exist in fresh-crater morphology. The continuous ejecta blankets of the large, fresh, Ganymedean craters in Fig. 12 have two distinct components, a textured facies and a more extensive facies that apparently only thinly mantles existing topography. The textured facies probably represents a sufficient amount of material that, if fluidized, could flow to form a pedestal scarp as a flow front. The other facies, despite its thinness, could still carry the geochemical signature of the ejecta; it may owe its relatively large extent to vapor-assisted transport. The lobate deposits surrounding Osiris (Fig. 12, top) are stronger evidence for ejecta fluidization and flow (again in analogy with similar Martian craters). Ejecta flow is probably a common, if not ubiquitous, aspect of cratering on Ganymede and Callisto, but conclusive evidence lies below the resolution of Voyager images.

Ejecta blankets are difficult but not impossible to identify on the Saturnian satellites. Plescia (1983) notes that Aeneas (150 km diameter) and Dido (125 km diameter), the best-imaged large craters on Dione, have rim deposits that extend about one crater radius from the rim crest, and that the smaller Creusa (30 km diameter) has a bright deposit of similar relative extent. Rim deposits can also be detected around larger fresh craters on Rhea (Fig. 13) and around Telemachus (90 km diameter) on Tethys (see Voyager 2 frame FDS 44003.57). The apparent paucity of ejecta blankets on the Saturnian satellites is probably due to low resolution and a general lack of albedo contrast with surrounding terrain. Bright ejecta deposits either do not form easily on these intrinsically bright objects, or any enchanced brightness is efficiently erased, or both. The theoretical arguments above do not support the notion that the continuous cjccta is significantly spread out by the low gravity of these satellites (see, e.g., Smith et al. 1981; Plescia 1983), and limited observations show the ejecta deposits are probably there.

Fields of secondary craters surrounding large craters and crater palimpsests (see Sec. II.E) on Ganymede are well documented (Passey and Shoemaker 1982; Shoemaker et al. 1982). Equivalent features are difficult to discriminate on Callisto's almost ubiquitously heavily cratered surface, but the radial alignment of some craters suggests that they are secondaries of nearby large craters. Crater chains, or *catena*, associated with the Gilgamesh and western equatorial basins on Ganymede are well developed, with individual crater diameters ranging up to 10 to 15 km (Passey and Shoemaker 1982). The Gilgamesh crater chains, in particular, resemble those of Orientale, and are probably due to the impact of multikilometer-sized, icy spall fragments. Numerous catena exist on Callisto as well, composed of craters up to 35 km in diameter, but they do not occur in association with each other, and the basin of origin is not obvious in most cases.

By the usual criteria, secondary craters have not been identified on the icy Saturnian satellites, even around large craters formed in or near plains units. A few crater chains exist on Dione, but they may be related to past endogenic activity (Moore 1984). As noted above, secondary crater fields due to spalls should expand in radial extent, measured in units of primary crater radius, with lower gravity. In fact, the gravitational fields of the Saturnian satellites are so low (< 30 cm s^{-2}) that secondaries should be distributed on a satellite-wide basis, and escape velocities are so low (< 700 m s^{-1}) that the high-velocity fraction of secondary-forming fragments should be ejected into Saturnocentric orbit, especially from the smaller satellites. These will be rapidly swept up on a generally satellite-wide basis and constitute an extended secondary tail (cf. Smith et al. 1981; see Sec. IV).

Numerous craters with bright ray and rim deposits are known on Ganymede and Callisto (e.g., Fig. 12), although the exact mechanism that accounts

Fig. 13. Voyager 1 image of heavily cratered terrain on Rhea. The identification of ejecta blankets is difficult, as in the lunar highlands, but rim units of the large fresh craters Agunua (A) and Yu-ti (Y) are topographically subdued and have lower small crater densities compared to surrounding terrain. Yu-ti ejecta appear textured at this resolution (1.3 km/pixel) and apparently has modified the interior morphology of the adjacent triangular crater. Arrows mark possible lobate edge to blanket of a 35 km diameter crater near Agunua. (Frame FDS 34953.03, centered at 62°N, 80°W; north towards lower right.)

for the relative brightness of rays on icy bodies has not been established. Bright materials on Ganymede and Callisto are also the least reddish (Johnson et al. 1983). In general, albedo and color are controlled by both grain size and composition (cf. chapter by Veverka et al.). The optical surface layers of Ganymede and Callisto are nearly pure water ice (> 90 wt. %), with minor contaminants acting as coloring agents, or *chromophores* (Clark 1980,1982; chapter by Clark et al.; Johnson et al. 1983; P. Lucey, personal communication, 1984). The reduction in grain size that accompanies ray formation may be responsible for the ray color and albedo. Provided the sizes of both contaminant and ice are large compared with visible wavelengths, the probability of an incoming photon striking reddish contaminating particles would be reduced if ice grains are preferentially comminuted. Another possible mechanism for producing bright, less reddish rays is increased wavelength-dependent scattering by very fine-grained frost.

A peculiar class of ray craters exists on Ganymede: those with dark rays. They are characterized by distinctive, very low-albedo ejecta deposits (Fig. 14a) and are the darkest and regionally reddest units on Ganymede (Conca 1981; Schenk and McKinnon 1985). The similarity between the albedos of these dark rays and that of Callisto's surface prevents easy recognition of similar features there. Poscolieri and Schultz (1980) and Poscolieri (1982) believe that dark-ray craters formed nearly exclusively in cratered terrain and within a restricted diameter range. Hartmann (1980b) used the latter fact to conclude that dark-ray ejecta came from specific stratigraphic horizons. Conca (1981) showed, however, that the findings of Poscolieri were erroneous: dark-ray craters are equally abundant on cratered and grooved terrain; they are observed at sizes down to the lower resolution limit of the Voyager cameras; and the upper size limit probably reflects their small total number (< 1% of all ray craters; Passey and Shoemaker 1982). These facts led Conca to conclude that dark rays were most likely due to projectile contamination, and that the enhanced number and lower albedos of dark-ray craters in the trailing hemisphere were due to magnetospheric sputtering. Shoemaker et al. (1982) point out that insolation-driven sublimation may also be important in the evolution of dark-ray deposits. Dark-ray craters are not well understood, but are probably important clues to the sources and history of contamination of the surfaces of Ganymede and Callisto.

Retention times for bright rays on Ganymede and Callisto may exceed 10^9 yr (Passey and Shoemaker 1982) although absolute time scales are very model-dependent (see Sec. IV). In contrast, there are only two observed ray craters on the Saturnian satellites: Cassandra on Dione (Fig. 14b) and a very bright pattern on Rhea's leading hemisphere (Smith et al. 1981). There are many possible reasons for this difference. First, ray-forming ejecta may generally move at velocities too high to be retained on the smaller satellites. Second, the optical surface layers of the Saturnian satellites are essentially pure water ice (> 99 wt. %) (Clark and Owensby 1981; Clark et al. 1984; chapter

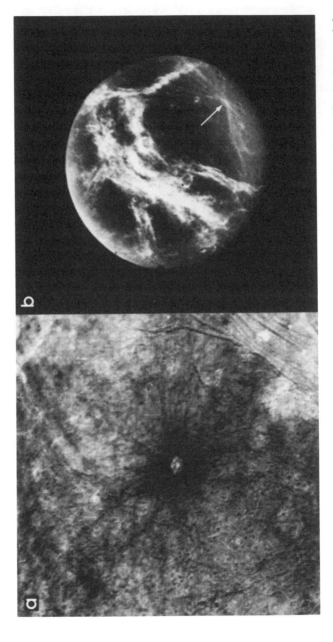

Fig. 14. Two rare types of ray craters: a dark ray crater on Ganymede and any type of ray crater on a Saturnian satellite. (a) Antum, a dark ray crater of ~ 20 km diameter on the trailing hemisphere of Ganymede (near 5°N, 220°W). Continuous dark deposits extend more than a crater diameter from the rim, and the longest rays are ~ 200 km. Portion of an older dark ray pattern is seen at top. (Voyager 2 frame FDS 20635.45; north at left.) (b) Cassandra, an oblique impact and the only bright ray crater on the trailing hemisphere of Dione (1120 km diameter). Arrow indicates probable direction of the impact trajectory. (Voyager 1 frame FDS 34933.38; north towards top.)

by Clark et al.), lacking the contaminating agents of Ganymede and Callisto (observed albedo differences on these satellites are possibly due to magnetospheric ions). Third, enhanced micrometeoroid bombardment of leading hemispheres (i.e., those centered on the apex of orbital motion; see Shoemaker and Wolfe [1982]) effectively erases albedo patterns there (Smith et al. 1981). Since the accretional efficiency of even small projectiles is strongly negative (i.e., erosion predominates), the optical surface layer is continually vaporized and/or ejected into orbit around Saturn. Ejected small particles become charged and are eventually sputtered away; the resulting ions are swept down the magnetotail and lost from the Saturnian system (McKinnon 1981c). The greater gravity of Ganymede and Callisto prevents this from happening there. The process is less relevant for the trailing hemispheres of Tethys, Dione and Rhea, where it is dominated by processes that yield the observed darkening. Of course, the optical surface layer of Enceladus may be affected by endogenic activity (see the chapter by Morrison et al.).

E. Morphology

The physics responsible for the diversity of observed crater morphologies is one of the less well-understood aspects of cratering mechanics. This section focuses on icy satellite craters; we do not attempt to summarize the voluminous, older literature concerning the Moon (the Kaula et al. chapter treats this in some detail). Indeed, to first order, the morphology of craters on icy satellites resembles that of craters on the Moon and terrestrial planets. Small craters are simple bowl shapes with raised rims; larger craters have collapsed: floors are uplifted, rim areas have slumped downward (possibly creating sets of terraces), and central peaks may have formed. Very large craters may be surrounded by one or more concentric tectonic structures (e.g., inward- and outward-facing scarps, and graben) or *rings*. The ensemble of crater and rings is termed a *multiringed basin* (or multiringed structure if a central depression does not exist). There are important morphological differences, however, between craters on icy satellites and on terrestrial planets. The general morphological sequence seen with increasing scale on the terrestrial planets, bowl-shaped craters → central peak craters → peak ring basins → multiringed basins (Howard 1974), is apparently lacking the peak ring form on the icy satellites. In its place are *central pit craters*, a type seen only in limited numbers and in preferred terrains on Mars (Wood et al. 1978; Hodges et al. 1980). The multiringed basins on icy satellites also exhibit more diverse morphologies than those observed on the terrestrial planets; these may be accounted for theoretically in terms of icy satellite rheologic structure. One of the most striking new crater forms on the icy satellites is the *palimpsest*— an albedo trace of a crater and its rim deposit that survives even after the original crater topography has been erased by processes such as viscous relaxation (Smith et al. 1979b). There are also many structures intermediate in appearance between craters and palimpsests, termed *penepalimpsests* in the comprehensive study of Passey and Shoemaker (1982). Below we briefly consider these various morphologies, especially with regard to intersatellite comparison.

Central-Peak Craters. The simple-to-complex crater transition occurs near 15 km diameter on the Moon (Pike 1977). Crater depth is proportional to diameter for simple craters; for complex craters it is nearly constant, varying from ~ 2.5 km to ~ 5.0 km as D varies from 15 km to 200 km (Pike 1974a). The inflection in depth-diameter statistics occurs near 11 km. Central peak occurrence reaches 50% near 20 km diameter (Cintala et al. 1977; Pike 1980a), and a number of other quantitative shape parameters change over the 10 to 30 km diameter range (Pike 1977,1980a).

On Ganymede, the transition to the central peak morphology is apparently reached by 5 km diameter (Greeley et al. 1982; Passey and Shoemaker 1982); discrimination of crater morphology at smaller sizes is severely limited by Voyager image resolution. Photoclinometrically determined crater depths (Passey 1982) are ~ 1 km for 10 km diameter craters. If the depth/diameter ratio is ~ 1/5 for small simple craters on Ganymede, then there would be an inflection in the depth/diameter ratio for Ganymedean craters below ~ 5 km diameter. The Callisto data have a higher resolution cut-off (at ~ 10 km diameter craters), apparently above the simple-to-complex crater transition (Passey and Shoemaker 1982).

No depth measurements of Saturnian satellite craters have yet been made, but central-peak statistics have been collected by M. Leake and one of us (WBM), reported by McKinnon (1982a). Figure 15 is a histogram of morphological classifications of craters on Mimas; the central-peak transition is somewhere in the 20 km to 35 km diameter range. A total of 1830 craters on Mimas, Enceladus, Tethys, Dione and Rhea have been classified, all in the morphologically freshest crater classes (C_1–C_3), as defined in the Lunar and Planetary Laboratory (LPL) system (cf. Wood and Andersson 1978; Leake 1981). Estimates for central-peak transitions (50% occurrence level) for these five satellites are plotted in Fig. 16, along with an estimate for Ganymede.

We interpret these data in terms of a viscoplastic (or Bingham material; see Malvern 1969, chapter 6) model of crater collapse and central-peak formation. In general, it is difficult to make a case for the driving force of collapse and central rebound to be anything other than gravity. The large compressions necessary to drive an elastic rebound are simply not available at the end of the excavation stage (McKinnon 1982b). Incidentally, the unique peak-forming flow fields seen in numerical calculations of extremely low-density projectile impacts (O'Keefe and Ahrens 1982b) [and probably associated with certain zero depth-of-burst and surface-tangent explosion trials (Jones 1977; Roddy 1977; also see Ullrich et al. 1977)] are not applicable to the general problem (McKinnon 1980). The role of gravity has been advocated for some time (see, e.g., Gilbert 1893; Shoemaker 1962; Quaide et al. 1965; Dence 1968; Gault

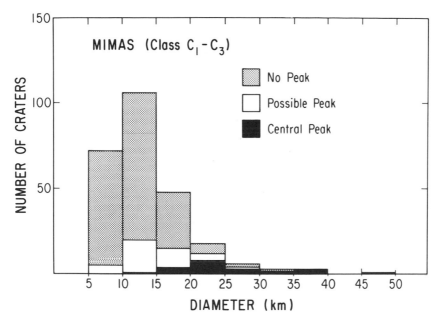

Fig. 15. Size-frequency histograms of craters on Mimas, classified according to presence of central peaks. The decrease in the number of craters < 10 km in diameter is due to the spread in resolution of the various images. Determinations for craters < 5 km in diameter are not considered reliable.

et al. 1968; Hartmann 1972*b*; Gault et al. 1975; Dence et al. 1977; Grieve et al. 1977,1981). We parameterize collapse and rebound in terms of an effective cohesion c^* and effective viscosity η^*, where these material parameters apply to the shock-weakened region surrounding the crater (see Sec. II.B). The cohesion determines when collapse can occur (Melosh 1977*b*; McKinnon 1978); once collapse occurs, the viscosity determines the degree of rebound or central-peak development (Melosh 1982*a*). Using the other relevant physical variables (ρ, g, H), two dimensionless parameters can be formed: the strength parameter, $\rho g H / c^*$, and the flow parameter, $\propto \rho g^{1/2} H^{3/2} / \eta^*$. Collapse occurs for those transient craters large enough that their strength parameters exceed a critical value (~ 7; McKinnon 1978; Melosh 1982*a*). For similar targets with similar values of c^*, the critical depth (or diameter) should scale inversely with g. This is best estimated from the inflection in the depth-diameter relation between simple and complex craters. If these data are unavailable, then it can be estimated from the central peak transition diameter. This diameter need not be inversely proportional to gravity, however; it depends on the scale dependence of η^* (Melosh 1982*a*; Melosh and Gaffney 1983). Unless η^* increases faster than H, the peak transition diameter becomes more similar to the critical collapse diameter as g decreases.

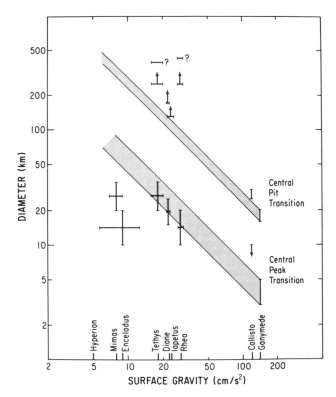

Fig. 16. Central peak and central pit transition diameters for craters on icy satellites derived from size-frequency analyses of morphology such as in Fig. 15. The central peak transition diameters (50% occurrence level) for Ganymede, Rhea, Dione, Tethys, Enceladus and Mimas are 3–5, 10–20, 15–25, 20–35, 10–20, and 20–35 km, respectively. Only an upper limit of 10 km is given for Callisto, due to limited resolution. A trend of inverse correlation with gravity is apparent, except for Mimas and Enceladus. The central pit transition diameters (50% occurrence level) for Ganymede and Callisto are 16 to 20 and 25 to 30 km, respectively (see Passey and Shoemaker 1982). Lower limits for central pit transition diameters on the Saturnian satellites are actually the largest definitely morphologically classifiable craters on each satellite, in each case a central peak crater. These lower limits lie above an inverse gravity trend extrapolated from the Ganymede transition. Also plotted are the largest craters on Tethys and Rhea; their morphological classifications are ambiguous (see text). Surface gravities are based on mass and radius data in Smith et al. (1979b,1982; cf. Table IV of the introductory chapter).

The Saturnian satellite data in Fig. 16 are scattered as a group, but lie at significantly larger diameters than the Ganymede transition. The shaded zone has a slope of − 1, indicating transition diameters inversely proportional to gravity. Overall, the correlation is reasonable, at least for Ganymede, Rhea, Dione and Tethys, consistent with theory provided the surfaces are all icy. Mimas has a distinctly lower transition diameter, but variations of a factor of two in effective cohesion are known for terrestrial planet targets (McKinnon

1980; Pike 1980a). Enceladus is even more anomalous, and it is possible that its crustal composition is different from water ice. The anomaly exists even though only the freshest crater classes (C_1 and C_2) were used in making the Enceladus estimate; this minimizes any effect of the pervasive viscous relaxation on Enceladus (Passey 1983). Effective cohesions during crater collapse for silicate targets are a few MPa (Melosh 1977b,1982a; McKinnon 1978, 1980); based on Fig. 16, the equivalent figure for water ice targets appears to be a few 0.1 MPa, and Enceladus appears to be weaker still.

From Fig. 16 it is apparent that if Hyperion were an icy satellite, then its large craters ($\sim$ 100 km diameter) should have central peaks. In Fig. 17, peaks are observable in craters of 80 and 140 km diameter. They would not occur if Hyperion were silicate, due to the greater effective strength of cratered rock debris as pointed out above. On satellites smaller than Hyperion, the simple crater form should persist up to sizes approaching that which would cause catastrophic fragmentation. For example, the great depths of the two largest craters on Amalthea (Thomas and Veverka 1982) suggest that they are simple.

Finally, we note that collapsed terrace walls are associated with craters having central peaks on Ganymede (Passey and Shoemaker 1982) and on the Saturnian satellites (see, e.g., Plescia 1983). Voyager resolution is generally insufficient to discriminate individual terraces, although Rhea images are an exception (e.g., the crater Yu-ti in Fig. 13). Each terrace is a slump block composed of a section of the original transient rim (Mackin 1969; Malin and Dzurisin 1978; Settle and Head 1979). While the transient rim must subside in some manner as the floor rebounds, the organized slumps and terraces indicate that the rim material is acting cohesively. For well-studied lunar terraces, derived cohesion estimates from terrace widths are in good agreement with estimates from the critical strength parameter (Melosh 1977b; McKinnon 1980).

Central-Pit Craters. The central-peak craters just discussed above are observed on Ganymede and Callisto only within a restricted diameter range. Instead, the ubiquitous central form at large sizes is a pit or depression, which is often rimmed. On Ganymede and Callisto, the smallest such central-pit craters are 16 and 18 km in diameter, respectively, and pits are dominant in craters greater than 20 and 30 km, respectively (Passey and Shoemaker 1982). In contrast, no confirmed central-pit craters are seen on the Saturnian satellites other than chance superpositions. In Fig. 16, the central-pit transition diameters for Ganymede and Callisto are plotted along with a g^{-1} extrapolation from the Ganymede estimate. Also given are lower limits for the central-pit transition diameter on Tethys, Dione, Iapetus and Rhea derived from the largest central-peak crater observed. Two possible central-pit craters are Odysseus (390 km diameter) on Tethys, whose irregular central-peak complex may indicate an incomplete transition to the pit form, and an unnamed large

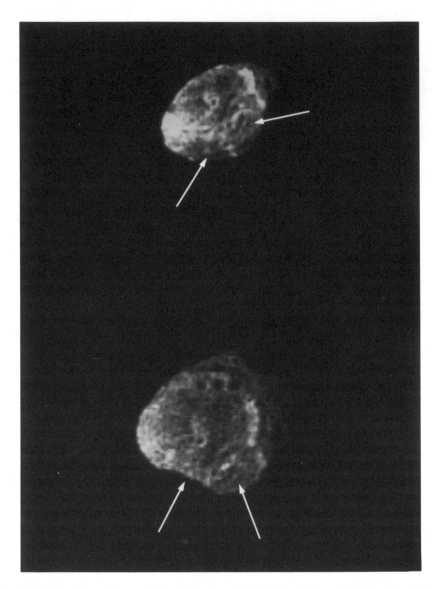

Fig. 17. Two views of the small, irregularly shaped, chaotically rotating satellite Hyperion (350 × 240 × 200 km; Thomas et al. 1983a). Two possible large craters (~ 140 and 80 km in diameter) with central peaks are identified, consistent with an ice-dominated bulk composition. The larger of the two craters is seen in profile in the bottom view. (Mosaic of Voyager 2 frames FDS 43959.06 and 43968.02.)

crater on Rhea ($\sim$ 435 km diameter, centered at 45°N, 140°W), which, when viewed at the terminator (Voyager 1 frame FDS 34869.55), exhibits a partial inner ring or pit rim. These two craters are plotted with question marks.

The lack of obvious central-pit craters on the Saturnian satellites suggests that the central-pit transition is a stronger function of gravity than g^{-1} (Fig. 16). Passey and Shoemaker (1982) argue that central pits form from the collapse of central peaks, and that peak material is fluidized by liquid water produced by shock and shear melting during transient crater excavation and collapse. In the context of the model of Melosh (1982a), the flow parameter governs whether the central peak is itself transient and whether a pit or inner ring can form. Perhaps η^* increases sufficiently rapidly with scale so that the central peak-to-pit transition is suppressed on the small icy satellites. However, it is not clear that this will happen if the fluidization mechanism is acoustic noise, as Melosh hypothesizes (Sec. II.C), nor is it obvious how acoustic fluidization could simultaneously account for the apparent fact (Fig. 16) that central-peak transitions are a weaker function of gravity than g^{-1}. On the other hand, if fluidization is accomplished by liquid water, the colder crustal temperatures of the Saturnian satellites and especially the longer peak rebound times due to lower gravity ($\tau_{\mathrm{rebound}} \sim 7\sqrt{H/g} \propto g^{-1}$ at the same gravity-scaled size) may cause slowly rising central peaks on the Saturnian satellites to freeze (effectively raising η^*) before they can collapse and form a pit (McKinnon 1982a). We expect that lithospheric temperatures influence pit formation. The stratigraphically uplifted peak is derived from warmer material, and hence may be more fluid. This is supported by the general correlation of increasing pit diameter (relative to crater diameter) with greater crater age and thus steeper lithospheric thermal profiles and higher internal temperatures.

Another speculative model of pit formation is that of Croft (1983), who proposes that initial strong zonation in shock-induced material properties (produced during coupling), due to strong thermal gradients in the target, manifests itself as a central molten zone that survives the growth of the crater and is incorporated into a central peak during the collapse phase; then the central pit forms from the collapse of this essentially liquid water peak. In this model, the lack of central-pit craters on the Saturnian satellites is due to their gentle lithospheric temperature gradients. A somewhat analogous mechanism of pit formation has been proposed by Greeley et al. (1982).

Palimpsests and Penepalimpsests.   The definition of palimpsest has led to disagreement among various workers (cf. Woronow and Strom 1981; Passey 1982). We prefer to restrict the definition to a vestigial albedo feature within which the topographic crater rim can no longer be identified. Such a definition should not be resolution-dependent, so palimpsest identifications from medium-resolution images should be regarded as provisional. All identified palimpsests on Ganymede lie wholly or partially within the dark, heavily cratered terrain; their albedos are similar to the relatively bright grooved and smooth terrains, and are thus not difficult to discriminate. Albedo contrasts on Callisto are more subtle (except for the center of the Valhalla multiringed structure), but numerous examples may be found there as well (Passey 1982). Palimpsests in grooved terrain on Ganymede would be difficult to identify on the basis of albedo alone, and in practice all large degraded crater forms in the grooved terrain have been classified as penepalimpsests.

Little, if any, remnant crater topography is apparent at Voyager resolutions within palimpsests on Ganymede. Fields of secondary craters may still surround a palimpsest, however, and this is one argument Passey and Shoemaker (1982) use to conclude that the original crater lies well within the albedo boundary of the palimpsest; in this interpretation, the boundary corresponds to the limit of relatively bright continuous ejecta. We note that this boundary is also remarkably circular in many instances. Hartmann (1984) questions this interpretation, preferring to equate the albedo boundary with the original crater rim. He cites the apparent presence of a (at least) partly encircling ridge at the outer albedo boundary, in some cases. Passey and Shoemaker (1982) and Shoemaker et al. (1982) point out, however, that where the resolution is sufficient, a remnant crater rim may be identified. Such very subdued craters are termed penepalimpsests. In these cases, the crater lies well within the palimpsestlike albedo boundary, as it should if penepalimpsests form a transitional morphologic sequence between craters and palimpsests.

The form of palimpsests and penepalimpsests probably derives from a combination of prompt collapse and later, slower viscous relaxation. The evidence for viscous relaxation in the icy crusts of Ganymede and Callisto is relatively straightforward. Topographic amplitudes are subdued planet-wide. Also long-wavelength topographic components are preferentially lost, leading, for example, to the observed bowed-up crater floors (Passey 1982; Shoemaker et al. 1982) and raised graben rims (see McKinnon and Melosh 1980; Shoemaker et al. 1982). There are some recognizable topographic features within some palimpsests (e.g., smooth central depressions and semiconcentric ridges or lineaments). It is difficult to find analogs for some of these features in fresh craters, although comparisons with central-pit craters (Croft 1983) and ringed basins (McKinnon 1981a; Passey and Shoemaker 1982) have been made. (Five very large, highly relaxed craters have especially unusual morphology; they are termed type II penepalimpsests by Passey and Shoemaker [1982].) Possibly, ancient craters, now recognized only as palimpsests, formed with different morphologies from those of more recent craters. Casacchia and Strom (1984) go so far as to suggest that palimpsest topography is primarily a manifestation of post-collapse morphology, which they ascribe to differences in the thermal states of the crusts and upper mantles of Ganymede and Callisto at the time of crater formation. We note that steeper temperature gradients and higher temperatures overall also enhance viscous relaxation effects, so it is difficult to argue for one process to the exclusion of the other.

Multiringed Basins and Structures.   Various types of multiringed structures on Ganymede and Callisto were initially recognized by the Voyager Imaging Team (Smith et al. 1979*a,b*). They ranged from the Valhalla and Asgard systems on Callisto, composed of numerous roughly concentric rings extending several basin radii away from the basin center, to the younger Gilgamesh on Ganymede, whose morphology is more similar to a terrestrial-planet multiringed basin. The arcuate-rimmed furrow systems that stretch across major units of ancient, heavily cratered terrain (Galileo, Marius and Nicholson Regio) were likened to sections of the Valhalla system, with most of the Ganymede ring systems and presumed craters of origin no longer visible, having been replaced by grooved terrain. An important qualitative interpretation was that the different ring forms were due to different planetary thermal (and hence rheologic) structures.

These interpretations were buttressed by the extension of the analytic model of ringed-basin formation (Melosh and McKinnon 1978) from thick to thin lithospheres (McKinnon and Melosh 1980; Melosh 1982*b*). In this model, ring formation is due to collapse of the transient basin cavity when the transient depth and lithospheric thickness are comparable. The lithosphere must be defined appropriately with respect to the time scale and stress level of basin collapse (see McKinnon 1981*a*), so that large craters on the Moon as well as on large, icy satellites can satisfy the mechanical requirements for ring formation. The thickness and strength of the lithosphere and the viscosity of the underlying asthenosphere determine whether the rings are catastrophic and tsunamilike (thin lithospheres and truly fluid asthenospheres), multiply concentric of the Valhalla type (thin, weak lithospheres and generally subsolidus asthenospheres), concentric of the Cordilleran type (thick, stronger lithospheres), or absent entirely (very thick lithospheres). The type of ring system observed today depends on the relative timing of those impacts that are preserved with respect to the thermal evolution of the target body.

The model of ringed-basin mechanics (ring tectonics) quantifies the relationship between ring structure and planetary or satellite thermal evolution. Estimates of lithospheric thickness and hence heat flow can be extracted from observations. None of the ring systems observed on the Moon or the icy satellites are of the tsunami type, so it is very unlikely that their mantles were predominantly liquid while their cratering records were being established. Valhalla-type patterns are seen around Valhalla ($\sim$ 4000 km diameter), Asgard ($\sim$ 1600 km diameter), near Asgard ($\sim$ 500 km diameter), and near Adlinda ($\sim$ 800 km diameter) on Callisto, and around near Osiris ($\sim$ 500 km diameter) on Ganymede (see Passey and Shoemaker [1982] for more detailed morphology). Along with the rimmed furrow patterns in the dark terrain, these offer dramatic evidence for the thinness of the lithospheres of Ganymede and Callisto $\sim$ 4 Gyr ago, and demonstrate that the tectonic effects of a single impact may be felt over an entire hemisphere. A comparison of Galileo Regio and portions of Valhalla shows that although Valhalla is older, the lithosphere of

Callisto was 1.5 to 2.0 times as thick at the time of formation. In addition, there are many smaller less well-studied ringed basins on Ganymede and Callisto (McKinnon 1981*a*; Passey and Shoemaker 1982).

In contrast with the above ring systems, the youngest basin on Ganymede, Gilgamesh, formed when the lithosphere had cooled and thickened sufficiently to yield only a few scarplike rings. A substantial lithospheric thickening is, therefore, inferred for Ganymede by the time Gilgamesh formed. Head and Solomon (1980) have shown that on the Moon, lithospheric thickness variations implied at the time of ring formation (all of the Cordilleran type) can be broadly correlated with elastic lithospheric thicknesses determined from the tectonic response to later mascon loading. The present lithospheric thicknesses of the Moon, Ganymede and Callisto are probably all too great to allow the formation of new rings, even if a basin-forming impact occurred. (Note that this does not preclude present-day formation of peak rings, which are a different type [see Melosh 1982*a*].)

Ring formation on the small Saturnian satellites is not impossible, but the rheological requirements are so extreme that if the satellites ever were warm enough, it is unlikely that such conditions could have lasted for long. The absence of clearly identifiable ring systems was successfully predicted (McKinnon 1981*a*), although Moore et al. (1985) believe that they have located extremely degraded examples of rings on Rhea.

Circumferential zones of graben are expected in Valhalla-type ring systems. Melosh (1982*b*) showed that between the graben zone and the remote, unfractured lithosphere (where stresses are elastic) there can occur a zone of outward-facing normal faults. In this zone, the lithosphere is in radial extension and the asthenosphere is stagnant or flowing weakly outward, leading to traction on lithospheric blocks opposite to that close to the crater. Smith et al. (1979*b*) originally identified the outer rings of Valhalla as ridges, but careful mapping by Remsberg (1981,1982) showed that they are actually outward-facing scarps of substantial relief ($\sim$ 1 to 2 km). (The interpretation was confused by the albedo effects of accumulated ice on northward facing slopes [Spencer and Maloney 1984].) Further discussion of multiringed basins on Ganymede and Callisto can be found in the chapter by McKinnon and Parmentier. While not all workers accept the basin model which we have discussed, no alternative theory as yet explains the diversity of structures seen.

F. Collisions, Breakup and Reaccretion, and Satellite Spins

So far in this chapter we have considered cratering impacts in which one body (the projectile) strikes another (the target) that is so large as to be considered a semiinfinite surface. Since planetary satellites are relatively small and some interplanetary projectiles can be quite large, there is another collisional regime to be considered for the largest impacts. If the scale of the transient crater, calculated according to the relations described earlier, approaches the scale of the target body, then the physics changes. Fracture planes may penetrate through the body and tensile waves reflected from the entire surface of the target may spall large parts of the target body away, especially near the collision's antapex. If the target body is spherically layered with sufficient strength contrast, these spherical layers may be stripped away. If the impact is large enough in scale, the target body may be *catastrophically fragmented* throughout. A fraction of the projectile's kinetic energy will be partitioned into kinetic energy of the resulting fragments, which will therefore be launched away from the target's original center of mass. For a very large collision, fragments representing more than half of the target mass may be launched at greater than escape velocity, in which case the body is said to have been *catastrophically disrupted*. (If $>> 50\%$ of the target mass is expelled, the collision is termed *supercatastrophic*.) Fragments traveling at less than escape velocity will reaccrete on a time scale of minutes to hours. Even if the body is entirely disrupted, if it is in orbit about a massive planet, the fragments will be tightly constrained in short-period orbits around the planet and will be subject to fairly rapid reaccretion. If a collision is off-center, the target body's spin may be greatly affected.

The subject of satellite fragmentation, disruption and reaccretion is now of great interest. Although Fujiwara et al. (1977) some years ago considered the implications of their collisional experiments for the possible breakup of the Martian satellites, it was the Voyager pictures of Saturn's satellites (including the retinue of small satellites and rings) that inspired serious consideration of the possible dramatic importance of satellite breakup and reaccretion in planetary orbit. While the thought seemed novel at first to some researchers, the groundwork for understanding these processes had already been established. A considerable body of literature exists concerning accretion of small bodies in circular orbits about a primary. Although the usual application is to the accretion of planetesimals into planets, or to the accretion of the Moon in Earth orbit, Soter (1971) long ago treated the reaccretion of ejecta from Phobos and Deimos onto those satellites.

The analogy is often made about such reaccretion being a kind of extended secondary cratering. While this is true for escaped and reaccreted crater ejecta, there are important differences in treating a disrupted body. Numerous experiments have been conducted by Fujiwara and his colleagues and by Hartmann (1980a) concerning the fragmentation of finite spheres and quasi-spheres. The size distribution of the fragments is not like that for primary crater ejecta for cases near the transition between the giant-crater case and the supercatastrophic-fragmentation case. The whole subject is of particular interest for understanding the origin of the asteroid size distribution, the origin of asteroid Hirayama families, and the origin of asteroid spin distributions. See, for example, the review by Davis et al. (1979). Recent publications on this topic by the three main research groups studying asteroid fragmentation are Davis et al. (1985), Zappalà et al. (1984), and Fujiwara (1982).

Catastrophic disruption depends on low-pressure material properties (such as tensile strength), which dominate for small satellites. The gravitational self-compression of larger bodies works against the propagation of fractures, so their effective tensile strength is increased. If an initially strong body is fragmented but reaccretes, it forms into a loosely cohesive rubble pile, which may respond to subsequent impacts very differently. Although the concept of a gravitationally bound rubble pile is widely discussed in an asteroidal context, some researchers believe that small bodies are usually strong and cohesive. Veverka and his coworkers have reported morphological features on Phobos, Amalthea and other satellites which suggest to them that the bodies are physically cohesive and strong (cf. Thomas and Veverka 1982). Others believe that virtually any morphological feature can be maintained on the surface of a rubble pile, just as angles-of-repose maintain topography in a sand box. An important question for further research is the degree of cohesiveness of satellites and asteroids of various sizes. If it is true that some satellites have been fragmented, disrupted and reaccreted in orbit, then their preexisting crater distributions will have been destroyed. Upon reaccretion, their surfaces may then display craters formed by the very low-velocity impacts of their own debris. Dynamical evolution studies (see, e.g., Farinella et al. 1986) seem to show that debris from one satellite cannot migrate with much efficiency to another satellite in a different orbit (for example, from Mimas to Enceladus). Even low-efficiency migration could be important, however (see Sec. IV).

Spin State.   A planetary satellite generally rotates about its axis of maximum moment of inertia (c-axis). Due to dissipation within the satellite, this axis is usually perpendicular to the orbit plane and the rotation rate is synchronous with the mean motion of the satellite about the primary. The axis of minimum moment of inertia (a-axis) is oriented toward the primary (see Peale [1977] for a full discussion). The dynamical and real figures of the satellite, dominated by its spin and tidal elongation in the gravity field of the primary, are nearly triaxial (with semimajor axes a, b and c). The satellite does not come into synchronous lock at any arbitrary position, but selects the lowest energy state in the primary's gravity field on the basis of its internal density configuration, which for any real body is nonuniform.

Any impact that adds spin angular momentum to the satellite induces a wobble or libration about this equilibrium condition, but this wobble is quickly damped. Sufficiently energetic impacts may induce, say, librational excursions about the c-axis which exceed 90°. In this case, the synchronous rotational lock is broken. Torques on the satellite due to the primary will speed up or slow down the satellite's spin as necessary in order to reestablish synchronism. Nevertheless since the dynamical figure is essentially symmetric with respect to 180° rotations about the c-axis, the hemispheres containing the apex and antapex of orbital motion may interchange. This may be one reason why the marked differences in crater densities on these hemispheres predicted to result from external bombardment (see, e.g., Shoemaker and Wolfe 1981,1982) are not observed. The stochastic exchange of apex and antapex hemispheres during establishment of the cratering record would tend to equilibrate crater densities (McKinnon 1981d; Plescia and Boyce 1982).

Melosh (1975) first pointed out the possibility of impact-induced exchange of the apex and antapex hemispheres of the Moon. (Very energetic impacts could even cause exchange of the northern and southern hemispheres of a satellite.) Examining solely the case of reorientations about the c-axis, the polar spin angular momentum added by an impact ΔL is related to the change in rotational kinetic energy ΔE by

$$\Delta L = (2I\Delta E)^{1/2} \tag{29}$$

where I is the polar moment of inertia (assumed to be that of a uniform sphere). (WBM wishes to thank J. Lissauer for pointing out a previous error in this stage of the analysis.)

The energy difference necessary for unlocking, corresponding to the potential energy difference of a 90° rotation about the c-axis, is

$$\Delta E = \frac{6Gm_s m_p}{a_s}\left(\frac{R_s}{a_s}\right)^2 C_{22} \tag{30}$$

where m_p is the primary's mass, and m_s, R_s, a_s and C_{22} are, respectively, the satellite's mass, radius, semimajor axis, and relevant gravitational potential coefficient (specifically, that of the unnormalized second-degree sectorial harmonic). Tides raised and dissipated during asynchronous rotation have been neglected. C_{22} is given by the sum of two terms:

$$C_{22} = \frac{3}{8}\left(\frac{R_s}{a_s}\right)^3 \frac{m_p}{m_s} + \lambda\mu C_{22}^{\mathbb{C}} \tag{31}$$

where λ and μ are coefficients explained below and $C_{22}^{\mathbb{C}}$ is the lunar value. The first term is due to the equilibrium hydrostatic tidal bulge, with the simplifying assumption that the satellite is of uniform density throughout. The second is due to intrinsic density anomalies and is scaled to the Moon.

$$\lambda = (\sigma_{max}^s/\sigma_{max}^{\mathbb{C}})\,(g_{\mathbb{C}}/g_s)^2 \tag{32}$$

is derived from the hypothesis that the anomaly magnitude is determined by long-term support of stress differences (Kaula 1968; Phillips and Lambeck 1980); $g_{\mathbb{C}}$ and g_s are the lunar and satellite surface gravities, and $\sigma_{max}^{\mathbb{C}}$ and σ_{max}^s are the maximum stress differences supportable on each body. For icy satellites, $(\sigma_{max}^s/\sigma_{max}^{\mathbb{C}}) \sim 10^{-1}$ is reasonable. We assume for simplicity that the density anomalies arise at similarly scaled radii in each satellite. Finally, μ

TABLE III
Gravitational Potential Coefficients, Potential Energy Barriers to
Break Synchronous Tidal Lock, and Crater Diameters Produced
by the Requisite Impacts

Satellite	C_{22} Hydrostatic	C_{22} Intrinsic	E^a (J)	$\langle E \rangle^b$ (J)	Crater[c] Diameter (km)
Mimas	5.5×10^{-3}	4.9×10^{-4}	3.8×10^{20}	3.4×10^{24}	230–810
Enceladus	2.9×10^{-3}	3.8×10^{-4}	2.9×10^{20}	2.7×10^{24}	210–750
Tethys	1.6×10^{-3}	9.6×10^{-5}	3.3×10^{21}	2.6×10^{25}	360–1200
Dione	6.6×10^{-4}	6.2×10^{-5}	1.1×10^{21}	1.6×10^{25}	310–990
Rhea	2.6×10^{-4}	3.8×10^{-5}	7.0×10^{20}	2.7×10^{25}	350–1100
Iapetus	9.8×10^{-7}	5.6×10^{-5}	2.9×10^{17}	3.6×10^{23}	120–370

[a] From Eq. (30) in text.
[b] Kinetic energy of projectile in representative unlocking case; see text.
[c] Range is based on limiting scaling laws with $\delta = 2.5$ g cm^{-3}; values near lower estimate are preferred.

is a factor ($\simeq 0.52$) that accounts for a portion of C_{22}^{C} being due to a fossil tidal bulge, and not to density anomalies (Lambeck and Pullan 1979).

Both contributions to C_{22} and the corresponding energies for apex-antapex exchange are calculated from Eqs. (30–32) for the classical Saturnian satellites (except Titan and Hyperion) and are given in Table III. For all but Iapetus, the tidal potential dominates the intrinsic potential, generally by an order of magnitude.

Let us consider a representative impact at $\pm 30°$ lat, with a $\theta = 45°$ trajectory striking N 45° E or N 45° W. Using Eqs. (29) and (30) to fix the amount of spin momentum about the c-axis contributed by the projectile, the mass of the projectile is calculated using velocities tabulated by (or interpolated from) Smith et al. (1981). A critical assumption here is the conservation of angular momentum before and after the impact. This requires that ejecta escaping the satellite leave isotropically in the satellite's center-of-mass rest frame (cf. Dobrovolskis and Burns 1984). However, this is almost certainly not true; a qualitative inference from the experiments of Gault and Wedekind (1978) that momentum transfer is incomplete (i.e., high-speed, escaping ejecta tend to be concentrated downrange) would increase the diameter estimates below.

Given mass and velocity estimates, the size of the resulting crater is calculated according to two limiting scaling laws: water-like scaling and sand-like scaling (Table I); a specific projectile density is also assumed, but crater size is only weakly dependent on this choice. The model crater geometry is a paraboloid of revolution with a depth/diameter ratio of 1/5. The crater diameters, for the two scaling laws, of sufficient magnitude to unlock each satellite are given in Table III. It is clear that only very large cratering events (> 200 km in diameter) can result in unlocking, except in the case of Iapetus (cf. Lissauer 1985). Even if the most realistic scaling is closer to the sandlike law (see Fig. 4), the diameter estimates in Table III do not account for widening by slumping, so only the very largest craters (e.g., Odysseus) could have unlocked and reoriented these satellites.

The kinetic energies of the representative projectiles are also given in Table III; note that the transfer of kinetic energy to a much more massive body is very inefficient. The gravitational binding energy of a satellite is $3/5$ G m_s^2/R_s; values for Mimas and Tethys, for example, are 4.2×10^{23} J and 4.3×10^{25} J, respectively. According to our limited understanding of energy partition in oblique impacts (Sec. II.B), the energies required to unlock Mimas and Enceladus exceed their gravitational binding energies. In general, major satellite-wide structural damage is expected to accompany unlocking.

The special case of Iapetus deserves comment. It is easily reorientable by crater formation in the > 100 to 200 km diameter range, and this must be factored into any discussion of the origin of its leading-trailing hemisphere albedo asymmetry. In addition, the predicted value of C_{22} due to density anomalies, $\sim 6 \times 10^{-5}$, exceeds that estimated by Peale (1977) for a lunar-like Iapetus, 1×10^{-5}. Peale has pointed out that in this case Iapetus should be in "Cassini state 1" with its spin axis nearly normal to its orbital plane, as opposed to an icy Iapetus in "Cassini state 2" with an obliquity exceeding $8°$. Iapetus has recently been shown to be in Cassini state 1 (Davies and Katayama 1984), which is fully consistent with an ice-dominated bulk composition; problems however exist because Iapetus' orbit is inclined by $8°$ to its Laplace plane (see Mignard et al. 1986).

G. Intersatellite Diameter Scaling

As a bridge to the section on crater statistics, we now develop a formalism for taking scaling into account when comparing crater populations on different satellites. For simplicity, we initially assume that all projectiles have the same composition. We attempt to relate the size of the crater produced to target properties and to the projectile velocity distribution at each target. From Fig. 7 it is apparent that most observed craters (> 1 km in diameter) formed in the gravity regime. Thus for simple and transient craters, which we assume to be geometrically similar, crater diameter should scale (from Eq. 4) as

$$D \propto \rho^{-1/3} \, g^{-\alpha/3} \, u^{2\alpha/3} \qquad (33)$$

for fixed projectile mass. Thus, by applying the proper normalization, one can directly compare the mass distributions of projectiles striking the satellites. This is strictly true only if all projectiles have the same density, but is

TABLE IV

Crater Diameters on Different Satellites for Physically Similar Projectiles

Satellite	External Projectiles				Internal Projectiles[a]	
	Impact Velocity[b] (km s⁻¹)	Transient Diameter[c] (km)	Simple-to-Complex Transition Diameter[d] (km)	Collapsed Diameter[e] (km)	Transient Diameter[c] (km)	Collapsed Diameter[e] (km)
Moon	20	1	~15	1	1	1
Phobos	13	3.2	unobserved			
Ganymede	17.4	1.4	~4	3.2	0.8	1.5
Callisto	14.5	1.4	(f)		0.9	
Rhea	15.2	1.9	~15	2.6	1.2	1.3
Dione	17.1	2.1	~20	2.6	1.2	1.2
Tethys	18.3	2.2	~27.5	2.5	1.3	1.1
Mimas	21.9	2.9	~27.5	3.6	1.5	1.4

[a] A modal velocity of 5 km s⁻¹ is assumed for Jovian and Saturnian satellites; comparison is to 20 km s⁻¹ lunar impacts.
[b] Weighted mean values; velocities for Jovian and Saturnian satellites based on Shoemaker and Wolfe (1981,1982).
[c] Transient craters are assumed to be geometrically similar.
[d] From Fig. 16.
[e] Crater volume is modeled as a right cylinder, with depth ratios taken from simple-to-complex transition diameters.
[f] Below resolution limit.

still approximately true if the average density of projectiles striking each satellite is different.

Table IV gives the diameters of (transient) craters formed on the icy satellites compared with the Moon ($D \equiv 1$) for projectile populations external and internal to the satellite systems; a comparison with Phobos is also included, although it is unlikely that any craters gravity-scale there. Icy satellite gravities are based on Smith et al. (1979b,1982); $g = 162$ and 0.6 cm s^{-2} on the Moon and Phobos, respectively. Heliocentric (or external) projectiles striking the icy satellites are predominantly comets (Shoemaker and Wolfe 1981,1982), while lunar craters are created by some combination of "asteroids" and "comets" (see Sec. IV). Assumed densities for the Moon, Phobos and the icy satellites are 3, 2.2, and 1 g cm^{-3}, respectively. The modal velocity for icy satellite planetocentric (or *internal*) projectiles is plausibly set at ~ 5 km s^{-1} (S. J. Weidenschilling, personal communication, 1984).

In general, external projectiles striking the icy satellites create larger simple or transient craters, due to lower satellite gravity and surface density. For internal impactors, crater sizes are more or less similar to those produced on the Moon by external sources. On the other hand, to compare planetocentric internal projectile cratering on icy satellites with that produced on the Moon by secondary impacts (e.g., $u \sim 1$ km s^{-1}), icy satellite crater diameters in Table IV should be increased by a factor of $(20)^{0.43} = 3.65$.

Complex craters are not geometrically similar, however, so we model collapsed craters as right cylinders with a limiting depth set at the simple-to-complex crater transition. This model is obviously oversimplified. It does not account for rim topography, possible subsidence that does not widen the crater, bulking, or the fact that complex crater depths slowly increase with diameter. As such, it severely overestimates the amount of collapse, but it does serve as an upper limit for the collapsed diameter. In this case, and for fixed projectile mass

$$D \propto H_c^{-1/2} \rho^{-1/2} g^{-\alpha/2} u^{\alpha} \tag{34}$$

where H_c is the limiting depth. For comparisons, however, only ratios of limiting depths are necessary; these are approximated by the ratios of the central peak transition diameters (Fig. 16).

Applying Eq. 34, external projectiles should create even larger collapsed craters on icy satellites compared with the Moon than they do transient craters. While the values in Table IV may be extreme, the same projectile should create a crater on Ganymede or Mimas, for example, that is at least twice as large as its lunar counterpart. Moreover, the higher transition diameters on the Saturnian satellites compared with Ganymede mitigate the effect of lower gravities on the smaller satellites; Saturnian and Jovian craters due to external projectiles scale more similarly than simple gravity scaling (Eq. 33) would imply.

We point out that each column in Table IV should be used independently; columns cannot be cross-compared because each is normalized to a lunar crater of a different geometric form. Also, each column implies a domain of applicability. Small craters on any two bodies should be compared using transient scaling, while large ones require collapsed scaling. At intermediate sizes, however, simple craters on one object may scale into the size range of complex craters on the other. In this range, the scaled diameter changes continuously from the transient to collapsed value. This table will be utilized in greater detail in Secs. III and IV.

We remark that other attempts to scale craters on the icy satellites have used modified Gault or Shoemaker energy-diameter scaling (see, e.g., Shoemaker and Wolfe 1982; Horedt and Neukum 1984a,b). A general problem with these approaches is the lack of explicit momentum dependence, which leads to an underestimate of the size of craters produced by low-velocity (i.e., internal) projectiles compared with high-velocity projectiles. More severe is the lack of accounting for the degree of collapse of complex craters on different objects. Shoemaker and Wolfe (1982) apply a constant enlargement factor to both lunar and Ganymedean craters, and Horedt and Neukum (1984b) ignore this issue altogether. In defense of these workers, however, we point out that the main thrust of their work was orbital dynamics, not crater scaling. In contrast, we have simplified consideration of orbital mechanics by using mean values from such calculations.

III. SATELLITE CRATER STATISTICS

The size-frequency distribution of craters on the surface of a planet or satellite is an observable characteristic of the body that has great potential for elucidating the geologic history of the body and the evolution of small-body populations in space in the vicinity of the body. Voyager made a major advance in this area by imaging crater populations on most of the larger satellites of Jupiter and Saturn. In combination with Viking data for Phobos and Deimos, the outer-planet satellite crater statistics greatly augment the preexisting crater data bases concerning Mercury, the Moon, Earth and Mars. Provided that the cratering record on a body has not been largely altered by the various exogenic and endogenic processes that degrade and erase craters, then it is possible to use our understanding of the probable original, pristine craterforms discussed in Sec. II of this chapter to enable us to deduce the production function, the size distribution of craters originally formed on the body's surface (without regard to crater erasure processes). The production function reflects the population characteristics of the asteroids, comets or other projectiles responsible for the cratering. Deviations from the production function in an observable crater record provide information about various physical and geologic processes that have affected the body's surface topography. The purpose of this section is to assess the observed crater distributions on the terrestrial planets and planetary satellites in order to establish constraints on the impacting populations and on the geologic histories of the satellites.

It is necessary to introduce terms related to the graphical representation of crater frequency data. The standard "differential size-frequency distribution plot" of incremental crater counts is presented in log-log format, with log diameter (log D) increasing to the right and log incremental frequency (log N) increasing upwards. The plotted frequency is the number of craters counted per km^2 divided by the width of the increment in diameter (in km). For typical crater populations, the counts lie approximately along a line, extending from the lower right to the upper left, of slope b. This slope is the exponent of a power law: $dN/dD = a D^b$. Although b is typically about -3, neither this value nor the power law approximation have a fundamental physical significance. Nevertheless, a slope of -3 has the characteristic that craters in equal logarithmic intervals occupy the same fraction of the surface area. Therefore, a more compact and useful graphical display—termed the relative size-frequency distribution plot (or R-plot)—has been adopted (Crater Analysis Techniques Working Group 1979) in which the log of the counts in increments D_{min} to D_{max} are plotted against log $(D_{min} D_{max})^{1/2}$ *relative to a* D^{-3} *power law* [the counts are divided by $(D_{min} D_{max})^{-3/2}$]. Thus, a -3 slope plots horizontally, and deviations above or below a horizontal line reflect greater or lesser crater densities. We employ the R-plot throughout, but we will refer to b (not $b + 3$) as the power law slope.

Figure 18 shows the crater size-frequency relations for various planets and satellites (or major units of those bodies). This is a schematic representation of available data; later we provide somewhat more specific discussion of the data sets and their uncertainties. It can be seen that some of the crater distributions are qualitatively similar to each other but that there are many wide differences from body to body and between different units on the same body. In the case of the Moon, the gross difference between the heavily cratered highlands and the sparsely cratered lunar maria populations reflects the destruction of preexisting craters in the maria by the basin-forming events and by subsequent volcanism. In a sense, the slate has been wiped clean and the mare surfaces (see "Lunar Maria" in Fig. 18) record the production function integrated over post-mare epochs. At diameters smaller than a few kilometers (and, especially, smaller than a few hundred meters) the interpretation of the post-mare crater population is rendered more difficult and controversial due to crater superposition (saturation); also there are competing hypotheses about the origin of the craters (whether they are primaries or secondaries, for example). But at larger sizes, it is widely agreed that the post-mare craters reflect the mix of asteroids and comets that have impacted the Moon during the past 2.5 to 3.9 Gyr (the ages of the maria). A critical question in planetary

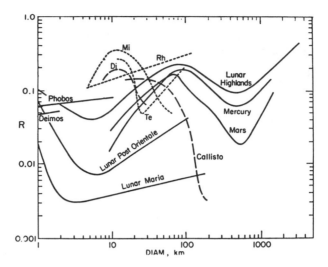

Fig. 18. Characteristic *R*-plot diameter-frequency relations for various satellites and terrestrial planets. These are approximate average relations, generally for heavily cratered terrains (see text). Mi, Rh, Di and Te identify Mimas, Rhea, Dione and Tethys, respectively.

science has been whether this production function has varied with time or with location in the solar system.

The crater population on the lunar highlands must necessarily include its share of craters of post-mare age, but it also includes many times that number of pre-mare craters. So many craters have formed in the lunar highlands that many investigators believe that the traits of the observable crater population essentially reflect the results of crater-upon-crater saturation, which would render interpretation in terms of a production function more difficult or impossible. Others believe the observed highland craters are not actually saturated and that the production function is similar to the observed population, or at least that it can be discerned by making corrections for the observed crater overlap.

The early Mariner missions to Mars and Mercury revealed crater populations on those planets that bear some similarities but also have important differences compared with the lunar crater populations. The crater numbers and morphologies for the more heavily cratered Martian terrains have generally been interpreted as reflecting severe alteration by a variety of endogenic processes (e.g., filling by dust, volcanism, water erosion, etc.; see references in Chapman and Jones 1977). Woronow, Gurnis (1981), and others (cf. Woronow 1977a) have argued, however, that some Martian highlands crater populations are very similar to crater populations on the lunar highlands and that both essentially reflect the same production function. The Mercurian crater population has generally been thought to show small but real differences from

the crater population on the lunar highlands; such differences have been ascribed to differences in the cratering process due, for example, to different
diameter scaling and to the effects of the process(es) that formed the intercrater plains on Mercury (cf. Strom 1979). The predominant view has been
that the highlands crater populations are in some sense saturated on all three
bodies and differ due to different endogenic processes; however, Strom, Woronow, Gurnis and their associates (cf. Strom et al. 1981) are impressed by the
similarity of the curves and apply theoretical arguments (discussed below) to
argue that they are essentially production populations. Everyone agrees that
the more lightly cratered terrains on all three bodies represent a production
population due to comets and asteroids in the inner solar system during recent
aeons. (The sparse crater population on Earth is greatly modified by rapid
erosion and other endogenic processes.) The difference in shape of the curves
for lightly cratered terrain compared with cratered terrain is subject to the
poor statistics of the former, especially at large diameters. But to the degree
the difference is real, it is ascribed by most investigators chiefly to the combined modifications of the cratered terrain due to crater saturation and ancient
endogenic processes, while Strom et al. believe it reflects a real change in the
nature of the projectile population between ancient and current epochs.

The Strom group also claims that some units intermediate in age between
highlands and post-mare populations show the highlands curve shape. These
units (for example, plains surrounding Orientale that are dated as post-
Orientale) are older, and more heavily cratered, than post-mare units but are
not so heavily cratered that anyone would claim that there are saturation
effects. This logically powerful argument of Strom et al. that the shape of the
production function has changed with time is weakened by doubts concerning
the selection of the counted craters (including identification of basin secondaries) and by the poor statistics due to the limited areas of the intermediate-
age units.

Voyager pictures of crater populations on the Galilean satellites and on
some of the larger satellites of Saturn have greatly enriched our data base. As
shown in Fig. 18, the observed crater populations do not appear to be the
same as any of those observed in the inner solar system, nor are they all similar to each other. In particular, Callisto (and Ganymede—not included here)
appear to have a relative lack of craters $\gtrsim$ 50 km diameter, compared with the
terrestrial planets. There is also a variety of populations observed on the satellites of Saturn (including overall crater densities at diameters near 10 km that
exceed those observed on any terrestrial body), leading the Voyager experi-
menters (Smith et al. 1981,1982) to propose that there are two populations of
bodies that have impacted the satellites in different proportions and/or at different times.

We will outline briefly the alternate perspectives on the implications of
the satellite crater data before delving into the differences in methodology and
interpretation that lead to different conclusions. One suite of hypotheses has

attempted to view the projectile populations in the solar system as being fixed, having the same population characteristics throughout solar system history and in both the inner and outer solar system, with the only major change being the decline in overall impact flux since early epochs of heavy bombardment. Neukum has been the most consistent advocate of this perspective, although he has recently advocated (Horedt and Neukum 1984b) that many of the craters on outer planet satellites were probably formed from low-velocity impacts, implying circumplanetary orbits rather than the heliocentric orbits that pertain to the inner planet projectile population. Shoemaker and his colleagues have emphasized the role of comets in cratering planetary and satellite surfaces in both ancient and recent epochs, and he has sought to explain any discrepancies with this hypothesis by invoking endogenic crater-obliteration processes (e.g., viscous relaxation of larger craters on Ganymede and Callisto) and the minimum possible number of additional source populations for cratering. Hartmann (1985) has emphasized the gross similarity of most observed crater populations and also is inclined to invoke endogenic crater-obliteration processes to explain any differences. Although there are major differences between the models of these and some other investigators, there has been, until very recently, general agreement that:

a. most observed crater populations represent the natural outcome of a cometary/asteroidal population that has been widespread throughout the solar system;
b. the only other projectile population is the so-called population II in the Saturn system (probably pieces of broken-up satellites or ejecta from small satellites reimpacting on the satellite surfaces, or some remnant of ring-body populations);
c. the cratering has been very extensive;
d. the differences in crater populations are chiefly due to a variety of active geologic processes that have affected planetary and satellite surfaces in ancient epochs (and recent epochs for the most lightly cratered surfaces).

The most extreme alternative view is that of Strom, Woronow, Gurnis and their associates. They are inclined to believe that the observed crater populations are production functions. Therefore, they infer from the differences in the crater populations that there were/are at least five distinct populations of projectiles:

1. An early terrestrial planet population, responsible for cratering highlands on the Moon, Mars and Mercury;
2. A recent terrestrial planet population, responsible for post-mare cratering;
3. A population deficient in larger projectiles, responsible for cratering Ganymede and Callisto, especially in early epochs;
4. Population I in the Saturn system, similar in shape to the recent ter-

restrial planet population, but responsible for the *early* cratering of several Saturnian satellites, such as Rhea;

5.   Population II in the Saturn system, evidently a circum-Saturnian population responsible for later cratering on some satellites.

[Plescia and Boyce (1985) have recently rejected the planetocentric origin for population II, but we find their scenario for Saturn-system cratering to be difficult to sustain (see Sec. IV.B below).]

We will show that there is a reasonable middle ground between the extremes that nonetheless admits of quite a few possibilities until certain questions are answered with greater confidence than is now possible. It is particularly important, for example, to clarify the role of crater-modification processes that could affect the shape of the frequency distribution of craters in the range 5 to 50 km; is it *possible*, for example, that all of the observed crater distributions in this range (except for Saturn population II) could be due to the same production population, but modified by various understandable endogenic processes? It is also important to identify the statistical significance of the crater count data that seem to show that the recent terrestrial planet population and Saturn population I lack the fall-off in craters > 50 km diameter characteristic of Ganymede and Callisto. We will present our reasons for believing that the early terrestrial planet population probably differs from that of Ganymede and Callisto at these larger sizes, despite the arguments of Passey and Shoemaker (1982) and others that the deficit of larger craters on the Galilean satellites is due to viscous relaxation.

A.   The Issue of Saturation

Since the early work of Marcus (1970), there has been considerable debate about what happens when a planetary surface is sufficiently cratered that craters start to overlap each other more and more. Eventually a state of saturation is achieved in which, on average, subsequent impacts erase preexisting craters as rapidly as they form new ones. A casual look at the most heavily cratered parts of the lunar highlands convinces the observer that it is approaching saturation, at least by craters of the predominant sizes. But the casual look cannot distinguish between the hypothesis that it is *nearly* saturated and the hypothesis that it is *completely* saturated, many times over. Two related questions concern (1) the differences between a production population and the resulting observed population of craters as saturation is approached or actually reached, and (2) whether or not endogenic processes have affected crater populations in certain heavily cratered terrains (e.g., what is the nature of the intercrater plains on the Moon and Mercury).

Most early studies of crater saturation concerned the situation where the slope of the incremental diameter-frequency production function was steeper than − 3, applicable to subkilometer-scale craters on the lunar maria. For this case, it is agreed that at small diameters, an equilibrium crater density is reached with a slope of -3. This case has some applicability to outer-planet satellite populations at larger sizes. The observed populations of craters ($>$ a few km diameter) on the Moon and terrestrial planets, however, have slopes averaging somewhat shallower than -3. This case has been studied by Woronow (1978); one of us (CRC) has extended this work using logically equivalent numerical models (see this chapter's Appendix). There are several important conclusions from these studies that are at variance with some common intuitive expectations:

(1) Such production functions reach only a kind of quasi-equilibrium because the surface area, and hence obliteration of preexisting craters, is dominated by the very largest craters, which are formed very infrequently. When a very large crater is formed that approaches the area of the region in which crater counts are made, it may destroy most preexisting craters, dropping the average crater density markedly; the surface then begins to be progressively recratered by smaller craters until the next giant impact suddenly starts the process all over again. Thus, there is no equilibrium density; rather the observed crater density bounces about temporally and varies spatially, depending on the proximity to recent giant impacts.

(2) The observed crater population resembles the production function, even after the surface has been saturated many times over. This follows from the previous conclusion because of the periodic recratering of obliterated areas by the production function. (Of course, in this case the production function inferred from observed craters is biased, in time, to recent epochs.)

(3) The typical crater density reached under this type of saturation depends on the ratio of the size(s) of craters under consideration to the largest crater that forms in (or affects) the area under consideration. The greater this dynamical range, the lower is the typical saturation density. This conclusion is particularly applicable to some early numerical simulations that employed a relatively small dynamical range; in being applied to interpret real planetary crater populations, these simulations effectively ignored the effects of basin-forming impacts on preexisting populations. For large bodies, like the terrestrial planets, basin-forming impacts must be predominantly responsible for affecting global crater densities. The precise mechanism whereby they manifest their effects is a subject of dispute; opinions vary widely about the extent to which basin ejecta can obliterate preexisting craters and possibly form intercrater plains. For small bodies, like small satellites and asteroids, the largest impacts may actually spall off large fragments or shatter the body itself. A particularly extreme way to reinitialize the surface of a body is for the fragments of such a shattered body to reaccumulate into a body with a new surface.

(4) If there is a very small dynamical range in the production population (e.g., if most craters are within a factor of a few of a predominant size), then the saturation density of craters *of those sizes* may exceed by many times the saturation density of craters of the same size produced by a broader produc-

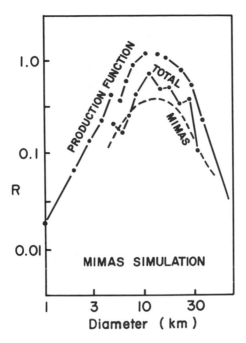

Fig. 19. Simulation with a narrow production function yielding a Mimas-like final crater population (labeled "Total") with a relatively high peak density. The dashed line is the observed crater population on Mimas.

tion population on another body. This is a likely explanation for why 20 km craters on Mimas are several times more numerous than equivalent-sized craters on other such supposedly saturated surfaces as pure lunar highlands: evidently the production function of Saturn's population II is lacking in projectiles much smaller or much larger than those that created the observable craters on Mimas. Figure 19 shows a simulation illustrating this effect. A Mimaslike production function yields a final curve of the same shape, which peaks out at an appreciably higher density than the results of a broader production function (compare with Fig. 22; also see this chapter's Appendix for a description of the simulation program).

A contentious issue is whether or not many of the most heavily cratered terrains on planetary satellites and on the terrestrial planets are saturated. Hartmann (1984) argues that the heavily cratered terrains on many bodies have crater distributions falling within factors of a few of what he terms an empirical saturation curve over at least part of the diameter range of this distribution. He argues that it would be a remarkable coincidence, if so-called true saturation density were several times higher, for cratering to have approached but never reached saturation on so many bodies. Hartmann has not, however, demonstrated that any particular crater distribution is necessarily saturated nor has he attempted to explain many of the significant variations of factors of a few within the envelope.

Woronow (1977a,b,1978) performed numerical simulations of saturation and concluded that saturation is not reached, for any of several production functions he studied, until crater densities reach at least several times higher than the observed crater densities on heavily cratered terrains of the terrestrial planets. He concluded that the only way to generate lower densities at saturation is to use parameters in his model that are unrealistic geologically. For example, he can generate lunar highland crater densities by assuming that a fresh crater's ejecta blanket destroys preexisting craters within a surrounding area 10 times the area of the fresh crater itself. He has referred to pictures of overlapping lunar and Mercurian craters that show ejecta blankets to be much less effective in obliterative power.

In our view, numerical models cannot be used for definitively proving the level of saturation densities to within a factor of a few, for two reasons: (1) the correspondence between model parameters and actual large-scale geologic processes is too uncertain; and (2) there is subjectivity in the recognition of highly degraded craters by different individuals who compile crater counts (sometimes approaching a factor of 2) which cannot be dealt with in the models. Nevertheless, it is useful to examine whether Woronow's model and his choices of parameters tend to overestimate or underestimate saturation densities. In several ways, we believe his past work has yielded unreasonably high predicted saturation densities, which may plausibly explain why the observed densities on cratered terrains are lower than predicted by his model. Furthermore, our own models (see this chapter's Appendix) demonstrate that there is no unique saturation density, but rather that the observed differences from body to body are understandable in terms of different production functions, combined with plausible differences in planetary geology, for different bodies. These same differences are also consistent with Woronow's modeling.

Woronow (1978) simulates a crater using 8 rim points. Apparently (Woronow, personal communication, 1983) a crater is not considered obliterated by subsequent impacts until all 8 rim points have been covered. In practice, we doubt that many craters would be counted that are missing more than about 2/3 of their rims; although the value 2/3 is itself subjective, we believe that Woronow's choice tends to retain craters in his simulations long after they would cease to be recognized in most counts of real, cratered surfaces on planets and satellites. (Woronow [1984] has recently considered this aspect of his models and concludes that it may be important.) Also, Woronow adopts a parameter F referring to the fractional size of the smallest crater that can obliterate a part of the rim of a larger crater. Woronow (1978) chose a value of $F = 0.354$ meaning that craters smaller than about 1/3 the diameter of another crater cannot degrade it and help to obliterate it. We believe, however, that a value of 0.1 or even less would be more appropriate, for two reasons:

(1) for the craters $\gtrsim$ 10 km diameter chiefly under discussion here, rim to-pography is typically within a factor of a few of 1 km and not strongly dependent on diameter (see Sec. II) so that, for example, a 15 km crater should be quite capable of penetrating through the topography of the rim of a 150 km crater; and (2) crater counters rely heavily on the presence of the narrow rim-crest of a crater, not on the full width of a rim, in recognizing a crater. If we are correct that Woronow's value for F underestimates the degree of obliteration by small craters, this will be most important for segments of the production function that have slopes steeper than -3.

Woronow's earlier studies (see, e.g., Woronow 1977a,b) employed a range of crater sizes (dynamical range) of a factor of 16. This artificially omitted the effects of very large craters obliterating very small ones. For example, the model could not deal with the effects of basin-forming impacts on the numbers of 10 to 30 km craters on the Moon. Woronow (1978) increased the dynamical range to 64, which is nearly sufficient but still misses the effects of the largest basins on the smallest craters (8 km) that are in the data set of observed lunar craters. In two simulations, Woronow used a dynamical range of 128 but states that computer resource limitations prevented him from running the simulations much beyond the observed lunar highland crater densities (actually he may have misplotted the crater data in his Figs. 8 and 9, in which case the runs proceeded further than he believed).

A major question concerns the efficacy of ejecta blankets in obliterating craters. As mentioned above, Woronow believes that they are very ineffective and has given some evidence to support his view. In his models, he has sometimes employed ejecta blanket parameters that he feels are too active. Some workers believe that major basin-forming events can be very destructive at large distances, as evidenced by the sculpturing of frontside lunar topography by the Imbrium event and by some interpretations of the nature of several highlands plains units (or consider Valhalla on Callisto, which has reduced small crater densities below the background level for distances up to ~ 1000 km from the original crater rim [Passey 1982]). Since the very largest impacts (the basins) dominate the saturation densities of small craters for shallow-sloped production functions (like the observed crater distribution on the terrestrial planets), the nature of basin ejecta blankets is critical to assessing the expected saturation level. On balance, we feel basins are likely to be quite effective in obliterating craters of the size under discussion.

Another point must be considered in comparing observed crater counts with simulations: are the counts appropriately restricted to areas where saturation cratering is the only obliterative process? The plots for average lunar highlands and Mercurian highlands shown in Fig. 18 are those of Strom et al. (Woronow et al. 1982) and Trask (1975), respectively, and include broad regions on both bodies. In particular, the counted area on Mercury includes the pervasive intercrater plains on that planet, which are widely (though not universally) believed to be due to volcanism. The lunar highlands data (except for the largest craters and basins) are based on frontside highlands only, which is an area with a larger percentage of lunar intercrater plains than the backside; the origin of the intercrater plains is more controversial on the Moon than on Mercury, but some of them may be of volcanic origin. Even if they are due to basin ejecta, the restriction of crater data to the frontside may bias the statistics. In short, it is possible that crater counts from more restricted parts of the highlands on the Moon and Mercury may better reflect the empirical level of saturation. Indeed, Hartmann (1985) shows lunar backside counts that have higher densities (especially for craters < 20 km diameter) than the counts of Strom et al. which include many intercrater plains. On the other hand, it would be inappropriate to consider counts from a small region of apparently "pure highlands" as "typical," if the region was selected by criteria that were biased against the real influences of large craters or basins. One must be especially careful not to mix counts from a variety of different geologic units, especially if the units are due to endogenic processes, in order to compare with numerical cratering models. Some of the early disputes concerning Voyager data on Saturn satellite craters were related to questions of how to separate appropriately well-defined units.

B. Production Functions Implied By Cratered Terrain Statistics

This section considers scaling considerations for crater production functions; these considerations relate such production functions to projectile populations, as applied in Sec. IV.

Several illustrations in this chapter present the results of our numerical simulations that may be compared with observed crater populations on various planets and satellites. While in the general sense we cannot claim uniqueness for the hypothesized production function that yields an end result similar to what is observed, we will describe the range of plausible production functions that could yield the observed crater populations, based on our experiments with a variety of production functions. (Naturally, there remains the possibility that endogenic processes might have modified the observed crater population, in addition to the effects of the cratering process itself. This possibility can be tested eventually by studies of crater counts distinguished by morphological class and by other photogeological techniques.) The cratering model is described briefly in this chapter's Appendix. We have employed parameters that we believe are more realistic than those of Woronow, although they are subject to large uncertainties. Compared with Woronow's model, this model has the following characteristics: (a) it employs a large dynamical range; (b) it defines crater rims by a complete circle of points, generally many more than the 8 rim points of Woronow; (c) it considers a crater obliterated when < 30% of the rim points remain; (d) it permits obliteration of a rim point by a subsequent impact of a crater of any size; and (e) it adopts a relatively narrow width for the crater obliteration zone of an ejecta blanket (ejecta blanket diameter = 1.1 times crater diameter for simulations reported here).

The Moon and Mercury.   The observed lunar highlands crater population can be interpreted as a saturated population only if it was produced by a
production function having approximately the same shape as the observed
population. (Compare counts in Fig. 18 with simulation in Fig. 22 [see the
chapter's Appendix]. Alternative production functions with steeper slopes do
not evolve into the observed shallow slopes.) We find that a true saturated
population should probably lie at a slightly higher density than is observed in
the frontside-dominated statistics of Strom. It is plausible to us that an endogenic process, or more effective obliteration by nearby basins than we have
modeled, is responsible for the slight discrepancy. The relatively small differences between crater curves for the more heavily cratered terrains on Mercury
and the Moon are most plausibly explained as being due to the endogenic process that created the intercrater plains on Mercury (see review by Strom
1979). But this endogenic process probably cannot explain the much larger
differences between the observed lunar highlands curve and any hypothesized
lunar production function having the same shape as the post-mare curve (Woronow, personal communication, 1983); the fact that the very obvious intercrater plains of Mercury have a population of observable small craters that is
reduced by only a factor of 2 compared with the Moon implies that the much
less obvious intercrater plains-forming process on the Moon could not produce
the much larger decrepancy in *shape* between pre- and post-mare lunar crater
populations on the highlands. It seems most reasonable to us to believe that
the two lunar populations were different. The error bars on the lunar mare
crater counts (cf. Strom et al. 1981) are large enough and there are sufficient
possibilities of systematic errors that it is difficult to rule out totally Neukum's
contention (G. Neukum, personal communication, 1984) that the two lunar
populations are identical; yet, they do seem to be markedly different. The
highlands production population could be somewhat more like the post-mare
population (which Hartmann [1984] believes he has observed in some backside highlands locations) if a size-dependent endogenic process has erased
many smaller frontside highland craters. But, as we have just discussed, the
required obliteration seems very large and also it does not seem to be reflected
in statistics of lunar crater morphologies (Chapman 1974). We feel that
Hartmann's data of lunar backside areas lacking the decrease in small highland
craters is most reasonably interpreted as being due to contamination of the
production function by secondary craters from proximate young basins rather
than an old highlands production function having a post-mare shape. We conclude that it is difficult to explain the fall off of smaller highlands craters except as a characteristic of the production function. Although these are our preferred interpretations of the lunar and Mercurian data, we reiterate that it
would be premature to be dogmatic about it.

Ganymede and Callisto.   Figure 20 shows the result of a production
function fairly similar in shape to the observed Callisto crater population; it

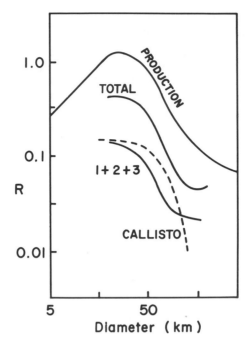

Fig. 20. Simulation of Callisto's crater population by a production function with relatively few
large craters. The observed Callisto crater distribution, shown dashed, agrees best with the
resulting curve for classes 1 + 2 + 3, but omitting class 4; these classes are described in the
appendix.

reaches saturation somewhat above the observed curve. Such a population is
capable of exceeding the lunar saturation curve at small sizes due to the lack
of large craters, for reasons analogous to the Mimas case (Fig. 19). The reason Callisto's craters do not reach densities quite so large as in the simulation
may be due to exceptionally destructive effects of very large basins, such as
Valhalla, and/or due to undercounting caused by the less than ideal sun-angles
on Voyager images. (Note that there is much better agreement with class 4
omitted; also Neukum [1985] has counted more craters than given by our observed curve, which is based on Woronow et al. [1982]; Neukum also disputes
the sharpness of large-crater fall off, which we take here to be real.) A matter
of great controversy is whether or not the decrease in frequency of craters
larger than several tens of km (essentially a total absence of craters larger than
150 km) is due to lack of such impacts in the production function or due to
relaxation and disappearance of such craters caused by the rheology and thermal history of the crustal layers of these icy satellites. Shoemaker et al. (1982)
and Passey and Shoemaker (1982) have argued the latter, but we find the arguments of Woronow and Strom (1981) and Gurnis (1985) convincing, i.e., that
anything approaching the lunar highlands production function would yield

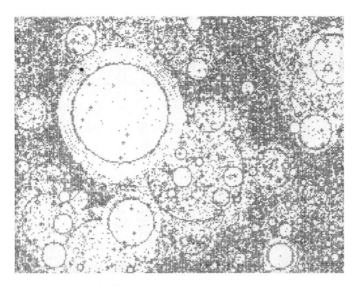

Fig. 21. A simulation dramatically showing the spatial nonuniformity of small craters caused by recent impacts of large craters. This simulation involves a fairly steep production function and wide ejecta blankets; since the viscous relaxation simulation is turned off, the rims of the large craters are still visible. Unlike most simulations (see the chapter's Appendix), the smallest craters are shown in this map. The spatial nonuniformity effects for Callisto and Ganymede are less dramatic than shown here, but are visible to the eye and readily demonstrable by statistical tests, as shown by Gurnis (1986).

holes in the spatial distribution of the smaller craters to a degree that is not seen. We have been able to reproduce the observed Callisto crater population using a lunar-highlands production function by incorporating diameter-dependent crater erasure (simulating viscous relaxation). But the resulting spatial distribution exhibits prominent holes where relatively recent large craters once existed (a qualitatively analogous effect is shown in Fig. 21).

Ganymede's older terrains have crater populations similar to those on Callisto. Hartmann has made counts on Ganymede's cratered terrain that restore palimpsests to the observed curve by considering the outer diameters of the albedo patches to be equivalent to the crater diameter; even with this extreme assumption (most workers believe that remnant topography within some palimpsests indicate that the equivalent diameters should be smaller; see Sec. II.E), little of the discrepancy between the Galilean satellites and the lunar highlands is removed. With regard to Ganymede's younger surfaces, available counts from the grooved terrain suggest that the production function deficient in larger craters remained similar into later stages of Ganymede's geologic history, although the statistical precision of the counts is much poorer.

As Strom et al. state, the shape of the observed diameter distribution on the Galilean satellites is distinctly different from the one observed on the terrestrial planet highlands. In fact, the obvious differences between the Ganymede/Callisto curves and the inner planet curves are magnified once scaling differences are taken into account, following the arguments developed in Sec. II (see below).

Satellites of Saturn. The Voyager images of the Saturnian satellites have been the basis for crater counts by Strom (1981), Plescia and Boyce (1982,1983), Plescia (1983) and Hartmann (1984). We consider here counts on Mimas, Enceladus, Tethys, Dione and Rhea. There are broad units on all of these satellites (although they are less obvious for Mimas) that reflect different crater populations. The crater data, in association with our numerical simulations, largely confirm the original view of Smith et al. (1981,1982) that the surfaces of these satellites chiefly reflect two populations of projectiles that impacted the satellites sequentially. Population I has a major proportion of large craters and is expressed on several satellites, including the heavily cratered terrains (generally not well observed by the Voyager spacecraft) of Dione and Tethys. More recent cratering by population II, which has a deficiency of large craters, is expressed on the younger surfaces of Dione and Tethys, from which population I has been partially or completely erased. Population II also appears to dominate Mimas' surface and is the population visible on those portions of Enceladus that have any craters at all.

Rhea presents a slight enigma. It has a substantial proportion of large craters (> 30 km diameter), especially on the leading hemisphere of its north polar region, which is generally considered to be the best expression of population I. (It is interesting to note that the *shape* of the population I distribution is approximately the same as the lunar post-mare distribution, although the *density* of craters is much higher, even exceeding the lunar highlands crater densities for diameters < 20 km.) The trailing hemisphere of Rhea's north polar region seems to show a crater population intermediate between population I and population II: there are fewer large craters and more smaller ones. A first-cut interpretation is that the saturated population I distribution was modified in this region and that a sprinkling of population II was superimposed on the partially remaining population I distribution. Some reason must be found to explain why the expression of population II on Rhea is so much reduced compared with the other Saturnian satellites (either lesser focusing by Saturn of an external projectile population on relatively distant Rhea, or localization of Saturnicentric sources of projectiles). A more difficult problem relates to the fact that the supposedly later population II craters have a *higher* spatial density on the supposedly *younger* trailing-hemisphere surface. One would expect the older surface to record the superposition of the younger population on top of the older one, which can hardly yield a lower crater density than is exhibited by the younger one alone. (Leading/trailing side focusing processes do not seem to be a promising way to explain this difference in expression of population II; we are concerned with near-polar regions and the differences are in the wrong sense anyway.)

The data are probably not at fault for the seeming paradox that the crater curves for Rhea's two polar regions seem to cross, because the counts of both Strom, and Plescia and Boyce agree in this respect. There could be a pervasive crater-degradation process operating in the north polar regions of the leading hemisphere that has degraded craters over the recent epochs during which population II has been emplaced; of course, it was the *other* hemisphere that suffered the original losses of population I craters. Moore et al. (1985) have discussed some other explanations including emplacement of a more erodable mantle on the apparently older surface, perhaps related to the mantling discussed by Plescia and Boyce (1982).

The presence of different units on the Saturnian satellites implies that they all (with the possible exception of Mimas) experienced widespread modification processes, presumably of endogenic origin, midway during the epochs of cratering (cf. Plescia 1983; Moore 1984). The exact form of the population I crater distribution is difficult to ascertain because it is obscured by population II at small diameters on all the satellites, except Rhea (perhaps); at larger diameters, the statistics are poor and in any case the population may have been modified. Therefore, it is difficult to place much weight on the otherwise intriguing apparent similarity of population I to the post-mare cratering of the terrestrial planets. If one discounts Enceladus as being uniquely subject to recent endogenic geologic activity, then there appears to be a tendency for population II to be more thoroughly expressed the closer a satellite is to Saturn. Our numerical simulations suggest that population II must be near the true saturation limit on Mimas, but that it is undersaturated on the other satellites. The predominant opinion is that population II is due to sweepup of Saturnicentric debris, perhaps formed by disruption of one or more satellites. The varying expression of this population on the different satellites may provide a clue to the origin of this debris; on the other hand, and notwithstanding the assertions about chronology by Plescia (1983) and Plescia and Boyce (1985), there is no independent constraint on the epoch of endogenic activity on the various satellites, so it may simply be the case that endogenic activity persisted longer on Dione and Rhea (as it most obviously did on Enceladus) than, for example, it did on Mimas.

Phobos and Deimos.   Thomas and Veverka (1980a) analyzed the crater populations on the Martian satellites. Lines through their counts are shown in Fig. 18. They noted the similarity of the curve shapes and heights to "saturation equilibrium" curves, such as those of Hartmann. They also noted that an extrapolation of the crater curves to larger sizes implied a significant chance of an even larger impact that could disrupt the satellites. The appearance of Stickney and its fracture pattern has been widely interpreted as representing a nearly disruptive impact (see the chapter by Thomas et al.). Relying on earlier arguments that Phobos retains appreciable internal strength and cannot be a rubble pile, Thomas and Veverka concluded that the Martian satellites are old and have recorded a cumulative crater flux since they were formed that has nearly reached or only barely exceeded saturation.

One of us (CRC; cf. Thomas et al. 1979b) has considered the Martian satellite crater data and reached somewhat different conclusions. The observed crater counts are actually appreciably lower than Hartmann's lunar saturation equilibrium curve. Indeed, they are lower for Deimos than for Phobos. This almost certainly implies that the crater populations are *not* saturated on these surfaces. As our earlier discussion showed, low-saturation equilibrium densities are explicable only if the dynamical range of impacting craters is large. For example, the lunar equilibrium curve is depressed below that advocated by Woronow only because of the effects of giant, basin-sized impacts. Since there are no apparent basin-analogs on the Martian satellites (especially on Deimos), they might be expected to exhibit higher crater densities at equilibrium. The fact that they do not is probably due to the would be basin-forming projectiles, which would probably break up objects as small as Phobos and Deimos.

Both the late heavy bombardment cratering population and the post-mare cratering populations, as expressed on the terrestrial planets, including Mars, have a relatively high proportion of large craters so that there is appreciable probability that the observed population would be accompanied by projectiles large enough to shatter the satellites. Once reassembled in Mars orbit (a rapid process, akin to the dust-belt process discussed by Soter [1971; see also Burns et al. 1984]), the satellite's pristine surface would then reaccumulate impacts up to approximately the levels observed today. Thus we propose that the observed crater populations on Phobos and Deimos represent a quasi-equilibrium density in a process of punctuated breakup and reassembly. The probable epoch of last breakup and reassembly of these satellites can be computed only by a chain of rather tenuous steps (including assumptions about the strengths of these bodies). The epoch could be well back into the epoch of late heavy bombardment or as recent as perhaps 1 to 2 Gyr ago.

IV. PROJECTILE POPULATIONS

As we have seen, there are appreciable differences among the crater distributions visible on different satellites and terrestrial planets. In this section, we consider the different projectile populations that can produce craters, fold in considerations of scaling based on Sec. II, and consider what constraints can be placed on the origin of observed craters.

Early studies of terrestrial-planet cratering identified three kinds of craters: primaries, secondaries and endogenic craters. Primary craters are formed by the impacts of comets and asteroids. Secondaries are craters formed by the comparatively low-velocity impacts of ejecta from primary craters; they often have recognizable spatial relationships to their primaries and they sometimes exhibit characteristic morphologies. Endogenic craters are generally rare on planets and satellites and, in any case, are often recognizable as being of internal origin by their morphologies and spatial relationships (e.g., crater chains). The extension of cratering statistics to very small satellites and to satellites in the outer solar system forces us to consider other subdivisions of the potential primary projectile population.

Comets include the populations of both new and short-period objects that penetrate the inner solar system. The outer solar system satellites are also affected by those comets in intermediate orbits that have been captured by outer planets like Saturn. The short-period comet population, which is captured in orbits with aphelia inside Jupiter's orbit, has little propensity for cratering outer-planet satellites. The craters produced by short-period comets may reflect population characteristics modified from the remaining populations of comets by their ablation, evaporation and disintegration in the inner solar system. Thus, there may not be a single comet population with a unique size distribution. Furthermore, the velocities at which comets in the various kinds of orbits impact different satellites are markedly different, and due account must be taken of the different resulting crater diameters.

Planetocentrically orbiting bodies constitute another kind of projectile population. In general, the time scales for dynamical evolution and sweepup of circumplanetary debris are very short. But, in principle, projectiles may remain in circumplanetary orbits long after the end of accretionary epochs, and some fraction of them may be slowly perturbed into satellite-crossing orbits, perhaps by subtle resonance-related processes. Other planetocentric projectile populations may be created by crater ejecta escaped from small satellites or by the catastrophic disruption of a small satellite, perhaps resulting from an impact by a large comet. It has also been suggested (cf. Wetherill 1977) that a low-velocity planet-grazing body could be tidally disrupted, leaving a portion of its debris in circumplanetary orbits; recently, however, this process has been discounted (Mizuno and Boss 1985). Anyway, the numerous small satellites and the ring system of Saturn testify to the plausibility of planetocentric projectiles, however they were formed.

The size distribution of such internal bodies could be different from external (cometary or asteroidal) populations if different processes gave rise to such populations. For example, some projectiles may be created as ejecta from large craters which have escaped into independent planetocentric orbits due to the low escape velocities of the small cratered satellites; they might yield crater populations with the steep-sloped (or narrow, unimodal) shapes of secondary crater populations on the terrestrial planets. Other projectiles might be formed by the shattering and disruption (by tensile failure) of a small satellite with minimal gravitational binding; experiments by Fujiwara et al.

(1977) and Hartmann (1980*a*) suggest a broader size distribution with a shallow slope (smaller population index) for such debris.

Impact velocities of projectile populations, both external and internal, are affected by gravitational focusing by the primary. In general, impact velocities are higher on satellites closer to the primary. There are also marked asymmetries in the longitudes on which craters form, including enhancement at the apex for external populations and enhancement at the antapex for planetocentric bodies having orbits with large semimajor axes and large eccentricities. Whether or not such asymmetries remain visible on the surfaces depends on the degree to which asymmetries might be masked by regional endogenic geologic activity and by the degree to which the satellites maintain synchronous rotation. Satellites may be jarred loose from synchronous lock by very large impacts (see Sec. II.F). We doubt that this can be accomplished by impacts less sizeable than those that would destroy most surface topography. Therefore we believe that the lack of obvious asymmetries on Saturn's satellites reflects saturation effects, especially for population I, and planetocentric projectiles for population II.

A. How Many Populations? Implications from Scaling

A major issue raised by the Voyager cratering data for the Galilean and Saturnian satellites concerns the number of distinct projectile populations required to explain the observations. If many of the crater populations on separate satellites and planets are due to one or a few populations, then there would be the possibility for linking the geologic histories of these bodies in time. Particularly if the cratering histories could all be linked to the lunar cratering history, which has been relatively well calibrated by the lunar rock ages, then we would have the answer to the long-standing question of the interplanetary correlation of geologic time.

There was a long-standing hypothesis, first developed in the 1960s after the early encounters with Mars, that the cratering history on the Earth and Moon could be extrapolated to the other terrestrial bodies. In particular, it was hypothesized that the late heavy bombardment (circa 4 Gyr ago) on the Moon occurred simultaneously on the other terrestrial bodies (cf. Murray et al. 1975). Furthermore, the subsequent cratering of the lunar maria, which continues today, is thought to have been similar on the other terrestrial bodies (after making appropriate small corrections for diameter scaling and accounting for plausible variations in the proportions of comets and asteroids from Mars into Mercury), mainly because of dynamical calculations that demonstrate that most populations of small bodies known to exist (or hypothesized to exist) in the inner solar system must be gravitationally scattered throughout the inner solar system and thus must be expressed on all of the bodies (cf. Hartmann et al. 1981).

The extrapolation of this concept to the Jupiter and Saturn systems was a natural jump as an analogy, but much less well founded on dynamical terms.

In particular, the influence of the asteroids, which is substantial and perhaps predominant in the inner solar system, is probably minimal for the Galilean satellites and absent for the Saturnian system. The role of comets and other ice-rich bodies that might not survive in, or be readily observable from, the inner solar system is problematical. Also, there is much more presumptive evidence that planetocentric populations might have played a major role in cratering the outer planet satellites, especially those in the Saturn system.

The Voyager team (Smith et al. 1981,1982) concluded that there were two populations of craters expressed on the surfaces of the Saturnian satellites. As we have already discussed (see Sec. III), Strom and his colleagues (cf. Woronow et al. 1982; Strom 1981) argue that at least 5 distinct projectile populations are required to explain all the observed crater data (the two Saturnian populations, a separate population required for Ganymede and Callisto, and both early and late populations in the inner solar system). Other researchers have attempted to minimize the required populations to two at most. Certainly before insisting that there are multiple projectile populations and/or that their characteristics have changed over solar system history, it is important to enquire whether some of the apparently different crater populations might, in fact, be due to a single projectile population. In the following subsections we consider whether scaling differences or other understandable processes can reconcile apparently *different* crater curves (to within their errors) as being possibly due to the *same* projectile population.

Could the Jovian and One of the Terrestrial Planet Populations be the Same?   One of the major mysteries concerns the fact that the crater populations on Ganymede and Callisto are different from both the inner planet cratering populations and the populations on the Saturnian satellites. The most obvious trait of the Jovian-system craters is the sharp drop-off at diameters $\gtrsim 60$ km. Although this has been ascribed by some to the effects of viscous relaxation of the larger craters, we have above agreed with Woronow and Strom (1981) and Gurnis (1985) that the drop-off is substantially real (i.e., reflects an equivalent sharp drop-off in the number of large projectiles). The largest craters in the Jovian system, above the scaled lunar transition diameter (perhaps 2 times 15 km), follow a very steep distribution which is completely inconsistent with the oppositely sloped lunar *highlands* curve at appropriately scaled sizes (about 2 times smaller).

Gurnis' analysis admits the possibility of there being some relaxation of large craters and a correspondingly less precipitous drop required of the Jovian projectile population. Since the statistics of the large, *post-mare* terrestrial-body craters are uncertain, it is possible that the Jovian population could be reconciled with the more recent terrestrial-planet production function for the *larger craters*, even though the same projectile would produce larger craters on Ganymede than on the Moon (see Table IV). Horedt and Neukum (1984b) have attempted to circumvent this problem by invoking much lower impact velocities on the Galilean satellites from a planetocentric source.

We believe the problems with reconciling the Jovian population with the terrestrial-planet populations are even worse for the smaller craters, although they have been less discussed in the literature. If we assume Ganymedean craters are 2 times larger than their lunar counterparts, only the largest Ganymedean craters may be compared directly with collapsed lunar craters larger than the 15 km simple/complex transition observed for the Moon. Smaller Ganymedean craters are equivalent to simple lunar craters, so the scaling varies from 2 (to perhaps as much as 3) down to 1.4 as one moves down to the smaller Ganymedean craters near the 5 km transition for that body. These craters, which trend along a nearly level slope on the R-plot, would continue sloping upward if translated to the Moon. Thus the effect of the scaling relationship is to worsen the already rather disparate comparison of slope of the post-mare distribution in the range 5 to 50 km. The comparison with the highlands terrestrial-planet population is still worse. In conclusion, we doubt that the Jovian crater populations could have been produced by the modern terrestrial planet projectile population and it is totally incompatible with the earlier population.

Is the Jovian Population Equivalent to Saturn Population I?   Strom and his colleagues have noted the similarity in shape between population I at Saturn and the post-mare terrestrial planet populations. We have previously discussed the possibility that the cratered-terrain populations on the terrestrial planets may not be a production function, but rather be the result of modification processes, in which case the *post*-mare population would be more generally reflective of cratering in the inner solar system throughout most of solar system history. To the degree that population I in the Saturn system is equivalent to the terrestrial planet population, then the arguments in the previous paragraph rule out a similarity with the Jovian population. On the other hand, population I is very poorly sampled in the Saturn system. It is chiefly expressed only on Rhea, on the leading side of its north polar region. There is minimal sampling of the population at diameters > 100 km. Furthermore, the nature of the population at smaller diameters is rendered uncertain by the question of the degree to which the possibly subsequent population II might be mixed in. Population I is evident on most other Saturnian satellites only by the presence of a few large craters. Since the external-impactor scaling between Saturnian and Jovian satellites is similar, the presence of 100 km scale craters on Saturnian satellites would seem to be incompatible with the absence of such large craters on Ganymede and Callisto, but the inadequacy of the statistics and scaling uncertainties do not permit a definite rejection of a similarity.

Population II in the Saturn System.   Could population II at Saturn be of external origin, and somehow be compatible with the Jovian population? We note that such Saturnian craters are relatively small, all within the simple crater domain. Hence craters on Mimas should be about 1.3 times as large as craters on Tethys, 1.4 times those on Dione, and 1.5 times those on Rhea. This scaling trend tends to be evident in the crater counts, although it does not distinguish between external and internal impactors because the same trend is applicable to both. Population II would make transient craters on Ganymede or Callisto only $1/2$ to $2/3$ as large as on the Saturnian satellites; such craters might then collapse and widen the final craters, in some cases, to be at least as large as the observable Saturnian-system craters, but not larger. Thus there is no way to reconcile the fall-off in population II craters near 20 to 30 km diameter with the somewhat analogous fall-off observed on Ganymede and Callisto at diameters at least twice as large. We thus must rule out the possibility of these two populations being identical if both are external impactors. Of course, if the Saturn population II were external, there would be a question about why that population is not manifested on the Jovian satellites, given that Saturn's population II is a relatively late population and Saturn would be an effective scatterer of any projectiles in its vicinity (this argument adversely affects the conclusions of Plescia and Boyce [1985]; see Sec. IV.B below).

Saturn population II is widely regarded as due to planetocentrically orbiting projectiles. Two possible sources for such bodies have been discussed: ejecta from some giant craters or fragments from a completely disrupted small satellite. A comparison of Saturnian simple craters formed at 5 km s^{-1} with secondary lunar craters landing at 1 to 1.5 km s^{-1} shows that the Saturnian craters would be 2 to 3 times bigger than their lunar secondary counterparts. The observed population II craters on the Saturnian satellites are much too big to be matched with the increasing numbers of small craters on the Moon, less than 2 to 3 km diameter, which some investigators believe are background secondaries. However, they are roughly similar in size to the apparent secondaries surrounding some larger craters (e.g., Copernicus) and are even too small to match the observed secondaries around major basins. To the degree that the impacts that produced the largest lunar craters and the mare basins would actually disrupt a small Saturnian satellite, it is possible that such disrupted fragments are compatible with the traits of population II. The mechanics of disruption of a body are different from those of cratering into a semi-infinite target, so it is by no means certain that the size distribution of fragments from a disrupted body would resemble that for ejecta fragments from a crater.

How Many Populations? A Summary.   We have summarized many arguments about the number of populations in Sec. III and so far in this section; some are air tight, but others are quite tenuous. In summary, it seems likely to us that at least four and perhaps all five of Strom's populations are in evidence on the outer-planet satellites and the terrestrial planets. Because the terrestrial planets are not our chief subject, we have not elaborated on the arguments that distinguish the cratered terrain populations from the post-mare populations on the Moon, Mars and Mercury, but those we have given seem reasonably compelling. It is conceivable that some of the comparisons and arguments could fall victim to possible biases in the data, especially given the statistical uncertainties of counts on less heavily cratered surfaces. In that case, an even greater role would have to be accepted on all the terrestrial planets, including the Moon, for widespread endogenic obliteration of smaller craters.

Phobos and Deimos exhibit crater populations similar in slope to the post-mare crater populations on the other terrestrial planets, provided the upturn at very small diameters for the latter is discounted, which is reasonable since it is widely attributed to secondary cratering. Our concept of the Martian satellites as being recratered following disruption and possible reassembly is consistent with the expression on their surfaces of a post-marelike production function, especially if the last reaccretion of the satellites occurred after the late heavy bombardment. Although the Phobos/Deimos data are at smaller diameters than those for which the older population is well characterized, an extrapolation of the 5 to 50 km curve into the size range applicable to the Martian satellite craters would give a different slope than is observed.

The production function evident on Ganymede and Callisto is a third population. There is no evidence that its characteristics changed with time. We have argued above that this population certainly cannot be reconciled with the old terrestrial planet population and probably not with the post-mare population, either. Earlier discussions of this issue in the literature focused on the absence of large craters on the Galilean satellites. Although we doubt that viscous relaxation can account for more than a part of the deficit, the statistics for large post-mare craters are not sufficient to rule out a similarity at such large sizes. But we believe our arguments about the scaling of smaller craters yield an iron-clad difference between the Galilean satellite crater population and both inner planet populations. It will be important in later discussions that we have *disconnected cratering in the Jupiter system from that in the inner solar system on the basis of population characteristics alone*.

A fourth population is population II on the Saturnian satellites, the generally later population responsible for the smaller craters. Its peaked size distribution is quite unlike the inner planet populations and we have shown that, as an external population, it cannot be reconciled with the craters on Jupiter's satellites. This population must be of planetocentric origin, although (as discussed in Sec. IV.B) Plescia and Boyce (1985) believe otherwise. Shoemaker and Wolfe (1981) and Smith et al. (1982) believe that population II represents secondary debris from disrupted satellites, which is consistent with the fact that the population shares traits with secondary crater populations formed, for example, by lunar basin impacts.

Saturn's population I may be a fifth distinct population. Saturnian satellites certainly were cratered by some population distinct from population II; it formed the largest craters (presumably including impacts large enough to disrupt the satellites) and it lacks the preponderance of smaller craters of population II. Unfortunately, because population I is largely masked by, or overwhelmed by, population II cratering, it is not well characterized. We have argued above that population I probably is inconsistent with the Galilean satellite population, but that is not statistically certain. The differences can be narrowed further by invoking a substantial amount of viscous relaxation of the larger craters on Ganymede and Callisto. Population I may resemble the inner-planet cratering populations, but that similarity would probably have no physical significance if we took the position that it is not expressed on the Galilean satellites, which lie physically in between the inner planets and the Saturn system. If possible population I should be better characterized.

B. Can Population II Provide an Interplanetary Cratering History Link?

Plescia and Boyce (1985) have presented an interpretation of Saturnian satellite cratering that addresses not only the question of projectile populations responsible for populations I and II, but also attempts to derive a model for the absolute cratering chronology in the Saturn system. At the outset, we must state that the model chronology is premature, for two chief reasons: (1) there is very little independent knowledge about the projectile flux history in the outer solar system; indeed even the *present* projectile population can be estimated only within very crude limits because, unlike the near-Earth asteroids, we cannot observe small bodies at Saturn's distance, and our knowledge of the size-distribution of new comets on eccentric orbits (which affect both Saturn's satellites and the inner solar system) is still quite poor; and (2) despite previously published assertions that the late heavy bombardment in the inner solar system may have occurred contemporaneously ($\sim$ 4 Gyr ago) at Jupiter, and despite Plescia and Boyce's extension of this idea to the Saturn system, there is no secure evidence linking this inner solar system event to outer-planet satellites. There are several ideas about the origin of the late heavy bombardment of the terrestrial planets, some of which provide no reason for expecting a contemporaneous bombardment on the satellites of Jupiter or Saturn (see further discussion below).

Plescia and Boyce (1985) believe that population I projectiles are substantially equivalent in identity and flux history to the projectiles that caused highlands cratering in the inner solar system; however, they acknowledge that there may be some contribution to satellite cratering from fragments of disrupted small satellites and that the inner planets were additionally cratered by asteroids, which would not have affected Saturnian satellites. Plescia and Boyce's most startling conclusion is that population II craters are not due to Saturnicentric bodies, but rather to the continuing bombardment by long-lived heliocentrically orbiting projectiles (long-period and perhaps other kinds of comets). They attempt to establish their chronology by identifying the onset of population II cratering as being linked to the onset of post-mare cratering in the inner solar system. As we will see, a flaw in their argument is the failure to account for the absence of population II crater distributions on other planets. In arriving at their conclusions, Plescia and Boyce make several ancillary assumptions or inferences with which we disagree. For example, they believe that crater densities are probably undersaturated on most satellites. They also believe that they can ignore the lack of apex/antapex crater density differences due to the ease of collisional spinup, whereas we believe collisional spinup is possible, but more difficult, requiring near-catastrophic impacts.

Plescia and Boyce (1985) seem to rule out the planetocentric projectile model for population II by invoking two chief points: they assert (a) that population I cratering must have ended before population II was expressed, and (b) that locally derived projectiles for population II must have been created just at the end of population I bombardment and must (for dynamical reasons) have been swept up on a geologically instantaneous time scale. These requirements lead to improbable timing coincidences.

We disagree, however, with both points. There is no reason to believe that population II must be sharply disconnected (temporally) from population I; because of the shape of the production function for II, it predominates at small sizes even while I predominates at large sizes. Both populations, for example, are clearly expressed on Tethys (see Fig. 18). While it is true that most debris from a large disrupted satellite will reaccumulate very rapidly, debris from small disrupted satellites may not reaccumulate at all and some fraction of especially high-velocity debris from larger disrupted bodies may be perturbed into different orbits and migrate, eventually to impact on other satellites over much longer time scales. The efficiency of migration need not be very high to account for population II craters. Recall that the volume of debris created by a disruptive event (or just a cratering event) is orders of magnitude greater than the volume of the incoming interplanetary projectile. Depending on the impact velocity of this debris on another satellite, migration efficiencies under 1% may nevertheless cause population II to dominate over direct population I cratering. Farinella et al. (1986) have studied the possible migration of fragments from Hyperion; while Hyperion fragments alone may be inadequate to account for population II cratering throughout the Saturn system, there are many other potential sources of debris. Of course, some population II craters may have formed by rapid reaccumulation onto some satellites (e.g., Mimas).

A major problem with the conclusions of Plescia and Boyce (1985), which they do not address, is why the continuing population II cratering by heliocentric bodies is not manifest on the Galilean satellites. (Above we have shown that the Galilean satellite population cannot be reconciled with population II.) Strangely, Plescia and Boyce believe they have found "a tie between the inner solar system and the outer solar system" flux curves in the "change in crater populations," by which they mean population I → II for Saturn and highlands → post-mare for the terrestrial planets; but population II in no way resembles the post-mare projectile population, as we have shown above, nor the spatially intermediate Jovian population, so there can be no physical identities among the populations nor their changes. We share the yearning of Plescia and Boyce for an interplanetary cratering chronology, but given the very different dynamical time scales that apply between the inner and outer solar systems and the different sources of cratering projectiles (including asteroids in one case and satellite fragments in the other), there appears to be no way to derive the absolute cratering chronology for the Saturn system through a connection with the inner solar system via the unique population II. We caution readers against using cratering ages based on the arbitrary and hypothetical cratering flux history proposed by Plescia and Boyce (1985, their Fig. 1); it could be right, but there is presently no basis for distinguishing it from many other potential flux histories.

Other researchers have addressed the relative roles of heliocentric versus planetocentric impactors. Horedt and Neukum (1984b) straddle the issue by concluding that the projectiles are predominantly of heliocentric origin, but that many of them may have impacted only after being captured into lower-velocity, planetocentric orbits. They downplay planetocentric projectiles produced by satellite disruption. The two chief observational constraints that compel Horedt and Neukum toward planetocentric projectiles, despite Neukum's long belief in the elegance of having all planets and satellites being cratered by a single heliocentric population, are: (1) the only slightly decreasing crater densities on the Saturnian satellites as a function of distance from Saturn; and (2) the lack of apex-antapex asymmetries. We regard many of the crater populations to be saturated, so the predicted variations with distance from Saturn would not generally be observable due to supersaturation. Point (2) may be partly explicable by saturation as well. We agree with Horedt and Neukum that many craters must have been formed by planetocentric populations, but for different reasons. In particular, our comparisons of crater populations on different planets, taking scaling into account, do not permit a reconciliation of population II with any other population. One of Horedt and Neukum's goals in turning to planetocentric populations was to lower impact velocities in ways that they thought would render the crater size distributions produced by the same projectile population on different planets more comparable. The efficiency of capturing heliocentric bodies into planetocentric orbits is low, at best. We prefer the virtually inevitable creation of planetocentric bodies by disruptive impact as being more probable and more consistent with the crater statistics and scaling laws.

C. Do Comets Provide an Interplanetary Cratering History Link?

By far the most detailed attempt to understand the relationship between the observed cratering record and projectile populations in the solar system is the work of Shoemaker and his colleagues. The most definitive discussion of this work is by Shoemaker and Wolfe (1982), concerning the Galilean satellite

craters. Unfortunately, the extrapolation of the approach to the Saturn satellites has been published in only abbreviated form (Shoemaker and Wolfe
1981; Smith et al. 1982) and Shoemaker has not discussed some more recent
relevant data in print. He has, however, graciously commented on an earlier
version of this discussion and critique, which we have since updated by including his observations.

Shoemaker and Wolfe's goal is ultimately to link the cratering chronology that has been established in the inner solar system (principally for the
Moon) to the crater populations observed in the satellite systems of the giant
planets. For the various types of projectile populations we have discussed
(comets, asteroids and planetocentric bodies), only comets actually spend
much time crossing the orbits of both the giant planets and the inner planets. If
a link can be established at all, it will be by understanding comets and any
other bodies in cometlike orbits. Unfortunately, despite the celestial grandeur
of comets, their nuclei—the part responsible for cratering—are still very
poorly characterized, although accelerating cometary research programs (including spacecraft missions) may soon improve our knowledge.

Shoemaker and Wolfe do the best they can to establish the size-frequency
relationships for long-period comets, short-period comets, and extinct comets. They also consider, but reject as negligible, bodies originating in the asteroid belt as possible cratering projectiles for the outer planet satellites. They
calculate plausible impact velocities, employ standard crater-scaling relationships, and deduce the modern cratering rates on the satellites. They find those
rates inadequate to produce the observed crater populations, in most cases, so
they invoke a late heavy bombardment (of comets), analogous in some ways
to the late heavy bombardment in the inner solar system. In the case of the
Saturn satellites, the late heavy bombardment is sufficient to fragment and
catastrophically disrupt many of them, some several times. The conclusion is
that comets (both ancient and modern) are responsible for essentially all of the
craters on the satellites of Jupiter and Saturn. In the case of the Galilean satellites, long-period comets are responsible for some craters, but short-period
comets (especially extinct ones) predominate. A hypothesized population of
fairly short-period "Saturn-family" comets are found to dominate cratering of
the Saturnian satellites, with some contribution by long-period comets, and
very minor contribution by true short-period comets. Insofar as the absolute
chronology of the cratering is concerned, Shoemaker and Wolfe interpret that
the onset and decline of the late heavy bombardment occurred on the satellites
of Saturn and Jupiter at the same time that it has been dated on the Moon,
ending just after 4 Gyr ago. They further conclude that the modern cratering
rates have been low, but sufficient so that all old satellite surfaces should be
visibly cratered, suggesting that the nearly craterless surfaces of Io, Europa
and portions of Enceladus imply comparatively recent endogenic resurfacing
processes on those bodies.

It is difficult to find logical flaws in the lengthy chain of arguments put forward by Shoemaker and Wolfe. But some links in the chain are based on rather tenuous data or on disputed assumptions. The heart of the problem is to establish the size-frequency relationships for comets. Other crucial steps in the argument involve the linkage between the terrestrial planets and the outer planet satellites. Let us consider these weak points in Shoemaker and Wolfe's argument.

In order to determine the size of a comet nucleus, one must know how bright the nucleus is and its albedo. With very few exceptions, albedos of comet nuclei are completely unknown. Also, it is very rare that an observer can be certain that the measured brightness refers just to the nucleus. A few Earth-approaching asteroids, with estimated sizes and in cometlike orbits, are good candidates for being extinct comets, but it is difficult to be sure of the cometary origin of any specific body (the Geminid parent body is the best candidate). Having a calibration on the size of one or two comet nuclei is just the beginning of the solution, however. In order to compare with observed crater populations, one must determine the numbers of comets of the sizes responsible for the observed craters. This is very difficult, indeed, for there are very serious questions of completeness that plague determination of size-frequency relations for comets. For example, short-period comet statistics are complete only for comets of 16 mag and brighter; on the Galilean satellites, such comets would form craters 60 km in diameter and larger, according to Shoemaker and Wolfe's scaling relations (which are adequate for this purpose), where the observed crater populations on Ganymede and Callisto are rapidly vanishing into statistical nothingness (either such craters rarely formed or they relaxed away). Reasonable completeness exists only for those long-period comets that would produce basins 300 km in diameter and up. For the extinct comets, which Shoemaker and Wolfe believe dominate the Galilean satellite cratering, there was only one known example at the time they published their analysis: the asteroid 944 Hidalgo. Shoemaker and Wolfe make plausibility arguments about how complete this sampling is, and about how big Hidalgo is. Their "statistics-of-one" arguments have been justified, in part, by the subsequent discovery of two new Earth-approaching Jupiter-crossers. Shoemaker and Wolfe arbitrarily adopt the identical slope for the population index of all comet populations (long-period, short-period, extinct—both ancient and modern). The extrapolations of comet numbers down into the size range of observed craters is, perhaps, the best that can be done, but it is without much justification. The situation is even worse for application to the Saturn system; the existence of the Saturn family of comets, while plausible, is based on a single, giant cometlike body, Chiron.

Shoemaker and Wolfe's preliminary estimates of the cratering efficacy of comets led to a prediction of a higher cratering rate on Earth and Moon than is observed. They use this discrepancy to decrease the comet flux by a factor of 8 (which implies either that comet photometry statistics are erroneous or that comets have higher albedos than were assumed [0.03]). However, some fraction of cratering in the Earth-Moon system is thought to be due to Earth-crossing objects derived from the asteroid belt rather than from the comet population. If noncometary objects dominate the terrestrial planet cratering rate, then the connection of the inner planet chronology to the outer solar system is weakened. It is by no means certain what fraction of cratering impacts on the terrestrial bodies is due to comets, active or extinct.

Whatever the correspondence is between the modern cratering populations in the solar system, it is very difficult to know what happened in the Jupiter and Saturn systems during the time of the late heavy bombardment in the inner solar system. As Shoemaker and Wolfe point out, if the late heavy bombardment projectile population had orbits resembling the modern Earth-crossers, they would have had negligible effect on the outer planet satellites (even including the fraction that is perturbed into transient Jupiter crossing orbits). While the existence of a late heavy bombardment in the inner solar system is generally accepted, its characteristics remain the subject of debate, and there is a variety of hypotheses about the origin of the bombardment. One hypothesis (Wetherill 1975) invokes the tidal breakup of a late-arriving planetesimal from the Uranus-Neptune zone by close passage near the Earth or Venus; the tidal breakup model now seems physically doubtful (Mizuno and Boss 1985). More recently, Wetherill (1981b) proposed that the late heavy bombardment was due to a long-lived decay of Earth-Venus zone planetesimals temporarily stored in purely Mars-crossing orbits before being perturbed again into other planet-crossing orbits. Although Wetherill again relied on tidal breakup to account for some ancillary details of the bombardment, the essential story remains valid independent of the validity of tidal breakup. For our purposes, the important point is that this bombardment would *not* affect the satellites of Jupiter or Saturn.

Shoemaker and Wolfe adopt a hypothesis that is simpler than Wetherill's, but not necessarily correct: the majority of outer-planet satellite cratering is due to a swarm of projectiles having a cometlike size distribution (that artifice of completeness calculations and extrapolations) but a temporal history identical to the late heavy bombardment in the inner solar system. This scenario could be true, but it is equivalent to neither Wetherill's original nor later hypothesis, nor others of which we are aware. Although an early "cometary" bombardment of outer-planet satellites is plausible, the form and even existence of any *late* heavy bombardment on the outer-planet satellites depends on Shoemaker and Wolfe's hypothesis being correct. The inferred geologic chronologies of Plescia and Boyce (1985) and others in turn depend on this unproven 4 Gyr ago late heavy bombardment in the outer solar system.

Our own arguments for the existence of four or five distinct populations do not seem to be consistent with the Shoemaker and Wolfe scenario and calculations, to the degree that they involve a single size-frequency distribution for comets and cometlike bodies that have struck the inner planets and outer-planet satellites in both ancient and modern times. Only the Saturn population

II is a physically separate population in the Shoemaker scenario. We believe, despite some of our earlier caveats, that it would be extremely difficult to reconcile all the other four populations as being identical to each other. Shoemaker (personal communication, 1984) grants the point that populations of cometary bodies might have different forms from original inhabitants of the Oort cloud due to cometary evolutionary processes, such as ablation, collisions, splitting and disintegration. But to the degree that he insists on a temporal and dynamical connection between different populations, he would still be in difficulty because his quantitative calculations are based on extrapolations of a single size-frequency relation for all such comets.

D. Conclusions

One way to reconcile the identification of four or five different cratering populations with the asteroid-comet populations in the solar system is as follows. We can identify the Ganymede-Callisto population as being due to comets. To the degree the inner-planet post-mare population is different, it reflects the augmented contribution of asteroids and asteroid fragments from the main asteroid belt, plus physically evolved captured short-period comets, in the inner solar system. The old highlands craters on the terrestrial planets reflect the circumstances (perhaps special?) that created the late heavy bombardment population. The dominant, or comparatively recent, bombardment of Saturnian satellites by the planetocentric population II is unique to Saturn; those craters overwhelm or obliterate much of the population I cratering record. If poorly-characterized population I is actually different from the comet population represented on Ganymede and Callisto, it could reflect a different epoch of ancient crater formation (e.g., since wiped clean on Ganymede and Callisto) or a different size-distribution of comets in the Saturn family from those extinct, perhaps partially evolved objects in the Jupiter family of short-period comets. Of course, the modern comet population (population I, which may or may not be similar to the Ganymedean crater population) must continue to express itself on the surfaces of the satellites of Saturn; but, in this picture, we suggest that it is overwhelmed by population II.

If our conclusions are correct, then the absolute interplanetary correlation of geologic time cannot be inferred reliably following the Shoemaker and Wolfe arguments. The most secure statement that can be made is that the Earth-Moon cratering record establishes an upper limit to the possible cratering rates on the Galilean satellites by comets. If our suggestion that it is the cometary population that is responsible for Galilean satellite cratering is incorrect, then no connections are secure. Despite this gloomy conclusion, however, the identification of several distinct cratering populations holds promise for further insights about the evolution in space and time of small-body populations in the solar system. For example, planetocentric populations are widely thought to be responsible for the Saturn system cratering and cannot be absolutely ruled out for the Jupiter system. If such debris is actually created by the catastrophic disruption of large satellites, we have potentially significant evidence about the outcomes of large-body breakups with potential application to the evolution of Phobos and Deimos, the asteroids, and possibly even the early comet population. On the other hand, a planetocentric population could be some kind of dynamical or collisional remnant of a primordial population of planetesimals or satellites. Could there be dynamically stable populations of bodies in Jupiter orbit beyond the Galilean satellites from which cratering projectiles could be derived by fragmentation and long time scale dynamical evolution? Farinella et al. (1986) have studied the question of cratering inner Saturnian satellites by debris from Hyperion, perturbed by Titan. Early Voyager Imaging Team results for Uranus suggest a mix of population I and II craters, similar to that exhibited in the Saturn system.

Although we find that the best published scenario linking planetary and satellite cratering (the work of Shoemaker) is not yet sufficient basis for establishing a paradigm for the interplanetary correlation of geologic time, we do not regard the situation as hopeless. Further observational and theoretical work on small-body populations (comets and asteroids) will improve our understanding of their collisional and orbital evolution and, in particular, of the physical attributes of comets. Also, the more detailed study of satellite cratering populations—including application of the powerful methodology of studying craters distinguished by morphological classes—will greatly augment the preliminary, rather coarse generalizations available in the literature and summarized in this chapter.

Acknowledgments. This research was supported by a grant and a contract from the National Aeronautics and Space Administration. We sincerely thank J. A. Burns, W. K. Hartmann, J. J. Lissauer, J. M. Moore, R. M. Schmidt, E. M. Shoemaker, L. A. Soderblom, P. M. Spudis, and R. J. Strom for reviews and comments on all or part of this long-gestating manuscript. The Planetary Science Institute is a division of Science Applications International Corporation.

APPENDIX

Cratering Simulations

We give here a brief description of a computerized cratering simulation model that is used in Sec. III for testing various hypotheses about the origin of observed crater distributions on planetary satellites. The code is written in Basic, with machine-language modules, for a small microcomputer. It takes an input production function, specified by a series of power law segments, and generates craters which are centered randomly within a rectangular grid of 96,000 pixels. Craters are defined as black circular rims. Any pixel falling within the circle is reset (obliterated) as are most pixels falling within the width of an ejecta blanket, which is a specified fraction of the crater diameter.

Generated craters larger than 8 pixels in diameter are recorded in two ways: as a circle on the 96,000 pixel map area and as an entry in a table, with specified diameter, x,y coordinate, and approximate relative time of creation. Smaller craters are recorded only as circles on the map. Craters smaller than 4 pixels in diameter are not shown as circles (and hence are not recorded) but they do have the effect of obliterating previously generated crater rims. Craters are generated with diameters spanning the range 1 to 200 pixels diameter. The program has an option for degrading craters by a process other than crater or ejecta-blanket overlap; craters may be degraded and obliterated at a rate that is a specified function of crater diameter.

At periodic intervals, statistical analysis may be performed on the evolving crater population. The computer program analyzes the crater map to determine what fraction of rim-pixels remain for each generated crater with a diameter of 8 pixels or more. A crater class is determined and tabulated, based on the percentage of rim which remains: class 1 = $>95\%$ of rim remaining; class 2 = 85 to 95%; class 3 = 50 to 85%; class 4 = 30 to 50%; class 5 = 10 to 30%. While class 5 craters may occasionally be recognizable, in general they are not and they are lumped together with craters with $< 10\%$ of rims remaining as obliterated craters. The sum of classes 1 to 4 is regarded as the total remaining crater population. The same criteria are, in principle, applied to craters with diameters $4 < D < 8$ pixels, which are identified in the map by a pattern-recognition scheme. However, it is not generally possible to recognize highly degraded (class 3 and 4) craters if they are very small; this limitation applies also to crater counts from spacecraft imagery. The smallest craters, for which we do not even generate black rims, cannot be counted in the statistics, but their numbers would be inaccurate anyway because the effects on them of craters < 1 pixel cannot be modeled. In many ways, then, this model involves a dynamical range of 200; there is a complete tabulation of craters, including highly degraded ones, over only a factor of 25 in diameter. Another limitation of the model is the fairly small map area. If the density of craters is substantially below 0.1 on the R-plot, there is often considerable statistical uncertainty. Most of the conclusions discussed in the text result from multiple runs, which were done to beat down the statistical noise.

The efficient use of computer memory for recording craters [the use of a table and a picture-map-memory for the less numerous longer-lived larger craters but only picture-map-memory for the numerous smaller (but short-lived) craters] combined with the virtually limitless, free use of a microcomputer help to overcome some of the problems encountered by Woronow (1978) in his somewhat similar cratering model which was written for a mainframe computer.

Figure 22 shows the results of a run intended to simulate saturation cratering by a production function of the same shape as the observed crater population on the lunar highlands (the ejecta-blanket obliteration circle is 1.1 times the crater diameter in this simulation). The production function, represented

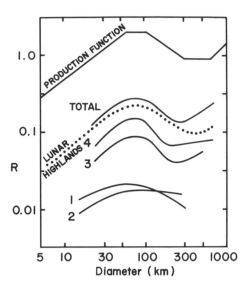

Fig. 22. *R*-plot diameter-frequency curves for a lunar-highlandslike production function. Curves are shown at the completion of the simulation for the four morphological classes of craters defined in the text as well as for their total. The dotted line is the *R* plot observed for the lunar highlands, as shown in Fig. 18.

here by five power law segments, generated sufficient craters to reach about 2.0 on the *R*-plot (the cumulative area of all emplaced craters was about 14 times the area of the map). The resulting curve at the end of the simulation (labeled Total) falls everywhere below 0.3; most simulations with this production function yield curve shapes and *R*-plot levels similar to Total after the cumulative area of emplaced craters reaches only 2 or 3 times the area of the map. This simulation confirms Woronow's (1978) finding that, even in saturation, the observed crater population resembles the production function (if it is like a lunar highlands population). However, the simulation contradicts Woronow's conclusion that the observed level on the lunar highlands lies far below the saturation level; counts in pure lunar highlands are almost identical to the curve Total in Fig. 22. As discussed in Sec. III, Woronow's model has limitations, or makes assumptions, that we feel are less realistic than those employed here.

The absolute levels of the curves in Fig. 22 for classes 1 to 4 reflect the class definitions given above. Few crater counts are available yet in the literature for different classes of craters on the outer-planet satellites. Such statistics can be very helpful, however, for distinguishing between different models for the evolution and degradation of a crater population. Before applying this model to such counts, when they become available, it may be necessary to adopt different definitions of rim completeness to match the definitions employed by those who count craters.

12. THE MOON

W. M. KAULA
University of California, Los Angeles

MICHAEL J. DRAKE
University of Arizona

and

JAMES W. HEAD
Brown University

The Moon is the fifth greatest of the natural satellites in absolute mass, and by far the largest in ratio to its primary if we disregard Charon. Its interior has evolved mainly through a cooling of its outer parts after extensive differentiation of a crust > 4.4 Gyr ago, so that impacts are the main shapers of the surface. The Moon's orbit has undergone considerable evolution because of tidal friction in the Earth, which has a 1/Q, averaged over time, at least 1000 times those of the major planets. The differences from other terrestrial bodies in bulk composition, as well as the high mass ratio to the primary, greatly complicate models for the Moon's origin.

I. INTRODUCTION

The Earth's Moon is anomalous in its mass compared to its primary; the ratio $1/81.30$ is the largest in the solar system by a factor of ten or more among the true planets (we exclude Pluto, now known to be only one-fifth as massive as the Earth's Moon). The Moon is also unusual in that (1) it has a mean density significantly lower than the densities of the terrestrial planets (allowing for compression): 3344 kg m^{-3}; (2) its petrology and seismology indicate an extraordinary dryness; (3) its oxygen isotope ratios constitute a tight group, overlapping ultramafic rocks of the Earth (in contrast to nearly all meteorites measured); and (4) it is quite distant from its primary: 60.39 mean planetary radii. The last of these anomalous properties is well understood as a consequence of exceptional dissipative character of the Earth, because of the oceans undulating between continents in response to the Moon's tidal attraction. The dryness strongly suggests that the Moon's low density comes from a deficiency in iron (by a factor of two-thirds) rather than an excess of volatiles. These two marked differences in bulk composition, together with the other anomalous properties listed above, make the origin of the Moon a special problem, inviting *ad hoc* treatment.

However, concerning the history of the Moon since its origin, a rather strong consensus has developed as to its nature and its evolution. The principal elemental constituents of the Moon are similar to those of the Earth's mantle plus crust, the main difference being the deficiency of volatiles. There might be a complementary enhancement in refractory elements (principally aluminum and calcium), but this is not absolutely required by the data. Most of the Moon is covered by aluminous rocks, but their character is quite different from continental rocks on Earth; the lunar terrae are clearly the consequence of large-scale meltings early in lunar history, not a several-staged processing throughout its evolution.

The different bulk composition and comparatively simple differentiation history have led to some major differences of lunar rocks from Earth rocks. The leading differences in rock characteristics are:

1. A lack of volatiles: no perceptible H_2O and lower levels of K_2O and Na_2O;
2. Low oxidation level: no Fe_2O_3, in addition to lesser indicators, such as Cr^{2+} and Eu^{2+};
3. Lower silica content: SiO_2 usually $< 48\%$;
4. Higher content of refractory siderophiles, such as Al_2O_3;
5. Higher mafic components: FeO, MgO, and TiO_2.

The early differentiation of the bulk of the lunar crust, > 4.4 Gyr ago, together with the Moon's smallness, led to an internal evolution which has been much less vigorous than those of the terrestrial planets (even Mercury underwent the early differentiation of a large core), and dwindled rapidly. Consequently, the main shaper of the surface, for at least 4 Gyr, has been impacts, which even controlled the locales of the relatively slight volcanism (compared to the Earth's); the basaltic maria are all situated in basins created by earlier impacts. It also is quite clear that most of this volcanism was completed more than 3 Gyr ago.

As consequences of this history, the Moon has a stratigraphy dependent mainly on cratering; a dichotomous petrology and geochemistry; some rocks reflecting early large-scale differentiations (much metamorphosed by impacts) and others indicating later partial melts at depth of small percentage; and a geophysics which indicates a moribund interior. Aside from some surprisingly strong remanent magnetism, the main puzzles, both dynamical and cosmochemical, arise from the origin of the Moon: in particular, how the protolunar material lost its iron and volatiles, and why so much matter remained in orbit around the Earth rather than being drawn in as the planet grew.

Study of the Moon has contributed to study of the other natural satellites mainly by stimulating techniques and insight concerning spectral properties, reflectivity, impacts, rotational dynamics, and orbital evolution. It is likely that Peale and colleagues would not have scored their coup of predicting the thermal behavior of Io (Peale et al. 1979) if they had not been frustrated in showing that tidal dissipation in the Moon was ever significant (Peale and Cassen 1978). However, even in this problem the key realization that Io has an orbital eccentricity forced by the resonance with Europa did not come from studying the Moon (chapter by Peale). Io and Europa are the two satellites that are by far the most similar to the Moon in size and mean density. However, their existence as members of the Galilean system around the giant planet Jupiter and the considerable volatiles in their composition indicate that the circumstances of their origins were probably quite different than that of the Moon, while the absence of craters and abundant presence of endogenic features manifest that their recent evolutions have also been quite distinct from the Moon's.

The few leading bulk properties of the Moon are given in Table I (cf. Tables III and IV of the Introduction). The next three sections of this chapter are in order of the decreasing accessibility of data: geology (Sec. II), petrology and geochemistry (Sec. III), and geophysics of the Moon (Sec. IV). The concluding briefer sections discuss the bulk composition (Sec. V), orbital evolution (Sec. VI), and origin of the Moon (Sec. VII).

TABLE I
Properties of the Earth's Moon[a]

Property	Symbol	Absolute Units	Ratios to Earth Parameters
Mass	M	7.347×10^{22} kg	$0.0123003\ M_{\oplus}$
Mean radius	$\bar{R}$	1.7375×10^{6} m	$0.275153\ R_{\oplus}$
Mean density	$\bar{\rho}$	3.344×10^{3} kg m^{-3}	$0.59045\ \rho_{\oplus}$
Moment-of-inertia ratio	I/MR^2	0.3904	$1.1812\ (I/MR^2)_{\oplus}$
Mean motion & mean rotation	$n, \bar{\omega}$	2.6617072×10^{-6} s	$0.036601\ \omega_{\oplus}$
Mean distance	a	3.84747×10^{5} km	$60.3904\ R_{\oplus}$

[a] Based mainly on Bills et al. (1980).

II. LUNAR GEOLOGY

A. Overview

Visual observations of the lunar surface and analysis of photographic images allow the major geologic processes operating there to be identified and understood. They also relate these processes to the stratigraphy and history of the Moon (Baldwin 1949,1963; Shoemaker and Hackman 1962; Wilhelms 1970,1984). Such analyses indicate that three fundamental processes—impact cratering, volcanism, and tectonism—have been dominant in lunar history. A knowledge of geologic processes and history, when combined with geochemical and geophysical data, permits significant inferences to be made concerning the origin and evolution of the Moon as a planetary body. The unique combination of data derived from lunar exploration (Earth-based and orbital remote sensing, manned reconnaissance, *in situ* geophysical stations, and sample return) makes the Moon a primary body for the study and understanding of a variety of general problems for which data are inadequate or unavailable elsewhere in the solar system. Thus, in the context of planetary satellites and planets, the Moon is not so much a Rosetta Stone, but rather a reminder of the range of information required to understand fully the fundamental aspects of planetary bodies. A second lesson from the Moon is that, although extensive lunar exploration has provided data presently unavailable for most other planetary bodies, major gaps exist in our data base and knowledge of the Moon (no farside gravity coverage, lack of global topography and geochemistry, etc.). For this reason, we outline both our current understanding as well as some major unanswered questions. First we describe the major geologic processes operating on the Moon in order to provide a context for our coming discussion of lunar chemistry, petrology, physics, and structure.

B. Impact Cratering

Early observations of the Moon revealed the widespread significance of cratering as a process; later the Apollo exploration provided evidence for their impact origin. Craters ranging from microscopic pits on returned samples to impact basins > 1000 km in diameter are observed on the Moon. The large number of craters, combined with low degradation rates and the availability of high-resolution images, allows fresh crater morphology to be studied and complements the information derived by investigating partially eroded terrestrial craters (Grieve and Head 1981).

On the Earth, there is a transition in crater structure from simple, at small crater diameters, to complex at larger crater diameters (Dence 1968). Observations of fresh lunar crater forms show that simple craters are generally < 15 km diameter and are characterized by a bowl-shaped form, an elevated rim capped by ejecta, and a depth/diameter relationship of about 1 to 5 (Pike 1977) (Fig. 1). Crater forms become complex over the range of 15 to 20 km in diameter. In this transitional range, the circular rim becomes more

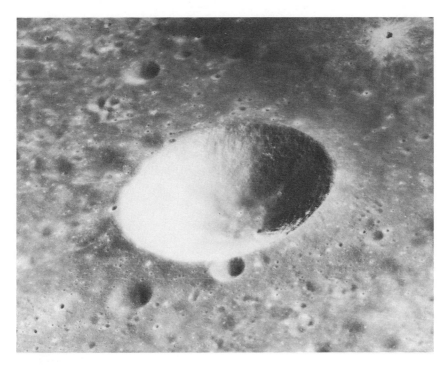

Fig. 1. A typical simple crater (Isidorus D) about 15 km in diameter with circular rims, smooth walls, and a bowl shape.

cuspate and scallop deposits occur at the base of the crater walls, indicating failure and inward slumping of the cavity walls. Complex craters (Fig. 2) are characterized by flat floors, well-developed wall terraces, central peaks, and a much lower depth/diameter ratio. The processes responsible for complex crater information can be assessed by combining the lunar data base with terrestrial data on crater substructure. For complex crater diameters of < 70 km, simple restoration of observed wall terraces to rim positions (pre-slumped) will recover the appropriate depth/diameter ratio (Settle and Head 1979). At larger diameters, slump processes alone are insufficient to account for observed volume variations and other processes—such as uplift and rebound of the cavity floor or basic changes in the cratering process—may be more important.

A second transition occurs from complex craters to basins. Between about 140 and 175 km diameter, some lunar craters display a fragmentary ring surrounding the central peak. At diameters greater than 175 km the central peak disappears and the structure is termed a peak-ring basin (Wood and Head 1976). Additional rings develop at diameters greater than 350 km; hence, these features are known as multiringed basins. The Orientale Basin is the

Fig. 2. The crater Aristarchus, approximately 40 km in diameter. The figure shows the terraced walls, flat floor, and central peak typical of a fresh complex crater.

most well-preserved example (Fig. 3), and contains three conspicuous rings: (1) The outermost Cordillera Mountain ring, defined by a 900 km diameter inward-facing circular scarp; (2) the intermediate Outer Rook ring, approximately 620 km in diameter and characterized by a continuous ring of massifs; and (3) the innermost Inner Rook ring, 480 km in diameter, comprised of a series of isolated mountain peaks, analogous to the peak rings in smaller peak-ring basins. A fourth feature, an inner depression of about 320 km in diameter, may be related to the thermal stresses associated with basin collapse and mantle uplift (Church et al. 1982). Although the origin of individual rings is controversial (Head 1974; Wilhelms et al. 1977; Croft 1981b), it is clear

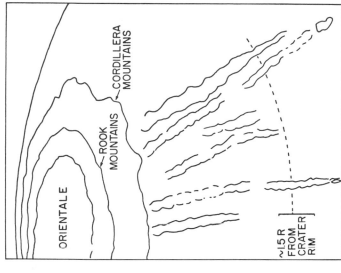

Fig. 3. The lunar Orientale Basin and ejecta deposits (LO IV-193M), looking north over the southeast portion of basin ejecta deposits to the basin. The Cordillera Mountain ring diameter is 900 km; the outer Rook Mountain ring is at 620 km. The sketch map shows major crater chains in the southeast. The dashed line about 1.5 R (crater radii) from the crater rim indicates radial position on Orientale deposits analogous to the distance of the Apollo 14 site from the Imbrium Crater rim.

TABLE II
Effects of Impact Cratering on the Lunar Crust[a]

Topography	Topographic differences of ≤ 8 km are created and the surface dichotomy between highlands and mare produced.
Gravity	Highs up to 200 mgal are associated with mascons within multiring basins; local lows are associated with ejecta and brecciation.
Magnetism	Local high-intensity anomalies may be associated with high-temperature impact deposits.
Seismic activity	Low seismic velocities in upper 20 to 25 km of crust are caused by fracturing and megaregolith development.
Distribution	Geologic units are related to major ejecta deposits, many recovered samples have been transported from their original locations.
Physical nature	90% of samples are impact products: breccias or melt rocks; many show the effects of shock.
Chemical nature	Bulk chemistry of samples is often a mixture of preexisting rock types, commonly enriched in meteoritic siderophile elements and with reset ages.

[a]Table after Grieve and Head (1981).

that the scale of these basin-forming events has strongly influenced many other aspects of lunar history and processes (Table II).

These basic characteristics of fresh crater forms, combined with terrestrial experimental and observational data and returned lunar samples, provide a general understanding of the physics of impact (see discussion in chapter by Chapman and McKinnon). The impacting body, traveling at perhaps 15 to 25 km s^{-1}, penetrates 2 to 3 times its radius and begins transferring the majority of its kinetic energy to the target rocks. An exponentially decaying compressional shock wave propagates radially into the planetary body with initial particle velocities measured in km s^{-1} and maximum pressures on the order of several megabars. The complex interaction of compressional and rarefaction waves deflects the original radial motions of particles upward and outward, beginning the excavation and ejection of target material to produce the transient cavity. Materials nearest the point of impact are subject to the highest pressures and are excavated first, followed by successive volumes of progressively less shocked material which land closer to the cavity rim (Oberbeck 1975). A volume of highly shocked and impact-melted material remains in the cavity and settles to the crater floor. The transient cavity is produced very rapidly (minutes even for craters measured in tens of km) and, for large craters undergoes rapid modification to produce the final crater form. Modification processes are poorly understood but involve massive uplift of the transient cavity floor through some combination of post-shock rebound of the cavity floor, and collapse of the unstable rim area (Grieve et al. 1977; Settle and Head 1979; Melosh 1981*a*).

Knowledge of the physics of cratering and the geology of the lunar surface permits one to assess the influence of this process (Table II). The flux of impacting projectiles and its variation with time dominates the degradation of surfaces by impact comminution and soil layer (regolith) production (see chapter by Veverka et al.). Younger lunar surface units (maria) are covered with a regolith of a few meters depth while the more ancient highlands are blanketed by a more coarsely fragmented layer of several km depth termed a megaregolith (Hartmann 1973). The *lateral* movement of material in the ejection process and the mixing of material during ejecta emplacement (Oberbeck 1975) has important bearing on the chemical and physical heterogeneity of the lunar crust and the actual origin of materials sampled at different locations on the Moon (Fig. 4). Similarly, the impact process is a significant contributor to *vertical* redistribution of crust and mantle materials. For example, Pieters (1982) has shown that the central peaks of Copernicus are characterized by the presence of olivine, a mineral uncommon in the surrounding highlands crust, but likely brought up from great depths by the modification stage of the cratering event. The significance of vertical and lateral redistribution of material is even more striking when considering the influence of a single impact basin on the lunar surface. The kinetic energy associated with the formation

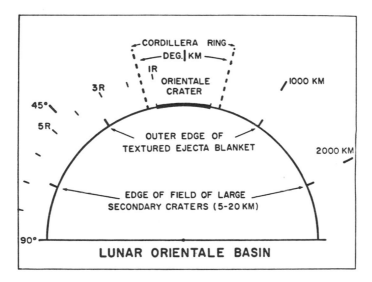

Fig. 4. Lunar Orientale Basin plotted on a lunar hemisphere showing the extent of various facies in kilometers, degrees, and crater radii (R). For the Imbrium Basin, the edge of the field of large secondary craters would extent to 180°.

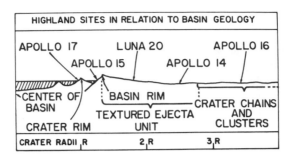

Fig. 5. Apollo highland sample return sites in relation to a composite hypothetical basin (figure from Head and Settle 1976).

of Orientale is estimated to be 10^3 to 10^4 times the present annual output of internal energy from the Earth. The basin-forming processes excavated 10^6 to 10^7 km^3 of material from a cavity whose original depth may have exceeded the thickness of the crust (Bratt et al. 1984) and redistributed this material over a lunar hemisphere, considerably modifying the surface geology of much of the Moon. Cavity collapse caused structural modifications and ring formation and approximately 2×10^5 km^3 of impact melt deposits settled to the basin floor (Head 1974). Analysis of the significance of samples collected at the various Apollo and Luna sites must take into account the location of the sampling sites relative to regional craters and basins (Fig. 5). In addition to implications for the *modification* of the lunar crust, impact processes appear to be a major source of energy for melting and crustal formation in the terminal stages of lunar accretion. Subsequent to the formation of the crust, impact basin formation also had a major influence on the nature and location of geophysical and thermal anomalies. Impact-basin formation caused crustal and lithospheric thinning, while transient cavity collapse caused major unwarping of isotherms and produced regional thermal anomalies (Bratt et al. 1981). Once formed, impact basins helped to localize the accumulation of mare lavas and the tectonically deformed lunar lithosphere (Solomon and Head 1979).

Many significant questions related to impact cratering on the Moon remain: e.g., the detailed nature of energy partitioning; the nature of impacting projectiles and their relationship to crater morphology; the role of substrate characteristics in crater morphology; the relationship of lunar crater morphology to terrestrial crater structure (Grieve and Head 1983); the relationship of breccia types to crater deposits; a more detailed understanding of the cratering process to lead to a quantitative assessment of craters as probes of crustal composition and structure; the depth of excavation and origin of rings associated with impact basins; and the nature and implication of extremely large impact basins (such as proposed by Whitaker [1981] and Wilhelms [1984]).

C. Volcanism

The lunar maria represent the most obvious and well-documented example of extrusive volcanism. The early recognition of lobate flow fronts followed by later sample collection established the volcanic and basaltic nature of the lunar maria. Lunar mare deposits, identified on the basis of their low-albedo and spectral-reflectance characteristics (Pieters 1978) cover approximately 17% of the surface (Head 1976), with the vast majority occurring on the Earth-facing hemisphere (Fig. 6). Mare deposits (Fig. 7) occur primarily in low areas produced by impact crater and basin formation (Head 1976) and have thicknesses ranging from meters to several kilometers (DeHon 1974; Head 1982).

Mare volcanic structures and features provide data from which eruption conditions can be inferred. Major features (Schultz 1976; Head 1976) include *flow fronts* which are relatively uncommon but which reach 30 to 35 m height in Mare Imbrium (Schaber 1973). *Sinuous rilles* are meandering troughs ranging from a few to several hundred km in length and up to 3 km in width. These enigmatic features are larger than lava channels found on Earth and often resemble terrestrial river channels with extreme meandering and occasional oxbows. Originating in linear to circular depressions at the edge of the lunar maria, these structures wind their way downslope, often toward central mare deposits. *Mare domes*, comparable to small terrestrial shield volcanoes in size and morphology (Head and Gifford 1980; Smith 1973), occur in various parts of the lunar maria and appear to represent local source vents for plains-type volcanism (Greeley 1977). A number of *cones* similar to terrestrial cinder cones have also been observed on the Moon (McCauley 1967). Numerous *dark-halo craters* with probable volcanic origins have been mapped on the Moon. These craters are surrounded by a halo whose albedo is comparable to dark mare materials, rather than the bright ejecta associated with most impact craters. *Dark-mantle deposits* are extensive units of low-albedo material located in uplands adjacent to several maria. They are relatively thin (tens to hundreds of meters) and appear to blanket surrounding terrain. Early workers considered these deposits to be of pyroclastic origin (Wilhelms 1970), but the locations of source vents and their mode of emplacement were not understood. Other features include *mare ridges* (see Fig. 8), which occur in the mare but are generally believed to be largely of tectonic origin (Lucchitta 1977a; Sharpton and Head 1981a); *craters* of possible volcanic origin ranging from small collapse craters on lunar mare surfaces (Schultz 1976) to larger features such as the Crater Kopff (McCauley 1968); and several *volcanic complexes* (Whitford-Stark and Head 1977) located primarily in Oceanus Procellarum (e.g., Marius Hills [McCauley 1967], Aristarchus-Harbinger region [Zisk et al. 1977], and the Rumker Hills [Scott and Eggleton 1973]).

Lunar volcanic features may be analyzed and lunar eruption conditions

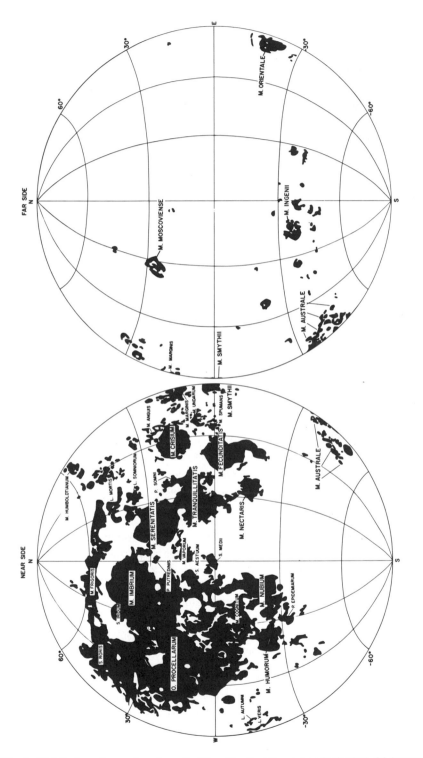

Fig. 6. Distribution of mare deposits on the Moon (Lambert Equal Area Projection). M denotes mare; O, Oceanus; S, Sinus; P. Palus; L, Lacus (figure from Head 1976). Compare to the lunar map shown in the Map Section.

Fig. 7. The lunar mare. This view shows relatively young mare lavas in southwestern Mare Imbrium lapping up onto the ejecta deposits of the crater Euler (A515-1701M).

inferred in terms of the major factors analyzed involved in the ascent and
eruption of lava on the Earth and Moon (Wilson and Head 1981,1983a). The
high mass eruption rates suggested by the extreme length of some lunar lava
flows and sinuous rilles require fissure systems with widths up to about 10 m.
This appears to be consistent with a lunar tectonic regime dominated by fractures associated with major impact basins and an extensional state of lithospheric stress correlated with the heating associated with the period of mare
volcanism. Calculations show that lower effusion rates could produce cinder
cones and spatter ridges up to a km wide along fissures, and central vents
could produce domes and cone deposits with diameters up to 5 km. If eruption
rates exceed about 10^7 kg s^{-1}, turbulence and thermal erosion will occur, producing a linear or circular depression around the vent area and eroding a sinuous rille downslope from the vent (Carr 1974; Head and Wilson 1980). The

absence of an atmosphere on the Moon insures that the disruption of magma into pyroclasts was common, even though volatile content of magmas was low. As the gas clouds expanded, pyroclasts were accelerated away from the vent. As gas density decreased, pyroclastics became separated from the gas flow and followed ballistic trajectories out to several hundred km. The dark mantle deposits, comprised of widespread areas of submillimeter-sized pyroclastic droplets, are believed to have formed this way. Isolated dark-halo craters, on the other hand, are thought to have developed by processes more closely related to terrestrial vulcanian eruptions (Head and Wilson 1979).

Photogeologic and remote-sensing data can be used to subdivide the lunar maria and to establish mare stratigraphic units in order to assess volcanic flux and style in space and time. There is strong evidence that mare volcanism began prior to the last major impact basins (Ryder and Taylor 1976; Schultz and Spudis 1983), but the accelerated impact flux obscures the details of timing, style, and abundance. Stratigraphic studies of mare deposits post-dating the Imbrium Basin have shown that the earliest deposits (3.5–3.8 Gyr) are the most significant (Head 1974; Howard et al. 1974; Boyce 1976; Whitford-Stark and Head 1980; Basaltic Volcanism Study Project 1981). Although younger deposits cover larger areas, they are relatively thin and less voluminous. Recent studies suggest that small surface eruptions could have occurred as late as 1 Gyr ago (Schultz and Spudis 1983). The general decrease in volume as a function of time is consistent with models of thermal evolution and the changing state of stress in the lithosphere (Solomon 1978; Solomon and Head 1980).

The high impact flux characteristic of the first 600 Myr of lunar history has obscured morphologic evidence of premare and nonmare volcanism. In addition to remote sensing data that may provide evidence for the location of rock types (such as KREEP: see Sec. III) characteristic of nonmare volcanic deposits (see, e.g., Spudis 1978; Hawke and Head 1978), there is some morphologic evidence for volcanic deposits of nonmare composition that may be synchronous with early mare deposits. Malin (1974) and Head and McCord (1978) have described a series of spectrally distinct domes of probable volcanic origin that are morphologically different from lunar mare domes. These features, exemplified by the Gruithuisen and Mairan Domes in northern Oceanus Procellarum, are similar to features associated with more viscous and explosive terrestrial magmas. Their relationship to known rock types is not established.

Sample and remote-sensing data suggest that mare basalt volcanism varied in composition and abundance. Early mare volcanism was concentrated in the eastern half of the Earth-facing hemisphere while the younger deposits are primarily in the central and western sides of this hemisphere. There is still considerable uncertainty, however, in the composition of lunar maria in space and time and the implications of these correlations for source area homogeneity or heterogeneity. Additional questions relate to the significance of volcanic complexes, the nature and extent of nonmare volcanism, the relationship of impact-basin formation and volcanism (particularly in reference to large features such as the Procellarum Basin), and the mechanisms of magma generation and ascent through the crust and lithosphere.

D. Tectonism

The diversity of tectonic features displayed by the Moon is narrow compared to the Earth. There is no evidence for large-scale strike-slip deformation on the Moon. Most tectonic features are local in scale and associated with mare basins. Linear rilles (Fig. 8) are flat-floored troughs ranging up to several km in width and extending in straight or arcuate patterns for tens to hundreds of km in length. These features are analogous to terrestrial graben and are believed to be extensional in origin (McGill 1971). Mare ridges (or wrinkle ridges) consist of linear segments that merge, overlap and are arranged *en echelon* in complex patterns that extend for hundreds of kilometers. The structure consists of a broad arch as much as 10 to 20 km in width and

Fig. 8. A view of the western edge of Mare Serenitatis showing linear rilles (lower center and right) and mare ridges (center and right). Note that linear rilles are flooded by later mare basalts which in turn are deformed by mare ridges (A17-0953M).

several hundred meters in height, and a narrow superposed ridge that meanders across, but is generally parallel to the broad arch. Mare ridges generally occur in systems that are subconcentric and subradial to the maria (Strom 1972; Lucchitta 1977a; Sharpton and Head 1981a). Although mare ridges have been attributed to both volcanism and tectonism, recent studies of surface and subsurface structure (Lucchitta 1977a; Sharpton and Head 1981a,b) strongly support a tectonic origin related to subsidence, which produces vertical and compressional stresses in the lunar maria. Linear rilles occur in relatively old mare units and appear to have formed prior to 3.6 (± 0.2) Gyr ago (Lucchitta and Watkins 1978). Mare ridges disrupt the youngest mare deposits and may have continued to form well into the last half of solar system history.

The lack of major crustal recycling on the Moon led Solomon (1977) to describe the Moon as a one-plate planet, characterized by a single unsegmented global lithospheric plate. The spatial distribution of rilles and ridges, their chronological sequence, the sequence and distribution of major volcanic units, and the topography of the present surfaces of the lunar maria, all suggest that lunar tectonism was dominated by loading, vertical movement and deformation localized in the lunar maria, but simultaneously related to the overall characteristics of thermal evolution. Specifically, the spatial and temporal distributions of linear rilles and mare ridges appear to be the product of two superposed stress systems: the local stress due to lithospheric loading by basaltic lava infilling of mare basins, and a global thermal stress associated with warming and cooling of the lunar interior (Solomon and Head 1980). Lunar thermal history models (chapter by Schubert et al.) that are consistent with the lack of global-scale extensional or compressional features predict an early period of modest global expansion and lithospheric extension during the first billion years of lunar history, followed by an extended period of modest planetary contraction and lithospheric compression lasting to the present.

Using plate-flexure theory, these observations can yield estimates of the thickness of the elastic lithosphere at the time when tectonic features were formed. Solomon and Head (1980) have outlined evidence for an increase in the effective thickness of the elastic lithosphere between about 100 Myr after basin formation and the time when volcanic infilling terminated. Lithospheric thickness was sufficient subsequent to mare volcanism to have maintained much of the final load, as indicated by the positive gravity anomalies or mascons (see Sec. IV). Major lateral heterogeneity of lithospheric thickness existed 3.6 to 3.8 Gyr ago as indicated by the varying tectonic response to imposed loads. Regions of thin lithosphere are correlated with regions of anomalous volcanic activity.

The Moon shows little evidence in its geological features for global stresses of sufficient magnitude to cause widespread lithospheric failure (cf. chapter by Squyres and Croft). Strom (1964) and others have suggested grid-like arrangements of lineaments which may represent fractures perhaps caused by tidal or thermal stresses early in lunar history. The Moon has been a one-plate planet at least since the formation of its thick global crust. Its early tectonic history was dominated by structural features and lithospheric thinning associated with impact crater and basin formation. The observed tectonic features (graben followed by mare ridges) are primarily the result of stresses produced by local loading by mare basalts superposed on global stresses related to internal heating and cooling. These local mare loads caused vertical tectonic movement (downwarping) of the globally continuous lithospheric shell. Variations in lithospheric thickness in space and time varied the style and extent of vertical tectonics (Head and Solomon 1981).

Major unresolved questions about the tectonics of the Moon include the nature of early heterogeneities in the thickness of the lithosphere, the relationship of impact energy and local thermal anomalies, the origin and significance of floor-fractured craters, and the role and extent of viscous relaxation in early lunar history.

E. History

Stratigraphic studies of the Moon have provided a framework for lunar history (Shoemaker and Hackman 1962; Wilhelms 1970,1984), the main subdivisions of which are shown in Fig. 9. From a geological point of view, crustal deformation, impact cratering, and basin formation dominated the first 600 to 700 Myr of lunar history, and obscured the role of nonmare and early mare volcanism. Basin formation appears to have been a major factor in crustal and lithospheric thinning and in the production of the first-order topography of the lunar surface. As the rate of impact cratering decreased, mare volcanism dominated the surface geology, decreasing in significance as a function of time. Although the age of the youngest flows is uncertain (Schultz and Spudis 1983), surface volcanism essentially ceased by about 2.5 Gyr ago. The tectonic history of the Moon is linked to its thermal evolution. Subse-

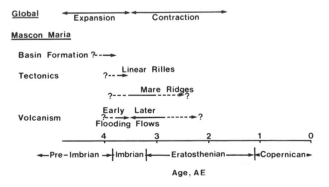

Fig. 9. Summary of lunar chronology (figure adapted from Wilhelms 1984, and Solomon and Head 1980).

quent to the formation of a globally continuous crust, the Moon was a one-plate planet with regional lithospheric thickness variations. Before about 3.6 Gyr, the Moon underwent heating and mild expansion resulting in an extensional state of stress in the lithosphere. These conditions were favorable to basaltic magma production and the ascent and emplacement of lavas, which collected in the impact basins. Lithospheric loading and flexure in the mare basins formed grabens (linear rilles). As the lunar interior cooled, basaltic magma production decreased, the Moon began mild contraction, the state of stress in the lithosphere became compressional, and the elastic lithosphere thickened. Smaller volcanic loads were applied to the surface, deformation was primarily compressional (mare ridges), and the lithosphere thickened sufficiently to support the later volcanic loads, contributing to the positive gravity anomalies (mascons). In the last one-half of solar system history, geologic activity on the lunar surface was limited primarily to exogenic bombardment.

III. LUNAR PETROLOGY AND GEOCHEMISTRY

A. Overview

Lunar rocks may be divided into three basic categories:

1. Terrae which formed by accumulation of minerals differing in density from their parent silicate material—cumulate rocks which are predominantly felsic and associated with early lunar differentiation;
2. Other terrae types identified primarily as components of soil and breccia which are rare among returned rock samples;
3. Mare basalts—lavas with ages significantly less than the age of the Moon.

Typical bulk compositions of these rock types are given in Table III. Felsic cumulate rocks and other terrae types are often substantially modified due to the early period of intense lunar bombardment. Mare basalts, on the other hand, are relatively unaffected by impact processes.

The three basic categories of lunar rocks may be related to one another through a model involving global lunar differentiation, possibly in a magma ocean, commencing contemporaneously with lunar formation. During this event, plagioclase effectively separated from mafic phases due to its greater buoyancy. At least three types of felsic cumulate rock groups exist: ferroan anorthosites, Mg-norites, and Mg-gabbronorites (Table III). One of these groups, possibly ferroan anorthosite, may be a remnant of the original lunar crust, while the others may have formed as igneous intrusions into that original crust. Complementary mafic cumulate rocks comprise the upper mantle of the Moon and most workers agree that they are involved in the subsequent genesis of mare basalts, although the nature of that involvement remains controversial. KREEP, a component enriched in lithophile-incompatible elements including potassium (K), the rare earth elements (REE) and phosphorus (P),

TABLE III
Bulk Compositions of Typical Lunar Rock Types[a]

	Ferroan Anorthosite 15415	Norite 77075	Troctolite 76535	Dunite 72417	Gabbronorite 61224
SiO_2	44.1	51.2	42.9	39.8	50.7
TiO_2	0.02	0.33	0.05	0.03	0.4
Al_2O_3	35.5	15.0	20.7	1.3	13.2
FeO	0.23	10.7	5.0	11.9	9.91
MnO	—	0.17	0.07	0.11	0.16
MgO	0.09	12.9	19.1	45.4	12.77
CaO	19.7	8.82	11.4	1.1	11.6
Na_2O	0.34	0.38	0.20	0.01	0.91
K_2O	—	0.18	0.03	—	0.02
Cr_2O_3	—	0.39	0.11	0.34	0.29
P_2O_5	0.01	—	0.03	—	—
TOTAL	99.99	100.1	99.6	100.0	99.96

	LKFM 14310	MKFM 15386	KREEP 15405	Granite 12013 Felsite	High-Ti 75055	Low-Ti 15455	VLT 24109	High-Al 14053
						Mare Basalts		
SiO_2	47.2	50.8	56.67	75.1	40.6	45.2	45.2	46.4
TiO_2	1.24	2.23	1.85	0.62	10.8	2.41	0.89	2.64
Al_2O_3	20.1	14.8	12.57	12.24	9.67	8.59	13.8	13.6
FeO	8.38	10.6	13.49	1.44	18.0	22.2	20.5	16.8
MnO	0.11	0.16	—	—	0.29	0.30	0.27	0.26
MgO	7.87	8.17	3.61	0.72	7.05	10.3	6.35	8.48
CaO	12.3	9.71	8.97	1.44	12.4	9.82	12.7	11.2
Na_2O	0.63	0.73	0.89	1.44	0.43	0.31	0.24	—
K_2O	0.49	0.67	1.93	7.00	0.08	0.04	0.01	0.10
Cr_2O_3	0.18	0.35	—	—	0.27	0.68	0.19	—
P_2O_5	0.34	0.70	—	—	—	—	—	—
TOTAL	98.8	99.0	99.98	100.00	99.5	99.8	100.2	99.5

[a] Data from S. R. Taylor (1982) and James and Flohr (1983). The text identifies the abbreviations used for rock types.

may have been derived from the last residue of this global differentiation event. Arguments supporting this model are presented in Sec. III.

An issue which remains unresolved is the question of whether the global differentiation event was an encircling magma ocean or a series of magma bodies which were smaller, shorter-lived, and not necessarily contemporaneous. If a magma ocean did exist, it is unclear to what depth the Moon was melted. There is some constraint from thermal history considerations, but the main problem is inadequate understanding of lunar origin and mechanism of assembly. The reader is directed to Wood et al. (1970), Wood (1975), Herbert et al. (1977a,1978), Solomon and Chaiken (1976), Solomon and Longhi (1977), Longhi (1980), Binder and Lange (1980), and Walker (1983) and the chapter by Schubert et al. for further reading.

B. Terrae Cumulate Rocks

The presence of small, plagioclase-rich fragments in the Apollo 11 regolith led to recognition that the lunar terrae were dominated by rocks in the anorthosite-norite-troctolite clan (Smith et al. 1970; Wood et al. 1970). However, the identification of distinct rock suites in the lunar terrae was hampered by the fact that most lunar terrae samples are polymict breccias, mixtures of unrelated rocks resulting from intense bombardment during the first half billion years of lunar history. A major advance occurred with the development of criteria for the identification of pristine rocks (Warren and Wasson 1977), providing a rational basis for the recognition of terrae rocks or rock fragments which have survived bombardment to preserve chemical information, even though isotopic systematics may be disturbed. When only pristine rocks are considered, two terrae cumulate groups immediately become apparent: the ferroan anorthosite trend and the Mg-suite trend (Fig. 10).

These two groups are disjunct, and it is not readily apparent that they could both form in the same global differentiation event. Clues to their geneses are provided by the study of the Stillwater Intrusion in Montana, a 2.7 Gyr old, layered terrestrial igneous complex with more than 5 km of exposed section, which is a possible analog of the lunar crust. Raedeke and McCallum (1980) have shown that rocks from the upper and lower banded zone of the Stillwater Intrusion mimic the lunar Mg-suite trend while rocks from the middle-banded zone mimic the lunar ferroan trend (Fig. 11). Note that the two trends overlap in the Stillwater but not in the lunar terrae. Detailed calculations (Raedeke and McCallum 1980) show that the lunar trends cannot be generated in the same way as the Stillwater trends. However, the Stillwater middle-banded zone is comprised largely of anorthosite, as is the lunar ferroan anorthosite suite. This correspondence suggests that the anorthosite trend results from fractional crystallization of mafic phases from melt trapped in the interstices of a feldspathic cumulate. Viewed in this context the ferroan anorthosites are most likely to represent original lunar crust, while the Mg-suite trend may represent one or more intrusions into that crust (James 1980).

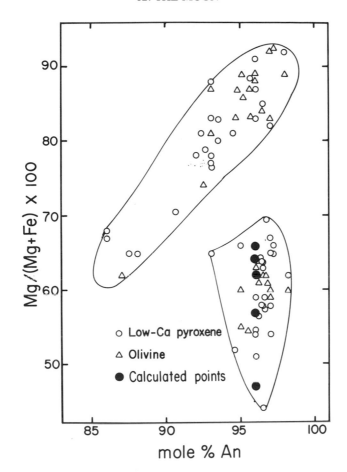

Fig. 10. Plot of mole percent anorthosite (An) in plagioclase versus mole percent Mg/(Mg + Fe) in coexisting mafic minerals for pristine lunar highlands samples (figure from Raedeke and McCallum [1980]; their paper gives details concerning calculated points).

Detailed studies suggest that the Mg-suite consists of at least two distinct groups, the Mg-norites and the Mg-gabbronites (James and Flohr 1983). The two groups are distinguished petrographically by the latter containing greater amounts of Ca-rich clinopyroxene, petrologically by the latter having somewhat less calcic feldspars at a given mafic phase composition (Fig. 12), and geochemically by the latter being distinct in a variety of trace elements (e.g., Fig. 13). This work suggests that our knowledge of lunar terrae rock groups may be restricted simply as a result of limited sampling. Lunar rocks are so complicated that, more than ten years after the last Apollo sample return, major scientific results are still forthcoming as a result of detailed studies.

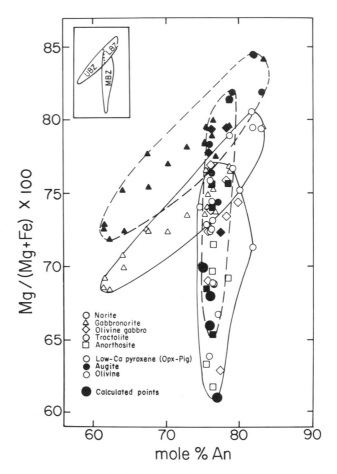

Fig. 11. Plot of mole percent anorthosite (An) in plagioclase versus mole percent Mg/(Mg + Fe) in coexisting mafic minerals for samples from the banded zone of the Stillwater Complex. Dashed line encloses augite/plagioclase data and solid line encloses olivine-orthopyroxene/ plagioclase data (figure from Raedeke and McCallum [1980]; their paper gives details concerning calculated points).

If the Mg-suite rocks really were intrusions into a ferroan anorthosite crust, direct dating might be expected to show that the ferroan anorthosites predate the Mg-suite samples. Unfortunately, such time resolution is not possible. Isotopic systematics have been disturbed for most systems as a result of intense bombardment. In addition, the isotope systems most likely to "see through" intense bombardment (Rb-Sr, Sm-Nd) are exceedingly difficult to apply to essentially monomineralic ferroan anorthosites. Finally, the ages deduced for lunar rocks might reflect isotopic closure after slow cooling rather than record the time of emplacement. All available mineral isochron dates for

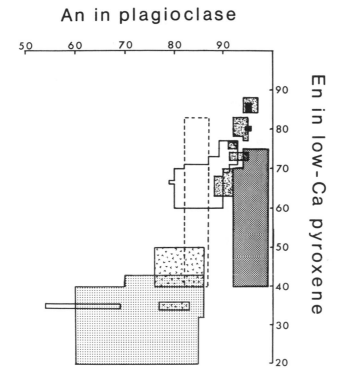

Fig. 12. Anorthosite (An) in plagioclase versus enstatite (En) in low-Ca pyroxene in Mg-gabbronorites (no pattern) and Mg-norites (irregular stipple), ferroan-anorthosites (gray), Apollo 14 troctolites (black), and alkali anorthosites (v's). KREEP is shown by dashed outline (figure after James and Flohr 1983).

pristine terrae rocks are given in Table IV. These rocks define a "total rock" Rb/Sr isochron of 4.38 (± 0.06) Gyr with an initial Sr isotopic composition indistinguishable from the basaltic achondrite best initial (BABI), an estimate of the initial solar system value (Carlson 1982).

Selected trace-element data for some lunar terrae rocks are shown in Fig. 14. Note that the ferroan anorthosite and the estimate of the average terrae composition (Taylor 1982) both have positive Eu anomalies, indicating that these rocks are close to the iron-iron oxide buffer. A clue to the scale of early lunar differentiation is that mare basalts have complementary rare earth element patterns that are most readily interpreted as having "seen" the formation of a plagioclase-rich crust. We shall return to this point in Sec. III.D.

The time of formation of the earliest terrae cumulate rocks cannot be unambiguously interpreted, but it is clear from ^{244}Pu/Xe and Rb/Sr isochronism for one troctolite (76535) that lunar differentiation occurred at least as early as 4.5×10^9 yr ago (Caffee et al. 1981, their Table 9). A younger

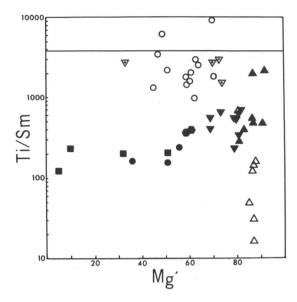

Fig. 13. Ti/Sm versus Mg* (100 Mg/Mg + Fe, molar) in highlands cumulates and KREEP-rich igneous rocks. The horizontal reference line indicates the value in chondrites. Open circles: ferroan anorthosites; filled circles: alkali anorthosites; filled upward-pointing triangles: troctolites and spinel troctolites of Mg-norite group; open upward-pointing triangles: Apollo 14 troctolites; filled downward-pointing triangles: Mg-norites; downward-pointing triangles with centered dot: Mg-gabbronorites; filled squares: evolved rocks; filled hexagons: KREEP basalts (figure after James and Flohr 1983).

TABLE IV
Rb-Sr and Sm-Nd Isotopic Results for Pristine Highland Rocks[a]

Sample	Type	Age (Sr)[b] (Gyr)	Age (Nd)[c] (Gyr)
15455c	Anorthositic norite	4.48 ± 0.12	—
60025	Anorthosite	3.85 ± 0.02	—
67667	Feldspathic lherzolite	—	4.18 ± 0.07
72255c	Norite	4.08 ± 0.05	—
72417	Dunite	4.45 ± 0.10	—
73255,27	Norite	—	4.23 ± 0.05
76535	Troctolite	4.51 ± 0.07	4.26 ± 0.06
77215	Norite	4.33 ± 0.04	4.37 ± 0.07
78236	Norite	4.29 ± 0.02	4.34 ± 0.04

[a]Table after Carlson (1982).
[b]Rb-Sr ages are calculated using $\lambda_{Rb} = 1.42 \times 10^{-11}$ yr^{-1}.
[c]Sm-Nd ages are calculated using $\lambda_{Sm} = 6.54 \times 10^{-12}$ yr^{-1}.

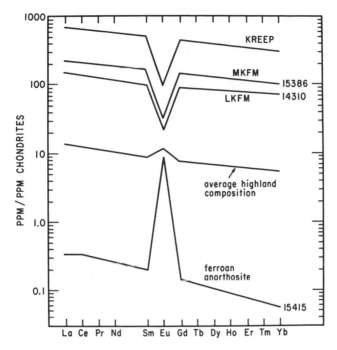

Fig. 14. Rare earth element abundances in terrae rock samples (figure after S. R. Taylor 1982). See text for abbreviations used.

Sm/Nd age probably represents metamorphic resetting of that isotope system. Additional direct chronological evidence of early differentiation has been presented by Oberli et al. (1978). By plotting $^{238}U/^{206}Pb$ versus $^{207}Pb/^{206}Pb$, these authors demonstrated that terrae rocks plot along two arrays, the lower intercept with concordia being interpreted as isotopic resetting during basin forming impacts. The upper and lower intersections with concordia for the two arrays are 4.47 and 3.86 Gyr, and 4.51 and 4.17 Gyr, respectively, again indicating differentiation occurring as early as 4.5 Gyr ago. The duration of this global differentiation event is unclear. Ages as young as 3.85 and 4.08 × 10^9 Gyr (Table III) are preferably interpreted as pertaining to younger intrusions or to isotopic resetting by impacts. The "age" of KREEP may be our best estimate of the end of global differentiation as discussed in the next section.

C. Other Terrae Volcanic Rock Types

Several compositions recognized on chemical grounds are common. These compositions are low-K Fra Mauro (LKFM), medium-K Fra Mauro (MKFM), and KREEP (Fig. 14). LKFM and MKFM occur in breccias but also as glasses. KREEP exists as a component in breccias and impact melts, and in two apparently volcanic rocks which may nevertheless have an impact origin. It is unclear whether significant terrae volcanism did occur. If it did, its records have been effectively obscured by the early period of intense bombardment.

We shall consider these rock types only in the context of global differentiation and will concentrate on KREEP. The orbital gamma-ray data obtained on Apollo 15 and 16 indicate that KREEP is principally restricted to the western Earth-facing hemisphere of the Moon, with high concentrations of U, Th, and K being detected over Oceanus Procellarium and Mare Imbrium (Metzger et al. 1979). A small occurrence on the lunar farside is associated with the crater Van der Graaf.

Unmodified samples of KREEP are rare in the Apollo and Luna collections. The two samples of KREEP which may be true volcanic basalts, 15382 and 15386, are 3.95 Gyr old (Nyquist et al. 1975). However, Rb-Sr model ages for KREEP-rich samples generally cluster around 4.4 Gyr, and Lugmair and Carlson (1978) have shown that Sm-Nd ages are consistent with establishment of a global KREEP reservoir at a time 4.36 ± 0.06 Gyr ago. If KREEP (or its precursor) represents the last residue of a global differentiation event such as a magma ocean, then termination of this event was at approximately 4.36 Gyr ago. Thus, early global differentiation spanned a period of approximately 150 Myr, from at least 4.51 Gyr to 4.36 Gyr ago.

D. Mare Basalts

Mare basalts are volumetrically insignificant but cover wide areas on the Earth-facing hemisphere of the Moon, with minor occurrences on the farside. Groundbased spectrophotometry (Pieters 1978) suggests that many varieties of mare basalt exist that are not present in the returned samples (see Color Plates 8, 9). Petrological data for the major sampled varieties of mare basalts are summarized in Fig. 15. Detailed mineral chemistry may be found in *Basaltic Volcanism on the Terrestrial Planets* (1981, Sec. 1.2.9). Rare earth element abundances are summarized here in Fig. 16. Ages of mare basalts are summarized in Table V. Our purpose in this chapter is not to discuss the detailed properties of mare basalts. Rather, it is to use mare basalts as petrological and geochemical probes of the lunar interior and of lunar differentiation processes.

There is a consensus that mare basalts were produced by melting of the lunar interior. Experimental petrological studies indicate that, with few exceptions, plagioclase is not a residual phase after melting of the lunar interior. Yet, almost all mare basalts have negative Eu anomalies (Fig. 16). In the absence of residual plagioclase, this feature must be a property of the lunar interior. (Plagioclase is the only major mineral that significantly concentrates divalent Eu in its structure relative to the other trivalent REE.) The complementary nature of the positive Eu anomaly in average terrae composition (Fig. 14) suggests that the lunar interior has "seen" the formation of the ter-

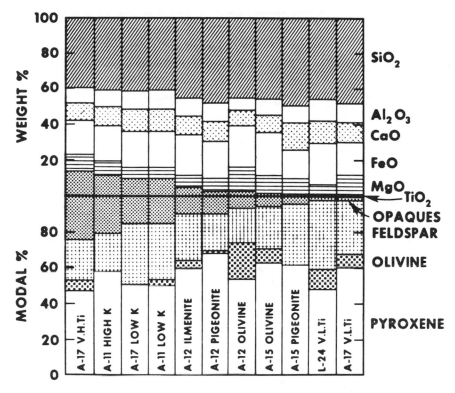

Fig. 15. Diagram showing correlation between major-element chemistry and modal mineralogy for mare basalts. Histogram is ordered from left to right by decreasing TiO_2 (or opaque oxide) content. Figure from *Basaltic Volcanism Study Project on the Terrestrial Planets* (1981).

rae. This observation is most readily understood if terrae and mare basalt source regions were established during the global differentiation event. Neodymium isotope systematics (Lugmair and Marti 1978) and most Rb-Sr model ages also indicate formation of mare basalt source regions approximately 4.4 Gyr ago (Fig. 17).

Most workers agree that mafic cumulates produced during global differentiation were involved in some manner in subsequent mare basalt petrogenesis. There is considerable controversy over the nature of that involvement, however. Nyquist et al. (1976–1979) in a series of papers have developed an internally consistent model for formation of a variety of mare basalts by remelting of mafic cumulates produced during global differentiation. Ringwood and Kesson (1976) advocate a dynamic model involving sinking of late-stage, ilmenite-rich cumulates of the postulated global differentiation event into a fertile, primitive interior of the Moon which was at or close to the solidus. Subsequently produced mare basalt magmas are hybrid prod-

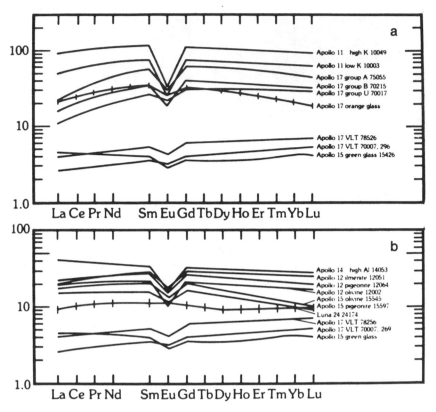

Fig. 16. Two REE patterns for various types of mare basalts (figure from S. R. Taylor 1982).

ucts with a chemical and isotopic memory of primitive and differentiated materials. Delano (1980) advocates sinking of rafts of mafic cumulates produced during the postulated global differentiation event to a depth of 400 to 500 km to create a heterogeneous mantle. These materials are then remelted to produce mare basalt magmas. This model is essentially a geometrical revision of the cumulate remelting model.

The problem in choosing between these competing hypotheses lies in the ambiguity of the depth of origin of mare basalts. In principle, if a magma is produced by equilibrium partial melting with two or more minerals remaining in the source region after extraction of magma, and if the magma ascends to the surface and erupts without modification due to crystallization or assimilation of wall-rock material, the depth of origin may be inferred from experimental petrology by searching for multiply saturated regions in pressure-temperature space. In practice, it is impossible to establish unambiguously that a mare basalt meets these criteria. The general trend of a transition from

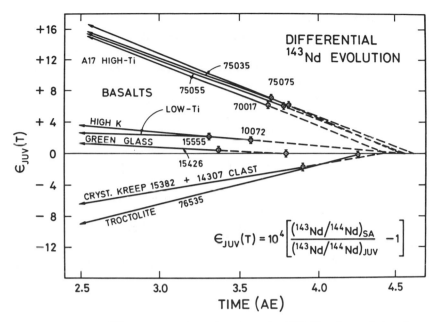

Fig. 17. Differential ^{143}Nd evolution for highland samples (KREEP and troctolite) and mare basalts relative to the Juvena meteorite standard. This diagram clearly shows the early differentiation of the Moon and the development of separate source reservoirs for mare basalts. The time of early differentiation is not well resolved by the Sm-Nd systematics, except for the fact that it is early (figure from Lugmair and Marti 1978).

high-Ti to low-Ti basalts with time (Table V), which has been interpreted by some authors to imply melting at greater depth as a function of time, may be an artifact of sampling. For example, unsampled high-Ti basalts in Mare Imbrium may be as young as 2.5 Gyr old (Boyce 1976), an observation more consistent with the views of Delano (1980) concerning mare basalt petrogenesis. Certainly the interior of the Moon was a dynamic place during early lunar history (Herbert 1980), and it is likely that some combination of cumulate foundering and remelting, coupled with crystallization and assimilation of wall-rock, occurred during production of mare basalt magmas.

IV. LUNAR PHYSICS AND STRUCTURE

In this section we present principal geophysical data, the geodesy, seismology, heat flow, and magnetism of the Moon and then use them to infer radial structure as well as any lateral inhomogeneities. Finally, we attempt to constrain lunar evolution from the constraints of all data types: geologic, compositional, and physical.

A. Shape and Gravity

Knowledge of the Moon's geometry and gravity depends primarily on observations from circumlunar spacecraft and secondarily on observations from the Earth. Useful global estimates of elevations and gravity potential exist, but accurate topographic maps are available only for regions covered by Apollo mapping cameras in sunlight, between longitudes 70°W and 180°E and latitudes ± 30°; accurate gravity fields are available only for the nearside within ~ latitudes ± 30°, because of dependence on tracking from the Earth, low spacecraft altitude, and variety of orbital inclination.

The leading parameters are given in Table I. The mean density of 3344 (± 2) kg m^{-3} from mass M and radius R is the most important constraint on the Moon's composition, limiting its iron content to less than 10%, or one third the Earth's content. The moment-of-inertia of 0.3904 MR^2, close to that for a homogeneous body, further limits the portion of the iron which can be separated into a core. The mean gravity of 162 cm s^{-2}, one-sixth that of the Earth, in turn implies one-sixth the near-surface stresses for a surface loading of the same mass and shape.

The mean radius of 1737.5 km is determined from a combination of Apollo spacecraft altimetry, photogrammetry, and landmark tracking with Earth-based photogrammetry and limb-profile measurements. Another inference from the same data is an offset of the center-of-figure from the center-of-mass: 2.0 km in the direction 19°S, 194°E. Like the Earth and Mars, the variability in the first degree harmonic of the lunar topography is appreciably larger than in higher harmonics (Bills and Ferrari 1977; Phillips and Lambeck 1980). The persistence of a first degree harmonic in the topography is related to the low stress levels associated therewith. The cause of the Moon's offset is probably associated with asymmetries primarily in the original crustal thickness and secondarily in subsequent basin formation.

The broad variations of the lunar topography appear to be positively correlated with age of geologic province, but not so simply related to gravity. The Moon's topography and gravity (Ferrari 1977; Ananda 1977; Bills and Ferrari 1980) can be characterized by half wavelength S, corresponding to spherical harmonic wave number ℓ by $S = \pi R / \ell$.

1. If $S > 1200$ km, the topography averages about one-seventh and the gravity about two-fifths the magnitude predicted by equal-stress extrapolation from the terrestrial. The gravity magnitude is about half that calculated from the topography, in contrast to a ratio of one-sixth for the Earth value, and the correlation is markedly positive, rather than oscillating from degree to degree.

2. If 1200 km $> S > 700$ km, the topography averages about one-seventh, but gravity averages about 1.05 times that predicted from the Earth value. The gravity is comparable to that calculated from the topography, but the correlation is somewhat negative.

3. If 700 km $> S >$ 400 km, the topography averages one-third the equal-
    stress prediction from the Earth's topography. The gravity averages about
    as much as predicted from the Earth values. The correlation is negligible.

The positive correlations and observed/calculated ratios for the half-
wavelengths > 1200 km are consistent with isostatic compensation at depths
150 to 350 km. However, the sharp drop in correlation for $S < 1200$ km indicates much shallower compensation and finite strength supporting the topography, as well as mass irregularities that are not positively correlated with topography. Most prominent among the latter are the mascons, mass excesses
on the nearside which correspond rather closely with the deepest features, the
mare-filled multiringed basins.

A significant inference from tracking of low-altitude Apollo 15 and 16
subsatellites is that mascon-associated maria have a high-density layer near
the surface, from the "edge effect" on the Doppler residual. This shallow
plate does not preclude a deeper dense intrusion from also contributing to the
mascon. Other features of the strongly constrained near-side, low-latitude
data are:

1. A broad negative gravity anomaly around the Orientale mascon;
2. No marked negative anomalies around other mascons;
3. Correlation of negative anomalies with craters of diameter less than
    ~ 300 km;
4. A broad positive anomaly associated with the highlands (Sjogren et al.
    1974; Kaula 1975; Sjogren 1977; Phillips and Lambeck 1980).

In general, the Moon appears to be closer to isostatic equilibrium than
the Earth: for the nearside, by a factor of about three, in the sense of stress
implication. The farside probably is farther from equilibrium, and probably
has a more positive correlation of gravity with topography.

An important application of the topographic and gravimetric data is to
extrapolate from the crustal thickness determined by seismology, D_s (see below), to obtain a global estimate of crustal thickness, D_m. This extrapolation
is based on the observation of broad isostatic compensation: as indicated by
small observed/calculated ratio of gravity anomalies and small correlation coefficients for the long wavelengths. Hence an isostatic balance equation can be
written between the seismometry region and the global average

$$\rho_C D_s + \rho_M [D_m + h_s - D_s] = \rho_C D_m \qquad (1)$$

where ρ_C is crustal density, ρ_M the mantle density, and h_s the altitude of the
seismometry region with respect to the global mean sphere. The essential parameter from the altimetry (Bills and Ferrari 1977) is $h_s = -1.5$ km. This
value, combined with $D_s = 56$ km, $\rho_C = 3.0$ g cm^{-3}, and $\rho_M = 3.4$ g cm^{-3}
obtains $D_m = 69$ km. However, the quantity which is more severely con-

strained is $(\rho_M - \rho_C)D_m \approx 27$ km $\times$ g cm^{-3}. This value represents a global shallow density deficiency which must be accounted for by a combination of alumina content, megaregolith porosity, and variations in magnesium/iron ratio (Wasson and Warren 1980).

B. Seismology

Seismometers operated until 1 October 1977, at four Apollo sites: 12, 14, 15, and 16. In addition, arrays of 3 or 4 geophones and ~ 100 m extent were laid out at Apollo sites 14, 16, and 17.

Recorded lunar seismic events can be classified in six types, as follows:

Local artificial sources. These include ascent stage thrust, thumpers, and explosive charges, all less than $\sim 2 \times 10^{13}$ erg and within 3 km of the geophone arrays (Cooper et al. 1974).

Spacecraft impacts. Four of $\sim 3 \times 10^{16}$ erg and five of $\sim 5 \times 10^{17}$ erg kinetic energy have been recorded, all but one of accurately known time and location. Seven of the nine were within 270 km of the midpoint between the Apollo 12 and 14 seismometers ($\sim 3°$S, 20°W in Mare Cognitum), and two were within 100 km of the Apollo 16 and 17 sites, respectively (Lammlein et al. 1974; Toksöz et al. 1974).

Meteoroid impacts. More than 150 natural impacts were identified, the largest of $\sim 10^{19}$ erg kinetic energy on the farside, at $\sim 35°$N, 130°E (Dorman et al. 1978). The portion of explosive or impact energy converted to seismic is quite minor (probably $< 10^{-4}$) so it is difficult to make a direct comparison in terms of energy to internally generated events.

Thermal moonquakes. Small events close to the seismometers are distinguished from meteoroid impacts by a faster rise time in the signal, a strong diurnal correlation, and several sets of events of similar waveform, suggesting repeated motions at the same sources under the heating effect of the Sun.

Shallow moonquakes. A few events at random times and locations are inferred to have sources near the surface but with rise times short compared to meteoroid impacts. The seismic energy received is comparable to that from all but the largest meteoroid impacts. The total energy release is estimated to be 2×10^{17} erg yr^{-1}, but this is unsure because of the paucity of events (Nakamura 1980; Goins et al. 1981a).

Deep moonquakes. Several thousand events have now been detected as coming from sources 700 to 1100 km deep in the Moon. The Richter magnitudes of these events are 0.5 to 1.3, equivalent to 10^7 to 10^9 erg. The total energy release rate is $\sim 10^{14}$ erg yr^{-1} (Goins et al. 1981a). Near-identity of waveforms indicates that there are repeated events at the same sites. The frequency of occurrence of deep shocks reveals periodicities of 13.6, 27.2, and 206 days, strongly suggestive of tidal triggering (Lammlein et al. 1974; Dainty et al. 1976).

It is convenient to discuss the seismic structure and activity in five parts.

1. Megaregolith: ≲ 2 km deep. The seismic refraction experiments obtained similar velocities at all three sites, but with varying thicknesses of velocity layers, consistent with an inverse correlation of the velocity gradient dV/dz, with age at each site. The 3 to 13 m thick regolith has P velocities of 100 to 115 m s^{-1}. At all sites, the velocity increases rapidly with depth. At the deepest-sounded site, Apollo 17, the velocity jumps to 1.0 km s^{-1} at 0.4 km and 4.7 km s^{-1} at 1.4 km depth (Cooper et al. 1974). Variations in thickness of this very low velocity layer are strongly indicated by the number of events detected by the Apollo seismometers; thus, Apollo 16 detected about five times as many events per year as Apollo 12 (Lammlein et al. 1974).

2. Crust: ≲ 60 km deep. Spacecraft impacts are most useful for this depth range. Because they are clustered around the Apollo 12 and 14 sites, the results are representative of the Mare Cognitum region. A long reverberation time for signals from impacts indicates that the dissipation factor $1/Q$ in the Moon is very small. The apparent randomness of phases in the signal indicates that there are many reflecting surfaces in the Moon. If these surfaces are randomly located and isotropically scatter the energy, then the transport of seismic energy is governed by the damped wave equation

$$\frac{\partial E}{\partial t} = \kappa \nabla^2 E - \omega E / Q \tag{2}$$

where E is energy density, κ is diffusivity and ω is frequency.

Solutions to body-wave travel times from artificial and meteoroid impacts find an increase of velocity to a crustal base 55 to 60 km deep (Goins et al. 1978; Lammlein et al. 1974), which is plausibly interpreted as a decrease of porosity with depth. However, the compressive velocity (V_p) of 6.8 km s^{-1}, which appears to prevail at depths 20 to 56 km, is significantly lower than measured for nonporous anorthosite, 7.3 km s^{-1}, or a plausible gabbro ($Al_2O_3 \sim 17\%$), 7.5 km s^{-1} (Liebermann and Ringwood 1976). It is, however, about equal to the typical velocity in the lower oceanic crust of the Earth.

3. Upper mantle: 60 to ~ 400 km deep. To obtain velocities in this region, meteoroid impacts and shallow moonquakes must be used. This region has highest velocities of 7.71 km s^{-1} for V_p and 4.5 km s^{-1} for the shear velocity V_s (Goins et al. 1981b; Nakamura 1983). These values are consistent with likely olivine-pyroxene compositions. The upper mantle also has, like the crust, extremely small dissipation factors ($1/Q$); values of 1/4800 for $1/Q_p$ (Dainty et al. 1976) and 1/4000 for $1/Q_s$ (Nakamura et al. 1976a; Nakamura and Koyama 1982) have been obtained. The principal uncertainty appears to be the depth of the bottom of the upper mantle, functionally defined as the depth at which there is an increase in the dissipation factor $1/Q$

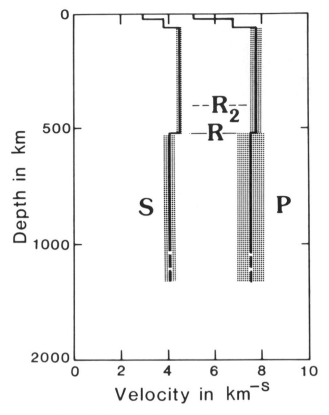

Fig. 18. Seismic velocities in the Moon. The levels R and R_2 are possible reflecting sources. The shading indicates the uncertainties in the velocities (figure from Goins et al. 1978).

and a drop in the V_s/V_p ratio (there may also be a drop in V_p, but as indicated by the shading on Fig. 18, this is unsure). Solutions by the two active groups are 300 and 520 km, respectively. Also uncertain is whether there is a reflecting layer associated with the transition; Goins et al. (1981b) come to a cautiously positive conclusion, while Nakamura's (1983) is negative.

4. *Middle mantle:* ∼ *400–1000 km deep.* In this zone, deep moonquakes must be used as sources. Within the 400 to 480 km zone, V_s appears to drop to ∼ 4.2 km s^{-1} and V_p to ∼ 7.6 km s^{-1} (Goins et al. 1981b; Nakamura 1983). The $1/Q$ increases markedly to something like 1/1500 for $1/Q_p$. Goins et al. (1981b) find these low velocities to prevail throughout the middle mantle, but Nakamura (1983) infers $V_p = 8.3$ km s^{-1} and $V_s = 4.7$ km s^{-1} for depths of 500 to 1000 km. The deep moonquakes appear to occur near the

lower boundary of this region, as though there was an interaction between a stiffer lithosphere and a softer asthenosphere.

5. Asthenosphere: ~ 1000–1740 km deep. No shear waves have been detected as having passed through this region. Even *P* waves are debatable; certainly there is a further increase in $1/Q_p$.

The seismic evidence of the Moon thus is consistent with an outermost layer $\lesssim 20$ km thick which has been fractured and broken up; a crust of low density ~ 60 km thick under Oceanus Procellarum; a dry lithosphere ~ 800 to 1000 km thick; and a central asthenosphere. The question of an iron core is unresolved. Application of the geodetic and seismological data to the problem of lunar composition leads to inferences that Mg/(Mg + Fe) of the Moon's upper mantle is about 0.80, significantly less than the Earth's upper mantle, but that the Al_2O_3 content could be similar (Buck and Toksöz 1980). The higher velocities for the middle mantle found by Nakamura (1983) suggest an appreciably higher MgO content than in the upper mantle.

C. Thermal Regime

Measured heat flows (after thermal equilibrium was obtained) are 21 erg cm^{-2} s^{-1} at Apollo 15, and 16 erg cm^{-2} s^{-1} at Apollo 17 (see the chapter by Schubert et al.). Correction for topography reduces the latter to 14 erg cm^{-2} s^{-1}. Extrapolation of a global mean naturally requires several assumptions.

1. Assume the total heat flux to be a combination of three parts, $Q = Q_M + Q_C + Q_A$, where Q_M is from the mantle, Q_C is from the crust, and Q_A is an added amount arising from the excavation of KREEP-like material.
2. Assume Q_M to be the same everywhere.
3. Assume Q_C to be proportional to crustal thickness, as inferred from topography plus isostasy (see Eq. 1).
4. Assume Q_A to be proportional to the excess of gamma ray counts over normal for the elevation.
5. Assume the heat-flow sites to be representative of their regional elevations and gamma ray counts.

With these assumptions one obtains 18 erg cm^{-2} s^{-1} for a global mean Q, 4 erg cm^{-2} s^{-1} for Q_M, and ~ 0.3 ppm mean *U* content of the crust (Langseth et al. 1976).

If steady state is assumed for both the Earth and the Moon, then the value of 18 erg cm^{-2} s^{-1} compared to 84 erg cm^{-2} s^{-1} in the Earth implies a 50% higher proportion of refractory lithophiles in the Moon than in the Earth (allowing for the differing K/U ratios). The Moon is probably closer to steady state than the Earth because a greater portion of its heat is likely to be shallow crustal sources, so that the enrichment should be somewhat greater than 50% (as also is indicated by the significantly lower density coupled with lower volatiles).

A more sensitive indicator of internal temperatures than the seismic $1/Q$ or V_s is the electrical conductivity, as inferred from the response of the Moon to variations in the solar wind. The essential physics of inferring $\sigma(r)$, the electrical conductivity as a function of radius, is reviewed by Sonett (1982). The applicable data for the Moon are simultaneous magnetometer readings on the surface and on a spacecraft in the general vicinity of the Moon, but far enough from it not to be influenced magnetically. The surface magnetometers are at Apollo 12, 14, 15, and 16 sites; the spacecraft magnetometer is on Explorer 35, a lunar satellite with 7759 km apocenter altitude. Inferences of conductivity have been made by both harmonic and transient analysis techniques.

The translation of conductivity into temperature also involves technical difficulties because of the extremely low oxygen fugacity. The conductivity-temperature relationship differs significantly between olivine and pyroxene; furthermore, it is sensitive to the chromium and aluminum contents of the rock. If the lunar mantle is orthopyroxene with small concentrations ($1-2\%$) of Al_2O_3 and Cr_2O_3, then the electrical conductivities indicate temperatures several hundred degrees Celsius below the solidus for the outer 1200 km (Huebner et al. 1979).

An important constraint on lunar thermal history is the absence of marked net global tension or compression (see corresponding discussion in the chapter by Squyres and Croft). Define

$$\Delta T(r) = T(r, t_2) - T(r, t_1) \tag{3}$$

where T is temperature, r is the radial coordinate, and t is time: $t_2 > t_1$. Then the corresponding change in total radius R is

$$\Delta R = \frac{1}{R^2} \int_0^R \alpha(r)\Delta T(r)r^2 \, dr \tag{4}$$

where $\alpha(r)$ is the coefficient of thermal expansion at r. If $\alpha(r)$ is taken as constant, then the volume-averaged temperature change

$$\overline{\Delta T} = \frac{3}{R^3} \int \Delta T r^2 \, dr = \frac{3\Delta R}{R\alpha} \tag{5}$$

or for $\alpha = 4 \times 10^{-5} \, ^\circ C^{-1}$, $|\Delta R| < 1$ km, $\overline{\Delta T} < 43$ K, a rather small change. If the Moon started with a hot outer shell as appears to be required by the petrological data, then to satisfy $\overline{\Delta T} \simeq 0$ requires starting with a colder center. The coldness of the outer few 100 km indicated by the seismic $1/Q$ and the electrical conductivity σ imply a temperature drop of 400° in this layer. Plausible lunar thermal histories indicate that the interior heats no more than 320°, whence the depth separating heating and cooling spheres is ≤ 300 km (Solomon and Chaikin 1976). The surface structure of the Moon also indicates significant regional-scale thermal evolution associated with maria basins (Solomon and Head 1980; Golombek and McGill 1983).

D. Remanent Magnetism

Significant magnetic intensity intrinsic to the Moon has been measured by (1) laboratory experiments on lunar rocks; (2) magnetometers at the Apollo 12, 14, 15, and 16 sites; and (3) magnetometers and electron detectors on the Apollo 15 and 16 subsatellites.

In contrast to terrestrial rocks, the stable magnetization of lunar samples is carried mainly by metallic iron particles, whose mass concentration ranges from a few times 0.01% in mare basalts to as much as 1% in some soils and breccias. The natural remanent magnetism (NRM) of lunar rocks covers a great range, from 10^{-7} to 10^{-3} emu g^{-1}, only partly correlated with the iron content. A possible mechanism is shock remanent magnetization (SRM). The questions are: (1) to what extent does hypervelocity impact heating lead to magnetization of rocks in an ambient magnetic field and (2) to what extent is the field itself enhanced by the impact. Experiments indicate that both effects are present; stable magnetization has been obtained from 14 km s^{-1} impacts in a 100 Gauss ambient field greater by a factor of only two than that obtained in 0.5 Gauss ambient field. However, other shock experiments have obtained significant demagnetization of rocks already magnetized. In the lunar samples themselves, shock effects are suggested by (1) a great range of NRM intensity among rocks of the same age at the same site (Apollo 15); and (2) some inverse correlation of NRM intensity with petrological indication of shock. Inferring the paleointensities involves technical difficulties; the results range from 0.2 to 3.0 Gauss (Fuller 1974; Cisowski et al. 1977; Srnka et al. 1978).

The remanent magnetic field at the Apollo sites, determined when external and induced fields are negligible, varies appreciably not only among sites, but over distances on the order of a kilometer at a site. At Apollo 15, 6 γ were measured; on the Apollo 16 Rover traverse of $\sim$ 6 kilometers, extremes of 120 and 310 γ were measured (Fuller 1974).

The Apollo subsatellites orbited at altitudes of $\sim$ 100 kilometers, so that the fields sensed by the magnetometers are averaged over comparable horizontal distances. The residual field attributable to lunar sources is on the order of $\pm$ 0.3 γ, with features of $\sim$ 300 km in extent. Such a weak field is consistent with the local gradients of $\sim$ 30 γ km^{-1} indicated at the landing sites. Some correlation with geologic age has been suggested: a decrease in intensity until Nectarian time, an increase to Imbrian time, and a decrease again since then. In the most recent analyses, the most pronounced magnetic anomalies appear to be associated with swirls of brighter material, which may result from the deflection of solar energetic particles which darken the surface (Hood et al. 1979; Hood and Schubert 1980; Hood 1981).

The small-magnitude and short-wavelength character of magnetic field variations lead to only a small component being attributable to a dipole: $\sim 1.3 \times 10^{18}$ Gauss-cm^3 for the moment, in the orbital plane of Apollo 15. If the source of the ~ 1 Gauss impressing field indicated by the samples was a deep internal dynamo, then a moment of $\sim 5 \times 10^{24}$ Gauss-cm^3 is required. However, if the now-removed impressing field was entirely internal, and if the shell remaining below the Curie point (and hence magnetized) is uniform and homogeneous, then no external magnetic field should remain. On the other hand, if the impressing field was external, there should be a measurable remanent field. These considerations lead to the suggestion that the Moon had a dynamo at one time, and hence has an iron core. The mean density, moment-of-inertia, seismic travel times, and electrical conductivity inferred from the induced variations in the field all could tolerate a core of ≤ 400 km radius. Problems for a lunar dynamo are the small size of this core, the energy source, and reconciling the dying out with the continued heating of the deep interior implied by the lack of net surface tension and compression (Eqs. 3–5) (Wiskerchen and Sonett 1977; Fuller 1974; Runcorn 1978,1983; Sonett 1982; Stevenson et al. 1983).

E. Evolution of the Moon's Interior

The generally dwindling level of internal activity suggested by the geologic and compositional records can be matched by thermal evolution models allowing for solid state convection and upward differentiation of K, U, Th, provided that an appropriate profile is assumed. Initial melting must be fairly deep in order to have partial melting persisting for 1.5 Gyr, so as to provide for mare basalt differentiation (see Fig. 19). The essential course of such a history is that inner-sphere heating balances the outer-shell cooling for the first 1.5 Gyr, but thereafter there is an overall cooling (Toksöz et al. 1978; see also Chapter 5 by Schubert et al.). The resultant average change $\overline{\Delta T}$ is about $-280°$ implying a ΔR of ~ 6 km, by Eq. (4). Also, a net heating persists over the innermost 400 to 500 km, rather than a cooling as appears to be required by the extinguished-dynamo hypothesis.

Most near-surface phenomena appear consistent with the foregoing standard thermal history. A rather deep magma ocean would be provided for crustal differentiation, and convection therein would have been vigorous enough to lead to the variation in crustal thickness accounting for the offset of center-of-figure from center-of-mass. The ultimate cause of mass concentrations associated with the ringed maria could have been the net heating of the inner part of the Moon relative to the outer. The pressure generated by thermal expansion would then be limited to its tectonic effect at a few locations of weakened lithosphere, i.e., the ringed basins. The thickening of this lithosphere with the temperature drop of the outer shell is sufficient to support the mascons with plausible stresses.

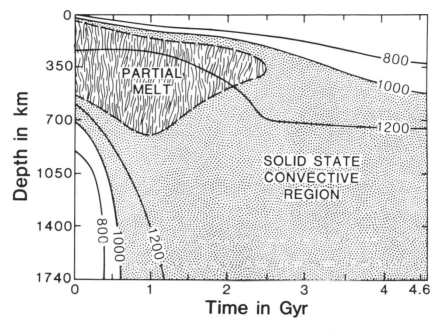

Fig. 19. Thermal evolution of the Moon's interior as a function of time. This evolution is inferred from an asteroid starting-temperature curve. The computational model simulates convective heat transfer and upward differentiation of heat sources as functions of temperature (figure from Toksöz et al. 1978). An asteroid starting temperature increases with radius due to increase in impact energy to a shallow depth above which heat is removed by impact stirring.

F. Summary

Although lunar composition, structure, and evolution are better understood than they were prior to the Apollo Project, several outstanding problems remain. Among these are the following:

1. The physical effects of impacts: ejecta distribution, melting, magnetization, etc.;
2. The identification of what were the igneous processes responsible for the formation of the gabbroic-anorthositic crust;
3. The history of crustal differentiation in the first ~ 0.5 Gyr;
4. The extent to which mare basalt differentiation differed from the sampled sites;
5. The true mean composition of the crust, in view of the V_p of ~ 6.8 km s^{-1};
6. The present mean temperature profile of the interior;
7. The correct implications of the apparent small change in radius for thermal evolution;

8. The source of the impressing field for the remanent magnetism of the surface;
9. The related questions of the existence of an iron core and a dynamo in the past.

The ultimate solution of some of these matters may depend on the initial composition and thermal state of the Moon. Since the Moon is small compared to the terrestrial planets and unavoidably anomalous in bulk composition, it may have formed quite inhomogeneously in several respects.

V. BULK COMPOSITION OF MOON

The bulk composition of the Moon is poorly known. Two principal approaches have been taken to its determination: geophysical approaches taking into account models of petrological and geochemical evolution; and geochemical approaches using geophysical measurements to fix some elemental abundances. These approaches can lead to quite different estimates of lunar bulk composition, as is illustrated below.

A. Constraints from Geophysics

Input data for most geophysical approaches are mass, density, upper and lower mantle seismic velocities, moment of inertia, and presumed lunar structure deduced from geophysical techniques. Using such input data, Buck (1982) concludes that the Mg/Si ratio in the Moon is closer to that of ordinary chondrites (0.8) than it is to the Earth (0.95), that the Mg/(Mg + Fe) ratio is approximately 0.8, that Al_2O_3 weight percent is in the range 3 to 5 wt. %, and that a small iron core of 1 to 2 wt. % of lunar mass is required. A range of possible bulk compositions fit these constraints. A composition preferred by Buck and Toksöz (1980) is given in Table VI.

B. Constraints from Geochemistry

The assumption is made that the Moon is constructed from the same components that were assembled to make the chondrites. This assumption permits a wide variety of compositions because the chondrites are comprised of many components; it is difficult to fix the relative and absolute abundances of most of these components unambiguously. Attempts to construct a lunar bulk composition are guided by three observations: (1) lunar rocks are depleted in volatile elements; (2) lunar rocks are enriched in refractory elements; and (3) the Moon is impoverished in metallic iron.

Refractory element concentrations can be estimated from global heat flow as discussed in Sec. IV. As mentioned there, this estimate will be an upper limit, due to the lag of heat loss behind heat generation (Schubert et al. 1980; see also Chapter 5). Allowing for the difference in K/U ratio, the global

TABLE V
Summary of Ages of Lunar Mare Basalts[a]

Mission	Basalt Type	Age Range (Gyr)
Apollo 11	Low K, high Ti	3.69–3.92
Apollo 11	Olivine, low Ti	3.21–3.29
	Pigeonite, low Ti	3.11–3.26
	Ilmenite, interm. Ti	3.08–3.23
	Feldspathic	3.08
Apollo 14	Feldspathic	3.85–3.96
Apollo 15	Olivine, low Ti	3.26
	Pigeonite, low Ti	3.30–3.37
Apollo 17	High Ti	3.59–3.79
Luna 16	Feldspathic	3.35–3.41
Luna 24	Very low Ti	3.28

[a]Table after S. R. Taylor (1982).

TABLE VI
Two Estimates of Lunar Bulk Composition[a]

	Buck & Toksöz (1980)	Taylor (1982)
SiO_2	48.37	44.47
TiO_2	0.40	0.31
Al_2O_3	5.00	6.15
FeO	12.90	10.96
MgO	29.02	32.79
CaO	3.83	4.61
Na_2O	0.15	0.09
Cr_2O_3	0.30	0.61

[a]Estimates exclude any core.

mean of 18 erg cm^{-2} s^{-1} (see Sec. IV) gives an upper limit to the refractory content of more than three times chondritic and a value 50% greater than that for the Earth's mantle. From an estimate of U concentration, abundances of other element groups may be estimated from well-behaved element ratios, e.g., U/K. Petrological and mass-balance constraints may also be used. A composition obtained in this general manner by Taylor (1982) is given in Table VI.

VI. INTERACTION OF THE MOON WITH THE EARTH

The orbit of the Moon differs from those of most satellites of major planets in that it has unusually large inclination, eccentricity, and semimajor axis (in ratio to the radius of its primary). The last two of these three properties arise from the tidal evolution of the Moon, which is dominated by dissipation in the Earth and negligibly affected by dissipation in the Moon (see chapter by Burns). However, the large inclination must result from circumstances of origin.

The rate at which the Moon is now moving away from the Earth has been analyzed in several papers over the decades, but observationally determined to two-digit accuracy only since a dozen years of laser ranging from Earth to Moon have accumulated (Dickey et al. 1983a; Burns' chapter discusses the implications). This definitive determination confirms the repeated suggestion (see, e.g., Kaula 1971) that, on the average, tidal dissipation must have been much less in the past due to different continent-ocean configurations. The most recent implementation of this suggestion has been made by Hansen (1982), who calculated the oceanic tidal dissipation associated with simplified continent configurations. While simple geometric forms tend to underestimate the amount of dissipation (because any irregularity enhances transfer of tidal energy from the second- to higher-degree harmonics), the analysis shows that the dissipation would be significantly less if the continents were concentrated in one block (particularly if polar), as has prevailed through most of geologic time. Hansen (1982) also gives a general review of tidal friction analyses; see also the chapter by Burns.

The tidal evolution of the Moon's orbit is unique because, among the planets, the Earth appears to be remarkable in the magnitude and variability of its dissipative character. The principal inference from this evolution is that the Moon was significantly closer to the Earth early in its history, and hence could have had an origin closely associated with the Earth. However, as the rate of change of the semimajor axis has the proportionality $\dot{a} \propto 1/a^{11/2}$, the Moon has spent nearly all its history rather far from the Earth. See the chapter by Burns for further discussion.

The Moon's rotation is locked to synchronism with its orbit, because of the attraction by the Earth for the Moon's irregular shape (see Fig. 1 in the chapter by Burns). This resonant lock affects the orientation, as well as the rate, of the Moon's spin. Meteoroid impact rates, together with plausible damping in the Moon, may lead to barely observable variations about the synchronism (Peale 1975,1976). A major impact could break the Moon's rotation free from the synchronism, but it would be reestablished quickly, on a geologic time scale. This reestablished coupling could be a somewhat different orientation with respect to the lunar body, if the impact was big enough to redistribute appreciable mass. Polar wander due to internal causes (Runcorn 1983) is also possible over long durations of geologic time.

Despite the commensurate rotation, there is tidal dissipation in the Moon because of the eccentricity and inclination of its orbit. It has sometimes been suggested that this dissipation caused appreciable heating in the Moon. However, a definitive analysis by Peale and Cassen (1978) demonstrated that this effect has been negligible throughout nearly all lunar history. Essentially, a body as small, as cold, and as homogeneous as the Moon holds itself together very tightly against the distorting effects of tidal attractions. Thus, while the origin of the Moon was closely bound up with the origin of the Earth, the subsequent evolution of the Moon has been only slightly affected by the fact that it is a satellite.

VII. ORIGIN OF THE MOON

A. Dynamical Considerations

The Moon is one of five sizeable rocky bodies within two AU (3×10^8 km) of the Sun. The next largest body within this region is probably $< 10^{-6}$ as massive as the Moon. The formation of these terrestrial bodies is very dependent on the formation of Jupiter ($\sim$ 300 times as massive as the Earth) and the Sun ($\sim$ 300,000 times as massive), two unsolved problems. However, it is generally held that the terrestrial planets grew by the agglomeration of smaller solid bodies (see the chapter by Safronov et al.). The strongest observational basis for this belief is probably that the planets have the same extreme depletions in inert gases (He, Ne, Ar, Kr, Xe) as chondritic meteorites. Hence a capturable gaseous nebula was probably lost when the protoplanetary material was in bodies as small as the meteorite parent bodies. Hypotheses for the origin of the Moon involve variants on capture, binary accretion and fission hypotheses (see the chapter by Stevenson et al.). Simple capture seems unlikely because the only energy dissipation mechanism associated therewith, tidal friction (Kaula 1971; Kaula and Harris 1975; chapter by Burns) is inadequate. Disruptive capture (Mitler 1975; Smith 1982) requires that a large number of objects of total mass much greater than the Moon must differentiate, be injected in Earth-intersecting orbits, and be disrupted inside the Roche limit. However, tidal disruption requires elastic behavior upon collision (Mizuno and Boss 1985) and it is questionable that it would devolatilize the material as thoroughly as required for the Moon.

Given a swarm of planetesimals in orbit about the Sun, current theoretical analyses and computer modelings of their dynamical interaction produce several dozen lunar-sized bodies, rather than a few terrestrial planets (Wetherill 1980). Ways to get around this difficulty have been suggested (Kaula 1983; chapter by Stevenson et al.). However, the essential implication for the origin of the Moon is that before there were five sizable terrestrial bodies, there were six; before there were six, there were seven; before seven, eight; and so forth: what might be called the ten-little-Indians hypothesis. Regardless of hypothesis, there are at least nine stages since the solar system became a separate entity which may have affected the composition of lunar samples:

1. Aggregation of planetesimals in the heliocentric nebula;
2. Intraplanetesimal differentation;
3. Planet growth;
4. Intraplanet differentiation;
5. Planet collision;
6. Planet disruption;
7. Aggregation of protosatellites in a geocentric accretion disk;
8. Lunar growth;
9. Intralunar differentiation.

Chondritic meteorites are direct evidence of processes at stage 1; differentiated meteorites, of stage 2; Earth rocks, of stage 4; and lunar samples, of stage 9. Unfortunately, there is little evidence relating directly to the stages making the bulk composition of the Moon different from that of the Earth: 5 and 7, the former of which would have entailed by far the greatest conversion of kinetic to thermal energy. Consequently, it is customary to assume either that stage 5 did not happen, or that its effects were the same as stages 1 and 7, i.e., selection by volatility, despite the great differences in physical circumstances.

Currently debated hypotheses of lunar origin can be said to concern stage 7: the source of material for a geocentric accretion disk. The fission hypothesis (see, e.g., Ringwood 1979) conjectures that the material came mainly from the Earth, and hence that stage 5, planet collision, was important (cf. Hartmann et al. 1986). The binary accretion hypothesis (see, e.g., Ruskol 1977; chapter by Safronov et al.) conjectures that the material came mainly by infall from outside the Earth-Moon system.

Dynamical evidences of the fission hypothesis are the lack of satellites around the other terrestrial planets, and the slow spin of Venus: i.e., the statistics of small numbers apply. However, in saying so it must be acknowledged that all terrestrial planets could have had accretion disks, but that of Mars did not evolve into satellites because even then it had a much smaller tidal dissipation factor $1/Q$, while Venus's and Mercury's did not because of their slow spins. The appreciable inclination of the Moon's orbit suggests a single major event, such as a planetary collision, rather than an accumulation of many small events. A dynamical objection to the fission hypothesis is that close passages, leading to tidal disruption, were more probable than collision. However, tidal disruption upon close passage has not been examined for the case of bodies comparable in size.

The great dynamical difficulty of the binary accretion hypothesis is a mechanism to differentiate material in the geocentric accretion disk efficiently enough to achieve the 2/3 reduction in iron content. Usually stage 2, intraplanetesimal differentiation, followed by planetesimal breakup leading to small silicate-rich and large iron-rich pieces is conjectured. The smaller pieces would then be more easily stopped by protosatellites orbiting the Earth. The principal dynamical objection to this hypothesis is that the much shorter geocentric time scale would lead to most of the Earth-orbiting material being collected in one body, and hence insensitive to the sizes of infalling planetesimals. This rapid aggregation into one body is also important to tidal friction keeping the growing Earth from pulling in most of the orbiting matter.

Further discussion of alternative hypotheses appears in the chapter by Stevenson et al.

B. Geochemical Considerations

It is clear that the Earth and Moon formed from fundamentally the same parental material. The strongest support of this conclusion comes from the similarity of oxygen isotopic compositions in the Earth and Moon (Clayton and Mayeda 1975) and their distinction from all other sampled solar system bodies except the enstatite chondrites and enstatite achondrites (Clayton et al. 1976; Clayton and Mayeda 1983). It is also clear that the Moon formed from condensed solar system material, as the absence of Eu and Yb anomalies mitigates against vaporization or selective condensation hypotheses (S. R. Taylor 1982). As discussed in the preceding section (VII.A), binary accretion and fission hypotheses are dynamically the most plausible. They are also susceptible to geochemical scrutiny. If the Earth and Moon formed as separate planetary bodies with the Moon assembled from material in geocentric orbit, the abundances of refractory siderophile elements inferred for the lunar mantle should be consistent with plausible lunar differentiation processes such as core formation. If, on the other hand, the Moon formed by fission of the Earth, the abundances of siderophile elements inferred for the mantles of the Earth and Moon 4.4 Gyr ago should have been the same.

Drake (1983) has shown that of the siderophile elements which may have been present in initially chondritic proportions in the Moon and whose geochemical behavior is predictable such that their abundances in the lunar mantle 4.4 Gyr ago have not been masked by subsequent igneous differentiation (W, Re, Mo, P), only W is consistent with the pure fission hypothesis for the origin of the Moon as advocated by Ringwood (1979). Rhenium, Mo and P are each inconsistent with the pure fission hypothesis (Figs. 20 and 21). The hybrid fission hypothesis of Wanke et al. (1983), that involves formation of a small core in the Moon after fission, circumvents many of the problems of the pure fission hypothesis. However, it is not readily reconcilable with the higher FeO content inferred for the lunar mantle compared to the terrestrial mantle. The abundances of siderophile elements in the Moon may be accounted for by a geophysically permissible core, and hence be consistent with the binary accretion hypothesis as advanced by Newsom and Drake (1982a,1983).

Chronological information also bears on the question of lunar origin. The presence of excess ^{129}Xe in both the Earth (Staudacher and Allegre 1982)

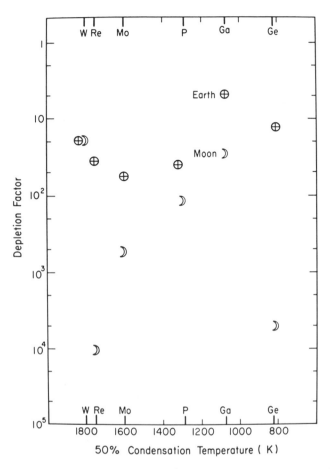

Fig. 20. Depletion of siderophile elements inferred for the mantles of the Earth and Moon relative to carbonaceous chondrites, versus a measure of volatility. Note here the lack of correlation of these parameters (figure from Drake 1983).

and Moon (Bernatowicz et al. 1980) implies that ^{129}I with a half-life of 17 Myr was alive during assembly of both objects. Thus, assembly of both Earth and Moon must have been complete within a few (probably < 10) half-lives following nucleosythesis of ^{129}I. The isochronism of ^{244}Pu/Xe and Rb/Sr chronometers for cumulate terrae troctolite 76535 (Caffee et al. 1981) requires that lunar assembly and some aspects of major differentiation be completed by 4.5 Gyr ago, essentially the age of the Earth. This requirement places severe time constraints on all hypotheses for lunar origin.

One cannot discriminate unambiguously on geochemical grounds between the hybrid fission hypothesis of Wanke et al. (1983) and the binary accretion hypothesis of Newsom and Drake (1982a,1983). The disparity be-

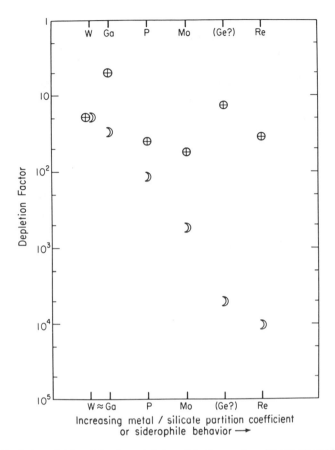

Fig. 21. Depletion of siderophile elements inferred for the mantles of the Earth and Moon relative to carbonaceous chondrites, versus of measure of siderophility. Note the correlation of
these parameters for the Moon (figure from Drake 1983).

tween lunar and terrestrial mantle FeO abundances, for example, may be
explained in an *ad hoc* way by continued growth of the Earth's core over geologic time (Wanke et al. 1983). However, isotopic evidence and geophysical
models both indicate rapid formation of the core. This impasse, coupled with
the observation that at least one lunar sample has been delivered to Earth as a
meteorite (see the special issue of *Geophysical Research Letters* [1983]),
raises the question of whether these two hypotheses should be considered as
end members of a continuum of models. Perhaps the material populating the
geocentric debris ring from which the Moon accumulated had two provenances: unprocessed virgin material, and material processed through a differentiating and still growing proto-Earth, which was ejected into geocentric or-

bit during impact events. The task of the dynamicist and the geochemist is to estimate the relative contributions of these two provenances.

Finally, one major inference may be drawn. The binary accretion and the hybrid fission hypotheses involve metal-silicate fractionation indigenous to the Moon, significantly enhancing the credibility of the existence of a small lunar core.

VIII. CONCLUSIONS

The Moon is probably the best understood object in the solar system because of the measurements made on and samples collected from its surface, as well as the simplicity of its behavior compared to the Earth's and the Sun's. Its history has largely been a warming of the center and a cooling of the outer parts, so that it now has a lithosphere several hundred kilometers thick. The preservation of a thick anorthositic crust differentiated soon after origin influences ideas about early crustal evolution of the terrestrial planets, while the Moon's size, composition, and orbit significantly constrain models of the origin of the Earth. On the other hand, the great differences in volatile content and dynamical circumstances has made the Moon a poor guide to understanding the bodies which it most resembles in the leading properties of mass and mean density: Io and Europa.

13. IO

DOUGLAS B. NASH
Jet Propulsion Laboratory

MICHAEL H. CARR
U.S. Geological Survey

JONATHAN GRADIE
Cornell University

DONALD M. HUNTEN
University of Arizona

and

CHARLES F. YODER
Jet Propulsion Laboratory

In this chapter we review the history of Io studies and present a comprehensive summary of what is currently known about this intriguing object. Io's surface color and spectra are dominated by sulfur compounds which may include various sulfur allotropes although their exact nature remains in doubt. A major question is whether high-temperature liquid sulfur forms can be quenched and retain their variegated colors on Io's surface. SO_2 is known as a condensed phase on Io, and SO_2 is concentrated equatorially on the leading hemisphere, while sulfur allotropes or other sulfur compounds may dominate the trailing hemisphere and polar regions. Silicates may be present beneath a veneer of sulfur and sodium-bearing compounds. Io's surface geology is dominated by volcanic processes yielding three kinds of surface features: vent regions, plains *and mountains up to 9 km high. Active plumes of volcanic debris reaching heights of 300 km are of two types: (1) small ($\sim$ 100 km height), long-duration (months to years) plumes, thought to be driven by low-temperature SO_2, and (2) large ($\sim$ 300 km height), short-duration (days to weeks) plumes, thought to be driven by high-temperature sulfur vapors. Io is resurfaced at an astoundingly high rate, perhaps as high as 10 cm per year, as a global average. Io's internal structure and the energy driving its volcanism are thought to be due to tidal interaction with Jupiter. The postulated internal structure inward from the surface is a quasi-rigid crust ($>$ 30 km thick), plastic convecting asthenosphere ($>$ 8 to 25 km thick), thin liquid shell ($>$ 1 km), thick mantle ($\sim$ 1000 km), and fluid core ($\sim$ 700 km). The steady state theoretical limit on tidal heat generation within or below a plastic asthenosphere, however, is about 0.2 W m^{-2}, set by the maximum convective heat transport in the asthenosphere. This value is exceeded by the presently observed heat flow radiated from hot spots on Io's surface ($\sim$ 1.0 to 2.0 W m^{-2}), suggesting that problems exist in understanding the relationship between tidal heat input and surface heat flow: Io's surface heat flow may be episodic and now unusually high. Io's atmosphere is poorly understood, with two conflicting but limiting models vying: (1) a thick equilibrium atmosphere with subsolar pressure of $\sim$ 10^{-7} bar buffered by solid SO_2 at the local surface temperature, and (2) a thin atmosphere controlled locally by venting and regionally by cold-trapping to $\sim$ 10^{-12} bar into a subsurface permafrost layer. The Io plasma torus, mostly ionized S and O, corotates with Jupiter's magnetic field in the plane of Jupiter's centrifugal equator centered at Io's orbital radius. The torus is supplied by material escaping from Io by some poorly understood process. Sputtering from the surface, atmosphere, or plume canopies may supply the sulfur, and thermal escape from Io may be appreciable for oxygen. The sodium cloud, composed of neutral sodium atoms, extends only partially around Jupiter along Io's orbital plane ahead of Io's orbit direction.*

I. INTRODUCTION

Io is the most active and bizarre of all known natural satellites. In the past decade, Io has been revealed to be a veritable wonderland of physics and chemistry, presenting a wide and rich array of phenomena for planetary scientists in many disciplines to attempt to unravel. Its prodigiously active volcanism and pervasive influence on Jupiter's giant magnetosphere are two dominant characteristics that make Io stand out as one of the most unusual objects in the solar system. Because of these unique features, studies of Io have been punctuated by more major turning points and surprise discoveries than for any other satellite.

A great deal of what we presently know about Io has been determined from the Voyager 1 and 2 spacecraft missions in 1979 and from Earth-based telescopic and theoretical studies since then. Thus we can break our discussion into pre-Voyager and post-Voyager periods. In this chapter, we first briefly review some of the early history of Io studies and then describe the current level of understanding of Io's physics, chemistry, geology, orbital dynamics, and geophysics, including its internal, surficial, atmospheric, plasma, and magnetospheric properties and how they interrelate.

Most of what we present in this chapter on Io is covered in more detail in key post-Voyager (special issue) journal papers, review articles and book chapters as follows: *Science*, vol. 204, June, 1979; *Nature*, vol. 280, August, 1979; *Science*, vol. 206, November, 1979; abstracts from IAU Colloq. 57, Kailua, Hawaii, May, 1980; *Icarus*, vol. 44, November, 1980; *Journal of Geophysical Research*, vol. 86, September, 1981; *Satellites of Jupiter*, ed. D. Morrison (Univ. of Arizona Press, 1982); *Physics of Jovian Magnetosphere*, ed. A. Dessler (Cambridge Univ. Press, 1983); and abstracts from IAU Colloq. 77, Ithaca, N.Y., July, 1983. A listing of the most pertinent physical properties of Io, derived from the above and other sources, is given here in Table I.

HISTORY OF PRE-VOYAGER IO STUDIES

Io was discovered in 1610 by Galileo (Galilei 1610) as the innermost of four bright objects forming a miniature solar system orbiting Jupiter. It became apparent by the end of the 18th century that the orbital spacings and motions of the inner three satellites were more regular than other planetary counterparts. Laplace (1805) demonstrated that the most distinctive feature of this miniature system was the special orbital configuration (the so-called Laplace resonance) which suggested a special dynamical relationship between the three inner satellites (see the chapter by Peale). Early 20th century work on the orbital motions of these bodies was done by Sampson (1921). The fundamental role that the orbital resonance has played in determining Io's major properties and thermal history was not discovered until recently (Peale et al. 1979).

Over the 300 years following Galileo's discovery, Io's character was not considered to be much different from the other Galilean satellites (in fact not too different from the Moon) until 1927 when photometric and colorimetric measurements (Stebbins 1927; Stebbins and Jacobsen 1928) showed Io to have a pronounced variation in brightness with orbital phase angle. For the next several decades, telescopic observations of Io provided gradual improvement of photometry and color data (summarized for that period by Harris 1961). These data showed that Io was the reddest object in the solar system and had an outstanding color variation with orbital phase angle.

Then, in 1964, Binder and Cruikshank (1964) reported an anomalous brightening of Io's surface as it emerged from eclipse. This first report of post-eclipse brightening and a possible atmosphere was the dawning of what in the next 15 yr would become a sequence of increasingly stunning surprises in our understanding of Io and its environs. In the same year, it was discovered by Bigg (1964) that bursts of decametric radio emission from Jupiter seemed to be controlled by Io's orbital position; this gave the first clue to an electrodynamic link between Io and the Jovian magnetosphere. Next came models of the electrodynamic interaction by Piddington and Drake (1968) and Goldreich

TABLE I
Io Global Physical Properties[a]

Radius	1815 ($\pm$5) km
Volume	2.52×10^{10} km^3 (2.52×10^{25} cm^3)
Surface area	4.15×10^7 km^2 (4.15×10^{17} cm^2)
Mass	8.92×10^{25} g
Relative mass (to Jupiter)	4.703 ($\pm$0.006) $\times$ 10^{-5}
Density	3.55 g cm^{-3}
Surface gravity	180 cm s^{-2}
Escape velocity	2.56 km s^{-1}
Orbital escape velocity	≥ 7.18 km s^{-1} (relative to Io surface)
Orbital radius (semimajor axis)	4.216×10^5 km; 5.95 R$_J$
Orbital period	42.456 hr
Mean orbital motion	203.2890 deg day^{-1}
Eccentricity[b]	0.0041 (forced); (1 $\pm$ 2) $\times$ 10^{-5} (free)
Inclination	0.027 deg
Orbital velocity	17.34 km s^{-1}
Magnetospheric velocity	56.8 km s^{-1}
Magnitude (V_0)	5.0
Albedo (V)	0.6
Heliocentric distance	5.20 AU (7.78×10^8 km)
Solar insolation	$\sim 3 \times 10^4$ erg cm^{-2} s^{-1}
Typical surface temperature	~ 135 K (subsolar)
Typical hotspot temperature	~ 300 K
Hotspot heat flow	$> 5 \times 10^{13}$ W
Global average heat flow	> 1.2 W m^{-2}
Jovian magnetic field (B) at Io	~ 2000 γ (or nT)
Corotation electric field at Io	0.113 V m^{-1} outward from Jupiter
Max potential across Io diameter	411 keV (500 keV if ionosphere included)
Pickup energy of torus ions	sulfur 540 eV, oxygen 270 eV

[a] Values given are derived from sources cited in the references to this chapter.
[b] The forced eccentricity is the dominant or actual eccentricity because it is so much larger than the free eccentricity.

and Lynden-Bell (1969) which addressed the coupling mechanism between Io's orbital motion and the inner magnetosphere.

In 1971 Io occulted a bright star and observations of this event provided the first accurate measure of Io's mean radius of 1818 km (subsequently revised to 1815 km; see Davies 1982) from which was derived (after the Pioneer-10 encounter in 1973 gave an accurate mass) an average global density of 3.55 g cm^{-3}, and an upper limit on Io's surface atmospheric pressure of 10^{-7} bar (Taylor et al. 1971; O'Leary and Van Flandern 1972; Smith and Smith 1972). About this time it was shown that Io has dark poles and relatively bright equatorial regions (Minton 1973) and a controversy developed over whether Io's reported post-eclipse brightening was real but sporadic (O'Leary and Veverka 1971; Fallon and Murphy 1971), or an artifact of observation resulting from light scattered off Jupiter (Franz and Millis 1971,1974). Several models were eventually put forth in attempts to explain the observations of post-eclipse brightening (Sinton 1973; Frey 1975; R. M. Nelson and B. W. Hapke 1978a) but no consensus was reached, and the question is still unsettled today (cf. chapter by Veverka et al.).

Also in the early 1970s, the techniques of measuring spectral reflectance were applied to Io in an attempt to identify its surface composition (Johnson and McCord 1970). Attention initially focused on the strong ultraviolet absorption and two kinds of material: (1) silicates were considered because of the relatively high bulk density of Io and the influence on thinking by the then current study of Apollo lunar rock samples; and (2) frosts were suggested by Io's high albedo, cold surface and proximity to gas-rich Jupiter. At the same time, early theoretical models of Io's formation (Lewis 1971a,b) tended to support the idea of a moon-like Io with a silicate crust containing surface frosts such as H_2O, NH_3, and various polymeric organic compounds that were modified by irradiation effects of Jupiter's magnetosphere (cf. Veverka 1971a; Lebofsky 1972). But also at this time several investigators were finding evidence against frost because the expected absorption bands were absent in Io's near-infrared spectrum (Moroz 1965; Gillett et al. 1970; Johnson and McCord 1971; Pilcher et al. 1972; Fink et al. 1973). One of these investigations (Johnson and McCord 1971) mentioned polysulfides as a possible surface coloring agent in the visible wavelengths. The idea of sulfur on Io's surface gained considerable momentum when Wamsteker (1973) pointed out, based on laboratory experiments, that free sulfur and some of its compounds closely matched the strong ultraviolet absorption and general reflectance spectrum of Io. Other laboratory spectral studies showed that many of the frosts that had been suggested earlier, such as H_2O, H_2S, NH_3, NH_4SH, CH_4 and CO_2, could not be present on Io's surface (Kieffer and Smythe 1974).

About this time, a new Io puzzle had developed when infrared photometry and radiometry measurements with groundbased telescopes showed Io to have discordant photometry at visible and infrared wavelengths and a higher brightness temperature at 10 μm than at 20 μm (Morrison et al. 1972; Hansen 1972; Sinton 1973). Equally puzzling were discordant eclipse measurements of Io's thermal inertia at various infrared wavelengths (Hansen 1973; Morrison and Cruikshank 1973) that prompted a rather $ad\ hoc$ two-layer inhomogeneous model for Io's surface (Hansen 1973). There were other baffling infrared observations (see, e.g., Hansen 1975) that were simply unexplainable in the context of that day's conception of Io. We now believe that these interpretive problems (summarized as they stood prior to Voyager by Morrison and Burns [1976], Morrison [1977], and Morrison and Morrison [1977]), were largely the result of an erroneous (but reasonable at that time) paradigm that Io was a cold, dead, insulating grey body like the Moon, receiving its surface energy only from incident sunlight.

Two events made 1973 an epochal year in the history of Io studies and focused much new attention on this satellite. First, Brown reported his discovery of sodium D line emission in the spectrum of Io (Brown 1974), which led to the mapping and characterization of the now well-known Io sodium cloud (Trafton et al. 1974; Trafton 1975a), and eventually to the discovery of a potassium cloud (Trafton 1975b). Second, the first spacecraft (Pioneer 10) visited the Jovian system and revealed from a telemetry occultation experiment that Io had an ionosphere and thus a thin atmosphere (Kliore et al. 1974), and from an ultraviolet spectrometer experiment that a torus of what was then thought to be hydrogen neutrals existed along Io's orbital path (Carlson and Judge 1974). Furthermore, energetic particle detectors on Pioneer 10 showed that Io was sweeping out a corridor of reduced particle density in Jupiter's radiation belt (Simpson et al. 1974).

These results promptly stimulated two theoretical models which were to influence strongly subsequent thinking about Io's surface composition and physics: (1) the evaporite hypothesis of Fanale et al. (1974,1977a,b) which argued for a salt/sulfur-rich surface composition enriched in sodium to account for the source of sodium to the cloud but which had trouble explaining the total absence of spectral evidence for bound water on the surface; and (2) the sputtering hypothesis of Matson et al. (1974) which provided a plausible ejection mechanism for supplying the sodium cloud by impact of magnetospheric ions, and which today remains the most viable theory. Resonant scattering of sunlight from neutral sodium atoms in an extended cloud was proposed by Trafton et al. (1974) and Matson et al. (1974) to account for the cloud emission, and later confirmed by systematic measurements of cloud brightness as function of orbital phase angle (Bergstralh et al. 1975,1977).

In 1975 Kupo et al. discovered ionized sulfur emission in the inner Jovian magnetosphere near Io's orbit but on the opposite side of Jupiter from the position of Io at the time (Kupo et al. 1976). Subsequent studies refined this picture of the sulfur emitting region from a semidisk (Kupo et al. 1976) to an annulus (Brown 1978) to a wedge-shaped ring (Nash 1979) to what we now know from detailed *in situ* measurements by Voyager 1 to be the Io plasma torus (Broadfoot et al. 1979; Sandel et al. 1979). Groundbased observations first revealed a high temperature of $\sim 10^4$ K and an electron density of $\sim 10^3$ cm^{-3} in the plasma (Brown 1976) (later confirmed by Voyager [Warwick et al. 1979; Eshleman et al. 1979]), the presence of ionized oxygen in the torus (Pilcher and Morgan 1979), and images of the wedge-shaped plasma ring (Pilcher 1980).

Meanwhile, spectral observations of Io's surface reflectance were expanding (Hansen 1975; Caldwell 1975; review by Johnson and Pilcher 1977; Nelson and Hapke 1978a) and extensive laboratory studies (Nash and Fanale 1977) were refining interpretations of the strong ultraviolet absorption feature and generally featureless high reflectance elsewhere in the visible, the prime evidence for sulfur on the surface. This work was joined by the discovery of a strong absorption band near 4 μm in Io's infrared spectrum simultaneously by several observing teams (Cruikshank et al. 1977,1978; Pollack et al. 1978a; Fink et al. 1978). This 4 μm feature is not characteristic of sulfur and it received considerable attention only to defy explanation (Fanale et al. 1977) until after Voyager 1 provided the clues that led to its identification as due to SO_2 on Io's surface.

Just prior to the March 1979 Voyager encounter, two events occurred that presaged the major and stunning surprise that was to follow. Witteborn et al. (1979) reported their observation of an intense temporary brightening of Io in the infrared from 2 to 5 μm. They explained it in the face of some skepticism as possibly an emission feature caused by a fraction of Io's surface having a brightness temperature of ~ 600 K, much hotter than the average daytime temperature of 130 K. Shortly after this report, and just a few days before the Voyager encounter, an epochal theoretical paper by Peale et al. (1979) appeared that addressed the dissipation of tidal energy in Io and predicted, ". . . widespread and recurrent volcanism. . . ."

The Voyager 1 flyby in March, 1979 spectacularly changed our way of thinking about Io. With the discovery of the absence of impact craters, the intense surface coloration (Smith et al. 1979a) and, most importantly, the presence of active volcanic plumes (Morabito et al. 1979), plus the detection of SO_2 gas and surface hotspots (Pearl et al. 1979), a new paradigm was born—Io was highly evolved and geophysically active, and its surface was young and chemically oxidizing.

III. SURFACE COMPOSITION AND TEXTURE

A. Composition

The surface composition of Io is unique among solar system bodies. Most of what we know about its surface composition is derived from groundbased disk-integrated spectral reflectance properties (see the chapter by Clark et al.) and *in situ* measurements by the Pioneer and Voyager spacecraft of ions in the Jovian magnetosphere (see the chapter by Cheng et al.) or inferred from the observed surface morphology and color differences seen in Voyager images. These observations have revealed a surface (see Color Plate 1) thought to be rich in sulfur, sulfur compounds (including SO_2), and alkali sulfides.

The composite spectral reflectance of Io obtained from groundbased observations is illustrated in Fig. 1. Besides the sharp drop in reflectance into the ultraviolet, the spectrum has two other striking features: the presence of a

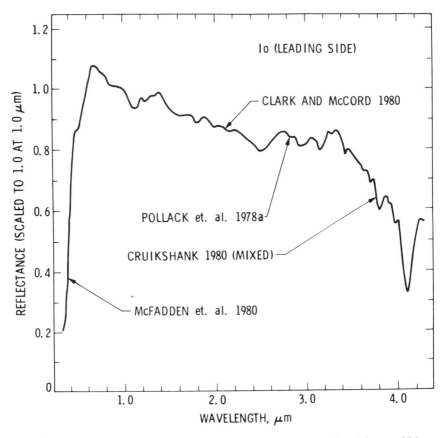

Fig. 1. The reflectance spectrum of Io's leading side (except data from 2.9 to 4.2 μm, which are from Cruikshank et al. [1978] and represent mixed orbital phase). The composite spectrum is from Clark and McCord (1980a) and contains data from McFadden et al. (1980) and Cruikshank (1980a). The sharp dropoff into the ultraviolet is the signature of sulfur or sulfur compounds. The absorption band at 4 μm is due to SO_2. The downward slope of the continuum from 0.6 to 4.3 μm is presently unexplained.

strong absorption near 4 μm and the absence of any absorption (in the 1.5 to 3.0 μm range) due to water. The strong infrared absorption at 4 μm observed by Cruikshank et al. (1978) and Pollack et al. (1978a) was identified from laboratory studies by Fanale et al. (1979), Smythe et al. (1979), and Hapke (1979) as SO_2 in the form of frost or adsorbate. Also, the visible and ultraviolet reflectance of SO_2 frost measured in the laboratory showed that SO_2 frost could be a component on Io's surface, but not a dominant one spectrally because of the frost's high brightness in the near-ultraviolet (Nash et al. 1980). The absence of water absorption bands meant that unlike the other Galilean satellites, Io has neither bound water nor free water ice on its surface.

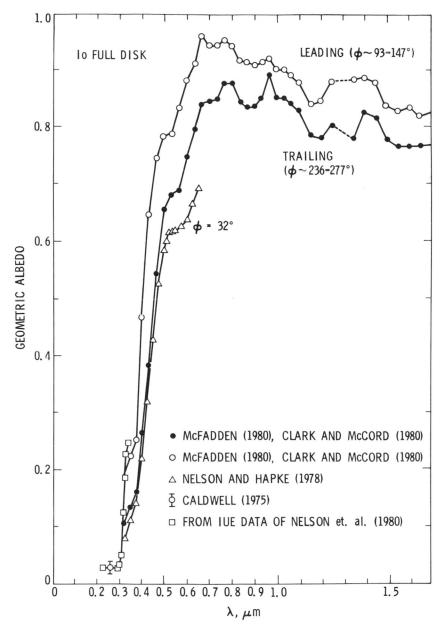

Fig. 2. Reflectance spectra of Io's full disk when Io is at different orbital phase angles (ϕ). Leading-side spectrum has been normalized to a geometric albedo of 0.90 at 1.0 μm, trailing-side spectrum to an albedo of 0.85 at 1.0 μm. The IUE data have been normalized to an albedo 0.029 at 0.26 μm. The steep absorption band edge at 0.31 μm is thought to be due to condensed SO_2, that at 0.4 to 0.5 μm due to sulfur or alkali sulfides. The band inflection at 0.55 μm and sloping shoulder between 0.6 and 0.7 μm are not well understood.

Sulfur. Early theories of the surface composition were based upon spectral reflectance observations in the range 0.4 to 1.0 μm where Io's full-disk spectrum (Fig. 2) shows a strong absorption edge from 0.4 to 0.6 μm and high flat continuum from 0.6 to 1.0 μm. Wamsteker et al. (1974) noted that the overall shape of the reflectance spectrum could be explained by a sulfur-rich surface. Later studies based on an extended spectral range into the near-ultraviolet and infrared concluded that sulfur allotropes as well as sulfur compounds and allotropes were important. Fanale et al. (1974) and Nash and Fanale (1977) proposed variations of the so-called evaporite model in which elemental sulfur mixed with sulfates of sodium, magnesium, and iron was deposited on the surface during the evaporation of outgassed volatiles. In an effort to explain subtle spectral features near 0.35 μm, Nelson and Hapke (1978*b*) proposed that some of the sulfur near postulated fumaroles may have undergone extensive thermal modification.

Soderblom et al. (1980) showed that the spectral reflectances of five spectrally distinguishable classes of surface areas on Io defined by Voyager (color filters at 0.35, 0.41, 0.48, 0.54, and 0.59 μm) were consistent with laboratory spectra measured for the various sulfur allotropes found in quenched sulfur (or sulfur glasses). They also noted that the ultraviolet spectral properties of the highly reflecting white areas on Io were similar to those of mixtures of pure S_8 and SO_2 frost.

Gradie et al. (1982), Gradie and Veverka (1984*a,b*), and Veverka et al. (1982*a*,1984) have compared the photometric properties of sulfur with those of Io. The photometric properties of sulfur depend on photometric geometry, particle size, contaminants, thermal history and temperature (see chapter by Veverka et al.). The temperature dependence of the long wavelength edge of the strong ultraviolet absorption is diagnostic of the presence of sulfur. When the temperature of powdered sulfur is lowered, the absorption edge (when observed by reflected light) moves toward the ultraviolet at about -1.6 Å K^{-1} as shown in Fig. 3. Veverka et al. (1981*a*) have examined Voyager images of Io emerging from eclipse in the Jovian shadow and report no anomalous brightness change at the 10% level expected for a sulfur surface. Veverka et al. (1982*a*,1984) report no temperature-induced spectral changes for regions of identical spectral properties when viewed near local noon and the terminator. One conclusion is that elemental sulfur is not ubiquitous on the surface. Hammel et al. (1985) suggest that no more than 50% of the surface could be covered with S_8 based on narrowband photometric observations of Io's full-disk albedo following eclipse reappearance. Another possibility is that the actual thermal conditions on Io's surface have not been adequately modeled in the laboratory. This perplexing topic is receiving further study.

Sagan (1979) pointed out that the color diversity of sulfur allotropes (such as S_4 and S_3) could provide an attractive explanation for the vivid reddish coloration of Io's surface displayed in the original Voyager images (Smith et al. 1979*a*). Since that time, an interesting and important controversy has

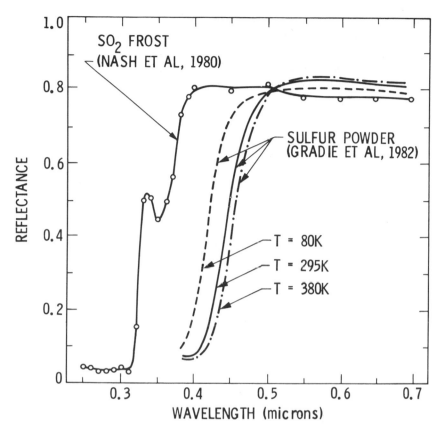

Fig. 3. Laboratory reflectance spectra of sulfur powder and sulfur dioxide frost. Both materials have a steep absorption band edge in the ultraviolet and are otherwise bright and featureless in the visible and near infrared. The absorption edge position of sulfur is temperature sensitive, shifting to shorter wavelength as temperature decreases; at typical Io surface temperature (135 K) ordinary yellow sulfur appears white. In the ultraviolet near 0.25 to 0.30 μm, SO_2 frost is less reflective than sulfur by $\sim 1/2$.

arisen, largely at the prodding of Young (1984), regarding what the real color of Io is. After discussion of this issue with Young and others, McEwen carefully reprocessed the Voyager image data and prepared a "natural color" image of Io (see Color Plate 4) and the following description of the situation: "The Voyager imaging systems include five color filters with the narrow-angle cameras: orange [0.59 ($\pm$0.03) μm], green [0.54 ($\pm$0.04) μm], blue [0.48 ($\pm$0.04) μm], violet [0.41 ($\pm$0.03) μm], and ultraviolet [0.35 ($\pm$0.03) μm]. Since the green filter was not used for many of the high-resolution color sequences, color images are commonly constructed with the orange, blue, and violet filter, displayed as red, green, and blue, respectively. Therefore, these filters are shifted to shorter wavelengths than what we see with our eyes. Most planets and satellites have fairly linear reflection spectra through the visible, so the orange-blue-violet filter color pictures are fair representations of natural color. However, the visible spectrum of Io [see Fig. 2] is markedly nonlinear, with a shoulder at ~ 0.5 μm, so the use of the orange, blue, and violet filters results in a redder image than what we would see with our eyes. The "natural color" image of Io is an attempt to correct this by using the orange filter for red, an average of the orange and blue filters for green, and an average of the blue and violet filters for blue. There remains a shift to shorter wavelengths, but Voyager did not return images at longer wavelengths than the orange, so this seems to be the best approximation. Io is predominantly pastel yellow, with orange, green and grey tints" (A. McEwen, personal communication, 1984). This is in agreement with Young's conclusion that Io's color is "greenish-yellow, not red" (Young 1984).

Sagan (1979), however, made the important suggestion that if the color of high-temperature allotropes of sulfur is preserved on quenching as Nelson and Hapke (1978b) suggested, then a morphology-color relationship should exist because of sulfur's unusual temperature/viscosity characteristics. Sill and Clark (1982; cf. chapter by Clark et al.) provide a discussion of the spectral reflectance properties of sulfur allotropes and associated sulfur compounds. These coloring agents are most likely to form on Io in and around the active volcanic regions and wherever molten sulfur can be quenched to < 150 K. It is thought by some that during the quenching process high-temperature allotropes that are stable in the melt but not at lower temperatures may be prevented from reverting to S_8. Pieri et al. (1984) make use of this assumed color diversity with temperature and viscosity to explain flow features in the Ra Patera region in terms of S_8 and other allotropic forms of sulfur. Fink et al. (1983) examined the behavior of liquid sulfur flows and concluded that surface folding instabilities and density inversions are likely to develop at the upper surface of Io sulfur flows, severely complicating any attempts to interpret surface colors.

Considerable controversy has arisen as to whether high-temperature sulfur allotropes would be stable on Io in spite of the low ambient surface temperature (< 135 K). Young (1984), Gradie and Moses (1983a), Fink et al. (1983), and Gradie et al. (1984) point out that quenching of molten, colored allotropes from sulfur is difficult to achieve, particularly in flows and ponds. Young (1984), in an instructive paper on the properties of sulfur, argues compellingly that "either the volcanic flows on Io are not sulfur, or some mechanism other than quenching is required to produce colored forms of sulfur in them." Gradie et al. (1984) note that droplets 10 to 100 μm in diameter, which quench by radiative (or more likely, evaporative) cooling, are more likely candidates for the form of sulfur on Io. Much smaller particles could quench by conduction to the cold surface of Io if their flight time is sufficiently short to keep them molten. Collins (1981) estimates that most of the mass in the plumes is ejected in particles with radius between 10^{-2} and 10^{-3} μm. However, the flight time of these particles (15 min; Johnson et al. 1980) is too long for them to be molten on impact.

Other mechanisms for color retention in molten sulfur are possible. Gradie and Moses (1983a) suggest that baking of the top layer in flows would not necessarily prevent the retention of high-temperature allotropes. Their laboratory experiments with boiled sulfur which is cooled slowly shows evidence that some yet unidentified high-temperature allotropes can remain in the solid for considerable time. It is not clear whether this property is intrinsic to the sulfur or caused by impurities in the sulfur sample.

Sulfur and Silicates.   The diverse morphologies on Io provide arguments in favor of the presence of silicate materials. As Carr et al. (1979) and Schaber (1982) have pointed out, the scarps and mountains, which rise as much as 10 km above the mean surface, cannot be supported by sulfur alone. Nor can the observed 2 km deep calderas form or survive in a material with a combination of thermal and strength properties similar to sulfur (Clow and Carr 1980). A mixture of sulfur and silicate, if not silicates alone, is required. Spectral data from Voyager and groundbased telescopes do not show any evidence for exposed silicate rocks. It must be remembered, however, that the Voyager data are limited by coverage in a spectral region not particularly diagnostic of silicates and the groundbased data are limited by spatial resolution. Moreover, only a thin veneer of sulfur-rich materials is needed to mask completely the silicate signature and any fresh silicate exposure would be rapidly coated with sulfur compounds as a result of plume and fumarolic activity.

Gradie and Moses (1983b) have examined the spectral reflectance properties of various mixtures of basalt and sulfur heated to 700 K. Gradie et al. (1984) have studied the spectral reflectance properties of basalt-sulfur mixtures heated to 1400 K. It can be concluded from both these studies that some combination of sulfur, basalt and temperature can be found that will match the spectral properties of some regions on Io in the limited spectral range available in the Voyager images. Figure 4 shows a comparison of some of the data of Gradie and Moses (1983b) with the spectral reflectance properties of Io surface materials associated with the two types of volcanic plumes defined by McEwen and Soderblom (1983). Unfortunately, disk-resolved spectral data from Io in the spectral region 0.7 to 2.5 μm, which would be diagnostic of some silicates such as pyroxene, olivine, and plagioclase, are not available. Groundbased full-disk spectral data indicate that, if present, the areal coverage on Io of these silicates is small.

Sulfur Dioxide.   Of all the surface materials proposed, only SO_2 has been clearly demonstrated to be present from the identification of the $\nu_1 + \nu_3$ combination band of a condensed SO_2 phase in Io's infrared spectrum (Fig. 5). Further support for SO_2 comes from ultraviolet spectral data; as shown in

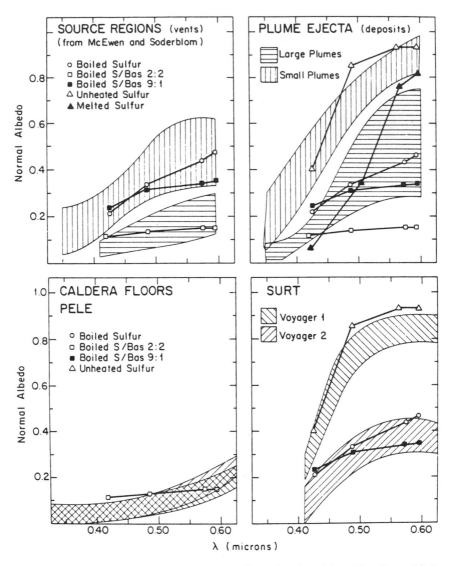

Fig. 4. Reflectance spectra of vent source regions, plume deposits and the caldera floors of Pele and Surt (from McEwen and Soderblom 1983) compared with laboratory spectra of various heated sulfur and sulfur/basalt mixtures (from Gradie and Moses 1983b). All laboratory data were convolved with the Voyager narrow angle camera response functions. The source regions of the small plumes appear consistent with boiled sulfur (ultra-pure sulfur heated to 717 K and cooled slowly to room temperature, then further cooled to 77 K) at ~ 120 K. The large plume source regions (the darkest pixels in each caldera) appear to have a larger opaque content, perhaps basalt, such as the boiled sulfur/basalt 2:2 mixture (ultra-pure sulfur plus dehydrated basalt heated to 717 K and cooled slowly to room temperatures, then further cooled to 77 K) at ~ 120 K. The small plume ejecta are qualitatively similar to pure (unheated) sulfur at T ~ 120 K or perhaps melted sulfur (ultra-pure sulfur heated to 433 K and cooled slowly to room temperature, then cooled to 77 K) mixed with highly reflecting, spectrally neutral (0.4 to 0.6 μm) SO_2 frost. The large plume ejecta are more consistent with the boiled sulfur. The caldera floor of Pele is consistent with a heated sulfur/basalt mixture. At the time of Voyager 1 the spectral reflectance of the caldera floor of Surt was matched by unheated sulfur but during the time of Voyager 2 the boiled sulfur or heated mixture of sulfur and a few percent basalt provide a better match.

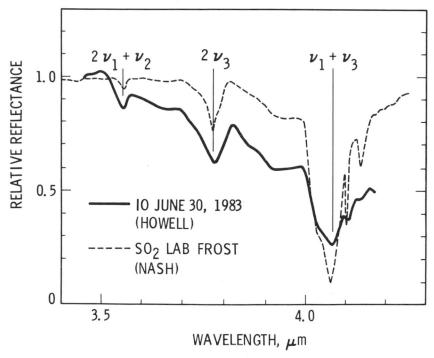

Fig. 5. Comparison of recent Io spectrum near 4 μm of Howell (Howell et al. 1984) with laboratory reflectance spectrum of SO$_2$ frost (Nash 1983) at approximately the same resolution. The
two spectra are scaled to 1.0 at 3.45 μm; the "absolute" reflectivity of the SO$_2$ frost relative to
a gold standard is ~0.95 at 3.45 μm. The principal band identifications in SO$_2$ are shown.
The two sharp bands on the high wavelength wing of the $\nu_1 + \nu_3$ band are identified as due to
heavy isotope fractions of SO$_2$ as follows: $^{34}S^{16}O_2$ at 4.108 μm; $^{32}S^{16}O^{18}O$ at 4.144 μm.
The Io spectrum was obtained at Mauna Kea on 30 June 1983 when Io was at an orbital phase
of ~150°.

Fig. 3, SO$_2$ frost has: (1) a steep absorption edge at 0.31 μm, and (2) very
low reflectivity at wavelengths below 0.31 μm. Io's spectrum has both of these
features (as shown in Fig. 2).

An important debate has centered on whether the stable SO$_2$ phase is ice
or frost, or adsorbate. An early widespread assumption was that the stable
phase was solid SO$_2$ as frost or ice. Yet, as argued by Nash (1981,1983), the
equilibrium vapor pressure required to sustain solid SO$_2$ (10^{-8} to 10^{-7} bar) is
much more than the average diurnal subsolar SO$_2$ pressure on Io. As noted by
Matson and Nash (1983), the lack of permanent polar caps, lack of regular
nighttime frost, and lack of regular post-eclipse brightening argue against a
stable solid SO$_2$ phase in equilibrium with an SO$_2$ atmosphere.

In the adsorbate model, summarized by Nash (1983), SO$_2$ is found as an adsorbed gas that coats the uppermost surface layer. SO_2 gas molecules adhere to the surfaces of grains in at least a partial monolayer by way of van der Waals forces or chemical bonding. The depth of the $\nu_1 + \nu_3$ spectral band observed at 4.05 μm is produced by multiple transmission through the SO_2 coatings on grains transparent at 4 μm (e.g., sulfur). Matson and Nash (1983) argue that a highly porous surface produced by fluffy, fine-grained volcanic ash is ideal for gas adsorption. High-resolution Fourier transform spectroscopy of Io by Fink et al. (1980b) placed the position of the Io band minimum at 4.054 μm, almost exactly at the position of the measured band for adsorbed SO_2 and significantly away from the earlier-reported position of the band for SO_2 frost and ice (4.07 to 4.08 μm) (Fanale et al. 1979; Smyth et al. 1979). However, the close match between band positions in more recent observations of Io's 4 μm band (Howell et al. 1984) and in a new laboratory spectrum of SO_2 frost (Nash 1983) indicate that the band on Io is primarily due to SO_2 frost, not adsorbate. Howell et al. (1984) suggest that SO_2 is more widespread on Io than earlier thought (cf. Nash et al. 1980); yet this estimate is made uncertain by the unknown degree to which the 4 μm reflectance band is contaminated by thermal emission from hotspots (D. Matson, personal communication, 1984). The controversy over frost vs. adsorbate as the SO_2 phase responsible for the strong 4 μm band seems at this writing to be settling toward the frost. Minor bands detected in Io's spectrum at 4.108 μm and 4.144 μm by Fink et al. (1980b) have been interpreted by Nash (1981) as due to heavy-isotope SO_2 on Io's surface.

Synoptic observations of Io by Nelson et al. (1980a) in the mid-ultraviolet (0.22 to 0.33 μm) with Earth-orbiting International Ultraviolet Explorer (IUE) spacecraft (typical data shown in Fig. 2), combined with laboratory spectral data on sulfur and SO_2 (e.g., Fig. 3) and groundbased infrared spectral data of Io (Fig. 5) suggest that the distribution of SO_2 varies longitudinally across Io's surface, with the SO_2 being most abundant on the visually bright leading hemisphere (72 to 137° longitude) and least abundant on the darker, red, trailing hemisphere (250 to 323° longitude). Synoptic observations of the 4 μm band by Howell et al. (1984) confirm this longitudinal variation in SO_2 but show no significant temporal variation between 1976 and 1981. They also conclude that significant amounts ($> 50\%$) of SO_2 exist in the nonwhite areas where they are intimately mixed with other phases such as sulfur. According to Baloga et al. (1983), the bright lateral surface markings or auras seen around flow features near Ra Patera and other regions on Io are probably caused by outgassed SO_2 recondensing on the adjacent cooler surface. In this interpretation, the SO_2 is liberated from the regolith by the heat of the nearby flow and follows a ballistic trajectory back to the surface. In their model, molecules such as H_2S_2, Na_2S_4 and S_8 do not fit the observations. The presence of large amounts of solid SO_2 on Io's equatorial surface regolith has important implications for Io's atmosphere and absence of polar caps which still remain to be adequately understood.

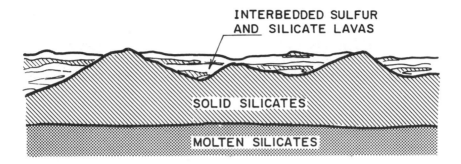

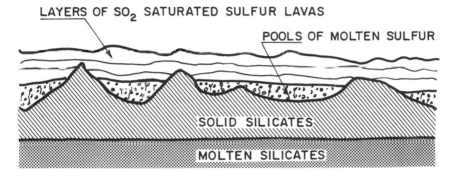

Fig. 6. Two models of the upper few kilometers of the crust of Io (after Keiffer 1982). In the model at top, interbedding of sulfur-silicate lavas is produced by silicate and sulfur volcanism. According to this model, landforms comprised of silicate and sulfur should be visible on the surface. In the lower model, sulfur and sulfur compounds overlay the hot silicates. The landforms in this model would be predominantly composed of sulfur and sulfur compounds.

Surface Composition Models.   The discovery of active volcanism on Io has resulted in two rather similar surface composition models for Io (Fig. 6). The lack of spectral evidence for water (either as frost or bound in minerals) requires that other volatiles be involved in the current volcanic activity. The spectral evidence and thermodynamic considerations (Smith et al. 1979*c*; Kieffer 1982) strongly suggest sulfur and SO_2 are the dominant volatiles. Current volcanic models predict that the surface will be composed almost entirely of sulfur and sulfur compounds or will be a mixture of interbedded silicate and sulfur (or sulfur derived) lavas. In both models the sulfur-rich surface layer is only a thin veneer overlying the silicate bulk of the planet. In any model some compound is required that can supply at least a few percent of Na and K since the presence of a neutral Na cloud around Io requires a source of Na at the surface (Matson et al. 1974; Nash and Nelson 1979; Fanale et al.

1982). Possible candidates are Na_2S and K_2S. Another important consideration is the source material for the sulfur. Hapke (1979) has argued that sulfides of iron, particularly troilite (FeS) brought to the crust from the core by vigorous mantle convection could provide the sulfur. Lewis (1982) counters that argument with geochemical evidence that FeS would not be the stable phase of the sulfide so that pyrite (FeS_2) and sulfates are the probable progenitors of sulfur and SO_2.

B. Texture

Our ideas concerning the microstructure and physical properties of Io's surface (see the chapter by Veverka et al.) are based upon: (1) optical observations of the photometric phase curve; (2) infrared observations of the thermophysical properties; and (3) the morphology and characteristics of the volcanic plumes.

Pang et al. (1981) have modeled the surface texture using groundbased and Voyager photopolarimeter full-disk observations of Io from 0 to 40° phase. The parameters of the best-fit model lead them to conclude that the bulk density of the topmost surface layer is about 0.7 to 0.8 g cm^{-3}. They conclude also that sulfur, if it exists in pure form, does not reside as terrestrial flowers or in colloidal form but as aggregates or as pollen-like structures. Their contention that SO_2 is not present as an optically-thick layer must be tempered with the fact that disk-integrated photometric properties are not necessarily germane to the photometric properties of specific regions. However, the photometric evidence for the lack of a widespread optically thick SO_2 layer is consistent with the conclusion of Howell et al. (1984) that the SO_2 must be mixed intimately with some other material.

Matson and Nash (1983) conclude from thermal eclipse measurements and calculations of the resurfacing rate (> 1 mm yr^{-1}) that at least the upper few centimeters of Io (Fig. 7) have a bulk density of ~ 0.3 g cm^{-3} and porosity as high as 90%. They liken this fluffy layer to the fairy-castle structure used to describe the uppermost layer of the lunar regolith. Their model for the upper 1 km of a sulfur regolith is described in Table II. Nelson et al. (1984) measured the ultraviolet albedo variation with low solar-phase angles ($< 6°$) for Io and Europa. They conclude that Io's uppermost surface layer is very porous and Europa's is very compact.

The typical surface temperature of Io's average surface near the noontime subsolar region is ~ 135 K (see, e.g., Simonelli and Veverka 1984a). The thermal conductivity of the upper few centimeters of this average surface in the Matson and Nash model is low enough to make the subsurface temperature of this layer significantly lower than the surface temperature at noon (Fig. 8). As Matson and Nash note, this low subsurface temperature, combined with modeled high porosity and permeability, could make this region an ideal subsurface cold-trap for gases. The bulk density calculated by Pang et al. (1981) is 2.5 times larger than that calculated by Matson and Nash for the

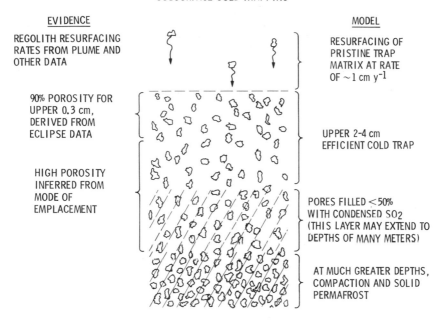

EVIDENCE

REGOLITH RESURFACING
RATES FROM PLUME AND
OTHER DATA

90% POROSITY FOR
UPPER 0.3 cm,
DERIVED FROM
ECLIPSE DATA

HIGH POROSITY
INFERRED FROM
MODE OF
EMPLACEMENT

MODEL

RESURFACING OF
PRISTINE TRAP
MATRIX AT RATE
OF ~1 cm y^{-1}

UPPER 2-4 cm
EFFICIENT COLD TRAP

PORES FILLED <50%
WITH CONDENSED SO_2
(THIS LAYER MAY EXTEND TO
DEPTHS OF MANY METERS)

AT MUCH GREATER DEPTHS,
COMPACTION AND SOLID
PERMAFROST

Fig. 7. Schematic vertical profile through Io's surface illustrating the essential elements of the subsurface cold-trapping model. The surface particles arrive by fallout from active plume fountains (figure from Matson and Nash 1983).

TABLE II
Estimated Average Surface Physical Properties of Io to
Depth of 1 km in Particulate Surface Regolith[a]

Depth (m)	Porosity (%)	Bulk Density (g cm^{-3})	Permeability	Thermal Conductivity (W cm^{-1} K^{-1})	Tempature at Base of Layer (K)
0–0.1	90	0.3	high	5×10^{-6}	97
0.1–1.0	60	0.8	medium	1×10^{-5}	107
1–10	40	1.2	low	5×10^{-5}	128
10–100	30	1.4	impermeable	5×10^{-4}	148
100–1000	0	2.0	impermeable	2×10^{-3}	198

[a]Table from Matson and Nash (1983).

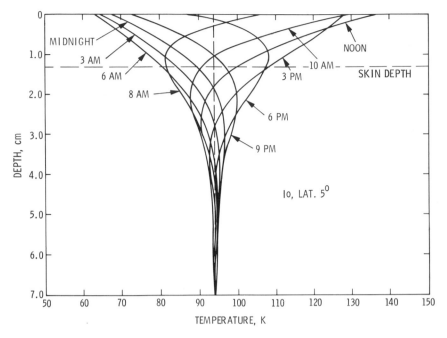

Fig. 8. Profiles of temperature vs. depth for Io's subsurface at different times of day. These curves were calculated for a latitude of 5°, but can be taken as typical of the equatorial region. As such, they constitute the highest-temperature-contrast case for the subsurface trapping model. Temperatures at greater depth can be found in Table II (Matson and Nash 1983).

upper 10 cm in their model. If this higher bulk density and thus lower porosity is more applicable, then the subsurface gas cold-trap will be less efficient.

IV. SURFACE GEOLOGY

Prior to arrival of the two Voyager spacecraft, mapping of Io's surface geology was virtually impossible. Data on Io came from Earth-based tele-scopic observations that were barely able to resolve even the gross surface markings. Io's 3630 km diameter from Earth is ~ 1 arcsec and the best observing conditions and equipment allowed albedo features of only ~ 0.2 arcsec, 1/5 of Io's diameter, to be distinguished. Thus, it was that the best pre-Voyager drawings and maps of Io (Minton 1973; Dollfus and Murray 1974; Murray 1975; Nash and Johnson 1979) showed only the most general features: dark polar regions, and bright equatorial regions with the leading hemisphere brighter than the trailing.

Voyager 1 opened a new window on Io by coming within 20,000 km of its surface and systematically photographing features on all scales down to as small as several km in its south polar region (Smith et al. 1979a). This quantum leap in information revealed Io's most stunning feature—active volcanism on a prodigious scale (Morabito et al. 1979)—and allowed geologists to map its associated surface features in relatively great detail (Masursky et al. 1979). They found nine active volcanic plumes ejecting material as high as 300 km above the surface and blanketing the surface with highly colored surface deposits (see Color Plate 5), some extending over 1000 km in diameter (Smith et al. 1979a; Strom et al. 1979). In addition, hundreds of pits, and calderas up to 200 km diameter, have been identified. Many are only recognizable as dark spots; they occur (see Fig. 9 foldout) in every region of Io's surface (Carr et al. 1979). Dark spots and calderas include the vents of active plumes and appear to mark places on Io's surface that are sources of both plumes and extensive surface flows (Carr et al. 1979; McCauley et al. 1979; Schaber 1980,1982).

An overriding character of Io's surface is its extreme youthfulness. There are no recognizable impact features down to the limiting image resolution of 600 m, and volcanic resurfacing rates were initially estimated to be as great as 0.1 cm yr^{-1} (Johnson et al. 1979); later, after a detailed study of plume particle-size mass distribution (Collins 1981), the rates were revised upward to as high as 10 cm yr^{-1} (Johnson and Soderblom 1982).

A. Landforms and Albedo Features

The structure and appearance of the Ionian surface is largely controlled by volcanic processes, which have produced a wide array of landforms and albedo features; these are classifiable into three broad categories: vent regions, plains, and mountains (Smith et al. 1979a; Carr et al. 1979; Schaber 1982); examples of all of these features can be seen in Color Plate 5.

Vent regions include (1) crater-like depressions, (2) dark, roughly circular markings, (3) the sources of radially directed flows, and (4) centers of bright halos. Vents are fairly evenly distributed across the surface, although Schaber (1982) has suggested that there may be a slight bias toward the equatorial regions. They also have a log-normal size distribution similar to impact craters on other planetary bodies. Where depressions are visible, they strongly resemble terrestrial calderas, having arcuate scalloped walls, floors at different levels, and smooth rims (Fig. 10). Most of the calderas are simply inset into the plains surface; they are rarely at the summit of a perceptible edifice, although two low circular shields at 17°S, 350°W are exceptions. However, even where there is no identifiable shield, radial flows commonly indicate that the source vent is at the summit of a low rise. Many caldera floors are completely or partly covered with dark material, which appears to be recent lava. The question of whether any of these dark lava flows could have been liquid sulfur at the time the Voyager images were obtained has been examined by Nelson et al. (1983) who found from laboratory studies that areas of liquid sulfur at any temperature on Io's surface would have appeared black in Voyager images. Bright halos which seem to be condensed vapor deposits around and partly within the calderas are also common. While many dark circular

0   100

km

Fig. 10. The 32 km diameter caldera Maasaw Patera at 40°S, 341°W. The caldera resembles terrestrial calderas in having smoothly arcuate walls and a floor with different levels. The deepest
part of the floor is 2.1 km below the rim. See also Color Plate 5.

markings are within observable calderas, most have no visible caldera, either because none exists or more probably, because none was visible due to poor illumination. The volcanic center is identifiable simply as an albedo marking or as the source of flows or other deposits. Some vents have been correlated with anomalously high temperatures (Pearl and Sinton 1982). The temperatures show a bimodal distribution with peaks at 650 K and 250 K. McEwen and Soderblom (1983) suggested that each peak temperature is associated with a different type of plume and that they result from the bimodal viscosity of liquid sulfur with temperature.

Plains constitute a second and the most extensive class of surface feature. They have numerous markings that range in color (in the exaggerated color images) from yellow to orange to dark red-brown. The most comprehensible are linear markings that are dark in images taken with the clear filter. The markings are clearly flows that radiate from vents, as for example Ra Patera (Fig. 11). Other dark flows that emanate from vents are broad and almost equidimensional in outline. They are recognizable as fluid flows only by their lobate margins. The color of the linear flows at Ra Patera changes radially outward from dark brown at the vent to orange distally, a color pattern that might be expected from sulfur cooling through its various allotropes (Sagan 1979; Picri et al. 1984), but, as noted above, this idea has been strongly challenged (Young 1984). Many of the flows are surrounded by a bright diffuse aura (Fig. 11). Work by Baloga et al. (1986) suggests that the auras are the result of heating of the ground by the flows and subsequent evolution of bright material from beneath the flows. On clear-filter images most flows appear darker than the surroundings, and the younger flows generally appear darker than older ones. But bright flows are present, as for example, at 2°S, 295°W. In general, the flow patterns are not clear from the albedo patterns. Dark flows, light flows and auras are so complexly superposed that individual flows are difficult to recognize in many places and, even if recognizable, cannot be traced for long distances. In addition, the continual rain of debris from the plumes may completely mask patterns intrinsic to the surface near plume vents and partly veil patterns elsewhere.

In several areas the plains are clearly layered. The layers are particularly visible at high southern latitudes where illuminated under low-sun angle. Escarpments 150 to 1700 m high (Schaber 1982) outline tabular, smooth-topped plateaus. Some of the escarpments are smoothly arcuate, as though formed by faulting; others are jagged as though eroded. Tabular units may be superposed so that the bounding escarpment of the upper unit transects that of the lower. The origin of plateaus is unclear. Some appear to start at vents such as Creidne Patera (Fig. 12 and Color Plate 5) but their extreme thickness and wide areal extent makes origin as a single discrete flow unlikely. Moreover, isolated plateaus, seemingly unrelated to vents, are also present, as at 80°S, 0°W. In places the escarpments seem to have been eroded to form alcoves separated by cuspate promontories, and to isolate mesa-like mountains from the main body

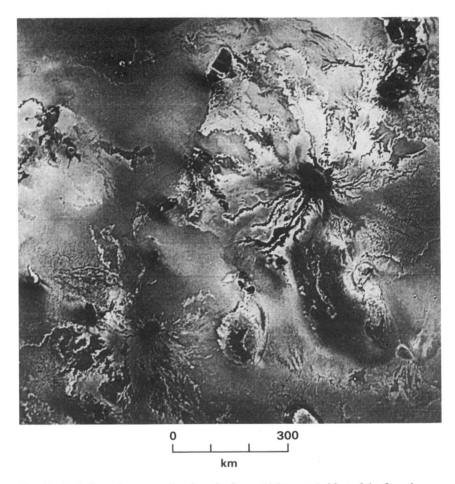

0   300

km

Fig. 11. Dark linear flows extending from Ra Patera (right center). Most of the flows have a
bright halo or aura. Numerous vents visible as dark marks have little discernible relief. The
longest flows are about 250 km long.

of the plateau. Diffuse bright marks occasionally occur along the escarpments, indicating fumarolic activity. McCauley et al. (1979) suggested that
the plateau units act as an aquifer for SO_2 which escapes along the escarpments, thus forming the observed brightenings. They suggested further that
escape of the SO_2 caused sapping and undermining of the escarpment to account for the erosion. However, the plateau materials themselves must also be
volatilized, or removed in some other, as yet undetermined, way to produce
the observed patterns.

A third major class of surface feature are the mountains. These mostly
have irregular, roughly equant outlines, and rugged, seemingly fractured, sur-

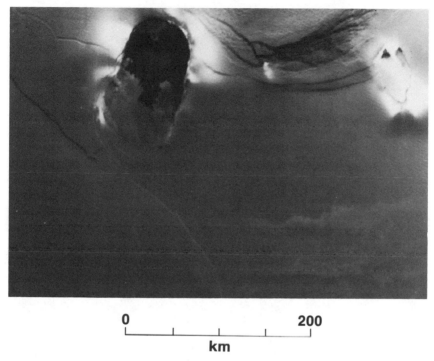

0 ⌞___⌞___⌞___⌞___⌟ 200

km

Fig. 12. Layered deposits near the caldera Creidne Patera at 52°S, 345°W. The heights of the escarpments range from 300 m to over 1000 m. Part of Creidne Patera, which was a hotspot (230 K) during Voyager 1 encounter, appears filled with recent dark lava. Sublimates have condensed around the northern half of the caldera to produce a bright irregular halo. See also Color Plate 5. (Voyager frame 0145J1+000, reprocessed by A. McEwen, T. Rock, L. Soderblom.)

faces. They range in size up to 200 km across and 9 km high. Several have characteristics that suggest a volcanic origin; some, such as Haemus Mons (e.g., lower-left corner Plate 5), are surrounded by a bright aureole, others partly surround a central depression, and the source of the prominent plume Pele is within an uplifted and fractured structure (see center of Fig. 13b) that resembles some of the other mountainous features. The mountains are not all necessarily volcanic. Smith et al. (1979c) interpreted some mountains as exposures of a silicate lithosphere that is almost everywhere else covered with sulfur-rich materials. Although the quality of the global photographic coverage available is very uneven, the mountains appear to be preferentially located in a 120° wide longitude band centered on 300°W (the same locale of large plumes and maximum thermal emission; see Sec. IV.B below), which suggests that Io's crust is asymmetric, since the silicate lithosphere is closer to the surface in this region (McEwen and Soderblom 1983).

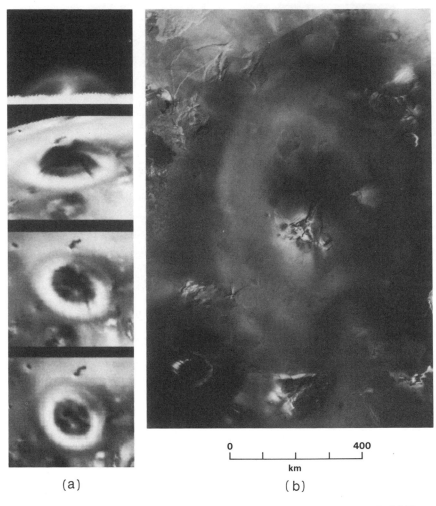

0   400
km

(a)   (b)

Fig. 13. Type examples of two classes of plumes and their deposits recognized on Io (McEwen and Soderblom 1983): (a) Prometheus, from various view angles. Smaller, lower velocity eruptions, thought to be driven by SO_2, reaching heights of ~ 100 km, are probably of long duration (over 2 yr). They form bright ringed deposits ~ 250 km in diameter and occur mostly in the equatorial belt. (b) Pele, from directly above. Larger, higher-velocity eruptions, of probably short duration (few days to few weeks), reaching altitudes of ~ 300 km, form large concentric orange-brown deposits ~ 1400 km in diameter. They occur in the longitude region near 300°W, and are thought to be sulfur-rich, SO_2-poor. (All images are from Voyager 1 data, reprocessed by A. McEwen, T. Rock, L. Soderblom. Image (b) includes some mosaicing lines.)

B. Plumes

The most spectacular indicators of active volcanism are the plumes (Fig. 13), umbrella-shaped fountains of fine-grained volcanic debris reaching to heights up to 300 km above the surface (Strom et al. 1979). The color of the plume deposits suggests that most of the ejected material is either sulfur or SO_2 (Fanale et al. 1982). Nine plumes were observed by Voyager 1 and eight were still active four months later during the Voyager 2 encounter (Strom et al. 1979; Strom and Schneider 1982).

Two types of plume have been recognized (McEwen and Soderblom 1983). The first type, exemplified by Prometheus (3°S, 153°W) (shown in Fig. 13a) includes all eight of the plumes that were active during both Voyager encounters. They are mostly from 50 to 200 km high and 150 to 550 km across at their base (Strom and Schneider 1982), which indicates ejection velocities of around 0.5 km s^{-1}. These plumes deposit bright whitish materials around their sources, and are concentrated at low latitudes but well spread in longitude. The fact that eight of the nine plumes remained active over a four-month time interval suggests that this type of plume has a lifetime measured in years.

The second type is exemplified by Pele (19°S, 257°W); its surface deposit is shown in Fig. 13b. This is the only plume of the second type that has been observed in eruption, but temporal changes in the albedo elsewhere on Io's surface indicate that two additional plumes of this type, Surt (45°N, 338°W) and Aten (48°S, 311°W) were active between the two Voyager encounters. The intermittent activity suggests that the Pele-type plumes have lifetimes typically of days or weeks, rather than years. Pele-type plumes are larger than the Prometheus-type. Pele was 300 km high and 1200 km across at its base, indicating ejection velocities of around 1 km s^{-1}, and the diameters of the deposits at Surt and Aten suggest plumes of similar dimensions. These larger plumes leave orange to red-black deposits around their vents, in contrast to the whitish deposits around the smaller, longer-lived type. They are located in a range of latitudes in a north-south zone centered near longitude 310°W. Thermal infrared measurements and inferred color temperatures for quenched liquid sulfur suggest vent temperatures of < 400 K for the small plumes and > 650 K for the large plumes.

Two types of models have been proposed to explain the plumes (Kieffer 1982). Smith et al. (1979c) proposed that the plumes are driven by SO_2 which is heated to the sulfur liquidus temperature (395 K) at depths of about 1.5 km. Formation of SO_2 vapor causes upward movement of the SO_2 liquid and vapor. As the gas-liquid mixture rises, more SO_2 vaporizes to accelerate the column further, and solid materials, such as sulfur, may be entrained by erosion of the conduit walls. Ultimately the triple point of SO_2 is reached, close to the surface, and any further expansion results in condensation of SO_2 snow. Support for SO_2 as the driver is the detection of gaseous SO_2 over the active plume Loki (Pearl et al. 1979). Reynolds et al. (1980), however, pointed to the difficulty of obtaining ejection velocities of 1 km s^{-1} required for Pele with such a low-temperature system. They, together with Consolmagno (1979), Hapke (1979), and Sinton (1980), proposed alternatively that the plumes are driven by sulfur heated to temperatures of 1000 to 1500 K at depths of several kilometers by hot, possibly molten, silicates. McEwen and Soderblom (1983) suggested that both SO$_2$- and S-driven plumes may be present. They argue that plumes of the Prometheus type are driven by SO$_2$ heated at the source by sulfur in the 400 to 450 K temperature range, where it has a viscosity minimum. In contrast, Pele-type plumes may be driven by sulfur heated to at least 650 K, by contact with hot silicates.

C. Resurfacing Rate

Io is being continually resurfaced by debris from the plumes and flows. Johnson et al. (1979) and Johnson and Soderblom (1982) calculated how much resurfacing could be attributed to plume activity. The main difficulty is in estimating how the mass of material within the plume is distributed by particle size. If most of the particles are around 1 μm in diameter, resurfacing rates of 10^{-3} to 10^{-4} cm yr^{-1} are derived from the optical depths of the plumes. However, Collins (1981) showed from scattering measurements of the Loki plumes that the particles are in the 1 μm size range only close to the vent; in the outer parts of the plume, the particles are mostly in the 0.01 to 0.1 μm size range. If most of the plume particles are in the 0.01 to 0.1 μm range, as Collins suggested, depositional rates may be as high as 10^{-2} to 1 cm yr^{-1}.

However, resurfacing rates can also be derived from the absence of impact craters. Current populations of impacting objects (asteroids and comets) suggest that impact rates in the Jovian system are within an order of magnitude of those in the inner solar system (Shoemaker and Wolfe 1982), and gravity-focusing by Jupiter should enhance impact rates on Io, bringing them close to inner solar system rates. Assuming lunar impact rates, Johnson and Soderblom (1982) estimated that Ionian resurfacing rates must be at least 10^{-1} cm yr^{-1}. Heat loss simply from bringing internal material to the surface at this rate is 0.02 to 0.2 W m^{-2} s^{-1}, well within the estimated global average heat loss of 1.2 W m^{-2} s^{-1} (Johnson et al. 1984). If all this internal heat loss were from resurfacing, the resurfacing rate would be 10 cm yr^{-1} (Johnson and Soderblom 1982).

D. Sulfur vs. Silicate Volcanism

A major uncertainty concerning the geology of the Ionian surface is the relative roles of sulfur and silicate volcanism. Smith et al. (1979c) proposed that an upper crust consisting largely of elemental sulfur and SO$_2$ overlies a layer of molten sulfur (sulfur ocean), possibly several kilometers thick (Fig. 6). The sulfur ocean rests on a silicate subcrust, which may be as thin as 30 km, which in turn may rest on a molten silicate layer that augments the tidal flexing of the crust. Estimates of the thickness of the various upper layers depend on what fraction of the total heat lost by the body is lost through conduction vs. convective overturn or volcanic activity. Measurements indicate that the total globally averaged heat flux is at least to 1.2 W m^{-2} (Johnson et al. 1984). A large fraction of this is probably lost by transport of materials to the surface (Reynolds et al. 1980; Matson et al. 1981; O'Reilly and Davies 1980). Smith et al. (1979c), assuming a conductive heat loss of 0.075 W m^{-2}, which represents only 6% of the total heat loss, estimated that the melting temperature of SO$_2$ (198 K) should be reached at depths of 0.5 to 1 km, and that melting temperatures of sulfur (400 K) should be reached at depths of about 1.5 km.

Carr et al. (1979) and Clow and Carr (1980) noted that such a crustal model is difficult to reconcile with the relief observed on the surface and proposed that silicates play a more prominent role in the evolution of the surface than Smith et al. originally envisaged. Several escarpments and calderas on the supposed sulfur crust have relief in the 1 to 3 km range (Arthur 1981; Schaber 1982). The deep calderas are particularly difficult to make in a sulfur crust. They form by brittle failure at vents where heat flows are atypically high. Even with the conservative assumption that only 6% of the heat is lost by conduction, the floors of the calderas are at depths where sulfur is molten and well below the depth where sulfur behaves ductilely (Clow and Carr 1980). Under these circumstances, a caldera could not form. Carr et al. (1979), and Clow and Carr (1980), therefore, proposed alternatively that the near-surface consists of interbedded sulfur and silicate materials (Fig. 6, top). They suggested that sulfur is recycled relatively rapidly near the surface but that sufficient silicates are intruded or erupted so that they significantly increase the structural strength of the near-surface rocks and, in addition, increase their thermal conductivity, thereby decreasing the thermal gradient. Another possibility is that impurities in the sulfur, such as Na$_2$S and K$_2$S, significantly increase the material's thermal conductivity and strength. Greeley et al. (1984) point out that, given the evidence for sulfur compounds and the likelihood of silicate volcanism on Io, secondary deposits of sulfur occurring in areas of silicate volcanism would be heated to sulfur melting temperatures and mobilized into flows that may travel long distances on Io and form relatively thin veneers over other surface features.

V. GEOPHYSICS AND THERMAL HISTORY

Io's internal physical structure is largely arrived at by analogy with more fully studied bodies such as the Earth and Moon. Io is similar to the Moon in its size, density, and apparent depletion of volatiles. On the other hand, Io is more like Earth than Moon in its level of surface activity. Our understanding of the mechanisms responsible for volcanism and resurfacing on Earth, how materials behave near the melting point, how the Earth convects heat upward (Stevenson and Turner 1979), and even the specifics of tidal dissipation on Earth may be applied to Io. The energy presently driving the volcanism on Io appears, from theoretical studies spurred by the Voyager encounter, to be predominantly derived from tidal dissipation of orbital energy; this is brought about by the forced eccentricity of Io's orbital motion coupled with enormous tides induced by Jupiter (Peale et al. 1979; Yoder 1979b; Lin and Papaloizou 1979b; Yoder and Peale 1981; Greenberg 1982a; Henrard 1983; chapter by Peale). Io's bulk chemical composition is inferred to be largely rocky with perhaps a small iron core (Consolmagno 1981a,b) based on (1) its known density, (2) the composition of primitive solar system condensates found in meteorites, and (3) models of the early Jovian nebula (Fanale et al. 1977a; Pollack et al. 1976; Prentiss and ter Haar 1979a). Its volatile depletion may be the result of the ongoing tidal heating of Io (Pollack and Witteborn 1980; Consolmagno 1981b). We refer the readers to the review by Pollack and Fanale (1982) for more details on this subject.

The key to understanding Io's internal geophysics is in resolving the details of the tidal friction heating mechanism and how this heat is removed. We shall first give a simple, if not compelling, picture of Io's physical structure and then follow this with a discussion of how heat is transported upward through the outer shell and end with a discussion of the various ways tidal flexing may deposit energy. Several useful physical properties are listed in Table III. The primary source of information below (Sec. V.A) is Schubert et al. (1981), Cassen et al. (1982), and especially Yoder and Faulkner (1984); also see chapters by Schubert et al. and by Peale in this book.

A. Internal Structure

The internal structure of Io is almost certainly affected by its thermal history and tidal friction in addition to its material constitution. Figure 14 shows a plausible structure (Schubert et al. 1981; Consolmagno 1981a,b; Cassen et al. 1982; cf. Fig. 2 in the chapter by Schubert et al.). The upper unit consists of a quasi-rigid crust overlying a more plastic asthenosphere. In this model the outer shell consisting of crust and asthenosphere is tidally decoupled from the interior by a liquid layer of uncertain depth and perhaps different composition. Below the liquid shell lies a hot but solid mantle. Also a substantial Fe-S liquid core may exist. If the global liquid layer is absent then the model presented here must be profoundly modified as to where tidal heating occurs and how the heat is transported to the surface.

The minimum thickness d_c of the crust can be estimated (at least locally) from the height of mountain-like terrain observed by Voyager 1 (Smith et al. 1979c). If the high terrain is supported isostatically (i.e., rises above the mean surface because the mountain units are of lower density than the overall crust), then we expect from the 5 to 10 km mountain elevations that $d_c > 30$ km, given a plausible density contrast of 0.5 g cm^{-3}. Below the quasi-rigid

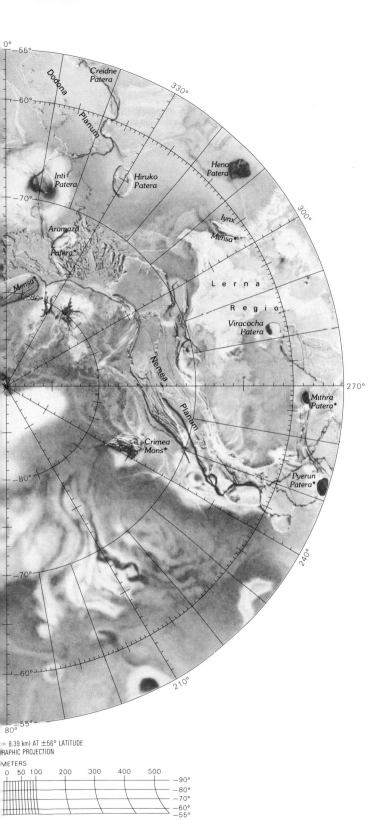

0°
−55°

Dodona

Creidne
Patera

330°

Planum

Inti
Patera

Hiruko
Patera

Heno
Patera

−60°

−70°

300°

Iynx
Mensa*

Aramazd
Patera*

Lerna

Mensa*

Regio

−80°

Viracocha
Patera

Nemea

270°

Mithra
Patera*

Planum

Crimea
Mons*

−80°

Pyerun
Patera*

240°

−70°

−60°

210°

−55°

80°

= 8.39 km) AT ±56° LATITUDE
RAPHIC PROJECTION

METERS

0 50 100   200   300   400   500

−90°
−80°
−70°
−60°
−55°

LAR REGION

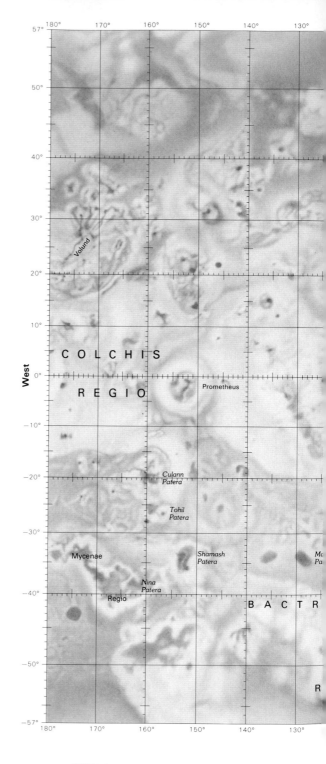

FIGURE 9A

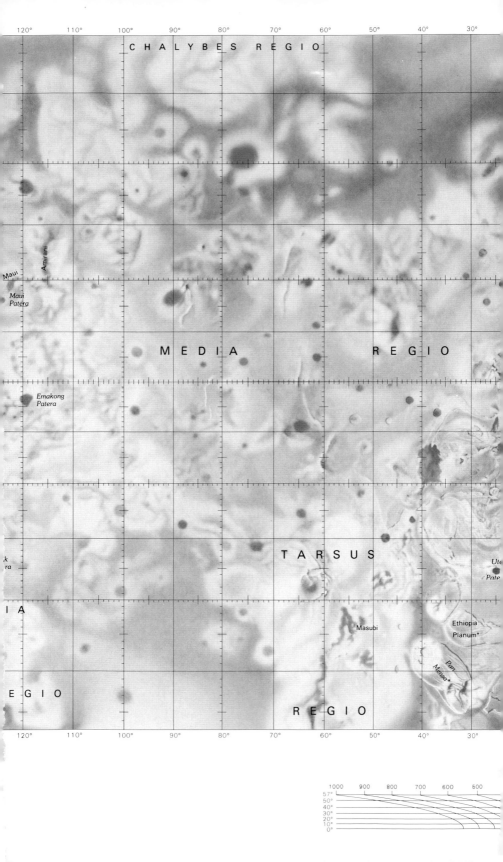

CHALYBES REGIO

120° 110° 100° 90° 80° 70° 60° 50° 40° 30°

Maui

Amirani

Maui
Patera

MEDIA   REGIO

Emakong
Patera

k
ra

TARSUS

Ute
Pate

I A

Masubi

Ethiopia
Planum

Pan
Mensa

EGIO

REGIO

120° 110° 100° 90° 80° 70° 60° 50° 40° 30°

1000 900 800 700 600 500
57°
50°
40°
30°
20°
10°
0°

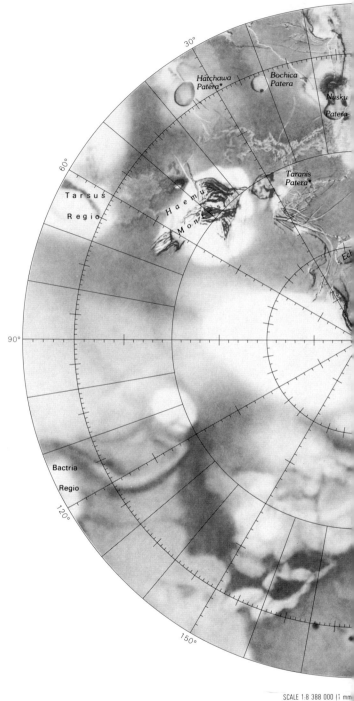

30°

Hatchawa
Patera*

Bochica
Patera

Nusku
Patera

60°

Taranis
Patera*

Tarsus
Regio

H a e m u s

M o n t e s

Ec

90°

Bactria

Regio

120°

150°

SCALE 1:8 388 000 (1 mm
POLAR STEREO

KIL
500   400   300   200   100 50
−90°
−80°
−70°
−60°
−55°

SOUTH P

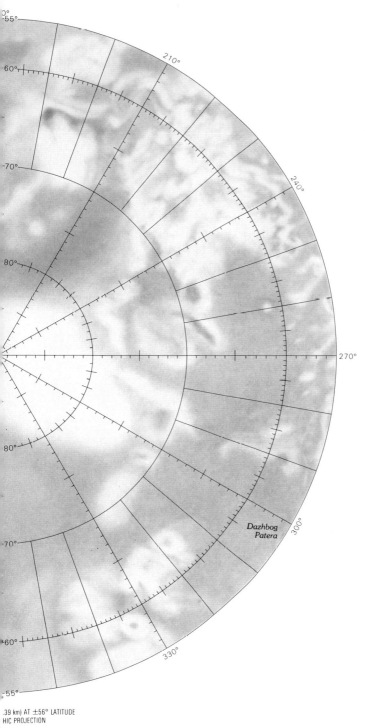

0°
−55°

60°

210°

−70°

240°

−80°

270°

80°

Dazhbog
Patera

300°

−70°

330°

−60°

−55°

.39 km) AT ±56° LATITUDE
HIC PROJECTION

TERS
50 100   200   300   400   500
—90°
—80°
—70°
—60°
—55°

AR REGION

North

20° 10° 0° 350° 340° 330° 320° 310° 300°

Dazhbog Patera

Surt

Manua Patera

Amaterasu Patera

Fuchi Patera

Loki

Loki Patera

Khalla Patera*

Reshet Catena*

Carancho Patera*

Tol-ava Patera*

Nyambe Patera

Sed Patera*

Purgine Patera*

Rata Patera

Dinjir Patera*

Marda Catena

Mama Patera*

Apis Tholus

Horus Patera

Ra Patera

Kibero Patera*

Ilmarinen Patera*

Tung Yo Fluctus*

Inachus Tholus

Huo Shen Patera

Ninurta Patera*

Mihr Patera*

Tung Yo Patera*

Kava Patera*

Podja Patera*

Angpeiu Patera*

Sui Jen Patera*

Shoshu Patera

Masaya Patera

Cataquil Patera*

Talus Patera

Vahagn Patera

Uta Fluctus*

Mbali Patera

Taw Patera

Menahka Patera*

Pautiwa Patera*

Iopolis Planum*

Sengen Patera

Lu Huo Patera*

Maasaw Patera

Agni Patera*

Kane Patera

Euboea Fluctus*

Argos Planum*

Aten Patera

Päive Patera*

Euboea Montes*

Sium Patera*

Hybristes Planum*

20° 10° 0° 350° 340° 330° 320° 310° 300°

South

SCALE 1:15 000 000 (1 mm = 15 km) AT 0° LATITUDE
MERCATOR PROJECTION

KILOMETERS

00 300   200   100 50 0 50 100   200   300   400   500   600   700   800   900   1000

57°
50°
40°
30°
20°
10°
0°

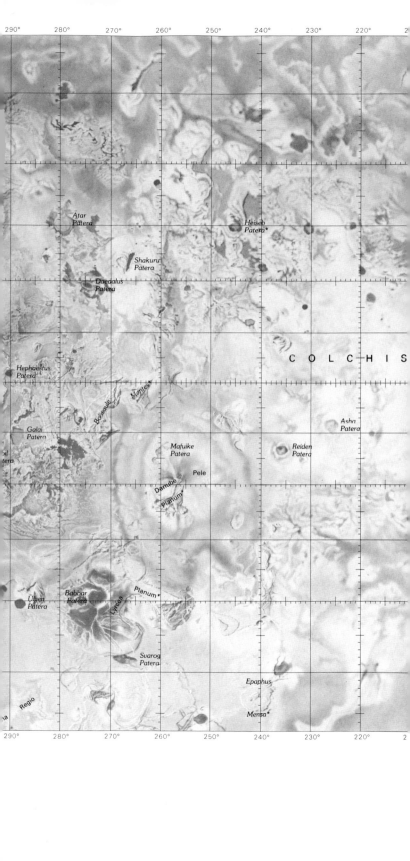

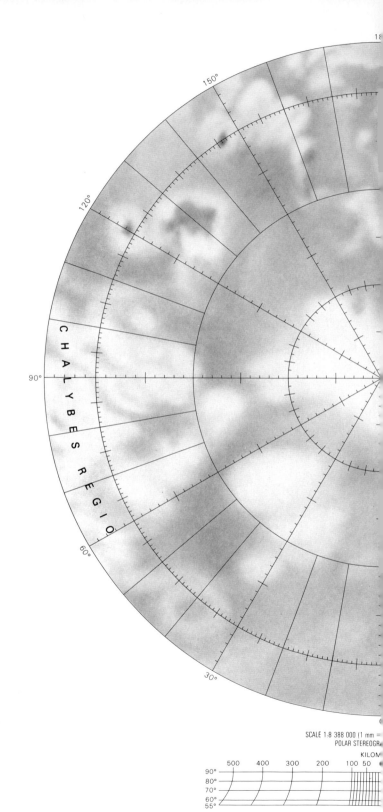

150°

18

120°

90°

C H A L Y B E S   R E G I O

60°

30°

SCALE 1:8 388 000 (1 mm =
POLAR STEREOGR

KILOM

500   400   300   200   100 50

90°
80°
70°
60°
55°

NORTH POL

FIGURE 9B

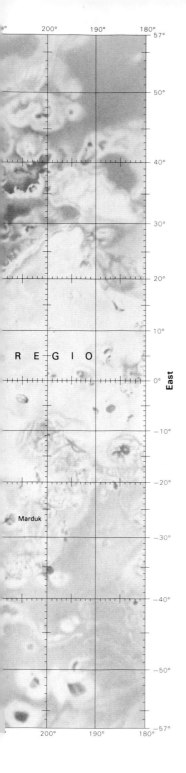

Fig. 9. Pictorial map of Io's surface based on Voyager 1 and 2 images, USGS Map No. I-1713, printing date April 1986. Map prepared by R. M. Batson and P. M. Bridges, USGS, Flagstaff. (a) Mercator projection, scale 1:15,000,000 at 0° latitude. (b) Polar stereographic projections, scale 1:8,388,000 at ±56° latitude. See also Map Section, pp. 890−891.

TABLE III
Material Properties for Solids that may Exist on and within Io

		Silicates	Sulfur	SO$_2$
Heat of fusion: (erg g^{-1})	H_{fusion}	1×10^{10}	4.4×10^8	$\sim 1 \times 10^9$
Heat capacity: (erg g^{-1} K^{-1})	C_p	1.2×10^7	0.7×10^7	$\sim 1 \times 10^7$
Density, solid:	ρ	3	2	1.9
liquid: (g cm^{-3})		2.7	< 1.8	1.5
Rigidity: (dyne cm^{-2})	μ	5×10^{10}	2×10^{10}	
Melting temperatures: (K)	T_m	1500	400	198
Thermal conductivity: (W cm^{-1} K^{-1})	k	4×10^{-2}	2×10^{-3}	
Thermal expansion: (K^{-1})	α	3×10^{-5}		
Viscosity of solid near melting point: (poise)	η_M	10^{16} to 10^{17}		
Viscosity of liquid near melting point: (poise)		10^4 to 10^5	$\sim 10^{-1}$ (at 400 K) $\sim 10^2$ (at 600 K)	
Electrical conductivity, solid: liquid: (ohm^{-1} cm^{-1})		$\sim 10^{-6}$ to 10^{-5} ~ 10	$\sim 10^{-18}$ $\geq 10^{-12}$	
Tensile strength at ~ 300 K: (kPa) (psi)		24 ~ 3500	1.4 ~ 200	
Compressive strength at ~ 300 K: (kPa) (psi)		207 $\sim 30,000$	21 ~ 3000	

crust, there may exist an asthenosphere of perhaps different composition. The dominant characteristics of this asthenospheric zone are that (1) it is relatively plastic and contains some partial melt because of its elevated temperature; (2) solid friction in the zone may also be substantially higher than in the crust; and (3) the heat generated here and deeper down is transported upwards by convection. The minimum thickness of the asthenospheric layer is about 8

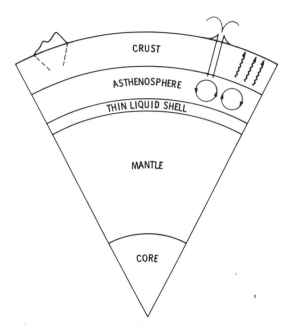

Fig. 14. Hypothetical model of Io's internal structure, not to scale. The major structural units include an outer shell consisting of the crust and asthenosphere, a liquid layer, a lower mantle and iron-rich core. Convection currents carry tidal-friction heat from lower to upper asthenosphere; eruptions and conduction carry heat to surface. Mountains exist on the crust with possible isostatically compensated roots.

km, based on the hypothesis that convection driven by tidal heating operates in this zone (Yoder and Faulkner 1984). The maximum thickness of the total outer shell is less certain, but could theoretically be estimated from the balance between heat production by tidal friction and upward heat transport (the subject of the following section).

The upper crust/asthenosphere units may be separated from the underlying mantle by a liquid layer of uncertain depth and composition. Schubert et al. (1981) and Consolmagno (1981a,b) argue that this layer could not be sustained indefinitely if the composition of the fluid layer and mantle are uniform. There are several ways to frustrate freezing of the liquid shell. Schubert et al. (1981; see chapter herein) argue that the liquid shell can be maintained if it is compositionally similar to the overlying asthenosphere, while the underlying mantle is chemically different with a higher solidus temperature. We shall argue later that viscous dissipation in this liquid shell is likely to produce enough heat to prevent its freezing if its thickness falls below 1 km.

The solid mantle is the largest unit with $d_m \simeq 1000$ km. By analogy with the Earth's Moon, the mean density of this region is expected to be only a few tenths g cm^{-3} higher than in the units above it. Since the tidal deformation of the mantle is roughly proportional to the density contrast at the fluid-core/mantle boundary, tidal flexing (hence dissipation) is negligible here. Io's mean global density of 3.55 g cm^{-3} is slightly higher than the lunar density of 3.34 g cm^{-3}. A substantial iron-rich core with 700 km radius in Io may be inferred if Io is otherwise compositionally similar to the Moon, which is suggested by their similar size and density.

However, Consolmagno (1981b) argues that the primordial composition of Io and Moon were quite dissimilar, Io being richer in volatiles and iron than the Moon. He estimates that Io's iron core may be as large as 1000 km in radius and account for as much as 20% of Io's mass. Such a large core could significantly alter the nonspherical, hydrostatic shape of Io caused by tidal deformation and synchronous rotation. The predicted difference in the three axes (R_1, R_2, R_3) defining the surface mean ellipsoid are

$$R_1 - R_3 \simeq \frac{5}{2} fR(R/a)^3 = 21.6 f \, \text{km} \tag{1}$$

$$R_1 - R_2 \simeq \frac{1}{4} (R_1 - R_3) = 5.4 f \, \text{km} \tag{2}$$

where

$$f = \frac{1}{\left[\dfrac{5}{2} - \dfrac{15}{4}\left(\dfrac{I}{MR^2}\right)\right]}. \tag{3}$$

If Io has uniform composition, then its mean moment of inertia $I = 0.40 \, MR^2$ (and the factor $f = 1$); here M is Io's mass and R its mean radius. But if Io has an iron-rich core of radius 1000 km, then I is reduced to $0.35 \, MR^2$ and $f = 0.84$.

Synnott et al. (1984) find $(R_1 - R_3) = 13$ to 16 km and $(R_1 - R_2) = 6$ to 9 km from measurements of Io's limb profiles obtained from Voyager images. Clearly, this data cannot yet be used to deduce core size given that the present uncertainties of these deviations from sphericity are 3 to 4 km. Davies (1983) obtains significantly smaller deviations of order $(R_1 - R_3) \sim 8$ (± 5) km and $(R_1 - R_2) \sim 0 \, (\pm 5)$ km. These results, if accurate, would imply that Io's shape is nonhydrostatic. Davies' technique also utilizes Voyager images, but results are obtained by tracking the position of surface features on Io. Factors such as mantle convection, lateral variations in the distribution of material (perhaps related to the ongoing resurfacing), nonsynchronous rotation (Greenberg and Weidenschilling 1984) or rapid polar wander (again related to the mass redistribution caused by volcanic activity) might explain significant deviations of Io's shape from its predicted, hydrostatic values. Measurements of Io's low-order, nonspherical gravity field by the planned Galileo mission, together with new measurements of its shape may resolve these issues.

B. Heat Loss Mechanisms

The complex balance between heat loss at Io's surface and the internal heat generated primarily by tidal friction is undoubtedly a major factor affecting Io's internal structure and past thermal history. The major heat loss mechanism at the surface is apparently thermal radiation to space from hotspots which comprise $\lesssim 3\%$ of surface area and which radiate heat equivalent to a mean heat flow of 1 to 2 W m^{-2} or total radiated power of (4 to 8) $\times$ 10^{13} W (Matson et al. 1981; Morrison and Telesco 1980; Sinton 1981; Pearl and Sinton 1982). Johnson et al. (1984) show that there is a strong longitude bias in the distribution of hotspots with about twice as much energy radiated on Io's trailing than on its leading hemisphere (see Fig. 15). Accounting for this bias in the data reduces the total hotspot radiated power to (4 to 6) $\times$ 10^{13} W. However, if there are significant hotspots near the poles, not seen from Earth, the total radiated power may be even greater.

The heat H_c conducted through the nonhotspot crust is proportional to the thermal conductivity k and temperature difference ΔT_c between lower and upper boundaries separated by distance d_c. Thus

$$H_c \simeq -k\,\frac{\Delta T_c}{d_c}.\tag{4}$$

D. Matson (personal communication, 1982) has established from analysis of thermal emission eclipse data of Io that the upper bound of heat flow from nonhotspot country rock surface of Io is < 0.5 W m^{-2}. Using this value for H_c we find that for $\Delta T_c = 1000$ K and $k = 4$ W m^{-1} K^{-1}, $d_c \geq 8$ km. On the other hand, the ability of Io to support mountainous terrain suggests that d_c is ≥ 30 km and H_c is < 0.1 W m^{-2} and this argument suggests that most of the heat flow at the surface is radiated at hotspots, a conclusion that may also be independently deduced from analysis of the radiation spectrum (Matson et al. 1981).

The heat flow driven by the temperature difference ΔT_ℓ across the warmer, plastic lithospheric layer of density ρ and thickness d_ℓ is primarily by convection (see the chapter by Schubert et al.). However, convection dominates over conduction only if the Rayleigh number Ra is greater than a critical value of Ra_{cr} of order 10^3.

$$Ra = \frac{\rho g \alpha \Delta T_\ell d_\ell^3}{\kappa \eta_0}\tag{5}$$

where g is surface gravity, α is the thermal expansion coefficient, κ is the thermal diffusivity, and η_0 is the mean viscosity. The onset of convection can be used to set a lower bound on d_ℓ.

Many materials behave like a Maxwell solid and deform plastically near

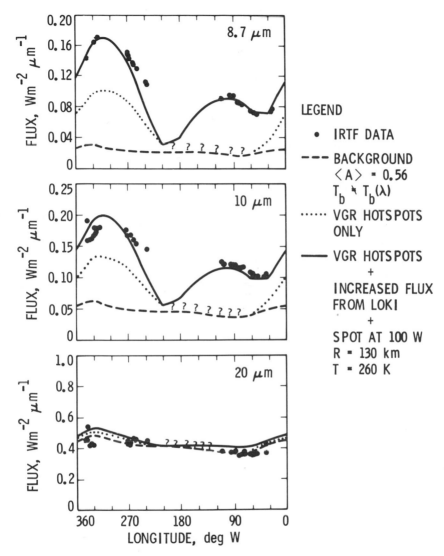

Fig. 15. Variation with longitude in surface heat flow from Io's interior as represented by disk-averaged infrared emission measured from an Earth-based infrared telescope (IRTF) at three wavelengths. The 8.7 μm and 10 μm data measure primarily emission from hotspots, the 20 μm data measure primarily emission from the nonhotspot average surface. Curves show results of model calculations for emission levels expected from background insolation heating, Voyager-observed hotspots, and Voyager hotspots plus increased flux from Loki plus a new hotspot at longitude 100 W (figure from Johnson et al. 1984).

their melting point. The creep viscosity $\eta(T)$ of such a solid near melting is exponentially sensitive to temperature T,

$$\eta(T) \simeq \eta_M \exp A\left(\frac{T_M}{T} - 1\right). \tag{6}$$

The viscosity may also depend on stress and pressure. The precise values of the creep viscosity parameters η_M and A in Eq. (6) are not well determined (Schubert et al. 1981; Stevenson 1981). The expected range for mantle materials is $10^{16} \leq \eta_M \leq 10^{17}$ poise and $25 \leq A \leq 35$. Booker and Stengel (1978) find that the appropriate viscosity η_0 to use in calculating the Rayleigh number in Eq. (5) corresponds to its value at the mean of the boundary temperatures. Since $T = T_M$ at the bottom of the asthenosphere, we find η_0 $(T_M - 1/2\ \Delta T_\ell) \simeq (20 \text{ to } 100)\ \eta_M$ for a plausible $\Delta T_\ell \sim 300$ K across the asthenosphere. The minimum thickness of the convecting layer is obtained from $Ra = Ra_{cr}$ and is 10 to 30 km. Obviously, this convecting layer beneath the crust need not exist if the decoupled upper mantle unit is sufficiently thin; i.e., the crust could be in direct contact with the liquid layer.

The transport of heat by convection H_ℓ is proportional to Ra^β, where $1/4 < \beta < 1/3$ (Sleep and Langan 1981). Booker and Stengel find for a variable viscosity fluid that

$$H_\ell \simeq -1.5\ k\frac{\Delta T_\ell}{d_\ell}\ (Ra/Ra_{cr})^{0.28} \tag{7}$$

while $Ra_{cr} \simeq 1700\ [\eta_M(T - \Delta T_\ell)/\eta]^{1/9}$ and $Ra \lesssim 10^6$. Since the viscosity increases rapidly with increasing ΔT_ℓ, H_ℓ, which is proportional to $\Delta T_\ell/\eta_0(\Delta T_\ell)$, must reach a maximum determined by

$$\partial H_\ell/\partial\Delta T_\ell = 0. \tag{8}$$

The corresponding temperature contrast which maximizes H_ℓ is of order 230 to 300 K for $25 \lesssim A \lesssim 35$. The maximum heat flow lies in the range

$$0.05\ \text{W m}^{-2} \leq H_\ell(\text{max}) \leq 0.2\ \text{W m}^{-2} \tag{9}$$

for the expected range for the creep viscosity parameters A and η_M given $d_\ell \sim$ 10 to 30 km. This limiting H_ℓ (max) suggests that the plastic asthenospheric zone may be unstable, if the heat generated within or below this layer is substantially more than can be convected upward, and melting within the asthenosphere will occur. There seem to be two possible outcomes if this limit is exceeded. Either the asthenosphere will thin out to reduce tidal heating and increase heat transport, or it will episodically melt and refreeze, perhaps producing periods of high and low activity on the surface (Consolmagno

1981a,b). Ojakangas and Stevenson (1985) have examined more fully the stability of tidal heat transport by convection using a simple parameterization which couples the effect of orbit evolution and resonance-pumping of the orbital eccentricity with a temperature-sensitive heating mechanism. Typically, they find that runaway tidal heating occurs once every 10^7 yr with each episode lasting about 10^6 yr (cf. chapter by Schubert et al.).

The above description changes drastically if the asthenosphere contains a significant fraction of partial melt. An interconnected liquid matrix which transports heat via thermal currents could substantially increase H_ℓ. This process could inhibit further melting of the asthenosphere and perhaps impede the formation of a global liquid layer.

C. Internal Heating Mechanisms

Although anelastic tidal flexing almost certainly heats Io, this dissipation would rapidly circularize Io's orbit, reducing the heating rate to zero in about 10^5 yr if other forces did not prevent it. The present consensus (cf. Greenberg 1982a) is that Io's orbital eccentricity e is maintained by a complex balance of several effects. First and foremost, the tidal torque arising from the tide Io raises on Jupiter acts to expand Io's orbit. Next, the resonance lock amongst the three inner satellites drives Io's eccentricity and transfers angular momentum from one satellite to another. Finally, the tide on Io, as depicted in Fig. 16 and described in the caption, provides a mechanism for tapping the mechanical energy of the orbital system and stabilizing the orbital resonance locks (see Yoder 1979b).

The tidal heating rate of Io is directly related, at least in a time average sense, to the orbital expansion rate while this latter rate is inversely proportional to the Jovian dissipation factor Q_J; the tidal Q equals 2 times the ratio of the peak tidal energy to the energy dissipated per flexing cycle. Estimates of Q_J derived from theoretical calculations of dissipation in Jupiter range from 10^9 (Goldreich and Nicholson 1977) to 10^5 or less (Stevenson 1983). A reliable Q_J will eventually be inferred (if $Q_J \leq 10^6$) from measurement of the secular change in Io's mean motion obtained from analysis by Lieske (1983) of eclipse timings and other astrometric data dating back to 1652. A global average heating rate of 1 W m^{-2} inferred from the hotspot heat flow observations implies $Q_J = 6 \times 10^4$. The lower bound of Q_J imposed by the finite mechanical energy in the orbital system is also near 6×10^4. This poses a serious but not insurmountable barrier to understanding the relationship between heat input and surface heat flow. As on Earth, the volcanic activity and resurfacing could be episodic and the presently observed heat flow greater than its mean. An alternative is that tidal friction heating within Io is also episodic. The third choice is that the Jovian Q_J has been reduced to near or below 10^5 comparatively recently because of evolutionary changes within Jupiter which have increased dissipation (Stevenson 1983).

Before we proceed with a discussion of solid friction in Io, let us con-

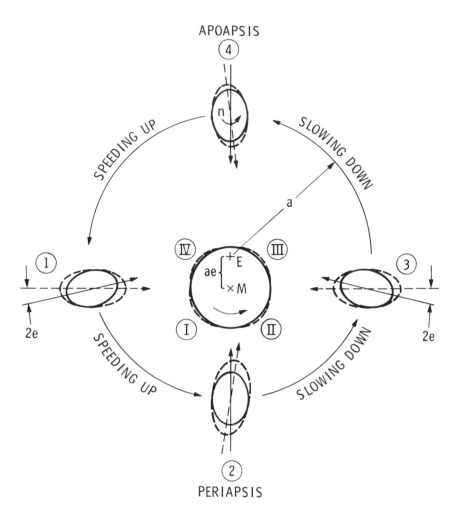

APOAPSIS
④

PERIAPSIS
②

Fig. 16. Schematic representation of Io's tidal interactions with Jupiter. View is of Io orbiting around Jupiter's center of mass M at average radius (semimajor axis) a about center of ellipse E with orbit eccentricity e, from above north pole, and not to scale. Io's permanent bulge (solid ellipse) caused by gravity gradient librates (relative to the sub-Io point on Jupiter) an amount $2e$ due to combination of uneven orbital velocity (from the eccentric orbit offset an amount ae) and uniform rotation rate n as Io orbits Jupiter. A tidal bulge (dashed ellipse) on Io results from uneven torque on the permanent bulge from Jupiter during Io's orbit. Dissipation of this inelastic tidal bulge in Io causes periodic displacement of the maximum tidal bulge which alternately leads (position 1) and trails (position 3) the permanent bulge; if no other forces were involved this dissipation would lead to a decay of Io's orbital eccentricity. Tide raised on Jupiter by Io lags behind Jupiter's rotation such that it always leads Io in its orbit; e.g., when Io is at position 2, the tide it produces on Jupiter is at II (and IV). Torque from this Jupiter tide would cause Io's orbit to expand and its eccentricity to diminish if it were not for the 2:1 resonant lock between Io and Europa; the lock prevents Io's orbit from expanding and forces Io's eccentricity to have a value greater than it otherwise would. This forced eccentricity maintains a high tidal dissipation rate—maximum near periapsis, minimum near apoapsis—and internal heating in Io (figure after Yoder 1979b).

sider briefly the contribution to the internal heat budget of some of the other mechanisms which may have significantly affected Io's thermal history: (1) radioactive decay, (2) accretional heating, (3) core formation, (4) electrical heating, and (5) tidal dissipation within a thin liquid layer or ocean.

Although the complement of radiogenic material within Io is unknown, about 6×10^{11} W is liberated at present by analogy with the Moon (Cassen and Reynolds 1974), and this is equivalent to a mean heat flow of 0.02 W m^{-2} at Io's surface. This source alone could be expected to raise Io's central temperature to near melting or higher. The impact energy buried by accretion is estimated to be about $1/2$ of the gravitational self-energy $E_g = 3/5 \, gMR$, which is $\simeq 1 \times 10^{36}$ erg (Kaula 1979) and is comparable to the total heat released by radioactive decay since Io's formation. The gravitational energy released by core formation $E_c \simeq (3/10) \, (\Delta\rho_c/\rho_c) \, M_c gR$, where M_c, ρ_c and $\Delta\rho_c$ are core mass, density, and core-mantle density contrast, respectively, is at least an order of magnitude smaller than accretional heating.

An electrical current through Io provides an additional source of internal heating (Goldreich and Lynden-Bell 1969; Ness et al. 1979; Gold 1979; Colburn 1980). A $\mathbf{v} \times \mathbf{B}_J$ voltage drop of ~ 411 keV across Io's diameter is generated by Io's orbital velocity $\mathbf{v}$ [$\simeq (\Omega - n)a$] relative to Jupiter's rotating magnetic field $\mathbf{B}_J$ where Ω = Jovian rotation rate, n = orbital mean angular velocity or mean motion, and a = mean orbital radius or semimajor axis. The charged plasma environment near Jupiter permits a current I to flow between Jupiter and Io, exiting near Jupiter's poles and flowing in two flux-tube loops along field lines to connect with Io near the sub-Jovian and anti-Jovian points on Io's surface (see Fig. 17). The electric current inferred from Voyager 1 flyby of Io's flux tube is sensitive to the existence or absence of an intrinsic magnetic field within Io (Kivelson et al. 1979). If such a field is absent, then the inferred current is 2.7×10^6 amp (Acuña et al. 1981). The presence of an intrinsic Ionian magnetic field reduces I to 7×10^5 amp (Southwood et al. 1980). Groundbased measurements of Io's radio emission (de Pater et al. 1982) have not resolved the controversy of whether or not Io has an intrinsic magnetic field. The maximum electrical heating rate $E_{\mathrm{EMF}} \simeq IR_J \mathbf{B}_J a$ $(\Omega - n) \simeq 2 \times 10^{11}$ W (Goldreich and Lynden-Bell 1969) where $\mathbf{B}_J = 0.02$ Gauss at Io's orbit and R_J is Jupiter's radius. However, either an intrinsic magnetic field (Kivelson et al. 1979) or an ionosphere (Colburn 1980; Herbert and Lichtenstein 1980) is likely to shunt the induced dc current around Io. The heat generated by an alternating current caused by the apparent wobbling of Jupiter's $\mathbf{B}_J$ field as seen by Io is also negligible (Colburn 1980). The relative motion as well induces an electromagnetic torque $N_{\mathrm{EMF}} = E_{\mathrm{EMF}}/(\Omega - n)$ which acts to expand Io's orbit even if the current is shunted through the atmosphere. However, the tidal torque expanding Io's orbit is greater than N_{EMF} if the tidal $Q_J < 7 \times 10^6$, as is likely.

Tidal dissipation on the Earth occurs primarily in the oceans (Lambeck 1980; Burns' chapter). The dominant drag force affecting tidal currents is

IO - A UNIPOLAR INDUCTOR

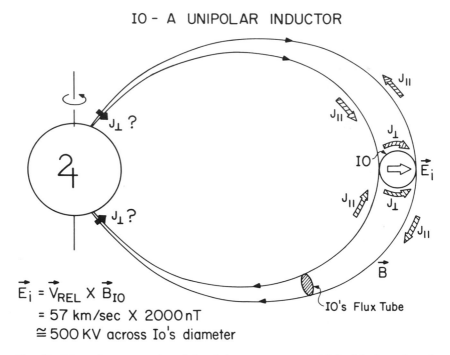

$$\vec{E}_i = \vec{V}_{REL} \times \vec{B}_{IO}$$
$$= 57 \text{ km/sec} \times 2000 \text{ nT}$$
$$\cong 500 \text{ KV across Io's diameter}$$

Fig. 17. Schematic representation of electrical current system model for Io's electrodynamic (unipolar inductor) interaction with Jupiter's magnetosphere. As Jupiter's magnetic field lines sweep past Io, currents as high as 10^6 amp flow approximately parallel ($J_{\parallel}$) to the magnetic field lines linking Io to Jupiter's ionosphere. The circuit is closed by transverse ($J_{\perp}$) current flow in the Jupiter ionosphere and in the ionosphere or interior of Io. These currents are thought to be responsible for the strong Io-controlled decametric radio emissions from Jupiter as well as minor heating of Io's ionosphere and/or interior. A total power of $\sim 10^{12}$ W is dissipated in the system (figure from Acuña et al. 1983).

thought to be bottom friction over shallow continental shelves. A similar mechanism will operate in a thin liquid shell in Io and will be a sensitive function of the mean depth d_f. The surface stress caused by the periodic tidal current is $\sim v_f(\eta_f \rho_f n/2)^{1/2}$ for laminar flow where v_f is the tidal velocity, and ρ_f and η_f are the magma ocean density and viscosity, respectively. However, the flow becomes turbulent for sufficiently small viscosity or "large" surface roughness. The onset of turbulence for smooth surfaces occurs when $(\eta_f n/\rho_f v_f^2)^{1/2} \sim 10^{-3}$, and the turbulent surface stress $\sim 10^{-3} \rho_f v_f^2$ is independent of viscosity. The rate of dissipation H_f per unit area at the upper and lower boundaries can be estimated using the skin friction approximation for turbulent stress ($\simeq 0.002 \rho_f v_f^2$), with the result that $H_f \simeq 0.002 \rho_f v_f^3$ (Lamb 1932). The mean tidal velocity v_f is $\simeq 0.2 (h/d_f) nR$, where h is the tidal amplitude (Sagan and Dermott 1982). The amplitude is $h \simeq 3eR (M_J/M)$

$(R/a)^3 \Delta h_2$, where e is eccentricity, and Δh_2 is the difference of the displacement Love numbers at the top and bottom boundaries. This estimate of h omits the effect of fluid inertia and drag which tend to decrease h. Fluid inertia becomes important if $d \lesssim 1$ km. If the decoupled outer shell is thin, then $\Delta h_2 \simeq 2.5/(1 + 60\mu \, (d_c + d_\ell)/11\rho g R^2) \simeq 1$ for $(d_c + d_\ell) = 100$ km, and tidal amplitude $h \simeq 30$ m. Thus, v_f could be as large as ~ 40 cm s^{-1} for $d_f = 1$ km. Also, $H_f \simeq 0.2 \, d_f^{-3}$ km^3 W m^{-2} at each surface, while the total $dE_f/dt = 8 \, \pi \, R^2 H_f$ is (Yoder and Faulkner 1984)

$$\frac{dE_f}{dt} \simeq 1 \times 10^{13} \, d_f^{-3} \text{ km}^3 \text{ W}. \tag{10}$$

The above expression represents a lower limit on the heating rate for a given depth of d_f. The same heating rate as in Eq. (10) can be obtained with larger d_f if the viscosity $\eta_f \gtrsim 10^{-6} \, \rho_f v_f^2/n \sim 10^2 \, d_f^{-1}$ poise-km and the flow is laminar. The viscosity of a silicate magma near its melting point is of order 10^4 to 10^5 poise and decreases with increasing temperature. The heating rate for laminar flow, which is proportional to $\eta^{1/2} \, d_f^{-2}$, can attain 10^{13} W for $d_f = 6$ km and $\eta_f = 10^5$ poise. The same heating rate occurs for turbulent flow ($\eta_f \ll 10^2$ poise) when $d_f = 1$ km.

Clearly, such a mechanism guarantees that once a thin liquid layer forms, dissipation will strongly inhibit its freezing. The silicate ocean can self-regulate its heat production by adjusting its depth by melting or freezing of material at the boundaries in response to the heat flow at the top of the layer. The maximum tidal currents (hence heating rate) occur at mid-latitudes rather than at Io's poles or equator if the ocean depth is everywhere uniform. However, this mechanism might achieve nearly uniform heating of the bottom of the outer shell by locally adjusting its depth (increase d_f where dissipation is high and decrease d_f where dissipation is low) and by redistribution via thermally driven fluid currents. Whether or not this mechanism provides significant tidal heating depends on the actual depth, which is expected to be a function of composition, the magnitude of the tidal heating occurring within the solid shell, and the rate that this heat together with the heat generated by tidal currents is transported to the surface.

Except for the last mechanism, it would appear by elimination that solid friction is the major source of internal heating. We shall now describe a model for solid friction within a tidally decoupled shell and discuss factors which affect its viability and efficiency. This model should not be taken too literally and does not necessarily represent even a consensus view. It serves only as a means to illustrate those issues which similar models must address.

The rate of global tidal heating dE/dt on a planetary satellite is a function of orbital parameters such as semimajor axis a, mean notion n and eccentricity e which determine the magnitude and frequency of the tidal force. The important material properties which affect dE/dt are mass M, material stiffness (encoded in the Love number k_2), and anelasticity which is described by a Q factor. The simplest formula for dE/dt is

$$\frac{dE_t}{dt} = \frac{21}{2} Mn^3a^2\left(\frac{R}{a}\right)^5 \frac{k_2}{Q} e^2 \tag{11}$$

$$= 6.2 \times 10^{15} (k_2/Q) \ W$$

and is adequate for estimating the general magnitude of dissipation, but the essential and poorly understood part of the equation is the k_2/Q factor. The amplitude of the tidal deformation is determined by the Love number k_2 which depends on the ratio of the body force proportional to $\rho g R$, promoting deformation, to the rigidity μ, inhibiting deformation. For a uniform spherical solid, $k_2 = (3/2)/(1 + 19\mu/2\rho g R)$ and equals 0.035 if we adopt $\mu = 5 \times 10^{11}$ dyne cm^{-2} for Io. Decoupling of a thin shell by a liquid layer can substantially increase shell deformation, and k_2 (shell) $\simeq 3/2(1 + 60\mu(d_c + d_\ell)/11 \ \rho g R^2)$.

Frictional dissipation is primarily confined to the rigid outer shell since tidal deformation of the lower mantle is strongly inhibited by the counterbalancing load provided by the fluid layer. Furthermore, dissipation in a solid is proportional to the square of the stress or deformation of the outer shell, times the shell volume V(shell); that is, the effective k_2/Q in Eq. (11) is roughly proportional to $[k_2(\text{shell})/k_2(\text{solid})]^2[V(\text{shell})/V(\text{total})]/Q(\text{shell})$, which reaches a maximum and decreases to zero as V(shell) decreases toward zero volume. The calculations by Peale et al. (1979) indicate that the maximum effective $k_2 \simeq 10 \ k_2$(solid) for a shell thickness of 90 km and falls to about 5 k_2(solid) for a shell thickness of 30 km.

If we characterize the solid dissipation of Io as that of a Maxwell body with uniform rigidity μ and viscosity η, then the local dissipation rate $d\varepsilon/dt$ is

$$\frac{d\varepsilon}{dt} = n\mu q^{-1} \sum e_{ij}^2 \tag{12}$$

$$e_{ij} = \frac{1}{2}\left(\frac{\partial u_i}{\partial x_j} + \frac{\partial u_j}{\partial x_i}\right) \tag{13}$$

$$q^{-1} = \frac{(n\eta/\mu)}{1 + (n\eta/\mu)^2}. \tag{14}$$

Here e_{ij} is the stress tensor, u_i and u_j are the Cartesian components of tidal displacement, and are roughly proportional to k_2 (shell), while q is a local dissipation function. Peale and Cassen (1978) obtained expressions for e_{ij} within a shell. The total energy production rate (Eq. 11) is obtained by integrating Eq. (12) over the volume of the shell. The creep viscosity characterizing convection may be as small as 10^{16} to 10^{17} cm^2 s^{-1} near melting

(Stevenson 1981). Although the creep viscosity for convection may not be appropriate (nor the only) mechanism for solid friction, it is suggestive that $(n\eta/\mu) \simeq 1$ to 10 for Io, supporting the idea that the major source of dissipation is flexing of the plastic asthenosphere. The Q(shell) is roughly $q[V(\text{shell})/V(\text{asthenosphere})] \simeq 5$ to 50. Thus, values for k_2/Q as large as 10^{-2} seem plausible and could account for the observed mean heat flow of ~ 1 W m^{-2}.

Tidal friction in a uniform shell tends to produce about 3 times more heat at the poles than near the equator. However, the geometry of a realistic shell (i.e., crust and asthenosphere) can also affect lateral variations in tidal heating. If the shell is comparatively thick (say about 100 km) compared to the plastic asthenosphere at its base (say 10 to 30 km), then the asthenosphere could locally self-regulate its thickness (hence its heating rate) so that the heat generated in the asthenosphere by solid friction and conducted or volcanically transported through the crust is about the same everywhere. Thus, in Io the asthenosphere may be thinner at the poles than at the equator.

What kind of evidence is required to determine whether solid friction in the asthenosphere or ocean-like dissipation in the underlying liquid layer is the dominant energy source? Solid friction in the asthenosphere will be the primary heat source if the overall shell thickness is greater than about 30 km. If for no other reason, the low vertical thermal gradient at its base would inhibit the transport of heat generated by ocean-like tidal currents and thereby inhibit ocean-like dissipation. If, however, the shell thickness is $\lesssim 30$ km, then dissipation caused by tidal currents would probably dominate over solid friction because of the absence of an asthenosphere to buffer it. Unfortunately, the resolution of this problem (let alone the more fundamental issue relating to the existence of a tidally decoupled outer shell) may require placement of a lander on Io.

Ross and Schubert (1986) have recently proposed a different model for tidal dissipation which bears some similarities with ocean dissipation but in which the low-viscosity liquid layer is missing. Instead of a thin liquid layer sandwiched between solid outer shell and mantle, they argue that the mantle itself is a highly viscous magma with high crystal fraction. The magma viscosity η is determined by crystal size D and crystal fraction χ. They adopt Sherman's (1968) empirical equation for the viscosity η for a crystal-rich partial melt

$$\eta = \eta_0 \exp\left[\frac{\delta D}{(\chi^{-1/3} - 1)} - 0.15\right] \tag{15}$$

where δ is an empirical constant of order 10 mm^{-1} and η_0 is the viscosity of the liquid component. Their preferred values of D and χ and η_0 are 0.5 mm and 0.4 and 10^5 poise, respectively. The resulting mantle viscosity is of order 10^{13} poise and leads to a heating rate comparable to the observed heat flow.

The plausibility of such a model depends on the absence of a low viscosity zone near the top of the mantle, since such a zone would tend to decouple the mantle beneath it from tidal flexing, much as an ocean-like layer suppresses tidal flexing of an underlying solid mantle. In order for the mantle slurry to remain well mixed, the influence of pressure on crystal fraction and size must be modest. Also the vertical convective velocity of order $0.1\ R\ [\alpha g H/\eta C_p]^{1/2} \simeq 1$ km yr^{-1} (Ginzberg et al. 1979) (where C_p is the heat capacity), must be much larger than the crystal settling velocity (i.e., Stokes velocity $= 2\Delta\rho g D^2/\eta$) driven by the effect of gravity on the difference in density $\Delta\rho$ of crystal and melt. This latter condition is probably satisfied since the maximum settling velocity is of order 1 m yr^{-1} for $\Delta\rho \simeq 0.1$ g cm^{-3}, $D \simeq 1$ mm, and $\eta \simeq 10^5$ poise.

What data could possibly be obtained which could discriminate between these various models? One major difference between dissipation in the highly viscous asthenosphere and ocean-like dissipation in either a thin liquid layer or thick mantle slurry is that the former probably leads to episodic frictional heating while the latter models predict uniform frictional heating with time. Episodic dissipation causes episodic excursions of the satellite orbits from their equilibrium configuration. Detection of a slight change in the near 2 to 1 ratio of the orbital mean motions of Io with respect to Europa (and thereby each satellite's forced eccentricity e) might indicate that an intense frictional heating episode is presently occurring (if $de/dt < 0$) or was recently completed (if $de/dt > 0$).

VI. ATMOSPHERE, TORUS, SODIUM CLOUD AND MAGNETOSPHERE

The subject matter of this section is given detailed discussion in four chapters in Morrison (1982b): Fanale et al. (1982) on Io's surface and its interaction with the atmosphere; Kumar and Hunten (1982) on the atmosphere and ionosphere; Pilcher and Strobel (1982) on the physics of the torus; and Sullivan and Siscoe (1982) on the torus as observed *in situ* by the Voyager plasma instruments. Most of the chapters in Dessler (1983) on the Jovian magnetosphere involve the Io torus to some degree (see, e.g., Brown et al. [1983a]); the topic was earlier reviewed by Brown and Yung (1976). Outer solar system atmospheres are reviewed by Trafton (1981).

A. Atmosphere

There is still very little consensus on the basic nature and amount of Io's atmosphere. An ionosphere was observed in 1973 at both limbs by the Pioneer 10 radio occultation experiment (Kliore et al. 1974,1975), and SO_2 was seen in 1979 over a hotspot and possibly through the plume of Loki by the Voyager 1 infrared spectrometer (Pearl et al. 1979). Straightforward explanations of these two observations require an atmospheric pressure on Io of 10^{-9} to 10^{-7}

bar (number densities 10^{11} to 10^{13} cm^{-3}; i.e., a "thick" atmosphere). However, sputtering from the surface is still the most attractive explanation for the large fluxes of fast sodium atoms that are seen near Io, and is viable only if the atmospheric pressure does not exceed 10^{-11} bar (i.e., a "thin" atmosphere) (Haff et al. 1981). An attempt to detect SO_2 gas absorption by the International Ultraviolet Explorer (IUE; Butterworth et al. 1980) sets an upper limit of 0.008 cm atm for the global average column abundance, slightly below that required to explain the Pioneer 10 ionospheric data and about 1/4 that expected if the Voyager-1 infrared interferometer spectrometer (IRIS) measurement represented an equilibrium thick atmosphere. However, there are uncertainties much bigger than a factor of 4 about the curve of growth for the ultraviolet band that was used in the IUE measurements (Belton 1982). The thin atmosphere interpretation of the IUE data was also questioned by Fanale et al. (1982) who pointed out that regional cold-trapping on Io's surface may reduce the global average pressure expected for an equilibrium atmosphere. Two additional conflicting views of Io's atmosphere have been put forth. One model is by Kumar and Hunten (1982) who advocate an equilibrium SO_2 atmosphere modified by photochemical production of O_2 that ends up as a major gas in the nightside atmosphere and which retards the transport of SO_2 to Io's cold polar regions. The other is by Matson and Nash (1983) who argue for a nonequilibrium (with the surface) SO_2 atmosphere whose surface pressure is greatly lowered from the equilibrium case by subsurface cold-trapping in a highly porous upper regolith layer.

The model atmosphere that has received the most extensive study is one of SO_2 in vapor-pressure equilibrium with surface deposits of SO_2 frost, the thick atmosphere model. Contours of the vapor pressure for this model have been given by Fanale et al. (1982) for two cases (Fig. 18): the SO_2 frost is assumed to occur (a) all over the surface, or (b) only in areas of higher albedo and therefore somewhat lower temperature. The actual atmospheric pressure contours would be different because strong winds must blow away from the subsolar region; although this difficult problem has been mentioned in an abstract (Ingersoll et al. 1986), no treatment has been published. Kumar (1979) presented computations of the dayside temperature profile, finding an unexpectedly large exospheric temperature of a little over 1000 K, because SO_2 is a much more efficient absorber of solar energy than the more familiar gases. His computed electron-density profiles (Kumar 1980) were in excellent agreement with the observed ones. For a brief summary, see Kumar (1982).

Aeronomical (or photochemical) computations for the thick atmosphere model require a number of assumptions about conditions at the upper boundary and the surface. Upward fluxes of S and O are simply postulated, without physical justification, as having values suitable to populate the Io plasma torus. At the surface, various plausible assumptions can be made, and some of them have been explored by Kumar (1982,1984). In these models (Fig. 19) there is a substantial buildup of O_2 as a photochemical by-product. Oxygen,

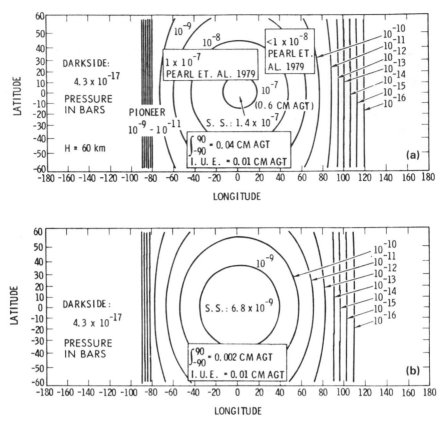

Fig. 18. Theoretical gas pressure distributions about the subsolar point on Io for an hypothetical SO$_2$ atmosphere: (a) for an equilibrium atmosphere everywhere buffered by SO$_2$ frost at local ground-surface temperature that results from average surface albedo and follows a cosine law; (b) for a regionally cold-trapped atmosphere buffered by SO$_2$ frost at surface temperature that follows variation in local surface albedo (figure from Fanale et al. 1982).

being noncondensable at Ionian temperatures, behaves very differently from SO$_2$ and can give a substantial pressure ($\sim 10^{-10}$ bar) on the night side. On the day side it can amount to $\sim 10\%$ of the SO$_2$. Pressures this large would have a profound effect on the possible SO$_2$ wind field. Above 300 km the dominant gas is O, whether O$_2$ is abundant or not.

Since the Sun-Io-Earth phase angle is always small ($< 12°$), both Pioneer 10 occultation events were near the terminator. On entry (downstream, sunlit) the peak electron density was 6×10^4 cm^{-3}; on exit (upstream, dark) it was 9×10^3 cm^{-3} and showed a much smaller vertical extent. Early models (see Kumar and Hunten 1982) assumed photoionization as the source and had difficulty reaching a high enough electron density. Cloutier et al. (1978) pointed out the potentially large effect of the magnetospheric plasma flowing by,

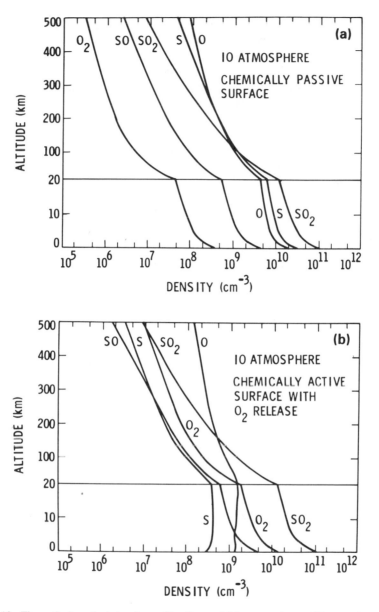

Fig. 19. Theoretical vertical density profiles for model Io atmospheres: (Note vertical scale stretch between 0 and 20 km elevation range.) (a) density of atmospheric constituents on Io for a model where the surface is assumed to be chemically passive, and photolysis products of SO_2 are allowed to escape; (b) Density of atmospheric constituents on Io for a model where the surface is assumed to be chemically active and serves as an efficient sink of S and O; however, O is converted to O_2 and returned to the atmosphere (figure from Kumar 1982).

which is still an unresolved question. Models based on an SO_2 atmosphere, and ignoring this cometary effect, were obtained by Kumar (1980). The principal source of ionization is a flux of low-energy electrons; the preferred energy is 100 eV. The amount of SO_2 required is consistent with the vapor pressure expected over the terminator. Another complication in interpreting the Pioneer 10 radio occultation data is a question raised by Matson et al. (1981b) as to whether or not any of the radio signals passed through erupting plumes on Io. If the number, size, and location of plumes seen by Voyager are typical, then the probability of at least one of the Pioneer occultations passing through a plume is high (T. V. Johnson and D. L. Matson, personal communication, 1982). Electron densities expected for a plume occultation have not yet been worked out.

A more recent thin atmosphere model for the basal pressure of Io's atmosphere has been developed by Matson and Nash (1983). This model takes account of the very high porosity ($\sim 90\%$) of Io's surface (Fig. 7). This porosity results in a steep temperature gradient (Fig. 8) in the upper few centimeters of the surface, conceivably allowing efficient subsurface cold-trapping of atmospheric gases such as SO_2. In the absence of local plume venting this tends to keep ambient surface pressures very low, as low as 10^{-12} bar in the subsolar region as shown in Fig. 20. In this model, the atmosphere, when not affected by active venting of gas, is in equilibrium with a subsurface permafrost rather than a much warmer surface frost. As pointed out by Matson and Nash (1983), it is important to keep in mind that the various observations pertaining to Io's atmosphere (Pioneer, Voyager, IUE, etc.) are not simultaneous but

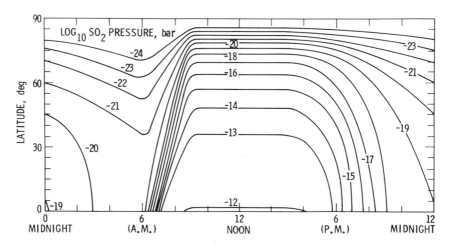

Fig. 20. Theoretical map of SO_2 gas pressure distribution at Io's surface assuming that subsurface cold-trapping is the only process actively controlling atmospheric pressure (figure from Matson and Nash 1983).

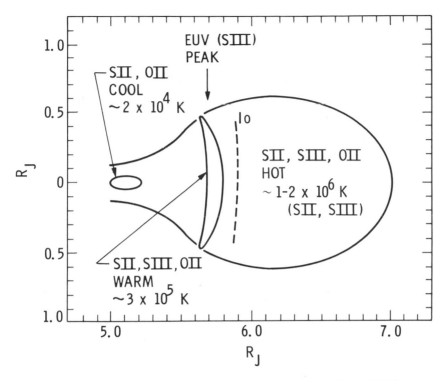

Fig. 21. Schematic cross section of Io torus emission in centrifugal coordinates. This is a representation of the torus when it is at its brightest, but it varies longitudinally and temporally. The outer part of the torus is much hotter than the inner part, and this difference is reflected in the greater vertical extent. Because of the tilt of the centrifugal equator with respect to Io's orbital plane, the position of Io appears to oscillate along the dashed line. (Sketch courtesy of C. Pilcher and J. Morgan, personal communication, 1984.)

obtained at times differing by more than a decade. With the now known high variability of volcanic activity on Io, the atmospheric data sets may each be strongly aliased and not representative of any single time or place.

B. Torus

Broadly speaking, the torus is a region enclosing Io's orbit at 5.9 Jovian radii (R_J) and with a thickness of about 1 R_J (Fig. 21). It contains neutral atoms (Na, K, O, S) and a plasma dominated by ions of S and O whose characteristics are summarized in Table IV, after Brown et al. (1983). The electron temperature is ~ 4 eV or 50,000 K, with a small admixture of a hotter component at ~ 600 eV. Inside 5.7 R_J the ion temperature is much lower, as low as 0.5 eV. Johnson and Strobel (1982) present a somewhat different picture, with properties varying with distance; the electron temperature is 2 eV at 5.1 R_J and 9.5 eV at 7.1 R_J. At 5.9 R_J, where comparison with Table IV is relevant,

TABLE IV
Composition of the Hot Plasma Torus and Number Densities[a]

Ion or atom	n (cm^{-3})	Ion or atom	n (cm^{-3})
O = O I	24	S = S I	7
O^+ = O II	1570	S^+ = S II	640
O^{+2} = O III	45	S^{+2} = S III	1040
O^{+3} = O IV	4	S^{+3} = S IV	190
e	5000	S^{+4} = S V	8

[a]Table after Brown et al. (1983b).

the neutral densities are similar, but the densities of electrons and ions are typically 40% as great.

A schematic cross section of the torus is shown in Fig. 21, and Color Plate 6 shows a groundbased telescopic image of the sulfur emissions. The symmetry plane of the torus is Jupiter's centrifugal equator, the locus of points where the magnetic field lines are farthest from the rotational axis. The plasma radiates intensely in the region 600 to 1200 Å where the Voyager ultraviolet spectrometers are most sensitive.

The signature of ionized SO_2^+ in the torus is seen in the Voyager ultraviolet data (Belcher 1983), and that of condensed SO_2 frost is seen in Io's surface reflection spectrum (Fanale et al. 1979; Smythe et al. 1979) (e.g., see Fig. 5). The conclusion is irresistible that all the material of the torus comes from Io. How it leaves Io is unknown, although some of the suggested mechanisms are discussed below and in Kumar and Hunten (1982). The power input necessary to drive the ultraviolet emissions observed by Voyager in 1979 is 2×10^{12} W (Broadfoot et al. 1979; Shemansky 1980) and the total ionization rate is 1 to 3×10^{28} s^{-1} (cf. Hill 1980; Brown et al. 1983). According to Smyth and Shemansky (1983), about 10% of this power comes from the velocity of newly formed ions relative to the magnetic field, which corotates with Jupiter. The source of the remainder might conceivably be some sort of plasma wave, or perhaps very energetic electrons from farther out in the magnetosphere, or perhaps it does come from corotation energy after all (see discussion in the chapter by Cheng et al.). Major temporal variations in the torus may occur as indicated by its apparent absence or much weaker ion density and radiation intensity during the Pioneer encounters in 1973–74.

Brown et al. (1983b) have presented an elaborate computation of the atomic physics of the torus, incorporating not only the usual processes of ionization by electron impact and recombination, but also charge exchange and loss from the region, both of which are very important. The time constant for the loss (assumed to be by diffusion) is about 100 days. A 10-day time is not consistent with the measured charge composition. Such a short time is required if the corotation energy of fresh ions is to be sufficient to supply all the necessary energy to the torus. Their conclusion, already mentioned above, is that some other energy source is necessary. The importance of charge exchange has also been stressed by Johnson and Strobel (1982).

The ultraviolet emission from the torus shows a distinct asymmetry, fixed in local time with the maximum around 1900 hr (Sandel and Broadfoot 1982; Shemansky and Sandel 1982). The average ratio of maximum to minimum is 1.53. Analysis of the spectra demonstrates that it is the electron energy that varies, not the electron or ion density. Any other result would have been most surprising in view of the 100-day characteristic time. An energy source concentrated near noon local time is required, and it must supply roughly half the total energy. It has also been suggested that the electron temperature is reversibly modulated by the agency of an electric field directed from dawn to dusk within the torus (Barbosa and Kivelson 1983; Ip and Goertz 1983).

C. Sodium Cloud

Neutral sodium atoms were the first substance to be discovered above Io's surface (Brown 1974). The sodium cloud does not in fact form a complete torus. Its rapid destruction by electron-impact ionization does not permit it to spread all the way around the orbit, although a much fainter cloud of Na permeates the whole Jovian magnetosphere system. It is concentrated in a banana-shaped cloud, the major part of which is ahead of Io and interior to its orbit (Fig. 22).

Groundbased studies of the Na cloud have concentrated on capturing its image and monitoring its distribution, shape, intensity, and temporal variations (Matson et al. 1978a; Goldberg et al. 1980; Pilcher 1980b; Brown and Schneider 1981; Pilcher et al. 1984). These observations (e.g., Color Plate 7) have shown that the cloud shape has directional extensions and other features that vary on time scales as short as minutes (Goldberg et al. 1984; Pilcher et al. 1984), but since 1976, when it was first imaged, the overall shape and character of the cloud has remained stable (Goldberg et al. 1984). The first celestial mechanical models for the distribution of sodium in Io's vicinity showed that Io is the source of the sodium and that it may be coming predominantly from the Jupiter-facing hemisphere of Io (Smyth and McElroy 1978; Carlson et al. 1978; Matson et al. 1978a). These models assumed a constant mean lifetime for the sodium atoms, and found a best fit to the data for a value of 20 hr (see also Macy and Trafton 1980); the source strength was about 3×10^{26} atoms s^{-1}, or 7×10^8 cm^{-2} s^{-1} if averaged over Io's surface (Carlson 1980). It is now clear (Smyth 1983) that the assumption of a constant lifetime is not realistic, and all these results require at least some revision. Figure 23 shows the neutral sodium lifetimes obtained for the actual plasma conditions; the lifetime is longer in the outer, than in the inner, part of the plasma torus due to variations in electron temperature there. When this effect is included, a broader sodium source distribution is acceptable; in particular,

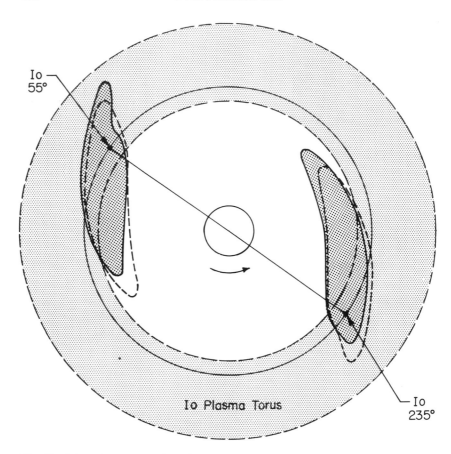

Fig. 22. Schematic representation of the Io torus and sodium cloud as viewed from above
Jupiter's north pole and at two solar phase angle positions of Io in its orbit. The torus plane
(large dashed circles) is inclined ∼ 7° to Io's orbit plane and wobbles relative to it (large solid
circle) as Jupiter rotates. The banana-shaped sodium cloud is in Io's orbit plane and skewed
toward Jupiter ∼ 35° from the instantaneous Io orbit direction. The cloud leads Io in its orbit.
The observed cloud shape (solid line, heavy shading) is perturbed by solar radiation pressure
(coming from the 180° Io phase direction), and thus is offset from its unperturbed position
(dashed shaped) relative to Io. Cloud shape shown represents the envelope of neutral sodium
atoms after 20 hr of flight time (figure from Smyth 1983).

there is no longer any need to assume that the sodium source is confined to
the inner hemisphere; either hemispheric or isotropic emission is permitted
(Smyth 1983).

Other theoretical studies have shown that the east-west asymmetry of the
cloud intensity and shape as it orbits Jupiter (discovered by Bergstralh et al.
1975, 1977; Goldberg et al. 1978) is apparently caused by solar radiation pressure, compressing the cloud near east elongation and expanding it near west

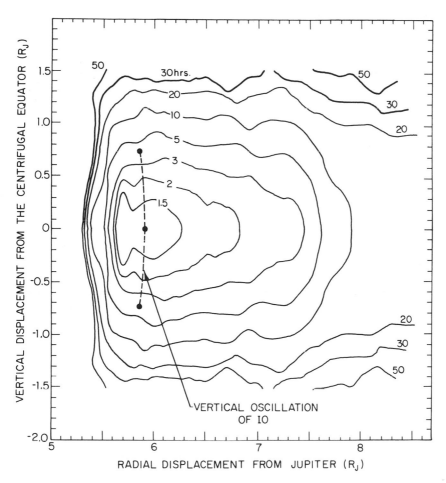

Fig. 23. Neutral sodium lifetime against electron impact ionization in the Io plasma torus. The lifetime calculated is based on experimental cross sections and Voyager 1 plasma density and temperature data. Dashed line represents north-south excursion zone of Io as it orbits within the tilted and rotating plasma torus. See Smyth (1983) for details.

elongation (Smyth 1979,1983). It was suggested by Trafton (1980b) that the Io plasma torus, the plane of which is inclined to Io's orbital plane, may be responsible for the north-south asymmetry in intensity of Io's sodium cloud because the cloud must regularly pass through the torus and be subject to ionization loss of neutral sodium atoms as noted above.

Other neutral atoms have been found about Io. The difficult detection of neutral O atoms was finally accomplished by Brown (1981), but there is little information on its distribution except that it is seen well away from Io. A model by Smyth and Shemansky (1983) shows a fairly uniform filling of the orbit along with a considerable enhancement near Io and leading it in the orbit. The source of O ions is therefore similarly distributed, and not confined to a very small volume around Io, as already concluded from the absence of any accompanying ultraviolet glow (Shemansky 1980). The detection of neutral S atoms has been announced by Durrance et al. (1983). Their measurement suggests an extensive neutral atomic cloud associated with the Io torus region.

D. Supply to the Torus and Sodium Cloud

Because Io's orbital corridor is within the torus volume, and the composition of ions in the torus is the same as neutrals or compounds thought to be on the surface and emanating from Io's volcanoes, there can be no doubt that the material in the torus has escaped from Io. The nature of the process or processes, however, is still obscure. The rate at which Io supplies mass to the torus has been estimated to be 1000 kg s^{-1}; thus only 0.1% of Io's mass has been lost throughout geologic time to the Jovian magnetosphere (Dessler 1980). Smyth and Shemansky (1983) and Brown et al. (1983a) find that the flux of O to the torus is 1.2×10^{10} cm^{-2} s^{-1} ($\sim 5 \times 10^{27}$ s^{-1}) and assume that the S flux is half as great. Taken together, these fluxes amount to 150 kg s^{-1}. The discrepancy with Dessler's number reflects the continuing difficulty, discussed above, in reconciling the atomic physics of the torus with the energy supply.

Thermal escape may be appreciable for O (Kumar and Hunten 1982), but could hardly apply to S. Ionization close to Io, followed with pickup by the corotating magnetic field, is ruled out by the absence of detectable ultraviolet emission (Shemansky 1980). Kumar (1984) has refined his photochemical and diffusive equilibrium model of Io's SO$_2$ atmosphere and argues that a lower limit atmosphere of 1.4×10^{-10} bar (column density $\sim 10^{16}$ cm^{-2}) is necessary to provide the required supply rates of O and S to the torus in its condition during the Voyager 1 encounter.

Sputtering, or impact by fast heavy ions, was suggested by Matson et al. (1974), originally to explain the presence and high velocity of sodium atoms. It is still the preferred mechanism, but in its simplest form is incompatible with the presence of a thick atmosphere, as mentioned above. On the other hand, thin atmosphere models encounter serious difficulties in achieving adequate supply rates of sulfur and oxygen to the torus (Kumar 1984; Cheng 1984a). Most of the material must be supplied as neutrals (Eviatar and Mekler 1984; Cheng 1984a; cf. chapter by Cheng et al.).

The supply mechanism must be very nearly the same on day and night sides, at least for sodium which shows very similar clouds at the two elongations (Bergstralh et al. 1977; Goldberg et al. 1980). Pilcher et al. (1984) estimate that the supply rate of sodium atoms must be 1 to 2×10^{26} s^{-1} to sustain the observed sodium cloud. Kumar and Hunten (1982) suggested a two-stage process, in which sodium is sputtered from the surface to the atmosphere by high-energy ions, and then from the exosphere by low-energy ions, which are much more abundant. Summers et al. (1983) have examined this scheme in detail, using ion fluxes observed by Voyager 1 and assuming that Na and K are present in cosmic proportions relative to S. They find escape rates in the required range for O as well as for Na and K. They also point out that the inner (Jupiter-facing) hemisphere of Io is favored because the flux of low-energy ions is greater there. Such an asymmetry can be part of the explanation of the geometry of the sodium cloud (Smyth and McElroy 1978), although the asymmetric lifetimes shown in Fig. 23 are involved. Loss of S is not discussed, doubtless because its density at high altitudes is very uncertain. Figure 19a suggests that it can be quite abundant, while it is relatively scarce in Fig. 19b. Beyond doubt, these two figures do not span the entire likely range. Prospects for a substantial escape rate of S therefore seem reasonable.

Sputtering of the top of Io's atmosphere is thought by Sieveka and Johnson (1984) to explain the directional fast-sodium features observed in Io's sodium cloud. Pilcher et al. (1984) propose a so-called magnetospheric wind-driven gas escape mechanism to explain the directional features outside Io's orbit. The atoms must be ejected with high velocity at right angles to Io's motion. Magnetospheric particles are stated to be the source; ejection at right angles is probable in collisions with a large impact parameter, which are usually effective for alkali metals (Brown et al. 1983a, their Sec. 6.3). Pilcher et al. (1984) also deduced that the sodium flux from Io may be higher from the equatorial regions than from the polar regions.

E. Io and the Jovian Magnetosphere

The effects of Io and its escaping atmosphere on the Jupiter magnetosphere are substantial and are discussed in the chapter by Cheng et al. and in Dessler (1983). Ions of S and O dominate much of the region; H, H_2, H_3, and Na ions are also detected. The S, O, and Na are thought to come from Io, whereas the molecular hydrogen ions presumably come from Jupiter or the other satellites, and H^+ mainly from the solar wind. Dense plasma forms a rotating disk originating from and extending out from Io's orbit and diffusing into the magnetosphere. Some of this material (at least sulfur) is thought to end up being deposited on the surfaces of other satellites immersed in the magnetosphere such as Europa and Amalthea (Lane et al. 1981; Eviatar et al. 1981; Gradie et al. 1980; Thomas and Veverka 1982; see the chapters by Veverka et al. and by Malin and Pieri).

Elsewhere in the magnetosphere, trapped ions and electrons are accelerated to 10 MeV or more. The fluxes of these high-energy particles generally increase with decreasing distance from Jupiter except that Io and its torus sweep out a corridor of lower particle density along their orbital path in the magnetosphere (Divine and Garrett 1983). However, the phase space density of heavy ions around 100 MeV *decreases* inwards and shows a precipitous drop between 12 and 8 R_J (Gehrels and Stone 1983). Undoubtedly, they are scattered into pitch angles that permit them to precipitate into Jupiter's high-latitude atmosphere (cf. Gurnett and Goertz 1983). The corresponding X-ray aurora has been reported by Metzger et al. (1983). Some of the energy input into the Io plasma torus may also be associated with these energetic ions.

Intense radio emissions from the inner magnetosphere and plasma waves within it are observed, some associated with Io's magnetic flux tube (Fig. 17) and some not. Io controls nearly all decametric (~ 5 to 39 MHz) radio emission either directly or through its torus by affecting the plasma conditions that result in the radio emissions; as summarized by Menietti et al. (1984), the emission is thought to be caused by electrons precipitating at the local gyrofrequency and occurs in conical sheets nested along field lines in the Io flux tube near Jupiter. The emission sheets sweep past the observer as Jupiter rotates, resulting in the apparent emission bursts that have been observed from Earth for several decades. In the case of lower-frequency kilometric radio emission discovered by Voyager, a shadow zone is cast by the Io plasma torus whose density is high enough to refract the radio waves (Kurth et al. 1980). This large topic on Jupiter radio emission, outside the scope of the present review, is discussed by Carr et al. (1983).

The effects of magnetospheric ion and electron bombardment on Io's surface have been considered by numerous investigators (see discussions in chapters by Veverka et al. and Cheng et al.). The incident ion flux is thought to have both high-energy and low-energy components. The typical energies and fluxes measured in the vicinity of Io's orbital corridor (by Pioneer and Voyager detectors) are as follows (data from summary by Divine and Garrett [1983]):

Ions

S^+, S^{++}	(~ 500 eV)	(2 to 6) $\times$ 10^9 cm^{-2} s^{-1}
O^+, O^{++}	(~ 250 eV)	(0.6 to 2) $\times$ 10^9 cm^{-2} s^{-1}
H^+	(30 KeV)	7×10^8 cm^{-2} s^{-1} (model-dependent estimate)
H^+	(1–20 MeV)	10^5 to 10^6 cm^{-2} s^{-1}
Electrons	(10–40 eV)	4×10^{11} to 1×10^{12} cm^{-2} s^{-1}
"	(0.1 MeV)	3×10^8 cm^{-2} s^{-1}
"	(3 MeV)	3×10^7 cm^{-2} s^{-1}
"	(21 MeV)	4×10^6 cm^{-2} s^{-1}

The low-energy (eV) ions and electrons are those composing the Io plasma torus. The higher-energy (MeV) protons and electrons are the conventional trapped magnetospheric particles. Because of the torus corotational velocity (57 km s^{-1} greater than Io's orbital motion), the low-energy particles would tend preferentially to impact Io's trailing hemisphere. In contrast, for the high-energy ions, there is no longitudinal dependence in their flux on Io because of their large gyroradii and cyclotron motion. The high-energy ion distributions near Io's orbit peak at pitch angles near 90° (Lanzerotti et al. 1981), thus Io's equatorial regions will experience larger fluxes of high-energy ions than will the polar regions.

The possibility of shielding Io's surface from Jovian magnetospheric particles, especially low-energy ions, by any intrinsic magnetic field of Io has been suggested (Kivelson et al. 1979; Southwood et al. 1980). Also, an atmosphere on Io may shield the surface; Haff et al. (1981) estimated that a surface atmospheric SO_2 pressure $> 10^{-11}$ bar will inhibit sputtering of surface atoms to space by magnetospheric particles. On the other hand, Watson (1981) calculated that, if there is no intrinsic atmosphere, surface sputtering by sulfur and oxygen plasma can result in an SO_2 corona that under certain circumstances may be the dominant atmospheric component on Io. He also demonstrated that most sputtered material will be redeposited on Io's surface, not lost to the magnetosphere as earlier suggested by Lanzerotti et al. (1978) and Pollack and Witteborn (1980).

Sputtering and erosion of SO_2 ice on Io's surface by low-energy S and O_2 and other ions has been investigated in the laboratory by Lanzerotti et al. (1982) and Johnson et al. (1981). They find very high sputtering rates and production rates for sputter products such as O_2, S, SO, S_3 and S_2O as well as SO_2. They confirm the earlier suggestion of Matson et al. (1974) that sputtering may be an important process for the supply of material to Io's exospheric region.

Chemical reactions producing SO_3 from SO_2 on Io's surface by the effects of heavy ions and high-energy (MeV) protons and low-energy (keV) ions have been suggested by Boring et al. (1983), Moore (1984) and Johnson et al. (1984a). But the maximum predicted rates of SO_3 production (1 to 2% concentration per year in the topmost 10 μm layer) are insignificant compared to the volcanic turnover rates of (mm to cm yr^{-1}) estimated from Voyager data (Johnson and Soderblom 1982).

Surface luminescence effects and color changes from ion bombardment on Io have been investigated with laboratory tests; Nash and Fanale (1977) found that sulfur powder did not luminesce nor change in color under low-energy (keV) proton bombardment. Moore (1984) determined that solid SO_2 irradiated with MeV protons yielded thermoluminescence behavior but that its intensity was very weak and such a process could not be invoked to explain Io's post-eclipse brightening enigma. She also found qualitatively that the irradiation produced some coloration of the solid SO_2 but that the concentration levels of similar coloration on Io would be very low.

Fast-ion sputter transport and redistribution of surface SO_2 from Io's equator to its poles has been considered (Watson 1981; Sieveka and Johnson 1982), but the absence of obvious polar SO_2 frost deposits suggests that continually active volcanic processes erase any long-term sputter redistribution effects (Sieveka and Johnson 1982). The presence of distinctly darkened polar regions on Io's surface, however, has long suggested the possibility of irradiation-darkening effects from bombardment by magnetospheric ions or electrons making up the Io flux tube that electrically couples Io's surface or ionosphere to the ionosphere of Jupiter. Any accumulated radiation darkening at the poles could also result from the fact that the volcanic resurfacing rate may be larger near the equator than near the poles, so the polar surfaces have a longer exposure time to incident charged particle (and solar ultraviolet) irradiation.

The current general picture of the effects of magnetospheric particles on Io's surface can be summarized as follows: (1) sputtering from the surface or atmosphere is the most plausible supply mechanism for the neutral sodium cloud and perhaps the supply to the ionized sulfur torus; (2) the cumulative irradiation effects on surface optical or chemical properties are probably insignificant compared with the supply of fresh material by inferred volcanic processes that overturn and replenish the surface at very high rates, erasing magnetospheric irradiation effects except in areas of low volcanic activity.

VII. MAJOR QUESTIONS AND FUTURE EXPECTATIONS

There are many outstanding questions and controversies at present regarding the way things really are in Io's interior, on its surface, and in its atmosphere and exosphere. We consider the major questions today are as follows:

1. Atmosphere: Is it thick or thin? What are its components besides SO_2? What is its temporal structure and composition?
2. Plumes: Are they driven by S, or SO_2 vapor phase(s), or both? Are other vapor phases involved?
3. Volcanism: Are the plains flows composed predominantly of sulfur, or silicate with a sulfur surface component?
4. Mountains and calderas: What are they composed of, what causes their relief and what is their depth of compensation?
5. Internal thermal conditions: How do we reconcile measured high surface heat flow with tidal theory? Is the heat vented episodically, the tidal heating itself episodic, or could the Jovian tidal torque (which is the ultimate source of heating on Io) have suddenly turned on due to, e.g., the recent onset of helium rainout within Jupiter's atmosphere? Does the location of hotspots change significantly over time, and why are mostly equatorial plume eruptions now active?
6. Tidal heating mechanisms: Is the primary energy source solid friction within a tidally decoupled shell, ocean currents in an exceptionally thin (~ 1 to 10 km) liquid layer beneath that shell, or viscous friction in a very soft mantle?
7. Rate of tidal expansion of Io's orbit: What is the present value of Q for

Jupiter, and what does it imply about the time-average heating rate of Io over geologic time compared to presently observed surface heat flow?

8. Magnetic field: Does Io have a large conducting core which acts like a hydromagnetic dynamo as on Earth?

9. Bulk composition: What is Io's bulk interior composition? How does this relate to the terrestrial bodies such as the Moon?

10. Surface composition: Is sulfur present in its stable elemental form, as metastable allotropes, or as compounds? Is condensed SO_2 present as solid ice, thick or thin frost, ubiquitous adsorbate, or all these forms; is it mixed with other materials or in discrete patches? What is the source of sodium (that feeds the sodium cloud) and how is it replenished? What silicate or other phases are present?

11. Plasma torus: How is material supplied to the torus? How is the torus related to surface volcanism? How do torus ions cool so rapidly?

12. Sodium cloud: How does sodium get off the surface? Why is the cloud so stable? How does it interact with the plasma torus?

13. Radio emission: How does Io control the decametric radio emissions from Jupiter?

Our expectations for finding the answers to some of these questions are good. Continued work by many investigators using Voyager data, obtaining new groundbased and Earth-orbiting telescopic data, conducting laboratory experiments on materials and processes, and carrying out theoretical studies will seek answers to the above questions and no doubt pose many new ones. Much of this work will be done in anticipation of the Galileo mission to Jupiter. In a few years the single Galileo spacecraft is scheduled to go into orbit at Jupiter and perform an intensive tour of the Galilean satellites. The planned mission includes a single close (1000 km) flyby of Io, but due to charged-particle radiation hazards in Jupiter's inner magnetosphere (1/3 of the total mission dose occurring at Io), no further close encounters with Io are scheduled during the spacecraft's planned two-year lifetime. The science objectives include imaging Io's entire surface at moderate resolution ($\sim$ 2 km) and sampling selected areas at resolution as high as 100 m, acquiring moderate-resolution spectral reflectance data of the entire surface, making synoptic observations of Io volcanism, hotspots, and other surface changes, and carrying out a radio occultation of Io, and at least one radio occultation of the Io torus. The history of Io studies portends that this sustained closeup look will reveal many more of Io's exciting secrets.

Acknowledgments. We thank D. Matson, W. Smyth, T. V. Johnson, R. Howell, J. Trauger, A. McEwen, B. Goldberg, R. Nelson, D. Pieri, C. Pilcher, W. Sinton, M. Acuña, W. Ward, A. Young, N. Divine and D. Stevenson for helpful discussions, comments, and assistance. Special thanks go to R. Howell for providing an infrared spectrum of Io in advance of publication,

D. B. NASH ET AL.

to C. Pilcher for a schematic drawing of the Io torus characteristics, to A. McEwen for color images of Io, to R. Batson for the map of Io, and to S. Prather for preparing the manuscript. Portions of this work represent the results of one phase of research carried out at the Jet Propulsion Laboratory, California Institute of Technology, under contract with the National Aeronautics and Space Administration.

14. EUROPA

MICHAEL C. MALIN
Arizona State University

and

DAVID C. PIERI
Jet Propulsion Laboratory

Europa, the second major satellite outward from Jupiter, is a lunar-sized object in synchronous rotation about that planet. Its high albedo and spectral characteristics indicate the presence of surface water ice and/or frost. Its density suggests a substantial silicate component of at least 85% by volume. A limited variety of landforms is seen in Voyager images, most likely because of poor spatial resolution of these data. The absence of numerous impact craters is interpreted as indicating a youthful surface and/or surface regenerating processes. A planet-wide lineament system has been interpreted to result from tectonic stresses induced by evolution of Europa's orbit and interior. Present data do not permit unambiguous tests of such interpretations. Greater insight may come from analytical or numerical models and future spacecraft observations.

Europa is the most enigmatic of the Galilean satellites of Jupiter, for it presents contradictory aspects that act to hide its true nature. Its albedo and spectral reflectance characteristics indicate the presence of water ice on its surface more clearly than any of the other satellites, and its density implies a silicate body not unlike Io or the Earth's Moon. Observations by the Voyager spacecraft are of the quality typical of Earth-based telescopic views of the Moon, but cover only a fraction of the surface (see Fig. 1, Color Plate 10 and Map Section). In addition to our Table I, the introductory chapter also lists the global properties of Europa.

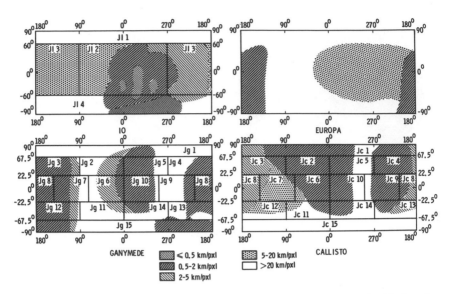

Fig. 1. Resolution/coverage of Voyager observations of Europa as compared with the other Galilean satellites.

TABLE I
Global Properties of Europa

Radius	1569 ($\pm$ 10) km
Mass	4.87 $\times$ 10 kg
	(39,000 reciprocal Jovian masses)
Density	3.04 g cm^{-3}
Orbital semimajor axis	670,900 km
Orbital period	3.551 days
Orbital eccentricity	0.0003
Orbital inclination	0°28ʹ1

Earth-based telescopic data, laboratory analysis, spacecraft observations, and theoretical models are all important to the study of Europa. In Sec. I, Earth-based telescopic and laboratory studies relevant to Europa's global spectral properties are presented. In Sec. II, spacecraft observations made by the Voyager imaging experiment are summarized. These observations address geomorphology with a lesser emphasis on composition. Tectonic models developed to describe and explain the Voyager observations are treated in Sec. III, followed by discussion of whole-planet compositional models to which they are closely related. In Secs. IV and V potential tests of alternative models are examined, deficiencies in the present observational data base are noted, and the remaining, outstanding questions unanswered by past and current research are summarized.

I. GLOBAL AND REGIONAL STUDIES OF SPECTRAL CHARACTERISTICS OF THE SURFACE

It has long been suspected that Europa is covered with frost, mainly of water composition. Kuiper (1957) first suggested the possible presence of water frost by showing, along with Moroz (1961), that both Europa and Ganymede have lower surface reflectivities at 1 μm than at 2 μm, as do many ices. From his polarization measurements, Veverka (1971a) proposed that Europa, Ganymede, and Callisto all had low-opacity surface layers, possibly of snow. Johnson and McCord (1971), from their observations at 1.6 μm, suggested the presence of ammonia frost. Lewis (1981a,b) proposed on theoretical grounds that the crusts of the Galilean satellites were composed of nearly pure ice. From the amount of infrared adsorption between 1.25 and 4 μm, Pilcher et al. (1972) estimated that between 50 and 100% of the hemisphere that faces in the direction of Europa's revolution around Jupiter (i.e., its leading hemisphere) was covered by water frost, and that, as a global average, about 70% of Europa's surface was covered by such frost, with Europa showing greater differences between its two hemispheres than did Ganymede (Pilcher et al. 1972). The strong albedo asymmetry of the leading and trailing hemispheres has been noted by many Earth-based investigators (see, e.g., Johnson 1970,1971; Morrison et al. 1974; McFadden et al. 1980). McFadden et al. (1980) also noted from reflectance measurements between 0.33 and 1.10 μm that a color asymmetry also exists such that Europa and Io are redder on their trailing hemispheres.

Other observations are also consistent with an interpretation of significant ice on Europa. For example, Fink and Larson (1975) estimated a brightness temperature of about 95 K at about 1.6 μm, lower than the actual temperature and in accord with a water adsorption band at that wavelength. At radar wavelengths, Europa appears anomalous (Ostro 1982) in that it possesses a strongly depolarized return and a very high radar albedo, the most geologically reasonable interpretation of which is a two-component icy regolith (e.g., clean ice/dirty ice or snow/solid ice interfaces; cf. Hagfors et al. 1985).

Recent spacecraft observations of the global spectral properties of Europa, including International Ultraviolet Explorer (IUE) and Voyager data, tend to support the interpretations of groundbased observations, although Voyager spectra show a steeper fall-off at short wavelengths than found for any of the groundbased data (Buratti 1983). Johnson et al. (1983), working with Voyager data, note that Europa's average global albedo is consistent with the pres-

TABLE II
Summary of Photometric Properties of Europa [a]

	Voyager (Clear Filter)	Earth-based (V Filter)
Phase coefficient, β (mag/deg)	0.0129 ± 0.0003	0.013 ± 0.001 [b]
Geometric albedo, p	$0.72 \pm$ (L) 0.62 ± 0.02 (T)	0.74 ± 0.03 (L) [c] 0.60 ± 0.03 (T)
Phase integral, q	1.09 ± 0.11	1.0 ± 0.1 [d]
Average normal reflectance, r_n (Voyager)	0.71 all bright plains 0.60 darker mottled terrain ($33° < \theta < 124°$) [e] 0.48 darker mottled terrain ($\theta \sim 340°$)	
Bond albedo, A_B	0.62 ± 0.14	0.58 ± 0.14 [d]

[a] Table from Buratti (1983).
[b] Average of leading (L) and trailing (T) hemisphere values (Millis and Thompson 1975).
[c] From Millis and Thompson (1975).
[d] Radiometric value from Morrison (1977).
[e] θ = subspacecraft longitude.

ence of water ice on the surface. Buratti's (1983) global and regional photometric parameters, derived from Voyager images, appear consistent with those derived from Earth-based observations (Table II).

Europa's surface can be divided into two major subdivisions on the basis of morphology, albedo, and characteristic structural style: dark mottled terrain and lineated plains of high and low albedo (Lucchitta and Soderblom 1982; Pieri and Hiller 1984; Pieri et al. 1985). Intricate local and global lineament systems of a variety of geometries are seen throughout all of these terrain units, as are dark irregular spots and patches. It appears, however, that Voyager filter colors of the mottled terrains and the bright and dark plains units are nearly indistinguishable as illustrated in Figs. 2 and 3 (Buratti and Golombek 1985). Furthermore, it appears that another group of features—brown spots, wedge-shaped lineaments, and gray bands—are spectrally distinct. This spectral dichotomy, which cuts across the terrain dichotomy, was suggested by Johnson et al. (1983) when they identified both high (HUVM) and low (LUVM) ultraviolet reflectance ratio subdivisions within the mottled and bright plains units (Fig. 4). Areas having Voyager orange albedos in the 0.65 to 0.70 range and ultraviolet albedos at about 0.27 were labeled HUVM, while areas having orange albedos of about 0.50 and ultraviolet albedos of about 0.19 to 0.15 were labeled LUVM (Fig. 4).

Johnson et al. (1983) attempted to synthesize the ultraviolet character of

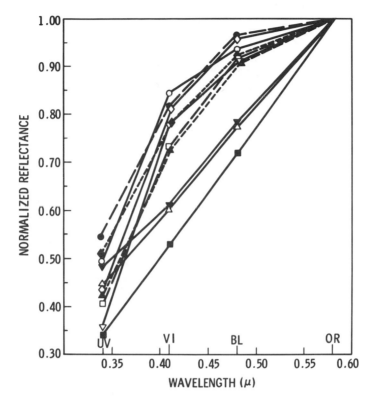

Fig. 2. Plots of normalized reflection as a function of wavelength for ten types of features on Europa as follow: ○, bright plains; ◆, plains; ▽, fractured plains; ◇, dark plains; ●, mottled gray terrain; □, mottled brown terrain; ▲, triple bands; ▼, brown spots; △, wedge-shaped bands; ■, gray bands. (Figure from Buratti and Golombek 1985.)

these areas and suggest that the best fit to the LUVM response is a 50%–50% mixture of HUVM and dark material similar to that found on Ganymede. They note that all Galilean satellites have dark materials roughly similar in spectral character and thus suggest the presence of a common silicate component (Fig. 5). Because of studies of possible sulfur implantation by migration of material from Io's orbit (see below and Sec. IV; also chapters by Veverka et al. and by Cheng et al.), a linear mixing of sulfur with HUVM was also tested by Johnson and coworkers, but this was shown not to match the LUVM spectrum. Laboratory studies of the mixing of different materials with different opacities (see, e.g., Johnson and Fanale 1973; Nash and Fanale 1977; Clark 1980; Gradie et al. 1980) have shown strong nonlinearities, the existence of which argues for a cautious attitude toward this type of "mixing model." Nevertheless, these results, combined with the more recent work of Buratti and Golombek (1985) suggest the presence of only two spectrally distinct

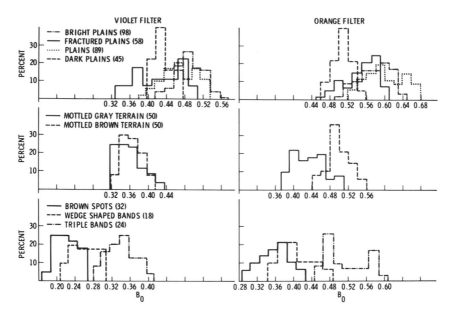

Fig. 3. Histograms of data groupings for Voyager violet and orange filters for nine Europa terrain types (figure from Buratti and Golombek 1985).

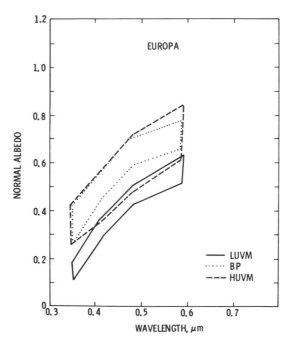

Fig. 4. Normal albedo for three different terrains on Europa (LUVM: low-ultraviolet materials; HUVM: high-ultraviolet materials; and BP: bright plains materials) shown as a function of wavelength (figure from T. V. Johnson et al. 1983).

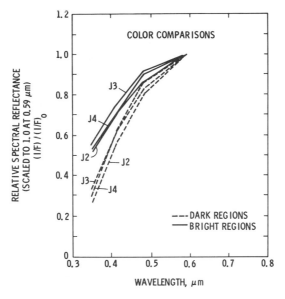

Fig. 5. Relative spectral reflectance (scaled to unity at 0.59 μm) for typical light (solid lines) and dark (dashed lines) regions of each of the icy satellites of Jupiter (figure from T. V. Johnson et al. 1983).

kinds of materials, despite a broad range of albedos. Lane et al. (1981) have dealt directly with the question of whether sulfur has been implanted on Europa's surface. They cite IUE observations of Europa that show a definite absorption band centered between 280 and 290 nm, that is not present in similar data for Ganymede. Lane and colleagues suggest that on Europa this spectral feature is caused by deeply implanted sulfur, swept from the magnetosphere, interacting with oxygen within the water ice lattice. They argue that the spectral signal does not result from the presence of sulfur dioxide frost, as no complementary infrared SO_2 band has been detected. Implantation of exogenic material as suggested by Lane et al. (1981) is also invoked by Nelson et al. (1980b) as a darkening mechanism to help explain the leading/trailing hemispheric asymmetry. Most recently, Ockert et al. (1985) report IUE observations of an SO_2 spectral feature on Europa's trailing side which further supports the earlier work. Darkening of large areas of Europa by exogenic implantation may compete with intrinsic or endogenic darkening of large tracts of the surfaces of Ganymede and Callisto. In the latter pair, groundbased photometric data can be explained by the location of large regions of old crustal material, as seen in Voyager data (see the chapter by McKinnon and Parmentier). On Europa, the lack of suitable high-resolution images makes the case for a similar situation less clear (Johnson et al. 1983), although low-resolution global coverage suggests that large areas of old, dark crust do not exist.

Spectral analyses of high-resolution Voyager images available for Europa have been sparse, as they have been for all of the Galilean satellites. Meier (1981) has employed a ternary color diagram for the spectral classification of particular features on Europa, using data derived from Voyager images exposed through violet, blue, and orange color filters. He finds two basic kinds of spectal signatures. One class of features (e.g., Type 1 lineaments; Pieri 1981) has a relatively narrow range of spectral variation and is thus clustered in his ternary diagram. The other class (represented by, e.g., Type 3 lineaments; Pieri 1981) has spectral signatures scattered across the ternary color field. Meier advances the theory that features of a more homogeneous composition have spectral signatures which clustered in his ternary color space, and that other features which show more spectral variation within the same geomorphic type are compositionally more diverse.

The most recent spectrophotometric analysis of high-resolution Voyager imaging data of Europa is presented by Buratti and Golombek (1985). They analyze over 500 data points using calibrated I/F data from the ultraviolet, violet, blue, and orange Voyager filters. Their data (Fig. 2) show only two major spectral groupings, which may themselves be members of a continuum (Fig. 3). Following the descriptive designations of Lucchitta and Soderblom (1982), these are: (1) light blue spectra—mottled brown and gray terrain, triple-band lineaments, bright plains, fractured plains, and dark plains; and (2) dark red spectra—brown spots, wedge-shaped lineaments, and gray bands. They hypothesize that the fractures in plains units have been infilled with dark material similar to that making up the brown spots and wedge-shaped bands tending to decrease the overall albedo of those units (e.g., mottled brown terrain and fractured plains). Triple bands appear lighter than brown spots and wedge-shaped bands due to their bright medial stripe. Detailed scans (Figs. 6 and 7) suggest that the central stripe of some triple bands is similar in spectral character to surrounding bright plains materials, and scans of some wedge-shaped lineaments show a brightening similar to that of the triple bands, possibly suggesting a similar emplacement mechanism.

Despite relatively low spatial resolution (≤ 4 km/line pair), Voyager multispectral observations of Europa provide clues to outstanding questions on the composition and history of its surface. It is clear from both Earth-based and spacecraft observations that the high global albedo of Europa is almost certainly due to the presence of substantial amounts of water ice across its surface. The precise texture and form of the icy subsurface is unknown, although Buratti (1983) suggests that photometry is consistent with low-density icy surface materials (see the chapter by Veverka et al.) and Squyres et al. (1983a) and Cook et al. (1982) have both suggested the presence of surface water frosts, possibly as a result of water volcanism.

It is significant that although the major morphology dichotomy exists be-

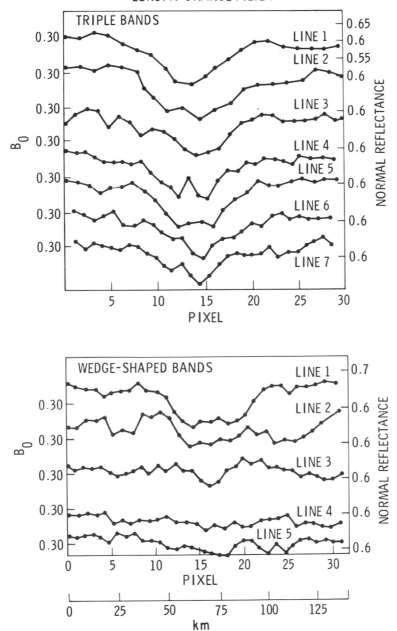

Fig. 6. Scans of B_0 and normalized reflectance across seven triple bands and five wedge-shaped lineaments (figure from Buratti and Golombek 1985). B_0 = average normal reflectance (r_n) for 0 phase angle.

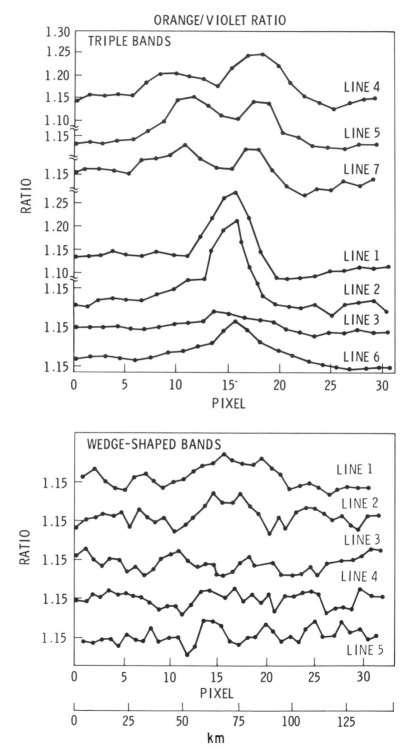

Fig. 7. Ratios for the data from Voyager orange and violet narrow-angle camera filters, some of which are shown above in Fig. 4.

tween plains and the mottled terrains, the major spectral dichotomy appears not between these units, but between the materials constituting the lineaments and dark spots, and everything else. Buratti and Golombek (1985) suggest that such a situation is consistent with overall albedo being controlled by the surface density of fractures. In their scenario, the mottled terrains, while having a spectral character identical to the plains units, appear darker because they have a higher abundance of fractures infilled with "dark red (possibly silicate-rich) material," but are of similar age. Lineaments, however, while they can be detected within the mottled terrains, can be seen only faintly, and may be less abundant there (Pieri 1981). In contrast, the highly fractured terrains around the anti-Jove point are clearly darkened by the increased surface density of lineaments, although they still appear smooth. However, another possibility is that the rough appearance of the mottled terrains is due to processes restricted in time or space, i.e., either they have endured a longer exposure to degradative processes, or such processes were more pronounced in those regions. Thus, the mottled terrains could be older than the plains units, and therefore were subjected to a longer period of impact or charged particle bombardment, which may have abetted excavative and/or sublimative processes. Alternatively, sublimative processes in these rough-appearing mottled regions may have been enhanced due to regionally higher heat flow.

II. SPACECRAFT OBSERVATIONS OF LANDFORMS

Landforms on Europa may be grouped by albedo, plan form, and relief (Figs. 8, 9 and 10). Linear, arcuate, and sinuous features are typically long, narrow, of relatively constant albedo and/or color, and with positive or little relief. Equidimensional forms are typically circular or irregular and may have positive or negative relief. Many features are identified exclusively on the basis of albedo, and have been treated as landforms even though they show no specific topographic expression. No quantitative information of surface topography is presently available for landforms on Europa.

There are four demonstrable landforms on Europa:

1. Craters (presumably of impact origin);
2. Pits and other irregular depressions (possibly endogenic);
3. Mounds, domes and other irregular features of positive relief (probably endogenic);
4. Linear, arcuate and sinuous ridges (endogenic).

In addition, two classes of features display ambiguous evidence of topographic expression: these include several large, arcuate lineations that may be shallow depressions, and several large, linear or arcuate bands that may be broad, low, asymmetric ridges.

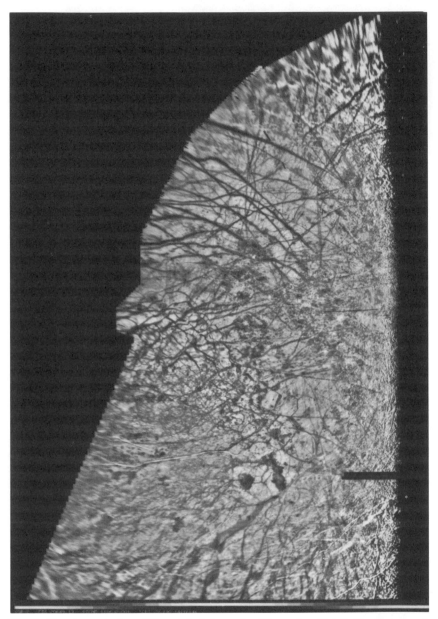

Fig. 8. Mercator projection of Voyager 2 photomosaic of Europa (see Map Section for an equal-area Lambert projection).

Fig. 9. Mercator airbrush map of equatorial Europa (see Map Section).

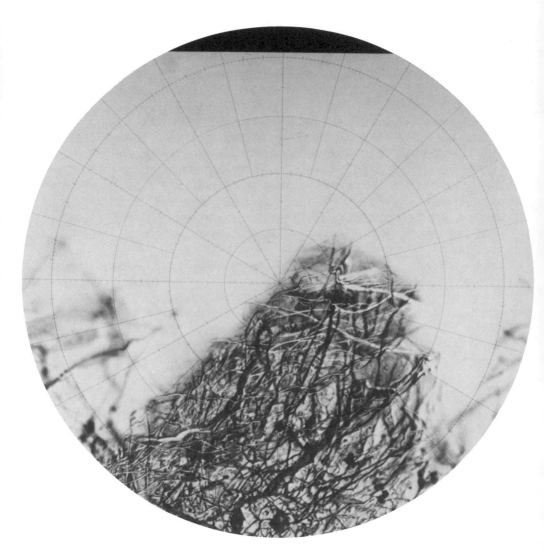

Fig. 10. Polar stereographic airbrush map of Europa's south pole (see Map Section).

Craters. Craters are not plentiful on Europa's surface. Observed crater densities are at least two orders of magnitude less than on Ganymede and Callisto (Lucchitta and Soderblom 1982; chapter by Chapman and McKinnon). These authors describe two main types of craters—palimpsests and fresh craters. Figures 11A and B show examples of each type. Cumulative size/frequency curves compiled by Lucchitta and Soderblom emphasize the paucity of impact features, but examination of near-terminator portions of Voyager images reveals some additional candidate impact structures (Fig. 11C). Many circular albedo features, possible craters after the interpretation of Lucchitta and Soderblom (1982, p. 541), complicate the story derived from crater statistical analyses. Image resolution is too poor to resolve clearly morphologic features useful in discriminating the effects of crater modification processes.

Impact craters are often used in attempts to determine relative ages of surface units. Europa's deficiency in impact craters larger than the limit of photographic resolution (about 1 km), even with large uncertainties, indicates that its surface is youthful relative to the surfaces of Ganymede and Callisto, and that some processes have been responsible for reshaping its surface in the relatively recent past. If such processes include general resurfacing, as is considered the case on Io, then the young age of the craters may also be extended to other surface features to be discussed in subsequent paragraphs. Thus, tectonic and geomorphic processes may have to be currently or recently active if the landforms they create are to survive the resurfacing process.

Pits. Irregular pits and other depressions are commonly visible along the terminator. They are characterized by relatively steep walls and no observable positive relief on their rims (Fig. 11D). Although some pits and depressions appear elongate, show directional trends, or are otherwise associated with other depressions, most are more randomly distributed.

Areas often termed mottled (see, e.g., Lucchitta and Soderblom 1982, p. 525) probably owe part of their apparent albedo and texture to pitting; the abundance of resolvable pits is much higher in these areas than on the plains units.

Irregular Features of Positive Relief. Mounds, domes and other irregular features of positive topographic expression are also visible along the terminator. They are less abundant than pits, but more abundant than clearly demonstrable impact craters. Often they occur at the plexus of several intersecting ridges (Fig. 11C) or astride a single ridge. Their relief, like that of the pits, is relatively low; they are seen clearly only within about 10° of the terminator. As much of the available high-resolution photographic coverage at low Sun elevation angles (high incidence angles) is within mottled terrain, most domes appear to occur in that terrain. This may well be an artifact created by limitations in the available data, as domes within plains units are more numer-

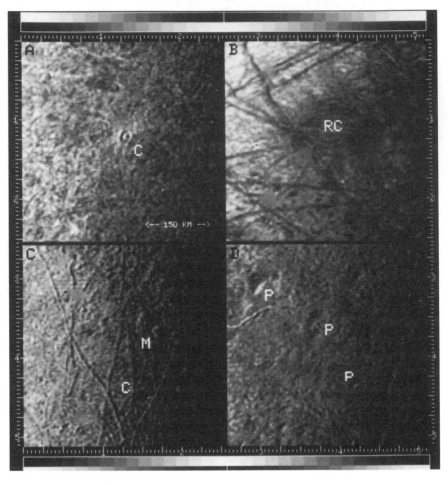

Fig. 11. Geomorphic features on Europa: (A) impact crater (c); (B) relic impact crater (rc); (C) crater (c) and mounds (m); (D) pits (p).

ous in higher-resolution images than in lower-resolution views of the same area. Domes range from ≤ 1 km to > 30 km in diameter.

Ridges.   Ridges are by far the most abundant and easily distinguishable landform on Europa. Although best seen along and parallel to the terminator, many ridges appear to transform at higher Sun elevation angles into thin, dark lineaments against light backgrounds. Their relief is probably small—no more than a few hundred meters—so their high visibility most likely results from rather abrupt and perhaps steep slopes. Other features, to be discussed below, may have greater relief, but appear less well defined, owing to ex-

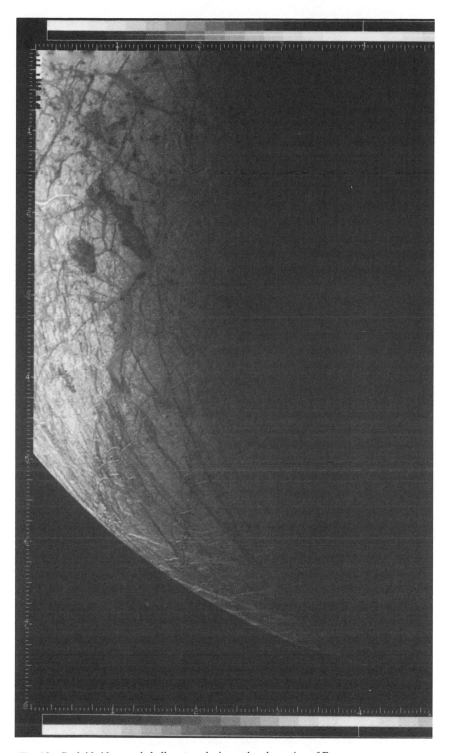

Fig. 12. Cycloid ridges and shallow troughs in south polar region of Europa.

tremely gentle slopes. Ridges are planimetrically complex. Most follow a generally arcuate path, with irregularities of direction superimposed on the general trend. Many ridges are segmented, with branches forming obtuse angles. A few have truly arcuate segments; these typically form a pattern resembling a cycloid (i.e., repetitively arcuate). Most of the cycloid ridges are found southward of 45° to 50°, although this apparent distribution may result from poor photographic coverage elsewhere. The wavelengths of the cycloids are generally 10 times their amplitudes; both measures are remarkably constant through numerous cycles. Cycloid ridges often cross one another; cycloids open southward north of 65° S latitude, but open northward south of that latitude. Figure 12 shows good examples of these types of ridges.

Only two other features on Europa display possible relief. One type appears to be associated with the largest, arcuate, dark lineaments, called linea. Figure 13 shows a significant change in the appearance of Minos Linea. This change, from a clearly visible dark curvilinear streak to a barely perceptible faint lineation, occurred during a period of about 2 hr, when viewing and illumination geometry angles changed by about 15°. Although it is possible that the change in brightness results from an unusual photometric function associated with the surface materials, it is more likely that it is a slope or topographic effect, suggesting a broad, gentle but possibly deep, linear depression. A similar depression is seen at more normal viewing and low Sun angles in Fig. 12. The second type of feature displaying apparent relief includes a pair of broad bands, arcuate in general plan form, but with cuspate margins, found near 55° S, 180° W and 60° S, 150° W (Figs. 8 and 12). Their variation in brightness as a function of azimuth suggests very gentle but definite slopes to the northwest.

The preceding discussion addresses only a small fraction of the features observed on the surface of Europa. The vast majority are seen solely as albedo markings that display apparent organization but no relief. These features have been the subject of several detailed accounts (Pieri 1981; Lucchitta and Soderblom 1982; Lucchitta et al. 1981; Helfenstein and Parmentier 1980); these accounts will not be repeated here. Typically, investigators have distinguished features first on the basis of pattern (wedge-shaped, linear, polygonal) and, second, on the basis of relative albedo (e.g., the term triple band used to denote a dark band with a central bright streak, or gray bands). In some cases, members of one class of linear feature merge into members of a different class. For example, bright forms streaking the centers of large, dark arcuate bands in many instances change into ridges as they approach the terminator. In other cases, the relationship between geographic coordinates and the direction of illumination suggests that such distinctions within the classification scheme are artifacts. For example, the northeast trend of ridges in the mid-latitude northern hemisphere, and the north-south trend of ridges in the equatorial terminator area may reflect the direction of illumination and not sub-Jovian stresses. The single most important result of the application of

Fig. 13. Changes seen in Minos Linea during a 2 hr period. The left image shows Minos Linea (B) at higher incidence and lower phase and viewing angles than seen in the right image. The prominent dark streak that makes up the linea can be seen to fade significantly with the relatively small shifts in viewing geometry.

these classification schemes is the strong geographical affinity of different types of lineaments. These patterns have been interpreted to arise from a variety of global or regional stresses. Theoretical considerations of these stresses will be discussed in Sec. III and in the chapter by Squyres and Croft.

Much of Europa's surface displays little or no organized pattern or form. These surfaces appear planar and differ on the basis of albedo and abundance of resolvable low-albedo spots. Lucchitta and Soderblom (1982) describe two types of terrains, high albedo plains and lower albedo, brown-hued mottled terrain. They interpret these units as primary or pristine crust (presumably ice or extremely ice-rich) and altered crust, respectively, modified principally by tectonism and intrusion of more silicate-rich materials from depth. Lucchitta and Soderblom argue against extrusive activity and for intrusion and/or tectonic deformation on the basis of the absence of clearly discernible superposition, transection, truncation, or embayment relationships between the two types of terrain. The poor resolution of the best photography make these arguments difficult to evaluate.

Europa has a very limited number of types of landforms at the resolution scales presently available to us. In a way, it resembles the Earth-based telescopic view of the Earth's Moon, which displays only two types of surfaces, mare and terra, and for the most part two topographically expressed landforms, impact craters and rilles. Detailed inspection of the Moon at many different Sun elevation angles permitted the addition of domes, mare ridges, and lava flows (in Mare Imbrium) to the list of surface features, but for the most part the Moon shows much less diversity in landform type than, for example, the Earth or Mars. In this comparison, then, Europa could be said to have had a geomorphic diversity not significantly different from the Moon's. However, the number of members of each type of landform *is* very different, and has severe implications concerning Europa's resurfacing history. As with the lunar maria, the details of such resurfacing are crucial to our understanding of the evolution of the satellite. The absence of images with the quality of Lunar Orbiter and Apollo photographs greatly limits our ability to evaluate Europa's recent geologic past.

III. GLOBAL TECTONICS

Europa possesses a global lineament system striking in its planet-wide extent as well as in its complexity and variety. A major unsolved question in the study of Europa involves these lineaments: how did this remarkably interconnected global system form and what does it tell us about the history of Europa's surface and interior? Additionally, we may ask whether or not the lineaments represent stresses related to the dynamical evolution of the entire Jovian satellite system. Much of the post-Voyager research about Europa concerns its lineament systems. In geologic experience, a lineament system is most often associated with tectonic processes, and thus most of the geologic speculation about the origin of the lineaments has centered on debate over the nature and origin of the stresses that may have created them (cf. chapter by Squyres and Croft).

Europa's lineament system has been studied by means of geographical pattern analysis and structural geologic analysis in the context of extrinsic and intrinsic stress. Extrinsic stress is defined here as stress imposed from without, such as tidal stress resulting from forced eccentricity, orbital recession, or differential (possibly asynchronous) rotation between crust and interior (Helfenstein and Parmentier 1980,1983,1985; Greenberg and Weidenschilling 1984). Intrinsic stress is defined here as stress imposed from within, such as expansion due to internal dehydration (Finnerty et al. 1981) or phase changes, or contraction due to cooling (Helfenstein and Parmentier 1983; chapter by Schubert et al.). Both extrinsic and intrinsic stress regimes can induce a range of crustal strain responses including faulting by compression, tension, and shear. Perhaps the most outstanding and difficult problem with regard to Europa's lineaments is to define the nature of the imposed stress, because combinations and permutations of various stress-process sequences appear to produce nonunique patterns at global scale (Helfenstein and Parmentier 1983,1985; Pieri 1981). However, even given the present state of knowledge about Europa, the situation is not totally hopeless. Despite the limited areal and resolution coverage of Voyager images ($<$ 25% of the surface at .4 km per line pair), certain observations can be made. First, there are discernible (e.g., albedo, morphological) appearance groupings among the lineaments (Pieri 1981; Lucchitta and Soderblom 1982); some of the lineaments exhibit a remarkable global continuity. Particularly at high ($\geq +45°$) latitudes, triple-band (Type 3; Pieri 1981) lineaments are remarkable in their continuous smoothly curvilinear paths, seemingly following great circles. Some of these lineaments show positive relief, while others manifest themselves as shallow depressions aligned with a dark lineament (see Sec. II). As continuous and long as the high-latitude lineaments are, lineaments seen at the anti-Jove position are quite short appearing at the limit of image resolution, giving that zone a fragmented appearance, as well as lowering the regional albedo. Lineaments which occur between the equator and $+45°$ latitude are intermediate between these two general types. In addition, other lineaments exist called flexi which have little albedo expression; these appear as narrow ridges and in plan view are broadly concentric to the anti-Jove position with a superimposed high spatial frequency cycloid pattern.

Thus, while Europa's lineaments as a group offer an astonishing complexity of morphology, albedo, size, and orientation, there are global and regional systematics in all of these parameters which may provide clues to the nature and timing of Europa's surface and interior stress regimes, as well as surface processes.

Several studies have tried to capitalize on grouping the lineaments by pattern and morphology. Pieri (1981) divided the lineaments into eight classes and analyzed the frequency distribution of sides of polygons formed by the intersections of the lineaments. He concluded that distinct pattern groupings of lineaments exist, some local and others global, and that the character of the groupings is probably related to the stress fields in which they formed. Pieri reports that patterns seen at the anti-Jove zone of Europa imply formation under isotropic (probably tensile) stress, while lineament patterns away from the anti-Jove point may imply a more directed stress field. Lucchitta and Soderblom (1982) and Lucchitta et al. (1981) developed a lineament classification scheme based on morphological and albedo characteristics of the lineaments, and attempted to put them into a geologic context. Their scheme included curvilinear dark bands, wedge-shaped dark bands, triple bands, and central stripes and ridges (in order of their inferred formation sequence). Schenk (1983) focused on the arrangements of wedge-shaped lineaments, inferring that they represent openings along fractures as the result of rotation and slippage of plates of ice. In evaluating these schemes, however, it remains difficult to confirm superposition and/or transectional relationships between lineaments or to infer their formation process by morphological analysis because of generally poor resolution of the image data.

Acknowledging the clear limitations of the data with which they were working, Helfenstein and Parmentier (1983) attempted to constrain models of Europa's interior and its response to tidal stress and stresses due to global volume changes. They have investigated two general cases, one assuming synchronous rotation of the body with respect to Jupiter (Helfenstein and Parmentier 1983) (Fig. 14) and more recently (Helfenstein and Parmentier 1985), the case of nonsynchronous rotation with respect to Jupiter (Fig. 15). They

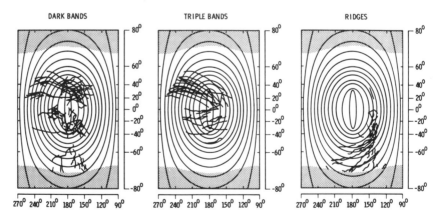

Fig. 14. Mercator projections of stress trajectories (curved, solid lines) parallel to the tangential horizontal stress σ_t arising from tidal distortion (orbital recession and/or orbital forced eccentricity) for the case of synchronous rotation (figure after Helfenstein and Parmentier 1983). Superimposed are the traces of dark bands, triple bands, and cycloid ridges (solid lines).

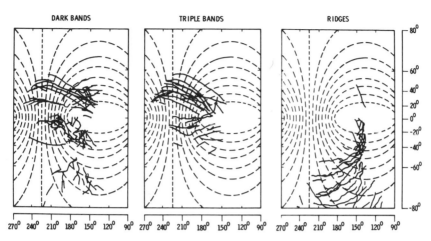

DARK BANDS   TRIPLE BANDS   RIDGES

Fig. 15. Stress trajectories (dashed) for the case of nonsynchronous rotation and lineaments (solid) digitized from Voyager 2 images. Stress trajectories are in the direction of the most tensile increment of horizontal surface stress due to an infinitesimal displacement of the tidal axis in the direction of increasing longitude. The instantaneous tidal axis is at 0° or 180° longitude and 0° latitude. The orientation of lineaments relative to stress trajectories provides a test of the hypothesis that lineaments are fractures produced by tidal stresses due to nonsynchronous rotation (figure after Helfenstein and Parmentier 1985).

condensed and combined the lineament classification schemes of Pieri (1981) and Lucchitta and Soderblom (1982), and use a scheme with three main groups—dark bands, triple bands, and cuspate ridges—and have attempted to determine whether the orientations and locations of these features are consistent with a tidal distortion model and volume change stresses. Their approach is to correlate surface tectonic patterns with the distribution of surface stresses that would arise from changes in the global figure of a planetary body. The amplitude of the distortion depends on the interior structure and mechanical properties of the deforming body, and the response may be short (elastic) or long (hydrostatic due to plastic or viscoelastic creep).

For the case of synchronous rotation there may be two end-member situations for Europa: Model 1, a thin shell *mechanically decoupled* from an elastic interior so that it deforms to nearly hydrostatic shape, corresponding to the thin icy shell–liquid mantle model of Squyres et al. (1983*a*) and; model 2, a thin shell *mechanically coupled* to an elastic interior corresponding to the model of Finnerty et al. (1981) that consists of a thin icy shell coupled to a rocky hydrated interior. This latter model predicts tectonic features produced by surface stresses on the basis of the state of stress and the failure criterion of brittle surface materials under synchronous rotation. It allows for (1) thrust and normal faulting (two horizontal principal stresses are of the same sign); for (2) tensional tectonics (tensional vs. strike-slip is controlled by ratio of tensile to shear strength and thus depends on depth); and for (3) change in the internal volume of the planet which causes additional isotropic stresses at the surface.

In the model assuming synchronous rotation, Helfenstein and Parmentier (1983) conclude that the orientations of the long, arcuate dark bands are consistent with formation stresses induced by an increase in semimajor axis of Europa's orbit (Fig. 14). Triple-band orientations suggest formation during planetary volume increase, apparently forming under conditions of lower mean normal stress than the dark bands unless both types formed concurrently (Lucchitta et al. 1981). Alternatively, dark bands and triple bands may have been formed at the same time, but stresses involved in triple-band formation may have been relatively shallow, consistent with a lower lithostatic stress (lower mean normal stress).

Helfenstein and Parmentier (1983) present several additional arguments specific to the question of the internal structure of Europa. One is that, for high latitudes, under stress associated with the increase in orbit semimajor axis, normal faults or tensional cracks will not occur for a constant volume model 1, but will occur for model 2. Patterns attributed to tensional stress appear to be strongly confined to the equatorial, anti-Jove zone. This observation seems to support the constant-volume, coupled-shell structure. However, Voyager imaging coverage of high northern and southern latitudes was restricted in area and only to very high phase angles, and arguments based on these data are inconclusive. Another tenet of the synchronous rotation model of Helfenstein and Parmentier is that the zone of polygonal fractures in the anti-Jove region, attributed to tensional stress, appears most likely to have formed *before* the accumulation of stresses associated with the increase in orbital semimajor axis. If these tensional features were formed after an increase in the size of the orbit, then orbital eccentricity variations greater than present values by an order of magnitude would be needed to induce such fracturing. Present orbital variations would produce tensional cracking at the anti-Jove region if the failure occurred before accumulation of the stress induced by the increase in the size of the orbit. In either case (i.e., tensional fracturing before or after stress accumulation), a small (0.01%) increase in planetary volume is required in the case of a completely solid (coupled-shell) Europa. For a decoupled-shell, no planetary volume increase is necessary. A volume decrease of about 0.2%, combined with stresses induced by the orbital semimajor axis increase, is needed by Helfenstein and Parmentier's model to explain the cuspate ridges found in Europa's more polar latitudes. Such a volume decrease can be achieved by a cooling of Europa by about 40° C in the case of the solid, coupled-shell model, or about 70° C in the case of the thin, decoupled-shell model.

In summary, Helfenstein and Parmentier (1983) conclude that the observed global fracture pattern on Europa, to the extent that it can be discerned in Voyager images, does not appear to be consistent with stresses due to tidal despinning; however, it apparently could be produced by a combination of stresses due to orbital recession, eccentricity, and internal contraction.

Stimulated by the work of Greenberg and Weidenschilling (1984), Helfenstein and Parmentier (1985) also examined patterns of triple bands, dark bands, and cycloid ridges produced under constant-volume, nonsynchronous rotation (Fig. 15). Implications of that model are several. Dark band and triple bands at high latitudes must have formed in a region where the principal stresses are of opposite sign (σ_1 and σ_2 are tensile and compressive, respectively) and thus must be tension cracks. Some of these features extend into the zone where both principal stresses are tensile. Here these lineaments indeed do appear shorter and appear to be arrayed polygonally as if the result of isotropic tension (Pieri 1981), also consistent with this model if the tidal axis had been shifted slightly to the east of its present position. Ridges have been interpreted as compressive (Finnerty et al. 1981) or as the result of the dike infilling of tensional fractures (Lucchitta and Soderblom 1982). When stress trajectories under this model are compared to ridge orientations, it is seen that they occur in zones of predicted tensional and compressive stress, and are generally consistent with a westward displacement of the tidal axis.

Since a moving tidal distortion would repeatedly fracture the surface and produce tensional features with a range of orientations within the same area, it is important to determine whether the observed lineament patterns reflect the accumulated scars of a finite amount of tidal axis motion, or only the last small bit. Ridges are probably relatively young because they still retain their topographic expression and thus Helfenstein and Parmentier suggest that under a nonsynchronous rotation scenario, they may be indicators of the most recent accumulated stress regime. Dark bands and triple bands, conversely, may be older; their eastward ends appear consistent with a more westward tidal axis orientation, and their westward extensions appear to be consistent with the current tidal axis orientation. They conclude that either nonsynchronous rotation or orbital recession and eccentricity combined with internal contraction would produce the observed lineament patterns.

The work by Helfenstein and Parmentier (1983,1985) is important because it presents specific statements about the orientations and sequence of formation of lineaments on Europa. With regard to the synchronous rotation model, if it is correct, for example, that the polygonal terrain in the anti-Jovian region formed before Europa's orbital distance from Jupiter increased, and if the essentially crater-free surface of Europa implies a young exposure age, then it would appear that Europa's orbital resonance with Io and the evolution of its orbit occurred relatively late in its history. If, however, crater obliteration occurs mainly by viscous relaxation, then the surface may be much older than its apparent exposure age, and tectonism may be much older as well. If, in addition, Europa's spin and orbit resonances with Jupiter and its sister satellites is primordial (Greenberg 1981), then the polygonal fracturing may be much older. With regard to the nonsynchronous vs. the synchronous rotation models, the Helfenstein and Parmentier work poses several interesting tests. For instance, if the long linear dark bands (peripheral to the anti-Jove zone) formed in response to orbital recession, then they must be interpreted as strike-slip faults, while they would be tension fractures under the nonsynchronous interpretation. While Schenk (1983) reports shear offset across these bands, the evidence is ambiguous. Also, tensional fracturing could be consistent with progressive shear offset as the tidal axis migrates westward. Whether ridges are tensional or compressional is of special interest because Helfenstein and Parmentier claim this would most clearly discriminate between the two models, as well as provide a test of the Finnerty et al. (1981) hypothesis. If viscous relaxation dissipates stress on about the same time scale as it accumulates owing to orbital recession, the nonsynchronous rotation model would clearly be favored, because it builds stress very rapidly, and with greater amplitude (Helfenstein and Parmentier 1985). Finally, if resurfacing mechanisms are rapid, as have been proposed by several authors (see, e.g., Helfenstein and Cook 1984; Squyres et al. 1983a; Eviatar et al. 1981), then the lineaments may be also relatively young. This situation would argue for a nonsynchronous rotation model, more than for orbital recession or planetary volume change.

As can be seen from the foregoing discussion, an understanding of the sequence of lineament formation and the cratering history of Europa are thus of critical importance to understanding its orbital evolution. Unfortunately, over 75% of Europa has not been seen at a resolution appropriate for the kind of analyses that could help in deciphering key lineament structural relationships. Models of lineament formation must for the present remain in the realm of informed speculation.

IV. SPUTTERING

The surfaces of the Galilean satellites interact directly with the intense Jovian magnetosphere (cf. chapter by Cheng et al.). Wolff and Mendis (1983) note that the plasma environment in the vicinity of the icy Galilean satellites is composed primarily of heavy ions, most likely a combination of various ionization states of oxygen and sulfur, as well as electrons, with velocities of the order of 100 to 200 km s^{-1}, as compared to satellite orbital velocities of about 10 km^{-1}. Thus, the trailing side of each satellite acts as a target for enhanced bombardment by particles moving in the Jovian magnetosphere. Wolff and Mendis (1983) predict substantial observable trailing hemisphere effects in the ultraviolet from ion implantation of sulfur and oxygen particles of energies < 10 keV (higher energy particles bury themselves too deeply and are more globally isotropic). They predict that such effects should be far greater on Europa than on either Ganymede or Callisto owing to the higher plasma densities at Europa. Any effect could be masked by micrometeorite gardening as well as by subliming of the water surface. Johnson and Pilcher

(1977) and McFadden et al. (1980) show that the trailing sides of Europa and Ganymede are darker at shorter wavelengths than their corresponding leading sides, with the effect being more pronounced on Europa. These observations are further corroborated by the ultraviolet observations reported by Nelson et al. (1980b) and Lane et al. (1981).

Intense heavy ion bombardment has been suggested not only as a mechanism of surface darkening but also as a mechanism for surface ablation. Lanzerotti et al. (1978) inferred a global ablation rate of about 1 vertical kilometer per Gyr owing to sputtering, while Johnson et al. (1981) calculate a rate of 450 meters per Gyr. Lane et al. (1981) suggest that sputtering is capable of removing on the order of 20 meters per Gyr. Eviatar et al. (1981) suggest a much more conservative rate of about 2 meters per 4.5 Gyr by ablation from incoming sulfur ejected from Io, resulting in a sulfur-enriched/water-depleted surface. This considerable range (2.5 orders of magnitude), based on essentially the same type of modeling, makes it difficult to evaluate the true potential of sputtering as a geomorphic agent.

V. COMPOSITIONAL MODELS AND THERMAL HISTORIES: HOW THICK IS THE ICE?

Pioneering models of the formation and thermal evolution of the Galilean satellites by Lewis (1971a,b), Consolmagno and Lewis (1978), and Fanale et al. (1977) concentrated on the implications of the satellite densities, sizes and distances from Jupiter. Most of these works highlight the role of water both as a density-depressing constituent and as a thermal energy transport medium. Later models examined the question of the existence of water or water ice layers covering Europa (see, e.g., Cassen et al. 1979a,1980b; Squyres et al. 1983a), although the conclusions of these studies seem disarmingly ephemeral. These earlier works are not reviewed here; see instead the chapter by Schubert et al. Rather, this section examines a single aspect of these models, namely, the nature of the outermost H_2O-rich layer, and the implications of past and potential observations in distinguishing between genetic models.

Evidence that the outer portion of Europa is covered by water ice seems irrefutable. As already noted, Europa's reflected light spectrum is dominated by water ice adsorptions (see the chapter by Clark et al.). Its high density (~ 3 g cm^{-3}) argues for a body composed mostly of silicates, although this value is sufficiently lower than most meteorites and Europa's Jovian neighbor, Io, to require a modest component ($\sim 5\%$ by mass; $\sim 15\%$ by volume) of a low-density volatile (presumably water) (cf. Fanale et al. 1977). The final state of this volatile component appears insensitive to its initial state. Accretion models that separate silicate phases from ice phases either nucleate ice on the Moon-sized silicate core or evolve through differentiation driven by radiogenic heat to a similar end. Models with silicates and volatiles in intimate mixture, as in hydrated mineral phases, similarly evolve to a dehydrated interior (amounting to about 60% in volume), covered by a layer of intensely hydrated materials perhaps a few hundred kilometers thick, covered in turn by a layer of ice (Ransford et al. 1981; Finnerty et al. 1981). Thus, in all cases, a final ice layer covers Europa, varying as a function of model only in thickness.

As the only major observable characteristic of this evolution is the ice surface, and as it is not diagnostic of any given model of Europa's thermal evolution, secondary clues, such as the pattern of lineaments, must be used to distinguish between alternative models. If these lineaments reflect stresses induced in the upper layers of Europa, it is unlikely that such stresses can be transmitted across a liquid interface (see the discussion by Squyres and Croft). Thus, models requiring a liquid layer beneath the ice crust and above the silicate core by necessity also require shallow stresses and strains. Conversely, those who believe in a thick ice crust must have stresses only within that portion of the crust that deforms as a brittle solid (~ 40 km; Finnerty et al. 1981), or else viscous deformation and dissipation would greatly reduce the surface expression of strain. Much of the interest in determining the origin of stresses in Europa's lithosphere stems from a desire to constrain this problem. As described in Sec. III, models for the origin of the stresses include tidally-induced differential torques (shallow), thermally-driven expansion/ contraction (deep seated), and thermal or compositional convection (depth unknown).

Other features may also place limits on the thickness of the ice crust. The bright ridges that occur in the centers of some dark lineaments may be extrusions or exposed intrusions of clean ice, following a kimberlite model of lineament formation favored by Finnerty et al. (1981). The pressure, temperature and stress relationships implied by such a model are very restrictive, but observations presently available cannot demonstrate its true applicability. Similarly, the tectonic, structural and thermal relationships permitting surficial extrusion of water (as noted by Wilson and Head [1984], among others) severely constrain models of Europa's near-surface stratigraphy. Unfortunately, lack of suitable, unambiguous observations of such phenomena again limits the evaluation of such models.

Age relationships are especially critical to deciphering the history of Europa. The paucity of craters indicates that processes capable of removing craters have been active in the recent past (i.e., < 500 Myr; Shoemaker and Wolff 1982). Unless a process can be found that removes craters but leaves lineaments and ridges intact, these latter features must also be relatively young. Thus, stresses necessary to create these lineaments must also be either recent, or at least capable of activating old patterns of structural weakness.

How, then, can the models of planetary composition, thermal evolution and planetary stress history be distinguished? Some tests may be forthcoming from higher-resolution Galileo images, recent or active water extrusions or intrusions with surface manifestations being obvious possibilities. Constraints on the relative timing of tectonic events may also permit some models to be distinguished. For example, current tidal stresses can be determined with much greater confidence than can be those active billions of years ago, but their effects are incalculable without knowledge or assumptions of the strength of Europa's lithospheric materials. Higher-resolution observations showing relationships favoring contemporary activity could thus be used to confirm such computations and thus provide support for determinations of crustal properties. However, without geophysical observations, most likely in the form of deviations of surface loads from gravitational isostasy, most observations will permit little advance beyond current speculations. Seismic experiments, both passive and active, will probably be required, as on the Earth and Moon, before a reasonably reliable interpretation of the bulk structure and composition of Europa can be obtained. Such observations are not likely to be made in the foreseeable future.

Acknowledgments. We thank P. Thomas for a critical review and J. A. Burns for his continuing encouragement. MCM acknowledges support of this research by a grant from NASA; DCP's work is one aspect of research conducted under a NASA grant to the Jet Propulsion Laboratory.

15. GANYMEDE AND CALLISTO

WILLIAM B. McKINNON
Washington University

and

E. M. PARMENTIER
Brown University

A fundamental issue for comparative planetology is that Ganymede and Callisto, although of similar size and bulk composition and sharing adjacent orbits, are very different. Moreover, volcanic, tectonic, impact and surface processes on these icy worlds challenge the understandings developed from the study of geology and geophysics of the terrestrial planets. We explore these differences and processes, attempting to quantify the dissimilarity, if any, between their internal structures that might explain the relatively vigorous history of geologic activity on Ganymede and the apparent lack of internal activity on Callisto. Groundbased visible, near-infrared, and active and passive microwave observations imply that, globally averaged, the optical surfaces and regoliths of both bodies are very ice-rich, and thus unlikely to be remnants of a crust created by homogeneous accretion. Ganymede's mantle is also relatively clean, and was the source region for the water, slush, or ice that resurfaced more than half the satellite. This resurfacing material is structurally confined in broad rifts or troughs, and is often (and characteristically) fractured to form grooved terrain. The detailed tectonic mechanism is unknown, but components of local, regional and global extension probably contribute. In contrast, the evidence for differentiation of Callisto's mantle is equivocal. It is usually argued that Callisto is undifferentiated, but models proposed so far to substantiate this all contain restrictive assumptions.

I. INTRODUCTION

In a cyan abyss who can detect the authority of the
Fashioner of Mutations—
Which condensed the frost and congealed the snow
to create her unique allure?
—from Moon, Hsü Yin

This fragment of a tenth-century Taoist poem reflects the then common belief that the Earth's Moon was both aqueous and crystalline, a lovely but frigid orb, sublimely yin. Despite the modern view of the Moon as an anhydrous, silicate planet, this fanciful conception of an ice world finds a ready home among the satellites of giant planets (see Schafer 1976).

Among the icy satellites, the large Galilean satellites Ganymede and Callisto occupy prominent positions. Ganymede is, of course, the largest satellite in the solar system, even bigger than the planet Mercury. Its surface appears both unusual and familiar; it is divided into two terrain types, dark and bright, has polar frost deposits, and is heavily tectonized and covered with abundant impact craters. Callisto is comparable in size to Mercury and, relative to Ganymede, appears more uniformly dark and very heavily cratered. No endogenic tectonic patterns have yet been identified on Callisto, but this fact does not make Callisto any less interesting, for it is precisely this lack of observed internal activity that resonates with studies of Ganymede. The basic dichotomy between Ganymede and Callisto, two worlds of similar size and bulk composition and sharing adjacent orbits, is a fundamental issue in comparative planetology.

The dichotomy is not one of mere appearance. Surface expression is related to internal structure and evolution, and these are ultimately related to the accretional state and conditions in the proto-Jovian nebula. As emphasized in the chapter by Stevenson et al., the issue of satellite accretion and its relationship to the general problem of planetary accretion remain among the most profound in planetary science. Coupled to questions of accretion are those of solar system bombardment history. The fact that Ganymede and Callisto have ancient surfaces of various ages is significant; they record the impact of various projectile populations such as comets and planetesimals from the zones of the terrestrial planets and Uranus and Neptune (cf. chapter by Chapman and McKinnon). The identification and time evolution of these bombardment sources constrain both satellite geologic history and planetary accretion dynamics.

Ganymede and Callisto are also important because many aspects of their geology are new; they extend the ruling paradigms of tectonism, volcanism and impact cratering developed from the study of the terrestrial planets. The structurally confined, resurfaced units on Ganymede, which when heavily tectonized are identified as grooved terrains, are unique in the solar system. The response of the lithospheres of both satellites to regional stress is remarkable in its geometric simplicity. These icy shells appear to have shorter tectonic memories than the lithospheres of the terrestrial planets; it is perhaps more appropriate to consider them as high-grade metamorphic terrains. Volcanism on Ganymede and Callisto is unique indeed, for the volcanic fluid is water. It is not yet clear whether this water is emplaced as liquid, slush or hot, "glacial" ice. Finally, impact craters exhibit novel morphologies: central pits instead of peaks, highly flattened forms known as palimpsests, and numerous rings of differing structure surrounding the largest impacts.

It is the intent of this chapter to focus on the Ganymede/Callisto dichotomy and its implications. In doing so, we attempt a synthesis of what we truly know of these fascinating objects and what we still need to learn. It is not our intent to duplicate recent comprehensive treatises, some of which may be found in Morrison (1982b), but to face squarely the serious unresolved questions concerning the regolith, surface geology, interior and tectonics of each of these worlds. We pose the following questions and then try to provide answers or at least to assess the state of our uncertainty.

1. What is the regolith made of? Can we estimate the relative fractions and distributions of ice and silicate phases?
2. How deep does resurfacing of the bright terrains and silicate contamination of the dark terrains extend? What is the source of contamination? What constraints are provided by studies of crater floors, ejecta blankets and rays?
3. How was the resurfacing accomplished? Did emplacement occur as a fluid or viscous liquid, or in the solid state?
4. To what extent are either Ganymede or Callisto differentiated? Can their divergent surface evolution be ascribed to different internal structures?
5. How do such planetary interiors transfer internal energy? Under what circumstances can water ice melt be produced?
6. What is the present-day heat flow and can it be estimated for previous epochs? How do estimates from crater relaxation studies compare to other methods?
7. How were the multiringed structures and furrows created? Are the furrows in the old cratered terrains of Ganymede due to large impacts?
8. How are the grooved and smooth terrains formed? Are volume changes and thermal stress sufficient? What are the roles of (a) primordial ocean closure, (b) very large impact, and (c) boundary layer instability? Why does plate tectonics not play a role?
9. Why are Ganymede and Callisto different?

II. THE PROTO-JOVIAN NEBULA

A simple, if not trivial, explanation for resurfacing and tectonism on Ganymede and its near total absence on Callisto is that Ganymede is at least partially differentiated while Callisto is undifferentiated (see, e.g., Schubert et al. 1981; see also the chapter by Schubert et al.). Certainly, the process of resurfacing may imply differentiation at some depth within a planet (and we shall argue in this chapter that this is the case for Ganymede), but the converse is not necessarily true. This explanation sidesteps the question of how a large, differentiated ice-rock satellite produces grooved terrain in the first place. Moreover, it does not seriously consider that such a satellite might have evolved to a geologically inactive state before its cratering record (observed today) was established.

To determine whether such a scenario is possible or likely, we must return to the accretional state. Ganymede and Callisto were born in a nebula of gas and dust which, according to best current understanding, formed when a portion of the protosolar nebula collapsed hydrodynamically onto a massive proto-Jovian core (on the order of 10 $M_\oplus$) (Mizuno 1980; Stevenson 1982b; chapter by Stevenson et al.). Net prograde angular momentum and viscous dissipation allowed the gas and dust to flatten into a disk. Temperatures in at least the outer portion of this proto-Jovian nebula were low enough to permit the condensation of water ice which, along with silicate, was then available for planetesimal accretion and satellite growth (Bodenheimer et al. 1980; Prinn and Fegley 1981; and references therein).

Because of the short orbital periods of material circulating about Jupiter, satellite formation is extremely rapid in gas-free accretion models (cf. chapter by Safronov et al.); accretion times are generally $< 10^5$ yr for Ganymede and Callisto (Coradini et al. 1982b; Shoberg 1982; Weidenschilling 1982). Thus, it is extremely difficult to maintain that either satellite could escape accretional melting of its outer layers. Consider the maximum temperature rise possible, given by the ratio of the gravitational binding energy of a homogeneous satellite to its heat capacity, $\Delta T = 3GM/5RC_p$, where C_p is the specific heat, and G, M and R have their usual meanings. With $C_p = 1.4 \times 10^7$ erg $g^{-1} K^{-1}$, appropriate to a $\sim 50/50$ ice/rock composition by mass, and values for GM and R from Table I, $\Delta T \sim 1300$ to 1600 K for Callisto and Ganymede (cf. Table II in chapter by Schubert et al.). Actually, this amount of accretional energy is sufficient to vaporize the ice in such an ice-rock satellite. A more realistic scenario accounts for the deposition of accretional energy as the satellite grows, with the natural outcome being a preferential heating of the satellite's outer layers. It is common to characterize the fraction of accretional energy retained at any radius as a factor h. Schubert et al. (1981) show that for accretional melting to be avoided for either of these two bodies requires $h \lesssim 0.05$. Values of $h > 0.1$ are usually considered appropriate (Schubert et al. 1981), and an explicit calculation of heat deposition and radiation during Ganymede and Callisto accretion by Coradini et al. (1982b) finds $h \sim 0.4$. We note that this is consistent with the time scale for radiating the total accretional energy to space, $\sim 2 \times 10^5$ (300 K/T)^4 yr for a 2500 km radius surface at temperature T.

TABLE I
Properties of Ganymede and Callisto [a]

	Ganymede	Callisto
Radius, R	2631 ($\pm$10) km	2400 ($\pm$10) km
Mass, M	1490 ($\pm$6) $\times 10^{23}$ g	1075 ($\pm$4) $\times 10^{23}$ g
GM	9940 ($\pm$100) $\times 10^{15}$ cm^3 s^{-2}	7100 ($\pm$215) $\times 10^{15}$ cm^3 s^{-2}
Surface gravity	144 ($\pm$2) cm s^{-2}	124 ($\pm$4) cm s^{-2}
Mean density	1.93 g cm^{-3}	1.83 g cm^{-3}
Percent silicate by mass	49–59	47–56
Distance from Jupiter	1.070 $\times 10^6$ km = 15.1 R$_J$	1.880 $\times 10^6$ km = 26.6 R$_J$
from Sun	5.203 AU	5.203 AU
Orbit period	7.155 day	16.689 day
around Sun	11.86 yr	11.86 yr
Eccentricity	0.001 (variable)	0.007
Inclination	0°.183	0°.253
Obliquity (average)	3°.08	3°.08
Visual albedo (p_V)	0.43	0.19
Bond albedo	0.35	0.11
Solar flux	3.7% $\times$ Earth	3.7% $\times$ Earth
Subsolar temperature	156 K	168 K
Equatorial subsurface temperature	117 K	126 K

[a] Sources: Morrison (1977,1982a); Mueller and McKinnon (1986); Squyres and Veverka (1981,1982); see introductory chapter by Burns.

If gas-free accretion is a valid model for their growth, Ganymede and Callisto should have melted to a depth of several hundred kilometers. Melting extends deeper than 1000 km in the models of Coradini et al. (1982*b*). Ganymede suffers greater melting due to its shorter formation time and larger mass, but it is only a matter of degree. There is no qualitative distinction between Ganymede and Callisto. The only way to make the distinction qualitative, i.e., to create an undifferentiated Callisto, is to require the accreting planetesimals to be so small that they cannot bury their accretional heat. For the proper choice of planetesimal size ($\lesssim$ 1 m across, roughly the depth that is in conductive communication with the surface at any time during accretion), Ganymede is minimally melted and Callisto unmelted. The planetesimals must not be too small, however, or *both* satellites are left undifferentiated. This theme will reappear several times in this discussion. Most scenarios do not naturally lead to divergent internal structures, but in nearly all there exists a special set of circumstances or restrictive assumptions that may. Clearly, the above range in planetesimal masses is restrictive, and wholly unrealistic for the accretion of 10^{26} g objects.

The accretional state of a deeply melted Ganymede or Callisto is pictured in Fig. 1. An undifferentiated ice-rock core is overlain by a differentiated mantle. Hydrated silicates form a lower mantle, having precipitated

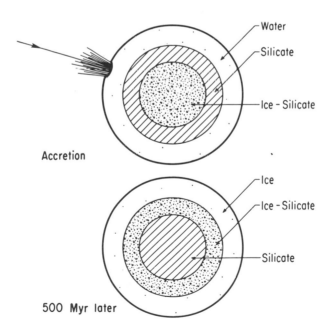

Fig. 1. Schematic internal structures for either Ganymede or Callisto during the accretional epoch, $\sim$ 4.5 Gyr ago (top), and approximately 500 Myr later (bottom), the formation epoch of the preserved cratering record according to conventional wisdom.

from a convecting, water upper mantle. This deep primordial ocean contains small particles in suspension and salts as well as possibly ammonia in solution (Lunine and Stevenson 1982a). The three-layer accretional state is a general representation; end members are completely differentiated and undifferentiated.

The preceding discussion has emphasized gas-free accretion, partly for illustrative purposes. Accretion time scales are small compared to nebular cooling times and to the overall disk lifetime (Prinn and Fegley 1981; Weidenschilling 1982; Pollack et al. 1976; see also Pollack and Consolmagno 1984), so the presence of gas cannot be realistically ignored. A first-order effect is probably to decrease accretion times further, but the effects on satellite composition and structure are potentially more profound. Lunine and Stevenson (1982a) showed that both Ganymede and Callisto develop thick water-saturated hydrogen-helium atmospheres as they accrete within the proto-Jovian nebula. The surface temperature in this case is not regulated by radiation to space but by convective transport to the nebula. It was originally hoped that the convective adiabat in the cooler region of Callisto's accretion zone might lead to an atmospheric buffering of Callisto's surface temperature below the melting point of ice throughout its accretion, but detailed analysis showed that both Callisto and Ganymede would be deeply melted. As in the gas-free case, Callisto is less differentiated. Lunine and Stevenson argue that aerodynamic effects may preferentially select for planetesimals in the 10 cm to 10 m radius range. In this case they are small enough to deposit their kinetic energy in the atmosphere, but large enough so that they do not substantially melt and disseminate in transit. Thus, they land cold (i.e., at the nebular ambient temperature). Lunine and Stevenson further maintain that disequilibrium between the hot (accretionally heated) atmosphere and the cold surface is sustainable, and that in this case it *is* possible to create a deeply differentiated Ganymede and an undifferentiated Callisto. The aerodynamic and disequilibrium surface models are based on restrictive assumptions, however, and we shall argue below (see end of Sec. VII) that the assumptions are, as in the gas-free case, not entirely realistic. Lunine and Stevenson in fact state as their best estimate that Callisto will be differentiated in its outer 200 km or so.

We note at this point that the problems of satellite accretion are far from solved (cf. chapters by Stevenson et al. and by Safronov et al.), so the basic assumptions of the above discussion are open to question. The present positions of the Galilean satellites may be far from the sites of their formation, due to the complex transfer of angular momentum between the nebula and the satellites (Lunine and Stevenson 1982a), so constructing nebular models on the basis of present-day satellite composition and position carries an element of risk. Shoemaker (1984) has suggested that all the Galilean satellites started ice-rich, but that very large impacts have preferentially stripped away primordial, icy mantles from Io and Europa. Io and Europa are the most affected, because the gravitational focusing of heliocentric projectiles towards Jupiter increases both projectile flux and projectile velocities (Shoemaker and Wolfe 1982). If this were true, it would require wholesale revision of our concept of the proto-Jovian nebula.

Cassen et al. (1982) remarked that the approximately < 50/50 ice/silicate mass ratios they calculated for Ganymede and Callisto were substantially less than an ice-rich solar composition. A mass balance calculation from the abundances tabulated in Anders and Ebihara (1982) gives a solar ratio of ~ 60/40. Perhaps Shoemaker's suggestion is linked to this preferential loss of water ice. Ahrens and O'Keefe (1984) proposed that selective impact volatilization (in a gas-free scenario) is responsible.

So far, we have implicitly assumed that accretion is homogeneous, and that silicate-rich cores do not form directly. Weidenschilling (1982) modeled the accumulation of different generations of rock- and ice-rich planetesimals, but Lunine and Stevenson feel that several aspects of planetesimal formation mitigate against this possibility. This and other aspects of satellite accretion are discussed more fully in the chapter by Stevenson et al. Our viewpoint on the nebular models or scenarios discussed here is that none is necessarily correct. We find, however, most of the detailed arguments of Lunine and Stevenson (1982a) to be physically powerful and sensible. The alternatives mentioned serve as cautions or caveats, to be seriously considered if the evidence warrants.

The general accretional structure in Fig. 1 is appropriate in both the gas-free and gas-rich cases. What, then, can we expect today? Internal evolution is treated in more detail in Sec. VII and in the chapter by Schubert et al., but some first-order statements are needed to guide the present discussion. The post-accretion water upper mantle will freeze; due to the efficient heat transfer by solid-state convection in water ice, freezing should be essentially complete by 4 Gyr ago (see, e.g., Reynolds and Cassen 1979). The intermediate silicate shell is manifestly unstable. While not yet modeled in detail, it seems plausible that some combination of processes (such as Rayleigh-Taylor instabilities, ice melting at the shell-core boundary, hydrofracturing, stoping and subduction) will cause the silicates to form a core on a similar time scale. In most cases the new intermediate ice-rock layer (Fig. 1) will not be susceptible to further melting and differentiation. (Its viscosity is dominated by water ice and thus solid-state convection in this layer is also efficient.) The observational question becomes to what degree are Ganymede and Callisto differentiated and how can this be determined. The following sections examine the evidence, starting at the surface and moving inward.

III. THE OPTICAL SURFACE

Voyager images of Ganymede and Callisto (e.g., Figs. 2 and 3), which are often highly processed, guide the perception, or misperception, of these objects. The chapters by Veverka et al. and by Clark et al. describe the physical properties of the optical surface and the composition of the surface layers, respectively. Portions of Ganymede and nearly all of Callisto are termed dark, but it is useful to remember that the dark terrains of both Ganymede and Callisto are brighter than the lunar highlands; normal visual albedos are ~ 0.3, ~ 0.2, and 0.11, respectively (Wildey 1977; Squyres and Veverka 1981; Johnson et al. 1983). This immediately indicates that icy compositions may be involved. The albedos of bright terrain on Ganymede and bright crater materials on Ganymede and Callisto are more reflective still, ~ 0.4, ~ 0.6 and ~ 0.4, respectively (Squyres and Veverka 1981; Johnson et al. 1983), making this inference a necessity. Figure 4, adapted from Johnson et al. (1983), illustrates the range in albedo values determined for various units by Voyager (cf. Fig. 7 in Squyres and Veverka [1981]). Ice as a major bulk component has been inferred for decades based on mean density (Table I), an argument that goes back to Kuiper (1952).

The presence of water ice on Ganymede and Callisto was confirmed by the detection of near-infrared absorption bands (Johnson and Pilcher 1977, and references therein; the chapter by Clark et al.). Furthermore, given sufficiently high resolution spectroscopy of these bands, the composition and state of the optical surface can be modeled. The optical surface is taken here to mean the portion of the surface that interacts with ultraviolet, visible and near-infrared solar radiation. Water ice has a fundamental absorption band, due to a combination of stretching and bending modes of the O—H covalent bond, near $3.0~\mu$m; overtones in the infrared occur at 2.02, 1.65, 1.55, 1.04, 0.9 and 0.81 μm (Clark 1981a). As demonstrated in the chapter by Clark et al. measurements of continuum levels and band positions, depths and shapes allow an estimate of ice-silicate mixing ratios and grain sizes (Clark 1981b, 1982; Clark and Lucey 1984). The higher and weaker overtones are more diagnostic of nonice contamination in ice-rich surfaces, while the strong fundamental is best for detecting small amounts of ice.

A primary result of laboratory studies is that small volumetric amounts of dark contaminant, if widely dispersed in ice grains so as to dominate scattering, can drastically lower spectral reflectance (Clark 1981b, 1982). On this basis and under the assumption of intimate mixtures, Clark (1980,1982) determined the globally averaged optical surfaces of Ganymede and Callisto to be ~ 90 wt.% and 30 to 90 wt.% water ice, respectively. The amount of bound water was found to be small for both ($< 5 \pm 5$ wt.%). More recent work limits the amount of dark grains in Ganymede's surface to $\lesssim 3$ wt.% and that in Callisto's to an astonishing $\lesssim 7$ to 8 wt.% (P. Lucey, personal communication, 1984; cf. Clark and Lucey 1984). The inference here is that despite appearances, the optical surfaces of these two satellites are close to pure water ice in composition, with darker units being more contaminated.

This conclusion, while not necessarily implying either satellite is significantly differentiated, is an argument against Callisto's surface being a primor-

Fig. 2. Global and regional views of Ganymede taken by Voyager 2 at ranges of 1,200,000 and 86,000 km, respectively; north is up. Ganymede is divided into dark, ancient, heavily cratered terrains and brighter, younger, grooved and smooth terrains. Also visible are the polar frost deposits, and young, bright rim and ray craters. Circular patches and parallel lineaments within the large region of dark terrain, Galileo Regio, are old crater palimpsests and elements of a rimmed furrow system, respectively. The lane of grooved terrain in the foreground image is Uruk Sulcus. (Frames FDS 20608.11 [background] and 20638.32 [foreground].) See also the Map Section and Color Plate 3d.

Fig. 3. Global and regional views of Callisto taken by Voyagers 2 and 1 at ranges of 1,100,000 and 320,000 km, respectively; north is ~ up. Callisto is heavily cratered. The bright region in the background image is the Asgard double multiringed system, and the northeast portion of the great Valhalla multiringed system is seen in the foreground image. (Frames FDS 20583.17 [background] and 16419.00 [foreground].) See also the Map Section and Color Plate 3e.

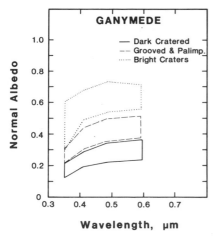 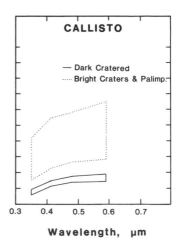

Wavelength, μm   Wavelength, μm

Fig. 4. Comparisons of normal spectral albedos for the major terrain types on Ganymede and Callisto. This figure is from Johnson et al. (1983) with correction factors of 1.30, 1.12, 1.0 and 0.9 for the ultraviolet, violet, blue and orange filters, respectively; these corrections to the calibration of Danielson et al. (1981) are based on comparisons between Voyager and ground-based data for Europa, Dione, Rhea, and Tethys (B. Buratti, personal communication, 1984). Other data show that a degree of overlap exists between the albedos of dark and bright (grooved) terrain (Squyres and Veverka 1981).

dial ice-rock mixture. A few caveats are in order, however. The modeling of Clark and coworkers, as stated above, assumes an intimate mixture of ice and contaminating particles. Any areal segregation of ice and dark material could modify the results. There is, of course, evidence for segregation at scales resolved by Voyager, e.g., the terrain dichotomy (and polar frost) on Ganymede, ice patches on north-facing slopes on Callisto (Spencer and Maloney 1984), and crater deposits on both satellites. Segregation below the limit of Voyager resolution may exist as well. From Voyager 2 IRIS (infrared interferometer spectrometer) data, Spencer (1983) presented evidence for large (up to 40 K) thermal contrasts on Ganymede, which he interpreted in terms of ice-rich and ice-poor patches. More interesting perhaps is the evidence for ice-free units on Callisto (Spencer 1984); Callisto's 3 μm albedo is 4% (L. Lebofsky and M. Feierberg, personal communication, 1984), and even a 1 μm ice film should cause this band to saturate. Accordingly, Spencer argues for a ice-poor surface. At this time no one has attempted to model the ground-based visible and near-infrared spectrum in terms of the major albedo units identified by Voyager (i.e., bright terrain, dark terrain, bright craters, and bright polar frost). Ultimately the NIMS (near-infrared mapping spectrometer) experiment on the Galileo Orbiter will be needed to resolve these issues.

We accept, at least for Ganymede, the interpretation of an optical surface that is nearly pure water ice. What then is the nature and source of the contaminant? Part of the answer lies in the colors of the surface materials (see the chapter by Veverka et al.). All albedo units on Ganymede (save the brightest craters) are spectrally red, meaning here to have an increasing reflectance from the ultraviolet to orange wavelengths as seen by Voyager, and the photometrically mature units (i.e., those exclusive of bright rim and ray craters) are the reddest (Squyres and Veverka 1982; Johnson et al. 1983). In addition, dark terrain materials appear somewhat redder than bright terrains (Johnson et al. 1983; Schenk and McKinnon 1985).

There also exist spectral differences between the leading and trailing hemispheres of orbital motion (chapters by Clark et al., Veverka et al. and Cheng et al.). On the trailing hemisphere, infrared band depths are deeper (Clark 1980; Clark et al. 1983a), visible albedos are lower, and ultraviolet albedos are especially low (Johnson and Pilcher 1977; Morrison and Morrison 1977; Johnson et al. 1983; Nelson et al. 1983; and references therein). These asymmetries, which also affect Europa but more strongly, are plausibly ascribed to a combination of enhanced magnetospheric sputtering on the trailing side (see the chapter by Cheng et al.), which preferentially destroys small ice grains and thus lowers visual reflectance and increases infrared band depth (Clark et al. 1983a), and enhanced implantation of magnetospheric sulfur on the trailing side, which could account for the lower ultraviolet spectral reflectance (Morrison and Morrison 1977; Lane et al. 1981; chapter by Malin and Pieri). That the ultraviolet asymmetry is global in nature and not sensitive to the underlying geology is further support for this interpretation (Nelson et al. 1983). It will be an interesting test for NIMS to determine if the infrared asymmetry behaves in the same manner.

On this basis, we conclude that ice on Ganymede's surface, even pure ice, will become reddened as it comes into photometric equilibrium with the space environment (referring to the steady-state appearance of an optical surface after prolonged exposure to solar radiation, magnetospheric ions, and micrometeoroid bombardment; that a steady state exists is a conjecture for the surfaces of Ganymede and other icy satellites, but its plausibility lies in analogy with the known optical maturation of the lunar regolith). This reddening is relevant because bright terrain material must be nearly pure ice in composition; i.e., if the globally averaged amount of dark grains is 3%, then the amount in the optical surface of the bright terrains must be smaller still. Adding a low-albedo contaminant to produce dark terrain and not flattening the spectra requires that the contaminant be reddish as well as low-albedo. It is tempting, considering the low overall concentrations implied by the infrared data, to assign an exogenic origin to this dark, reddish contaminant. We shall return to this point in Sec. V.

The correlation of low albedo with reddish color holds true for Callisto's surface as well, both for units on Callisto itself and in comparison with units on Ganymede (Johnson et al. 1983). Interestingly, Clark et al. (1983a) find little hemispheric asymmetry in infrared band depth for Callisto. This is consistent with the extremely low level of ion bombardment at Callisto compared to that at Ganymede (Sieveka and Johnson 1982), if sputtering is the cause of such asymmetries. In addition, Wolff and Mendis (1983) have proposed that an extremely tenuous O_2 atmosphere, derived from sublimation or sputtering, stands off most of the flux of magnetospheric particles, such as it is. No such atmosphere has been detected yet, however. Callisto does have a leading/trailing hemisphere asymmetry in ultraviolet and visible reflectance, but it is in a sense opposite to that of Ganymede (and Europa) (Johnson and Pilcher 1977; Morrison and Morrison 1977; Nelson et al. 1983; chapter by Veverka et al.). The large, bright, multiringed impact features are on the slightly darker, leading hemisphere, making this all the more intriguing. As to an explanation, if magnetospheric interactions are deemed insignificant, then one is left by default with the asymmetric effects of micrometeoroid bombardment (i.e., gardening and accumulation). This is enhanced on the leading side and could be responsible for the extra darkening. It may also be responsible for the well-known hemispheric asymmetry in visual polarization properties (Dollfus 1975; Mandeville et al. 1980; see also Veverka 1977b). Dollfus and his colleagues interpreted pre-Voyager measurements in terms of purely silicate surfaces; Pang et al. (1983) have reinterpreted the data in terms of differing microstructures in an ice-rock optical surface, but the physical explanation remains obscure.

Redistribution of water molecules by sputtering, coupled with trapping at cold, high latitudes, has been invoked as the cause of the polar frost caps, or shrouds, on Ganymede (Sieveka and Johnson 1982). Purves and Pilcher (1980) showed that sublimation is an inadequate redistribution mechanism as mean free paths are too short. Shaya and Pilcher (1984) went on to describe a model for deposition of a global ice layer (due to boiling of water-rich volcanics) that ultimately results in an enhancement in the iciness of the polar regolith (see Sec. IV below); impact is another possible redistribution mechanism. Johnson (1985) takes a different approach, arguing that frostiness is primarily a result of the unique microstructure created by sputtering. This microstructure is not stable in regions such as the equatorial latitudes of Ganymede, where sublimation dominates reprocessing of the optical surface. The lack of polar frost on Callisto could then be due to the global dominance of sublimation over (yet again) very weak sputtering.

We need to understand more fully the exogenic processes that affect a satellite's optical surface, i.e., sublimation, sputtering and micrometeoroid gardening, as well as the moderating effects of tenuous atmospheres. Until then, generalizations about Ganymede's or Callisto's structure or evolution based on, say, the presence or absence of polar frost, seem premature. The preponderance of evidence does, however, favor very ice-rich optical surfaces for much of Callisto and nearly all of Ganymede. These optical surfaces are only on the order of a centimeter thick, however (the depth being determined by the competing effects of ice transparency, grain size and contaminant distribution). Since we can be misled by the optical surface, it is important to ask if these compositional inferences extend deeper, into the regolith.

IV. REGOLITH

Because the surfaces of Ganymede and Callisto are continually exposed to meteoroidal bombardment, it is logical to assume that an impact-generated soil, or regolith, exists. Models of satellite regoliths are presented in the chapter by Veverka et al. Most of our knowledge of regolith processes (and all of our *in situ* knowledge) comes from study of the lunar regolith; but it is unclear how applicable the lunar model is to icy satellites. Fortunately, the structure and composition of the Ganymede and Callisto regoliths can be probed by Earth-based radar and passive microwave measurements. In fact, the microwave window is potentially of tremendous value, because water ice is very transparent at these wavelengths and becomes more transparent with decreasing temperature. The extinction index (i.e., the imaginary part of the complex index of refaction), n_{im}, of solid water ice at the surface temperatures of these satellites varies from $\sim 10^{-3}$ to $\sim 10^{-5}$ over the 1 mm to 1 m wavelength (λ) range (Warren 1984); this translates into absorption lengths, given by $\lambda/4\pi n_{im}$, of hundreds to thousands of wavelengths. Thus, if the regolith is ice rich, radar signals and microwave emission can probe down several meters, if not hundreds of meters (even in the absence of a regolith's natural porosity). This behavior is in contrast with solid silicates, whose absorption lengths are on the order of a few wavelengths (Campbell and Ulrichs 1969). Thus, not only are radar and microwaves deep probes, but they are potentially extremely sensitive to variations in the ice/silicate mix.

The radar data taken to date show very high geometric albedos compared with the terrestrial planets (~ 0.3 to 0.4, ~ 0.16 and 0.025 for Ganymede, Callisto and the terrestrial planets, respectively [Ostro 1982]). They also show inverted circular polarization ratios, i.e., for a circularly polarized transmission, the handedness of the returned signal is more in the sense of that transmitted. Usually, the handedness is reversed upon reflection or depolarized by scattering (see, e.g., Ostro 1982). In addition, linearly polarized signals are returned essentially depolarized. Detailed models have been proposed to explain these observations; they invoke multiple subsurface reflections from either randomly oriented ice-vacuum interfaces (Goldstein and Green 1980) or planar veins of contrasting electromagnetic properties (Ostro 1982). A more recent proposal involves not reflection, but a gentle refraction of the radar beam (due to inhomogeneous dielectric structure) back through the surface (Hagfors et al. 1985). See also Eshleman (1986).

The essence of all these models is that very little signal is returned by direct surface reflection. Rather, multiple reflections or refraction at depth are necessary. Furthermore, they all require a very long pathlength to work, which would seem to rule out anything approaching a 50/50 silicate/ice mix-

ture, even for Callisto. Both geometric albedo and the anomalous circular polarization ratio decrease for Ganymede viewing geometries that include large
units of dark terrain and more so for Callisto (Ostro 1982). Hence, there is a
correlation between surface albedo and radar properties, and the contaminant
inferences from the last section may apply to the regolith as a whole. Unfortunately, none of the models so far investigated has attempted to constrain the
ice/silicate ratio, so these conclusions are not definitive (especially for Callisto). That the deep regoliths of Ganymede and Callisto are not silicate rich
(i.e., have ice/silicate ratios that approach the satellite average) cannot be
conclusively rejected.

Microwave measurements in the centimeter range, especially high-
precision VLA (Very Large Array) measurements, are consistent with the
high radar reflectivities (dePater et al. 1984; Muhleman et al. 1984a; Ulich et
al. 1984). In particular, the disk-averaged emissivities of Ganymede and Callisto are very close to one minus their respective radar albedos, as they should
be. Millimeter wavelength observations are especially interesting as they can,
by virtue of their shorter wavelengths, probe the upper regolith. Unfortunately, the millimeter wavelength picture is confused, with conflicting results
reported (Muhleman et al. 1984a; Ulich et al. 1984). Still, some interesting
differences between Ganymede and Callisto are apparent. Callisto's brightness
temperature is greater than that of Ganymede, and its disk-averaged emissivity may be essentially unity, indicating either that the rather novel structures advocated above to explain the radar observations do not apply at these
wavelengths or at shallower levels in Callisto's regolith, or that these measurements are sensing the diurnal surface temperature variation. Further interpretation, i.e., unscrambling the effects of temperature, composition and
structure, will be difficult.

We cannot overemphasize the great potential of radar and radio studies.
We note that the Galilean satellites will again be observable by the Arecibo
telescope starting in 1987 and the Goldstone radar is always available. This
kind of data and the fascinating interpretations it provokes will not be super-
seded by the Galileo mission, although bistatic sounding of satellite surfaces
by Galileo during Earth occultation is a possibility.

At this time the structure of the regolith is unknown. The models proposed to explain the radar returns are all interesting (see, e.g., Goldstein and
Green's [1980] conception of Ganymede's icy surface as "crazed and fissured
and covered by jagged ice boulders"), but the connection of these models to
geological reality is difficult to judge. On the theoretical end, Shoemaker et
al. (1982) have applied their model of lunar regolith growth to Ganymede, and
these authors predict regolith thicknesses of a few tens of meters today. The
model given in the chapter by Veverka et al., Fig. 7d, predicts more variable and
generally greater thickness. Note that radar and microwave techniques have the
capacity to sound an entire such regolith, if ice rich. Shoemaker et al. (1982)
also point out that the upper portion should have a very low thermal conduc-

tivity (as does the lunar regolith), a fact attested to by thermal infrared measurements (see Morrison 1977; Hanel et al. 1979b). The possibility also exists that the regolith anneals itself. For lack of a better estimate, the 130 K isotherm is often taken as the annealing temperature. While related to the vapor pressure curve of ice, the annealing isotherm will in fact depend on pore size distribution and planetary thermal history. Smoluchowski and McWilliam's (1984) calculations predict that pores will migrate up the temperature gradient (i.e., downward), coalesce, and eventually be partly crushed if the porosity exceeds a close-packing percentage of ~ 40%. Further densification in regions of gentle thermal gradient is accomplished by creep. Unfortunately, no detailed modeling has yet been attempted on this important surface layer that controls lithospheric temperature and hence crustal and thermal evolution.

The model of polar frost formation of Shaya and Pilcher (1984) deserves comment here. Their premise is that a clean ice layer, one to several meters thick, once completely covered Ganymede. Thermal migration then caused the ice in the equatorial to mid-latitudes to migrate poleward over geologic time. Eventually, nearly all this ice was removed from a broad equatorial belt, leaving behind thermally stable polar ice caps with all of the transported low-latitude ice piled up at the cap edges. Simultaneous gardening and regolith formation mixed the deposited ice with the subjacent "original" surface materials, the ultimate result being an enhancement in the iciness of the polar regolith and optical surface. One problem with this model is that there are no observed bright polar collars. Furthermore, thermal migration should continue from the equatorial latitudes, resulting in a continual decrease in albedo from the cap edges to the equator. This is not observed. In fact, if we accept that both dark and bright terrain regoliths are very ice rich, then the existence of a planetary ice sheet is irrelevant. The migration of water to form a darker equatorial band should occur in any case. Shaya and Pilcher's (implicit) explanation for the lack of latitudinal albedo gradient for low latitudes on Ganymede is that thermal migration does not occur at points where the cap edge has since retreated poleward. They justify this by asserting that thermal migration has not created polar caps on Callisto. But by this they assume the process that they mean to demonstrate (i.e., thermal migration does not affect Ganymede's low latitudes because it does not). Despite this model's intriguing premise that a planetary ice sheet resulted from the eruption of liquid water during bright terrain formation, we find it difficult to support.

It is possible that somewhere beneath the optical surfaces and regoliths of Ganymede and Callisto, there exist silicate-rich layers or accumulations. It is also possible that silicate-rich units are being sensed by infrared and radio observations, but are being lost in the globally averaged signal. These silicate-rich units may in fact be more representative of the bulk composition of the satellite surfaces. Nevertheless, at this time there is simply no evidence that suggests that the regoliths and optical surfaces of various terrains on Ganymede are anything but ice rich. The case for Callisto, we freely admit, is less clear, but so far no data compels us to identify any large-scale silicate enrichment of Callisto's surface either.

V. CRUST AND CRATERING RECORD

We have identified geochemical contrasts between the various terrains on Ganymede and Callisto. The question then becomes: how deep do they extend? That is, can we define a "crust" that is geochemically distinct from the underlying "mantle?" The answer is beyond the reach of electromagnetic sounding at present, so we employ impact craters as probes. We note the use of the term crust here should not be confused with lithosphere, which is defined mechanically.

Impact craters are ubiquitous on Ganymede and Callisto, a fact that attests to the great geologic age of their surfaces. Increasing crater density is correlated with decreasing terrain albedo, implying that the Ganymede bright terrain, Ganymede dark terrain, and Callisto's surface form a sequence of geologic units of increasing relative age. The *absolute* ages are highly uncertain due to the unknown flux history of impacting bodies (see the chapter by Chapman and McKinnon). The dark terrains, however, are believed to date from a period of intense bombardment early in the geologic history of Ganymede and Callisto, while some areas of bright terrain may be comparable in age to the lunar maria (3.5 to 4 Gyr old) (Smith et al. 1979a,b). Some regions of smooth, relatively uncratered bright terrain may be much younger.

More precise estimates have been provided by the impact flux history model of Shoemaker and Wolfe (1982). In particular, the age of Callisto's surface is given as $\gtrsim 4.0$ Gyr, that of the Ganymede dark terrains span ~ 3.8 to 4.0 Gyr, and Ganymede bright grooved and smooth terrains are $\lesssim 3.8$ Gyr old (Shoemaker and Wolfe 1982; Passey and Shoemaker 1982a; Shoemaker et al. 1982). The most heavily cratered bright terrain on Ganymede, near the south pole, has a comparable crater density to the least cratered dark terrain (Shoemaker ct al. 1982), so the ages of dark and bright terrain on Ganymede overlap.

The Shoemaker and Wolfe model assumes a lunar-like variation of projectile flux with time. There is, however, a dearth of craters > 60 km in diameter on Ganymede and Callisto compared with the preserved terrestrial planet record (Strom et al. 1981). If these large craters originally formed but were largely erased by processes such as viscous relaxation (see below), they would have left large holes in the spatial distribution of craters (Woronow and Strom 1982). A rigorous study of the spatial distribution statistics of such a process shows that while a degree of erasure of large craters can be tolerated, a lunar-like crater population cannot be reconciled with that expressed on Ganymede and Callisto (Gurnis 1986). This difference becomes more acute when crater scaling is taken into account, as projectiles of equal mass create larger craters on Ganymede and Callisto (by at least a factor of 1.5 in diameter) than on the

Moon (see the chapter by Chapman and McKinnon). This implies a projectile population in the Jovian region different from that in the region of the terrestrial planets. Because projectile populations that affect Jupiter also substantially affect the terrestrial planets while populations indigenous to the region of the terrestrial planets cannot significantly affect the Jovian satellites, the ancient Jovian bombardment was either contemporaneous with, or earlier than, the late heavy bombardment of the terrestrial planets. Thus, it is plausible that all of the terrains are older than given above.

We first address the thickness of the bright terrains on Ganymede. Because they are younger than the dark terrains, it is possible that dark terrain underlies grooved and smooth terrain. Crater rims are the highest topographic features in the dark terrain, but flooded and embayed remnants of these rims are not visible in the bright terrain, even where craters are sharply truncated at a bright/dark terrain contact. Because crater rims are generally under 1 km in height, as determined photoclinometrically by Passey (1982), this height can serve as a rough lower limit to grooved terrain thickness. Some distinctive patchy, smooth terrains (discussed in Sec. VI, below) are conceivably thinner. Shoemaker et al. (1982) stated that there is no evidence for craters even as large as 50 km in diameter excavating dark material from depth in the bright terrain. This suggested to them that in many areas the thickness of bright deposits could exceed 10 km. Recently, however, dark halo craters in bright terrain have been recognized as the sought-after evidence of dark terrain excavation (Schenk and McKinnon 1985). In appropriately imaged regions of bright terrain, such as Uruk Sulcus (Fig. 5), there are three types of craters: large craters with a low-albedo annulus or halo, smaller craters without such halos, and bright rim and ray craters of all sizes. The implication is that as the bright ejecta blankets of fresh craters evolve toward photometric equilibrium with the space environment, the larger ones become darker than the surrounding bright terrain. Halo radius is linearly correlated with crater radius, implying that the halo is related to the continuous ejecta blanket and is not erased with time. Furthermore, the albedos and colors of the halos are intermediate between those of surrounding bright terrain and nearby dark terrain. The interpretation of Schenk and McKinnon (1985) is that craters above a certain size are able to excavate dark terrain material, which forms a stratigraphic horizon buried at depth. The estimate for bright terrain thickness in Uruk Sulcus is 1.0 to 1.6 km. Bright terrain thicknesses elsewhere on Ganymede are inferred to be similar, and probably do not exceed a few kilometers.

We note that the very rare dark-ray craters, which have been implicated in the past in the excavation of silicate-rich units (Hartmann 1980b; Poscolieri 1982), are now judged to be due to projectile contamination (cf. Conca 1981; Shoemaker et al. 1982; chapter by Chapman and McKinnon). Ray ejecta, with their uncertain provenance (see the chapter by Chapman and McKinnon), is a poorly controlled stratigraphic probe compared with the continuous ejecta blanket in any case.

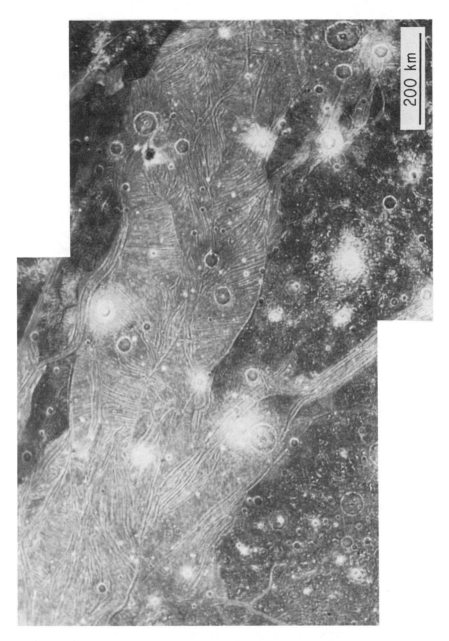

Fig. 5. High-resolution image (centered at 1° S, 161° W) of a lane of bright grooved and smooth terrain, Uruk Sulcus (see Fig. 2). Noteworthy are multiple groove sets and superposed dark halo craters; the latter are arguably indicative of dark terrain excavation (see text). Portions of Voyager 2 frames FDS 20637.17 and 20637.20 were mosaicked and contrast stretched. Scene is approximately 1350 km across with north at top.

Dark terrain on Ganymede also appears to have a limited vertical extent. Shoemaker et al. (1982) noted that most large craters (those $\gtrsim$ 50 km in diameter) on the dark terrain have persistent bright albedo deposits. Some of these "craters" are so highly flattened that recognition of familiar topographic features (such as rims) is all but impossible at Voyager resolutions (e.g., Fig. 6a). These vestigial albedo traces are termed palimpsests (see Passey and Shoemaker 1982a). The bright material may be ejecta differentiated by the heat of the impact, cleaner material excavated from deeper stratigraphic levels, or later flooding by clean water or ice as suggested by Hartmann (1984). Shoemaker et al. (1982) chose the second alternative, citing as evidence (smaller) dark-floored craters superposed on the outer parts of palimpsests and the lack of dark floors for craters superposed on palimpsest centers (see Fig. 6a); that is, the clean ice in the outer portions of palimpsests is interpreted to be a thin layer, while that in the inner portions is significantly deeper (perhaps indefinitely deeper, forming a window into the mantle). The excavation depth of a 50 km diameter crater is roughly 5 km (see the discussion in Schenk and McKinnon [1985]), so these observations point to a cleaner mantle at depths greater than this.

Our favored interpretation is that Ganymede truly has a crust. It is defined with respect to a cleaner mantle by the presence of a dark contaminant. It is not the product of geochemical differentiation, as on the Earth or Moon, and thus does not have a distinct lower boundary like Earth's "Moho." Grooved and smooth terrain are resurfaced units within Ganymede's crust. The situation for Callisto is (as usual) not clear. There, palimpsests are not bright except for Valhalla, the largest (Fig. 6b). This particular case is equivocal in terms of a mantle window. Superposed dark-floored craters are observed in the outer part of the Valhalla palimpsest, but no constraint exists on the thickness of the central "bright" material. The one conspicuous crater superposed near the center still has a bright rim, so no compositional inferences can be made about its floor or blanket from Voyager images. Such photometric immaturity would not hamper interpretation of high-quality infrared spectra, however. This crater and the entire Valhalla palimpsest is a prime target for NIMS. Overall, the lack of persistently bright albedo features around large craters on Callisto implies that if Callisto has a definable crust, it is relatively thick.

The correlation of crater density with albedo (and redness) on both satellites suggests (as in Sec. III) that meteoroidal contamination is responsible (Pollack et al. 1978a; Hartmann 1980b; Johnson et al. 1983). But, because certain units of bright and dark terrain on Ganymede have comparable crater densities, this contamination must predate the observed crater population in large measure. Inconsistent estimates for the rate of darkening of various terrains on Ganymede and Callisto by infalling debris (Johnson et al. 1983) imply the same. The dark surfaces probably accumulated material for a period of time before they became stable enough to retain craters. The contaminant is

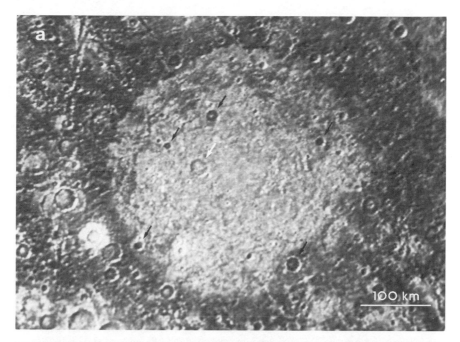

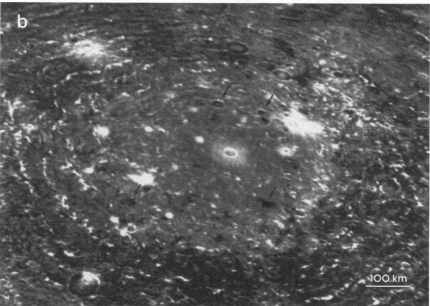

Fig. 6. Crater palimpsests on Ganymede and Callisto. (a) Memphis Facula, Galileo Regio, ~ 350 km across (centered at 15°5 N, 132°5 W with north at top); (b) Valhalla, ~ 600 km across (centered at 15° N, 57° W with north towards lower right). The presence of dark floored craters within the peripheries of both is evidence for the surficial nature of the outer bright materials. Centrally located superposed craters remain as bright as their surroundings on Ganymede palimpsests; the case for Valhalla is equivocal. Palimpsests on Callisto are not initially, or do not remain, bright compared to surrounding terrain. (Frames FDS 20538.29, rectified [a], and 16422.11 [b].)

apparently reddish and thus may be more indicative of reddish D-type carbonaceous material than spectrally neutral C-type carbonaceous material (Schenk and McKinnon 1984; see also Hartmann et al. 1982). A definitive spectral identification of D-type material, however, requires a rising reflectance into the near-infrared, and is thus another job for NIMS. Pollack et al. (1978a) hypothesized that dust particles blasted off the outer Jovian satellites by small impacts spiral inwards under the action of Poynting-Robertson drag and are responsible for the contamination of the surfaces of Ganymede and Callisto. It is interesting to note that the recent photometric work of Tholen and Zellner (1984) indicates that the outer Jovian satellites look like C-type asteroids. (They are generally considered to be captured; see discussions in the chapters by Burns and by Stevenson et al.) There is an obvious and intriguing parallel to the hypothesis that links dust from Phoebe (in the Saturn system) to the dark leading hemisphere of Iapetus (see also Bell 1984).

Multiring and Furrow Systems

Impact craters also provide constraints on thermal and rheologic history. The largest impacts can create systems of concentric faults, or rings, surrounding the crater. The type and extent of the fault pattern depends on the scale of the impact and the rheologic structure of the satellite (see the chapter by Chapman and McKinnon). Such multiringed basins or structures are found on both Ganymede and Callisto. Very large impacts are also probably responsible for the ancient systems of furrows that cross the dark terrains. The most prominent furrows are rimmed and form arcuate parallel and nearly parallel sets of great length ($>$ 1000 km in many instances; see Fig. 7). As discussed in McKinnon and Melosh (1980; see chapter by Squyres and Croft), the furrows are best interpreted as graben formed in an extensional and largely axially symmetric tectonic regime; the rims can be accounted for by later viscous relaxation (also advocated by Shoemaker et al. 1982). The logical source of such stress is the prompt collapse of a large impact crater (there is a dearth but not total absence of these). On Ganymede, however, widespread resurfacing (see next section) has hidden or eliminated the presumed craters of origin.

There are at least four furrow systems on Ganymede: one that crosses units of Marius Regio and Galileo Regio (there called the Lakhmu Fossae) (Fig. 7); a second, the Zu Fossae, that are best expressed at the northwestern boundary of Galileo Regio and appear to predate the main set (McKinnon 1981b; Shoemaker et al. 1982; Casacchia and Strom 1984); one in Nicholson Regio (see, e.g., Voyager 1 frame FDS 16405.12); and a set of high curvature in northern Perrine Regio (see Voyager 1 frame FDS 16402.02).

An impact origin for the furrows (or graben) in Galileo Regio has been questioned by Casacchia and Strom (1984). Their most cogent argument is that interfurrow spacing perpendicular to strike remains nearly constant over 60° of arc, as opposed to increasing with distance from the center of curvature (a prediction they quote from McKinnon and Melosh [1980]). A more telling

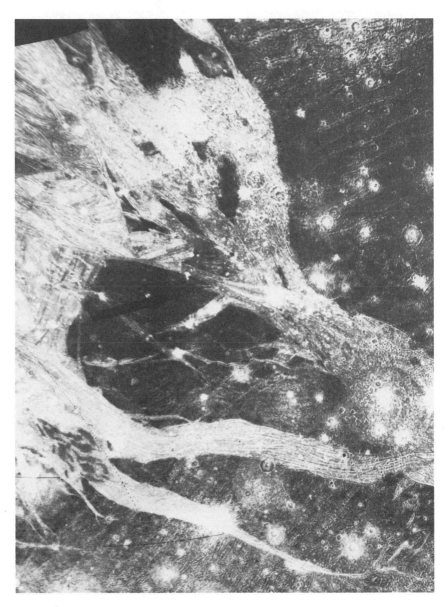

Fig. 7. Computer mosaic of Marius Regio quadrangle (Jg-4) and surroundings. Arcuate furrow systems in Galileo Regio (upper left) and Marius Regio (center and bottom) may have originally been part of a single eccentric system. The bright, grooved terrain boundary of Galileo Regio is a continuation of Uruk Sulcus (see Figs. 2 and 5 as well as the Map Section). Marius Regio is transected by two narrower lanes of grooved terrain, and contains a large number of palimpsests and highly flattened craters. The mosaic is a Lambert conformal projection (centered near 40° N, 195° W); width varies from ~ 2000 km (top) to ~ 1600 km (bottom); north is approximately up.

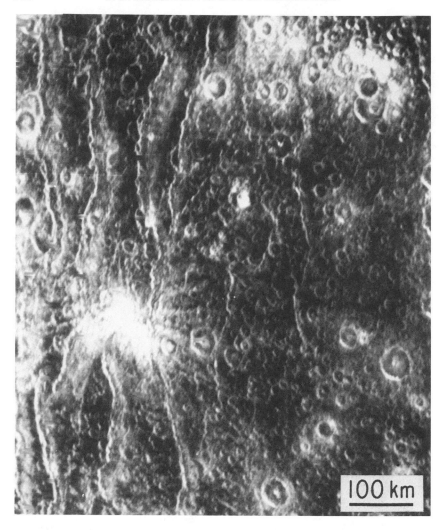

Fig. 8. Outward-facing ring scarps of the Valhalla multiringed system on Callisto. Brighter materials at scarp bases may be due to flooding. Crater counts on these units are consistent with that on the Valhalla palimpsest, indicating emplacement at or near the time of impact (Remsberg 1982). Scene is approximately 675 km across; north is towards the upper right. (Voyager 1 frame FDS 16424.46 [centered near 38° N, 28° W].)

question would be to ask why are all the rings graben; that is, why is there no transition from graben to outward-facing normal faults away from the center of curvature, as observed around Valhalla on Callisto (Fig. 8; also see Melosh 1982*b*)? The answers to both questions need to be resolved; they may be related and may be important clues that the rheologic structure of the Ganymedean mantle differed in some way from that of Callisto. In addition, careful study of the geometry of the furrows (Zuber and Parmentier 1984*a*) shows that even furrows within the unbroken Galileo Regio do not form a circular concentric system. The width of the furrows in Marius Regio ($\sim$ 6 km) and Galileo Regio ($\sim$ 10 km) also changes abruptly across the intervening bright terrain. This raises doubts about whether the furrows in these areas ever formed a continuous system. At the very least, a significant deviation from axial symmetry is indicated. We note, however, that portions of the Valhalla system also deviate noticeably from circularity.

The morphology of the observed ring systems (and furrow systems, if an impact origin is correct) implies something important about the mantles of Ganymede and Callisto at the time of impact: they were both predominantly solid to the depths affected by large impact (on the order of one crater diameter, or several hundred km). Thus, it is very unlikely that Ganymede and Callisto possessed thin (e.g., < 50 km thick) ice crusts, floating on liquid water, at the time of the ring forming impacts. This conclusion is especially important for Ganymede, because the furrows predate nearly the entire cratering record, and are thus the earliest tectonic structures preserved on the satellite. Note, however, that this does not preclude the presence of liquid water at depth.

Multiring systems are not, unfortunately, very sensitive to the silicate content of the mantle or interior of either satellite. Ring formation depends on rheologic structure; hence the modest increase in viscosity, perhaps an order of magnitude (McKinnon 1982*c*; Friedson and Stevenson 1983; but see Kirk and Stevenson 1986), caused by even a 50/50 ice/silicate mass ratio can be offset by an equally modest temperature increase (of order 10 K).

Heat Flow

Furrows on Ganymede and rings of similar morphology on Callisto, when interpreted as graben, can be used to infer directly surface heat flow, regionally averaged, at the time of furrow or ring formation. The method involved assumes that the inward-dipping master faults of the graben penetrate the satellite's lithosphere and intersect at the top of the asthenosphere. (The lithosphere is defined as the relatively stiff outer shell of a satellite that responds in brittle and/or plastic fashion to stress, and the asthenosphere is the more deformable layer below that responds ductilely; the contrast in rheological behavior reflects the strong dependence of [ice] viscosity on temperature.) This style of graben and rift formation is known on the Earth in regions where the entire lithosphere or some other mechanically competent layer is in extension (e.g., the Canyonlands graben, the Baikal and Rhinegraben rifts). On Ganymede and Callisto, however, the lithosphere-asthenosphere boundary must be carefully defined with respect to the time and stress scale of the source of the extension, here considered to be basin collapse (see McKinnon 1981*a*). A model for the temperature profile created by solid-state convection in an icy mantle or interior is also needed for the heat-flow estimate, but such

TABLE II

Lithospheric Thickness and Heat-Flow Estimates from Ring and Furrow
Systems on Callisto and Ganymede

	T_s (K)	H^a (km)	q (mW m^{-2})	Age[b] (Gyr)
Valhalla	120	15–20	~20	3.92
Galileo Regio	110	10	~40	3.87
Marius Regio	110	6	~65	3.87
Gilgamesh	110	30–100	~5–15	3.2–3.5

[a]From McKinnon and Melosh (1980) and Zuber and Parmentier (1984a).
[b]From Passey (1982), Passey and Shoemaker (1982a) and Shoemaker et al. (1982), and adapted to the mean-age heat-flow model of Passey (1982).

a model (i.e., parameterized convection; see Sec. VII on internal evolution) is subject to large uncertainties. Although detailed heat-flow calculations have been presented (McKinnon 1982c), rough estimates can be obtained by simply assuming that furrow (or graben) width is equal to lithospheric thickness (reasonable for faults of 60° dip), and that the heat flow q through the lithosphere is given by a conductive temperature profile

$$q = 567/H \ln(T_b/T_s) \text{ W m}^{-1} \tag{1}$$

where H is lithosphere thickness, T_b and T_s are the temperatures at the base and top of the lithosphere, respectively, and the thermal conductivity of solid ice is given by $567/T$ W m^{-1} (Klinger 1980). Estimates are presented in Table II for Valhalla, Galieo and Marius Regio, and the younger Gilgamesh basin on Ganymede, assuming $T_b \sim 218$ K (0.8 of the melting temperature, which is reasonable for basin collapse stresses).

The evolution of heat flow on Ganymede and Callisto can also be constrained by systematic study of the viscous relaxation of impact craters, provided that relaxation occurs, of course. This phenomenon, predicted by Johnson and McGetchin (1973) and treated theoretically by Phillips and Malin (1980) and Parmentier and Head (1981), is plausible given the rheology of ice at the lithospheric temperatures of these satellites (Goodman et al. 1981; Weertman 1983; Durham et al. 1983). Passey (1982) showed, by matching photoclinometrically derived topographic profiles to theoretical ones, that viscous relaxation has affected craters on both satellites. Furthermore, he quantified the degree of relaxation as a function of crater size, and constrained the viscosity evolution of the lithospheres of both satellites using the cratering flux history model of Shoemaker and Wolfe (1982). The e-folding time for the lithospheric viscosity gradient was determined to range between 2×10^8 and 10^9 yr for both the grooved and cratered terrain on Ganymede

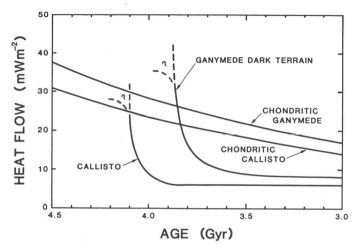

Fig. 9. Model heat flows from Ganymede and Callisto as a function of age. The Ganymede dark terrain and Callisto curves are derived from the crater viscous relaxation study of Passey (1982). The high ends of the heat flow curves, which correspond to the earliest crater retention ages, are not well constrained. In the study of Passey (1982), Ganymede bright terrain formation begins at 3.8 Gyr, and has a heat flow ~ 5 mW m^{-2} greater than the dark terrain at any time. Plotted for comparison is the heat flow that would be produced in equilibrium with the decay of a chondritic abundance of long-lived radioactive elements. These curves (from Mueller and McKinnon 1986) are maximum values, calculated for fully differentiated models of Ganymede and Callisto. Undifferentiated models give values $\sim 15\%$ less.

and between 10^8 and 10^9 yr for Callisto. This was translated into an e-folding time scale for the heat flow, about 10^8 yr for each terrain type (Fig. 9). As is immediately apparent from Fig. 9, the sharp dropoffs in modeled heat flow do not correspond well with the gentle decline predicted if heat flow is in equilibrium with internal heating by long-lived radioactive elements (the decay time for a chondritic abundance of U, Th and K is $\sim 2 \times 10^9$ yr). The long-term heat flow estimates are not in sharp disagreement, nor are the ancient peak values; it is the time histories that are difficult to reconcile. Passey (1982) notes that the e-folding time of the heat flow is tied to the assumed e-folding time of the cratering flux, such that if the cratering flux decays with a time constant of 2×10^8 yr, then the heat flow will as well. The dependence on bombardment time scale of the model cannot rationalize the divergent heat flow histories, however.

A comparison of Fig. 9 with the heat flow values in Table II is provocative. While it would be unwise to overinterpret these rough estimates, it does appear that the regional heat flows in Galileo and Marius Regio significantly exceeded that expected from the decay of long-lived radioisotopes alone. We note that Passey (1982) could not determine the history of the heat flow earlier than a certain time on the Ganymede dark terrain or on Callisto. In particular, it is not known whether the heat flows on these older terrains continue to increase with increasing age or whether they plateau at certain values (see the question marks in Fig. 9); if they plateau, the levels at which they do so are undetermined (Passey 1982).

These remarks concerning heat flow will undoubtedly undergo revision in the next few years. Viscous relaxation studies and parameterized convection models have assumed that the ice viscosity applicable to these processes is Newtonian, that is, not explicitly stress dependent. This assumption was reasonable at the time given the stress levels involved and the available knowledge of ice rheology. New experimental work on low-temperature ice deformation (Durham et al. 1983; Kirby et al. 1985) implies that topographic relaxation and solid-state convection may be inherently non-Newtonian. The premises of parameterized convection in a strongly temperature-dependent viscosity fluid are also being reconsidered (see, e.g., Christensen 1985).

The evidence for viscous relaxation of craters as small as 10 km in diameter on the cratered terrains of Ganymede and Callisto raises the possibility that craters of this size have been lost from the observable record. Thus, the ages quoted here (in Table II and Fig. 9) should be properly thought of as *crater retention* ages (Passey and Shoemaker 1982a). This is a concept distinct from that applied to the heavily cratered terrains of the terrestrial planets, where age usually means that of igneous crystallization, or for example, the emplacement of an ejecta blanket. The crater retention age is more properly thought of as a metamorphic age, and the old terrains as structural equivalents of terrestrial high-grade metamorphic terrains. This leads to a natural hypothesis for the timing of the contamination of the older crusts of each satellite; the epochs that predate the crater retention ages of various dark terrain units but that postdate the time the various lithospheric units could resist recycling into the interior. Whether or not such recycling ever existed and the form it might have taken (e.g., convection or impact mixing) are not crucial.

There is some evidence that lithospheric recycling may have occurred, at least on Ganymede. The persistence of relatively high heat flow in Marius Regio may imply that the lithosphere there stabilized relatively late and that, compared to Galileo Regio, crustal contamination was volumetrically less and did not extend as deep on average. The consequences could be its higher albedo compared to that of Galileo Regio and a relatively easier formation and preservation of palimpsests. Palimpsests occupy nearly half the terrain on Marius Regio, compared with one-quarter on Galileo Regio (Shoemaker et al. 1982).

Further constraints on the internal structures and degrees of differentiation of Ganymede and Callisto are more difficult to develop. In principle, though, these can be provided by the careful interpretation of resurfacing, endogenic tectonism, or the absence of either.

VI. TECTONICS AND RESURFACING

Ganymede's surface has undergone a complex and active geologic evolution (see, e.g., Smith et al. 1979*a*,*b*). Approximately 40% of the surface is composed of dark, heavily cratered terrain. The dark terrain is divided into polygonal regions separated by less heavily cratered bright terrain, the bright terrain often forming swaths or lanes (sulci). Wedge-shaped areas of bright terrain also penetrate and partially divide dark regions. The bright terrain frequently contains long, linear topographic depressions called grooves, and this terrain has been accordingly called grooved terrain.

Several studies, including the chapter by Squyres and Croft, have examined the style of global tectonics associated with the emplacement of bright terrain. This terrain truncates topographic features in the older dark terrain, and the terrain boundaries are often sharply defined even at the highest Voyager resolution (600 m per pixel). This is particularly true for narrow bands of bright terrain and suggests that bright terrain material was tectonically or structurally confined in many areas (Lucchitta 1980; Shoemaker et al. 1982; Parmentier et al. 1982). In some irregularly shaped areas, bright terrain has a distinctive dark mottling suggestive of a thin layer of bright material overlying dark terrain (e.g., Fig. 10). In such areas the contacts between bright and dark terrain can be diffuse, and craters appear to be partially flooded by bright material. But in most areas, sharp curvilinear contacts between bright and dark terrain argue against flooding of dark terrain as the main mechanism for the emplacement of bright terrain.

If the tectonic emplacement of bright terrain were accommodated by the complete separation of adjacent areas of dark terrain, then large lateral movements of lithospheric plates would be required. In at least some areas, this would necessitate lateral displacements of adjacent areas of dark terrain with displacements comparable to the width of bright terrain bands. Although some features can be interpreted as evidence for lateral shear (Lucchitta 1980), large lateral strike-slip displacements have not been demonstrated. Shoemaker et al. (1982) cite the nonconcentricity of the ancient furrow system in Marius and Galileo Regiones as the best evidence for large lateral displacement, but as discussed in the previous section, the furrow system was probably nonconcentric initially. The clearest evidence of shear displacement is the lateral offset of a few crater rims across transecting linear features, but in these cases the offset is at most a few kilometers (Lucchitta 1980).

If complete separation of areas of dark terrain had occurred, as in continental drift on the Earth, it should be possible to fit these areas back together. The boundaries of some adjacent dark regions are roughly parallel, but detailed examination shows that they do not fit back together very well (Parmentier et al. 1982). Furthermore, if simple drifting apart had occurred, features such as the two parts of an impact crater transected by a bright band should be visible on both sides of the band. For bright terrain bands that are wider

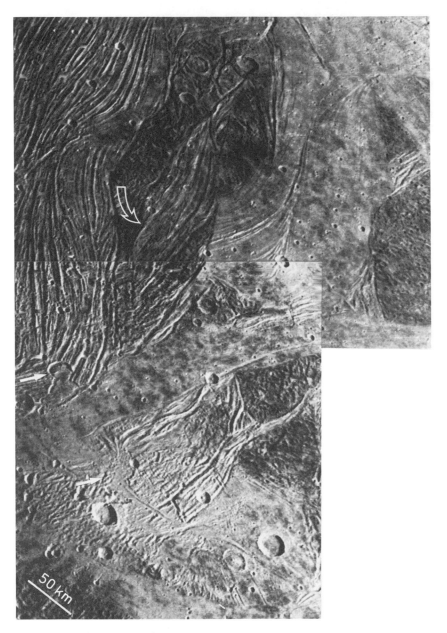

Fig. 10. Grooved and smooth terrain on Ganymede. The smooth terrains exhibit a distinctive patchy or mottled appearance; circular bright patches may be barely flooded craters. Also seen are grooved, dark terrain (large arrow) and two flooded crater complexes (small white arrows); the upper complex appears flooded by material significantly more viscous than water or slush. Taken near the south pole by Voyager 2, these are among the highest resolution images available; north is at bottom. (Mosaic of frames FDS 20640.27 [5a] and 20640.25 [5b] [centered near 75° S, 170° W].)

than the crater transected, this is never observed. The evidence for excavation of dark terrain material from beneath bright terrain by craters, as discussed in the previous section, also implies that large separations between dark terrain units are unnecessary.

All available geologic evidence is consistent with the formation of bright terrain by rifting, i.e., finite lithospheric extension concentrated in relatively narrow, linear zones. As in terrestrial rift zones, extension by normal faulting and thinning of cold, dense lithosphere would lead to surface subsidence and a tectonic depression in which bright terrain material (clean ice) can be confined (cf. Parmentier and Head 1984). Beneath the bright terrain, older dark terrain may have simply subsided, as in terrestrial rift zones, or if it were more dense, it may have foundered into the interior. Thus, the interpretation of dark halo craters as evidence for shallow resurfacing, if accepted, becomes an important constraint on the mechanics of rifting process. We also note that the presence of small, isolated blocks of dark terrain in many areas of bright terrain (e.g., Fig. 11) argues against the complete foundering of the dark terrain.

The emplacement mechanism of bright terrain and the physical state of bright terrain material at the time of emplacement are unknown. We have inferred earlier in this chapter that bright terrain material is less contaminated with dark particles than dark terrain, and that bright terrain is close to pure water ice in composition. (The possible presence of ammonia hydrate ices, neglected so far, will be considered in the next section.) The source of this clean ice is the Ganymedean mantle, and this is direct evidence for differentiation of the mantle at some depth. Bright terrain may have been emplaced as liquid water, partially crystallized water (i.e., slush), or as ice. Direct evidence for any of these emplacement modes has not yet been identified. One topographic dome that may be of volcanic origin has been described (Squyres 1980c). Features such as lava flows and volcanic constructs, however, cannot be identified unequivocally at the resolution of Voyager images. The highest-resolution images (shown in Fig. 10) are nevertheless provocative. Smooth units give an appearance of thinness easily compatible with liquid water or slush, while possible flow fronts are suggestive of the termini of what would be essentially hot "glacial" ice moving viscously across a much colder and more rigid surface. It appears that Voyager was on the threshold of making definitive morphological observations; the much improved resolution of Galileo images should prove decisive.

The question of the physical state of bright terrain material during emplacement is an important one from the standpoint of the resurfacing mechanism and its implications for Ganymede's internal evolution. Shoemaker et al. (1982) favor liquid water, arguing that an isostatic column of liquid water cannot rise through a less dense ice crust to flood the surface, but can fill stratigraphic depressions. They cite the confined nature of the resurfacing and stipulate that the dark terrain is not so contaminated with dark nonice (and

Fig. 11. Elevated outlier of dark terrain, ~ 120 km across. The northernmost extension of
Galileo Regio is in the background; north pole is off to the right. The scene is approximately
800 km across. (Voyager 1 frame 16405.56 [centered near 70° N, 25° W].)

presumably silicaceous) particles as to be denser than liquid water. The problem with this "freeboard" argument is that there is no obvious way to get liquid water to rise through a less dense ice crust or lithosphere. Crevasses created by tension could partly penetrate the lithosphere, but deeper extensional
faults (such as those that create graben) are still under net compression, and
will not be used as conduits if the magma is more dense than the surrounding
material (the hydrostatic driving force for such a fluid-filled crack is downward). If the crust is sufficiently contaminated ($\gtrsim 12\%$ silicates by mass), so
that buoyancy-driven water-filled cracks rise through it, there is still the difficulty of getting the water to rise through the cleaner ice mantle below. The
problem is exacerbated if the water is denser due to dissolved salts or other
solutes. Arguments for emplacement in the solid state fare no better. Although
there is no density problem in getting hot ice to rise through the mantle or
crust, the large viscosity of the ice should more naturally lead to diapirism
(see the chapter by Squyres and Croft). Such diapirism has yet to be identified
in connection with the bright terrains.

Grooves

Tectonic features, generically termed grooves, are abundant on the surface of Ganymede and are most strongly concentrated in the bright terrain.
Grooves have widely varying morphologies and may be the expression of
a variety of tectonic processes (see chapter by Squyres and Croft, Fig. 4).

Where grooves occur singly, they have the form of narrow linear surface depressions with widths ranging from a few kilometers to down to the limit of Voyager resolution. These features have been interpreted as tension cracks or narrow graben (Shoemaker et al. 1982; Parmentier et al. 1982; Squyres 1982). Other grooves occur in well-defined pairs of two individual grooves separated by up to 10 km. Based on theoretical and laboratory studies of the viscous relaxation of topography, these groove pairs are most simply interpreted as viscously relaxed graben (Parmentier et al. 1982). Other grooves frequently occur in sets of many parallel grooves (e.g., Fig. 5). Crest-to-trough topographic amplitudes are on the order of a few hundred meters (Squyres 1981b). The spacing of grooves varies from 3.5 to 17 km with mean of about 8 km; the spacing within an individual groove set is more uniform and there is no correlation with latitude or longitude (Grimm and Squyres 1985). These features may be tension cracks, grabens or extensional, ductile flow features due to boudinage of a strong near-surface layer (Fink and Fletcher 1981; Parmentier et al. 1982). In the latter case, the spacing of grooves would form at the dominant wavelength of extensional flow instabilities, about 3 to 4 times the thickness of the strong surface layer. Thus, a number of investigators favor the interpretation of groove sets as extensional tectonic features. It has been pointed out, however, that based on their morphology alone, groove sets could be trains of folds produced by horizontal compressional deformation (Parmentier et al. 1982).

The relationship of grooves to the emplacement of bright terrain is not fully understood. Grooves in the bright terrain, except for a few examples (see Shoemaker et al. 1982), do not disrupt impact craters. Impact craters in the bright terrain appear to be at least as circular as those in the dark terrain (Zuber and Parmentier 1984a). Therefore, grooves must be produced almost contemporaneously with the emplacement of bright terrain. This precludes the possibility that grooves are compressional features related to long-term cooling and contraction of the interior (Zuber and Parmentier 1984b). Grooves may be the surface expression of faults or tension cracks at depth in underlying dark terrain or may develop solely at shallow depth due to continuing extension of the bright terrain. In long, narrow bands of bright terrain, grooves form sets that are nearly parallel to the boundaries of the band. However, in wide, more complex areas of bright terrain, for example in Uruk Sulcus, grooves have more widely varying morphology and occur in a variety of orientations. Golombek and Allison (1981) suggested a model in which grooves of varying orientation and morphology are related to the sequential emplacement of bright terrain (see Fig. 12); complex areas of bright terrain may form by sequentially developing structural cells (cf. Shoemaker et al. 1982). Thus grooves may be produced by deformation related to the tectonics of bright terrain emplacement. Mechanisms of groove formation related to the solidification and cooling of water-rich magma have also been suggested, however (Parmentier and Head 1984).

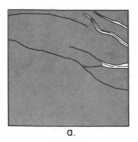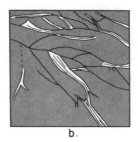

a.   b.   c.

Fig. 12. Stages of development of grooved terrain in the model of Golombek and Allison (1981). Formation of master grooves is followed by the formation of secondary grooves and resurfacing. Tertiary grooves form sets (D and E) within structurally isolated polygons. Smooth terrain polygon (northwest of A) is unfractured. This sketch map covers a section of Uruk Sulcus that overlaps Fig. 5; north is up.

The emplacement of bright terrain in rift zones may be related to internal expansion of Ganymede. Squyres (1980a; see chapter by Squyres and Croft) has shown that an increase in surface area of 5 to 7% is possible due to differentiation of a homogeneously accreted body. Several estimates of the amount of extension associated with bright terrain emplacement have been attempted. Assuming that bands of bright terrain are rift zones bounded by 60° dipping normal faults, Parmentier et al. (1982) estimated that only about 1% internal volume change would be required to produce 5 globe-circling rift zones filled with a 10 km thickness of bright material. Assuming that individual grooves in Uruk Sulcus are simple graben bounded by normal faults with dips greater than 60°, Golombek (1982) estimated a 1% upper limit on Ganymede's expansion. McKinnon (1981b) calculated the magnitude of elastic membrane stresses in the large unbroken lithospheric shell of Galileo Regio. For a lithosphere strength on the order of 10 MPa (100 bar), an upper limit of 1% expansion can occur; this strength is probably a severe overestimate for a lithosphere in tension (cf. Brace and Kohlstedt 1980). Parmentier et al. (1981) considered the case of a viscoelastic shell by modifying the solution of Turcotte (1974) for a shallow elastic shell. A larger amount of expansion is possible if it occurs over times longer than the relaxation time for elastic stresses. This time (the Maxwell time) is given by η/μ, where η is the viscosity and μ is the shear modulus of the shell. Passey (1982) estimated from crater relaxation studies, discussed in the last section, that the mean surface viscosity of Ganymede's lithosphere was $1.0 (\pm 0.5) \times 10^{25}$ Pa s (1 Pa s = 10 poise). A near-surface viscosity of this order is consistent with the preservation of furrows in the dark terrain over a few Gyr as well, even in the presence of an initially steep thermal gradient. This gives a Maxwell time of order 10^8 yr, assuming an ice shear modulus of 3.5 GPa (Gammon et al. 1983) for the surface. Therefore, large amounts of internal expansion associated with differentiation on time scales of 10^8 yr or longer may be permitted.

Whether expansion accommodated by ductile strain leads to tectonic structures similar to those observed is uncertain. Even the extensional instability failure model of Fink and Fletcher (1981) ultimately results in surface fracturing. Thus, to restate, the limits of McKinnon (1981a,b) and Golombek (1982) apply to the accumulated elastic stress and strain that lead to brittle failure. If the tectonics of Ganymede are primarily a manifestation of brittle failure, which is the predominant view, then these limits apply to the expansion associated with bright terrain formation. Some global expansion associated with differentiation may, of course, have occurred before bright terrain formation (e.g., during accretion).

The concentration of groove formation in the bright terrains is not obviously consistent with global expansion as the sole process producing stress; that is, initial fracturing in bright-terrain regions should relieve expansion stress there. Thus it is now generally recognized that the response of the lithosphere to stress was highly nonuniform spatially, that additional regional and local stresses were important, or both (McKinnon 1981b; Parmentier et al. 1982; Zuber and Parmentier 1984b; Grimm and Squyres 1985). An example of a possible regional stress source would be convective upwelling (see the chapter by Squyres and Croft); a local source would be the cooling of bright terrain material itself (as noted above).

Kirk and Stevenson (1983) have outlined a specific mechanical model for grooved terrain formation. It assumes a deeply melted and differentiated Ganymede, as discussed in the beginning of this chapter. The liquid-water upper mantle freezes until just a residual liquid layer remains at the depth of minimum melting temperature for ice (the ice I-ice III-water triple point, 251.2 K and 207 MPa; Eisenberg and Kauzmann 1969; see also Fig. 8 in the chapter by Squyres and Croft). At this stage the lower ice III layer has a significantly larger potential temperature than the ice I layer above, and this serves to drive an extremely vigorous convective exchange between the two layers. It may also provide both extensional stress sufficient to fracture the lithosphere and ice in the solid state for resurfacing. This model is related to the general phenomenon of convective upwelling mentioned above, but its extreme vigor implies that global expansion is probably not necessary for it to operate; it would probably operate in the presence of global contraction.

Other mechanisms have been suggested in connection with grooved terrain formation. One is the possible role of giant impact (T. V. Johnson, personal communication, 1983). This is attractive from the standpoint of furrows in the dark terrain predating the creation of bright terrain. Hence, if furrows are the surviving sections of giant impact rings, there is a circumstantial link in time to the formation of the oldest bright terrain. On the other hand, giant impacts did not cause the formation of grooved terrain on Callisto (to our knowledge), nor were the impacts which formed furrows (if furrows are impact-related) the largest ones on Ganymede. Presumably, even larger impact events predate furrow formation. Thus, a giant impact is not likely to directly cause grooved terrain formation, but if the physical circumstances are right, then it may act as a trigger. Kirk and Stevenson (1983), in fact, invoke giant impact to initiate the convective instability in their model. Still other mechanisms for grooved terrain formation, such as diapirs originating as boundary layer instabilities at the bottom of an ice/silicate lower mantle (McKinnon 1981*a,b*) or changing mantle convective patterns brought about by the different viscosities of various water ice polymorphs (Shoemaker et al. 1982), remain unevaluated and speculative.

One mechanism not responsible for grooved terrain is a Ganymedean analog to terrestrial plate tectonics. We have noted that little evidence exists for large-scale crustal extension or strike-slip motion. Similarly, evidence for large-scale lithospheric turnover is meager. But this is not to say that plate tectonics cannot exist on an ice-mantle planet such as Ganymede; this is a topic for further research. Its manifestation would probably differ from that on the Earth, though, if only because the temperature and pressure dependence of the viscosity of ice I differs markedly from that of typical terrestrial mantle silicates.

Although the origin of bright terrain formation and grooved terrain tectonics is not well understood, some points do stand out.

1. Extensional tectonics are probably dominant. A more definite interpretation is hampered by the resolution of Voyager images and the possible effects of viscous relaxation and mass wasting (mainly impact degradation).
2. The material that resurfaced the bright terrains is differentiated. This key conclusion cannot be overlooked in interpreting the internal evolution of Ganymede.
3. Stresses concentrate in regions of groove formation, and resurfacing and groove formation are causally connected. Far from being trivial, this connection suggests that the same internal process is responsible for the tectonics and the volcanism.

The mechanical behavior of Ganymede's lithosphere deserves comment here. Both in furrow and groove formation, the response of the lithosphere is often remarkable in its large-scale geometric simplicity. Two conditions appear necessary for this to be achieved; the patterns of driving stress must be themselves geometrically simple, and the lithosphere must be fairly homogeneous mechanically. The latter is probably realized on Ganymede (and on Callisto) by virtue of the "metamorphic" nature of the lithosphere discussed in the previous section; that is, in at least the lower lithosphere, fractures can anneal and the tectonic memory of the lithosphere should be limited (compared to that of the Moon or Mars, for example).

Resurfacing on Callisto

Callisto has not figured in the discussion of this section simply because no tectonic structures other than those related to large impacts have been identified there. Some evidence exists for resurfacing, however; brighter units, apparently topographically confined, exist at the bases of the outward-facing scarps of the Valhalla system (see Fig. 8). While probably related in time to the impact (Remsberg 1982), these units, unlike the brighter material of the Valhalla palimpsest, cannot be impact melt. They may be evidence that cleaner ice exists in Callisto's interior, at depths exceeding 15 to 20 km (the thickness of the lithosphere at the time of ring formation).

VII. INTERNAL EVOLUTION

General Considerations

The studies of Lewis (1971a,b) were the first to consider the chemical composition, structure and evolution of ice-silicate planetary bodies like Ganymede and Callisto. Later investigations by Consolmagno and Lewis (1976,1977,1978) developed more detailed models of internal evolution for a range of chemical models. Both equilibrium and disequilibrium condensation were considered as well as homogeneous and inhomogeneous accretion. The evolution of any planetary body is governed by its internal sources of heat and the mechanisms by which heat is transferred from the interior to the surface. The Consolmagno and Lewis models included radioactivity due to chondritic abundances of U, Th and K in the silicate component and, for homogeneous accretion, the heat resulting from differentiation of a silicate core. Ice-silicate differentiation was assumed to occur as the icy component in any particular region of the interior melted. Because of uncertainties in the viscosity or strength of solid icy materials, the models did not consider heat transfer due to solid-state convection or diapiric rise of melt. In this case radioactive heating in a body the size of Ganymede or Callisto is sufficient to cause large amounts of melting of the icy component and complete ice-silicate differentiation. The models predicted that, in the absence of solid-state convection, the icy mantles of Ganymede and Callisto would at present be largely in the liquid state.

As the chapter by Schubert et al. discusses, several subsequent studies showed that solid-state deformation of the icy component could significantly affect the internal evolution of ice/silicate bodies. Thermally activated solid state creep can be described by a temperature-dependent viscosity

$$\eta = \eta_m \exp [A(T_m/T - 1)] \tag{2}$$

where η_m is the viscosity at the melting temperature T_m, and A is a constant proportional to the creep activation energy. Based on estimates of η_m and A from laboratory measurements for ice I (see, e.g., Glen 1975) and studies of terrestrial ice sheets (Hughes 1976), Reynolds and Cassen (1979) showed that a solid ice lithosphere, overlying an internal liquid water layer as predicted by the Consolmagno and Lewis models, would be unstable to thermal convection. Based on convective heat transfer relationships derived from laboratory experiments and boundary layer theory, Reynolds and Cassen predicted that solid-state convection in an ice lithosphere would refreeze an initially liquid water mantle in a time on the order of 10^8 yr. For possibly higher values of η_m, the refreezing time could be longer than 10^9 yr, but it appears unlikely that chondritic amounts of radioactivity in a differentiated body with the size and mass of Ganymede or Callisto could prevent refreezing of a liquid water mantle that was melted during accretion or differentiation.

Parmentier and Head (1979a,b) calculated the magnitude of viscosity required to cause extensive melting in an undifferentiated ice-silicate body containing radioactive heat sources. These estimates were also based on convective heat transfer relationships derived from theoretical and experimental studies of convection. The viscosity of the interior beneath the near-surface thermal boundary layer, which controls the rate of convective overturn, must exceed about 10^{15} Pa s (10^{16} poise) for melting to occur near the depth of minimum ice-melting temperature. This is greater than the melting temperature viscosity of both low-pressure and high-pressure ice polymorphs (Poirier et al. 1981; Sotin et al. 1985), implying that chondritic heating in an undifferentiated Ganymede or Callisto-like body is not sufficient to cause large amounts of melting. This, however, may not preclude small amounts of melting in a layer near the minimum water ice melting temperature, at depths of about 100 km in an undifferentiated body. Differentiation of ice and silicate by settling of silicate particles through ice in the solid state is slow. If radioactive heating is not sufficient to cause melting, then Ganymede and Callisto, assuming they accreted homogeneously and unmelted, may remain undifferentiated.

Parmentier and Head (1979a,b) also pointed out that an ice-silicate lithosphere may sink diapirically into an underlying liquid layer. Estimates based on the growth time of Rayleigh-Taylor instabilities suggest that diapirs could form quickly, in times on the order of a few years to a few hundred years, and thus contribute significantly to refreezing of an initially liquid water mantle. If the viscosity of ice in the interior were high enough that solid-state thermal convection could not prevent melting, diapirs of liquid water could form in a relatively short time, on the order of 10^2 yr or less.

We point out that the viscosity relationship (Eq. 2) applies to a single deformation mechanism that is either Newtonian (no explicit stress dependence) or non-Newtonian at a fixed stress level. Ice Ih is known or suspected to flow by several competing mechanisms, each of which is dominant in a certain temperature and stress range (see Goodman et al. 1981; Durham et al. 1983; and Weertman 1983). Recent measurements have empirically defined these different mechanisms for ice Ih at temperatures relevant to icy satellite interiors, but at rather high differential stresses (Durham et al. 1983; Kirby et al. 1985). In addition, flow laws have been determined for ice II, III, and VI (Durham et al. 1985; Sotin et al. 1985), and preliminary measurements have been made on ice V (Durham et al. 1985). Extrapolation of the ice Ih laws to the low stress levels appropriate to the convecting interiors of Ganymede and Callisto (~ 1 bar; Schubert et al. 1981) shows that the interiors are less viscous than Eq. (2) indicates if A and η_m are high temperature values. Thus the conclusions discussed here with regard to likelihood of melting or maintaining a deep liquid layer are reinforced. The effects of the higher-pressure rheologies are not yet clear.

Relation to the Ganymede/Callisto Dichotomy

The formation of bright terrain on Ganymede has been frequently considered a product of internal melting, usually a direct product. Therefore thermal evolution models have attempted to explain why more melting might have occurred within Ganymede than within Callisto. Based on the multiple lines of evidence presented in this chapter, however, bright terrain does not represent clean ice emplaced in a primordial ice-silicate crust created by homogeneous accretion. Rather, the evidence favors that at least the upper mantle of Ganymede is differentiated; bright terrain material represents an additional episode of differentiation in which clean water, slush, or ice was emplaced in a crust defined by meteoroidal contamination. The depth of the differentiated upper mantle of Ganymede is not known. The source region for bright terrain material could conceivably underlie undifferentiated material. Callisto may have a differentiated upper mantle as well, but this conclusion rests primarily on evidence that Callisto's optical surface and regolith, globally averaged, are ice-rich; evidence for cleaner ice at depth, based on impact-excavation and resurfacing, is tantalizing but tentative. Thus Callisto may be only marginally differentiated.

Did tectonic (and volcanic) activity on Callisto, similar to that responsible for the formation of bright terrain on Ganymede, ever occur? If so, it must have taken place early enough that the high flux of impacting bodies eradicated all evidence. Has Callisto remained tectonically inactive throughout its evolution, or do the observed differences between Ganymede and Callisto indicate that tectonic activity persisted for longer times on Ganymede? These questions cannot be answered with any degree of certainty, but it is important to consider the predictions of model studies in deciding what may be the most likely explanation for the observed differences. Although bright terrain may not be the immediate result of melting, it is still pertinent to consider models that predict greater melting within Ganymede than within Callisto.

A number of studies have examined more detailed thermal evolution models of an ice-silicate body with radioactive heating. As in earlier efforts, a heat transfer-Rayleigh number relationship was assumed. With given initial conditions, the average temperature of the interior can then be calculated as a function of time (Kawakami and Mizutani 1980; Thurber et al. 1980; Zuber and Parmentier 1984b). Other studies have assumed a thermal steady state in which heat flux from the interior equals the heat generated internally (Cassen et al. 1980*a*; Schubert et al. 1981; Friedson and Stevenson 1983). This would be a good approximation if the interior temperature does not change rapidly. The formulation of these models involves a number of uncertainties and they differ in several important respects. Since the heat transfer-Rayleigh number relationships employed are derived for convecting fluids of constant viscosity, the temperature at which the viscosity appearing in the Rayleigh number should be evaluated is uncertain. For a viscosity that is strongly temperature dependent, the exact choice of this temperature is important. The question of whether or not convective motions penetrate ice phase transitions is not fully resolved. The models of Thurber et al. (1980) include estimates of heat transfer due to diapiric rise of liquid. The others do not. Despite these differences and uncertainties, all the models appear to agree with the general conclusion that chondritic abundances of U, Th and K are not sufficient to cause appreciable amounts of melting of a pure water ice interior if deformation is dominated by the viscosity of ice. If extensive melting and differentiation occur during accretion, greater radioactive heat sources in Ganymede may delay refreezing of a liquid water mantle by as much as 5×10^8 yr (Cassen et al. 1980*a*). In the evolution of a homogeneously accreted and undifferentiated body, the amount of melting that occurs would depend on the initial temperature following accretion, but this melting would be confined to a layer near the depth of minimum melting temperature and would occur only during the earliest portion of the evolution. If Ganymede accreted at a higher temperature than Callisto, this may be a possible explanation of the observed dichotomy. However, differences based solely on varying amounts of radioactive heating due to dissimilarities in their silicate content do not appear to be sufficient.

The differences in silicate content of the two bodies may influence the viscosity of ice-silicate mixtures in their interiors. This has been pointed out by Thurber et al. (1980), Friedson and Stevenson (1983) and Zuber and Parmentier (1984*b*). The higher silicate content of Ganymede would result in higher viscosity and therefore favor greater amounts of melting. Greater amounts of melting in Ganymede could have resulted in more rapid or more complete differentiation. The principal difficulty in evaluating this hypothesis is the uncertainty in estimating the viscosity of ice-silicate mixtures for relatively high silicate volume fractions. The volume fraction of silicates, estimated from the size and mass of the body, also depends on the density of the silicates (see Table I; also Table I of the chapter by Schubert et al.). This is nevertheless a factor that would favor higher temperatures and greater amounts of melting in Ganymede and cannot be excluded as an explanation for the observed differences between Ganymede and Callisto. Small variations in silicate fraction (or the size-frequency distribution of silicate particles) could certainly affect viscosity much more than they would affect radioactive heat source strength.

Cassen et al. (1980*a*) studied the magnitude of tidal heating to determine whether this heat source is capable of maintaining a layer of liquid water near the depth of minimum melting temperature. The importance of tidal heating depends on the orbital eccentricity and the dissipation factor Q. For $Q = 100$, a reasonable value (cf. chapters by Burns and by Peale), the orbit of Callisto must be highly eccentric while for Ganymede a more modest eccentricity is required to maintain a partially molten layer. This difference is primarily a consequence of the smaller mean orbital radius of Ganymede. In contrast to Io and Europa, however, Ganymede experiences an insignificant forced eccentricity due to orbital resonance, and Callisto experiences none. Even if the orbital eccentricity of Ganymede were initially high enough to cause melting, its free orbital eccentricity would decay in a time on the order of 10^8 yr. Even if the youngest terrain on Ganymede is as old as the lunar maria, 3.5 to 4 $\times$ 10^9 yr, tidal heating alone does not appear to be a satisfactory explanation for its origin. The uncertainties in the impact flux leave open the possibility that the bright terrain is older, however.

Accretion and differentiation represent potentially large sources of internal energy. As discussed at the beginning of this chapter and in the chapters by Safronov et al. and Schubert et al., the amount of energy converted to heat and retained in the interior of a growing planet depends strongly on both the rate of accretion and the depth to which individual impacts heat the interior. A number of workers (see, e.g., Parmentier and Head 1979b; Cassen et al. 1980a; Coradini et al. 1982b) suggest or predict greater amounts of accretional melting for Ganymede than for Callisto, but the amounts are not grossly dissimilar.

Thus we are led to the hypothesis that none of the above differences between Ganymede and Callisto are by themselves sufficient to account for the divergent evolution of the two satellites, but because radioactive heating, higher viscosity due to silicates, and tidal and accretional heating all favor Ganymede, they may act in concert. Differentiation may only occur or may be much more complete on Ganymede, and a liquid water layer may persist there for an extended period of time compared to Callisto.

Scenarios for Divergence

An alternate point of view is that Ganymede and Callisto have evolved in radically divergent ways. Several investigators have sought the necessary mechanisms, but as noted earlier in this chapter, special circumstances or assumptions are usually required. Friedson and Stevenson (1983) determined that certain combinations of ice viscosity and silicate packing fraction could lead to a breakdown in the self-regulation of internal temperature by solid-state convection, manifested as melting and differentiation at the depth of the ice minimum melting temperature. It is possible, for a homogeneously accreted but undifferentiated Ganymede and Callisto, that conditions were just right for melting to occur in Ganymede and not in Callisto; but as the authors state, there is no way to determine this *a priori*. Perhaps the most significant point made by Friedson and Stevenson (1983) is that the gravitational potential energy released during initial melting and unmixing is sufficient to drive further differentiation, and may lead to runaway (and possibly complete) differentiation. A related point is put forth by Mueller and McKinnon (1986), who studied the thermal structure in a three-layer Ganymede, specifically a clean ice upper mantle, mixed ice-silicate lower mantle, and silicate core (this structure is assumed to be due ultimately to accretional differentiation, as discussed earlier in this chapter). If the upper mantle is sufficiently shallow (approximately, at pressures above the ice III-ice V transition of ≈ 344 MPa), then it is unlikely that melting can be avoided in the upper thermal boundary layer of the lower mantle. Unless the upper clean ice layer is so thin that it remains conductive, differentiation down to at least the ice III-ice V phase boundary is possible. An internal structure of this sort can be said to possess an accretional trigger.

Lunine and Stevenson's (1982a) model for a deeply melted Ganymede and minimally melted Callisto (discussed in Sec. II) depends on two critical assumptions: the unconfined compressive strengths of ice-silicate planetesimals range between 1 and 100 kPa (so that they break up into a specific range of sizes), and the impact of the planetesimals with the satellite surface does not allow planetesimal fragments to equilibrate thermally with the hot overlying atmosphere (and melt their ice fractions). The latter may not be reasonable if a thin liquid layer is present at the surface (which the authors consider) and the planetesimals are as weak as they suppose. In an impact of this sort, even if at low velocity, the planetesimals should break up and be dispersed in a very steep watery ejecta cone. The planetesimal strength range is ostensibly based on that of carbonaceous chondrites (Baldwin and Shaeffer [1971] quoted in Pollack et al. [1979]). Baldwin and Shaeffer actually estimated ~ 100 kPa (1 bar), and it could be argued that the strength of a cold ice-silicate mixture is greater. The meteorites probably represent regolith samples of their respective parent bodies, after all. The essential point here is that a critical parameter is unknown.

Greenberg (1982a; cf. chapters by Burns and by Peale) proposed that Ganymede, Europa and Io have until recently been in a deeply resonant configuration, possibly episodically so. A concomitant large forced eccentricity for Ganymede would result in a large supply of tidal heat, while Callisto would not be affected at all. Although there is no tectonic evidence on Ganymede for tidally induced fracture within the past billion years, such evidence has not been explicitly searched for. Recent calculations by Greenberg (personal communication, 1985) suggest that deep resonance is dynamically possible but that Ganymede's forced eccentricity never exceeds a few times 10^{-3}. At this level the tidal heating, calculated from relationships in Cassen et al. (1982), is negligible.

Finally, the incorporation of ammonia into Ganymede or Callisto may have had a major effect on their internal evolution. If incorporated as a solid hydrate in a homogeneously accreted but undifferentiated Ganymede or Callisto, a (low melting point, ~ 200 K but dependent on pressure) peritectic or eutectic melt should form. This melt is buoyant and of low viscosity, so resurfacing should occur. Callisto is much more likely to incorporate ammonia (di)hydrate, due to its position in a cooler part of the proto-Jovian nebula (see, e.g., Lunine and Stevenson 1982a), so preferential resurfacing on Ganymede is not an obvious outcome of this scenario. Lunine and Stevenson (1982a) also point out that in the more likely case of accretional melting, ammonia from the nebula may go directly into solution in the initial liquid water upper mantle, even if no ammonia hydrates condense out directly. As such a mantle freezes, ammonia is concentrated in a residual water layer lying either at the depth of the ice minimum melting temperature or, if melting does not initially (or later) extend to this depth, at the top of the unmelted interior. The satellite interior only cools so much over time, however, and due to the combined freezing point depression provided by the ammonia and other solutes, the residual layer is not likely to freeze (Kirk and Stevenson 1983). Whether or not this liquid participates in resurfacing is very uncertain. Whatever reduction in density the ammonia provides is probably offset by salts and other heavier solutes. On the other hand, the vigor of the instability proposed by Kirk and Stevenson (1983) raises the possibility of liquid entrainment in the ascending ice. Detection of ammonia-water ice on any satellite surface has proven to be notoriously difficult (see Lanzerotti et al. 1984; McKinnon 1985), but its possible identification on the surface of Europa (R. H. Brown, personal communication, 1985) makes the presence of ammonia within Ganymede much more tenable.

VIII. CODA

At the end of the Introduction to this chapter the question "Why are Ganymede and Callisto different?" was asked. We have arrived at no certain answer, but below we outline a number of possible pathways or scenarios for divergent internal evolution. Each assumes that accretion is homogeneous rather than heterogeneous and, with the exception of scenario (6), that the initial differences between Ganymede and Callisto are not great.

1. Ganymede and Callisto accrete undifferentiated or minimally differentiated. Runaway differentiation is then initiated in Ganymede only, possibly dependent on major early tidal heating. Evolution of a differentiated Ganymede results in bright grooved and smooth terrain. Early eruption of ammonia-water melt may occur on Callisto, but any record of this is obliterated by cratering. Otherwise, Callisto remains tectonically and volcanically inactive.

2. Ganymede is partially differentiated during accretion, and Callisto is undifferentiated or minimally differentiated. Further differentiation of Ganymede is accretionally triggered, and may go to completion. Implications for both satellites are the same as for scenario (1) above.

TABLE III
Some Galileo Measurements and the Questions They Address

NIMS spectra of various albedo units (bright terrain, dark terrain, etc.)	Composition of optical surfaces; degree of contamination of different terrains
NIMS spectra as a function of latitude and longitude on Ganymede and Callisto	Nature of magnetospheric and micrometeoroid control on surface photometric properties
NIMS spectra of dark halo craters on Ganymede	Confirmation of hypothesis of dark terrain excavation; if confirmed, extension of bright terrain thickness estimates satellite-wide
NIMS spectra of dark ray craters on Ganymede	Nature and source of dark contaminant
NIMS spectra of Valhalla palimpsest and bright units at ring-scarp bases	Determination of composition and possible thickness of bright units on Callisto; relationship to interior
High-resolution imaging of smooth terrain on Ganymede	Emplacement of bright terrain as water, slush or solid ice; identification of source vents
High-resolution imaging of grooved terrain	Origin of multiple grooves as horst and graben, crevasses, or extensional or compressional folds
High-resolution imaging of ring and furrow systems; completion of reconnaissance imaging	Relationship between furrow systems and impact rings; source basins for crater chains on Callisto
Bistatic radar sounding of regoliths and crusts during Earth occultation	Regolith and crust composition and structure
Second-degree gravitational potential harmonic coefficients	Internal structures

3. Ganymede and Callisto are both partially but not deeply differentiated, with Ganymede more deeply melted. In this case, both satellites possess accretional triggers and may proceed to complete differentiation. Both satellites go through an episode of bright terrain volcanism and tectonism, but Callisto's is early enough to be masked later by heavy cratering. The timing requirements of this scenario are severe, but have not been adequately tested.

4. Ganymede and Callisto are both partially differentiated, with melting extending to relatively great depth on Ganymede. In this case, Ganymede is stable against further differentiation (except in its silicate core), while Callisto possesses an accretional trigger and thus becomes the more differentiated of the two satellites. Intuitively, this predicts the wrong re-

spective levels of tectonism and volcanism, but cannot be dismissed out of hand.

5. Ganymede and Callisto are both deeply differentiated during accretion. Neither satellite possesses an accretional trigger and both are stable against further noncore differentiation. Both evolve similarly, but Callisto evolves faster, as in scenario (3). Questions of timing of surface activity and cratering are also similar to scenario (3).

6. Ganymede is deeply differentiated during accretion, and Callisto is minimally melted. Although considered a less likely initial configuration in this chapter, long-term evolution should be along the lines of scenarios (1) or (2).

Some of these possibilities should be eliminated when Galileo reaches the Jupiter system, as the degree of central condensation of both satellites will hopefully be constrained by measurements of their respective second-degree gravitational potential harmonic coefficients, C_{20} (or J_2) and C_{22}. These and other diagnostic Galileo measurements, mostly culled from this chapter, are summarized in Table III. With a successful Galileo mission, the authority of the fashioner of mutations may yet be detected.

Acknowledgments. We would like to thank N. Teichman for her oriental scholarship, P. Schenk for his assistance with the figures, and S. Squyres for a helpful review. This work was supported by various grants from the NASA Planetary Geology and Geophysics Program.

16. THE SATELLITES OF SATURN

DAVID MORRISON
University of Hawaii

TOBIAS OWEN
SUNY/Stony Brook

and

LAURENCE A. SODERBLOM
U.S. Geological Survey

The satellites of Saturn, as revealed by the two Voyager encounters and continu-
ing Earth-based studies, are surprisingly heterogeneous, in spite of the regu-
larity of the satellite system and their apparently similar bulk compositions. All
are composed at least half of water ice, but there may be considerable variation
in more volatile trace constituents, which may have played an important role in
their evolution. Titan is one of the largest satellites in the planetary system, but
since little is known about the surface, it cannot be compared geologically with
Ganymede or Callisto, which it resembles in size and bulk composition. In this
review, we concentrate on the atmospheric chemistry of Titan. The other Saturn
satellites are all more or less heavily cratered, but most also show clear evi-
dence of substantial endogenic modification and resurfacing. In this chapter, we
review the current knowledge of the geology of these satellites, and of what they
can tell us of the geologic and impact history of the Saturn system.

I. INTRODUCTION

The largest satellite of Saturn, Titan, was discovered in 1655 by C.
Huygens, with four additional objects found by J. D. Cassini two decades

later. By the beginning of the 20th century, telescopic observers had found a total of nine Saturn satellites, constituting the largest known regular satellite system. In 1944, G. P. Kuiper discovered the atmosphere of Titan, and during the 1970s considerable progress was made in determining the basic physical properties of the satellites, although fundamental questions, such as the mass and composition of Titan's atmosphere, remained beyond the reach of telescopic studies. Most of what is now known about these satellites is the direct result of the two Voyager spacecraft encounters in 1980 and 1981 (Morrison 1982; Smith et al. 1982).

Immediately following the Voyager encounters, mainly first-order interpretation of the geologic and photometric properties of the satellites of Saturn was carried out. Half a dozen new icy worlds were suddenly available for analysis, including comparisons with the four Galilean satellites of Jupiter revealed by Voyager in 1979. Titan was an object of even greater fascination, with attention focused on the evolution of its massive, organically rich atmosphere. The spacecraft studies have continued to be supplemented by ground-based observations, primarily carried out with large telescopes. The data from the satellites of Saturn have also played a major part in the synthesis of a theory of the impact history of the planetary system (see the chapter by Chapman and McKinnon), which can now be extended from the inner planets out to 10 AU from the Sun.

Many of the initial post-Voyager results were summarized in three chapters of the University of Arizona book, *Saturn* (Gehrels and Matthews 1984), which went to press at almost the same time as the Natural Satellites Meeting (IAU Colloquium #77) that forms the basis for much of the present volume. These three chapters covered the geologic perspective on the satellites (Morrison et al. 1984), the optical properties (Cruikshank et al. 1984) and Titan (Hunten et al. 1984). Because relatively little new information has become available since these summaries were completed, it has not seemed appropriate to include chapters at the same level of detail in the present volume. Rather, this book reflects a swing of the pendulum back from the satellites of Saturn toward the Jovian satellites. With the forthcoming Galileo mission to Jupiter, we may anticipate that the Jovian satellites will continue to be a major focus of interest into the 1990s.

The present chapter is in large part an edited and updated version of Morrison et al. (1984), with the addition of a major new section on the atmosphere of Titan emphasizing work done since the completion of the review by Hunten et al. (1984). The interested reader is referred to the satellite chapters in *Saturn* for additional background.

II. SURVEY OF THE SATELLITES

Table I summarizes the basic physical data for the 17 known satellites of Saturn; comparable data are tabulated in the book's introductory chapter. This

TABLE I
Satellites of Saturn[a]

Satellite		a(R$_S$)	P(day)	e	i(°)	$V(1,0)$[d]	p_v[d]	R(km)[e]	$M(10^{23}$ g$)$[f]	ρ(g cm^{-3})
SXVII	Atlas	2.276	0.602	0.002	0.3	9.5	0.4	20 × ? × 10	—	—
SXVI	Prometheus	2.310	0.613	0.004	0.0	6.2	0.6	70 × 50 × 37	—	—
SXV	Pandora	2.349	0.629	0.004	0.1	6.9	0.5	55 × 45 × 33	—	—
SX	Janus	2.51[b]	0.69[b]	—[c]	—[c]	4.9	0.6	110 × 95 × 80	—	—
SXI	Epimetheus					6.1	0.5	70 × 58 × 50	—	—
SI	Mimas	3.08	0.94	0.020	1.5	3.3	0.6	197 ± 3	0.46 ± 0.05	1.4 ± 0.2
SII	Enceladus	3.95	1.37	0.004	0.0	2.2	1.0	251 ± 5	0.8 ± 0.3	1.2 ± 0.5
SIII	Tethys	4.88	1.89	0.000	1.1	0.7	0.8	530 ± 10	7.6 ± 0.9	1.2 ± 0.1
SXIII	Telesto	4.88	1.89	—	—	9.1	~1.0	15 × 10 × 8	—	—
SXIV	Calypso	4.88	1.89	—	—	9.4	0.7	12 × 11 × 11	—	—
SIV	Dione	6.26	2.74	0.002	0.0	0.8	0.6	560 ± 5	10.5 ± 0.3	1.4 ± 0.1
SXII	Helene	6.26	2.74	0.005	0.2	8.8	0.6	17 × 16 × 15	—	—
SV	Rhea	8.73	4.52	0.001	0.4	0.1	0.6	765 ± 5	24.9 ± 1.5	1.2 ± 0.1
SVI	Titan	20.3	15.95	0.029	0.3	−1.2	0.2	2575 ± 2	1345.7 ± 0.3	1.88
SVII	Hyperion	24.6	21.28	0.104	0.4	4.8	0.3	205 × 130 × 110	—	—
SVIII	Iapetus	59	79.3	0.028	14.7[c]	0.7–2.5	0.4–0.08	718 ± 18	18.8 ± 1.2	1.2 ± 0.1
SIX	Phoebe	215	550	0.163	150	6.8	0.06	110 ± 10	—	—

[a]Table adapted from Morrison et al. (1984); see also Tables III and IV in the introductory chapter.
[b]No orbital determinations have been made for these coorbital satellites since their 1982 "orbital exchange."
[c]Variable.
[d]Thomas et al. (1983a); Morrison (1982b); Smith et al. (1982); Cruikshank et al. (1984).
[e]Smith et al. (1982); Thomas et al. (1983a,b,c); Davies and Katayama (1983,1984). Uncertainties for the tri-axial dimensions of the small satellites are generally ~ 10% (Thomas et al. 1983a,c).
[f]Tyler et al. (1982), except Enceladus from Kozai (1976).

TABLE II
Photometric and Radiometric Properties of Saturn's Satellites[a]

Satellite	Voyager Phase Angle Coverage	Voyager Clear Filter: $\lambda \sim 0.47\ \mu m$ [b]			Voyager Infrared (IRIS) Results [c]	
		Geometric Albedo p (± 0.1)	Phase Integral q (± 0.1)	Bond Albedo $p \cdot q$ (± 0.1)	Subsolar Point Temperature (K)	Bolometric Bond Albedo A_{bol}
Rhea	2–68°	0.65	0.70	0.45	100 ± 2	0.65 ± 0.03
Dione	7–74°	0.55	0.80	0.45	—	—
Tethys	10–42°	0.80	0.75	0.60	93 ± 4	0.73 ± 0.05
Enceladus	12–42°	1.00	0.85	0.9	75 ± 3	0.89 ± 0.02
Mimas	12–75°	0.75	0.80	0.60	—	—

[a] Table taken from Cruikshank et al. (1984).
[b] After Buratti et al. (1982) and Buratti (1983).
[c] After Hanel et al. (1981a,1982).

system includes: Titan, the one object in the Saturnian system that is comparable to the large Galilean satellites; the 8 additional icy satellites that are readily seen from Earth and for which Voyager provided a basic geologic survey, and 8 smaller satellites discovered either by Voyager or by telescopic observations made around the time when the rings were seen edge-on from Earth. Table II, taken from Cruikshank et al. (1984), provides a similar summary of photometric data for the larger satellites; see also Figs. 10, 11, 15 and 16 of the chapter by Veverka et al. Thermal models of the intermediate satellites of Saturn are presented in Figs. 11–15 of the chapter by Schubert et al. with interior models given in Fig. 4 of the same chapter.

Recent work on Titan is discussed in Sec. III. For the photometric properties of the smaller satellites we refer primarily to the review by Cruikshank et al. (1984), although some recent work is specifically mentioned. Primarily, however, Sec. II deals with the geologic perspective provided by Voyager for the middle-sized satellites of Saturn; sketches of most of these satellites are shown in the Map Section.

Iapetus

Perhaps the most mysterious of the Saturn satellites, Iapetus is unique in its range of surface albedos, from ~ 0.5 (a value typical of icy objects) to 0.05 in the interior parts of its leading hemisphere. Before Voyager, ground-based observations had established the size and approximate albedo range for Iapetus, as well as the virtually perfect longitudinal symmetry of the dark material with respect to direction of orbital motion. The diameter of Iapetus from Voyager data is 1436 ($\pm$ 36) km (Davies and Katayama 1984).

Voyager 1 provided images at a maximum resolution of 50 km/line-pair, primarily showing the Saturn-facing hemisphere and the boundary between the leading (dark) and trailing (light) sides. At this resolution, little topographic detail could be seen, but a large equatorial dark ring ~ 300 km in diameter centered near longitude 300° (i.e., extending into the light side) strikingly interrupted the symmetry of the albedo boundary. Voyager 2 passed closer (909,000 km) and imaged primarily the complementary (anti-Saturn) hemisphere. The highest-resolution images (17 to 20 km/line-pair) were of the region stretching from the north pole to the equator and from longitude 190° to 290° and showed primarily the high albedo surface (Fig. 1). These images revealed large numbers of craters and helped to clarify the character of the boundary of the dark region.

The airbrush map of Iapetus (Fig. 2) obtained from Voyager images shows that the dark material is strongly concentrated not just on the leading hemisphere, but distributed symmetrically about the apex of orbital motion. Photometric analyses (Squyres and Sagan 1982; Goguen et al. 1983b; Squyres et al. 1984) indicate that the reflectance of the surface is strikingly bimodal, with values in the range 0.4 to 0.5 for the bright side and 0.02 to 0.04 for the dark side. The albedo is lowest near the apex, but the brightest area is not at

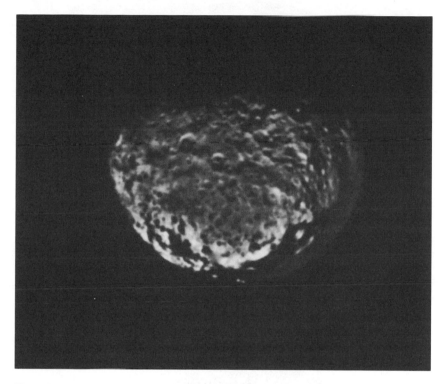

Fig. 1. Best Voyager 2 view of Iapetus. This image shows the northern hemisphere, primarily on
the trailing (bright) side, at a resolution of 19 km/line-pair. Filtering has emphasized the heavily cratered topography of the bright side. (Voyager/JPL image 260-1477.)

the antapex. In that part of the boundary imaged by Voyager 2, the width of
the transition zone from bright to dark surface material is only a few hundred
kilometers.

The highest-resolution Voyager images reveal that the bright side (and
particularly the north polar region) is heavily cratered, with a surface density
of 205 ($\pm$ 16) craters ($D > 30$ km) per 10^6 km^2 (Plescia and Boyce 1983).
Extrapolated to 10 km diameter, this density corresponds to > 2000 craters ($D
> 10$ km) per 10^6 km^2, a value comparable to that on other densely cratered
objects, such as Mercury, Callisto or the lunar highlands. Unfortunately, due
to low signal to noise in the images, no craters or other topographic detail can
be seen unambiguously in the dark part of the surface, so we cannot establish
whether this large crater density extends over both hemispheres of Iapetus.

A distinctive feature of the dark-light boundary on Iapetus is the presence of a number of dark-floored craters in the brighter material, but the absence of light-floored or halo craters (or other white spots) within the dark
material. The dark-floored craters and other dark spots below the resolution of
the pictures extend near the equator to longitude 270°, the antapex point.

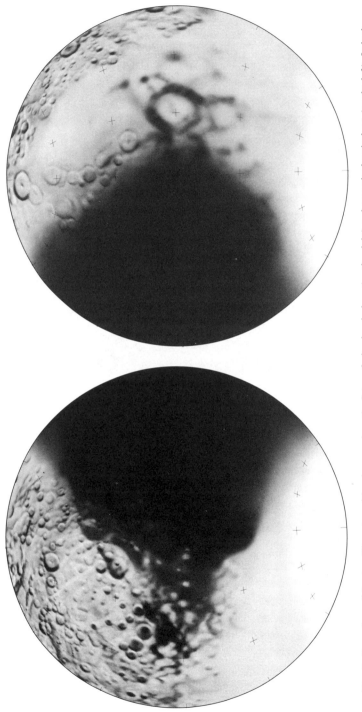

Fig. 2. Two views of Iapetus. These two airbrush maps show the Saturn-facing hemisphere (on the right) and opposite hemisphere (on the left). In both cases north is at the top. The dark region on the leading hemisphere faces in the direction of the satellite's motion about Saturn. These U.S. Geological Survey maps are presented in an equal area projection (maps prepared by J. L. Inge and E. M. Lee under the direction of R. M. Batson).

There are some craters (for instance, near 200°) that exhibit dark areas suggestive of deposits that face the boundary. These oriented dark deposits and the appearance of the contact regions generally suggest that the dark material is superimposed on the bright terrain and is therefore younger. The absence of bright craters within the dark material indicates that the deposit is either thick (> 10 km) or that freshly exposed material from below is darkened or buried on a time scale short relative to the interval between major cratering events (perhaps $\sim 10^8$ yr).

An important additional clue to the relationship between dark and light materials is provided by the density of Iapetus, 1.16 ($\pm$ 0.09) g cm^{-3} (Campbell and Anderson 1985). This value is similar to that of the other icy satellites of Saturn, and consistent with models in which water ice is the primary bulk constituent.

In view of Voyager's poor imaging coverage of the leading hemisphere, the nature and origin of the dark material on Iapetus remains problematical. The strong symmetry with respect to the direction of orbital motion virtually demands some external control, if not an external origin, for the dark material. Yet the topographic relationship of dark deposits in the floors of bright-side craters suggests an internal origin for at least these deposits. It may be that two or more mechanisms are involved.

Bell et al. (1985) have carried out the most extensive analysis of the possible composition of the dark material. Using a set of high-accuracy spectrophotometric data obtained at Mauna Kea in the range from 0.3 to 2.6 μm, they were able to isolate the spectral reflectance of the dark material by removing the effects of the residual water ice that remains even under optimum geometry. They find (Fig. 3) that the isolated dark-unit material is very red in the visible and near infrared, but that it gradually flattens near 2.0 μm (see the chapter by Clark et al., in particular Fig. 17). Bell et al. were able to match this spectrum with a laboratory mixture of simulated meteoritic organic polymers (10%) and hydrated silicates (90%).

If the dark deposits are primarily externally controlled, a mechanism is required for their production. It has been suggested that this dark material is a coating of dust eroded from the surface of Phoebe (or some undiscovered other external satellites in a retrograde orbit) and is drifting inward under Poynting-Robertson drag. Alternatively, impacts of material from Phoebe or elsewhere may modify the surface of the leading hemisphere with the result that low-albedo material becomes concentrated at the surface; in such a scenario the dark material is endogenic, but exogenic impacts concentrate it. Bell et al. (1985) favor this exogenic hypothesis and suggest that this dark material is a very primitive component of the original condensate that formed Iapetus.

Rhea

A near twin of Iapetus in size, but without its dark material, Rhea may represent a relatively uncomplicated archetype of the icy satellites of the outer

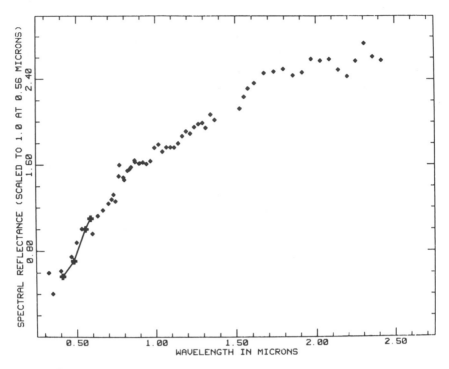

Fig. 3. Visible and infrared spectral reflectance of the Iapetus dark material obtained by Bell et al. from telescopic observations by means of a linear mixing model. (Figure from Bell et al. 1985.)

solar system; its spectrum is shown in Fig. 16 of the chapter by Clark et al. Rhea has a diameter of 1530 km and a density of 1.24 ($\pm$ 0.05) g cm^{-3} (Campbell and Anderson 1985). Its geometric albedo is 0.6, similar to that of the poles and trailing hemisphere of Iapetus. Although in no way mimicking the albedo difference of Iapetus, Rhea does exhibit its own type of longitudinal asymmetry, which is paralleled by its neighbor Dione.

On a hemispheric scale, the asymmetry of Rhea shows up as a brightness variation of 0.2 mag well aligned with orbital position, but having the opposite phase as Iapetus, with a broad maximum brightness on the leading side and a sharper minimum on the trailing (orbital longitude 270° $\pm$ 5°) (Noland et al. 1974; Cruikshank 1979; see also the chapters by Clark et al. and Veverka et al.). Even at low resolution (100 to 200 km/line-pair), Voyager pictures also showed the asymmetry: the leading side was fairly uniform in brightness, while the trailing side appeared mottled with spots or patches of different albedo, leading to an overall lower value.

The best resolution of much of the trailing hemisphere was only ~ 50 km/line-pair. The highest albedo areas are wispy in form, giving the impression of cloud bands or broad irregular rays of ejecta superimposed on a darker, cratered terrain. Although roughly similar to some of the bright streaks on Ganymede, these Rhea features do not appear in radial patterns, nor is there any indication that they originate at impact centers. There are too few well-defined streaks for a statistical analysis of their orientations, but their breadth and irregular curvilinear shapes are not diagnostic of either external or internal processes. There are no comparable features on the surfaces of the Galilean satellites.

The highest-resolution images of any of the icy Saturnian satellites were obtained of Rhea by Voyager 1 ($\sim$ 1 km/line-pair; Fig. 4). The orbital position of the satellite was such that primarily the bright leading hemisphere was illuminated, but these images include the transition to darker terrains at about longitude 315°. Throughout both terrains, the surface is dominated by well-formed impact craters and resembles the highland provinces of the Moon and Mercury. These craters are notably different from the flattened, relaxed forms common on Ganymede and Callisto, apparently because of the stiffer ice crust and lower surface gravity on Rhea (cf. Chapman and McKinnon's chapter). Also apparently absent are compact ejecta blankets, possibly another consequence of low surface gravity. The most unusual feature of these craters is the presence on some crater walls of curious bright patches. These may be exposures of relatively fresh ice, but the exact mechanism of emplacement is unclear.

Plescia and Boyce (1982) have analyzed the crater size-frequency distributions in three areas of Rhea: two polar regions and an equatorial area near longitude 30° in the leading hemisphere. They find that the demarcation between the polar terrain that has retained large craters and that in which they are absent is very near 0° (the sub-Saturn longitude). The demarkation apparently defines the boundary of a major resurfacing event that took place near the end of the period of heavy bombardment of this satellite. However, Lissauer (1985) has more recently concluded that some of these apparent variations are an artifact of the lighting geometry in the Voyager images, and that the crater density is more uniform than previously thought.

All the areas of Rhea in which crater counts have been carried out are very heavily cratered for sizes below $\sim$ 30 km (see Fig. 13 in the chapter by Chapman and McKinnon). Moore et al. (1986) have studied the craters and related landforms on Rhea in some detail, concluding that there has been considerable ejecta mantling on parts of the surface, perhaps associated with two large multiringed basins. They suggest several distinct terrain ages, with impact cratering interrupted by periods of endogenic resurfacing. However, in view of the questions raised by Lissauer (1985b), it is probably premature to derive a detailed geologic history for Rhea until agreement is reached on the interpretation of the Voyager images.

Fig. 4. Mosaic of the north polar region of Rhea. These images were acquired during the close approach of Voyager 1 to the satellite from a distance of roughly 50,000 km. Due to the high angular velocity of Rhea relative to the spacecraft, image motion had to be compensated for by slewing the scan platform during the exposure of each image, resulting in an effective resolution of ~ 1 km/line-pair. The north pole is roughly centered along the terminator. (Voyager/JPL image P-23177.)

Dione

Although smaller ($D = 1120$ km) than Rhea, Dione exhibits most of the same general features, plus a few new wrinkles of its own. It has the highest well-determined density among the inner satellites: 1.43 (± 0.04) g cm^{-3}. Like Rhea, it has a bright icy surface (albedo ~ 0.5), is brighter on its leading hemisphere than its trailing, and has a mottled trailing hemisphere with brighter, wispy terrain superimposed on a darker background. The lightcurve amplitude is 0.6 mag, the greatest of any large satellite except Iapetus. Dione also shows considerably more evidence of internal activity than does Rhea.

Buratti et al. (1984) find from an analysis of disk-resolved Voyager photometry that both Rhea and Dione have little or no limb darkening, with photometric functions similar to that of the Moon. Thus, in spite of their high albedos, these two satellites do not display the photometric signature of multiple scattering. In contrast, there is substantial limb darkening on Mimas, Tethys and Enceladus.

Although the Voyager imaging resolution (~ 2 km/line-pair) is not as great as on Rhea, the coverage is more complete, thanks to Dione's more rapid

Fig. 5. Varied terrains on Dione. This Voyager 1 image at a resolution of ~ 4 km/line-pair shows a portion of the Saturn-facing hemisphere. Near the terminator in the leading hemisphere is a heavily cratered region including many large craters, while some of the enigmatic wispy terrain can be seen on the right. (Voyager/JPL image P-23101.)

rotation (Fig. 5). Craters can be clearly identified in the trailing hemisphere, and while none is as large as the biggest on the leading side, there is no demonstratably significant difference in the cratering population. The most striking features of this side of Dione are, however, those that make up the wispy terrain. Imaged at a resolution of a few km, their contrast exceeds a factor of 3, with reflectivities ranging from 0.2 for parts of the background to 0.7 in the bright streaks (Buratti 1983,1984). One set of markings appears radial to a large ($D \sim 50$ km) crater, but generally they are curvilinear and complex and do not resemble crater rays.

Where they can be traced westward into higher-resolution images, the wispy bright markings appear more in the nature of relatively narrow, bright lines, some of which extend into the leading hemisphere, where they lose contrast against the brighter background. A few of these narrow streaks are associated with topographic features—narrow linear troughs and ridges—cutting across the leading hemisphere. Some of these features, particularly in Palatine Chasma in the south polar region, strongly resemble grabens and horsts (Moore 1984). This association suggests that many, if not all, of the bright features may be controlled by a global tectonic system of fractures or faults. They may consist of surface deposits of ice outgassed along the fracture system, perhaps carried up from the interior by more volatile substances such as ammonia or methane.

Dione has a low overall crater population relative to Rhea (see Fig. 18 in Chapman and McKinnon's chapter, or Fig. 6d in the chapter by Veverka et al.). It also particularly lacks craters with $D > 30$ km, suggesting that the impacting flux recorded on its surface may be the same as that responsible for cratering the apparently younger terrains on Rhea. In addition, the crater densities are much more varied on Dione, indicating a more active geologic history. Terrain types in the leading hemisphere have been described as ranging from heavily cratered to smooth, lightly cratered plains.

Plescia and Boyce (1982) have analyzed the crater size-frequency distributions for three areas: a rough terrain with numerous large craters, centered at 50°S, 20° longitude; a plains unit with intermediate crater density centered on the equator at 50° longitude; and a smooth plains unit on the equator at 65° longitude. Even the heavily cratered area has a 20 km cumulative density of 270 ($\pm$ 50) craters per 10^6 km^2, significantly less than the least-cratered regions investigated on Rhea. The 20 km densities for the intermediate and lightly cratered plains are both 45 ($\pm$ 25) craters per 10^6 km^2, but these curves rise steeply, and the three units may not be distinguishable for crater diameters below 10 km. The slopes of the curves are about -2 for the heavily cratered terrain and perhaps as high as -4 for the plains unit, suggesting to Plescia and Boyce that the two surfaces may reflect different populations of impacting bodies as well as different ages.

The presence of extensive resurfaced plains units on Dione as well as the existence of troughs or valleys hundreds of kilometers in length indicate a relatively high level of post-accretional alteration of the surface. In spite of its small size, Dione's level of endogenic activity exceeded in intensity and duration that of the larger Rhea. Moore (1984) has synthesized a history for Dione that is probably tectonic and volcanic, involving a global expansion that produced large-scale lineaments near the end of the heavy bombardment, followed by eruption (probably of an NH_3-H_2O melt) to form the less-cratered plains units. Moore then suggests a long period of interior cooling, global shrinkage and horizontal compression of the crust.

Tethys

With a diameter of 1060 ($\pm$ 20) km Tethys is a near twin of Dione. Its density, unfortunately, is not well determined. Its albedo (0.8) is notably higher, however, than any satellite encountered so far in this review (Table II), and its trailing hemisphere lacks the darker regions and conspicuous wispy terrain of Rhea and Dione. As shown by Buratti et al. (1984), Tethys also has a photometric function substantially different from that of Dione. All parts of the surface of Tethys are densely cratered. Voyager 1 imaged primarily the Saturn-facing hemisphere with a resolution of $\sim$ 15 km/line-pair, while Voyager 2 viewed principally the opposite hemisphere at resolutions $\leq$ 5 km/line-pair (Fig. 6), with a single fractional frame at 2 km resolution.

Two outstanding topographic features dominate the images of Tethys (see the Map Section): a giant crater and a trench or valley of global proportions. The crater, named Odysseus, is 400 km in diameter ($\sim$ 40% of the diameter of the satellite) and lies in the leading hemisphere centered at 30°N, 130° longitude. Odysseus is the largest crater in the Saturnian system, with only the dark ring on the Saturn-facing side of Iapetus of even comparable dimensions.

Images of Odysseus on the limb of Tethys show that the floor has rebounded by tens of kilometers, so that today it presents a convex face with about the same radius of curvature as the rest of the surface, rather than the concave shape associated with smaller craters. Such a rebound by creep or viscous flow in the icy lithosphere is not unexpected, with a thermal gradient of $\sim$ 0.1 K km^{-1} sufficient for ice to flow in a body of this size (Passey 1983).

Ithaca Chasma, the great trough or valley system of Tethys, stretches around at least 3/4 of the circumference along a great circle centered roughly on Odysseus and passing near the north pole. Its width is $\sim$ 100 km, and its total area amounts to nearly 10% of the surface of Tethys. Over much of its length, Ithaca Chasma is complex, with multiple subparallel walls and troughs. Rough estimates suggest that the walls are several kilometers high (see the discussion in the chapter by Squyres and Croft; also McKinnon 1985b).

The best global views of Tethys show that it, like Dione and Rhea, exhibits terrains of different geologic ages. The well-imaged region north of the equator near longitude 60° is rough, hilly and densely cratered (cumulative density for $D < 20$ km of 500 ($\pm$ 92) craters per 10^6 km^2), with most of the

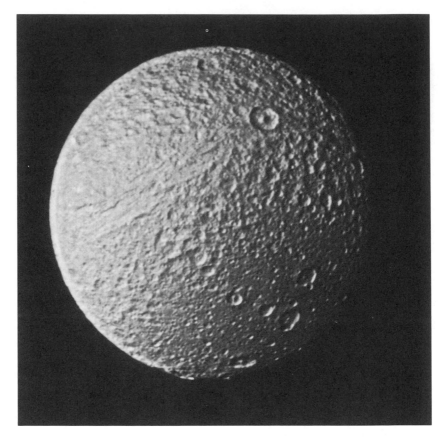

Fig. 6. Best global view of Tethys. This Voyager 2 image has a resolution of ~ 4 km/line-pair, similar to that of Fig. 5. The boundary between a heavily cratered unit (top right) and a more lightly cratered plain (bottom right) is easily distinguished. Ithaca Chasma stretches diagonally across the upper part of the image, with a large fresh-looking crater superimposed on it. (Voyager/JPL image P-24065.)

larger craters highly degraded (Plescia and Boyce 1983). Part of the trailing hemisphere is less rugged and is best described as a lightly cratered plains unit. (crater density 800 ($\pm$ 66) craters per 10^6 km^2 for $D > 10$ km; Plescia and Boyce 1983). As with Dione, it appears that episodes of surface flooding persisted after the accretionary era.

Because of its large area, it has been possible for Plescia and Boyce (1983) to measure crater densities inside Ithaca Chasma. The values are similar to those in the adjacent plains, with 20 km crater numbers of 120 ($\pm$ 50) craters per 10^6 km^2. Thus, while Ithaca Chasma is the least cratered and possibly the youngest major unit identified on Tethys, its age is not dramatically less than that associated with the plains deposits on either Tethys or Dione.

Enceladus

Mimas and Enceladus, the innermost of the classical satellites, are both small, with well-determined diameters of 394 ($\pm$ 6) km and 502 ($\pm$ 10) km, respectively (Davies and Katayama 1983a). However, they are not twins; in fact, the striking differences between them provide one of the major mysteries of the Saturn system.

Even before the Voyager missions it was apparent that Enceladus had a remarkably high albedo, and the spacecraft images quickly established that the surface was strongly backscattering, with normal reflectance > 1.1, substantially greater than that of any common natural surface, such as freshly fallen snow. The mean geometric albedo is 1.04 ± 0.15, and the Bond albedo is 0.89 ± 0.02 (Veverka et al. 1985), resulting in the lowest surface temperature among the Saturn satellites (Cruikshank et al. 1984). The corresponding single-scattering albedo for surface grains is a remarkably high 0.99.

Voyager 2 imaged the northern half of the trailing hemisphere at a resolution of 2 km/line-pair. This picture (Fig. 7) revealed a surface quite unlike anything else seen in the Saturnian system, with striking indications of large-scale endogenic activity.

Enceladus has a wide diversity of terrains, ranging from substantially cratered (cumulative density for $D > 10$ km of 800 ($\pm$ 300) craters per 10^6 km^2, similar to the intermediate regions of Dione and Tethys) down to regions in which all impact craters have been erased (density for $D > 10$ km of < 100 craters per 10^6 km^2). In addition, the craters themselves display a range of forms indicating various degrees of crustal relaxation, and a system of peculiar curvilinear ridges up to 1 km in height is a prominent feature of the crater-free areas.

At least five distinct terrains have been identified on the basis of crater populations and other landforms; ct$_1$ with highly flattened craters 10 to 20 km in diameter; ct$_2$ with well-preserved craters in the same size range; cp with bowl-shaped craters 5 to 10 km in diameter but at a lower density; sp$_1$, a lightly cratered smooth plains unit with a rectilinear groove pattern; and finally the youngest units of crater-free smooth plains (sp$_2$) and ridged plains (rp). The locations of these units are shown in a geologic sketch map (Fig. 8). In unit rp, the spacing between ridges ranges from 7 to 15 km, while the ridge heights determined from photoclinometry (Passey 1983) range from a few hundred meters to 1.5 km. (For comparison, the groove spacing on Ganymede is 3 to 10 km, and the relief is < 1 km.)

Where the ridged terrain stretches in a corridor nearly to the north pole, its contacts with the older cratered plains units reveal a complex history. Evidently the sequence of events included replacement of the older units, formation and subsequent flattening of craters on the new surface, and later emplacement of the ridged terrain. Additional intermediate stages in the evolution of the crust are suggested by some of the details of the contact between

Fig. 7. High-resolution mosaic of Enceladus. These are the best Voyager 2 images of this satellite, with a resolution of 2 km/line-pair. The varied terrain types are delineated on a geologic
   sketch map (see Fig. 8). (Voyager/JPL image P-23956.)

the younger and older terrains. While absolute ages cannot be assigned, the
crater-free terrains are almost certainly younger than 10^9 yr, and may be
younger than 10^8 yr.

Passey (1983) has used photoclinometry to measure the profiles of large
craters on Enceladus. On the basis of these profiles, he distinguishes different
degradation histories even for regions of similar crater density (see the chapter

by Chapman and McKinnon). For example, in one heavily cratered region all of the larger craters are extremely flattened and have bowed-up floors, indicative of relaxation by lithospheric creep. In another part of ct_1, $\sim 75\%$ of the craters with $D > 8$ km are flattened, while 25% appear fresh and unrelaxed. Passey concludes that the viscosity of the lithosphere in the heavily cratered regions has been between 10^{24} and 10^{25} P. The exact time scale for viscous relaxation is unknown, but Passey argues that it should be between 10^8 and 4

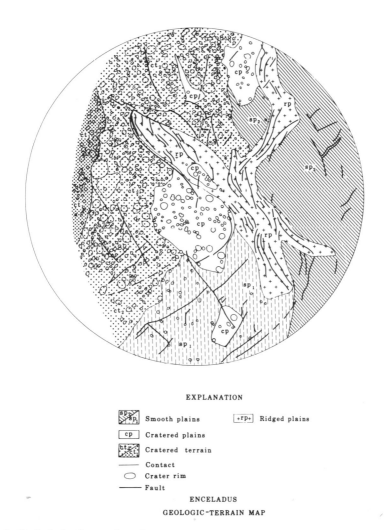

EXPLANATION

sp₂ / sp₁	Smooth plains	+rp+	Ridged plains
cp	Cratered plains		
ct₂ / ct₁	Cratered terrain		
——	Contact		
◯	Crater rim		
——	Fault		

ENCELADUS

GEOLOGIC-TERRAIN MAP

Fig. 8. Geologic sketch map of terrains on Enceladus. Compiled from images like that in Fig. 7, the units are identified on the basis of the abundance of superposed impact craters or sinuous ridges where impact craters are absent.

$\times\ 10^9$ yr. The variations in degree of crater relaxation from one region to another suggest that the heat flow has varied from one terrain unit to the next.

In spite of the great range of surface ages indicated by the varied topography, the photometric properties of the surface of Enceladus are remarkably uniform. Buratti et al. (1983) have noted that this uniformity applies to color and scattering properties as well as albedo. The total range of albedo over the surface is $\sim 20\%$, with the trailing hemisphere lighter, but even these small differences are not correlated with the ages of the major trailing-side terrain units discussed above, which differ in albedo by $\leq 2\%$. As Buratti et al. argue, both the unusual nature of the surface optical properties and the uniformity of their distribution suggest that the optical properties are determined by a ubiquitous surface layer of relatively recent age (see the chapter by Veverka et al.).

A further intriguing clue to the unique history of Enceladus may lie in the relationship to the tenuous E ring of Saturn, which has a peak in its intensity near the orbit of the satellite (Baum et al. 1981). Photometric studies (Pang et al. 1982c) indicate that the E ring is composed of spherical, micrometer-sized particles, which should have lifetimes of $< 10^4$ yr (Haff et al. 1983; Johnson et al. 1984b). It is therefore suggested that the E ring is a recent phenomenon associated with Enceladus. Possibly the same event that produced the uniform, uncontaminated optical surface of the satellite also sprayed out ice particles to form the E ring, although both sputtering and continued micrometeoroid erosion may also be important. Whatever the nature of this event, it probably occurred within the past few thousand years.

The extraordinary nature of Enceladus requires an explanation, presumably involving an unexpectedly large internal heat source. Since the mass of the satellite is not well determined from either groundbased or spacecraft data, its density remains relatively uncertain, but there is no reason to expect an excess of radionucleides, certainly not at the level required to maintain the satellite in a molten state to the present time. Tidal heating has been suggested (Yoder 1979b), but Poirier et al. (1983) have argued that tidal effects could not provide the required heating unless the past eccentricity of the orbit of Enceladus was substantially greater than seems likely (cf. Lissauer et al. 1984). Thus, the problem of the internal activity of Enceladus remains unresolved (see also the chapters by Peale, by Schubert et al. and by Stevenson et al.).

Mimas

Mimas, with a diameter of 394 ($\pm$ 6) km, is the smallest and innermost of the classical, regular satellites of Saturn. It was imaged best by Voyager 1, which provided coverage primarily of the southern hemisphere at a resolution of ~ 2 km/line-pair. Its surface is exceedingly heavily cratered, with little indication of major endogenic modifications. The albedo is fairly uniform, and the lightcurve amplitude is < 0.1 mag (Buratti 1983).

The most striking feature of Mimas is a very well-preserved 130 km

Fig. 9. Leading hemisphere of Mimas. The 125 km diameter crater Herschel is prominent in a heavily cratered surface. Resolution is ~ 4 km/line-pair. (Voyager/JPL image P-23210.)

crater, Herschel, nearly centered on the leading hemisphere (Fig. 9). The crater walls are ~ 5 km high, and parts of the floor may be ~ 10 km deep. A large central peak has a base 20 to 30 km across and rises to ~ 6 km above the crater floor. Unlike the large crater Odysseus on Tethys, Herschel does not appear significantly modified by crustal creep or relaxation.

The next largest crater on Mimas is less than half the size of Herschel, and most are < 50 km in diameter. The cratering, although everywhere heavy, is not uniform. On the leading hemisphere, west of Herschel, there are many craters with $D > 40$ km, while the south polar and adjacent areas are deficient in craters with $D > 20$ km and are devoid of craters with $D > 30$ km. Plescia and Boyce (1982) measured a cumulative density ($D > 10$ km) of 1250 ($\pm$ 490) craters per 10^6 km^2 near the south pole and suggested that the values in the leading hemisphere near Herschel might be several times higher yet. They note an inflection in the size-frequency distribution curves between $D = 15$ and $D = 40$ km which suggests a mixing of two populations of craters, perhaps the result of resurfacing. Evidence exists, therefore, for some modification of the surface during the period of heavy bombardment, perhaps similar to the processes seen more clearly on Rhea, Dione and Tethys.

Small Satellites

The small satellites of Saturn are defined here to include 2 classical satellites, Hyperion and Phoebe, as well as 10 objects in the inner part of the system, between the orbit of Dione and the outer edge of the A ring. These include Atlas, the F ring shepherds, the coorbitals, several Lagrangian satellites and two satellites in ring gaps that are inferred from the waves they induce even though at this writing their images have not yet been identified (Showalter et al. 1985a; Marouf and Tyler 1985). Some properties of these small satellites are summarized in Table I (see also the chapter by Thomas et al.).

Phoebe, the outermost satellite, orbits Saturn at a distance of 215 R$_S$ in a retrograde sense. Voyager 2 observed Phoebe from a range of $\sim 2 \times 10^6$ km, providing images over a 24 hr period at a resolution of ~ 40 km/line-pair. Our discussion of these images is based on the analysis of Thomas et al. (1983b).

Phoebe is a dark, approximately spherical object with a diameter of ~ 220 km and a geometric albedo of 0.05 to 0.06. Its rotation is nonsynchronous, with a period of 9.4 (± 0.2) hr (prograde) determined from both the disk-integrated Voyager lightcurve and by tracking individual markings. The most prominent surface markings are scattered brighter patches with contrast up to 50% relative to the darker, bland areas. The Voyager color data agree with groundbased observations in showing a flatter spectrum than that of the dark side of Iapetus, excluding Phoebe-type dust as the sole darkening agent for Iapetus (Cruikshank et al. 1983; Bell et al. 1985).

At the Voyager resolution no craters or other topography can be identified unambiguously. This observation does not mean that Phoebe is uncratered; with only 11 pixels across the satellite image and the low phase angles of the Voyager images, Thomas et al. argue that even the largest expected craters ($D \sim 100$ km) would be only marginally visible. Generally, its dark surface and retrograde orbit suggest that Phoebe is a captured object with primitive surface composition (see the chapter by Burns), but there is little direct information in the Voyager observations concerning either its origin or its geologic history.

Hyperion is the largest of the small satellites, occupying a moderately eccentric orbit between Titan and Iapetus. Voyager 2 images at resolutions of ~ 10 km/line-pair reveal it to be an irregularly shaped object $\sim 370 \times 280 \times 225$ km, with angular features and facets (Thomas and Veverka 1985). While nearly as large as Mimas or Enceladus, its rugged profile is in dramatic contrast to their smooth, spherical shapes. It is surely the fragmented remnant of a larger parent body. Hyperion has a dirty ice surface, with a lower albedo and weaker ice bands than those of the inner, regular satellites (Tholen and Zellner 1982; Cruikshank et al. 1984; see also the chapters by Cruikshank and Brown and by Clark et al., specifically Fig. 18). Hyperion's surface includes one crater 120 km in diameter with ~ 10 km vertical relief. Other deep craters 40 to 50 km across can be recognized, and several major scarps (one nearly

300 km long) are also present (see Fig. 17 of the chapter by Chapman and McKinnon). Cratering and spallation (see, e.g., Thomas et al. 1981, 1984) appear to have been the dominant processes in its geologic history.

A major controversy for Hyperion concerns its rotation period, which is apparently not synchronous. Both the high-resolution Voyager images and the analysis of the lightcurve from low-resolution images yield a 13 day periodicity with a spin vector lying close to its orbital plane (Thomas et al. 1981,1984; Thomas and Veverka 1985). A similar 13 day period was derived from 1983 groundbased photometry by J. Goguen. Dynamical analysis, however, suggests that this period may not be stable, and that the rotation of Hyperion should be chaotic (Wisdom et al. 1983; see also Peale's chapter); indeed 1985 data do not seem to be consistent with the 13 day period (Binzel et al. 1985). In such a situation, the satellite is expected to be tumbling irregularly, with a period that could change by tens of percent in a few weeks.

The inner, small satellites (Table I; see the chapter by Thomas et al. as well as the introductory chapter to the book) all have dimensions smaller than Hyperion or Phoebe. The largest are the two coorbitals (Janus: 220 × 190 × 160 km; Epimetheus: 140 × 115 × 100 km). Next are the two F-ring shepherds ($\sim$ 100 km), while the others do not exceed 40 km in major diameters. With imaging resolutions that generally amount to only a few pixels per diameter, the degree of geologic information that can be gleaned for these objects is clearly limited (Thomas et al. 1983a).

Photometrically, the analysis of Thomas et al. (1983a) shows that all of these inner satellites have bright, presumably ice surfaces, with geometric albedos ranging from $\sim$ 0.4 up to perhaps as high as 1.0 for Telesto, the leading Lagrangian of Tethys. The irregular shapes and cratered surfaces of these satellites suggest a violent past, and it seems likely that all are remnants of once larger bodies.

The two coorbitals are large enough to permit crater counts. Thomas et al. (1983b) have identified 12 craters on Janus and 16 on Epimetheus. Within the (rather large) statistical uncertainties, the crater densities for these two satellites agree with the values for heavily cratered surfaces ($\sim$ 1000 per 10^6 km^2 for $D \geq 10$ km) on the other Saturnian satellites. Yoder and Synnott (1984; see also Peale's chapter) have carried out a dynamical analysis to estimate the masses of Janus and Epimetheus, and their results are surprisingly low, suggesting densities of < 1.0 g cm^{-3}.

All of the small inner satellites are in orbits that are remarkable in some way: as coorbitals, shepherds or Lagrangians. Various scenarios have been suggested for their origin, generally involving the breakup of larger parent objects (see the introductory chapter). We will return to some of these ideas in the final sections of this chapter.

III. TITAN

Introduction

Titan has similar size and bulk composition to the Galilean satellites Ganymede and Callisto, but unlike them it has formed and retained a major atmosphere, with surface pressure greater than that of the Earth. This satellite has been of special interest since the 1944 discovery of its atmosphere. By the early 1970s, the existence of an opaque aerosol layer had been demonstrated, and spectroscopy had revealed the presence of some hydrocarbons. Following the two Voyager encounters, Titan has remained an object of study, as well as being identified as a high-priority target for future spacecraft exploration.

The properties of Titan and its atmosphere were reviewed in detail by Hunten et al. (1984). The basic physical data are summarized in Table III, taken from Hunten et al. Unfortunately, we have no information on the nature of this satellite's surface, and no discussion is possible of either its geology or

TABLE III
Properties of Titan[a]

Surface radius, R_T	2575 km
Mass, M	1.346×10^{26} g $= 0.022 \times$ Earth's
GM	8.976×10^{18} cm^3 s^{-2}
Surface gravity, g_s	135 cm s^{-2}
Mean density	1.881 g cm^{-3}
Rock/ice ratio (by mass)	$\sim 52:48$
Distance from Saturn	1.226×10^6 km $= 20$ R$_S$
from Sun	9.546 AU
Orbit period	15.95 day
around Sun	30 yr
Obliquity	26°7 (assumed)
Temperature, K [b]	
Surface	94
Effective	86
Tropopause (42 km)	71.4
Stratopause (200 km)	170
Exobase (1600 km)	186 ± 20
Bond albedo	0.20
Solar flux	1.1% $\times$ Earth's
Surface pressure	1496 ± 20 mbar

[a] Table taken from Hunten et al. (1984).
[b] Atmospheric temperatures below 200 km assume mean molecular weight 28 and scale with this value (Lindal et al. 1983).

surface chemistry. Until other missions are flown, our focus of study will stay with the atmosphere.

In this chapter, we discuss only the atmosphere of Titan and the implications that can be drawn for the chemical evolution of the satellite. First, we outline the main points of progress prior to Voyager. Pre-Voyager knowledge of the chemistry of the atmosphere has been extensively reviewed by Trafton (1974). Next, we discuss the Voyager results and, finally, we conclude with a description of a model for the origin of Titan's atmosphere.

Pre-Voyager Observations

Methane. In Kuiper's (1944) pioneering attempt to determine the abundance of methane in Titan's atmosphere, he concluded that the vertical column abundance was $\alpha = 200$ meter amagats (m-am). This estimate did not take into account curve-of-growth or temperature effects and assumed a clear atmosphere above a reflecting surface.

The next step came 20 years later, when Trafton (1972,1974) succeeded in observing the $3\nu_3$ methane band near 1.1 μm. For an atmosphere of pure methane, an abundance of 2 km-am was required, a factor of 10 larger than Kuiper's estimate. On the other hand, the observations could also be explained if a large amount of some undetected gas were present to broaden the methane bands, producing the same absorption with less methane. In this case, methane would be a minor atmospheric constituent.

Trafton's several papers, as summarized in Trafton (1981), emphasized the unique features of methane absorption on Titan compared with the other outer planets. The explanation for these peculiarities was suggested to lie in the existence of a thick aerosol layer in the satellite's atmosphere. Nevertheless, a simple scattering model used by Hunten (1978) also indicated that methane by itself could not account for the observed intensities of the various bands, suggesting, but not demonstrating, the presence of other major constituents.

Other Constituents. A concerted effort to search for ammonia in Titan's atmosphere produced steadily decreasing upper limits. The lowest published value is that of Fink and Larson (1979), who set $\eta\alpha < 2$ cm-am (η is the airmass factor that accounts for the geometry of the observation and any scattering in the atmosphere). This is consistent with the expectation that ammonia will be readily photodissociated with subsequent escape of hydrogen, or that ammonia is present but frozen out on Titan's surface.

The other gases that were thought to be present on Titan were identified on the basis of unresolved bands detected in emission in the 8 to 14 μm region of the spectrum. The discovery by Low (1965) that Titan was radiating an anomalously high thermal flux led to the suggestion by Danielson et al. (1973) of a thermal inversion in Titan's upper atmosphere. Danielson et al. identified

C_2H_6 as the gas responsible for the strong emission peak at 12.2 μm, and Gillett (1975) suggested the presence of CH_3D, C_2H_4 and C_2H_2.

Observations of the 2000 to 3000 Å region of Titan's ultraviolet spectrum by the IUE (International Ultraviolet Explorer) showed no evidence of discrete, gaseous absorbers. The satellite's reflectivity remains near 3% throughout this region. The absence of Rayleigh scattering indicated that the aerosol suspected from spectroscopy and polarimetry must be relatively high in Titan's atmosphere and essentially ubiquitous (Caldwell et al. 1981).

Conclusions. Prior to the Voyager 1 encounter, two extreme models for the atmosphere of Titan were being seriously considered. Caldwell (1977) favored a model in which the atmosphere was essentially pure methane, in equilibrium with solid methane at a surface with a temperature of ~ 86 K. The surface pressure of this atmosphere was close to 20 mbar. Hunten (1978) favored an atmosphere consisting of 20 bar of nitrogen, in which methane was merely a minor constituent. The surface temperature in this model was ~ 200 K, maintained by the greenhouse effect from pressure-induced absorption by the thick nitrogen atmosphere. Observations of Titan with the VLA (Very Large Array) in January 1980 by Jaffe et al. (1980) showed that the high-temperature model was inadmissible; the measured surface temperature was 87 ($\pm$ 9) K. Nevertheless, a predominantly nitrogen atmosphere with a surface pressure of 2 to 3 bar was compatible with these results.

From arguments based on cosmic abundances and models for atmospheric evolution, Cess and Owen (1973) had suggested the presence of substantial quantities of neon and/or argon. Clearly some spectroscopically undetectable component had to be present in substantial quantities if all of the previously described observations of methane were correct. However, there was enough disagreement among the various workers in this field to prevent a well-supported consensus from developing for any particular model atmosphere.

The Voyager Encounters

Global Appearance. The pictures returned by the Voyager 1 cameras in the fall of 1980 (Smith et al. 1981) confirmed the variety of previous observations that suggested the satellite's atmosphere must be filled with aerosols. No evidence of any surface detail could be found. The only visible markings were a dark north polar hood and a difference in the reflectivity and color of the northern and southern hemispheres, with a boundary just at Titan's equator. The northern hemisphere appeared slightly browner and darker than the southern at this season (onset of spring in the northern hemisphere). Above this smog-filled atmosphere lay another haze layer, apparently at the same elevation as the north polar hood and extending entirely around the satellite. Pictures obtained nine months later by Voyager 2 (Smith et al. 1982) showed little change, although the north polar hood now appeared as an annulus.

The reasons for this north-south asymmetry are not yet entirely clear. Sromovsky et al. (1981) suggested that we are observing the effect of a long dynamical time constant in Titan's atmosphere. Although the local season of the Voyager encounter was spring in Titan's northern hemisphere, Sromovsky et al. argued that the response of the atmosphere to solar radiation was delayed by one full season (about 7.5 yr) so that Voyager was seeing a winter-summer asymmetry. This model can be tested by groundbased observations of the satellite's brightness, since its reflectivity has been observed to vary (Lockwood 1977). The observations available to date (October 1985) appear to support at least this aspect of the Sromovsky et al. model.

Atmospheric Composition.   Information about atmospheric composition was quickly forthcoming from other experiments. Voyager's ultraviolet spectrometer detected emission lines from both molecular and atomic nitrogen in Titan's upper atmosphere (Broadfoot et al. 1981). When this result was combined with Voyager 1 radio occultation data, it became apparent that the atmosphere was in fact predominantly nitrogen, with methane a minor constituent. The temperature and pressure at Titan's surface were found to be 94 ($\pm$ 2) K and 1.5 bar, respectively (Tyler et al. 1981; Hanel et al. 1981; Lindal et al. 1983). Thus, Hunten's model was essentially correct, if the surface were just taken at the 1.5 bar level.

Under these conditions, methane should be in the liquid state at Titan's surface. However, a careful study of the radio occultation temperature profile shows no departure from a dry adiabatic lapse rate. This finding rules out a global methane ocean and restricts the presence of methane seas to high latitudes, away from the points of occultation immersion or emersion. On the other hand, ethane is one of the main products of methane photochemistry in Titan's atmosphere, and this gas would also form a liquid at the surface conditions on Titan. Lunine et al. (1983) have raised the possibility that the entire satellite might be covered by a global ocean of ethane, in which some methane, nitrogen and other atmospheric constituents could be dissolved. The bottom of this ocean would be covered by deposits of other organic materials forming in the atmosphere, excluding solid acetylene (Matteson 1984).

The Voyager infrared spectrometer confirmed the presence of CH_4, C_2H_2, C_2H_4 and C_2H_6 and immediately added HCN (Hanel et al. 1981). Subsequent analyses have identified a number of other gases, and carbon monoxide was added to this list as a result of groundbased observations. A summary of current identifications and abundances for minor constituents is given in Table IV (Samuelson et al. 1981; Maguire et al. 1981; Kunde et al. 1981; Samuelson et al. 1983; Lutz et al. 1983; Muhleman et al. 1984). The photochemical reactions required to produce these compounds have been recently described by Strobel (1982) and Yung et al. (1984). Laboratory simulations of reactions in Titan's atmosphere have been carried out by Scattergood

TABLE IV
Atmospheric Composition of Titan below 0.1 mbar

Gas	Band	Position (cm^{-1})	Mole Fraction Inferred Indirectly	Mole Fraction Measured Directly
Nitrogen	N_2	—	—	0.65–0.98	
Argon(?)	Ar	—	—	0–0.25	
Methane	CH_4	ν_4	1304	0.02–0.10	
Hydrogen	H_2	S_0	360		2×10^{-3} [a]
		S_1	600		
Carbon monoxide	CO	(3–0) $X^1\Sigma^+$	6350		6×10^{-5}—1.5×10^{-4} [b]
Ethane	C_2H_6	ν_9	821		2×10^{-5} [c]
Propane	C_3H_8	ν_{21}	748		4×10^{-6} [d]
Acetylene	C_2H_2	ν_5	729		2×10^{-6} [c]
Ethylene	C_2H_4	ν_7	950		4×10^{-7} [c]
Hydrogen cyanide	HCN	ν_2	712		2×10^{-7} [c]
Methyl acetylene	C_3H_4	ν_9	633		3×10^{-8} [c]
		ν_{10}	328		
Diacetylene	C_4H_2	ν_8	628		$\sim 10^{-8}$—10^{-7} [c]
		ν_9	220		
Cyano-acetylene	HC_3N	ν_5	663		$\sim 10^{-8}$—10^{-7} [c]
		ν_6	499		
Cyanogen	C_2N_2	ν_5	233		$\sim 10^{-8}$—10^{-7} [c]
Carbon dioxide	CO_2	ν_2	667		1.5×10^{-9} [e]

[a] Samuelson et al. 1981; [b] Lutz et al. 1983; [c] Kunde et al. 1981; [d] W. C. Maguire, personal communication; [e] Samuelson et al. 1983b.

and Owen (1977), Bar-Nun and Podolak (1979), Khare et al. (1982), and Cerceau et al. (1985).

In addition to these gases, two independent studies of the abundance of CH_3D on Titan have been carried out. A preliminary investigation of the 8.6 μm emission band in the Voyager spectra led Kim and Caldwell (1982) to derive a value of D/H = $4.2^{+2}_{-1.5} \times 10^{-4}$. More recently, de Bergh et al. (1986) obtained D/H = $1.65^{+1.65}_{-0.8} \times 10^{-4}$ from a study of groundbased observations of the $3\nu_2$ band at 1.6 μm. In fact, corrections to the temperature-pressure profile of Titan's atmosphere and new broadening coefficients for the ν_6 band at 8.6 μm reduce the original Kim and Caldwell value to 2×10^{-4}, in good agreement with de Bergh et al..

Table IV also includes the main constituents, which are nitrogen, methane and possibly argon. The nitrogen and methane have both been detected spectroscopically, as mentioned above. However, it has proved difficult to determine their relative abundances. The nitrogen detection refers only to the extreme upper atmosphere. It is deduced to be the main atmospheric constituent because the mean molecular weight derived from the radio occultation experiment is very close to 28.

Methane is a problem because it can condense in Titan's atmosphere; hence the column abundance measured at some point in the atmosphere cannot be extrapolated unambiguously to the surface. Samuelson et al. (1981) point out the possibility that methane could be present with a constant mixing ratio of 0.027 above the level where the atmosphere reaches its minimum temperature of 75 K. This is at a pressure of 200 mbar, which would suggest a column abundance of 0.31 km-am. Below the temperature minimum, the methane mixing ratio can increase with temperature. Samuelson et al. adopt a value of 0.06 as a mean, which would lead to a surface column abundance of 6 km-am.

These figures suggest that the groundbased near-infrared spectroscopic measurements probably never correspond to an optical path that reaches the satellite's surface. Instead, scattering occurs from the aerosol and whatever methane clouds may be present at and below the tropopause. It will now take some careful modeling to reconcile the observations of bands of various strengths at differing regions of the spectrum. Information currently being gained on the properties of the particles making up the aerosol should greatly aid this process.

The presence of argon is even more problematic. Broadfoot et al. (1981) set an upper limit of 6% on the argon abundance from the absence of the 1048 Å emission. But this refers to the upper layers of the atmosphere, where diffusive separation will tend to deplete the argon abundance. Samuelson et al. (1981) suggested a mixing ratio as high as 12% (assuming ^{40}Ar), in order to obtain the mean molecular weight of 28.6 derived from radio occultation measurements at that time. Subsequent analysis of the occultation data by Lindal et al. (1983) allows the possibility of a molecular weight of 28.0, so that no argon at all might be present. The uncertainties in the existing data set are simply too great to allow the determination of a meaningful abundance for this noble gas.

The Origin of the Atmosphere

If Titan had captured its atmosphere from the solar nebula (or from an unfractionated proto-Saturnian nebula), one would expect to find a large amount of neon in the atmosphere. The cosmic abundances of neon and nitrogen are nearly equal, implying an atmospheric composition of 67% Ne and 33% N_2 by number. The fact that the derived mean molecular weight of Titan's atmosphere is equal to or slightly larger than 28 proves that this is not the case.

Evidently the atmosphere represents a devolatilization of the ices and rocky material that accreted to form the planet. But what was that material? The mean density of 1.86 g cm^{-3} derived for Titan indicates that the satellite may be slightly enriched in silicates compared with an object that condensed from an unfractionated portion of the solar nebula. The bulk composition appears to be 52% rock vs 40% for a solar abundance model.

Twenty-five years ago, Miller (1961) suggested that the ice that makes up most of the mass of the satellites of Saturn is actually methane hydrate, $CH_4 \cdot 7H_2O$. According to Lewis (1973), methane hydrate formation in the cooling solar nebula goes to completion with the exhaustion of H_2O at $T = 60$ K. At this temperature, the saturation vapor pressure of Ne is above 40 bar, so formation of Ne hydrate is out of the question. Indeed, no trapping of neon by ice is found even at temperatures as low as 25 K (Bar-Nun et al. 1985). The absence of Ne in Titan's atmosphere is thus easily explained.

It is commonly assumed that the nitrogen one finds in the outer solar system was present in the solar nebula as ammonia. But Lewis and Prinn (1980) have pointed out that the kinetics of the formation of ammonia from nitrogen are such that nitrogen is more likely to be present as N_2 if it entered the nebula in that form. On the other hand, Prinn and Fegley (1981) have demonstrated that the thermal gradient in the nebula surrounding a forming planet can lead to rapid conversion of N_2 to NH_3 if temperatures are sufficiently high. Thus, we cannot tell if the nitrogen we now see in Titan's atmosphere was originally in the form of N_2 or NH_3. If nitrogen were incorporated by Titan in the form of NH_3, Atreya et al. (1978) have shown that Titan must have been considerably warmer in the past. Otherwise an insufficient amount of ammonia would get into the atmosphere where photodissociation could gradually produce the presently observed quantity of N_2. Atreya et al. require temperatures in excess of 150 K during the early phases of Titan's history. Such temperatures require a significant greenhouse effect, but ammonia is an excellent infrared absorber, capable of sustaining the necessary thermal balance.

A clear indication of the presence of clathrate hydrates in the ices that formed Titan would be provided if argon could be unequivocally detected in the satellite's atmosphere. If the atmosphere of Titan actually contains several percent of argon as originally suggested by Samuelson et al. (1981), this would have to be primordial [36]Ar and [38]Ar trapped in the ices that formed the satellite along with the methane. The degassing of [40]Ar from the decay of radioactive [40]K in the satellite's rocky core could only produce a mole fraction of 0.01% of the atmosphere, about one thousand times too small. According to Cameron's (1982) list of cosmic abundances of the elements, we expect N/Ar = 22 in the solar nebula. This leads to an expected ratio of N_2/Ar = 11 in Titan's atmosphere. All we can say at present is that this ratio is somewhere between infinity (no primordial argon) and 6 (the upper limit for argon set by Lindal et al. [1983] from the maximum mean molecular weight compatible with the radio occultation observations). According to Strobel (1985), the present atmospheric abundance of nitrogen is about 20% less than the amount produced by Titan during the history of the solar system, thanks to escape and chemical deposition. If Lunine et al. (1983) are correct and a global ethane ocean is present, about 50% of the present atmospheric abundance of nitrogen could be dissolved in this ocean.

The conclusion of this exercise is that an atmosphere containing several percent of argon is indeed consistent with what we know about cosmic abundances and what we can deduce about conditions in Titan's atmosphere. Consistency is very different from necessity, however, and we still lack any firm evidence proving the existence of argon on Titan. Once such evidence (or a definitive lack thereof) is established, we shall have an excellent test of the models that have been proposed for atmosphere production. The presence of a large amount of argon would require the incorporation of this gas by low-temperature adsorption or clathrate production, processes that would capture methane, nitrogen and/or ammonia as well.

The presence of two oxygen-containing compounds in this reducing environment invites special discussion. In the clathrate hydrate model, it is assumed that CO is incorporated in the ices along with the N_2, Ar and CH_4. CO would then be released into the atmosphere with these other gases. Reactions with OH then produce CO_2. Alternatively, OH can produce CO from CH_4 directly (Samuelson et al. 1983). In either case, the source of the OH is thought to be particles of ice bombarding Titan as the satellite orbits Saturn.

One can imagine the outgassing that produces the atmosphere to occur early, during the accretion of the object, and also during its subsequent history, as a result of internal heat. For example, Lupo and Lewis (1979) have produced a model for Titan with internal temperatures above 273 K. The argon, methane, nitrogen and carbon monoxide released from the ices at depth would then migrate to the surface, perhaps through the kinds of cracks and fissures that we see now on the ancient surfaces of Dione and Tethys. An attractive feature of this evolutionary model is that there is no requirement for a higher surface temperature on Titan in the past than the value we observe today, unlike the case that starts with nitrogen in the form of NH_3.

Finally, the high value of D/H measured in the methane in Titan's atmosphere seems to be telling us something about the ways in which hydrogen-containing compounds were incorporated in planets and satellites. Pinto et al. (1986) have shown that the maximum enrichment of D/H in CH_3D that could occur from the sum of a variety of processes on Titan is only 4.4. If we eliminate the unlikely process of catalysis by metallic grains at 500 K from this sum, the enrichment factor drops to 2.2. This is significantly smaller than the observed factor of 8^{+8}_{-4} (de Bergh et al. 1986). Evidently the enrichment occurred prior to formation of the satellite. This is a much higher value of D/H than is found in the atmospheres of the major planets, leading Owen et al. (1986) to suggest that the solar system formed with at least two deuterium reservoirs: one in hydrogen gas and a second in hydrogen-containing compounds trapped in ices. The latter seems to be represented in meteorites as well as in the terrestrial oceans. To be certain of this conclusion, it is necessary to obtain D/H on Titan with greater precision and to evaluate this important ratio in the gases produced by comets.

Titan, Triton and Pluto

Both Triton and Pluto (also discussed in the chapter by Cruikshank and Brown) are icy bodies sufficiently massive that we might *a priori* expect them to have atmospheres like Titan. But it is dangerous to extrapolate from a body formed in the special conditions of the proto-Saturnian nebula to two objects with quite different histories. Once again we are interested in N_2, NH_3, CH_4, CO and Ar. Again, neon is highly improbable. Of these gases, only CH_4 is likely to be detectable by remote measurements, unless very large quantities of CO are present. Observations to date suggest that CH_4 is present on both Pluto and Triton, although in what form is not yet completely clear (Cruikshank et al. 1976; Cruikshank and Silvaggio 1979,1980). An absorption band at 2.15 μm in the spectrum of Triton (Fig. 21 of the chapter by Clark et al.) has been attributed to the presence of liquid nitrogen on this satellite's surface (Cruikshank 1984; cf. Rieke et al. 1985). This feature has not yet been identified in spectra of Pluto. No trace of NH_3 frozen on the surface of either body has yet been reported.

Thus, an underlying similarity with Titan may well exist, but just how close the histories have been remains to be determined. One thing is already clear: the atmospheres of Pluto and Triton do not contain a thick, ubiquitous, Titan-style aerosol; nor have they produced one in the past, unless it is thoroughly buried under a new surface coating. Is this because of the absence of magnetospheric electrons, the weaker ultraviolet flux from the Sun, the lack of sufficient atmospheric N_2, or simply the thinness of the atmospheres in each case?

We are presently in a situation with respect to our knowledge about these two bodies that is very similar to the pre-Voyager picture of Titan. For Triton, we will have the answers after August 1989 when Voyager 2 flies through the Neptune system. However interesting these results are in and of themselves, we can also be certain that they will increase our understanding of Titan as well. This is particularly true for the significance of local environments for the incorporation of CH_4/CO, NH_3/N_2 in ices accreting in the outer solar nebula. Pluto is not yet in anyone's budget, so the answers here will come much more slowly.

IV. GEOLOGIC HISTORIES

Although any attempt to trace a detailed geologic history of the Saturnian satellite system is premature, it is instructive to compare these satellites on the basis of existing data and to sketch some of the processes that might have influenced their development. In this section, we discuss such processes under three headings: origin, exogenic processes and endogenic processes.

Origins

The regular satellite system of Saturn presumably had a common origin in a circum-Saturnian nebula. The ubiquitous presence of water ice in the ring and satellite system indicates that temperatures throughout the nebula were cold enough to permit the condensation of water, unlike the Jovian case. The presence of both methane and nitrogen in the atmosphere of Titan suggests that ammonia and methane were incorporated, perhaps as clathrates, at least in the outer part of the Saturnian satellite system. If these two volatiles are present on Titan, they were probably also included in the building blocks of Hyperion and Iapetus, and perhaps even in those of the inner satellites.

Some constraints on bulk composition of Saturn's satellites are provided by their densities, which are for the most part adequately determined after the Voyager missions (Fig. 10; see also the discussion of satellite masses in the introductory chapter). The mass for Mimas has not been directly measured by Voyager, but the mass ratio Mimas/Tethys is known from resonance studies, and therefore Mimas can be scaled to the Tethys value. The mass of Enceladus, unfortunately, remains poorly determined, but the values for the other measured objects are all consistent to within the stated errors with a uniform density of 1.3 (± 0.1) g cm^{-3}. This density is consistent with a composition of 60 to 70% ice if the remaining fraction is chondritic (Lupo and Lewis 1979). As shown in Fig. 10, this ice/rock proportion is also consistent with the compressed density for Titan of 1.89 g cm^{-3}. Although minor stochastic differences may exist among the compositions of the measured satellites of Saturn, the contrast with the regular progression in densities displayed by the Galilean satellites of Jupiter is striking. Either the Saturn proto-satellite nebula was fairly uniform in composition, or else the process of accretion involved sufficient mixing to homogenize the satellites as they formed (cf. the chapter by Stevenson et al.).

The satellites could have been heated and melted by accretional energy, tidal despinning or radioactive energy from their presumed chondritic fraction (chapter by Schubert et al.). Because of the very low melting temperatures of either pure H$_2$O ice or various clathrates, it seems nearly certain that the larger bodies must have differentiated early in their history. However, smaller satellites may still retain their primitive composition and structure.

As will be discussed in the next section, the accretionary and post-accretionary bombardments of the inner satellites must have been intense and probably led to many cases of fragmentation and even disintegration of some of the smaller early satellites. Presumably the various coorbital, shepherd and Lagrangian satellites assumed their present identities at that time. The rings themselves could have been formed as the consequence of similar bombardment-triggered disruptive events.

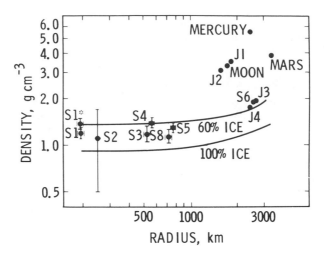

Fig. 10. Density versus radius trends for the icy Saturnian satellites. Taken from Smith et al. (1982), these data are compared with model curves for comparison of rock and ice mixtures calculated by Lupo and Lewis (1979). Two values are shown for S1 (Mimas). The lower is the classical value based on mass determinations from Earth-based astronomical observations; the upper from Voyager radio science measurements (Tyler et al. 1982).

Finally, we would guess (primarily on dynamical grounds) that Phoebe is a captured object. Its moderately regular shape suggests that it was captured without catastrophic collisions, probably at a time when gas drag from the proto-satellite nebula was still significant (see, e.g., Pollack et al. 1979).

Exogenic Processes

The heavily cratered surfaces seen throughout the Saturnian system testify to the role of impacts in satellite evolution. The observed crater densities are summarized in Table V. Regolith depth models are given in Fig. 7d of the chapter by Veverka et al. However, in order to interpret the impacts, we must consider the effects of Saturn and the satellites themselves in determining the flux of bombarding particles. These effects were first emphasized for the Galilean satellites by Shoemaker and Wolfe (1982), and have been further developed for the Saturnian system by Smith et al. (1981,1982); they are further described in the chapter by Chapman and McKinnon.

The number of impacts and their energy for sources of projectiles outside the Saturnian system are strongly influenced by the gravitational attraction of the planet, and will be significantly enhanced for satellites close to the planet. The present cratering rate on Rhea from such sources should be about twice that on Iapetus, and for Mimas the rate is 20 times that on Iapetus. The cratering rates are also affected by the masses of the satellites themselves, but for objects as small as Saturn's satellites, the dominant effect is that of the deep

TABLE V
Crater Density Cumulative Values per 10^6 km² [a]

Satellite	Terrain	Crater Diameter	
		10 km	**20 km**
Iapetus	Bright side (average)	[2000] [b]	740 ± 30
Rhea	North polar	1100 ± 20	300 ± 100
	Equatorial (unmantled)	700 ± 150	200 ± 100
	Equatorial (mantled)	160 ± 45	80 ± 40
Dione	Heavily cratered	[1000]	270 ± 50
	Intermediate	700 ± 100	45 ± 25
	Lightly cratered	260 ± 50	45 ± 25
Tethys	Heavily cratered	[2000]	500 ± 92
	Intermediate	690 ± 72	170 ± 26
	Plains	800 ± 66	140 ± 28
Enceladus	Heavily cratered	800 ± 300	[300]
	Fractured plains	200 ± 100	[50]
	Smooth plains	< 100	—
Mimas	South polar	1200 ± 500	< 100

[a] Adapted from Plescia and Boyce (1982,1983). Table taken from Morrison et al. (1984).
[b] Brackets in each case indicate extrapolation.

gravitational well of the planet. An additional effect for synchronous satellites is the enhancement of cratering rates on the leading hemisphere relative to the trailing. The variation from apex to antapex increases with orbital velocity: at Rhea the factor is 6, and at Mimas it is nearly 20. However, we note that no dependence of crater density upon longitude has been seen for any Saturn satellite (Plescia and Boyce 1983), suggesting that they may not have been in synchronous rotation during periods of heavy bombardment or, alternatively, that the impacting bodies were in planetocentric rather than heliocentric orbits (Horedt and Neukum 1984; cf. Chapman and McKinnon's chapter).

Present cratering rates due to long-period and Saturn-family comets have been estimated to range from ~ 1 (cumulative crater count $D > 10$ km per 10^6 km² per 10^9 yr) for Iapetus to ~ 16 in the same units for Mimas. If these calculations correctly represent the cratering rates that have applied, they can be used to estimate surface ages. For instance, the rate for Enceladus is ~ 10, equal to the observed upper limit for the crater-free terrain, yielding an age for that terrain < 10^9 yr. Such calculations are, however, probably at best adequate to give an order-of-magnitude upper limit to surface ages. In fact, the Saturnian system presumably experienced much higher impact fluxes in the past, as shown by the fact that at estimated current rates it would require ~ 10^{12} yr to produce the crater density observed on Iapetus.

A less assumption-dependent analysis of the significance of cratering statistics can be obtained by adopting Iapetus as a fiducial mark. The bright side of Iapetus has a cumulative crater density of ~ 2000 per 10^6 km^2 for $D >$ 10 km. On the assumption that these craters were produced by projectiles from an external source, Smith et al. (1982) calculated the cumulative impacts for the other Saturn satellites over the same period, and thus obtained a lower limit to the total crater density for each object. In all cases the calculated cratering exceeds the observed value, implying saturation of the surface or resurfacing subsequent to the stabilization of Iapetus' surface. Furthermore, the cratering rates extrapolated in this way from Iapetus are so high in the inner part of the system as to imply impact disruption sometime in the past for all satellites interior to Rhea. This exercise indicates, for instance, that Enceladus and Mimas should each have been disrupted and reaccreted four or five times during the same time period in which Iapetus was developing its current density of craters, if the Iapetus craters indeed represent an impact flux external to the Saturn system. These arguments are less compelling, however, if a major source of impactors in planetocentric orbits is considered (Horedt and Neukum 1984).

In a sense, these two exercises bracket the probable cratering history of the satellites. So long as we assume that all the satellites were impacted by the same population of externally derived projectiles that cratered the bright side of Iapetus, we are led to the conclusion that disruptive impacts were common for the inner satellites. However, the implied cratering rates are orders of magnitude higher than the currently estimated rates for the known external source, i.e., the comets. An alternative supposition is that a major part of the satellite cratering, including that of Iapetus, might have been due to some other population of debris more closely associated with the Saturnian system. In that case, catastrophic impacts might have been less common.

As the chapter by Chapman and McKinnon details, there is evidence in the cratering record for different populations of impacting bodies. At the very least there seem to be two: population 1, representing the tail-off of a post-accretional heavy bombardment, and population 2, similar to the secondary populations on the terrestrial planets and possibly resulting from debris generated by collisions within the system. Both of these populations are, of course, distinct from the external, cometary population causing impacts today.

The Voyager observations indicate that population 1 craters are found in abundance on Rhea, on the older parts of Dione, and in the heavily cratered terrain of Tethys. Population 2 characterizes the younger terrains on all three of these satellites, plus most of the surface of Mimas, and all of Enceladus.

Although none of the satellites shows an apex-to-antapex crater gradient of the sort predicted for the flux on a synchronous satellite, this focusing of impacts by orbital motion may still have played an important role in the surface evolution of Rhea and Dione (as well as, possibly, Iapetus). Suppose that those satellites, and perhaps Tethys as well, developed a global wispy terrain as a result of an early episode of internal activity. It is then reasonable that subsequent impacts reworked the leading hemispheres of Rhea and Dione, and all of Tethys, to the point where the evidence of endogenic activity was erased, whereas the wispy terrain was preserved on the protected trailing hemispheres of Rhea and Dione. This scenario seems quite reasonable for the inner satellites, but in this simple form it does not appear to be consistent with the lack of longitudinal control of densities of craters of $D > 10$ km as counted in the Voyager images.

A very recent impact event in the Saturnian satellite system has been suggested by McKinnon (1983) as the source of the E ring and the remarkable photometric properties of Enceladus. McKinnon argues that the sphericity and narrow size distribution of the E-ring particles (Pang et al. 1984a) are indicative of condensation from a liquid, and that a water cloud would most readily be generated from the impact melt associated with a crater-forming event on Enceladus. He calculates a current mass for the E ring of a few times 10^{10} g, equivalent to an ice sphere of ~ 20 m radius. This is not an unreasonable mass for even a small cometary impact, such as might have occurred during the past 10^3 to 10^4 yr. In this hypothesis, the optical properties of Enceladus result from reaccretion onto the satellite of these micrometer-sized ice particles, and are unrelated to the unique geologic history of this satellite. Such "E rings" would be produced periodically in the Saturnian system, and it is coincidence that we find one now in association with Enceladus.

We should not leave the topic of exogenic processes without again raising the issue of the dark material on Iapetus. None of the Voyager data have altered the conclusion based on the symmetry of the lightcurve that the albedo distribution on this satellite is externally controlled. If anything, the spacecraft images have strengthened the case by demonstrating that the albedo contours approximately follow calculated contours of equal impacts for a synchronous satellite.

Endogenic Processes

The thermal evolution of small bodies, and particularly of small icy bodies, is a topic that owes its recent interest to the Voyager observations of the Jovian and Saturnian satellite systems; it is considered in detail by Schubert et al. in their chapter. Early discussions by Lewis (1971a) and Consolmagno and Lewis (1976,1977) dwelt primarily with the Galilean satellites, but are also applicable to Saturn's system, especially their suggestions of the possible role of small quantities of methane and ammonia in stimulating early melting and differentiation. These two volatiles may also be implicated in the outgassing of icy satellites as they evolve.

One of the clearest cases for thermal evolution from among Saturn's satellites is provided by Tethys, with its fracture system, Ithaca Chasma. If the satellite were once a ball of liquid water covered by a thin frozen crust, freezing of the interior would have produced expansion of the surface comparable to the area of the chasm (see the chapter by Squyres and Croft). It is not clear, however, why the extension would be concentrated in a single narrow lane rather than distributed among multiple faults over the surface. The smaller-scale valleys or graben of Dione, similar in size and density to Tethys, might also be due to expansion of an ice object as it freezes. There are suggestions of similar grooves on Mimas, but these have been nearly destroyed by subsequent impacts.

All of the inner satellites also show evidence of several stages of partial resurfacing, presumably by liquid escaping from the interior. In the absence of unexpected internal heat sources, it has been suggested that low-temperature melts might have been provided by a water-ammonia eutectic, which melts at 170 K. Similar explanations may apply to the putative outgassing along fracture lines that may have produced the wispy terrain of Rhea and Dione. Clearly several different styles of internal activity are indicated, none of them very well understood. Studies of thermal evolution and of volcanism on icy bodies are still in their infancy.

Enceladus provides the strongest evidence for internal activity, with its multiple localized resurfacing events and the plastic deformation of the crust, probably extending to the current geologic epoch (i.e., $< 10^9$ yr old). This dramatic and continuing evolution appears to require an internal heat source orders of magnitude larger than would be expected from known causes.

The most detailed analysis of lithospheric processes on Enceladus has been made by Passey (1983). His studies of the flattening of craters and the bowing up of their floors suggest a history of viscous relaxation of topography. The modified craters are located in distinct zones adjacent to other units, with similar crater densities, where craters have not been flattened. These modifications of craters appear to require viscosities one to two orders of magnitude less than for Ganymede or Callisto. Passey concludes that such a large difference in viscosity requires a difference in lithospheric material, as well as a substantial heat source on Enceladus. The presence of ammonia as well as water ice in the lithosphere of Enceladus would satisfy this requirement.

Even more dramatic than the alteration of craters is the formation on Enceladus of the plains terrains, which are nearly crater-free and may have ages $< 10^9$ yr. The contacts between cratered regions and the younger plains are often linear and abrupt, truncating preexisting craters. The curvilinear grooves that characterize part of these plains are morphologically similar to those on Ganymede. Squyres et al. (1983b; see the chapter by Squyres and Croft) believe that the grooves resulted from open extension fractures in a relatively brittle crust. Widespread flooding of parts of the surface with water magma probably occurred at about the same time.

At some of the contacts between younger and older terrain, craters appear to have been distorted by a horizontal strain associated with extensional faulting. Kargel (1983) has also noted examples where there appears to have been strike-slip faulting with ~ 20 km of offset, and he argues for past plate tectonic activity, including a possible analogy between the grooves near longitude 0° and terrestrial midocean ridges. Along spreading centers he expects flooding and even pyroclastic release of water in the form of large volcanic fountains. These ideas are intriguing, but perhaps go beyond what can firmly be supported on the basis of current data. Even the distortion of craters near geologic contacts is debatable, and Passey (1983) has suggested that a fortuitous superposition of craters seen near the limit of resolution provides an acceptable alternative interpretation.

Although the time scale for geologic activity on Enceladus is not known, it is clear that some terrains on this object are much younger than any seen elsewhere among Saturn's satellites, indicating a much more persistent activity. Such long-lived activity on so small an object requires a heat source considerably greater than that expected from radioactive decay in a presumed silicate core, which is only $\sim 10^8$ W, two orders of magnitude below that required for melting. If tidal heating is the explanation of the unique activity level of Enceladus, it implies a past forced eccentricity much higher than can be maintained or episodically generated by the Dione resonance (Lissauer et al. 1984).

More generally, the thermal evolution of bodies of the size and composition of the Saturnian satellites is a topic of great interest and no little complexity. The Voyager data indicating that substantial evolution has taken place will surely focus interest on this topic. A critical unknown for the development of quantitative models is the exact chemistry of the satellite interiors.

Although the mean densities are consistent with comparable mixtures of water ice and chondritic rock, we do not know the actual density of the rocky component, and therefore we cannot establish the relative proportions of rock and ice. Further, we have very little data on the presence of non-H_2O ices, even a small admixture of which can greatly alter the rheology and phase changes of the mixture. Spectrometric data on surface composition cannot resolve this question in a quantitative way, because even at 90 K such species as ammonium clathrate and methane clathrate are less stable than H_2O and could be selectively depleted in the regolith. Thus, while it seems clear that the presence of these ices may be implicated in the more active geologic histories of some of Saturn's satellites, this speculation cannot be demonstrated based on present data.

17. SMALL SATELLITES

P. THOMAS, J. VEVERKA and S. DERMOTT
Cornell University

*Satellites smaller than Mimas (r = 195 km) are distinguished by irregular over-
all shapes and by rough limb topography. Material properties and impact cratering dominate the shaping of these objects. Long fragmentation histories can
produce a variety of internal structures, but so far there is no direct evidence
that any small satellite is an equilibrium ellipsoid made up of noncohesive,
gravitationally bound rubble. On many bodies that orbit close to their primary,
the tidal and rotational components of surface gravity strongly affect the directions of local g and thereby affect the redistribution of regolith by mass wasting.
Downslope movement of regolith is extensive on Deimos, and is probably effective on many other small satellites. It is shown that in some cases observed patterns of downslope mass wasting could produce useful constraints on the satellite's mean density. The diversity of features seen in the few high-resolution
images of small satellites currently available suggests that these objects have
undergone complex histories of cratering, fragmentation, and regolith evolution.*

I. INTRODUCTION

What are small satellites? To some extent the answer is a matter of definition. Certainly Deimos is a small satellite, but what about Dione? We will see
that a useful definition of a small satellite is one small enough that collisions
with other bodies have played a dominant role in the evolution of its shape.
Such satellites can be expected to have irregular shapes and heavily cratered
surfaces. As will be shown below, a useful demarcation line between large and
small satellites can be made at an object like Mimas, whose radius is 195 km.
Of the 44 known satellites in the solar system, the 24 smaller than Mimas are
the subject of this chapter (Table I). In composition as well as in provenance

TABLE I
Physical Properties[a]

	Diameter (km) a	b	c	a/c[b]	Best Resolution[c] (km/ℓp)	Coverage[d] %	Crater Density[f]	Visible Albedo	Composition	Density (g cm⁻³)
Phobos	27,	22,	18	1.5	0.006	95	LH	0.06	Carbonaceous	1.9
Deimos	15,	12,	11	1.5	0.003	80	~LH	0.06	Carbonaceous	2.1
Metis	40,	—,	40	1	20	S[e]	—	0.05–0.1	Rock?	—
Adrastea	25,	20,	15	1.7	20	S[e]	—	0.05–0.1	"	—
Amalthea	270,	166,	150	1.8	13	60	~LH	0.06	"	—
Thebe	110,	—,	90	1.2	20	S[e]	—	0.05–0.1	"	—
Leda		10		—	—	—	—	—	"	—
Himalia		180		—	—	—	—	0.03?	"	—
Lysithea		20		—	—	—	—	—	"	—
Elara		80		—	—	—	—	0.03?	"	—
Ananke		20		—	—	—	—	—	"	—
Carme		30		—	—	—	—	—	"	—
Pasiphae		40		—	—	—	—	—	"	—
Sinope		30		—	—	—	—	—	"	—

TABLE I (continued)

	Diameter (km) a	b	c	a/c[b]	Best Resolution[c] (km/ℓp)	Coverage[d] %	Crater Density[f]	Visible Albedo	Composition	Density (g cm^{-3})
Atlas	37,	—,	27	1.4	13.4	S[e]	—	0.5	Ice?	—
Prometheus	140,	100,	75	1.9	6.8	S[e]	—	0.5	"	—
Pandora	110,	85,	65	1.7	7.6	S[e]	—	0.5	"	—
Janus	220,	190,	160	1.4	6.2	50	~LH	0.5	"	—
Epimetheus	140,	115,	100	1.4	3.2	40	~LH	0.5	"	—
Telesto	—,	24,	22	—	12.4	S[e]	—	0.6	"	—
Calypso	30,	25,	15	2.0	4.6	S[e]	—	0.9	"	—
Helene	35,	—,	29	1.2	6.4	S[e]	—	0.6	"	—
Hyperion	350,	240,	200	1.7	9.0	40	~1/4 LH?	0.25	"	—
Phoebe	230,	220,	210	1.1	22	S[e]	—	0.06	Rock?	—
									Carbonaceous?	

[a]Compiled from Cruikshank et al. (1982), Synnott (personal communication), Thomas et al. (1983a,b), Veverka and Burns (1980), and Veverka et al. (1981b). See also Table IV of the introductory chapter.
[b]Ratio of long and short axes.
[c]The best resolution images in km/lp.
[d]The percent of surface usefully imaged for surface features.
[e]S = size data only.
[f]Crater densities are referenced to the approximate density on lunar highlands (LH) (Thomas and Veverka 1980a).

TABLE II
Orbital Elements of Small Satellites[a]

	Period	a (km $\times$ 10^3)	e	i^b (deg)
Mars				
Phobos	7.65 h	9.378	0.015	1.02
Deimos	30.29 h	23.46	0.00052	1.82
Jupiter				
Metis	7.075 h	127.96	~ 0	~ 0
Adrastea	7.159 h	128.98	~ 0	~ 0
Amalthea	11.95 h	181.3	0.003	0.45
Thebe	16.20 h	221.9	0.013	0.9
Leda	240 d	11,110	0.147	26.7
Himalia	251 d	11,470	0.158	27.6
Lysithea	260 d	11,710	0.130	29.0
Elara	260 d	11,740	0.207	24.8
Ananke	617 d	20,700	0.170	147
Carme	692 d	22,350	0.21	164
Pasiphae	735 d	23,300	0.38	145
Sinope	758 d	23,700	0.28	153
Saturn				
Atlas	14.46 h	137.67	0.002	0.3
Prometheus	14.712 h	139.35	0.004	0
Pandora	15.085 h	141.7	0.004	0.1
Janus	16.664 h	151.42	0.009	0.3
Epimetheus	16.672 h	151.472	0.007	0.1
Telesto	45.307 h	294.67	—	—
Calypso	45.307 h	294.67	—	—
Helene	2.74 h	378.06	0.005	0.2
Hyperion	21.28 h	3,560	0.104	0.4
Phoebe	550 d	13,210	0.163	186

[a] Compiled from Veverka and Burns (1980), Morrison (1982a), and B. A. Smith et al. (1982). Cf. Table III of the introductory chapter.
[b] Relative to Laplace plane.

they form a diverse group. In terms of origin, some may be related to asteroids, others perhaps even to the nuclei of extinct comets; many may be fragments of larger precursor bodies (cf. introductory chapter).

Observed characteristics of small satellites vary widely. Many have "regular" orbits which are prograde, almost circular, and of low inclination (Table II). Some have "irregular" orbits which are highly inclined (in some cases retrograde) and markedly eccentric. Orbital periods range from 7.1 hr (Metis) to 758 d (Sinope). While many of the small satellites are in synchronous spin states, a significant fraction are not (Sec. VI). Limited spectral reflectance data show that surface materials range from carbonaceous rocks to water ice (Table I); observed albedos vary between 0.05 and at least 0.60.

At very small diameters one must distinguish between small satellites and ring particles. The smallest of the tabulated satellites, Deimos, is clearly not a ring particle, but future spacecraft missions may image rings sufficiently well to raise such an ambiguity. Davis et al. (1984) have suggested that very small ephemeral satellites may exist in Saturn's rings. They envisage clumps of small ring particles some meters across which are stable on time scales of days to weeks. It is not known whether such clumps really exist or whether they could reach sizes large enough to be detectable by spacecraft.

Our list of small satellites is undoubtedly incomplete (see also the discussion on completeness in the introductory chapter). However, it is unlikely that any small satellites will be found in association with the three inner planets (Kuiper 1961), nor should we expect that Mars has any satellites other than Phobos and Deimos (Duxbury et al. 1982). In the case of Jupiter, our inventory is probably incomplete, as indicated both by the Voyager experience and recent telescopic discoveries of outer satellites. This remark applies even more forcefully to Saturn, where the current inventory is almost certainly incomplete. Certain models of Uranus' rings (Goldreich and Tremaine 1979) imply that at least several small shepherding satellites remain to be discovered by Voyager 2 during its flyby in 1986 [see Table IV of the introductory chapter]. There probably are also undiscovered satellites of Neptune.

The regular orbits of most large satellites have led to the conviction that such bodies formed close to their present locations as part of the same general process that produced the primary (see chapter by Stevenson et al.). In some cases, most notably Jupiter's Galilean satellites, their inferred composition has been used to constrain the early chemical/physical conditions in the neighborhood of the protoplanet. While there has been considerable success in understanding the compositional gradient of the Galilean satellites in terms of the early superluminous phase of Jupiter (Pollack and Fanale 1982), our understanding of the compositional characteristics of the Saturn satellite system is less advanced (possibly because of a greater role of fragmentation and mixing of debris in the case of these smaller bodies).

For some small satellites, capture is a viable hypothesis of origin. For example, the carbonaceous-like composition of Phobos and Deimos has led to various proposals in which these objects are small asteroid-like bodies that were captured in the very early history of Mars when the protoplanet may have had an extended atmosphere (Hunten 1979). Less likely schemes do not rely on the early atmosphere to produce the drag needed to effect capture, but postulate the capture of a much larger satellite (large enough for tidal braking to be effective) and its subsequent destruction by collisions (Veverka and Burns 1980). However, Yoder (1982) concludes that the orbits of Phobos and Deimos in fact are most compatible with a noncapture origin. (The chapter by Burns elaborates on these issues.)

Capture may also account for the outer satellites of Jupiter (Pollack et al. 1979). On the other hand, the inner satellites of Jupiter may not be captured bodies; more likely, they and the ring are the collisional remains of larger bodies subjected to intense bombardment and fragmentation, along the lines suggested for the rings and small satellites of Saturn by Shoemaker (Smith et al. 1982). In some scenarios Amalthea could represent the largest fragment or a leftover core, in which case one might expect a refractory composition for this satellite (Gradie et al. 1980).

It has been suggested that the inner satellites of Saturn are collisionally derived objects (Smith et al. 1982); they may be pieces of material that formed close to Saturn. However, reading information about conditions around early Saturn from inner satellite makeup is complicated (even if we knew their exact compositions) due to possible differentiation of the parent bodies before fragmentations, and subsequent mixing of the debris. An important issue concerning the small inner satellites of Saturn is whether some are reaccumulations of fragments, or whether the observed satellites are the surviving largest fragments. Because of its retrograde orbit, Saturn's outer satellite Phoebe is usually assumed to be a captured object (Pollack et al. 1979).

The hypothesized shepherding satellites within the rings of Uranus would have an origin common with that of the ring particles: collisional fragmentation of small satellites deep in the potential well of a planet. In this context, it should be noted that the mode of formation of the larger regular satellites of Uranus remains enigmatic due to the high tilt of the planet's spin axis (98°) (Singer 1975). Nereid, the outer satellite of Neptune, is commonly considered to be a captured object because of the high eccentricity of its orbit (0.75). As in the case of Phoebe, there is no information on where it came from or how it was captured (cf. Pollack et al. 1979).

In terms of available data, our set of information for small satellites is much more limited than for their large counterparts. As Table I indicates, in most cases many essential parameters are uncertain or even unknown. Only for Phobos and Deimos do we have fairly complete information. For some small satellites we know almost nothing about their physical properties. The situation could improve somewhat given vigorous telescopic observing programs; yet, imaging of surface detail, accurate shape determinations and mean density measurements require spacecraft missions.

To date no spacecraft has been launched specifically to study small satellites, although plans for a Soviet Phobos mission for the late 1980s have been mentioned. Phobos and Deimos were studied in detail by the Mariner 9 and Viking Orbiter spacecraft as part of the overall investigations of Mars; in fact, the Viking Orbiters returned higher resolution images of these satellites than they did of the Martian surface. Many small satellites of Jupiter and Saturn were investigated by the Voyager spacecraft in 1979–1981, although the trajectories did not allow comprehensive imaging of all satellites. The inner and outer Jovian satellites and some of the very small Saturnian satellites remain the most poorly studied. Of these, the inner satellites of Jupiter will be scrutinized by Galileo in the late 1980s; the outer Jovian ones are most amenable to careful Earth-based observations.

In this chapter we review some of the characteristics of the small satellites, especially their shapes and surface features, and comment on possible future studies of these objects, representative examples of which are shown in Fig. 1.

II. SIZES AND SHAPES

Can small satellites be segregated into distinct groups on the basis of size and shape? In Fig. 2 we have plotted the volumes and axial ratios (a/c) of the satellites of Mars, Jupiter, and Saturn in ascending order of size, realizing that in some cases (e.g., outer satellites of Jupiter) current estimates of these parameters are only approximate. The overall plot shows little segregation by size; the largest gap by volume (nearly a factor of 10) occurs between Enceladus and Tethys. The continuity in the distribution of satellite volumes evident in Fig. 2 is largely due to the characteristics of the Saturn system of satellites. By contrast, a dichotomy between small and large satellites exists in the Jovian system: the ratio of volume of Europa (the smallest Galilean satellite) is more than 4000 times that of Amalthea (the largest "small" satellite).

The most striking feature of Fig. 2 is that satellites the size of Mimas or larger ($r = 195$ km) are all nearly spherical, while smaller satellites, with the possible exception of Phoebe ($a/c = 1.1 \pm 0.1$), are significantly nonspherical. To first order, satellites can be divided into two categories based on size and shape: the larger satellites tend to be nearly spherical, the smaller ones tend to have irregular shapes. Outlines (silhouettes) of a number of small irregularly shaped satellites are shown in Fig. 3, all to a common scale.

The significance of observed axial ratios of small satellites has been debated for years (Soter and Harris 1977; Hartmann 1979; Morrison and Burns 1976; Davis et al. 1981; Dermott 1979,1984c). Some authors have hypothesized that satellites and asteroids small enough to have suffered nearly catastrophic impacts may be gravitationally bound rubble piles (Davis et al. 1979; Farinella et al. 1983). A further hypothesis is that such loose agglomerations of debris will approximate equilibrium ellipsoids. For a homogeneous satel-

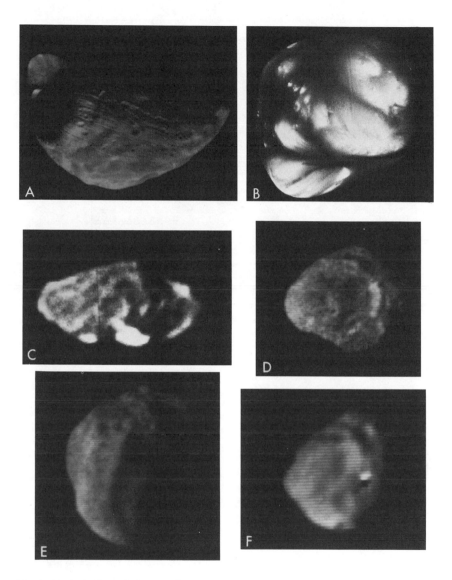

Fig. 1. Typical views of some small satellites: A: Phobos. 27 km across (Viking Orbiter image 357A64); B: Deimos. 12 km across, taken at $\alpha = 5°7$. Bright albedo features trend downslope (Viking Orbiter image 507A01); C: Amalthea. View is 275 km across, $\alpha = 35°$, centered at 92°W, north is at top (Voyager image FDS = 16377.34); D: Hyperion. $\alpha = 56°$, view is 300 km across (FDS = 43968.02); E: Epimetheus. $\alpha = 55°$, 140 km top to bottom (FDS = 34942.29); F: Janus. $\alpha = 25°$, 200 km top to bottom (FDS = 34936.11). Crude sketch maps of these satellites are available in the map section.

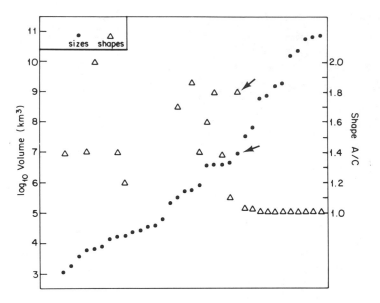

Fig. 2. Sizes and shapes of all the satellites of Mars, Jupiter and Saturn. Triangles designate the ratio of longest to shortest axes for those objects with adequate image coverage. Solid circles indicate volumes. Data are arranged in order of satellite volume. Data for Hyperion, the largest irregularly shaped satellite, are identified by arrows.

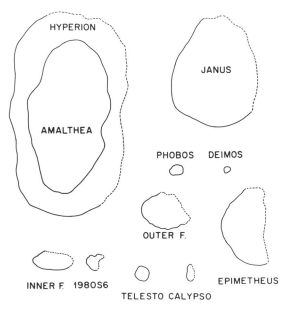

Fig. 3. Silhouettes of small satellites plotted at the same scale. Solid lines are limbs, dashed lines are terminators.

lite in a particular orbit, the axial ratios of the equilibrium ellipsoid depend solely on the object's density. For small satellites that produce no measurable perturbations on their neighbors and for which masses have not been determined by spacecraft flybys, this technique, if applicable, could provide our only information on density.

This technique is of some importance in the cases of Mimas and Enceladus, because the masses of these satellites are considerably uncertain. The mean density of Mimas, as determined from radio measurements of the mass of Tethys by Voyager 2 and from the observed ratio of the libration amplitudes of Mimas and Tethys, is 1.42 ($\pm$ 0.18) g cm^{-3} (Tyler et al. 1982). This is at odds with the classical determination of the satellite masses from the libration amplitudes and from the libration period of Mimas which results in a value of 1.16 ($\pm$ 0.06) g cm^{-3} (Dermott 1984c). In the case of Enceladus, although this satellite is in a resonance with Dione, the amplitude of libration is poorly determined; the density of the satellite is 1.27 ($\pm$ 0.45) g cm^{-3}. The comparatively smooth surface of Enceladus, in particular, suggests that its shape must be close to that of an equilibrium ellipsoid and thus that a determination of the shape would result in an estimate of the mean density. However, the very fact that its surface appears to have been molten in the past suggests that the satellite may be differentiated, in which case the satellite would not be homogeneous and its shape would also depend on its moment of inertia. In these circumstances, Dermott (1984c) has pointed out that, if a satellite has a deep mantle of known material (water ice, for example), then a determination of a satellite's shape could result in useful estimates of its mean density, mass and moment of inertia.

To test the usefulness of the equilibrium-ellipsoid hypothesis we should compare directly measured mean densities with those inferred from the shapes of these bodies under the equilibrium-ellipsoid model. Unfortunately, measured mean densities are available only for Phobos and Deimos; however, useful constraints can also be obtained from the measured shapes of Amalthea and Hyperion. The results are summarized in Table III. Using the latest shapes for Phobos and Deimos determined by Duxbury and Callahan (1982),

TABLE III
Tests of Equilibrium-Ellipsoid Models

	a/c Observed	a/c Equilibrium (based on ρ)	ρ Observed	ρ Inferred a/c	ρ Inferred a/b
Phobos	1.43	unstable	2.2	3.5	5.6
Deimos	1.39	1.04	1.7	0.24	0.32
Amalthea	1.78	—	—	1.1	1.2
Hyperion	1.8	—	—	<0.01	

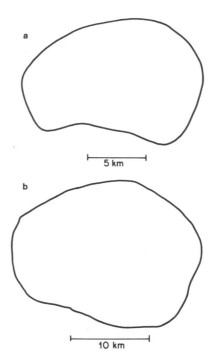

Fig. 4. Silhouettes of Phobos and Deimos. (a) Deimos, from composite of images, $A - C$ plane. Mars is to the right. (b) Phobos, from image 111A03, centered at: 3°S, 311°W.

we see that the equilibrium-ellipsoid model is not realistic. While the densities of both satellites are in the neighborhood of 2 g cm^{-3}, the equilibrium-ellipsoid model gives 3.5 g cm^{-3} for Phobos and 0.2 g cm^{-3} for Deimos. In fact, in the case of Phobos there is no stable rubble pile ellipsoid with as low a density as the satellite actually has. Thus, at least in the cases of Phobos and Deimos, the rubble-pile equilibrium-ellipsoid model fails badly. This result is not surprising, considering their silhouettes (Fig. 4).

Of the small satellites of Saturn only Hyperion provides a useful test because of its very irregular shape and its slow spin rate (Thomas et al. 1984). Hyperion's calculated mean density turns out to be less than 0.01 g cm^{-3}. Although we lack a measured value of the density, it is safe to say that here again the model is not realistic. In this context it should be stated that Hyperion may not provide a fair test, because dynamical arguments have been advanced by Farinella et al. (1983) that this particular satellite cannot be a rubble pile, but should rather be considered a residual core that has survived from the catastrophic breakup of a larger precursor.

For Amalthea, one derives a model density of about 1 g cm^{-3} using the axial ratios quoted by Veverka et al. (1981). On general grounds, such a value seems too low (Gradie et al. 1980). However, it is essential to realize that the axial ratios commonly quoted for small satellites cannot be meaningfully applied to such calculations unless the shapes really are well approximated by ellipsoids. In the case of Amalthea, for instance, the a/c cross section is only approximately an ellipse, and the best-fit ellipse to this cross section corresponds to an ellipsoid that is unstable for any assumed density.

Even if the satellites are claimed to be ellipsoidal, to obtain useful values of the mean density from the above model, one requires more accurate determinations of the axial ratios than are usually available. To illustrate, we can consider the case of Janus. The best estimates of a/c and a/b are 1.4 ± 0.15 and 1.2 ± 0.09, respectively; corresponding model densities are 0.7 to 1.1 g cm^{-3} (from a/c) and 0.9 to 2.3 g cm^{-3} (from a/b). Given the likely icy composition of Janus, the range of values is too wide to be useful (Yoder, personal communication, 1984).

We conclude that at the present time there is sufficient evidence to show that at least four small satellites (Phobos, Deimos, Hyperion and Amalthea) cannot be considered equilibrium-ellipsoid rubble piles. In the remaining cases, data on shapes are sufficiently uncertain that meaningful model calculations cannot be made.

III. LIMB TOPOGRAPHY

Two methods are available for determining the detailed shapes of small satellites. The more comprehensive method, involving the determination of control points (latitude, longitude, radius) has been applied only to Phobos and Deimos (Duxbury and Callahan 1982). In fact, Phobos is the only small satellite for which estimated topographic contours have been published (Turner 1978). Data on the other small satellites are less detailed, and one must rely on axial ratios and limb topography for shape information. Limbs are effectively skylines that hide depressions and thus provide only lower limits to the topographic variations along particular sections. Compared to nearly spherical objects such as Mimas and Enceladus, nonspherical small satellites display considerable limb topography. In Figs. 5 and 6 we show some examples plotted as deviations from the best-fit ellipse for a particular section. Data for all satellites are plotted on the same vertical scale (deviation from ellipse expressed as a fraction of the mean radius); the horizontal scales vary. The contrast in limb topography between objects such as Phobos, Deimos, Amalthea, Janus, and Hyperion, and that of Enceladus and Mimas, is striking.

To include other small satellites in this discussion, we must resort to even lower-resolution limb data. To incorporate such data, we simply measure the linear dimensions of concavities on the limbs and tabulate the largest for each object. For some objects such as Phobos and Deimos, the coverage is so complete that our measurement almost certainly represents the largest concavity on the body; for other satellites the measured concavity could be significantly

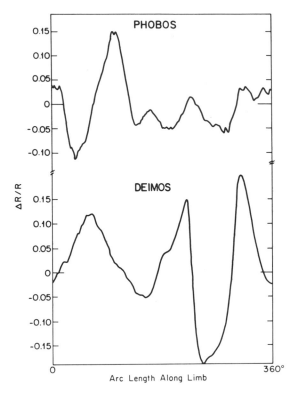

Fig. 5. Limb topography on small satellites. Deviation of limb profile from a best-fit ellipse is plotted as a fraction of average satellite radius. For Phobos and Deimos the profiles are 360° of arc (cf. Fig. 6).

less than the maximum present because of limited coverage. In Fig. 7 we plot the dimension of the largest concavity against the mean radius of the satellite. Mimas marks an abrupt change in the trend.

From various aspects of the above discussion it is evident that there are two distinct classes of small satellites: those that are more or less spherical and show little limb topography, and those that are irregularly shaped and have very rugged limb profiles. The traditional interpretation is that the smaller and irregular ones were shaped by fragmentation processes whereas the shapes of the larger ones are controlled by gravity. Johnson and Mc-Getchin (1973) discussed the shapes of satellites and asteroids in terms of material strength and the ability to support topographic loads. Their considerations show that other factors being equal, smaller satellites can support larger relative topography.

The energetics of changing a satellite's shape may be more important than just support of topography. Because fragmentation energy varies as r^3

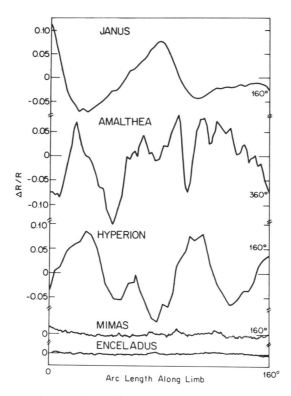

Fig. 6. Limb topography on small satellites. Deviation of limb profile from a best-fit ellipse is plotted as a function of average satellite radius. For Amalthea the profile is 360° of arc; for the others the profiles are about 160° of arc (cf. Fig. 5).

and depends on material strength, whereas gravitational binding varies as r^5 (Pollack et al. 1973), the conclusion is inescapable that for a particular material the r^5/r^3 ratio between gravitational binding energy and fragmentation energy implies a rapid transition with increase in radius from dominantly strength-controlled processes to those dominated by gravitational ones (Thomas et al. 1983a). This transition is probably documented by the significant differences in topographic and shape data of Hyperion and Mimas shown in Figs. 5, 6 and 7. The actual topographic loads on Hyperion are at least the order of those on Mimas, and could be greater because of the much greater topography.

IV. CRATERING, SHAPING BY IMPACTS, AND FRAGMENTATION OF SMALL SATELLITES

Shaping and fragmentation by impacts have undoubtedly been important processes in the evolution of small satellites and asteroids (see, e.g., Chap-

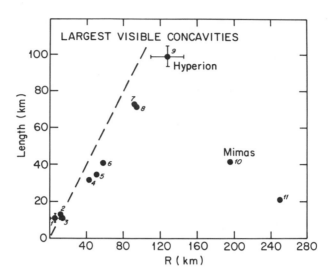

Fig. 7. Lengths of largest visible concavities on small satellites plotted vs. the average satellite radius. Dashed line shows length=R. 1: Deimos; 2: Phobos; 3: Calypso; 4: Outer F-ring shepherd; 5: Inner F-ring shepherd; 6: Epimetheus; 7: Janus; 8: Amalthea; 9: Hyperion; 10: Mimas; 11: Enceladus.

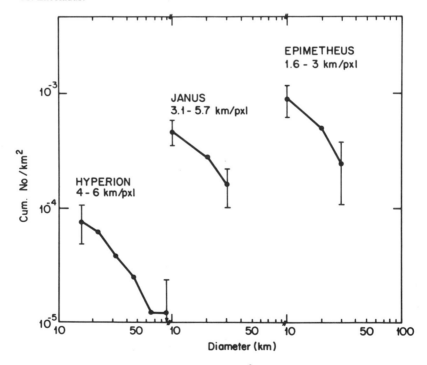

Fig. 8. Cumulative crater densities (number > D/km^2) on three Saturnian satellites. Also indicated is the linear resolution of the images used in km/pixel.

man and Davis 1975; Davis et al. 1979; Fujiwara 1980; Farinella et al. 1982; Smith et al. 1982). All small satellites imaged at resolutions better than 25 line pairs/diameter appear to be heavily cratered. The maximum crater densities on Phobos and Deimos are near the highest densities observed on the Moon (Thomas and Veverka 1980*a*). Craters are prominent on the other small satellites, but due to the limited resolution of available images, the tabulated densities depend heavily on only a few discernible craters; thus, they have large uncertainties. Examples of crater counts for three small Saturn satellites are shown in Fig. 8. These counts overlap the independent estimates by Plescia and Boyce (1983) only at the largest crater diameters, illustrating that in dealing with such low-resolution images, personal judgment plays a role in determining what is counted as a crater (generally Plescia and Boyce count more craters than we do for the three satellites in Fig. 8).

In many instances the available resolution is not sufficient to attempt to answer key questions, such as whether there exist leading/trailing side asymmetries in crater densities. In the case of Epimetheus and Janus, the data are also insufficient to search for uncratered or sparsely cratered areas (possible evidence of the disruption of the putative coorbital parent body).

Crater densities equivalent to those observed on Phobos or Iapetus imply that the largest crater on small objects should have a diameter which approaches the satellite radius (Thomas and Veverka 1980*a*). Impacts corresponding to somewhat larger craters are presumed to lead to massive fragmentation of the parent body (cf. Fujiwara et al. 1977). Smith et al. (1982) have presented a model of cratering in the Saturn system which implies cratering rates at Janus and Epimetheus over 30 times as great as those at Iapetus. In this scheme the fragmentation of small satellites close to Saturn would be a common process, and indeed Smith et al. suggest that all Saturn satellites within the orbit of Rhea have been fragmented and reaccreted at least once in the past 4.5 Gyr. Unfortunately we do not know if the small satellites are single fragments or a reaccumulation of debris. Similarly, in the Jovian satellite system, the high crater density on Callisto suggests very high cratering rates within the orbit of Io (Shoemaker and Wolfe 1982) and the possibility that the inner small satellites of Jupiter are collisionally shaped and derived fragments.

Although the largest concavities discussed above (Fig. 7) were not included in the crater counts, their presence suggests strongly the occurrence of noncatastrophic impacts more severe than those which formed the largest well-defined craters. The generally irregular outlines of the small satellites (Fig. 3), the concavity data (Fig. 7), and predictions of extremely high crater production rates, all suggest that impact-related phenomena played a major role in the shaping of these objects.

What are the possible configurations resulting from severe impacts? Many laboratory studies have shown the progression of damage with increasing impact energy per target mass: simple craters, craters and fracturing, craters and spallation, and finally the destruction of the target body (see, e.g., Fujiwara et al. 1977). Consequences for asteroids (and to some extent for small satellites) have been explored by Davis et al. (1979) and Farinella et al. (1981,1982). Considerable emphasis has been placed on impacts that fragment but do not disperse objects, thereby yielding rubble piles whose mechanical properties are very different from those of the original objects (Barks 1960; Farinella et al. 1982).

Some schematic configurations of heavily battered satellites are shown in Fig. 9. The simplest configuration results from the chipping of small fragments from a rigid core. If the debris is ultimately lost, a situation that is less likely for satellites than for asteroids, the object will evolve into an irregular, fractured core. If debris is retained (directly or by reaccretion), the buildup of thick regolith layers could smooth out the shape of the object. However, as noted by Goguen et al. (1979), the volume of material excavated by a crater density similar to that on Phobos is only 5% of the object's volume. Thus, a cratering history more extensive than that observed on Phobos is necessary to change significantly the shape of the parent body. Phobos's crater Stickney, which has a diameter nearly equal to the satellite's average radius, is not the largest concavity on the satellite. A large saddle associated with Kepler's ridge (Pollack et al. 1973b) and a major asymmetry along the a axis dominate the satellite's shape. Not surprisingly, the impact history preserved on a satellite's surface in terms of well-defined craters does not play a dominant role in determining the present shape of the body.

Very severe impacts will fragment the parent body and, in the case of a satellite orbiting close to its primary, significant reaccumulation of debris could occur (Burns and Showalter 1983). Such fragments might be reaccreted in gravitationally stable groups (Fig. 9b); in such cases the shape of the satellite is determined by the shape and size distribution of the fragments, not by the satellite's density. The latter situation arises only if one assumes that the fragmentation debris is all very fine in size, an assumption which on the basis of extrapolation from small-scale fragmentation experiments appears unjustified.

The reaccretion of large fragments as depicted in Fig. 9b produces essentially the same result as the low-velocity collisions postulated to be responsible for the shape of Hektor by Hartmann (1979). Hartmann has pointed out that collisional accretion of objects could yield irregularly shaped asteroids and that this may be an alternative process to forming irregular shapes by collisional fragmentation.

Further impacts may eventually reduce all parts of the satellite shown in Fig. 9b to fragments small compared to the average radius. Theoretically, these pieces could reaccrete into a gravitationally-shaped ellipsoid, whose shape would be determined solely by its density and orbit, if the satellite is homogeneous. However, even loose material can support slopes; angles of repose for such material are in the range of 30 to 35°, and theoretically are inde-

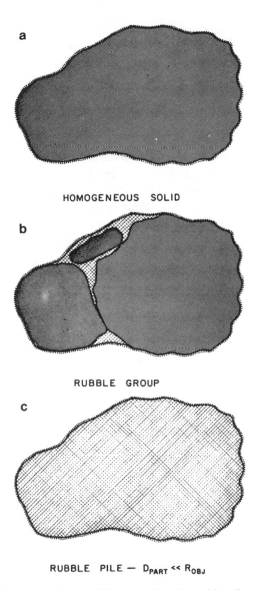

a

HOMOGENEOUS SOLID

b

RUBBLE GROUP

c

RUBBLE PILE — $D_{PART} \ll R_{OBJ}$

Fig. 9. Schematic cross sections of different configurations of heavily cratered satellites: (a) structurally continuous and homogeneous solid with thin regolith; (b) reaccumulated large fragments and some small fragments that slightly smooth the resulting shape; (c) rubble pile of particles that are small compared to the satellite's radius. Topography results from non-catastrophic impacts subsequent to major reaccretion.

pendent of g. Thus, the configuration in Fig. 9c is a true rubble pile, and could be a natural satellite. Its shape, however, need not contain useful density information. The shape of a true rubble pile is presumably the result of cratering and redistribution of debris, and will always be somewhat irregular.

Objects that are postulated to have been repeatedly battered, broken, dispersed and reaccreted are often termed rubble piles, piles of rocks, megaregolith objects, etc. Brecciated objects may be a preferable description, because this term does not leave the impression that such objects must lack cohesion. Meteorites and lunar impact breccias include much weak and crumbly material, but they do show that the impact process also can cause welding and cementation of fragments. Additionally, hydrothermal processes could be important in cementing fragments of many objects, and might be especially effective on icy bodies.

V. INTERNAL STRUCTURE OF SMALL SATELLITES

There is little direct information on the internal structure of small satellites. Even in the best of cases we only know the mean density, but have no information on the mass distribution within the object. Satellites could presumably be nearly homogeneous, or have a central concentration of mass, or have irregular distribution of fragments of different densities. While methods of measuring the moments of inertia of small satellites can be conceived (precession, libration, in addition to close flybys by spacecraft), only in the case of Phobos has any information been obtained. Using a control network of 98 craters on 43 Viking Orbiter Phobos pictures, Duxbury and Callahan (1982) obtained a value of 0.8 ($\pm$ 0.2) deg for the observed forced rotational libration of that satellite. The theoretical value (Peale 1977) is

$$\delta = 6e\left[\left(\frac{C}{B-A}\right) - 3\right]^{-1} \qquad (1)$$

where e is orbital eccentricity and A, B, C are principal moments of inertia. If we assume that the composition of the satellite is homogeneous, then from the observed shape we obtain $(B - A)/C \approx 0.2$ and the expected value of δ is three times larger than the observed value (Duxbury and Callahan 1982). The observed value of δ requires that $(B - A)/C \approx 0.1$. This could be reconciled with the observed shape if the satellite is not homogeneous but consists of a comparatively dense core surrounded by a deep, low-density regolith.

Equally few direct data are available about the effective mechanical strength of small satellites. While very weak structures have been postulated in rubble-pile models, the few indirect observables that bear upon this question suggest that most small satellites are made of material that has significant mechanical strength. For example, on Phobos, patterns of grooves and ridges suggest a moderately competent internal structure. The grooves define a series of nearly parallel planes, most of which are also approximately parallel to the satellite's intermediate axis (see Map Section). Although the pattern strongly suggests fracture planes, a completely satisfactory explanation for their origin does not exist and the regularity of the pattern on such an irregularly shaped object remains puzzling (Fujiwara and Asada 1983). The pattern of grooves does not appear compatible with fractures that might result from tidal stresses alone (Dobrovolskis 1982). The depth and width of these inferred fractures are unknown. What is clear is that the orientations of the groove planes and the satellite's axes are related; it is not known if it is a cause and effect relation or whether both derive from some primitive anisotropy in the Phobos material.

Other prominent structural features on the surface of Phobos are very long ridges, some of which form approximately parallel sets. Their pattern does not correspond to that of the grooves, and their origin is unclear. Some internal fabric is indicated by the nearly rectangular shape of crater Hall ($d = 5$ km) whose sides align with many nearby ridges. The nature of the anisotropy or fabric is unknown, but it must extend to scales of at least 9 km.

Prominent ridges are also observed on Deimos (Fig. 1). These divide the satellite into three major regions. The smooth areas between the ridges are slopes in which significant downhill regolith movement occurs. Deimos may be an object that is being gradually worn down by impact erosion and these ridges may be the expression of underlying bedrock. The ridges on Deimos are difficult to explain as fortuitous alignments of fragments and impossible to explain as sedimentary features because of the observed downslope movement of regolith away from the ridges. Although we do not know the strength of the material underlying the ridges, the continuity of the ridges suggests structural continuity for much of the length of the satellite.

Long straight ridges indicative of structural continuity are also found on Amalthea, and on the outer F shepherd where a prominent spine exists in the northern hemisphere. In general, however, small satellites other than Phobos and Deimos have not been imaged sufficiently well to detect reliably systems of ridges, grooves, or other possible evidence of internal structure.

Pressures within small bodies are low. For a spherical body of uniform density, the pressure P at the center is given by

$$P \simeq 1.4 \times 10^{-3} \rho^2 r^2 \text{ bar} \qquad (2)$$

where $\rho(\text{g cm}^{-3})$ is the density and r (km) is the radius. Thus central pressures reach only ~ 15 bar inside Janus, ~ 75 bar inside Mimas, and a mere 0.5 bar inside Phobos.

Conventional arguments show that the thermal history involving heating by long-lived radioactive nuclides alone would be dull for such small objects (see chapter by Schubert et al.). Consolmagno and Lewis (1978) calculate that if long-lived radioactivity is the only significant source of energy, bodies smaller than 500 km in radius will not be heated enough to melt and differentiate if made of water ice and rock in "cosmic proportions." Although the presence of other ices such as NH_3 may help (a mixture of ammonium hydrate and water melts at 173 K at 1 bar), even in this instance bodies as small as considered here will have uneventful thermal histories. Equilibrium thermal gradients of about 3 K km^{-1} apply for a rocky body ($\rho \sim 3$ g cm^{-3}) of chondritic composition, and of about 1 K km^{-1} for an ice/rock body of Ganymede's composition (Lewis 1971). Thus, due to heat sources of long-lived radioactive nuclides alone, central temperatures will never be higher than 300 K above the surface temperature for a body with a radius of 100 km. That it is difficult to melt bodies even as large as Enceladus by conventional methods is illustrated by detailed calculations such as those of Ellsworth and Schubert (1983), Consolmagno and Lewis (1978), and others (see also chapter by Schubert et al.).

It would, however, be unwise to conclude that small satellites, even those which are not fragments of significantly larger parent bodies, have had uneventful thermal histories. For some satellites tidal heating could have been important (cf. Enceladus) (Squyres et al. 1983b). More generally, many meteorites bear witness that their parent bodies, some no bigger than the larger small satellites discussed here, had thermal histories hot enough to melt and differentiate them (Dodd 1981). Other meteorites show definite evidence of lithification and even of hydrothermal alteration. Gas-rich meteorites are commonly interpreted as lithified regolith breccias (Anders 1978), while Kerridge and Bunch (1979) have reviewed extensive evidence for the aqueous alteration of carbonaceous meteorites.

VI. DYNAMICAL CONSIDERATIONS

A. Rotation States

The theory of satellite rotation states has been reviewed by Peale (1977) and others. A satellite is expected to be in a synchronous rotation state, spinning about a principal axis of inertia, the latter oriented perpendicular to the orbital plane. The characteristic time to reach synchronous rotation for close satellites is very much less than the age of the solar system (see Table 6.1 of Peale 1977). Indeed, synchronous spin states prevail among small satellites about which we have information on rotation. Known exceptions include Phoebe, Hyperion, and some of the outer satellites of Jupiter (e.g., JVI and JVII). It is likely that the spin state of Nereid is also nonsynchronous, but there are as yet no observations.

The rotation of Hyperion has attracted special attention; Wisdom et al. (1984) have predicted that this satellite should have a chaotic spin state. The average torque that a planet exerts on a satellite increases with $(B - A)/C$, where A, B, C are the principal moments of inertia of the satellite, and with the eccentricity of the satellite's orbit. For Hyperion, because both of these quantities are exceptionally large, the changes in the spin rate of the satellite that can occur over one libration cycle are large enough to change the spin from one resonant state to the next, and it has been shown that the resultant motion is chaotic (Wisdom et al. 1984). Thomas et al. (1984; Thomas and Veverka 1985) found that during the Voyager 2 encounter Hyperion was spinning with a period of 13 d about an axis lying close to Hyperion's orbital plane. This rotation, over 61 d, is apparently consistent with the theory of Wisdom et al. (1984). Tidal dissipation could reduce the eccentricity of Hyperion's orbit and reduce the perturbing torques. However, gravitational perturbations due to Titan force Hyperion's eccentricity to remain at the comparatively high value of 0.104.

B. Coorbital Satellites

Some of the smaller satellites in the solar system share their orbits with larger companions. A major puzzle of the distribution of these satellites is that all the known coorbitals are associated with Saturnian satellites. The Jovian system has been equally well observed but no coorbitals have been discovered. The status of the Uranian system in this respect will probably remain undetermined until the arrival of the Voyager 2 spacecraft in 1986.

The groundbased and spacecraft observations of Saturn coorbitals have been summarized by Aksnes and Franklin (1978), Marsden (1980), Sinclair (1984), and chapter by Peale. 1980 S6 orbits close to the leading Lagrangian equilibrium point $L4$ of Dione. Telesto and Calypso orbit close to the leading $L4$ and the lagging $L5$ Lagrangian equilibrium points, respectively, of Tethys. Enceladus appears to be devoid of coorbitals, although Baum et al. (1981) observe a slight bunching of E-ring material, or possibly a faint satellite, close to the $L5$ point of Enceladus. Chenette and Stone (1983) consider that the electron absorption signature observed by the cosmic ray system on Voyager 2 and the electron absorption features observed by Pioneer 11 could be due to the presence of localized absorbing material in the orbit of Mimas. The latter interpretation of the Pioneer 11 data supports that of Simpson et al. (1980). Other interpretations of the data have been given by Van Allen (1982). Synnott and Terrile (1982; Synnott 1986) report that Voyager 2 images reveal a satellite (1981 S12) of diameter ~ 10 km very near the orbit of Mimas, but trailing Mimas by 108°. The most interesting case is that of Janus and Epimetheus. These satellites have comparable masses and move in horseshoe paths about a common mean orbit.

The dynamics of coorbital satellite systems have been discussed by Dermott et al. (1979,1980,1984a), Dermott and Murray (1981a,b), Yoder et al. (1983), Sinclair (1984) and chapter by Peale. Dermott et al. (1979,1980) pointed out two reasons why we might expect horseshoe orbits in the solar system to be associated only with very small satellites. First, the ratio of the radial widths of those regions where, respectively, tadpole alone and tadpole and horseshoe orbits are possible is $\sim (m/M)^{1/6}$, where m is the mass of the larger of the two coorbitals and M is the mass of the planet. It follows that the horseshoe orbit region is only dominant if $(m/M)^{1/6} \ll 1$. The second reason obtains from considerations of orbital stability. Dermott et al. (1980); Dermott and Murray (1981a) have suggested that the horseshoe paths may be imperfectly periodic and that random changes in the semimajor axes of the satellites of magnitude $\delta a/a \sim \pm\ m/M$ may occur on each encounter of the coorbitals. The horseshoe configuration would then have a lifetime, $L \sim T(m/M)^{-5/3}$, where T is the orbital period of the satellites. For $L > 10^8$ yr, we require $m/M < 10^{-7}$.

On encounter, satellites in horseshoe paths exchange energy and angular momentum and experience near-symmetric changes with respect to their common mean orbit: while one satellite moves from the inside to the outside of its horseshoe path, the other satellite moves from the outside to the inside of its own path. Observations of the changes in the mean motions that occur on encounter lead to an estimate of the ratio of the satellite masses, while the time between encounters depends both on the difference of their semimajor axes and on the sum of their masses (Dermott and Murray 1981a,b). The available observations have been analysed by Yoder and Synnott (1984; cf. chapter by Peale), who estimates that the ratio of the masses of Janus and Epimetheus is ~ 3.5. From the estimate of the sum of the masses and the observed volumes, the density of Janus and Epimetheus is estimated as 0.65 ($\pm$ 0.4) g cm^{-3} (Yoder, personal communication, 1984).

Because Janus and Epimetheus are of comparable size and have irregular shapes, it is natural to suggest that they are collision fragments from a single body (Smith et al. 1982). However, the relative velocities of the satellites are much smaller than their mutual escape velocity, and formation by disruption would require very special circumstances (Dermott and Murray 1981b; Yoder et al. 1983). Dermott and Murray (1981b) consider that if collisions have had a role in the formation of coorbital satellites, then the formation of a narrow ring of coorbital debris out of which the coorbital satellites then accrete is a necessary intermediate step. However, it is possible that drag forces on the satellites have acted to reduce their relative velocities (Yoder et al. 1983; Dermott 1984a,b). In particular, any force that acts to push the satellites away from the planet will reduce the relative velocity of the satellites and drive the smaller of the two satellites towards one of the Lagrangian equilibrium points of the larger satellite. This transformation of the orbits will occur on a time scale comparable with that required to produce significant changes in the orbital radii of the satellites. Both tidal forces and ring torques need to be considered.

Drag forces may also have had a role in the early evolution of coorbital satellite systems. Dermott (1984b) and Dermott et al. (1985) have suggested that those drag forces that operated during planet and satellite formation may have acted to disrupt all the primordial coorbital systems. In that case, the Saturnian coorbitals may exist because they are not primordial but the results of later disruptions caused by cometary impacts (Smith et al. 1982). That being the case, Dermott (1984b) argues that Uranian coorbitals may exist only if the Uranian satellites have suffered disruptions since the time of planet formation. Small satellites, such as Miranda, that are close to the planet, are the most likely candidates.

VII. EFFECTIVE GRAVITATIONAL ACCELERATION AND SLOPES ON SMALL SATELLITES

The directions of effective slopes on some irregularly shaped satellites may not be intuitively obvious. Combination of the gravitational attraction of a nonspherical body with tidal and rotational accelerations can yield net accelerations in directions that are far from the surface point's radius vector. An extreme case occurs when the acceleration causes loss of material from the satellite (cf. Dobrovolskis and Burns 1980). The result is that on a satellite such as Phobos or Amalthea it is not evident by simply looking at a picture what the relationship of the surface normal **n** and gravitational acceleration **g** is at any point. Accurate knowledge of this relationship is, however, essential to morphologic studies, because the local direction of slope angles influences ejecta emplacement, regolith migration, and mass wasting in general. If one can observe the effects of downslope movement (e.g., brighter streamers on the surface of Deimos), then one can derive data on local slope angles which in some cases depend strongly on the mean density of the satellite. For some irregularly shaped small satellites and asteroids this technique could be useful in the inference of mean densities.

Assuming that satellites are ellipsoids, the calculation of slope angles for various satellites that are strongly affected by tides have been made by Davis et al. (1981). Although such an approach reveals general trends, such as the relative potentials of the poles and equatorial points, it says little of the real topography.

We have calculated **g** and effective slope angles at individual points (pixels) of well-determined limb profiles for several small satellites. The scheme of calculation is as follows: the satellite is assumed to have elliptical cross sections normal to the limb profile. One axis is defined by the limb profile (axis 1, Fig. 10). Because we rarely have good data on limb profiles that are orthogonal to one another, the other axis must be estimated (axis 2, Fig. 10). Reasonable estimates are got simply by scaling the first axis by the similar axial ratio of the whole satellite (B/C in Fig. 10). This estimate works well because it is measured where the axes 1 and 2 are widest, i.e., the most massive part (it is most poorly constrained at the smaller parts). Note that for Fig. 10, the north and south profiles are calculated separately with slightly different C-axes. This scheme imposes some unrealistic symmetry, but allows a calculation of an accuracy appropriate to our data. For Deimos, the satellite is so asymmetric that additional point sources of mass were added to account for the two high points in the southern hemisphere (i.e., the slabs were not chosen as ellipses).

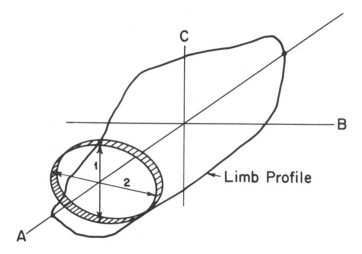

Fig. 10. Approximation of satellite shapes for calculation of gravitational accelerations by elliptical slabs perpendicular to well-determined limb profile. Axis 1 of thin slab is determined by
the limb profile; axis 2 is scaled by the B/C ratio of the whole satellite.

For a homogeneous satellite, surface accelerations due to tidal, rotational and gravitational forces are determined by the satellite and primary
masses, and by the size of the satellite and its orbit; they were calculated using
the expressions of Dobrovolskis and Burns (1980). Because we are not concerned with trajectories of ejecta, the Coriolis effects treated by Dobrovolskis
and Burns are not pertinent to our problem.

Using the slope angles calculated at each point along the limb we can
construct a profile, by integrating the slope angles between points along
the limb profile. Such profiles, which should probably be termed quasi-
topography, are useful in that they indicate the direction in which loose surface material would tend to move at a particular point on the limb.

The usefulness of the technique can be illustrated by applying it to
Amalthea, a satellite for which no direct estimate of mean density exists. We
can assume two rather different values of ρ: first, 1.1 g cm^{-3} which corresponds to the equilibrium-ellipsoid value (see above), and second, 3.5 g cm^{-3}
(similar to Io's value).

The calculated magnitude of g varies greatly over Amalthea. For $\rho = 1.1$
g cm^{-3}, g varies from 1.4 to 2.8 cm s^{-2} from 180°W, 0°N to near the north
pole. For $\rho = 3.5$ g cm^{-3}, the variation from 0°N, 180°W to the pole is from
5.5 to 8.3 cm s^{-2}. The variation is due both to the change in the tidal component as well as to changes in distance to the greater part of the satellite's mass.

The quasi-topographic profiles are quite different for the two densities
assumed (Fig. 11a,b). Except in the vicinity of the crater Gaea, slope angles
for $\rho = 1.1$ g cm^{-3} are < 10° (at an image resolution of 6.5 km/pixel). Slope

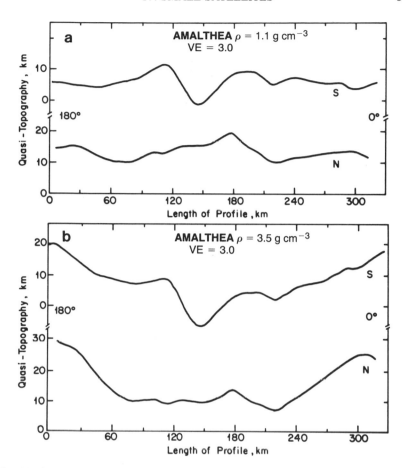

Fig. 11. Calculated effective topography on Amalthea. The calculated topography is the integrated slope angle along the profiles. The slope angles depend upon **g** which varies with the assumed density; hence, the calculated quasi-topography varies with assumed densities. *VE* is vertical exaggeration. (a) Calculated topography along 0° and 180° longitudes on Amalthea, assuming that density is 1.1 g cm^{-3}. Sub-Jupiter point is on right; S = southern profile, N = northern profile. Compare to limb outline in Fig. 3. (b) Same as plot (a), but for density of 3.5 g cm^{-3}.

angles are larger (8 to 15°) if ρ = 3.5 g cm^{-3}. Slope angles within the crater Gaea are little affected by the assumed satellite density, because the crater is near the pole where rotational components are vanishingly small and tidal accelerations are nearly parallel to the gravitational components. Effective slope angles outside of Gaea depend strongly on the assumed value of ρ. Thus, a determination by high-resolution imaging of the direction of regolith movement in this area would provide a valuable constraint on the mean density of the satellite.

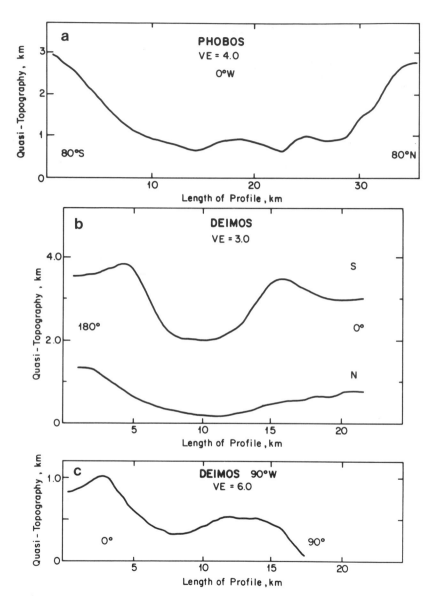

Fig. 12. Calculated effective topography on Phobos and Deimos. The calculated topography is the integrated slope angle along the profiles. Slope angles depend upon **g** which varies with the assumed density; hence, the calculated quasi-topography varies with different assumed densities. Density is 2.0 g cm^{-3}. *VE* is vertical exaggeration. (a) Calculated topography for Phobos along the 0° longitude from 80°N to 80°S from limb profile approximated from Viking image 126A83. The sub-Mars point is to the right. (b) Calculated topography on Deimos along 0° and 180° longitudes. Sub-Mars point is to the right; S = southern profile; N = northern profile. See Fig. 4 for satellite profile. (c) Calculated topography at 90°W on Deimos from 15°S (left) to 90°N.

A quasi-topographic profile calculated along the 0° longitude section of Phobos is shown in Fig. 12a. For this satellite we know that the mean density is close to 2 g cm^{-3} (Table I). The huge depression in the quasi-topographic profile occurs at the sub-Mars point, as already noted by Dobrovolskis and Burns (1980) and by Davis et al. (1981).

Profiles calculated for Deimos (density $\sim$ 2 g cm^{-3}; Table I), certainly one of the most irregularly shaped satellites yet seen, are shown in Fig. 12b,c. Here, as for Amalthea, the asymmetry in shape along the A axis will displace the rotation axis slightly toward the 0° longitude point and cause the 180°W, 0°N point to be at a slightly higher effective elevation than the sub-Mars point. The large saddle in the southern hemisphere of Deimos has slope angles as great as 34°. Slopes away from the ridge at low latitudes are roughly 15° over a 2 km distance. The magnitude of g varies from 0.28 to 0.34 cm s^{-2} over the surface of the satellite. Compared to Phobos, where tidal and rotational components make up 40% of the total effective gravitational acceleration, these two effects are very small on Deimos because of its much greater distance from Mars.

VIII. REGOLITH CHARACTERISTICS

The lunar surface remains the only satellite regolith studied directly and, as such, provides much of the basis for models of regoliths on other satellites and asteroids (cf. the discussion in the chapter by Veverka et al.). For most small satellites data on regoliths are limited to such superficial parameters as albedos, photometric phase functions, some spectral reflectance information, and very limited thermal data.

Generation of impact debris on small satellites may differ from that on larger bodies because of low gravity and possibly because of different material properties. The lower gravity may lead to larger craters and thus more debris, but the debris may be more widely spread over the object. The small gravitational accelerations (most much less than 2% that of the Moon) may favor the formation of very low-density regoliths. The rapid increase in the density of lunar regolith in the top centimeters may be a function of self compaction of highly stirred material. Seismic shaking and impact pressures may dominate compaction of the deeper lunar regolith (Carrier et al. 1973; Houston et al. 1974). While the upper centimeter or so of a small satellite regolith may be much fluffier than that of the Moon (consistent with thermal inertia data for Martian satellites [Lunine et al. 1982]), if the impact and sedimentation dynamics control regolith densities, then at depth, small satellite regoliths may not be very different from that of the Moon.

Phobos and Deimos are the only small satellites for which spacecraft data can resolve spatial variations in regolith properties (Thomas 1979; Thomas and Veverka 1980b; Lunine et al. 1982). The presence of substantial loose material on Phobos is demonstrated by the morphology of the grooves: smooth concave walls and slump features up to 100 m in relief. Layering is exposed in at least two crater walls. Some craters have flat or concentric floors that may also indicate layering. Infrared radiometry (Lunine et al. 1982) shows a surface of very low thermal inertia; the model-dependent particle sizes are $< 100 \mu$m. On both Phobos and Deimos the depth of the loose surface layer must be at least 1 cm to satisfy the radiometry. The Phobos data are consistent with a small exposure ($\sim 5\%$) of higher-inertia (blocky) material. The Deimos radiometry is not precise enough to place constraints on the presence or absence of a similar small blocky fraction.

The Deimos regolith has at least two principal components: brighter (albedo $\sim 8\%$) material that originates at ridge crests and crater rims, and material of albedo $\sim 6\%$ that appears to have been widely dispersed and subsequently concentrated in low areas (including crater floors). Deimos' surface is radically different in appearance from that of Phobos (see Fig. 1 and Sec. VII). Up to 12 m of debris has been accumulated locally within craters; accumulations in low regional areas could exceed 100 m (Thomas and Veverka 1980b).

Blocks are present on both satellites; they range in size from ~ 3 to 5 m (the resolution limits) to ~ 150 m across, and are evidently a component of crater ejecta (Lee et al. 1986). Such blocks are not numerous enough to account for the high thermal inertia component detected by Lunine et al. (1982) on Phobos.

Disk-resolved photometry of Phobos and Deimos has been carried out by Noland and Veverka (1977a,b), Goguen et al. (1978), French (1980), and Klaasen et al. (1979). Klaasen et al. (1979) found the normal reflectance of Phobos and Deimos through the Viking clear filter ($\lambda \sim 0.55 \mu$m) to be essentially identical (0.066 ± 0.006). Phobos' phase coefficient is slightly higher than that of Deimos, perhaps as a result of greater macroscopic roughness (Klaasen et al. 1979; French 1980). Darker areas within and near some fresh craters on Phobos are apparently the result of locally higher phase coefficients (Goguen et al. 1978). Goguen et al. speculated that these areas might contain impact-generated vesicular glass. French (1980) examined the bright markings on Deimos and concluded that their higher albedo but similar color indicate comminution of the background dark material. The material is believed to be similar to that in carbonaceous chondrites (Pang et al. 1978; Pollack et al. 1978b).

Available data on the regoliths of other small satellites are essentially albedos and phase coefficients. Voyager infrared temperature measurements of Amalthea (Simonelli 1983) are best explained in terms of a surface of low thermal inertia, rather than a surface of high thermal inertia rocks.

The regoliths of the small (icy?) objects orbiting Saturn might hold many surprises if the unusual radar properties of Ganymede and Callisto (Ostro 1982) are general to impact processes on icy satellites.

IX. REGOLITH DISTRIBUTION

Phobos and Deimos present interesting contrasts in the distribution of regoliths over their surfaces which may be used to discuss related phenomena on other small objects. Figure 1 and maps and photographs in Thomas and Veverka (1980*b*) demonstrate the long-distance transport of regolith on Deimos. This transport can be in extremely straight lines, as is shown by dark "shadows" in the bright material (Fig. 1). Directions of transport are consistent with the slopes calculated (Figs. 11,12) and with the general outline of the satellite. The bright material, coming from ridges and crater rims, is probably not voluminous (Thomas and Veverka 1980*b*). Its unimpeded movement over 30 to 90° of arcs implies a very smooth substrate. Indeed, images of 2 m resolution show extremely smooth surfaces with some craters and blocks, but most craters are ghosts even where incidence angles are 80°. The fill inside craters implies that layers of ~ 7 m total thickness have been spread widely over Deimos and then have been concentrated by downslope movement. Calculations show that visible large craters can supply ~ 12 m of debris over all of Deimos. Such quantities of debris are sufficient to fill small craters, but certainly do not significantly change the overall shape of Deimos. If the smooth surfaces are indeed sedimentary, we can make two observations: (1) they are not equipotential surfaces and could be subject to further movement, and (2) debris would have to have come from many large craters not included in present mapping. If the 11 km concavity in the southern hemisphere is a crater, its ejecta would constitute a significant fraction of the volume of the whole satellite. Such an amount of material could form the large smooth surfaces. If the smooth surfaces are not sedimentary, then we have a very interesting and completely unresolved problem.

Phobos presents an entirely different appearance: there are no regional albedo features that trend downslope nor are there other indications of significant regolith transport. The surface is rough at all observable scales. The lack of prominent regional downslope movement is easily explained by the roughness of the surface. The topographic profiles such as displayed in Fig. 12a also emphasize the difficulty of looking for sediment concentrations near the sub-Mars point; they would be trapped in local depressions over a very wide area.

Features interpreted as an ejecta blanket around crater Stickney by Thomas (1979) and the distribution of blocks on Phobos and Deimos suggest that crater ejecta on Phobos are retained close to the source craters but are widely dispersed on Deimos (Lee et al. 1986). This postulated difference could account for some of the differences in the two objects, but probably does not account for all the smoothing of Deimos' topography. The differences could arise if Phobos' craters were gravity-scaled while those on Deimos are strength-scaled. The exact relationship of gravity and material properties that could provide for this difference is poorly constrained, given the almost total lack of data on the mechanical properties of these objects.

The dynamical environment of Phobos and Deimos has been investigated by Dobrovolskis and Burns (1980) and Davis et al. (1981). They note that the rotation of Phobos (and small satellites in similar short-period orbits: Metis, Amalthea, etc.) is sufficiently rapid to cause the escape velocity (body reference frame) of material launched prograde to be significantly less than material launched retrograde. The major asymmetries in the extent of ejecta blankets that may result apply only to a small fraction of Phobos ejecta if the interpretation of Lee et al. (1986) on ejecta blankets is correct. Dobrovolskis and Burns found that some ejecta from Deimos could enter temporary orbits about that satellite before reimpacting it. One important result of the calculations of Davis et al. (1981) is that streams of ejecta cannot match the pattern of grooves on Phobos, either now or earlier when the satellite was in a longer-period orbit.

Our knowledge of regolith distribution on other small satellites is necessarily less detailed than for Phobos and Deimos. Amalthea has bright albedo markings on ridge crests and crater walls that in some respects (but *not* in terms of spectral reflectance [Thomas et al. 1981]) are like those on Deimos, and may be related to local mass movement. Bright markings on Saturn's inner F ring shepherd are even more difficult to characterize, but could be ice with a less-dark contaminant than for other areas. Phoebe has locally brighter spots, whose nature is unresolved in the long-range images (Thomas et al. 1983*b*). Hyperion has diffusely bounded albedo variations of $\sim 15\%$ (Thomas and Veverka 1985). Other small satellites show only hints of albedo markings or other possible evidence of real differences in surface materials. Disk-integrated studies of the outer Jovian satellites have not demonstrated heterogeneities in composition with rotation aspect.

X. DIVERSE SURFACE FEATURES

A. Morphology of Impact Craters

Crater morphology on small objects has been investigated by Cintala et al. (1978), Thomas and Veverka (1980*b*), and Pike (1980). Fresh, bowl-shaped craters (diameter D) on Phobos and Deimos appear to have depths $\sim 0.2\ D$, approximately the lunar value. Some flat-floored and concentric-floored craters occur on these satellites, possibly indicating the effects of substrate layering.

The transition from simple to complex crater morphology (central peaks, etc.) might not occur on small satellites if gravity scaling were the only control (Pike 1980). Two poorly resolved features (diameters of 40 and 90 km) on Amalthea and Hyperion may be craters with central peaks. However, because it is difficult even to count large craters on most of these objects, discussion of crater morphology is premature.

B. Grooves

The grooves on Phobos have been discussed in detail by many investigators (Thomas et al. 1979; Davis et al. 1981; Weidenschilling 1979; Head and Cintala 1979; and Dobrovolskis 1982). We believe there is strong circumstantial evidence that these grooves are due to fractures caused by the impact which produced Stickney. It is possible that tidal stresses played a complementary role as suggested by Weidenschilling (1979). There is little evidence in the distribution and morphology of grooves to support any scheme in which the grooves were produced by impact ejecta (Thomas et al. 1979a). The view endorsed here is that they represent a loss of regolith over fractures in the more rigid body of Phobos. The exact mechanism of regolith loss (ejection, drainage?) and the actual cause of the fractures remain controversial.

If the mechanism of loss of regolith were by expulsion, raised rims on the grooves might be expected. Although inspection of all the Phobos images revealed a few possible instances of raised rims (Thomas 1978), they are not common. Photoclinometry has been used to search in more detail for raised rims on grooves. The method (Thomas 1978) leaves much uncertainty in the calculated slope angles, but gives reasonable indications for reversal of slopes relative to a particular reference surface. Four photoclinometric profiles across grooves are shown in Fig. 13. The grooves in 13a and b may have raised rims, but the local topography is complex, and subtle nearby grooves make the interpretation of rims superposed on flat areas somewhat uncertain. In areas near the crater Stickney (Fig. 13c,d) large grooves have complex cross sections and there is no certain method for detecting constructional rims in this area. From the difficulty of finding examples of raised rims, we conclude that most grooves formed by methods that did not deposit significant material nearby as rims. The data give support to models of origin by drainage of regolith rather than expulsion. Thomas and Veverka (1979) note that if the grooves are impact generated, they may be common features on other small solar system objects, including asteroids. Models such as those of Weidenschilling (1979) that require the presence of tidal stresses would imply that no grooves will be found on small asteroids. The only other satellite imaged at a resolution sufficient to reveal grooves is Deimos; none exist on its surface.

XI. CONCLUSIONS

Satellites smaller than Mimas are distinguished by their elongate shapes and rough-limb topography compared to larger satellites. This division probably results from a dominance of gravitational effects in the shaping processes on the larger objects, whereas strength effects are more important in the case of the smaller ones.

Predicted high impact rates, observed high crater densities, and the irregular shapes of small satellites suggest a major role for impact fragmenta-

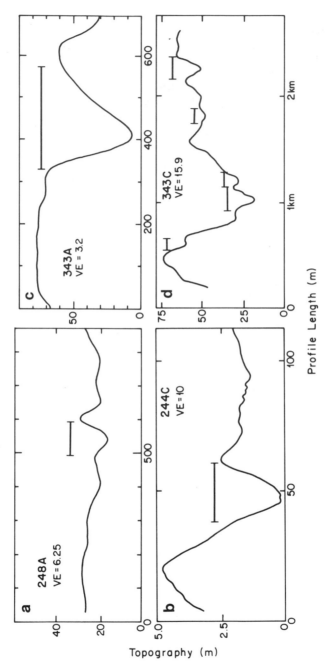

Fig. 13. Approximate topography across grooves (marked by horizontal lines) on Phobos determined by photoclinometry. *VE* is vertical exaggeration. Topography is complex and the removal of general Phobos curvature is uncertain. Only profiles (a) and (b) show possible raised rims on a flat surface. Profile lengths are measured in m except for (d) which is in km.

tion in the shaping of small satellites. Progressive fragmentation can lead to many different structural states of satellites: solid or fractured residual cores, groups of a few large fragments, or rubble piles of loose particles. Distinction between these models on the basis of available imaging data is usually impossible. Because satellites can be expected to retain well-developed regoliths, such a distinction based on remote sensing techniques may always remain difficult.

Images of small satellites obtained have so far revealed no objects for which "gravitationally-bound equilibrium-ellipsoid" is an appropriate description. Even if some objects do consist of loose, reaccumulated material, the observed irregular shapes preclude the use of axial ratios to estimate their mean densities.

Directions of the effective gravity on small satellites that rotate rapidly and are subject to significant tidal effects depend strongly on the object's mean density. Thus, patterns of downslope movement might constrain the mean densities of some bodies.

The only two small satellites that have been imaged in detail, Phobos and Deimos, are very different: one shows widespread downslope transport of regolith, the other has much evidence of internal structures (grooves). Thus, a wide diversity of surface phenomena should be expected even on small bodies.

Many fundamental questions concerning small satellites remain unanswered: Why do Saturn's small satellites differ in composition from those of Jupiter? How do satellites get into coorbital and Lagrangian orbits? Why are there no Lagrangian satellites in the Jovian system? What is the primordial origin of any captured small satellites? How are such satellites captured? How much have the small satellites evolved collisionally?

Some of these problems might be resolved by Earth-based observations and theoretical investigations. Others require detailed high-resolution study by spacecraft, something which is not likely in the near future for many of these bodies. Important clues could come, however, from spacecraft investigations of related bodies.

Acknowledgments. Preparation of this chapter was supported in part by grants from the National Aeronautics and Space Administration. We wish to thank J. Gradie, P. Nicholson, and J. Burns for helpful discussions and M. Roth for technical help.

18. SATELLITES OF URANUS AND NEPTUNE, AND THE PLUTO-CHARON SYSTEM

DALE P. CRUIKSHANK
University of Hawaii

and

ROBERT HAMILTON BROWN
Jet Propulsion Laboratory

Recent work on the Uranian satellites has revealed many of their basic physical properties. Radiometric measurements have shown that Ariel, Umbriel, Titania and Oberon have diameters which range from 1630 to 1110 km and albedos which range from 0.30 to 0.18. Spectrophotometric observations of Miranda suggest that it may have the highest albedo of the known Uranian satellites and a diameter of about 500 km. All five known satellites of Uranus have surfaces composed of water ice contaminated with small amounts of dark material. The dark material on the surfaces of Ariel, Umbriel, Titania and Oberon has spectral similarities to carbon black, charcoal, carbonaceous chondritic material and other dark, spectrally neutral materials. Large density differences among Ariel, Umbriel, Titania and Oberon, with density increasing with distance from Uranus have been proposed. Of the two confirmed satellites of Neptune, only Triton is sufficiently bright to permit studies of its physical characteristics. While the diameter and mass are uncertain, spectroscopy has indicated methane and possibly nitrogen on Triton's surface. An atmosphere must surround Triton, although strong seasonal effects will profoundly influence the volatile distribution on the surface and in the atmosphere. Pluto and Charon constitute a unique planet-satellite pair in the solar system. Pluto is sublunar in size and has a surface with solid methane distributed irregularly upon it. The low mean density of Pluto suggests a predominantly volatile bulk composition. A tenuous atmosphere consisting of methane and perhaps other gases envelopes Pluto.

[836]

In 1977 when *Planetary Satellites* (Burns 1977*a*) was published, very little was known about the physical properties or the surface compositions of the satellites of Uranus and Neptune, and the satellite of Pluto had not yet been discovered. This state of ignorance has since changed markedly with improvements in the technology of electro-optical detector systems, especially detectors optimized for the near-infrared. The situation can be expected to improve even more dramatically when the Voyager 2 spacecraft makes its flybys through the Uranian and Neptunian systems, with observations of the satellites, Triton in particular. [Voyager images of Uranian moons are shown later.]

Since the 1973 review of planetary satellites by Newburn and Gulkis, and that in 1974 by Morrison and Cruikshank, a most useful review of outer solar system planets and satellites has been published by Trafton (1981). A major review of Uranus, Neptune, and their rings and satellites, current as of mid 1984, has been produced under the editorship of Bergstralh (1984). In this chapter we also consider both Pluto and its satellite, since both are satellite-size bodies.

I. SATELLITES OF URANUS

The satellites of Uranus comprise the most distant regular satellite system in the solar system. Regular satellites are those whose orbits are very nearly circular and whose orbital planes are nearly coincident with the equatorial plane of the central body (they are discussed in the introductory chapter). The Uranian satellite system represents such a system, having five known satellites and nine known rings, all of whose orbits fit the criteria for a regular system (Miranda could be considered a slightly irregular satellite owing to its inclination of $4°22$). In order of distance from Uranus, the five satellites are UV Miranda, UI Ariel, UII Umbriel, UIII Titania and UIV Oberon. A compilation of the known properties of the satellites of Uranus prior to 1982 has been published by Cruikshank (1982); therefore this section will concentrate primarily on work done since that review.

A. Orbital Properties

Studies of the dynamics of satellite systems have three general goals: first, a complete characterization of the orbital properties of the satellites in order to predict their positions accurately; second, a determination of the mass and the gravitational moments of the central body from the observed orbital parameters; and third, the determination of the satellite masses from observation of those orbital parameters which are modified by the gravitational interactions of the individual satellites. Several studies have been made of the dynamics of the Uranian satellite system and thorough discussions of the theory and observations prior to 1979 have been published by Greenberg (1975*a*, 1976, 1979*a*). A review of recent studies of the masses of the Uranian

TABLE I

Orbital Properties of the Satellites of Uranus and Neptune[a]

	Sidereal Period[a] P (days)	Semimajor Axis a (km)	Eccentricity e	Inclination i (deg)
Satellites of Uranus[b]				
Miranda	1.41347925 (14)	129390 ± 135	0.0027 ± 0.0006	4.22 ± 0.16
Ariel	2.52037935 (10)	191020 ± 90	0.0034 ± 0.0003	0.31 ± 0.11
Umbriel	4.1441772 (2)	266300 ± 90	0.0050 ± 0.0003	0.36 ± 0.08
Titania	8.7058717 (3)	435920 ± 75	0.0022 ± 0.0001	0.142 ± 0.031
Oberon	13.4622389 (5)	583530 ± 60	0.0008 ± 0.0001	0.101 ± 0.024
Satellites of Neptune				
Triton[c]	5.87686 (1)	354290 ± 200	0.00	21.0 ± 1.5(R)[e]
Nereid[d]	360.14	5517000	0.75612	27.522

[a] Uncertainties (one standard deviation of the mean) in the last 1 or 2 digits are given in parentheses for the sidereal periods.
[b] Data adapted from Veillet (1983a; cf. Dermott and Nicholson 1986).
[c] Data from A. W. Harris (1984b).
[d] Data are the osculating elements referred to the 1950 equator, epoch 1981.0, from Veillet (1982).
[e] R = retrograde motion. Compare to Table III of the introductory chapter.

satellites has also been published by Greenberg (1984a). An important study of the dynamical properties of the Uranian satellites has been completed by Veillet (1983a) in which most of the satellite astrometric observations made since their discovery have been analyzed. Using earlier data from various observers and his own data obtained during the current epoch when the satellite orbits are viewed nearly pole-on, Veillet has determined the masses of Miranda, Ariel, Umbriel, Titania and Oberon from measurements of their mutual gravitational perturbations (this work is questioned by Dermott and Nicholson [1986]). In Table I, adapted from Veillet's work, it can be seen that all the orbits have low eccentricities and all but that of Miranda have essentially zero inclination. The regularity of the Uranian system is in contrast to the Jovian and Saturnian satellite systems, in which the exterior members have highly irregular orbits. It should be noted, however, that Uranus may have a few irregular satellites which are too faint to be observed from the ground. A recent search for outer satellites to magnitude $V = 23$ by Cruikshank [in preparation] has revealed no objects exterior to Oberon.

One especially interesting dynamical aspect of the Uranian system is the near-coincidence of the rotation axis of Uranus and the plane of its orbit and the ecliptic. Uranus' axial inclination of 98° with respect to the ecliptic pole is the most extreme example of axial tilt among the planets. The present (1985) aspect as seen from the Earth is essentially polar; this has simplified the derivation of some orbital parameters from astrometric observations, as well as some observational work, such as photometry. The polar aspect does, however, frustrate groundbased observations intended to determine orbital inclinations of the satellites or to search for such properties as albedo asymmetries with respect to the satellites' leading and trailing sides. The current orbits of the Uranian satellites strongly constrain the events that are responsible for the observed axial tilt of Uranus (see, e.g., Singer 1975), which in turn may have implications for the origin of the satellites.

Squyres et al. (1985) have concluded from an analysis of the eccentricities of the Uranian satellites that the inner three could not have maintained nonzero eccentricities over the age of the solar system unless their internal properties are more like rock than ice, or unless some presently inactive process has recently increased their eccentricities. They suggest that a recent passage of Miranda, Ariel and Umbriel through an exact resonance (their orbital periods are presently close to, but not exactly in resonance) may be responsible for their residual eccentricities (cf. Dermott and Nicholson 1986). If this is the case, Squyres et al. conclude that Ariel may have recently suffered significant tidal heating.

B. Surface Compositions

Most of what is known about the surface compositions of satellites in the outer solar system is derived from observations of their reflectance spectra (see chapter by Clark et al.). Absorption features characteristic of the surface mineralogy of a planetary body can be observed over the entire region of the solar spectrum whenever there is a detectable amount of reflected light (approximately 0.3 to 5 μm for groundbased work, depending on the temperature of the object). Observations of the near-infrared (0.8 to 5 μm) reflectance of icy bodies are particularly diagnostic of surface composition because several cosmochemically important molecules (e.g., H_2O, NH_3 and CH_4) have vibrational transitions resulting in absorptions seen in spectra of their diffuse reflectance. Near-infrared spectrophotometry has recently been applied to the Uranian satellites and has resulted in a reasonable characterization of their surface compositions.

The first study of the low-resolution, near-infrared spectral reflectance of Titania and Oberon was published by Cruikshank (1980b). His spectra show absorptions at 1.5 and 2.0 μm characteristic of the presence of water ice or frost on the surfaces of these two satellites. In a follow-up study of Ariel and Umbriel, Cruikshank and Brown (1981) found similar absorptions in these satellites' reflectance spectra. Due to its faintness and proximity to Uranus, the reflectance spectrum of Miranda is extremely difficult to observe; however, Brown and Clark (1984) succeeded in obtaining a spectrum of Miranda in the 1.6 to 2.4 μm spectral region that clearly shows a deep absorption at 2.0 μm characteristic of water ice (Fig. 1). With the Brown and Clark observations it is now known that all five Uranian satellites have water ice surfaces. Soifer et al. (1981) published spectrophotometry of Umbriel, Titania and

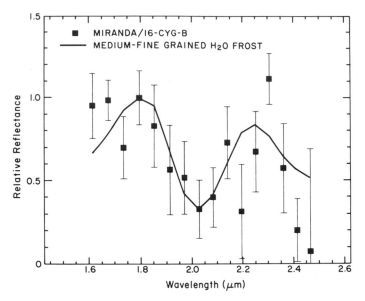

Fig. 1. The reflectance spectrum of Miranda from Brown and Clark (1984). The Miranda spectrum is normalized to 1.0 at 1.79 μm and is overlaid with a spectrum of fine-grained water frost from Clark (1981a). The laboratory frost spectrum has been convolved to the resolution of the Miranda spectrum and also normalized to 1.0 at 1.79 μm.

TABLE II
Radii, Masses, Densities and Albedos

	Radius[a] (km)	Mass[b] (10^{23} g)	Density[b] (g cm^{-3})	p_V [a]
Satellites of Uranus				
Miranda	250 ± 110	1.7 ± 1.7	~ 3	~ 0.5
Ariel	665 ± 65	15.6 ± 3.5	1.3 ± 0.5	0.30 ± 0.06
Umbriel	555 ± 50	10.0 ± 4.2	1.4 ± 0.7	0.19 ± 0.04
Titania	800 ± 60	59.0 ± 7.0	2.7 ± 0.7	0.23 ± 0.04
Oberon	815 ± 70	60.0 ± 7.0	2.6 ± 0.8	0.18 ± 0.04
Satellites of Neptune				
Triton[c]	1750 ± 250	—	—	~ 0.4
Nereid	?	?	?	?

[a] Radii and visual geometric albedos (p_V) are from Brown et al. (1982a) except for the radius and albedo of Miranda which are from Brown and Clark (1984).
[b] Masses and densities are from Veillet (1983a; cf. Dermott and Nicholson 1986).
[c] See text for caveats concerning the values of parameters for Triton, and the discussion of Table IV in the introductory chapter.

Oberon which confirmed the discovery of water ice absorptions and raised the possibility that the surfaces of these three satellites have lower albedos than are characteristic of pure water ice surfaces heavily gardened by meteoritic infall. That some of the Uranian satellites have relatively low albedos was firmly established by the radiometric measurements of Brown et al. (1982b) who found that the visual geometric albedos (p_V) of Ariel, Umbriel, Titania and Oberon are intermediate to the visual geometric albedo range for most solar system bodies (approximately 0.03 to 0.7). Table II lists the albedos determined by Brown et al. for the four brightest Uranian satellites. With their p_V range of 0.2 to 0.3, the nearest albedo analogs for the Uranian satellites among other icy satellites are Callisto and Hyperion whose p_V are 0.19 and 0.28, respectively (Squyres and Veverka 1981; Cruikshank and Brown 1982). As we shall see below, Hyperion may be similar to Ariel in surface composition as well as in albedo.

Because the relatively low albedos of the Uranian satellites suggest the presence of a dark contaminant on their surfaces, some recent studies of the near-infrared reflectance of the Uranian satellites have concentrated on the identification of this material. Brown and Cruikshank (1983) and Brown (1983) have obtained reflectance spectra of Ariel, Umbriel, Titania and Oberon in the 0.8 to 2.6 μm spectral region which indicate that the nonwater component of their surfaces has a bland spectral reflectance with spectral similarities to such substances as charcoal, carbon black, carbonaceous chondritic material and other neutrally colored, low-reflectance materials. Brown (1983) has combined the near-infrared reflectance spectra of the Uranian satellites with photovisual data to give the composite spectra shown in Fig. 2. All spectra in Fig. 2 show the strong H_2O absorption at 2.0 μm and some show the 1.5 μm absorption as well. A laboratory spectrum of a sample of fine-grained water frost obtained by Clark (1981a) is given in Fig. 3 (see other plots in the chapter by Clark et al.) for comparison.

Areal mixtures of water frost and isolated patches of dark, opaque, spectrally neutral material have also been investigated as possible analogs to the surfaces of Ariel, Umbriel, Titania and Oberon (Brown 1983). Figure 4 displays the results of a spectral matching process which involves the linear superposition of two laboratory spectra: that of fine-grained water frost, and that of an intimate mixture of 30 wt % charcoal and 70 wt % water ice. The relatively close (though nonunique) matches produced demonstrate the consistency of the Uranian satellite spectra with areal mixtures of water frost and isolated patches of dark spectrally neutral materials. It should be noted, however, that with available data it is not possible to determine conclusively whether the dominant state of dispersal of the dark component of the Uranian satellite surfaces is voluminal or areal. If the dark component of the Uranian satellite surfaces exists in a predominantly intimate mix, its intrinsic albedo may be higher than that of charcoal or carbon black (Brown 1983).

While searching for spectral analogs for the Uranian satellites, Brown

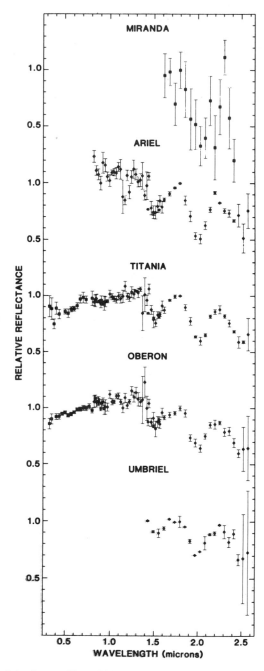

Fig. 2. Spectra of the five satellites of Uranus in rough order of the strength of their 2.0 μm water ice absorption. The Miranda data are from Brown and Clark (1984); the 0.8 to 1.6 μm data for Ariel, Umbriel and Titania are from Brown (1982,1983); the 1.4 to 2.6 μm data for Ariel, Umbriel, Titania and Oberon are from Brown and Cruikshank (1983); the 0.3 to 1.1 μm data for Titania and Oberon are from Bell et al. (1979).

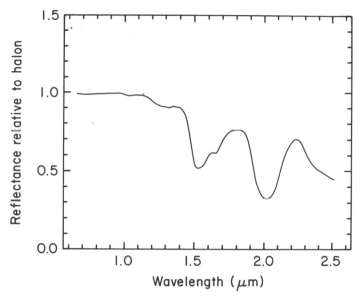

Fig. 3. A spectrum of fine-grained water frost from R. N. Clark (1981*a*).

and Cruikshank (1983) noticed that the 1.5 to 2.6 μm reflectance spectrum of
Saturn's satellite Hyperion (Cruikshank and Brown 1982) is very similar to
that of Ariel. This point is illustrated in Fig. 5 where the 1.5 to 2.6 μm spectra of Ariel and Hyperion have been normalized to 1.0 at 1.79 μm and overlaid. The reflectance similarity of the two bodies is further demonstrated in
Fig. 4 and by the similarity of the two bodies' visual geometric albedos (0.30
for Ariel and 0.28 for Hyperion). Nevertheless, though it is interesting, this
comparable spectral signature need not be a compositional similarity of the
dark components, and may only result from a similarity in the distribution
(areal vs. voluminal) of the dark components (Brown 1983).

Many theories of the formation of bodies in the outer solar system predict the incorporation of volatiles such as ammonia, methane and carbon
monoxide into the surfaces of the Uranian satellites (Cameron 1973*a,b*;
Lewis 1971*a*,1972,1973; Lunine and Stevenson 1983; see also chapter by Stevenson et al. and references therein). Conclusive evidence for the presence of
volatiles other than water ice in the Uranian satellite surfaces has not yet been
found, but there are some interesting features of the reflectance spectra of
Ariel and Hyperion which might indicate the presence of NH_3, CH_4 or CO.
The spectral feature in question is subtle, but amounts to a depression of the
continuum at 2.25 μm in the Ariel and Hyperion spectra relative to that of
pure water frost (see Figs. 3, 4 and 5). The depression results in the spectra of
Ariel and Hyperion peaking at 2.19 μm instead of at 2.25 μm as is normal for

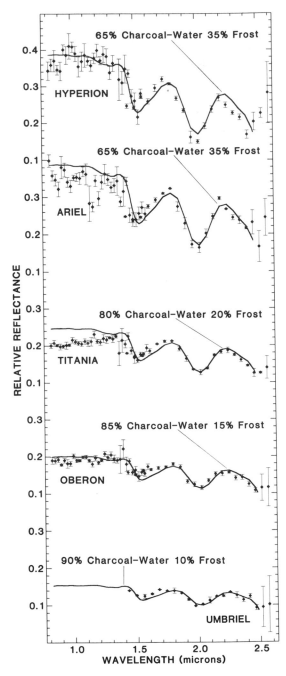

Fig. 4. The spectra of Hyperion, Ariel, Titania and Oberon overlaid with spectra constructed from a linear superposition of the spectra of fine-grained water frost and an intimate mixture of 30 wt % charcoal and 70 wt % water ice. The data are from Brown (1983) and demonstrate one class of spectral analogs for the surfaces of these four satellites. The laboratory data have been convolved to the same resolution as the telescopic data. Note that the normal reflectances of the laboratory spectra approximately match the satellites' geometric albedos.

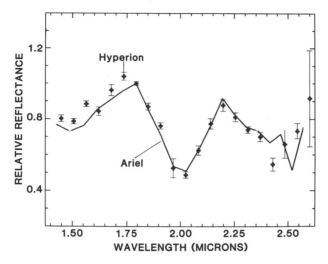

Fig. 5. The spectra of Ariel and Hyperion, both normalized to 1.0 at 1.79 μm. The data are from
Brown and Cruikshank (1983). To facilitate comparison, no error bars are shown for the Ariel
spectrum although they are comparable in size to those of the Hyperion spectrum.

pure water ice or a mixture of water ice and small amounts of spectrally neutral material. Spectra of Ariel obtained during the 1983 apparition of Uranus
(Brown and Clark, unpublished) confirm this spectral feature. Ammonia,
methane and carbon monoxide all have strong absorptions in the 2.2 to
2.3 μm region, but the low-resolution and precision of the existing telescopic
data prevents a positive identification as to which of these compounds, if any,
is present on the surfaces of Ariel or Hyperion.

C. Opposition Surges

A nonlinear increase in logarithmic brightness at a solar phase angle near
0° (opposition surge) has been found for many solar system bodies, but recent
observations indicate that the opposition surges for the Uranian satellites are
unusually strong. Brown and Cruikshank (1983) have observed the near-
infrared opposition surges of the Uranian satellites and found that they are the
largest known (over the 3° of solar phase angle which can be observed from
Earth). In Fig. 6 we show the near-infrared opposition surges of Ariel, Umbriel, Titania and Oberon as well as the visual opposition surge of Saturn's
rings, which until recently was thought to be the largest known. As can be
seen from Fig. 6, the opposition surges of at least three of the Uranian satellites are 0.5 mag or more. A follow-up study of the broadband visual (*V*) opposition surges of the Uranian satellites (Goguen et al. 1984) tentatively
shows comparable results, but more observations are required to characterize
fully their visual phase curves. It is not clear what surface properties of the

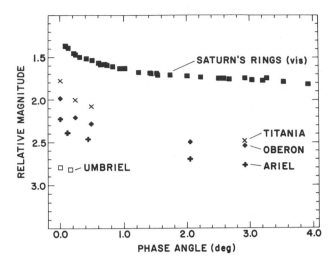

Fig. 6. The near-infrared opposition brightness surges of Ariel, Umbriel, Titania and Oberon. The data are from Brown and Cruikshank (1983). Also plotted are the visual opposition surge data for Saturn's rings from Franklin and Cook (1965). The data for the Uranian satellites contain an arbitrary offset to facilitate comparison to the data for Saturn's rings.

Uranian satellites might be responsible for the large opposition surges, but recent work points to surfaces composed of particles with highly backscattering phase functions (Pang and Rhoads 1983). Hapke (1983) has been able to model the very large opposition surges of the Uranian satellites by requiring that the density of the surface layer of scattering particles increase with depth from near zero to close-packed over a distance of about 30 times the mean particle size. Veverka and Gradie (1983; cf. chapter by Veverka et al.) argue that grain size distribution, grain shape and packing texture are some of several mechanisms that can be invoked to explain why the surfaces of icy satellites are, in general, not Lambert scatterers.

D. Masses, Radii and Densities

Mean density, as deduced from mass and radius measurements, is an important physical property related to a satellite's origin. Until recently, however, this quantity could only be estimated for the satellites of Uranus. The distance from Earth and the faintness of these satellites have made direct determinations of their radii using groundbased techniques impossible. For the same reasons, observations from which accurate masses of the Uranian satellites can be derived are very difficult. Nevertheless, recent studies by Brown et al. (1982b), Brown and Clark (1984), and Veillet (1983a) have made progress in the determination of the radii and masses. Brown et al. used the photometric/radiometric technique to determine the radii of Ariel, Umbriel, Titania and Oberon. They measured the 20 μm thermal fluxes from the satellites, combined them with measurements of the satellites' broadband visual fluxes, and interpreted them with a version of the standard radiometric model (Morrison 1973; Morrison and Lebofsky 1979; Brown et al. 1982*a*). As can be seen in Table II, the satellites of Uranus are comparable in size to the largest of Saturn's icy satellites Dione, Iapetus and Rhea, which have radii of 660, 730 and 765 km, respectively (Smith et al. 1982; see Table IV of the introductory chapter; chapter by Morrison et al.). The Uranian satellites are therefore among the largest satellites in the solar system, but are considerably smaller than the giant satellites Ganymede, Callisto, and Titan, whose radii are approximately 2500 km.

As mentioned above, Veillet (1983*a*) derived masses for all the Uranian satellites from observations of their mutual orbital perturbations. Combining his mass measurements with the radii measured by Brown et al. (1982*b*), Veillet computed mean densities for Ariel, Umbriel, Titania and Oberon (Table II). The density of Miranda listed in Table II was derived from the diameter estimate by Brown and Clark (1984) and the mass estimate by Veillet. As can be seen in this table, the Uranian satellites seem to form two distinct density groups (although the error bars nearly overlap). Given that all the Uranian satellites are known to have water ice surfaces, and that the densities of Ariel and Umbriel are comparable to those of several of Saturn's satellites whose bulk compositions are thought to be about 40% silicates and 60% water ice by weight, one can reasonably surmise that the bulk compositions of Ariel and Umbriel may be similar. In contrast, Titania and Oberon have densities which indicate that a much larger fraction of their bulk compositions is of high-density materials like silicates. Because they are large enough to have undergone melting and at least partial differentiation (Lewis 1971*b*; chapter by Stevenson et al.), densities of 2.6 to 2.7 g cm^{-3} for Titania and Oberon suggest that they may each have a large core which is mostly rock with a thin skin of water ice comprising their crusts and mantles. If the apparent density variations are in fact real (and as yet this is uncertain; Dermott and Nicholson 1986), then we have a system whose density gradient is *opposite* to that expected if primordial heat from the accretion and contraction of Uranus determined what materials were available for incorporation into the satellites. This might be seen to favor scenarios for the origin of the Uranian satellites which are closely connected to the catastrophic events hypothesized to be responsible for the present axial orientation of the Uranian system (Singer 1975). Nevertheless, large uncertainties in the densities of the Uranian satellites counsel restraint with regard to speculations about the satellites' origins based on their apparent density differences. This is particularly true in view of the observations by Dermott and Nicholson (1986) that Veillet's dynamical analysis for the masses of the satellites did not include secular variations in the orbital eccentricities, inclinations and precession rates caused by mutual perturbations. A preliminary analysis by Dermott and Nicholson suggests that the mass of Ariel, for example, might be very much less than that given by

Veillet, resulting in a mean density substantially less than 1.0 g cm^{-3}. Further thoughts regarding the origins of the Uranian satellites can be found in the chapter by Stevenson et al.

E. Summary of the Uranian System

The Uranian satellites comprise a system of five regular satellites, all having water ice surfaces. Ariel, Umbriel, Titania and Oberon have opposition surges which are among the largest in the solar system and have low albedos compared to those typical of pure, heavily gardened water ice surfaces. Present with the water ice on the surfaces of Ariel, Umbriel, Titania and Oberon is a dark, spectrally neutral component, which has spectral characteristics similar to those of carbon black, charcoal, carbonaceous chondritic material, and other neutrally colored, low-albedo materials. Compounds more volatile than water ice (e.g., methane, ammonia and carbon monoxide) have not yet been conclusively shown to exist on any surface in the Uranian system, although there is as yet an unidentified spectral feature in the spectrum of Ariel which may result from the presence of the hydrate of one of these compounds. The Uranian satellites are comparable in size to the largest of Saturn's icy satellites while density measurements suggest that the bulk compositions of Ariel and Umbriel might be different from those of Titania and Oberon. Ariel and Umbriel seem to have densities similar to those of the icy Saturnian satellites and may have similar bulk compositions. Titania and Oberon have densities which suggest that they may have a larger bulk fraction of silicates or other high-density materials than do Ariel and Umbriel.

II. SATELLITES OF NEPTUNE

A. Triton

1. Orbital Properties and Dynamics. Triton is the largest satellite of Neptune. It was seen first by William Lassell in 1846, less than three weeks after Neptune itself was discovered by Leverrier, Adams, Galle and d'Arrest, and was soon found to lie in a retrograde orbit with a sidereal period of 5^{d} 21.044^{h}. The retrograde motion has given rise to interesting speculation on the origin and dynamical future of Triton (chapter by Burns).

The orbit of a retrograde satellite about a planet in prograde rotation will decay at a rate dependent upon the tidal dissipation in the planet (Burns 1977b,1982). In an early study of the Triton-Neptune system, McCord (1966) showed that Triton's orbit would decay completely in some 10 to 100 million years if the tidal dissipation factor of Neptune Q_N lies in the range 10^2 to 10^3. Values of Q for the other giant planets have been derived from studies of the limited outward evolution of their prograde satellites (Goldreich and Soter 1966), and are found to exceed about 10^4. In the specific case of Uranus, a planet of similar size and apparently similar composition to that of Neptune, Q_U exceeds a few times 10^4, as derived from the present orbit of the satellite

Miranda (Goldreich and Soter 1966). If Neptune's internal structure is similar to that of Uranus, with a correspondingly high dissipation factor, the decay time of Triton's orbit is long, perhaps $> 10^{10}$ yr (Harris 1984c).

The origin of Triton is unknown. An early suggestion by Lyttleton (1936) tied Triton's origin to that of Pluto; a similar scenario has been considered more recently by Harrington and Van Flandern (1979), who proposed that both Pluto and the irregular satellite system of Neptune might have resulted from a single encounter of a massive body with Neptune. The capture of Triton by Neptune in its present retrograde orbit is favored by Farinella et al. (1980) and McKinnon (1984) as the simplest mechanism, although most details of the capture process are still highly uncertain. Pluto may still play a role in the capture scheme, although McKinnon (1984) suggests that they are dynamically unrelated (see also the discussions in the chapters by Burns and by Peale). Below, we discuss similarities in the dimensions and chemical compositions of Pluto and Triton which are tantalizing, but do not lead us along an obvious path toward a solution to the problems of the origin of these bodies.

The orbital parameters of Triton given in Table I represent recent determinations by Harris (1983,1984b). An additional parameter of interest to us here is the period of precession of the node (637 ± 40 yr), and the acceleration of the mean motion ($n < 2°$/century). Harris finds that the spin axis of Triton is inclined $\sim 0°.9$ from the orbit normal in a direction away from the pole of Neptune. The latitude on Triton of the subsolar point is $+ 40°$ and is increasing so that at the present time the south pole is in permanent darkness. As a consequence of the precession of the orbit of Triton about the equatorial plane of Neptune and the motion of Neptune around the Sun, there is a periodic variation of the latitude of the subsolar point on the satellite which also varies in amplitude. Figure 7 shows this phenomenon, in which the strongest seasonal effects come at intervals of about 600 yr. Seasonal effects with long-term cold traps at dark poles are important in considerations of the surface composition and dynamics of surface-atmosphere interactions.

2. Surface and Atmosphere of Triton.   The spectrum of Triton has been explored with relatively low spectral resolution from 0.3 to 2.5 μm. From 0.3 to 0.6 μm, the spectrum is red and without any distinct absorption fea-

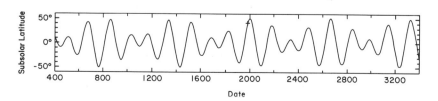

Fig. 7. The latitude of the subsolar point on Triton, computed by Harris (1984b).

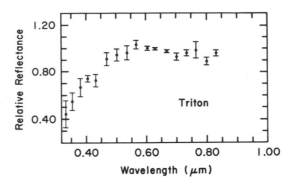

Fig. 8. The relative reflectance of Triton in the photovisual spectral region, observed with a multifilter photometer (previously unpublished data from Bell et al. 1979). Error bars are ± 1 σ in this and the following figures.

tures, as shown in Fig. 8, which combines photographic spectrophotometry (Cruikshank et al. 1979) and multifilter photometry (Bell et al. 1979); the data sets are in agreement. Longward of 0.8 μm, absorption bands appear, as shown in the composite spectrum of Triton in Fig. 9, in which the 0.8 to 2.5 μm data are from Cruikshank and Apt (1984) and Cruikshank et al. (1984a). The majority of these bands are identified as due to methane on the basis of comparisons with synthetic laboratory spectra of gaseous methane calculated for the low-temperature conditions on Triton (Cruikshank and Apt 1984) and with laboratory spectra of methane ice. The 2.3 μm absorption band was first identified as methane by Cruikshank and Silvaggio (1979) from spectra of lower resolution. In high-resolution spectra, Apt et al. (1983) have found the 0.89 μm band of methane and have noted that it is weaker than the same band in the spectrum of Pluto.

In addition to the six bands in the Triton spectrum that match the methane spectrum, there is an additional feature, centered at 2.16 μm, located on the steep slope of the 2.3 μm methane band that cannot itself be attributed to methane either in the gaseous or solid state. This may be the same feature reported at 2.10 μm by Rieke et al. (1981). Cruikshank et al. (1984a) have tentatively identified this band as the density-induced (2–0) absorption band of molecular nitrogen. The identification is regarded as tentative because it is based upon the presence of one band alone, and the apparent coincidence of the central wavelengths of N_2 observed in the laboratory and the band on Triton. From a consideration of the expected temperature of Triton's surface and the phase equilibrium of nitrogen, Cruikshank et al. (1983b,1984a) showed that, in order for the observed spectral band to be produced in gaseous nitrogen, the surface pressure would exceed that at which condensation would occur at the relevant temperature. Thus, the nitrogen should exist as a liquid or solid, depending upon the exact temperature. Laboratory observations show

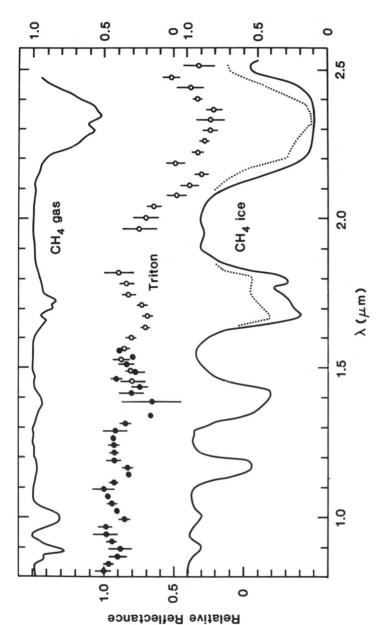

Fig. 9. Composite spectrum of Triton in the range 0.8 to 2.5 μm. Solid dots are data obtained on 18.6 May 1982; open circles are data from 2.4 June 1980. The synthetic methane gas spectrum was calculated by Apt for $T = 55$ K and 50 m-Am density at 10^{-3} bar. The methane ice data are diffuse reflection spectra obtained by Brown and Clark. The dashed segments of the line represent spectra of a sample of methane ice mixed with carbon black (figure from Cruikshank and Apt 1984).

that the 2.16 μm feature also occurs in liquid nitrogen. From a determination of the absorption coefficient in the band, Cruikshank et al. calculated that the spectral feature on Triton is produced by absorption through a few tens of centimeters of liquid nitrogen on the surface of the satellite (see Rieke et al. 1985).

This result raises interesting possibilities about the atmosphere of Triton, its interaction with liquid nitrogen and solid methane, and the nature of the satellite's red color. Estimates of Triton's temperature are also complicated because of the unknown effects of the atmosphere. It is probable that the atmosphere itself has not yet been observed because it is mostly nitrogen at a surface pressure regulated by the temperature of the liquid or solid; the equilibrium vapor pressure of nitrogen at $T = 64$ K is 0.13 atm. The vapor pressure of methane is less by about 10^3, so that this gas is likely a minor constituent of the satellite's atmosphere.

Additional observational results show that the strengths of the methane bands on Triton vary with its orbital position (Cruikshank and Apt 1984). Because Triton is in locked synchronous rotation in its orbit around Neptune, the orbital variability reflects a nonuniform distribution of methane on the surface of the satellite. There are presently no conclusions about the possible variability of the strength of the nitrogen band. Triton does not show a pronounced photometric variability with its orbital position. Franz (1981) found an amplitude of about 0.06 mag (at $\lambda = 0.56$ μm), with the maximum at western elongation (leading hemisphere). The variability of the methane band strength appears to be in the sense that the strongest bands are also found near western elongation, although this observation requires confirmation.

Methane is highly soluble in liquid nitrogen and has the effect of raising the freezing point of the mixture. The implications of this for Triton have been explored in a very preliminary way by Cruikshank et al. (1984a); it appears that the colder portions of the satellite, particularly those near the pole in extended darkness at the present season, are sufficiently cold to permit the nitrogen sea to freeze. Whether or not there is a diurnal freeze-thaw cycle depends on the heat capacity of the sea and its global extent, as well as the possibility that other materials are dissolved in it, which may lower the freezing point.

In attempts to model the infrared spectrum of Triton with laboratory observations of methane ice and liquid nitrogen, Cruikshank et al. (1983b, 1984a) found that an additional component was necessary to match the shape of the continuum at various wavelengths. The absorption spectrum of water ice provides the additional component needed to fit the Triton spectrum to a precision commensurate with the quality of the telescopic data for the satellite. The best-fit model for Triton's spectrum, together with the individual components shown separately, is given in Fig. 10.

While Cruikshank et al. (1984a) noted various reasons why the proposed nitrogen component of the infrared spectrum was best explained by liquid, Lunine and Stevenson (1985) reconsidered the state of volatiles on the satellite

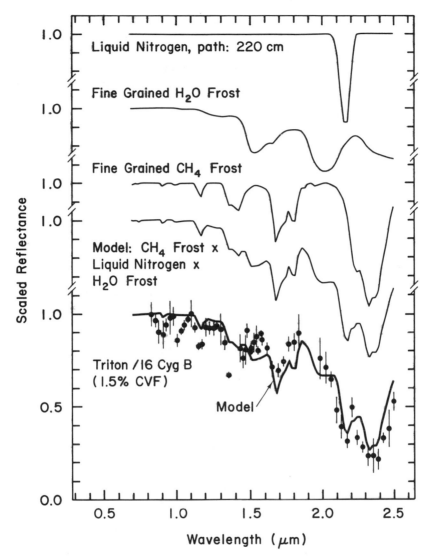

Fig. 10. A model of the Triton spectrum with resolution 1.5%, together with the individual components used in the calculation. In trying to fit the data (solid points with error bars), emphasis was placed on the region 1.8 to 2.5 μm. The model calculations were made by Clark using laboratory spectra that he obtained in collaboration with the authors of this chapter (figure from Cruikshank et al. 1984a).

from the point of view of phase equilibrium between the solid and liquid components. They concluded that the simplest configuration of the volatiles is with both N_2 and CH_4 in the solid states, possibly in a microscopic mixture or a disequilibrium assemblage with a nonuniform distribution across the satellite's surface. They found that a nitrogen ocean requires more restrictive assumptions about the temperature regime on Triton.

While a detailed study of the short-wavelength end of the spectrum where Triton shows a distinct reddish color has not yet been accomplished, interesting possibilities arise from a consideration of the fact that other methane-bearing bodies in the outer solar system also have reddish hues. In the case of Saturn's satellite Titan, the red color is probably that of the aerosol photochemically produced from the methane in the upper atmosphere. Pluto is also red. Photochemistry on Triton may produce reddish organic solid matter from methane and nitrogen. If there is methane dissolved in the liquid nitrogen, red organic matter may be suspended in the liquid. Delitsky (1983) has given some consideration to the complex organic chemistry that may occur in a nitrogen sea of Triton.

The equilibrium temperature of the subsolar point on Triton can be calculated from

$$T_{ss} = 1.523 \times 10^9 \left(\frac{1 - A}{R^2 \varepsilon_b} \right)^{1/4} \tag{1}$$

where A is the bolometric Bond albedo, R is the heliocentric distance in centimeters, and ε_b is the bolometric emissivity (Table II of the introductory chapter tabulates T_{ss}). Estimates of A, as we note below, give $A \sim 0.4$. For unit emissivity, $\varepsilon_b = 1$, $T_{ss} = 60$ K. The disk-averaged temperature, assuming a temperature distribution proportional to $\cos^{1/4}\theta$ where θ is the angle between the insolation and surface normal, is

$$T_{avg} = \left(\frac{2}{\pi} \right)^{1/4} T_{ss} = 57 \text{ K}. \tag{2}$$

As noted, however, the actual temperature near the subsolar point may be moderated by the heat capacity of the atmosphere and the possible shallow liquid nitrogen sea covering some fraction of the surface.

In summary, the surface of Triton is covered in part by solid methane, and nitrogen may also be present in a liquid or frozen state. Reddish photochemical products may give the surface a slight coloration, and water ice may occur as crystals suspended in the liquid nitrogen or as a solid mixed with the methane frost on expanses of solid surface (spectral modeling favors a suspension of fine crystals in the liquid). In this scenario, the satellite has an atmosphere of nitrogen with other possible minor constituents; the surface pressure may be regulated by the vapor pressure at the local temperature, but strong diurnal and seasonal effects are probable (see Trafton 1983). This model is subject to test by the planned occultation of the Voyager 2 spacecraft by Triton in 1989.

We note that Cruikshank et al. (1979) suggested that Triton has a rocky surface, a conclusion based on the red slope of the photovisual spectral reflectance. Here, following Cruikshank et al. (1984a), that same red slope is interpreted as a possible organic contaminant of the surface ices. The possibility that it is instead related to a red silicate contaminant should not be neglected, however, because the nature of the red contaminant is not yet clearly established.

3. Radius and Mass of Triton. Lebofsky et al. (1982b) obtained a weak detection of the 22.5 μm thermal radiation of Triton, thus permitting the radius and bolometric geometric albedo to be determined through the photometric/radiometric technique widely used for determinations of these parameters for asteroids and planetary satellites (Morrison 1973; Morrison and Lebofsky 1979). Nearly simultaneously, Morrison et al. (1982) attempted the same measurement but derived only an upper limit to the thermal flux. Both groups of investigators obtained essentially the same result, namely, the radius of Triton is 1750 ± 250 km, and the bolometric geometric albedo is ∼ 0.4.

It is important to inject a cautionary note before these results are taken as certain, however. The thermal model upon which the radius and albedo are calculated from the infrared flux assumes that the surface of the satellite is in instantaneous thermal equilibrium with the solar insolation and that thermal radiation from surface areas not in sunlight is negligible. For a liquid surface or one covered with a substantial atmosphere, the large heat capacity would violate the assumption of instantaneous thermal equilibrium. In the case of Triton, the degree of validity of the model, and hence the quality of the calculated radius and albedo, depends on the true nature of the surface, in particular, how much is solid and how much might be covered by liquid. In the extreme, it is possible that the radius given above is underestimated by $\sqrt{2}$ and that the albedo is a factor of 2 too high (Cruikshank et al. 1984a).

As an historical aside, it is interesting to recall Kuiper's (1954) direct measurement of the diameter of Triton with a diskmeter attached to the Hale 5 m telescope. He found the diameter to be 0.173 arcsec (Triton was 30.1 AU from Earth), corresponding to a radius of 1900 km, or within the errors of the determination discussed above. Large systematic errors in Kuiper's technique, especially for the smaller objects he measured (such as Pluto), show that this agreement is fortuitous. For comparison, the largest planetary satellite is Ganymede with $r = 2631$ km; for Titan, $r = 2575$ km; for Io, $r = 1815$ km; and for the Moon, $r = 1738$ km.

The mass of Triton has been estimated from its effect on the motion of Neptune in separate studies by Nicholson et al. (1931) and Alden (1940, 1943). The values obtained are, respectively $(3.5 \pm 1.4) \times 10^{-3}$ M_N and

$(1.34 \pm 0.3) \times 10^{-3} M_N$. Alden's mass value is based upon observations during a three-year period with a telescope of astrometic precision, while the observations made by Nicholson et al. were much less extensive and were obtained with a reflecting telescope. On these bases, the mass value by Alden appears preferable. The errors quoted for Alden's mass is a probable error in the weighted mean of three separate determinations in successive years. An updated determination reported by T. C. Van Flandern (personal communication, 1982) gives $1.7 \times 10^{-3} M_N$, consistent with the Alden determinations. The mass value often quoted (see, e.g., Morrison and Cruikshank 1974) is that given in the compilation by Duncombe et al. (1974), which is an equally weighted mean of the Nicholson et al. and Alden (1940,1943) determinations. This mean value, when taken with $r = 1750$ km, as discussed above, gives a mean density that appears unrealistic. Even the Alden determination alone, which is much smaller than that of Nicholson, gives a mean density on the order of 6 g cm^{-3} and should be viewed with caution. Our experience with other bodies in the outer solar system suggests that for these volatile-dominated objects the mean densities should lie in the range 1 to 3 g cm^{-3}. In a sense, anguish over the mass and radius of Triton is an idle exercise, because the planned flyby of Voyager 2 through the Neptune system in 1989 will yield these quantities with far greater precision than can be achieved by groundbased observations.

B. Nereid

Nereid is a 19th magnitude satellite of Neptune discovered in 1948 by Kuiper. It lies at a mean distance of 0.037 AU from Neptune in an orbit of high eccentricity (0.76) and high inclination (27°5), with a period of 360.14 d. The most recent study of the orbit, based on new observations, is that by Veillet (1982).

There is virtually no physical information on Nereid apart from the photographically determined brightness. From plausible assumptions about color (the difference between the V magnitude and the photographic magnitude), Morrison and Cruikshank (1974) found $V(1,0) = +4.0$. If the geometric albedo is 0.04, the radius is 525 km; if it is 0.4, the radius is 165 km.

The inclined, eccentric (direct) orbit suggests an origin by capture, in which case Nereid could be representative of the small, icy satellites, or could be a dark asteroid of mainly silicate composition. There is no information to support either interpretation, but Nereid will soon be within reach of multicolor photometric systems, and observations at a range of wavelengths will help discern between a surface predominantly composed of rocky or carbonaceous material and one of ice.

C. A Third Satellite of Neptune?

Neptune was monitored photometrically from two locations separated by 6 km during the close approach of the planet to a star in May 1981. Both stations recorded a drop in signal lasting 8.1 s. Reitsema et al. (1982) have interpreted their observations as the occultation of the star by a previously unknown satellite of Neptune. If the object lies in Neptune's orbital plane, it orbits at a distance of 3 Neptune radii from the planet's center and has a minimum radius of 90 km. Attempts to image this suspected object have thus far failed. This "satellite" may in fact be a Neptunian ring (Hubbard et al. 1986; cf. Hubbard 1986).

III. PLUTO AND CHARON

A. Orbital Properties and Rotation

We consider Pluto and its satellite Charon together in this chapter because both are satellite-size bodies with compositional characteristics comparable to other planetary satellites. It is not necessary to review the orbital properties and dynamics of Pluto itself since this subject has been treated in detail in recent studies; the elements and the history of their determination have been reviewed by Seidelmann et al. (1980).

The elements of Charon, determined from astrometric observations by photographic and speckle interferometric techniques, and from eclipse observations, are given in Table III (adapted from Tholen [1985]). The period of 6.38723 ($\pm$ 0.00027) d is important because it corresponds to the diurnal rotation period of Pluto itself, as determined by the periodicity of the lightcurve measured photometrically (see, e.g., Tedesco and Tholen 1980). Charon and Pluto thus appear to be locked in synchronous revolution/rotation in a situation apparently unique among the bodies of intermediate size in the solar system (some asteroids may have satellites which share this property; this dynamical state is considered in the chapters by Burns and by Peale). We discuss the diurnal period of Pluto and its photometric lightcurve below.

Other important aspects of the orbit of Charon are the high inclination to the ecliptic and the radius of the satellite's orbit, $a = 19,130$ ($\pm$ 460) km. With the currently accepted radius of Pluto ($R_P = 1500$ km, discussed below), the orbital radius of Charon corresponds to 13 R_P.

Pluto's rotational axis is similarly inclined, as required by the dynamical end state of tidal evolution, and confirmed by the increasing lightcurve amplitude as we approach an equatorial viewing aspect. Tidal evolution has erased any inclination between the initial spin and orbit of Pluto and its satellite, and has also circularized the orbit, making it impossible to classify Charon as a "regular" or "irregular" satellite based on its present orbit (Harris, personal communication, 1984).

Photometric observations in 1954–1955 (Walker and Hardie 1955) showed a lightcurve of Pluto with amplitude 0.1 mag (V wavelength) and with a period of 6.390 ($\pm$ 0.0003) d. Observations in 1964 allowed Hardie (1965) to refine the period to 6.3860 ($\pm$ 0.0003) d, but he also suspected that the amplitude of the lightcurve had increased. By 1980 the amplitude had in-

TABLE III
The Pluto-Charon System

Orbit of Charon[a]	
Orbital radius	$a = 19{,}130\ (\pm\ 460)$ km
Period	$P = 6.38723\ (\pm\ 0.00027)$ d
Inclination to celestial equator	$I = 94°3 \pm 1°5$
Node on equator	$\Omega = 223°7 \pm 1°4$
True anomaly (31 Jan 1982 $= $ JD 2445000.5)	$f = 78°6 \pm 1°9$
Mass of Pluto + Charon[a]	$(1.35 \pm 0.1) \times 10^{25}$ g $= 0.0023\ M_{\cdot \oplus} = 0.184\ M_{moon}$
Rotation period of Pluto[b]	$P = 6.38726\ (\pm\ 0.00007)$ d (synodic) $P = 6.38755$ d (sidereal)
Radii[c]	
Pluto	$R_p = 1500\ (\pm\ ^{200}_{300})$ km
Charon	$R_C = 650\ (\pm\ 100)$ km

[a] From Tholen (1985).
[b] From D. J. Tholen (personal communication). The synodic period is that determined from the lightcurve in 1980–1983. The sidereal period is the best estimate, dependent on an assumption of the orientation of the planet's pole.
[c] Calculated from considerations of mean density and the physical observations discussed in the text.

creased to 0.3 mag and the mean opposition magnitude had *decreased* by 0.3 mag compared to the 1954–1955 value (Tedesco and Tholen 1980). The secular decrease in brightness at maximum light is thus 0.008 mag yr^{-1} over the interval noted, and the decrease in *mean* light is 0.012 mag yr^{-1}. The observing geometry (aspect) of Pluto has changed by about 62° during this 25 year period, while the distance from the Sun has changed from 34.9 AU to 29.9 AU (at opposition dates).

The secular change in brightness may be caused either by the changing distance from the Sun or by the changing aspect, or by some combination of the two. The increasing solar insolation could result in volatilization of surface ices, normally high in albedo, with larger surface exposures of otherwise covered or included dark soil. Because the aspect is changing from a polar to an equatorial view of Pluto, it appears more likely that the decrease in brightness results from darker surface areas progressively moving into view. This interpretation is favored because the lightcurve study of Tedesco and Tholen shows that the minimum light is fading at twice the rate of maximum light. While the lightcurve shows no significant color change shortward of 0.55 μm, reflectances at 0.70 and 0.86 μm are 0.02 and 0.03 mag higher, respectively, at minimum light than at maximum. This suggests that the darker material moving into view is reddish in color, a point to which we shall return below. The ongoing series of eclipses/occultations will be valuable in resolving some of these issues (Binzel et al. 1985*a*).

Pluto's elliptical orbit brings it within the orbit of Neptune near perihelion. In its current orbit, Pluto is closer to the Sun than Neptune in the interval from 21 January 1979 to 14 March 1999; perihelion passage will occur on 12 September 1989 at $r = 29.64730$ AU (Meeus 1983).

B. Surface Composition

The first direct information on the surface composition of Pluto came from a study by Cruikshank et al. (1976), who found evidence for solid methane from measurements in infrared filter bands selected to distinguish among the ices of water, methane, and ammonia; a neutral, rocky surface would have also been evident from the data obtained. Spectrophotometric observations (Cruikshank and Silvaggio 1980; Soifer et al. 1980) subsequently confirmed the strong bands of methane in the near infrared, although there had been an earlier tentative confirmation by Lebofsky et al. (1979). In Fig. 11 we show a composite spectrum of Pluto from 0.8 to 2.5 μm in which six distinct methane bands are apparent. Comparison with synthetic methane spectra computed by J. Apt for the low-temperature conditions on Pluto, and with laboratory spectra of methane gas and ice, suggest that the Pluto spectrum is best matched by the ice spectrum. In further support of the ice interpretation is the strength of the infrared bands which are variable with the planet's rotation, indicating a nonuniform distribution of methane ice across the surface (Cruikshank and Brown, in preparation).

Observations of the spectrum of Pluto in the region 0.6 to 1.0 μm at higher spectral resolution than the near-infrared spectrophotometry show additional methane bands (Fink et al. 1980*a*; Apt et al. 1983). Structure in the bands, particularly that at 0.89 μm, indicates that some of the absorption is due to methane gas. Buie and Fink (1983) find that the strengths of the bands in this spectral region are variable in phase with Pluto's lightcurve, indicating that a component of the absorption is attributable to methane ice having bands coincident with the gas phase bands.

Pluto has a distinct reddish color in the photovisual spectral region (0.3 to 1.0 μm), as seen in the data from several observers (Harris 1961; Fix et al. 1970; Lane et al. 1976; Bell et al. 1979; Barker et al. 1980; Apt et al. 1983). Apt et al. have a diagram with many of these results; here we show in Fig. 12 the data of Bell et al. (1979) and those of Barker et al. (1980) for comparison. The discrepancy between the data sets in the extreme violet end of the graph is probably not significant.

The reddish color is not characteristic of methane ice and therefore represents direct evidence for a second spectral component on Pluto's surface. The initial five-color photometric observations made in the early 1950s (Harris

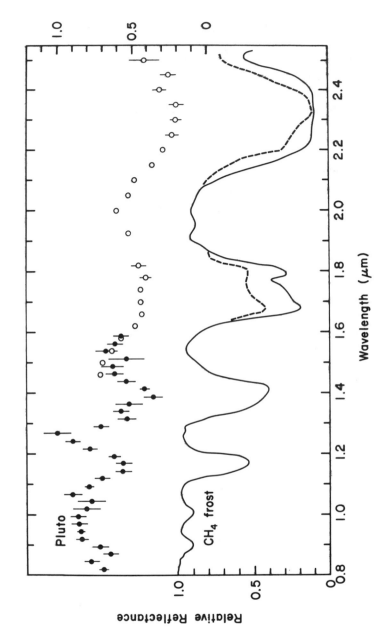

Fig. 11. Composite spectrum of Pluto. The open circles are from Soifer et al. (1980), while the solid points were obtained by Cruikshank and Brown in 1982. The methane ice spectrum is the same as that shown in Fig. 9.

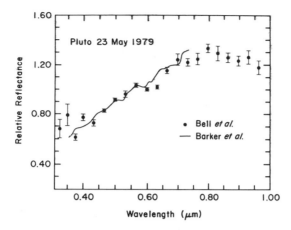

Fig. 12. Spectral reflectance of Pluto in the photovisual spectral region. The Bell et al. (1979) data from 23 May 1979 are previously unpublished; they were obtained with a multifilter photometer. The solid line is from Barker et al. (1980).

1961) show the reddish color clearly; when the early observations are compared with those of Bell et al. (1979) there is no clear indication of change. Therefore, we cannot say with certainty that the surface component progressively coming into view at the current epoch is alone responsible for the overall reddish color. There may be a colored surface constituent mixed intimately with the methane ice. Photometry with high precision may help resolve the question of the surface distribution of components of different colors and albedos during the current epoch of mutual eclipses and occultations of Pluto and Charon (Binzel et al. 1985a).

As we have already noted, Triton is also reddish in the photovisual spectral range, as is the atmosphere of Titan. The fact that Pluto and the other two objects are rich in methane in one phase or another, as well as laboratory results which show that tholins (hydrocarbon derivatives of methane made under planetary conditions [Kharc et al. 1983]) are red, suggest some common causal thread of connection. The reddish colors of Pluto and Triton may indicate the presence of surface deposits of complex hydrocarbons of the general type that cause the photochemical smog in the atmosphere of Titan.

C. The Atmosphere of Pluto

Although the atmosphere of Pluto is not central to this review, it is important to consider it in the context of possible atmospheres on other bodies comparable in size to Pluto. Fink et al. (1980a) found the first direct spectroscopic evidence for a methane atmosphere on Pluto, while Cruikshank and Silvaggio (1980) speculated that there must be a tenuous atmosphere because of the high vapor pressure of CH_4. Fink et al. (1980a) found structure in the

0.89 μm methane band and, by comparison with synthethic spectra, inferred an atmosphere with 27 ($\pm$7) m-Am of methane, giving a surface pressure of 0.15 mbar. This derived abundance does not take into account coincident absorption by methane ice, so the derived gas abundance should be less than 27 m-Am.

Trafton (1980a) showed that, for an isothermal atmosphere in contact with the surface, the hydrodynamic escape of methane is such that an amount equal to the mass Pluto would be lost in less than the age of the solar system. He proposed that the loss of methane would be slowed by diffusion through a heavier and undetected atmospheric constituent. Hunten and Watson (1982) have pointed out, however, that the adiabatic expansion of an escaping atmosphere will produce a cold trap unless there is sufficient heat input in the high atmosphere. Plausible sources of heat are quantitatively insufficient in the case of Pluto (and perhaps also in the case of Triton), and the escape rate should be slowed by the cold trap to a value consistent with the presence of solid methane on the surface of the planet. The escape parameter defined by Hunten and Watson (1982) differs only by a factor of 3 for Charon, thus indicating that only a modest amount of methane will be lost from the satellite as well as from the planet.

D. Masses and Radii of Pluto and Charon

Since the discovery of Pluto in 1930, the determination of its diameter and mass, and hence its mean density, has vexed astronomers anxious to ascertain whether the planet falls into the general categories of terrestrial or Jovian objects. Duncombe and Seidelmann (1980) have reviewed the mass determinations, while the radius has been discussed by numerous authors. With a relatively high surface geometric albedo assured by the discovery of methane ice on Pluto, coupled with the likelihood of a largely volatile bulk composition consistent with other small bodies in the outer solar system, Cruikshank et al. (1976) suggested that Pluto is sublunar in size and that it has a mean density representative of a mixture of common volatiles. The discovery of Pluto's satellite established that the mass is very small, 1.48×10^{25} g (Pluto plus Charon) (Harrington and Christy 1981), or 0.0025 M$_\oplus$. The determination of the radius (Lebofsky et al. 1982b; Morrison et al. 1982; see Table IV) has remained a problem, although various considerations of plausible mean density and the range of observed values of the radius make it possible to infer a value with a high degree of confidence.

A number of recent determinations of the dimensions of the Pluto-Charon system are given in Table IV. The original diskmeter measurement by Kuiper (1954) is not listed because it is known to be spurious. If the mass of Pluto represents 7/8 of the mass of the Pluto-Charon system, a mean density of 1.0 g cm^{-3} is equivalent to a radius of 1460 km. Mean densities much less than 1.0 g cm^{-3} are implausible for bodies with a substantial fraction of water ice and perhaps some silicate minerals as well. For a radius of 1700 km, the

TABLE IV

Dimensions of Pluto-Charon System

Observations	Maximum Separation Pluto-Charon (arcsec)	Radius of Pluto (km)	Radius of Charon (km)	Δm (P-C) (stellar magnitudes)[c]	Source
Photographs	0.84 ± 0.03	—	—	1.7 ± 0.1 (B)	Thomsen and Ables (1978)
Speckle	~ 0.8	1500 ± 200[a] 1800 ± 200[b]	—	—	Arnold et al. (1979)
Occultation	—	—	≥ 600	—	Walker (1980)
Speckle	0.85 ± 0.01	2000 ± 200	1000 ± 100	1.6 ± 0.2 (V)	Bonneau and Foy (1980)
Speckle	—	1500 ± 200	550 ± 300	2.2 ± 0.5 (V)	Hege et al. (1982)
Infrared	—	≤ 1700	—	—	Morrison et al. (1982)
Infrared	—	≤ 1300	—	—	Lebofsky et al. (1982b)
CCD Images	—	—	—	1.85 (V)	Reitsema et al. (1983)
Eclipses/Transits	—	1545-1660	600-850	2.1	Tholen (1985)

[a] No limb darkening.

[b] Limb darkening by cosine law.

[c] B = blue filter band, $\lambda = 0.44$ μm; V = visual filter band, $\lambda = 0.56$ μm.

mean density becomes 0.6 g cm^{-3} which is marginally possible for the unlikely case of a planet of pure methane ice.

Mutual eclipses of Pluto and Charon began in 1985 (Binzel et al. 1985a), as had been predicted shortly after the discovery of the satellite in 1978. Tholen (1985) has used observations of the first few eclipse events to refine the orbit of Charon and to make a preliminary estimate of the dimensions of both the satellite and Pluto itself; his results are given in Table IV. As the mutual events continue, it will become possible to refine the dimensions of both bodies, as well as to measure the spectral reflectance of the satellite.

It will be seen in Table IV that there is a general convergence of results on the dimensions of both Pluto and Charon, showing that the satellite is very large in comparison to its primary, and that the size of Pluto is comparable to that of the Moon. Both factors are of extreme interest in the context of the formation and subsequent evolution of the bodies in the outer solar system.

E. Conclusion

As we consider together the satellites of Uranus and Neptune and the Pluto-Charon pair, we see many interesting similarities and differences. It is clear that Pluto and Triton have some important common characteristics, but great differences as well. In that their interiors are expected to consist primarily of water ice, as is found on the surfaces of the Uranian satellites, it is of uncommon interest that the surfaces of Pluto and Triton have large quantities of solid methane rather than water ice.

The differences and similarities are clues to the mystery of the formation and evolution of these bodies, and provide a compelling challenge to the observers who use modern telescopes and spacecraft, and to theorists who try to see the complete picture in a puzzle with many missing pieces.

Note added in proof: *Preliminary Results of the Voyager 2 Encounter with the Satellites of Uranus.* During the encounter of the Voyager 2 spacecraft with the Uranian system in January 1986, many outstanding new results were obtained for the satellites (Smith et al. 1986). In addition to the discovery of ten new satellites, all interior to Miranda, the diameters and masses of the five larger known satellites were determined. Images of moderate resolution were obtained for Oberon, Titania and Umbriel, and images of high resolution were obtained for Ariel and Miranda. The largest of the newly discovered satellites, 1985 U1, was imaged with sufficient resolution to show a few craters and a slightly irregular shape. Both Ariel and Miranda were measured with the infrared instrument (IRIS) on the spacecraft; the observed temperatures are apparently in accord with those predicted on the basis of the geometric albedos of the satellites determined from groundbased photometry and the radii established by the imaging experiment on the spacecraft (Hanel et al. 1986). In this note we give some preliminary results of the Imaging Science satellite investigations in progress; some additional material is described in an appendix to the book's introductory chapter.

Physical Parameters. In planning for the Voyager 2 encounter with the Uranian system, groundbased data on the albedos and radii of the satellites were used to gauge exposure times and field-filling of the imaging system. These values proved to be quite close to the radii and surface reflectances determined by the spacecraft. Optical navigation images taken before and during the encounter, plus radio science data, permitted the determination of the masses of the five large

TABLE A-1

Satellite	Semimajor Axis (km)	Radius (km)	Mean Density (g cm^{-3})	Geometric Albedo[a]	Normal Reflectance[b]
1985 U1	86000	85 ± 5	—	—	0.07
Miranda	129783±66	242 ± 5	1.26 ± 0.39	0.22	0.34
Ariel	191239±57	580 ± 5	1.65 ± 0.30	0.38	0.40
Umbriel	265969±48	595 ± 10	1.44 ± 0.23	0.16	0.19
Titania	435844±86	805 ± 5	1.59 ± 0.09	0.23	0.28
Oberon	582596±71	775 ± 10	1.50 ± 0.10	0.20	0.24

[a]The geometric albedos in the table are derived from those given by Brown et al. (1982) as scaled for the satellite radii measured on Voyager 2 images. Brown et al. used groundbased photometric values of V(1,0) taken from Cruikshank (1982) and unpublished observations by Brown and J. Goguen.

[b]The normal reflectance for 1985 U1 was calculated from a Voyager 2 photometrically calibrated image and the assumption of an asteroidal phase function. The normal reflectances of the other satellites were derived from Voyager images and a strongly peaked phase function as measured on images of Titania.

TABLE A-2
The Newly Discovered Satellites of Uranus

Satellite	Semimajor[a] Axis (km)	Approximate Radius (km)	Geometric Albedo
1986 U1	66100	40	0.09
1986 U2	64600	40	0.06
1986 U3	61800	30	0.04
1986 U4	69900	60	0.04
1986 U5	75300	30	0.05
1986 U6	62700	30	0.04
1986 U7	49700	25	0.05
1986 U8	53800	25	0.05
1986 U9	59200	25	0.05

[a]The semimajor axes of the satellite orbits were calculated from the revolution periods of the satellites derived from Voyager data. In Table A-1, the orbit dimensions of the larger satellites were derived from their known periods and a new value of the mass of Uranus determined from Voyager data. Data rounded off to the nearest 100 km.

satellites; these quantities together with the radii give the mean densities with much greater certainty than had been possible with the groundbased data alone (in which the masses were the least certain).

In Table A-1 we give the physical parameters determined from a preliminary analysis of the Voyager data for the largest of the satellites; some of these data can be compared with those given in the main text of this chapter (Tables I and II).

Geometric Parameters.   The preliminary orbital semimajor axes for the newly discovered Uranian satellites are given in Table A-2 (Smith et al. 1986).

Images of the Satellites.   The Voyager spacecraft produced stunning images of the five previously known satellites of Uranus (Smith et al. 1986), and in this section we reproduce a small selection of those with the highest surface resolution.

1.   *Oberon.* The best picture of this satellite was obtained at a distance of 660,000 km and with a resolution of about 12 km on the surface (Fig. A-1). Craters appear prominently, in many cases surrounded by bright rays similar to those found on Callisto. Near the center of this image is a large crater with a bright central peak and a variegated floor. Surface topographic features include the mountain peak on the lower left limb in this image. There are no obvious global structural features on Oberon.

2.   *Titania.* The highest-resolution picture of Titania was obtained at a distance of 369,000 km giving a resolution of about 13 km. The contrast-enhanced and spatially filtered image in Fig. A-2 is a composite of two images through the clear filter of the narrow-angle camera. It shows an abundance of impact craters and prominent global tectonic features exemplified by the large fault valleys extending some 1500 km in length and 75 km in width. Near the

Fig. A-1.   Voyager 2 image of Oberon (Jet Propulsion Laboratory photograph P-29501C).

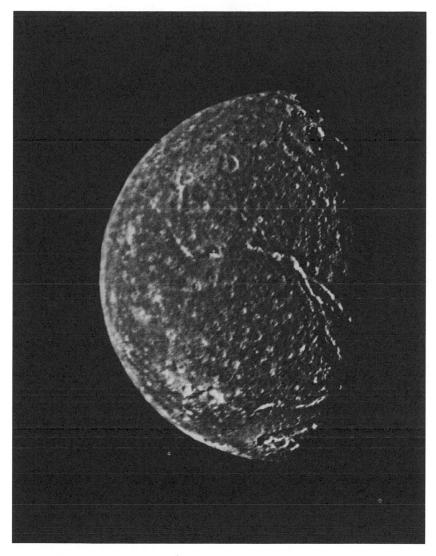

Fig. A-2. Voyager 2 composite image of Titania (Jet Propulsion Laboratory photograph P-29522 B/W).

bottom of the image is an impact crater of more than 200 km diameter through which a fault valley cuts. Near the top of the image is a large crater of diameter about 300 km.

3. *Umbriel.* Heavily cratered Umbriel was imaged with a resolution about 10 km from a distance of 557,000 km through the Voyager 2 clear filter on the narrow-angle camera (Fig. A-3). The uniform distribution of craters and the apparent absence of major global tectonic features suggests that this body has experienced a low level of geologic activity. The low geometric albedo of the surface is maintained by an absence of bright rays associ ated with the impact craters. The prominent crater on the upper right terminator is about

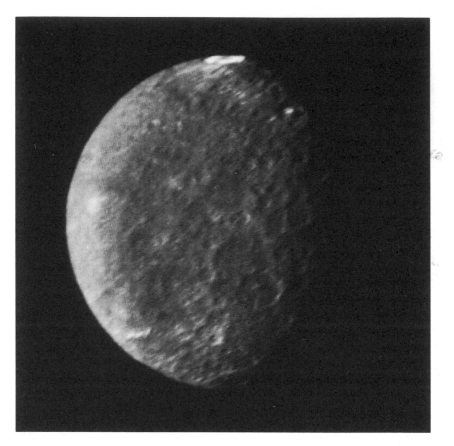

Fig. A-3. Voyager 2 image of Umbriel (Jet Propulsion Laboratory photograph P-29521 B/W).

110 km in diameter and has a bright central peak. At the very top of the image is a bright ring-shaped feature which is the most reflective part of the surface seen in Voyager images. The ring is near the satellite's equator and is about 140 km in diameter, and its reflectivity is comparable to surface material found on some of the other, higher-albedo satellites.

4. *Ariel*. The relatively close approach of the spacecraft to Ariel permitted pictures with a surface resolution of about 2.4 km. Fig. A-4 is a mosaic of the four highest-resolution Ariel images obtained with the clear filter of the narrow-angle camera from a distance of about 130,000 km. Numerous craters are seen in the mosaic down to the threshold of resolution. The highly pitted terrain is crossed by several valleys and fault scarps which may have formed as a result of expansion and stretching of the satellite's crust. The largest fault valleys, near the terminator at right, as well as the smooth region near the center of Fig. A-4, have been partially filled with smooth younger deposits. The deposits have themselves been incised by narrow sinuous features, possibly resulting from the flow of fluids.

5. *Miranda*. The geometry of the Voyager 2 encounter with Uranus and its satellites was dictated by a number of constraints, including that of directing the spacecraft on to Neptune for a 1989 encounter. As a consequence of this requirement plus the unusual orientation of the Uranian system, it was possible to fly very close to only one of the five previously known satellites, Miranda. As can be seen in the main text of this chapter, there was extremely little pre-encounter knowledge of Miranda. The images obtained with the Voyager

Fig. A-4. Voyager 2 image of Ariel (Jet Propulsion Laboratory photograph P-29520 B/W).

cameras have revealed a more startling surface than was imagined possible. We reproduce here a selection of the highest-resolution views obtained with the narrow-angle camera and a clear filter from a distance of 26,000 to 36,000 km. The best surface resolution is about 0.6 km.

Miranda is an asteroid-size satellite with an extraordinary diversity of surface structures. Textural and structural provinces are assembled in a highly complex pattern of cratered terrain, layered expanses, faulted and compressed regions, and a range of albedos that are in some cases matched to the topography and in others apparently unrelated. Fig. A-5 shows a chevron-shaped structural and albedo feature that lies in sharp contrast to the

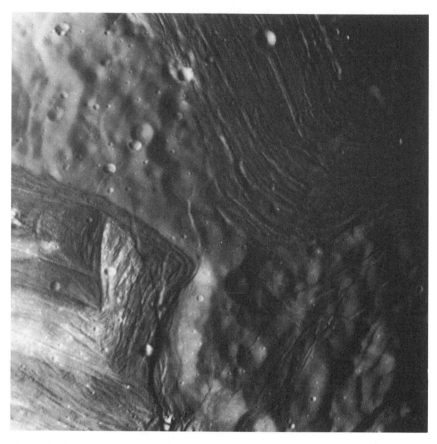

Fig. A-5. Voyager 2 image of Miranda (Jet Propulsion Laboratory photograph P-29515 B/W).

smooth cratered terrain. The width of this picture is about 220 km. In Fig. A-6, another section of the satellite is seen with the bright chevron in the upper portion of the frame. Graben-like faults cross the terrain in this portion of the surface, interrupting impact craters and the tectonically produced superstructure. A major fault scarp in the lower right catches the sunlight near the terminator. An astonishing array of topographic types is seen in Fig. A-7, in which fractures, grooves, and craters fill this narrow-angle clear filter view. The grooves and troughs reach depths of a few kilometers and expose materials of different albedos in the subsurface. Seen at their intersection with the upper limb, the grooves reveal an *en echelon* structure. The best of the narrow-angle, high-resolution images of Miranda have been assembled in a mosaic (see frontispiece to book) showing most of the surface visible to the spacecraft. The individual frames, which represent some of the highest resolution views of any solar system body obtained by the Voyager spacecraft, were obtained under especially difficult conditions of low light levels, with consequent long exposures, and rapid spacecraft motion past the target.

6.   *Satellite 1985 U1*. The first satellite discovered at Uranus by Voyager 2 is a body of some 170 km diameter and low (about 7%) surface albedo. It was found early enough in the mission to permit targeting of the narrow-angle camera at close approach so as to produce the single image in Fig. A-8 taken from a distance of about 500,000 km and having a surface resolution of about 10 km. Irregular topography, especially near the terminator

Fig. A-6. Voyager 2 image of Miranda (Jet Propulsion Laboratory photograph P-29512 B/W).

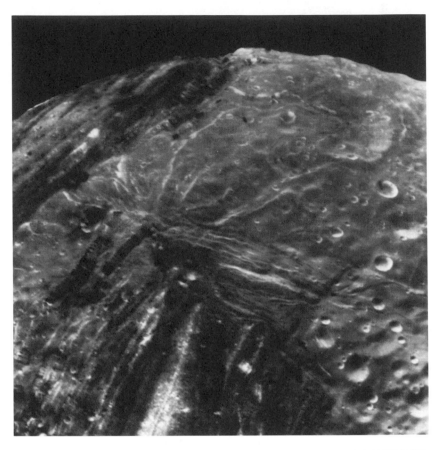

Fig. A-7. Voyager 2 image of Miranda (Jet Propulsion Laboratory photograph P-29513 B/W).

and limb, reveals three or four rather large craters. The general shape of the satellite is spherical.

In spite of the unfavorable geometry for studies of the satellites of Uranus during the Voyager 2 encounter, tremendous strides were made in opening these remote worlds to close inspection and study. The diversity among the six satellites imaged is a window on their histories as individual bodies and as members of the enigmatic Uranian system consisting of a tipped planet, strangely oriented magnetic field, unusual rings, and large family of satellites. It now remains for the detailed study of the images and other satellite data to permit us to make some strides in understanding the origin and evolution of Uranus and its family.

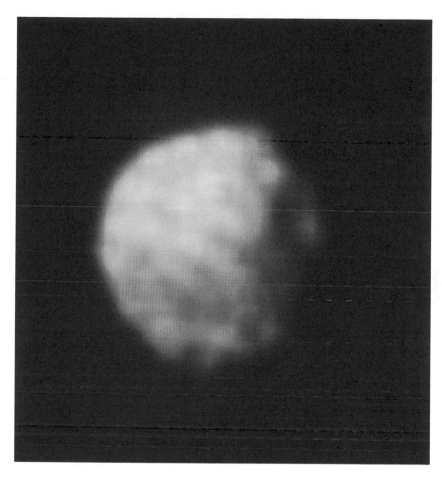

Fig. A-8. Voyager 2 image of satellite 1985 U1 (Jet Propulsion Laboratory photograph P-29519 B/W).

Color Section

Color Plate 1. Voyager photos of each Galilean satellite. Clockwise from top left are: Io, Europa, Callisto and Ganymede. Note the general reflectance levels and colors of each satellite. Io's reds, oranges, yellows and whites may be indicative of sulfur dioxide and sulfur allotropes (see Plate 4), while the other three satellites are composed of ice with some rock. The surface of Callisto has the most rock and dust and is the darkest, while that of Europa has the least rock and is the brightest. Ganymede is intermediate. (See, e.g., chapter by Clark et al.)

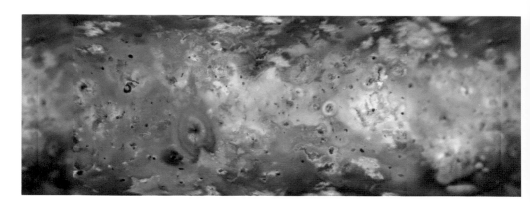

a

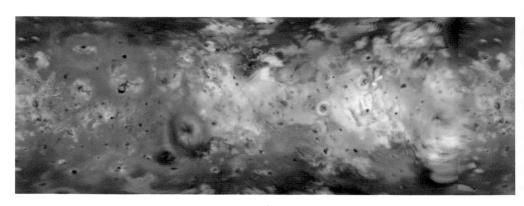

b

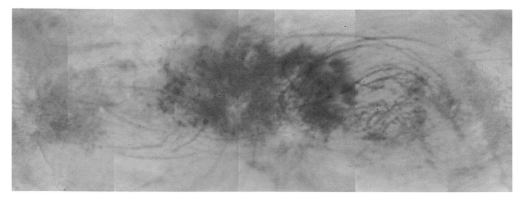

c

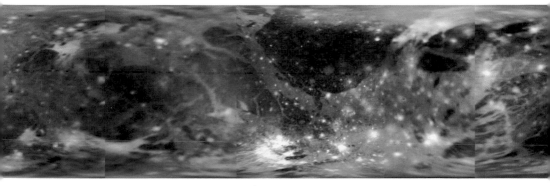

d

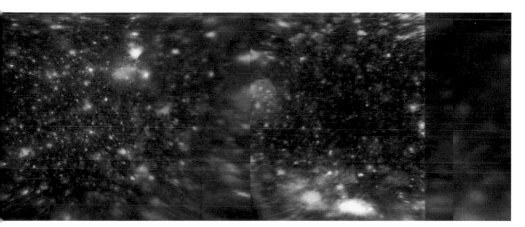

e

Color Plates 2 and 3. Simple cylindrical global mosiacs of the Galilean satellites: (a) and (b) Voyager 1 and 2 images of Io constructed from images taken through the orange, blue and violet filters; (c) true-color image of Europa; (d) false-color image of Ganymede constructed from orange, blue and ultraviolet filters; (e) false-color image of Callisto utilizing the clear, violet and ultraviolet filters. Color reconstruction courtesy of A. McEwen and L. Soderblom. (See Burns' introductory chapter.)

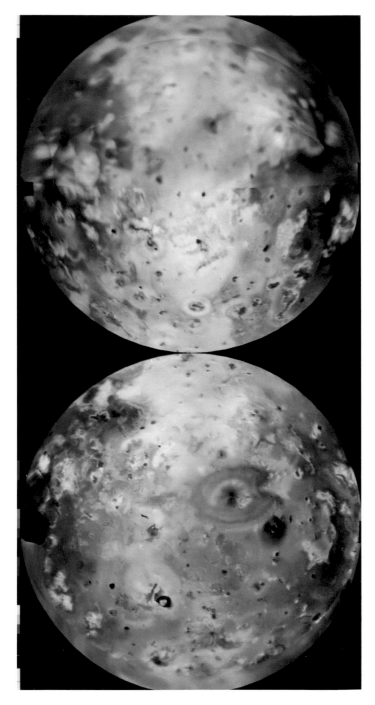

Color Plate 4. Approximate natural-color image of Io. These are mosaics made from Voyager data rendered in equal-area Lambert projection centered on the trailing (right) and leading (left) hemispheres. Image processing by A. McEwen and L. Soderblom, USGS, Flagstaff. (See chapter by Nash et al.)

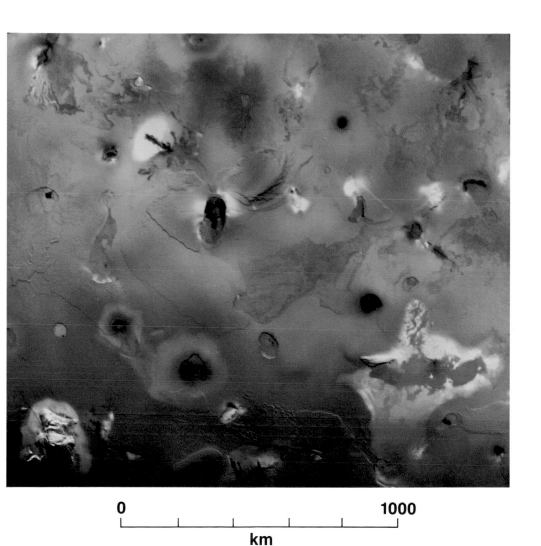

0   1000

km

Color Plate 5. Voyager color image of Io's surface in the high-southern latitude region. Center of the image is near lat 55°S and long 338°W. South pole is at the bottom center. East-west dimension of view is ~ 1800 km. Caldera Massaw Patera (shown in Fig. 10 of text) is at upper (north) center of view; Creidne Patera (Fig. 12 in text) is just below it. (See chapter by Nash et al.)

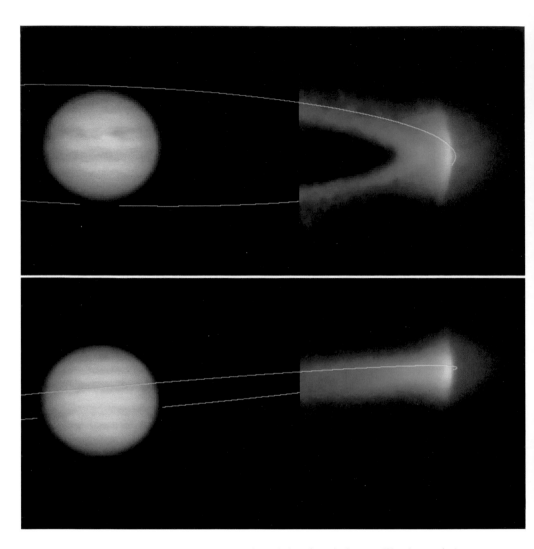

Color Plate 6. False-color images of the sulfur emissions from the Io torus (West Ansa region) at two rotational phases separated by 3 hr, obtained from Earth-based observations by Trauger at Las Campanas Observatory, Chile, on 4 May 1983. Regions of the torus populated predominantly in S^{++} appear with red hues, those predominantly S^+ appear green. Yellow curve marks the centrifugal confinement equator associated with the orbital path of Io (see chapter by Nash et al. and Trauger 1984 for details).

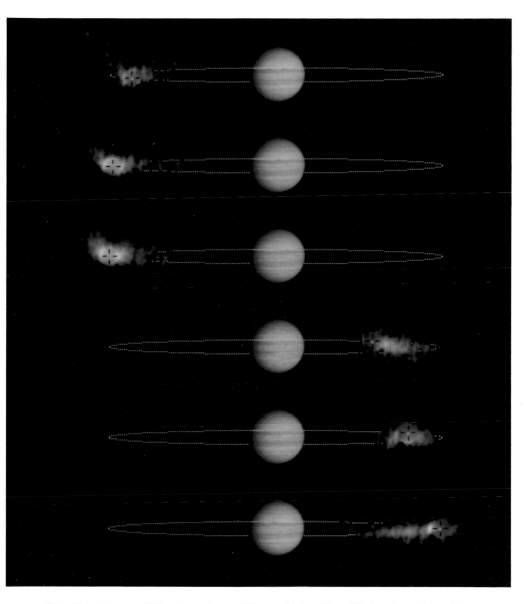

Color Plate 7. Images of Io's sodium cloud at different orbital positions of Io. Location and size of Io are indicated by the dot within the crossbars; its orbit by dotted ellipse. Emissions (shown here in yellow) are from neutral sodium atoms excited by resonant scattering of sunlight in the sodium D_2 line at wavelength of 5890 Å. Brightness of the cloud shown is on the order of 1 kilorayleigh, approximately that of the Earth's auroral displays. These images were obtained by Goldberg from Earth-based observations on 4–5 May 1982 at JPL's Table Mountain Observatory (see chapter by Nash et al. and Goldberg et al. 1984 for details).

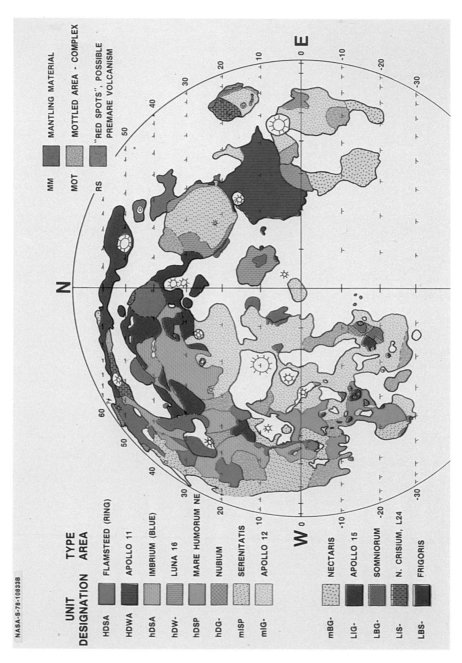

Color Plate 8. Major basalt types for the frontside of the Moon as derived from groundbased telescopic spectral reflectance data (from Pieters 1978; see chapter by Kaula et al.).

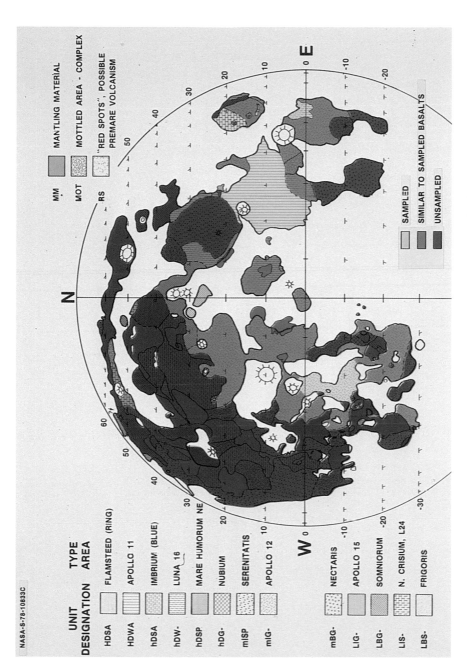

Color Plate 9. Sampled and unsampled lunar basalt types on spectral reflectance data (from Pieters 1978; see chapter by Kaula et al.).

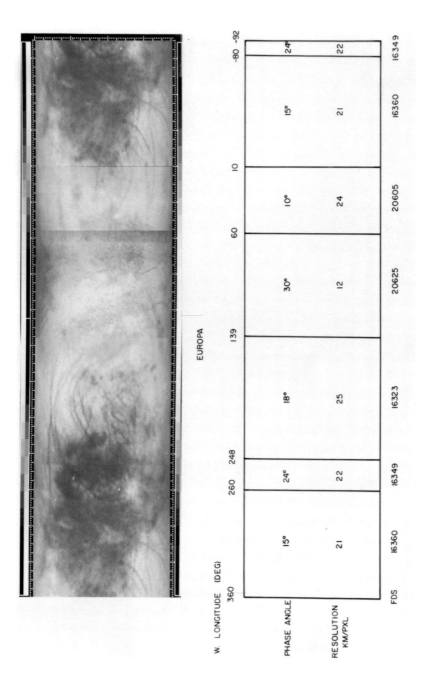

Color Plate 10. Color mosaic (top) and map of image reference data (bottom) of Voyager's global color observations of Europa (Johnson et al. 1983). Top to bottom, the image covers from 60°N to 60°S latitude in this cylindrical projection; the left edge is at about 80°W longitude and the rightmost sector of the image is a repeat of the leftmost sector (over 360° is thus covered). The dark region on the left corresponds to the trailing hemisphere of Europa, thought to be darkened by ion implantation. The higher albedo leading hemisphere is seen in the center of the mosaic. (See chapter by Malin and Pieri.)

Map Section

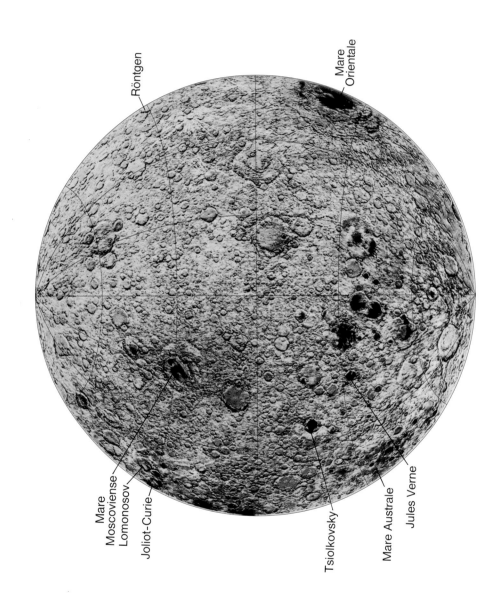

Röntgen

Mare
Orientale

Mare
Moscoviense
Lomonosov

Joliot-Curie

Tsiolkovsky

Mare Australe

Jules Verne

MOON
FAR SIDE

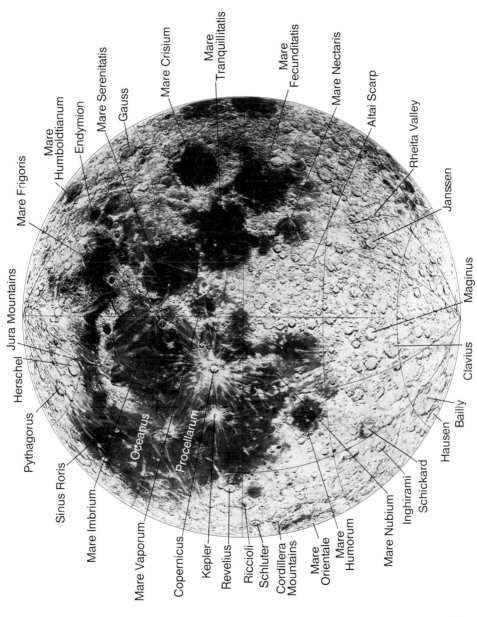

MOON
NEAR SIDE

Rutwa Patera

0°

Tarsus Regio

Uta Patera

Masubi

Kane Patera

Bochica Patera

Nusku Patera

Haemus Mons

CHALYBES REGIO

MEDIA
REGIO

Amirani

Maui

Maui Patera

Volund

COLCHIS
REGIO

180°

Prometheus

Emakong Patera

Culann Patera

Tohil Patera

Mycenae Regio

Nina Patera

Shamash Patera

Malik Patera

Bactria Regio

IO
LEADING SIDE

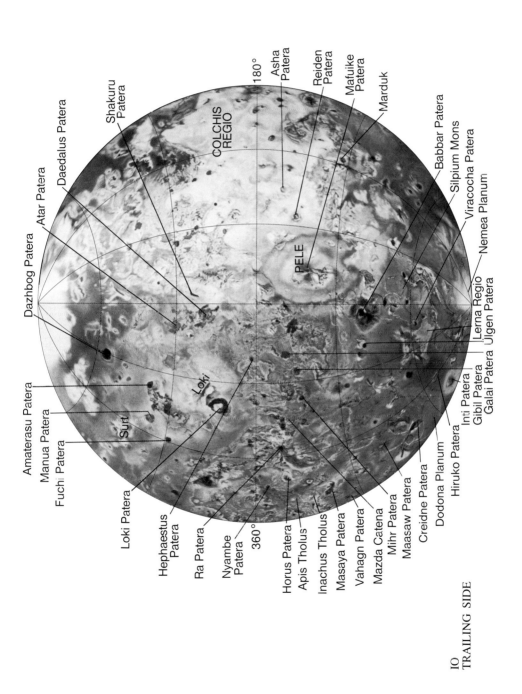

IO
TRAILING SIDE

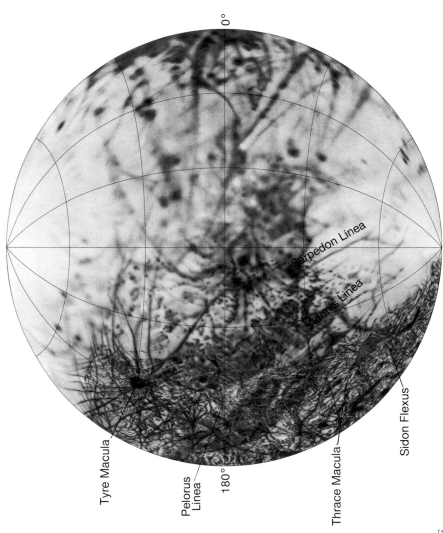

EUROPA
LEADING SIDE

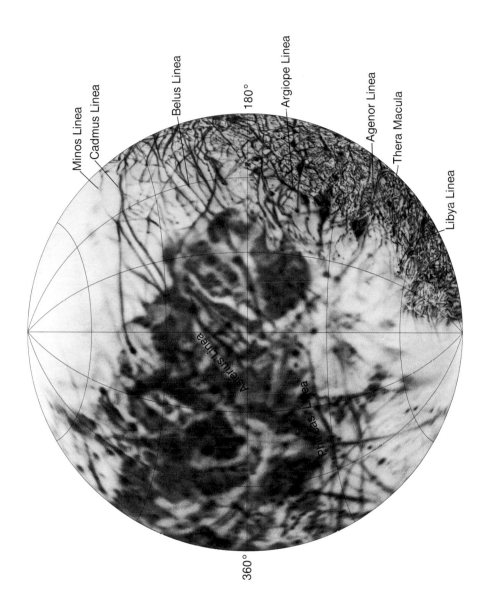

EUROPA
TRAILING SIDE

GANYMEDE
LEADING SIDE

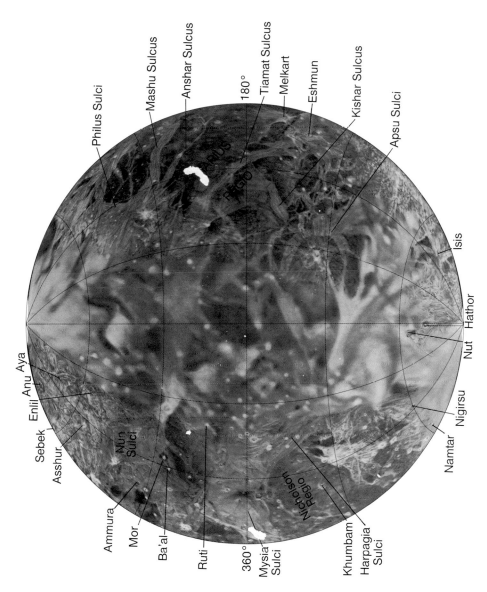

GANYMEDE
TRAILING SIDE

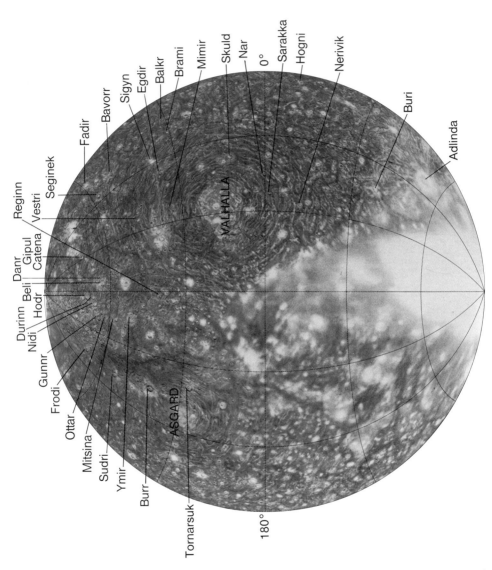

CALLISTO
LEADING SIDE

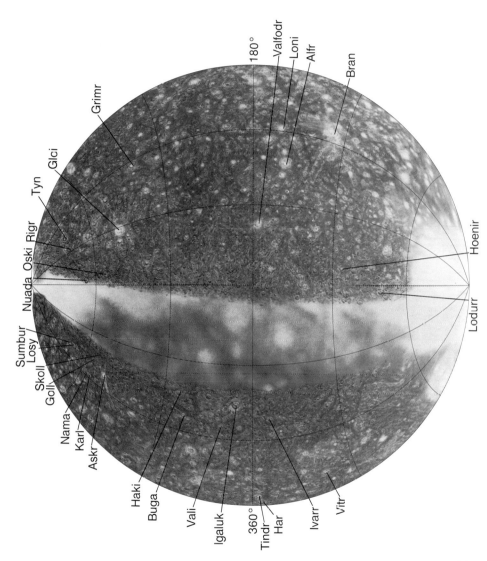

CALLISTO
TRAILING SIDE

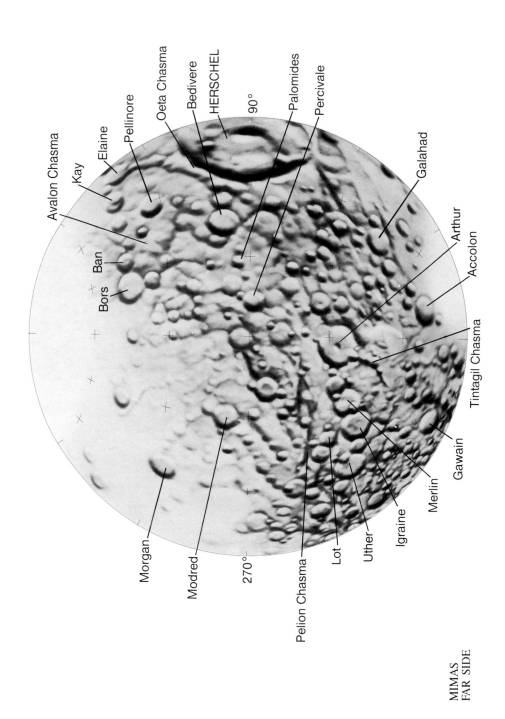

MIMAS
FAR SIDE

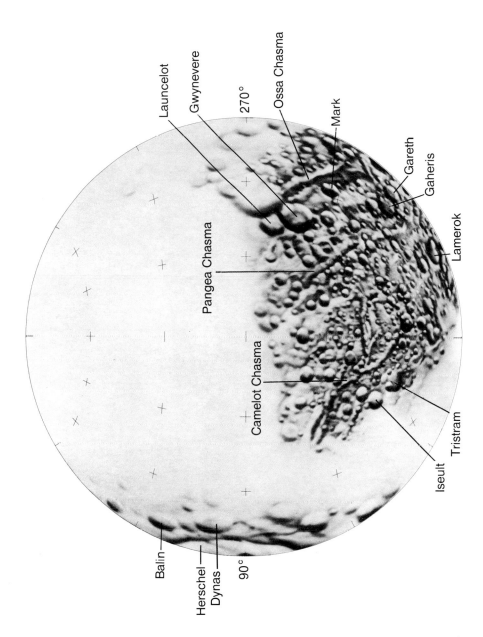

MIMAS
NEAR SIDE

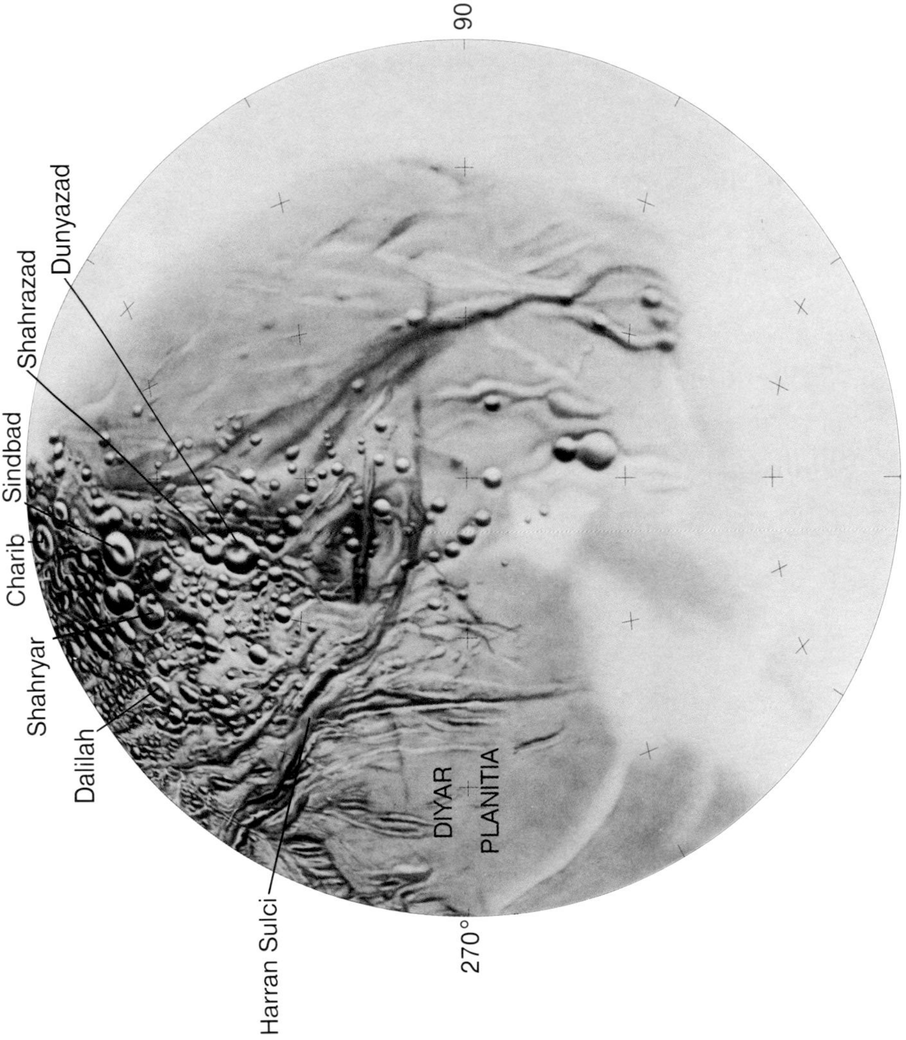

90°

Dunyazad

Shahrazad

Sindbad

Charib

Shahryar

Dalilah

Harran Sulci

DIYAR
PLANITIA

270°

ENCELADUS
FAR SIDE

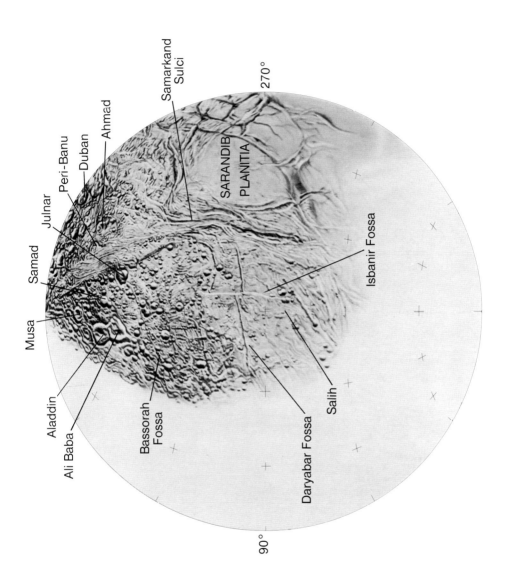

ENCELADUS
NEAR SIDE

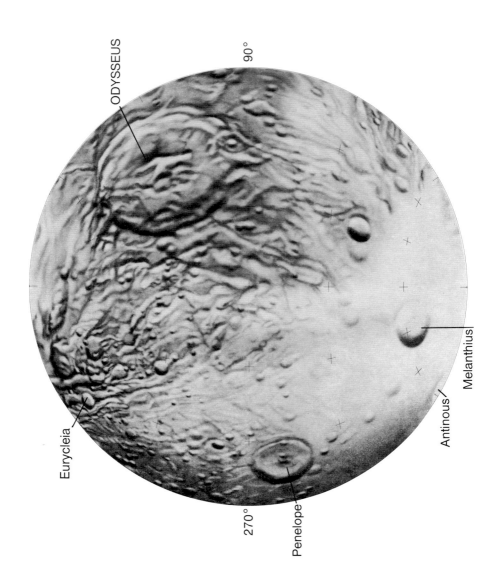

TETHYS
FAR SIDE

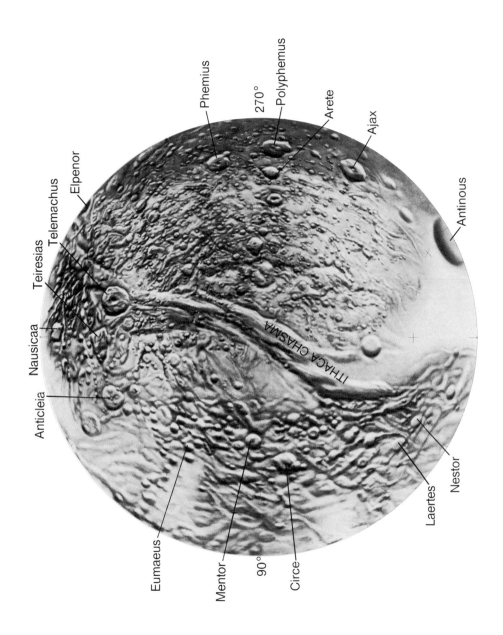

TETHYS
NEAR SIDE

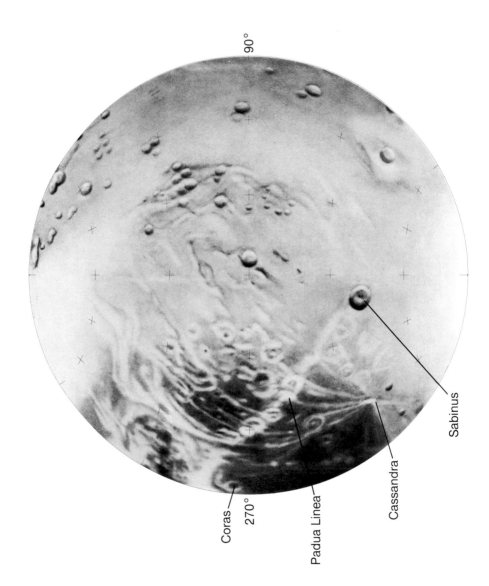

DIONE
FAR SIDE

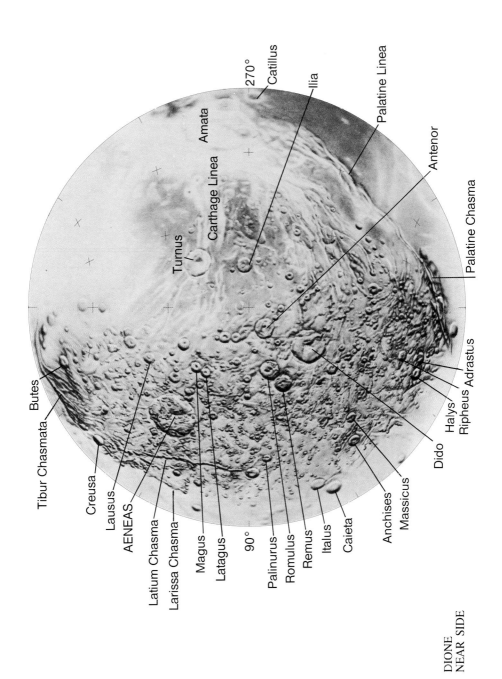

270°
Catillus
Amata
Ilia
Palatine Linea
Carthage Linea
Antenor
Turnus
Palatine Chasma
Butes
Tibur Chasmata
Creusa
Lausus
AENEAS
Latium Chasma
Larissa Chasma
Magus
Latagus
90°
Palinurus
Romulus
Remus
Italus
Caieta
Anchises
Massicus
Dido
Halys
Ripheus Adrastus

DIONE
NEAR SIDE

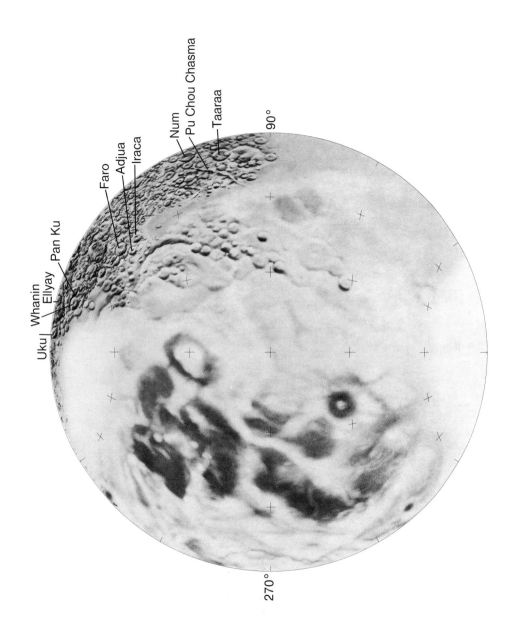

RHEA
FAR SIDE

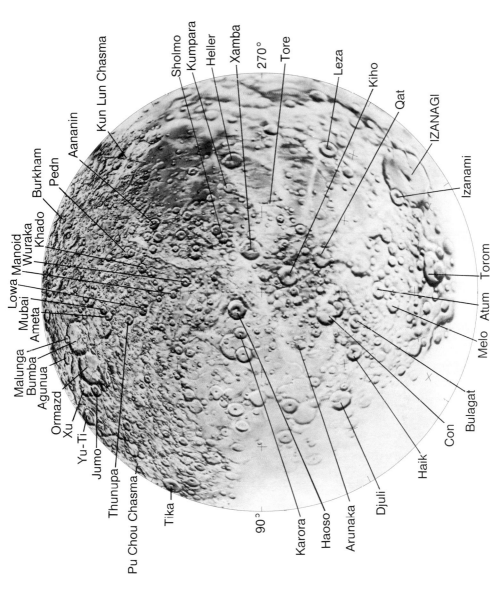

RHEA
NEAR SIDE

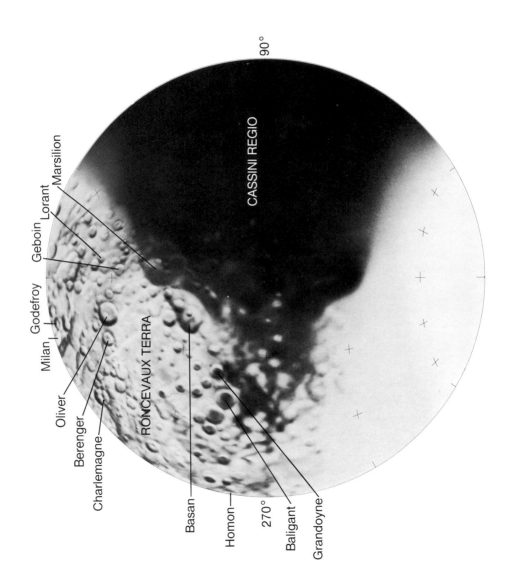

90°

CASSINI REGIO

Marsilion

Lorant

Geboin

Godefroy

Milan

Oliver

Berenger

Charlemagne

RONCEVAUX TERRA

Basan

Homon

270°

Baligant

Grandoyne

IAPETUS
FAR SIDE

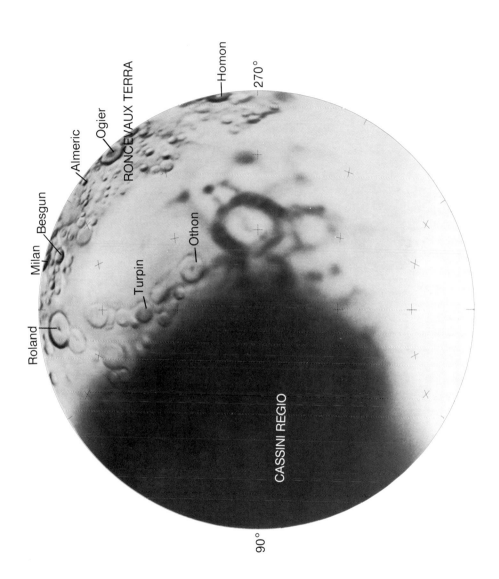

IAPETUS
NEAR SIDE

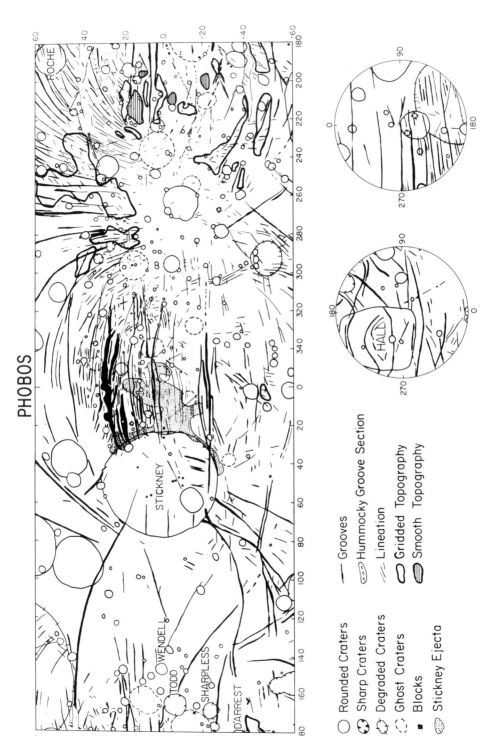

PHOBOS

ROCHE

STICKNEY

WENDELL

TODD

SHARPLESS

D'ARREST

HALL

○ Rounded Craters
✪ Sharp Craters
◌ Degraded Craters
◌ Ghost Craters
▪ Blocks
⬮ Stickney Ejecta

— Grooves
〰 Hummocky Groove Section
╱╱ Lineation
◯ Gridded Topography
⬭ Smooth Topography

PHOBOS

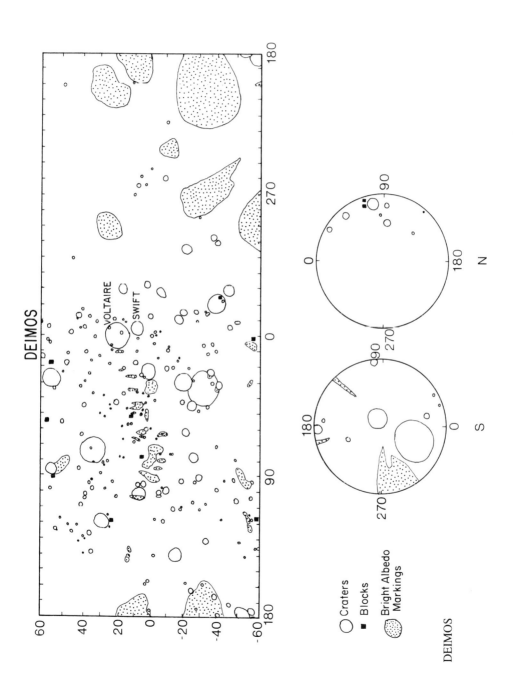

DEIMOS

○ Craters
■ Blocks
▨ Bright Albedo
   Markings

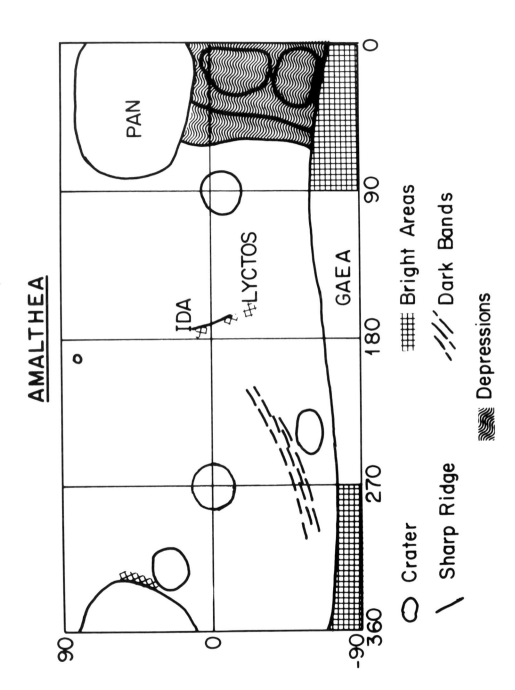

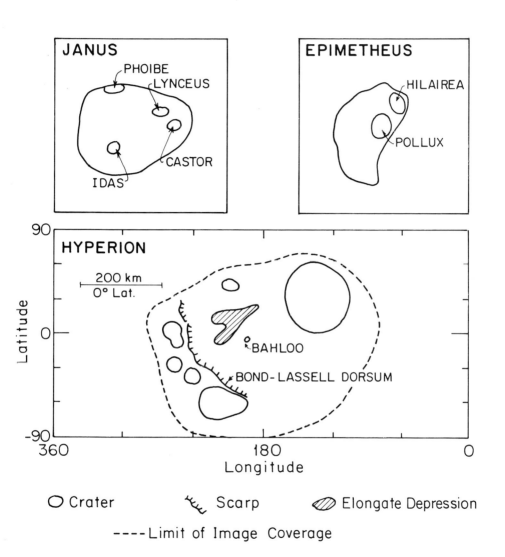

Bibliography

BIBLIOGRAPHY

Compiled by Melanie Magisos

Ables, H. D., and Thomsen, B. 1979. Quoted in Arnold, S. J., Boksenberg, A., and Sargent, W. L. W. 1979. Measurements of the diameter of Pluto by speckle interferometry. *Astrophys. J.* 234:L159–L163.

Acuña, M. H., Neubauer, F. M., and Ness, N. F. 1981. Standing Alfvén wave current system at Io: Voyager 1 observations. *J. Geophys. Res.* 86:8513–8522.

Acuña, M. H., Behannon, K. W., and Connerney, J. E. P. 1983. Jupiter's magnetic field and magnetosphere. In *Physics of the Jovian Magnetosphere,* ed. A. J. Dessler (Cambridge: Cambridge Univ. Press), pp. 1–50.

Adams, J. B. 1968. Lunar and Martian surface: Petrologic significance of absorption bands in the near infrared. *Science* 159:1453–1455.

Adams, J. B., and McCord, T. B. 1973. Vitrification darkening in the lunar highlands and identification of Descartes material at the Apollo 16 site. *Proc. Lunar Sci. Conf.* 4:163–177.

Adler, I., and Trombka, J. I. 1977. Orbital chemistry: Lunar surface analysis from the x-ray and gamma ray remote sensing experiments. *Phys. Chem. Earth* 10:17–43.

Adler, I., Trombka, J., Schmadebeck, R., Lowman, P., Blodget, N., Yin, L., and Eller, E. 1973. Results of the Apollo 15 and 16 x-ray experiment. *Proc. Lunar Sci. Conf.* 4:2783–2791.

Ahrens, T. J., and O'Keefe, J. D. 1972. Shock melting and vaporization of lunar rocks and minerals. *The Moon* 4:214–249.

Ahrens, T. J., and O'Keefe, J. D. 1977. Equations of state and impact-induced shock-wave attenuation on the moon. In *Impact and Explosion Cratering,* eds. D. J. Roddy, R. O. Pepin, and R. B. Merrill (New York: Pergamon), pp. 639–656.

Ahrens, T. J., and O'Keefe, J. D. 1978. Energy and mass distributions of impact ejecta blankets on the moon and Mercury. *Proc. Lunar Planet. Sci. Conf.* 9:3787–3802.

Ahrens, T. J., and O'Keefe, J. D. 1984. Shock vaporization and the accretion of the icy satellites of Jupiter and Saturn. *Lunar Planet. Sci.* XV:3–4 (abstract).

Alden, H. L. 1940. Mass of the satellite of Neptune. *Astron. J.* 49:71–72.

Alden, H. L. 1943. Observations of the satellite of Neptune. *Astron. J.* 50:110–111.

Alfvén, H. 1963. The early history of the Moon and the Earth. *Icarus* 1:357–363.

Allan, R. R. 1969. Evolution of Mimas-Tethys commensurability. *Astron. J.* 74:497–506.

Allen, C. C. 1977. Rayed craters on the moon and Mercury. *Earth Planet. Sci. Lett.* 15:179–188.

Allen, C. W. 1973. *Astrophysical Quantities,* 3rd ed. (London: Athlone Press).

Allen, D. A. 1971. The method of determining infrared diameters. In *Physical Studies of Minor Planets,* ed. T. Gehrels, NASA SP-267, pp. 41–44.

Allison, M. L., Head, J. W., and Parmentier, E. M. 1980. High albedo terrain on Ganymede: Origin as flooded graben. *The Satellites of Jupiter, IAU Coll. 57,* abstract booklet, 6–7.

Ananda, M. R. 1977. Lunar gravity: A mass point model. *J. Geophys. Res.* 82:3049–3064.

Anders, E. 1975. Do stony meteorites come from comets? *Icarus* 24:363–371.

Anders, E. 1978. Most stony meteorites come from the asteroid belt. In *Asteroids: An Exploration Assessment,* NASA CP-2053, pp. 57–75.

Anders, E., and Ebihara, M. 1982. Solar-system abundances of the elements. *Geochim. Cosmochim. Acta* 46:277–339.

Anderson, H. N., and Zeigler, J. F. 1977. *Hydrogen: Stopping Powers and Ranges in All Elements* (New York: Pergamon).

Andrews, R. J. 1977. Characteristics of debris from small-scale cratering experiments. In *Impact and Explosion Cratering,* eds. D. J. Roddy, R. O. Pepin, and R. B. Merrill (New York: Pergamon), pp. 1089–1100.

Apt, J., Carleton, N. P., and MacKay, C. D. 1983. Methane on Triton and Pluto: New CCD spectra. *Astrophys. J.* 270:342–350.

Arnold, J. R., Peterson, L. E., Metzger, A. E., and Trombka, J. I. 1972. Gamma-ray spectrometer experiment. In *Apollo 15 Preliminary Science Report,* NASA SP-289, 16-1–16-6.

Arnold, S. J., Boksenberg, A., and Sargent, W. L. W. 1979. Measurement of the diameter of Pluto by speckle interferometry. *Astrophys. J.* 234:L159–L163.

Arthur, D. W. G. 1981. *Vertical Dimensions of the Galilean Satellites.* NASA TM-81776.

Arvidson, R., Drozd, R., Guinness, E., Hohenberg, C., and Morgan, C. 1976. Cosmic ray exposure ages of Apollo 17 samples and the age of Tycho. *Proc. Lunar Sci. Conf.* 7:2817–2832.

Ashby, M. F., and Verrall, R. A. 1978. Micromechanisms of flow and fracture, and their relevance to the rheology of the upper mantle. *Phil. Trans. Roy. Soc. London* A288:59–93.

Asknes, K. 1977. Properties of satellite orbits: Ephemerides, dynamical constants and satellite phenomena. In *Planetary Satellites,* ed. J. A. Burns (Tucson: Univ. of Arizona Press), pp. 27–42.

Asknes, K. 1985. The tiny satellites of Jupiter and Saturn and their interactions with the rings. In *Stability of the Solar System and its Minor Natural and Artificial Bodies,* ed. V. Szebehely (Dordrecht: Reidel), pp. 3–16.

Asknes, K., and Franklin, F. A. 1978. The evidence for faint satellites of Saturn reexamined. *Icarus* 36:107–118.

Astronomical Almanac. 1985. (Washington: U.S. Government Printing Office).

Atreya, S. K., Donahue, T. M., and Kuhn, W. R. 1978. Evolution of a nitrogen atmosphere on Titan. *Science* 201:611–613.

Aumann, H. H., Gillett, F. C., Beichman, C. A., DeJong, T., Houck, J. R., Low, F. J., Neugebauer, G., Walker, R. G., and Wesselius, P. R. 1984. Discovery of a shell around Alpha Lyrae. *Astrophys. J.* 278:L23–L27.

Baedecker, P. A., Chou, C. L., and Wasson, J. T. 1972. The extralunar component in lunar soils and breccias. *Proc. Lunar Sci. Conf.* 3:1343–1359.

Bagenal, F., and Sullivan, J. 1981. Direct plasma measurements in the Io torus and inner magnetosphere of Jupiter. *J. Geophys. Res.* 86:8447–8466.

Baker, R. W., and Gerberich, W. W. 1979. The effect of crystal size and dispersed solid inclusions on the activation energy for creep of ice. *J. Glaciol.* 24:179–194.

Baldwin, B., and Shaeffer, Y. 1971. Ablation and break-up of large meteoroids during atmospheric entry. *J. Geophys. Res.* 76:4653–4668.

Baldwin, R. B. 1949. *The Face of the Moon* (Chicago: Univ. of Chicago Press).

Baldwin, R. B. 1963. *The Measure of the Moon* (Chicago: Univ. of Chicago Press).

Baloga, S. M., Pieri, D. C., and Matson, D. L. 1986. The auras of volcanic flows on Io. Submitted to *J. Geophys. Res.*

Banfi, V. 1984. Future dynamical evolution of the Neptune-Triton system. A new synthetic method of analysis. *Earth, Moon and Planets* 30:43–52.

Barbosa, D. D., and Kivelson, M. G. 1983. Dawn-dusk electric field asymmetry of the Io plasma torus. *Geophys. Res. Lett.* 10:210–213.

Barker, E. S., Cochran, W. D., and Cochran, A. L. 1980. Spectrophotometry of Pluto from 3500 to 7350 Å. *Icarus* 44:43–52.

Barks, C. 1960. Island in the sky. *Uncle Scrooge* 29:1–18.

Barton, L. A. 1983. Magnetosphere Ion Erosions of the Icy Satellites of Saturn. M.S. Thesis, Univ. of Virginia.

Basaltic Volcanism Study Project. 1981. Basaltic Volcanism on the Terrestrial Planets (New York: Pergamon Press).

Bates, R. L., and Jackson, J. A. 1980. *Glossary of Geology* (Falls Church: Amer. Geol. Inst.).

Baum, W. A., Kreidl, T., Westphal, J. A., Danielson, G. E., Seidelmann, P. K., Pascu, D., and Currie, D. G. 1981. Saturn's E-ring. I. CCD observations of March 1980. *Icarus* 47:84–96.

Beatty, J. K., O'Leary, B., and Chaikin, A., eds. 1982. *The New Solar System,* 2nd ed. (Cambridge, MA: Sky Publishing Corp.).

Belcher, J. 1983. Low-energy plasma in the Jovian magnetosphere. In *Physics of the Jovian Magnetosphere,* ed. A. J. Dessler (Cambridge: Cambridge Univ. Press), pp. 68–105.

Bell, J. F. 1984a. A Search for Ultraprimitive Material in the Solar System. Ph.D. Thesis, University of Hawaii, Honolulu.

Bell, J. F. 1984b. Callisto: Jupiter's Iapetus. Lunar Planet. Sci. XV:44–45 (abstract).

Bell, J. F., and Hawke, B. R. 1982. Spectral studies of lunar dark halo craters. Bull. Amer. Astron. Soc. 14:754 (abstract).

Bell, J. F., and Hawke, B. R. 1984. Lunar dark-haloed impact craters: Origin and implications for early mare volcanism. J. Geophys. Res. 89:6899–6910.

Bell, J. F., Clark, R. N., McCord, T. B., and Cruikshank, D. P. 1979. Reflection spectra of Pluto and three distant satellites. Bull. Amer. Astron. Soc. 11:570 (abstract).

Bell, J. F., Cruikshank, D. P., and Gaffey, M. J. 1985. The composition and origin of the Iapetus dark material. Icarus 61:192–207.

Belton, M. J. S. 1982. An interpretation of the near-ultraviolet absorption spectrum of SO_2: Implications for Venus, Io, and laboratory measurements. Icarus 52:149–165.

Belton, M. J. S., and Terrile, R. 1984. Rotational properties of Uranus and Neptune. In Uranus and Neptune, ed. J. T. Bergstralh, NASA CP-2330, pp. 327–347.

Benz, W., Slattery, W. L., and Cameron, A. G. W. 1986. The origin of the Moon and the single-impact hypothesis I. Icarus 66. In press.

Berge, G. L., and Muhleman, D. O. 1975. Callisto: Disk temperature at 3.71 cm wavelength. Science 187:441–443.

Bergstralh, J. T., ed. 1984. Uranus and Neptune, NASA CP-2330.

Bergstralh, J. T., Matson, D. L., and Johnson, T. V. 1975. Sodium D-line emission from Io: Synoptic observations from Table Mountain Observatory. Astrophys. J. 195:L131–L133.

Bergstralh, J. T., Young, J. W., Matson, D. L., and Johnson, T. V. 1977. Sodium D line emission from Io: A second year of synoptic observations from Table Mountain Observatory. Astrophys. J. 211:L51–L55.

Bernatowicz, T. J., Hohenberg, C. M., Hudson, B., Kennedy, B. M., Laul, J. C., and Podosek, F. A. 1980. Noble gas component organization in 14301. Proc. Lunar Planet. Sci. Conf. 11:629–668.

Bibring, J. P., Borg, J., Burlingame, A. L., Langevin, Y., Maurette, M., and Vassent, B. 1975. Solar wind and solar flare maturation of the lunar regolith. Proc. Lunar Sci. Conf. 6:3471–3493.

Biggs, E. K. 1964. Influence of the satellite Io on Jupiter's decametric emission. Nature 203:1008–1010.

Bills, B. G., and Ferrari, A. J. 1977. A harmonic analysis of lunar topography. Icarus 31:244–259.

Bills, B. G., and Ferrari, A. J. 1980. A harmonic analysis of lunar gravity. J. Geophys. Res. 85:1013–1025.

Binder, A. B. 1974. On the origin of the moon by rotational fission. The Moon 11:53–76.

Binder, A. B. 1982. Post-Imbrium global lunar tectonism: Evidence for an initially totally molten moon. Moon and Planets 26:117–133.

Binder, A. B., and Cruikshank, D. P. 1964. Evidence for an atmosphere on Io. Icarus 3:299–305.

Binder, A. B., and Lange, M. A. 1980. On the thermal history, thermal state, and related tectonism of a moon of fission origin. J. Geophys. Res. 85:3194–3208.

Binzel, R. P., Tholen, D. J., Tedesco, E. F., Buratti, B. J., and Nelson, R. M. 1985. The detection of eclipses in the Pluto-Charon system. Science 228:1193–1195.

Birch, F. 1954. Heat from radioactivity. In Nuclear Geology, ed. H. Faul (New York: Wiley), pp. 148–174.

Birkhoff, G., MacDougall, D. P., Pugh, E. M., and Taylor, G. 1948. Explosives with lined cavities. J. Appl. Phys. 19:563–582.

Bjorkman, M. D. 1984. Feasibility of determining impact conditions from total crater melt. Lunar Planet. Sci. XV:64–65 (abstract).

Black, D. C., and Matthews, M. S., eds. 1985. Protostars & Planets II (Tucson: Univ. of Arizona Press).

Blair, G. N., and Owen, F. W. 1974. The UBV orbital phase curves of Rhea, Dione, and Tethys. Icarus 22:224–229.

Blamont, J. 1974. The atmosphere of the rings of Saturn. In The Rings of Saturn, eds. F. Palluconi and G. Pettengill, NASA SP-343, pp. 197–199.

Bodenheimer, P., and Pollack, J. B. 1984. Calculations of the origins of the giant planets. *Bull. Amer. Astron. Soc.* 16:702 (abstract) and submitted to *Icarus,* 1986.

Bodenheimer, P., Grossman, A. S., DeCampli, W. M., Marcy, G., and Pollack, J. B. 1980. Calculations of the evolution of the giant planets. *Icarus* 41:293–308.

Bondi, H., and Hoyle, F. 1944. On the mechanism of accretion by stars. *Mon. Not. Roy. Astron. Soc.* 168:603–637.

Bonneau, D., and Foy, R. 1980. Interferometry au 3.60 m CFH I: Résolution du systeme Pluto-Charon. *Astron. Astrophys.* 92:L1–L4.

Booker, J. R., and Stengal, J. C. 1978. Further thoughts on convective heat transport in a variable-viscosity fluid. *J. Fluid Mech.* 84:289–291.

Borderies, N., and Goldreich, P. 1984. A simple derivation of capture probabilities for the J + 1: J and J + 2: J orbit-orbit resonance problems. *Celestial Mech.* 32:127–136.

Borderies, N., Goldreich, P., and Tremaine, S. 1984. Unsolved problems in planetary ring dynamics. In *Planetary Rings,* eds. R. Greenberg and A. Brahic (Tucson: Univ. of Arizona Press), pp. 713–734.

Boring, J. W., Garrett, J. W., Sieveka, E., and Johnson, R. E. 1983. Sputter-induced atmosphere and molecular ejection on Io. *Lunar Planet. Sci.* XIV:61–62 (abstract).

Boss, A. P. 1986. The origin of the Moon. *Science* 231:341–345.

Boss, A. P., and Mizuno, H. 1985. Dynamic fission instability of dissipative protoplanets. *Icarus* 63:134–152.

Boss, A. P., and Peale, S. J. 1986. Dynamical constraints on the origin of the Moon. In *The Origin of the Moon,* eds. W. K. Hartmann, R. J. Phillips, and G. J. Taylor (Houston: Lunar Planet. Inst.). In press.

Bowell, E., Chapman, C. R., Gradie, J. C., Morrison, D., and Zellner, B. 1978. Taxonomy of asteroids. *Icarus* 35:313–335.

Boyce, J. M. 1976. Ages of flow units in lunar nearside maria based on Lunar Orbiter 4 photographs. *Proc. Lunar Sci. Conf.* 7:2712–2728.

Brace, W. F., and Kohlstedt, D. L. 1980. Limits on lithospheric stress imposed by laboratory experiment. *J. Geophys. Res.* 85:6248–6252.

Bratt, S. R., Solomon, S. C., and Head, J. W. 1981. The evolution of multi-ringed basins: Cooling, subsidence, and thermal stress. *Lunar Planet. Sci.* IX:109 (abstract).

Bratt, S. R., Solomon, S. C., Head, J. W., and Thurber, C. H. 1984. Mantle uplift beneath lunar basins: Clues to the understanding of basin formation. *Lunar Planet. Sci.* XV:88–89 (abstract).

Bridge, H. S., Belcher, J. W., Lazarus, A. J., Sullivan, J. D., McNutt, R. L., Bagenal, F., Scudder, J. D., Sittler, E. C., Siscoe, G. L., Vasylunias, V. M., Goertz, C. K., Yeates, C. M. 1979. Plasma observations near Jupiter: Initial results from Voyager 1. *Science* 204:987–990.

Bridge, H. S., Bagenal, F., Belcher, J. W., Lazarus, A. J., McNutt, R. L., Sullivan, J. D., Gazis, P. R., Hartle, R. E., Ogilvie, K. W., Scudder, J. D., Sittler, E. C., Eviatar, A., Siscoe, G. L., Goertz, C. K., and Vasylunias, V. M. 1982. Plasma observations near Saturn: Initial results from Voyager 2. *Science* 215:563–570.

Broadfoot, A., Belton, M. J. S., Takacs, P. Z., Sandel, B. R., Shemansky, D. E., Holberg, J. B., Ajello, J. M., Atreya, S. K., Donahue, T. M., Moos, H. W., Bertaux, J. L., Blamont, J. E., Strobel, D. F., McConnell, J. C., Dalgarno, A., Goody, R., and McElroy, M. B. 1979. Extreme ultraviolet observations from Voyager 1 encounter with Jupiter. *Science* 204:979–982.

Broadfoot, A. L., Sandel, B. R., Shemansky, P. E., Holberg, J. B., Smith, G. R., Strobel, D. F., McConnell, J. C., Kumar, S., Hunten, D. M., Atreya, S. K., Donahue, T. M., Moos, H. W., Bertaux, J. L., Blamont, J. E., Pomphrey, R. B., Linick, S. 1981. Extreme ultraviolet observations from Voyager 1 encounter with Saturn. *Science* 212:206–211.

Brosch, N., Mendelson, H., and Ibbetson, P. A. 1986. The substantial atmosphere of Pluto. Submitted to *Icarus.*

Brosche, P., and Sündermann, J., eds. 1978. *Tidal Friction and the Earth's Rotation* (New York: Springer-Verlag).

Brosche, P., and Sündermann, J., eds. 1982. *Tidal Friction and the Earth's Rotation II* (New York: Springer-Verlag).

Brouwer, D., and Clemence, G. M. 1961. *Methods of Celestial Mechanics* (New York: Academic).

Brown, R. A. 1973. Optical line emission from Io. Paper presented at Copernicus Symp. IV, IAU Symp. 65, Torun, Poland, September.

Brown, R. A. 1974. Optical line emission from Io. In *Exploration of the Planetary System, IAU Symp. 65*, eds. A. Woszczyk and C. I. Waniszewska (Dordrecht: Reidel), pp. 527–531.

Brown, R. A. 1976. A model of Jupiter's sulfur nebula. *Astrophys. J.* 206:L179–L183.

Brown, R. A. 1978. Measurements of SII optical emission from the thermal plasma of Jupiter. *Astrophys. J.* 224:L97–L98.

Brown, R. A. 1981. The Jupiter hot plasma torus: Observed electron temperature and energy flows. *Astrophys. J.* 244:1072–1080.

Brown, R. A., and Ip, W.-H. 1981. Atomic clouds as a distributed source for the Io plasma torus. *Science* 213:1493–1495.

Brown, R. A., and Schneider, N. 1981. Sodium remote from Io. *Icarus* 48:519–535.

Brown, R. A., and Yung, Y. L. 1976. Io: Its atmosphere and optic emissions. In *Jupiter*, ed. T. Gehrels (Tucson: Univ. of Arizona Press), pp. 1102–1145.

Brown, R. A., Pilcher, C., and Strobel, D. 1983a. Spectrophotometric studies of the Io torus. In *Physics of the Jovian Magnetosphere*, ed. A. J. Dessler (Cambridge: Cambridge Univ. Press), pp. 197–225.

Brown, R. A., Shemansky, D. E., and Johnson, R. E. 1983b. A deficiency of OIII in the Io plasma torus. *Astrophys. J.* 264:309–323.

Brown, R. H. 1982. The Satellites of Uranus: Spectrophotometric and Radiometric Studies of Their Surface Properties and Diameters. Ph.D. Thesis, Univ. of Hawaii, Honolulu.

Brown, R. H. 1983. The Uranian satellites and Hyperion: New spectrophotometry and compositional implications. *Icarus* 56:414–425.

Brown, R. H. 1984. Physical properties of the Uranian satellites. In *Uranus and Neptune*, ed. J. T. Bergstralh, NASA CP-2330, pp. 437–462.

Brown, R. H., and Clark, R. N. 1984. Surface of Miranda: Identification of water ice. *Icarus* 58:288–292.

Brown, R. H., and Cruikshank, D. P. 1983. The Uranian satellites: Surface compositions and opposition brightness surges. *Icarus* 55:83–92.

Brown, R. H., and Cruikshank, D. P. 1985. The moons of Uranus, Neptune and Pluto. *Sci. Amer.* 253(1):38–48.

Brown, R. H., Morrison, D., Telesco, C. M., and Brunk, W. E. 1982a. Calibration of the radiometric asteroid scale using occultation diameters. *Icarus* 52:188–195.

Brown, R. H., Cruikshank, D. P., and Morrison, D. 1982b. Diameters and albedos of satellites of Uranus. *Nature* 300:423–425.

Brown, W. L., Lanzerotti, L. J., Poate, J. M., and Augustyniak, W. M. 1978. Sputtering of ice by MeV light ions. *Phys. Rev. Lett.* 40:1027–1030.

Brown, W. L., Augustyniak, W. M., Brody, E., Cooper, B., Lanzerotti, L. J., Ramirez, A., Evatt, R., and Johnson, R. E. 1980a. Energy dependence of the erosion of H_2O ice films by H and He ions. *Nucl. Instr. Meth.* 170:321–325.

Brown, W. L., Augustyniak, W. M., Lanzerotti, L. J., Johnson, R. E., and Evatt, R. 1980b. Linear and nonlinear processes in the erosion of H_2O ice by fast light ions. *Phys. Rev. Lett.* 45:1632.

Brown, W. L., Lanzerotti, L. J., and Johnson, R. E. 1982a. Fast ion bombardment of ices and its astrophysical implications. *Science* 218:525–531.

Brown, W. L., Augustyniak, W. M., Simmons, E., Marcantonio, K. J., Lanzerotti, L. J., Johnson, R. E., Boring, J. W., Reimann, C. T., Foti, G., and Pirronello, V. 1982b. Erosion and molecule formation in condensed gas films by electronic energy loss of fast ions. *Nucl. Inst. Meth.* 198:1–8.

Bryan, J. B., Burton, D. E., Cunningham, M. E., and Lettis, L. A. Jr. 1978. A two-dimensional computer simulation of hypervelocity impact cratering: Some preliminary results for Meteor Crater, Arizona. *Proc. Lunar Planet. Sci. Conf.* 9:3931–3964.

Buck, W. R. 1982. Geophysical constraints on lunar composition. *Workshop on Pristine Lunar Highlands Rocks and the Early History of the Moon* (Houston: Lunar and Planetary Inst.), pp. 22–25 (abstract).

Buck, W. R., and Toksöz, M. N. 1980. The bulk composition of the Moon based on geophysical constraints. *Proc. Lunar Planet. Sci. Conf.* 11:2043–2058.

Buhl, D. 1971. Lunar rocks and thermal anomalies. *J. Geophys. Res.* 76:3384–3390.

Buie, M. W. 1984. Lightcurve CCD Spectrophotometry of Pluto. Ph.D. Thesis, Univ. of Arizona.

Buie, M. W., and Fink, U. 1983. Lightcurve spectrophotometry of methane absorptions on Pluto. *Bull. Amer. Astron. Soc.* 15:860 (abstract).

Buie, M. W., and Fink, U. 1984. Methane frost on Pluto: Model implications from spectrophotometry. *Bull. Amer. Astron. Soc.* 16:651 (abstract).

Bullen, K. E. 1967. The origin of the Moon. In *Mantles of the Earth and Terrestrial Planets,* ed. S. K. Runcorn (New York: Wiley), pp. 261–264.

Buratti, B. J. 1983. Photometric Properties of Europa and the Icy Satellites of Saturn. Ph.D. Thesis, Cornell Univ.

Buratti, B. 1984. Voyager disk resolved photometry of the Saturnian satellites. *Icarus* 59:392–405.

Buratti, B. 1985. Application of a radiative transfer model to bright icy satellites. *Icarus* 61:208–217.

Buratti, B. J., and Golombeck, M. P. 1985. Europa: Geologic implications of spectrophotometry. In preparation for *Icarus.*

Buratti, B., and Veverka, J. 1983. Voyager photometry of Europa. *Icarus* 55:93–110.

Buratti, B., and Veverka, J. 1984. Voyager photometry of Rhea, Dione, Tethys, Enceladus and Mimas. *Icarus* 58:254–264.

Buratti, B. J., Veverka, J., and Thomas, P. 1982. Voyager photometry of Saturn's satellites. NASA TM-85127, pp. 41–43.

Burns, J. A. 1968. Jupiter's decametric radio emission and the radiation belts of its Galilean satellites. *Science* 159:971–972.

Burns, J. A. 1972. The dynamical characteristics of Phobos and Deimos. *Rev. Geophys. Space Phys.* 10:462–483.

Burns, J. A. 1973. Where are the satellites of the inner planets? *Nature* 242:23–25.

Burns, J. A. 1975. The angular momentum of solar system bodies: Implications for asteroid strengths. *Icarus* 25:545–554.

Burns, J. A. 1976a. The consequences of the tidal slowing of Mercury. *Icarus* 28:453–458.

Burns, J. A. 1976b. An elementary derivation of the perturbation equations of celestial mechanics. *Amer. J. Phys.* 44:944–949.

Burns, J. A., ed. 1977a. *Planetary Satellites* (Tucson: Univ. of Arizona Press).

Burns, J. A. 1977b. Orbital evolution. In *Planetary Satellites,* ed. J. A. Burns (Tucson: Univ. of Arizona Press), pp. 113–156.

Burns, J. A. 1978. On the orbital evolution and origin of the Martian moons. *Vistas in Astronomy* 22:193–210.

Burns, J. A. 1982. The dynamical evolution of the solar system. In *Formation of Planetary Systems,* ed. A. Brahic (Paris: Editions Cepadues), pp. 404–501.

Burns, J. A., and Cuzzi, J. N. 1985. Phantoms of Saturn's F ring. *Bull. Amer. Astron. Soc.* 17:922 (abstract).

Burns, J. A., and Showalter, M. 1983. Dust rings from satellites. *Bull. Amer. Astron. Soc.* 15:814–815 (abstract).

Burns, J. A., Lamy, P. L., and Soter, S. 1979. Radiation forces on small particles in the solar system. *Icarus* 40:1–48.

Burns, J. A., Showalter, M. R., Cuzzi, J. N., and Pollack, J. B. 1980. Physical processes in Jupiter's ring? Clues to its origin by Jove! *Icarus* 44:339–360.

Burns, J. A., Showalter, M. R., and Morfill, G. E. 1984. The ethereal rings of Jupiter and Saturn. In *Planetary Rings,* eds. R. Greenberg and A. Brahic (Tucson: Univ. of Arizona Press), pp. 200–272.

Butterworth, P. S., Caldwell, J., Moore, V., Owen, T., Rwolo, A. R., and Lane, A. L. 1980. An upper limit to the global SO_2 abundance on Io. *Nature* 235:308–309.

Byl, J., and Ovendon, M. W. 1975. On the satellite capture problem. *Mon. Not. Roy. Astron. Soc.* 173:579–584.

Cadogan, P. H. 1981. *The Moon: Our Sister Planet* (Cambridge: Cambridge Univ. Press).

Caffee, M., Hohenberg, C. M., and Hudson, B. 1981. Troclolite 76535: A study in the preservation of early isotopic records. *Proc. Lunar Planet. Sci. Conf.* 12:99–115.

Calcagno, L., Strazzulla, G., Fichera, M., and Foti, G. 1983a. Ion-induced polymerization in benzene frozen films. *Rad. Eff. Lett.* 76:143–148.

Calcagno, L., Strazzulla, G., and Foti, G. 1983b. Carbon build-up by fast protons on silicon substrate in CH_4 atmosphere. *Lett. Nuovo Cim.* 37:303–306.

Caldwell, J. 1975. Ultraviolet observations of small bodies in the solar system by OAO-2. *Icarus* 25:384–396.

Caldwell, J. 1977. Thermal radiation from Titan's atmosphere. In *Planetary Satellites*, ed. J. A. Burns (Tucson: Univ. of Arizona Press), pp. 438–450.

Caldwell, J., Owen, T., Rivolo, A. R., Moore, V., Hunt, G. E., and Butterworth, P. S. 1981. Observations of Uranus, Neptune, and Titan by the International Ultraviolet Explorer. *Astron. J.* 86:298–305.

Cameron, A. G. W. 1973*a*. Formation of the outer planets. *Space Sci. Rev.* 14:383–391.

Cameron, A. G. W. 1973*b*. Elemental and isotopic abundances of the volatile elements in the outer planets. *Space Sci. Rev.* 14:392–400.

Cameron, A. G. W. 1975. Cosmogonical considerations regarding Uranus. *Icarus* 24:280–284.

Cameron, A. G. W. 1978. Physics of the primitive solar nebula and of giant gaseous protoplanets. In *Protostars and Planets*, ed. T. Gehrels (Tucson: Univ. of Arizona Press), pp. 453–487.

Cameron, A. G. W. 1983*a*. Dissipation of thick accretion disks. *Astrophys. Space Sci.* 93: 295–303.

Cameron, A. G. W. 1983*b*. Origin of the atmospheres of the terrestrial planets. *Icarus* 56: 195–201.

Cameron, A. G. W. 1985. Formation of the prelunar accretion disk. *Icarus* 62:319–327.

Cameron, A. G. W., and Pollack, J. B. 1976. On the origin of the solar system and of Jupiter and its satellites. In *Jupiter*, ed. T. Gehrels (Tucson: Univ. of Arizona Press), pp. 61–84.

Cameron, A. G. W., and Ward, W. R. 1976. The origin of the moon. *Lunar Sci.* VII:120–122 (abstract).

Cameron, A. G. W., DeCampli, W. M., and Bodenheimer, P. 1982. Evolution of giant, gaseous protoplanets embedded in the primitive solar nebula. *Icarus* 49:298–312.

Campbell, J. K., and Anderson, J. D. 1985. Gravity field of the Saturnian system from Pioneer and Voyager tracking data. *Bull. Amer. Astron. Soc.* 17:697–698 (abstract).

Campbell, J. K., and Synnott, S. P. 1985. Gravity field of the Jovian system from Pioneer and Voyager tracking data. *Astron. J.* 90:364–372.

Campbell, M. J., and Ulrichs, J. 1969. Electrical properties of rocks and their significance for lunar radar observations. *J. Geophys. Res.* 74:5867–5881.

Canuto, V. M., Goldman, I., and Hubickyj, O. 1984. A formula for the Shakura-Sunyaev turbulent viscosity parameter. *Astrophys. J.* 280:L55–L58.

Cappalo, R. J., Counselman, C. C. III, King, R. W., and Shapiro, I. I. 1981. Tidal dissipation in the Moon. *J. Geophys. Res.* 86:7180–7184.

Carlson, R. W. 1980. Photosputtering of ice and hydrogen around Saturn's rings. *Nature* 283:461.

Carlson, R. W. 1982. Chronologic and isotopic systematics of lunar highland rocks. *Workshop on Pristine Lunar Highland Rocks and the Early History of the Moon* (Houston: Lunar and Planetary Inst.), pp. 31–35 (abstract).

Carlson, R. W., and Judge, D. L. 1974. Pioneer 10 photometric observations at Jupiter encounter. *J. Geophys. Res.* 79:3623–3633.

Carlson, R. W., Matson, D. L., Johnson, T. V., and Bergstralh, J. T. 1978. Sodium D-line emission from Io: Comparison of observed and theoretical line profiles. *Astrophys. J.* 223: 1082–1086.

Carr, M. H. 1974. The role of lava erosion in the formation of lunar rilles and Martian channels. *Icarus* 22:1–23.

Carr, M. H. 1986. Silicate volcanism on Io. *J. Geophys. Res.* 91:3521–3532.

Carr, M. H., Masursky, H., Strom, R. G., and Terrile, R. J. 1979. Volcanic features of Io. *Nature* 280:729–733.

Carr, T. D., Desch, M. D., and Alexander, J. K. 1983. Phenomenology of magnetospheric radio emissions. In *Physics of the Jovian Atmosphere*, ed. A. J. Dessler (New York: Cambridge Univ. Press), pp. 226–284.

Carrier, W. C. III, Mitchell, J. K., and Mahmood, A. 1973. The relative density of lunar soil. *Proc. Lunar Sci. Conf.* 4:2403–2411.

Carter, N. L. 1976. Steady state flow of rocks. *Rev. Geophys. Space Phys.* 14:301–360.

Carusi, A., and Valsecchi, G. B. 1979. Numerical simulations of close encounters between Jupiter and minor bodies. In *Asteroids*, ed. T. Gehrels (Tucson: Univ. of Arizona Press), pp. 391–416.

Casacchia, R., and Strom, R. G. 1984. Geologic evolution of Galileo Regio, Ganymede. *Proc. Lunar Planet. Sci. Conf. 14* in *J. Geophys. Res. Suppl.* 88:B419–B428.

Cashore, J., and Woronow, A. 1984. A new Monte Carlo model of lunar megaregolith development. *Proc. Lunar Planet. Sci. Conf. 15* in *J. Geophys. Res. Suppl.* 90:C811–C816.

Cassen, P., and Pettibone, D. 1976. Steady accretion of a rotating fluid. *Astrophys. J.* 208: 500–511.

Cassen, P., and Reynolds, R. T. 1973. The role of convection in the Moon. *J. Geophys. Res.* 78:3203–3215.

Cassen, P., and Reynolds, R. T. 1974. Convection in the Moon: Effect of variable viscosity. *J. Geophys. Res.* 79:2937–2944.

Cassen, P., and Summers, A. 1983. Models of the formation of the solar nebula. *Icarus* 53: 26–40.

Cassen, P., Reynolds, R. T., and Peale, S. J. 1979a. Is there liquid water on Europa? *Geophys. Res. Lett.* 6:731–734.

Cassen, P., Reynolds, R. T., Graziani, F., Summers, A., McNellis, J., and Blalock, L. 1979b. Convection and lunar thermal history. *Phys. Earth Planet. Int.* 19:183–196.

Cassen, P., Peale, S. J., and Reynolds, R. T. 1980a. On the comparative evolution of Ganymede and Callisto. *Icarus* 41:232–239.

Cassen, P., Peale, S. J., and Reynolds, R. T. 1980b. Tidal dissipation in Europa: A correction. *Geophys. Res. Lett.* 7:987–988.

Cassen, P., Smith, B., Miller, R., and Reynolds, R. 1981. Numerical experiments on the stability of preplanetary disks. *Icarus* 48:377–392.

Cassen, P. M., Peale, S. J., and Reynolds, R. T. 1982. Structure and thermal evolution of the Galilean satellites. In *Satellites of Jupiter,* ed. D. Morrison (Tucson: Univ. of Arizona Press), pp. 93–128.

Cayley, A. 1859. Tables of the developments of functions in the theory of elliptic motion. *Mem. Roy. Astron. Soc.* 29:191–306.

Cazenave, A. 1982. Tidal friction parameters from satellite observations. In *Tidal Friction and the Earth's Rotation II,* eds. P. Brosche and J. Sündermann (New York: Springer-Verlag), pp. 4–18.

Cazenave, A., Dobrovolskis, A., and Lago, B. 1980. Orbital history of the Martian satellites with inferences on their origin. *Icarus* 44:730–744.

Cellino, A., Pannunzio, R., Zappalà, V., Farinella, P., and Paolicchi, P. 1985. Do we observe light curves of binary asteroids? *Astron. Astrophys.* 144:355–362.

Chacko, S., and DeBremaecker, J. C. 1982. The evolution of the Moon: A finite element approach. *Moon and Planets* 27:467–492.

Chandrasekhar, S. 1960. *Radiative Transfer* (New York: Dover).

Chandrasekhar, S. 1961. *Hydrodynamic and Hydromagnetic Stability* (Oxford: Clarendon Press).

Chapman, C. R. 1974. Cratering on Mars. I. Cratering and obliteration history. *Icarus* 22: 272–291.

Chapman, C. R., and Davis, D. R. 1975. Asteroidal collisional evolution: Evidence for a much larger earlier population. *Science* 190:553–556.

Chapman, C. R., and Jones, K. L. 1977. Cratering and obliteration history of Mars. *Ann. Rev. Earth Planet. Sci.* 5:515–540.

Chenette, D. L., and Stone, E. C. 1983. The Mimas ghost revisited: An analysis of the electron flux and electron microsignatures observed in the vicinity of Mimas and Saturn. *J. Geophys. Res.* 88:8755–8764.

Cheng, A. F. 1980. Effects of Io's volcanoes on the plasma torus and Jupiter's magnetosphere. *Astrophys. J.* 242:812–827.

Cheng, A. F. 1982. SO_2 ionization and dissociation in the Io plasma torus. *J. Geophys. Res.* 87:5301–5304.

Cheng, A. F. 1984a. Escape of sulfur and oxygen from Io. *J. Geophys. Res.* 89:3939–3944.

Cheng, A. F. 1984b. Magnetosphere, rings, and moons of Uranus. In *Uranus and Neptune,* ed. J. T. Bergstralh, NASA CP-2330, pp. 541–556.

Cheng, A. F., and Lanzerotti, L. J. 1978. Ice sputtering by radiation belt protons and the rings of Saturn and Uranus. *J. Geophys. Res.* 83:2597–2602.

Cheng, A. F., Lanzerotti, L. J., and Pirronello, V. 1982. Charged particle sputtering of ice surfaces in Saturn's magnetosphere. *J. Geophys. Res.* 87:4567–4570.

Cheng, A. F., Maclennan, C. G., Lanzerotti, L. J., Paonessa, M., and Armstrong, T. 1983. Energetic ion losses near Io's orbit. *J. Geophys. Res.* 88:3936–3944.

Chirikov, B. V. 1979. A universal instability of many-dimensioned oscillator systems. *Phys. Rep.* 52:263–379.

Christensen, U. 1984. Convection with pressure and temperature dependent non-Newtonian rheology. *Geophys. J. Roy. Astron. Soc.* 77:343–384.

Christensen, U. R. 1985. Thermal evolution models for the earth. *J. Geophys. Res.* 90:2995–3007.

Christy, J. W., and Harrington, R. S. 1978. The satellite of Pluto. *Astron. J.* 83:1005–1008.

Christy, J. W., and Harrington, R. S. 1980. The discovery and orbit of Charon. *Icarus* 44:38–40.

Church, S., Head, J. W., and Solomon, S. C. 1982. Multi-ringed basin interiors: Structure and early evolution of Orientale. *Lunar Planet. Sci* XIII:98 (abstract).

Cintala, M. J., Wood, C. A., and Head, J. W. 1977. The effect of target characteristics on fresh crater morphology: Preliminary results for the moon and Mercury. *Proc. Lunar Sci. Conf.* 8:3409–3425.

Cintala, M. J., Head, J. W., and Veverka, J. 1978. Characteristics of the cratering process on small satellites and asteroids. *Proc. Lunar Planet. Sci. Conf.* 9:3803–3930.

Cisowski, S. M., Hale, C., and Fuller, M. 1977. On the intensity of ancient lunar fields. *Proc. Lunar Sci. Conf.* 8:725–750.

Cisowski, S. M., Collinson, D. W., Runcorn, S. K., and Stephenson, A. 1983. A review of lunar paleointensity data and implications for the origin of lunar magnetism. *Proc. Lunar Planet. Sci. Conf. 13* in *J. Geophys. Res. Suppl.* 88:A691–A704.

Clancy, R. T., and Danielson, G. E. 1981. High resolution albedo measurements of Io from Voyager 1. *J. Geophys. Res.* 86:8627–8634.

Clark, R. N. 1980. Ganymede, Europa, Callisto, and Saturn's rings: Compositional analysis from reflectance spectroscopy. *Icarus* 44:388–409.

Clark, R. N. 1981a. Water frost and ice: The near-infrared spectral reflectance 0.65–2.5 μm. *J. Geophys. Res.* 86:3087–3096.

Clark, R. N. 1981b. The spectral reflectance of water-mineral mixtures at low temperatures. *J. Geophys. Res.* 86:3074–3086.

Clark, R. N. 1982. Implications of using broadband photometry for compositional remote sensing of icy objects. *Icarus* 49:244–257.

Clark, R. N., and Lucey, P. G. 1984. Spectral properties of ice-particulate mixtures and implications for remote sensing. 1. Intimate mixtures. *J. Geophys. Res.* 89:6341–6348.

Clark, R. N., and McCord, T. B. 1980a. The Galilean satellites: New near-infrared reflectance measurements (0.65–2.5 μm) and a 0.325–5 μm summary. *Icarus* 41:323–339.

Clark, R. N., and McCord, T. B. 1980b. The rings of Saturn: New infrared reflectance measurements and a 0.326–4.08 μm summary. *Icarus* 43:161–168.

Clark, R. N., and Owensby, P. D. 1981. The infrared spectrum of Rhea. *Icarus* 46:354–360.

Clark, R. N., and Roush, T. L. 1984. Reflectance spectroscopy: Quantitative analysis techniques for remote sensing applications. *J. Geophys. Res.* 89:6329–6340.

Clark, R. N., Singer, R. B., Owensby, P. D., and Fanale, F. P. 1980. Galilean satellites: High precision near-infrared spectrophotometry (0.65–2.5μm) of the leading and trailing sides. *Bull. Amer. Astron. Soc.* 12:713–714 (abstract).

Clark, R. N., Brown, R. H., Nelson, M. L., and Hayashi, J. 1983a. Surface composition of Enceladus. *Bull. Amer. Astron. Soc.* 15:853 (abstract).

Clark, R. N., Fanale, F. P., and Zent, A. P. 1983b. Frost grain metamorphism: Implications for remote sensing of planetary surfaces. *Icarus* 56:233–245.

Clark, R. N., Brown, R. H., Owensby, P. D., and Steele, A. 1984. Saturn's satellites: Near-infrared spectrophotometry (0.65–2.5 μm) of the leading and trailing sides and compositional implications. *Icarus* 58:265–281.

Clark, S. P. Jr. 1966. Viscosity. *Geol. Soc. Amer. Memoir* 96:291–300.

Clarke, J. T. 1982. Detection of auroral H Ly-α emission from Uranus. *Astrophys. J.* 263:L105–L109.

Clarke, J. T., Moos, H. W., Atreya, S., and Lane, A. 1981. IUE detection of bursts of H L-α emission from Saturn. *Nature* 290:226–227.

Clayton, R. N., and Mayeda, T. K. 1975. Genetic relations between the Moon and meteorites. *Proc. Lunar Sci. Conf.* 6:1761–1769.

Clayton, R. N., and Mayeda, T. K. 1983. Oxygen isotopes in encrites, shergottites, nakhlites, and chassignites. *Earth Planet. Sci. Lett.* 62:1–6.

Clayton, R. N., Onuma, N., and Mayeda, T. K. 1976. A classification of meteorites based on oxygen isotopes. *Earth Planet. Sci. Lett.* 30:10–18.

Cloutier, P. A., Daniell, R. E. Jr., Dessler, A. J., and Hill, T. W. 1978. A cometary ionosphere model for Io. *Astrophys. Space Sci.* 55:93–112.

Clow, G. D., and Carr, M. H. 1980. Stability of sulfur slopes on Io. *Icarus* 44:268–279.

Colburn, D. S. 1980. Electromagnetic heating of Io. *J. Geophys. Res.* 85:7257–7261.

Colburn, D. S., and Reynolds, R. T. 1985. Electrolytic currents in Europa. *Icarus* 63:39–44.

Collins, S. A. 1981. Spatial color variations in the volcanic plume at Loki on Io. *J. Geophys. Res.* 86:8621–8626.

Collins, S. A., Cook, A. F. II, Cuzzi, J. N., Danielson, G. E., Hunt, G. E., Johnson, T. V., Morrison, D., Owen, T., Pollack, J. B., Smith, B. A., and Terrile, R. J. 1980. First Voyager view of the rings of Saturn. *Nature* 228:439–443.

Colombo, G., and Franklin, F. A. 1971. On the formation of the outer satellite groups of Jupiter. *Icarus* 15:186–191.

Colombo, G., and Shapiro, I. I. 1966. The rotation of the planet Mercury. *Astrophys. J.* 145:296–305.

Conca, J. 1981. Dark-ray craters on Ganymede. *Proc. Lunar Planet. Sci. Conf.* 12:1599–1606.

Conel, J. E. 1969. Infrared emissivities of silicates: Experimental results and a cloudy atmosphere model of spectral emission from condensed particulate mediums. *J. Geophys. Res.* 74:1614–1634.

Consolmagno, G. J. 1979. Sulfur volcanoes on Io. *Science* 205:397–398.

Consolmagno, G. J. 1981a. An Io thermal model with intermittent volcanism. *Proc. Lunar Planet. Sci. Conf.* 12:175–177.

Consolmagno, G. J. 1981b. Io: Thermal models and chemical evolution. *Icarus* 47:36–45.

Consolmagno, G. J. 1985. Resurfacing Saturn's satellites: Models of partial differentiation and expansion. *Icarus* 64:401–413.

Consolmagno, G. J., and Lewis, J. S. 1976. Structural and thermal models of icy Galilean satellites. In *Jupiter,* ed. T. Gehrels (Tucson: Univ. of Arizona Press), pp. 1035–1051.

Consolmagno, G. J., and Lewis, J. S. 1977. Preliminary thermal history models of icy satellites. In *Planetary Satellites,* ed. J. A. Burns (Tucson: Univ. of Arizona Press), pp. 492–500.

Consolmagno, G. J., and Lewis, J. S. 1978. The evolution of icy satellite interiors and surfaces. *Icarus* 34:280–293.

Consolmagno, G. J., and Lewis, J. S. 1980. The chemical thermal evolution of Io. *The Satellites of Jupiter, IAU Coll. 57,* Univ. of Hawaii, abstract booklet, 7–11.

Consolmagno, G. J., and Reiche, H. A. T. O. 1981. Pronouncing the names of the moons of Saturn (or pulling teeth from Tethys). *EOS Trans. AGU* 63:145–147.

Conway, B. A. 1982. On the history of the lunar orbit. *Icarus* 51:610–622.

Conway, B. A. 1986. Stability and evolution of primeval lunar satellite orbits. *Icarus* 66. In press.

Cook, A. F., and Franklin, F. A. 1970. An explanation of the light curve of Iapetus. *Icarus* 13:282–291.

Cook, A. F., Shoemaker, E. M., Soderblom, L. A., Mullins, K. F., and Fiedler, R. 1982. Volcanism in ice on Europa. *Bull. Amer. Astron. Soc.* 14:736–737 (abstract).

Cook, J. G., and Leaist, D. G. 1983. An exploratory study of the thermal conductivity of methane hydrate. *Geophys. Res. Lett.* 10:397–399.

Cooper, B. H., and Tombrello, T. A. 1984. Enhanced erosion of frozen H_2O films by high energy ^{19}F ions. *Rad. Eff.* 80:203–221.

Cooper, H. F. 1977. A summary of explosion cratering phenomena relevant to meteor impact events. In *Impact and Explosion Cratering,* eds. D. J. Roddy, R. O. Pepin, and R. B. Merrill (New York: Pergamon), pp. 11–44.

Cooper, M. R., Konrath, R. L., and Watkins, J. S. 1974. Lunar near-surface structure. *Rev. Geophys. Space Phys.* 12:291–308.

Coradini, A., Federico, C., and Magni, G. 1981. Gravitational instabilities in satellite disks and formation of regular satellites. *Astron. Astrophys.* 99:255–261.

Coradini, A., Federico, C., and Lanciano, P. 1982. Ganymede and Callisto: Accumulation heat content. In *The Comparative Study of the Planets*, eds. A. Coradini and M. Fulchignoni (Dordrecht: Reidel), pp. 61–70.

Counselman, C. C. 1973. Outcomes of tidal evolution. *Astrophys. J.* 180:307–314.

Courtin, R., Gautier, D., Marten, A., and Bezard, B. 1983. The composition of Saturn's atmosphere at northern latitudes from Voyager IRIS spectra. *Bull. Amer. Astron. Soc.* 15:831 (abstract).

Crater Analysis Techniques Work Group. 1979. Standard techniques for presentation and analysis of crater size-frequency data. *Icarus* 37:467–474.

Croft, S. K. 1980. Cratering flow fields: Implications for the excavation and transient expansion stages of crater formation. *Proc. Lunar Planet. Sci. Conf.* 11:2347–2378.

Croft, S. K. 1981*a*. The excavation stage of basin formation: A qualitative model. In *Multi-Ring Basins: Formation and Evolution, Proc. Lunar Planet. Sci. 12A* (New York: Pergamon), pp. 207–225.

Croft, S. K. 1981*b*. Modification stage of basin formation. In *Multi-Ring Basins: Formation and Evolution, Proc. Lunar Planet. Sci. 12A* (New York: Pergamon), pp. 227–257.

Croft, S. K. 1981*c*. Impacts in ice and snow: Implications for crater scaling on icy satellites. *Lunar Planet. Sci.* XII:135–136 (abstract).

Croft, S. K. 1981*d*. On the origin of pit craters. *Lunar Planet. Sci.* XII:196–198 (abstract).

Croft, S. K. 1983. A proposed origin for palimpsests and anomalous pit craters on Ganymede and Callisto. *Proc. Lunar Planet. Sci. Conf. 14* in *J. Geophys. Res. Suppl.* 88:B71–B89.

Croft, S. K. 1984. Effects of ice creep relations on thermal histories and viscous relaxation for icy satellites. Unpublished.

Croft, S. K. 1986. Cratering on Ganymede. Submitted to *Icarus*.

Cruikshank, D. P. 1977. Radii and albedos of four Trojan asteroids and Jovian satellites 6 and 7. *Icarus* 30:224–230.

Cruikshank, D. P. 1979. The surfaces and interiors of Saturn's satellites. *Rev. Geophys. Space Phys.* 17:165–176.

Cruikshank, D. P. 1980*a*. The infrared spectrum of Io, 2.8–5.2 μm. *Icarus* 41:240–245.

Cruikshank, D. P. 1980*b*. Near-infrared studies of the satellites of Saturn and Uranus. *Icarus* 41:246–258.

Cruikshank, D. P. 1982. The satellites of Uranus. In *Uranus and the Outer Planets*, ed. G. E. Hunt (Cambridge: Cambridge Univ. Press), pp. 193–216.

Cruikshank, D. P. 1984. Physical properties of the satellites of Neptune. In *Uranus and Neptune*, ed. J. T. Bergstralh, NASA CP-2330, pp. 425–436.

Cruikshank, D. P., and Apt, J. 1984. Methane on Triton: Physical state and distribution. *Icarus* 58:306–311.

Cruikshank, D. P., and Brown, R. H. 1981. The Uranian satellites: Water ice on Ariel and Umbriel. *Icarus* 45:607–611.

Cruikshank, D. P., and Brown, R. H. 1982. Surface composition and radius of Hyperion. *Icarus* 50:82–87.

Cruikshank, D. P., and Silvaggio, P. 1979. Triton: A satellite with an atmosphere. *Astrophys. J.* 233:1016–1020.

Cruikshank, D. P., and Silvaggio, P. 1980. The surface and atmosphere of Pluto. *Icarus* 41:96–102.

Cruikshank, D. P., Pilcher, C. B., and Morrison, D. 1976. Pluto: Evidence for methane frost. *Science* 194:835–837.

Cruikshank, D. P., Jones, T. J., and Pilcher, C. B. 1977. Absorption bands in the spectrum of Io. *Bull. Amer. Astron. Soc.* 9:465 (abstract).

Cruikshank, D. P., Jones, T. J., and Pilcher, C. B. 1978. Absorption bands in the spectrum of Io. *Astrophys. J.* 225:L89–L92.

Cruikshank, D. P., Stockton, A., Dyck, H. M., Becklin, E. E., and Macy, W. 1979. The diameter and reflectance of Triton. *Icarus* 40:104–114.

Cruikshank, D. P., Degewij, J., and Zellner, B. H. 1982. The outer satellites of Jupiter. In *Satellites of Jupiter*, ed. D. Morrison (Tucson: Univ. of Arizona Press), pp. 129–146.

Cruikshank, D. P., Bell, J. F., Gaffey, M. J., Brown, R. H., Howell, R., Beerman, C., and Rognstad, M. 1983*a*. The dark side of Iapetus. *Icarus* 53:90–104.

Cruikshank, D. P., Brown, R. H., and Clark, R. N. 1983*b*. Nitrogen on Triton. *Bull. Amer. Astron. Soc.* 15:857 (abstract).

Cruikshank, D. P., Brown, R. H., and Clark, R. N. 1984a. Nitrogen on Triton. *Icarus* 58: 293–305.
Cruikshank, D. P., Veverka, J., and Lebofsky, L. A. 1984b. Satellites of Saturn: The optical properties. In *Saturn*, eds. T. Gehrels and M. S. Matthews (Tucson: Univ. of Arizona Press), pp. 640–667.
Curran, D. R., Shockey, D. A., Seaman, L., and Austin, M. 1977. Mechanisms and models of cratering in earth media. In *Impact and Explosion Cratering*, eds. D. J. Roddy, R. O. Pepin, and R. B. Merrill (New York: Pergamon), pp. 1057–1087.
Cuzzi, J. N., and Scargle, J. D. 1985. Wavy edges suggest moonlet in Encke's group. *Astrophys. J.* 292:276–290.
Cuzzi, J. N., Pollack, J. B., and Summers, A. L. 1980. Saturn's rings: Particle composition and size distribution as constrained by observations at microwave wavelengths. *Icarus* 44: 683–705.
Cuzzi, J. N., Lissauer, J. J., Esposito, L. W., Holberg, J. B., Marouf, E. A., Tyler, G. L., and Boischot, A. 1984. Saturn's rings: Properties and processes. In *Planetary Rings*, eds. R. Greenberg and A. Brahic (Tucson: Univ. of Arizona Press), pp. 73–199.
Dainty, A. M., Toksöz, M. N., and Stein, S. 1976. Seismic investigation of the lunar interior. *Proc. Lunar Sci. Conf.* 7:3057–3076.
Danby, J. M. A. 1962. *Fundamentals of Celestial Mechanics* (New York: Macmillan).
Danielson, G. E., Kupferman, P. N., Johnson, T. V., and Soderblom, L. A. 1981. Radiometric performance of the Voyager cameras. *J. Geophys. Res.* 86:8683–8689.
Darwin, G. H. 1880. On the secular change in elements of the orbit of a satellite revolving around a tidally distorted planet. *Phil. Trans. Roy. Soc. London* 171:713–891.
Darwin, G. H. 1908. *Tidal Friction and Cosmogony*, vol. 2, *Scientific Papers* (London: Cambridge Univ. Press).
Davies, M. E. 1982. Cartography and nomenclature for the Galilean satellites. In *Satellites of Jupiter*, ed. D. Morrison (Tucson: Univ. of Arizona Press), pp. 911–914.
Davies, M. E. 1983. The shape of Io. *Natural Satellites, IAU Coll. 77*, abstract booklet, Cornell Univ., p. 14.
Davies, M. E., and Katayama, F. Y. 1981. Coordinates of features on the Galilean satellites. *J. Geophys. Res.* 86:8635–8657.
Davies, M. E., and Katayama, F. Y. 1983a. The control networks of Mimas and Enceladus. *Icarus* 53:332–340.
Davies, M. E., and Katayama, F. Y. 1983b. The control networks of Tethys and Dione. *J. Geophys. Res.* 88:8729–8735.
Davies, M. E., and Katayama, F. Y. 1983c. The control network of Rhea. *Icarus* 56:603–610.
Davies, M. E., and Katayama, F. Y. 1984. The control network of Iapetus. *Icarus* 59:199–204.
Davies, M. E., Abalakin, V. K., Lieske, J. H., Seidelmann, P. K., Sinclair, A. T., Sinzi, A. M., Smith, B. A., and Tjuflin, Y. S. 1983. Report of the IAU working group on cartographic coordinates and rotational elements of the planets and satellites: 1982. *Celestial Mech.* 29: 309–321.
Davis, D. R., Chapman, C. R., Greenberg, R., Weidenschilling, S. J., and Harris, A. W. 1979. Collisional evolution of asteroids: Populations, rotations, and velocities. In *Asteroids*, ed. T. Gehrels (Tucson: Univ. of Arizona Press), pp. 528–557.
Davis, D. R., Housen, K. R., and Greenberg, R. 1981. The unusual dynamical environment of Phobos and Deimos. *Icarus* 47:220–233.
Davis, D. R., Weidenschilling, S. J., Chapman, C. R., and Greenberg, R. 1984. Saturn ring particles as dynamic ephemeral bodies. *Science* 224:744–746.
Davis, D. R., Chapman, C. R., Weidenschilling, S. J., and Greenberg, R. J. 1985. Collisional history of asteroids: Evidence from Vesta and the Hirayama families. *Icarus* 62:30–53.
Degewij, J., and van Houten, C. J. 1979. Distant asteroids and outer Jovian satellites. In *Asteroids*, eds. T. Gehrels (Tucson: Univ. of Arizona Press), pp. 417–435.
Degewij, J., Andersson, L. E., and Zellner, B. 1980a. Photometric properties of outer planetary satellites. *Icarus* 44:520–540.
Degewij, J., Cruikshank, D. P., and Hartmann, W. K. 1980b. Near-infrared colorimetry of J6 Himalia and S9 Phoebe: A summary of 0.3–2.2 μm reflectances. *Icarus* 44:541–547.
Delano, J. W. 1980. Chemistry and liquidus phase relations of Apollo 15 red glass: Implications for the deep lunar interior. *Proc. Lunar Planet. Sci. Conf.* 11:251–288.

Delitsky, M. L. 1983. Chemistry of Triton's ocean. *Bull. Amer. Astron. Soc.* 15:857 (abstract).

Delton, R. A. 1974. Thickness of mare material in the Tranquillitatis and Nectaris basins. *Proc. Lunar Sci. Conf.* 5:53–59.

Dence, M. R. 1968. Shock zoning at Canadian craters: Petrography and structural implications. In *Shock Metamorphism of Natural Materials,* eds. B. M. French and N. M. Short (Baltimore: Mono Book Corp.), pp. 168–184.

Dence, M. R., Grieve, R. A. F., and Robertson, P. B. 1977. Terrestrial impact structures: Principal characteristics and energy considerations. In *Impact and Explosion Cratering,* eds. D. J. Roddy, R. O. Pepin, and R. B. Merrill (New York: Pergamon), pp. 247–275.

de Pater, I., Jaffe, W. J., Brown, R. A., and Berge, G. L. 1982. Radio emission from Io. *Astrophys. J.* 261:396–401.

de Pater, I., Brown, R. A., and Dickel, J. R. 1984. VLA observations of the Galilean satellites. *Icarus* 57:93–101.

Dermott, S. F. 1978. Pluto, Herculina, Mercury and Venus: Their real and imaginary satellites. Unpublished.

Dermott, S. F. 1979*a*. Tidal dissipation in the solid cores of the major planets. *Icarus* 37:310–321.

Dermott, S. F. 1979*b*. Shapes and gravitational moments of satellites and asteroids. *Icarus* 37:575–586.

Dermott, S. F. 1984*a*. Dynamics of narrow rings. In *Planetary Rings,* eds. R. Greenberg and A. Brahic (Tucson: Univ. of Arizona Press), pp. 589–637.

Dermott, S. F. 1984*b*. Origin and evolution of the Uranian and Neptunian satellites: Some dynamical considerations. In *Uranus and Neptune,* ed. J. T. Bergstralh, NASA CP-2330, pp. 377–404.

Dermott, S. F. 1984*c*. Rotation and the internal structures of the major planets and their inner satellites. *Phil. Trans. Roy. Soc. London* A313:123–138.

Dermott, S. F., and Murray, C. D. 1981*a*. The dynamics of tadpole and horseshoe orbits. I. Theory. *Icarus* 48:1–11.

Dermott, S. F., and Murray, C. D. 1981*b*. The dynamics of tadpole and horseshoe orbits. II. The coorbital satellites of Saturn. *Icarus* 48:12–22.

Dermott, S. F., and Nicholson, P. D. 1986. The masses of the Uranian satellites. *Nature* 319:115–120.

Dermott, S. F., Gold, T., and Sinclair, A. T. 1979. The rings of Uranus: Nature and origin. *Astron. J.* 84:1225–1234.

Dermott, S. F., Murray, C. D., and Sinclair, A. T. 1980. The narrow rings of Jupiter, Saturn and Uranus. *Nature* 284:309–313.

Dermott, S. F., Murray, C. D., Fox, K., and Williams, I. P. 1986. Drag forces in the three-body problem and the formation of coorbital satellites. In preparation.

Dessler, A. J. 1980. Mass injection rate from Io into the Io plasma torus. *Icarus* 44:291–295.

Dessler, A. J., ed. 1983. *Physics of the Jovian Magnetosphere* (Cambridge: Cambridge Univ. Press).

Dickey, J. O., Newhall, X. X., Williams, J. G., and Yoder, C. F. 1983*a*. Modulation of the lunar tidal acceleration. *EOS Trans. AGU* 64:204 (abstract).

Dickey, J. O., Williams, J. G., Newhall, X. X., and Yoder, C. F. 1983*b*. Geophysical applications of lunar laser ranging. *Proc. IAG Symp. IUGG 18 General Assembly Hamburg FRG,* August, vol. 2 (Columbus, OH: Dept. of Geodetic Science and Surveying, Ohio State Univ.), pp. 509–521.

Divine, N., and Garrett, H. B. 1983. Charged particle distribution in Jupiter's magnetosphere. *J. Geophys. Res.* 88:6889–6903.

Dobrovolskis, A. 1982. Internal stresses in Phobos and other triaxial bodies. *Icarus* 52:136–148.

Dobrovolskis, A. R., and Burns, J. A. 1980. Life near the Roche limit: Behavior of ejecta from satellites close to planets. *Icarus* 42:422–441.

Dobrovolskis, A. R., and Burns, J. A. 1984. Angular momentum drain: A mechanism for despinning asteroids. *Icarus* 57:464–476.

Dodd, R. T. 1981. *Meteorites: A Petrologic-Chemical Synthesis* (Cambridge: Cambridge Univ. Press).

Dollfus, A. 1961. Polarization studies of planets. In *Planets and Satellites,* ed. G. P. Kuiper (Chicago: Univ. of Chicago Press), pp. 343–399.

Dollfus, A. 1971. Physical studies of asteroids by polarization of the light. In *Physical Studies of Minor Planets*, ed. T. Gehrels, NASA SP-267, pp. 95–116.

Dollfus, A. 1975. Optical polarimetry of the Galilean satellites of Jupiter. *Icarus* 25:416–431.

Dollfus, A., and Geake, J. E. 1975. Polarimetric properties of the lunar surface and its interpretation. Part 7. Other solar system bodies. *Proc. Lunar Sci. Conf.* 6:2749–2768.

Dollfus, A., and Murray, J. 1974. La rotation, la cartographie et la photometrie des satellites de Jupiter. In *Exploration of the Planetary System, IAU Symp. 65*, eds. A. Woszczyk and C. Iwaniszewska (Dordrecht: Reidel), pp. 513–525.

Dollfus, A., and Zellner, B. 1979. Optical polarimetry of asteroids and laboratory samples. In *Asteroids*, ed. T. Gehrels (Tucson: Univ. of Arizona Press), pp. 170–183.

Donnison, J. R. 1978. The escape of natural satellites from Mercury and Venus. *Astrophys. Space Sci.* 59:499–501.

Dorman, H. J., Evans, S., Nakamura, Y., and Latham, G. 1978. On the time-varying properties of the lunar seismic meteoroid population. *Proc. Lunar Planet. Sci. Conf.* 9:3615–3626.

Dormand, J. R., and Woolfson, M. M. 1980. The origin of Pluto. *Mon. Not. Roy. Astron. Soc.* 193:171–174.

Drake, M. J. 1983. Geochemical constraints on the origin of the Moon. *Geochim. Cosmochim. Acta* 47:1759–1767.

Drobyshevski, E. M. 1979. Magnetic field of Jupiter and the volcanism and rotation of the Galilean satellites. *Nature* 282:811–813.

Duncombe, R. L., and Seidelmann, P. K. 1980. A history of the determination of Pluto's mass. *Icarus* 44:12–18.

Duncombe, R. L., Klepczynski, W. J., and Seidelmann, P. K. 1974. The masses of the planters, satellites, and asteroids. *Fund. Cos. Phys.* 1:119–165.

Durham, W. B., Heard, H. C., and Kirby, S. H. 1983. Experimental deformation of crystalline (H_2O) ice at high pressure and low temperature: Preliminary results. *Proc. Lunar Planet. Sci. Conf. 14* in *J. Geophys. Res. Suppl.* 89:B377–B392.

Durham, W. B., Kirby, S. H., and Heard, H. C. 1985. Rheology of the high pressure H_2O ices II, III, and V. *Lunar Planet. Sci.* XVI:198–199 (abstract).

Durisen, R. H., and Scott, E. H. 1984. Implications of recent numerical calculations for the fission theory of the origin of the Moon. *Icarus* 58:153–158.

Durrance, S. T., and Moos, H. W. 1982. Intense Ly-α emission from Uranus. *Nature* 299:428–429.

Durrance, S. T., Feldman, P., and Weaver, H. 1982. Rocket detection of ultraviolet emission from neutral oxygen and sulfur in the Io torus. *Astrophys. J.* 267:L125–L129.

Duxbury, T. C., and Callahan, J. 1982. Phobos and Deimos cartography. *Lunar Planet. Sci.* XIII:190 (abstract).

Duxbury, T. C., Ocampo, A. C., and Doyle, R. J. 1982. Martian satellite search frm Viking 2. *Lunar Planet. Sci.* XIII:192 (abstract).

Dyal, P., Parkin, C. W., and Daily, W. D. 1974. Magnetism of the interior of the Moon. *Rev. Geophys. Space Phys.* 12:568–591.

Dyal, P., Parkin, C. W., and Daily, W. D. 1976. Structure of the lunar interior from magnetic field measurements. *Proc. Lunar Sci. Conf.* 7:3077–3095.

Eischens, R. P., and Pliski, W. A. 1958. Infrared spectra of adsorbed molecules. In *Advances in Catalysis* (New York: Academic), pp. 1–56.

Eisenberg, D., and Kauzmann, W. 1969. *The Structure and Properties of Water* (Oxford: Oxford Univ. Press).

Elliot, J. L., and Nicholson, P. D. 1984. The rings of Uranus. In *Planetary Rings*, eds. R. J. Greenberg and A. Brahic (Tucson: Univ. of Arizona Press), pp. 25–72.

Ellsworth, K., and Schubert, G. 1983. Saturn's icy satellites: Thermal and structural models. *Icarus* 54:490–510.

Epstein, E. E., Janssen, M. A., and Cuzzi, J. N. 1984. Saturn's rings: 3-mm low inclination observations and derived properties. *Icarus* 58:403–411.

Eshelman, V. R. 1986. Mode decoupling during retrorefraction as an explanation for bizarre radar echoes from icy moons. *Nature* 319:755–757.

Eshelman, V. R., Tyler, G. L., Wood, G. E., Lindal, G. F., Anderson, J. D., Levy, G. S., and Croft, T. A. 1979. Radio science with Voyager at Jupiter: Initial Voyager 2 results and a Voyager 1 measure of the Io torus. *Science* 206:959–962.

Eshelman, V. R., Lindal, G. F., and Tyler, G. L. 1983. Is Titan wet or dry? *Science* 221:53–55.

Eviatar, A., and Mekler, Y. 1984. Aperiodic ion temperature variations in the Io plasma torus. *J. Geophys. Res.* 89:1496–1500.

Eviatar, A., and Podolak, M. 1983. Titan's gas and plasma torus. *J. Geophys. Res.* 88:833–840.

Eviatar, A., and Siscoe, G. 1980. Limit on rotational energy available to excite Jovian aurora. *Geophys. Res. Lett.* 7:1085–1088.

Eviatar, A., Siscoe, G. L., Johnson, T. V., and Matson, D. L. 1981. Effects of Io ejecta on Europa. *Icarus* 47:75–83.

Eviatar, A., Bar-Nun, A., and Podolak, M. 1985. Europan surface phenomena. *Icarus* 61: 185–191.

Fallon, F. W., and Murphy, R. E. 1971. Absence of post-eclipse brightening of Io and Europa in 1971. *Icarus* 15:492–496.

Fanale, F. P., Johnson, T. V., and Matson, D. L. 1974. Io: A surface evaporite deposit? *Science* 186:922–924.

Fanale, F. P., Johnson, T. V., and Matson, D. L. 1977a. Io's surface and the histories of the Galilean satellites. In *Planetary Satellites*, ed. J. A. Burns, (Tucson: Univ of Arizona Press), pp. 379–405.

Fanale, F. P., Johnson, T. V., and Matson, D. L. 1977b. Io's surface composition: Observational constraints and theoretical considerations. *Geophys. Res. Lett.* 4:303–306.

Fanale, F. P., Brown, R. H., Cruikshank, D. P., and Clark, R. N. 1979. Significance of absorption features in Io's IR reflectance spectrum. *Nature* 280:760–763.

Fanale, F. P., Banerdt, W. B., Elson, L. S., Johnson, T. V., and Zurek, R. W. 1982. Io's surface: Its phase composition and influence on Io's atmosphere and Jupiter's magnetosphere. In *Satellites of Jupiter*, ed. D. Morrison (Tucson: Univ. of Arizona Press), pp. 756–781.

Farinella, P., Milani, A., Nobili, A. M., and Valsecchi, G. B. 1979. Tidal evolution of the Pluto-Charon system. *Moon and Planets* 20:415–421.

Farinella, P., Milani, A., Nobili, A. M., and Valsecchi, G. B. 1980. Some remarks on the capture of Triton and the origin of Pluto. *Icarus* 44:810–812.

Farinella, P., Paolicchi, P., Tedesco, E. F., and Zappalà, V. 1981. Triaxial equilibrium ellipsoids among the asteroids? *Icarus* 46:114–123.

Farinella, P., Paolicchi, P., and Zappalà, V. 1982. The asteroids as outcomes of catastrophic collisions. *Icarus* 52:409–433.

Farinella, P., Milani, A., Nobili, A. M., Paolicchi, P., and Zappalà, V. 1983. Hyperion: Collisional disruption of a resonant satellite. *Icarus* 54:353–360.

Farinella, P., Paolicchi, P., Strom, R., and Zappalà, V. 1986. The fate of Hyperion's fragments. Submitted to *Icarus*.

Federico, C., and Lanciano, P. 1983. Thermal and structural evolution of four satellites of Saturn. *Ann. Geophys.* 1:469–476.

Feierberg, M. A., Lebofsky, L. A., and Larson, H. P. 1981. Spectroscopic evidence for aqueous alteration products on the surfaces of low-albedo asteroids. *Geochim. Cosmochim. Acta* 45:971–981.

Ferrari, A. J. 1977. Lunar gravity: A harmonic analysis. *J. Geophys. Res.* 82:3065–3084.

Ferrari, A. J., Sinclair, W. S., Sjogren, W. L., Williams, J. G., and Yoder, C. F. 1980. Geophysical parameters of the Earth-Moon system. *J. Geophys. Res.* 85:3939–3951.

Fiedler, G., ed. 1971. *Geology and Physics of the Moon* (Amsterdam: Elsevier).

Fink, J. H., and Fletcher, R. C. 1981. A mechanical analysis of extensional instability on Ganymede. In *Reports of Planetary Geology Program–1981*, NASA TM-84211, pp. 51–53.

Fink, J. H., Park, S. O., and Greeley, R. 1983. Cooling and deformation of sulfur flows. *Icarus* 56:38–50.

Fink, J., Gault, D., and Greeley, R. 1984. The effect of viscosity on impact cratering and possible application to the icy satellites of Saturn and Jupiter. *J. Geophys. Res.* 89:417–423.

Fink, U., and Larson, H. 1975. Temperature dependence of the water-ice spectrum between 1 and 4 μm: Application to Europa, Ganymede, and Saturn's rings. *Icarus* 24:411–420.

Fink, U., and Larson, H. P. 1979. The infrared spectra of Uranus, Neptune, and Titan from 0.8–2.5 microns. *Astrophys. J.* 233:1021–1040.

Fink, U., Dekkers, N. H., and Larson, H. P. 1973. Infrared spectra of the Galilean satellites of Jupiter. *Astrophys. J.* 179:L155–L159.

Fink, U., Larson, H. P., Gautier, T. N., and Treffers, R. R. 1976. Infrared spectra of the satellites of Saturn: Identification of water ice on Iapetus, Rhea, Dione, and Tethys. *Astrophys. J.* 207:L63–L67.

Fink, U., Larson, H. P., Lebofsky, L. A., Feierberg, M., and Smith, H. 1978. The 2–4 μm spectrum of Io. *Bull. Amer. Astron. Soc.* 10:580 (abstract).

Fink, U., Smith, B. A., Benner, C. D., Johnson, J. R., Reitsema, H. J., and Westphal, J. A. 1980a. Detection of CH_4 atmosphere on Pluto. *Icarus* 44:62–71.

Fink, U., Larson, H. P., Lebofsky, L. A., Frierberg, M., and Smith, H. 1980b. High resolution spectrum of Io from 2–4 μm. *The Satellites of Jupiter, IAU Coll. 57,* abstract booklet.

Finnerty, A. A., Ransford, G. A., Pieri, D. C., and Collerson, K. D. 1981. Is Europa's surface cracking due to thermal evolution? *Nature* 289:24–27.

Fish, R. A., Goles, G. G., and Anders, E. 1960. The record in meteorites. III. On the development of meteorites in asteroidal bodies. *Astrophys. J.* 132:243–258.

Fix, J. D., Neff, J. S., and Kelsey, L. A. 1970. Spectrophotometry of Pluto. *Astron. J.* 75:895–896.

Flasar, F. M. 1983. Oceans on Titan? *Science* 221:55–57.

Fletcher, R. C., and Hallet, B. 1983. Unstable extension of the lithosphere: A mechanical model for basin-and-range structure. *J. Geophys. Res.* 88:7457–7466.

Foti, G., Calcagno, L., Sheng, K. L., and Strazzulla, G. 1984. Micrometer-sized polymer layers synthesized by MeV ions impinging on frozen methane. *Nature* 310:126–128.

Francis, P. W., Thorpe, R. S., Brown, G. C., and Glasscock, J. 1980. Pyroclastic sulfur eruption and Poas Volcano, Costa Rica. *Nature* 283:754–755.

Frank, L. A., Burek, B. G., Ackerson, K. L., Wolfe, J. H., and Mihalov, J. D. 1980. Plasmas in Saturn's magnetosphere. *J. Geophys. Res.* 85:5695–5708.

Franklin, F. A., and Cook, A. F. 1965. Optical properties of Saturn's rings. II. Two-color phase curves of the two bright rings. *Astron. J.* 70:704–720.

Franz, O. G. 1978. UBV photometry of Triton. *Bull. Amer. Astron. Soc.* 10:585 (abstract).

Franz, O. G. 1981. UBV photometry of Triton. *Icarus* 45:602–606.

Franz, O. G., and Millis, R. L. 1971. A search for an anomalous brightening on Io after eclipse. *Icarus* 14:13–15.

Franz, O. G., and Millis, R. L. 1974. A search for posteclipse brightening on Io in 1973. II. *Icarus* 23:431–436.

Franz, O. G., and Millis, R. L. 1975. Photometry of Dione, Tethys, and Enceladus on the UBV system. *Icarus* 24:433–442.

French, L. 1980. *Photometric Properties of Carbonaceous Chondrites and Related Materials.* Ph.D. Thesis, Cornell Univ.

French, R. G. 1984. Oblatenesses of Uranus and Neptune. In *Uranus and Neptune,* ed. J. T. Bergstralh, NASA CP-2330, pp. 349–355.

French, R. G., Elliot, J. L., and Levine, S. E. 1986. Structure of the Uranian rings. II. Perturbations of the ring orbits and widths. *Icarus* 66. In press.

Frey, H. 1975. Posteclipse brightening and non-brightening of Io. *Icarus* 25:439–446.

Fricker, P. E., Reynolds, R. T., and Summers, A. L. 1967. On the thermal history of the Moon. *J. Geophys. Res.* 72:2649–2663.

Friedson, A. J., and Stevenson, D. J. 1983. Viscosity of rock-ice mixtures and applications to the evolution of icy satellites. *Icarus* 56:1–14.

Frondel, J. W. 1975. *Lunar Mineralogy* (New York: Wiley).

Fujiwara, A. 1980. On the mechanism of catastrophic destruction of minor planets by high-velocity impacts. *Icarus* 41:356–364.

Fujiwara, A. 1982. Complete fragmentation of the parent bodies of Themis, Eos, and Koronis families. *Icarus* 52:434–443.

Fujiwara, A., and Asada, N. 1983. Impact fracture patterns on Phobos ellipsoids. *Icarus* 56:590–602.

Fujiwara, A., Kamimoto, G., and Tsukamoto, A. 1977. Destruction of basaltic bodies by high velocity impact. *Icarus* 31:277–288.

Fuller, M. 1974. Lunar magnetism. *Rev. Geophys. Space Phys.* 12:23–70.

Gaffey, M. J. 1978. Mineralogical characterizations of asteroid surface materials: Evidence for unsampled meteorite types. *Meteoritics* 13:471–472.

Gaffney, E. S. 1984. Effect of strain-rate dependent yield strength on crater scaling relations. *Geophys. Res. Lett.* 11:121–123.

Galilei, G. 1610. *Sidereus Nuncius,* Venice, Italy.

Gammon, P. H., Klefte, H., and Clouter, M. J. 1983. Elastic constants of ice samples by Brillouin spectroscopy. *J. Phys. Chem.* 87:4025–4029.

Ganapathy, R., Keays, R. R., Laub, J. C., and Anders, E. 1970. Trace elements in Apollo 11 lunar rocks: Implications for meteorite influx and origin of the moon. *Geochim. Cosmochim. Acta Suppl.* 1:1117–1142.

Gardner, C. S. 1959. Adiabatic invariants of periodic classical systems. *Phys. Rev.* 115: 791–794.

Gary, B. L., and Kheim, S. J. 1978. Interpretation of ground-based microwave measurements of the moon using a detailed regolith properties model. *Proc. Lunar Planet. Sci. Conf.* 9: 2885–2900.

Gatley, T., Kieffer, H., Miner, E., and Neugebauer, G. 1974. Infrared observations of Phobos from Mariner 9. *Astrophys. J.* 190:497–503.

Gault, D. E. 1978. Experimental impact "craters" formed in water: Gravity scaling realized. *EOS Trans. AGU* 59:1121 (abstract).

Gault, D. E., and Heitowit, E. D. 1963. The partition of energy for hypervelocity impact craters formed in rock. *Proc. 6th Hypervelocity Impact Symp.,* vol. 2 (Cleveland, OH: Firestone Rubber Co.), pp. 419–456.

Gault, D. E., and Sonett, C. P. 1982. Laboratory simulations of pelagic asteroidal impact: Atmospheric injection, benthic topography, and the surface wave radiation field. *Geol. Soc. Amer. Spec. Paper 190,* pp. 69–92.

Gault, D. E., and Wedekind, J. A. 1977. Experimental hypervelocity impacts into quartz sand. II. Effects of gravitational acceleration. In *Impact and Explosion Cratering,* eds. D. J. Roddy, R. O. Pepin, and R. B. Merrill (New York: Pergamon), pp. 1231–1244.

Gault, D. E., and Wedekind, J. A. 1978. Experimental studies of oblique impact. *Proc. Lunar Planet. Sci. Conf.* 9:3843–3875.

Gault, D. E., Shoemaker, E. M., and Moore, H. J. 1963. Spray ejected from the lunar surface by meteroid impact. NASA Tech. Note D-1767.

Gault, D. E., Quaide, W. L., and Oberbeck, V. R. 1968. Impact cratering mechanics and structures. In *Shock Metamorphism of Natural Materials,* eds. B. M. French and N. M. Short (Baltimore: Mono Book Corp.), pp. 87–99.

Gault, D. E., Horz, F., Brownlee, D. E., and Hartung, J. B. 1974. Mixing of the lunar regolith. *Proc. Lunar Sci. Conf.* 5:2365–2386.

Gault, D. E., Guest, J. E., Murray, J. B., Dzurisin, D., and Malin, M. C. 1975. Some comparisons of impact craters on Mercury and the Moon. *J. Geophys. Res.* 80:2444–2460.

Gault, D. E., Burns, J. A., Cassen, P., and Strom, R. G. 1977. Mercury. *Ann. Rev. Astron. Astrophys.* 15:97–126.

Gautier, D., Bezard, B., Marten, A., Baluteau, J. P., Scott, N., Chedin, A., Kunde, V., and Hanel, R. 1982. The C/H ratio in Jupiter from the Voyager infrared investigation. *Astrophys. J.* 257:901–912.

Gavrilov, S. V., and Zharkov, V. N. 1977. Love numbers of the giant planets. *Icarus* 32:443–449.

Gehrels, N., and Stone, E. 1983. Energetic oxygen and sulfur ions in the Jovian magnetosphere and their contribution to the auroral excitation. *J. Geophys. Res.* 88:5537–5550.

Gehrels, T., ed. 1978. *Protostars and Planets* (Tucson: Univ. of Arizona Press).

Gehrels, T., and Matthews, M. S., eds. 1984. *Saturn* (Tucson: Univ. of Arizona Press).

Gerstenkorn, H. 1955. Über Gezeitenreibung beim Zweikörpenproblem. *Z. Astrophys.* 36: 245–274.

Gerstenkorn, H. 1969. The earliest past of the Earth-Moon system. *Icarus* 11:189–207.

Gilbert, G. K. 1893. The moon's face: A study of the origin of its features. *Bull. Phil. Soc. Washington* 12:241–292.

Gillett, F. C. 1975. Further observations of 8–13 micron observations of Titan. *Astrophys. J.* 201:L41–L43.

Gillett, F. C., Merrill, K. M., and Stein, W. A. 1970. Albedo and thermal emission of Jovian satellites. I–IV. *Astrophys. Lett.* 6:247–249.

Ginzberg, A. I., Golitzen, G. S., and Fedorov, K. N. 1979. Measurements of the time scale of convection in the liquid cooled from the surface. *Izvestiya, Atmospheric and Ocean Physics* 15(3):227–228.

Glass, B. P. 1982. *Introduction to Planetary Geology* (Cambridge: Cambridge Univ. Press).

Glen, J. W. 1975. The mechanics of ice. *Cold. Reg. Res. Eng. Lab. Monogr.* II-C2b, Hanover, NH.

Goertz, C. K. 1980. Io's interaction with the plasma torus. *J. Geophys. Res.* 85:2949–2956.

Goetze, C., and Kohlstedt, D. L. 1973. Laboratory study of dislocation climb and diffusion in olivine. *J. Geophys. Res.* 78:5961–5971.

Goguen, J. D. 1981. A Theoretical and Experimental Investigation of the Photometric Functions of Particulate Surfaces. Ph.D. Thesis, Cornell Univ.

Goguen, J. D., and Sinton, W. M. 1985. Characterization of Io's volcanic activity by infrared polarimetry. *Science* 230:65–69.

Goguen, J., Veverka, J., Thomas, P., and Duxbury, T. C. 1978. Phobos: Photometry and origin of dark markings on crater floors. *Geophys. Res. Lett.* 5:981–984.

Goguen, J., Veverka, J., and Duxbury, T. 1979. Marsshine on Phobos. *Icarus* 37:377–388.

Goguen, J., Trippico, M., and Morrison, D. 1983a. Voyager feature-fixed photometry of the surface of Iapetus. *Bull. Amer. Astron. Soc.* 15:855 (abstract).

Goguen, J., Cruikshank, D. P., Hammel, H., and Hartmann, W. K. 1983b. The rotational lightcurve of Hyperion during 1983. *Bull. Amer. Astron. Soc.* 15:854 (abstract).

Goguen, J., Hammel, H., and Brown, R. H. 1986. Photometry of Titania, Oberon and Triton. Submitted to *Icarus*.

Goins, M. R., Toksöz, M. N., and Dainty, A. M. 1978. Seismic structure of the lunar mantle: An overview. *Proc. Lunar Planet. Sci. Conf.* 9:3575–3588.

Goins, N. R., Dainty, A. M., and Toksöz, M. N. 1981a. Seismic energy release of the Moon. *J. Geophys. Res.* 86:378–388.

Goins, N. R., Dainty, A. M., and Toksöz, M. N. 1981b. Lunar seismology: The internal structure of the Moon. *J. Geophys. Res.* 86:5061–5074.

Gold, L. W. 1958. Some observations on the dependence of strain rate on stress ice. *Can. J. Phys.* 36:1265–1275.

Gold, L. W. 1977. Engineering properties of fresh-water ice. *J. Glaciol.* 19:197–212.

Gold, T. 1979. Electrical origin of the outbursts on Io. *Science* 206:1071–1073.

Goldberg, B. A., Carlson, R. W., Matson, D. L., and Johnson, T. V. 1978. Io's sodium cloud. An orbital phase dependent asymmetry. *EOS Trans. AGU* 59:1122 (abstract).

Goldberg, B. A., Mekler, Y. U., Carlson, R. W., Johnson, T. V., and Matson, D. L. 1980. Io's sodium emission cloud and the Voyager 1 encounter. *Icarus* 44:305–317.

Goldberg, B. A., Garneau, G. W., and LaVoie, S. K. 1984. Io's sodium cloud. *Science* 226:512–516.

Golden, L. M. 1979. The effect of surface roughness on the transmission of microwave radiation through a planetary surface. *Icarus* 38:451–455.

Goldreich, P. 1963. On the eccentricity of satellite orbits in the solar system. *Mon. Not. Roy. Astron. Soc.* 126:257–268.

Goldreich, P. 1965a. An explanation of the frequent occurrence of commensurable motions in the solar system. *Mon. Not. Roy. Astron. Soc.* 130:159–181.

Goldreich, P. 1965b. Inclination of satellite orbits about an oblate precessing planet. *Astron. J.* 70:5–9.

Goldreich, P. 1966. History of the lunar orbit. *Rev. Geophys.* 4:411–439.

Goldreich, P., and Lynden-Bell, D. 1969. Io: A Jovian unipolar inductor. *Astrophys. J.* 156:59–78.

Goldreich, P., and Nicholson, P. D. 1977. Turbulent viscosity and Jupiter's tidal Q. *Icarus* 30:301–304.

Goldreich, P., and Peale, S. J. 1966. Spin-orbit coupling in the solar system. *Astron. J.* 71:425–438.

Goldreich, P., and Peale, S. J. 1970. The obliquity of Venus. *Astron. J.* 75:273–284.

Goldreich, P., and Soter, S. 1966. Q in the solar system. *Icarus* 5:375–389.

Goldreich, P., and Tremaine, S. 1978. The velocity dispersion in Saturn's rings. *Icarus* 34:227–239.

Goldreich, P., and Tremaine, S. 1979. Towards a theory for the Uranian rings. *Nature* 277:97–99.

Goldreich, P., and Tremaine, S. 1982. The dynamics of planetary rings. *Ann. Rev. Astron. Astrophys.* 20:249–283.

Goldreich, P., and Ward, W. R. 1973. The formation of planetesimals. *Astrophys. J.* 183: 1051–1061.

Goldstein, H. 1980. *Classical Mechanics* (Reading, MA: Addison-Wesley).

Goldstein, R. M., and Green, R. R. 1980. Ganymede: Radar surface characteristics. *Science* 207:179–180.

Golombek, M. P. 1982. Constraints on the expansion of Ganymede and the thickness of the lithosphere. *Proc. Lunar Planet. Sci. Conf. 13* in *J. Geophys. Res. Suppl.* 87:A77–A83.

Golombek, M. P., and Allison, M. L. 1981. Sequential development of grooved terrain and polygons on Ganymede. *Geophys. Res. Lett.* 8:1139–1142.

Golombek, M. P., and Bruckenthal, E. 1983. Origin of triple bands on Europa. *Lunar Planet. Sci.* XIV:251–252 (abstract).

Golombek, M. P., and McGill, G. E. 1983. Grabens, basin tectonics, and maximum total expansion of the Moon. *J. Geophys. Res.* 88:3563–3578.

Gooding, J. 1975. A High Temperature Study on the Vaporation of Alkalis from Motton Basutto under High Vacuum: A Model for Lunar Volcanism. M.S. Thesis, Univ. of Hawaii.

Goodman, D. J., Frost, H. J., and Ashby, M. F. 1981. The plasticity of polycrystalline ice. *Phil. Mag.* A43:665–695.

Goody, R. M., and Walker, J. C. G. 1972. *Atmospheres* (Englewood Cliffs: Prentice Hall).

Gradie, J., and Moses, J. 1983a. Spectral reflectance of unquenched sulfur. *Lunar Planet. Sci.* XIV:255–256 (abstract).

Gradie, J., and Moses, J. 1983b. Spectral reflectance properties of basalt/sulfur mixtures. *Bull. Amer. Astron. Soc.* 15:850 (abstract).

Gradie, J., and Tedesco, E. 1982. Compositional structure of the asteroid belt. *Science* 216: 1405–1407.

Gradie, J., and Veverka, J. 1984a. Photometric properties of powdered sulfur. *Icarus* 58:227–245.

Gradie, J., and Veverka, J. 1984b. Wavelength dependence of phase coefficients. *Icarus* 66. In press.

Gradie, J., Thomas, P., and Veverka, J. 1980. The surface composition of Amalthea. *Icarus* 44:373–387.

Gradie, J., Ostro, S. J., Thomas, P., and Veverka, J. 1982. Sulfur on Io: Laboratory measurements of spectral properties. *Lunar Planet. Sci.* XIII:275–276 (abstract).

Gradie, J., Ostro, S. J., Thomas, P., and Veverka, J. 1984. Glass on the surface of Io and Amalthea. *J. Noncryst. Solids* 67:421–432.

Graps, A. L., Lane, A. L., Horn, L. J., and Simmons, K. E. 1984. Evidence for material between Saturn's A and F rings from the Voyager 2 photopolarimeter experiment. *Icarus* 60:409–415.

Greeley, R. 1977. Basaltic "plains" volcanism. In *Volcanism of the Eastern Snake River Plain, Idaho,* eds. R. Greeley and J. S. King, NASA CR-154621, pp. 23–44.

Greeley, R., Fink, J., Gault, D. E., Snyder, D. B., Guest, J. E., and Schultz, P. H. 1980. Impact cratering in viscous targets: Laboratory experiments. *Proc. Lunar Planet. Sci. Conf.* 11: 2075–2097.

Greeley, R., Fink, J. H., Gault, D. E., and Guest, J. E. 1982. Experimental simulation of impact cratering on icy satellites. In *Satellites of Jupiter,* ed. D. Morrison (Tucson: Univ. of Arizona Press), pp. 340–378.

Greeley, R., Thelig, E., and Christensen, P. 1984. The Mauna Loa sulfur flow as an analogy to secondary sulfur flows(?) on Io. *Icarus* 60:189–199.

Greenberg, J. M. 1982. What are comets made of? A model based on interstellar dust. In *Comets,* ed. L. L. Wilkening (Tucson: Univ. of Arizona Press), pp. 131–163.

Greenberg, R. 1973a. Evolution of satellite resonances by tidal dissipation. *Astron. J.* 78: 338–346.

Greenberg, R. 1973b. The inclination-type resonance of Mimas and Tethys. *Mon. Not. Roy. Astron. Soc.* 165:305–311.

Greenberg, R. 1975a. The dynamics of Uranus' satellites. *Icarus* 24:325–332.

Greenberg, R. 1975b. On the Laplace relation among the satellites of Uranus. *Mon. Not. Roy. Astron. Soc.* 173:121–129.

Greenberg, R. 1976. The Laplace relation and the masses of Uranus' satellites. *Icarus* 29: 427–433.

Greenberg, R. 1977. Orbit-orbit resonances in the solar system: Varieties and similarities. *Vistas in Astron.* 21:209–239.

Greenberg, R. 1979*a*. The motions of Uranus' satellites: Theory and application. In *Dynamics of the Solar System*, ed. R. Duncombe (Dordrecht: Reidel), pp. 177–180.

Greenberg, R. 1979*b*. Growth of large, late-stage planetesimals. *Icarus* 39:141–150.

Greenberg, R. 1981. Tidal evolution of the Galilean satellites: A linearized theory. *Icarus* 46:415–423.

Greenberg, R. 1982*a*. Orbital evolution of the Galilean satellites. In *Satellites of Jupiter*, ed. D. Morrison (Tucson: Univ. of Arizona Press), pp. 65–92.

Greenberg, R. 1982*b*. Planetesimals to planets. In *Formation of the Planetary Systems*, ed. A. Brahic (Toulouse, France: Editions Cepadues), pp. 515–569.

Greenberg, R. 1984*a*. Satellite masses in the Uranus and Neptune systems. In *Uranus and Neptune*, ed. J. T. Bergstralh, NASA CP-2330, pp. 463–480.

Greenberg, R. 1984*b*. Stability of the Laplace relation in deep resonance. *Bull. Amer. Astron. Soc.* 16:687 (abstract).

Greenberg, R. 1984*c*. Orbital resonances among Saturn's satellites. In *Saturn*, eds. T. Gehrels and M. S. Matthews (Tucson: Univ. of Arizona Press), pp. 593–608.

Greenberg, R. 1985. Evolution paths for Galilean satellites. *Bull. Amer. Astron. Soc.* 17: 921–922 (abstract).

Greenberg, R., and Brahic, A., eds. 1984. *Planetary Rings* (Tucson: Univ. of Arizona Press).

Greenberg, R., and Weidenschilling, S. J. 1984. How fast do Galilean satellites spin? *Icarus* 58:186–196.

Greenberg, R., Counselman, C. C., and Shapiro, I. I. 1972. Orbit-orbit resonance capture in the solar system. *Science* 178:747–749.

Greenberg, R., Davis, D. R., Hartmann, W. K., and Chapman, C. R. 1977. Size distribution of particles in planetary rings. *Icarus* 30:769–779.

Greenberg, R., Weidenschilling, S. J., Chapman, C. R., and Davis, D. R. 1984. From icy planetesimals to outer planets and comets. *Icarus* 59:87–113.

Grieve, R. A. F., and Cintala, M. J. 1981. A method for estimating the initial impact conditions of terrestrial cratering events, exemplified by its application to Brent Crater, Ontario. *Proc. Lunar Planet. Sci. Conf.* 12:1607–1621.

Grieve, R. A. F., and Head, J. W. 1981. Impact cratering: A geological process on the planets. *Episodes* 1981(2):3–9.

Grieve, R. A. F., and Head, J. W. 1983. The Manicouagan impact structure: An analysis of its original dimensions and form. *J. Geophys. Res.* 88:A807–A818.

Grieve, R. A. F., Dence, M. R., and Robertson, P. B. 1977. Cratering processes: As interpreted from the occurrence of impact melts. In *Impact and Explosion Cratering*, eds. D. J. Roddy, R. O. Pepin and R. B. Merrill (New York: Pergamon), pp. 791–814.

Grieve, R. A. F., Palme, H., and Plant, A. G. 1980. Siderophile-rich particles in the melt rocks at E. Clearwater impact structure, Quebec: Their characteristics and relationship to the impacting body. *Contrib. Min. Petro.* 75:187–198.

Grieve, R. A. F., Robertson, P. B., and Dence, M. R. 1981. Constraints on the formation of ring impact structures, based on terrestrial data. In *Proc. Lunar Planet. Sci. Conf.* 12A:37–57.

Grimm, R. E., and Squyres, S. W. 1985. Spectral analysis of groove spacing on Ganymede. *J. Geophys. Res.* 90:2013–2021.

Grossman, L., and Larimer, J. 1974. Early chemical history of the solar system. *Rev. Geophys. Space Phys.* 12:71–101.

Guest, J. E., and Greeley, R. 1977. *Geology on the Moon* (London: Wykeham Publications).

Gurnett, D. A., and Goertz, C. K. 1983. Ion cyclotron waves in the Io plasma torus: Polarization reversal of whistler mode noise. *Geophys. Res. Lett.* 10:587–590.

Gurnis, M. 1981. Martian cratering revisited: Implications for early geologic evolution. *Icarus* 48:62–75.

Gurnis, M. 1986. Spatial distribution of craters on Callisto and limits on the crater production population of the Jovian system. Submitted to *Icarus*.

Gussow, W. C. 1968. Salt diapirism: Implications of temperature, and energy source of emplacement. In *Diapirism and Diapirs,* eds. J. Braunstein and G. D. O'Brian (Tulsa, OK: Assoc. Petrol. Geol.), pp. 16–52.

Haff, P. K. 1980. A model for the formation of thin films in dirty ice targets by sputtering: Application to the satellites of Jupiter. *Proc. Symp. on Thin Film Interfaces and Interactions,* 80–2, eds. J. E. E. Baglin and J. M. Poate (Princeton: Electrochemical Society), pp. 21–28.

Haff, P. K., and Watson, C. C. 1979. The erosion of planetary and satellite atmospheres by energetic atomic particles. *J. Geophys. Res.* 84:8436–8442.

Haff, P. K., Watson, C. C., and Yung, Y. L. 1981. Ejection of matter from Io. *J. Geophys. Res.* 86:6933–6938.

Haff, P. K., Eviatar, A., and Siscoe, G. L. 1983. Ring and plasma: The enigmae of Enceladus. *Icarus* 56:426–438.

Hafner, W. 1951. Stress distribution and faulting. *Geol. Soc. Amer. Bull.* 62:373–398.

Hagfors, T., Gold, T., and Ierkic, H. M. 1985. Refraction scattering as origin of the anomalous radar returns of Jupiter's satellites. *Nature* 315:637–640.

Hammel, H. B., Goguen, J. D., Sinton, W. M., and Cruikshank, D. P. 1985. Observational test for sulfur allotropes on Io. *Icarus* 64:125–132.

Handin, J. 1966. Strength and ductility. *Geol. Soc. Amer. Memoir* 97:223–289.

Handin, J., and Logan, J. M. 1981. Experimental tectonophysics. *Geophys. Res. Lett.* 7:647–650.

Hanel, R., Conrath, B., Flasar, F. M., Kunde, V., Lowman, P., Maguire, W., Pearl, J., Pirraglia, J., Samuelson, R., Gautier, D., Gierasch, P., Kumar, S., and Ponnamperuma, C., 1979a. Infrared observations of the Jovian system from Voyager 1. *Science* 204:972–976.

Hanel, R., Conrath, B., Flaser, F. M., Herath, L., Kunde, V., Lowman, P., Maguire, W., Pearl, J., Pirraglia, J., Samuelson, R., Gautier, D., Gierasch, P., Horn, L., Kumar, S., and Ponnamperuma, C. 1979b. Infrared observations of the Jovian system from Voyager 2. *Science* 206:952–956.

Hanel, R., Conrath, B., Flasar, F. M., Kunde, V., Maguire, W., Pearl, J., Pirraglia, J., Samuelson, R., Herath, L., Allison, M., Cruikshank, D., Gautier, D., Gierasch, P., Horn, L., Koppany, R., and Ponnamperuma, C. 1981. Infrared observations of the Saturnian system from Voyager 1. *Science* 212:192–200.

Hanel, R., Conrath, B., Flasar, F. M., Kunde, V., Maguire, W., Pearl, J., Pirraglia, J., Samuelson, R., Cruikshank, D., Gautier, D., Gierasch, P., Horn, L., and Schulte, P. 1986. Infrared observations of the Uranian system. *Science.* In press.

Hanks, T. C., and Anderson, D. L. 1972. Origin, evolution and present thermal state of the moon. *Phys. Earth Planet. Int.* 5:409–425.

Hansen, K. S. 1982. Secular effects of oceanic tidal dissipation on the Moon's orbit and the Earth's rotation. *Rev. Geophys. Space Phys.* 20:457–480.

Hansen, O. L. 1972. Thermal Radiation from the Galilean Satellites Measured at 10 and 20 Microns. Ph.D. thesis, California Inst. of Tech.

Hansen, O. L. 1973. Ten micron eclipse observations of Io, Europa, and Ganymede. *Icarus* 18:237–246.

Hansen, O. L. 1975. Infrared albedos and rotation curves of the Galilean satellites. *Icarus* 26:24–29.

Hanson, R. B., Jones, B. F., and Lin, D. N. C. 1983. The astrometric position of T Tauri and the nature of its companion. *Astrophys. J.* 270:L27–L30.

Hapke, B. 1963. A theoretical photometric function for the lunar surface. *J. Geophys. Res.* 68:4571–4586.

Hapke, B. 1971. Inferences from optical properties concerning the surface texture and composition of asteroids. In *Physical Studies of Minor Planets,* NASA SP-267, pp. 67–77.

Hapke, B. 1979. Io's surface and environs: A magmatic-volatile model. *Geophys. Res. Lett.* 6:799–802.

Hapke, B. 1981. Bidirectional reflectance spectroscopy 1. Theory. *J. Geophys. Res.* 86:3039–3054.

Hapke, B. 1983. The opposition effect. *Bull. Amer. Astron. Soc.* 15:853 (abstract).

Hapke, B. 1984. Bidirectional reflectance spectroscopy. 3. Correction for macroscopic roughness. *Icarus* 59:41–59.

Hapke, B., and Cassidy, W. 1978. Is the moon really as smooth as a billiard ball? Remarks concerning recent models of sputter fractionation on the lunar surface. *Geophys. Res. Lett.* 5:297–300.

Hapke, B. W., and VanHorn, H. 1963. Photometric studies of complex surfaces, with applications to the Moon. *J. Geophys. Res.* 68:4545–4570.

Hapke, B., and Wells, E. 1981. Bidirectional reflectance spectroscopy. 2. Experiments and observations. *J. Geophys. Res.* 86:3055–3060.

Hardie, R. 1965. A re-examination of the light variation of Pluto. *Astron. J.* 70:140 (abstract).

Haring, R. A., Haring, A., Klein, F. S., Kummel, A. C., and de Vries, A. E. 1983. Reaction sputtering of simple condensed gases by keV heavy ion bombardment. *Nucl. Instr. Meth.* 211:529–533.

Harrington, R. S., and Christy, J. W. 1980. The satellite of Pluto. II. *Astron. J.* 85:168.

Harrington, R. S., and Christy, J. W. 1981. The satellite of Pluto. III. *Astron. J.* 86:442–443.

Harrington, R. S., and Seidelmann, P. K. 1981. The dynamics of the Saturnian satellites, 1980 S1 and 1980 S3. *Icarus* 47:97–99.

Harrington, R. S., and van Flandern, T. C. 1979. The satellites of Neptune and the origin of Pluto. *Icarus* 39:131–136.

Harris, A. W. 1977. An analytical theory of planetary rotation rates. *Icarus* 31:168–174.

Harris, A. W. 1978*a*. The formation of the outer planets. *Lunar Planet. Sci.* IX:459–461.

Harris, A. W. 1978*b*. Satellite formation. II. *Icarus* 34:128–145.

Harris, A. W. 1979. Asteroid rotation rates. II. A theory for the collisional evolution of rotation rates. *Icarus* 40:145–153.

Harris, A. W. 1983. Physical characteristics of Neptune and Triton inferred from the orbital motion of Triton. *Natural Satellites, IAU Coll. 77,* abstract booklet, Cornell Univ.

Harris, A. S. 1984*a*. The origin and evolution of planetary rings. In *Planetary Rings,* ed. R. Greenberg and A. Brahic (Tucson: Univ. of Arizona Press), pp. 641–659.

Harris, A. W. 1984*b*. Physical properties of Neptune and Triton inferred from the orbit of Triton. In *Uranus and Neptune,* ed. J. Bergstralh, NASA CP-2330, pp. 357–376.

Harris, A. W. 1984*c*. Letter to the editor. *Sky and Telescope* 67:108.

Harris, A. W., and Kaula, W. M. 1975. A co-accretional model of satellite formation. *Icarus* 24:516–524.

Harris, A. W., and Ward, W. R. 1982. Dynamical constraints on the formation and evolution of planetary bodies. *Ann. Rev. Earth Planet. Sci.* 10:61–108.

Harris, D. L. 1961. Photometry and colorimetry of planets and satellites. In *Planets and Satellites,* eds. G. P. Kuiper and B. M. Middlehurst (Chicago: Univ. of Chicago Press), pp. 272–342.

Harrison, M., and Schoen, R. 1967. Evaporation of ice in space: Saturn's rings. *Science* 157:1175.

Hartmann, W. K. 1972*a*. Paleocratering the Moon: Review of post-Apollo data. *Astron. Space Sci.* 18:48.

Hartmann, W. K. 1972*b*. Interplanet variations in the scale of crater morphology—Earth, Mars, Moon. *Icarus* 17:707–713.

Hartmann, W. K. 1973. Ancient lunar mega-regolith and subsurface structure. *Icarus* 18:634–636.

Hartmann, W. K. 1979. Diverse puzzling asteroids and a possible unified explanation. In *Asteroids,* ed. T. Gehrels (Tucson: Univ. of Arizona Press), pp. 466–479.

Hartmann, W. K. 1980*a*. Continued low-velocity impact experiments at Ames Vertical Gun Facility: Miscellaneous results. *Lunar Planet. Sci.* XI:404–406 (abstract).

Hartmann, W. K. 1980*b*. Surface evolution of two-component stone/ice bodies in the Jupiter region. *Icarus* 44:441–453.

Hartmann, W. K. 1983. *Moon and Planets,* 2nd ed. (Belmont, CA: Wadsworth Publishing).

Hartmann, W. K. 1984. Does crater "saturation equilibrium" occur in the solar system. *Icarus* 60:56–74.

Hartmann, W. K. 1985. Impact experiments. I. Ejecta velocity distributions and related results from regolith targets. *Icarus* 63:69–98.

Hartmann, W. K., and Davis, D. R. 1975. Satellite-sized planetesimals. *Icarus* 24:504–515.

Hartmann, W. K., Strom, R. G., Grieve, R. A. F., Weidenschilling, S. J., Diaz, J., Blasius, K. R., Chapman, C. R., Woronow, A., Shoemaker, E. M., Dence, M. R., and Jones, K. L. 1981. Chronology of planetary volcanism by comparative studies of planetary cratering. In

Basaltic Volcanism of the Terrestrial Planets, Basaltic Volcanism Project (Houston: Lunar Planet. Inst.), pp. 1049–1127.

Hartmann, W. K., Cruikshank, D. P., and Degewij, J. 1982. Remote comets and related bodies: VJHK colorimetry and surface materials. *Icarus* 52:377–408.

Hartmann, W. K., Phillips, R. J., and Taylor, G. J., eds. 1986. *Origin of the Moon* (Houston: Lunar Planet. Inst.). In press.

Hawke, B. R., and Head, J. W. 1978. Lunar KREEP volcanism: Geological evidence for history and mode of emplacement. *Proc. Lunar Planet. Sci. Conf.* 9:3285–3309.

Hawkes, I., and Mellor, M. 1972. Deformation and fracture of ice under uniaxial stress. *J. Glaciol.* 11:103–131.

Hayashi, C. 1981. Formation of the planets. In *Fundamental Problems in the Theory of Stellar Evolution, IAU Symp. 93,* eds. D. Sugimoto, D. Q. Lamb, and D. N. Schramm (Dordrecht: Reidel), pp. 113–128.

Hayashi, C., Nakazawa, K., and Adachi, I. 1977. Long term behavior of planetesimals and the formation of planets. *Publ. Astron. Soc. Japan* 29:163–196.

Hayashi, C., Nakazawa, K., and Mizuno, H. 1979. Earth's melting due to the blanketing effect of the primordial dense atmosphere. *Earth Planet. Sci. Lett.* 43:22–28.

Hayashi, C., Nakazawa, K., and Nakagawa, Y. 1985. Formation of the solar system. In *Protostars & Planets II,* eds. D. C. Black and M. S. Matthews (Tucson: Univ. of Arizona Press), pp. 1100–1153.

Haynes, J. 1978. *Effect of Temperature on the Strength of Snow-Ice,* USA CRREL Report 78–27.

Head, J. W. 1974. Orientale multi-ring basin interior and implications for the petrogenesis of lunar highland samples. *Moon* 11:327–356.

Head, J. W. 1976. Lunar volcanism in space and time. *Rev. Geophys. Space Phys.* 14:265–300.

Head, J. W. 1982. Lava flooding of ancient planetary crusts: Geometry, thickness, and volumes of flooded lunar impact basins. *Moon and Planets* 26:61–88.

Head, J. W., and Cintala, M. J. 1978. Grooves on Phobos: Evidence for possible secondary cratering origin. NASA TM-80339, pp. 19–21.

Head, J. W., and Gifford, A. 1980. Lunar mare domes: Classification and mode of origin. *Moon and Planets* 22:235–258.

Head, J. W., and McCord, T. B. 1978. Imbrian-age highland volcanism on the Moon: The Gruithuisen and Mairan domes. *Science* 199:1433–1436.

Head, J. W., and Settle, M. 1976. Relation of lunar highland sample sites to the Imbrium Basin deposits. In *Interdisciplinary Studies by the Imbrium Consortium,* vol. 1 (Cambridge: Center for Astrophysics), pp. 5–14.

Head, J. W., and Solomon, S. C. 1980. Lunar basin structure: Possible influence of variations in lithospheric thickness. *Lunar Planet. Sci.* XI:421–423 (abstract).

Head, J. W., and Solomon, S. C. 1981. Tectonic evolution of the terrestrial planets. *Science* 213:62–76.

Head, J. W., and Wilson, L. 1979. Alphonsus-type dark-halo craters: Morphology, morphometry and eruption conditions. *Proc. Lunar Planet. Sci. Conf.* 10:2861–2897.

Head, J. W., and Wilson, L. 1980. The formation of eroded depressions around the sources of lunar sinuous rilles: Observations. *Lunar Planet. Sci.* XI:426–428 (abstract).

Hege, K. K., Hubbard, E., Drummond, J. D., Strittmatter, P. A., Worden, S. P., and Lauer, T. 1981. Speckle interferometric observations of Pluto and Charon. *Icarus* 50:72–81.

Helfenstein, P., and Cook, A. F. 1984. Active venting of Europa? An analysis of a transient bright surface feature. *Lunar Planet. Sci.* XV:354–355 (abstract).

Helfenstein, P., and Parmentier, E. M. 1980. Fractures on Europa: Possible response of an ice crust to tidal deformation. *Proc. Lunar Planet. Conf.* 11:1987–1998.

Helfenstein, P., and Parmentier, E. M. 1983. Patterns of fracture and tidal stress on Europa. *Icarus* 53:415–430.

Helfenstein, P., and Parmentier, E. M. 1985. Patterns of fracture and tidal stresses due to non-synchronous rotation: Implications for fracturing on Europa. *Icarus* 61:175–185.

Helfenstein, P., Wilson, L., and Walker, N. 1984. Photometric classification of terrain units on Ganymede and implications for the Galileo mission. *Proc. Lunar Planet Sci. Conf.* 15:265–300.

Henrard, J. 1982a. Capture into resonance: An extension of the use of adiabatic invariants. *Celestial Mech.* 27:3–22.

Henrard, J. 1982b. The adiabatic invariant: Its use in celestial mechanics. In *Applications of Modern Dynamics to Celestial Mechanics and Astrodynamics,* ed. V. Szebehely (Dordrecht: Reidel), pp. 153–171.

Henrard, J. 1983. Orbital evolution of the Galilean satellites: Capture into resonance. *Icarus* 53:55–67.

Henrard, J., and Lemaître, A. 1983. A second fundamental model for resonance. *Celestial Mech.* 30:197–218.

Heppenheimer, T. A., and Porco, C. 1977. New contributions to the problem of capture. *Icarus* 30:385–401.

Herbert, F. L. 1980. Time-dependent lunar density models. *Proc. Lunar Planet. Sci. Conf.* 11:2015–2030.

Herbert, F. L., and Lichtenstein, B. R. 1980. Joule heating of Io's ionosphere by unipolar induction currents. *Icarus* 44:296–304.

Herbert, F. L., Drake, M. J., Sonett, C. P., and Wiskerchen, M. J. 1977a. Some constraints on the thermal history of the lunar magma ocean. *Proc. Lunar Sci. Conf.* 8:573–582.

Herbert, F. L., Sonett, C. P., and Wiskerchen, M. J. 1977b. Model "zero-age" lunar thermal profiles resulting from electrical induction. *J. Geophys. Res.* 82:2054–2060.

Herbert, F. L., Drake, M. J., and Sonett, C. P. 1978. Geophysical and geochemical evolution of the lunar magma ocean. *Proc. Lunar Planet. Sci. Conf.* 9:249–262.

Herbst, E. 1978. The current state of interstellar chemistry of dense clouds. In *Protostars and Planets,* ed. T. Gehrels (Tucson: Univ. of Arizona Press), pp. 88–99.

Herkenhoff, K. E., and Stevenson, D. J. 1984. Formation of Saturn's E ring by evaporation of liquid from the surface of Enceladus. *Lunar Planet. Sci.* XV:361–362 (abstract).

Hill, T. W. 1980. Corotation in lag in Jupiter's magnetosphere: Comparison of observation and theory. *Science* 207:301–302.

Hill, T. W., and Michel, F. C. 1976. Heavy ions from the Galilean satellites and the centrifugal distortion of the Jovian magnetosphere. *J. Geophys. Res.* 81:4561–4565.

Hill, T. W., Dessler, A. J., and Goertz, C. K. 1983a. Magnetospheric models. In *Physics of the Jovian Magnetosphere,* ed. A. J. Dessler (Cambridge: Cambridge Univ. Press), pp. 353–394.

Hill, T. W., Dessler, A. J., and Tassbach, M. E. 1983b. Aurora on Uranus: A Faraday disk dynamo mechanism. *Planet. Space Sci.* 31:1187–1198.

Hobbs, P. V. 1974. *Ice Physics* (London: Oxford Univ. Press).

Hodges, C. A., Shew, N. B., and Clow, G. 1980. Distribution of central pit craters on Mars. *Lunar Planet. Sci.* XI:450–452 (abstract).

Holsapple, K. A. 1979. Impact experiments with ambient atmospheric pressure. *EOS Trans. AGU* 60:871 (abstract).

Holsapple, K. A., and Bjorkman, M. D. 1983. On the scaling of impact crater growth. *EOS Trans. AGU* 64:747 (abstract).

Holsapple, K. A., and Schmidt, R. M. 1979. A material-strength model for apparent crater volume. *Proc. Lunar Planet. Sci. Conf.* 10:2757–2777.

Holsapple, K. A., and Schmidt, R. M. 1982. On the scaling of crater dimensions. 2. Impact processes. *J. Geophys. Res.* 87:1849–1870.

Hood, L. L. 1981. Sources of lunar magnetic anomalies and their bulk directions of magnetization: Additional evidence from Apollo orbital data. *Proc. Lunar Planet. Sci. Conf.* 12:817–830.

Hood, L. L., and Schubert, G. 1980. Lunar magnetic anomalies and surface optical properties. *Science* 208:49–51.

Hood, L. L., Coleman, P. J. Jr., and Wilhelms, D. E. 1979. The Moon: Sources of crustal magnetic anomalies. *Science* 204:53–57.

Hood, L. L., Russell, C. T., and Coleman, P. J. Jr. 1981. Contour maps of lunar remanent magnetic fields. *J. Geophys. Res.* 86:1055–1069.

Hood, L. L., Herbert, F., and Sonett, C. P. 1982. The deep lunar electrical conductivity profile: Structural and thermal inferences. *J. Geophys. Res.* 87:5311–5326.

Hooke, R. L. 1981. Flow law for polycrystalline ice in glaciers: Comparison of theoretical predictions, laboratory data, and field measurements. *Rev. Geophys. Space Phys.* 19:664–672.

Hooke, R. L., Dahlin, B. B., and Kauper, M. T. 1972. Creep of ice containing dispersed fine sand. *J. Glaciol.* 11:327–336.

Horedt, G. P. 1980. Accretional heating as the major cause of compositional differences among meteoritic parent bodies, the Moon and Earth. *Icarus* 43:215–221.

Horedt, G. P., and Neukum, G. 1984*a*. Comparison of six crater-scaling laws. *Earth, Moon and Planets* 31:265–269.

Horedt, G. P., and Neukum, G. 1984*b*. Planetocentric versus heliocentric impacts on the Jovian and Saturnian satellite system. *J. Geophys. Res.* 89:10405–10410.

Horedt, G. P., and Neukum, G. 1984*c*. Cratering rate over the surface of a synchronous satellite. *Icarus* 60:710–717.

Horner, V. M., and Greeley, R. 1982. Pedestal craters on Ganymede. *Icarus* 51:549–562.

Hörz, F., Gall, H., Gibbons, R. E., Hill, R. E., and Gault, D. E. 1976. Large scale cratering of the lunar highlands: Some Monte Carlo model considerations. *Proc. Lunar Sci. Conf.* 7:2931–2945.

Hörz, F., Ostertag, R., and Rainey, D. A. 1983. Bunte breccia of the Ries: Continuous deposits of large impact craters. *Rev. Geophys. Space Phys.* 21:1667–1725.

Houghton, J. T. 1977. *The Physics of Atmospheres* (Cambridge: Univ. of Cambridge Press).

Housen, K. R. 1976. A model of regolith formation on asteroids. *Meteoritics* 11:300–301.

Housen, K. R., and Wilkening, L. L. 1982. Regoliths on small bodies in the solar system. *Ann. Rev. Planet. Sci.* 10:355–376.

Housen, K. R., Wilkening, L. L., Chapman, C. R., and Greenberg, R. 1979. Asteroidal regoliths. *Icarus* 39:317–351.

Housen, K. R., Schmidt, R. M., and Holsapple, K. A. 1983. Crater ejecta scaling laws: Fundamental forms based on dimensional analysis. *J. Geophys. Res.* 88:2485–2499.

Houston, W. N., Mitchell, J. K., and Carrier, W. D. 1974. Lunar soil density and porosity. *Proc. Lunar Planet. Sci. Conf.* 5:2361–2364.

Howard, K. A. 1974. Fresh lunar impact craters: Review of variation with size. *Proc. Lunar Sci. Conf.* 5:67–69.

Howard, K. A., Wilhelms, D. E., and Scott, D. H. 1974. Lunar basin formation and highland stratigraphy. *Rev. Geophys. Space Phys.* 12:309–327.

Howell, R. R., and McGinn, M. T. 1985. Infrared speckle observations of Io: An eruption in the Loki region. *Science* 230:63–65.

Howell, R. R., Cruikshank, D. P., and Fanale, F. P. 1984. Sulfur dioxide on Io: Spatial distribution and physical state. *Icarus* 57:83–92.

Huang, T.-Y., and Innanen, K. A. 1983. The gravitational escape/capture of planetary satellites. *Astron. J.* 88:1537–1548.

Hubbard, W. B. 1968. Tides in the giant planets. *Icarus* 23:42–50.

Hubbard, W. B. 1986. 1981NI: A Neptune arc? *Science* 231:1276–1278.

Hubbard, W. B., and Stevenson, D. J. 1984. Interior structure of Saturn. In *Saturn*, eds. T. Gehrels and M. S. Matthews (Tucson: Univ. of Arizona Press), pp. 47–87.

Hubbard, W. B., Brahic, A., Sicardy, B., Elicer, L.-R., Roques, F., and Vilas, F. 1986. Occultation detection of a Neptunian ring-like arc. *Nature* 319:636–640.

Huebner, J. S., Duba, A., and Wiggins, L. B. 1979. Electrical conductivity of pyroxene which contains trivalent cations: Laboratory measurements and the lunar profile. *J. Geophys. Res.* 84:4652–4656.

Hughes, T. J. 1976. The theory of thermal convection in polar ice sheets. *J. Glaciol.* 16:41–71.

Hunt, G. R. 1977. Spectral signatures of particulate minerals, in the visible and near-infrared. *Geophys.* 42:501–513.

Hunt, G. R., and Salisbury, J. W. 1970. Visible and near infrared spectra of minerals and rocks. I. Silicate minerals. *Mod. Geol.* 1:283–300.

Hunt, G. R., and Salisbury, J. W. 1971. Visible and near infrared spectra of minerals and rocks. II. Carbonates. *Mod. Geol.* 2:23–30.

Hunt, G. R., and Vincent, R. K. 1968. The behavior of spectral features in the infrared emission from particulate surfaces of various grain sizes. *J. Geophys. Res.* 73:6039–6046.

Hunt, G. R., Salisbury, J. W., and Lenhoff, C. J. 1971*a*. Visible and near infrared spectra of minerals and rocks. III. Oxides and hydroxides. *Mod. Geol.* 2:195–205.

Hunt, G. R., Salisbury, J. W., and Lenhoff, C. J. 1971*b*. Visible and near infrared spectra of minerals and rocks. IV. Sulphides and sulphates. *Mod. Geol.* 3:1–14.

Hunt, G. R., Salisbury, J. W., and Lenhoff, C. J. 1972. Visible and near infrared spectra of minerals and rocks. V. Halides, arsenates, vanadates, and borates. *Mod. Geol.* 3:121–132.

Hunt, G. R., Salisbury, J. W., and Lenhoff, C. J. 1973. Visible and near infrared spectra of minerals and rocks. VI. Additional silicates. *Mod. Geol.* 4:85–106.

Hunten, D. M. 1977. Titan's atmosphere and surface. In *Planetary Satellites*, ed. J. A. Burns (Tucson: Univ. of Arizona Press), pp. 430–437.

Hunten, D. M. 1978. A Titan atmosphere with a surface temperature of 200 K. In *The Saturn System*, eds. D. M. Hunten and D. Morrison, NASA CP-2068, pp. 127–140.

Hunten, D. M. 1979. Capture of Phobos and Deimos by protoatmospheric drag. *Icarus* 37: 113–123.

Hunten, D. M. 1985. Blow-off of an atmosphere and possible application to Io. *Geophys. Res. Lett.* 12:271–273.

Hunten, D. M., and Morrison, D., eds. 1978. *The Saturn System*, NASA CP-2068.

Hunten, D. M., and Watson, A. J. 1982. Stability of Pluto's atmosphere. *Icarus* 51:665–667.

Hunten, D. M., Tomasko, M. G., Flasar, F. M., Samuelson, R. E., Strobel, D. F., and Stevenson, D. J. 1984. Titan. In *Saturn*, ed. T. Gehrels and M. S. Matthews (Tucson: Univ. of Arizona Press), pp. 671–759.

Illies, J. H. 1970. Graben tectonics as related to crust-mantle interaction. In *Graben Problems*, ed. J. H. Illies and S. Mueller (Stuttgart: Schweitzerbart), pp. 4–27.

Ingersoll, P. A., Summers, M. E., and Schlipf, S. G. 1985. Supersonic meteorology of Io: Sublimation-driven flow of SO_2. *Icarus* 64:375–390.

Ip, W.-H. On plasma transport in the vicinity of the rings of Saturn: A siphon flow mechanism. *J. Geophys. Res.* 88:819–822.

Ip, W.-H., and Goertz, C. K. 1983. An interpretation of the dawn-dusk asymmetry of UV emission from the Io plasma torus. *Nature* 302:232–233.

Irvine, W. M. 1966. The shadowing effect in diffuse reflection. *J. Geophys. Res.* 71:2931–2937.

Irvine, W. M., and Lane, A. P. 1973. Photometric properties of Saturn's rings. *Icarus* 18: 171–176.

Irvine, W. M., and Pollack, J. B. 1968. Infrared optical properties of water and ice spheres. *Icarus* 8:324–360.

Jacobs, J. A. 1975. *The Earth's Core* (New York: Academic Press).

Jaffe, W., Caldwell, J. J., and Owen, T. 1980. Radius and brightness temperature observations of Titan at centimeter wavelengths by the Very Large Array. *Astrophys. J.* 242:806–811.

James, O. B. 1980. Rock of the early lunar crust. *Proc. Lunar Planet. Sci. Conf.* 11:365–393.

James, O. B., and Flohr, M. K. 1983. Subdivision of the Mg-suite noritic rocks into Mg-gabbronorites and Mg-norites. *J. Geophys. Res.* 88:A603–A614.

Jeanloz, R., and Ahrens, T. J. 1980. Release adiabat measurements on minerals: The effect of viscosity. *J. Geophys. Res.* 84:7545–7548.

Jeffreys, H. 1920. Tidal friction in shallow seas. *Phil. Trans. Roy. Soc. London* A221:239–264.

Jeffreys, H. 1952. *The Earth*, 3rd ed. (Cambridge: Cambridge Univ. Press).

Jeffreys, H. 1961. The effects of tidal friction on eccentricity and inclination. *Mon. Not. Roy. Astron. Soc.* 122:339–343.

Johnson, R. E. 1985. Polar frost formation on Ganymede. *Icarus* 62:344–347.

Johnson, R. E., and Strobel, D. 1982. Charge exchange in the Io torus and exosphere. *J. Geophys. Res.* 87:10385–10393.

Johnson, R. E., Lanzerotti, L. J., Brown, W. L., and Armstrong, T. P. 1981. Erosion of Galilean satellite surfaces by Jovian magnetosphere particles. *Science* 212:1027–1030.

Johnson, R. E., Boring, J., Reimann, C., Barton, L., Sieveka, E., Garrett, J., Farmer, K., Brown, W., and Lanzerotti, L. J. 1983*a*. Plasma ion induced molecular ejection on the Galilean satellites: Energies of ejected molecules. *Geophys. Res. Lett.* 10:892–895.

Johnson, R. E., Lanzerotti, L. J., and Brown, W. L. 1983*b*. Charged particle bombardment of ice grains and comets. *Astron. Astrophys.* 123:343–346.

Johnson, R. E., Garrett, J., Boring, J., Barton, L., and Brown, W. 1984*a*. Erosion and modification of SO_2 ice by ion bombardment of the surface of Io. *Proc. Lunar Planet. Sci. Conf. 14* in *J. Geophys. Res. Suppl.* 89:B711–B715.

Johnson, R. E., Sieveka, E. M., Taylor, G., and Lanzerotti, L. J. 1984*b*. Sputtering of the icy satellites of Saturn: A source of a neutral torus and magnetospheric plasma. *Bull. Amer. Astron. Soc.* 16:661 (abstract).

Johnson, R. E., Lanzerotti, L. J., and Brown, W. L. 1984*c*. Sputtering processes: Erosion and chemical change. *Adv. Space. Res.* 4:41–51.

Johnson, R. E., Barton, L. A., Boring, J. W., Jesser, W. A., Brown, W. L., and Lanzerotti, L. J. 1985. Charged particle modification of ices in the Saturnian and Jovian systems. In *Ices in the Solar System,* eds. J. Klinger, D. Benest, A. Dollfus, and R. Smoluchowski (Dordrecht: Reidel), pp. 301–316.

Johnson, T. V. 1970. Albedo and Spectral Reflectance of the Galilean Satellites of Jupiter. Ph.D. Thesis, California Inst. of Technology.

Johnson, T. V. 1971. Galilean satellites: Narrowband photometry 0.30 to 1.10 μm. *Icarus* 14:94–111.

Johnson, T. V. 1978. The Galilean satellites of Jupiter: Four worlds. *Ann. Rev. Earth Planet. Sci.* 6:93–125.

Johnson, T. V. 1981. The Galilean satellites. In *The New Solar System,* ed. J. K. Beatty, B. O'Leary, and A. Chaikin (Cambridge: Sky Pub. Corp.), pp. 143–160.

Johnson, T. V., and Fanale, F. P. 1973. Optical properties of carbonaceous chondrites and their relationship to asteroids. *J. Geophys. Res.* 78:8507–8578.

Johnson, T. V., and McCord, R. B. 1970. Galilean satellites: The spectral reflectivity 0.30–1.10 μm. *Icarus* 13:37–42.

Johnson, T. V., and McCord, R. B. 1971. Spectral geometric albedo of the Galilean satellites, 0.3 to 2.5 μm. *Astrophys. J.* 169:589–594.

Johnson, T. V., and McGetchin, T. R. 1973. Topography on satellite surfaces and the shape of asteroids. *Icarus* 18:612–620.

Johnson, T. V., and Pilcher, C. B. 1977. Satellite spectrophotometry and surface compositions. In *Planetary Satellites,* ed. J. A. Burns (Tucson: Univ. of Arizona Press), pp. 232–268.

Johnson, T. V., and Soderblom, L. A. 1982. Volcanic eruptions on Io: Implications for surface evolution and mass loss. In *Satellites of Jupiter,* ed. D. Morrison (Tucson: Univ. of Arizona Press), pp. 634–646.

Johnson, T. V., Veeder, G. J., and Matson, D. L. 1975. Evidence for frost on Rhea's surface. *Icarus* 24:428–432.

Johnson, T. V., Saunders, R. S., Matson, D. L., and Mosher, J. A. 1977. A TiO$_2$ abundance map of the northern maria. *Proc. Lunar Sci. Conf.* 8:1029–1036.

Johnson, T. V., Cook, A. F. II, Sagan, C., and Soderblom, L. A. 1979. Volcanic resurfacing rates and implications for volatiles on Io. *Nature* 280:746–750.

Johnson, T. V., Morfill, G., and Grün, E. 1980. Dust in Jupiter's magnetosphere: An Io source? *Geophys. Res. Lett.* 7:305–308.

Johnson, T. V., Lanzerotti, L. J., Brown, W. L., and Armstrong, T. P. 1981. Erosion of Galilean satellite surfaces by Jovian magnetospheric particles. *Science* 212:1027–1030.

Johnson, T. V., Soderblom, L. A., Mosher, J. A., Danielson, G. E., Cook, A. F., and Kupfer-man, P. 1983. Global multispectral mosaics of the icy Galilean satellites. *J. Geophys. Res.* 88:5789–5805.

Johnson, T. V., Morrison, D., Matson, D. L., Veeder, G. J., Brown, R. H., and Nelson, R. M. 1984. Volcanic hot spots on Io: Stability and longitudinal distribution. *Science* 226:134–137.

Jones, E. M., and Kodis, J. W. 1982. Atmospheric effects of large body impacts: The first few minutes. *Geol. Soc. Amer. Special Paper 190,* pp. 175–186.

Jones, G. H. S. 1977. Complex craters in alluvium. In *Impact and Explosion Cratering,* eds. D. J. Roddy, R. O. Pepin, and R. B. Merrill (New York: Pergamon), pp. 163–183.

Judge, D., Wu, F., and Carlson, R. 1980. Ultraviolet observations of the Saturn system. *Science* 209:431–434.

Kane, T. R. 1965. Attitude stability of earth-pointing satellites. *AIAA J.* 3:726–731.

Kaula, W. M. 1963. Tidal dissipation in the Moon. *J. Geophys. Res.* 68:4959–4965.

Kaula, W. M. 1964. Tidal dissipation by solid friction and the resulting orbital evolution. *Rev. Geophys.* 2:661–685.

Kaula, W. M. 1968. *An Introduction to Planetary Physics: The Terrestrial Planets* (New York: Wiley).

Kaula, W. M. 1971. Dynamical aspects of lunar origin. *Rev. Geophys. Space Phys.* 9:217–238.

Kaula, W. M. 1975. The gravity and shape of the Moon. *EOS Trans. AGU* 56:304–316.

Kaula, W. M. 1977. On the origin of the moon, with emphasis on bulk composition. *Proc. Lunar Sci. Conf.* 8:321–331.

Kaula, W. M. 1979. Thermal evolution of earth and moon growing by planetesimal impacts. *J. Geophys. Res.* 84:999–1008.

Kaula, W. M. 1980. The beginning of the Earth's thermal evolution. In *The Continental Crust and its Mineral Deposits*, ed. D. W. Strangway (Waterloo: Geol. Assoc. of Canada Special Paper), pp. 25–34.

Kaula, W. M. 1986. Formation of the Sun and its planets. In *Physics of the Sun*, vol. 3, ed. P. A. Sturrock (Dordrecht: Reidel), pp. 1–32.

Kaula, W. M., and Beachey, A. E. 1984. Mechanical models of close approaches and collisions of large protoplanets. *Conf. Origin of Moon*, Kona, HI, October, p. 59 (abstract).

Kaula, W. M., and Harris, A. W. 1973. Dynamically plausible hypotheses of lunar origin. *Nature* 245:367–369.

Kaula, W. M., and Harris, A. W. 1975. Dynamics of lunar origin and orbital evolution. *Rev. Geophys. Space Phys.* 13:363–371.

Kaula, W. M., and Yoder, C. F. 1976. Lunar orbit evolution and tidal heating of the Moon. *Proc. Lunar Sci. Conf.* VII:440–442 (abstract).

Kaula, W. M., Schubert, G., Lingenfelter, R. E., Sjogren, W. L., and Wolenhaupt, W. R. 1974. Apollo laser altimetry and inferences as to lunar structure. *Proc. Lunar Sci. Conf.* 5:3049–3058.

Kawakami, S., and Mizutani, H. 1980. Ganymede and Callisto: Thermal and structural evolution model. *Proc. 13th ISAS Lunar Planet. Symp.*, Univ. of Tokyo, pp. 330–345.

Kawakami, S., Mizutani, H., Takagi, Y., Kto, M., and Kumazawa, M. 1983. Impact experiments on ice. *J. Geophys. Res.* 88:5806–5814.

Keenan. D. W. 1981. Galactic tidal limits on star clusters. II. Tidal radius and outer dynamical structure. *Astron. Astrophys.* 95:340–348.

Kelly, A., Cook, A. H., and Greenwood, G. W., eds. 1978. Creep of engineering materials and of the Earth. *Phil. Trans. Roy. Soc. London* A288.

Kerridge, J. F., and Bunch, T. E. 1979. Aqueous activity on asteroids: Evidence from carbonaceous meteorites. In *Asteroids*, ed. T. Gehrels (Tucson: Univ. of Arizona Press), pp. 745–764.

Khare, B. N., Sagan, C., Zumberge, J. F., Sklarew, D. S., and Nagy, B. 1982. Organic solids produced by electrical discharge in reducing atmospheres: Tholin molecular analysis. *Icarus* 48:290–297.

Khare, B. N., Sagan, C., Arakawa, E. T., Suits, F., Callicott, T. A., and Williams, M. W. 1983. Optical constants of Titan tholin aerosols. *Bull. Amer. Astron. Soc.* 15:842–843 (abstract).

Kheim, S. J., and Langseth, M. G. 1975. Microwave emission spectrum of the Moon: Mean global heat flow and average depth of regolith. *Science* 187:64–66.

Kieffer, H. H., and Smythe, W. D. 1974. Frost spectra: Comparison with Jupiter's satellites. *Icarus* 21:506–512.

Kieffer, S. W. 1977. Impact conditions required for formation of melt by jetting in silicates. In *Impact and Explosion Cratering*, eds. D. J. Roddy, R. O. Pepin, and R. B. Merrill (New York: Pergamon), pp. 751–769.

Kieffer, S. W. 1982. Dynamics and thermodynamics of volcanic eruptions: Implications for the plumes on Io. In *Satellites of Jupiter*, ed. D. Morrison (Tucson: Univ. of Arizona Press), pp. 647–723.

Kieffer, S. W., and Simonds, C. H. 1980. The role of volatiles and lithology in the impact cratering process. *Rev. Geophys. Space Phys.* 18:143–181.

Kim, S. J., and Caldwell, J. J. 1982. The abundance of CH_3D in the atmosphere of Titan, derived from 8- to 14-μm thermal emission. *Icarus* 52:473–482.

King, E. A. 1976. *Space Geology: An Introduction* (New York: J. Wiley).

Kirby, S. H., Durham, W. B., and Heard, H. C. 1985. Rheologies of H_2O ices I, II, and III at high pressures: A progress report. In *Ices in the Solar System*, eds. J. Klinger, D. Benest, A. Dollfus, and R. Smoluchowski (Dordrecht: Reidel), pp. 89–107.

Kirk, R. L., and Stevenson, D. J. 1983. Thermal evolution of a differentiated Ganymede and implications for surface features. *Lunar Planet. Sci.* XIV:373–374 (abstract).

Kirk, R. L., and Stevenson, D. J. 1986. Thermal evolution of a differentiated Ganymede and implications for surface features. Submitted to *Icarus*.

Kivelson, M., Slavin, J., and Southwood, D. 1979. Magnetospheres of the Galilean satellites. *Science* 205:491–493.

Klassen, K. P., Duxbury, T. C., and Veverka, J. 1979. Photometry of Phobos and Deimos from Viking Orbiter images. *J. Geophys. Res.* 84:8478–8496.

Klinger, J. 1980. Influence of a phase transition of ice on the heat and mass balance of comets. *Science* 209:271–272.

Klinger, J., Benest, D., Dollfus, A., and Smoluchowski, R., eds. *Ices in the Solar System* (Dordrecht: Reidel).

Kliore, A. J., Caine, D. L., Fjeldbo, G., Seidel, B. L., and Rasool, S. I. 1974. Preliminary results on the atmospheres of Jupiter and Io from the Pioneer 10 S-band occultation experiment. *Science* 183:323.

Kliore, A., Fjeldbo, G., Seidel, B., Sweetman, D., Sesplaukis, T., and Woiceshyn, P. 1975. Atmosphere of Io from Pioneer 10 radio occultation measurements. *Icarus* 24:407–410.

Knopoff, L. 1964. Q. *Rev. Geophys.* 2:625–660.

Kohlstedt, D. L., Nichols, H. P. K., and Hornack, P. 1980. The effect of pressure on the rate of dislocation recovery in olivine. *J. Geophys. Res.* 85:3122–3130.

Kovach, R. L., Watkins, J. S., and Talwani, P. 1973. The properties of the lunar shallow crust: An overview from Apollo 14, 16, and 17. *Lunar Sci.* IV:444–445 (abstract).

Kovalevsky, J., and Sagnier, J.-L. 1977. Motions of natural satellites. In *Planetary Satellites,* ed. J. A. Burns (Tucson: Univ. of Arizona Press), pp. 43–62.

Kowal, C. T. 1975. Probable new satellite of Jupiter. *IAU Circ. 2855.*

Krimigis, S. M., Armstrong, T. P., Axford, W. I., Bostrom, C. D., Fan, C. Y., Gloeckler, G., and Lanzerotti, L. J. 1977. The low energy charged particle experiment (LECP) on the Voyager spacecraft. *Space Sci. Rev.* 21:329–354.

Krimigis, S. M., Armstrong, T. P., Axford, W. I., Bostrom, C. O., Fan, C. Y., Gloeckler, G., Lanzerotti, L. J., Keath, E. P., Zwickly, R. D., Corbury, J. F., and Hamilton, D. C. 1979. Hot plasma environment of Jupiter: Voyager 2 results. *Science* 201:977–984.

Krimigis, S. M., Carbary, J. F., Keath, E. P., Bostrom, C. O., Axford, W. I., Gloeckler, G., Lanzerotti, L. J., and Armstrong, T. P. 1981a. Characteristics of hot plasma in the Jovian magnetosphere: Results from the Voyager spacecraft. *J. Geophys. Res.* 86:8227–8257.

Krimigis, S. M., Armstrong, T. P., Axford, W. I., Bostrom, C. O., Gloeckler, G., Keath, E. P., Lanzerotti, L. J., Carbary, J. F., Hamilton, D. C., and Roelof, E. C. 1981b. Low energy charged particles in Saturn's magnetosphere: Results from Voyager 1. *Science* 212:225–231.

Krimigis, S. M., Carbary, J. F., Keath, E. P., Armstrong, T. P., Lanzerotti, L. J., and Gloeckler, G. 1983. General characteristics of hot plasma and energetic particles in the Saturnian magnetosphere: Results from the Voyager spacecraft. *J. Geophys. Res.* 88:8871–8892.

Krohn, J., and Sündermann, J. 1982. Paleotides before the Permian. In *Tidal Friction and the Earth's Rotation II,* eds. P. Brosche and J. Sündermann (New York: Springer-Verlag), pp. 190–209.

Kuiper, G. P. 1944. Titan: A satellite with an atmosphere. *Astrophys. J.* 100:378–383.

Kuiper, G. P. 1951. On the origin of the irregular satellites. *Proc. Natl. Acad. Sci. USA* 37:717–721.

Kuiper, G. P. 1952. Planetary atmospheres and their origin. In *The Atmospheres of the Earth and the Planets,* Rev. Ed., ed. G. P. Kuiper (Chicago: Univ. of Chicago Press), pp. 306–405.

Kuiper, G. P. 1954. Report of the commission for physical observations of planets and satellites. *Trans. IAU* 9:250.

Kuiper, G. P. 1957. Infrared observations of planets and satellites. *Astron. J.* 62:295 (abstract).

Kuiper, G. P. 1961. Limits of completeness. In *Planets and Satellites,* eds. G. P. Kuiper and B. M. Middlehurst (Chicago: Univ. of Chicago Press), pp. 575–591.

Kuiper, G. P., Cruikshank, D. P., and Fink, U. 1970. *Cited in* The composition of Saturn's rings. *Sky and Telescope* 39:14.

Kumar, S. S. 1977. The escape of natural satellites from Mercury and Venus. *Astrophys. Space Sci.* 51:235–238.

Kumar, S. 1979. The stability of an SO_2 atmosphere on Io. *Nature* 280:758–760.

Kumar, S. 1980. A model of the SO_2 atmosphere and ionosphere of Io. *Geophys. Res. Lett.* 7:9–12.

Kumar, S. 1982. Photochemistry of SO_2 in the atmosphere of Io and implications on atmospheric escape. *J. Geophys. Res.* 87:1677–1684.

Kumar, S. 1984. Sulfur and oxygen escape from Io and a lower limit to atmospheric SO_2 at Voyager 1 encounter. *J. Geophys. Res.* 89:7399–7406.

Kumar, S., and Hunten, D. 1982. The atmospheres of Io and other satellites. In *Satellites of Jupiter,* ed. D. Morrison (Tucson: Univ. of Arizona Press), pp. 782–806.

Kunde, V. G., Aikin, A. C., Hanel, R. A., Jennings, D. E., Maguire, W. C., and Samuelson, R. E. 1981. C_4H_2, HC_3N, and C_2N_2 in Titan's atmosphere. *Nature* 292:686–688.

Kupo, I., Mekler, Yu., and Eviatar, A. 1976. Detection of ionized sulfur in the Jovian magnetosphere. *Astrophys. J.* 205:L51–L53.

Kurth, W. S., Gurnett, D. A., and Scarf, F. L. 1980. Spatial and temporal studies of Jovian kilometric radiation. *Geophys. Res. Lett.* 7:61–64.

Kusaka, T., Nakano, T., and Hayashi, C. 1970. Growth of solid particles in the primordial solar nebula. *Prog. Theor. Phys.* 44:1580–1595.

Labotka, T. C., Kempa, M. J., White, C., Papike, J. J., and Laul, J. C. 1980. The lunar regolith: Comparative petrology of the Apollo sites. *Proc. Lunar Planet. Sci. Conf.* 11:1285–1305.

Lamb, H. 1932. *Hydrodynamics* (New York: Dover).

Lambe, T. W., and Whitman, R. V. 1979. *Soil Mechanics, SI Version* (New York: Wiley).

Lambeck, K. 1975. Effects of tidal dissipation in the oceans on the Moon's orbit and the Earth's rotation. *J. Geophys. Res.* 80:2917–2925.

Lambeck, K. 1977. Tidal dissipation in the oceans: Astronomical, geophysical and oceanographic consequences. *Phil. Trans. Roy. Soc. London* A287:545–594.

Lambeck, K. 1979. On the orbital evolution of the Martian satellites. *J. Geophys. Res.* 84:5651–5658.

Lambeck, K. 1980. *The Earth's Variable Rotation: Geophysical Causes and Consequences* (Cambridge: Cambridge Univ. Press).

Lambeck, K., and Pullan, S. 1980. The lunar fossil bulge hypothesis revisited. *Phys. Earth Planet. Int.* 22:29–35.

Lammlein, D. R., Latham, G. V., Dorman, J., Nakamura, Y., and Ewing, M. 1974. Lunar seismicity, structure and tectonics. *Rev. Geophys. Space Phys.* 12:1–22.

Lanciano, P., Federico, C., and Coradini, A. 1981. Primordial thermal history of growing planetary objects. *Lunar Planet. Sci.* XII:586–588 (abstract).

Landau, L. D., and Lifshitz, E. M. 1960. *Mechanics* (London: Pergamon).

Lane, A. L., Nelson, R. M., and Matson, D. L. 1981. Evidence for sulphur implantation in Europa's UV absorption band. *Nature* 292:38–39.

Lane, W. A., Neff, J. S., and Fix, J. D. 1976. A measurement of the relative reflectance of Pluto at 0.86 μm. *Publ. Astron. Soc. Pacific* 88:77–79.

Lange, M. A., and Ahrens, T. J. 1981. Fragmentation of ice by low velocity impact. *Proc. Lunar Planet. Sci. Conf.* 12:1667–1687.

Lange, M. A., and Ahrens, T. J. 1983. The dynamic tensile strength of ice and ice-silicate mixtures. *J. Geophys. Res.* 88:1197–1208.

Langevin, Y., and Arnold, J. R. 1977. The evolution of the lunar regolith. *Ann. Rev. Earth Planet. Sci.* 5:449–489.

Langevin, Y., and Maurette, M. 1980. A model for small body regolith evolution: The critical parameters. *Lunar Planet. Sci.* XI:602–604 (abstract).

Langseth, M. G., Keihm, S. J., and Peters, K. 1976. Revised lunar heat-flow values. *Proc. Lunar Sci. Conf.* 7:3143–3171.

Lanzerotti, L. J., and Brown, W. L. 1983. Supply of SO_2 for the atmosphere of Io. *J. Geophys. Res.* 88:989–990.

Lanzerotti, L. J., Robbins, M. F., Tolk, N. H., and Neff, S. H. 1975. Scans of Io, Europa, and Ganymede in the NaD region. *Pub. Astron. Soc. Pacific* 87:449–452.

Lanzerotti, L. J., Brown, W. L., Poate, J. M., and Augustyniak, W. M. 1978. On the contributions of water products from Galilean satellites to the Jovian magnetosphere. *Geophys. Res. Lett.* 5:155–158.

Lanzerotti, L. J., Maclennan, C. G., Armstrong, T. P., Krimigis, S. M., Lepping, R. P., and Ness, N. F. 1981. Ion and electron angular distributions in the Io torus region of the Jovian magnetosphere. *J. Geophys. Res.* 86:8491–8496.

Lanzerotti, L. J., Brown, W. L., Augustyniak, W. M., Johnson, R. E., and Armstrong, T. P. 1982. Laboratory studies of charged particle erosion of SO_2 ice and applications to the frosts of Io. *Astrophys. J.* 259:920–929.

Lanzerotti, L. J., Maclennan, C. G., Brown, W. L., Johnson, R. E., Barton, L. A., Reimann, C. T., Garrett, J. W., and Boring, J. W. 1983. Implications of Voyager data for energetic ion erosion of the icy satellites of Saturn. *J. Geophys. Res.* 88:8765–8770.

Lanzerotti, L. J., Brown, W. L., Marcantonio, K. J., and Johnson, R. E. 1984. Production of ammonia-depleted surface layers on the Saturnian satellites by ion sputtering. *Nature* 312:139–140.

Lanzerotti, L. J., Brown, W. L., and Johnson, R. E. 1985. Laboratory studies of ion irradiations of water, sulfur dioxide, and methane ices. In *Ices in the Solar System*, eds. J. Klinger, D. Benest, A. Dollfus and R. Smoluchowski (Dordrecht: Reidel), pp. 317–336.

Laplace, P. S. 1805. *Mecanique Celeste,* vol. 4 (Paris: Coureier). Trans. by N. Bowditch, rpt. 1966, New York: Chelsea.

Leake, M. A. 1981. The Intercrater Plains of Mercury and the Moon: Their Nature, Origin, and Role in Terrestrial Planet Evolution. Ph.D. thesis, Univ. of Arizona.

Lebofsky, L. A. 1972. Reflectivities of ammonium hydrosulfides: Application to the interpretation of reflection spectra of outer solar system bodies. *Bull. Amer. Astron. Soc.* 4:362 (abstract).

Lebofsky, L. A. 1973. Chemical Composition of Saturn's Rings and Icy Satellites. Ph.D. Thesis, Massachusetts Institute of Technology.

Lebofsky, L. A. 1975. Stability of frosts in the solar system. *Icarus* 25:205–217.

Lebofsky, L. A. 1977. Identification of water frost on Callisto. *Nature* 269:785–787.

Lebofsky, L. A., and Fegley, M. B. Jr. 1976. Laboratory reflection spectra for the determination of chemical composition of icy bodies. *Icarus* 28:379–387.

Lebofsky, L. A., Rieke, G. H., and Lebofsky, M. J. 1979. Surface composition of Pluto. *Icarus* 37:554–558.

Lebofsky, L. A., Feierberg, M. A., and Tokunaga, A. T. 1982*a.* Infrared observations of the dark side of Iapetus. *Icarus* 49:382–386.

Lebofsky, L. A., Rieke, G. H., and Lebofsky, M. 1982*b.* The radii and albedos of Triton and Pluto. *Bull. Amer. Astron. Soc.* 14:766 (abstract).

Lee, S., Veverka, J., and Thomas, P. 1986. Phobos, Deimos, and the Moon: Size and distribution of crater ejecta blocks. Submitted to *Icarus.*

Lee, W. H. 1968. Effects of selective fusion on the thermal history of the Earth's mantle. *Earth Planet. Sci. Lett.* 4:270–276.

Lenard, A. 1959. Adiabatic invariance to all orders. *Ann. Phys.* 6:261–276.

LePoire, D. J., Cooper, B. H., Melcher, C. L., and Tombrello, T. A. 1983. Sputtering of SO_2 by high energy ions. *Rad. Eff.* 71:245–259.

Levin, B. J. 1962. Thermal history of the Moon. In *The Moon,* ed. Z. Kopal and Z. K. Mikhailov (New York: Academic Press), pp. 157–167.

Lewis, J. S. 1971*a.* Satellites of the outer planets: Their physical and chemical nature. *Icarus* 15:174–185.

Lewis, J. S. 1971*b.* Satellites of the outer planets: Thermal models. *Science* 172:1127–1128.

Lewis, J. S. 1972. Low temperature condensation from the solar nebula. *Icarus* 16:241–252.

Lewis, J. S. 1973. Chemistry of the outer solar system. *Space Sci. Rev.* 14:401–411.

Lewis, J. S. 1974. The temperature gradient in the solar nebula. *Science* 186:440–443.

Lewis, J. S. 1982. Io: Geochemistry of sulfur. *Icarus* 50:103–114.

Lewis, J. S., and Prinn, R. G. 1980. Kinetic inhibition of CO and N_2 reduction in the solar nebula. *Astrophys. J.* 238:357–364.

Liebermann, R. C., and Ringwood, A. F. 1976. Elastic properties of anorthite and the nature of the lunar crust. *Earth Planet. Sci. Lett.* 31:69–74.

Lieske, J. H. 1980. Improved ephemerides of the Galilean satellites. *Astron. Astrophys.* 82:340–348.

Lieske, J. H. 1983. A collection of Galilean satellite observations 1652–1982. In *Dynamical Trapping and Evolution in the Solar System*, eds. V. V. Markellos and Y. Kozai (Dordrecht: Reidel), pp. 51–59.

Lin, C. C., and Shu, F. H. 1966. On the spiral arms of disk galaxies. II. Outline of a theory of density waves. *Proc. Natl. Acad. Sci. U.S.A.* 55:229–234.

Lin, D. N. C. 1981. On the origin of the Pluto-Charon system. *Mon. Not. Roy. Astron. Soc.* 197:1081–1085.

Lin, D. N. C., and Bodenheimer, P. 1982. On the evolution of convective accretion disk models of the primordial solar nebula. *Astrophys. J.* 262:768–779.

Lin, D. N. C., and Papaloizou, J. 1979*a.* Tidal torques on accretion disks in binary systems with extreme mass ratios. *Mon. Not. Roy. Astron. Soc.* 186:799–812.

Lin, D. N. C., and Papaloizou, J. 1979*b*. On the structure of circumbinary accretion discs and the tidal evolution of commensurable satellites. *Mon. Not. Roy. Astron. Soc.* 188:191–201.

Lindal, G. F., Wood, G. E., Hotz, H. B., Sweetnam, D. N., Eshleman, V. R., and Tyler, G. L. 1983. The atmosphere of Titan: An analysis of the Voyager 1 radio occultation measurements. *Icarus* 53:348–363.

Lindholm, U. S., Yeakley, L. M., and Magy, A. 1974. The dynamic strength and fracture properties of Dresser basalt. *Int. J. Rock Mech. Min. Sci. Geomech. Abstr.* 11:181–191.

Lindsay, J. F. 1976. *Lunar Stratigraphy and Sedimentology* (New York: Elsevier).

Linker, J. A., Kivelson, M., Moreno, M., and Walker, R. 1985. Explanation of the inward displacement of Io's hot plasma torus and consequences for sputtering sources. *Nature* 315:373–378.

Lionel, W., and Head, J. W. 1983. Water volcanism. *Natural Satellites, IAU Coll. 77,* abstract booklet, p. 21.

Lissauer, J. J. 1985. Can cometary bombardment disrupt synchronous rotation of planetary satellites? *J. Geophys. Res.* 90:11,289–11,293.

Lissauer, J. J., and Cuzzi, J. N. 1985. Rings and moons: Clues to understanding the solar nebula. In *Protostars & Planets II,* eds. D. C. Black and M. S. Matthews (Tucson: Univ. of Arizona Press), pp. 920–956.

Lissauer, J. J., Peale, S. J., and Cuzzi, J. N. 1984. Ring torque on Janus and the melting of Enceladus. *Icarus* 58:159–168.

Lissauer, J. J., Goldreich, P., and Tremaine, S. 1985*a*. Evolution of the Janus-Epimetheus coorbital resonance due to torque from Saturn's rings. *Icarus* 64:425–434.

Lissauer, J. J., Squyres, S. W., Hartmann, W. K., and Lin, K. 1985*b*. Bombardment history of the Saturn system. *Bull. Amer. Astron. Soc.* 17:738 (abstract).

Livingston, C. W. 1960. *Explosions in Ice,* U.S. Army Snow, Ice and Permafrost Research Establishment Tech. Rept. 75.

Lockwood, G. W. 1977. Secular brightness increases of Titan, Uranus, and Neptune, 1972–1976. *Icarus* 32:413–430.

Lockwood, G. W., Thompson, D. T., and Lumme, K. 1980. A possible detection of solar variability from photometry of Io, Europa, Callisto, and Rhea, 1976–1979. *Astron. J.* 85:961–967.

Lockwood, G. W., Lutz, B. L., Thompson, D. T., and Bus, E. S. 1986. The albedo of Titan. *Astrophys. J.* 303:511–520.

Longaretti, P.-Y., and Borderies, N. 1986. Non-linear study of the Mimas 5:3 density wave. *Icarus.* In press.

Longhi, J. 1980. A model of early lunar differentiation. *Proc. Lunar Planet. Sci. Conf.* 11:289–315.

Love, A. E. H. 1944. *A Treatise on the Mathematical Theory of Elasticity,* 4th ed. (New York: Dover).

Low, F. J. 1965. Planetary radiation at infrared and millimeter wavelengths. *Lowell Obs. Bull.* 6:184–187.

Lucchitta, B. K. 1976. Mare ridges and related highland scarps: Result of vertical tectonism? *Proc. Lunar Sci. Conf.* 7:2761–2782.

Lucchitta, B. K. 1977*a*. Topography, structure, and mare ridges in southern Mare Imbrium and northern Oceanus Procellarum. *Proc. Lunar Sci. Conf.* 8:2691–2703.

Lucchitta, B. K. 1977*b*. Crater clusters and light mantle at the Apollo 17 site: A result of secondary impact from Tycho. *Icarus* 30:80–96.

Lucchitta, B. K. 1980. Grooved terrain on Ganymede. *Icarus* 44:481–501.

Lucchitta, B. K., and Soderblom, L. A. 1982. The geology of Europa. In *Satellites of Jupiter,* ed. D. Morrison (Tucson: Univ. of Arizona Press), pp. 521–555.

Lucchitta, B. K., and Watkins, J. A. 1978. Age of graben systems on the Moon. *Proc. Lunar Planet. Sci. Conf.* 9:3459–3472.

Lucchitta, B. K., Soderblom, L. A., and Ferguson, H. M. 1981. Structures on Europa. *Proc. Lunar Planet. Sci. Conf.* 12:1555–1567.

Lugmair, G. W., and Carlson, R. W. 1978. The Sm-Nd history of KREEP. *Proc. Lunar Planet. Sci. Conf.* 9:689–704.

Lugmair, G. W., and Marti, K. 1978. Lunar initial ^{143}Nd/^{144}Nd: Differential evolution of the lunar crust and mantle. *Earth Planet. Sci. Lett.* 39:349–357.

Lumme, K., and Bowell, E. 1981. Radiative transfer in the surfaces of atmosphereless bodies. I. Theory. *Astron. J.* 86:1694–1704.

Lunine, J. I. 1985. Volatiles in the Outer Solar System: I. Thermodynamics of Clathrate Hydrates. II. Ethane Ocean on Titan. III. Evolution of Primordial Titan Atmosphere. Ph.D. Thesis, California Inst. of Technology.

Lunine, J. I., and Stevenson, D. J. 1982*a*. Formation of the Galilean satellites in a gaseous nebula. *Icarus* 52:14–39.

Lunine, J. I., and Stevenson, D. J. 1982*b*. Post accretional evolution of Titan's surface and atmosphere. *Bull. Amer. Astron. Soc.* 14:713 (abstract).

Lunine, J. I., and Stevenson, D. J. 1983. The role of clathrates in the formation and evolution of icy satellites. *Natural Satellites, IAU Coll. 77*, abstract booklet.

Lunine, J. I., and Stevenson, D. J. 1985*a*. Thermodynamics of clathrate hydrate at low and high pressures with application to the outer solar system. *Astrophys. J. Suppl.* 58:493–531.

Lunine, J. I., and Stevenson, D. J. 1985*b*. Physics and chemistry of sulphur lakes on Io. *Icarus* 64:345–367.

Lunine, J. I., Neugebauer, G., and Jakosky, B. M. 1982. Infrared observations of Phobos and Deimos from Viking. *J. Geophys. Res.* 87:10297–10305.

Lunine, J. I., Stevenson, D. J., and Yung, Y. L. 1983. Ethane ocean on Titan. *Science* 222:1229–1230.

Lupo, M. J. 1982. Mass-radius relationships in icy satellites after Voyager. *Icarus* 52:40–53.

Lupo, M. J., and Lewis, J. S. 1979. Mass-radius relationships in icy satellites. *Icarus* 40:125–135.

Lutz, B. L., de Bergh, C., and Owen, T. 1983. Titan: Discovery of carbon monoxide in its atmosphere. *Science* 220:1374–1375.

Lynden-Bell, D., and Pringle, J. E. 1974. The evolution of viscous disks and the origin of nebular variables. *Mon. Not. Roy. Astron. Soc.* 168:603–637.

Lyot, B. 1929. Studies of the polarization of planets. *Ann. Obs. Meudon* 8:1.

Lyttleton, R. A. 1936. On the possible results of an encounter of Pluto with the Neptunian system. *Mon. Not. Roy. Astron. Soc.* 97:108–115.

MacDonald, G. J. F. 1960. Stress history of the Moon. *Planet. Space Sci.* 2:249–255.

MacDonald, G. J. F. 1961. Interior of the Moon. *Science* 133:1045–1050.

MacDonald, G. J. F. 1964. Tidal friction. *Rev. Geophys.* 2:467–541.

MacDougall, D., Rajan, R. S., and Price, P. B. 1974. Gas-rich meteorites: Possible evidence for origin on a regolith. *Science* 183:73–74.

Mackin, J. H. 1969. Origin of lunar maria. *Geol. Soc. Amer. Bull.* 80:735–748.

Macy, W., and Trafton, L. 1980. The distribution of sodium on Io's cloud: Implications. *Icarus* 41:131–141.

Maguire, W. C., Hanel, R. A., Jennings, D. E., Kunde, V. G., and Samuelson, R. E. 1981. C_3H_8 and C_3H_4 in Titan's atmosphere. *Nature* 292:683–686.

Malin, M. C. 1974. Lunar red spots: Possible pre-mare materials. *Earth Planet. Sci. Lett.* 21:331–341.

Malin, M. C. 1980. Morphology of lineaments on Europa. In *The Satellites of Jupiter, IAU Coll. 57*, abstract booklet, 7–2.

Malin, M. C., and Dzurisin, D. 1978. Modifications of crater landforms: Evidence from the moon and Mercury. *J. Geophys. Res.* 83:233–243.

Malvern, L. E. 1969. *Introduction to the Mechanics of a Continuous Medium* (New Jersey: Prentice-Hall).

Mandeville, J. C., Geake, J. E., and Dollfus, A. 1980. Reflectance polarimetry of Callisto and the evolution of the Galilean satellites. *Icarus* 41:343–355.

Marcus, A. H. 1969. Speculations on mass loss by meteoroid impact and formation of the planets. *Icarus* 11:76–87.

Marcus, A. H. 1970. Comparison of equilibrium size distributions for lunar craters. *J. Geophys. Res.* 75:4977–4984.

Markellos, V. V., and Roy, A. E. 1981. Hill stability of satellite orbits. *Celestial Mech.* 23:269–275.

Marouf, E. A., and Tyler, G. L. 1985. Identification of two moonlets embedded in Saturn's rings. *Bull. Amer. Astron. Soc.* 17:717 (abstract). Submitted to *Nature*.

Marouf, E. A., Tyler, G. L., Zebker, H. A., and Eshleman, V. R. 1983. Particle-size distributions in Saturn's rings from Voyager 1 radio occultation. *Icarus* 54:189–211.

Marsden, B. G. 1980. Planets and satellites galore. *Icarus* 44:29–37.

Marvin, U. B. 1973. The moon after Apollo. *Technology Rev.* 75:12–23.

Marvin, U. B., Wood, J. A., Taylor, G. J., Reid, J. B., Powell, B. N., Dickey, J. S., and Bower, J. F. 1971. Relative proportions and probable sources of rock fragments in the Apollo 12 soil samples. *Proc. Lunar Sci. Conf.* 2:679–699.

Mason, B. 1971. *Handbook of Elemental Abundances in Meteorites* (New York: Gordon and Breach).

Mason, R., Guest, J. E., and Cooke, G. N. 1976. An Imbrium pattern of graben on the Moon. *Proc. Geol. Assoc. (London)* 87(2):161–168.

Masursky, H., Schaber, G. G., Soderblom, L. A., and Strom, R. G. 1979. Preliminary geological mapping of Io. *Nature* 280:725–729.

Matson, D. L. 1971. Infrared observations of asteroids. In *Physical Studies of Minor Planets*, ed. T. Gehrels, NASA SP-267, pp. 41–44.

Matson, D. M., and Nash, D. B. 1981. Io's atmosphere: Pressure control by subsurface regolith cold trapping. *Lunar Planet Sci.* XII:664–666 (abstract).

Matson, D. L., and Nash, D. 1983. Io's atmosphere: Pressure control by regolith cold trapping and surface venting. *J. Geophys. Res.* 88:4771–4783.

Matson, D. L., Johnson, T. V., and Fanale, F. P. 1974. Sodium D-line emission from Io: Sputtering and resonant scattering hypothesis. *Astrophys. J.* 192:L43–L46.

Matson, D. L., Johnson, T. V., and Veeder, G. J. 1977. Soil maturity and planetary regoliths: The Moon, Mercury, and the asteroids. *Proc. Lunar Sci. Conf.* 8:1001–1011.

Matson, D. L., Goldberg, B. A., Johnson, T. V., and Carlson, R. W. 1978a. Images of Io's sodium cloud. *Science* 199:531–533.

Matson, D. L., Veeder, G. J., and Lebofsky, L. A. 1978b. Infrared observations of asteroids from earth and space. In *Asteroids: An Exploration Assessment*, eds. D. Morrison and W. Wells, NASA CP-2053, pp. 127–144.

Matson, D. L., Ransford, G. A., and Johnson, T. V. 1981a. Heat flow from Io (J1). *J. Geophys. Res.* 86:1664–1672.

Matson, D. L., Johnson, T. V., and Nash, D. B. 1981b. Io's atmosphere: Regolith cold trapping hypothesis and the interpretation of the Pioneer 10 radio occultation. Paper presented at "4th Conference on Physics of Jovian and Saturnian Magnetospheres," Johns Hopkins Univ., Laurel, MD.

Matson, D. L., Johnson, T. V., Veeder, G. J., Nelson, R. M., Morrison, D., Brown, R. H., and Tokunaga, A. T. 1983. The mystery of the Galilean satellites' brightness temperatures. *Bull. Amer. Astron. Soc.* 15:852 (abstract).

Maurette, M., and Price, P. B. 1975. Electron microscopy of irradiation effects in space. *Science* 187:121–129.

Maxwell, D. E. 1977. Simple Z model of cratering, ejection, and the overturned flap. In *Impact and Explosion Cratering*, eds. D. J. Roddy, R. O. Pepin, and R. B. Merrill (New York: Pergamon), pp. 1003–1008.

Maxwell, D. E., and Seifert, K. 1974. Modeling of cratering, close-in displacement, and ejecta. *Report DNA 3628F* (Washington: Defense Nuclear Agency).

Maxwell, T. A., and El-Baz, F. 1978. The nature of rays and sources of highland material in Mare Crisium. In *Mare Crisium: The View from Luna 24*, eds. R. B. Merrill and J. J. Papike (New York: Pergamon), pp. 89–103.

Mayeva, S. V. 1965. Some calculations of the thermal history of Mars and the Moon. *Soviet Phys.–Doklady* 9:945–948. In Russian.

McCarthy, D. W. Jr., Probst, R. G., and Low, F. J. 1985. Infrared detection of a close cool companion to Van Biesbroeck 8. *Astrophys. J.* 290:L9–L13.

McCauley, J. F. 1967. Geological map of the Hevelius region of the Moon. *U.S. Geol. Surv. Misc. Inv. Ser.*, Map I–491.

McCauley, J. F. 1968. Geological results from lunar precursor problems. *Amer. Inst. Aeronautics and Astronautics J.* 6:1991–1996.

McCauley, J. F., Smith, B. A., and Soderblom, L. A. 1979. Erosional scarps on Io. *Nature* 280:736–738.

McConnell, R. K. Jr., McClaine, L. A., Lee, D. W., Aronson, J. R., and Allen, R. V. 1967. A model for planetary igneous differentiation. *Rev. Geophys.* 5:121–172.

McCord, T. B. 1966. Dynamical evolution of the Neptune system. *Astron. J.* 71:585–590.

McCord, T. B. 1968. The loss of retrograde satellites in the solar system. *J. Geophys. Res.* 73:1497–1500.

McCord, T. B., Clark, R. N., Hawke, B. R., McFadden, L. A., Owensby, P. D., Pieters, C. M., and Adams, J. B. 1981. Moon: Near-infrared reflectance, a first good look. *J. Geophys. Res.* 86:10833–10892.

McDonough, T., and Brice, N. 1973. A Saturnian gas ring and the recycling of Titan's atmosphere. *Icarus* 20:136–145.

McEwen, A. S., and Soderblom, L. A. 1983. Two classes of volcanic plumes on Io. *Icarus* 55:191–218.

McEwen, A. S., and Soderblom, L. A. 1984. Multiple image photometric solutions for the Galilean satellites, NASA TM-86246, pp. 261–262.

McEwen, A. S., Matson, D. L., Johnson, T. V., and Soderblom, L. A. 1985. Volcanic hot spots on Io: Correlation with low-albedo calderas. *J. Geophys. Res.* 90:12345–12377.

McFadden, L. A., Bell, J. F., and McCord, T. B. 1980. Visible spectral reflectance measurements (0.33–1.10 μm) of the Galilean satellites at many orbital phase angles. *Icarus* 44:410–430.

McGetchin, T. R., Settle, M., and Head, J. W. 1973. Radial thickness variation in impact crater ejecta: Implications for lunar basin deposits. *Earth Planet. Sci. Lett.* 20:226–236.

McGill, G. E. 1971. Altitude of fractures bounding straight and arcuate lunar rilles. *Icarus* 14:53–58.

McGregor, R. K., and Emery, A. F. 1969. Free convection through vertical plane layers: Moderate and high Prandtl number fluids. *J. Heat Trans.* 91:391–403.

McKay, D. S., Morrison, D. A., Clanton, U.S., Ladle, G. H., and Lindsay, J. F. 1971. Apollo 12 soil and breccia. *Proc. Lunar Sci. Conf.* 2:755–773.

McKinnon, W. B. 1978. An investigation into the role of plastic failure in crater modification. *Proc. Lunar Planet. Sci. Conf.* 9:3965–3973.

McKinnon, W. B. 1980. Large impact craters and basins: Mechanics of Syngenetic and Postgenetic Modification. Ph.D. thesis, California Inst. of Technology.

McKinnon, W. B. 1981a. Application of ring tectonic theory to Mercury and other solar system bodies. In *Multi-Ring Basins: Formation and Evolution, Proc. Lunar Planet. Sci. 12A* (New York: Pergamon), pp. 259–273.

McKinnon, W. B. 1981b. Tectonic deformation of Galileo Regio and limits to the planetary expansion of Ganymede. *Proc. Lunar Planet. Sci. Conf.* 12B:1585–1597.

McKinnon, W. B. 1981c. EJECTION! Vapor entrainment during cratering, and erosion of the Saturnian satellites. *Bull. Amer. Astron. Soc.* 13:741 (abstract).

McKinnon, W. B. 1981d. Reorientation of Ganymede and Callisto by impact, and interpretation of the cratering record. *EOS Trans. AGU* 62:318 (abstract).

McKinnon, W. B. 1982a. Cratering the icy satellites of Saturn: Effects myriad and sundry. *Saturn*, abstract booklet, Univ. of Arizona.

McKinnon, W. B. 1982b. Impact into the Earth's ocean floor: Preliminary experiments, a planetary model, and possibilities for detection. *Geol. Soc. Amer. Spec. Paper 190*, pp. 129–142.

McKinnon, W. B. 1982c. Problems pertaining to the internal structures of Ganymede and Callisto. *Lunar Planet. Sci.* XIII:499–500 (abstract).

McKinnon, W. B. 1983a. Origin of the E-ring: Condensation of impact vapor or boiling of impact melt? *Lunar Planet. Sci.* XIV:487–488 (abstract).

McKinnon, W. B. 1983b. The strength-gravity transition in cratering. *EOS Trans. AGU* 64:747 (abstract).

McKinnon, W. B. 1983c. Consequences of a capture origin for Triton. *Bull. Amer. Astron. Soc.* 15:857 (abstract).

McKinnon, W. B. 1984. On the origin of Triton and Pluto. *Nature* 311:355–358.

McKinnon, W. B. 1985. Geology of icy satellites. In *Proc. of NATO Workshop on Ices in the Solar System*, eds. J. Klinger, D. Benest, A. Dollfus, and R. Smoluchowski (Dordrecht: Reidel), pp. 829–856.

McKinnon, W. B. 1986. The strength-gravity and other transitions in cratering. Submitted to *Geophys. Res. Lett.*

McKinnon, W. B., and Melosh, H. J. 1980. Evolution of planetary lithospheres: Evidence from multiring basins on Ganymede and Callisto. *Icarus* 44:454–471.

Meador, W. E., and Weaver, W. R. 1975. A photometric function for diffuse reflection by particulate materials, NASA TN-D-7903.

Meeus, J. 1983. *Astronomical Tables of the Sun, Moon, and Planets* (Richmond, VA: Willman-Bell, Inc.).

Meier, T. A. 1981. Color distribution fields of geomorphic features on Europa: Initial results from a new technique 47–47. *Repts. of the Planet. Geol. Prog. 1981*, ed. H. Holt, NASA TM-84211, pp. 47–49.

Mekler, Y., and Eviatar, A. 1984. Spectroscopic observations of Io. *Astrophys. J.* 193: L151–L152.

Melcher, C. L., LePoire, D. J., Cooper, B. H., and Tombrello, T. A. 1982. Erosion of frozen sulfur dioxide by ion bombardment: Applications to Io. *Geophys. Res. Lett.* 9:1151–1154.

Melosh, H. J. 1975. Large impact craters and the moon's orientation. *Earth Planet. Sci. Lett.* 26:353–360.

Melosh, H. J. 1977a. Global tectonics of a despun planet. *Icarus* 31:221–243.

Melosh, H. J. 1977b. Crater modification by gravity: A mechanical analysis of slumping. In *Impact and Explosion Cratering*, eds. D. J. Roddy, R. O. Pepin, and R. B. Merrill (New York: Pergamon), pp. 1245–1260.

Melosh, H. J. 1979. Acoustic fluidization: A new geologic process? *J. Geophys. Res.* 84: 7513–7520.

Melosh, H. J. 1980. Cratering mechanics: Observational, experimental, and theoretical. *Ann. Rev. Earth Planet. Sci.* 8:65–93.

Melosh, H. J. 1981a. A Bingham plastic model of crater collapse. *Lunar Planet. Sci.* XII: 702–704 (abstract).

Melosh, H. J. 1981b. Atmospheric break-up of terrestrial impactors. In *Multi-Ring Basins: Formation and Evolution, Proc. Lunar Planet. Sci. 12A* (New York: Pergamon), pp. 29–35.

Melosh, H. J. 1982a. A schematic model of crater modification by gravity. *J. Geophys. Res.* 87:371–380.

Melosh, H. J. 1982b. A simple mechanical model of Valhalla Basin, Callisto. *J. Geophys. Res.* 87:1880–1890.

Melosh, H. J. 1984. Impact ejection, spallation, and the origin of meteorites. *Icarus* 59:234–260.

Melosh, H. J., and Gaffney, E. S. 1983. Acoustic fluidization and the scale dependence of impact crater morphology. *Proc. Lunar Planet. Sci. Conf. 13* in *J. Geophys. Res. Suppl.* 88:A830–A834.

Melosh, H. J., and McKinnon, W. B. 1978. The mechanics of ringed basin formation. *Geophys. Res. Lett.* 5:985–988.

Mendis, D. A., and Axford, W. I. 1974. Satellites and magnetospheres of the outer planets. *Ann. Rev. Earth Planet. Sci.* 2:419–474.

Menietti, J. D., Green, J. L., Gulkis, S., and Six, F. 1984. Three-dimensional ray tracing of the Jovian magnetosphere in the low-frequency range. *J. Geophys. Res.* 89:1489–1495.

Metzger, A. E., Trombka, J. I., Peterson, L. E., Reedy, J. C., and Arnold, J. R. 1973. Lunar surface radioactivity: Preliminary results of the Apollo 15 and 16 gamma ray spectrometer experiment. *Science* 179:800–803.

Metzger, A. E., Haines, E. L., Etchegaray-Ramirez, M. I., and Hawke, B. R. 1979. Thorium concentrations in the lunar surface. III. Deconvolution of the Appenninus region. *Proc. Lunar Planet. Sci. Conf.* 10:1701–1718.

Metzger, A. E., Gilman, D. A., Luthey, J. L., Luthey, K. C., Schnopper, H. W., Sweard, F. D., and Sullivan, J. D. 1983. The detecton of x rays from Jupiter. *J. Geophys. Res.* 88:7731–7741.

Mignard, F. 1979. The evolution of the lunar orbit revisited, I. *Moon and Planets* 22:301–315.

Mignard, F. 1980. The evolution of the lunar orbit revisited, II. *Moon and Planets* 23:185–201.

Mignard, F. 1981a. On a possible origin of Charon. *Astron. Astrophys.* 96:L1–L2.

Mignard, F. 1981b. The lunar orbit revisited, III. *Moon and Planets* 24:189–207.

Mignard, F. 1981c. The mean elements of Nereid. *Astron. J.* 86:1728–1729.

Mignard, F. 1981d. Evolution of the Martian satellites. *Mon. Not. Roy. Astron. Soc.* 194: 365–379.

Mignard, F. 1982a. Radiation pressure and dust particle dynamics. *Icarus* 49:347–366.

Mignard, F. 1982b. Long time integration of the Moon's orbit. In *Tidal Friction and the Earth's Rotation II,* eds. P. Brosche and J. Sündermann (New York: Springer-Verlag), pp. 67–91.

Mignard, F. 1985. Tidal and non-tidal acceleration of the Earth's rotation. Preprint.

Mignard, F., Soter, S., and Burns, J. A. 1986. Phoebe dust and Iapetus: A critical assessment. In preparation.

Miki, S. 1982. The gaseous flow around a protoplanet in the primordial solar nebula. *Prog. Theor. Phys.* 67:1053–1067.

Miller, S. L. 1961. The occurrence of gas hydrates in the solar system. *Proc. Natl. Acad. Sci. U.S.A.* 47:1798–1808.

Millis, R. L. 1977. UBV photometry of Iapetus: Results from five apparitions. *Icarus* 31:81–88.

Millis, R. L. 1978. Photoelectric photometry of JV. *Icarus* 33:319–321.

Millis, R. L., and Thompson, D. T. 1975. UBV photometry of the Galilean satellites. *Icarus* 26:408–419.

Minear, J. W., and Fletcher, C. R. 1978. Crystallization of a lunar magma ocean. *Proc. Lunar Planet. Conf.* 9:263–283.

Minnaert, M. 1961. Photometry of the Moon. In *Planets and Satellites,* vol. 3, *The Solar System,* eds. G. P. Kuiper and B. M. Middlehurst (Chicago: Univ. of Chicago Press), pp. 213–248.

Minton, R. B. 1973. The red polar caps of Io. *Comm. Lunar Planet. Lab.* 10:35–39.

Mitler, H. E. 1975. Formation of an iron-poor Moon by partial capture, or: Yet another exotic theory of lunar origin. *Icarus* 24:256–268.

Mizuno, H. 1980. Formation of the giant planets. *Prog. Theor. Phys.* 64:544–557.

Mizuno, H., and Boss, A. P. 1985. Tidal disruption of dissipative planetesimals. *Icarus* 63:109–133.

Mizuno, H., and Wetherill, G. W. 1984. Grain abundance in the primordial atmosphere of the earth. *Icarus* 59:74–86.

Mizuno, H., Nakazawa, K., and Hayashi, C. 1978. Instability of a gaseous envelope surrounding a planetary core and formation of giant planets. *Prog. Theor. Phys.* 60:699–710.

Moore, H. J., and Gault, D. E. 1963. Relations between dimensions of impact craters and properties of rock targets and projectiles. *Astrogeologic Studies Annual Progress Report, Part B: Crater Investigations* (Washington: U.S. Geological Survey), pp. 38–79.

Moore, H. J., Hodges, C. A., and Scott, D. H. 1974. Multiringed basins—illustrated by Orientale and associated features. *Proc. Lunar Sci. Conf.* 5:71–100.

Moore, J. M. 1984. The tectonic and volcanic history of Dione. *Icarus* 59:205–220.

Moore, J. M., and Ahearn, J. L. 1983. The geology of Tethys. *J. Geophys. Res.* 88:A577–A584.

Moore, J. M., Horner, V. M., and Greeley, R. 1984. The geomorphology of Rhea. NASA TM-87563, pp. 376–378.

Moore, J. M., Horner, V. M., and Greeley, R. 1985. The geomorphology of Rhea: Implications for geologic history and surface processes. *Proc. Lunar Planet. Sci. Conf. 15* in *J. Geophys. Res. Suppl.* 90:C785–C795.

Moore, M. H. 1984. Studies of proton-irradiated SO₂ at low temperatures: Implications for Io. *Icarus* 59:114–128.

Moore, M. H., and Donn, B. 1982. The infrared spectrum of a laboratory synthesized residue: Implications for the 3.4 micron interstellar absorption feature. *Astrophys. J.* 257:L47–L50.

Moore, M. H., Donn, B., Khanna, R., and A'Hearn, M. 1983. Studies of the proton irradiation of cometary-type ice mixtures. *Icarus* 54:388–405.

Morabito, L. A., Synnott, S. P., Kupferman, P., and Collins, S. A. 1979. Discovery of currently active extraterrestrial volcanism. *Science* 204:972.

Morfill, G. 1982. Interplanetary meteoroids as the source of the ring atmosphere and plasma. *Saturn,* abstract booklet, Univ. of Arizona.

Morfill, G. E., Fechtig, H., Grün, E., and Goertz, C. K. 1983a. Some consequences of meteoroid impacts on Saturn's rings. *Icarus* 55:439–447.

Morfill, G., Grün, E., and Johnson, T. V. 1983b. Saturn's E, G, and F rings: Modulated by the plasma sheet? *J. Geophys. Res.* 88:5573–5579.

Morgan, J. P., Coleman, J. M., and Gagliano, S. M. 1968. Mudlumps: Diapiric structures in Mississippi delta sediments. In *Diapirism and Diapirs,* eds. J. Braunstein and G. D. O'Brian (Tulsa, OK: Amer. Assoc. Petrol. Geol.), pp. 145–161.

Moroz, V. I. 1961. On the infra-red spectrum of Jupiter and Saturn (0.9–2.5 micrometers). *Astron. Z.* 38:1080–1081. In Russian.

Moroz, V. I. 1965. Infrared spectrophotometry of the Moon and the Galilean satellites of Jupiter. *Astron. Z.* 42:1287–1295. In Russian. Trans. *Soviet. Astron. A.J.* 9:999–1006.

Morrison, D. 1973. Determination of radii of satellites and asteroids from radiometry and photometry. *Icarus* 19:1–14.

Morrison, D. 1977. Radiometry of satellites and of the rings of Saturn. In *Planetary Satellites,* ed. J. A. Burns (Tucson: Univ. of Arizona), pp. 269–301.

Morrison, D. 1982*a*. Introduction to the satellites of Jupiter. In *Satellites of Jupiter,* ed. D. Morrison (Tucson: Univ. of Arizona Press), pp. 3–43.

Morrison, D., ed. 1982*b*. *Satellites of Jupiter* (Tucson: Univ. of Arizona Press).

Morrison, D. 1982*c*. The satellites of Jupiter and Saturn. *Ann. Rev. Astron. Astrophys.* 20:469–495.

Morrison, D. 1983. Outer planets satellites. *Rev. Geophys. Space Phys.* 21:151–159.

Morrison, D., and Burns, J. A. 1976. The Jovian satellites. In *Jupiter,* ed. T. Gehrels (Tucson: Univ. of Arizona Press), pp. 991–1034.

Morrison, D., and Cruikshank, D. P. 1973. Thermal properties of the Galilean satellites. *Icarus* 18:224–236.

Morrison, D., and Cruikshank, D. P. 1974. Physical properties of the natural satellites. *Space Sci. Rev.* 15:641–739.

Morrison, D., and Lebofsky, L. A. 1979. Radiometry of asteroids. In *Asteroids,* ed. T. Gehrels (Tucson: Univ. of Arizona Press), pp. 184–205.

Morrison, D., and Morrison, N. D. 1977. Photometry of the Galilean satellites. In *Planetary Satellites,* ed. J. A. Burns (Tucson: Univ. of Arizona Press), pp. 363–378.

Morrison, D., and Telesco, C. M. 1980. Io: Observational constraints on the internal energy and thermodynamics of the surface. *Icarus* 44:226–233.

Morrison, D., Cruikshank, D. P., and Murphy, R. E. 1972. Temperatures of Titan and the Galilean satellites at 20 microns. *Astrophys. J.* 173:L142–L146.

Morrison, D., Morrison, N. D., and Lazarewicz, A. R. 1974. Four color photometry of the Galilean satellites. *Icarus* 23:399–416.

Morrison, D., Jones, T. J., Cruikshank, D. P., and Murphy, R. E. 1975. The two faces of Iapetus. *Icarus* 24:157–171.

Morrison, D., Cruikshank, D. P., Pilcher, C. B., and Rieke, G. H. 1976. Surface compositions of the satellites of Saturn from infrared photometry. *Astrophys. J.* 207:L213–L216.

Morrison, D., Cruikshank, D. P., and Burns, J. A. 1977. Introducing the satellites. In *Planetary Satellites,* ed. J. A. Burns (Tucson: Univ. of Arizona Press), pp. 3–17.

Morrison, D., Cruikshank, D. P., and Brown, R. H. 1982. Diameters of Triton and Pluto. *Nature* 300:425–427.

Morrison, D., Johnson, T. V., Shoemaker, E. M., Soderblom, L. A., Thomas, P., Veverka, J., and Smith, B. A. 1984. Satellites of Saturn: Geological perspective. In *Saturn,* eds. T. Gehrels and M. S. Matthews (Tucson: Univ. of Arizona Press), pp. 609–639.

Morrison, L. V., and Ward, C. C. 1975. An analysis of the transits of Mercury: 1677–1973. *Mon. Not. Roy. Astron. Soc.* 173:183–206.

Morrison, R. H., and Oberbeck, V. R. 1975. Geomorphology of crater and basin deposits—emplacement of the Fra Mauro formation. *Proc. Lunar Sci. Conf.* 6:2503–2530.

Morrison, R. H., and Oberbeck, V. R. 1978. A composition and thickness model for lunar impact crater and basin deposits. *Proc. Lunar Planet. Sci. Conf.* 9:3763–3785.

Moshev, V. V. 1979. Viscosity relations for heavily filled suspensions. *Fluid Mech. Soviet Res.* 8:88–96.

Mueller, S. W., and McKinnon, W. B. 1986. Three-layer models of Ganymede and Callisto: Structure and evolution. Submitted to *Icarus.*

Muhleman, D., Berge, G., and Rudy, D. 1984*a*. Microwave emission from Titan and the Galilean satellites. *Bull. Amer. Astron. Soc.* 16:686 (abstract).

Muhleman, D. O., Berge, G. L., and Clancy, R. T. 1984*b*. Microwave measurements of carbon monoxide on Titan. *Science* 223:393–396.

Mulholland, J. D. 1980. Scientific achievements from ten years of lunar laser ranging. *Rev. Geophys. Space Phys.* 18:549–564.

Munk, W. H., and MacDonald, G. J. F. 1960. *The Rotation of the Earth* (Cambridge: Cambridge Univ. Press).

Murase, T., and McBirney, A. R. 1973. Properties of some common igneous rocks and their melts at high temperatures. *Geol. Soc. Amer. Bull.* 84:3563–3592.

Murcray, F. J., and Goody, R. 1978. Pictures of the Io sodium cloud. *Astrophys. J.* 226:327–335.

Murphy, R. E., Cruikshank, D. P., and Morrison, D. 1972. Radii, albedos, and the 20-micron brightness temperature of Iapetus and Rhea. *Astrophys. J.* 177:L93–L96.

Murray, B. C., Strom, R. G., Trask, N. J., and Gault, D. E. 1975. Surface history of Mercury: Implications for terrestrial planets. *J. Geophys. Res.* 80:2508–2514.

Murray, B., Malin, M. C., and Greeley, R. 1981. *Earthlike Planets* (San Francisco: W. Freeman).

Murray, G. E. 1968. Salt structures of Gulf of Mexico basin: A review. In *Diapirism and Diapirs,* eds. J. Braunstein and G. D. O'Brian (Tulsa, OK: Amer. Assoc. Petrol. Geol.), pp. 99–121.

Murray, J. B. 1975. New observations of surface markings on Jupiter's satellites. *Icarus* 25:397–404.

Mutch, T. A. 1970. *Geology of the Moon: A Stratigraphic View* (Princeton: Princeton Univ. Press).

Nakagawa, Y. 1978. Statistical behavior of planetesimals in the primitive solar system. *Prog. Theor. Phys.* 59:1834–1851.

Nakagawa, Y., Nakazawa, K., and Hayashi, C. 1981. Growth and sedimentation of dust grains in the primordial solar nebula. *Icarus* 45:517–528.

Nakagawa, Y., Hayashi, C., and Nakazawa, K. 1983. Accumulation of planetesimals in the solar nebula. *Icarus* 54:361–376.

Nakamura, Y. 1980. Shallow moonquakes: How they compare with earthquakes. *Proc. Lunar Planet. Sci. Conf.* 11:1847–1854.

Nakamura, Y. 1983. Seismic velocity structure of the lunar mantle. *J. Geophys. Res.* 88:677–686.

Nakamura, Y., and Koyama, J. 1982. Seismic Q of the lunar upper mantle. *J. Geophys. Res.* 87:4855–4861.

Nakamura, Y., Latham, G., Lammlein, D., Ewing, M., Duennebier, F., and Dorman, J. 1974. Deep lunar interior inferred from recent seismic data. *Geophys. Res. Lett.* 1:137–140.

Nakamura, Y., Duennebier, F. K., Latham, G. V., and Dorman, H. J. 1976a. Structure of the lunar mantle. *J. Geophys. Res.* 81:4818–4824.

Nakamura, Y., Latham, G. V., and Dorman, H. J. 1976b. Seismic structure of the Moon. *Proc. Lunar Sci. Conf.* 7:602–603.

Nakamura, Y., Latham, G. V., Dorman, H. J., and Duennebier, F. K. 1976c. Seismic structure of the Moon: A summary of current status. *Proc. Lunar Sci. Conf.* 7:3113–3121.

Nakamura, Y., Latham, G. V., and Dorman, H. J. 1982. Apollo lunar seismic experiment: Final summary. *Proc. Lunar Planet. Sci. Conf. 13* in *J. Geophys. Res. Suppl.* 87:A117–A123.

Nakazawa, K., and Hayashi, C. 1984. Tidal disruption in binary encounter of protoplanetary bodies. In preparation. (Results described in Hayashi et al. 1985.)

Nakazawa, K., and Nakagawa, Y. 1981. Origin of the solar system: Planetary growth in the gaseous nebula. *Suppl. Prog. Theor. Phys.* 70:11–34.

Nakazawa, K., Komuro, T., and Hayashi, C. 1983. Origin of the Moon: Capture by gas drag of the Earth's primordial atmosphere. *Moon and Planets* 28:311–327.

Nash, D. B. 1979. Jupiter sulfur plasma ring. *EOS Trans. AGU* 60:307 (abstract).

Nash, D. B. 1981. Io's 4 μm IR bands: Evidence for normal- and heavy-isotope SO_2 adsorbed on surface particulates. *EOS Trans. AGU* 62:316 (abstract).

Nash, D. B. 1983. Io's 4 μm band and the role of adsorbed SO_2. *Icarus* 54:511–523.

Nash, D. B., and Fanale, F. P. 1977. Io: Surface composition model based on reflectance spectra of sulfur/salt mixtures and proton irradiation experiments. *Icarus* 31:40–80.

Nash, D. B., and Johnson, T. V. 1979. Albedo distribution of Io's surface. *Icarus* 38:69–74.

Nash, D. B., and Nelson, R. M. 1979. Spectral evidence for sublimates and absorbates on Io. *Nature* 280:763–766.

Nash, D., Fanale, F., and Nelson, R. 1980. SO_2 frost: UV-visible reflectivity and limits on Io surface coverage. *Geophys. Res. Lett.* 7:665–668.

Neff, J. S., Lane, W. A., and Fix, J. D. 1974. An investigation of the rotational period of the planet Pluto. *Pub. Astron. Soc. Pacific* 86:225–230.

Nelson, M. L., McCord, T. B., Clark, R. N., Johnson, T. V., Matson, D. 1983. Spectral evidence for magnetospheric interactions with the surfaces of the icy Galilean satellites. *Lunar Planet. Sci.* XIV:554–555 (abstract).

Nelson, M. L., McCord, T. B., Clark, R. N., Johnson, T. V., Matson, D. L., Mosher, J. A., and Soderblom, L. A. 1986. Europa: Characterization and interpretation of global spectral surface units. *Icarus* 65:129–151.

Nelson, R. M., and Hapke, B. W. 1978*a*. Spectral reflectivities of the Galilean satellites and Titan 0.32 to 0.86 micrometers. *Icarus* 36:304–329.

Nelson, R. M., and Hapke, B. W. 1978*b*. Possible correlation of Io's posteclipse brightening with major solar flares. *Icarus* 33:203–209.

Nelson, R. M., Lane, A. L., Matson, D. L., Fanale, F. P., Nash, D. B., and Johnson, T. V. 1980*a*. Io longitudinal distribution of sulfur dioxide frost. *Science* 210:784–786.

Nelson, R. M., Matson, D. L., Lane, A. L., Motteler, F. C., and Ockert, M. E. 1980*b*. Ultraviolet albedos of the Galilean satellites with IUE. *Bull. Amer. Astron. Soc.* 12:713 (abstract).

Nelson, R. M., Pieri, D., Baloga, S., Nash, D., and Sagan, C. 1983. The reflection spectrum of liquid sulfur: Implications for Io. *Icarus* 56:409–413.

Ness, N. F., Acũna, M. H., Lepping, R. P., Burlaga, L. F., Behannon, K. W., and Neubauer, F. M. 1979. Magnetic field studies at Jupiter by Voyager 1: Preliminary results. *Science* 204:982–987.

Neubauer, F. M. 1980. Nonlinear standing Alfvén wave current system at Io: Theory. *J. Geophys. Res.* 85:1171–1178.

Neubauer, F. M., Gurnett, D. A., Scudder, J. D., and Hartle, R. E. 1984. Titan's magnetospheric interaction. In *Saturn*, eds. T. Gehrels and M. S. Matthews (Tucson: Univ. of Arizona Press), pp. 760–787.

Neukum, G. 1985. Cratering records of the satellites of Jupiter and Saturn. *Adv. Space Res.* 5(8):107–116.

Newburn, R. L. Jr., and Gulkis, S. 1973. A survey of the outer planets: Jupiter, Saturn, Uranus, Neptune, Pluto, and their satellites. *Space Sci. Rev.* 14:179–271.

Newsom, H. E., and Drake, M. J. 1982*a*. Constraints on the Moon's origin from the partitioning behavior of tungsten. *Nature* 297:210–212.

Newsom, H. E., and Drake, M. J. 1982*b*. Geochemical arguments against an origin of the Moon by fission from the Earth. *Meteoritics* 17:259–260.

Newsom, H. E., and Drake, M. J. 1983. Experimental investigation of the partitioning of phosphorus between metal and silicate phases: Implications for the Earth, Moon, and encrite parent body. *Geochim. Cosmochim. Acta* 47:93–100.

Nicholson, P. D., and Jones, J. J. 1980. Two-micron spectrophotometry of Uranus and its rings. *Icarus* 42:54–67.

Nicholson, S. B., van Maanen, A., and Willis, H. C. 1931. A preliminary determination of the mass of Neptune's satellite. *Pub. Astron. Soc. Pacific* 43:261–262.

Noland, M., and Veverka, J. 1977*a*. Photometric functions of Phobos and Deimos. II. Surface photometry of Phobos. *Icarus* 30:200–211.

Noland, M., and Veverka, J. 1977*b*. Photometric functions of Phobos and Deimos. III. Surface photometry of Deimos. *Icarus* 30:212–223.

Noland, M., Veverka, J., Morrison, D., Cruikshank, D. P., Lazarewicz, A. R., Morrison, N. D., Elliot, J. L., Goguen, J., and Burns, J. A. 1974. Six-color photometry of Iapetus, Titan, Rhea, Dione, and Tethys. *Icarus* 23:334–354.

Nyquist, L. E., Bansal, B. M., and Wiesmann, H. 1975. Rb-Sr ages and intial $^{87}Sr/^{86}Sr$ for Apollo 17 basalts and KREEP basalt 15386. *Proc. Lunar Sci. Conf.* 6:1445–1465.

Nyquist, L. E., Bansal, B. M., and Wiesmann, H. 1976. Sr-isotopic constraints on the petrogenesis of Apollo 17 mare basalts. *Proc. Lunar Sci. Conf.* 7:1507–1528.

Nyquist, L. E., Bansal, B. M., Wooden, J. L., and Wiesmann, H. 1977. Sr-isotopic constraints on the petrogenesis of Apollo 12 mare basalts. *Proc. Lunar Sci. Conf.* 8:1383–1415.

Nyquist, L. E., Wiesmann, H., Bansal, B., Wooden, J., and McKay, G. 1978. Chemical and Sr-isotopic characteristics of the Luna 24 samples. In *Mare Crisium: The View from Luna 24*, eds. R. B. Merrill and J. J. Papike (New York: Pergamon), pp. 631–656.

Nyquist, L. E., Shih, C.-Y., Wooden, J. L., Bansal, B. M., and Wiesmann, H. 1979. The Sr and Nd isotopic record of Apollo 12 basalts: Implications for lunar geochemical evolution. Proc. Lunar Planet. Sci. Conf. 9:77–114.

Oberbeck, V. R. 1971. A mechanism for the production of lunar crater rays. *The Moon* 2: 263–278.

Oberbeck, V. R. 1975. The role of ballistic erosion and sedimentation in lunar stratigraphy. *Rev. Geophys. Space Phys.* 13:337–362.

Oberbeck, V. R. 1977. Application of high explosion cratering data to planetary problems. In *Impact and Explosion Cratering*, eds. D. J. Roddy, R. O. Pepin, and R. B. Merrill (New York: Pergamon), pp. 45–65.

Oberbeck, V. R., and Morrison, R. H. 1974. Laboratory simulation of the herringbone pattern associated with lunar secondary crater chains. *The Moon* 9:415–455.

Oberbeck, V. R., and Morrison, R. H. 1976. Candidate areas for *in situ* ancient lunar materials. *Proc. Lunar Sci. Conf.* 7:2983–3006.

Oberbeck, V. R., Quaide, W. L., Mahan, M., and Paulson, J. 1973. Monte Carlo calculations of lunar regolith thickness distributions. *Icarus* 19:87–107.

Oberbeck, V. R., Morrison, R. H., Hörz, F., Quaide, W. L., and Gault, D. E. 1974. Smooth plains and continuous deposits of craters and basins. *Proc. Lunar Sci. Conf.* 5:111–136.

Oberlin, F., McCulloch, M. R., Tera, F., Papanastassiou, D. A., and Wasserburg, G. J. 1978. Early lunar differentiation constraints from U-Th-Pb, Sm-Nd, and Rb-Sr model ages. *Lunar Planet. Sci.* IX:832–834 (abstract).

O'Brian, G. D. 1968. Survey of diapirs and diapirism. In *Diapirism and Diapirs*, eds. J. Braunstein and G. D. O'Brian (Tulsa, OK: Amer. Assoc. Petrol. Geol.), pp. 1–9.

Officer, C. B. 1974. *Introduction to Theoretical Geophysics* (Berlin: Springer-Verlag).

Ojakangas, G. W., and Stevenson, D. J. 1986. Episodic volcanism of tidally heated satellites with application to Io. *Icarus* 66. In press.

O'Keefe, J. A. 1970. The origin of the Moon. *J. Geophys. Res.* 75:6565–6574.

O'Keefe, J. A., and Sullivan, F. 1978. Fission origin of the Moon: Cause and timing. *Icarus* 35:272–283.

O'Keefe, J. D., and Ahrens, T. J. 1975. Shock effects from a large impact on the moon. *Proc. Lunar Sci. Conf.* 6:2831–2844.

O'Keefe, J. D., and Ahrens, T. J. 1976. Impact ejecta on the moon. *Proc. Lunar Sci. Conf.* 7:3007–3025.

O'Keefe, J. D., and Ahrens, T. J. 1977. Impact-induced energy partitioning, melting and vaporization on terrestrial planets. *Proc. Lunar Sci. Conf.* 8:3357–3374.

O'Keefe, J. D., and Ahrens, T. J. 1982*a*. The interaction of the Cretaceous/Tertiary extinction bolide with the atmosphere, ocean, and solid Earth. *Geol. Soc. Amer. Special Paper 190*, pp. 103–120.

O'Keefe, J. D., and Ahrens, T. J. 1982*b*. Cometary and meteorite swarm impact on planetary surfaces. *J. Geophys. Res.* 87:6668–6680.

O'Leary, B. T., and van Flandern, T. C. 1972. Io's triaxial figure. *Icarus* 17:209–215.

O'Leary, B. T., and Veverka, J. 1971. On the anomalous brightening of Io after eclipse. *Icarus* 14:265–268.

Ollerhead, R. W., Bøttiger, J., Davies, J. A., L'Ecuyer, J., Haugen, H. K., and Matsunami, N. 1980. Evidence for a thermal spike mechanism in the erosion of frozen xenon. *Rad. Eff.* 49:203–212.

O'Nions, R. K., Evenson, N. M., and Hamilton, P. J. 1979. Geochemical modelling of mantle differentiation and crustal growth. *J. Geophys. Res.* 84:6091–6101.

Onorato, P. I. K., Uhlmann, D. R., and Simonds, C. H. 1978. The thermal history of the Manicouagan impact melt sheet, Quebec. *J. Geophys. Res.* 83:2789–2798.

Öpik, E. J. 1972. Comments on lunar origin. *Irish Astron. J.* 10:190–238.

O'Reilly, T. C., and Davies, G. F. 1980. Magma transport of heat on Io: A mechanism allowing a thick lithosphere. *Geophys. Res. Lett.* 8:313–316.

Orient, O. J., and Srivastava, S. 1984. Mass spectrometric determination of partial and total electron impact ionization cross-sections of SO_2 from threshold to 200 eV. *J. Chem. Phys.* 80:140–143.

Ornatskaya, O. J., Alber, Y. I., and Ryanzontseva, I. L. 1977. Calculations of the Moon's thermal history at different concentrations of radioactive elements, taking into account differentiation on melting. In *The Soviet-American Conference on Cosmochemistry of the Moon and Planets*, NASA SP-300, pp. 347–366.

Orphal, D. L. 1977. Calculations of explosion cratering. II. Cratering mechanics and phenomenology. In *Impact and Explosion Cratering*, eds. D. J. Roddy, R. O. Pepin, and R. B. Merrill (New York: Pergamon), pp. 484–509.

Orphal, D. L., Borden, W. F., Larson, S. A., and Schultz, P. H. 1980. Impact melt generation and transport. *Proc. Lunar Planet. Sci. Conf.* 11:2309–2323.

Ostro, S. J. 1982. Radar properties of Europa, Ganymede, and Callisto. In *Satellites of Jupiter,* ed. D. Morrison (Tucson: Univ. of Arizona Press), pp. 213–236.

Ostro, S. J. 1983. Planetary radar astronomy. *Rev. Geophys. Space Phys.* 21:186–196.

Owen, T. 1982. The composition and origin of Titan's atmosphere. *Planet. Space Sci.* 30: 833–838.

Owen, T., Lutz, B. L., and de Bergh, C. 1986. Deuterium in the outer solar system: Evidence for two distinct reservoirs. *Nature* 320:244–246.

Pang, K. D. 1983. Planetary astronomy in ancient and modern China. *Bull. Amer. Astron. Soc.* 15:840 (abstract).

Pang, K. D., and Nicholson, P. D. 1986. Spectral measurements of the Uranian rings. *J. Geophys. Res.* In press.

Pang, K. D., and Rhoads, J. W. 1983. Interpretation of disk-integrated photometry of the Uranian satellites. *Natural Satellites, IAU Coll. 77,* abstract booklet, Cornell Univ.

Pang, K. D., Pollack, J. B., Veverka, J., Lane, A. L., and Ajello, J. M. 1978. The composition of Phobos: Evidence for carbonaceous chondrite surface from spectral analysis. *Science* 199:64–66.

Pang, K. D., Lumme, K., and Bowell, E. 1981. Microstructure and particulate properties of Io and Ganymede: Comparison with other solar system bodies. *Proc. Lunar Planet. Sci. Conf.* 12:1543–1553.

Pang, K. D., Voge, C. C., and Rhoads, J. W. 1982. Macrostructure and microphysics of Saturn's E ring. Proc. "Planetary Rings/Anneaux des Planètes Conf.," Toulouse, France, August (abstract).

Pang, K. D., Ajello, J., Lumme, K., and Bowell, E. 1983. Interpretation of integrated disk photometry of Callisto and Ganymede. *Proc. Lunar Planet. Sci. Conf.* 13 in *J. Geophys. Res. Suppl.* 88:A569–A576.

Pang, K. D., Voge, C. C., Rhoads, J. W., and Ajello, J. M. 1984a. The E ring of Saturn and satellite Enceladus. *J. Geophys. Res.* 89:9459–9470.

Pang, K. D., Voge, C. C., and Rhoads, J. W. 1984b. Macrostructure and microphysics of Saturn's E-ring. In *Planetary Rings / Anneaux des Planètes*, ed. A. Brahic Toulouse: Editions Cepadues), pp. 607–613.

Papadakos, D. N., and Williams, I. P. 1983. Escape of ejecta from cratered solar system satellites. *Mon. Not. Roy. Astron. Soc.* 204:635–645.

Papaloizou, J., and Lin, D. N. C. 1984. On the tidal interaction between protoplanets and the primordial solar nebula. I. Linear calculation of the role of angular momentum exchange. *Astrophys. J.* 285:818–834.

Papike, J. J., Simon, S. B., and Carl, J. C. 1982. The lunar regolith: Chemistry, mineralogy, and petrology. *Rev. Geophys. Space Phys.* 20:761–826.

Parameswaran, V. R., and Jones, S. J. 1975. Brittle fracture of ice at 77 K. *J. Glaciol.* 14: 305–315.

Parmentier, E. M., and Head, J. W. 1979a. Internal processes affecting surfaces of low-density satellites: Ganymede and Callisto. *J. Geophys. Res.* 84:6263–6276.

Parmentier, E. M., and Head, J. W. 1979b. Some possible effects of solid-state deformation on the thermal evolution of ice-silicate planetary bodies. *Proc. Lunar Planet. Sci. Conf.* 10:2403–2419.

Parmentier, E. M., and Head, J. W. 1981. Viscous relaxation of impact craters on icy planetary surfaces: Determination of viscosity variation with depth. *Icarus* 47:100–111.

Parmentier, E. M., and Head, J. W. 1984. A model for the emplacement and solidification of bright terrain on Ganymede. *Lunar Planet. Sci.* XV:633–634 (abstract).

Parmentier, E. M., Zuber, M. T., and Head, J. W. 1981. Ganymede tectonics: Global rifting due to planetary expansion? *Conf. on the Processes of Planetary Rifting* (Houston: Lunar Planet. Inst.), pp. 28–30 (abstract).

Parmentier, E. M., Squyres, S. W., Head, J. W., and Allison, M. L. 1982. The tectonics of Ganymede. *Nature* 295:290–293.

Passey, Q. R. 1982. Viscosity Structure of the Lithospheres of Ganymede, Callisto, and Enceladus, and of the Earth's Mantle. Ph.D. Thesis, California Inst. of Technology.

Passey, Q. R. 1983. Viscosity of the lithosphere of Enceladus. *Icarus* 53:105–120.

Passey, Q. R., and Shoemaker, E. M. 1982*a*. Craters and basins on Ganymede and Callisto: Morphological indicators of crustal evolution. In *Satellites of Jupiter,* ed. D. Morrison (Tucson: Univ. of Arizona Press), pp. 379–434.

Passey, Q. R., and Shoemaker, E. M. 1982*b*. Early thermal histories of Ganymede and Callisto. *Lunar Planet. Sci.* XIII:619–620 (abstract).

Peale, S. J. 1975. Dynamical consequences of meteorite impacts on the Moon. *J. Geophys. Res.* 80:4939–4946.

Peale, S. J. 1976*a*. Excitation and relaxation of the wobble, precession, and libration of the Moon. *J. Geophys. Res.* 81:1813–1827.

Peale, S. J. 1976*b*. Orbital resonances in the solar system. *Ann. Rev. Astron. Astrophys.* 14:215–245.

Peale, S. J. 1977. Rotation histories of the natural satellites. In *Planetary Satellites,* ed. J. A. Burns (Tucson: Univ. Arizona Press), pp. 87–112.

Peale, S. J. 1978. An observational test for the origin of the Titan-Hyperion orbital resonance. *Icarus* 36:240–244.

Peale, S. J. 1984. The rotation of Hyperion. *Phil. Trans. Roy. Soc. London* A313:147–156.

Peale, S. J., and Cassen, P. 1978. Contribution of tidal dissipation to lunar thermal history. *Icarus* 36:245–269.

Peale, S. J., and Greenberg, R. 1980. On the Q of Jupiter. *Lunar Planet. Sci.* XI:871–873 (abstract).

Peale, S. J., and Wisdom, J. 1984. Do current observations support the hypothesis of chaotic rotation for Hyperion? *Bull. Amer. Astron. Soc.* 16:686 (abstract).

Peale, S. J., Cassen, P., and Reynolds, R. T. 1979. Melting of Io by tidal dissipation. *Science* 203:892–894.

Peale, S. J., Cassen, P. M., and Reynolds, R. T. 1980. Tidal dissipation, orbital evolution, and the nature of Saturn's inner satellites. *Icarus* 43:65–72.

Pearl, J. C., and Sinton, W. M. 1982. Hot spots of Io. In *Satellites of Jupiter,* ed. D. Morrison (Tucson: Univ. of Arizona Press), pp. 724–755.

Pearl, J., Hanel, R., Kunde, V., Maguire, W., Fox, K., Gupta, S., Ponnamperuma, C., and Raulin, F. 1979. Identification of gaseous SO_2 and new upper limits for other gases on Io. *Nature* 280:757–758.

Pearl, J., Hanel, R., Ospina, M., Samuelson, R., Jere, G., and Khanna, R. 1983. Implications of Voyager infrared spectra for the crustal state of Io. *Natural Satellies, IAU Coll.* 77, abstract booklet, Cornell Univ., p. 18.

Pechernikova, G. V., Majeva, S. V., and Vitjazev, A. V. 1984. The dynamics of circumplanetary swarms. *Sov. Astron. J. Lett.* 10:703–710.

Peltier, W. R. 1985. The Lageos constraint on deep mantle viscosity: Results from a new normal mode method for the inversion of viscoelastic relaxation spectra. *J. Geophys. Res.* 90:9411–9421.

Pendleton, Y. J., and Black, D. C. 1983. Further studies on criteria for the onset of dynamical instability in general three-body systems. *Astron. J.* 88:1415–1419.

Perri, F., and Cameron, A. G. W. 1974. Hydrodynamic instability of the solar nebula in the presence of a planetary core. *Icarus* 22:416–425.

Pettit, E. 1961. Planetary temperature measurements. In *Planets and Satellites,* vol. 3, *The Solar System,* eds. G. P. Kuiper and B. M. Middlehurst (Chicago: Univ. of Chicago Press), pp. 400–428.

Pettit, E., and Nicholson, S. B. 1930. Lunar radiation and temperatures. *Astrophys. J.* 71:102–135.

Phillips, R. J., and Ivins, E. R. 1979. Geophysical observations pertaining to solid state convection in the terrestrial planets. *Phys. Earth Planet. Int.* 19:107–148.

Phillips, R. J., and Lambeck, K. 1980. Gravity fields of the terrestrial planets: Long-wavelength anomalies and tectonics. *Rev. Geophys. Space Phys.* 18:27–76.

Phillips, R. J., and Malin, M. C. 1980. Ganymede: A relationship between thermal history and crater statistics. *Science* 210:185–188.

Phinney, R. A., and Anderson, D. L. 1967. Present knowledge about the thermal history of the Moon. In *Physics of the Moon*, ed. S. F. Singer (Tarzana, CA: Amer. Astron. Soc.), pp. 161–170.

Phinney, W. C., and Simonds, C. H. 1977. Dynamical implications of the petrology and distribution of impact melt rocks. In *Impact and Explosion Cratering*, eds. D. J. Roddy, R. O. Pepin, and R. B. Merrill (New York: Pergamon), pp. 771–790.

Piddington, J. H., and Drake, J. F. 1968. Electrodynamic effects of Jupiter's satellite Io. *Nature* 217:935–937.

Piekutowski, A. J. 1977. Cratering mechanisms observed in laboratory-scale high-explosive experiments. In *Impact and Explosion Cratering*, eds. D. J. Roddy, R. O. Pepin, R. B. Merrill (New York: Pergamon), pp. 1245–1260.

Pieri, D. C. 1981. Lineament and polygon patterns on Europa. *Nature* 289:17–21.

Pieri, D. C., and Hiller, K. 1984. J3 Quadrangle, Europa: Preliminary geologic designations. In *Reports Planet. Geol. Prog. 1983,* ed. H. Holt, NASA TM-86246, pp. 318–320.

Pieri, D. C., Baloga, S. M., Nelson, R. M., and Sagan, C. 1984. The sulfur flows of Ra Patera, Io. *Icarus* 60:685–700.

Pieri, D. C., Hiller, K., and Rudnyk, R. 1985. Geological map, J3 quadrangle of Europa. Submitted to NASA/USGS Galilean Satellite Mapping Prog.

Pieters, C. M. 1977. Characterization and Distribution of Lunar Basalt Types Using Remote Sensing Techniques. Ph.D. Thesis, Massachusetts Inst. of Technology.

Pieters, C. M. 1978. Mare basalt types on the front side of the Moon: A summary of spectral reflectance data. *Proc. Lunar Planet. Sci. Conf.* 9:2825–2849.

Pieters, C. M. 1982. Copernicus crater central peak: Lunar mountain of unique composition. *Science* 215:59–61.

Pieters, C. M., Adams, J. B., Head, J. W., McCord, T. B., and Zisk, S. H. 1982. Primary ejecta in crater rays: The Copernicus example. *Lunar Planet. Sci.* XIII:623–624 (abstract).

Pieters, C. M., Adams, J. B., Mouginis-Mark, P. J., Zisk, S. H., Smith, M. O., Head, J. W., McCord, T. B. 1985. The nature of crater rays: Copernicus example. *J. Geophys. Res.* 90:12,393–12,413.

Pike, R. J. 1974*a*. Depth/diameter relations of fresh lunar craters: Revision from spacecraft data. *Geophys. Res. Lett.* 1:291–294.

Pike, R. J. 1974*b*. Ejecta from large craters on the moon: Comments on the geometric model of McGetchin et al. *Earth Planet. Sci. Lett.* 23:265–271.

Pike, R. J. 1977. Size dependence in the shape of fresh impact craters on the moon. In *Impact and Explosion Cratering*, eds. D. J. Roddy, R. O. Pepin, and R. B. Merrill (New York: Pergamon), pp. 484–509.

Pike, R. J. 1980*a*. Control of crater morphology by gravity and target type: Mars, Earth, Moon. *Proc. Lunar Planet. Sci. Conf.* 11:2159–2189.

Pike, R. J. 1980*b*. Formation of complex impact craters: Evidence from Mars and other planets. *Icarus* 43:1–19.

Pilcher, C. B. 1979. The stability of water on Io. *Icarus* 37:559–574.

Pilcher, C. B. 1980*a*. Images of Jupiter's sulfur ring. *Science* 207:181–183.

Pilcher, C. B. 1980*b*. Transient sodium ejection from Io. *Bull. Amer. Astron. Soc.* 12:675 (abstract).

Pilcher, C. B., and Morgan, J. 1979. Detection of singly ionized oxygen around Jupiter. *Science* 205:297–298.

Pilcher, C. B., and Strobel, D. F. 1982. Emission from neutrals and ions in the Jovian magnetosphere. In *Satellites of Jupiter,* ed. D. Morrison (Tucson: Univ. of Arizona Press), pp. 807–845.

Pilcher, C. B., Chapman, C. R., Lebofsky, L. A., and Kieffer, H. H. 1970. Saturn's rings: Identification of water frost. *Science* 178:1087–1089.

Pilcher, C. B., Ridgway, S. T., and McCord, T. B. 1972. Galilean satellites: Identification of water frost. *Science* 178:1087–1089.

Pilcher, C. B., Smyth, W. H., Combi, M. R., and Fertel, J. H. 1984. Io's sodium directional features: Evidence for a magnetospheric-wind-driven gas escape mechanism. *Astrophys. J.* 287:427–444.

Pinto, J. P., Lunine, J. I., Kim, S. J., and Yung, Y. L. 1985. The D to H ratio and the origin and evolution of Titan's atmosphere. *Nature* 319:388–390.

Pirronello, V., Strazzulla, G., Foti, G., and Rimini, E. 1981. MeV helium erosion yield of nitrogen frozen gas. *Nucl. Instr. Meth.* 182/183:315–317.

Pirronello, V., Brown, W. L., Lanzerotti, L. J., Simmons, E., and Marcantonio, K. J. 1982. Formaldehyde formation in a H_2O/CO_2 ice mixture under irradiation by gas ions. *Astrophys. J.* 262:636–640.

Platzman, G. W., Curtis, G. A., Hansen, K. S., and Slater, R. D. 1981. Normal modes of the world ocean. Part II. Description of modes in the period range 8 to 80 hours. *J. Phys. Oceanography* 11:579–603.

Plescia, J. B. 1983. The geology of Dione. *Icarus* 56:255–277.

Plescia, J. B., and Boyce, J. M. 1982. Crater densities and geological histories of Rhea, Dione, Mimas, and Tethys. *Nature* 295:285–290.

Plescia, J. B., and Boyce, J. M. 1983. Crater numbers and geological histories of Iapetus, Enceladus, Tethys and Hyperion. *Nature* 301:666–670.

Plescia, J. B., and Boyce, J. M. 1985. Impact cratering history of the Saturnian satellites. *J. Geophys. Res.* 90:2029–2037.

Plummer, H. C. 1918, rpt. 1960. *Introductory Treatise on Dynamical Astronomy* (New York: Dover).

Podolak, M., and Bar-Nun, A. 1979. A constraint on the distribution of Titan's atmospheric aerosol. *Icarus* 39:272–276.

Podolak, M., Noy, N., and Bar-Nun, A. 1979. Photochemical aerosols in Titan's atmosphere. *Icarus* 40:193–198.

Poincaré, H. 1892. *Les Methods Nouvelles de la Mecanique Celeste* (Paris: Gauthier-Villars).

Poirier, J. P. 1982. Rheology of ices: A key to the tectonics of the ice moons of Jupiter and Saturn. *Nature* 299:683–688.

Poirier, J. P., Sotin, C., and Peyronneau, J. 1981. Viscosity of high-pressure ice VI and evolution and dynamics of Ganymede. *Nature* 292:225–227.

Poirier, J. P., Boloh, L., and Chambon, P. 1983. Tidal dissipation in small viscoelastic ice moons: The case of Enceladus. *Icarus* 55:218–230.

Pollack, J. B. 1977. Phobos and Deimos. In *Planetary Satellites*, ed. J. A. Burns (Tucson: Univ. of Arizona Press), pp. 319–345.

Pollack, J. B., and Consolmagno, G. 1984. Origin and evolution of the Saturn system. In *Saturn*, eds. T. Gehrels and M. S. Matthews (Tucson: Univ. of Arizona Press), pp. 811–866.

Pollack, J. B., and Fanale, F. 1982. Origin and evolution of the Jupiter satellite system. In *Satellites of Jupiter*, ed. D. Morrison (Tucson: Univ. of Arizona Press), pp. 872–910.

Pollack, J. B., and Reynolds, R. T. 1974. Implications of Jupiter's early contraction history for the composition of the Galilean satellites. *Icarus* 21:248–253.

Pollack, J. B., and Witteborn, F. C. 1980. Evolution of Io's volatile inventory. *Icarus* 44:249–267.

Pollack, J. B., Summers, A., and Baldwin, B. 1973a. Estimates of the size of particles in the rings of Saturn and their cosmogonic implications. *Icarus* 20:263–278.

Pollack, J. B., Veverka, J., Noland, M., Sagan, C., Duxbury, T. C., Acton, C. H. Jr., Born, G. H., Hartmann, W. K., and Smith, B. A. 1973b. Mariner 9 television observations of Phobos and Deimos. II. *J. Geophys. Res.* 78:4313–4326.

Pollack, J. B., Grossman, A. S., Moore, R., and Graboske, H. C. Jr. 1976. The formation of Saturn's satellites and rings as influenced by Saturn's contraction history. *Icarus* 29:35–48.

Pollack, J. B., Grossman, A. S., Moore, R., and Graboske, H. C. Jr. 1977. A calculation of Saturn's gravitational contraction history. *Icarus* 30:111–128.

Pollack, J. B., Witteborn, F. C., Erickson, E. F., Strecker, D. W., Baldwin, B. J., and Bunch, T. E. 1978a. Near-infrared spectra of the Galilean satellites: Observations and compositional implications. *Icarus* 36:271–303.

Pollack, J. B., Veverka, J., Pang, K., Colburn, D., Lane, A. L., and Ajello, J. M. 1978b. Multicolor observations of Phobos with the Viking Lander cameras: Evidence for a carbonaceous chondrite composition. *Science* 199:66–69.

Pollack, J. B., Burns, J. A., and Tauber, M. E. 1979. Gas drag in primordial circumplanetary envelopes: A mechanism for satellite capture. *Icarus* 37:587–611.

Poscolieri, M. 1982. Stratigraphic relationships among the upper layers of the outer Galilean satellites, inferred from the investigation of their ray systems. In *The Comparative Study of the Planets*, eds. A. Coradini and M. Fulchignoni (Dordrecht: Reidel), pp. 485–494.

Poscolieri, M., and Schultz, P. H. 1980. Crater rays on Ganymede and Callisto. *The Satellites of Jupiter, IAU Coll. 57,* abstract booklet, Univ. of Hawaii.

Prentice, A. J. R., and Ter Haar, D. 1979*a*. Origin of the Jovian ring and the Galilean satellites. *Nature* 280:300–302.

Prentice, A. J. R., and Ter Haar, D. 1979*b*. Formation of the regular satellite systems and rings of the major planets. *Moon and Planets* 21:43–62.

Prinn, R. G., and Fegley, B. Jr. 1981. Kinetic inhibition of CO and N_2 reduction in circumplanetary nebulae: Implications for satellite composition. *Astrophys. J.* 249:308–317.

Puetter, R. C., and Russell, R. W. 1977. The 2–4 μm spectrum of Saturn's rings. *Icarus* 32: 37–40.

Pullan, S., and Lambeck, K. 1980. On constraining lunar mantle temperatures from gravity data. *Proc. Lunar Planet. Sci. Conf.* 11:2031–2041.

Purves, N. G., and Pilcher, C. B. 1980. Thermal migration of water on the Galilean satellites. *Icarus* 43:51–55.

Qiu, Y., Griffith, J. E., Meng, W. J., and Tombrello, T. A. 1983. Sputtering of silicon and its compounds in the electronic stopping region. *Rad. Eff.* 70:231–236.

Quaide, W. L., and Oberbeck, V. R. 1968. Thickness determinations of the lunar surface layer from lunar impact craters. *J. Geophys. Res.* 73:5247–5270.

Quaide, W. L., Gault, D. E., and Schmidt, R. A. 1965. Gravitative effects on lunar impact structure. *Ann. N.Y. Acad. Sci.* 123:563–572.

Raedeke, L. D., and McCallum, I. S. 1980. A comparison of fractionation trends in the lunar crust and the Stillwater Complex. *Proc. Conf. Lunar Highlands Crust* (New York: Pergamon), pp. 133–153.

Rajan, R. S. 1974. On the irradiation history and origin of gas-rich meteorites. *Geochim. Cosmochim. Acta* 38:777–788.

Ramberg, H. 1981. *Gravity, Deformation and the Earth's Crust.* (London: Academic Press).

Ransford, G. A., Finnerty, A. A., and Collerson, K. D. 1981. Europa's petrological thermal history. *Nature* 289:21–24.

Reid, M. 1973. The tidal loss of satellite-orbiting objects and its implications for the lunar surface. *Icarus* 20:240–248.

Reitsema, H. J., Hubbard, W. B., Lebofsky, L. A., and Tholen, D. J. 1982. Occultation by a possible third satellite of Neptune. *Science* 215:289–291.

Reitsema, H. J., Vilas, F., and Smith, B. A. 1983. A charge-coupled device observation of Charon. *Icarus* 56:75–79.

Remsberg, A. R. 1981. A structural analysis of Valhalla Basin, Callisto. *Lunar Planet. Sci.* XII:874–876 (abstract).

Remsberg, A. R. 1982. Tectonics of Valhalla Basin, Callisto. M.S. Thesis, State Univ. of New York at Stony Brook.

Reynolds, R. T., and Cassen, P. 1979. On the internal structure of the major satellites of the outer planets. *Geophys. Res. Lett.* 6:121–124.

Reynolds, R. T., Peale, S. J., and Cassen, P. 1980. Energy constraints and plume volcanism. *Icarus* 44:234–239.

Rieke, G. H. 1975. The temerature of Amalthea. *Icarus* 25:333–334.

Rieke, G. H., Lebofsky, L. A., Lebofsky, M. J., and Montgomery, E. F. 1981. Unidentified features in the spectrum of Triton. *Nature* 294:59–60.

Rieke, G. H., Lebofsky, L. A., and Lebofsky, M. J. 1985. A search for nitrogen on Triton. *Icarus* 64:153–155.

Rigden, S. M., and Ahrens, T. J. 1981. Impact vaporization and lunar origin. *Lunar Planet. Sci.* XII:885–887 (abstract).

Rinehart, J. S. 1968. Intense destructive stresses resulting from stress wave interactions. In *Shock Metamorphism of Natural Materials,* eds. B. M. French and N. M. Short (Baltimore: Mono Book Corp.), pp. 31–42.

Ringwood, A. E. 1960. Some aspects of the thermal evolution of the Earth. *Geochim. Cosmochim. Acta* 20:241–249.

Ringwood, A. E. 1970. Origin of the Moon: The precipitation hypothesis. *Earth Planet. Sci. Lett.* 8:131–140.

Ringwood, A. E. 1979. *Origin of the Earth and Moon* (Berlin: Springer-Verlag).

Ringwood, A. E., and Kesson, S. E. 1976. A dynamic model for mare basalt petrogenesis. *Proc. Lunar Sci. Conf.* 7:1697–1722.

Roddy, D. J. 1977. Large-scale impact and explosion craters: Comparisons of morphological and structural analogs. In *Impact and Explosion Cratering,* eds. D. J. Roddy, R. O. Pepin, and R. B. Merrill (New York: Pergamon), pp. 185–246.

Roddy, D. J., Boyce, J. M., Colton, G. W., and Dial, A. L. 1975. Meteor Crater, Arizona, rim drilling with thickness, structural uplift, depth, volume, and mass-balance calculations. *Proc. Lunar Sci. Conf.* 6:2621–2644.

Roscoe, R. 1952. The viscosity of suspensions of rigid spheres. *British J. Appl. Phys.* 3:267–269.

Ross, J. V., Avé Lallement, H. G., and Carter, N. L. 1979. Activation volume for creep in the upper mantle. *Science* 203:261–263.

Ross, M. N., and Schubert, G. 1985. Tidally forced viscous heating in a partially molten Io. *Icarus* 64:391–400.

Ross, M., and Schubert, G. 1986. Tidal dissipation in a viscoelastic planet. *J. Geophys. Res.* 91:D447–D452.

Ross, R. G., and Anderson, P. 1982. Clathrate and other solid phases in the tetrahydrofuran-water system: Thermal conductivity and heat capacity under pressure. *Canadian J. Chem.* 60:881–892.

Rosse, Lord. 1868. On the radiation of heat from the Moon. *Proc. Roy. Soc.* 17:436–443.

Roy, A. E., and Ovenden, M. W. 1954. On the occurrence of commensurable mean motions in the solar system. *Mon. Not. Roy. Astron. Soc.* 114:232–241.

Roy, A. E., Carusi, A., Valsecchi, G. B., and Walker, I. W. 1984. The use of the energy and angular momentum integrals to obtain a stability criterion in the general hierarchical three-body problem. *Astron. Astrophys.* 141:25–29.

Rubincam, D. P. 1975. Tidal friction and the early history of the Moon's orbit. *J. Geophys. Res.* 80:1537–1548.

Runcorn, S. K. 1962. Convection in the Moon. *Nature* 195:1150–1151.

Runcorn, S. K. 1977. Early melting of the Moon. *Proc. Lunar Sci. Conf.* 8:463–469.

Runcorn, S. K. 1978. The ancient lunar core dynamo. *Science* 194:771–773.

Runcorn, S. K. 1983. Lunar magnetism, polar displacements and primeval satellites in the Earth-Moon system. *Nature* 304:589–596.

Ruskol, E. L. 1960a. On the origin of protoplanets. In *Problems of Cosmogony,* vol. 7 (Moscow: Acad. Sci. USSR), pp. 8–14. In Russian.

Ruskol, E. L. 1960b. Origin of the Moon I. *Sov. Astron. A.J.* 4:657–668.

Ruskol, E. L. 1963. On the origin of the moon. II. The growth of the moon in the circumterrestrial swarm of satellites. *Sov. Astron. A.J.* 7:221–227.

Ruskol, E. L. 1967. The tidal history and origin of the Earth-Moon system. *Sov. Astron. A.J.* 10:659–665.

Ruskol, E. L. 1972a. On the initial distance of the Moon forming in the circumterrestrial swarm. In *The Moon, IAU Symp. 47,* eds. S. K. Runcorn and H. C. Urey (Dordrecht: Reidel), pp. 402–404.

Ruskol, E. L. 1972b. The role of the satellite's swarm in the origin of the rotation of the Earth. *Astron. Vestn.* 6:91–95. In Russian.

Ruskol, E. L. 1972c. The origin of the Moon. III. Some aspects of the dynamics of the circumterrestrial swarm. *Sov. Astron. A.J.* 15:646–654.

Ruskol, E. L. 1975. *The Origin of the Moon* (Moscow: Nauka). In Russian. Trans. NASA TT-F-16623.

Ruskol, E. L. 1977. The origin of the Moon. In *Proc. Soviet-Amer. Conf. Cosmochemistry of the Moon and Planets,* eds. J. Pomeroy and N. Hubbard, NASA SP-370, pp. 815–822.

Ruskol, E. L. 1981. Formation of planets. In *The Solar System and Its Exploration* (Paris: ESA), pp. 107–113.

Ruskol, E. L. 1982a. Origin of planetary satellites. *Izvestiya Earth Phys.* 18:425–433. In Russian.

Ruskol, E. L. 1982b. The origin of satellites. *Phys. Earth Acad. Sci. USSR* 6:40–51. In Russian.

Ryder, G., and Taylor, G. J. 1976. Did mare-type volcanism commence early in lunar history? *Proc. Lunar Sci. Conf.* 7:1741–1755.

Saari, D. G. 1984. From rotations and inclinations to zero configurational velocity surfaces. I. A natural rotating co-ordinate system. *Celestial Mech.* 33:299–318.

Saari, J. M., and Shorthill, R. W. 1963. Isotherms of crater regions on the illuminated and eclipsed Moon. *Icarus* 2:115–136.

Saari, J. M., and Shorthill, R. W. 1967. *Isothermal and Isophotic Atlas of the Moon*, NASA CP CR-855.

Safronov, V. S. 1958. On the turbulence in the protoplanetary cloud. *Rev. Mod. Phys.* 30: 1023–1024.

Safronov, V. S. 1969. *Evolution of the Protoplanetary Cloud and Formation of the Earth and Planets* (Moscow: Nauka). In Russian. Trans. NASA TT-F-677, 1972.

Safronov, V. S. 1978. The heating of the Earth during its formation. *Icarus* 33:3–12.

Safronov, V. S., and Ruskol, E. L. 1957. On the hypothesis of turbulence in the protoplanetary cloud. *Voprosy Kosmogonii* 5:22–46. In Russian.

Safronov, V. S., and Ruskol, E. L. 1977. Accumulation of satellites. In *Planetary Satellites,* ed. J. A. Burns (Tucson: Univ. of Arizona Press), pp. 501–512.

Safronov, V. S., and Ruskol, E. L. 1982. On the origin and initial temperature of Jupiter and Saturn. *Icarus* 49:284–296.

Sagan, C. 1979. Sulfur flows on Io. *Nature* 280:750–753.

Sagan, C., and Dermott, S. F. 1982. The tide in the seas of Titan. *Nature* 300:731–733.

Sagan, C., Khare, B. N., Lewis, J. S. 1984. Organic matter in the Saturn system. In *Saturn,* eds. T. Gehrels and M. S. Matthews (Tucson: Univ. of Arizona Press), pp. 788–807.

Sammis, C. G., Smith, J. C., Schubert, G., and Yuen, D. A. 1977. Viscosity-depth profile of the Earth's mantle: Effects of polymorphic phase transitions. *J. Geophys. Res.* 82:3747–3761.

Sammis, C. G., Smith, J. C., and Schubert, G. 1981. A critical assessment of estimation methods for activation volume. *J. Geophys. Res.* 86:10707–10718.

Sampson, R. A. 1921. Theory of the four great satellites of Jupiter. *Mem. Roy. Astron. Soc.* 63.

Samuelson, R. E., Hanel, R. A., Kunde, V. G., and Maguire, W. C. 1981. Mean molecular weight and hydrogen abundance of Titan's atmosphere. *Nature* 292:688–693.

Samuelson, R. E., Maguire, W. C., Hanel, R. A., Kunde, V. G., Jennings, D. E., Yung, Y. L., and Aikin, A. C. 1983. CO_2 on Titan. *J. Geophys. Res.* 88:8709–8715.

Sandel, B. R., and Broadfoot, A. L. 1982. Io's hot plasma torus: A synoptic view from Voyager. *J. Geophys. Res.* 87:212–218.

Sandel, B. R., Shemansky, D., Broadfoot, A., Bertaux, J., Blamont, J., Belton, M. J. S., Ajello, J. M., Holberg, J. B., Atreya, S. K., Donahue, T. M., Moos, H. W., Strobel, D. F., McConnell, J. C., Dalgarno, A., Goody, R., McElroy, M. B., and Takacs, P. Z. 1979. Extreme ultraviolet observations from Voyager 2 encounter with Jupiter. *Science* 206:962–966.

Sandel, B. R., Shemansky, D. E., Broadfoot, A. L., Holberg, J. B., Smith, G. R., McConnell, J. C., Strobel, D. F., Atreya, S. K., Donahue, T. M., Moos, H. W., Hunten, D. M., Pomphrey, R. B., and Linick, S. 1982. Extreme ultraviolet observations from the Voyager 2 encounter with Saturn. *Science* 215:548–553.

Scattergood, T., and Owen, T. 1977. On the sources of ultraviolet absorption in spectra of Titan and the outer planets. *Icarus* 30:780–788.

Schaber, G. G. 1973. Lava flows in Mare Imbrium: Geologic evidence from Apollo orbital photography. *Proc. Lunar Sci. Conf.* 4:73–92.

Schaber, G. G. 1980. The surface of Io: Geologic units, morphology and tectonics. *Icarus* 43:302–333.

Schaber, G. G. 1982. The geology of Io. In *Satellites of Jupiter,* ed. D. Morrison (Tucson: Univ. of Arizona Press), pp. 556–597.

Schafer, E. H. 1976. A trip to the moon. *J. Amer. Oriental Soc.* 96(1):27–37.

Schenk, P. M. 1983. The crustal morphology and structure of Europa. *Natural Satellites, IAU Coll. 77,* abstract booklet, Cornell Univ., 4–A.

Schenk, P. M. 1984. The crustal tectonics and history of Europa: A structural, morphological, and comparative study. In *Advances in Planetary Geology,* NASA TM-86247, pp. 3–111.

Schenk, P. M., and McKinnon, W. B. 1984. Dark ray craters on Ganymede: Evidence for impact of D-type asteroids? *Bull. Amer. Astron. Soc.* 16:683 (abstract).

Schenk, P. M., and McKinnon, W. B. 1985. Dark halo craters and the thickness of grooved terrain on Ganymede. *Proc. Lunar Planet. Sci. Conf. 16* in *J. Geophys. Res. Suppl.* 90: C775–C783.

Schmidt, O. Yu. 1950. Genesis of planets and satellites. *Izv. Acad. Nauk SSSR* 14:29–45. In Russian.

Schmidt, O. Yu. 1957. *Four Lectures on the Theory of the Earth's Origin,* 3rd ed. (Moscow: Izdatelstvo AN SSSR). In Russian.

Schmidt, R. M. 1980. Meteor Crater: Energy of formation—implications of centrifuge scaling. *Proc. Lunar Planet. Sci. Conf.* 11:2099–2128.

Schmidt, R. M. 1981. Scaling crater time-of-formation. *EOS Trans. AGU* 62:944 (abstract).

Schmidt, R. M. 1983. Strength-gravity transition for impact craters in wet sand. *Lunar. Planet. Sci.* XIV:666–667 (abstract).

Schmidt, R. M. 1984. Transient crater motions: Saturated sand centrifuge experiments. *Lunar Planet. Sci.* XV:722–723 (abstract).

Schmidt, R. M., and Holsapple, K. A. 1982. Estimates of crater size for large-body impact: Gravity-scaling results. *Geol. Soc. Amer. Special Paper 190,* pp. 93–102.

Schoenberg, E. 1925. Investigations concerning theories of the illumination of the Moon based on photometric measurements. *Acta Soc. Scien. Fennicae* 50(9).

Schonfeld, E. 1982. Organic chemistry on Europa? *Lunar Planet. Sci.* XIII:691 (abstract).

Schubert, G., and Lichtenstein, B. R. 1974. Observations on Moon-plasma interactions by orbital and surface experiments. *Rev. Geophys. Space Phys.* 12:592–626.

Schubert, G., Turcotte, D. L., and Oxburgh, E. R. 1969. Stability of planetary interiors. *Geophys. J. Roy. Astron. Soc.* 18:441–460.

Schubert, G., Yuen, D. A., and Turcotte, D. L. 1975. Role of phase transitions in a dynamic mantle. *Geophys. J. Roy. Astron. Soc.* 42:705–735.

Schubert, G., Young, R. E., and Cassen, P. 1977. Solid state convection models of the lunar internal temperature. *Phil. Trans. Roy. Soc. London* A285:523–536.

Schubert, G., Cassen, P., and Young, R. E. 1979. Subsolidus convective cooling histories of terrestrial planets. *Icarus* 38:192–211.

Schubert, G., Stevenson, D. J., and Cassen, P. 1980. Whole planet cooling and the radiogenic heat source contents of the Earth and Moon. *J. Geophys. Res.* 85:2531–2538.

Schubert, G., Stevenson, D. J., and Ellsworth, K. 1981. Internal structures of the Galilean satellites. *Icarus* 47:46–59.

Schultz, P. H. 1976. *Moon Morphology* (Austin: Univ. of Texas Press).

Schultz, P. H., and Gault, D. E. 1979. Atmospheric effects on Martian ejecta emplacement. *J. Geophys. Res.* 84:7669–7687.

Schultz, P. H., and Gault, D. E. 1982. Impact ejecta dynamics in an atmosphere: Experimental results and extrapolations. *Geol. Soc. Amer. Special Paper 190,* pp. 153–174.

Schultz, P. H., and Gault, D. E. 1983. High-velocity clustered impacts: Experimental results. *Lunar Planet. Sci.* XIV:674–675 (abstract).

Schultz, P. H., and Lutz-Garihan, A. B. 1981. Ancient polar locations on Mars: Evidence and implications. *Papers Presented to the Third International Colloquium on Mars* (Houston: Lunar Planet. Inst.), pp. 229–231 (abstract).

Schultz, P. H., and Mendell, W. 1978. Orbital infrared observations of lunar craters and possible implications for impact ejecta emplacement. *Proc. Lunar Planet. Sci. Conf.* 9:2857–2883.

Schultz, P. H., and Spudis, P. D. 1979. Evidence for ancient mare volcanism. *Proc. Lunar Planet. Sci. Conf.* 10:2899–2918.

Schultz, P. H., and Spudis, P. O. 1983. Beginning and end of lunar mare volcanism. *Nature* 302:233–236.

Schulz, M., and Eviatar, A. 1977. Charged particle absorption by Io. *Astrophys. J.* 211: L149–L154.

Scott, D. H., and Eggleton, R. E. 1973. Geological map of the Rümker quadrangle of the Moon. *U.S. Geol. Surv. Misc. Inv. Ser.,* Map I–805.

Scott, D. H., Diaz, J. M., and Watkins, J. A. 1975. The geologic evaluation and regional synthesis of metric and panoramic photographs. *Proc. Lunar Sci. Conf.* 6:2531–2540.

Scrutton, C. T. 1978. Periodic growth features in fossil organisms and the length of the day and month. In *Tidal Friction and the Earth's Rotation,* eds. P. Brosche and J. Sündermann (New York: Springer-Verlag), pp. 154–196.

Seiberling, L. E., Meins, C. K., Cooper, B. H., Griffith, J. E., Mendenhall, M. H., and Tombrello, T. A. 1982. The sputtering of insulating materials by fast heavy ions. *Nucl. Instr. Meth.* 198:17–25.

Seidelmann, P. K., Kaplan, G. H., Pulkkiner, K. F., Santoro, E. J., and van Flandern, T. C. 1980. Ephemeris of Pluto. *Icarus* 44:19–28.

Seidelmann, P. K., Harrington, R. S., Pascu, D., Baum, W. A., Currie, D. G., Westphal, J. A., and Danielson, G. E. 1981. Saturn satellite observations and orbits from the 1980 ring plane crossing. *Icarus* 47:282–287.

Sekiya, M. 1983. Gravitational instabilities in a dust-gas layer and formation of planetesimals in the solar nebula. *Prog. Theor. Phys.* 69:1116–1130.

Sekiya, M., Hayashi, C., and Nakazawa, K. 1981. Dissipation of the primordial terrestrial atmosphere due to irradiation of the solar far-UV during T Tauri stage. *Prog. Theor. Phys.* 66:1301–1316.

Settle, M., and Head, J. W. 1977. Radial variation of lunar crater rim topography. *Icarus* 31:123–135.

Settle, M., and Head, J. W. 1979. The role of rim slumping in the modification of lunar impact craters. *J. Geophys. Res.* 84:3081–3096.

Settle, M., Head, J. W., and McGetchin, T. R. 1974. Ejecta from large craters on the Moon: Discussion. *Earth Planet. Sci. Lett.* 23:271–274.

Shakura, N. I., and Sunyaev, R. A. 1973. Black holes in binary systems: Observational appearances. *Astron. Astropys.* 24:337–355.

Sharpe, H. N., and Peltier, W. R. 1978. Parameterized mantle convection and the Earth's thermal history. *Geophys. Res. Lett.* 5:737–740.

Sharpton, V. L., and Head, J. W. 1981. The origin of mare ridges: Evidence from basalt stratigraphy and substructure in Mare Serenitatis. *Lunar Planet. Sci.* XII:961–963 (abstract).

Sharpton, V. L., and Head, J. W. 1982. Stratigraphy and structural evolution of southern Mare Serenitatis: A reinterpretation based on Apollo lunar sounder experiment data. *J. Geophys. Res.* 87:10983–10998.

Shaya, E. J., and Pilcher, C. B. 1984. Polar cap formation on Ganymede. *Icarus* 58:74–80.

Shemansky, D. 1980. Mass loading and the diffusion loss rate of the Io plasma torus. *Astrophys. J.* 242:1266–1277.

Shemansky, D. E., and Sandel, B. R. 1982. The injection of energy into the Io plasma torus. *J. Geophys. Res.* 87:219–229.

Sherman, P. 1968. *Emusion Science* (New York: Academic).

Shoberg, T. G. 1982. On the Accretion of the Terrestrial Planets with Possible Applications to the Satellites of the Outer Planets. M.A. Thesis, Washington Univ., St. Louis.

Shoemaker, E. M. 1962. Interpretation of lunar craters. In *Physics and Astronomy of the Moon,* ed. Z. Kopal (New York: Academic), pp. 283–357.

Shoemaker, E. M. 1963. Impact mechanics at Meteor Crater, Arizona. In *The Moon, Meteorites, and Comets,* vol. 4, eds. B. M. Middlehurst and G. P. Kuiper (Chicago: Univ. of Chicago Press), pp. 301–336.

Shoemaker, E. M. 1984. Kuiper Prize Lecture, 16th DPS Meeting, Kona, HI.

Shoemaker, E. M., and Hackman, R. J. 1962. Stratigraphic basis for a lunar time scale. In *The Moon,* eds. Z. Kopal and Z. K. Mikhailov (New York: Academic), pp. 289–300.

Shoemaker, E. M., and Wolfe, R. F. 1981. Evolution of the Saturnian satellites: The role of impact. *Lunar Planet. Sci.* XII:1–3 (abstract).

Shoemaker, E. M., and Wolfe, R. F. 1982. Cratering time scales for the Galilean satellites. In *Satellites of Jupiter,* ed. D. Morrison (Tucson: Univ. of Arizona Press), pp. 277–339.

Shoemaker, E. M., Hackman, R. J., and Eggleton, R. E. 1963. Interplanetary correlation of geologic time. *Adv. Astronaut. Sci.* 88:70–89.

Shoemaker, E. M., Batson, R. M., Holt, H. E., Morris, E. C., Rennilson, J. J., and Whitaker, E. A. 1965. Television observations from Surveyor VII. In *Surveyor VII Mission Report, Part II, Science Results,* NASA-JPL TR-32-1264, pp. 9–76.

Shoemaker, E. M., Batson, R. M., Holt, H. E., Morris, E. C., Rennilson, J. J., and Whitaker, E. A. 1969. Observations of the lunar regolith and the earth from the television camera on Surveyor 7. *J. Geophys. Res.* 74:6081–6119.

Shoemaker, E. M., Hait, M. H., Swann, G. A., Schleicher, D. L., Dahlem, D. H., Schaber, G. G., and Sutton, R. L. 1970. Lunar regolith at Tranquility Base. *Science* 175:4697–4704.

Shoemaker, E. M., Lucchitta, B. K., Plescia, J. B., Squyres, S. W., and Wilhelms, D. E. 1982. The geology of Ganymede. In *Satellites of Jupiter,* ed. D. Morrison (Tucson: Univ. of Arizona Press), pp. 435–520.

Shor, V. A. 1975. The motion of the Martian satellites. *Celestial Mech.* 12:61–75.

Short, N. M. 1975. *Planetary Geology* (New Jersey: Prentice-Hall).

Shorthill, R. W., and Saari, J. M. 1965. Radiometric and photometric mapping of the Moon through a lunation. *Ann. N.Y. Acad. Sci.* 123:722–739.

Showalter, M. R., Cuzzi, J. N., Marouf, E. A., and Esposito, L. W. 1985a. Satellite wakes and the orbit of the Encke gap moonlet. *Bull. Amer. Astron. Soc.* 17:716 (abstract). Also *Icarus* 66. In press.

Showalter, M. R., Burns, J. A., Cuzzi, J. N., and Pollack, J. B. 1985b. Discovery of Jupiter's "gossamer" ring. *Nature* 316:526–528.

Showalter, M. R., Burns, J. A., Cuzzi, J. N., and Pollack, J. B. 1986. Jupiter's ring system: New results on ring structure and particle properties. Submitted to *Icarus*.

Shu, F. H., and Lubow, S. H. 1981. Mass, angular momentum, and energy transfer in close binary stars. *Ann. Rev. Astron. Astrophys.* 19:277–293.

Shu, F. H., Dones, L., Lissauer, J. J., Yuan, C., and Cuzzi, J. N. 1985. Non-linear spiral density waves: Viscous damping. *Astrophys. J.* 299:542–573.

Shvashtein, Z. I. 1973. Experimental studies in an ice-research laboratory. In *Ice Physics and Ice Engineering*, ed. G. N. Vakolev (Jerusalem: Israel Program for Scientific Translations).

Siegfried, R. W., and Solomon, S. C. 1974. Mercury: Internal structure and thermal evolution. *Icarus* 23:192–205.

Sieveka, E. 1983. Redistribution and Coronal Formation on the Galilean Satellites by Plasma Ion Sputtering of the Surfaces. Ph.D. Thesis, Univ. of Virginia.

Sieveka, E. M., and Johnson, R. E. 1982. Thermal- and plasma-induced molecular redistribution on the icy satellites. *Icarus* 51:528–548.

Sieveka, E. M., and Johnson, R. E. 1984. Ejection of atoms and molecules from Io by plasma-ion impact. *Astrophys. J.* 287:418–426.

Sieveka, E. M., and Johnson, R. E. 1985. Non-isotropic coronal atmosphere on Io. *J. Geophys. Res.* 90:5327–5331.

Sigmund, P. 1969. Theory of sputtering I: Sputtering yield of amorphous and polycrystalline targets. *Phys. Rev.* 184:383–416.

Sill, G. T., and Clark, R. N. 1982. Composition of the surfaces of the Galilean satellites. In *Satellites of Jupiter*, eds. D. Morrison (Tucson: Univ. of Arizona Press), pp. 174–212.

Simonds, C. H., Floran, R. J., McGee, P. E., Phinney, W. C., and Warner, J. L. 1978. Petrogenesis of melt rock, Manicougan impact structure, Quebec. *J. Geophys. Res.* 83:2773–2788.

Simonelli, D. P. 1983. Amalthea: Implications of the temperature observed by Voyager. *Icarus* 54:524–538.

Simonelli, D. P., and Veverka, J. 1983. Opposition effect of individual regions on Io. *Natural Satellites, IAU Coll. 77*, abstract booklet, Cornell Univ.

Simonelli, D. P., and Veverka, J. 1984a. Voyager disk-integrated photometry of Io. *Icarus* 59:406–425.

Simonelli, D. P., and Veverka, J. 1984b. Accurate albedos of the brightest regions on Io. *Bull. Amer. Astron. Soc.* 16:654 (abstract).

Simonelli, D. P., and Veverka, J. 1985. Disk-resolved photometry of Io. I. Near opposition limb darkening. *Icarus* 66. In press.

Simpson, J. A., Hamilton, D. C., McKibben, R. B., Mogro-Campero, A., Pyle, K. R., and Tuzzolino, A. J. 1974. The protons and electrons trapped in the Jovian dipole magnetic field region and their interaction with Io. *J. Geophys. Res.* 79:3522–3544.

Simpson, J. A., Bastian, T. S., Chenette, D. L., McKibben, R. B., and Pyle, K. R. 1980. The trapped radiations of Saturn and their absorption by satellites and rings. *J. Geophys. Res.* 85:5731–5762.

Sinclair, A. T. 1972. On the origin of the commensurabilities amongst the satellites of Saturn. *Mon. Not. Roy. Astron. Soc.* 160:169–187.

Sinclair, A. T. 1974. On the origin of the commensurabilities of the satellites of Saturn. II. *Mon. Not. Roy. Astron. Soc.* 166:165–179.

Sinclair, A. T. 1975. The orbital resonance amongst the Galilean satellites of Jupiter. *Mon. Not. Roy. Astron. Soc.* 171:59–72.

Sinclair, A. T. 1983. A reconsideration of the evolution hypothesis of the origin of the resonances among Saturn's satellites. In *Dynamical Trapping and Evolution in the Solar System*, eds. V. V. Markellos and Y. Kozai (Dordrecht: Reidel), pp. 19–25.

Sinclair, A. T. 1984. Perturbations on the orbits of companions of the satellites of Saturn. *Astron. Astrophys.* 136:161–166.

Singer, A. V. 1983. Acceleration and crushing of ejecta by an impact-generated gas cloud. *EOS Trans. AGU* 64:254 (abstract).

Singer, R. B. 1981. Near-infrared reflectance of mineral mixtures: Systematic combinations of pyroxenes, olivine, and iron oxides. *J. Geophys. Res.* 86:7967–7982.

Singer, S. F. 1968. The origin of the Moon and geophysical consequences. *Geophys. J.* 15: 205–226.

Singer, S. F. 1970. How did Venus lose its angular momentum? *Science* 170:1196–1198.

Singer, S. F. 1971. The Martian satellites. In *Physical Studies of Minor Planets*, ed. T. Gehrels, NASA SP-267, pp. 399–405.

Singer, S. F. 1975. When and where were the satellites of Uranus formed? *Icarus* 25:484–488.

Sinton, W. M. 1962. Temperatures on the lunar surface. In *Physics and Astronomy of the Moon*, ed. Z. Kopal (New York: Academic), pp. 407–428.

Sinton, W. M. 1972. A near-infrared view of the Uranus system. *Sky and Telescope* 44:304–305.

Sinton, W. M. 1973. Does Io have an ammonia atmosphere? *Icarus* 20:284–296.

Sinton, W. M. 1977. Uranus: The rings are black. *Science* 198:503–504.

Sinton, W. M. 1980. Io: Are vapor explosions responsible for the 5-μm outbursts? *Icarus* 43: 56–64.

Sinton, W. M. 1981. The thermal emission spectrum of Io and a determination of the heat flux from its hot spots. *J. Geophys. Res.* 86:3122–3128.

Sinton, W., Tokunaga. A., Becklin, E., Gatley, I., Lee, T., and Lonsdale, C. J. 1980. Io: Groundbased observations of hot spots. *Science* 210:1015–1017.

Siscoe, G. L. 1971. Two magnetic tail models for Uranus. *Planet. Space Sci.* 19:483–490.

Sjogren, W. L. 1977. Lunar gravity determinations and their implications. *Phil. Trans. Roy. Soc. London* A285:219–226.

Sjogren, W. L. 1983. Planetary geodesy. *Rev. Geophys. Space Phys.* 21:528–537.

Sjogren, W. L., R. N. Wimberly, and Wollenhumpt, W. R. 1978. Lunar gravity in the Apollo 15 and 16 subsatellites. *The Moon* 9:115–128.

Slattery, W. L., DeCampli, W. M., and Cameron, A. G. W. 1980. Protoplanetary core formation by rain-out of minerals. *Moon and Planets* 23:381–390.

Sleep, N. H., and Langan, R. T. 1981. Thermal evolution of the Earth: Some recent developments. *Adv. Geophys.* 23:1–23.

Slobodkin, L. S., Buyakov, I. F., Triput, N. S., Cess, R. D., Caldwell, J., and Owen, T. 1980. Spectra of SO_2 frost for application to emission observations of Io. *Nature* 285:211–212.

Smith, B. A. 1984. Near infrared imaging of Uranus and Neptune. In *Uranus and Neptune*, ed. J. T. Bergstralh, NASA CP-2330, pp. 213–223.

Smith, B. A., and Smith, S. A. 1972. Upper limits for an atmosphere on Io. *Icarus* 17:218–222.

Smith, B. A., and Terrile, R. J. 1984. A circumstellar disk around β Pictoris. *Science* 226: 1421–1424.

Smith, B. A., Soderblom, L. A., Johnson, T. V., Ingersoll, A. P., Collins, S. A., Shoemaker, E. M., Hunt, G. E., Masursky, H., Carr, M. H., Davies, M. E., Cook, A. F. II, Boyce, J., Danielson, G. E., Owen, T., Sagan, C., Beebe, R. F., Veverka, J., Strom, R. G., McCauley, J. F., Morrison, D., Briggs, G. A., and Suomi, V. E. 1979*a*. The Jupiter system through the eyes of Voyager 1. *Science* 204:951–972.

Smith, B. A., Soderblom, L. A., Beebe, R., Boyce, J., Briggs, G., Carr, M., Collins, S. A., Cook, A. F. II, Danielson, G. E., Davies, M. E., Hunt, G. E., Ingersoll, A., Johnson, T. V., Masursky, H., McCauley, J., Morrison, D., Owen, T., Sagan, C., Shoemaker, E. M., Strom, R., Suomi, V. E., and Veverka, J. 1979*b*. The Galilean satellites and Jupiter: Voyager 2 imaging science results. *Science* 206:927–950.

Smith, B. A., Shoemaker, E. M., Kieffer, S. W., and Cook, A. F. II. 1979*c*. The role of SO_2 in volcanism on Io. *Nature* 280:738–743.

Smith, B. A., Reitsema, H. J., Fountain, J. W., and Larson, S. M. 1980. Saturn's synergistic co-orbital satellites. *Bull. Amer. Astron. Soc.* 12:727 (abstract).

Smith, B. A., Soderblom, L., Beebe, R., Boyce, J., Briggs, G., Bunker, A., Collins, S. A., Hansen, C. J., Johnson, T. V., Mitchell, J. L., Terrile, R. J., Carr, M., Cook, A. F. II, Cuzzi, J., Pollack, J. B., Danielson, G. E., Ingersoll, A., Davies, M. E., Hunt, G. E., Masursky, H.,

Shoemaker, E., Morrison, D., Owen, T., Sagan, C., Veverka, J., Strom, R., and Suomi, V. E. 1981. Encounter with Saturn: Voyager 1 imaging science results. *Science* 212:163–191.

Smith, B. A., Soderblom, L., Batson, R., Bridges, P., Inge, J., Masursky, H., Shoemaker, E., Beebe, R., Boyce, J., Briggs, G., Bunker, A., Collins, S. A., Hansen, C. J., Johnson, T. V., Mitchell, J. L., Terrile, R. J., Cook, A. F. II, Cuzzi, J., Pollack, J. B., Danielson, G. E., Morrison, D., Owen, T., Sagan, C., Veverka, J., Strom, R., and Suomi, V. E. 1982. A new look at the Saturn system: The Voyager 2 images. *Science* 215:504–537.

Smith, B. A., Soderblom, L. A., Beebe, R., Bliss, D., Boyce, J. M., Brahic, A., Briggs, G. A., Brown, R. H., Collins, S. A., Cook, A. F. II, Croft, S. K., Cuzzi, J. N., Danielson, G. E., Davies, M. E., Dowling, T. E., Godfrey, D., Hansen, C. J., Harris, C., Hunt, G. E., Ingersoll, A. P., Johnson, T. V., Krauss, R. J., Masursky, H., Morrison, D., Owen, T., Plescia, J., Pollack, J. B., Porco, C. P., Rages, K., Sagan, C., Shoemaker, E. M., Sromovsky, L. A., Stoker, C., Strom, R. G., Suomi, V. E., Synnott, S. P., Terrile, R. J., Thomas, P., Thompson, W. R., and Veverka, J. 1986. Voyager 2 in the Uranian system: Imaging science results. Submitted to *Science*.

Smith, E. I. 1973. Identification, distribution, and significance of lunar volcanic domes. *The Moon* 6:3–31.

Smith, J. C., and Born, G. H. 1976. Secular acceleration of Phobos and Q of Mars. *Icarus* 27:52–54.

Smith, J. V. 1974. Origin of the Moon by disintegrative capture with chemical differentiation followed by sequential accretion. *Lunar Sci.* V:718–720 (abstract).

Smith, J. V. 1982. Heterogeneous growth of meteorites and planets, especially the Earth and Moon. *J. Geol.* 90:1–48.

Smith, J. V., Anderson, A. T., Newton, R. C., Olsen, E. J., Wyllie, P. J., Crewe, A. V., Isaacson, M. S., and Johnson, D. 1970. Petrologic history of the Moon inferred from petrography, mineralogy, and petrogenesis of Apollo 11 rocks. *Proc. Apollo 11 Lunar Sci. Conf.*, ed. A. A. Levinson (New York: Pergamon), pp. 897–925.

Smith, M. L., and Dahlen, F. A. 1981. The period and Q of the Chandler wobble. *Geophys. J. Roy. Astron. Soc.* 64:223–231.

Smoluchowski, R. 1983. Solar system ice: Amorphous or crystalline? *Science* 222:161–163.

Smoluchowski, R., and McWilliam, A. 1984. Structure of ices on satellites. *Icarus* 58:282–287.

Smrekar, S., and Pieters, C. M. 1984. Spectral similarities of impact melts from large lunar highland craters. *Lunar Planet. Sci.* XV:802–803 (abstract).

Smyth, W. H. 1979. Io's sodium cloud: Explanation of the east-west asymmetries. *Astrophys. J.* 234:1148–1153.

Smyth, W. H. 1983. Io's sodium cloud: Explanation of the east-west asymmetries. II. *Astrophys. J.* 264:708–725.

Smyth, W., and McElroy, M. 1978. Io's sodium cloud: Comparison of models and two-dimensional images. *Astrophys. J.* 226:336–346.

Smyth, W., and Shemansky, D. 1983. Escape and ionization of atomic oxygen from Io. *Astrophys. J.* 271:865–875.

Smythe, W. D., Nelson, R. M., and Nash, D. B. 1979. Spectral evidence for SO_2 frost or absorbate on Io's surface. *Nature* 280:766.

Soderblom, L., Johnson, T., Morrison, D., Danielson, G. E., Smith, B., Veverka, J., Cook, A., Sagan, C., Kupferman, P., Pieri, D., Mosher, J., Avis, C., Gradie, J., and Clancy, T. 1980. Spectrophotometry of Io: Preliminary Voyager 1 results. *Geophys. Res. Lett.* 7:963–966.

Soifer, B. T., Neugebauer, G., and Gatley, I. 1979. The near-infrared reflectivity of the dark and light faces of Iapetus. *Astron. J.* 84:1644–1646.

Soifer, B. T., Neugebauer, G., and Matthews, K. 1980. The 1.5-2.5 μm spectrum of Pluto. *Astron. J.* 85:166–167.

Soifer, B. T., Neugebauer, G., and Matthews, K. 1981. Near-infrared spectrophotometry of the satellites and rings of Uranus. *Icarus* 45:612–617.

Solomon, S. C. 1978. On the volcanism and thermal tectonics on one-plate planets. *Geophys. Res. Lett.* 5:461–464.

Solomon, S. C., and Chaiken, J. 1976. Thermal expansion and thermal stress in the Moon and terrestrial planets: Clues to early thermal history. *Proc. Lunar Sci. Conf.* 7:3229–3243.

Solomon, S. C., and Head, J. W. 1979. Vertical movement in mare basins: Relation to mare emplacement, basin tectonics, and lunar thermal history. *J. Geophys. Res.* 84:1667–1682.

Solomon, S. C., and Head, J. W. 1980. Lunar mascon basins: Lava filling, tectonics, and evolution of the lithosphere. *Rev. Geophys. Space Phys.* 18:107–141.

Solomon, S. C. and Longhi, J. 1977. Magma oceanography. I. Thermal evolution. *Proc. Lunar Sci. Conf.* 8:583–599.

Sonett, C. P. 1982. Electromagnetic induction in the Moon. *Rev. Geophys. Space Phys.* 20: 411–455.

Sonett, C. P., Smith, B. F., Colburn, D. S., Schubert, G., and Schwartz, K. 1972. The reduced magnetic field of the Moon: Conductivity profiles and inferred temperature. *Proc. Lunar Sci. Conf.* 3:2309–2336.

Sonett, C. P., Colburn, D. S., and Schwartz, K. 1975. Formation of the lunar crust: An electrical source of heating. *Icarus* 24:231–255.

Soter, S. 1971. The dust belts of Mars. CRSR Report No. 462, Cornell Univ.

Soter, S. 1974. Paper presented at IAU Planetary Satellites Conference, Cornell Univ., Ithaca, NY.

Soter, S., and Harris, A. W. 1977. The equilibrium figures of Phobos and other small bodies. *Icarus* 30:192–199.

Sotin, C., Gillet, P., and Poirier, J. P. 1985. Creep of high-pressure ice VI. *Ices in the Solar System*, eds. J. Klinger, D. Benest, A. Dollfus, and R. Smoluchowski (Dordrecht: Reidel), pp. 109–118.

Southwood, D. J., Kivelson, M. G., Walker, R. J., and Slavin, J. A. 1980. Io and its plasma environment. *J. Geophys. Res.* 85:5959–5968.

Spencer, J. R. 1982. Voyager Ganymede stellar occulation and surface ice temperatures. In *Rept. Planet. Geol. Prog.—1982*, NASA TM-85127, pp. 32–34 (abstract).

Spencer, J. R. 1983. Analysis of Voyager thermal infrared observations of Ganymede. *Natural Satellites, IAU Coll. 77*, abstract booklet, Cornell Univ., p. 25.

Spencer, J. R. 1984. An ice-poor interpretation of Callisto's reflectance spectrum. *Bull. Amer. Astron. Soc.* 16:685 (abstract).

Spencer, J. R., and Maloney, P. R. 1984. Mobility of water ice on Callisto: Evidence and implications. *Geophys. Res. Lett.* 11:1223–1226.

Spencer Jones, H. 1939. The rotation of the Earth and the secular accelerations of the Sun, Moon and planets. *Mon. Not. Roy. Astron. Soc.* 99:541–558.

Spudis, P. D. 1978. Composition and origin of the Apennine Bench Formation. *Proc. Lunar Planet. Sci. Conf.* 9:3379–3394.

Squyres, S. W. 1980a. Volume changes in Ganymede and Callisto and the origin of grooved terrain. *Geophys. Res. Lett.* 7:593–596.

Squyres, S. W. 1980b. Surface temperatures and retention of H_2O frost on Ganymede and Callisto. *Icarus* 44:502–510.

Squyres, S. W. 1980c. Topographic domes on Ganymede: Ice vulcanism or isostatic upwarping. *Icarus* 44:472–480.

Squyres, S. W. 1981a. The Morphology and Evolution of Ganymede and Callisto. Ph.D. Thesis, Cornell Univ.

Squyres, S. W. 1981b. The topography of Ganymede's grooved terrain. *Icarus* 46:156–168.

Squyres, S. W. 1981c. Ice volcanism in the outer solar system. *EOS Trans. AGU* 62:1080–1081 (abstract).

Squyres, S. W. 1982. The evolution of tectonic features on Ganymede. *Icarus* 52:545–559.

Squyres, S. W., and Reynolds, R. T. 1983. Tidal evolution of the Uranian satellites. *Bull. Amer. Astron. Soc.* 15:1014 (abstract). *See* Squyres et al. 1985.

Squyres, S. W., and Sagan, C. 1983. Albedo asymmetry of Iapetus. *Nature* 303:782–785.

Squyres, S. W., and Veverka, J. 1981. Voyager photometry of surface features on Ganymede and Callisto. *Icarus* 46:137–155.

Squyres, S. W., and Veverka, J. 1982. Color photometry of surface features on Ganymede and Callisto. *Icarus* 52:117–125.

Squyres, S. W., Reynolds, R. T., Cassen, P. M., and Peale, S. J. 1983a. Liquid water and active resurfacing on Europa. *Nature* 301:225–226.

Squyres, S. W., Reynolds, R. T., Cassen, P. M., and Peale, S. J. 1983b. The evolution of Enceladus. *Icarus* 53:319–331.

Squyres, S. W., Buratti, B., Veverka, J., and Sagan, C. 1984. Voyager photometry of Iapetus. *Icarus* 59:426–435.

Squyres, S. W., Reynolds, R. T., and Lissauer, J. J. 1985. The enigma of the Uranian satellites' orbital eccentricities. *Icarus* 61:218–223.

Srnka, L. J. 1978. Spontaneous magnetic field generation in hypervelocity impacts. *Proc. Lunar Planet. Sci. Conf.* 8:785–792.

Sromovsky, L. A., Suomi, V. E., Pollack, J. B., Krauss, R. J., Limaye, S. S., Owen, T., Revercomb, H. E., and Sagan, C. 1981. Titan brightness contrasts: Implications of Titan's north-south brightness asymmetry. *Nature* 292:698–702.

Stanyukovich, K. P. 1950. Elements of the physical theory of meteors and the formation of meteor craters. *Meteoritika* 7:39–62.

Staudacher, T., and Allegre, C. J. 1982. Terrestrial xenology. *Earth Planet. Sci. Lett.* 60: 389–406.

Stebbins, J. 1927. The light variations of the satellites of Jupiter and their applications to measures of the solar constant. *Lick Obs. Bull.* 13:1–11.

Stebbins, J., and Jacobsen, T. S. 1928. Further photometric measures of Jupiter's satellites and Uranus, with tests for the solar constant. *Lick Obs. Bull.* 13:180–195.

Steiger, R. H., and Jaeger, E. 1977. Subcommission on geochronology: Convention on the use of decay constants in geo- and cosmo-chronology. *Earth Planet. Sci. Lett.* 36:359–362.

Stevenson, D. J. 1981. Models of the Earth's core. *Science* 214:611–619.

Stevenson, D. J. 1982a. Interiors of the giant planets. *Ann. Rev. Earth Planet. Sci.* 10:257–295.

Stevenson, D. J. 1982b. Formation of the giant planets. *Planet. Space Sci.* 30:755–764.

Stevenson, D. J. 1982c. Volcanism and igneous processes in small icy satellites. *Nature* 298: 142–144.

Stevenson, D. J. 1982d. Migration of fluid-filled cracks: Applications to terrestrial and icy bodies. *Lunar Planet. Sci.* XIII:768–769 (abstract).

Stevenson, D. J. 1983. Anomalous bulk viscosity of two-phase fluids and implications for planetary interiors. *J. Geophys. Res.* 88:2445–2455.

Stevenson, D. J. 1984a. On forming the giant planets quickly (superganymedean puffballs). *Lunar Planet. Sci.* XV:822–823 (abstract).

Stevenson, D. J. 1984b. Composition, structure, and evolution of Uranian and Neptunian satellites. In *Uranus and Neptune*, NASA CP-2330, pp. 405–423.

Stevenson, D. J. 1985a. Lunar formation from impact on the Earth: Is it possible? In *Proc. Origin of the Moon Conf.*, Kona, HI, 14–16 October 1984.

Stevenson, D. J. 1985b. Implications of very large impacts for Earth accretion and lunar formation. *Lunar Planet. Sci.* XVI:819 (abstract).

Stevenson, D. J., and Anderson, A. 1981. Volcanism and igneous processes in small icy satellites. *EOS Trans. AGU* 62:1081 (abstract).

Stevenson, D. J., and Turner, T. S. 1979. Fluid models of mantle convection. In *The Earth: Its Origin, Structure, and Evolution*, ed. M. W. McElhinny (New York: Academic), pp. 227–263.

Stevenson, D. J., and Yoder, C. F. 1982. A fluid outer core for the Moon and its implications for lunar dissipation, free librations and magnetism. *Lunar Planet. Sci.* XII:1043–1046 (abstract).

Stevenson, D. J., Spohn, T., and Schubert, G. 1983. Magnetism and thermal evolution of the terrestrial planets. *Icarus* 54:466–489.

Stewart, G. R., and Kaula, W. M. 1980. A gravitational kinetic theory for planetesimals. *Icarus* 44:154–171.

Stocker, R. L., and Ashby, M. F. 1973. On the rheology of the upper mantle. *Rev. Geophys. Space Phys.* 11:391–426.

Stöffler, D., Gault, D. E., Wedekind, J. A., and Polkowski, G. 1975. Experimental hypervelocity impact into quartz sand: Distribution and shock metamorphism of ejecta. *J. Geophys. Res.* 80:4062–4077.

Strazzulla, G., Calcagno, L., and Foti, G. 1985. Dark material by fast protons on frozen methane. *Il Nuovo Cimento*. In press.

Strobel, D. F. 1982. Chemistry and evolution of Titan's atmosphere. *Planet. Space Sci.* 30: 839–848.

Strobel, D. F. 1985. Paper presented at ESA/ASSA Workshop on The Atmospheres of Saturn and Titan, Alpbach, Austria, September. Proceedings in press.

Strobel, D., and Shemansky, D. 1982. EUV emission from Titan's upper atmosphere: Voyager 1 encounter. *J. Geophys. Res.* 87:1361–1368.

Strom, R. G. 1964. Analysis of lunar lineaments, I: Tectonic maps of the moon. *Comm. Lunar Planet. Lab* 2(39):205–216.

Strom, R. G. 1972. Lunar mare ridges, rings, and volcanic ring complexes. *The Moon, IAU Symp. 47,* eds. S. K. Runcorn and H. Urey (Dordrecht: Reidel), pp. 187–215.

Strom, R. G. 1979. Mercury: A post-Mariner 10 assessment. *Space Sci. Rev.* 24:3–70.

Strom, R. G. 1981. Crater populations on Mimas, Dione, and Rhea. *Lunar Planet. Sci.* XII:7–9 (abstract).

Strom, R. G., and Schneider, N. M. 1982. Volcanic eruption plumes on Io. In *Satellites of Jupiter,* ed. D. Morrison (Tucson: Univ. of Arizona Press), pp. 598–633.

Strom, R. G., Terrile, R. J., Masursky, H., and Hansen, C. 1979. Volcanic eruption plumes on Io. *Nature* 280:733–736.

Strom, R. G., Woronow, A., and Gurnis, M. 1981. Crater populations on Ganymede and Callisto. *J. Geophys. Res.* 86:8659–8674.

Struve, G. 1933. *Veroff. Berlin-Babelsber* 6, part 4.

Sullivan, J. D., and Bagenal, F. 1979. In situ identification of various ionic species in Jupiter's magnetosphere. *Nature* 280:798–799.

Sullivan, J. D., and Siscoe, G. L. 1982. In situ observations of Io torus plasma. In *Satellites of Jupiter,* ed. D. Morrison (Tucson: Univ. of Arizona Press), pp. 846–871.

Summers, M. E., Yung, Y. L., and Haff, P. K. 1983. A two-stage mechanism for the escape of Na and K from Io. *Nature* 304:710–712.

Sündermann, J. 1982. The resonance behavior of the world ocean. In *Tidal Friction and the Earth's Rotation II,* eds. P. Brosche and J. Sündermann (New York: Springer-Verlag), pp. 165–174.

Sündermann, J., and Brosche, P. 1978. Numerical computation of tidal friction for present and ancient oceans. In *Tidal Friction and the Earth's Rotation,* eds. P. Brosche and J. Sündermann (New York: Springer-Verlag), pp. 124–144.

Switkowski, Z. E., Haff, P. K., Tombrello, T. A., and Burnett, D. S. 1977. Mass fractionation of the lunar surface by solar wind sputtering. *J. Geophys. Res.* 82:3797–3804.

Synnott, S. P. 1986. Evidence for the existence of additional small satellites of Saturn. Submitted to *Icarus.*

Synnott, S. P., and Terrille, R. J. 1982. Satellites of Saturn. *IAU Circ.* 3660.

Synnott, S. P., Peters, C. F., Smith, B. A., and Morabito, L. A. 1981. Orbits of the small satellites of Saturn. *Science* 212:191–192.

Synnott, S. P., Callahan, J. C., Riedel, J. E., and Donegan, A. T. 1984. Shape of Io. *Bull. Amer. Astron. Soc.* 16:657 (abstract).

Synnott, S. P., Riedel, J. E., and Gaskell, R. W. 1985. The shape of Io. *Bull. Amer. Astron. Soc.* 17:692 (abstract).

Szebehely, V. 1967. *Theory of Orbits* (New York: Academic).

Szeto, A. M. K. 1983. Orbital evolution and origin of the Martian satellites. *Icarus* 55:133–168.

Taylor, D. B. 1984. A comparison of the theory of the motion of Hyperion with observations made during 1967–1982. *Astron. Astrophys.* 141:151–158.

Taylor, G., O'Leary, B., van Flandern, T., Barthodi, P., Owen, F., Hubbard, W., Smith, B., Smith, S., Fallon, F., Devinney, E., and Oliver, J. 1971. Occultation of Beta Scorpii C by Io on May 14, 1971. *Nature* 234:405–406.

Taylor, S. R. 1975. *Lunar Science: A Post-Apollo View* (New York: Pergamon).

Taylor, S. R. 1982. *Planetary Science: A Lunar Perspective* (New York: Pergamon).

Tedesco, E. F., and Tholen, D. J. 1980. Photometric observations of Pluto in 1980. *Bull. Amer. Astron. Soc.* 12:729 (abstract).

Tholen, D. J. 1985. The orbit of Pluto's satellite. *Astron. J.* 90:2353–2359.

Tholen, D. J., and Zellner, B. 1983. Eight-color photometry of Hyperion, Iapetus, and Phoebe. *Icarus* 53:341–347.

Tholen, D. J., and Zellner, B. 1984. Multi-color photometry of outer Jovian satellites. *Icarus* 58:246–253.

Thomas, P. 1978. The Morphology of Phobos and Deimos. Ph.D. Thesis, Cornell Univ.

Thomas, P. 1979. Surface features of Phobos and Deimos. *Icarus* 40:223–243.

Thomas, P., and Dermott, S. F. 1985. The shape of Mimas. *Bull. Amer. Astron. Soc.* 17:738 (abstract).

Thomas, P., and Veverka, J. 1979. Grooves on asteroids: A prediction. *Icarus* 40:394–405.

Thomas, P., and Veverka, J. 1980*a*. Crater densities on the satellites of Mars. *Icarus* 41: 365–380.

Thomas, P., and Veverka, J. 1980*b*. Downslope movement of material on Deimos. *Icarus* 42: 234–250.

Thomas, P., and Veverka, J. 1982. Amalthea. In *Satellites of Jupiter*, ed. D. Morrison (Tucson: Univ. of Arizona Press), pp. 147–173.

Thomas, P., and Veverka, J. 1985. Hyperion: Analysis of Voyager observations. *Icarus* 64: 414–424.

Thomas, P., Veverka, J., Duxbury, T., and Bloom, A. 1979*a*. The grooves on Phobos: Distribution, morphology, and possible origin. *J. Geophys. Res.* 84:8457–8477.

Thomas, P., Veverka, J., and Chapman, C. R. 1979*b*. Crater populations on the satellites of Mars. NASA TM-80339, pp. 15–18.

Thomas, P., Veverka, J., Morrison, D., and Davies, M. 1981. Photometry and topography of Saturn's small satellites from Voyager data. *Bull. Amer. Astron. Soc.* 13:720–721 (abstract).

Thomas, P., Veverka, J., Morrison, D., Davies, M., and Johnson, T. V. 1983*a*. Saturn's small satellites: Voyager imaging results. *J. Geophys. Res.* 88:8743–8754.

Thomas, P., Veverka, J., Morrison, D., Davies, M., Johnson, T. V., and Smith, B. A. 1983*b*. Phoebe: Voyager 2 observations. *J. Geophys. Res.* 88:8736–8742.

Thomas, P., Veverka, J., Wenkert, D., Danielson, G. E., and Davies, M. E. 1984. Hyperion: 13-day rotation from Voyager data. *Nature* 307:716–717.

Thomas, P. J., and Schubert, G. 1986. Crater relaxation as a probe of Europa's interior. *Proc. Lunar Sci. Conf. 16* in *J. Geophys. Res. Suppl.* 91:D453–D459.

Thompson, A. C., and Stevenson, D. J. 1983. Two-phase gravitational instabilities in thin disks with application to the origin of the moon. *Lunar Planet. Sci.* XIV:787–788 (abstract).

Thompson, T. W., Cutts, J. A., Shorthill, R. W., and Zisk, S. H. 1980. Infrared and radar signatures of lunar craters: Implications about crater evolution. In *Proc. Conf. Lunar Highland Crust*, eds. J. J. Papike and R. B. Merrill (New York: Pergamon), pp. 483–489.

Thomsen, B., and Ables, H. D. 1978. Measurement of the angular separation and magnitude difference for the Pluto/Charon system. *Bull. Amer. Astron. Soc.* 10:586 (abstract).

Thomsen, J. M., Austin, M. G., Ruhl, S. F., Schultz, P. H., and Orphal, D. L. 1979. Calculational investigation of impact cratering dynamics: Early time material motions. *Proc. Lunar Planet. Sci. Conf.* 10:2741–2756.

Thurber, C. H., and Solomon, S. C. 1978. An assessment of crustal thickness variations on the lunar nearside: Models, uncertainties, and implications for crustal differentiation. *Proc. Lunar Sci. Conf.* 9:3481–3497.

Thurber, C. H., Hsui, A. T., and Toksöz, M. N. 1980. Thermal evolution of Ganymede and Callisto: Effects of solid-state convection and constraints from Voyager imagery. *Proc. Lunar Planet. Sci. Conf.* 11:1957–1977.

Toksöz, M. N., Dainty, H. M., Solomon, S. C., and Andersson, K. R. 1974. Structure of the Moon. *Rev. Geophys. Space Phys.* 12:539–567.

Toksöz, M. N., Hsui, A. T., and Johnston, D. H. 1978. Thermal evolutions of the terrestrial planets. *Moon and Planets* 18:281–320.

Tomasko, M. G. 1980. Preliminary results of polarimetry and photometry of Titan at large phase angles from Pioneer 11. *J. Geophys. Res.* 85:5937–5942.

Tomasko, M. G., and Smith, P. H. 1982. Photometry and polarimetry of Titan: Pioneer 11 observations and their implications for aerosol properties. *Icarus* 51:65–95.

Tombaugh, C. W., Robinson, J. C., Smith, B. A., and Murrell, A. S. 1959. *The Search for Small Natural Earth Satellites: Final Technical Report*, Physical Science Laboratory, New Mexico State Univ.

Tozer, D. 1965. Heat transfer and convection currents. *Phil. Trans. Roy. Soc. London* A258: 252–271.

Tozer, D. 1967. Towards a theory of thermal convection in the mantle. In *The Earth's Mantle*, ed. T. F. Gaskell (London: Academic), pp. 325–353.

Trafton, L. M. 1972. On the possible detection of H_2 in Titan's atmosphere. *Astrophys. J.* 175: 285–293.

Trafton, L. M. 1974. Titan: Unidentified strong absorptions in the photometric infrared. *Icarus* 21:175–187.

Trafton, L. 1975a. High-resolution spectra of Io's sodium emission. *Astrophys. J.* 202: L107–L112.

Trafton, L. 1975b. Detection of a potassium cloud near Io. *Nature* 258:690–692.

Trafton, L. 1976. A search for emission features in Io's extended cloud. *Icarus* 27:429–437.

Trafton, L. 1980a. Does Pluto have a substantial atmosphere? *Icarus* 44:53–61.

Trafton, L. 1980b. An explanation for the alternating north-south asymmetry of Io's sodium cloud. *Icarus* 44:318–325.

Trafton, L. 1981. The atmospheres of the outer planets and satellites. *Rev. Geophys. Space Phys.* 19:43–89.

Trafton, L. 1983. Does Triton's atmosphere undergo large seasonal variations? *Natural Satellites, IAU Coll. 77*, abstract booklet, Cornell Univ.

Trafton, L. M. 1984. Seasonal variations in Triton's atmospheric mass and composition. In *Uranus and Neptune*, ed. J. T. Bergstralh, NASA CP-2330, pp. 481–496.

Trafton, L., and Stern, S. A. 1981. On the global distribution of Pluto's atmosphere. *Astrophys. J.* 267:872–881.

Trafton, L., Parkinson, T., Macy, W. Jr. 1974. The spatial extent of sodium emission around Io. *Astrophys. J.* 190:L85–L89.

Trask, N. J. 1975. In *Proc. Internatl. Coll. Planet. Geol., Geol. Rom.* 15:471.

Trauger, J. 1984. The Jovian nebula: A post-Voyager perspective. *Science* 226:337–341.

Troitsky, V. S. 1965. Investigation of the surfaces of the Moon and planets by thermal radiation. *Radio Sci.* 69:1585–1612.

Tullis, J. A. 1980. High temperature deformation of rocks and minerals. *Rev. Geophys. Space Phys.* 17:1137–1154.

Turcotte, D. L. 1974. Membrane tectonics. *Geophys. J. Roy. Astron. Soc.* 36:33–42.

Turcotte, D. L., and Oxburgh, E. R. 1969. Implications of convection within the Moon. *Nature* 223:250–251.

Turcotte, D. L., and Oxburgh, E. R. 1970. Lunar convection. *J. Geophys. Res.* 75:6549–6552.

Turner, R. J. 1978. A model of Phobos. *Icarus* 33:116–140.

Tyler, G. L., Eshleman, V. R., Anderson, J. D., Levy, G. S., Lindal, G. F., Wood, G. E., and Croft, T. A. 1981. Radio science investigations of the Saturn system with Voyager 1: Preliminary results. *Science* 212:201–206.

Tyler, G. L., Eshleman, V. R., Anderson, J. D., Levy, G. S., Lindal, G. F., Wood, G. E., and Croft, T. A. 1982. Radio science with Voyager 2 at Saturn: Atmosphere and ionosphere and the masses of Mimas, Tethys, and Iapetus. *Science* 215:553–558.

Tyler, G. L., Sweetnam, D. N., Anderson, J. D., Campbell, J. C., Eshelman, V. R., Hinson, D. P., Levy, G. S., Lindal, G. F., Marouf, E. A., and Simpson, R. A. 1986. Radio-science observations of the Uranian system with Voyager 2: Properties of the atmosphere, rings and satellites. *Science*. In press.

Ulich, B. L., and Conklin, E. K. 1976. Observations of Ganymede, Callisto, Ceres, Uranus and Neptune at 3.33 mm wavelength. *Icarus* 27:183–189.

Ulich, B. L., Dickel, J. R., and dePater, I. 1984. Planetary observations at a wavelength of 1.32 mm. *Icarus* 60:590–598.

Ullrich, G. W., Roddy, D. J., and Simmons, G. 1977. Numerical simulations of a 20-ton TNT detonation on the Earth's surface and implications concerning the mechanics of central uplift formation. In *Impact and Explosion Cratering,* eds. D. J. Roddy, R. O. Pepin, and R. O. Merrill (New York: Pergamon), pp. 959–982.

Urey, H. C. 1951. The origin and development of the Earth and other terrestrial planets. *Geochim. Cosmochim. Acta* 1:209–277.

Urey, H. C. 1962. The origin of the Moon. In *Physics and Astronomy of the Moon*, ed. Zd. Kopal (New York: Academic), pp. 481–523.

Usselman, T. M. 1975. Experimental approach to the state of the core: Part 1. The liquid relations of the Fe-rich portion of the Fe-Ni-S system from 30 to 100 kb. *Amer. J. Sci.* 275:278–290.

Valdes, F., and Freitas, R. A. Jr. 1983. A search for objects near the Earth-Moon Lagrangian points. *Icarus* 53:453–457.

Van Allen, J. A., Thomsen, M. F., and Randall, B. A. 1980. The energetic charged particle absorption signature of Mimas. *J. Geophys. Res.* 85:5709–5718.

Van Arsdale, W. E. 1985. Orbital dynamics of a viscoelastic body. *J. Geophys. Res.* 90: 6887–6892.

Van Arsdale, W. E., and Burns, J. A. 1979. A self-consistent tidal theory for imperfectly elastic bodies. *Lunar Planet. Sci.* X:1259–1261 (abstract).

Van Flandern, T. C. 1981. Is the gravitational constant changing? *Astrophys. J.* 248:813–816.

Van Flandern, T. C., Tedesco, E. F., and Binzel, R. P. 1979. Satellites of asteroids. In *Asteroids*, ed. T. Gehrels (Tucson: Univ. of Arizona Press), pp. 443–465.

Veeder, G. J., and Matson, D. L. 1980. The relative reflectance of Iapetus at 1.6 and 2.2 μm. *Astron. J.* 85:969–972.

Veillet, C. 1982. Orbital elements of Nereid from new observations. *Astron. Astrophys.* 112:277–280.

Veillet, C. 1983a. De l'observation et du mouvement des satellites d'Uranus. Ph.D. Thesis, Université de Paris.

Veillet, C. 1983b. Observations of Miranda; New orbit and mass; Ariel and Umbriel. *Astron. Astrophys.* 118:211–216.

Veverka, J. 1971a. Polarization measurements of the Galilean satellites of Jupiter. *Icarus* 14:355–359.

Veverka, J. 1971b. Asteroid polarimetry: A progress report. In *Physical Studies of Minor Planets*, ed. T. Gehrels, NASA SP-267, pp. 91–94.

Veverka, J. 1977a. Photometry of satellite surfaces. In *Planetary Satellites*, ed. J. A. Burns (Tucson: Univ. of Arizona Press), pp. 171–209.

Veverka, J. 1977b. Polarimetry of satellite surfaces. In *Planetary Satellites*, ed. J. A. Burns (Tucson: Univ. of Arizona Press), pp. 210–231.

Veverka, J., and Burns, J. 1980. The moons of Mars. *Ann. Rev. Earth Planet. Sci.* 8:527–558.

Veverka, J., and Gradie, J. 1983. Why don't icy satellites scatter like model snow-covered planets? *Bull. Amer. Astron. Soc.* 15:856–857 (abstract).

Veverka, J., and Thomas, P. 1979. Phobos and Deimos: A preview of what asteroids are like? In *Asteroids*, ed. T. Gehrels (Tucson: Univ. of Arizona Press), pp. 628–651.

Veverka, J., Goguen, J., Yang, S., and Elliot, J. L. 1978. On matching the spectrum of Io: Variations in the photometric properties of sulfur-containing mixtures. *Icarus* 37:249–255.

Veverka, J., Simonelli, D., Thomas, P., Morrison, D., and Johnson, T. V. 1981a. Voyager search for posteclipse brightening on Io. *Icarus* 47:60–74.

Veverka, J., Thomas, P., Davies, M., and Morrison, D. 1981b. Amalthea: Voyager imaging results. *J. Geophys. Res.* 86:8675–8692.

Veverka, J., Gradie, J., Thomas, P., and Ostro, S. 1982a. How much S_8 (Cyclo-octasulfur) is there on the surface of Io? *Lunar Planet. Sci.* XIII:823–824 (abstract).

Veverka, J., Thomas, P., and Synnott, S. 1982b. The inner satellites of Jupiter. *Vistas in Astron.* 25:245–262.

Veverka, J., Gradie, J., and Thomas, P. 1984. Is there sulfur on Io? Personal communication.

Veverka, J., Thomas, P., Thompson, R. W., and Buratti, B. 1985. Enceladus, again. *Bull. Amer. Astron. Soc.* 17:739 (abstract).

Vickery, A. M. 1984. Physical constraints on the origin of the shergottites, nakhlites, and chassignites. Ph.D. Thesis, State Univ. of New York at Stony Brook.

Vilas, F., and McCord, T. B. 1976. Mercury: Spectral reflectance measurements (0.33–1.60 μm) 1974/5. *Icarus* 28:593–599.

Vitjazev, A. V., and Pechernikova, G. V. 1981. Solution of the problem of planets' rotation in the statistical theory of accumulation. *Sov. Astron. J. Lett.* 58:869–878. In Russian.

Vitjazev, A. V., and Pechernikova, G. V. 1982. Models of preplanetary disk around F-G stars. *Sov. Astron. J. Lett.* 8:371–377. In Russian.

Voight, B., ed. 1978. *Rockslides and Avalanches* (New York: Elsevier).

Voight, G. H., Hill, T. W., and Dessler, A. J. 1983. The magnetosphere of Uranus: Plasma sources, convection, and field configuration. *Astrophys. J.* 266:390–401.

Walker, A. R. 1980. An occultation by Charon. *Mon. Not. Roy. Astron. Soc.* 192:47P–50P.

Walker, D. 1983. Lunar and terrestrial crust formation. *Proc. Lunar Planet. Sci. Conf.* 14 in *J. Geophys. Res. Suppl.* 88:B17–B25.

Walker, M. F., and Hardie, R. 1955. A photometric determination of the rotation period of Pluto. *Pub. Astron. Soc. Pacific* 67:224–231.

Wamsteker, W. 1973. Narrow-band photometry of the Galilean satellites. *Comm. Lunar Planet. Lab.* 9:171–177.

Wamsteker, W., Kroes, R. L., and Fountain, J. A. 1974. On the surface composition of Io. *Icarus* 10:1–7.

Wänke, H., Dreibus, G., Palme, H., Rammensee, W., and Weckwerth, G. 1983. Geochemical evidence for the formation of the Moon from material of the Earth's mantle. *Lunar Planet. Sci.* XIV:818–819 (abstract).

Ward, W. R. 1976. Some remarks on the accretion problem. In *Fisica e Geologia Planetaria* (Rome: Atti dei Convegni Lincea), 25:22–239.

Ward, W. R. 1981. Orbital inclination of Iapetus and the rotation of the Laplacian plane. *Icarus* 46:97–107.

Ward, W. R. 1982. Comments on the long-term stability of the Earth's obliquity. *Icarus* 50:444–448.

Ward, W. R., and Cameron, A. G. W. 1978. Disc evolution within the Roche limit. *Lunar Planet. Sci.* IX:1205–1207 (abstract).

Ward, W. R., and Reid, M. J. 1973. Solar tidal friction and satellite loss. *Mon. Not. Roy. Astron. Soc.* 164:21–32.

Warren, P. H., and Wasson, J. T. 1977. Pristine nonmare rocks and the nature of the lunar crust. *Proc. Lunar Sci. Conf.* 8:2215–2235.

Warren, S. J. 1984. Optical constants of ice from the ultraviolet to the microwave. *Appl. Opt.* 23:1206–1225.

Warwick, J. W., Pearce, J., Riddle, A., Alexander, J., and Desch, M. 1979. Voyager 1 planetary radio astronomy observations near Jupiter. *Science* 204:995–998.

Wasserburg, G. J., Papanastassiou, D. A., Tera, F., and Huneke, J. C. 1977. The accumulation and bulk composition of the moon: Outline of a lunar chronology. *Phil. Trans. Roy. Soc. London* A285:7–22.

Wasson, J. T., and Warren, P. H. 1980. Contribution of the mantle to lunar asymmetry. *Icarus* 44:752–771.

Watkins, J. S., and Kovach, R. L. 1972. Apollo 14 active seismic experiment. *Science* 175:1244–1245.

Watson, C. C. 1981. The sputter-generation of planetary coronae: Galilean satellites of Jupiter. *Proc. Lunar Planet. Sci. Conf.* 12:1569–1583.

Webb, D. J. 1980. Tides and tidal friction in a hemispherical ocean centered at the equator. *Geophys. J. Roy. Astron. Soc.* 61:573–600.

Webb, D. J. 1982a. Tides and the evolution of the Earth-Moon system. *Geophys. J. Roy. Astron. Soc.* 70:261–271.

Webb, D. J. 1982b. On the reduction in tidal dissipation produced by increases in the Earth's rotation rate and its effect on the long-term history of the Moon's orbit. In *Tidal Friction and the Earth's Rotation II*, eds. P. Brosche and J. Sündermann (New York: Springer-Verlag), pp. 210–221.

Wechsler, A. E., Glaser, P. E., and Little, A. D. 1972. Thermal properties of granulated materials. *Thermal Characteristics of the Moon. Prog. Astronaut. Aeronaut.* 28:215–241.

Weertman, J. 1971. Velocity at which liquid-filled cracks move in the Earth's crust or in glaciers. *J. Geophys. Res.* 76:8544–8553.

Weertman, J. 1973. Creep of ice. In *Physics and Chemistry of Ice*, eds. E. Whalley, S. J. Jones, and L. W. Gold (Ottawa: Roy. Soc. Canada), pp. 320–337.

Weertman, J. 1983. Creep deformation of ice. *Ann. Rev. Earth Planet. Sci.* 11:215–240.

Weertman, J., and Weertman, J. R. 1975. High temperature creep of rock and mantle viscosity. *Ann. Rev. Earth Planet. Sci.* 3:293–315.

Wehner, E. K., KenKnight, C. E., and Rosenberg, D. L. 1963. Modification of the lunar surface by solar wind bombardment. *Planet. Space Sci.* 11:1257–1261.

Weidenschilling, S. J. 1977. Aerodynamics of solid bodies in the solar nebula. *Mon. Not. Roy. Astron. Soc.* 180:57–70.

Weidenschilling, S. J. 1979. A possible origin for the grooves of Phobos. *Nature* 282:697–698.

Weidenschilling, S. J. 1980. Dust to planetesimals: Settling and coagulation in the solar nebula. *Icarus* 44:172–189.

Weidenschilling, S. J. 1981. How fast can an asteroid spin? *Icarus* 46:124–126.

Weidenschilling, S. J. 1982. Origin of regular satellites. In *The Comparative Study of the Planets*, eds. A. Coradini and M. Fulchignoni (Dordrecht: Reidel), pp. 49–59.

Weidenschilling, S. J., Chapman, C. R., Davis, D. R., and Greenberg, R. 1984. Ring particles:

Collisional interactions and physical nature. In *Planetary Rings,* eds. R. Greenberg and A. Brahic (Tucson: Univ. of Arizona Press), pp. 367–415.

Weiser, H., Vitz, R. C., and Moos, H. 1977. Detection of Lyman α emission of the Saturnian disk and from the ring system. *Science* 197:755–757.

Wesselink, A. J. 1948. Heat conductivity and nature of the lunar surface material. *Bull. Astron. Inst. Netherlands* 10:351–390.

Wetherill, G. 1975. Late heavy bombardment of the Moon and terestrial planets. *Proc. Lunar Sci. Conf.* 6:1539–1561.

Wetherill, G. W. 1976. The role of large bodies in the formation of the Earth. *Proc. Lunar Sci. Conf.* 7:3245–3257.

Wetherill, G. 1977. Evolution of the Earth's planetesimal swarm subsequent to the formation of the Earth and Moon. *Proc. Lunar Sci. Conf.* 8:1–16.

Wetherill, G. W. 1980*a.* Formation of the terrestrial planets. *Ann. Rev. Astron. Astrophys.* 18: 77–113

Wetherill, G. W. 1980*b.* Numerical calculations relevant to the accumulation of the terrestrial planets. In *The Continental Crust and Its Mineral Deposits,* ed. D. W. Strangway, Geol. Assoc. Canada Special Paper 20, pp. 3–24.

Wetherill, G. W. 1981*a.* Solar wind origin of ^{36}Ar on Venus. *Icarus* 46:70–80.

Wetherill, G. W. 1981*b.* Nature and origin of basin-forming projectiles. *Proc. Lunar Planet. Sci. Conf.* 12:1–18.

Whipple, F. L. 1972. On certain aerodynamic processes for asteroids and comets. In *From Plasma to Planet: Proc. Nobel Symp. No. 21,* ed. A. Elvius (New York: Wiley), pp. 211–232.

Whitaker, E. A. 1981. The lunar Procellarum basin. In *Multi-Ring Basins: Evolution and Formation, Proc. Lunar Planet. Sci. 12A* (New York: Pergamon), pp. 105–111.

Whitehead, J. A., and Luther, D. S. 1975. Dynamics of laboratory diapir and plume models. *J. Geophys. Res.* 80:705–717.

Whitford-Stark, J. L., and Head, J. W. 1977. The Procellarum volcanic complexes: Contrasting styles of volcanism. *Proc. Lunar Sci. Conf.* 8:2705–2724.

Whitford-Stark, J. L., and Head, J. W. 1980. Stratigraphy of Oceanus Procellarum basalts: Sources and styles of emplacement. *J. Geophys. Res.* 85:6579–6609.

Whyte, A. J. 1980. *The Planet Pluto* (Toronto: Pergamon Press).

Wildey, R. L. 1977. A digital file of the lunar normal albedo. *The Moon* 16:231–277.

Wilhelms, D. E. 1970. Summary of lunar stratigraphy-telescopic observations. U.S. Geol. Surv. Prof. Paper 599-F.

Wilhelms, D. E. 1980. Stratigraphy of part of the lunar near side. U.S. Geol. Surv. Prof. Paper 1046-A.

Wilhelms, D. E. 1984. Moon. In *The Geology of the Terrestrial Planets,* ed. M. H. Carr, NASA SP-469, pp. 106–205.

Wilhelms, D. E., Hodges, C. A., and Pike, R. J. 1977. Nested crater model of lunar ringed basins. In *Impact and Explosion Cratering,* eds. D. J. Roddy, R. O. Pepin, and R. B. Merrill (New York: Pergamon), pp. 539–562.

Wilhelms, D. E., Oberbeck, V. R., and Aggarwal, H. R. 1978. Size-frequency distributions of primary and secondary lunar impact craters. *Proc. Lunar Planet. Sci. Conf.* 9:3735–3762.

Wilkening, L. L. 1971. *Particle Track Studies and the Origin of Gas-Rich Meteorites* (Tempe: Center for Meteorite Studies, Arizona State Univ.).

Wilson, L., and Head, J. W. 1981. Ascent and eruption of basaltic magma on the Earth and Moon. *J. Geophys. Res.* 86:2971–3001.

Wilson, L., and Head, J. W. 1983*a.* A comparison of volcanic eruption processes on Earth, Moon, Mars, Io and Venus. *Nature* 302:663–669.

Wilson, L., and Head, J. W. 1983*b.* Water volcanism. *Natural Satellites, IAU Colloquium 77,* abstract booklet, Cornell Univ., pp. 21.

Wilson, L., and Head, J. W. 1984. Aspects of water eruption on icy satellites. *Lunar Planet. Sci.* XV:924–925 (abstract).

Winter, D. F., and Krupp, J. A. 1971. Directional characteristics of infrared emission from the Moon. *The Moon* 2:279–292.

Wisdom, J. 1983. Chaotic behavior and the origin of the 3/1 Kirkwood gap. *Icarus* 56:51–74.

Wisdom, J., and Peale, S. J. 1984. The light curve of Hyperion. *Bull. Amer. Astron. Soc.* 16:707 (abstract).

Wisdom, J., Peale, S. J., and Mignard, F. 1984. The chaotic rotation of Hyperion. *Icarus* 58:137–152.

Wise, D. U. 1963. An origin of the moon by rotational fission during formation of the earth's core. *J. Geophys. Res.* 68:1547–1554.

Wise, D. U. 1969. Origin of the Moon from the Earth: Some new mechanisms and comparisons. *J. Geophys. Res.* 74:6034–6045.

Wiskerchen, M. J., and Sonett, C. P. 1977. A lunar metal core? *Proc. Lunar Sci. Conf.* 8: 515–535.

Witteborn, F. C., Bregman, J. D., and Pollack, J. P. 1979. Io: An intense brightening near 5 micrometers. *Science* 203:643–646.

Wolf, M. 1980. Theory and application of the polarization-albedo rules. *Icarus* 44:780–792.

Wolff, R. S., and Mendis, D. A. 1983. On the nature of the interaction of the Jovian magnetosphere with the icy Galilean satellites. *J. Geophys. Res.* 88:4749–4769.

Woltjer, J. 1928. The motion of Hyperion. *Ann. Sternw. Leiden* 16(3):1–139.

Wood, C. A., and Andersson, L. 1978. New morphometric data for fresh lunar craters. *Proc. Lunar Planet. Sci. Conf.* 9:3669–3689.

Wood, C. A., and Head, J. W. 1976. Comparison of impact basins on Mercury, Mars and the Moon. *Proc. Lunar Sci. Conf.* 7:3629–3651.

Wood, C. A., Head, J. W., and Cintala, M. J. 1978. Interior morphology of fresh Martian craters: The effects of target characteristics. *Proc. Lunar Planet. Sci. Conf.* 9:3691–3709.

Wood, J. A. 1975. Lunar petrogenesis in a well-stirred magma ocean. *Proc. Lunar Sci. Conf.* 6:1087–1102.

Wood, J. A. 1977. Origin of the Earth's moon. In *Planetary Satellites,* ed. J. A. Burns (Tucson: Univ. of Arizona Press), pp. 513–529.

Wood, J. A. 1986. Moon over Mauna Loa: A review of hypotheses of formation of Earth's Moon. In *Origin of the Moon,* eds. W. K. Hartmann, R. J. Phillips and G. J. Taylor (Houston: Lunar Planet. Inst.). In press.

Wood, J. A., and Mitler, H. E. 1974. Origin of the moon by a modified capture mechanism, or half a loaf is better than a whole one. *Lunar Sci.* V:851–853.

Wood, J. A., Dickey, J. S., Marvin, U. B., and Powell, B. N. 1970. Lunar anorthosites and a geophysical model of the Moon. *Proc. Apollo 11 Lunar Sci. Conf.,* ed. A. A. Levinson (New York: Pergamon), pp. 965–988.

Woronow, A. 1977a. A size-frequency study of large Martian craters. *J. Geophys. Res.* 82: 5807–5820.

Woronow, A. 1977b. Crater saturation and equilibrium: A Monte Carlo simulation. *J. Geophys. Res.* 82:2447–2456.

Woronow, A. 1978. A general cratering-history model and its implications for the lunar highlands. *Icarus* 34:76–88.

Woronow, A. 1984. Factors affecting saturation density in computer simulations. *Lunar Planet. Sci.* XV:939–940 (abstract). *See also* A Monte Carlo study of parameters affecting computer simulations of crater saturation density. *Proc. Lunar Planet. Sci. Conf. 15* in *J. Geophys. Res. Suppl.* 90:C817–C824.

Woronow, A., and Strom, R. G. 1982. Limits on large-crater production and obliteration on Callisto. *Geophys. Res. Lett.* 8:891–894.

Woronow, A., Strom, R., and Gurnis, M. 1982. Interpreting the cratering record: Mercury to Ganymede and Callisto. In *Satellites of Jupiter,* ed. D. Morrison (Tucson: Univ. of Arizona Press), pp. 237–276.

Yabushita, S. 1969. Stability analysis of Saturn's rings with differential rotation. *Mon. Not. Roy. Astron. Soc.* 133:247–263.

Yang, J., and Epstein, S. 1983. Interstellar organic matter in meteorites. *Geochim. Cosmochim. Acta* 47:2199–2215.

Yavorski, B. M., and Detlaf, A. A. 1968. *Handbook on physics for engineers and students,* 4th ed. (Moscow: Nauka). In Russian.

Yoder, C. F. 1973. On the Establishment and Evolution of Orbit-Orbit Resonances. Ph.D. Thesis, University of California, Santa Barbara.

Yoder, C. F. 1979a. Diagrammatic theory of transition of pendulumlike systems. *Celestial Mech.* 19:3–29.

Yoder, C. F. 1979b. How tidal heating in Io drives the Galilean orbital resonance locks. *Nature* 279:767–770.

Yoder, C. F. 1981a. Tidal friction and Enceladus' anomalous surface. *EOS Trans. AGU* 62:939 (abstract).

Yoder, C. F. 1981b. Free librations of a dissipative Moon. *Phil. Trans. Roy. Soc. London* A303: 327–338.

Yoder, C. F. 1982. Tidal rigidity of Phobos. *Icarus* 49:327–346.

Yoder, C. F., and Faulkner, J. 1984. Tidal heating mechanisms. Presented at Natural Satellites, IAU Coll. 77, Ithaca, NY.

Yoder, C. F., and Peale, S. J. 1981. The tides of Io. *Icarus* 47:1–35.

Yoder, C. F., and Synnott, S. P. 1984. Masses of Janus and Epimetheus. *Bull. Amer. Astron. Soc.* 16:687 (abstract).

Yoder, C. F., Williams, J. G., Dickey, J. O., Schutz, B. E., Eames, R. E., and Tapley, B. D. 1983. Secular variations of the Earth gravitational harmonic J_2 coefficient from Lageos and nontidal acceleration of the Earth's rotation. *Nature* 303:757–762.

Yoder, C. F., Colombo, G., Synnott, S. P., Yoder, K. A. 1983. Theory of motion of Saturn's coorbiting satellites. *Icarus* 53:431–443.

Yoder, C. F., Williams, J. G., Dickey, J. O., and Newhall, X. X. 1984. Tidal dissipation in the Earth and Moon from lunar laser ranging. *Conf. Origin of the Moon,* abstract booklet, Univ. of Hawaii, p. 31.

Young, A. T. 1984. No sulfur flows on Io. *Icarus* 58:197–226.

Yung, Y. L., Allen, M., and Pinto, J. P. 1984. Photochemistry of the atmosphere of Titan: Comparison between model and observations. *Astrophys. J. Suppl.* 55:465–506.

Zappalà, V., Farinella, P., Knezevic, Z., and Paolicchi, P. 1984. Collisional origin of the asteroid families: Mass and velocity distributions. *Icarus* 59:261–285.

Zeldovich, Y. B., and Raizer, Yu. P. 1967. *Physics of Shock Waves and High-Temperature Hydrodynamic Phenomena* (New York: Academic).

Zharkov, V. N., Kozenko, A. V., and Mayeva, S. V. 1984. Structure and origin of Martian satellites. *Astron. Vestnik* 18:83–99. In Russian.

Ziegler, J. F. 1977. *Helium: Stopping Powers and Ranges in All Elements* (New York: Pergamon).

Zisk, S., Hodges, C. A., Moore, H. J., Shorthill, R. W., Thompson, T. W., Whitaker, E. A., and Wilhelms, D. E. 1977. The Aristarchus-Harbinger region of the Moon: Surface geology and history from recent remote-sensing observations. *The Moon* 17:59–99.

Zuber, M. T., and Parmentier, E. M. 1984a. A geometric analysis of surface deformation: Implications for the tectonic evolution of Ganymede. *Icarus* 60:200–210.

Zuber, M. T., and Parmentier, E. M. 1984b. Lithospheric stresses due to radiogenic heating of an ice-silicate planetary body: Implications for Ganymede's tectonic evolution. *Proc. Lunar Planet. Sci. Conf. 14* in *J. Geophys. Res. Suppl.* 89:B429–B437.

Glossary

GLOSSARY*

Compiled by Melanie Magisos

accretion

the agglomeration of matter together to form larger bodies such as stars, planets and moons.

adiabatic lapse rate

in an atmosphere the rate of temperature decrease with altitude for a parcel that does not exchange heat with its surroundings.

aeon (AE)

10^9 yr; *see* Gyr.

albedo

reflectivity; geometric albedo: ratio of an object's brightness at zero phase angle to the brightness of a perfectly diffusing disk with the same position and apparent size as the planet; Bond albedo: fraction of the total incident light reflected in all directions by a spherical body.

allotrope

one of two or more solid, liquid or gaseous forms of an element, in one phase of matter.

amagat

a unit of molecular volume at 0°C and a pressure of 1 atmosphere. This unit varies slightly from one gas to another, but in general it corresponds to 2.24×10^4 cm^3. Also, a unit of density equal to 0.046 gram mole per liter

*We have used various definitions from *Glossary of Astronomy and Astrophysics* by J. Hopkins (by permission of the University of Chicago Press, copyright 1980 by the University of Chicago), from *Astrophysical Quantities* by W. W. Allen (London: Athlone Press, 1973) and from *Dictionary of Astronomy, Space, and Atmospheric Phenomena* by D. F. Tver (New York: Van Nostrand Reinhold Co., 1979).

at 1 atm pressure ($\sim$ 2.687 $\times$ 10^{19} molecules cm^{-3}) under standard conditions.

anorthosite   a granular plutonic igneous rock composed almost wholly of plagioclase.

apoapse   the orbital point farthest from the focus of attraction.

asthenosphere   a weak spherical shell located below the lithosphere, in which isostatic adjustments take place, magmas may be generated, and seismic waves are strongly attenuated.

bar   a unit of atmospheric pressure; 1 bar = 10^6 dyne cm^{-2} = 0.987 atm.

basalt   a general term for dark-colored mafic igneous rocks, commonly extrusive but locally intrusive (i.e., dikes), composed chiefly of calcic plagioclase and clinopyroxene; the fine-grained equivalent of gabbro.

blackbody   an idealized body which absorbs all radiation of all wavelengths incident on it. The radiation emitted by a blackbody is a function of temperature only. Because it is a perfect absorber, it is also a perfect emitter.

Boltzmann equation   an equation giving the distribution of particles among energy states for a group of atoms, ions and electrons in a gas of known temperature.

breccia   a rock composed of broken rock fragments cemented together by finer-grained material.

brightness temperature   the temperature that a blackbody would have to have in order to emit radiation of the observed intensity at a given wavelength. This quantity is particularly useful when the Rayleigh-Jeans approximation is valid (as it often is in radio astronomy) because in this approximation it is directly proportional to the specific intensity. It is useful whenever there is reason to believe that it corresponds to a physical temperature; in other cases it merely indicates the radiation's intensity at a given wavelength.

carbonaceous chondrite   a class of chemically primitive meteorites characterized by hydrated minerals and organic (carbon) compounds.

central peak   a central high area produced in an impact crater by inward and upward movement of underlying material.

Clapeyron curve   the equilibrium boundary between two phases of the same substance in a temperature-pressure diagram.

corotation   the configuration in which a plasma moves in a planetary magnetosphere at the same angular rate as that of the planetary magnetic field, resulting in rigid body rotation.

crust   the outermost solid layer of a planet or satellite, mostly consisting of crystalline rock and extending no more than a few kilometers from the surface.

Curie point   the temperature marking the transition between ferromagnetism and paramagnetism, or between the ferroelectric phase and the paraelectric phase.

DAM   decametric radiation.

diapirism   the process of piercing or rupturing of domed or uplifted rocks by mobile core material, by tectonic stresses as in anticlinal folds, by the effect of geostatic load in sedimentary strata as in salt domes and shale diapirs, or by igneous intrusion, forming diapiric structure such as plugs.

differentiation   processes by which planets and satellites develop concentric layers or zones of different chemical and mineralogical composition.

dike   the intrusion of magma into a crack that cuts across the existing crust.

eccentricity   a number which defines the shape of an ellipse. It is the ratio of the distance from center to focus to the semimajor axis.

eclogite   a granular rock composed essentially of garnet (almandine-pyrope) and pyroxene (omphacite).

eddy viscosity   the effective viscosity of a turbulent medium, usually enormously larger than the molecular viscosity.

ejecta blanket — the deposit surrounding an impact crater composed of material ejected from the crater during its formation.

emu — electromagnetic unit.

exobase — the critical altitude of an atmosphere above which collisions are so infrequent that typical particles moving above the escape velocity will escape on ballistic trajectories.

facies — the aspect, appearance and characteristics of a rock unit, usually reflecting the conditions of its origin; especially as differentiating the unit from adjacent or associated units.

fault — a fracture or zone of fractures along which the sides are displaced relative to one another, parallel to the fracture.

forsterite — a whitish or yellowish mineral of the olivine group: Mg_2SiO_4. It is isomorphous with fayalite, and occurs chiefly in metamorphosed dolomites and crystalline limestones.

fumarole — a volcanic vent from which gases are emitted.

gabbro — a plutonic rock consisting of calcic plagioclase (commonly labradorite) and clinopyroxene, with or without orthopyroxene and olivine. Apatite and magnetite or ilmenite are common accessories.

gamma — a unit of magnetic field strength, equal to 10^{-5} gauss.

graben — an elongate crustal depression bounded by normal faults on its long sides.

Griffith failure theory — a theory of failure in solids based on growth of microfractures.

grooves — curvilinear tectonic depressions found on some icy satellites.

Gyr — gigayear = 10^9 yr.

heat capacity
(specific heat)

that quantity of heat required to increase the temperature of a system by one degree at constant pressure and volume. It is usually expressed in calories per degree Celsius per gram.

hematite

a mineral, Fe_2O_3 (hexagonal rhombohedral), the principal ore of iron.

H-function

(the Chandrasekhar H-function) the solution to an integral equation that appears in radiative transfer problems (e.g., to compute the reflectance of a surface or atmosphere). The H-function describes the results of multiple-scattering in an atmosphere or surface, and depends on the angle of incident or emitted radiation, and on the single-scattering albedo.

Hill sphere

the approximately spherical region within which a planet, rather than the Sun, dominates the motion of particles.

horseshoe orbit

the motion of an orbiting particle that alternately nearly overtakes another body and then slows down so as to be nearly overtaken by the other body. In a reference frame rotating with the orbit of the other body's orbital motion, the particle follows a horseshoe-shaped path.

Hugoniot

the locus of points describing the pressure-volume-energy relations or states that may be achieved within a material by shocking it from a given initial state.

ice phases

the various crystalline forms, or polymorphs, of ice, each stable in its own range of temperature and pressure.

I/F

the intensity in the focal plane at a vidicon divided by the flux of the Sun divided by π. If a object scatters light according to Lambert's law and if the solar incidence angle is i, then I/F may be written in terms of $\bar{A}$ as follows: $I/F = \bar{A} \cos i$, where $\bar{A}$ is defined as the geometric albedo weighted with the system response and solar flux. For a more complete explanation, see Danielson et al. 1981, pp. 8683 and 8685.

IRIS

Infrared Interferometer Spectrometer, a Voyager instrument.

isostatic   a substance subject to equal pressure from every side, in
                     hydrostatic equilibrium.

IUE   the International Ultraviolet Explorer, an Earth-oribiting
                     observatory.

Jacobi ellipsoid   the triaxial ellipsoidal form that homogeneous, uni-
                     formly rotating, self-gravitating masses will assume if
                     they rotate sufficiently rapidly.

keV   kilo electron volt $= 10^3$eV.

Kolmogorov   a spectrum in a homogeneous and isotropic turbulent
spectrum   medium with energy continually transferred between
                     turbulent eddies of different sizes. The distribution of
                     energy cascading down the different scales is referred to
                     as the Kolmogorov spectrum.

KREEP   lunar basaltic material rich in radioactive elements (K
                     for potassium, REE for rare earth elements, P for
                     phosphorus).

Lagrangian   the five equilibrium points in the restricted three-body
points   problem. Two of the Lagrange points (L_4 and L_5) are lo-
                     cated at the vertices of equilateral triangles formed by
                     the two primaries (e.g., Sun and Saturn, or Saturn and
                     satellite) and are stable; the other three are unstable and
                     lie on the line connecting the two primaries.

Lagrangian orbit   an orbit in which a particle oscillating about one of the
                     stable Lagrangian equilibrium points defined by the re-
                     stricted three-body problem moves.

Lagrangian   a satellite moving in a Lagrangian orbit.
satellite

Lambert's law   a simple scattering law according to which the intensity
                     of scattered light is independent of the emission angle.
                     An ideal Lambert surface scatters light uniformly in all
                     directions (i.e., a diffuse scatterer).

Laplace   the coefficients appearing in the Fourier expanded form
coefficients   of the disturbing function which represent the gravita-
                     tional potential due to one orbiting perturbing body on

another orbiting perturbed body. They are functions of the ratio of the semimajor axes of the two orbiters.

Laplace plane — the orientation of the orbital plane of a satellite that is perturbed by the planetary oblateness, the Sun and nearby, massive satellites and that is not fixed in inertial space. The latter perturbations cause the pole of the satellite orbit to precess about the pole of the Laplacian plane.

LECP — Low Energy Charged Particle experiment on the Voyager spacecraft.

libration — a small oscillation around an equilibrium configuration, such as the angular change in the face that a synchronously rotating satellite presents towards the focus of its orbit.

lightcurve — the brightness values plotted as a function of time.

limb — the edge of the apparent disk of a celestial body, as of the Sun, the Moon, a planet or a satellite.

linea — elongate markings.

lithosphere — the stiff upper layer of a planetary body, including the crust and part of the upper mantle, lying above the weaker asthenosphere.

Love number — the parameter that describes the enhanced gravitational potential of a body due to the redistribution of mass in the body caused by the external perturbation; the value depends on internal structure and the nature of the perturbation.

lp — line pair of an image.

Lyman-α emission — a line in the far-ultraviolet part of the spectrum (1215 Å) strongly absorbed and easily emitted by hydrogen atoms. Lyman-β and Lyman-γ are others in the Lyman spectral series similar but at shorter wavelengths.

macula — a dark spot.

magma
: mobile or fluid rock material, lava, generalized to refer to any material that behaves like silicate magma in the Earth.

magnetic Bode's law
: the approximate, linear empirical relation between the rotational angular momentum of a planet and its dipole magnetic moment.

magnetic permeability
: a factor, characteristic of a material, that equals the magnetic induction produced in a material divided by the magnetic field strength; it is a tensor when these quantities are not parallel.

magnetic Reynolds number
: a dimensionless number, $R_m = lv/\nu_m$, where l is the length scale of motion, v is a velocity and ν_m is the magnetic viscosity ($\nu_m = c^2/4\pi\sigma_e$, where σ_e is electrical conductivity and c is the speed of light).

magnetism, remanent
: that component of a rock's magnetization whose direction is fixed relative to the rock and which is independent of moderate applied magnetic fields.

magnetosphere
: the region of space surrounding a planet in which the planet's magnetic field dominates that of the solar wind. In the specific case of the Earth, the outer region of its ionosphere, starting at $\sim$ 1000 km above Earth's surface and extending to $\sim$ 60,000 km (or considerably farther, at least 100 Earth radii in the magnetotail on the side away from the Sun).

magnetotail
: the far downstream portion of a magnetosphere embedded in a flowing medium. The magnetotail contains magnetic fields oriented mainly toward or away from the central body.

mantle
: the interior zone of a planet or satellite below the crust and above the core, which is divided into the upper mantle and the lower mantle with a transition zone between.

mare (pl., maria)
: an area on the moon or Mars that appears darker and smoother than its surroundings. Lunar maria are scattered basaltic flows.

mascons	the dozen or so large-scale gravity anomalies that are found on the Moon. Most are gravity enhancements and are associated with maria.
Maxwell solid	a viscoelastic material in which strain is the sum of an elastic and a viscoelastic part; hence such a solid deforms elastically for rapid load change and deforms plastically for low load changes (e.g., like "silly putty").
Maxwell time	the characteristic time for stresses in a viscoelastic material to relax, given by viscosity divided by shear modulus.
MeV	million electron volt $= 10^6$ eV.
mgal	an acceleration (usually gravitational) of 10^{-3} cm s^{-2}.
mol	1 mol $=$ 1 mole $= 6.023 \times 10^{23}$ atoms (or molecules).
MR	megarayleigh (R $= 10^6/4\pi$ photons cm^{-2} steradian^{-1}.
Myr	10^6 yr.
nm	nanometer $= 10^{-9}$m $= 10$ Å.
NIMS	Near Infrared Mapping Spectrometer on the Galileo spacecraft.
Nusselt number (Nu)	dimensionless ratio of heat flow with convection to that carried by conduction alone.
obliquity	the angle between an object's axis of rotation and the pole of its orbit.
occultation	the cutoff of light or radiation from a celestial body (or spacecraft) due to its passage behind another body.
olivine	the most abundant mineral in chondritic meteorites, $(Mg,Fe)_2SiO_4$.
opacity	a loosely defined term referring to the ability of a medium to extinguish radiation of any given wavelength. In various applications, opacity has been used to mean: (a) optical thickness divided by physical thickness;

(b) optical or radio thickness; or (c) mass extinction coefficient.

orbital commen-
surabilities

the pairs of orbital frequencies that are in a ratio of small whole numbers.

orbital elements

six quantities that fully describe an orbit; along with time, they specify the position of an orbiting body along its path. A typical set of orbital elements are: (1) semi-major axis a, the approximate mean distance from the gravitating central body; (2) eccentricity e, the departure from circularity; (3) inclination i, the angle between the orbit plane and some reference plane such as the ecliptic; (4) longitude of the ascending node, the angle between some line in the reference plane and the point where the body crosses the reference plane moving south to north; (5) argument of periapse, the angle along the orbit plane from the ascending node to the position where the body is closest to the central object; and (6) epoch T, the time of passage through periapse.

Pa

pascal, 1 Pa $= 10^{-5}$ bar.

palimpsest

a roughly circular albedo spot on Ganymede or Callisto that is presumed to mark the site of a former crater and its rim deposit. Most, if not all, of the topographic structure has disappeared, but the visual distinction from adjacent crust remains.

patera

a crater with irregular or scalloped edges; on Io, a volcanic vent surrounded by irregular flows.

periapse

the orbital point nearest to the focus of attraction.

phase angle

the angle between Earth, object and Sun.

photosputtering

the process whereby molecules on the surface of a body can be dissociated by absorption of ultraviolet photons, causing escape of dissociation fragments from the surface.

pit crater

an impact crater containing a central pit rather than a central peak.

plagioclase   a mineral group, formula $(Na,Ca)Al(Si,Al)Si_2O_8$; a solid solution series from $NaAlSi_3O_8$ (albite) to $CaAl_2Si_2O_8$ (anorthite), triclinic. It is one of the commonest rock-forming minerals.

planetcsimal   a hypothetical early body of intermediate (perhaps 1 to 100 m) size out of which all solar system members are presumed to have accumulated.

planum   a high plateau.

plasma   the completely ionized gas, the so-called fourth state of matter in which the temperature is too high for atoms, as such, to exist and which consists of free electrons and free atomic nuclei.

Poincaré (virial) theorem   a theorem stating that the total kinetic energy of all the stars in a cluster is equal to half the negative gravitational potential energy of the cluster.

poise   a unit of viscosity, defined as the shear stress (tangential force per unit area, dyne cm^{-2}) required to maintain a unit difference in velocity ($cm\ s^{-1}$) between two parallel planes separated by 1 cm of fluid. 1 poise = 1 dyne s cm^{-2} = 1 $g\ cm^{-1}s^{-1}$ in cgs units.

Poisson's ratio   a ratio of lateral to longitudinal elastic strain in a body that has been stressed longitudinally.

polarization   the action or process of affecting radiation, especially light, such that the vibrations assume some definite form. Light which has encountered an index of refraction boundary will have different reflection coefficients depending on the orientation of the electric vector. Polarization is defined as negative if the light reflected from a boundary is greater in the plane given by the scattering plane (source-boundary-observer) than in the perpendicular plane. If the light intensity is the same in both perpendicular and parallel directions, the light is unpolarized, and if it is greater in the perpendicular direction, the polarization is called positive.

Poynting-Robertson effect
: a drag that primarily affects small grains and is due to asymmetrical re-radiation of absorbed light; it causes grains to spiral inward toward the body which they orbit.

P-wave
: a longitudinal seismic wave whose movement is accompanied by compression and rarefaction. P-waves travel faster than S (shear)-waves and can penetrate fluid regions.

pyroxene
: a group of common rock-forming silicates which have ratios of metal oxides (MgO, FeO or CaO) to SiO_2 of $1:1$. These are called metasilicates. Pure members of this group are $MgSiO_3$ (enstatite), and $FeSiO_3$ (ferrosilite). Pure $CaSiO_3$ does not crystallize with the pyroxene structure. Ca does substitute for up to 50% of the Mg and Fe in the pyroxene structure.

Q
: the anelasticity parameter, defined as $2\pi E_0/\delta E$, where E_0 is energy stored in some periodic distortion and δE is energy dissipated during one cycle of distortion. Dissipation introduces delay in tidal response, and this has implications for orbital and rotational evolution. Note that a small Q implies a large dissipation.

Raman effect
: the change of wavelength on scattering. It occurs when incident radiation excites (or de-excites) atoms or molecules from their initial states.

Rayleigh number (Ra)
: a dimensionless combination of parameters involving the temperature gradient and the coefficients of thermal conductivity and kinematic viscosity, which determines when a fluid, under specified geometrical conditions, will become convectively unstable.

Rayleigh scattering
: the scattering at shorter wavelengths of light by molecules or small particles. The scattering is inversely proportional to the fourth power of the wavelength.

Rayleigh-Taylor instability
: the gravitational instability which results when a heavy fluid overlies a lighter fluid (e.g., a cold dense gas above a hot rarefied gas).

RD objects
: an old term used to describe asteroids with very low albedos and reddish spectra; now called D asteroids.

REE	rare earth elements, the lathanide series in the periodic table.
reflectance	the fraction of the total radiant flux incident upon something (like a planetary atmosphere or surface) that is reflected; in general it varies according to the wavelength distribution of the incident radiation. Reflectance is usually measured for a narrow range of wavelengths through a filter or other dispersive device.
reflectance spectrum	the reflectance of an object measured as a function of wavelength (*see also* reflectance).
regio	a large area of distinctive albedo markings.
regolith	a layer of fragmentary debris produced by meteoritic impact on the surface of any celestial object.
resonance	the enhanced response of any oscillating system to an external stimulus which has the same driving frequency as the natural frequency of the system; higher order resonances occur when these frequencies are commensurable.
Reynolds number	a dimensionless number ($R = Lv/\nu$), where L is a typical dimension of the system, v is a measure of the velocities that prevail, and ν is the kinematic viscosity) that governs the likelihood for the occurence of turbulence in fluids (*see also* magnetic Reynolds number).
rille	one of several trench-like or crack-like valleys, up to several hundred km long and 1 to 2 km wide, commonly occurring on the surface of the Moon and other satellites. Rilles appear to be relatively youthful features and many apparently represent fracture systems originating in the brittle material.
rms	root mean square.
Roche limit	the minimum distance at which a fluid satellite influenced by its own gravitation and that of a central mass can be in mechanical equilibrium. For a satellite of zero tensile strength, and the same mean density as its primary, in a circular orbit around its primary, this critical distance is 2.46 times the radius of the primary.

Roche (Hill's) lobe — the largest closed zero-velocity potential surface surrounding a secondary body (satellite) in orbit about a primary. This surface passes through the Lagrange L_2 point which lies between the primary and secondary bodies.

Rutherford scattering — the Coulomb scattering of ions by atomic nuclei.

saturation cratering — the condition of maximum possible density of impact features on a planetary or satellite surface, beyond which additional cratering does not alter the observed crater distribution.

scarps — the line of cliffs produced by faulting or erosion, or a cliff-like face or slope of considerable linear extent.

shield volcano — a volcanic mountain in the shape of a broad, flattened dome.

single-scattering albedo — the fraction of light that survives an interaction with a particle and is not absorbed.

solar nebula — the gas and dust from which the solar system formed; also, the nebula surrounding the protosun when the planets and smaller bodies were still accreting.

solar wind — the energetic charged particles that flow radially outward from the solar corona, carrying mass and angular momentum away from the Sun.

spallation — a nuclear reaction in which the energy of incident particles is so high that several particles are ejected from the target nucleus for each incident one; hence both the mass number and atomic number of the target nucleus are changed.

sputtering — the expulsion of atoms or atomic fragments from a solid, caused by impact of energetic particles.

Stokes velocity — the terminal velocity of a sphere falling through a viscous fluid under the influence of gravity.

sulcus — a complex area of subparallel furrows and ridges.

S-wave   in geophysics, a seismic shear wave in which components move transverse to the direction of propagation. *S*-waves cannot penetrate fluid regions. *S*-waves move more slowly than *P* (pressure)-waves. In atomic collision theory, a wave representing a collision between two particles with zero relative angular momentum.

tadpole orbit   an alternative term for a Lagrangian orbit.

terminator   the line of sunrise or sunset on a planet or satellite.

thermal conductivity   the proportionality constant that gives the amount of heat conducted through a unit cross section in unit time under the influence of unit heat gradient. In cgs units the thermal conductivity is expressed in calories per centimeter squared per second per degree Celsius.

thermal emission   the emission of electromagnetic radiation from a body due to its temperature and emissivity.

thermal emission spectrum   the thermal emission from a body measured as a function of wavelength (*see also* reflectance).

thermal inertia   a material parameter which indicates the rate at which a body's temperature responds to changing heat input. It is proportional to the square root of the product of thermal conductivity and volume heat capacity.

UVS   ultraviolet spectrometer, a Voyager instrument.

V_0 the mean opposition visual magnitude, measured at zero phase angle.

$V(1,0)$ the absolute visual magnitude, at unit distance and zero phase, defined by $V = V(1,0) + 5 \log R\Delta + F(\alpha)$ where R is the distance from the Sun and Δ from the Earth (both measured in AU) and $F(\alpha)$ is the phase function.

van der Waal forces   the relatively weak attractive forces operative between neutral atoms and molecules.

vapor   any substance in the gaseous state; a gasified liquid or solid.

VIMS   Visual and Infrared Mapping Spectrometer on the Galileo
                              spacecraft.

Young's modulus   the proportionality constant between stress (force per
                             unit area) and strain (change in length per unit length) for
                             an elastic material. In cgs units it is expressed in dynes
                             cm^{-2} or lbs ft^{-2}.

Acknowledgments to Funding Agencies and Referees

ACKNOWLEDGMENTS TO
FUNDING AGENCIES AND REFEREES

The editors acknowledge the support of the National Aeronautics and Space Administration for the preparation of the book. The following authors wish to acknowledge specific funds involved in supporting the preparation of their chapters.

Burns, J. A.: NASA Grants NAGW-310 and NAGW-521, and NSF Grant AST-8214651
Chapman, C. R.: NASA Contract NASW-3718
Cheng, A. F.: NASA Grant NAGW-154
Clark, R. N.: NASA Interagency Agreement W15805
Croft, S. K.: NASA Grant NSG-7146
Dermott, S.: NASA Grants NAGW-111, NAGW-392, and NSG-7156
Drake, M. J.: NASA Grant NAG-9-39
Gradie, J.: NASA Grant NAGW-193
Haff, P. K.: NASA Grant NAGW-202
Harris, A. W.: NASA Grant NAS7-918
Head, J. W. III: NASA Grant NGR-40-002-116
Johnson, R. E.: NASA Grant NAGW-186 and NSF Grant AST-82-00477
Johnson, T. V.: NASA Grant NAS7-918
Lunine, J. I.: NASA Grants NAGW-185 and NAGW-450
Malin, M. C.: NASA Grant NAGW-1
Matson, D. L.: NASA Grant NAS7-918
McKinnon, W. B.: NASA Grant NAGW-432
Nash, D. B.: NASA Grant NAS7-918
Owen, T.: NASA Grant NGR 33-015-141 and NASA Contract 953614
Parmentier, E. M.: NASA Grant NSG-7605
Peale, S. J.: NASA Grant NGR-05-010-062
Pieri, D. C.: NASA Grant NAS7-918
Schubert, G.: NASA Grant NSG-7315
Squyres, S. W.: NASA RTOP 151-05-60-08
Stevenson, D. J.: NASA Grants NAGW-185 and NAGW-450
Thomas, P.: NASA Grants NAGW-111, NAGW-392, and NSG-7156
Veverka, J.: NASA Grants NSG-7156 and NSG 7616
Yoder, C. F.: NASA Grant NAS7-918

The editors also wish to acknowledge the help provided by the following referees:

D. P. Cruikshank, A. R. Dobrovolskis, T. C. Duxbury, M. Golombeck, J. Gradie, R. Greeley, R. Greenberg, A. W. Harris, J. Henrard, H. H. Kieffer, K. Lambeck, L. A. Lebofsky, J. J. Lissauer, M. C. Malin, D. L. Matson, A. S. McEwen, W. B. McKinnon, E. M. Parmentier, S. J. Peale, C. Pilcher, J. B. Pollack, W. M. Sinton, G. L. Siscoe, L. A. Soderblom, S. C. Solomon, P. D. Spudis, S. W. Squyres, D. J. Stevenson, R. G. Strom, S. R. Taylor, P. Thomas, C. F. Yoder

Index

INDEX*

---

*Italicized page numbers indicate figures or tables.